AF880054

Functions Modeling Change
A Preparation for Calculus
Third Edition

Connally, Hughes-Hallett, Gleason, et al.

with additional material from

Finite Mathematics
Ninth Edition
Michael Sullivan

Calculus
Eight Edition
Howard Anton/Irl Bivens/Stephen Davis

Precalculus Algebra MAC 1140
Mathematics Department
Sante Fe Community College

WILEY
CUSTOM SERVICES

To order books or for customer service, please call 1(800)-CALL-WILEY (225-5945).

Printed in the United States of America.

ISBN 978-0-470-89581-8

10 9 8 7

FUNCTIONS MODELING CHANGE:
A Preparation for Calculus

Third Edition

Produced by the Calculus Consortium and initially funded by a National Science Foundation Grant.

Eric Connally
Harvard University Extension

Deborah Hughes-Hallett
University of Arizona

Andrew M. Gleason
Harvard University

Philip Cheifetz
Nassau Community College

Ann Davidian
Gen. Douglas MacArthur HS

Daniel E. Flath
Macalester College

Brigitte Lahme
Sonoma State University

Patti Frazer Lock
St. Lawrence University

Jerry Morris
Sonoma State University

Karen Rhea
University of Michigan

Ellen Schmierer
Nassau Community College

Pat Shure
University of Michigan

Carl Swenson
Seattle University

Katherine Yoshiwara
Los Angeles Pierce College

Elliot J. Marks

with the assistance of
Frank Avenoso
Nassau Community College

John Wiley & Sons, Inc.

Dedicated to Maria, Ben, Jonah, and Isabel

PUBLISHER	Laurie Rosatone
SENIOR ACQUISITIONS EDITOR	Angela Y. Battle
MARKETING MANAGER	Amy Sell
FREELANCE DEVELOPMENTAL EDITOR	Anne Scanlan-Rohrer
SENIOR PRODUCTION EDITOR	Ken Santor
ASSISTANT EDITOR	Shannon Corliss
MARKETING ASSISTANT	Tara Martinho
COVER DESIGNER	Hope Miller
COVER PHOTO	©Scott Berner/Picture Quest

This book was set in Times Roman by the Consortium using TeX, Mathematica, and the package AsTeX, which was written by Alex Kasman. It was printed and bound by Von Hoffmann Press. The cover was printed by Von Hoffmann Press. The process was managed by Elliot Marks.

This material is based upon work supported by the National Science Foundation under Grant No. DUE-9352905. Opinions expressed are those of the authors and not necessarily those of the Foundation.

ISBN-13 978-0-471-79303-8
ISBN-10 0471-79303-5

Printed in the United States of America

10 9 8 7 6 5 4 3 2 1

PREFACE

Mathematics has the extraordinary power to reduce complicated problems to simple rules and procedures. Therein lies the danger in teaching mathematics: it is possible to teach the subject as nothing but the rules and procedures—thereby losing sight of both the mathematics and of its practical value. The third edition of *Functions Modeling Change: A Preparation for Calculus* continues our effort to refocus the teaching of mathematics on concepts as well as procedures.

A Curriculum Balancing Skills and Conceptual Understanding

These materials stress conceptual understanding and multiple ways of representing mathematical ideas. Our goal is to provide students with a clear understanding of the ideas of functions as a solid foundation for subsequent courses in mathematics and other disciplines. When we designed this curriculum under an NSF grant, we started with a clean slate. We focused on the key concepts, emphasizing depth of understanding.

Skills are developed in the context of problems and reinforced in a variety of settings, thereby encouraging retention. This balance of skills and understanding enables students to realize the power of mathematics in modeling.

The Calculus Consortium for Higher Education

This book is the work of faculty at a consortium of ten institutions, generously supported by the National Science Foundation. It represents the first consensus between such a diverse group of faculty to have shaped a mainstream precalculus text. Bringing together the results of research and experience with the views of many users, this text is designed to be used in a wide range of institutions.

Guiding Principles: Varied Problems and the Rule of Four

Since students usually learn most when they are active, the exercises in a text are of central importance. In addition, we have found that multiple representations encourage students to reflect on the meaning of the material. Consequently, we have been guided by the following principles.

- Our problems are varied and some are challenging. Many cannot be done by following a template in the text.
- The Rule of Four, originally introduced by the consortium, promotes multiple representations. Each concept and function is represented symbolically, numerically, graphically, and verbally.
- The components of a precalculus curriculum should be tied together by clearly defined themes. Functions as models of change is our central theme, and algebra is integrated where appropriate.
- Fewer topics are introduced than is customary, but each topic is treated in greater depth. The core syllabus of precalculus should include only those topics that are essential to the study of calculus.
- Problems involving real data are included to prepare students to use mathematics in other fields.
- To use mathematics effectively, students need skill in both symbolic manipulation and the use of technology. The exact proportions of each may vary widely, depending on the preparation of the student and the wishes of the instructor.
- Materials for precalculus should allow for a broad range of teaching styles. They should be flexible enough to use in large lecture halls, small classes, or in group or lab settings.

The Third Edition: Increased Flexibility

The third edition retains the hallmarks of earlier editions and provides instructors with more flexibility.

- **Composition of functions** is introduced in Chapter 2 for instructors who want to introduce this topic early. However, instructors have the flexibility to delay this topic until Chapter 8.
- The former Section 3.4 has been split into two sections, the first on continuous growth and e and the second on compound interest. This gives instructors the flexibility to discuss the properties of e and compound interest together or separately.
- The **introduction of** e has been rewritten to increase understanding.
- **Limit notation** has been introduced in Chapters 1 and 3 for instructors who wish to include this.
- The material on **trigonometric identities** in Chapter 7 has been expanded, and new problems added.
- A new section on the **geometric properties of conics** has been added to Chapter 12.
- **Algebra review** is integrated throughout the text. More skill building has been emphasized wherever appropriate. The algebraic tools used in a chapter are summarized for easy reference in a "Tools" section.
- **Data and problems** have been updated and revised as appropriate. Many new problems have been added.
- **ConcepTests** for precalculus are now available for instructors looking for innovative ways to promote active learning in the classroom. Further information is provided under the Supplementary Materials on page viii.

What Student Background is Expected?

Students using this book should have successfully completed a course in intermediate algebra or high school algebra II. The book is thought-provoking for well-prepared students while still accessible to students with weaker backgrounds. Providing numerical and graphical approaches as well as the algebraic gives students another way of mastering the material. This approach encourages students to persist, thereby lowering failure rates.

Our Experiences

Previous editions of this book were used by hundreds of schools around the country. In this diverse group of schools, the first two editions were successfully used with many different types of students in semester and quarter systems, in large lectures and small classes, as well as in full year courses in secondary schools. They were used in computer labs, small groups, and traditional settings, and with a number of different technologies.

Content

The central theme of this course is functions as models of change. We emphasize that functions can be grouped into families and that functions can be used as models for real-world behavior. Because linear, exponential, power, and periodic functions are more frequently used to model physical phenomena, they are introduced before polynomial and rational functions. Once introduced, a family of functions is compared and contrasted with other families of functions.

A large number of the examples and problems that students see in this precalculus course are given in the context of real-world problems. Indeed, we hope that students will be able to create mathematical models that will help them understand the world in which they live. The inclusion of non-routine problems is intended to establish the idea that such problems are not only part of mathematics, but in some sense, are the point of mathematics.

The book does not require any specific software or technology. Instructors have used the material with graphing calculators and graphing software. Any technology with the ability to graph functions will suffice.

Chapter 1: Linear Functions and Change

This chapter introduces the concept of a function and the graphical, tabular, symbolic, and verbal representations of functions. The advantages and disadvantages of each representation are discussed. Rates of change are introduced and used to characterize linear functions. A section on fitting a linear function to data is included.

The **Tools** section for Chapter 1 reviews linear equations and the coordinate plane.

Chapter 2: Functions

Function notation is introduced in Chapter 1 and studied in more detail in this chapter. Domain and range, and the concepts of composite and inverse functions are introduced. The idea of concavity is investigated using rates of change. The chapter includes sections on piecewise and quadratic functions.

The **Tools** section for Chapter 2 reviews quadratic equations.

Chapter 3: Exponential Functions

This chapter introduces the family of exponential functions and the number e. Exponential and linear functions are compared, and exponential equations are solved graphically.

The **Tools** section for Chapter 3 reviews the properties of exponents.

Chapter 4: Logarithmic Functions

Logarithmic functions to base 10 and base e are introduced to solve exponential equations and to serve as inverses of exponential functions. After manipulations with logarithms, Chapter 4 focuses on modeling with exponential functions and logarithms. Logarithmic scales and a section on linearizing data conclude the chapter.

The **Tools** section for Chapter 4 reviews the properties of logarithms.

Chapter 5: Transformations of Functions and Their Graphs

This chapter investigates transformations—shifting, reflecting, and stretching. These ideas are applied to the family of quadratic functions.

The **Tools** section for Chapter 5 reviews completing the square.

Chapter 6: Trigonometric Functions

This chapter, which focuses on modeling periodic phenomena, introduces the trigonometric functions: the sine, cosine, tangent, and briefly, the secant, cosecant, and cotangent. The inverse trigonometric functions are introduced.

The **Tools** section for Chapter 6 reviews right triangle trigonometry.

Chapter 7: Trigonometry

This chapter focuses on triangles, identities, and polar coordinates. It includes a section on modeling using trigonometric functions. Complex numbers and polar coordinates are introduced.

Chapter 8: Compositions, Inverses, and Combinations of Functions

This chapter covers combinations of functions. Composite and inverse functions, which were introduced in Chapter 2, are investigated in more detail in this chapter.

Chapter 9: Polynomial and Rational Functions

This chapter covers power functions, polynomials, and rational functions. The chapter concludes by comparing several families of functions, including polynomial and exponential functions, and by fitting functions to data.

The **Tools** section for Chapter 9 reviews algebraic fractions.

Chapter 10: Vectors

This chapter contains material on vectors and operations involving vectors. An introduction to matrices is included in the last section.

Chapter 11: Sequences and Series

This chapter introduces arithmetic and geometric sequences and series and their applications.

Chapter 12: Parametric Equations and Conic Sections

The concluding chapter looks at parametric equations, implicit functions, the hyperbolic functions, and the conic sections: circles, ellipses, and hyperbolas. The chapter includes a section on the geometrical properties of the conic sections and their applications to orbits.

Supplementary Materials

The following supplementary materials are available for the Third Edition:

- **Instructor's Manual** contains teaching tips, lesson plans, syllabi, and worksheets. The Instructor's Manual has been expanded and revised to include worksheets, identification of technology-oriented problems, and new syllabi. (ISBN 0470-10819-3)
- **Printed Test Bank** contains test questions arranged by section. (ISBN 0470-10821-5)
- **Instructor's Solution Manual** with complete solutions to all problems. (ISBN 0470-10820-7)
- **Student Solution Manual** with complete solutions to half the odd-numbered problems. (ISBN 0470-10561-5)
- **Student Study Guide** includes study tips, learning objectives, practice problems, and solutions. The topics are tied directly to the book. (ISBN 0470-5-3)
- **Getting Started Graphing Calculator Manual** instructs students on how to utilize their TI-83 and TI-84 series calculators with the text. Contains samples, tips, and trouble shooting sections to answer students' questions. (ISBN 0470-10558-5)
- **Computerized Test Bank**: Available in both PC and Macintosh formats, the Computerized Test Bank allows instructors to create, customize, and print a test containing any combinations of questions from a large bank of questions. Instructors can also customize the questions or create their own. (ISBN 0470-10822-3)
- **Book Companion Site** The accompanying website contains all instructor supplements as well as web quizzes for student practice.
- **Wiley PLUS** This powerful and highly integrated suite of online teaching and learning resources provides course management options to instructors an students. Instructors can automate the process of assigning, delivering, and grading algorithmically generated homework exercises, hints and solutions while providing students with immediate feedback. Additionally student tutorials, an instructor gradebook, integrated links to the electronic version of the text and all of the text supplemental materials are provided. For more information, visit www.wiley.com/college/wileyplus or contact your local Wiley representative for more details.

- **The Faculty Resource Network** is a peer-to-peer network of academic faculty dedicated to the effective use of technology in the classroom. This group can help you apply innovative classroom techniques, implement specific software packages, and tailor the technology experience to the specific needs of each individual class. Ask your Wiley representative for more details.

ConcepTests

ConcepTests, modeled on the pioneering work of Harvard physicist Eric Mazur, are questions designed to promote active learning during class, particularly (but not exclusively) in large lectures. Our evaluation data show students taught with ConcepTests outperformed students taught by traditional lecture methods 73% versus 17% on conceptual questions, and 63% versus 54% on computational problems. ConcepTests arranged by section are available in print, PowerPoint, and Classroom Response System-ready formats from your Wiley representative. (ISBN 0470-10818-5)

Acknowledgments

We would like to thank the many people who made this book possible. First, we would like to thank the National Science Foundation for their trust and their support; we are particularly grateful to Jim Lightbourne and Spud Bradley.

We are also grateful to our Advisory Board for their guidance: Benita Albert, Lida Barrett, Simon Bernau, Robert Davis, Lovenia Deconge-Watson, John Dossey, Ronald Douglas, Eli Fromm, Bill Haver, Don Lewis, Seymour Parter, John Prados, and Stephen Rodi.

Working with Laurie Rosatone, Angela Battle, Amy Sell, Anne Scanlan-Rohrer, Ken Santor, Hope Miller, and Shannon Corliss at John Wiley is a pleasure. We appreciate their patience and imagination.

Many people have contributed significantly to this text. They include: Fahd Alshammari, David Arias, Tim Bean, Charlotte Bonner, Bill Bossert, Brian Bradie, Noah S. Brannen, Mike Brilleslyper, Donna Brouillette, Jo Cannon, Ray Cannon, Kenny Ching, Pierre Cressant, Laurie Delitsky, Bob Dobrow, Ian Dowker, Carolyn Edmond, Maryann Faller, Aidan Flanagan, Brendan Fry, Brad Garner, Carrie Garner, John Gerke, Christie Gilliland, Wynne Guy, Donnie Hallstone, David Halstead, Larry Henly, Dean Hickerson, Jo Ellen Hillyer, Bob Hoburg, Phil Hotchkiss, Mike Huffman, Mac Hyman, Rajini Jesudason, Loren Johnson, Mary Kilbride, Steve Kinholt, Kandace Kling, Rob LaQuaglia, Barbara Leasher, Richard Little, David Lovelock, Len Malinowski, Nancy Marcus, Bill McCallum, Kate McGivney, Gowri Meda, Bob Megginson, Deborah Moore, Eric Motylinski, Bill Mueller, Kyle Niedzwiecki, Kathryn Oswald, Igor Padure, Bridget Neale Paris, Janet Ray, Ken Richardson, Halip Saifi, Sharon Sanders, Ellen Schmierer, Mary Schumacher, Mike Seery, Mike Sherman, Donna Sherrill, Kanwal Singh, Fred Shure, Myra Snell, Natasha Speer, Sonya Stanley, Jim Stone, Peggy Tibbs, Jeff Taft, Elias Toubassi, Jerry Uhl, Pat Wagener, Dale Winter, and Xianbao Xu.

Reports from the following reviewers were most helpful in shaping the second and third editions: Victor Akatsa, Jeffrey Anderson, Ingrid Brown-Scott, Linda Casper, Kim Chudnick, Ray Collings, Monica Davis, Helen Doerr, Diane Downie, Peter Dragnev, Patricia Dueck, Jennifer Fowler, David Gillette, Donnie Hallstone, Jeff Hoherz, Majid Hosseini, Rick Hough, Pallavi Ketkar, William Kiele, Mile Krajcevski, John LaMaster, Phyllis Leonard, Daphne MacLean, Diane Mathios, Vince McGarry, Maria Miles, Laura Moore-Mueller, Dave Nolan, Linda O'Brien, Scott Perry, Mary Rack, Emily Roth, Barbara Shabell, Diana Staats, John Stadler, Mary Jane Sterling, Allison Sutton, John Thomason, Diane Van Nostrand, Linda Wagner, Nicole Williams, Jim Winston, and Vauhn Wittman-Grahler,

Special thanks are owed to "Suds" Sudholz for administering the project and to Alex Kasman for his software support.

Eric Connally	Deborah Hughes-Hallett	Andrew M. Gleason	Philip Cheifetz
Ann Davidian	Dan Flath	Brigitte Lahme	Patti Frazer Lock
Elliot Marks	Jerry Morris	Karen Rhea	Ellen Schmierer
Pat Shure	Carl Swenson	Katherine Yoshiwara	

To Students: How to Learn from this Book

- This book may be different from other math textbooks that you have used, so it may be helpful to know about some of the differences in advance. At every stage, this book emphasizes the *meaning* (in practical, graphical or numerical terms) of the symbols you are using. There is much less emphasis on "plug-and-chug" and using formulas, and much more emphasis on the interpretation of these formulas than you may expect. You will often be asked to explain your ideas in words or to explain an answer using graphs.
- The book contains the main ideas of precalculus in plain English. Success in using this book will depend on reading, questioning, and thinking hard about the ideas presented. It will be helpful to read the text in detail, not just the worked examples.
- There are few examples in the text that are exactly like the homework problems, so homework problems can't be done by searching for similar–looking "worked out" examples. Success with the homework will come by grappling with the ideas of precalculus.
- Many of the problems in the book are open-ended. This means that there is more than one correct approach and more than one correct solution. Sometimes, solving a problem relies on common sense ideas that are not stated in the problem explicitly but which you know from everyday life.
- This book assumes that you have access to a calculator or computer that can graph functions and find (approximate) roots of equations. There are many situations where you may not be able to find an exact solution to a problem, but can use a calculator or computer to get a reasonable approximation. An answer obtained this way can be as useful as an exact one. However, the problem does not always state that a calculator is required, so use your own judgement.
- This book attempts to give equal weight to four methods for describing functions: graphical (a picture), numerical (a table of values), algebraic (a formula) and verbal (words). Sometimes it's easier to translate a problem given in one form into another. For example, you might replace the graph of a parabola with its equation, or plot a table of values to see its behavior. It is important to be flexible about your approach: if one way of looking at a problem doesn't work, try another.
- Students using this book have found discussing these problems in small groups helpful. There are a great many problems which are not cut-and-dried; it can help to attack them with the other perspectives your colleagues can provide. If group work is not feasible, see if your instructor can organize a discussion session in which additional problems can be worked on.
- You are probably wondering what you'll get from the book. The answer is, if you put in a solid effort, you will get a real understanding of functions as well as a real sense of how mathematics is used in the age of technology.

Table of Contents

4 LOGARITHMIC FUNCTIONS 151

5 TRANSFORMATIONS OF FUNCTIONS AND THEIR GRAPHS 193

6 TRIGONOMETRIC FUNCTIONS 243

Chapter One

LINEAR FUNCTIONS AND CHANGE

A function describes how the value of one quantity depends on the value of another. A function can be represented by words, a graph, a formula, or a table of numbers. Section 1.1 gives examples of all four representations and introduces the notation used to represent a function. Section 1.2 introduces the idea of a rate of change.

Sections 1.3–1.6 investigate linear functions, whose rate of change is constant. Section 1.4 gives the equations for a line, and Section 1.5 focuses on parallel and perpendicular lines. In Section 1.6, we see how to approximate a set of data using linear regression.

The Tools Section on page 55 reviews linear equations and the coordinate plane.

1.1 FUNCTIONS AND FUNCTION NOTATION

In everyday language, the word *function* expresses the notion of dependence. For example, a person might say that election results are a function of the economy, meaning that the winner of an election is determined by how the economy is doing. Someone else might claim that car sales are a function of the weather, meaning that the number of cars sold on a given day is affected by the weather.

In mathematics, the meaning of the word *function* is more precise, but the basic idea is the same. A function is a relationship between two quantities. If the value of the first quantity determines exactly one value of the second quantity, we say the second quantity is a function of the first. We make the following definition:

> A **function** is a rule which takes certain numbers as inputs and assigns to each input number exactly one output number. The output is a function of the input.

The inputs and outputs are also called *variables*.

Representing Functions: Words, Tables, Graphs, and Formulas

A function can be described using words, data in a table, points on a graph, or a formula.

Example 1 It is a surprising biological fact that most crickets chirp at a rate that increases as the temperature increases. For the snowy tree cricket (*Oecanthus fultoni*), the relationship between temperature and chirp rate is so reliable that this type of cricket is called the thermometer cricket. We can estimate the temperature (in degrees Fahrenheit) by counting the number of times a snowy tree cricket chirps in 15 seconds and adding 40. For instance, if we count 20 chirps in 15 seconds, then a good estimate of the temperature is $20 + 40 = 60°\text{F}$.

The rule used to find the temperature T (in °F) from the chirp rate R (in chirps per minute) is an example of a function. The input is chirp rate and the output is temperature. Describe this function using words, a table, a graph, and a formula.

Solution

- **Words**: To estimate the temperature, we count the number of chirps in fifteen seconds and add forty. Alternatively, we can count R chirps per minute, divide R by four and add forty. This is because there are one-fourth as many chirps in fifteen seconds as there are in sixty seconds. For instance, 80 chirps per minute works out to $\frac{1}{4} \cdot 80 = 20$ chirps every 15 seconds, giving an estimated temperature of $20 + 40 = 60°\text{F}$.
- **Table**: Table 1.1 gives the estimated temperature, T, as a function of R, the number of chirps per minute. Notice the pattern in Table 1.1: each time the chirp rate, R, goes up by 20 chirps per minute, the temperature, T, goes up by 5°F.
- **Graph**: The data from Table 1.1 are plotted in Figure 1.1. For instance, the pair of values $R = 80$, $T = 60$ are plotted as the point P, which is 80 units along the horizontal axis and 60 units up the vertical axis. Data represented in this way are said to be plotted on the *Cartesian plane*. The precise position of P is shown by its coordinates, written $P = (80, 60)$.

Rule of Four:
Words
Formula
Graph
Table

Table 1.1 *Chirp rate and temperature*

R, chirp rate (chirps/minute)	T, predicted temperature (°F)
20	45
40	50
60	55
80	60
100	65
120	70
140	75
160	80

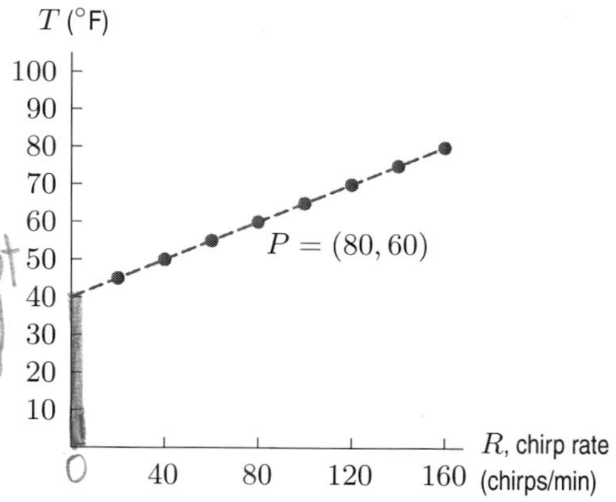

Figure 1.1: Chirp rate and temperature

- **Formula**: A formula is an equation giving T in terms of R. Dividing the chirp rate by four and adding forty gives the estimated temperature, so:

$$\underbrace{\text{Estimated temperature (in °F)}}_{T} = \frac{1}{4} \cdot \underbrace{\text{Chirp rate (in chirps/min)}}_{R} + 40.$$

Rewriting this using the variables T and R gives the formula:

$$T = \frac{1}{4}R + 40.$$

Let's check the formula. Substituting $R = 80$, we have

$$T = \frac{1}{4} \cdot 80 + 40 = 60$$

which agrees with point $P = (80, 60)$ in Figure 1.1. The formula $T = \frac{1}{4}R + 40$ also tells us that if $R = 0$, then $T = 40$. Thus, the dashed line in Figure 1.1 crosses (or intersects) the T-axis at $T = 40$; we say the T-*intercept* is 40.

All the descriptions given in Example 1 provide the same information, but each description has a different emphasis. A relationship between variables is often given in words, as at the beginning of Example 1. Table 1.1 is useful because it shows the predicted temperature for various chirp rates. Figure 1.1 is more suggestive of a trend than the table, although it is harder to read exact values of the function. For example, you might have noticed that every point in Figure 1.1 falls on a straight line that slopes up from left to right. In general, a graph can reveal a pattern that might otherwise go unnoticed. Finally, the formula has the advantage of being both compact and precise. However, this compactness can also be a disadvantage since it may be harder to gain as much insight from a formula as from a table or a graph.

Mathematical Models

When we use a function to describe an actual situation, the function is referred to as a **mathematical model**. The formula $T = \frac{1}{4}R + 40$ is a mathematical model of the relationship between the temperature and the cricket's chirp rate. Such models can be powerful tools for understanding phenomena and making predictions. For example, this model predicts that when the chirp rate is 80 chirps per minute, the temperature is 60°F. In addition, since $T = 40$ when $R = 0$, the model predicts that the chirp rate is 0 at 40°F. Whether the model's predictions are accurate for chirp rates down to 0 and temperatures as low as 40°F is a question that mathematics alone cannot answer; an understanding of the biology of crickets is needed. However, we can safely say that the model does not apply for temperatures below 40°F, because the chirp rate would then be negative. For the range of chirp rates and temperatures in Table 1.1, the model is remarkably accurate.

In everyday language, saying that T is a function of R suggests that making the cricket chirp faster would somehow make the temperature change. Clearly, the cricket's chirping does not cause the temperature to be what it is. In mathematics, saying that the temperature "depends" on the chirp rate means only that knowing the chirp rate is sufficient to tell us the temperature.

Function Notation

To indicate that a quantity Q is a function of a quantity t, we abbreviate

$$Q \text{ is a function of } t \quad \text{to} \quad Q \text{ equals "} f \text{ of } t \text{"}$$

and, using function notation, to

$$Q = f(t).$$

Thus, applying the rule f to the input value, t, gives the output value, $f(t)$. In other words, $f(t)$ represents a value of Q. Here Q is called the *dependent variable* and t is called the *independent variable*. Symbolically,

$$\text{Output} = f(\text{Input})$$

or

$$\text{Dependent} = f(\text{Independent}).$$

We could have used any letter, not just f, to represent the rule.

Example 2 The number of gallons of paint needed to paint a house depends on the size of the house. A gallon of paint typically covers 250 square feet. Thus, the number of gallons of paint, n, is a function of the area to be painted, A ft^2. We write $n = f(A)$.

(a) Find a formula for f.
(b) Explain in words what the statement $f(10{,}000) = 40$ tells us about painting houses.

Solution (a) If $A = 5000$ ft^2, then $n = 5000/250 = 20$ gallons of paint. In general, n and A are related by the formula

$$n = \frac{A}{250}.$$

(b) The input of the function $n = f(A)$ is an area and the output is an amount of paint. The statement $f(10{,}000) = 40$ tells us that an area of $A = 10{,}000$ ft^2 requires $n = 40$ gallons of paint.

The expressions "Q depends on t" or "Q is a function of t" do *not* imply a cause-and-effect relationship, as the snowy tree cricket example illustrates.

Example 3 Example 1 gives the following formula for estimating air temperature based on the chirp rate of the snowy tree cricket:

$$T = \frac{1}{4}R + 40.$$

In this formula, T depends on R. Writing $T = f(R)$ indicates that the relationship is a function.

Functions Don't Have to Be Defined by Formulas

People sometimes think that functions are always represented by formulas. However, the next example shows a function which is not given by a formula.

Example 4 The average monthly rainfall, R, at Chicago's O'Hare airport is given in Table 1.2, where time, t, is in months and $t = 1$ is January, $t = 2$ is February, and so on. The rainfall is a function of the month, so we write $R = f(t)$. However there is no equation that gives R when t is known. Evaluate $f(1)$ and $f(11)$. Explain what your answers mean.

Table 1.2 *Average monthly rainfall at Chicago's O'Hare airport*

Month, t	1	2	3	4	5	6	7	8	9	10	11	12
Rainfall, R (inches)	1.8	1.8	2.7	3.1	3.5	3.7	3.5	3.4	3.2	2.5	2.4	2.1

Solution The value of $f(1)$ is the average rainfall in inches at Chicago's O'Hare airport in a typical January. From the table, $f(1) = 1.8$. Similarly, $f(11) = 2.4$ means that in a typical November, there are 2.4 inches of rain at O'Hare.

When Is a Relationship Not a Function?

It is possible for two quantities to be related and yet for neither quantity to be a function of the other.

Example 5 A national park contains foxes that prey on rabbits. Table 1.3 gives the two populations, F and R, over a 12-month period, where $t = 0$ means January 1, $t = 1$ means February 1, and so on.

Table 1.3 *Number of foxes and rabbits in a national park, by month*

t, month	0	1	2	3	4	5	6	7	8	9	10	11
R, rabbits	1000	750	567	500	567	750	1000	1250	1433	1500	1433	1250
F, foxes	150	143	125	100	75	57	50	57	75	100	125	143

(a) Is F a function of t? Is R a function of t?
(b) Is F a function of R? Is R a function of F?

Solution

(a) Both F and R are functions of t. For each value of t, there is exactly one value of F and exactly one value of R. For example, Table 1.3 shows that if $t = 5$, then $R = 750$ and $F = 57$. This means that on June 1 there are 750 rabbits and 57 foxes in the park. If we write $R = f(t)$ and $F = g(t)$, then $f(5) = 750$ and $g(5) = 57$.

(b) No, F is not a function of R. For example, suppose $R = 750$, meaning there are 750 rabbits. This happens both at $t = 1$ (February 1) and at $t = 5$ (June 1). In the first instance, there are 143 foxes; in the second instance, there are 57 foxes. Since there are R-values which correspond to more than one F-value, F is not a function of R.

Similarly, R is not a function of F. At time $t = 5$, we have $R = 750$ when $F = 57$, while at time $t = 7$, we have $R = 1250$ when $F = 57$ again. Thus, the value of F does not uniquely determine the value of R.

How to Tell if a Graph Represents a Function: Vertical Line Test

What does it mean graphically for y to be a function of x? Look at the graph of y against x. For a function, each x-value corresponds to exactly one y-value. This means that the graph intersects any vertical line at most once. If a vertical line cuts the graph twice, the graph would contain two points with different y-values but the same x-value; this would violate the definition of a function. Thus, we have the following criterion:

> **Vertical Line Test**: If there is a vertical line which intersects a graph in more than one point, then the graph does not represent a function.

Example 6 In which of the graphs in Figures 1.2 and 1.3 could y be a function of x?

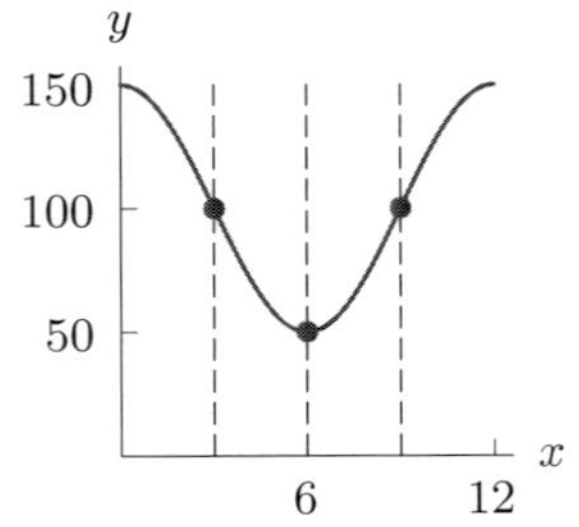

Figure 1.2: Since no vertical line intersects this curve at more than one point, y could be a function of x

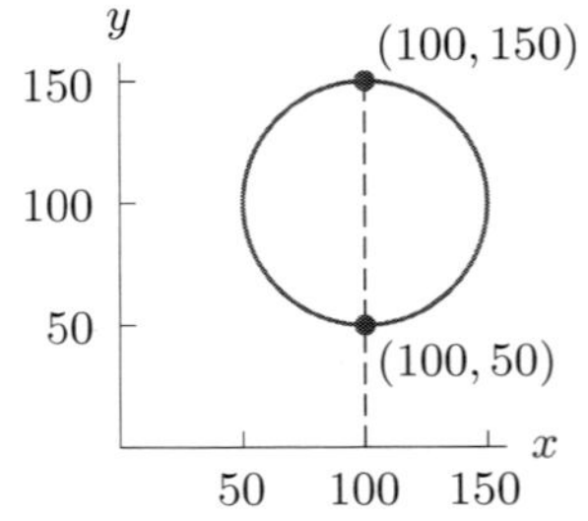

Figure 1.3: Since one vertical line intersects this curve at more than one point, y is not a function of x

Solution The graph in Figure 1.2 could represent y as a function of x because no vertical line intersects this curve in more than one point. The graph in Figure 1.3 does not represent a function because the vertical line shown intersects the curve at two points.

A graph fails the vertical line test if at least one vertical line cuts the graph more than once, as in Figure 1.3. However, if a graph represents a function, then *every* vertical line must intersect the graph at no more than one point.

Exercises and Problems for Section 1.1

Exercises

Exercises 1–4 use Figure 1.4.

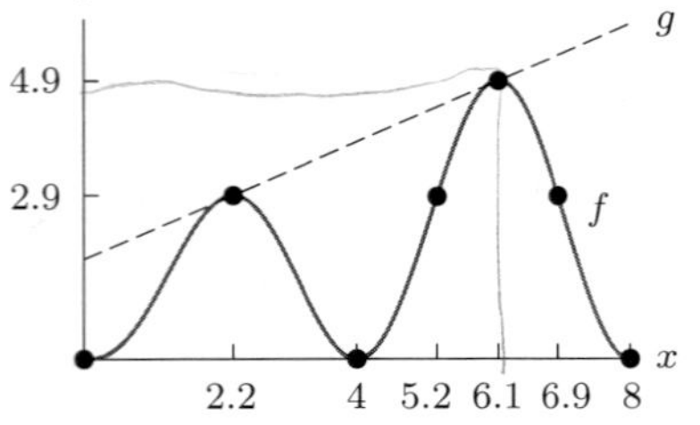

Figure 1.4

1. Find $f(6.9)$

2. Give the coordinates of two points on the graph of g.

3. Solve $f(x) = 0$ for x

4. Solve $f(x) = g(x)$ for x

In Exercises 5–6, write the relationship using function notation (i.e. y is a function of x is written $y = f(x)$).

5. Number of molecules, m, in a gas, is a function of the volume of the gas, v.

6. Weight, w, is a function of caloric intake, c.

7. **(a)** Which of the graphs in Figure 1.5 represent y as a function of x? (Note that an open circle indicates a point that is not included in the graph; a solid dot indicates a point that is included in the graph.)

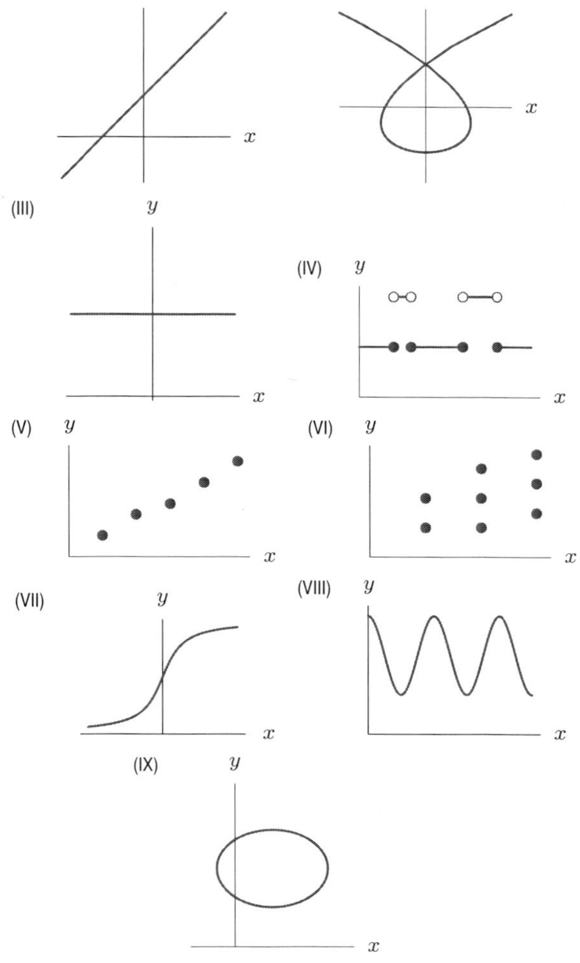

Figure 1.5

(b) Which of the graphs in Figure 1.5 could represent the following situations? Give reasons.

(i) SAT Math score versus SAT Verbal score for a small number of students.

(ii) Total number of daylight hours as a function of the day of the year, shown over a period of several years.

(c) Among graphs (I)–(IX) in Figure 1.5, find two which could give the cost of train fare as a function of the time of day. Explain the relationship between cost and time for both choices.

8. Using Table 1.4, graph $n = f(A)$, the number of gallons of paint needed to cover a house of area A. Identify the independent and dependent variables.

Table 1.4

A	0	250	500	750	1000	1250	1500
n	0	1	2	3	4	5	6

9. Use Figure 1.6 to fill in the missing values:

(a) $f(0) = ?$ **(b)** $f(?) = 0$

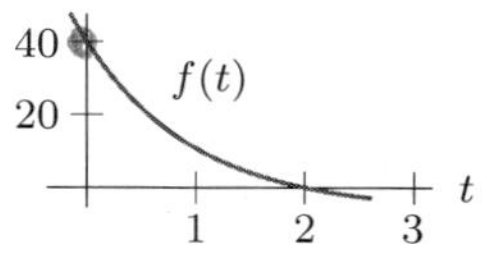

Figure 1.6

10. Use Table 1.5 to fill in the missing values. (There may be more than one answer.)

(a) $f(0) = ?$ **(b)** $f(?) = 0$
(c) $f(1) = ?$ **(d)** $f(?) = 1$

Table 1.5

x	0	1	2	3	4
$f(x)$	4	2	1	0	1

11. **(a)** You are going to graph $p = f(w)$. Which variable goes on the horizontal axis?

(b) If $10 = f(-4)$, give the coordinates of a point on the graph of f.

(c) If 6 is a solution of the equation $f(w) = 1$, give a point on the graph of f.

12. **(a)** Make a table of values for $f(x) = 10/(1 + x^2)$ for $x = 0, 1, 2, 3$.

(b) What x-value gives the largest $f(x)$ value in your table? How could you have predicted this before doing any calculations?

In Exercises 13–16, label the axes for a sketch to illustrate the given statement.

13. "Over the past century we have seen changes in the population, P (in millions), of the city. . ."

14. "Sketch a graph of the cost of manufacturing q items. . ."

15. "Graph the pressure, p, of a gas as a function of its volume, v, where p is in pounds per square inch and v is in cubic inches."

16. "Graph D in terms of y. . ."

Problems

17. **(a)** Ten inches of snow is equivalent to about one inch of rain.[1] Write an equation for the amount of precipitation, measured in inches of rain, $r = f(s)$, as a function of the number of inches of snow, s.
(b) Evaluate and interpret $f(5)$.
(c) Find s such that $f(s) = 5$ and interpret your result.

18. You are looking at the graph of y, a function of x.

(a) What is the maximum number of times that the graph can intersect the y-axis? Explain.
(b) Can the graph intersect the x-axis an infinite number of times? Explain.

19. Let $f(t)$ be the number of people, in millions, who own cell phones t years after 1990. Explain the meaning of the following statements.

(a) $f(10) = 100.3$ **(b)** $f(a) = 20$
(c) $f(20) = b$ **(d)** $n = f(t)$

In Problems 20–22, use Table 1.6, which gives values of $v = r(s)$, the eyewall wind profile of a typical hurricane.[2] The eyewall of a hurricane is the band of clouds that surrounds the eye of the storm. The eyewall wind speed v (in mph) is a function of the height above the ground s (in meters).

Table 1.6

s	0	100	200	300	400	500
v	90	110	116	120	121	122
s	600	700	800	900	1000	1100
v	121	119	118	117	116	115

20. Evaluate and interpret $r(300)$.

21. At what altitudes does the eyewall windspeed appear to equal or exceed 116 mph?

22. At what height is the eyewall wind speed greatest?

23. Table 1.7 shows the daily low temperature for a one-week period in New York City during July.

(a) What was the low temperature on July 19?
(b) When was the low temperature 73°F?
(c) Is the daily low temperature a function of the date?
(d) Is the date a function of the daily low temperature?

Table 1.7

Date	17	18	19	20	21	22	23
Low temp (°F)	73	77	69	73	75	75	70

24. Table 1.8 gives $A = f(d)$, the amount of money in bills of denomination d circulating in US currency in 2005.[3] For example, there were $60.2 billion worth of $50 bills in circulation.

(a) Find $f(100)$. What does this tell you about money?
(b) Are there more $1 bills or $5 bills in circulation?

Table 1.8

Denomination ($)	1	2	5	10	20	50	100
Circulation ($bn)	8.4	1.4	9.7	14.8	110.1	60.2	524.5

25. Use the data from Table 1.3 on page 5.

(a) Plot R on the vertical axis and t on the horizontal axis. Use this graph to explain why you believe that R is a function of t.
(b) Plot F on the vertical axis and t on the horizontal axis. Use this graph to explain why you believe that F is a function of t.
(c) Plot F on the vertical axis and R on the horizontal axis. From this graph show that F is not a function of R.
(d) Plot R on the vertical axis and F on the horizontal axis. From this graph show that R is not a function of F.

26. Since Roger Bannister broke the 4-minute mile on May 6, 1954, the record has been lowered by over sixteen seconds. Table 1.9 shows the year and times (as min:sec) of new world records for the one-mile run.[4]

(a) Is the time a function of the year? Explain.
(b) Is the year a function of the time? Explain.
(c) Let $y(r)$ be the year in which the world record, r, was set. Explain what is meant by the statement $y(3{:}47.33) = 1981$.
(d) Evaluate and interpret $y(3{:}51.1)$.

Table 1.9

Year	Time	Year	Time	Year	Time
1954	3:59.4	1966	3:51.3	1981	3:48.53
1954	3:58.0	1967	3:51.1	1981	3:48.40
1957	3:57.2	1975	3:51.0	1981	3:47.33
1958	3:54.5	1975	3:49.4	1985	3:46.32
1962	3:54.4	1979	3:49.0	1993	3:44.39
1964	3:54.1	1980	3:48.8	1999	3:43.13
1965	3:53.6				

[1] http://mo.water.usgs.gov/outreach/rain, accessed May 7, 2006.
[2] Data from the National Hurricane Center, www.nhc.noaa.gov/aboutwindprofile.shtml, last accessed October 7, 2004.
[3] *The World Almanac and Book of Facts*, 2006 (New York), p. 89.
[4] www.infoplease.com/ipsa/A0112924.html, accessed January 15, 2006.

27. Rebecca Latimer Felton of Georgia was the first woman to serve in the US Senate. She took the oath of office on November 22, 1922 and served for just two days. The first woman actually elected to the Senate was Hattie Wyatt Caraway of Arkansas. She was appointed to fill the vacancy caused by the death of her husband, then won election in 1932, was reelected in 1938, and served until 1945. Table 1.10 shows the number of female senators at the beginning of the first session of each Congress.[5]

(a) Is the number of female senators a function of the Congress's number, c? Explain.
(b) Is the Congress's number a function of the number of female senators? Explain.
(c) Let $S(c)$ represent the number of female senators serving in the cth Congress. What does the statement $S(104) = 8$ mean?
(d) Evaluate and interpret $S(108)$.

Table 1.10

Congress, c	96	98	100	102	104	106	108
Female senators	1	2	2	2	8	9	14

28. A bug starts out ten feet from a light, flies closer to the light, then farther away, then closer than before, then farther away. Finally the bug hits the bulb and flies off. Sketch the distance of the bug from the light as a function of time.

29. A light is turned off for several hours. It is then turned on. After a few hours it is turned off again. Sketch the light bulb's temperature as a function of time.

30. The sales tax on an item is 6%. Express the total cost, C, in terms of the price of the item, P.

31. A cylindrical can is closed at both ends and its height is twice its radius. Express its surface area, S, as a function of its radius, r. [Hint: The surface of a can consists of a rectangle plus two circular disks.]

32. According to Charles Osgood, CBS news commentator, it takes about one minute to read 15 double-spaced typewritten lines on the air.[6]

(a) Construct a table showing the time Charles Osgood is reading on the air in seconds as a function of the number of double-spaced lines read for $0, 1, 2, \ldots, 10$ lines. From your table, how long does it take Charles Osgood to read 9 lines?
(b) Plot this data on a graph with the number of lines on the horizontal axis.
(c) From your graph, estimate how long it takes Charles Osgood to read 9 lines. Estimate how many lines Charles Osgood can read in 30 seconds.
(d) Construct a formula which relates the time T to n, the number of lines read.

33. Match each story about a bike ride to one of the graphs (i)–(v), where d represents distance from home and t is time in hours since the start of the ride. (A graph may be used more than once.)

(a) Starts 5 miles from home and rides 5 miles per hour away from home.
(b) Starts 5 miles from home and rides 10 miles per hour away from home.
(c) Starts 10 miles from home and arrives home one hour later.
(d) Starts 10 miles from home and is halfway home after one hour.
(e) Starts 5 miles from home and is 10 miles from home after one hour.

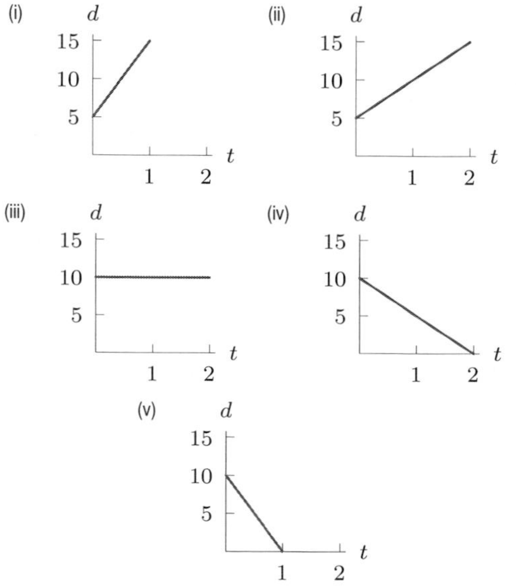

34. A chemical company spends \$2 million to buy machinery before it starts producing chemicals. Then it spends \$0.5 million on raw materials for each million liters of chemical produced.

(a) The number of liters produced ranges from 0 to 5 million. Make a table showing the relationship between the number of million liters produced, l, and the total cost, C, in millions of dollars, to produce that number of million liters.
(b) Find a formula that expresses C as a function of l.

35. The distance between Cambridge and Wellesley is 10 miles. A person walks part of the way at 5 miles per hour, then jogs the rest of the way at 8 mph. Find a formula that expresses the total amount of time for the trip, $T(d)$, as a function of d, the distance walked.

[5] www.senate.gov, accessed January 15, 2006.
[6] T. Parker, *Rules of Thumb*, (Boston: Houghton Mifflin, 1983).

36. A person leaves home and walks due west for a time and then walks due north.

(a) The person walks 10 miles in total. If w represents the (variable) distance west she walks, and D represents her (variable) distance from home at the end of her walk, is D a function of w? Why or why not?

(b) Suppose now that x is the distance that she walks in total. Is D a function of x? Why or why not?

1.2 RATE OF CHANGE

Sales of digital video disc (DVD) players have been increasing since they were introduced in early 1998. To measure how fast sales were increasing, we calculate a *rate of change* of the form

$$\frac{\text{Change in sales}}{\text{Change in time}}.$$

At the same time, sales of video cassette recorders (VCRs) have been decreasing. See Table 1.11.

Let us calculate the rate of change of DVD player and VCR sales between 1998 and 2003. Table 1.11 gives

$$\begin{array}{c}\text{Average rate of change of DVD}\\ \text{player sales from 1998 to 2003}\end{array} = \frac{\text{Change in DVD player sales}}{\text{Change in time}} = \frac{3050 - 421}{2003 - 1998} \approx 525.8 \begin{array}{c}\text{mn \$/}\\ \text{year.}\end{array}$$

Thus, DVD player sales increased on average by \$525.8 million per year between 1998 and 2003. See Figure 1.7. Similarly, Table 1.11 gives

$$\begin{array}{c}\text{Average rate of change of VCR sales}\\ \text{from 1998 to 2003}\end{array} = \frac{\text{Change in VCR sales}}{\text{Change in time}} = \frac{407 - 2409}{2003 - 1998} \approx -400.4 \begin{array}{c}\text{mn \$/}\\ \text{year.}\end{array}$$

Thus, VCR sales decreased on average by \$400.4 million per year between 1998 and 2003. See Figure 1.8.

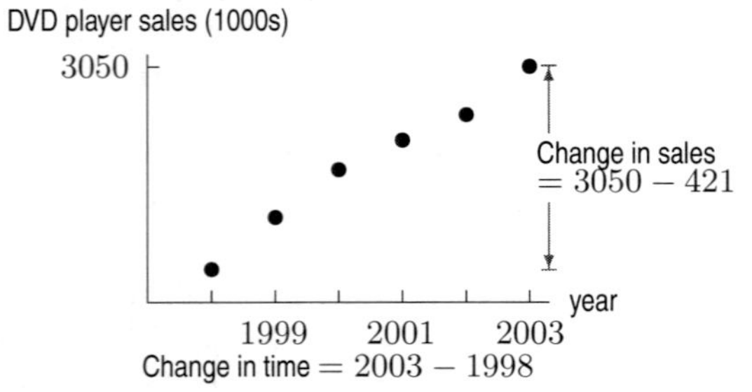

Figure 1.7: DVD player sales

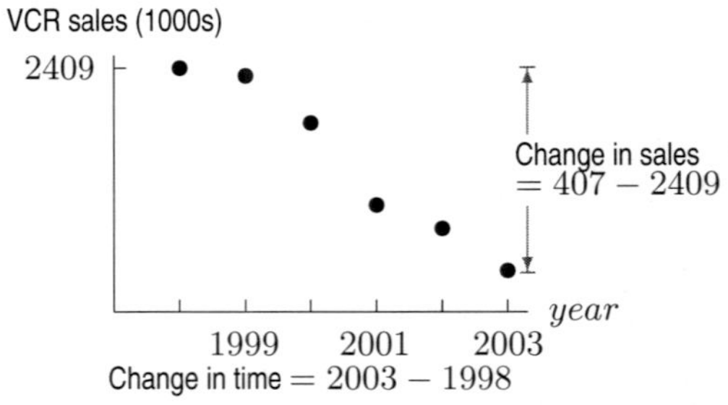

Figure 1.8: VCR sales

Table 1.11 *Annual sales of VCRs and DVD players in millions of dollars*[7]

Year	1998	1999	2000	2001	2002	2003
VCR sales (million \$)	2409	2333	1869	1058	826	407
DVD player sales (million \$)	421	1099	1717	2097	2427	3050

[7] www.census.gov/prod/2005pubs/06statab/manufact.pdf, accessed January 16, 2006.

Rate of Change of a Function

The rate of change of sales is an example of the rate of change of a function. In general, if $Q = f(t)$, we write ΔQ for a change in Q and Δt for a change in t. We define:[8]

> The **average rate of change**, or **rate of change**, of Q with respect to t over an interval is
>
> $$\begin{array}{c}\text{Average rate of change}\\ \text{over an interval}\end{array} = \frac{\text{Change in } Q}{\text{Change in } t} = \frac{\Delta Q}{\Delta t}.$$

The average rate of change of the function $Q = f(t)$ over an interval tells us how much Q changes, on average, for each unit change in t within that interval. On some parts of the interval, Q may be changing rapidly, while on other parts Q may be changing slowly. The average rate of change evens out these variations.

Increasing and Decreasing Functions

In the previous example, the average rate of change of DVD player sales is positive on the interval from 1998 to 2003 since sales of DVD players increased over this interval. Similarly, the average rate of change of VCR sales is negative on the same interval since sales of VCRs decreased over this interval. The annual sales of DVD players is an example of an *increasing function* and the annual sales of VCRs is an example of a *decreasing function*. In general we say the following:

> If $Q = f(t)$ for t in the interval $a \leq t \leq b$,
> - f is an **increasing function** if the values of f increase as t increases in this interval.
> - f is a **decreasing function** if the values of f decrease as t increases in this interval.

Looking at DVD player sales, we see that an increasing function has a positive rate of change. From the VCR sales, we see that a decreasing function has a negative rate of change. In general:

> If $Q = f(t)$,
> - If f is an increasing function, then the average rate of change of Q with respect to t is positive on every interval.
> - If f is a decreasing function, then the average rate of change of Q with respect to t is negative on every interval.

Example 1 The function $A = q(r) = \pi r^2$ gives the area, A, of a circle as a function of its radius, r. Graph q. Explain how the fact that q is an increasing function can be seen on the graph.

[8] The Greek letter Δ, delta, is often used in mathematics to represent change. In this book, we use rate of change to mean average rate of change across an interval. In calculus, rate of change means something called instantaneous rate of change.

Solution The area increases as the radius increases, so $A = q(r)$ is an increasing function. We can see this in Figure 1.9 because the graph climbs as we move from left to right and the average rate of change, $\Delta A/\Delta r$, is positive on every interval.

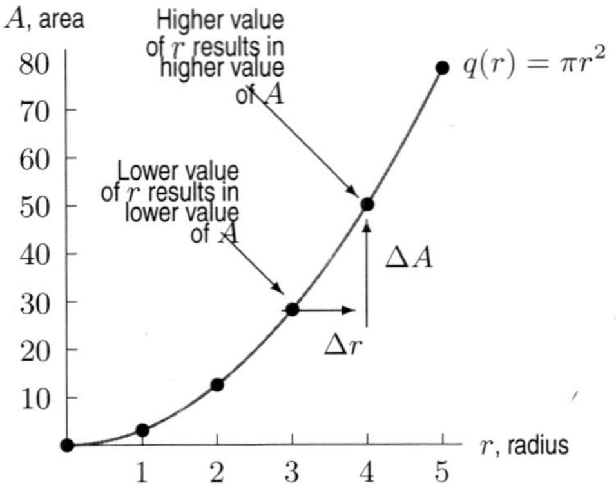

Figure 1.9: The graph of an increasing function, $A = q(r)$, rises when read from left to right

Example 2 Carbon-14 is a radioactive element that exists naturally in the atmosphere and is absorbed by living organisms. When an organism dies, the carbon-14 present at death begins to decay. Let $L = g(t)$ represent the quantity of carbon-14 (in micrograms, μg) in a tree t years after its death. See Table 1.12. Explain why we expect g to be a decreasing function of t. How is this represented on a graph?

Table 1.12 *Quantity of carbon-14 as a function of time*

t, time (years)	0	1000	2000	3000	4000	5000
L, quantity of carbon-14 (μg)	200	177	157	139	123	109

Solution Since the amount of carbon-14 is decaying over time, g is a decreasing function. In Figure 1.10, the graph falls as we move from left to right and the average rate of change in the level of carbon-14 with respect to time, $\Delta L/\Delta t$, is negative on every interval.

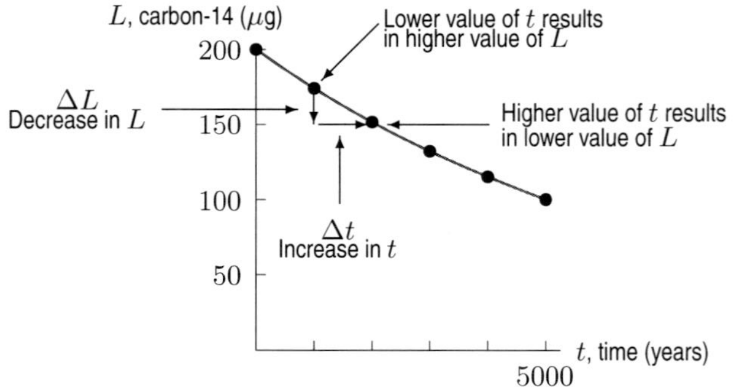

Figure 1.10: The graph of a decreasing function, $L = g(t)$, falls when read from left to right

In general, we can identify an increasing or decreasing function from its graph as follows:

- The graph of an increasing function rises when read from left to right.
- The graph of a decreasing function falls when read from left to right.

Many functions have some intervals on which they are increasing and other intervals on which they are decreasing. These intervals can often be identified from the graph.

Example 3 On what intervals is the function graphed in Figure 1.11 increasing? Decreasing?

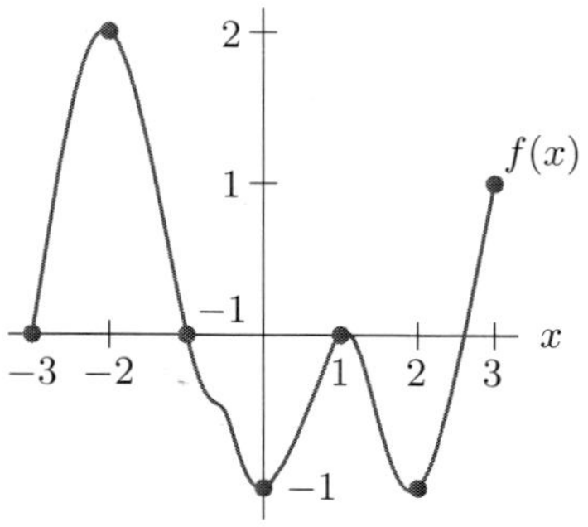

Figure 1.11: Graph of a function which is increasing on some intervals and decreasing on others

Solution The function appears to be increasing for values of x between -3 and -2, for x between 0 and 1, and for x between 2 and 3. The function appears to be decreasing for x between -2 and 0 and for x between 1 and 2. Using inequalities, we say that f is increasing for $-3 < x < -2$, for $0 < x < 1$, and for $2 < x < 3$. Similarly, f is decreasing for $-2 < x < 0$ and $1 < x < 2$.

Function Notation for the Average Rate of Change

Suppose we want to find the average rate of change of a function $Q = f(t)$ over the interval $a \le t \le b$. On this interval, the change in t is given by

$$\Delta t = b - a.$$

At $t = a$, the value of Q is $f(a)$, and at $t = b$, the value of Q is $f(b)$. Therefore, the change in Q is given by

$$\Delta Q = f(b) - f(a).$$

Using function notation, we express the average rate of change as follows:

$$\begin{array}{c}\text{Average rate of change of } Q = f(t) \\ \text{over the interval } a \le t \le b\end{array} = \frac{\text{Change in } Q}{\text{Change in } t} = \frac{\Delta Q}{\Delta t} = \frac{f(b) - f(a)}{b - a}.$$

In Figure 1.12, notice that the average rate of change is given by the ratio of the rise, $f(b)-f(a)$, to the run, $b - a$. This ratio is also called the *slope* of the dashed line segment.[9]

[9]See Section 1.3 for further discussion of slope.

In the future, we may drop the word "average" and talk about the rate of change over an interval.

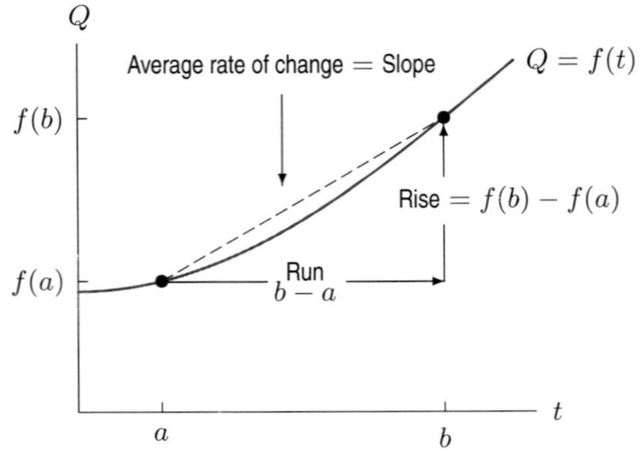

Figure 1.12: The average rate of change is the ratio Rise/Run

In previous examples we calculated the average rate of change from data. We now calculate average rates of change for functions given by formulas.

Example 4 Calculate the average rates of change of the function $f(x) = x^2$ between $x = 1$ and $x = 3$ and between $x = -2$ and $x = 1$. Show your results on a graph.

Solution Between $x = 1$ and $x = 3$, we have

$$\begin{aligned}\text{Average rate of change of } f(x) \text{ over the interval } 1 \le x \le 3 &= \frac{\text{Change in } f(x)}{\text{Change in } x} = \frac{f(3) - f(1)}{3 - 1} \\ &= \frac{3^2 - 1^2}{3 - 1} = \frac{9 - 1}{2} = 4.\end{aligned}$$

Between $x = -2$ and $x = 1$, we have

$$\begin{aligned}\text{Average rate of change of } f(x) \text{ over the interval } -2 \le x \le 1 &= \frac{\text{Change in } f(x)}{\text{Change in } x} = \frac{f(1) - f(-2)}{1 - (-2)} \\ &= \frac{1^2 - (-2)^2}{1 - (-2)} = \frac{1 - 4}{3} = -1.\end{aligned}$$

The average rate of change between $x = 1$ and $x = 3$ is positive because $f(x)$ is increasing on this interval. See Figure 1.13. However, on the interval from $x = -2$ and $x = 1$, the function is partly decreasing and partly increasing. The average rate of change on this interval is negative because the decrease on the interval is larger than the increase.

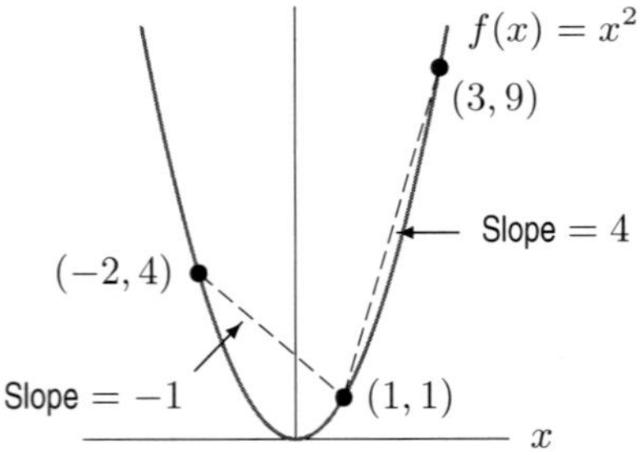

Figure 1.13: Average rate of change of $f(x)$ on an interval is slope of dashed line on that interval

---

Exercises and Problems for Section 1.2

Exercises

1. If G is an increasing function, what can you say about $G(3) - G(-1)$?

2. If F is a decreasing function, what can you say about $F(-2)$ compared to $F(2)$?

Exercises 3–7 use Figure 1.14.

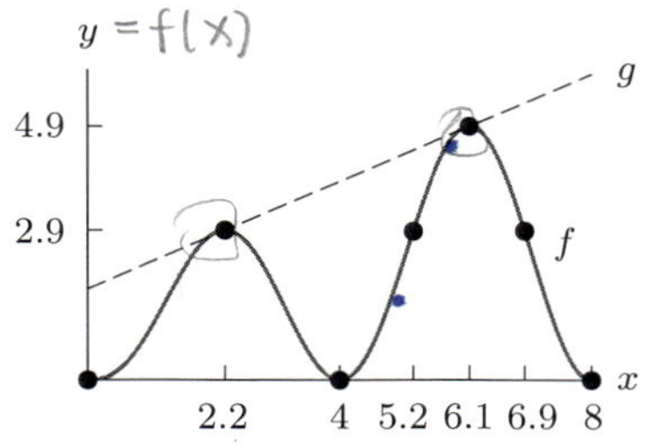

Figure 1.14

3. Find the average rate of change of f for $2.2 \leq x \leq 6.1$.

4. Give two different intervals on which $\Delta f(x)/\Delta x = 0$.

5. What is the average rate of change of g between $x = 2.2$ and $x = 6.1$?

6. What is the relation between the average rate of change of f and the average rate of change of g between $x = 2.2$ and $x = 6.1$?

7. Is the rate of change of f positive or negative on the following intervals?

(a) $2.2 \leq x \leq 4$ **(b)** $5 \leq x \leq 6$

8. Table 1.11 on page 10 gives the annual sales (in millions) of compact discs and vinyl long playing records. What was the average rate of change of annual sales of each of them between

(a) 1982 and 1984? **(b)** 1986 and 1988?

(c) Interpret these results in terms of sales.

9. Table 1.11 on page 10 shows that CD sales are a function of LP sales. Is it an increasing or decreasing function?

10. Figure 1.15 shows distance traveled as a function of time.

(a) Find ΔD and Δt between:

(i) $t = 2$ and $t = 5$ (ii) $t = 0.5$ and $t = 2.5$
(iii) $t = 1.5$ and $t = 3$

(b) Compute the rate of change, $\Delta D/\Delta t$, and interpret its meaning.

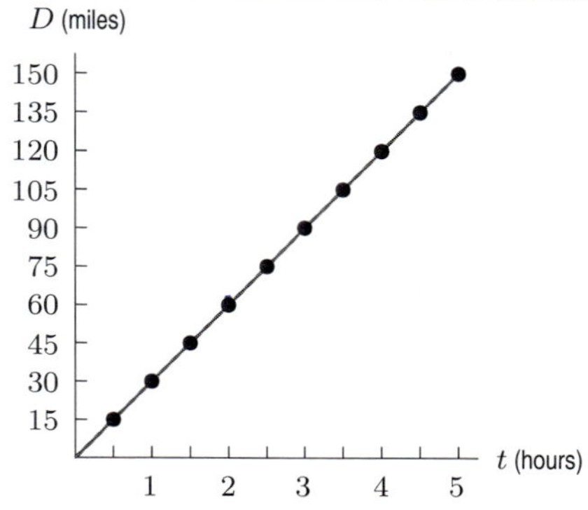

Figure 1.15

11. Table 1.13 shows data for two populations (in hundreds) for five different years. Find the average rate of change of each population over the following intervals.

(a) 1990 to 2000 **(b)** 1995 to 2007
(c) 1990 to 2007

Table 1.13

Year	1990	1992	1995	2000	2007
P_1	53	63	73	83	93
P_2	85	80	75	70	65

12. Table 1.14 gives the populations of two cities (in thousands) over a 17-year period.

(a) Find the average rate of change of each population on the following intervals:

(i) 1990 to 2000 (ii) 1990 to 2007
(iii) 1995 to 2007

(b) What do you notice about the average rate of change of each population? Explain what the average rate of change tells you about each population.

Table 1.14

Year	1990	1992	1995	2000	2007
P_1	42	46	52	62	76
P_2	82	80	77	72	65

Problems

13. Because scientists know how much carbon-14 a living organism should have in its tissues, they can measure the amount of carbon-14 present in the tissue of a fossil and then calculate how long it took for the original amount to decay to the current level, thus determining the time of the organism's death. A tree fossil is found to contain 130 μg of carbon-14, and scientists determine from the size of the tree that it would have contained 200 μg of carbon-14 at the time of its death. Using Table 1.12 on page 12, approximately how long ago did the tree die?

14. Table 1.15 shows the number of calories used per minute as a function of body weight for three sports.[10]

(a) Determine the number of calories that a 200-lb person uses in one half-hour of walking.
(b) Who uses more calories, a 120-lb person swimming for one hour or a 220-lb person bicycling for a half-hour?
(c) Does the number of calories used by a person walking increase or decrease as weight increases?

Table 1.15

Activity	100 lb	120 lb	150 lb	170 lb	200 lb	220 lb
Walking	2.7	3.2	4.0	4.6	5.4	5.9
Bicycling	5.4	6.5	8.1	9.2	10.8	11.9
Swimming	5.8	6.9	8.7	9.8	11.6	12.7

15. **(a)** What is the average rate of change of $g(x) = 2x - 3$ between the points $(-2, -7)$ and $(3, 3)$?
(b) Based on your answer to part (a), is g increasing or decreasing on the given interval? Explain.
(c) Graph the function and determine over what intervals g is increasing and over what intervals g is decreasing.

16. **(a)** Let $f(x) = 16 - x^2$. Compute each of the following expressions, and interpret each as an average rate of change.

(i) $\dfrac{f(2) - f(0)}{2 - 0}$ (ii) $\dfrac{f(4) - f(2)}{4 - 2}$

(iii) $\dfrac{f(4) - f(0)}{4 - 0}$

(b) Graph $f(x)$. Illustrate each ratio in part (a) by sketching the line segment with the given slope. Over which interval is the average rate of decrease the greatest?

[10] From *1993 World Almanac*.

17. Figure 1.16 shows the graph of the function $g(x)$.

(a) Estimate $\dfrac{g(4) - g(0)}{4 - 0}$.
(b) The ratio in part (a) is the slope of a line segment joining two points on the graph. Sketch this line segment on the graph.
(c) Estimate $\dfrac{g(b) - g(a)}{b - a}$ for $a = -9$ and $b = -1$.
(d) On the graph, sketch the line segment whose slope is given by the ratio in part (c).

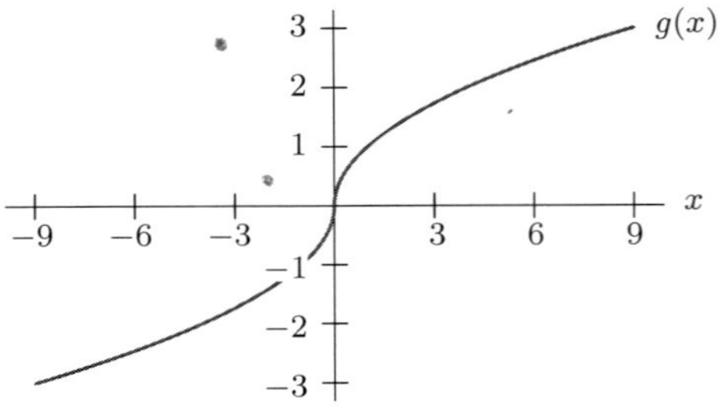

Figure 1.16

For the functions in Problems 18–20:

(a) Find the average rate of change between the points
(i) $(-1, f(-1))$ and $(3, f(3))$
(ii) $(a, f(a))$ and $(b, f(b))$
(iii) $(x, f(x))$ and $(x + h, f(x + h))$
(b) What pattern do you see in the average rate of change between the three pairs of points?

18. $f(x) = 5x - 4$

19. $f(x) = \frac{1}{2}x + \frac{5}{2}$

20. $f(x) = x^2 + 1$

21. Find the average rate of change of $f(x) = 3x^2 + 1$ between the points

(a) $(1, 4)$ and $(2, 13)$
(b) (j, k) and (m, n)
(c) $(x, f(x))$ and $(x + h, f(x + h))$

22. The surface of the sun has dark areas known as sunspots, which are cooler than the rest of the sun's surface. The number of sunspots fluctuates with time, as shown in Figure 1.17.

(a) Explain how you know the number of sunspots, s, in year t is a function of t.

(b) Approximate the time intervals on which s is an increasing function of t.

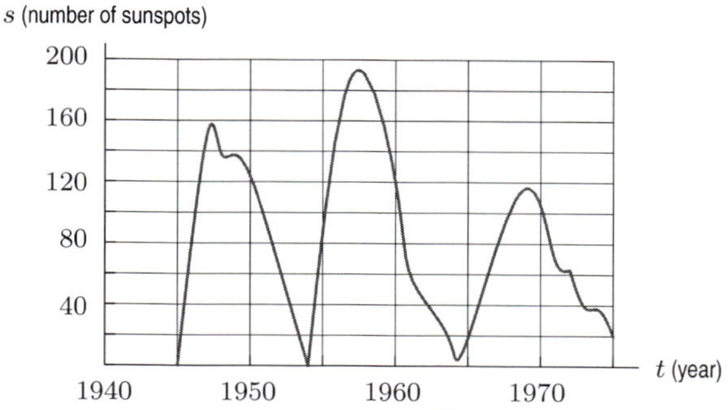

Figure 1.17

23. Table 1.16 gives the amount of garbage, G, in millions of tons, produced[11] in the US in year t.

(a) What is the value of Δt for consecutive entries in this table?

(b) Calculate the value of ΔG for each pair of consecutive entries in this table.

(c) Are all the values of ΔG you found in part (b) the same? What does this tell you?

Table 1.16

t	1960	1970	1980	1990	2000
G	90	120	150	205	234

24. Table 1.17 shows the times, t, in sec, achieved every 10 meters by Carl Lewis in the 100 meter final of the World Championship in Rome in 1987.[12] Distance, d, is in meters.

(a) For each successive time interval, calculate the average rate of change of distance. What is a common name for the average rate of change of distance?

(b) Where did Carl Lewis attain his maximum speed during this race? Some runners are running their fastest as they cross the finish line. Does that seem to be true in this case?

Table 1.17

t	0.00	1.94	2.96	3.91	4.78	5.64
d	0	10	20	30	40	50
t	6.50	7.36	8.22	9.07	9.93	
d	60	70	80	90	100	

25. Table 1.18 gives the average temperature, T, at a depth d, in a borehole in Belleterre, Quebec.[13] Evaluate $\Delta T/\Delta d$ on the the following intervals, and explain what your answers tell you about borehole temperature.

(a) $25 \leq d \leq 150$
(b) $25 \leq d \leq 75$
(c) $100 \leq d \leq 200$

Table 1.18

d, depth (m)	25	50	75	100
T, temp (°C)	5.50	5.20	5.10	5.10
d, depth (m)	125	150	175	200
T, temp (°C)	5.30	5.50	5.75	6.00
d, depth (m)	225	250	275	300
T, temp (°C)	6.25	6.50	6.75	7.00

1.3 LINEAR FUNCTIONS

Constant Rate of Change

In the previous section, we introduced the average rate of change of a function on an interval. For many functions, the average rate of change is different on different intervals. For the remainder of this chapter, we consider functions which have the same average rate of change on every interval. Such a function has a graph which is a line and is called *linear*.

[11] www.epa.gov/epaoswer/hon-hw/muncpl/pubs/MSW05rpt.pdf, accessed January 15, 2006.
[12] W. G. Pritchard, "Mathematical Models of Running", *SIAM Review.* 35, 1993, pp. 359–379.
[13] Hugo Beltrami of St. Francis Xavier University and David Chapman of the University of Utah posted this data at http://geophysics.stfx.ca/public/borehole/borehole.html http://geophysics.stfx.ca/public/borehole/borehole.html.

Population Growth

Mathematical models of population growth are used by city planners to project the growth of towns and states. Biologists model the growth of animal populations and physicians model the spread of an infection in the bloodstream. One possible model, a linear model, assumes that the population changes at the same average rate on every time interval.

Example 1 A town of 30,000 people grows by 2000 people every year. Since the population, P, is growing at the constant rate of 2000 people per year, P is a linear function of time, t, in years.

(a) What is the average rate of change of P over every time interval?
(b) Make a table that gives the town's population every five years over a 20-year period. Graph the population.
(c) Find a formula for P as a function of t.

Solution

(a) The average rate of change of population with respect to time is 2000 people per year.
(b) The initial population in year $t = 0$ is $P = 30{,}000$ people. Since the town grows by 2000 people every year, after five years it has grown by

$$\frac{2000 \text{ people}}{\text{year}} \cdot 5 \text{ years} = 10{,}000 \text{ people}.$$

Thus, in year $t = 5$ the population is given by

$$P = \text{Initial population} + \text{New population} = 30{,}000 + 10{,}000 = 40{,}000.$$

In year $t = 10$ the population is given by

$$P = 30{,}000 + \underbrace{2000 \text{ people/year} \cdot 10 \text{ years}}_{20{,}000 \text{ new people}} = 50{,}000.$$

Similar calculations for year $t = 15$ and year $t = 20$ give the values in Table 1.19. See Figure 1.18; the dashed line shows the trend in the data.

Table 1.19 *Population over 20 years*

t, years	P, population
0	30,000
5	40,000
10	50,000
15	60,000
20	70,000

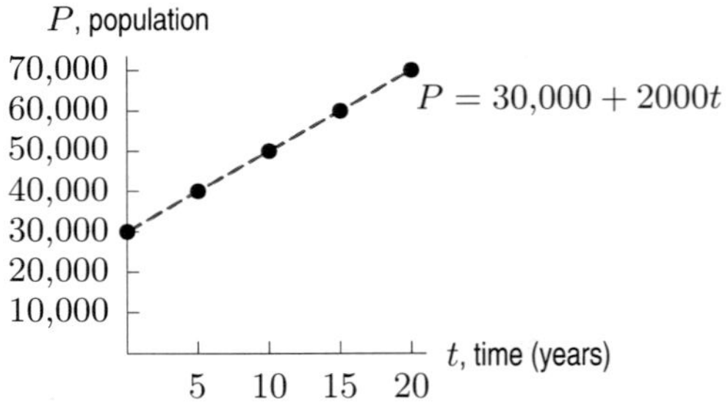

Figure 1.18: Town's population over 20 years

(c) From part (b), we see that the size of the population is given by

$$\begin{aligned} P &= \text{Initial population} + \text{Number of new people} \\ &= 30{,}000 + 2000 \text{ people/year} \cdot \text{Number of years}, \end{aligned}$$

so a formula for P in terms of t is

$$P = 30{,}000 + 2000t.$$

The graph of the population data in Figure 1.18 is a straight line. The average rate of change of the population over every interval is the same, namely 2000 people per year. Any linear function has the same average rate of change over every interval. Thus, we talk about *the* rate of change of a linear function. In general:

- A **linear function** has a constant rate of change.
- The graph of any linear function is a straight line.

Financial Models

Economists and accountants use linear functions for *straight-line depreciation.* For tax purposes, the value of certain equipment is considered to decrease, or depreciate, over time. For example, computer equipment may be state-of-the-art today, but after several years it is outdated. Straight-line depreciation assumes that the rate of change of value with respect to time is constant.

Example 2 A small business spends $\$20{,}000$ on new computer equipment and, for tax purposes, chooses to depreciate it to $\$0$ at a constant rate over a five-year period.

(a) Make a table and a graph showing the value of the equipment over the five-year period.
(b) Give a formula for value as a function of time.

Solution (a) After five years, the equipment is valued at $\$0$. If V is the value in dollars and t is the number of years, we see that

$$\begin{array}{c}\text{Rate of change of value} \\ \text{from } t=0 \text{ to } t=5\end{array} = \frac{\text{Change in value}}{\text{Change in time}} = \frac{\Delta V}{\Delta t} = \frac{-\$20{,}000}{5 \text{ years}} = -\$4000 \text{ per year.}$$

Thus, the value drops at the constant rate of $\$4000$ per year. (Notice that ΔV is negative because the value of the equipment decreases.) See Table 1.20 and Figure 1.19. Since V changes at a constant rate, $V = f(t)$ is a linear function and its graph is a straight line. The rate of change, $-\$4000$ per year, is negative because the function is decreasing and the graph slopes down.

Table 1.20 *Value of equipment depreciated over a 5-year period*

t, year	V, value ($)
0	20,000
1	16,000
2	12,000
3	8,000
4	4,000
5	0

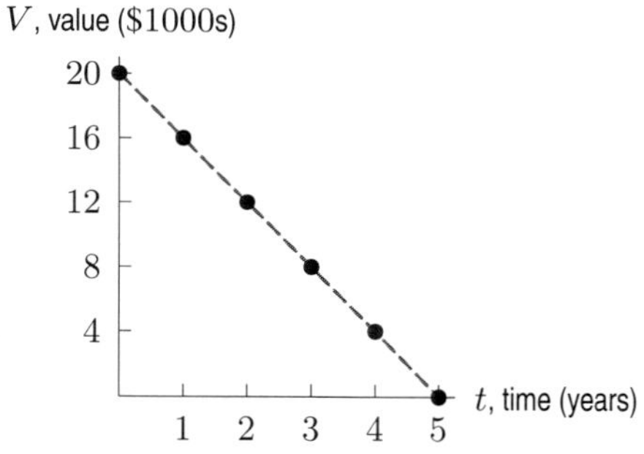

Figure 1.19: Value of equipment depreciated over a 5-year period

(b) After t years have elapsed,

$$\text{Decrease in value of equipment} = \$4000 \cdot \text{Number of years} = \$4000t.$$

The initial value of the equipment is $\$20{,}000$, so at time t,

$$V = 20{,}000 - 4000t.$$

The total cost of production is another application of linear functions in economics.

A General Formula for the Family of Linear Functions

Example 1 involved a town whose population is growing at a constant rate with formula

$$\text{Current population} = \underbrace{\text{Initial population}}_{30{,}000 \text{ people}} + \underbrace{\text{Growth rate}}_{2000 \text{ people per year}} \times \underbrace{\text{Number of years}}_{t}$$

so

$$P = 30{,}000 + 2000t.$$

In Example 2, the value, V, as a function of t is given by

$$\text{Total cost} = \underbrace{\text{Initial value}}_{\$20{,}000} + \underbrace{\text{Change per year}}_{-\$4000 \text{ per year}} \times \underbrace{\text{Number of years}}_{t}$$

so

$$V = 20{,}000 + (-4000)t.$$

Using the symbols x, y, b, m, we see formulas for both of these linear functions follow the same pattern:

$$\underbrace{\text{Output}}_{y} = \underbrace{\text{Initial value}}_{b} + \underbrace{\text{Rate of change}}_{m} \times \underbrace{\text{Input}}_{x}.$$

Summarizing, we get the following results:

If $y = f(x)$ is a linear function, then for some constants b and m:

$$y = b + mx.$$

- m is called the **slope**, and gives the rate of change of y with respect to x. Thus,

$$m = \frac{\Delta y}{\Delta x}.$$

If (x_0, y_0) and (x_1, y_1) are any two distinct points on the graph of f, then

$$m = \frac{\Delta y}{\Delta x} = \frac{y_1 - y_0}{x_1 - x_0}.$$

- b is called the **vertical intercept**, or **y-intercept**, and gives the value of y for $x = 0$. In mathematical models, b typically represents an initial, or starting, value of the output.

Every linear function can be written in the form $y = b + mx$. Different linear functions have different values for m and b. These constants are known as *parameters*.

Example 3 In Example 1, the population function, $P = 30{,}000 + 2000t$, has slope $m = 2000$ and vertical intercept $b = 30{,}000$. In Example 2, the value of the computer equipment, $V = 20{,}000 - 4000t$, has slope $m = -4000$ and vertical intercept $b = 20{,}000$.

Tables for Linear Functions

A table of values could represent a linear function if the rate of change is constant, for all pairs of points in the table; that is,

$$\text{Rate of change of linear function} = \frac{\text{Change in output}}{\text{Change in input}} = \text{Constant.}$$

Thus, if the value of x goes up by equal steps in a table for a linear function, then the value of y goes up (or down) by equal steps as well. We say that changes in the value of y are *proportional* to changes in the value of x.

Example 4 Table 1.21 gives values of two functions, p and q. Could either of these functions be linear?

Table 1.21 *Values of two functions p and q*

x	50	55	60	65	70
$p(x)$	0.10	0.11	0.12	0.13	0.14
$q(x)$	0.01	0.03	0.06	0.14	0.15

Solution The value of x goes up by equal steps of $\Delta x = 5$. The value of $p(x)$ also goes up by equal steps of $\Delta p = 0.01$, so $\Delta p/\Delta x$ is a constant. See Table 1.22. Thus, p could be a linear function.

Table 1.22 *Values of* $\Delta p/\Delta x$

x	$p(x)$	Δp	$\Delta p/\Delta x$
50	0.10		
		0.01	0.002
55	0.11		
		0.01	0.002
60	0.12		
		0.01	0.002
65	0.13		
		0.01	0.002
70	0.14		

Table 1.23 *Values of* $\Delta q/\Delta x$

x	$q(x)$	Δq	$\Delta q/\Delta x$
50	0.01		
		0.02	0.004
55	0.03		
		0.03	0.006
60	0.06		
		0.08	0.016
65	0.14		
		0.01	0.002
70	0.15		

In contrast, the value of $q(x)$ does not go up by equal steps. The value climbs by 0.02, then by 0.03, and so on. See Table 1.23. This means that $\Delta q/\Delta x$ is not constant. Thus, q could not be a linear function.

It is possible to have data from a linear function in which neither the x-values nor the y-values go up by equal steps. However the rate of change must be constant, as in the following example.

Example 5 The former Republic of Yugoslavia exported cars called Yugos to the US between 1985 and 1989. The car is now a collector's item.[14] Table 1.24 gives the quantity of Yugos sold, Q, and the price, p, for each year from 1985 to 1988.

(a) Using Table 1.24, explain why Q could be a linear function of p.
(b) What does the rate of change of this function tell you about Yugos?

Table 1.24 *Price and sales of Yugos in the US*

Year	Price in \$, p	Number sold, Q
1985	3990	49,000
1986	4110	43,000
1987	4200	38,500
1988	4330	32,000

Solution (a) We are interested in Q as a function of p, so we plot Q on the vertical axis and p on the horizontal axis. The data points in Figure 1.20 appear to lie on a straight line, suggesting a linear function.

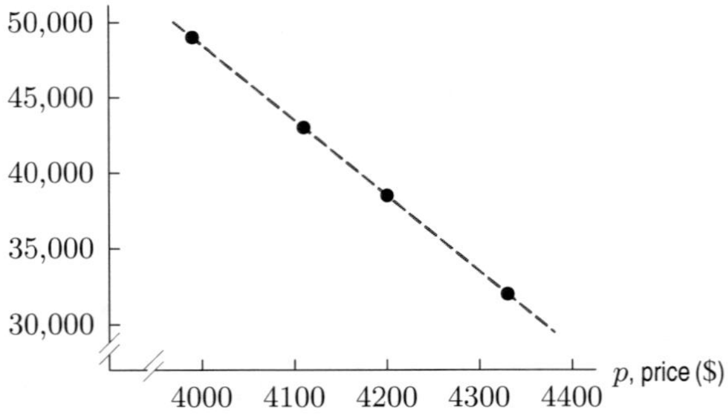

Figure 1.20: Since the data from Table 1.24 falls on a straight line, the table could represent a linear function

To provide further evidence that Q is a linear function, we check that the rate of change of Q with respect to p is constant for the points given. When the price of a Yugo rose from \$3990 to \$4110, sales fell from 49,000 to 43,000. Thus,

$$\Delta p = 4110 - 3990 = 120,$$

$$\Delta Q = 43{,}000 - 49{,}000 = -6000.$$

Since the number of Yugos sold decreased, ΔQ is negative. Thus, as the price increased from \$3990 to \$4110,

$$\text{Rate of change of quantity as price increases} = \frac{\Delta Q}{\Delta p} = \frac{-6000}{120} = -50 \text{ cars per dollar.}$$

Next, we calculate the rate of change as the price increased from \$4110 to \$4200 to see if the rate remains constant:

$$\text{Rate of change} = \frac{\Delta Q}{\Delta p} = \frac{38{,}500 - 43{,}000}{4200 - 4110} = \frac{-4500}{90} = -50 \text{ cars per dollar,}$$

[14]www.inet.hr/~pauric/epov.htm, accessed January 16, 2006.

and as the price increased from \$4200 to \$4330:

$$\text{Rate of change} = \frac{\Delta Q}{\Delta p} = \frac{32{,}000 - 38{,}500}{4330 - 4200} = \frac{-6500}{130} = -50 \text{ cars per dollar.}$$

Since the rate of change, -50, is constant, Q could be a linear function of p. Given additional data, $\Delta Q/\Delta p$ might not remain constant. However, based on the table, it appears that the function is linear.

(b) Since ΔQ is the change in the number of cars sold and Δp is the change in price, the rate of change is -50 cars per dollar. Thus the number of Yugos sold decreased by 50 each time the price increased by \$1.

Warning: Not All Graphs That Look Like Lines Represent Linear Functions

The graph of any linear function is a line. However, a function's graph can look like a line without actually being one. Consider the following example.

Example 6 The function $P = 100(1.02)^t$ approximates the population of Mexico in the early 2000s. Here P is the population (in millions) and t is the number of years since 2000. Table 1.25 and Figure 1.21 show values of P over a 5-year period. Is P a linear function of t?

Table 1.25 *Population of Mexico t years after 2000*

t (years)	P (millions)
0	100
1	102
2	104.04
3	106.12
4	108.24
5	110.41

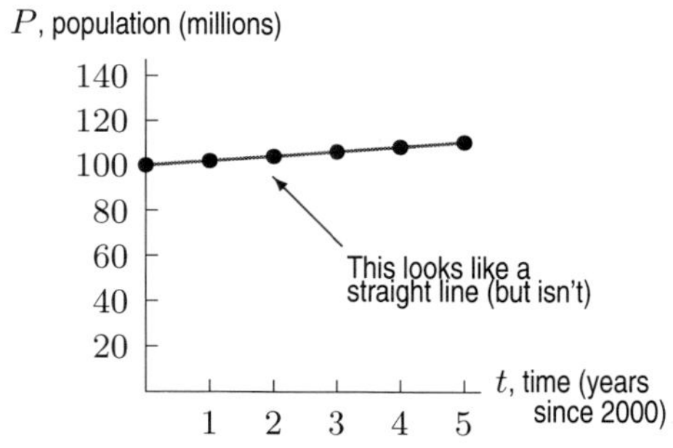

Figure 1.21: Graph of $P = 100(1.02)^t$ over 5-year period: Looks linear (but is not)

Solution The formula $P = 100(1.02)^t$ is not of the form $P = b + mt$, so P is not a linear function of t. However, the graph of P in Figure 1.21 appears to be a straight line. We check P's rate of change in Table 1.25. When $t = 0$, $P = 100$ and when $t = 1$, $P = 102$. Thus, between 2000 and 2001,

$$\text{Rate of change of population} = \frac{\Delta P}{\Delta t} = \frac{102 - 100}{1 - 0} = 2.$$

For the interval from 2001 to 2002, we have

$$\text{Rate of change} = \frac{\Delta P}{\Delta t} = \frac{104.04 - 102}{2 - 1} = 2.04,$$

and for the interval from 2004 to 2005, we have

$$\text{Rate of change} = \frac{\Delta P}{\Delta t} = \frac{110.41 - 108.24}{5 - 4} = 2.17.$$

Thus, P's rate of change is not constant. In fact, P appears to be increasing at a faster and faster rate. Table 1.26 and Figure 1.22 show values of P over a longer (60-year) period. On this scale, these points do not appear to fall on a straight line. However, the graph of P curves upward so gradually at first that over the short interval shown in Figure 1.21, it barely curves at all. The graphs of many nonlinear functions, when viewed on a small scale, appear to be linear.

Table 1.26 *Population over 60 years*

t (years since 2000)	P (millions)
0	100
10	121.90
20	148.59
30	181.14
40	220.80
50	269.16
60	328.10

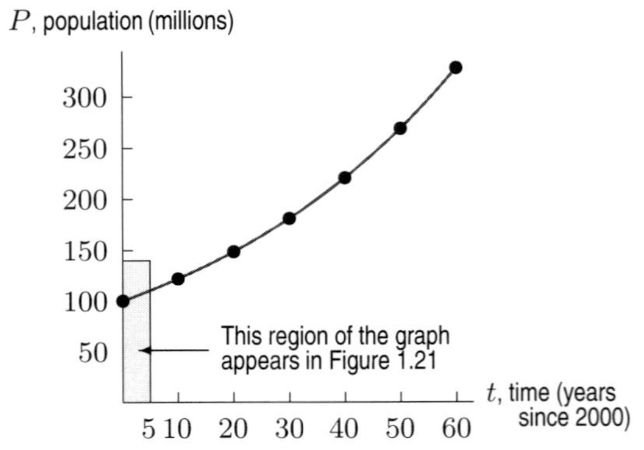

Figure 1.22: Graph of $P = 100(1.02)^t$ over 60 years: Not linear

Exercises and Problems for Section 1.3

Exercises

Which of the tables in Exercises 1–6 could represent a linear function?

1.

x	0	100	300	600
$g(x)$	50	100	150	200

2.

x	0	10	20	30
$h(x)$	20	40	50	55

3.

t	1	2	3	4	5
$g(t)$	5	4	5	4	5

4.

x	0	5	10	15
$f(x)$	10	20	30	40

5.

γ	9	8	7	6	5
$p(\gamma)$	42	52	62	72	82

6.

x	−3	−1	0	3
$j(x)$	5	1	−1	−7

In Exercises 7–10, identify the vertical intercept and the slope, and explain their meanings in practical terms.

7. The population of a town can be represented by the formula $P(t) = 54.25 - \frac{2}{7}t$, where $P(t)$ represents the population, in thousands, and t represents the time, in years, since 1970.

8. A stalactite grows according to the formula $L(t) = 17.75 + \frac{1}{250}t$, where $L(t)$ represents the length of the stalactite, in inches, and t represents the time, in years, since the stalactite was first measured.

9. The profit, in dollars, of selling n items is given by $P(n) = 0.98n - 3000$.

10. A phone company charges according to the formula: $C(n) = 29.99 + 0.05n$, where n is the number of minutes, and $C(n)$ is the monthly phone charge, in dollars.

Problems

11. In 2006, the population of a town was 18,310 and growing by 58 people per year. Find a formula for P, the town's population, in terms of t, the number of years since 2006.

12. The population, $P(t)$, in millions, of a country in year t, is given by the formula $P(t) = 22 + 0.3t$.

(a) Construct a table of values for $t = 0, 10, 20, \ldots, 50$.
(b) Plot the points you found in part (a).
(c) What is the country's initial population?
(d) What is the average rate of change of the population, in millions of people/year?

13. In 2003, the number, N, of cases of SARS (Severe Acute Respiratory Syndrome) reported in Hong Kong[15] was initially approximated by $N = 78.9 + 30.1t$, where t is the number of days since March 17. Interpret the constants 78.9 and 30.1.

14. Table 1.27 shows the cost C, in dollars, of selling x cups of coffee per day from a cart.

(a) Using the table, show that the relationship appears to be linear.
(b) Plot the data in the table.
(c) Find the slope of the line. Explain what this means in the context of the given situation.
(d) Why should it cost $50 to serve zero cups of coffee?

Table 1.27

x	0	5	10	50	100	200
C	50.00	51.25	52.50	62.50	75.00	100.00

15. A woodworker sells rocking horses. His start-up costs, including tools, plans, and advertising, total $5000. Labor and materials for each horse cost $350.

(a) Calculate the woodworker's total cost, C, to make 1, 2, 5, 10, and 20 rocking horses. Graph C against n, the number of rocking horses that he carves.
(b) Find a formula for C in terms of n.
(c) What is the rate of change of the function C? What does the rate of change tell us about the woodworker's expenses?

16. Table 1.28 gives the area and perimeter of a square as a function of the length of its side.

(a) From the table, decide if either area or perimeter could be a linear function of side length.
(b) From the data make two graphs, one showing area as a function of side length, the other showing perimeter as a function of side length. Connect the points.
(c) If you find a linear relationship, give its corresponding rate of change and interpret its significance.

Table 1.28

Length of side	0	1	2	3	4	5	6
Area of square	0	1	4	9	16	25	36
Perimeter of square	0	4	8	12	16	20	24

17. Make two tables, one comparing the radius of a circle to its area, the other comparing the radius of a circle to its circumference. Repeat parts (a), (b), and (c) from Problem 16, this time comparing radius with circumference, and radius with area.

18. Sri Lanka is an island which experienced approximately linear population growth from 1950 to 2000. On the other hand, Afghanistan was torn by warfare in the 1980s and did not experience linear nor near-linear growth.[16]

(a) Table 1.29 gives the population of these two countries, in millions. Which of these two countries is A and which is B? Explain.
(b) What is the approximate rate of change of the linear function? What does the rate of change represent in practical terms?
(c) Estimate the population of Sri Lanka in 1988.

Table 1.29

Year	1950	1960	1970	1980	1990	2000
Population of country A	8.2	9.8	12.4	15.1	14.7	23.9
Population of country B	7.5	9.9	12.5	14.9	17.2	19.2

19. In each case, graph a linear function with the given rate of change. Label and put scales on the axes.

(a) Increasing at 2.1 inches/day
(b) Decreasing at 1.3 gallons/mile

20. A new Toyota RAV4 costs $21,000. The car's value depreciates linearly to $10,500 in three years time. Write a formula which expresses its value, V, in terms of its age, t, in years.

[15] World Health Organisation, www.who.int/csr/sars/country/en
[16] www.census.gov/ipc/www/idbsusum.html, accessed January 12, 2006.

21. Table 1.18 on page 17 gives the temperature-depth profile, $T = f(d)$, in a borehole in Belleterre, Quebec, where T is the average temperature at a depth d.

(a) Could f be linear?
(b) Graph f. What do you notice about the graph for $d \geq 150$?
(c) What can you say about the average rate of change of f for $d \geq 150$?

22. Outside the US, temperature readings are usually given in degrees Celsius; inside the US, they are often given in degrees Fahrenheit. The exact conversion from Celsius, C, to Fahrenheit, F, uses the formula

$$F = \frac{9}{5}C + 32.$$

An approximate conversion is obtained by doubling the temperature in Celsius and adding 30° to get the equivalent Fahrenheit temperature.

(a) Write a formula using C and F to express the approximate conversion.
(b) How far off is the approximation if the Celsius temperature is $-5^\circ, 0^\circ, 15^\circ, 30^\circ$?
(c) For what temperature (in Celsius) does the approximation agree with the actual formula?

23. Tuition cost T (in dollars) for part-time students at Stonewall College is given by $T = 300 + 200C$, where C represents the number of credits taken.

(a) Find the tuition cost for eight credits.
(b) How many credits were taken if the tuition was \$1700?
(c) Make a table showing costs for taking from one to twelve credits. For each value of C, give both the tuition cost, T, and the cost per credit, T/C. Round to the nearest dollar.
(d) Which of these values of C has the smallest cost per credit?
(e) What does the 300 represent in the formula for T?
(f) What does the 200 represent in the formula for T?

24. A company finds that there is a linear relationship between the amount of money that it spends on advertising and the number of units it sells. If it spends no money on advertising, it sells 300 units. For each additional \$5000 spent, an additional 20 units are sold.

(a) If x is the amount of money that the company spends on advertising, find a formula for y, the number of units sold as a function of x.
(b) How many units does the firm sell if it spends \$25,000 on advertising? \$50,000?
(c) How much advertising money must be spent to sell 700 units?
(d) What is the slope of the line you found in part (a)? Give an interpretation of the slope that relates units sold and advertising costs.

25. A report by the US Geological Survey[17] indicates that glaciers in Glacier National Park, Montana, are shrinking. Recent estimates indicate that the area covered by glaciers has decreased from over 25.5 km^2 in 1850 to about 16.6 km^2 in 1995. Let $A = f(t)$ give the area t years after 2000, and assume that $f(t) = 16.2 - 0.062t$. Explain what your answers to the following questions tell you about glaciers.

(a) Give the slope and A-intercept.
(b) Evaluate $f(15)$.
(c) How much glacier area disappears in 15 years?
(d) Solve $f(t) = 12$.

26. Graph the following function in the window $-10 \leq x \leq 10, -10 \leq y \leq 10$. Is this graph a line? Explain.

$$y = -x\left(\frac{x - 1000}{900}\right)$$

27. Graph $y = 2x + 400$ using the window $-10 \leq x \leq 10, -10 \leq y \leq 10$. Describe what happens, and how you can fix it by using a better window.

28. Graph $y = 200x + 4$ using the window $-10 \leq x \leq 10, -10 \leq y \leq 10$. Describe what happens and how you can fix it by using a better window.

29. Figure 1.23 shows the graph of $y = x^2/1000 + 5$ in the window $-10 \leq x \leq 10, -10 \leq y \leq 10$. Discuss whether this is a linear function.

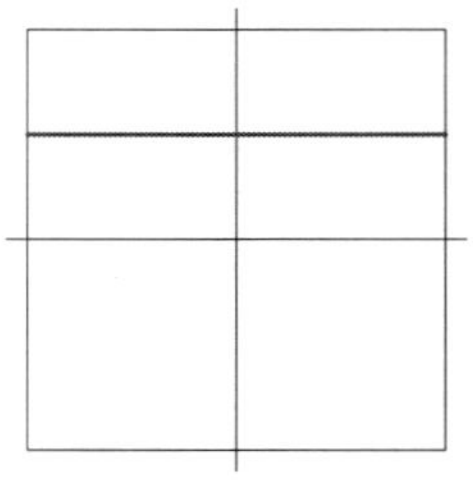

Figure 1.23

30. The cost of a cab ride is given by the function $C = 1.50 + 2d$, where d is the number of miles traveled and C is in dollars. Choose an appropriate window and graph the cost of a ride for a cab that travels no farther than a 10 mile radius from the center of the city.

[17] *Glacier Retreat in Glacier National Park, Montana*, http://www.nrmsc.usgs.gov/research/glacier_retreat.htm, accessed October 15, 2003.

31. The graph of a linear function $y = f(x)$ passes through the two points $(a, f(a))$ and $(b, f(b))$, where $a < b$ and $f(a) < f(b)$.

(a) Graph the function labeling the two points.
(b) Find the slope of the line in terms of f, a, and b.

32. Let $f(x) = 0.003 - (1.246x + 0.37)$.

(a) Calculate the following average rates of change:

(i) $\dfrac{f(2) - f(1)}{2 - 1}$ (ii) $\dfrac{f(1) - f(2)}{1 - 2}$

(iii) $\dfrac{f(3) - f(4)}{3 - 4}$

(b) Rewrite $f(x)$ in the form $f(x) = b + mx$.

1.4 FORMULAS FOR LINEAR FUNCTIONS

To find a formula for a linear function we find values for the slope, m, and the vertical intercept, b in the formula $y = b + mx$.

Finding a Formula for a Linear Function from a Table of Data

If a table of data represents a linear function, we first calculate m and then determine b.

Example 1 A grapefruit is thrown into the air. Its velocity, v, is a linear function of t, the time since it was thrown. A positive velocity indicates the grapefruit is rising and a negative velocity indicates it is falling. Check that the data in Table 1.30 corresponds to a linear function. Find a formula for v in terms of t.

Table 1.30 *Velocity of a grapefruit t seconds after being thrown into the air*

t, time (sec)	1	2	3	4
v, velocity (ft/sec)	48	16	−16	−48

Solution Figure 1.24 shows the data in Table 1.30. The points appear to fall on a line. To check that the velocity function is linear, calculate the rates of change of v and see that they are constant. From time $t = 1$ to $t = 2$, we have

$$\text{Rate of change of velocity with time} = \frac{\Delta v}{\Delta t} = \frac{16 - 48}{2 - 1} = -32.$$

For the next second, from $t = 2$ to $t = 3$, we have

$$\text{Rate of change} = \frac{\Delta v}{\Delta t} = \frac{-16 - 16}{3 - 2} = -32.$$

You can check that the rate of change from $t = 3$ to $t = 4$ is also -32.

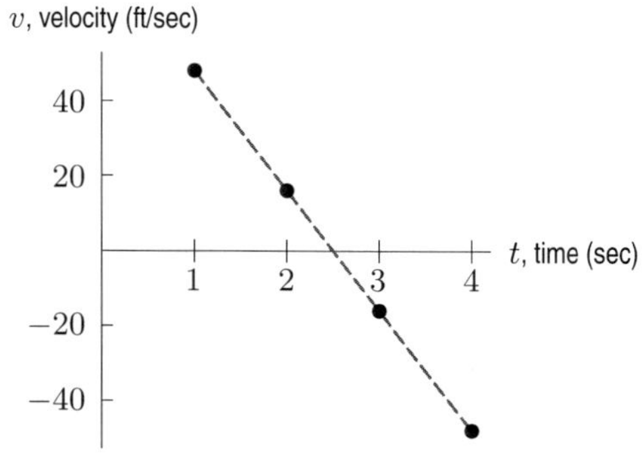

Figure 1.24: Velocity of a grapefruit is a linear function of time

A formula for v is of the form $v = b + mt$. Since m is the rate of change, we have $m = -32$ so $v = b - 32t$. The initial velocity (at $t = 0$) is represented by b. We are not given the value of v when $t = 0$, but we can use any data point to calculate b. For example, $v = 48$ when $t = 1$, so

$$48 = b - 32 \cdot 1,$$

which gives

$$b = 80.$$

Thus, a formula for the velocity is $v = 80 - 32t$.

What does the rate of change, m, in Example 1 tell us about the grapefruit? Think about the units:

$$m = \frac{\Delta v}{\Delta t} = \frac{\text{Change in velocity}}{\text{Change in time}} = \frac{-32 \text{ ft/sec}}{1 \text{ sec}} = -32 \text{ ft/sec per second.}$$

The value of m, -32 ft/sec per second, tells us that the grapefruit's velocity is decreasing by 32 ft/sec for every second that goes by. We say the grapefruit is accelerating at -32 ft/sec per second. (The units ft/sec per second are often written ft/sec^2. Negative acceleration is also called deceleration.)[18]

Finding a Formula for a Linear Function from a Graph

We can calculate the slope, m, of a linear function using two points on its graph. Having found m we can use either of the points to calculate b, the vertical intercept.

Example 2 Figure 1.25 shows oxygen consumption as a function of heart rate for two people.

(a) Assuming linearity, find formulas for these two functions.

(b) Interpret the slope of each graph in terms of oxygen consumption.

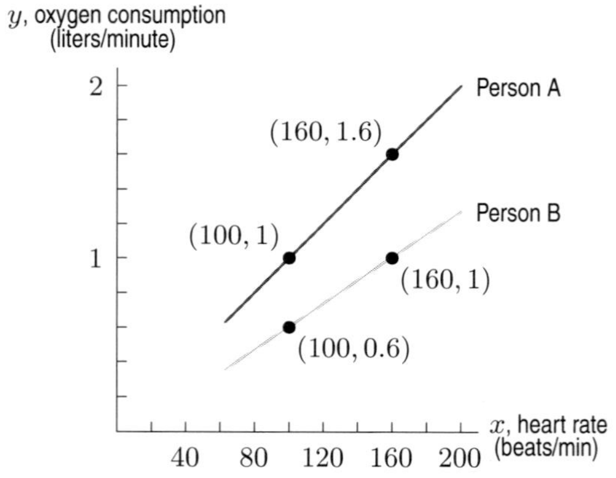

Figure 1.25: Oxygen consumption of two people running on treadmills

[18]The notation ft/sec^2 is shorthand for ft/sec per second; it does not mean a "square second" in the same way that areas are measured square feet or square meters.

Solution

(a) Let x be heart rate and let y be oxygen consumption. Since we are assuming linearity, $y = b + mx$. The two points on person A's line, $(100, 1)$ and $(160, 1.6)$, give

$$\text{Slope of } A\text{'s line } = m = \frac{\Delta y}{\Delta x} = \frac{1.6 - 1}{160 - 100} = 0.01.$$

Thus $y = b + 0.01x$. To find b, use the fact that $y = 1$ when $x = 100$:

$$\begin{aligned} 1 &= b + 0.01(100) \\ 1 &= b + 1 \\ b &= 0. \end{aligned}$$

Alternatively, b can be found using the fact that $x = 160$ if $y = 1.6$. Either way leads to the formula $y = 0.01x$.

For person B, we again begin with the formula $y = b + mx$. In Figure 1.25, two points on B's line are $(100, 0.6)$ and $(160, 1)$, so

$$\text{Slope of } B\text{'s line } = m = \frac{\Delta y}{\Delta x} = \frac{1 - 0.6}{160 - 100} = \frac{0.4}{60} \approx 0.0067.$$

To find b, use the fact that $y = 1$ when $x = 160$:

$$\begin{aligned} 1 &= b + (0.4/60) \cdot 160 \\ 1 &= b + 1.067 \\ b &= -0.067. \end{aligned}$$

Thus, for person B, we have $y = -0.067 + 0.0067x$.

(b) The slope for person A is $m = 0.01$, so

$$m = \frac{\text{Change in oxygen consumption}}{\text{Change in heart rate}} = \frac{\text{Change in liters/min}}{\text{Change in beats/min}} = 0.01 \frac{\text{liters}}{\text{heart beat}}.$$

Every additional heart beat (per minute) for person A translates to an additional 0.01 liters (per minute) of oxygen consumed.

The slope for person B is $m = 0.0067$. Thus, for every additional beat (per minute), person B consumes an additional 0.0067 liter of oxygen (per minute). Since the slope for person B is smaller than for person A, person B consumes less additional oxygen than person A for the same increase in pulse.

What do the y-intercepts of the functions in Example 2 say about oxygen consumption? Often the y-intercept of a function is a starting value. In this case, the y-intercept would be the oxygen consumption of a person whose pulse is zero (i.e. $x = 0$). Since a person running on a treadmill must have a pulse, in this case it makes no sense to interpret the y-intercept this way. The formula for oxygen consumption is useful only for realistic values of the pulse.

Finding a Formula for a Linear Function from a Verbal Description

Sometimes the verbal description of a linear function is less straightforward than those we saw in Section 1.3. Consider the following example.

Example 3 We have $24 to spend on soda and chips for a party. A six-pack of soda costs $3 and a bag of chips costs $2. The number of six-packs we can afford, y, is a function of the number of bags of chips we decide to buy, x.

(a) Find an equation relating x and y.

(b) Graph the equation. Interpret the intercepts and the slope in the context of the party.

Solution (a) If we spend all $24 on soda and chips, then we have the following equation:

$$\text{Amount spent on chips} + \text{Amount spent on soda} = \$24.$$

If we buy x bags of chips at $2 per bag, then the amount spent on chips is $\$2x$. Similarly, if we buy y six-packs of soda at $3 per six-pack, then the amount spent on soda is $\$3y$. Thus,

$$2x + 3y = 24.$$

We can solve for y, giving

$$3y = 24 - 2x$$
$$y = 8 - \frac{2}{3}x.$$

This is a linear function with slope $m = -2/3$ and y-intercept $b = 8$.

(b) The graph of this function is a discrete set of points, since the number of bags of chips and the number of six-packs of soda must be (nonnegative) integers.

To find the y-intercept, we set $x = 0$, giving

$$2 \cdot 0 + 3y = 24.$$

So $3y = 24$, giving $y = 8$.

Substituting $y = 0$ gives the x-intercept,

$$2x + 3 \cdot 0 = 24.$$

So $2x = 24$, giving $x = 12$. Thus the points $(0, 8)$ and $(12, 0)$ are on the graph.

The point $(0, 8)$ indicates that we can buy 8 six-packs of soda if we buy no chips. The point $(12, 0)$ indicates that we can buy 12 bags of chips if we buy no soda. The other points on the line describe affordable options between these two extremes. For example, the point $(6, 4)$ is on the line, because

$$2 \cdot 6 + 3 \cdot 4 = 24.$$

This means that if we buy 6 bags of chips, we can afford 4 six-packs of soda.

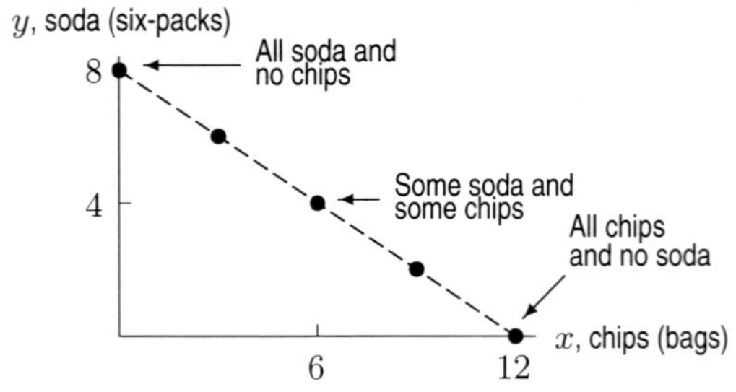

Figure 1.26: Relation between the number of six-packs, y, and the number of bags of chips, x

The points marked in Figure 1.26 represent affordable options. All affordable options lie on or below the line $2x + 3y = 24$. Not all points on the line are affordable options. For example, suppose we purchase one six-pack of soda for $\$3.00$. That leaves $\$21.00$ to spend on chips, meaning we would have to buy 10.5 bags of chips, which is not possible. Therefore, the point $(10.5, 1)$ is not an option, although it is a point on the line $2x + 3y = 24$.

To interpret the slope, notice that

$$m = \frac{\Delta y}{\Delta x} = \frac{\text{Change in number of six-packs}}{\text{Change in number of bags of chips}},$$

so the units of m are six-packs of soda per bags of chips. The fact that $m = -2/3$ means that for each additional 3 bags of chips purchased, we can purchase 2 fewer six-packs of soda. This occurs because 2 six-packs cost $\$6$, the same as 3 bags of chips. Thus, $m = -2/3$ is the rate at which the amount of soda we can buy decreases as we buy more chips.

Alternative Forms for the Equation of a Line

In Example 3, the equation $2x + 3y = 24$ represents a linear relationship between x and y even though the equation is not in the form $y = b + mx$. The following equations represent lines.

- The *slope-intercept form* is
  $y = b + mx$ where m is the slope and b is the y-intercept.
- The *point-slope form* is
  $y - y_0 = m(x - x_0)$ where m is the slope and (x_0, y_0) is a point on the line.
- The *standard form* is
  $Ax + By + C = 0$ where A, B, and C are constants.

If we know the slope of a line and the coordinates of a point on the line, it is often convenient to use the point-slope form of the equation.

Example 4 Use the point-slope form to find the equation of the line for the oxygen consumption of Person A in Example 2.

Solution In Example 2, we found the slope of person A's line to be $m = 0.01$. Since the point $(100, 1)$ lies on the line, the point-slope form gives the equation

$$y - 1 = 0.01(x - 100).$$

To check that this gives the same equation we got in Example 2, we multiply out and simplify:

$$\begin{aligned} y - 1 &= 0.01x - 1 \\ y &= 0.01x. \end{aligned}$$

Alternatively, we could have used the point $(160, 1.6)$ instead of $(100, 1)$, giving

$$y - 1.6 = 0.01(x - 160).$$

Multiplying out again gives $y = 0.01x$.

Exercises and Problems for Section 1.4

Exercises

If possible, rewrite the equations in Exercises 1–9 in slope-intercept form, $y = b + mx$.

1. $5(x + y) = 4$

2. $3x + 5y = 20$

3. $0.1y + x = 18$

4. $5x - 3y + 2 = 0$

5. $y - 0.7 = 5(x - 0.2)$

6. $y = 5$

7. $3x + 2y + 40 = x - y$

8. $x = 4$

9. $\frac{x + y}{7} = 3$

Is each function in Exercises 10–15 linear? If so, rewrite it the form $y = b + mx$.

10. $g(w) = -\frac{1 - 12w}{3}$

11. $F(P) = 13 - \frac{2^{-1}}{4}P$

12. $j(s) = 3s^{-1} + 7$

13. $C(r) = 2\pi r$

14. $h(x) = 3^x + 12$

15. $f(x) = m^2x + n^2$

Find formulas for the linear functions in Exercises 16–23.

16. Slope -4 and x-intercept 7

17. Slope 3 and y-intercept 8

18. Passes through the points $(-1, 5)$ and $(2, -1)$

19. Slope $2/3$ and passes through the point $(5, 7)$

20. Has x-intercept 3 and y-intercept -5

21. Slope 0.1, passes through $(-0.1, 0.02)$

22. Function f has $f(0.3) = 0.8$ and $f(0.8) = -0.4$

23. Function f has $f(-2) = 7$ and $f(3) = -3$.

Exercises 24–30 give data from a linear function. Find a formula for the function.

24.

Year, t	0	1	2
Value of computer, $\$V = f(t)$	2000	1500	1000

25.

Price per bottle, p ($)	0.50	0.75	1.00
Number of bottles sold, $q = f(p)$	1500	1000	500

26.

Temperature, $y = f(x)$ (°C)	0	5	20
Temperature, x (°F)	32	41	68

27.

Temperature, $y = f(x)$, (°R)	459.7	469.7	489.7
Temperature, x (°F)	0	10	30

28.

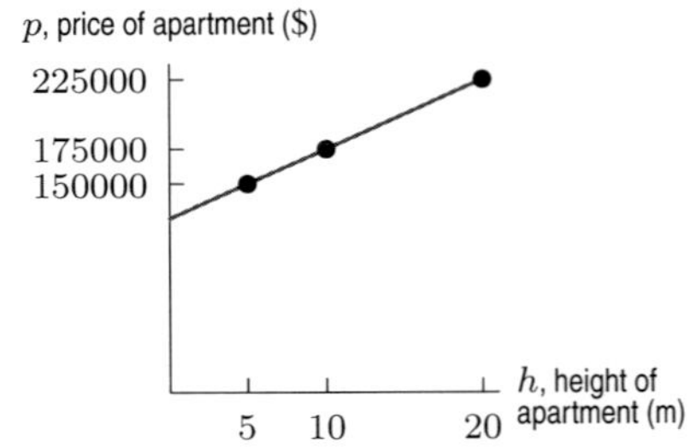

29.

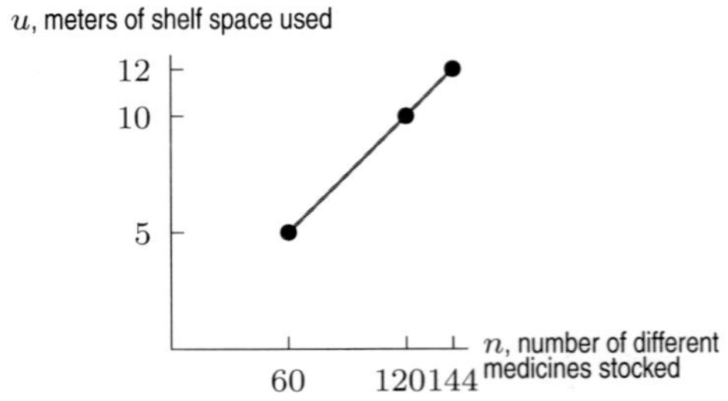

30.

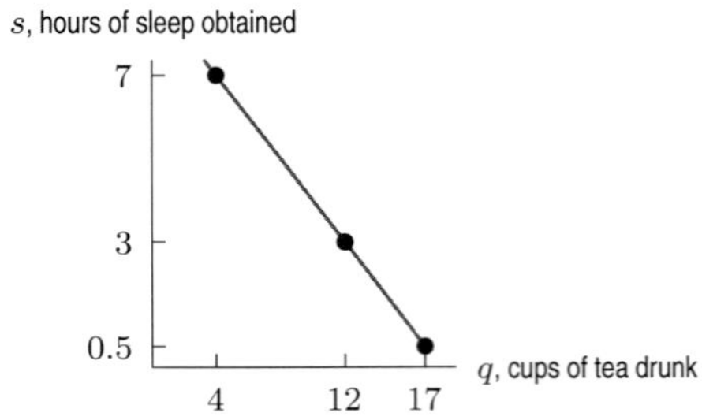

Problems

31. Find the equation of the line l, shown in Figure 1.27, if its slope is $m = 4$.

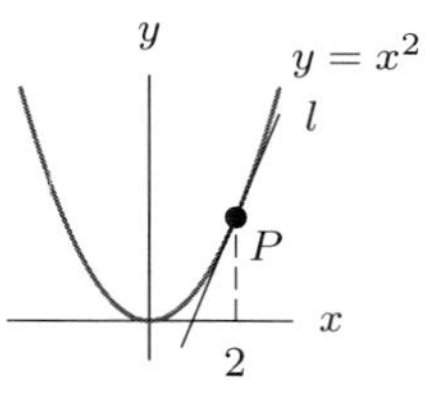

Figure 1.27

32. Find a formula for the line intersecting the graph of $f(x)$ at $x = 1$ and $x = 3$, where

$$f(x) = \frac{10}{x^2 + 1}.$$

33. Find a formula for the linear function $h(t)$ whose graph intersects the graph of $j(t) = 30(0.2)^t$ at $t = -2$ and $t = 1$.

34. Find the equation of the line l in Figure 1.28. The shapes under the line are squares.

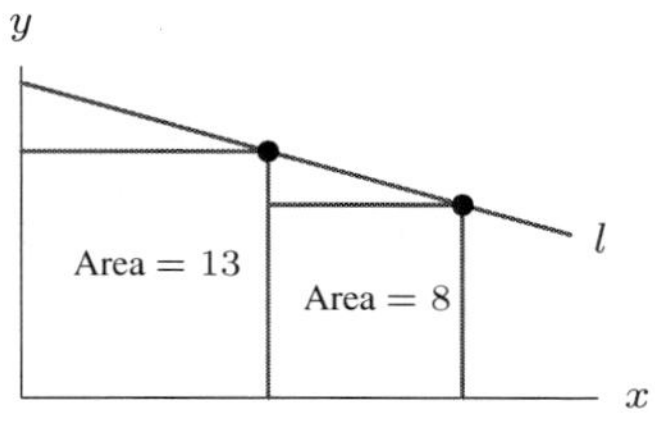

Figure 1.28

35. Find the equation of line l in Figure 1.29.

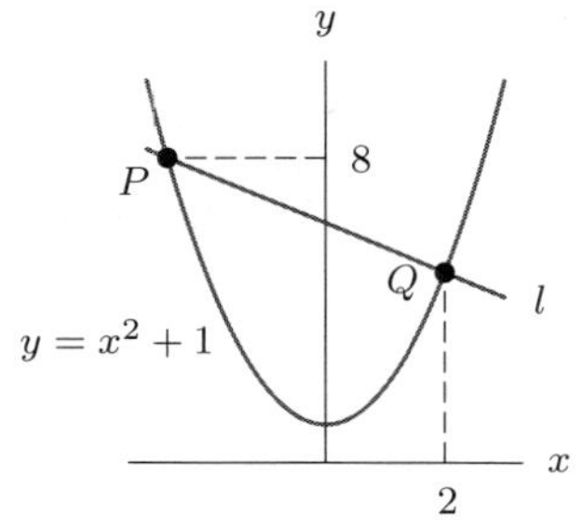

Figure 1.29

36. Find an equation for the line l in Figure 1.30 in terms of the constant A and values of the function f.

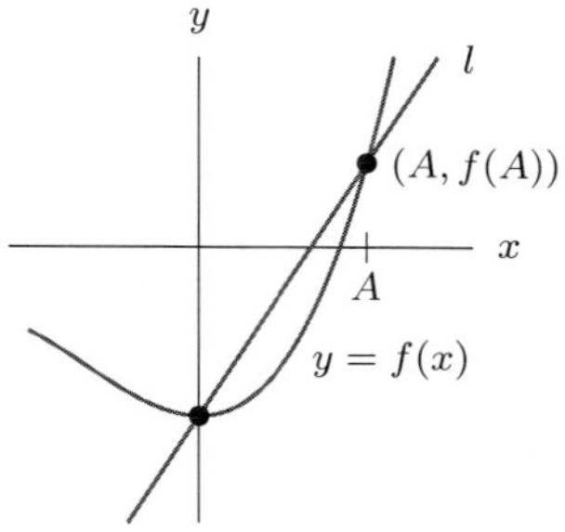

Figure 1.30

37. A bullet is shot straight up into the air from ground level. After t seconds, the velocity of the bullet, in meters per second, is approximated by the formula

$$v = f(t) = 1000 - 9.8t.$$

(a) Evaluate the following: $f(0)$, $f(1)$, $f(2)$, $f(3)$, $f(4)$. Compile your results in a table.

(b) Describe in words what is happening to the speed of the bullet. Discuss why you think this is happening.

(c) Evaluate and interpret the slope and both intercepts of $f(t)$.

(d) The gravitational field near the surface of Jupiter is stronger than that near the surface of the earth, which, in turn, is stronger than the field near the surface of the moon. How is the formula for $f(t)$ different for a bullet shot from Jupiter's surface? From the moon?

38. In a college meal plan you pay a membership fee; then all your meals are at a fixed price per meal.

(a) If 30 meals cost \$152.50 and 60 meals cost \$250, find the membership fee and the price per meal.

(b) Write a formula for the cost of a meal plan, C, in terms of the number of meals, n.

(c) Find the cost for 50 meals.

(d) Find n in terms of C.

(e) Use part (d) to determine the maximum number of meals you can buy on a budget of \$300.

39. John wants to buy a dozen rolls. The local bakery sells sesame and poppy seed rolls for the same price.

(a) Make a table of all the possible combinations of rolls if he buys a dozen, where s is the number of sesame seed rolls and p is the number of poppy seed rolls.

(b) Find a formula for p as a function of s.

(c) Graph this function.

40. A theater manager graphed weekly profits as a function of the number of patrons and found that the relationship was linear. One week the profit was $11,328 when 1324 patrons attended. Another week 1529 patrons produced a profit of $13,275.50.

(a) Find a formula for weekly profit, y, as a function of the number of patrons, x.
(b) Interpret the slope and the y-intercept.
(c) What is the break-even point (the number of patrons for which there is zero profit)?
(d) Find a formula for the number of patrons as a function of profit.
(e) If the weekly profit was $17,759.50, how many patrons attended the theater?

41. The demand for gasoline can be modeled as a linear function of price. If the price of gasoline is $p = \$2.10$ per gallon, the quantity demanded in a fixed period is $q = 65$ gallons. If the price rises to $2.50 per gallon, the quantity demanded falls to 45 gallons in that period.

(a) Find a formula for q in terms of p.
(b) Explain the economic significance of the slope of your formula.
(c) Explain the economic significance of the q-axis and p-axis intercepts.

42. An empty champagne bottle is tossed from a hot-air balloon. Its upward velocity is measured every second and recorded in Table 1.31.

(a) Describe the motion of the bottle in words. What do negative values of v represent?
(b) Find a formula for v in terms of t.
(c) Explain the physical significance of the slope of your formula.
(d) Explain the physical significance of the t-axis and v-axis intercepts.

Table 1.31

t (sec)	0	1	2	3	4	5
v (ft/sec)	40	8	−24	−56	−88	−120

43. A business consultant works 10 hours a day, 6 days a week. She divides her time between meetings with clients and meetings with co-workers. A client meeting requires 3 hours while a co-worker meeting requires 2 hours. Let x be the number of co-worker meetings the consultant holds during a given week. If y is the number of client meetings for which she has time remaining, then y is a function of x. Assume this relationship is linear and that meetings can be split up and continued on different days.

(a) Graph the relationship between y and x. [Hint: Consider the maximum number of client and co-worker meetings that can be held.]
(b) Find a formula for y as a function of x.
(c) Explain what the slope and the x- and y-intercepts represent in the context of the consultant's meeting schedule.
(d) A change is made so that co-worker meetings take 90 minutes instead of 2 hours. Graph this situation. Describe those features of this graph that have changed from the one sketched in part (a) and those that have remained the same.

44. The development time, t, of an organism is the number of days required for the organism to mature, and the development rate is defined as $r = 1/t$. In cold-blooded organisms such as insects, the development rate depends on temperature: the colder it is, the longer the organism takes to develop. For such organisms, the degree-day model[19] assumes that the development rate r is a linear function of temperature H (in °C):

$$r = b + kH.$$

(a) According to the degree-day model, there is a minimum temperature H_{min} below which an organism never matures. Find a formula for H_{min} in terms of the constants b and k.
(b) Define S as $S = (H - H_{\text{min}})t$, where S is the number of degree-days. That is, S is the number of days t times the number of degrees between H and H_{min}. Use the formula for r to show that S is a constant. In other words, find a formula for S that does not involve H. Your formula will involve k.
(c) A certain organism requires $t = 25$ days to develop at a constant temperature of $H = 20$°C and has $H_{\text{min}} = 15$°C. Using the fact that S is a constant, how many days does it take for this organism to develop at a temperature of 25°C?
(d) In part (c) we assumed that the temperature H is constant throughout development. If the temperature varies from day to day, the number of degree-days can be accumulated until they total S, at which point the organism completes development. For instance, suppose on the first day the temperature is $H = 20$°C and that on the next day it is $H = 22$°C. Then for these first two days

Total number of degree days

$$= (20 - 15) \cdot 1 + (22 - 15) \cdot 1 = 12.$$

[19]Information drawn from a web site created by Dr. Alexei A. Sharov at the Virginia Polytechnic Institute, http://www.ento.vt.edu/ sharov/PopEcol/popecol.html.

Based on Table 1.32, on what day does the organism reach maturity?

Table 1.32

Day	1	2	3	4	5	6	7	8	9	10	11	12
H (°C)	20	22	27	28	27	31	29	30	28	25	24	26

45. (Continuation of Problem 44.) Table 1.33 gives the development time t (in days) for an insect as a function of temperature H (in °C).

(a) Find a linear formula for r, the development rate, in terms of H.

(b) Find the value of S, the number of degree-days required for the organism to mature.

Table 1.33

H, °C	20	22	24	26	28	30
t, days	14.3	12.5	11.1	10.0	9.1	8.3

46. Describe a linear (or nearly linear) relationship that you have encountered outside the classroom. Determine the rate of change and interpret it in practical terms.

1.5 GEOMETRIC PROPERTIES OF LINEAR FUNCTIONS

Interpreting the Parameters of a Linear Function

The slope-intercept form for a linear function is $y = b + mx$, where b is the y-intercept and m is the slope. The parameters b and m can be used to compare linear functions.

Example 1 With time, t, in years, the populations of four towns, P_A, P_B, P_C and P_D, are given by the following formulas:

$$P_A = 20{,}000 + 1600t, \quad P_B = 50{,}000 - 300t, \quad P_C = 650t + 45{,}000, \quad P_D = 15{,}000(1.07)^t.$$

(a) Which populations are represented by linear functions?

(b) Describe in words what each linear model tells you about that town's population. Which town starts out with the most people? Which town is growing fastest?

Solution

(a) The populations of towns A, B, and C are represented by linear functions because they are written in the form $P = b + mt$. Town D's population does not grow linearly since its formula, $P_D = 15{,}000(1.07)^t$, cannot be expressed in the form $P_D = b + mt$.

(b) For town A, we have

$$P_A = \underbrace{20{,}000}_{b} + \underbrace{1600}_{m} \cdot t,$$

so $b = 20{,}000$ and $m = 1600$. This means that in year $t = 0$, town A has 20,000 people. It grows by 1600 people per year.

For town B, we have

$$P_B = \underbrace{50{,}000}_{b} + \underbrace{(-300)}_{m} \cdot t,$$

so $b = 50{,}000$ and $m = -300$. This means that town B starts with 50,000 people. The negative slope indicates that the population is decreasing at the rate of 300 people per year.

For town C, we have

$$P_C = \underbrace{45{,}000}_{b} + \underbrace{650}_{m} \cdot t,$$

so $b = 45{,}000$ and $m = 650$. This means that town C begins with 45,000 people and grows by 650 people per year.

Town B starts out with the most people, 50,000, but town A, with a rate of change of 1600 people per year, grows the fastest of the three towns that grow linearly.

The Effect of the Parameters on the Graph of a Linear Function

The graph of a linear function is a line. Changing the values of b and m gives different members of the family of linear functions. In summary:

> Let $y = b + mx$. Then the graph of y against x is a line.
> - The y-intercept, b, tells us where the line crosses the y-axis.
> - If the slope, m, is positive, the line climbs from left to right. If the slope, m, is negative, the line falls from left to right.
> - The slope, m, tells us how fast the line is climbing or falling.
> - The larger the magnitude of m (either positive or negative), the steeper the graph of f.

Example 2 (a) Graph the three linear functions P_A, P_B, P_C from Example 1 and show how to identify the values of b and m from the graph.
(b) Graph P_D from Example 1 and explain how the graph shows P_D is not a linear function.

Solution (a) Figure 1.31 gives graphs of the three functions:

$$P_A = 20{,}000 + 1600t, \qquad P_B = 50{,}000 - 300t, \quad \text{and} \quad P_C = 45{,}000 + 650t.$$

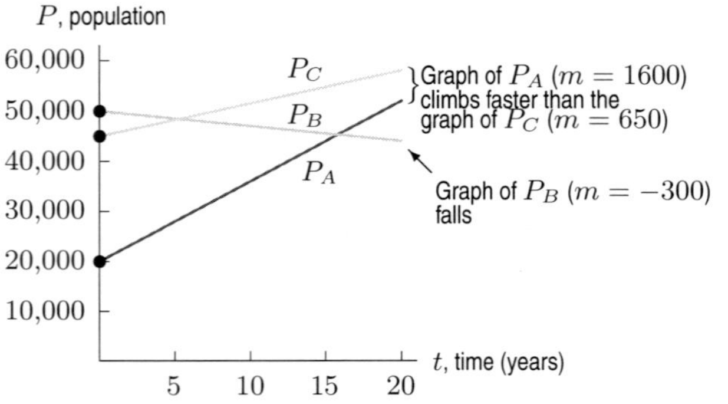

Figure 1.31: Graphs of three linear functions, P_A, P_B, and P_C, showing starting values and rates of climb

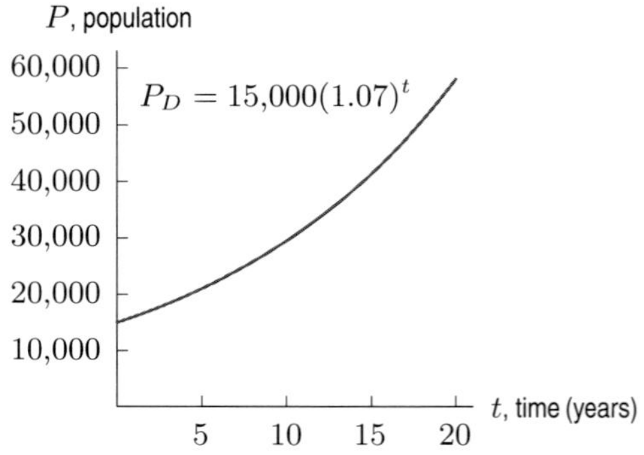

Figure 1.32: Graph of $P_D = 15{,}000(1.07)^t$ is not a line

The values of b identified in Example 1 tell us the vertical intercepts. Figure 1.31 shows that the graph of P_A crosses the P-axis at $P = 20{,}000$, the graph of P_B crosses at $P = 50{,}000$, and the graph of P_C crosses at $P = 45{,}000$.

Notice that the graphs of P_A and P_C are both climbing and that P_A climbs faster than P_C. This corresponds to the fact that the slopes of these two functions are positive ($m = 1600$ for P_A and $m = 650$ for P_C) and the slope of P_A is larger than the slope of P_C.

The graph of P_B falls when read from left to right, indicating that population decreases over time. This corresponds to the fact that the slope of P_C is negative ($m = -300$).

(b) Figure 1.32 gives a graph of P_D. Since it is not a line, P_D is not a linear function.

Intersection of Two Lines

To find the point at which two lines intersect, notice that the (x, y)-coordinates of such a point must satisfy the equations for both lines. Thus, in order to find the point of intersection algebraically, solve the equations simultaneously.[20]

[20]If you have questions about the algebra in this section, see the Tools Section on page 55.

If linear functions are modeling real quantities, their points of intersection often have practical significance. Consider the next example.

Example 3 The cost in dollars of renting a car for a day from three different rental agencies and driving it d miles is given by the following functions:

$$C_1 = 50 + 0.10d, \qquad C_2 = 30 + 0.20d, \qquad C_3 = 0.50d.$$

(a) Describe in words the daily rental arrangements made by each of these three agencies.
(b) Which agency is cheapest?

Solution

(a) Agency 1 charges \$50 plus \$0.10 per mile driven. Agency 2 charges \$30 plus \$0.20 per mile. Agency 3 charges \$0.50 per mile driven.

(b) The answer depends on how far we want to drive. If we are not driving far, agency 3 may be cheapest because it only charges for miles driven and has no other fees. If we want to drive a long way, agency 1 may be cheapest (even though it charges \$50 up front) because it has the lowest per-mile rate.

The three functions are graphed in Figure 1.33. The graph shows that for d up to 100 miles, the value of C_3 is less than C_1 and C_2 because its graph is below the other two. For d between 100 and 200 miles, the value of C_2 is less than C_1 and C_3. For d more than 200 miles, the value of C_1 is less than C_2 and C_3.

By graphing these three functions on a calculator, we can estimate the coordinates of the points of intersection by tracing. To find the exact coordinates, we solve simultaneous equations. Starting with the intersection of lines C_1 and C_2, we set the costs equal, $C_1 = C_2$, and solve for d:

$$\begin{aligned} 50 + 0.10d &= 30 + 0.20d \\ 20 &= 0.10d \\ d &= 200. \end{aligned}$$

Thus, the cost of driving 200 miles is the same for agencies 1 and 2. Solving $C_2 = C_3$ gives

$$\begin{aligned} 30 + 0.20d &= 0.50d \\ 0.30d &= 30 \\ d &= 100, \end{aligned}$$

which means the cost of driving 100 miles is the same for agencies 2 and 3.

Thus, agency 3 is cheapest up to 100 miles. Agency 1 is cheapest for more than 200 miles. Agency 2 is cheapest between 100 and 200 miles. See Figure 1.33. Notice that the point of intersection of C_1 and C_3, $(125, 62.5)$, does not influence our decision as to which agency is the cheapest.

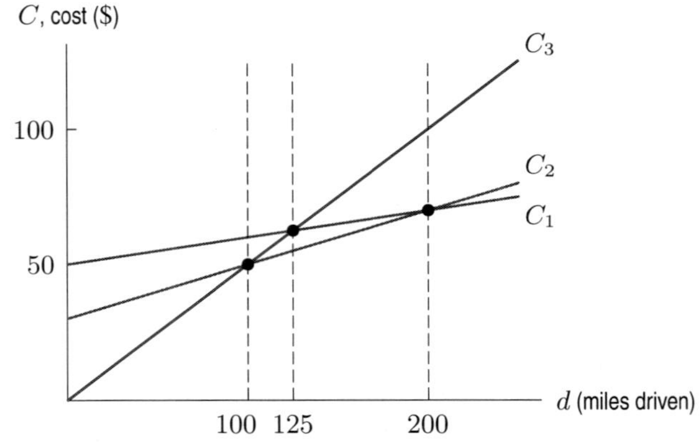

Figure 1.33: Cost of driving a car d miles when renting from three different agencies. Cheapest agency corresponds to the lowest graph for a given d value

Equations of Horizontal and Vertical Lines

An increasing linear function has positive slope and a decreasing linear function has negative slope. What about a line with slope $m = 0$? If the rate of change of a quantity is zero, then the quantity does not change. Thus, if the slope of a line is zero, the value of y must be constant. Such a line is horizontal.

Example 4 Explain why the equation $y = 4$ represents a horizontal line and the equation $x = 4$ represents a vertical line.

Solution The equation $y = 4$ represents a linear function with slope $m = 0$. To see this, notice that this equation can be rewritten as $y = 4 + 0 \cdot x$. Thus, the value of y is 4 no matter what the value of x is. See Figure 1.34. Similarly, the equation $x = 4$ means that x is 4 no matter what the value of y is. Every point on the line in Figure 1.35 has x equal to 4, so this line is the graph of $x = 4$.

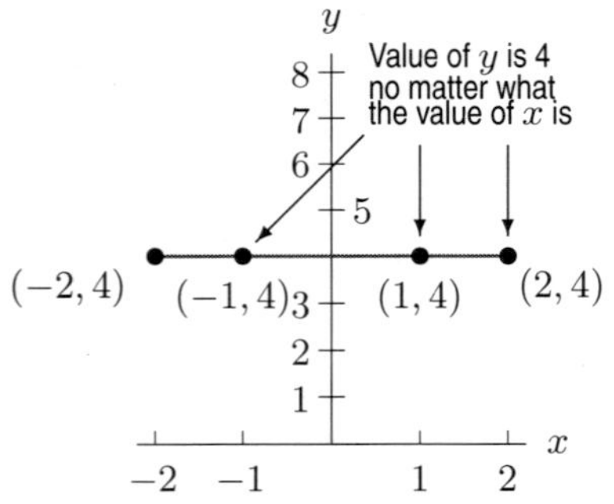

Figure 1.34: The horizontal line $y = 4$ has slope 0

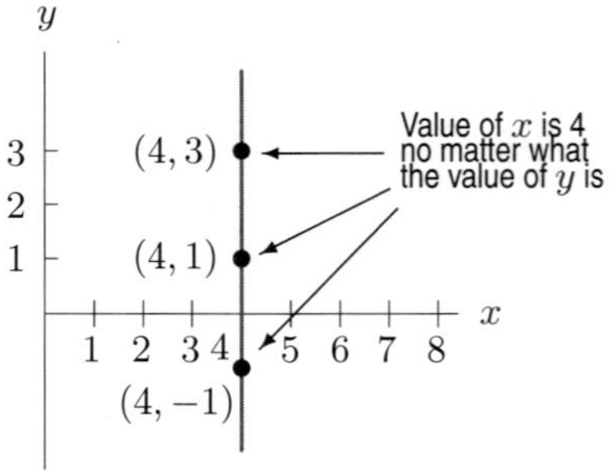

Figure 1.35: The vertical line $x = 4$ has an undefined slope

What is the slope of a vertical line? Figure 1.35 shows three points, $(4, -1)$, $(4, 1)$, and $(4, 3)$ on a vertical line. Calculating the slope, gives

$$m = \frac{\Delta y}{\Delta x} = \frac{3 - 1}{4 - 4} = \frac{2}{0}.$$

The slope is undefined because the denominator, Δx, is 0. The slope of every vertical line is undefined for the same reason. All the x-values on such a line are equal, so Δx is 0, and the denominator of the expression for the slope is 0. A vertical line is not the graph of a function, since it fails the vertical line test. It does not have an equation of the form $y = b + mx$.

In summary,

> For any constant k:
> - The graph of the equation $y = k$ is a horizontal line and its slope is zero.
> - The graph of the equation $x = k$ is a vertical line and its slope is undefined.

Slopes of Parallel and Perpendicular Lines

Figure 1.36 shows two parallel lines. These lines are parallel because they have equal slopes.

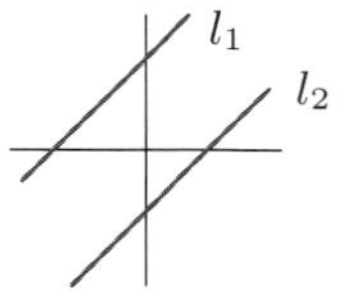

Figure 1.36: Parallel lines: l_1 and l_2 have equal slopes

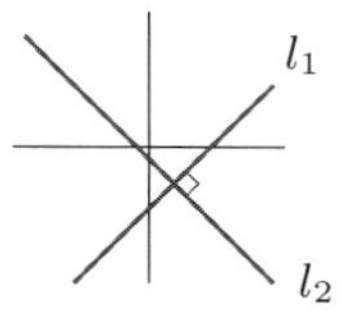

Figure 1.37: Perpendicular lines: l_1 has a positive slope and l_2 has a negative slope

What about perpendicular lines? Two perpendicular lines are graphed in Figure 1.37. We can see that if one line has a positive slope, then any perpendicular line must have a negative slope. Perpendicular lines have slopes with opposite signs.

We show that if l_1 and l_2 are two perpendicular lines with slopes, m_1 and m_2, then m_1 is the negative reciprocal of m_2. If m_1 and m_2 are not zero, we have the following result:

> Let l_1 and l_2 be two lines having slopes m_1 and m_2, respectively. Then:
> - These lines are parallel if and only if $m_1 = m_2$.
> - These lines are perpendicular if and only if $m_1 = -\dfrac{1}{m_2}$.

In addition, any two horizontal lines are parallel and $m_1 = m_2 = 0$. Any two vertical lines are parallel and m_1 and m_2 are undefined. A horizontal line is perpendicular to a vertical line. See Figures 1.38–1.40.

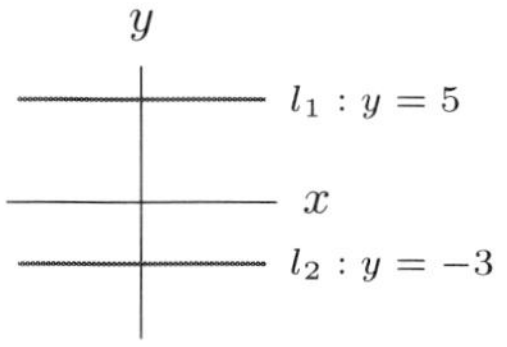

Figure 1.38: Any two horizontal lines are parallel

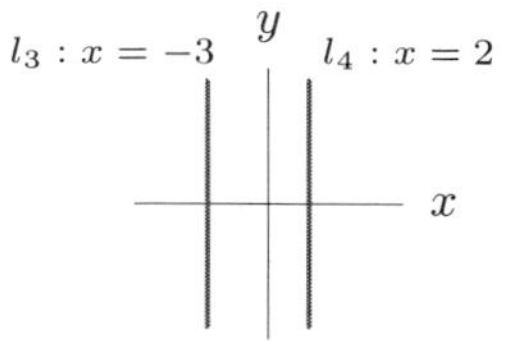

Figure 1.39: Any two vertical lines are parallel

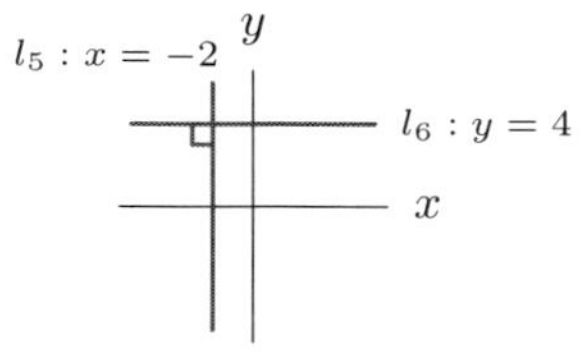

Figure 1.40: A horizontal line and a vertical line are perpendicular

Justification of Formula for Slopes of Perpendicular Lines

Figure 1.41 shows l_1 and l_2, two perpendicular lines with slope m_1 and m_2. Neither line is horizontal or vertical, so m_1 and m_2 are both defined and nonzero. We will show that

$$m_1 = -\frac{1}{m_2},$$

We use the two triangles, ΔPQR and ΔSPR. We show that ΔPQR and ΔSPR are similar by showing that corresponding angles have equal measure. The line PR is horizontal, so $\angle QRP = \angle SRP$ since both are right angles. Since ΔQPS is a right triangle, $\angle S$ is complementary to $\angle Q$ (that is, $\angle S$ and $\angle Q$ add to 90°). Since ΔQRP is a right triangle, $\angle QPR$ is complementary to $\angle Q$. Therefore $\angle S = \angle QPR$. Since two pairs of angles in ΔPQR and ΔSPR have equal measure, the third must be equal also; the triangles are similar.

Corresponding sides of similar triangles are proportional. (See Figure 1.42.) Therefore,

$$\frac{\|RS\|}{\|RP\|} = \frac{\|RP\|}{\|RQ\|},$$

where $\|RS\|$ means the length of side RS.

Next, we calculate m_1 using points S and P, and we calculate m_2 using points Q and P. In Figure 1.41, we see that

$$\Delta x = \|RP\|, \quad \Delta y_1 = \|RS\|, \quad \text{and} \quad \Delta y_2 = -\|RQ\|,$$

where Δy_2 is negative because y-values of points on l_2 decrease as x increases. Thus,

$$m_1 = \frac{\Delta y_1}{\Delta x} = \frac{\|RS\|}{\|RP\|} \quad \text{and} \quad m_2 = \frac{\Delta y_2}{\Delta x} = -\frac{\|RQ\|}{\|RP\|}.$$

Therefore, using the result obtained from the similar triangles, we have

$$m_1 = \frac{\|RS\|}{\|RP\|} = \frac{\|RP\|}{\|RQ\|} = -\frac{1}{m_2}.$$

Thus, $m_1 = -1/m_2$.

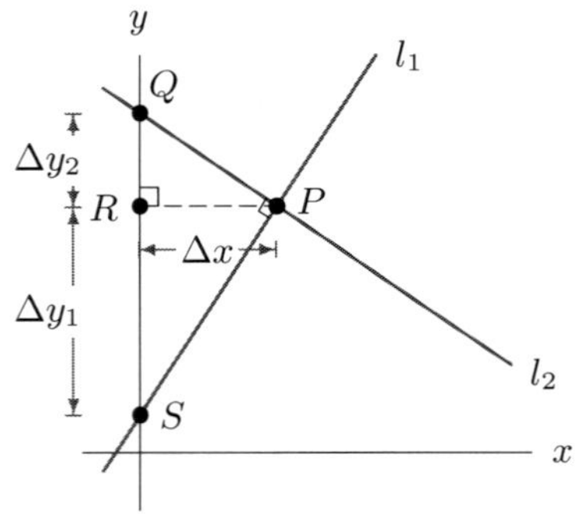

Figure 1.41

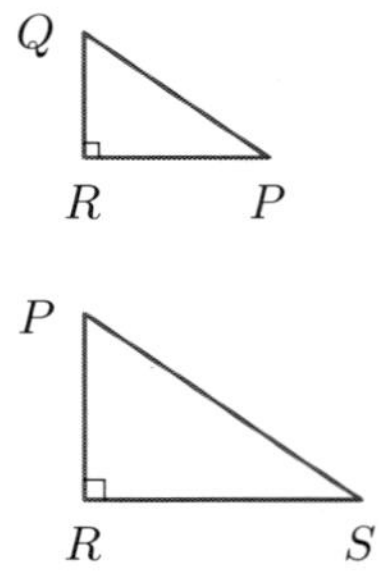

Figure 1.42

Exercises and Problems for Section 1.5

Exercises

1. Without a calculator, match the functions (a)–(c) to the graphs (i)–(iii).

(a) $f(x) = 3x + 1$ **(b)** $g(x) = -2x + 1$
(c) $h(x) = 1$

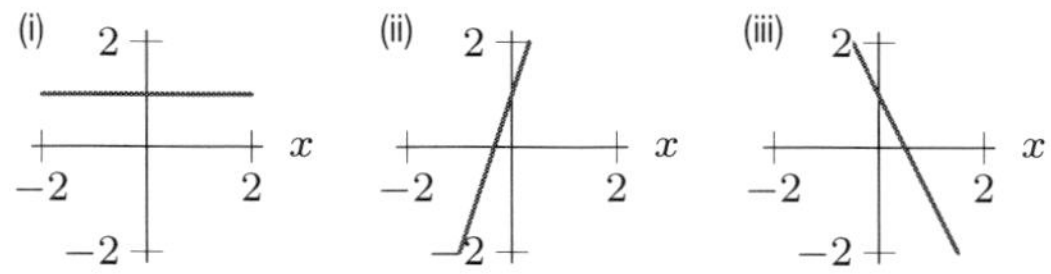

2. Without a calculator, match the equations (a)–(g) to the graphs (I)–(VII).

(a) $y = x - 5$ **(b)** $-3x + 4 = y$
(c) $5 = y$ **(d)** $y = -4x - 5$
(e) $y = x + 6$ **(f)** $y = x/2$
(g) $5 = x$

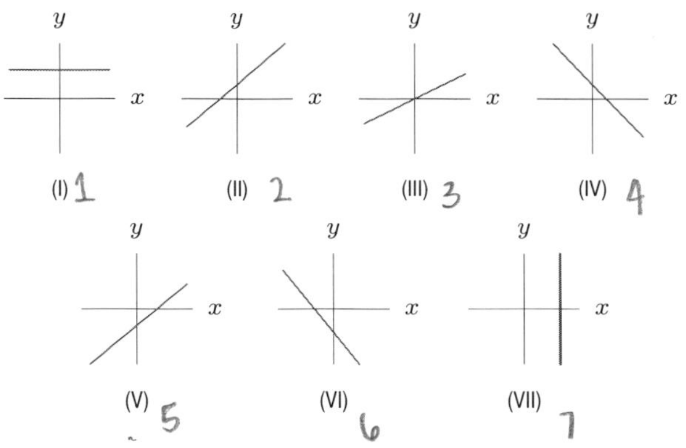

In Exercises 3–5, which line has the greater

(a) Slope? **(b)** y-intercept?

3. $y = -1 + 2x, \quad y = -2 + 3x$

4. $y = 3 + 4x, \quad y = 5 - 2x$

5. $y = \frac{1}{4}x, \quad y = 1 - 6x$

6. Figure 1.43 gives lines, A, B, C, D, and E. Without a calculator, match each line to f, g, h, u or v:

$f(x) = 20 + 2x$
$g(x) = 20 + 4x$
$h(x) = 2x - 30$
$u(x) = 60 - x$
$v(x) = 60 - 2x$

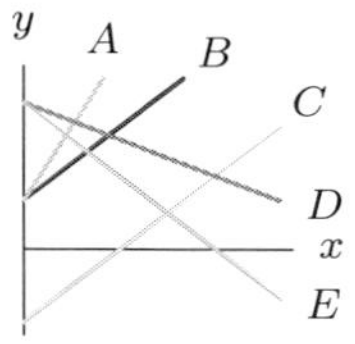

Figure 1.43

7. Without a calculator, match the following functions to the lines in Figure 1.44:

$f(x) = 5 + 2x$
$g(x) = -5 + 2x$
$h(x) = 5 + 3x$
$j(x) = 5 - 2x$
$k(x) = 5 - 3x$

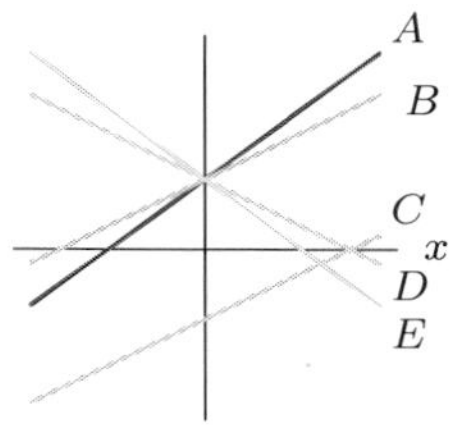

Figure 1.44

8. **(a)** By hand, graph $y = 3$ and $x = 3$.
(b) Can the equations in part (a) be written in slope-intercept form?

Are the lines in Exercises 9–14 perpendicular? Parallel? Neither?

9. $y = 5x - 7$; $y = 5x + 8$

10. $y = 4x + 3$; $y = 13 - \frac{1}{4}x$

11. $y = 2x + 3 \quad y = 2x - 7$

12. $y = 4x + 7 \quad y = \frac{1}{4}x - 2$

13. $f(q) = 12q + 7$; $g(q) = \frac{1}{12}q + 96$

14. $2y = 16 - x$; $4y = -8 - 2x$

Problems

15. Find a formula for the line parallel to the line $y = 20 - 4x$ and containing the point $(3, 12)$.

16. Find the equation of the linear function g whose graph is perpendicular to the line $5x - 3y = 6$; the two lines intersect at $x = 15$.

17. Line l is given by $y = 3 - \frac{2}{3}x$ and point P has coordinates $(6, 5)$.

(a) Find the equation of the line containing P and parallel to l.
(b) Find the equation of the line containing P and perpendicular to l.
(c) Graph the equations in parts (a) and (b).

18. Fill in the missing coordinates for the points in the following figures.

(a) The triangle in Figure 1.45.
(b) The parallelogram in Figure 1.46.

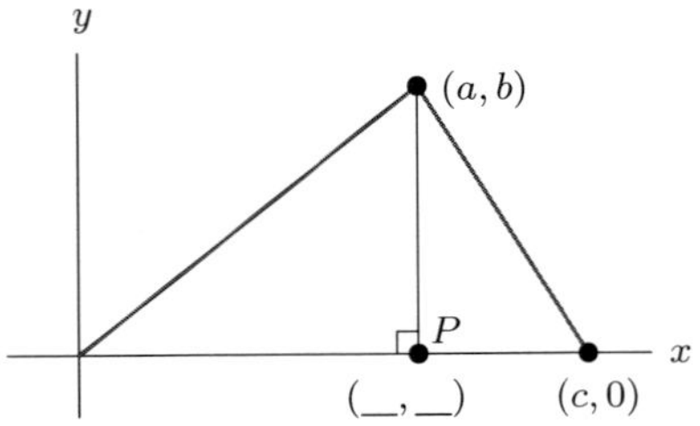

Figure 1.45

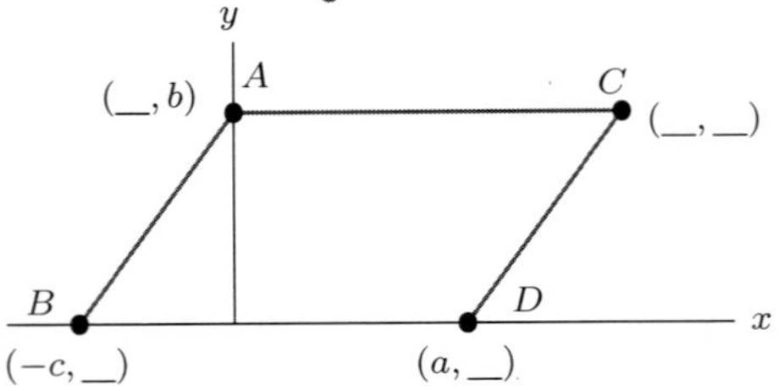

Figure 1.46

19. Using the window $-10 \le x \le 10, -10 \le y \le 10$, graph $y = x$, $y = 10x$, $y = 100x$, and $y = 1000x$.

(a) Explain what happens to the graphs of the lines as the slopes become large.
(b) Write an equation of a line that passes through the origin and is horizontal.

20. Graph $y = x + 1$, $y = x + 10$, and $y = x + 100$ in the window $-10 \le x \le 10, -10 \le y \le 10$.

(a) Explain what happens to the graph of a line, $y = b + mx$, as b becomes large.
(b) Write a linear equation whose graph cannot be seen in the window $-10 \le x \le 10, -10 \le y \le 10$ because all its y-values are less than the y-values shown.

21. The graphical interpretation of the slope is that it shows steepness. Using a calculator or a computer, graph the function $y = 2x - 3$ in the following windows:

(a) $-10 \le x \le 10$ by $-10 \le y \le 10$
(b) $-10 \le x \le 10$ by $-100 \le y \le 100$
(c) $-10 \le x \le 10$ by $-1000 \le y \le 1000$
(d) Write a sentence about how steepness is related to the window being used.

22. Find the coordinates of point P in Figure 1.47.

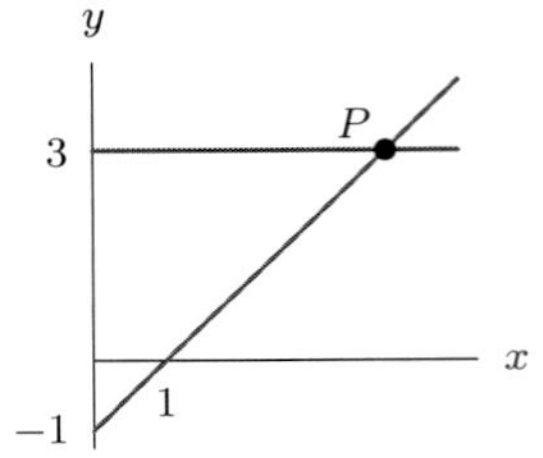

Figure 1.47

23. Estimate the slope of the line in Figure 1.48 and find an approximate equation for the line.

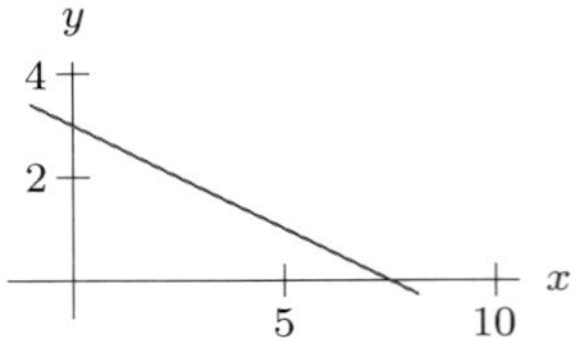

Figure 1.48

24. Line l in Figure 1.49 is parallel to the line $y = 2x + 1$. Find the coordinates of the point P.

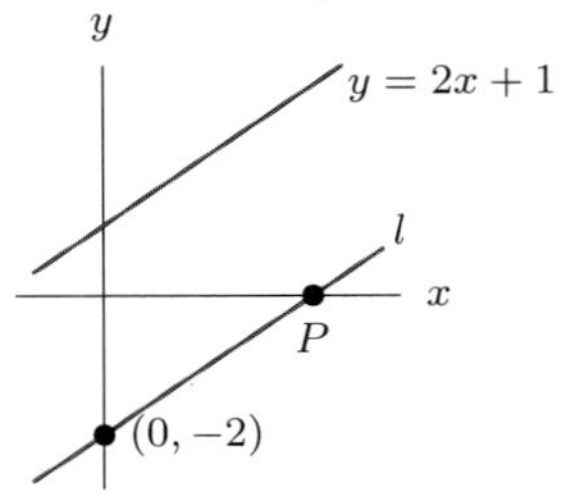

Figure 1.49

25. Find the equation of the line l_2 in Figure 1.50.

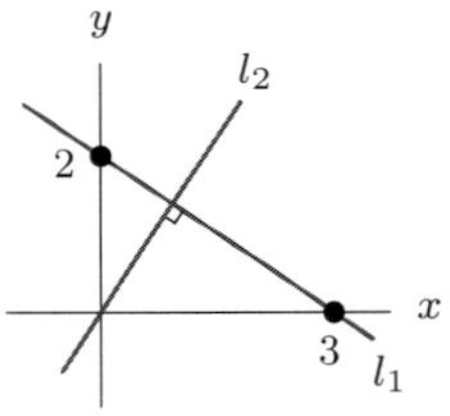

Figure 1.50

In Problems 26–27, what is true about the constant β in the following linear equation if its graph has the given property?

$$y = \frac{x}{\beta - 3} + \frac{1}{6 - \beta}.$$

26. Positive slope, positive y-intercept.

27. Perpendicular to the line $y = (\beta - 7)x - 3$.

28. A circle of radius 2 is centered at the origin and goes through the point $(-1, \sqrt{3})$.

(a) Find an equation for the line through the origin and the point $(-1, \sqrt{3})$.
(b) Find an equation for the tangent line to the circle at $(-1, \sqrt{3})$. [Hint: A tangent line is perpendicular to the radius at the point of tangency.]

29. Find an equation for the altitude through point A of the triangle ABC, where A is $(-4, 5)$, B is $(-3, 2)$, and C is $(9, 8)$. [Hint: The altitude of a triangle is perpendicular to the base.]

30. The cost of a Frigbox refrigerator is $950, and it depreciates $50 each year. The cost of an Arctic Air refrigerator is $1200, and it depreciates $100 per year.

(a) If a Frigbox and an Arctic Air are bought at the same time, when do the two refrigerators have equal value?
(b) If both refrigerators continue to depreciate at the same rates, what happens to the values of the refrigerators in 20 years time? What does this mean?

31. You need to rent a car and compare the charges of three different companies. Company A charges 20 cents per mile plus $20 per day. Company B charges 10 cents per mile plus $35 per day. Company C charges $70 per day with no mileage charge.

(a) Find formulas for the cost of driving cars rented from companies A, B, and C, in terms of x, the distance driven in miles in one day.
(b) Graph the costs for each company for $0 \leq x \leq 500$. Put all three graphs on the same set of axes.
(c) What do the slope and the vertical intercept tell you in this situation?
(d) Use the graph in part (b) to find under what circumstances company A is the cheapest? What about Company B? Company C? Explain why your results make sense.

32. You want to choose one long-distance telephone company from the following options.

- Company A charges $0.37 per minute.
- Company B charges $13.95 per month plus $0.22 per minute.
- Company C charges a fixed rate of $50 per month.

Let Y_A, Y_B, Y_C represent the monthly charges using Company A, B, and C, respectively. Let x be the number of minutes per month spent on long distance calls.

(a) Find formulas for Y_A, Y_B, Y_C as functions of x.
(b) Figure 1.51 gives the graphs of the functions in part (a). Which function corresponds to which graph?
(c) Find the x-values for which Company B is cheapest.

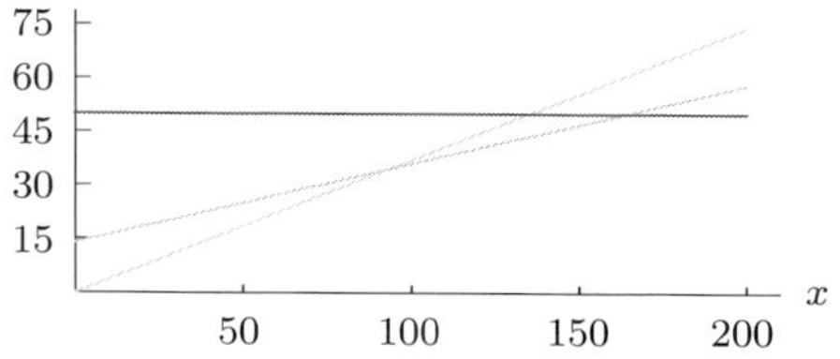

Figure 1.51

33. The solid waste generated each year in the cities of the US is increasing.[21] The solid waste generated, in millions of tons, was 88.1 in 1960 and 234 in 2000. The trend appears linear during this time.

(a) Construct a formula for the amount of municipal solid waste generated in the US by finding the equation of the line through these two points.
(b) Use this formula to predict the amount of municipal solid waste generated in the US, in millions of tons, in the year 2020.

34. Fill in the missing coordinates in Figure 1.52. Write an equation for the line connecting the two points. Check your answer by solving the system of two equations.

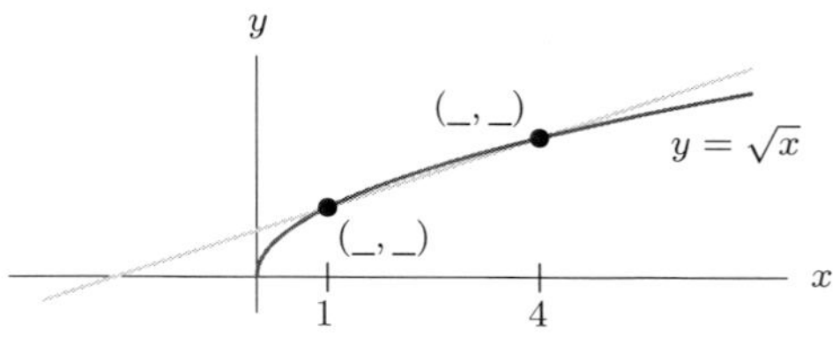

Figure 1.52

35. Two lines are given by $y = b_1 + m_1 x$ and $y = b_2 + m_2 x$, where b_1, b_2, m_1, and m_2 are constants.

(a) What conditions are imposed on b_1, b_2, m_1, and m_2 if the two lines have no points in common?
(b) What conditions are imposed on b_1, b_2, m_1, and m_2 if the two lines have all points in common?
(c) What conditions are imposed on b_1, b_2, m_1, and m_2 if the two lines have exactly one point in common?
(d) What conditions are imposed on b_1, b_2, m_1, and m_2 if the two lines have exactly two points in common?

[21] www.epa.gov/aeposwer/non-hw/muncpl/pubs/MSW05rpt.pdf, accessed January 10, 2006.

36. A commission is a payment made to an employee based on a percentage of sales made. For example, car salespeople earn commission on the selling price of a car. In parts (a)–(d), explain how to choose between the options for different levels of sales.

(a) A weekly salary of \$100 or a weekly salary of \$50 plus 10% commission.

(b) A weekly salary of \$175 plus 7% commission or a weekly salary of \$175 plus 8% commission.

(c) A weekly salary of \$145 plus 7% commission or a weekly salary of \$165 plus 7% commission.

(d) A weekly salary of \$225 plus 3% commission or a weekly salary of \$180 plus 6% commission.

1.6 FITTING LINEAR FUNCTIONS TO DATA

When real data are collected in the laboratory or the field, they are often subject to experimental error. Even if there is an underlying linear relationship between two quantities, real data may not fit this relationship perfectly. However, even if a data set does not perfectly conform to a linear function, we may still be able to use a linear function to help us analyze the data.

Laboratory Data: The Viscosity of Motor Oil

The viscosity of a liquid, or its resistance to flow, depends on the liquid's temperature. Pancake syrup is a familiar example: straight from the refrigerator, it pours very slowly. When warmed on the stove, its viscosity decreases and it becomes quite runny.

The viscosity of motor oil is a measure of its effectiveness as a lubricant in the engine of a car. Thus, the effect of engine temperature is an important determinant of motor-oil performance. Table 1.34 gives the viscosity, v, of motor oil as measured in the lab at different temperatures, T.

Table 1.34 *The measured viscosity, v, of motor oil as a function of the temperature, T*

T, temperature (°F)	v, viscosity (lbs·sec/in^2)
160	28
170	26
180	24
190	21
200	16
210	13
220	11
230	9

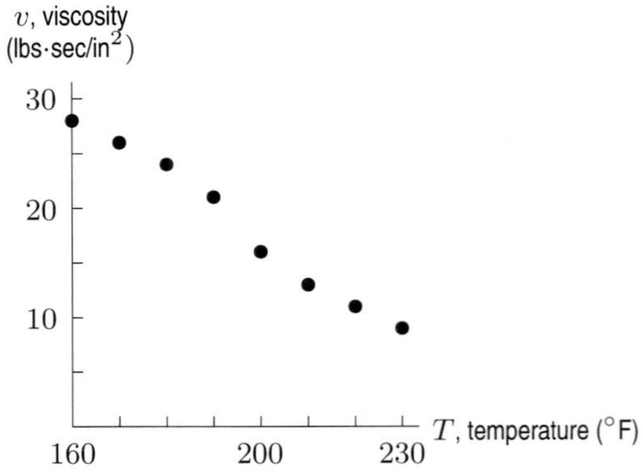

Figure 1.53: The viscosity data from Table 1.34

The *scatter plot* of the data in Figure 1.53 shows that the viscosity of motor oil decreases, approximately linearly, as its temperature rises. To find a formula relating viscosity and temperature, we fit a line to these data points.

Fitting the best line to a set of data is called *linear regression*. One way to fit a line is to draw a line "by eye." Alternatively, many computer programs and calculators compute regression lines. Figure 1.54 shows the data from Table 1.34 together with the computed regression line,

$$v = 75.6 - 0.293T.$$

Notice that none of the data points lie exactly on the regression line, although it fits the data well.

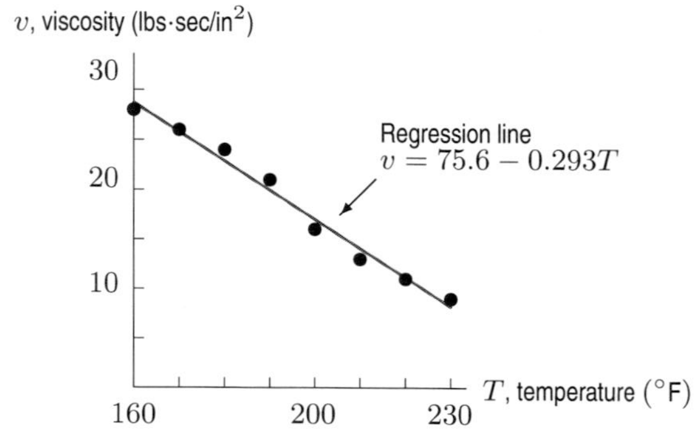

Figure 1.54: A graph of the viscosity data from Table 1.34, together with a regression line (provided by a calculator)

The Assumptions Involved In Finding a Regression Line

When we find a regression line for the data in Table 1.34, we are assuming that the value of v is related to the value of T. However, there may be experimental errors in our measurements. For example, if we measure viscosity twice at the same temperature, we may get two slightly different values. Alternatively, something besides engine temperature could be affecting the oil's viscosity (the oil pressure, for example). Thus, even if we assume that the temperature readings are exact, the viscosity readings include some degree of uncertainty.

Interpolation and Extrapolation

The formula for viscosity can be used to make predictions. Suppose we want to know the viscosity of motor oil at $T = 196°$F. The formula gives

$$v = 75.6 - 0.293 \cdot 196 \approx 18.2 \text{ lb} \cdot \text{sec/in}^2.$$

To see that this is a reasonable estimate, compare it to the entries in Table 1.34. At 190°F, the measured viscosity was 21, and at 200°F, it was 16; the predicted viscosity of 18.2 is between 16 and 21. See Figure 1.55. Of course, if we measured the viscosity at $T = 196°$F in the lab, we might not get exactly 18.2.

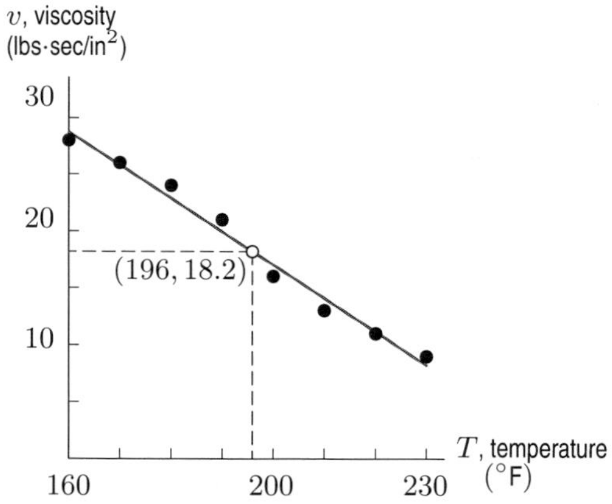

Figure 1.55: Regression line used to predict the viscosity at 196°

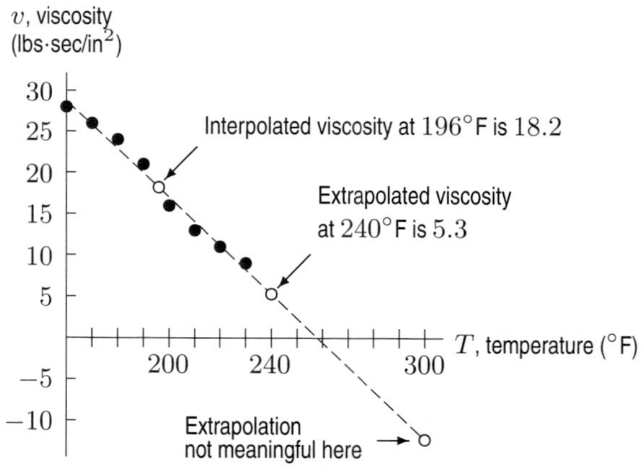

Figure 1.56: The data from Table 1.34 together with the predicted viscosity at $T = 196°$, $T = 240°$, and $T = 300°$

Since the temperature $T = 196°$ is between two temperatures for which v is known ($190°$ and $200°$), the estimate of 18.2 is said to be an *interpolation*. If instead we estimate the value of v at a temperature outside the values for T in Table 1.34, our estimate is called an *extrapolation*.

Example 1 Predict the viscosity of motor oil at $240°$F and at $300°$F.

Solution At $T = 240°$F, the formula for the regression line predicts that the viscosity of motor oil is

$$v = 75.6 - 0.293 \cdot 240 = 5.3 \text{ lb} \cdot \text{sec/in}^2.$$

This is reasonable. Figure 1.56 shows that the predicted point—represented by an open circle on the graph—is consistent with the trend in the data points from Table 1.34.

On the other hand, at $T = 300°$F the regression-line formula gives

$$v = 75.6 - 0.293 \cdot 300 = -12.3 \text{ lb} \cdot \text{sec/in}^2.$$

This is unreasonable because viscosity cannot be negative. To understand what went wrong, notice that in Figure 1.56, the open circle representing the point $(300, -12.3)$ is far from the plotted data points. By making a prediction at $300°$F, we have assumed—incorrectly—that the trend observed in laboratory data extended as far as $300°$F.

In general, interpolation tends to be more reliable than extrapolation because we are making a prediction on an interval we already know something about instead of making a prediction beyond the limits of our knowledge.

How Regression Works

How does a calculator or computer decide which line fits the data best? We assume that the value of y is related to the value of x, although other factors could influence y as well. Thus, we assume that we can pick the value of x exactly but that the value of y may be only partially determined by this x-value.

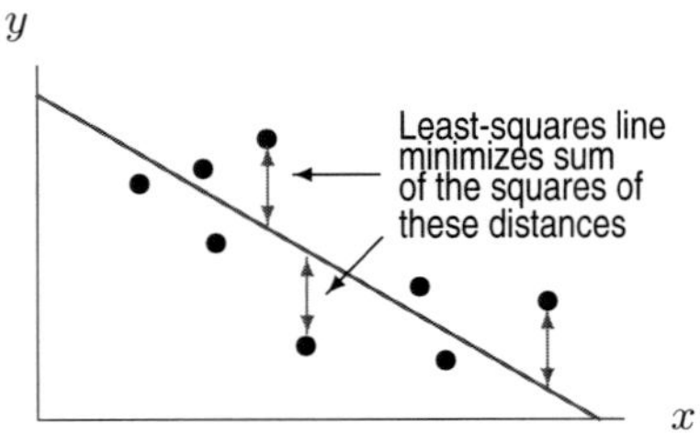

Figure 1.57: A given set of data and the corresponding least-squares regression line

One way to fit a line to the data is shown in Figure 1.57. The line shown was chosen to minimize the sum of the squares of the vertical distances between the data points and the line. Such a line is called a *least-squares line*. There are formulas which a calculator or computer uses to calculate the slope, m, and the y-intercept, b, of the least-squares line.

Correlation

When a computer or calculator calculates a regression line, it also gives a *correlation coefficient*, r. This number lies between -1 and $+1$ and measures how well a particular regression line fits the data. If $r = 1$, the data lie exactly on a line of positive slope. If $r = -1$, the data lie exactly on a line of negative slope. If r is close to 0, the data may be completely scattered, or there may be a non-linear relationship between the variables. (See Figure 1.58.)

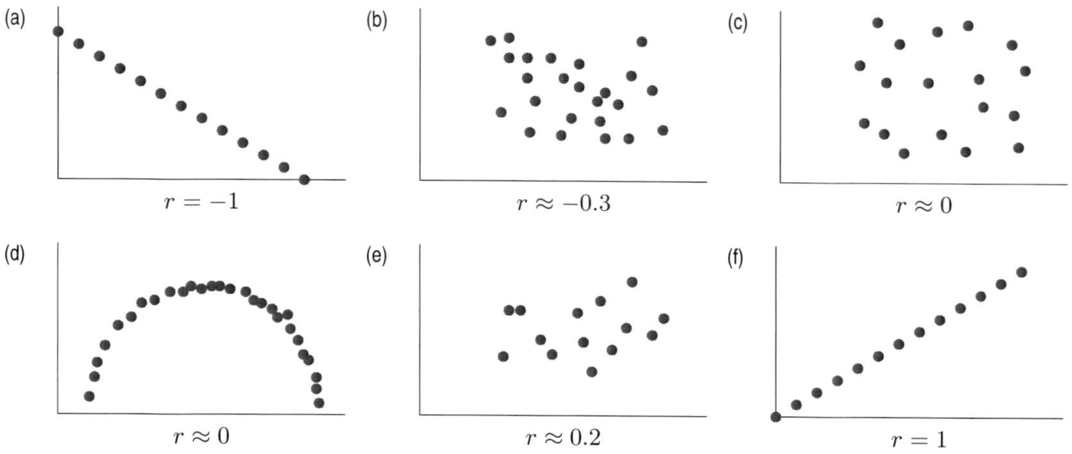

Figure 1.58: Various data sets and correlation coefficients

Example 2 The correlation coefficient for the viscosity data in Table 1.34 on page 44 is $r \approx -0.99$. The fact that r is negative tells us that the regression line has negative slope. The fact that r is close to -1 tells us that the regression line fits the data well.

The Difference between Relation, Correlation, and Causation

It is important to understand that a high correlation (either positive or negative) between two quantities does *not* imply causation. For example, there is a high correlation between children's reading level and shoe size.[22] However, large feet do not cause a child to read better (or vice versa). Larger feet and improved reading ability are both a consequence of growing older.

Notice also that a correlation of 0 does not imply that there is no relationship between x and y. For example, in Figure 1.58(d) there is a relationship between x and y-values, while Figure 1.58(c) exhibits no apparent relationship. Both data sets have a correlation coefficient of $r \approx 0$. Thus a correlation of $r = 0$ usually implies there is no linear relationship between x and y, but this does not mean there is no relationship at all.

[22]From *Statistics*, 2ed, by David Freedman. Robert Pisani, Roger Purves, Ani Adhikari, p. 142 (New York: W.W.Norton, 1991).

Exercises and Problems for Section 1.6

1. Match the r values with scatter plots in Figure 1.59.

$$r = -0.98, \quad r = -0.5, \quad r = -0.25,$$

$$r = 0, \quad r = 0.7, \quad r = 1.$$

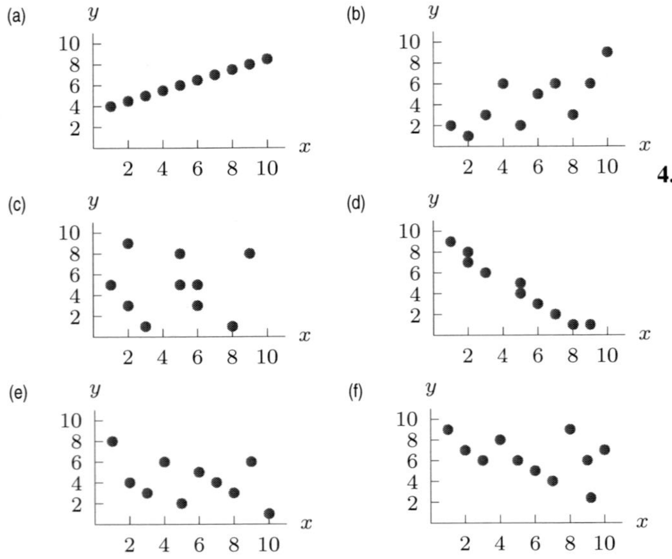

Figure 1.59

2. Table 1.35 shows the number of calories burned per minute by a person walking at 3 mph.

(a) Make a scatter plot of this data.
(b) Draw a regression line by eye.
(c) Roughly estimate the correlation coefficient by eye.

Table 1.35

Body weight (lb)	100	120	150	170	200	220
Calories	2.7	3.2	4.0	4.6	5.4	5.9

3. The rate of oxygen consumption for Colorado beetles increases with temperature. See Table 1.36.

(a) Make a scatter plot of this data.
(b) Draw an estimated regression line by eye.
(c) Use a calculator or computer to find the equation of the regression line. (Alternatively, find the equation of your line in part (b).) Round constants in the equation to the nearest integer.
(d) Interpret the slope and each intercept of the regression equation.
(e) Interpret the correlation between temperature and oxygen rate.

Table 1.36

°C	10	15	20	25	30
Oxygen consumption rate	90	125	200	300	375

4. Table 1.37 shows the IQ of ten students and the number of hours of TV each watches per week.

(a) Make a scatter plot of the data.
(b) By eye, make a rough estimate of the correlation coefficient.
(c) Use a calculator or computer to find the least squares regression line and the correlation coefficient. Your values should be correct to four decimal places.

Table 1.37

IQ	110	105	120	140	100	125	130	105	115	110
TV	10	12	8	2	12	10	5	6	13	3

5. An ecologist tracked 145 deer that were born in 1997. The number of deer, d, living each subsequent year is recorded in Table 1.38.

(a) Make a scatter plot of this data. Let $t = 0$ represent 1997.
(b) Draw by eye a good fitting line and estimate its equation. (Round the coefficients to integers.)
(c) Use a calculator or computer to find the equation of the least squares line. (Round the coefficients to integers.)
(d) Interpret the slope and each intercept of the line.
(e) Interpret the correlation between the year and the number of deer born in 1997 that are still alive.

Table 1.38

Year	1997	1998	1999	2000	2001	2002	2003	2004	2005
Deer	145	144	134	103	70	45	32	22	4

6. Table 1.39 gives the data on hand strength collected from college freshman using a grip meter.

(a) Make a scatter plot of these data treating the strength of the preferred hand as the independent variable.
(b) Draw a line on your scatter plot that is a good fit for these data and use it to find an approximate equation for the regression line.
(c) Using a graphing calculator or computer, find the equation of the least squares line.
(d) What would the predicted grip strength in the non-preferred hand be for a student with a preferred hand strength of 37?
(e) Discuss interpolation and extrapolation using specific examples in relation to this regression line.
(f) Discuss why r, the correlation coefficient, is both positive and close to 1.
(g) Why do the points tend to cluster into two groups on your scatter plot?

Table 1.39 *Hand strength for 20 students in kilograms*

Preferred	28	27	45	20	40	47	28	54	52	21
Nonpreferred	24	26	43	22	40	45	26	46	46	22
Preferred	53	52	49	45	39	26	25	32	30	32
Nonpreferred	47	47	41	44	33	20	27	30	29	29

7. Table 1.40 shows men's and women's world records for swimming distances from 50 meters to 1500 meters.[23]

(a) What values would you add to Table 1.40 to represent the time taken by both men and women to swim 0 meters?
(b) Plot men's time against distance, with time t in seconds on the vertical axis and distance d in meters on the horizontal axis. It is claimed that a straight line models this behavior well. What is the equation for that line? What does its slope represent? On the same graph, plot women's time against distance and find the equation of the straight line that models this behavior well. Is this line steeper or flatter than the men's line? What does that mean in terms of swimming? What are the values of the vertical intercepts? Do these values have a practical interpretation?
(c) On another graph plot the women's times against the men's times, with women's times, w, on the vertical axis and men's times, m, on the horizontal axis. It should look linear. How could you have predicted this linearity from the equations you found in part (b)? What is the slope of this line and how can it be interpreted? A newspaper reporter claims that the women's records are about 8% slower than the men's. Do the facts support this statement? What is the value of the vertical intercept? Does this value have a practical interpretation?

Table 1.40 *Men's and women's world swimming records*

Distance (m)	50	100	200	400	800	1500
Men (sec)	21.64	47.84	104.06	220.08	458.65	874.56
Women (sec)	24.13	53.62	116.64	243.85	496.22	952.10

8. In baseball, Henry Aaron holds the record for the greatest number of home-runs hit in the major leagues. Table 1.41 shows his cumulative yearly record[24] from the start of his career, 1954, until 1973.

(a) Plot Aaron's cumulative number of home runs H on the vertical axis, and the time t in years along the horizontal axis, where $t = 1$ corresponds to 1954.
(b) By eye draw a straight line that fits these data well and find its equation.
(c) Use a calculator or computer to find the equation of the regression line for these data. What is the correlation coefficient, r, to 4 decimal places? To 3 decimal places? What does this tell you?
(d) What does the slope of the regression line mean in terms of Henry Aaron's home-run record?
(e) From your answer to part (d), how many home-runs do you estimate Henry Aaron hit in each of the years 1974, 1975, 1976, and 1977? If you were told that Henry Aaron retired at the end of the 1976 season, would this affect your answers?

Table 1.41 *Henry Aaron's cumulative home-run record, H, from 1954 to 1973, with t in years since 1953*

t	1	2	3	4	5	6	7	8	9	10
H	13	40	66	110	140	179	219	253	298	342
t	11	12	13	14	15	16	17	18	19	20
H	366	398	442	481	510	554	592	639	673	713

[23]Data from "The World Almanac and Book of Facts: 2006," World Almanac Education Group, Inc., New York, 2006.
[24]Adapted from "Graphing Henry Aaron's home-run output" by H. Ringel, The Physics Teacher, January 1974, page 43.

CHAPTER SUMMARY

- **Functions**
  Definition: a rule which takes certain numbers as inputs and assigns to each input exactly one output number.
  Function notation, $y = f(x)$.
  Use of vertical line test.
- **Average rate of change**
  Average rate of change of $Q = f(t)$ on $[a, b]$ is
  $$\frac{\Delta Q}{\Delta t} = \frac{f(b) - f(a)}{b - a}.$$
  Increasing, decreasing functions; identifying from average rate of change.
- **Linear Functions**
  Value of y changes at constant rate.
- **Formulas for Linear Functions**
  Slope-intercept form: $y = b + mx$.
  Point-slope form: $y - y_0 = m(x - x_0)$.
  Standard form: $Ax + By + C = 0$.
- **Properties of Linear Functions**
  Interpretation of slope, vertical and horizontal intercepts.
  Intersection of lines: Solution of equations.
  Parallel lines: $m_1 = m_2$.
  Perpendicular lines: $m_1 = -\frac{1}{m_2}$.
- **Fitting Lines to Data**
  Linear regression; correlation. Interpolation, extrapolation; dangers of extrapolation.

REVIEW EXERCISES AND PROBLEMS FOR CHAPTER ONE

Exercises

In Exercises 1–5 a relationship is given between two quantities. Are both quantities functions of the other one, or is one or neither a function of the other? Explain.

1. $7w^2 + 5 = z^2$ **2.** $y = x^4 - 1$ **3.** $m = \sqrt{t}$

4.

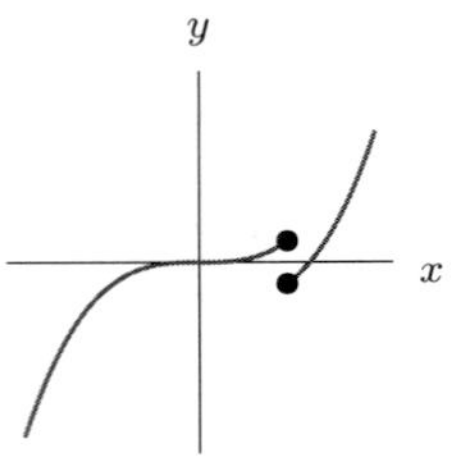

Figure 1.60

5. The number of gallons of gas, g, at \$2 per gallon and the number of pounds of coffee, c, at \$10 per pound that can be bought for a total of \$100.

6. In 2005, you have 40 CDs in your collection. In 2008, you have 120 CDs. In 2012, you have 40. What is the average rate of change in the size of your CD collection between

(a) 2005 and 2008? (b) 2008 and 2012?
(c) 2005 and 2012?

7. Find the average rate of change of $f(x) = 3x^2 + 1$ between the points

(a) $(1, 4)$ and $(2, 13)$ (b) (j, k) and (m, n)
(c) $(x, f(x))$ and $(x+h, f(x+h))$

In Exercises 8–9, could the table represent a linear function?

8.

λ	1	2	3	4	5
$q(\lambda)$	2	4	8	16	32

9.

t	3	6	9	12	15
$a(t)$	2	4	6	8	10

Problems 10–12 give data from a linear function. Find a formula for the function.

10.

x	200	230	300	320	400
$g(x)$	70	68.5	65	64	60

11.

t	1.2	1.3	1.4	1.5
$f(t)$	0.736	0.614	0.492	0.37

12.

t	5.2	5.3	5.4	5.5
$f(t)$	73.6	61.4	49.2	37

In Exercises 13–14, which line has the greater

(a) Slope? (b) y-intercept?

13. $y = 7 + 2x$, $y = 8 - 15x$

14. $y = 5 - 2x$, $y = 7 - 3x$

Are the lines in Exercises 15–18 perpendicular? Parallel? Neither?

15. $y = 5x + 2 \quad y = 2x + 5$

16. $y = 14x - 2 \quad y = -\frac{1}{14}x + 2$

17. $y = 3x + 3 \quad y = -\frac{1}{3}x + 3$

18. $7y = 8 + 21x;\ 9y = 77 - 3x$

Problems

Find formulas for the linear functions in Problems 19–22.

19. The graph of f contains $(-3, -8)$ and $(5, -20)$.

20. $g(100) = 2000$ and $g(400) = 3800$

21. $P = h(t)$ gives the size of a population that begins with 12,000 members and grows by 225 members each year.

22. The graph of h intersects the graph of $y = x^2$ at $x = -2$ and $x = 3$.

23. Find the equation of the line parallel to $3x + 5y = 6$ and passing through the point $(0, 6)$.

24. Find the equation of the line passing through the point $(2, 1)$ and perpendicular to the line $y = 5x - 3$.

25. Find the equations of the lines parallel to and perpendicular to the line $y + 4x = 7$, and through the point $(1, 5)$.

26. You have zero dollars now and the average rate of change in your net worth is \$5000 per year. How much money will you have in forty years?

27. A flight costs \$10,000 to operate, regardless of the number of passengers. Each ticket costs \$127. Express profit, π, as a linear function of the number of passengers, n, on the flight.

28. Table 1.42 gives the ranking r for three different names—Hannah, Alexis, and Madison. Of the three names, which was most popular and which was least popular in

(a) 1995? **(b)** 2004?

Table 1.42 *Ranking of names—Hannah (r_h), Alexis (r_a), and Madison (r_m)—for girls born between 1995 ($t = 0$) and 2004 ($t = 9$)*[25]

t	0	1	2	3	4	5	6	7	8	9
r_h	7	7	5	2	2	2	3	3	4	5
r_a	14	8	8	6	3	6	5	5	7	11
r_m	29	15	10	9	7	3	2	2	3	3

29. Table 1.42 gives information about the popularity of the names Hannah, Madison, and Alexis. Describe in words what your answers to parts (a)–(c) tell you about these names.

(a) Evaluate $r_m(0) - r_h(0)$.

(b) Evaluate $r_m(9) - r_h(9)$.

(c) Solve $r_m(t) < r_a(t)$.

30. Figure 1.61 gives the depth of the water at Montauk Point, New York, for a day in November.

(a) How many high tides took place on this day?

(b) How many low tides took place on this day?

(c) How much time elapsed in between high tides?

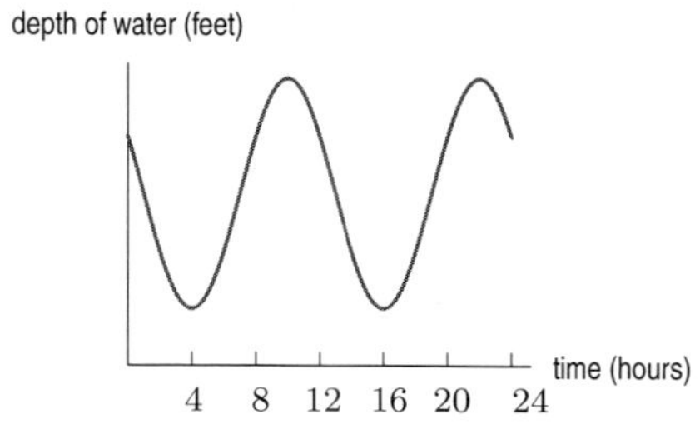

Figure 1.61

31. **(a)** Is the area, A, of a square a function of the length of one of its sides, s?

(b) Is the area, A, of a rectangle a function of the length of one of its sides, s?

32. A person's blood sugar level at a particular time of the day is partially determined by the time of the most recent meal. After a meal, blood sugar level increases rapidly, then slowly comes back down to a normal level. Sketch a person's blood sugar level as a function of time over the course of a day. Label the axes to indicate normal blood sugar level and the time of each meal.

[25] Data from the SSA website at www.ssa.gov, accessed January 12, 2006.

33. Many people think that hair growth is stimulated by haircuts. In fact, there is no difference in the rate hair grows after a haircut, but there *is* a difference in the rate at which hair's ends break off. A haircut eliminates dead and split ends, thereby slowing the rate at which hair breaks. However, even with regular haircuts, hair will not grow to an indefinite length. The average life cycle of human scalp hair is 3-5 years, after which the hair is shed.[26]

Judy trims her hair once a year, when its growth is slowed by split ends. She cuts off just enough to eliminate dead and split ends, and then lets it grow another year. After 5 years, she realizes her hair won't grow any longer. Graph the length of her hair as a function of time. Indicate when she receives her haircuts.

34. At the end of a semester, students' math grades are listed in a table which gives each student's ID number in the left column and the student's grade in the right column. Let N represent the ID number and the G represent the grade. Which quantity, N or G, must necessarily be a function of the other?

35. A price increases 5% due to inflation and is then reduced 10% for a sale. Express the final price as a function of the original price, P.

36. An 8-foot tall cylindrical water tank has a base of diameter 6 feet.

(a) How much water can the tank hold?
(b) How much water is in the tank if the water is 5 feet deep?
(c) Write a formula for the volume of water as a function of its depth in the tank.

37. Figure 1.62 shows the fuel consumption (in miles per gallon, mpg) of a car traveling at various speeds.

(a) How much gas is used on a 300 mile trip at 40 mph?
(b) How much gas is saved by traveling 60 mph instead of 70 mph on a 200 mile trip?
(c) According to this graph, what is the most fuel-efficient speed to travel? Explain.

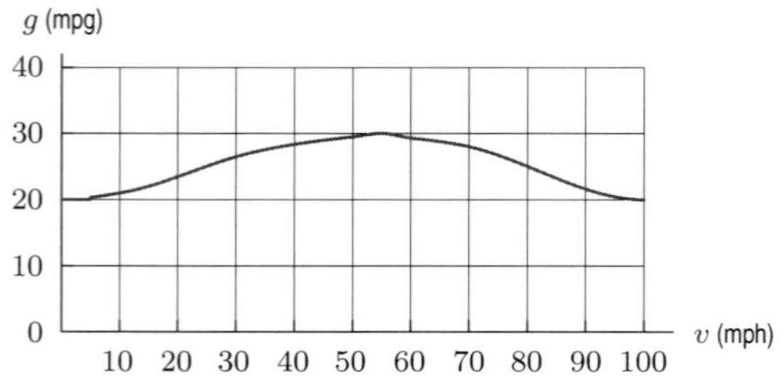

Figure 1.62

38. Academics have suggested that loss of worker productivity can result from sleep deprivation. An article in the Sunday, September 26, 1993, *New York Times* quotes David Poltrack, the senior vice president for planning and research at CBS, as saying that seven million Americans are staying up an hour later than usual to watch talk show host David Letterman. The article goes on to quote Timothy Monk, a professor at the University of Pittsburgh School of Medicine, as saying "... my hunch is that the effect [on productivity due to sleep deprivation among this group] would be in the area of a 10 percent decrement." The article next quotes Robert Solow, a Nobel prize-winning professor of economics at MIT, who suggests the following procedure to estimate the impact that this loss in productivity will have on the US economy—an impact he dubbed "the Letterman loss." First, Solow says, we find the percentage of the work force who watch the program. Next, we determine this group's contribution to the gross domestic product (GDP). Then we reduce the group's contribution by 10% to account for the loss in productivity due to sleep deprivation. The amount of this reduction is "the Letterman loss."

(a) The article estimated that the GDP is $\$6.325$ trillion, and that 7 million Americans watch the show. Assume that the nation's work force is 118 million people and that 75% of David Letterman's audience belongs to this group. What percentage of the work force is in Dave's audience?
(b) What percent of the GDP would be expected to come from David Letterman's audience? How much money would they have contributed if they had not watched the show?
(c) How big is "the Letterman Loss"?

39. There are x male job-applicants at a certain company and y female applicants. Suppose that 15% of the men are accepted and 18% of the women are accepted. Write an expression in terms of x and y representing each of the following quantities:

(a) The total number of applicants to the company.
(b) The total number of applicants accepted.
(c) The percentage of all applicants accepted.

40. You start 60 miles east of Pittsburgh and drive east at a constant speed of 50 miles per hour. (Assume that the road is straight and permits you to do this.) Find a formula for d, your distance from Pittsburgh as a function of t, the number of hours of travel.

[26] *Britannica Micropaedia* vol. 5. (Chicago: Encyclopaedia Britannica, Inc., 1989).

Table 1.43 gives the cost, $C(n)$, of producing a certain good as a linear function of n, the number of units produced. Use the table to answer Problems 41–43.

Table 1.43

n (units)	100	125	150	175
$C(n)$ (dollars)	11000	11125	11250	11375

41. Evaluate the following expressions. Give economic interpretations for each.

(a) $C(175)$ **(b)** $C(175) - C(150)$

(c) $\dfrac{C(175) - C(150)}{175 - 150}$

42. Estimate $C(0)$. What is the economic significance of this value?

43. The *fixed cost* of production is the cost incurred before any goods are produced. The *unit cost* is the cost of producing an additional unit. Find a formula for $C(n)$ in terms of n, given that

$$\text{Total cost} = \text{Fixed cost} + \text{Unit cost} \cdot \text{Number of units}$$

44. Sketch a family of functions $y = -2 - ax$ for five different values of x with $a < 0$.

45. Assume A, B, C are constants with $A \neq 0$, $B \neq 0$. Consider the equation

$$Ax + By = C.$$

(a) Show that $y = f(x)$ is linear. State the slope and the x- and y-intercepts of $f(x)$.

(b) Graph $y = f(x)$, labeling the x- and y-intercepts in terms of A, B, and C, assuming

(i) $A > 0, B > 0, C > 0$

(ii) $A > 0, B > 0, C < 0$

(iii) $A > 0, B < 0, C > 0$

CHECK YOUR UNDERSTANDING

Are the statements in Problems 1–54 true or false? Give an explanation for your answer.

1. $Q = f(t)$ means Q is equal to f times t.

2. A function must be defined by a formula.

3. If $P = f(x)$ then P is called the dependent variable.

4. Independent variables are always denoted by the letter x or t.

5. It is possible for two quantities to be related and yet neither be a function of the other.

6. A function is a rule that takes certain values as inputs and assigns to each input value exactly one output value.

7. It is possible for a table of values to represent a function.

8. If Q is a function of P, then P is a function of Q.

9. The graph of a circle is not the graph of a function.

10. If $n = f(A)$ is the number of angels that can dance on the head of a pin whose area is A square millimeters, then $f(10) = 100$ tells us that 10 angels can dance on the head of a pin whose area is 100 square millimeters.

11. Average speed can be computed by dividing the distance traveled by the time elapsed.

12. The average rate of change of a function Q with respect to t over an interval can be symbolically represented as $\dfrac{\Delta t}{\Delta Q}$.

13. If $y = f(x)$ and as x increases, y increases, then f is an increasing function.

14. If f is a decreasing function, then the average rate of change of f on any interval is negative.

15. The average rate of change of a function over an interval is the slope of a line connecting two points of the graph of the function.

16. The average rate of change of $y = 3x - 4$ between $x = 2$ and $x = 6$ is 7.

17. The average rate of change of $f(x) = 10 - x^2$ between $x = 1$ and $x = 2$ is the ratio $\dfrac{10 - 2^2 - 10 - 1^2}{2 - 1}$.

18. If $y = x^2$ then the slope of the line connecting the point $(2, 4)$ to the point $(3, 9)$ is the same as the slope of the line connecting the point $(-2, 4)$ to the point $(-3, 9)$.

19. A linear function can have different rates of change over different intervals.

20. The graph of a linear function is a straight line.

21. If a line has the equation $3x + 2y = 7$, then the slope of the line is 3.

22. A table of values represents a linear function if $\dfrac{\text{Change in output}}{\text{Change in input}} = \text{constant}$.

23. If a linear function is decreasing, then its slope is negative.

24. If $y = f(x)$ is linear and its slope is negative, then in the expression $\dfrac{\Delta y}{\Delta x}$ either Δx or Δy is negative, but not both.

25. A linear function can have a slope that is zero.

26. If a line has slope 2 and y-intercept -3, then its equation may be written $y = -3x + 2$.

27. The line $3x + 5y = 7$ has slope $3/5$.

28. A line that goes through the point $(-2, 3)$ and whose slope is 4 has the equation $y = 4x + 5$.

29. The line $4x + 3y = 52$ intersects the x-axis at $x = 13$.

30. If $f(x) = -2x + 7$ then $f(2) = 3$.

31. The line that passes through the points $(1, 2)$ and $(4, -10)$ has slope 4.

32. The linear equation $y - 5 = 4(x + 1)$ is equivalent to the equation $y = 4x + 6$.

33. The line $y - 4 = -2(x + 3)$ goes through the point $(4, -3)$.

34. The line whose equation is $y = 3 - 7x$ has slope -7.

35. The line $y = -5x + 8$ intersects the y-axis at $y = 8$.

36. The equation $y = -2 - \frac{2}{3}x$ represents a linear function.

37. The lines $y = 8 - 3x$ and $-2x + 16y = 8$ both cross the y-axis at $y = 8$.

38. The graph of $f(x) = 6$ is a line whose slope is six.

39. The lines $y = -\frac{4}{5}x + 7$ and $4x - 5y = 8$ are parallel.

40. The lines $y = 7 + 9x$ and $y - 4 = -\frac{1}{9}(x + 5)$ are perpendicular.

41. The lines $y = -2x + 5$ and $y = 6x - 3$ intersect at the point $(1, 3)$.

42. If two lines never intersect then their slopes are equal.

43. The equation of a line parallel to the y-axis could be $y = -\frac{3}{4}$.

44. A line parallel to the x-axis has slope zero.

45. The slope of a vertical line is undefined.

46. Fitting the best line to a set of data is called linear regression.

47. The process of estimating a value within the range for which we have data is called interpolation.

48. Extrapolation tends to be more reliable than interpolation.

49. If two quantities have a high correlation then one quantity causes the other.

50. If the correlation coefficient is zero, there is not a relationship between the two quantities.

51. A correlation coefficient can have a value of $-\frac{3}{7}$.

52. A value of a correlation coefficient is always between negative and positive one.

53. A correlation coefficient of one indicates that all the data points lie on a straight line.

54. A regression line is also referred to as a least squares line.

TOOLS FOR CHAPTER ONE: LINEAR EQUATIONS AND THE COORDINATE PLANE

Solving Linear Equations

To solve a linear equation, we isolate the variable.

Example 1 Solve $22 + 1.3t = 31.1$ for t.

Solution We subtract 22 from both sides. Since $31.1 - 22 = 9.1$, we have

$$1.3t = 9.1.$$

We divide both sides by 1.3, so

$$t = \frac{9.1}{1.3} = 7.$$

Example 2 Solve $3 - [5.4 + 2(4.3 - x)] = 2 - (0.3x - 0.8)$ for x.

Solution We begin by clearing the innermost parentheses on each side. Using the distributive law, this gives

$$3 - [5.4 + 8.6 - 2x] = 2 - 0.3x + 0.8.$$

Then

$$\begin{aligned} 3 - 14 + 2x &= 2 - 0.3x + 0.8 \\ 2.3x &= 13.8, \\ x &= 6. \end{aligned}$$

Example 3 Solve $ax = c + bx$ for x. Assume $a \neq b$.

Solution To solve for x, we first get all the terms involving x on the left side by subtracting bx from both sides

$$ax - bx = c.$$

Factoring on the left, $ax - bx = (a - b)x$, enables us to solve for x by dividing both sides by $(a - b)$:

$$\begin{aligned} x(a - b) &= c \\ x &= \frac{c}{(a - b)}. \end{aligned}$$

Since $a \neq b$, division by $(a - b)$ is possible.

Example 4 Solve for q if $p^2q + r(-q-1) = 4(p+r)$.

Solution We first collect all the terms containing q on the left side of the equation.

$$\begin{aligned} p^2q - rq - r &= 4p + 4r \\ p^2q - rq &= 4p + 5r. \end{aligned}$$

To solve for q, we factor and then divide by the coefficient of q.

$$\begin{aligned} q(p^2 - r) &= 4p + 5r \\ q &= \frac{4p+5r}{p^2 - r}. \end{aligned}$$

Solving Exactly Versus Solving Approximately

Some equations can be solved exactly, often by using algebra. For example, the equation $7x - 1 = 0$ has the exact solution $x = 1/7$. Other equations can be hard or even impossible to solve exactly. However, it is often possible, and sometimes easier, to find an approximate solution to an equation by using a graph or a numerical method on a calculator. The equation $7x-1 = 0$ has the approximate solution $x \approx 0.14$ (since $1/7 = 0.142857\ldots$). We use the sign $\approx$, meaning approximately equal, when we want to emphasize that we are making an approximation.

Systems of Linear Equations

To solve for two unknowns, we must have two equations—that is, two relationships between the unknowns. Similarly, three unknowns require three equations, and n unknowns (n an integer) require n equations. The group of equations is known as a *system* of equations. To solve the system, we find the *simultaneous* solutions to all equations in the system.

We can solve these equations either by *substitution* (see Example 5) or by *elimination* (see Example 6).

Example 5 Solve for x and y in the following system of equations using substitution.

$$\begin{cases} y + \dfrac{x}{2} = 3 \\ 2(x+y) = 1 - y \end{cases}$$

Solution Solving the first equation for y, we write $y = 3 - x/2$. Substituting for y in the second equation gives

$$2\left(x + \left(3 - \frac{x}{2}\right)\right) = 1 - \left(3 - \frac{x}{2}\right).$$

Then

$$\begin{aligned} 2x + 6 - x &= -2 + \frac{x}{2} \\ x + 6 &= -2 + \frac{x}{2} \\ 2x + 12 &= -4 + x \\ x &= -16. \end{aligned}$$

Using $x = -16$ in the first equation to find the corresponding y, we have

$$y - \frac{16}{2} = 3$$
$$y = 3 + 8 = 11.$$

Thus, the solution that simultaneously solves both equations is $x = -16$, $y = 11$.

Example 6 Solve for x and y in the following system of equations using elimination.

$$\begin{cases} 8x - 5y = 11 \\ -2x + 10y = -1. \end{cases}$$

Solution To eliminate y, we observe that if we multiply the first equation by 2, the coefficients of y are -10 and 10:

$$\begin{cases} 16x - 10y = 22 \\ -2x + 10y = -1. \end{cases}$$

Adding these two equations gives

$$14x = 21$$
$$x = 3/2.$$

We can substitute this value for x in either of the original equations to find y. For example

$$8\left(\frac{3}{2}\right) - 5y = 11$$
$$12 - 5y = 11$$
$$-5y = -1$$
$$y = 1/5$$

Thus, the solution is $x = 3/2$, $y = 1/5$,

Intersection of Two Lines

The coordinates of the point of intersection of two lines satisfies the equations of both lines. Thus, the point can be found by solving the equations simultaneously.

Example 7 Find the point of intersection of the lines $y = 3 - \frac{2}{3}x$ and $y = -4 + \frac{3}{2}x$.

Solution Since the y-values of the two lines are equal at the point of intersection, we have

$$-4+\frac{3}{2}x=3-\frac{2}{3}x.$$

Notice that we have converted a pair of equations into a single equation by eliminating one of the two variables. This equation can be simplified by multiplying both sides by 6:

$$\begin{aligned}6\left(-4+\frac{3}{2}x\right)&=6\left(3-\frac{2}{3}x\right)\\ -24+9x&=18-4x\\ 13x&=42\\ x&=\frac{42}{13}.\end{aligned}$$

We can evaluate either of the original equations at $x=\frac{42}{13}$ to find y. For example, $y=-4+\frac{3}{2}x$ gives

$$y=-4+\frac{3}{2}\left(\frac{42}{13}\right)=\frac{11}{13}.$$

Therefore, the point of intersection is $\left(\frac{42}{13},\frac{11}{13}\right)$. You can check that this point also satisfies the other equation. The lines and their point of intersection are shown in Figure 1.63.

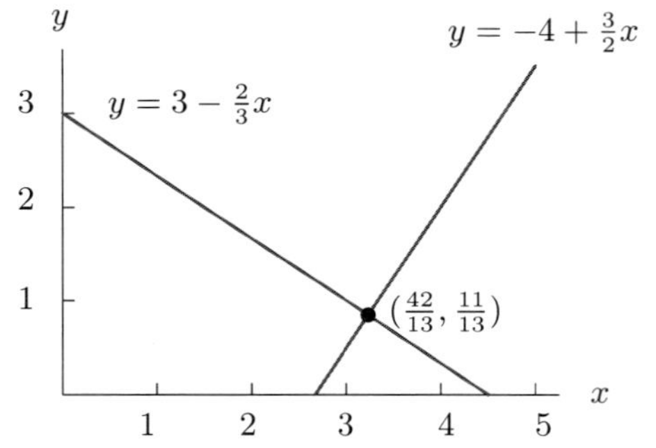

Figure 1.63: Intersection of lines is solution to simultaneous equations

Exercises on Tools for Chapter 1

Solve the equations in Exercises 1–17.

1. $3x=15$

2. $-2y=12$

3. $4z=22$

4. $x+3=10$

5. $y-5=21$

6. $w-23=-34$

7. $2x-5=13$

8. $7-3y=-14$

9. $13t+2=47$

10. $2x-5=4x-9$

11. $0.5x-3=7$

12. $17-28y=13y+24$

13. $\frac{5}{3}(y+2)=\frac{1}{2}-y$

14. $3t-\frac{2(t-1)}{3}=4$

15. $2(r+5)-3=3(r-8)+21$

16. $B-4[B-3(1-B)]=42$

17. $1.06s-0.01(248.4-s)=22.67s$

In Exercises 18–32, solve for the indicated variable.

18. $A=l\cdot w,$ for l.

19. $C=2\pi r,$ for r.

20. $I=Prt,$ for P.

21. $C=\frac{5}{9}(F-32),$ for F.

22. $l=l_0+\frac{k}{2}w$, for w.

23. $h=v_0t+\frac{1}{2}at^2$, for a.

24. $by-d=ay+c$, for y.

25. $ab+ax=c-ax$, for x.

26. $3xy+1=2y-5x$, for y.

27. $u(v+2)+w(v-3)=z(v-1)$, for v.

28. $S=\frac{rL-a}{r-1}$, for r.

29. $\frac{a-cx}{b+dx}+a=0$, for x.

30. $\frac{At-B}{C-B(1-2t)}=3$, for t.

31. $y'y^2+2xyy'=4y$, for y'.

32. $2x-(xy'+yy')+2yy'=0$, for y'.

Solve the systems of equations in Exercises 33–40.

33. $\begin{cases} x+y=3 \\ x-y=5 \end{cases}$

34. $\begin{cases} 2x-y=10 \\ x+2y=15 \end{cases}$

35. $\begin{cases} 3x-2y=6 \\ y=2x-5 \end{cases}$

36. $\begin{cases} x=7y-9 \\ 4x-15y=26 \end{cases}$

37. $\begin{cases} 2x+3y=7 \\ y=-\frac{3}{5}x+6 \end{cases}$

38. $\begin{cases} 3x-y=17 \\ -2x-3y=-4 \end{cases}$

39. $\begin{cases} 2(x+y)=3 \\ x=y+3(x-5) \end{cases}$

40. $\begin{cases} ax+y=2a \\ x+ay=1+a^2 \end{cases}$

Determine the points of intersection for Exercises 41–44.

41.

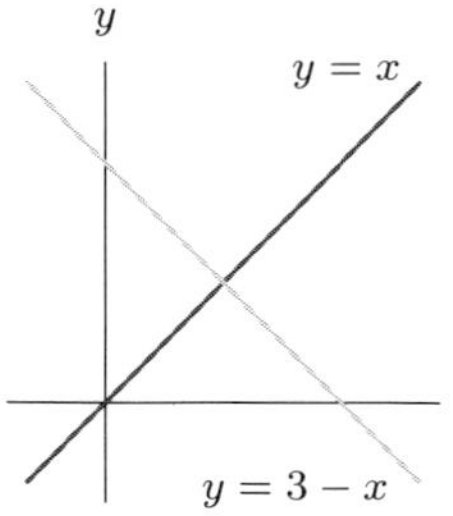

42.

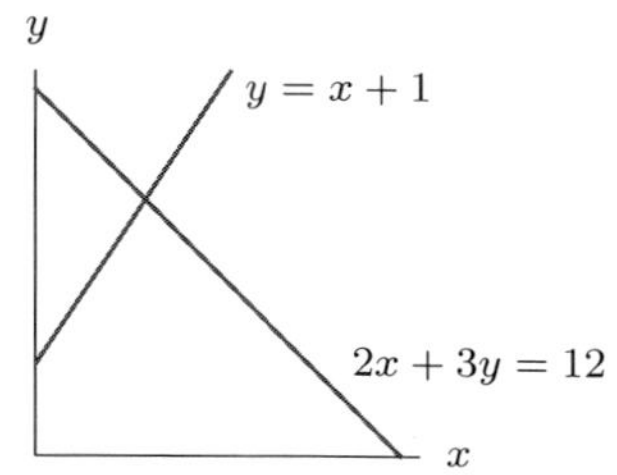

43.

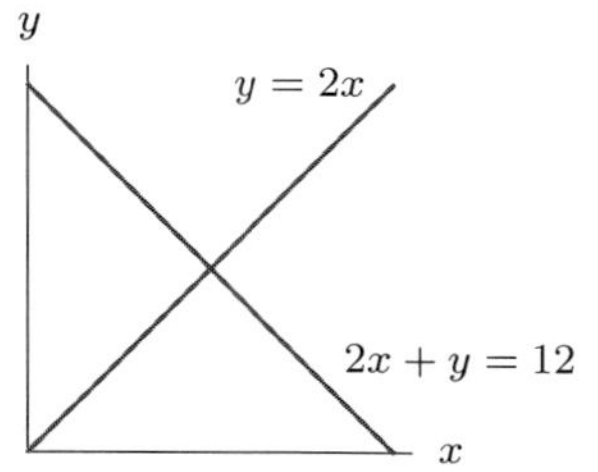

44.

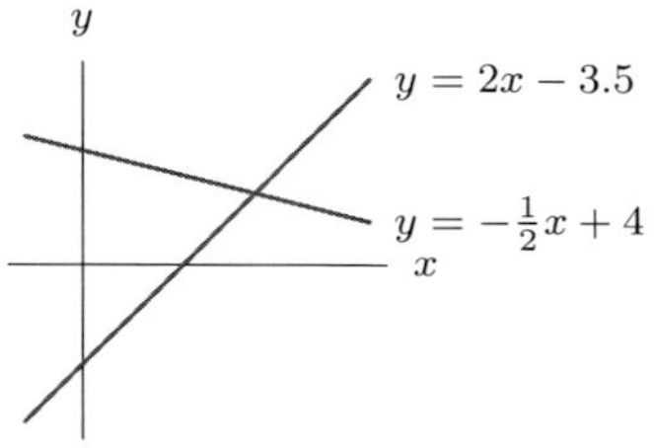

The figures in Problems 45–48 are parallelograms. Find the coordinates of the labeled point(s).

45.

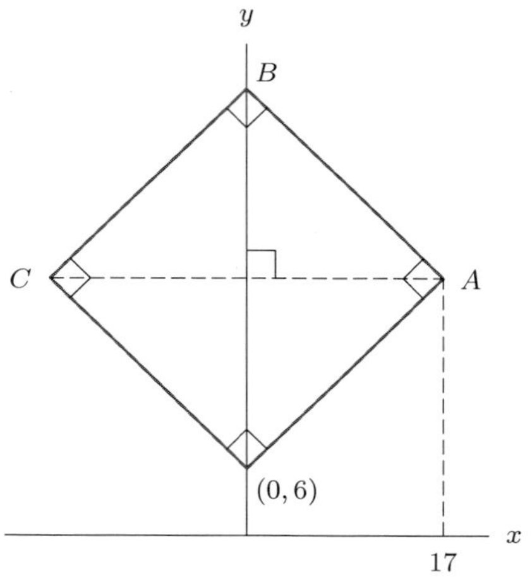

46.

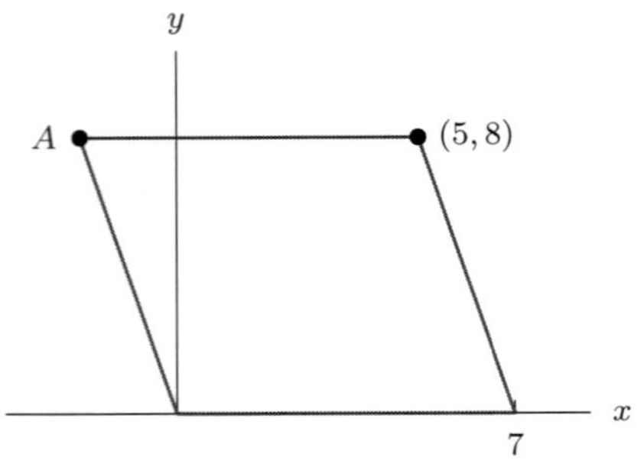

47.

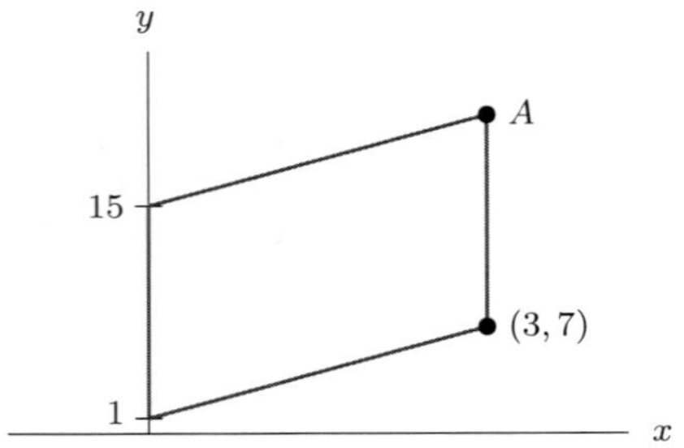

48.

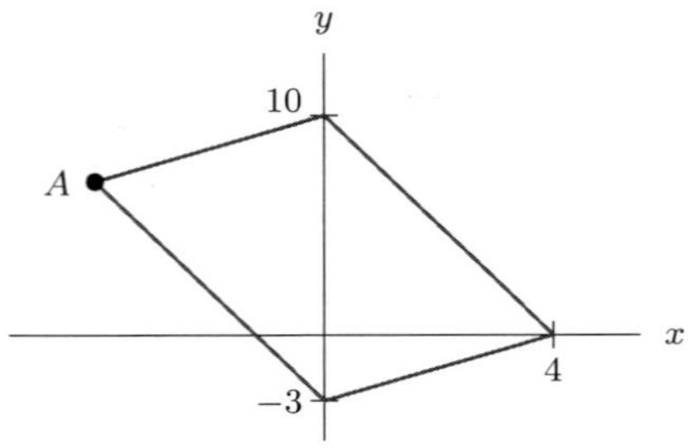

The figures in Problems 49–52 contain a semicircle with the center marked. Find the coordinates of A, a point on the diameter, and B, an extreme point (highest, lowest, or farthest to the right).

49.

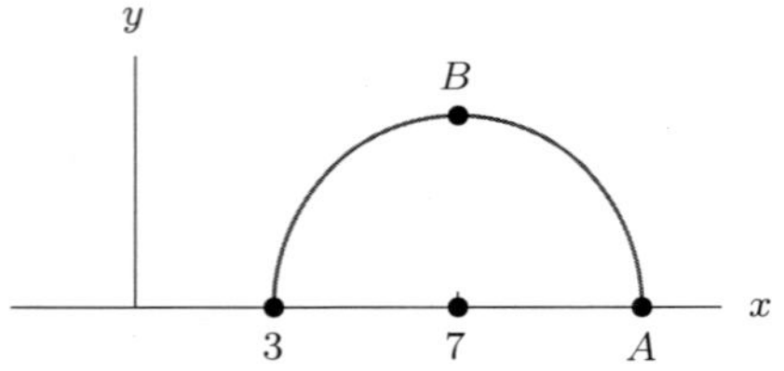

50.

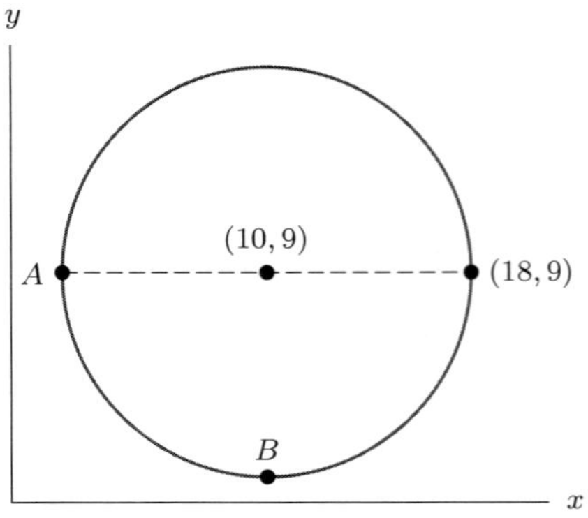

51.

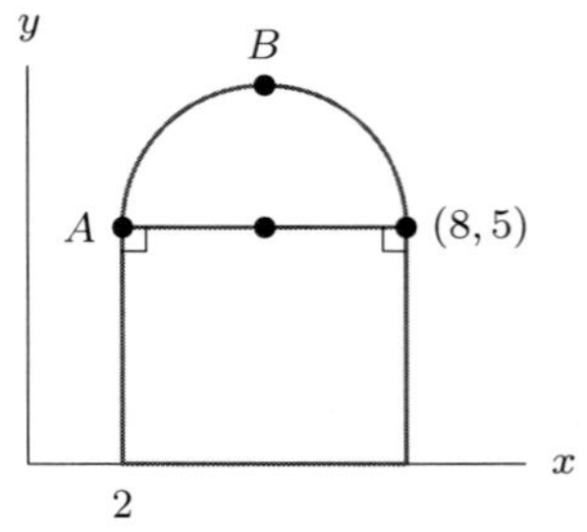

52.

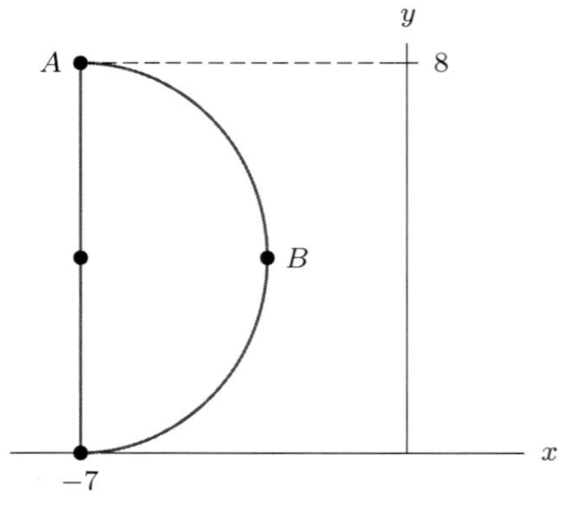

Chapter Two

FUNCTIONS

In this chapter, we investigate properties and notation common to all functions. We begin with a closer look at function notation, including an introduction to inverse functions. The ideas of domain, range, and piecewise defined functions are addressed. Concavity is then introduced and applied to quadratic functions.

The Tools Section on page 99 reviews quadratic equations.

2.1 INPUT AND OUTPUT

Finding Output Values: Evaluating a Function

Evaluating a function means calculating the value of a function's output from a particular value of the input.

In the housepainting example on page 4, the notation $n = f(A)$ indicates that n is a function of A. The expression $f(A)$ represents the output of the function—specifically, the amount of paint required to cover an area of A ft^2. For example $f(20{,}000)$ represents the number of gallons of paint required to cover a house of 20,000 ft^2.

Example 1 Using the fact that 1 gallon of paint covers 250 ft^2, evaluate the expression $f(20{,}000)$.

Solution To evaluate $f(20{,}000)$, calculate the number of gallons required to cover 20,000 ft^2:

$$f(20{,}000) = \frac{20{,}000 \text{ ft}^2}{250 \text{ ft}^2/\text{gallon}} = 80 \text{ gallons of paint.}$$

Evaluating a Function Using a Formula

If we have a formula for a function, we evaluate it by substituting the input value into the formula.

Example 2 The formula for the area of a circle of radius r is $A = q(r) = \pi r^2$. Use the formula to evaluate $q(10)$ and $q(20)$. What do your results tell you about circles?

Solution In the expression $q(10)$, the value of r is 10, so

$$q(10) = \pi \cdot 10^2 = 100\pi \approx 314.$$

Similarly, substituting $r = 20$, we have

$$q(20) = \pi \cdot 20^2 = 400\pi \approx 1257.$$

The statements $q(10) \approx 314$ and $q(20) \approx 1257$ tell us that a circle of radius 10 cm has an area of approximately 314 cm^2 and a circle of radius 20 cm has an area of approximately 1257 cm^2.

Example 3 Let $g(x) = \dfrac{x^2+1}{5+x}$. Evaluate the following expressions.

(a) $g(3)$ (b) $g(-1)$ (c) $g(a)$

Solution (a) To evaluate $g(3)$, replace every x in the formula with 3:

$$g(3) = \frac{3^2+1}{5+3} = \frac{10}{8} = 1.25.$$

(b) To evaluate $g(-1)$, replace every x in the formula with (-1):

$$g(-1) = \frac{(-1)^2+1}{5+(-1)} = \frac{2}{4} = 0.5.$$

(c) To evaluate $g(a)$, replace every x in the formula with a:

$$g(a) = \frac{a^2+1}{5+a}.$$

Evaluating a function may involve algebraic simplification, as the following example shows.

Example 4 Let $h(x) = x^2 - 3x + 5$. Evaluate and simplify the following expressions.

(a) $h(2)$ (b) $h(a-2)$ (c) $h(a) - 2$ (d) $h(a) - h(2)$

Solution Notice that x is the input and $h(x)$ is the output. It is helpful to rewrite the formula as

$$\text{Output} = h(\text{Input}) = (\text{Input})^2 - 3 \cdot (\text{Input}) + 5.$$

(a) For $h(2)$, we have Input $= 2$, so

$$h(2) = (2)^2 - 3 \cdot (2) + 5 = 3.$$

(b) In this case, Input $= a - 2$. We substitute and multiply out

$$\begin{aligned} h(a-2) &= (a-2)^2 - 3(a-2) + 5 \\ &= a^2 - 4a + 4 - 3a + 6 + 5 \\ &= a^2 - 7a + 15. \end{aligned}$$

(c) First input a, then subtract 2:

$$\begin{aligned} h(a) - 2 &= a^2 - 3a + 5 - 2 \\ &= a^2 - 3a + 3. \end{aligned}$$

(d) Since we found $h(2) = 3$ in part (a), we subtract from $h(a)$:

$$\begin{aligned} h(a) - h(2) &= a^2 - 3a + 5 - 3 \\ &= a^2 - 3a + 2. \end{aligned}$$

Finding Input Values: Solving Equations

Given an input, we evaluate the function to find the output. Sometimes the situation is reversed; we know the output and we want to find the corresponding input. If the function is given by a formula, the input values are solutions to an equation.

Example 5 Use the cricket function $T = \frac{1}{4}R + 40$, introduced on page 3, to find the rate, R, at which the snowy tree cricket chirps when the temperature, T, is 76°F.

Solution We want to find R when $T = 76$. Substitute $T = 76$ into the formula and solve the equation

$$\begin{aligned} 76 &= \frac{1}{4}R + 40 \\ 36 &= \frac{1}{4}R && \text{subtract 40 from both sides} \\ 144 &= R. && \text{multiply both sides by 4} \end{aligned}$$

The cricket chirps at a rate of 144 chirps per minute when the temperature is 76°F.

Example 6 Suppose $f(x) = \dfrac{1}{\sqrt{x-4}}$.

(a) Find an x-value that results in $f(x) = 2$.
(b) Is there an x-value that results in $f(x) = -2$?

Solution (a) To find an x-value that results in $f(x) = 2$, solve the equation

$$2 = \frac{1}{\sqrt{x-4}}.$$

Square both sides

$$4 = \frac{1}{x-4}.$$

Now multiply by $(x-4)$

$$\begin{aligned} 4(x-4) &= 1 \\ 4x - 16 &= 1 \\ x &= \frac{17}{4} = 4.25. \end{aligned}$$

The x-value is 4.25. (Note that the simplification $(x-4)/(x-4) = 1$ in the second step was valid because $x - 4 \neq 0$.)

(b) Since $\sqrt{x-4}$ is nonnegative if it is defined, its reciprocal, $f(x) = \dfrac{1}{\sqrt{x-4}}$ is also nonnegative if it is defined. Thus, $f(x)$ is not negative for any x input, so there is no x-value that results in $f(x) = -2$.

In the next example, we solve an equation for a quantity that is being used to model a physical quantity; we must choose the solutions that make sense in the context of the model.

Example 7 Let $A = q(r)$ be the area of a circle of radius r, where r is in cm. What is the radius of a circle whose area is 100 cm^2?

Solution The output $q(r)$ is an area. Solving the equation $q(r) = 100$ for r gives the radius of a circle whose area is 100 cm^2. Since the formula for the area of a circle is $q(r) = \pi r^2$, we solve

$$\begin{aligned} q(r) = \pi r^2 &= 100 \\ r^2 &= \frac{100}{\pi} \\ r &= \pm\sqrt{\frac{100}{\pi}} = \pm 5.642. \end{aligned}$$

We have two solutions for r, one positive and one negative. Since a circle cannot have a negative radius, we take $r = 5.642$ cm. A circle of area 100 cm^2 has a radius of 5.642 cm.

Finding Output and Input Values From Tables and Graphs

The following two examples use function notation with a table and a graph respectively.

Example 8 Table 2.1 shows the revenue, $R = f(t)$, received or expected, by the National Football League,[1] NFL, from network TV as a function of the year, t, since 1975.

(a) Evaluate and interpret $f(25)$. (b) Solve and interpret $f(t) = 1159$.

Table 2.1

Year, t (since 1975)	0	5	10	15	20	25	30
Revenue, R (million \$)	201	364	651	1075	1159	2200	2200

[1] *Newsweek*, January 26, 1998.

Solution (a) The table shows $f(25) = 2200$. Since $t = 25$ in the year 2000, we know that NFL's projected revenue from TV was \$2200 million in the year 2000.

(b) Solving $f(t) = 1159$ means finding the year in which TV revenues were \$1159 million; it is $t = 20$. In 1995, NFL's TV revenues were \$1159 million.

Example 9 A man drives from his home to a store and back. The entire trip takes 30 minutes. Figure 2.1 gives his velocity $v(t)$ (in mph) as a function of the time t (in minutes) since he left home. A negative velocity indicates that he is traveling away from the store back to his home.

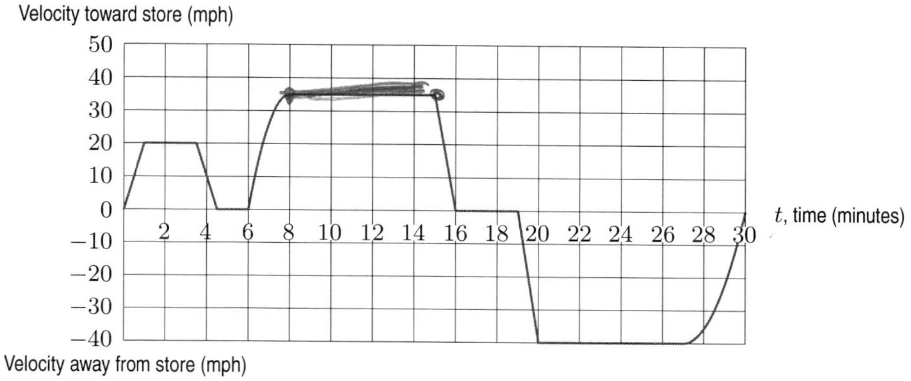

Figure 2.1: Velocity of a man on a trip to the store and back

Evaluate and interpret:

(a) $v(5)$ (b) $v(24)$ (c) $v(8) - v(6)$ (d) $v(-3)$

Solve for t and interpret:

(e) $v(t) = 15$ (f) $v(t) = -20$ (g) $v(t) = v(7)$

Solution (a) To evaluate $v(5)$, look on the graph where $t = 5$ minutes. Five minutes after he left home, his velocity is 0 mph. Thus, $v(5) = 0$. Perhaps he had to stop at a light.

(b) The graph shows that $v(24) = -40$ mph. After 24 minutes, he is traveling at 40 mph away from the store.

(c) From the graph, $v(8) = 35$ mph and $v(6) = 0$ mph. Thus, $v(8) - v(6) = 35 - 0 = 35$. This shows that the man's speed increased by 35 mph in the interval between $t = 6$ minutes and $t = 8$ minutes.

(d) The quantity $v(-3)$ is not defined since the graph only gives velocities for nonnegative times.

(e) To solve for t when $v(t) = 15$, look on the graph where the velocity is 15 mph. This occurs at $t \approx 0.75$ minute, 3.75 minutes, 6.5 minutes, and 15.5 minutes. At each of these four times the man's velocity was 15 mph.

(f) To solve $v(t) = -20$ for t, we see that the velocity is -20 mph at $t \approx 19.5$ and $t \approx 29$ minutes.

(g) First we evaluate $v(7) \approx 27$. To solve $v(t) = 27$, we look for the values of t making the velocity 27 mph. One such t is of course $t = 7$; the other t is $t \approx 15$ minutes.

Exercises and Problems for Section 2.1

Exercises

In Exercises 1–2, evaluate the function for $x = -7$.

1. $f(x) = x/2 - 1$

2. $f(x) = x^2 - 3$

For Exercises 3–6, calculate exactly the values of y when $y = f(4)$ and of x when $f(x) = 6$.

3. $f(x) = \dfrac{6}{2 - x^3}$

4. $f(x) = \sqrt{20 + 2x^2}$

5. $f(x) = 4x^{3/2}$

6. $f(x) = x^{-3/4} - 2$

7. If $f(x) = 2x+1$, (a) Find $f(0)$ (b) Solve $f(x) = 0$.

8. If $f(t) = t^2-4$, (a) Find $f(0)$ (b) Solve $f(t) = 0$.

9. If $g(x) = x^2-5x+6$, (a) Find $g(0)$ (b) Solve $g(x) = 0$.

10. If $g(t) = \dfrac{1}{t+2}-1$, (a) Find $g(0)$ (b) Solve $g(t) = 0$.

11. If $f(x) = \dfrac{x}{1-x^2}$, find $f(-2)$.

12. If $h(x) = ax^2 + bx + c$, find $h(0)$.

13. If $P(t) = 170 - 4t$, find $P(4) - P(2)$.

14. If $g(x) = -\frac{1}{2}x^{1/3}$, find $g(-27)$.

15. Let $h(x) = 1/x$. Find **(a)** $h(x+3)$ **(b)** $h(x) + h(3)$

16. Let $f(x) = \dfrac{2x+1}{x+1}$. For what value of x is $f(x) = 0.3$?

17. **(a)** Using Table 2.2, evaluate $f(1)$, $f(-1)$, and $-f(1)$.
(b) Solve $f(x) = 0$ for x.

Table 2.2

x	-1	0	1	2
$f(x)$	0	-1	2	1

18. **(a)** In Figure 2.2, estimate $f(0)$.
(b) For what x-value(s) is $f(x) = 0$?
(c) For what x-value(s) is $f(x) > 0$?

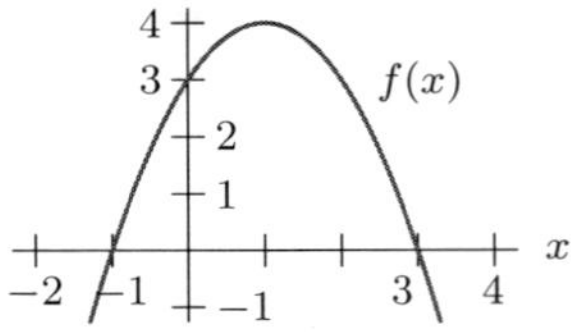

Figure 2.2

19. Let $F = g(t)$ be the number of foxes in a park as a function of t, the number of months since January 1. Evaluate $g(9)$ using Table 1.3 on page 5. What does this tell us about the fox population?

20. Let $F = g(t)$ be the number of foxes in month t in the national park described in Example 5 on page 5. Solve the equation $g(t) = 75$. What does your solution tell you about the fox population?

Problems

21. If $V = \frac{1}{3}\pi r^2 h$ gives the volume of a cylinder, what is the value of V when $r = 3$ inches and $h = 2$ inches? Give units.

22. Let $q(x) = 3 - x^2$. Evaluate and simplify:

(a) $q(5)$ **(b)** $q(a)$
(c) $q(a - 5)$ **(d)** $q(a) - 5$
(e) $q(a) - q(5)$

23. Let $p(x) = x^2 + x + 1$. Find $p(-1)$ and $-p(1)$. Are they equal?

24. Let $f(x) = 3 + 2x^2$. Find $f\left(\dfrac{1}{3}\right)$ and $\dfrac{f(1)}{f(3)}$. Are they equal?

25. Chicago's average monthly rainfall, $R = f(t)$ inches, is given as a function of month, t, in Table 2.3. (January is $t = 1$.) Solve and interpret:

(a) $f(t) = 3.7$ **(b)** $f(t) = f(2)$

Table 2.3

t	1	2	3	4	5	6	7	8
R	1.8	1.8	2.7	3.1	3.5	3.7	3.5	3.4

26. Let $g(x) = x^2 + x$. Find formulas for the following functions. Simplify your answers.

(a) $g(-3x)$ **(b)** $g(1-x)$ **(c)** $g(x+\pi)$
(d) $g(\sqrt{x})$ **(e)** $g(1/(x+1))$ **(f)** $g(x^2)$

27. Let $f(x) = \dfrac{x}{x-1}$.

(a) Find and simplify

(i) $f\left(\frac{1}{t}\right)$ (ii) $f\left(\frac{1}{t+1}\right)$

(b) Solve $f(x) = 3$.

28. (a) Find a point on the graph of $h(x) = \sqrt{x+4}$ whose x-coordinate is 5.

(b) Find a point on the graph whose y-coordinate is 5.

(c) Graph $h(x)$ and mark the points in parts (a) and (b).

(d) Let $p = 2$. Calculate $h(p+1) - h(p)$.

29. Use the graph of $f(x)$ in Figure 2.3 to estimate:

(a) $f(0)$ **(b)** $f(1)$ **(c)** $f(b)$ **(d)** $f(c)$ **(e)** $f(d)$

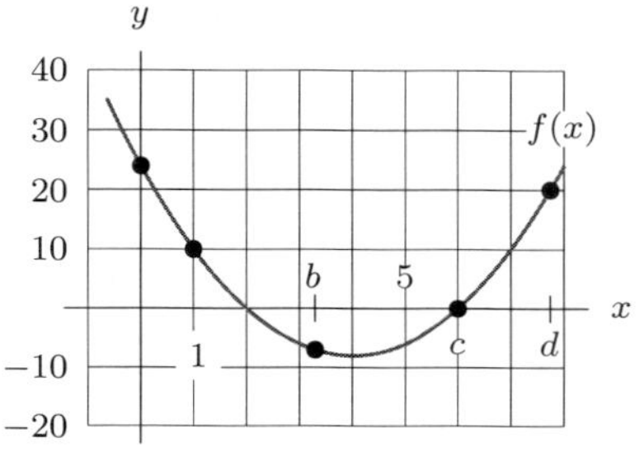

Figure 2.3

30. (a) Using Figure 2.4, fill in Table 2.4.

Table 2.4

x	-2	-1	0	1	2	3
$h(x)$						

(b) Evaluate $h(3) - h(1)$ (c) Evaluate $h(2) - h(0)$
(d) Evaluate $2h(0)$ (e) Evaluate $h(1) + 3$

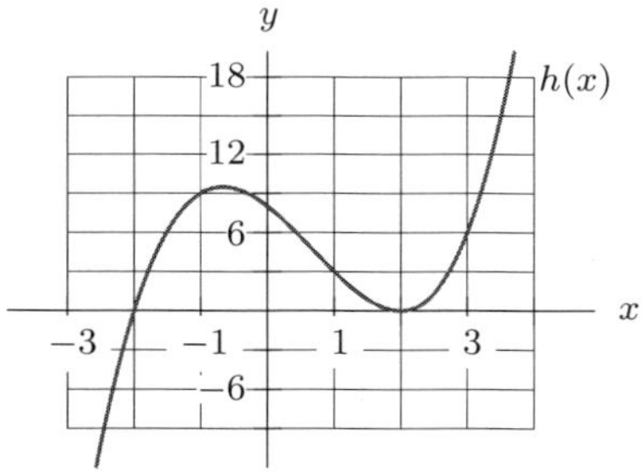

Figure 2.4

31. Let $f(x) = \sqrt{x^2 + 16} - 5$.

(a) Find $f(0)$

(b) For what values of x is $f(x)$ zero?

(c) Find $f(3)$

(d) What is the vertical intercept of the graph of $f(x)$?

(e) Where does the graph cross the x-axis?

32. Use the letters a, b, c, d, e, h in Figure 2.5 to answer the following questions.

(a) What are the coordinates of the points P and Q?

(b) Evaluate $f(b)$.

(c) Solve $f(x) = e$ for x.

(d) Suppose $c = f(z)$ and $z = f(x)$. What is x?

(e) Suppose $f(b) = -f(d)$. What additional information does this give you?

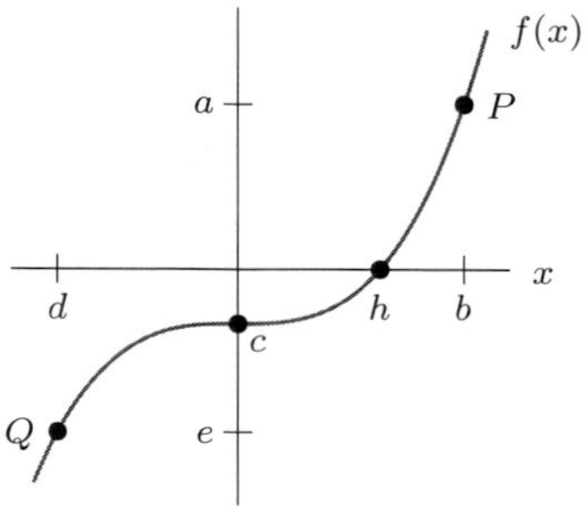

Figure 2.5

33. A ball is thrown up from the ground with initial velocity 64 ft/sec. Its height at time t is

$$h(t) = -16t^2 + 64t.$$

(a) Evaluate $h(1)$ and $h(3)$. What does this tell us about the height of the ball?

(b) Sketch this function. Using a graph, determine when the ball hits the ground and the maximum height of the ball.

34. Let $v(t) = t^2 - 2t$ be the velocity, in ft/sec, of an object at time t, in seconds.

(a) What is the initial velocity, $v(0)$?

(b) When does the object have a velocity of zero?

(c) What is the meaning of the quantity $v(3)$? What are its units?

35. Let $s(t) = 11t^2 + t + 100$ be the position, in miles, of a car driving on a straight road at time t, in hours. The car's velocity at any time t is given by $v(t) = 22t + 1$.

(a) Use function notation to express the car's position after 2 hours. Where is the car then?
(b) Use function notation to express the question, "When is the car going 65 mph?"
(c) Where is the car when it is going 67 mph?

36. New York state income tax is based on taxable income, which is part of a person's total income. The tax owed to the state is calculated using the taxable income (not total income). In 2005, for a single person with a taxable income between \$20,000 and \$100,000, the tax owed was \$973 plus 6.85% of the taxable income over \$20,000.

(a) Compute the tax owed by a lawyer whose taxable income is \$68,000.
(b) Consider a lawyer whose taxable income is 80% of her total income, $\$x$, where x is between \$85,000 and \$120,000. Write a formula for $T(x)$, the taxable income.
(c) Write a formula for $L(x)$, the amount of tax owed by the lawyer in part (b).
(d) Use $L(x)$ to evaluate the tax liability for $x = 85{,}000$ and compare your results to part (a).

37. **(a)** The Fibonacci sequence is a sequence of numbers that begins 1, 1, 2, 3, 5, ... Each term in the sequence is the sum of the two preceding terms. For example,

$$2 = 1 + 1, \quad 3 = 2 + 1, \quad 5 = 2 + 3, \ldots.$$

Based on this observation, complete the following table of values for $f(n)$, the nth term in the Fibonacci sequence.

n	1	2	3	4	5	6	7	8	9	10	11	12
$f(n)$	1	1	2	3	5							

(b) The table of values in part (a) can be completed even though we don't have a formula for $f(n)$. Does the fact that we don't have a formula mean that $f(n)$ is not a function?
(c) Are you able to evaluate the following expressions using parts (a) and (b)? If so, do so; if not, explain why not.

$$f(0), \quad f(-1), \quad f(-2), \quad f(0.5).$$

38. **(a)** Complete Table 2.5 using

$$f(x) = 2x(x-3) - x(x-5) \quad \text{and} \quad g(x) = x^2 - x.$$

What do you notice? Graph these two functions. Are the two functions the same? Explain.

(b) Complete Table 2.6 using

$$h(x) = x^5 - 5x^3 + 6x + 1 \quad \text{and} \quad j(x) = 2x + 1.$$

What do you notice? Graph these two functions. Are the two functions the same? Explain.

Table 2.5

x	-2	-1	0	1	2
$f(x)$					
$g(x)$					

Table 2.6

x	-2	-1	0	1	2
$h(x)$					
$j(x)$					

39. A psychologist conducts an experiment to determine the effect of sleep loss on job performance. Let $p = f(t)$ be the number of minutes it takes the average person to complete a particular task if they have lost t minutes of sleep, where $t = 0$ means they have had exactly 8 hours of sleep. For instance, $f(60)$ is the amount of time it takes the average person to complete the task after sleeping for only 7 hours. Let $p_0 = f(0)$ and let t_1, t_2, and t_3 be positive constants. Explain what the following statements tell you about sleep loss and job performance.

(a) $f(30) = p_0 + 5$
(b) $f(t_1) = 2p_0$
(c) $f(2t_1) = 1.5f(t_1)$
(d) $f(t_2 + 60) = f(t_2 + 30) + 10$

Problems 40–41 concern $v = r(s)$, the eyewall wind profile of a hurricane at landfall, where v is the eyewall wind speed (in mph) as a function of s, the height (in meters) above the ground. (The eyewall is the band of clouds that surrounds the eye of the storm.) Let s_0 be the height at which the wind speed is greatest, and let $v_0 = r(s_0)$. Interpret the following in terms of the hurricanes.

40. $r(0.5s_0)$

41. $r(s) = 0.75v_0$

2.2 DOMAIN AND RANGE

In Example 4 on page 5, we defined R to be the average monthly rainfall at Chicago's O'Hare airport in month t. Although R is a function of t, the value of R is not defined for every possible value of t. For instance, it makes no sense to consider the value of R for $t = -3$, or $t = 8.21$, or $t = 13$ (since a year has 12 months). Thus, although R is a function of t, this function is defined only for certain values of t. Notice also that R, the output value of this function, takes only the values $\{1.8, 2.1, 2.4, 2.5, 2.7, 3.1, 3.2, 3.4, 3.5, 3.7\}$.

A function is often defined only for certain values of the independent variable. Also, the dependent variable often takes on only certain values. This leads to the following definitions:

> If $Q = f(t)$, then
> - the **domain** of f is the set of input values, t, which yield an output value.
> - the **range** of f is the corresponding set of output values, Q.

Thus, the domain of a function is the set of input values, and the range is the set of output values.

If the domain of a function is not specified, we usually assume that it is as large as possible—that is, all numbers that make sense as inputs for the function. For example, if there are no restrictions, the domain of the function $f(x) = x^2$ is the set of all real numbers, because we can substitute any real number into the formula $f(x) = x^2$. Sometimes, however, we may restrict the domain to suit a particular application. If the function $f(x) = x^2$ is used to represent the area of a square of side x, we restrict the domain to positive numbers.

If a function is being used to model a real-world situation, the domain and range of the function are often determined by the constraints of the situation being modeled, as in the next example.

Example 1 The house painting function $n = f(A)$ in Example 2 on page 4 has domain $A > 0$ because all houses have some positive area. There is a practical upper limit to A because houses cannot be infinitely large, but in principle, A can be as large or as small as we like, as long as it is positive. Therefore we take the domain of f to be $A > 0$.

The range of this function is $n \geq 0$, because we cannot use a negative amount of paint.

Choosing Realistic Domains and Ranges

When a function is used to model a real situation, it may be necessary to modify the domain and range.

Example 2 Algebraically speaking, the formula

$$T = \frac{1}{4}R + 40$$

can be used for all values of R. If we know nothing more about this function than its formula, its domain is all real numbers. The formula for $T = \frac{1}{4}R + 40$ can return any value of T when we choose an appropriate R-value (See Figure 2.6.) Thus, the range of the function is also all real numbers. However, if we use this formula to represent the temperature, T, as a function of a cricket's chirp rate, R, as we did in Example 1 on page 2, some values of R cannot be used. For example, it does not make sense to talk about a negative chirp rate. Also, there is some maximum chirp rate R_{max}

that no cricket can physically exceed. Thus, to use this formula to express T as a function of R, we must restrict R to the interval $0 \leq R \leq R_{max}$ shown in Figure 2.7.

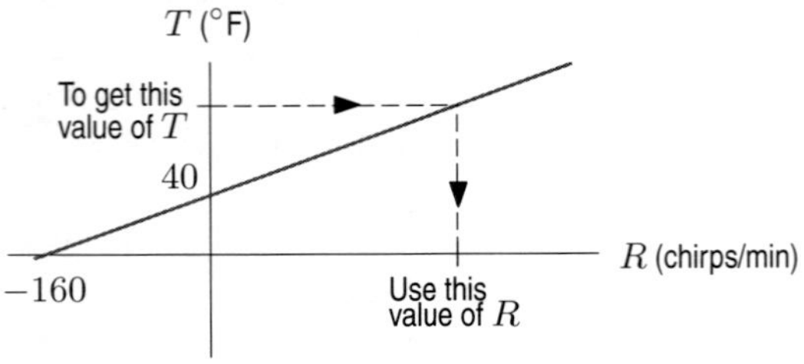

Figure 2.6: Graph showing that any T value can be obtained from some R value

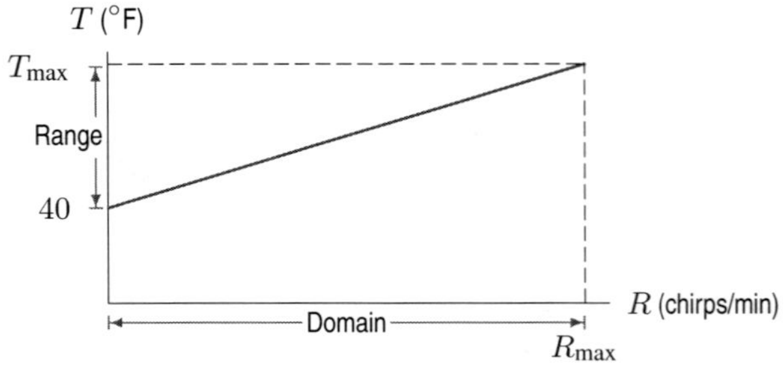

Figure 2.7: Graph showing that if $0 \leq R \leq R_{max}$, then $40 \leq T \leq T_{max}$

The range of the cricket function is also restricted. Since the chirp rate is nonnegative, the smallest value of T occurs when $R = 0$. This happens at $T = 40$. On the other hand, if the temperature gets too hot, the cricket will not be able to keep chirping faster. If the temperature T_{max} corresponds to the chirp rate R_{max}, then the values of T are restricted to the interval $40 \leq T \leq T_{max}$.

Using a Graph to Find the Domain and Range of a Function

A good way to estimate the domain and range of a function is to examine its graph. The domain is the set of input values on the horizontal axis which give rise to a point on the graph; the range is the corresponding set of output values on the vertical axis.

Example 3 A sunflower plant is measured every day t, for $t \geq 0$. The height, $h(t)$ centimeters, of the plant[2] can be modeled by using the *logistic function*

$$h(t) = \frac{260}{1 + 24(0.9)^t}.$$

(a) Using a graphing calculator or computer, graph the height over 80 days.
(b) What is the domain of this function? What is the range? What does this tell you about the height of the sunflower?

Solution (a) The logistic function is graphed in Figure 2.8.

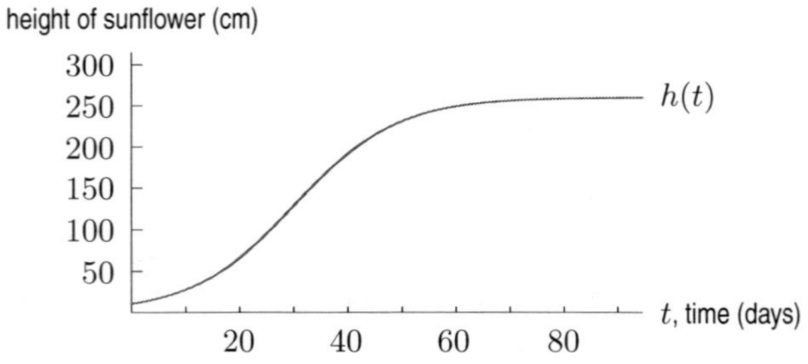

Figure 2.8: Height of sunflower as a function of time

[2]Adapted from H.S. Reed and R.H. Holland, "Growth of an Annual Plant Helianthus" *Proc. Nat. Acad. Sci.*, 5, 1919.

(b) The domain of this function is $t \geq 0$. If we consider the fact that the sunflower dies at some point, then there is an upper bound on the domain, $0 \leq t \leq T$, where T is the day on which the sunflower dies.

To find the range, notice that the smallest value of h occurs at $t = 0$. Evaluating gives $h(0) = 10.4$ cm. This means that the plant was 10.4 cm high when it was first measured on day $t = 0$. Tracing along the graph, $h(t)$ increases. As t-values get large, $h(t)$-values approach, but never reach, 260. This suggests that the range is $10.4 \leq h(t) < 260$. This information tells us that sunflowers typically grow to a height of about 260 cm.

Using a Formula to Find the Domain and Range of a Function

When a function is defined by a formula, its domain and range can often be determined by examining the formula algebraically.

Example 4 State the domain and range of g, where

$$g(x) = \frac{1}{x}.$$

Solution The domain is all real numbers except those which do not yield an output value. The expression $1/x$ is defined for any real number x except 0 (division by 0 is undefined). Therefore,

$$\text{Domain: all real } x, \quad x \neq 0.$$

The range is all real numbers that the formula can return as output values. It is not possible for $g(x)$ to equal zero, since 1 divided by a real number is never zero. All real numbers except 0 are possible output values, since all nonzero real numbers have reciprocals. Thus

$$\text{Range: all real values}, \quad g(x) \neq 0.$$

The graph in Figure 2.9 indicates agreement with these values for the domain and range.

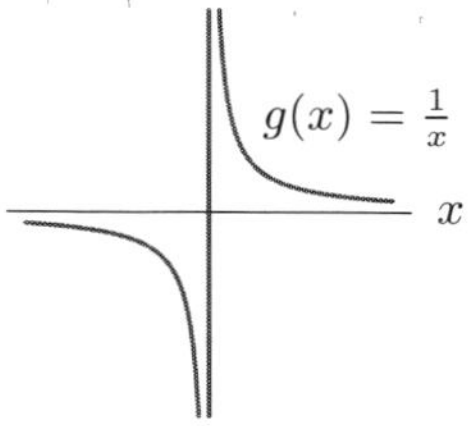

Figure 2.9: Domain and range of $g(x) = 1/x$

Example 5 Find the domain of the function $f(x) = \dfrac{1}{\sqrt{x-4}}$ by examining its formula.

Solution The domain is all real numbers except those for which the function is undefined. The square root of a negative number is undefined (if we restrict ourselves to real numbers), and so is division by zero. Therefore we need

$$x - 4 > 0.$$

Thus, the domain is all real numbers greater than 4.

$$\text{Domain:} \quad x > 4.$$

In Example 6 on page 63, we saw that for $f(x) = 1/\sqrt{x-4}$, the output, $f(x)$, cannot be negative. Note that $f(x)$ cannot be zero either. (Why?) The range of $f(x) = 1/\sqrt{x-4}$ is $f(x) > 0$. See Problem 16.

Exercises and Problems for Section 2.2

Exercises

In Exercises 1–4, use a graph to find the range of the function on the given domain.

1. $f(x) = \dfrac{1}{x}, \quad -2 \le x \le 2$

2. $f(x) = \dfrac{1}{x^2}, \quad -1 \le x \le 1$

3. $f(x) = x^2 - 4, \quad -2 \le x \le 3$

4. $f(x) = \sqrt{9 - x^2}, \quad -3 \le x \le 1$

Graph and give the domain and range of the functions in Exercises 5–12.

5. $f(x) = (x-4)^3$

6. $f(x) = x^2 - 4$

7. $f(x) = 9 - x^2$

8. $f(x) = x^3 + 2$

9. $f(x) = \sqrt{8 - x}$

10. $f(x) = \sqrt{x - 3}$

11. $f(x) = \dfrac{-1}{(x+1)^2}$

12. $f(x) = \dfrac{1}{x^2}$

In Exercises 13–14, estimate the domain and range of the function. Assume the entire graph is shown.

13.

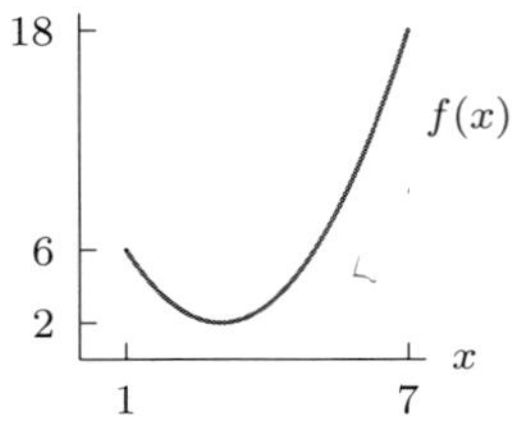

14.

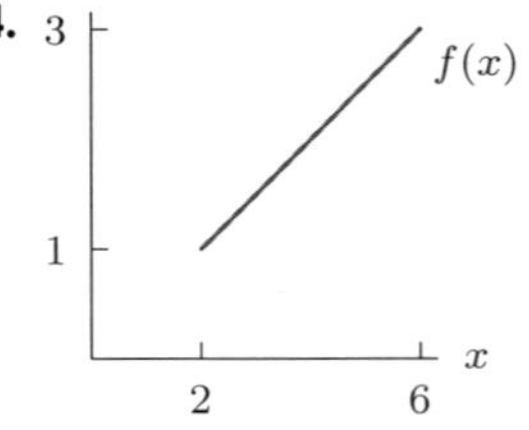

Problems

Find the domain and range of functions in Exercises 15–18 algebraically.

15. $q(x) = \sqrt{x^2 - 9}$

16. $f(x) = \dfrac{1}{\sqrt{x-4}}$

17. $m(x) = 9 - x$

18. $n(x) = 9 - x^4$

In Exercises 19–24, find the domain and range.

19. $f(x) = -x^2 + 7$

20. $f(x) = \sqrt[3]{x + 77}$

21. $f(x) = x^2 + 2$

22. $f(x) = 1/(x+1) + 3$

23. $f(x) = x - 3$

24. $f(x) = (x-3)^2 + 2$

25. Give a formula for a function whose domain is all non-negative values of x except $x = 3$.

26. Give a formula for a function that is undefined for $x = 8$ and for $x < 4$, but is defined everywhere else.

27. A restaurant is open from 2 pm to 2 am each day, and a maximum of 200 clients can fit inside. If $f(t)$ is the number of clients in the restaurant t hours after 2 pm each day, what are a reasonable domain and range for $f(t)$?

28. What is the domain of the function f giving average monthly rainfall at Chicago's O'Hare airport? (See Table 1.2 on page 5)

29. A movie theater seats 200 people. For any particular show, the amount of money the theater makes is a function of the number of people, n, in attendance. If a ticket costs \$4.00, find the domain and range of this function. Sketch its graph.

30. A car gets the best mileage at intermediate speeds. Graph the gas mileage as a function of speed. Determine a reasonable domain and range for the function and justify your reasoning.

31. **(a)** Use Table 2.7 to determine the number of calories that a person weighing 200 lb uses in a half–hour of walking.[3]

(b) Table 2.7 illustrates a relationship between the number of calories used per minute walking and a person's weight in pounds. Describe in words what is true about this relationship. Identify the dependent and independent variables. Specify whether it is an increasing or decreasing function.

(c) (i) Graph the linear function for walking, as described in part (b), and estimate its equation.

(ii) Interpret the meaning of the vertical intercept of the graph of the function.

(iii) Specify a meaningful domain and range for your function.

(iv) Use your function to determine how many calories per minute a person who weighs 135 lb uses per minute of walking.

Table 2.7 *Calories per minute as a function of weight*

Activity	100 lb	120 lb	150 lb	170 lb	200 lb	220 lb
Walking	2.7	3.2	4.0	4.6	5.4	5.9
Bicycling	5.4	6.5	8.1	9.2	10.8	11.9
Swimming	5.8	6.9	8.7	9.8	11.6	12.7

[3]Source: 1993 World Almanac. Speeds assumed are 3 mph for walking, 10 mph for bicycling, and 2 mph for swimming.

In Exercises 32–33, find the domain and range.

32. $g(x) = a + 1/x$, where a is a constant

33. $q(r) = (x - b)^{1/2} + 6$, where b is a constant

34. The last digit, d, of a phone number is a function of n, its position in the phone book. Table 2.8 gives d for the first 10 listings in the 1998 Boston telephone directory. The table shows that the last digit of the first listing is 3, the last digit of the second listing is 8, and so on. In principle we could use a phone book to figure out other values of d. For instance, if $n = 300$, we could count down to the 300th listing in order to determine d. So we write $d = f(n)$.

(a) What is the value of $f(6)$?
(b) Explain how you could use the phone book to find the domain of f.
(c) What is the range of f?

Table 2.8

n	1	2	3	4	5	6	7	8	9	10
d	3	8	4	0	1	8	0	4	3	5

35. In month $t = 0$, a small group of rabbits escapes from a ship onto an island where there are no rabbits. The island rabbit population, $p(t)$, in month t is given by

$$p(t) = \frac{1000}{1 + 19(0.9)^t}, \quad t \geq 0.$$

(a) Evaluate $p(0)$, $p(10)$, $p(50)$, and explain their meaning in terms of rabbits.
(b) Graph $p(t)$ for $0 \leq t \leq 100$. Describe the graph in words. Does it suggest the growth in population you would expect among rabbits on an island?
(c) Estimate the range of $p(t)$. What does this tell you about the rabbit population?
(d) Explain how you can find the range of $p(t)$ from its formula.

36. Bronze is an alloy or mixture of the metals copper and tin. The properties of bronze depend on the percentage of copper in the mix. A chemist decides to study the properties of a given alloy of bronze as the proportion of copper is varied. She starts with 9 kg of bronze that contain 3 kg of copper and 6 kg of tin and either adds or removes copper. Let $f(x)$ be the percentage of copper in the mix if x kg of copper are added ($x > 0$) or removed ($x < 0$).

(a) State the domain and range of f. What does your answer mean in the context of bronze?
(b) Find a formula in terms of x for $f(x)$.
(c) If the formula you found in part (b) was not intended to represent the percentage of copper in an alloy of bronze, but instead simply defined an abstract mathematical function, what would be the domain and range of this function?

2.3 PIECEWISE DEFINED FUNCTIONS

A function may employ different formulas on different parts of its domain. Such a function is said to be *piecewise defined*. For example, the function graphed in Figure 2.10 has the following formulas:

$$y = x^2 \text{ for } x \leq 2, \quad y = 6 - x \text{ for } x > 2 \qquad \text{or more compactly} \qquad y = \begin{cases} x^2 & \text{for } x \leq 2 \\ 6 - x & \text{for } x > 2. \end{cases}$$

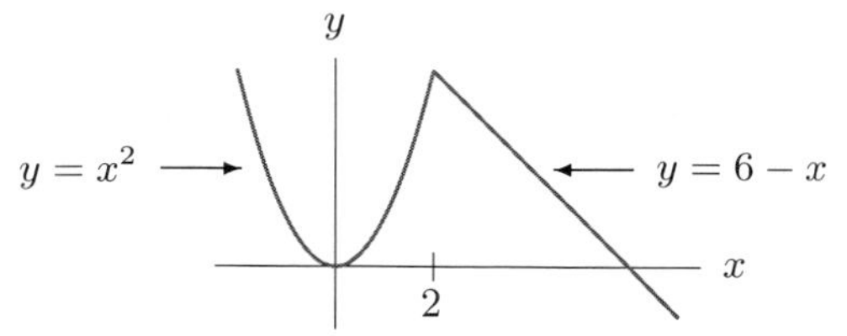

Figure 2.10: Piecewise defined function

Example 1 Graph the function $y = g(x)$ given by the following formulas:

$$g(x) = x + 1 \quad \text{for} \quad x \leq 2 \qquad \text{and} \qquad g(x) = 1 \quad \text{for} \quad x > 2.$$

Using bracket notation, this function is written:

$$g(x) = \begin{cases} x + 1 & \text{for } x \leq 2 \\ 1 & \text{for } x > 2. \end{cases}$$

Solution For $x \leq 2$, graph the line $y = x + 1$. The solid dot at the point $(2, 3)$ shows that it is included in the graph. For $x > 2$, graph the horizontal line $y = 1$. See Figure 2.11. The open circle at the point $(2, 1)$ shows that it is not included in the graph. (Note that $g(2) = 3$, and $g(2)$ cannot have more than one value.)

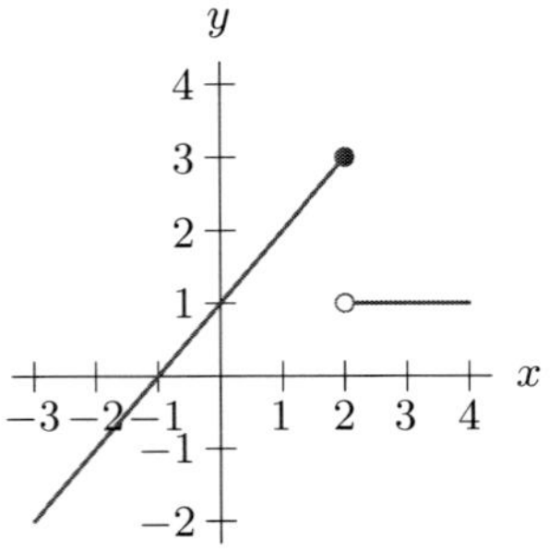

Figure 2.11: Graph of the piecewise defined function g

Example 2 A long-distance calling plan charges 99 cents for any call up to 20 minutes in length and 7 cents for each additional minute or part of a minute.

(a) Use bracket notation to write a formula for the cost, C, of a call as a function of its length t in minutes.
(b) Graph the function.
(c) State the domain and range of the function

Solution

(a) For $0 < t \leq 20$, the value of C is 99 cents. If $t > 20$, we subtract 20 to find the additional minutes and multiply by the rate 7 cents per minute.[4] The cost function in cents is thus

$$C = f(t) = \begin{cases} 99 & \text{for } 0 < t \leq 20 \\ 99 + 7(t - 20) & \text{for } t > 20, \end{cases}$$

or, after simplifying,

$$C = f(t) = \begin{cases} 99 & \text{for } 0 < t \leq 20 \\ 7t - 41 & \text{for } t > 20. \end{cases}$$

(b) See Figure 2.12.
(c) Because negative and zero call lengths do not make sense, the domain is $t > 0$. From the graph, we see that the range is $C \geq 99$.

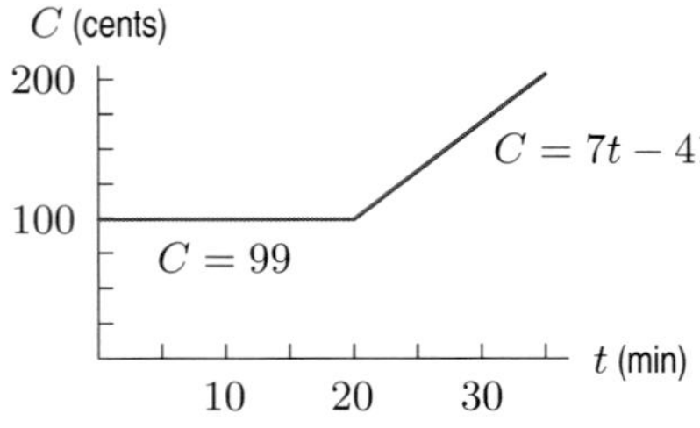

Figure 2.12: Cost of a long-distance phone call

[4] In actuality, most calling plans round the call length to whole minutes or specified fractions of a minute.

Example 3 The Ironman Triathlon is a race that consists of three parts: a 2.4 mile swim followed by a 112 mile bike race and then a 26.2 mile marathon. A participant swims steadily at 2 mph, cycles steadily at 20 mph, and then runs steadily at 9 mph.[5] Assuming that no time is lost during the transition from one stage to the next, find a formula for the distance d, covered in miles, as a function of the elapsed time t in hours, from the beginning of the race. Graph the function.

Solution For each leg of the race, we use the formula Distance = Rate · Time. First, we calculate how long it took for the participant to cover each of the three parts of the race. The first leg took $2.4/2 = 1.2$ hours, the second leg took $112/20 = 5.6$ hours, and the final leg took $26.2/9 \approx 2.91$ hours. Thus, the participant finished the race in $1.2 + 5.6 + 2.91 = 9.71$ hours.

During the first leg, $t \leq 1.2$ and the speed is 2 mph, so

$$d = 2t \qquad \text{for} \qquad 0 \leq t \leq 1.2.$$

During the second leg, $1.2 < t \leq 1.2 + 5.6 = 6.8$ and the speed is 20 mph. The length of time spent in the second leg is $(t - 1.2)$ hours. Thus, by time t,

$$\text{Distance covered in the second leg} = 20(t - 1.2) \quad \text{for } 1.2 < t \leq 6.8.$$

When the participant is in the second leg, the total distance covered is the sum of the distance covered in the first leg (2.4 miles) plus the part of the second leg that has been covered by time t.

$$\begin{aligned} d &= 2.4 + 20(t - 1.2) \\ &= 20t - 21.6 \qquad \text{for } 1.2 < t \leq 6.8. \end{aligned}$$

In the third leg, $6.8 < t \leq 9.71$ and the speed is 9 mph. Since 6.8 hours were spent on the first two parts of the race, the length of time spent on the third leg is $(t - 6.8)$ hours. Thus, by time t,

$$\text{Distance covered in the third leg} = 9(t - 6.8) \quad \text{for } 6.8 < t \leq 9.71.$$

When the participant is in the third leg, the total distance covered is the sum of the distances covered in the first leg (2.4 miles) and the second leg (112 miles), plus the part of the third leg that has been covered by time t:

$$\begin{aligned} d &= 2.4 + 112 + 9(t - 6.8) \\ &= 9t + 53.2 \qquad \text{for } 6.8 < t \leq 9.71. \end{aligned}$$

The formula for d is different on different intervals of t:

$$d = \begin{cases} 2t & \text{for} \quad 0 \leq t \leq 1.2 \\ 20t - 21.6 & \text{for} \quad 1.2 < t \leq 6.8 \\ 9t + 53.2 & \text{for} \quad 6.8 < t \leq 9.71. \end{cases}$$

Figure 2.13 gives a graph of the distance covered, d, as a function of time, t. Notice the three pieces.

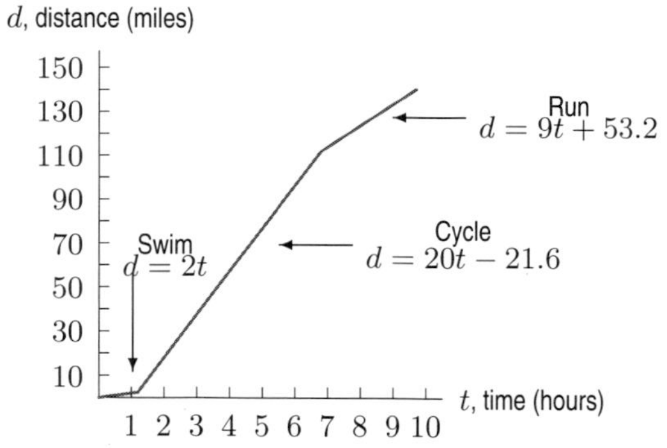

Figure 2.13: Ironman Triathlon: d as a function of t

[5] Data supplied by Susan Reid, Athletics Department, University of Arizona.

The Absolute Value Function

The absolute value of a x, written $|x|$, is defined piecewise

$$\text{For positive } x, \quad |x| = x.$$

$$\text{For negative } x, \quad |x| = -x.$$

(Remember that $-x$ is a positive number if x is a negative number.) For example, if $x = -3$, then

$$|-3| = -(-3) = 3.$$

For $x = 0$, we have $|0| = 0$. This leads to the following two-part definition:

> The **Absolute Value Function** is defined by
>
> $$f(x) = |x| = \begin{cases} x & \text{for} \quad x \geq 0 \\ -x & \text{for} \quad x < 0 \end{cases}.$$

Table 2.9 gives values of $f(x) = |x|$ and Figure 2.14 shows a graph of $f(x)$.

Table 2.9 *Absolute value function*

x	$\|x\|$
−3	3
−2	2
−1	1
0	0
1	1
2	2
3	3

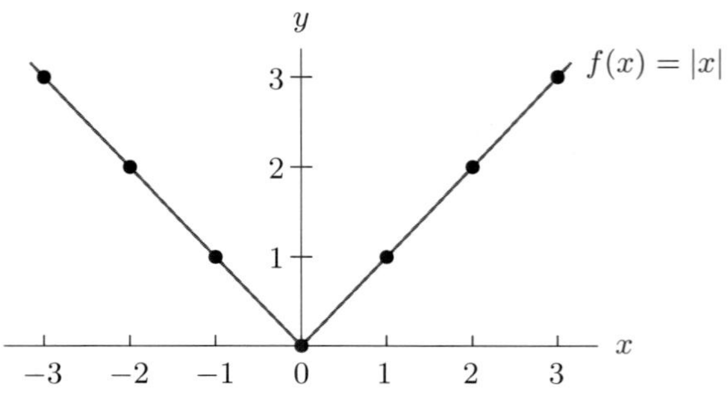

Figure 2.14: Graph of absolute value function

Exercises and Problems for Section 2.3

Exercises

Graph the piecewise defined functions in Exercises 1–4. Use an open circle to represent a point which is not included and a solid dot to indicate a point which is on the graph.

1. $f(x) = \begin{cases} -1, & -1 \leq x < 0 \\ 0, & 0 \leq x < 1 \\ 1, & 1 \leq x < 2 \end{cases}$

2. $f(x) = \begin{cases} x+1, & -2 \leq x < 0 \\ x-1, & 0 \leq x < 2 \\ x-3, & 2 \leq x < 4 \end{cases}$

3. $f(x) = \begin{cases} x+4, & x \leq -2 \\ 2, & -2 < x < 2 \\ 4-x, & x \geq 2 \end{cases}$

4. $f(x) = \begin{cases} x^2, & x \leq 0 \\ \sqrt{x}, & 0 < x < 4 \\ x/2, & x \geq 4 \end{cases}$

For Exercises 5–6, find the domain and range.

5. $G(x) = \begin{cases} x+1 & \text{for} \quad x < -1 \\ x^2+3 & \text{for} \quad x \geq -1 \end{cases}$

6. $F(x) = \begin{cases} x^3 & \text{for} \quad x \leq 1 \\ 1/x & \text{for} \quad x > 1 \end{cases}$

In Exercises 7–10, write formulas for the functions.

7.

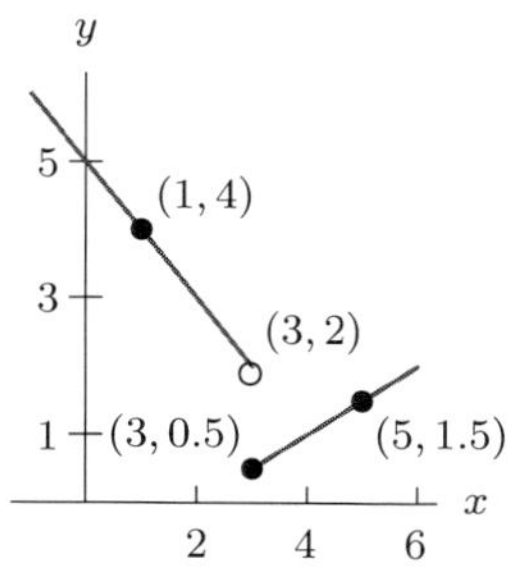

8.

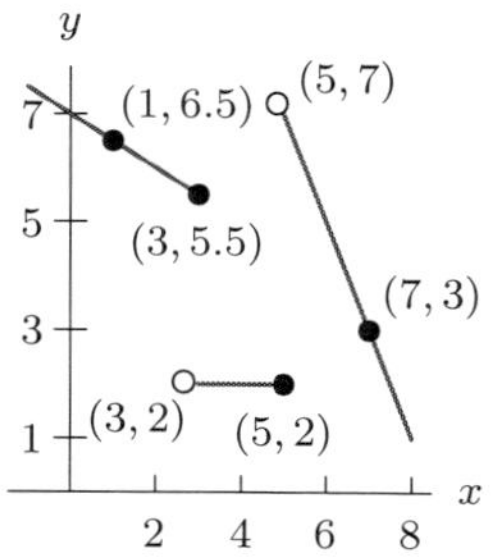

9.

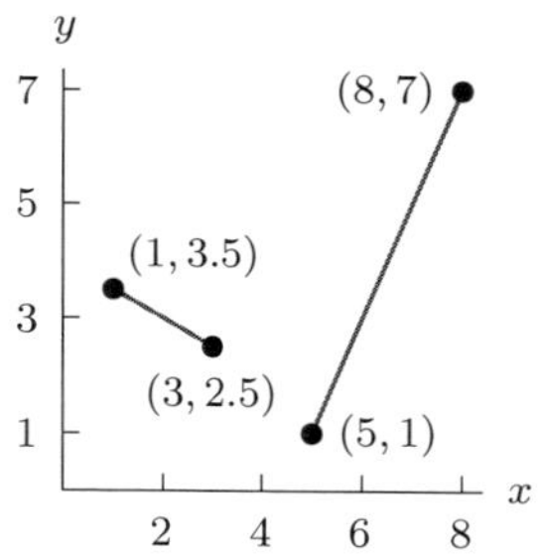

10.

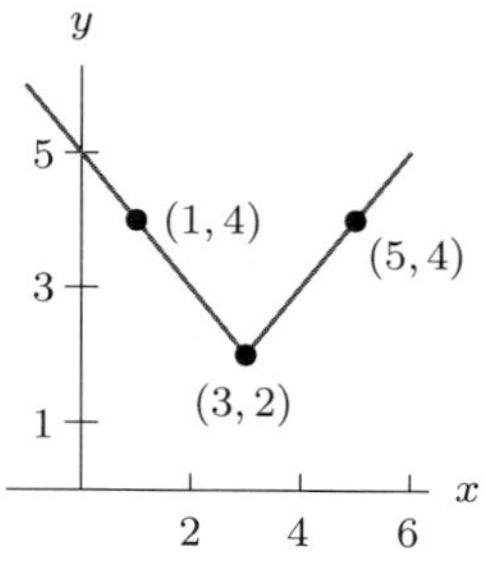

Problems

11. Consider the graph in Figure 2.15. An open circle represents a point which is not included.

(a) Is y a function of x? Explain.
(b) Is x a function of y? Explain.
(c) The domain of $y = f(x)$ is $0 \le x < 4$. What is the range of $y = f(x)$?

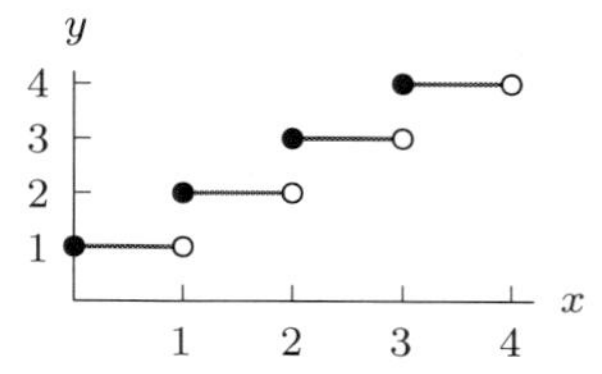

Figure 2.15

12. Many people believe that $\sqrt{x^2} = x$. We will investigate this claim graphically and numerically.

(a) Graph the two functions x and $\sqrt{x^2}$ in the window $-5 \le x \le 5$, $-5 \le y \le 5$. Based on what you see, do you believe that $\sqrt{x^2} = x$? What function does the graph of $\sqrt{x^2}$ remind you of?
(b) Complete Table 2.10. Based on this table, do you believe that $\sqrt{x^2} = x$? What function does the table for $\sqrt{x^2}$ remind you of? Is this the same function you found in part (a)?

Table 2.10

x	-5	-4	-3	-2	-1	0	1	2	3	4	5
$\sqrt{x^2}$											

(c) Explain how you know that $\sqrt{x^2}$ is the same as the function $|x|$.
(d) Graph the function $\sqrt{x^2} - |x|$ in the window $-5 \le x \le 5$, $-5 \le y \le 5$. Explain what you see.

13. **(a)** Graph $u(x) = |x|/x$ in the window $-5 \le x \le 5$, $-5 \le y \le 5$. Explain what you see.
(b) Complete Table 2.11. Does this table agree with what you found in part (a)?

Table 2.11

x	-5	-4	-3	-2	-1	0	1	2	3	4	5
$\lvert x\rvert/x$											

(c) Identify the domain and range of $u(x)$.
(d) Comment on the claim that $u(x)$ can be written as

$$u(x) = \begin{cases} -1 & \text{if } x < 0, \\ 0 & \text{if } x = 0, \\ 1 & \text{if } x > 0. \end{cases}$$

14. The charge for a taxi ride in New York City is \$2.50 for the first $1/4$ of a mile, and \$0.40 for each additional $1/4$ of a mile (rounded up to the nearest $1/4$ mile).

(a) Make a table showing the cost of a trip as a function of its length. Your table should start at zero and go up to two miles in $1/4$-mile intervals.
(b) What is the cost for a 1.25-mile trip?
(c) How far can you go for \$5.30?
(d) Graph the cost function in part (a).

15. A museum charges \$40 for a group of 10 or fewer people. A group of more than 10 people must, in addition to the \$40, pay \$2 per person for the number of people above 10. For example, a group of 12 pays \$44 and a group of 15 pays \$50. The maximum group size is 50.

(a) Draw a graph that represents this situation.
(b) What are the domain and range of the cost function?

16. A floor-refinishing company charges $\$1.83$ per square foot to strip and refinish a tile floor for up to 1000 square feet. There is an additional charge of $\$350$ for toxic waste disposal for any job which includes more than 150 square feet of tile.

(a) Express the cost, y, of refinishing a floor as a function of the number of square feet, x, to be refinished.
(b) Graph the function. Give the domain and range.

17. A contractor purchases gravel one cubic yard at a time.

(a) A gravel driveway L yards long and 6 yards wide is to be poured to a depth of 1 foot. Find a formula for $n(L)$, the number of cubic yards of gravel the contractor buys, assuming that he buys 10 more cubic yards of gravel than are needed (to be sure he'll have enough).
(b) Assuming no driveway is less than 5 yards long, state the domain and range of $n(L)$. Graph $n(L)$ showing the domain and range.
(c) If the function $n(L)$ did not represent an amount of gravel, but was a mathematical relationship defined by the formula in part (a), what is its domain and range?

18. At a supermarket checkout, a scanner records the prices of the foods you buy. In order to protect consumers, the state of Michigan passed a "scanning law" that says something similar to the following:

> If there is a discrepancy between the price marked on the item and the price recorded by the scanner, the consumer is entitled to receive 10 times the difference between those prices; this amount given must be at least \$1 and at most \$5. Also, the consumer will be given the difference between the prices, in addition to the amount calculated above.

For example: If the difference is 5¢, you should receive \$1 (since 10 times the difference is only 50¢ and you are to receive at least \$1), plus the difference of 5¢. Thus, the total you should receive is $\$1.00 + \$0.05 = \$1.05$,
If the difference is 25¢, you should receive 10 times the difference in addition to the difference, giving $(10)(0.25) + 0.25 = \$2.75$.
If the difference is 95¢, you should receive \$5 (because $10(.95) = \$9.50$ is more than \$5, the maximum penalty), plus 95¢, giving $5 + 0.95 = \$5.95$.

(a) What is the lowest possible refund?
(b) Suppose x is the difference between the price scanned and the price marked on the item, and y is the amount refunded to the customer. Write a formula for y in terms of x. (Hints: Look at the sample calculations.)
(c) What would the difference between the price scanned and the price marked have to be in order to obtain a $\$9.00$ refund?
(d) Graph y as a function of x.

19. Many printing presses are designed with large plates that print a fixed number of pages as a unit. Each unit is called a signature. A particular press prints signatures of 16 pages each. Suppose $C(p)$ is the cost of printing a book of p pages, assuming each signature printed costs $\$0.14$.

(a) What is the cost of printing a book of 128 pages? 129 pages? p pages?
(b) What are the domain and range of C?
(c) Graph $C(p)$ for $0 \leq p \leq 128$.

20. Gore Mountain is a ski resort in the Adirondack mountains in upstate New York. Table 2.12 shows the cost of a weekday ski-lift ticket for various ages and dates.

(a) Graph cost as a function of age for each time period given. (One graph will serve for times when rates are identical).
(b) For which age group does the date affect cost?
(c) Graph cost as a function of date for the age group mentioned in part (b).
(d) Why does the cost fluctuate as a function of date?

Table 2.12 *Ski-lift ticket prices at Gore Mountain, 1998–1999*[6]

Age	Opening-Dec 12	Dec 13-Dec 24	Dec 25-Jan 3	Jan 4-Jan 15	Jan 16-Jan 18
Up to 6	Free	Free	Free	Free	Free
7–12	\$19	\$19	\$19	\$19	\$19
13–69	\$29	\$34	\$39	\$34	\$39
70+	Free	Free	Free	Free	Free
Age	Jan 19-Feb 12	Feb 13-Feb 21	Feb 22-Mar 28	Mar 29-Closing	
Up to 6	Free	Free	Free	Free	
7–12	\$19	\$19	\$19	\$19	
13–69	\$34	\$39	\$34	\$29	
70+	Free	Free	Free	Free	

[6]The Olympic Regional Development Authority.

2.4 COMPOSITE AND INVERSE FUNCTIONS

Composition of Functions

Two functions may be connected by the fact that the output of one is the input of the other. For example, to find the cost, C, in dollars, to paint a room of area A square feet, we need to know the number, n, of gallons of paint required. Since one gallon covers 250 square feet, we have the function $n = f(A) = A/250$. If paint is \$30.50 a gallon, we have the function $C = g(n) = 30.5n$. We substitute $n = f(A)$ into $g(n)$ to find the cost C as a function of A.

Example 1 Find a formula for cost, C, as a function of area, A.

Solution Since we have

$$C = 30.5n \quad \text{and} \quad n = \frac{A}{250},$$

substituting for n in the formula for C gives

$$C = 30.5\frac{A}{250} = 0.122A.$$

We say that C is a "function of a function", or *composite function*. If the function giving C in terms of A is called h, so $C = h(A)$, then we write

$$C = h(A) = g(f(A)).$$

The function h is said to be the *composition* of the functions f and g. We say f is the *inside* function and g is the *outside* function. In this example, the composite function $C = h(A) = g(f(A))$ tells us the cost of painting an area of A square feet.

Example 2 The air temperature, T, in °F, is given in terms of the chirp rate, R, in chirps per minute, of a snowy tree cricket by the function

$$T = f(R) = \frac{1}{4}R + 40.$$

Suppose one night we record the chirp rate and find that it varies with time, x, according to the function

$$R = g(x) = 20 + x^2 \qquad \text{where } x \text{ is in hours since midnight and } 0 \le x \le 10.$$

Find how temperature varies with time by obtaining a formula for h, where $T = h(x)$.

Solution Since $f(R)$ is a function of R and $R = g(x)$, we see that g is the inside function and f is the outside function. Thus we substitute $R = g(x)$ into f:

$$T = f(R) = f(g(x)) = \frac{1}{4}g(x) + 40 = \frac{1}{4}(20 + x^2) + 40 = \frac{1}{4}x^2 + 45.$$

Thus, for $0 \le x \le 10$, we have

$$T = h(x) = \frac{1}{4}x^2 + 45.$$

Example 3 shows another example of composition.

Example 3 If $f(x) = x^2$ and $g(x) = 2x + 1$, find

(a) $f(g(x))$ (b) $g(f(x))$

Solution (a) We have

$$f(g(x)) = f(2x + 1) = (2x + 1)^2.$$

(b) We have

$$g(f(x)) = g(x^2) = 2(x^2) + 1 = 2x^2 + 1.$$

Notice that $f(g(x))$ is not equal to $g(f(x))$ in this case.

Inverse Functions

The roles of a function's input and output can sometimes be reversed. For example, the population, P, of birds is given, in thousands, by $P = f(t)$, where t is the number of years since 2007. In this function, t is the input and P is the output. If the population is increasing, knowing the year enables us to calculate the population. Thus we can define a new function, $t = g(P)$, which tells us the value of t given the value of P instead of the other way round. For this function, P is the input and t is the output.The functions f and g are called *inverses* of each other. A function which has an inverse is said to be *invertible*.

The fact that f and g are inverse functions means that they go in "opposite directions." The function f takes t as input and outputs P, while g takes P as input and outputs t.

Inverse Function Notation

In the preceding discussion, there was nothing about the names of the two functions that stressed their special relationship. If we want to emphasize that g is the inverse of f, we call it f^{-1} (read "f-inverse"). To express the fact that the population of birds, P, is a function of time, t, we write

$$P = f(t).$$

To express the fact that the time t is also determined by P, so that t is a function of P, we write

$$t = f^{-1}(P).$$

The symbol f^{-1} is used to represent the function that gives the output t for a given input P.

Warning: The -1 which appears in the symbol f^{-1} for the inverse function is not an exponent. Unfortunately, the notation $f^{-1}(x)$ might lead us to interpret it as $\frac{1}{f(x)}$. The two expressions are not the same in general: $f^{-1}(x)$ is the output when x is fed into the inverse of f, while $\frac{1}{f(x)}$ is the reciprocal of the number we get when x is fed into f.

Example 4 Using $P = f(t)$, where P represents the population, in thousands, of birds on an island and t is the number of years since 2007:

(a) What does $f(4)$ represent? (b) What does $f^{-1}(4)$ represent?

Solution

(a) The expression $f(4)$ is the bird population (in thousands) in the year 2011.
(b) Since f^{-1} is the inverse function, f^{-1} is a function which takes population as input and returns time as output. Therefore, $f^{-1}(4)$ is the number of years after 2007 at which there were 4,000 birds on the island.

Finding a Formula for the Inverse Function

In the next example, we find the formula for an inverse function.

Example 5 The cricket function, which gives temperature, T, in terms of chirp rate, R, is

$$T = f(R) = \frac{1}{4} \cdot R + 40.$$

Find a formula for the inverse function, $R = f^{-1}(T)$.

Solution The inverse function gives the chirp rate in terms of the temperature, so we solve the following equation for R:

$$T = \frac{1}{4} \cdot R + 40,$$

giving

$$T - 40 = \frac{1}{4} \cdot R$$
$$R = 4(T - 40).$$

Thus, $R = f^{-1}(T) = 4(T - 40)$.

Domain and Range of an Inverse Function

The input values of the inverse function f^{-1} are the output values of the function f. Thus, the domain of f^{-1} is the range of f. For the cricket function, $T = f(R) = \frac{1}{4}R + 40$, if a realistic domain is $0 \le R \le 160$, then the range of f is $40 \le T \le 80$. See Figure 2.16.

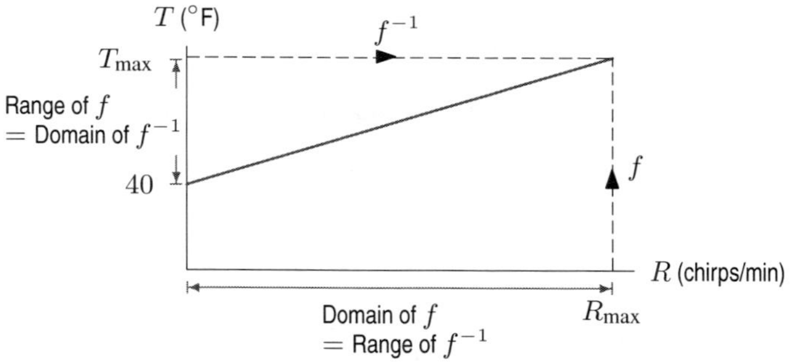

Figure 2.16

A Function and its Inverse Undo Each Other

Example 6 Calculate the composite functions $f^{-1}(f(R))$ and $f(f^{-1}(T))$ for the cricket example. Interpret the results.

Solution Since $f(R) = \frac{1}{4}R + 40$. and $f^{-1}(T) = 4(T - 40)$, we have

$$f^{-1}(f(R)) = f^{-1}\left(\frac{1}{4} \cdot R + 40\right) = 4\left(\left(\frac{1}{4} \cdot R + 40\right) - 40\right) = R.$$

$$f(f^{-1}(T)) = f(4(T - 40)) = \frac{1}{4}(4(T - 40)) + 40 = T.$$

To interpret these results, we use the fact that $f(R)$ gives the temperature corresponding to chirp rate R, and $f^{-1}(T)$ gives the chirp rate corresponding to temperature T. Thus $f^{-1}(f(R))$ gives the chirp rate at temperature $f(R)$, which is R. Similarly, $f(f^{-1}(T))$ gives the temperature at chirp rate $f^{-1}(T)$, which is T.

Example 6 illustrates the following result, which we see is true in general in Chapter 8.

The functions f and f^{-1} are called inverses because they "undo" each other when composed.

Exercises and Problems for Section 2.4

Exercises

In Exercises 1–4, give the meaning and units of the composite function.

1. $A(f(t))$, where $r = f(t)$ is the radius, in centimeters, of a circle at time t minutes, and $A(r)$ is the area, in square centimeters, of a circle of radius r centimeters.

2. $R(f(p))$, where $Q = f(p)$ is the number of barrels of oil sold by a company when the price is p dollars/barrel and $R(Q)$ is the revenue earned in millions of dollars.

3. $a(g(w))$, where $F = g(w)$ is the force, in newtons, on a rocket when the wind speed is w meters/sec and $a(F)$ is the acceleration, in meters/sec^2, when the force is F newtons.

4. $P(f(t))$, where $l = f(t)$ is the length, in centimeters, of a pendulum at time t minutes, and $P(l)$ is the period, in seconds, of a pendulum of length l.

In Exercises 5–12, use $f(x) = x^2 + 1$, $g(x) = 2x + 3$.

5. $f(g(0))$ **6.** $f(g(1))$ **7.** $g(f(0))$ **8.** $g(f(1))$

9. $f(g(x))$ **10.** $g(f(x))$ **11.** $f(f(x))$ **12.** $g(g(x))$

In Exercises 13–17, give the meaning and units of the inverse function. (Assume f is invertible.)

13. $P = f(t)$ is population in millions in year t.

14. $T = f(H)$ is time in minutes to bake a cake at H°F.

15. $N = f(t)$ number of inches of snow in the first t days of January.

16. $V = f(t)$ is the speed in km/hr of an accelerating car t seconds after starting.

17. $I = f(r)$ is the interest earned, in dollars, on a \$10,000 deposit at an interest rate of r% per year, compounded annually.

In Exercises 18–21, find the inverse function.

18. $y = f(t) = 2t + 3$ **19.** $Q = f(x) = x^3 + 3$

20. $y = g(t) = \sqrt{t} + 1$ **21.** $P = f(q) = 14q - 2$

22. Use the graph in Figure 2.17 to fill in the missing values:

(a) $f(0) = ?$ **(b)** $f(?) = 0$
(c) $f^{-1}(0) = ?$ **(d)** $f^{-1}(?) = 0$

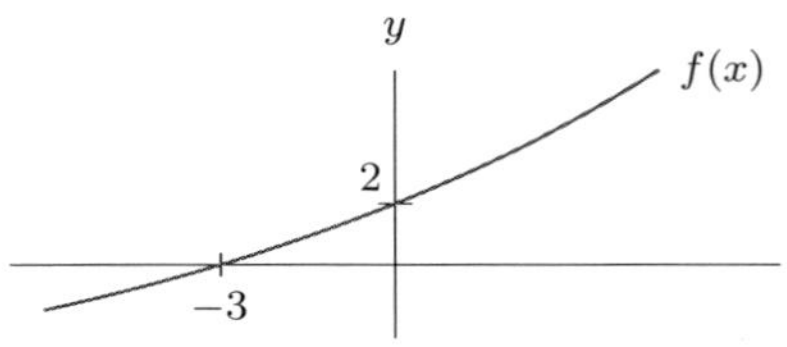

Figure 2.17

23. Use the graph in Figure 2.18 to fill in the missing values:

(a) $f(0) = ?$ **(b)** $f(?) = 0$
(c) $f^{-1}(0) = ?$ **(d)** $f^{-1}(?) = 0$

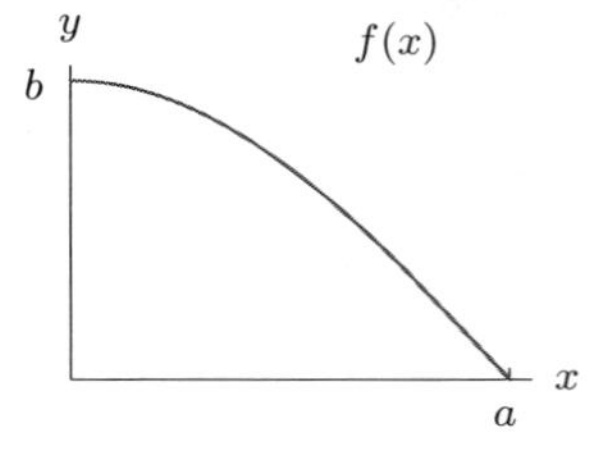

Figure 2.18

Problems

24. Let $n = f(A) = A/250$, where n is the number of gallons of paint and A is the area to be painted. Find a formula for the inverse function $A = g(n)$.

25. Calculate the composite functions $f^{-1}(f(A))$ and $f(f^{-1}(n))$ from Problem 24. Explain the results.

26. Table 2.13 gives values of an invertible function, f.

(a) Using the table, fill in the missing values:

(i) $f(0) = ?$ (ii) $f(?) = 0$
(iii) $f^{-1}(0) = ?$ (iv) $f^{-1}(?) = 0$

(b) How do the answers to (i)–(iv) in part (a) relate to one another? In particular, how could you have obtained the answers to (iii) and (iv) from the answers to (i) and (ii)?

Table 2.13

x	-2	-1	0	1	2
$f(x)$	5	4	2	0	-3

27. Use the values of the invertible function in Table 2.14 to find as many values of g^{-1} as possible.

Table 2.14

t	1	2	3	4	5
$y = g(t)$	7	12	13	19	22

28. The cost (in dollars) of producing x air conditioners is $C = g(x) = 600 + 45x$. Find a formula for the inverse function $g^{-1}(C)$.

29. The formula $V = f(r) = \frac{4}{3}\pi r^3$ gives the volume of a sphere of radius r. Find a formula for the inverse function, $f^{-1}(V)$, giving radius as a function of volume.

30. The cost, C, in thousands of dollars, of producing q kg of a chemical is given by $C = f(q) = 100 + 0.2q$. Find and interpret

(a) $f(10)$ **(b)** $f^{-1}(200)$ **(c)** $f^{-1}(C)$

31. The perimeter of a square of side s is given by $P = f(s) = 4s$. Find and interpret

(a) $f(3)$ **(b)** $f^{-1}(20)$ **(c)** $f^{-1}(P)$

32. The gross domestic product (GDP) of the US is given by $G(t)$ where t is the number of years since 1990 and the units of G are billions of dollars.

(a) What does $G(11)$ represent?
(b) What does $G^{-1}(9873)$ represent?

33. The formula for the volume of a cube with side s is $V = s^3$. The formula for the surface area of a cube is $A = 6s^2$.

(a) Find and interpret the formula for the function $s = f(A)$.
(b) If $V = g(s)$, find and interpret the formula for $g(f(A))$.

34. Interpret and evaluate $f(100)$ and $f^{-1}(100)$ for the house-painting function, $n = f(A) = A/250$. (See Problem 24.)

35. The cost of producing q thousand loaves of bread is $C(q)$ dollars. Interpret the following statements in terms of bread; give units.

(a) $C(5) = 653$
(b) $C^{-1}(80) = 0.62$
(c) The solution to $C(q) = 790$ is 6.3
(d) The solution to $C^{-1}(x) = 1.2$ is 150.

36. The area, $A = f(s)$ ft^2, of a square wooden deck is a function of the side s feet. A can of stain costs \$29.50 and covers 200 square feet of wood.

(a) Write the formula for $f(s)$.
(b) Find a formula for $C = g(A)$, the cost in dollars of staining an area of A ft^2.
(c) Find and interpret $C = g(f(s))$.
(d) Evaluate and interpret, giving units:

(i) $f(8)$ (ii) $g(80)$ (iii) $g(f(10))$

37. The radius, r, in centimeters, of a melting snowball is given by $r = 50 - 2.5t$, where t is time in hours. The snowball is spherical, with volume $V = \frac{4}{3}\pi r^3$ cm^3. Find a formula for $V = f(t)$, the volume of the snowball as a function of time.

38. A circular oil slick is expanding with radius, r in yards, at time t in hours given by $r = 2t - 0.1t^2$, for t in hours, $0 \le t \le 10$. Find a formula for the area in square yards, $A = f(t)$, as a function of time.

39. Carbon dioxide is one of the "greenhouse" gases that are believed to affect global warming. Between January 1998 and January 2003, the concentration of carbon dioxide in the earth's atmosphere increased steadily from 365 parts per million (ppm) to 375 ppm. Let $C(t)$ be the concentration in ppm of carbon dioxide t years after 1998.

(a) State the domain and range of $C(t)$.
(b) What is the practical meaning of $C(4)$?
(c) What does $C^{-1}(370)$ represent?

40. Table 2.15 shows the cost, $C(m)$, of a taxi ride as a function of the number of miles, m, traveled.

(a) Estimate and interpret $C(3.5)$ in practical terms.
(b) Assume C is invertible. What does $C^{-1}(3.5)$ mean in practical terms? Estimate $C^{-1}(3.5)$.

Table 2.15

m	0	1	2	3	4	5
$C(m)$	0	2.50	4.00	5.50	7.00	8.50

41. The perimeter, in meters, of a square whose side is s meters is given by $P = 4s$.

(a) Write this formula using function notation, where f is the name of the function.
(b) Evaluate $f(s + 4)$ and interpret its meaning.
(c) Evaluate $f(s) + 4$ and interpret its meaning.
(d) What are the units of $f^{-1}(6)$?

2.5 CONCAVITY

Concavity and Rates of Change

The graph of a linear function is a straight line because the average rate of change is a constant. However, not all graphs are straight lines; they may bend up or down. Consider the salary function $S(t)$ shown in Table 2.16 and Figure 2.19, where t is time in years since being hired. Since the rate of change increases with time, the slope of the graph increases as t increases, so the graph bends upward. We say such graphs are *concave up*.

Table 2.16 *Salary: Increasing rate of change*

t (years)	S (\$1000s)	Rate of change $\Delta S/\Delta t$
0	40	
		3.2
10	72	
		5.6
20	128	
		10.2
30	230	
		18.1
40	411	

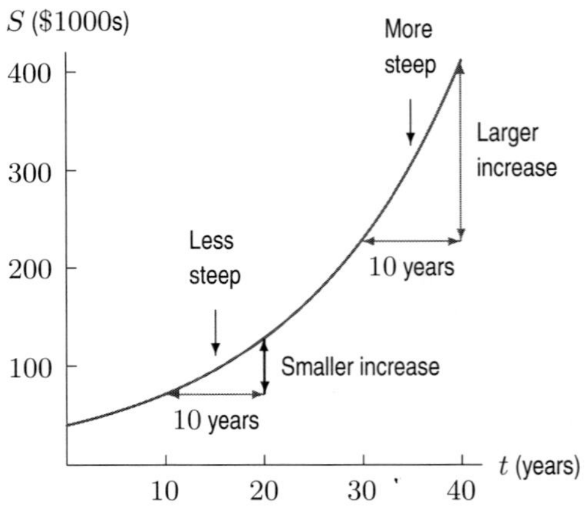

Figure 2.19: Graph of salary function is concave up because rate of change increases

The next example shows that a decreasing function can also be concave up.

Example 1 Table 2.17 shows Q, the quantity of carbon-14 (in μg) in a 200 μg sample remaining after t thousand years. We see from Figure 2.20 that Q is a decreasing function of t, so its rate of change is always negative. What can we say about the concavity of the graph, and what does this mean about the rate of change of the function?

Table 2.17 *Carbon-14: Increasing rate of change*

t (thousand years)	Q (μg)	Rate of change $\Delta Q/\Delta t$
0	200	
		-18.2
5	109	
		-9.8
10	60	
		-5.4
15	33	

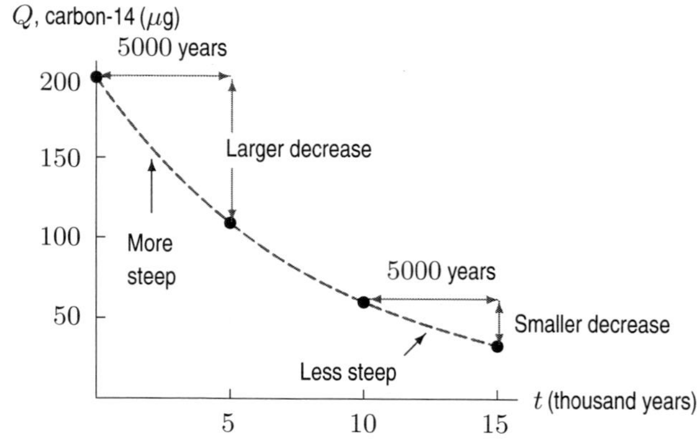

Figure 2.20: Graph of the quantity of carbon-14 is concave up

Solution The graph bends upward, so it is concave up. Table 2.18 shows that the rate of change of the function is increasing, because the rate is becoming less negative. Figure 2.20 shows how the increasing rate of change can be visualized on the graph: the slope is negative and increasing.

Graphs can bend downward; we call such graphs *concave down.*

Example 2 Table 2.18 gives the distance traveled by a cyclist, Karim, as a function of time. What is the concavity of the graph? Was Karim's speed (that is, the rate of change of distance with respect to time) increasing, decreasing, or constant?

Solution Table 2.18 shows Karim's speed was decreasing throughout the trip. Figure 2.21 shows how the decreasing speed leads to a decreasing slope and a graph which bends downward; thus the graph is concave down.

Table 2.18 *Karim's distance as a function of time, with the average speed for each hour*

t, time (hours)	d, distance (miles)	Average speed, $\Delta d/\Delta t$ (mph)
0	0	
		20 mph
1	20	
		15 mph
2	35	
		10 mph
3	45	
		7 mph
4	52	
		5 mph
5	57	

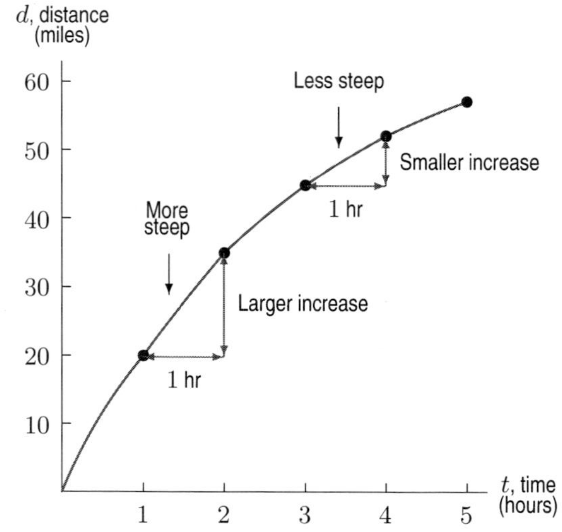

Figure 2.21: Karim's distance as a function of time

Summary: Increasing and Decreasing Functions; Concavity

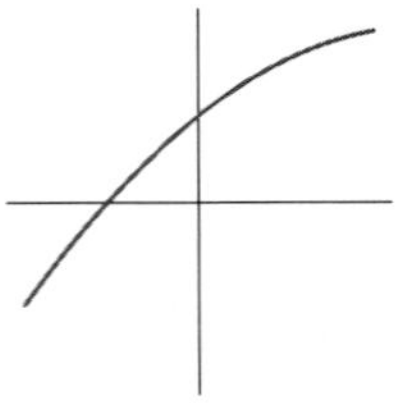

Figure 2.22: Increasing and concave down

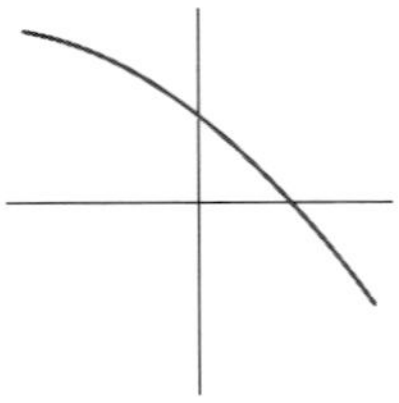

Figure 2.23: Decreasing and concave down

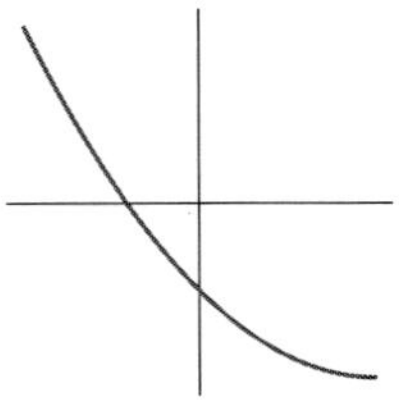

Figure 2.24: Decreasing and concave up

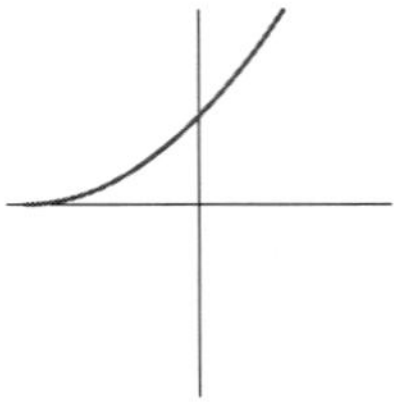

Figure 2.25: Increasing and concave up

Figures 2.22–2.25 reflect the following relationships between concavity and rate of change:

- If f is a function whose rate of change increases (gets less negative or more positive as we move from left to right[7]), then the graph of f is **concave up**. That is, the graph bends upward.
- If f is a function whose rate of change decreases (gets less positive or more negative as we move from left to right), then the graph of f is **concave down**. That is, the graph bends downward.

If a function has a constant rate of change, its graph is a line and it is neither concave up nor concave down.

Exercises and Problems for Section 2.5

Exercises

Do the graphs of the functions in Exercises 1–8 appear to be concave up, concave down, or neither?

1.

x	0	1	3	6
$f(x)$	1.0	1.3	1.7	2.2

2.

t	0	1	2	3	4
$f(t)$	20	10	6	3	1

3.

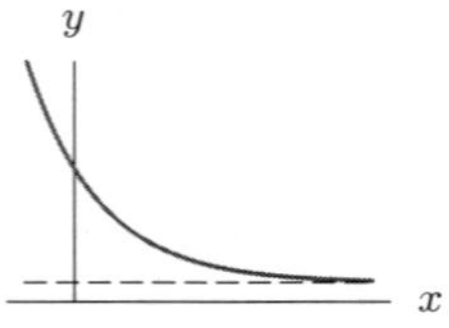

4.

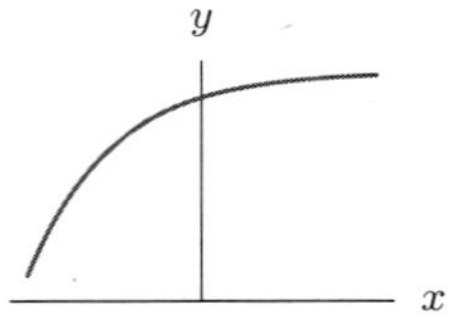

5. $y = x^2$

6. $y = -x^2$

7. $y = x^3, x > 0$

8. $y = x^3, x < 0$

9. Calculate successive rates of change for the function, $p(t)$, in Table 2.19 to decide whether you expect the graph of $p(t)$ to be concave up or concave down.

Table 2.19

t	0.2	0.4	0.6	0.8
$p(t)$	−3.19	−2.32	−1.50	−0.74

10. Calculate successive rates of change for the function, $H(x)$, in Table 2.20 to decide whether you expect the graph of $H(x)$ to be concave up or concave down.

Table 2.20

x	12	15	18	21
$H(x)$	21.40	21.53	21.75	22.02

11. Sketch a graph which is everywhere negative, increasing, and concave down.

12. Sketch a graph which is everywhere positive, increasing, and concave up.

[7]In fact, we need to take the average rate of change over an arbitrarily small interval.

Problems

Are the functions in Problems 13–17 increasing or decreasing? What does the scenario tell you about the concavity of the graph modeling it?

13. When money is deposited in the bank, the amount of money increases slowly at first. As the size of the account increases, the amount of money increases more rapidly, since the account is earning interest on the new interest, as well as on the original amount.

14. After a cup of hot chocolate is poured, the temperature cools off very rapidly at first, and then cools off more slowly, until the temperature of the hot chocolate eventually reaches room temperature.

15. When a rumor begins, the number of people who have heard the rumor increases slowly at first. As the rumor spreads, the rate of increase gets greater (as more people continue to tell their friends the rumor), and then slows down again (when almost everyone has heard the rumor).

16. When a drug is injected into a person's bloodstream, the amount of the drug present in the body increases rapidly at first. If the person receives daily injections, the body metabolizes the drug so that the amount of the drug present in the body continues to increase, but at a decreasing rate. Eventually, the quantity levels off at a saturation level.

17. When a new product is introduced, the number of people who use the product increases slowly at first, and then the rate of increase is faster (as more and more people learn about the product). Eventually, the rate of increase slows down again (when most people who are interested in the product are already using it).

18. Match each story with the table and graph which best represent it.

(a) When you study a foreign language, the number of new verbs you learn increases rapidly at first, but slows almost to a halt as you approach your saturation level.
(b) You board an airplane in Philadelphia heading west. Your distance from the Atlantic Ocean, in kilometers, increases at a constant rate.
(c) The interest on your savings plan is compounded annually. At first your balance grows slowly, but its rate of growth continues to increase.

(E)

x	0	5	10	15	20	25
y	20	275	360	390	395	399

(F)

x	0	5	10	15	20	25
y	20	36	66	120	220	400

(G)

x	0	5	10	15	20	25
y	20	95	170	245	320	395

(I)
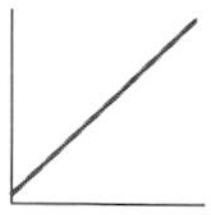
(II)

(III)

19. Match each of the following descriptions with an appropriate graph and table of values.

(a) The weight of your jumbo box of Fruity Flakes decreases by an equal amount every week.
(b) The machinery depreciated rapidly at first, but its value declined more slowly as time went on.
(c) In free fall, your distance from the ground decreases faster and faster.
(d) For a while it looked like the decline in profits was slowing down, but then they began declining ever more rapidly.

(E)

x	0	1	2	3	4	5
y	400	384	336	256	144	0

(F)

x	0	1	2	3	4	5
y	400	320	240	160	80	0

(G)

x	0	1	2	3	4	5
y	400	184	98	63	49	43

(H)

x	0	1	2	3	4	5
y	412	265	226	224	185	38

(I)

(II)
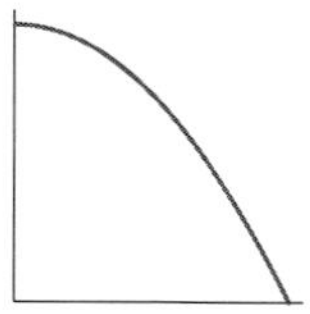
(III)
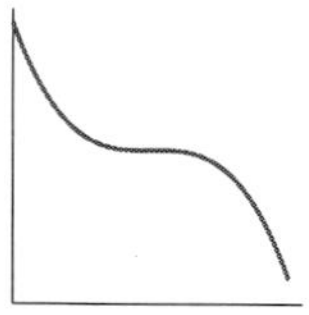
(IV)
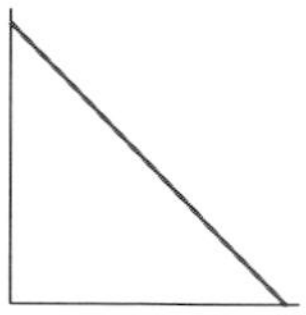

20. An incumbent politician running for reelection declared that the number of violent crimes is no longer rising and is presently under control. Does the graph shown in Figure 2.26 support this claim? Why or why not?

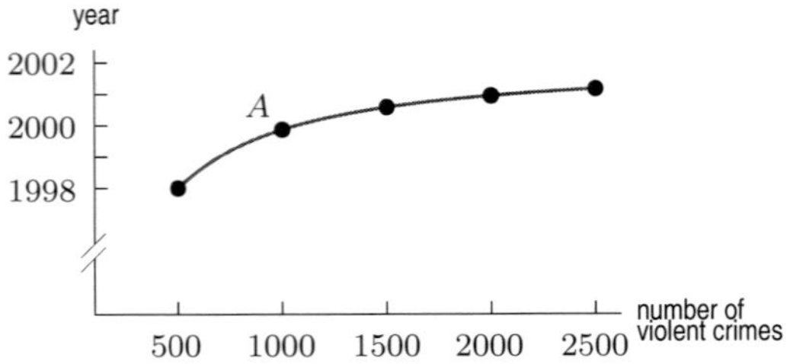

Figure 2.26

21. The rate at which water is entering a reservoir is given for time $t > 0$ by the graph in Figure 2.27. A negative rate means that water is leaving the reservoir. For each of the following statements, give the largest interval on which:

(a) The volume of water is increasing.
(b) The volume of water is constant.
(c) The volume of water is increasing fastest.
(d) The volume of water is decreasing.

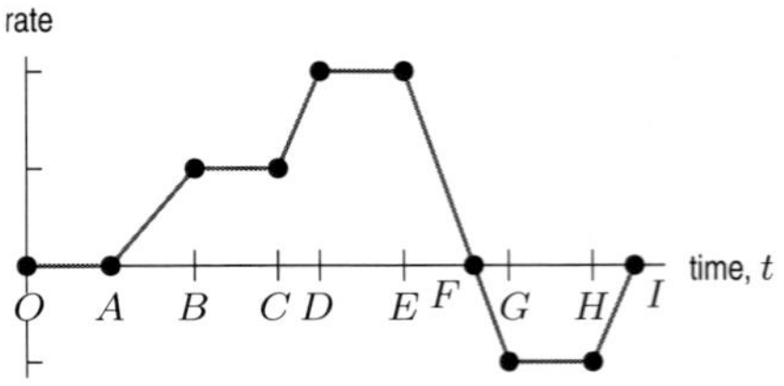

Figure 2.27

22. The relationship between the swimming speed U (in cm/sec) of a salmon to the length l of the salmon (in cm) is given by the function[8]

$$U = 19.5\sqrt{l}.$$

(a) If one salmon is 4 times the length of another salmon, how are their swimming speeds related?
(b) Graph the function $U = 19.5\sqrt{l}$. Describe the graph using words such as increasing, decreasing, concave up, concave down.
(c) Using a property that you described in part (b), answer the question "Do larger salmon swim faster than smaller ones?"
(d) Using a property that you described in part (b), answer the question "Imagine four salmon—two small and two large. The smaller salmon differ in length by 1 cm, as do the two larger. Is the difference in speed between the two smaller fish, greater than, equal to, or smaller than the difference in speed between the two larger fish?"

2.6 QUADRATIC FUNCTIONS

A baseball is "popped" straight up by a batter. The height of the ball above the ground is given by the function $y = f(t) = -16t^2 + 64t + 3$, where t is time in seconds after the ball leaves the bat and y is in feet. See Figure 2.28. Although the path of the ball is straight up and down, the graph of its height as a function of time is concave down. The ball goes up fast at first and then more slowly because of gravity.

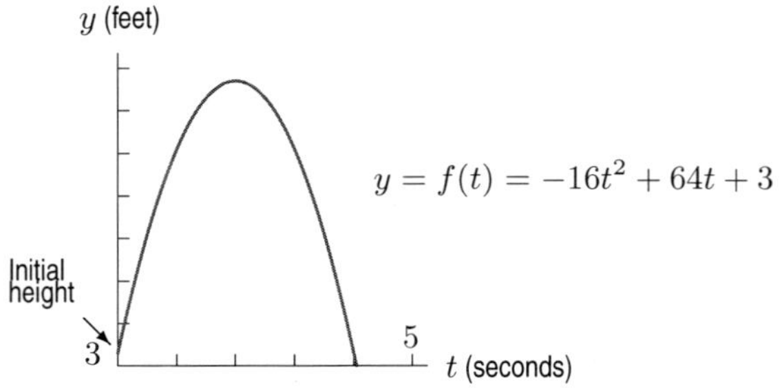

Figure 2.28: The height of a ball t seconds after being "popped up". (Note: This graph does not represent the ball's path)

[8]From K. Schmidt-Nielsen, *Scaling, Why is Animal Size so Important?* (Cambridge: CUP, 1984).

The baseball height function is an example of a *quadratic function*, whose general form is $y = ax^2 + bx + c$.

Finding the Zeros of a Quadratic Function

A natural question to ask is when the ball hits the ground. The graph suggests that $y = 0$ when $t \approx 4$. (See Problem 25 on page 92.) We can phrase the question symbolically: For what value of t does $f(t) = 0$? Input values of t which make the output $f(t) = 0$ are called *zeros* of f. It is easy to find the zeros of a quadratic function if its formula can be factored (see the Tools section to review factoring).

Example 1 Find the zeros of $f(x) = x^2 - x - 6$.

Solution To find the zeros, set $f(x) = 0$ and solve for x by factoring:

$$x^2 - x - 6 = 0$$
$$(x-3)(x+2) = 0.$$

Thus the zeros are $x = 3$ and $x = -2$.

Some quadratic functions can be expressed in *factored form*,

$$q(x) = a(x-r)(x-s),$$

where a, r, and s are constants, $a \neq 0$. Note that r and s are zeros of the function q. The factored form of the function f in Example 1 is $f(x) = (x-3)(x+2)$.

We can also find the zeros of a quadratic function by using the quadratic formula. (See the Tools section to review the quadratic formula.)

Example 2 Find the zeros of $f(x) = x^2 - x - 6$ by using the quadratic formula.

Solution We must solve the equation $x^2 - x - 6 = 0$. For this equation, $a = 1$, $b = -1$, and $c = -6$. Thus

$$x = \frac{-b \pm \sqrt{b^2 - 4ac}}{2a} = \frac{-(-1) \pm \sqrt{(-1)^2 - 4(1)(-6)}}{2(1)}$$
$$= \frac{1 \pm \sqrt{25}}{2} = 3 \text{ or } -2.$$

The zeros are $x = 3$ and $x = -2$, the same as we found by factoring.

The zeros of a function occur at the x-intercepts of its graph. Not every quadratic function has x-intercepts, as we see in the next example.

Example 3 Figure 2.29 shows a graph of $h(x) = -\frac{1}{2}x^2 - 2$. What happens if we try to use algebra to find its zeros?

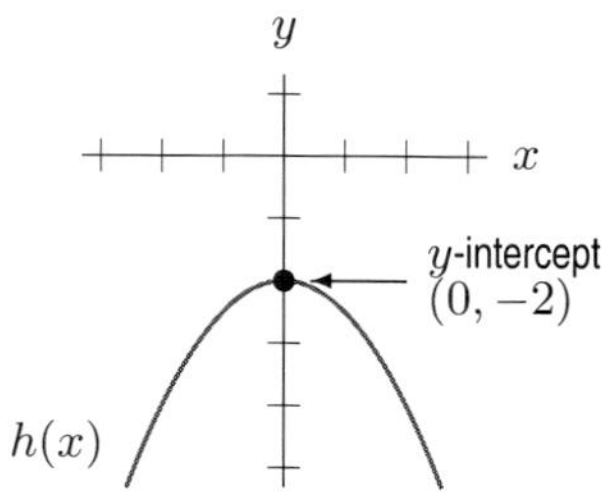

Figure 2.29

Solution To find the zeros, we solve the equation

$$
\begin{aligned}
-\frac{1}{2}x^2 - 2 &= 0 \\
-\frac{1}{2}x^2 &= 2 \\
x^2 &= -4 \\
x &= \pm\sqrt{-4}.
\end{aligned}
$$

Since $\sqrt{-4}$ is not a real number, there are no real solutions, so h has no real zeros. This corresponds to the fact that the graph of h in Figure 2.29 does not cross the x-axis.

Concavity and Quadratic Functions

Unlike a linear function, whose graph is a straight line, a quadratic function has a graph which is either concave up or concave down.

Example 4 Let $f(x) = x^2$. Find the average rate of change of f over the intervals of length 2 between $x = -4$ and $x = 4$. What do these rates tell you about the concavity of the graph of f?

Solution Between $x = -4$ and $x = -2$, we have

$$\text{Average rate of change of } f = \frac{f(-2) - f(-4)}{-2 - (-4)} = \frac{(-2)^2 - (-4)^2}{-2 + 4} = -6.$$

Between $x = -2$ and $x = 0$, we have

$$\text{Average rate of change of } f = \frac{f(0) - f(-2)}{0 - (-2)} = \frac{0^2 - (-2)^2}{0 + 2} = -2.$$

Between $x = 0$ and $x = 2$, we have

$$\text{Average rate of change of } f = \frac{f(2) - f(0)}{2 - 0} = \frac{2^2 - 0^2}{2 - 0} = 2.$$

Between $x = 2$ and $x = 4$, we have

$$\text{Average rate of change of } f = \frac{f(4) - f(2)}{4 - 2} = \frac{4^2 - 2^2}{4 - 2} = 6.$$

Since these rates are increasing, we expect the graph of f to be bending upward. Figure 2.30 confirms that the graph is concave up.

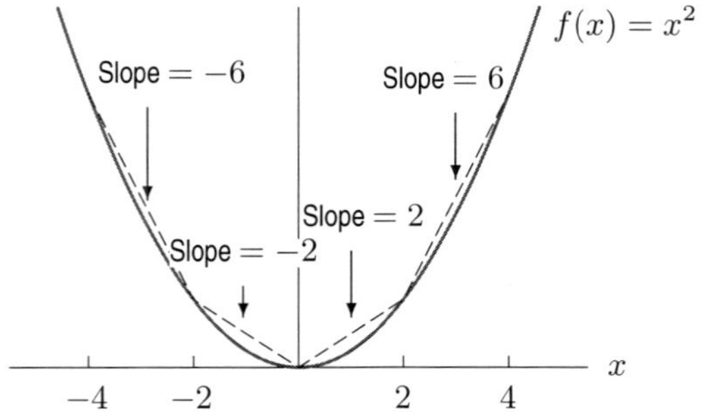

Figure 2.30

Example 5 A high diver jumps off a 10-meter springboard. For h in meters and t in seconds after the diver leaves the board, her height above the water is in Figure 2.31 and given by

$$h = f(t) = -4.9t^2 + 8t + 10.$$

(a) Find and interpret the domain and range of the function and the intercepts of the graph.
(b) Identify the concavity.

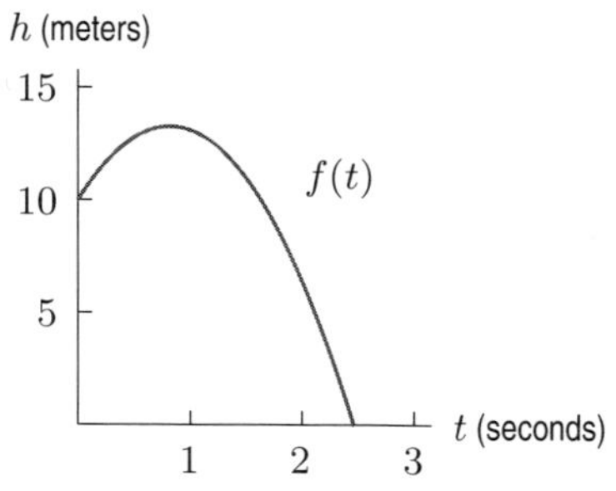

Figure 2.31: Height of diver as a function of time

Solution (a) The diver enters the water when her height is 0. This occurs when

$$h = f(t) = -4.9t^2 + 8t + 10 = 0.$$

Using the quadratic formula to solve this equation, we find $t = 2.462$ seconds. The domain is the interval of time the diver is in the air, namely $0 \leq t \leq 2.462$. To find the range of f, we look for the largest and smallest outputs for h. From the graph, the diver's maximum height appears to occur at about $t = 1$, so we estimate the largest output value for f to be

$$f(1) = -4.9 \cdot 1^2 + 8 \cdot 1 + 10 = 13.1 \text{ meters}$$

Thus, the range of f is approximately $0 \leq f(t) \leq 13.1$.

The vertical intercept of the graph is

$$f(0) = -4.9 \cdot 0^2 + 8 \cdot 0 + 10 = 10 \text{ meters}.$$

The diver's initial height is 10 meters (the height of the springboard). The horizontal intercept is the point where $f(t) = 0$, which we found in part (a). The diver enters the water approximately 2.462 seconds after leaving the springboard.

(b) In Figure 2.31, we see that the graph is bending downward over its entire domain, so it is concave down. This is confirmed by Table 2.21, which shows that the rate of change, $\Delta h/\Delta t$, is decreasing.

Table 2.21

t (sec)	h (meters)	Rate of change $\Delta h/\Delta t$
0	10	
		5.55
0.5	12.775	
		0.65
1.0	13.100	
		−4.25
1.5	10.975	
		−9.15
2.0	6.400	

Exercises and Problems for Section 2.6

Exercises

Are the functions in Exercises 1–7 quadratic? If so, write the function in the form $f(x) = ax^2 + bx + c$.

1. $f(x) = 2(7 - x)^2 + 1$
2. $L(P) = (P + 1)(1 - P)$
3. $g(m) = m(m^2 - 2m) + 3\left(14 - \dfrac{m^3}{3}\right) + \sqrt{3}m$
4. $h(t) = -16(t - 3)(t + 1)$
5. $R(q) = \dfrac{1}{q^2}(q^2 + 1)^2$
6. $K(x) = 13^2 + 13^x$
7. $T(n) = \sqrt{5} + \sqrt{3n^4} - \sqrt{\dfrac{n^4}{4}}$
8. Find the zeros of $Q(r) = 2r^2 - 6r - 36$ by factoring.
9. Find the zeros of $Q(x) = 5x - x^2 + 3$ using the quadratic formula.
10. Find two quadratic functions with zeros $x = 1$, $x = 2$.
11. Solve for x using the quadratic formula and demonstrate your solution graphically:
    (a) $6x - \frac{1}{3} = 3x^2$ **(b)** $2x^2 + 7.2 = 5.1x$
12. Without a calculator, graph $y = 3x^2 - 16x - 12$ by factoring and plotting zeros.

In Exercises 13–24, find the zeros (if any) of the function algebraically.

13. $y = (2 - x)(3 - 2x)$
14. $y = 2x^2 + 5x + 2$
15. $y = 4x^2 - 4x - 8$
16. $y = 7x^2 + 16x + 4$
17. $y = 9x^2 + 6x + 1$
18. $y = 6x^2 - 17x + 12$
19. $y = 5x^2 + 2x - 1$
20. $y = 3x^2 - 2x + 6$
21. $y = -17x^2 + 23x + 19$
22. $y = 89x^2 + 55x + 34$
23. $y = x^4 + 5x^2 + 6$
24. $y = x - \sqrt{x} - 12$

Problems

25. Use the quadratic formula to find the time at which the baseball in Figure 2.28 on page 88 hits the ground.
26. Is there a quadratic function with zeros $x = 1$, $x = 2$ and $x = 3$?
27. Determine the concavity of the graph of $f(x) = 4 - x^2$ between $x = -1$ and $x = 5$ by calculating average rates of change over intervals of length 2.
28. Graph a quadratic function which has all the following properties: concave up, y-intercept is -6, zeros at $x = -2$ and $x = 3$.
29. Without a calculator, graph the following function by factoring and plotting zeros:
$$y = -4cx + x^2 + 4c^2 \quad \text{for} \quad c > 0$$
30. A ball is thrown into the air. Its height (in feet) t seconds later is given by $h(t) = 80t - 16t^2$.
    (a) Evaluate and interpret $h(2)$.
    (b) Solve the equation $h(t) = 80$. Interpret your solutions and illustrate them on a graph of $h(t)$.
31. Let $V(t) = t^2 - 4t + 4$ represent the velocity of an object in meters per second.
    (a) What is the object's initial velocity?
    (b) When is the object not moving?
    (c) Identify the concavity of the velocity graph.
32. The percentage of schools with interactive videodisc players[9] each year from 1992 to 1996 is shown in Table 2.22. If x is in years since 1992, show that this data set can be approximated by the quadratic function $p(x) = -0.8x^2 + 8.8x + 7.2$. What does this model predict for the year 2004? How good is this model for predicting the future?

Table 2.22

Year	1992	1993	1994	1995	1996
Percentage	8	14	21	29.1	29.3

[9]Data from R. Famighetti, ed. *The World Almanac and Book of Facts: 1999.* (New Jersey: Funk and Wagnalls, 1998).

33. Let $f(x) = x^2$ and $g(x) = x^2 + 2x - 8$.

(a) Graph f and g in the window $-10 \le x \le 10$, $-10 \le y \le 10$. How are the two graphs similar? How are they different?

(b) Graph f and g in the window $-10 \le x \le 10$, $-10 \le y \le 100$. Why do the two graphs appear more similar on this window than on the window from part (a)?

(c) Graph f and g in the window $-20 \le x \le 20$, $-10 \le y \le 400$, the window $-50 \le x \le 50$, $-10 \le y \le 2500$, and the window $-500 \le x \le 500$, $-2500 \le y \le 250{,}000$. Describe the change in appearance of f and g on these three successive windows.

34. A relief package is dropped from a moving airplane. Since the package is initially released with a forward horizontal velocity, it follows a parabolic path (instead of dropping straight down). Figure 2.32 shows the height of the package, h, in km, as a function of the horizontal distance, d, in meters, it has traveled since it was dropped.

(a) From what height was the package released?

(b) How far away from the spot above which it was released does the package hit the ground?

(c) Write a formula for $h(d)$. [Hint: The package starts falling at the highest point on the parabola].

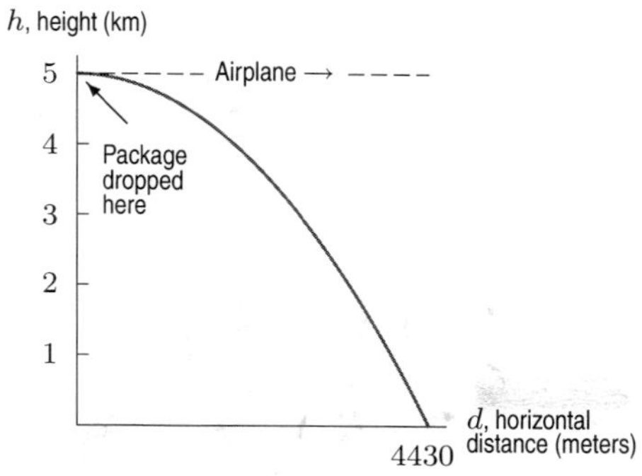

Figure 2.32

35. **(a)** Fit a quadratic function to the first three data points in Table 2.23. Use the fifth point as a check.

(b) Find the formula for a linear function that passes through the second two data points.

(c) Compare the value of the linear function at $x = 3$ to the value of the quadratic at $x = 3$.

(d) Compare the values of the linear and quadratic functions at $x = 50$.

(e) For approximately what x values do the quadratic and linear function values differ by less than 0.05? Using a calculator or computer, graph both functions on the same axes and estimate an answer.

Table 2.23

x	0	1	2	3	50
y	1.0	3.01	5.04	7.09	126.0

CHAPTER SUMMARY

- **Input and Output**
  Evaluating functions: finding $f(a)$ for given a.
  Solving equations: finding x if $f(x) = b$ for given b.
- **Domain and Range**
  Domain: set of input values.
  Range: set of output values
  Piecewise functions: different formulas on different intervals.
- **Inverse functions**
  If $y = f(x)$, then $f^{-1}(y) = x$.
  Evaluating $f^{-1}(b)$. Interpretation of $f^{-1}(b)$. Formula for $f^{-1}(y)$ given formula for $f(x)$.
- **Concavity**
  Concave up: increasing rate of change.
  Concave down: decreasing rate of change.
- **Quadratic Functions**
  Standard form for quadratic functions:
  $$f(x) = ax^2 + bx + c, \quad a \ne 0.$$
  Factored form gives zeros r, s of quadratic:
  $$f(x) = a(x - r)(x - s).$$
  Quadratic formula:
  $$x = \frac{-b \pm \sqrt{b^2 - 4ac}}{2a}.$$

REVIEW EXERCISES AND PROBLEMS FOR CHAPTER TWO

Exercises

If $p(r) = r^2 + 5$, evaluate the expressions in Exercises 1–2.

1. $p(7)$ **2.** $p(x) + p(8)$

3. Let $h(x) = x^2 + bx + c$. Evaluate and simplify:

(a) $h(1)$ **(b)** $h(b+1)$

4. If $g(x) = x\sqrt{x} + 100x$, evaluate without a calculator

(a) $g(100)$ **(b)** $g(4/25)$ **(c)** $g(1.21 \cdot 10^4)$

5. Find the zeros of $s(l) = 7l - l^2$.

6. **(a)** How can you tell from the graph of a function that an x-value is not in the domain? Sketch an example.
(b) How can you tell from the formula for a function that an x-value is not in the domain? Give an example.

In Exercises 7–10, state the domain and range.

7. $h(x) = x^2 + 8x$ **8.** $f(x) = \sqrt{x-4}$

9. $\dfrac{r(x)}{\sqrt{4-\sqrt{x-4}}} =$ **10.** $g(x) = \dfrac{4}{4+x^2}$

11. Let $g(x) = x^2 + x$. Evaluate and simplify the following.

(a) $-3g(x)$ **(b)** $g(1) - x$
(c) $g(x) + \pi$ **(d)** $\sqrt{g(x)}$
(e) $g(1)/(x+1)$ **(f)** $(g(x))^2$

12. Let $f(x) = 1 - x$. Evaluate and simplify the following.

(a) $2f(x)$ **(b)** $f(x) + 1$ **(c)** $f(1-x)$
(d) $(f(x))^2$ **(e)** $f(1)/x$ **(f)** $\sqrt{f(x)}$

In Exercises 13–14, let $f(x) = 3x - 7$, $g(x) = x^3 + 1$ to find a formula for the function.

13. $f(g(x))$ **14.** $g(f(x))$

In Exercises 15–16, find the inverse function.

15. $y = f(x) = 3x - 7$ **16.** $y = g(x) = x^3 + 1$

17. Use the graph in Figure 2.33 to fill in the missing values:

(a) $f(0) = ?$ **(b)** $f(?) = 0$
(c) $f^{-1}(0) = ?$ **(d)** $f^{-1}(?) = 0$

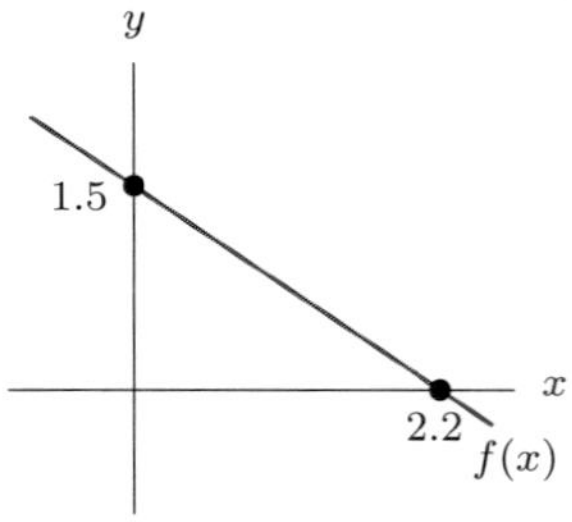

Figure 2.33

In Exercises 18–20, let $P = f(t)$ be the population, in millions, of a country at time t in years and let $E = g(P)$ be the daily electricity consumption, in megawatts, when the population is P. Give the meaning and units of the function. Assume both f and g are invertible.

18. $g(f(t))$ **19.** $f^{-1}(P)$ **20.** $g^{-1}(E)$

Problems

In Figure 2.34, show the coordinates of the point(s) representing the statements in Problems 21–24.

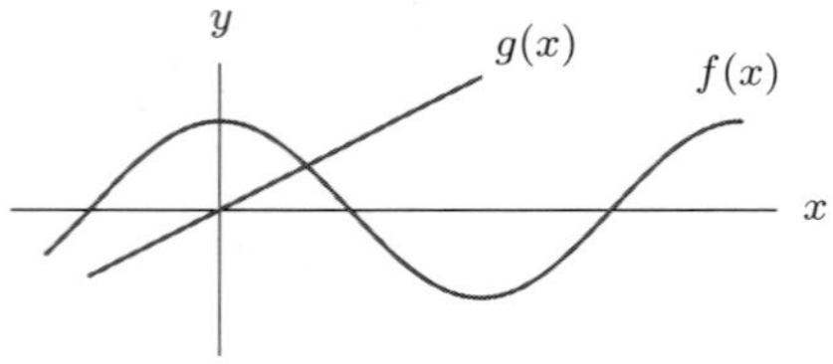

Figure 2.34

21. $f(0) = 2$

22. $f(-3) = f(3) = f(9) = 0$

23. $f(2) = g(2)$

24. $g(x) > f(x)$ for $x > 2$

In Problems 25–27, if $f(x) = \dfrac{ax}{a+x}$, find and simplify

25. $f(a)$ **26.** $f(1-a)$ **27.** $f\left(\dfrac{1}{1-a}\right)$

28. **(a)** Find the side, $s = f(d)$, of a square as function of its diagonal d.
(b) Find the area, $A = g(s)$, of a square as function of its side s.
(c) Find the area $A = h(d)$ as a function of d.
(d) What is the relation between f, g, and h?

29. The cost of producing q thousand loaves of bread is $C(q)$ dollars. Interpret the following statements in terms of bread; give units.

(a) $C(5) = 653$
(b) $C^{-1}(80) = 0.62$
(c) The solution to $C(q) = 790$ is 6.3
(d) The solution to $C^{-1}(x) = 1.2$ is 150.

In Exercises 30–32, let $H = f(t) = \frac{5}{9}(t - 32)$, where H is temperature in degrees Celsius and t is in degrees Fahrenheit.

30. Find and interpret the inverse function, $f^{-1}(H)$.

31. Using the results of Exercise 30, evaluate and interpret:

(a) $f(0)$ **(b)** $f^{-1}(0)$
(c) $f(100)$ **(d)** $f^{-1}(100)$

32. The temperature, $t = g(n) = 68 + 10 \cdot 2^{-n}$, in degrees Fahrenheit of a room is a function of the number, n, of hours that the air conditioner has been running. Find and interpret $f(g(n))$. Give units.

33. The period, T, of a pendulum of length l is given by $T = f(l) = 2\pi\sqrt{l/g}$, where g is a constant. Find a formula for $f^{-1}(T)$ and explain its meaning.

34. The area, in square centimeters, of a circle whose radius is r cm is given by $A = \pi r^2$.

(a) Write this formula using function notation, where f is the name of the function.
(b) Evaluate $f(0)$.
(c) Evaluate and interpret $f(r + 1)$.
(d) Evaluate and interpret $f(r) + 1$.
(e) What are the units of $f^{-1}(4)$?

35. An epidemic of influenza spreads through a city. Figure 2.35 is the graph of $I = f(w)$, where I is the number of individuals (in thousands) infected w weeks after the epidemic begins.

(a) Evaluate $f(2)$ and explain its meaning in terms of the epidemic.
(b) Approximately how many people were infected at the height of the epidemic? When did that occur? Write your answer in the form $f(a) = b$.
(c) Solve $f(w) = 4.5$ and explain what the solutions mean in terms of the epidemic.
(d) The graph used $f(w) = 6w(1.3)^{-w}$. Use the graph to estimate the solution of the inequality $6w(1.3)^{-w} \geq 6$. Explain what the solution means in terms of the epidemic.

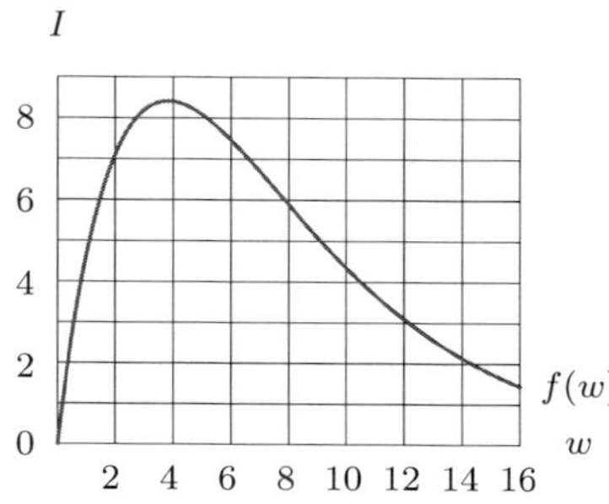

Figure 2.35

36. Let t be time in seconds and let $r(t)$ be the rate, in gallons/second, that water enters a reservoir:

$$r(t) = 800 - 40t.$$

(a) Evaluate the expressions $r(0)$, $r(15)$, $r(25)$, and explain their physical significance.
(b) Graph $y = r(t)$ for $0 \leq t \leq 30$, labeling the intercepts. What is the physical significance of the slope and the intercepts?
(c) For $0 \leq t \leq 30$, when does the reservoir have the most water? When does it have the least water?
(d) What are the domain and range of $r(t)$?

37. Suppose that $f(x)$ is invertible and that both f and f^{-1} are defined for all values of x. Let $f(2) = 3$ and $f^{-1}(5) = 4$. Evaluate the following expressions, or, if the given information is insufficient, write unknown.

(a) $f^{-1}(3)$ (b) $f^{-1}(4)$ (c) $f(4)$

38. Suppose that $j(x) = h^{-1}(x)$ and that both j and h are defined for all values of x. Let $h(4) = 2$ and $j(5) = -3$. Evaluate if possible:

(a) $j(h(4))$ **(b)** $j(4)$ **(c)** $h(j(4))$
(d) $j(2)$ **(e)** $h^{-1}(-3)$ **(f)** $j^{-1}(-3)$
(g) $h(5)$ **(h)** $(h(-3))^{-1}$ **(i)** $(h(2))^{-1}$

39. Values of f and g are given in Table 2.24.

(a) Evaluate $f(1)$ and $g(3)$.
(b) Describe in full sentences the patterns you see in the values for each function.
(c) Assuming that the patterns you observed in part (b) hold true for all values of x, calculate $f(5)$, $f(-2)$, $g(5)$, and $g(-2)$.
(d) Find possible formulas for $f(x)$ and $g(x)$.

Table 2.24

x	-1	0	1	2	3	4
$f(x)$	-4	-1	2	5	8	11
$g(x)$	4	1	0	1	4	9

40. Let $k(x) = 6 - x^2$.

(a) Find a point on the graph of $k(x)$ whose x-coordinate is -2.
(b) Find two points on the graph whose y-coordinates are -2.
(c) Graph $k(x)$ and locate the points in parts (a) and (b).
(d) Let $p = 2$. Calculate $k(p) - k(p-1)$.

41. Let $f(a)$ be the cost in dollars of a pounds of organic apples at the Gourmet Garage in New York City in January 2006. What do the following statements tell you? What are the units of each of the numbers?

(a) $f(2) = 4.98$ **(b)** $f(0.5) = 1.25$
(c) $f^{-1}(0.62) = 0.25$ **(d)** $f^{-1}(12.45) = 5$

42. Let $t(x)$ be the time required, in seconds, to melt 1 gram of a compound at x°C.

(a) Express the following statement as an equation using $t(x)$: It takes 272 seconds to melt 1 gram of the compound at 400°C.
(b) Explain the following equations in words:
(i) $t(800) = 136$ (ii) $t^{-1}(68) = 1600$
(c) Above a certain temperature, doubling the temperature, x, halves the melting time. Express this fact with an equation involving $t(x)$.

43. Table 2.25 shows $N(s)$, the number of sections of Economics 101, as a function of s, the number of students in the course. If s is between two numbers listed in the table, then $N(s)$ is the higher number of sections.

Table 2.25

s	50	75	100	125	150	175	200
$N(s)$	4	4	5	5	6	6	7

(a) Evaluate and interpret:
(i) $N(150)$ (ii) $N(80)$ (iii) $N(55.5)$
(b) Solve for s and interpret:
(i) $N(s) = 4$ (ii) $N(s) = N(125)$

Problems 44–45 concern studies which indicate that as carbon dioxide (CO_2) levels rise, hurricanes will become more intense.[10] Hurricane intensity is measured in terms of the minimum central pressure (in mb): the lower the pressure, the more powerful the storm. Since warm ocean waters fuel hurricanes, P is a decreasing function of H, sea surface temperature in °C. Let $P = n(H)$ be the hurricane-intensity function for present-day CO_2 levels, and let $P = N(H)$ be the hurricane-intensity function for future projected CO_2 levels. If H_0 is the average temperature in the Caribbean Sea, what do the following quantities tell you about hurricane intensity?

44. $N(H_0) - n(H_0)$ **45.** $n(H_0 + 1) - n(H_0)$

46. Table 2.26 shows the population, P, in millions, of Ireland[11] at various times between 1780 and 1910, with t in years since 1780.

(a) When was the population increasing? Decreasing?
(b) For each successive time interval, construct a table showing the average rate of change of the population.
(c) From the table you constructed in part (b), when is the graph of the population concave up? Concave down?
(d) When was the average rate of change of the population the greatest? The least? How is this related to part (c)? What does this mean in human terms?
(e) Graph the data in Table 2.26 and join the points by a curve to show the trend in the data. From this graph identify where the curve is increasing, decreasing, concave up and concave down. Compare your answers to those you got in parts (a) and (c). Identify the region you found in part (d).
(f) Something catastrophic happened in Ireland between 1780 and 1910. When? What happened in Ireland at that time to cause this catastrophe?

Table 2.26 *The population of Ireland from 1780 to 1910, where t = 0 corresponds to 1780*

t	0	20	40	60	70	90	110	130
P	4.0	5.2	6.7	8.3	6.9	5.4	4.7	4.4

[10] Journal of Climate, September 14, 2004, pages 3477–3495.
[11] Adapted from D. N. Burghes and A. D. Wood, Ellis Horwood, *Mathematical Models in the Social, Management and Life Science,* p. 104 (Ellis Horwood, 1980).

47. The surface area of a cylindrical aluminum can is a measure of how much aluminum the can requires. If the can has radius r and height h, its surface area A and its volume V are given by the equations:

$$A = 2\pi r^2 + 2\pi rh \quad \text{and} \quad V = \pi r^2 h.$$

(a) The volume, V, of a 12 oz cola can is 355 cm^3. A cola can is approximately cylindrical. Express its surface area A as a function of its radius r, where r is measured in centimeters. [Hint: First solve for h in terms of r.]

(b) Graph $A = s(r)$, the surface area of a cola can whose volume is 355 cm^3, for $0 \le r \le 10$.

(c) What is the domain of $s(r)$? Based on your graph, what, approximately, is the range of $s(r)$?

(d) The manufacturers wish to use the least amount of aluminum (in cm^2) necessary to make a 12 oz cola can. Use your answer in (c) to find the minimum amount of aluminum needed. State the values of r and h that minimize the amount of aluminum used.

(e) The radius of a real 12 oz cola can is about 3.25 cm. Show that real cola cans use more aluminum than necessary to hold 12 oz of cola. Why do you think real cola cans are made to this way?

CHECK YOUR UNDERSTANDING

Are the statements in Problems 1–51 true or false? Give an explanation for your answer.

1. If $f(t) = 3t^2 - 4$ then $f(2) = 0$.

2. If $f(x) = x^2 - 9x + 10$ then $f(b) = b^2 - 9b + 10$.

3. If $f(x) = x^2$ then $f(x + h) = x^2 + h^2$.

4. If $q = \dfrac{1}{\sqrt{z^2 + 5}}$ then the values of z that make $q = \frac{1}{3}$ are $z = \pm 2$.

5. If $W = \dfrac{t + 4}{t - 4}$ then when $t = 8$, $W = 1$.

6. If $f(t) = t^2 + 64$ then $f(0) = 64$.

7. If $f(x) = 0$ then $x = 0$.

8. If $f(x) = x^2 + 2x + 7$ then $f(-x) = f(x)$.

9. If $g(x) = \dfrac{3}{\sqrt{x^2 + 4}}$ then $g(x)$ can never be zero.

10. If $h(p) = -6p + 9$ then $h(3) + h(4) = h(7)$.

11. The domain of a function is the set of input values.

12. If a function is being used to model a real world situation, the domain and range are often determined by the constraints of the situation being modeled.

13. The domain of $f(x) = \dfrac{4}{x - 3}$ consists of all real numbers x, $x \neq 0$.

14. If $f(x) = \sqrt{2 - x}$, the domain of f consists of all real numbers $x \ge 2$.

15. The range of $f(x) = \dfrac{1}{x}$ is all real numbers.

16. The range of $y = 4 - \dfrac{1}{x}$ is $0 < y < 4$.

17. If $f(x) = \frac{2}{5}x + 6$ and its domain is $15 \le x \le 20$ then the range of f is $12 \le x \le 14$.

18. The domain of $f(x) = \dfrac{x}{\sqrt{x^2 + 1}}$ is all real numbers.

19. The graph of the absolute value function $y = |x|$ has a V shape.

20. The domain of $f(x) = |x|$ is all real numbers.

21. If $f(x) = |x|$ and $g(x) = |-x|$ then for all x, $f(x) = g(x)$.

22. If $f(x) = |x|$ and $g(x) = -|x|$ then for all x, $f(x) = g(x)$.

23. If $y = \dfrac{x}{|x|}$ then $y = 1$ for $x \neq 0$.

24. If $f(x) = \begin{cases} 3 & \text{if } x < 0 \\ x^2 & \text{if } 0 \le x \le 4, \\ 7 & \text{if } x > 4 \end{cases}$ then $f(3) = 0$.

25. Let $f(x) = \begin{cases} x & \text{if } x < 0 \\ x^2 & \text{if } 0 \le x \le 4 \\ -x & \text{if } x > 4 \end{cases}$ If $f(x) = 4$ then $x = 2$.

26. If $f(3) = 5$ and f is invertible, then $f^{-1}(3) = 1/5$.

27. If $h(7) = 4$ and h is invertible, then $h^{-1}(4) = 7$.

28. If $f(x) = \frac{3}{4}x - 6$ then $f^{-1}(8) = 0$.

29. If $R = f(S) = \frac{2}{3}S + 8$ then $S = f^{-1}(R) = \frac{3}{2}(R - 8)$.

30. In general $f^{-1}(x) = (f(x))^{-1}$.

31. If $f(x) = \dfrac{x}{x + 1}$ then $f(t^{-1}) = \dfrac{1/t}{1/t + 1}$.

32. The units of the output of a function are the same as the units of output of its inverse.

33. The functions $f(x) = 2x + 1$ and $g(x) = \frac{1}{2}x - 1$ are inverses.

34. If $q = f(x)$ is the quantity of rice in tons required to feed x million people for a year and $p = g(q)$ is the cost, in dollars, of q tons of rice, then $g(f(x))$ is the dollar cost of feeding x million people for a year.

35. If $f(t) = t+2$ and $g(t) = 3t$, then $g(f(t)) = 3(t+2) = 3t+6$.

36. A fireball has radius $r = f(t)$ meters t seconds after an explosion. The volume of the ball is $V = g(r)$ meter^3 when it has radius r meters. Then the units of measurement of $g(f(t))$ are meter^3/sec.

37. If the graph of a function is concave up, then the average rate of change of a function over an interval of length 1 increases as the interval moves from left to right.

38. The function f in the table could be concave up.

x	-2	0	2	4
$f(x)$	5	6	8	12

39. The function g in the table could be concave down.

t	-1	1	3	5
$g(t)$	9	8	6	3

40. A straight line is concave up.

41. A function can be both decreasing and concave down.

42. If a function is concave up, it must be increasing.

43. The quadratic function $f(x) = x(x+2)$ is in factored form.

44. If $f(x) = (x+1)(x+2)$, then the zeros of f are 1 and 2.

45. A quadratic function whose graph is concave up has a maximum.

46. All quadratic equations have the form $f(x) = ax^2$.

47. If the height above the ground of an object at time t is given by $s(t) = at^2 + bt + c$, then $s(0)$ tells us when the object hits the ground.

48. To find the zeros of $f(x) = ax^2 + bx + c$, solve the equation $ax^2 + bx + c = 0$ for x.

49. Every quadratic equation has two real solutions.

50. There is only one quadratic function with zeros at $x = -2$ and $x = 2$.

51. A quadratic function has exactly two zeros.

TOOLS FOR CHAPTER 2: QUADRATIC EQUATIONS

Expanding an Expression

The *distributive property* for real numbers a, b, and c tells us that

$$a(b+c) = ab + ac,$$

and

$$(b+c)a = ba + ca.$$

We use the distributive property and the rules of exponents to multiply algebraic expressions involving parentheses. This process is sometimes referred to as *expanding* the expression.

Example 1 Multiply the following expressions and simplify.

(a) $3x^2\left(x + \frac{1}{6}x^{-3}\right)$ (b) $\left((2t)^2 - 5\right)\sqrt{t}$

Solution

(a) $3x^2\left(x + \frac{1}{6}x^{-3}\right) = \left(3x^2\right)(x) + \left(3x^2\right)\left(\frac{1}{6}x^{-3}\right) = 3x^3 + \frac{1}{2}x^{-1}$

(b) $\left((2t)^2 - 5\right)\sqrt{t} = (2t)^2(\sqrt{t}) - 5\sqrt{t} = \left(4t^2\right)\left(t^{1/2}\right) - 5t^{1/2} = 4t^{5/2} - 5t^{1/2}$

If there are two terms in each factor, then there are four terms in the product:

$$(a+b)(c+d) = a(c+d) + b(c+d) = ac + ad + bc + bd.$$

The following special cases of the above product occur frequently. Learning to recognize their forms aids in factoring.

$$(a+b)(a-b) = a^2 - b^2$$
$$(a+b)^2 = a^2 + 2ab + b^2$$
$$(a-b)^2 = a^2 - 2ab + b^2$$

Example 2 Expand the following and simplify by gathering like terms.

(a) $(5x^2+2)(x-4)$ (b) $(2\sqrt{r}+2)(4\sqrt{r}-3)$ (c) $\left(3 - \frac{1}{2}x\right)^2$

Solution

(a) $\left(5x^2+2\right)(x-4) = \left(5x^2\right)(x) + \left(5x^2\right)(-4) + (2)(x) + (2)(-4) = 5x^3 - 20x^2 + 2x - 8$

(b) $(2\sqrt{r}+2)(4\sqrt{r}-3) = (2)(4)(\sqrt{r})^2 + (2)(-3)(\sqrt{r}) + (2)(4)(\sqrt{r}) + (2)(-3) = 8r + 2\sqrt{r} - 6$

(c) $\left(3 - \frac{1}{2}x\right)^2 = 3^2 - 2(3)\left(\frac{1}{2}x\right) + \left(-\frac{1}{2}x\right)^2 = 9 - 3x + \frac{1}{4}x^2$

Factoring

To write an expanded expression in factored form, we "un-multiply" the expression. Some techniques for factoring are given in this section. We can check factoring by remultiplying.

Removing a Common Factor

It is sometimes useful to factor out the same factor from each of the terms in an expression. This is basically the distributive law in reverse:

$$ab + ac = a(b + c).$$

One special case is removing a factor of -1, which gives

$$\boxed{-a - b = -(a + b)}$$

Another special case is

$$\boxed{(a - b) = -(b - a)}$$

Example 3 Factor the following:

(a) $\frac{2}{3}x^2y + \frac{4}{3}xy$ (b) $(2p + 1)p^3 - 3p(2p + 1)$ (c) $-\frac{s^2t}{8w} - \frac{st^2}{16w}$

Solution

(a) $\frac{2}{3}x^2y + \frac{4}{3}xy = \frac{2}{3}xy(x + 2)$

(b) $(2p + 1)p^3 - 3p(2p + 1) = (p^3 - 3p)(2p + 1) = p(p^2 - 3)(2p + 1)$

(Note that the expression $(2p + 1)$ was one of the factors common to both terms.)

(c) $-\frac{s^2t}{8w} - \frac{st^2}{16w} = -\frac{st}{8w}\left(s + \frac{t}{2}\right).$

Grouping Terms

Even though all the terms may not have a common factor, we can sometimes factor by first grouping the terms and then removing a common factor.

Example 4 Factor $x^2 - hx - x + h$.

Solution

$$x^2 - hx - x + h = (x^2 - hx) - (x - h) = x(x - h) - (x - h) = (x - h)(x - 1)$$

Factoring Quadratics

One way to factor quadratics is to mentally multiply out the possibilities.

Example 5 Factor $t^2 - 4t - 12$.

Solution If the quadratic factors, it will be of the form

$$t^2 - 4t - 12 = (t + ?)(t + ?).$$

We are looking for two numbers whose product is -12 and whose sum is -4. By trying combinations, we find

$$t^2 - 4t - 12 = (t - 6)(t + 2).$$

Example 6 Factor $4 - 2M - 6M^2$.

Solution By a similar method as in the previous example, we find $4 - 2M - 6M^2 = (2 - 3M)(2 + 2M)$.

Perfect Squares and the Difference of Squares

Recognition of the special products $(x + y)^2$, $(x - y)^2$ and $(x + y)(x - y)$ in expanded form is useful in factoring. Reversing the results in the last section, we have

$$a^2 + 2ab + b^2 = (a + b)^2,$$
$$a^2 - 2ab + b^2 = (a - b)^2,$$
$$a^2 - b^2 = (a - b)(a + b).$$

When we can see that terms in an expression we want to factor are squares, it often makes sense to look for one of these forms. The difference of squares identity (the third one listed above) is especially useful.

Example 7 Factor: (a) $16y^2 - 24y + 9$ (b) $25S^2R^4 - T^6$ (c) $x^2(x - 2) + 16(2 - x)$

Solution
(a) $16y^2 - 24y + 9 = (4y - 3)^2$
(b) $25S^2R^4 - T^6 = \left(5SR^2\right)^2 - \left(T^3\right)^2 = \left(5SR^2 - T^3\right)\left(5SR^2 + T^3\right)$
(c) $x^2(x-2) + 16(2-x) = x^2(x-2) - 16(x-2) = (x-2)\left(x^2 - 16\right) = (x-2)(x-4)(x+4)$

Solving Quadratic Equations

Example 8 Give exact and approximate solutions to $x^2 = 3$.

Solution The exact solutions are $x = \pm\sqrt{3}$; approximate ones are $x \approx \pm 1.73$, or $x \approx \pm 1.732$, or $x \approx \pm 1.73205$. (since $\sqrt{3} = 1.732050808\ldots$). Notice that the equation $x^2 = 3$ has only two exact solutions, but many possible approximate solutions, depending on how much accuracy is required.

Solving by Factoring

Some equations can be put into factored form such that the product of the factors is zero. Then we solve by using the fact that if $a \cdot b = 0$, then either a or b (or both) is zero.

Example 9 Solve $(x + 1)(x + 3) = 15$ for x.

Solution Do not make the mistake of setting $x + 1 = 15$ and $x + 3 = 15$. It is not true that $a \cdot b = 15$ means that $a = 15$ or $b = 15$ (or both). (Although it is true that if $a \cdot b = 0$, then $a = 0$ or $b = 0$, or both.) So, we must expand the left-hand side and set the equation equal to zero:

$$x^2 + 4x + 3 = 15,$$
$$x^2 + 4x - 12 = 0.$$

Then, factoring gives

$$(x - 2)(x + 6) = 0.$$

Thus $x = 2$ and $x = -6$ are solutions.

Example 10 Solve $2(x + 3)^2 = 5(x + 3)$.

Solution You might be tempted to divide both sides by $(x+3)$. However, if you do this you will overlook one of the solutions. Instead, write

$$\begin{aligned}2(x+3)^2-5(x+3)&=0\\(x+3)\,(2(x+3)-5)&=0\\(x+3)(2x+6-5)&=0\\(x+3)(2x+1)&=0.\end{aligned}$$

Thus, $x=-\dfrac{1}{2}$ and $x=-3$ are solutions.

Solving with the Quadratic Formula

Alternatively, we can solve the equation $ax^2+bx+c=0$ by using the quadratic formula:

$$x=\frac{-b\pm\sqrt{b^2-4ac}}{2a}.$$

The quadratic formula is derived by completing the square for $y=ax^2+bx+c$. See page 239 in Tools for Chapter 5.

Example 11 Solve $11+2x=x^2$.

Solution The equation is

$$-x^2+2x+11=0.$$

The expression on the left does not factor using integers, so we use

$$x=\frac{-2+\sqrt{4-4(-1)(11)}}{2(-1)}=\frac{-2+\sqrt{48}}{-2}=\frac{-2+\sqrt{16\cdot3}}{-2}=\frac{-2+4\sqrt{3}}{-2}=1-2\sqrt{3},$$

$$x=\frac{-2-\sqrt{4-4(-1)(11)}}{2(-1)}=\frac{-2-\sqrt{48}}{-2}=\frac{-2-\sqrt{16\cdot3}}{-2}=\frac{-2-4\sqrt{3}}{-2}=1+2\sqrt{3}.$$

The exact solutions are $x=1-2\sqrt{3}$ and $x=1+2\sqrt{3}$.

The decimal approximations to these numbers $x=1-2\sqrt{3}=-2.464$ and $x=1+2\sqrt{3}=4.464$ are approximate solutions to this equation. The approximate solutions could be found directly from a graph or calculator.

Exercises on Tools for Chapter 2

For Exercises 1–18, expand and simplify.

1. $3(x+2)$

2. $5(x-3)$

3. $2(3x-7)$

4. $-4(y+6)$

5. $12(x+y)$

6. $-7(5x-8y)$

7. $x(2x+5)$

8. $3z(2x-9z)$

9. $-10r(5r+6rs)$

10. $x(3x-8)+2(3x-8)$

11. $5z(x-2)-3(x-2)$

12. $(x+1)(x+3)$

13. $(x-2)(x+6)$

14. $(5x-1)(2x-3)$

15. $(x+2)(3x-8)$

16. $(y+1)(z+3)$

17. $(12y-5)(8w+7)$

18. $(5z-3)(x-2)$

Multiply and write the expressions in Problems 19–27 without parentheses. Gather like terms.

19. $-(x-3)-2(5-x)$

20. $(x-5)6-5(1-(2-x))$

21. $\left(3x-2x^2\right)4+(5+4x)(3x-4)$

22. $\left(t^2+1\right)50t-\left(25t^2+125\right)2t$

23. $P(p-3q)^2$

24. $\left(A^2-B^2\right)^2$

25. $4(x-3)^2+7$

26. $-\left(\sqrt{2x}+1\right)^2$

27. $u\left(u^{-1}+2^u\right)2^u$

For Exercises 28–76, factor completely if possible.

28. $2x+6$

29. $3y+15$

30. $5z-30$

31. $4t-6$

32. $10w-25$

33. u^2-2u

34. $3u^4-4u^3$

35. $3u^7+12u^2$

36. $12x^3y^2-18x$

37. $14r^4s^2-21rst$

38. x^2+3x+2

39. x^2+3x-2

40. x^2-3x+2

41. x^2-3x-2

42. x^2+2x+3

43. x^2-2x-3

44. x^2-2x+3

45. x^2+2x-3

46. $2x^2+5x+2$

47. $3x^2-x-4$

48. $2x^2-10x+12$

49. $x^2+3x-28$

50. x^3-2x^2-3x

51. x^3+2x^2-3x

52. $ac+ad+bc+bd$

53. $x^2+2xy+3xz+6yz$

54. $x^2-1.4x-3.92$

55. $a^2x^2-b^2$

56. $\pi r^2+2\pi rh$

57. $B^2-10B+24$

58. c^2+x^2-2cx

59. x^2+y^2

60. a^4-a^2-12

61. $(t+3)^2-16$

62. $x^2+4x+4-y^2$

63. a^3-2a^2+3a-6

64. $b^3-3b^2-9b+27$

65. $c^2d^2-25c^2-9d^2+225$

66. $hx^2+12-4hx-3x$

67. $r(r-s)-2(s-r)$

68. $y^2-3xy+2x^2$

69. $x^2e^{-3x}+2xe^{-3x}$

70. $t^2e^{5t}+3te^{5t}+2e^{5t}$

71. $(s+2t)^2-4p^2$

72. $P(1+r)^2+P(1+r)^2r$

73. $x^2-6x+9-4z^2$

74. $dk+2dm-3ek-6em$

75. $\pi r^2-2\pi r+3r-6$

76. $8gs-12hs+10gm-15hm$

Solve the equations in Exercises 77–108.

77. $x^2+7x+6=0$

78. $y^2-5y-6=0$

79. $2w^2+w-10=0$

80. $4s^2+3s-15=0$

81. $\dfrac{2}{x}+\dfrac{3}{2x}=8$

82. $\dfrac{3}{x-1}+1=5$

83. $\sqrt{y-1}=13$

84. $\sqrt{5y+3}=7$

85. $\sqrt{2x-1}+3=9$

86. $\dfrac{21}{z-5}-\dfrac{13}{z^2-5z}=3$

87. $-16t^2+96t+12=60$

88. $r^3-6r^2=5r-30$

89. $g^3-4g=3g^2-12$

90. $8 + 2x - 3x^2 = 0$

91. $2p^3 + p^2 - 18p - 9 = 0$

92. $N^2 - 2N - 3 = 2N(N - 3)$

93. $\frac{1}{64}t^3 = t$

94. $x^2 - 1 = 2x$

95. $4x^2 - 13x - 12 = 0$

96. $60 = -16t^2 + 96t + 12$

97. $n^5 + 80 = 5n^4 + 16n$

98. $5a^3 + 50a^2 = 4a + 40$

99. $y^2 + 4y - 2 = 0$

100. $\frac{2}{z-3} + \frac{7}{z^2 - 3z} = 0$

101. $\frac{x^2 + 1 - 2x^2}{(x^2+1)^2} = 0$

102. $4 - \frac{1}{L^2} = 0$

103. $2 + \frac{1}{q+1} - \frac{1}{q-1} = 0$

104. $\sqrt{r^2 + 24} = 7$

105. $\frac{1}{\sqrt[3]{x}} = -2$

106. $3\sqrt{x} = \frac{1}{2}x$

107. $10 = \sqrt{\frac{v}{7\pi}}$

108. $\frac{(3x+4)(x-2)}{(x-5)(x-1)} = 0$

In Exercises 109–112, solve for the indicated variable.

109. $T = 2\pi\sqrt{\frac{l}{g}}$, for l.

110. $Ab^5 = C$, for b.

111. $|2x + 1| = 7$, for x.

112. $\frac{x^2 - 5mx + 4m^2}{x - m} = 0$, for x

Solve the systems of equations in Exercises 113–117.

113. $\begin{cases} y = 2x - x^2 \\ y = -3 \end{cases}$

114. $\begin{cases} y = 1/x \\ y = 4x \end{cases}$

115. $\begin{cases} x^2 + y^2 = 36 \\ y = x - 3 \end{cases}$

116. $\begin{cases} y = 4 - x^2 \\ y - 2x = 1 \end{cases}$

117. $\begin{cases} y = x^3 - 1 \\ y = e^x \end{cases}$

118. Let ℓ be the line of slope 3 passing through the origin. Find the points of intersection of the line ℓ and the parabola whose equation is $y = x^2$. Sketch the line and the parabola, and label the points of intersection.

Determine the points of intersection for Problems 119–120.

119.

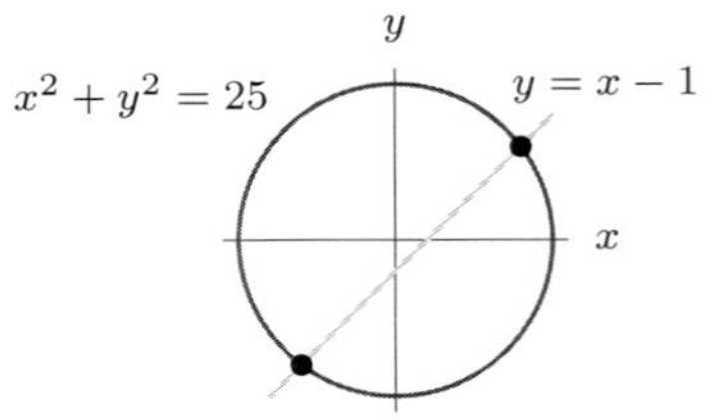

120.

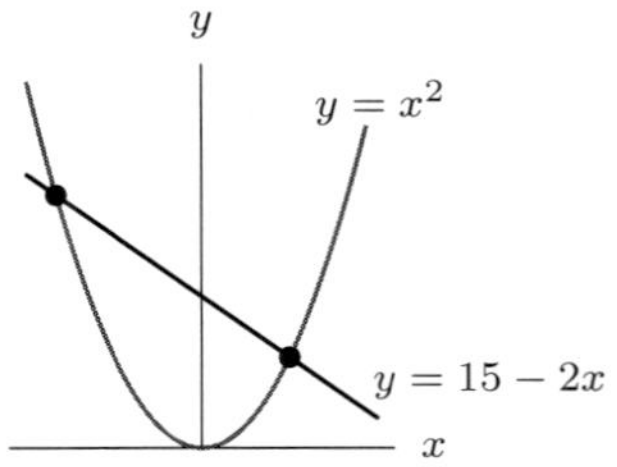

Chapter Three

EXPONENTIAL FUNCTIONS

Exponential functions represent quantities that increase or decrease at a constant percent rate. In contrast to linear functions, in which a constant amount is added per unit input, an exponential function involves multiplication by a constant factor for each unit increase in input value. Examples include the balance of a savings account, the size of some populations, and the quantity of a chemical that decays radioactively.

The Tools Section on page 146 reviews the properties of exponents.

3.1 INTRODUCTION TO THE FAMILY OF EXPONENTIAL FUNCTIONS

Growing at a Constant Percent Rate

Linear functions represent quantities that change at a constant rate. In this section we introduce functions that change at a constant *percent* rate, the *exponential functions*.

Salary Raises

Example 1 After graduation from college, you will probably be looking for a job. Suppose you are offered a job at a starting salary of \$40,000 per year. To strengthen the offer, the company promises annual raises of 6% per year for at least the first five years after you are hired. Let's compute your salary for the first few years.

If t represents the number of years since the beginning of your contract, then for $t = 0$, your salary is $\$40,000$. At the end of the first year, when $t = 1$, your salary increases by 6% so

$$\begin{aligned}\begin{array}{c}\text{Salary when}\\ t=1\end{array} &= \text{Original salary } + 6\% \text{ of Original salary}\\ &= 40000 + 0.06 \cdot 40000\\ &= 42400 \text{ dollars.}\end{aligned}$$

After the second year, your salary again increases by 6%, so

$$\begin{aligned}\begin{array}{c}\text{Salary when}\\ t=2\end{array} &= \text{Former salary } + 6\% \text{ of Former salary}\\ &= 42400 + 0.06 \cdot 42400\\ &= 44944 \text{ dollars.}\end{aligned}$$

Notice that your raise is higher in the second year than in the first since the second 6% increase applies both to the original \$40,000 salary and to the \$2400 raise given in the first year.

Salary calculations for four years have been rounded and recorded in Table 3.1. At the end of the third and fourth years your salary again increases by 6%, and your raise is larger each year. Not only are you given the 6% increase on your original salary, but your raises earn raises as well.

Table 3.1 *Raise amounts and resulting salaries for a person earning 6% annual salary increases*

Year	Raise amount (\$)	Salary (\$)
0		40000.00
1	2400.00	42400.00
2	2544.00	44944.00
3	2696.64	47640.64
4	2858.44	50499.08

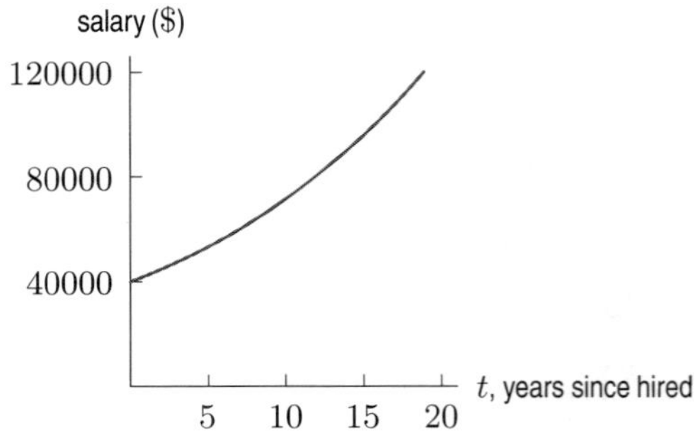

Figure 3.1: Salary over a 20-year period

Figure 3.1 shows salary over a 20-year period assuming that the annual increase remains 6%. Since the rate of change of your salary (in dollars per year) is not constant, the graph of this function is not a line. The salary increases at an increasing rate, giving the graph its upward curve.

Population Growth

Exponential functions provide a reasonable model for many growing populations.

Example 2 During the early 2000s, the population of Mexico increased at a constant annual percent rate of 2%. Since the population grew by the same percent each year, it can be modeled by an exponential function.

Let's calculate the population of Mexico for the first few years after 2000. In 2000, the population was 100 million. The population grew by 2%, so

$$\begin{aligned}\text{Population in 2001} &= \text{Population in 2000} + 2\% \text{ of Population in 2000}\\ &= 100 + 0.02(100)\\ &= 100 + 2 = 102 \text{ million.}\end{aligned}$$

Similarly,

$$\begin{aligned}\text{Population in 2002} &= \text{Population in 2001} + 2\% \text{ of Population in 2001}\\ &= 102 + 0.02(102)\\ &= 102 + 2.04 = 104.04 \text{ million.}\end{aligned}$$

The calculations for years 2000 through 2004 have been rounded and recorded in Table 3.2. The population of Mexico increased by slightly more each year than it did the year before, because each year the increase is 2% of a larger number.

Table 3.2 *Calculated values for the population of Mexico*

Year	ΔP, increase in population	P, population (millions)
2000	—	100
2001	2	102
2002	2.04	104.04
2003	2.08	106.12
2004	2.12	108.24

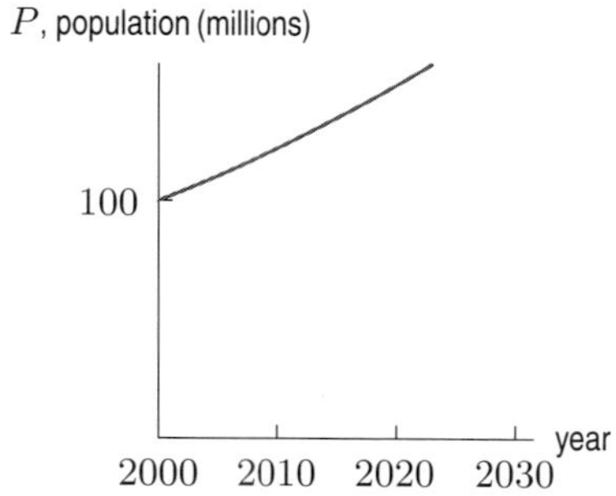

Figure 3.2: The projected population of Mexico, assuming 2% annual growth

Figure 3.2 gives a graph of the population of Mexico over a 30-year period, assuming a 2% annual growth rate. Notice that this graph curves upward like the graph in Figure 3.1.

Radioactive Decay

Exponential functions can also model decreasing quantities. A quantity which decreases at a constant percent rate is said to be decreasing exponentially.

Example 3 Carbon-14 is used to estimate the age of organic compounds. Over time, radioactive carbon-14 decays into a stable form. The decay rate is 11.4% every 1000 years. For example, if we begin with a 200 microgram (μg) sample of carbon-14 then

$$\begin{aligned}\text{Amount remaining after 1000 years} &= \text{Initial amount} - 11.4\% \text{ of Initial amount}\\ &= 200 - 0.114 \cdot 200\\ &= 177.2.\end{aligned}$$

Similarly,

$$\begin{aligned}\text{Amount remaining after 2000 years} &= \text{Amount remaining after 1000 years} - 11.4\% \text{ of } \text{Amount remaining after 1000 years}\\ &= 177.2 - 0.114(177.2) \approx 156.999,\end{aligned}$$

and

$$\begin{aligned}\text{Amount remaining after 3000 years} &= \text{Amount remaining after 2000 years} - 11.4\% \text{ of } \text{Amount remaining after 2000 years}\\ &= 156.999 - 0.114 \cdot 156.999 \approx 139.101.\end{aligned}$$

These calculations are recorded in Table 3.3. During each 1000-year period, the amount of carbon-14 that decays is smaller than in the previous period. This is because we take 11.4% of a smaller quantity each time.

Table 3.3 *The amount of carbon-14 remaining over time*

Years elapsed	Amount decayed (μg)	Amount remaining (μg)
0	—	200.0
1000	22.8	177.2
2000	20.2	156.999
3000	17.898	139.101

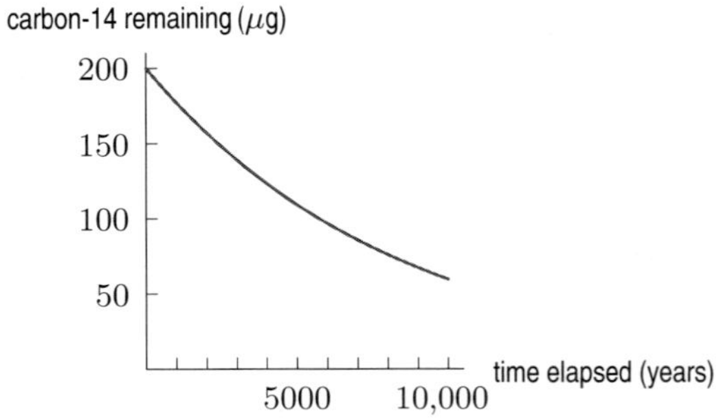

Figure 3.3: Amount of carbon-14 over 10,000 years

Figure 3.3 shows the amount of carbon-14 left from a 200 μg sample over 10,000 years. Because the amount decreases by a smaller amount over each successive time interval, the graph is not linear but bends upward.

Growth Factors and Percent Growth Rates

The Growth Factor of an Increasing Exponential Function

The salary in Example 1 increases by 6% every year. We say that the annual percent growth rate is 6%. But there is another way to think about the growth of this salary. We know that each year,

$$\text{New salary} = \text{Old salary} + 6\% \text{ of Old salary}.$$

We can rewrite this as follows:

$$\text{New salary} = 100\% \text{ of Old salary} + 6\% \text{ of Old salary}.$$

So

$$\text{New salary} = 106\% \text{ of Old salary}.$$

Since $106\% = 1.06$, we have

$$\text{New salary} = 1.06 \cdot \text{Old salary.}$$

We call the 1.06 the *annual growth factor*.

The Growth Factor of a Decreasing Exponential Function

In Example 3, the carbon-14 changes by -11.4% every 1000 years. The negative growth rate tells us that the quantity of carbon-14 decreases over time. We have

$$\text{New amount} = \text{Old amount} - 11.4\% \text{ of Old amount,}$$

which can be rewritten as

$$\text{New amount} = 100\% \text{ of Old amount} - 11.4\% \text{ of Old amount.}$$

So,

$$\text{New amount} = 88.6\% \text{ of Old amount.}$$

Since $88.6\% = 0.886$, we have

$$\text{New amount} = 0.886 \cdot \text{Old amount.}$$

Hence the growth factor is 0.886 per millennium. The fact that the growth factor is less than 1 indicates that the amount of carbon-14 is decreasing, since multiplying a quantity by a factor between 0 and 1 decreases the quantity.

Although it may sound strange to refer to the growth factor, rather than decay factor, of a decreasing quantity, we will use growth factor to describe both increasing and decreasing quantities.

A General Formula for the Family of Exponential Functions

Because it grows at a constant percentage rate each year, the salary, S, in Example 1 is an example of an exponential function. We want a formula for S in terms of t, the number of years since being hired. Since the annual growth factor is 1.06, we know that for each year,

$$\text{New salary} = \text{Previous salary} \cdot 1.06.$$

Thus, after one year, or when $t = 1$,

$$S = \underbrace{40{,}000}_{\text{Previous salary}} \cdot 1.06.$$

Similarly, when $t = 2$,

$$S = \underbrace{40{,}000(1.06)}_{\text{Previous salary}} \cdot 1.06 = 40{,}000(1.06)^2.$$

Here there are *two* factors of 1.06 because the salary has increased by 6% twice. When $t = 3$,

$$S = \underbrace{40,000(1.06)^2}_{\text{Previous salary}} \cdot 1.06 = 40{,}000(1.06)^3$$

and continues in this pattern so that after t years have elapsed,

$$S = 40{,}000 \underbrace{(1.06)(1.06)\ldots(1.06)}_{t \text{ factors of } 1.06} = 40{,}000(1.06)^t.$$

After t years the salary has increased by a factor of 1.06 a total of t times. Thus,

$$S = 40{,}000(1.06)^t.$$

These results, which are summarized in Table 3.4, are the same as in Table 3.1. Notice that in this formula we assume that t is an integer, $t \geq 0$, since the raises are given only once a year.

Table 3.4 *Salary after t years*

t (years)	S, salary (\$)
0	40,000
1	$40{,}000(1.06) = 42{,}400.00$
2	$40{,}000(1.06)^2 = 44{,}944.00$
3	$40{,}000(1.06)^3 = 47{,}640.64$
t	$40,000(1.06)^t$

This salary formula can be written as

$$S = \text{Initial salary} \cdot (\text{Growth factor})^t.$$

In general, we have:

An **exponential function** $Q = f(t)$ has the formula

$$f(t) = ab^t, \quad b > 0,$$

where a is the initial value of Q (at $t = 0$) and b, the base, is the growth factor: $b > 1$ gives exponential growth, $0 < b < 1$ gives exponential decay. The growth factor is given by

$$b = 1 + r$$

where r is the decimal representation of the percent rate of change.

The constants a and b are called *parameters*.. The base b is restricted to positive values because if $b < 0$ then b^t is undefined for some exponents t, for example, $t = 1/2$.

Every function in the form $f(t) = ab^t$ with the input, t, in the exponent is an exponential function, provided $a \neq 0$. Note that if $b = 1$, then $f(t) = a \cdot 1^t = a$ and $f(t)$ is a constant, so when $b = 1$, the function is generally not considered exponential. Graphs showing exponential growth and decay are in Figures 3.4 and 3.5. Notice that in both cases the graph is concave up.

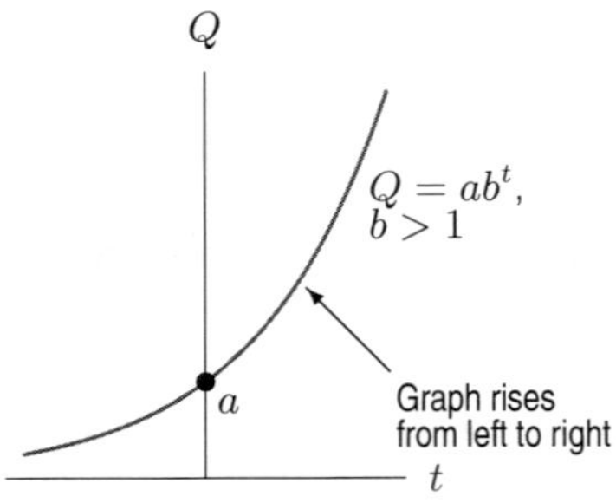

Figure 3.4: Exponential growth: $b > 1$

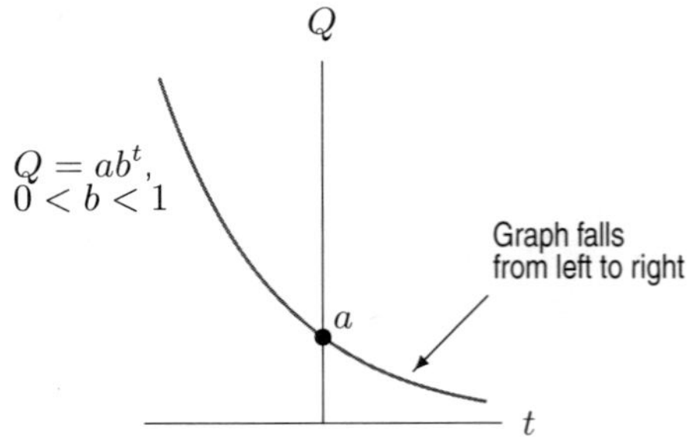

Figure 3.5: Exponential decay: $0 < b < 1$

Example 4 Use the formula $S = 40{,}000(1.06)^t$ to calculate your salary after 4 years, 12 years, and 40 years.

Solution After 4 years, $t = 4$, and we have

$$S = 40{,}000(1.06)^4 \approx \$50{,}499.08.$$

Notice that this agrees with Table 3.1 on page 106. After 12 years, $t = 12$, and we have

$$S = 40{,}000(1.06)^{12} \approx \$80{,}487.86.$$

After 12 years, the salary has more than doubled from the initial salary of $\$40,000$. When $t = 40$ we have

$$S = 40{,}000(1.06)^{40} \approx \$411{,}428.72.$$

Thus if you work for 40 years and consistently earn 6% annual raises, your salary will be over $\$400{,}000$ a year.

Example 5 Carbon-14 decays at a rate of 11.4% every 1000 years. Write a formula for the quantity, Q, of a 200 μg sample remaining as a function of time, t, in thousands of years.

Solution The growth factor of carbon-14 over 1000 years is $1 - 0.114 = 0.886$. Originally, there are 200 μg, so the quantity remaining after t thousand years is given by

$$Q = 200(0.886)^t.$$

Example 6 Using Example 2 on page 107, find a formula for P, the population of Mexico (in millions), in year t where $t = 0$ represents the year 2000.

Solution In 2000, the population of Mexico was 100 million, and it was growing at a constant 2% annual rate. The growth factor is $b = 1 + 0.02 = 1.02$, and $a = 100$, so

$$P = 100(1.02)^t.$$

Because the growth factor may change eventually, this formula may not give accurate results for large values of t.

Example 7 What does the formula $P = 100(1.02)^t$ predict when $t = 0$? When $t = -5$? What do these values tell you about the population of Mexico?

Solution If $t = 0$, then , since $(1.02)^0 = 1$, we have

$$P = 100(1.02)^0 = 100.$$

This makes sense because $t = 0$ stands for 2000, and in 2000 the population was 100 million. When $t = -5$ we have

$$P = 100(1.02)^{-5} \approx 90.573.$$

To make sense of this number, we must interpret the year $t = -5$ as five years before 2000; that is, as the year 1995. If the population of Mexico had been growing at a 2% annual rate from 1995 onward, then it was 90.573 million in 1995.

Example 8 On August 2, 1988, a US District Court judge imposed a fine on the city of Yonkers, New York, for defying a federal court order involving housing desegregation.[1] The fine started at $100 for the first day and was to double daily until the city chose to obey the court order.

(a) What was the daily percent growth rate of the fine?
(b) Find a formula for the fine as a function of t, the number of days since August 2, 1988.
(c) If Yonkers waited 30 days before obeying the court order, what would the fine have been?

[1] *The Boston Globe,* August 27, 1988.

Solution

(a) Since the fine increased each day by a factor of 2, the fine grew exponentially with growth factor $b = 2$. To find the percent growth rate, we set $b = 1 + r = 2$, from which we find $r = 1$, or 100%. Thus the daily percent growth rate is 100%. This makes sense because when a quantity increases by 100%, it doubles in size.

(b) If t is the number of days since August 2, the formula for the fine, P in dollars, is

$$P = 100 \cdot 2^t.$$

(c) After 30 days, the fine is $P = 100 \cdot 2^{30} \approx 1.074 \cdot 10^{11}$ dollars, or \$107,374,182,400.

Exercises and Problems for Section 3.1

Exercises

What is the growth factor in Exercises 1–4? Assume time is measured in the units given.

1. Water usage is increasing by 3% per year.

2. A city grows by 28% per decade.

3. A diamond mine is depleted by 1% per day.

4. A forest shrinks 80% per century.

In Exercises 5–10, you start with 500 items. How many do you have after the following change?

5. 10% increase

6. 100% increase

7. 1% decrease

8. 42% decrease

9. 42% increase followed by 42% decrease

10. 42% decrease followed by 42% increase

In Exercises 11–14, give the starting value a, the growth factor b, and the growth rate r if $Q = ab^t = a(1 + r)^t$.

11. $Q = 1750(1.593)^t$

12. $Q = 34.3(0.788)^t$

13. $Q = 79.2(1.002)^t$

14. $Q = 0.0022(2.31)^{-3t}$

15. Without a calculator or computer, match each exponential formula to one of the graphs I–VI.

(a) $10(1.2)^t$ **(b)** $10(1.5)^t$ **(c)** $20(1.2)^t$
(d) $30(0.85)^t$ **(e)** $30(0.95)^t$ **(f)** $30(1.05)^t$

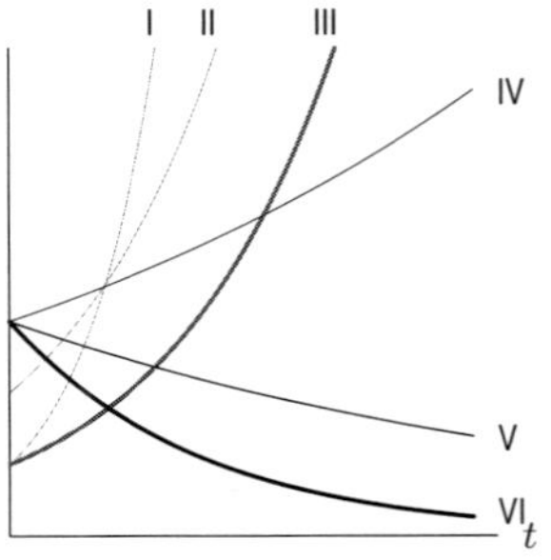

16. The populations, P, of six towns with time t in years are given by.

(i) $P = 1000(1.08)^t$ (ii) $P = 600(1.12)^t$
(iii) $P = 2500(0.9)^t$ (iv) $P = 1200(1.185)^t$
(v) $P = 800(0.78)^t$ (vi) $P = 2000(0.99)^t$

(a) Which towns are growing in size? Which are shrinking?
(b) Which town is growing the fastest? What is the annual percent growth rate for that town?
(c) Which town is shrinking the fastest? What is the annual percent "decay" rate for that town?
(d) Which town has the largest initial population (at $t = 0$)? Which town has the smallest?

17. A quantity increases from 10 to 12. By what percent has it increased? Now suppose that it had increased from 100 to 102. What is the percent increase in this case?

18. An investment decreases by 5% per year for 4 years. By what total percent does it decrease?

Problems

19. Find a formula for $P = f(t)$, the size of a population that begins in year $t = 0$ with 2200 members and decreases at a 3.2% annual rate.

20. The value \$$V$ of an investment in year t is given by $V = 2500(1.0325)^t$. Describe the investment in words.

21. In 2006, the cost of a train ticket from Boston to New York was \$73. Assume that the price rises by 10% per year. Make a table showing the price of tickets each year until 2010.

22. In 2006, the population of a country was 70 million and growing at a rate of 1.9% per year. Assuming the percentage growth rate remains constant, express the population, P, as a function of t, the number of years after 2006.

23. The mass, Q, of a sample of Tritium (a radioactive isotope of Hydrogen), decays at a rate of 5.626% per year. Write a function giving the mass of a 726 gram sample after a time, t, in years. Graph this decay function.

24. Every year, a lake becomes more polluted, and 2% fewer organisms can live in it. If in 2010 there are one million organisms, write an equation relating O, the number of organisms, to time, t, in years since 2010.

25. The value, V, of a \$100,000 investment that earns 3% annual interest is given by $V = f(t)$ where t is in years. How much is the investment worth in 3 years?

26. The amount (in milligrams) of a drug in the body t hours after taking a pill is given by $A(t) = 25(0.85)^t$.

(a) What is the initial dose given?
(b) What percent of the drug leaves the body each hour?
(c) What is the amount of drug left after 10 hours?
(d) After how many hours is there less than 1 milligram left in the body?

27. **(a)** The annual inflation rate is 3.5% per year. If a movie ticket costs \$7.50, find a formula for p, the price of the ticket t years from today, assuming that movie tickets keep up with inflation.
(b) According to your formula, how much will movie tickets cost in 20 years?

28. In the year 2004, a total of 7.9 million passengers took a cruise vacation.[2] The global cruise industry has been growing at approximately 10% per year for the last decade; assume that this growth rate continues.

(a) Write a formula to approximate the number, N, of cruise passengers (in millions) t years after 2004.
(b) How many cruise passengers are predicted in the year 2010? Approximately how many passengers went on a cruise in the year 2000?

29. A typical cup of coffee contains about 100 mg of caffeine and every hour approximately 16% of the amount of caffeine in the body is metabolized and eliminated.

(a) Write C, the amount of caffeine in the body in mg as a function of t, the number of hours since the coffee was consumed.
(b) How much caffeine is in the body after 5 hours?

30. Radioactive gallium-67 decays by 1.48% every hour; there are 100 milligrams initially.

(a) Find a formula for the amount of gallium-67 remaining after t hours.
(b) How many milligrams are left after 24 hours? After 1 week?

31. In 2005 the number of people infected by a virus was P_0. Due to a new vaccine, the number of infected people has decreased by 20% each year since 2005. In other words, only 80% as many people are infected each year as were infected the year before. Find a formula for $P = f(n)$, the number of infected people n years after 2005. Graph $f(n)$. Explain, in terms of the virus, why the graph has the shape it does.

32. You owe \$2000 on a credit card. The card charges 1.5% monthly interest on your balance, and requires a minimum monthly payment of 2.5% of your balance. All transactions (payments and interest charges) are recorded at the end of the month. You make only the minimum required payment every month and incur no additional debt.

(a) Complete Table 3.5 for a twelve-month period.
(b) What is your unpaid balance after one year has passed? At that time, how much of your debt have you paid off? How much money in interest charges have you paid your creditors?

Table 3.5

Month	Balance	Interest	Minimum payment
0	\$2000.00	\$30.00	\$50.00
1	\$1980.00	\$29.70	\$49.50
2	\$1960.20		
⋮			

33. Polluted water is passed through a series of filters. Each filter removes 85% of the remaining impurities. Initially, the untreated water contains impurities at a level of 420 parts per million (ppm). Find a formula for L, the remaining level of impurities, after the water has been passed through a series of n filters.

34. The UN Food and Agriculture Organization estimates that 4.2% of the world's natural forests existing in 1990 were gone by the end of the decade. In 1990, the world's forest cover stood at 3843 million hectares.[3]

(a) How many million hectares of natural forests were lost during the 1990s?

[2]The Worldwatch Institute, *Vital Signs* 2005 (New York: W.W. Norton & Company, 2002), p. 100.
[3]The Worldwatch Institute, *Vital Signs* 2002 (New York: W.W. Norton & Company, 2002), p. 104.

(b) How many million hectares of natural forests existed in the year 2000?
(c) Write an exponential formula approximating the number of million hectares of natural forest in the world t years after 1990.
(d) What was the annual percent decay rate during the 1990s?

35. The *Home* section of many Sunday newspapers includes a mortgage table similar to Table 3.6. The table gives the monthly payment per $1000 borrowed for loans at various interest rates and time periods. Determine the monthly payment on a

(a) $60,000 mortgage at 8% for fifteen years.
(b) $60,000 mortgage at 8% for thirty years.
(c) $60,000 mortgage at 10% for fifteen years.
(d) Over the life of the loan, how much money would be saved on a 15-year mortgage of $60,000 if the rate were 8% instead of 10%?
(e) Over the life of the loan, how much money would be saved on an 8% mortgage of $60,000 if the term of the loan was fifteen years rather than thirty years?

Table 3.6

Interest rate (%)	15-year loan	20-year loan	25-year loan	30-year loan
8.00	9.56	8.37	7.72	7.34
8.50	9.85	8.68	8.06	7.69
9.00	10.15	9.00	8.40	8.05
9.50	10.45	9.33	8.74	8.41
10.00	10.75	9.66	9.09	8.78
10.50	11.06	9.99	9.45	9.15
11.00	11.37	10.33	9.81	9.53
11.50	11.69	10.67	10.17	9.91

36. Every year, teams from 64 colleges qualify to compete in the NCAA basketball playoffs. For each round, every team is paired with an opponent. A team is eliminated from the tournament once it loses a round. So, at the end of a round, only one half the number of teams move on to the next round. Let $N(r)$ be the number of teams remaining in competition after r rounds of the tournament have been played.

(a) Find a formula for $N(r)$ and graph $y = N(r)$.
(b) How many rounds does it take to determine the winner of the tournament?

37. Figure 3.6 is the graph of $f(x) = 4 \cdot b^x$. Find the slope of the line segment PQ in terms of b.

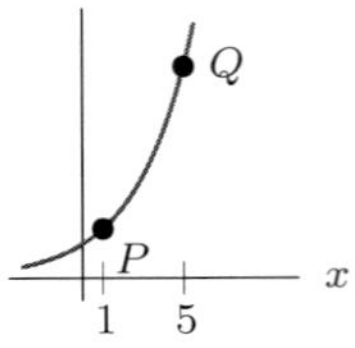

Figure 3.6

38. A one-page letter is folded into thirds to go into an envelope. If it were possible to repeat this kind of tri-fold 20 times, how many miles thick would the letter be? (A stack of 150 pieces of stationery is one inch thick; 1 mile = 5280 feet.)

39. Let P be the number of students in a school district, N be the size of the tax base (in households), and r be the average annual tax rate (in $/household).

(a) Find a formula for R, the total tax revenue, in terms of N and r.
(b) Find a formula for A, the average revenue per student.
(c) Suppose the tax base goes up by 2% and the tax rate is raised by 3%. Find formulas for the new tax base and tax rate in terms of N and r.
(d) Using your answer to part (c), find a formula for the new total tax revenue in terms of R. By what percent did R increase?
(e) Over the time period in part (c), the student population rises by 8%. Find a formula for the new average revenue in terms of A. Did the average revenue rise or fall? By how much?

Use Figure 3.7 in Problems 40–43.

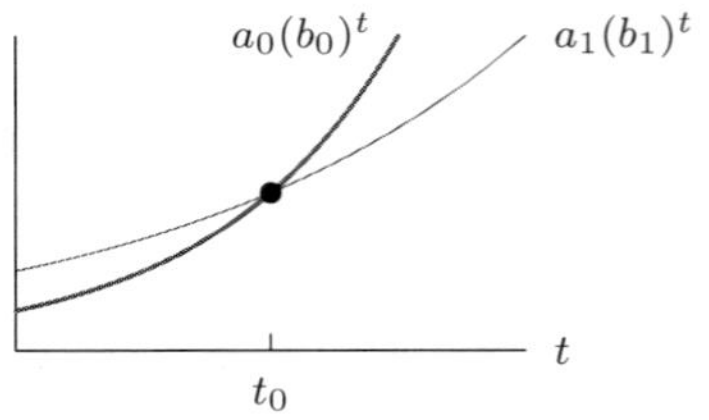

Figure 3.7

40. Which is greater, a_0 or a_1?

41. Which is greater, b_0 or b_1?

42. What happens to t_0 if a_0 is increased while the other quantities remain fixed?

43. What happens to t_0 if b_1 is decreased while the other quantities remain fixed?

3.2 COMPARING EXPONENTIAL AND LINEAR FUNCTIONS

The exponential function $Q = ab^t$ represents a quantity changing at a constant percent rate. In this section we compare exponential and linear models and we fit exponential models to data from tables and graphs.

Identifying Linear and Exponential Functions From a Table

Table 3.7 gives values of a linear and an exponential functions. Notice that the value of x changes by equal steps of $\Delta x = 5$. The function f is linear because the difference between consecutive values of $f(x)$ is constant: $f(x)$ increases by 15 each time x increases by 5.

Table 3.7 *Two functions, one linear and one exponential*

x	20	25	30	35	40	45
$f(x)$	30	45	60	75	90	105
$g(x)$	1000	1200	1440	1728	2073.6	2488.32

On the other hand, the difference between consecutive values of $g(x)$ is *not* constant:

$$1200 - 1000 = 200$$
$$1440 - 1200 = 240$$
$$1728 - 1440 = 288.$$

Thus, g is not linear. However, the *ratio* of consecutive values of $g(x)$ is constant:

$$\frac{1200}{1000} = 1.2, \quad \frac{1440}{1200} = 1.2, \quad \frac{1728}{1440} = 1.2,$$

and so on. Note that $1200 = 1.2(1000)$, $1440 = 1.2(1200)$, $1728 = 1.2(1440)$. Thus, each time x increases by 5, the value of $g(x)$ increases by a factor of 1.2. This pattern of constant ratios is indicative of exponential functions. In general:

> For a table of data that gives y as a function of x and in which Δx is constant:
> - If the *difference* of consecutive y-values is constant, the table could represent a linear function.
> - If the *ratio* of consecutive y-values is constant, the table could represent an exponential function.

Finding a Formula for an Exponential Function

To find a formula for the exponential function in Table 3.7, we must determine the values of a and b in the formula $g(x) = ab^x$. The table tells us that $ab^{20} = 1000$ and that $ab^{25} = 1200$. Taking the ratio gives

$$\frac{ab^{25}}{ab^{20}} = \frac{1200}{1000} = 1.2.$$

Notice that the value of a cancels in this ratio, so

$$\frac{ab^{25}}{ab^{20}} = b^5 = 1.2.$$

We solve for b by raising each side to the $(1/5)^{\text{th}}$ power:

$$(b^5)^{1/5} = b = 1.2^{1/5} \approx 1.03714.$$

Now that we have the value of b, we can solve for a. Since $g(20) = ab^{20} = 1000$, we have

$$a(1.03714)^{20} = 1000$$
$$a = \frac{1000}{1.03714^{20}} \approx 482.253.$$

Thus, a formula for g is $g(x) = 482.253(1.037)^x$. (Note: We could have used $g(25)$ or any other value from the table to find a.)

Modeling Linear and Exponential Growth Using Two Data Points

If we are given two data points, we can fit either a line or an exponential function to the points. The following example compares the predictions made by a linear model and an exponential model fitted to the same data.

Example 1 At time $t = 0$ years, a species of turtle is released into a wetland. When $t = 4$ years, a biologist estimates there are 300 turtles in the wetland. Three years later, the biologist estimates there are 450 turtles. Let P represent the size of the turtle population in year t.

(a) Find a formula for $P = f(t)$ assuming linear growth. Interpret the slope and P-intercept of your formula in terms of the turtle population.
(b) Now find a formula for $P = g(t)$ assuming exponential growth. Interpret the parameters of your formula in terms of the turtle population.
(c) In year $t = 12$, the biologist estimates that there are 900 turtles in the wetland. What does this indicate about the two population models?

Solution (a) Assuming linear growth, we have $P = f(t) = b + mt$, and

$$m = \frac{\Delta P}{\Delta t} = \frac{450 - 300}{7 - 4} = \frac{150}{3} = 50.$$

Calculating b gives

$$300 = b + 50 \cdot 4$$
$$b = 100,$$

so $P = f(t) = 100 + 50t$. This formula tells us that 100 turtles were originally released into the wetland and that the number of turtles increases at the constant rate of 50 turtles per year.

(b) Assuming exponential growth, we have $P = g(t) = ab^t$. The values of a and b are calculated from the ratio

$$\frac{ab^7}{ab^4} = \frac{450}{300},$$

so

$$b^3 = 1.5.$$

Thus,

$$b = (1.5)^{1/3} \approx 1.145.$$

Using the fact that $g(4) = ab^4 = 300$ to find a gives

$$a(1.145)^4 = 300$$
$$a = \frac{300}{1.145^4} \approx 175, \quad \text{Rounding to the nearest whole turtle}$$

so $P = g(t) = 175(1.145)^t$. This formula tells us that 175 turtles were originally released into the wetland and the number increases at about 14.5% per year.

(c) In year $t = 12$, there are approximately 900 turtles. The linear function from part (a) predicts

$$P = 100 + 50 \cdot 12 = 700 \text{ turtles.}$$

The exponential formula from part (b), however, predicts

$$P = 175(1.145)^{12} \approx 889 \text{ turtles.}$$

The fact that 889 is closer to the observed value of 900 turtles suggests that, during the first 12 years, exponential growth is a better model of the turtle population than linear growth. The two models are graphed in Figure 3.8.

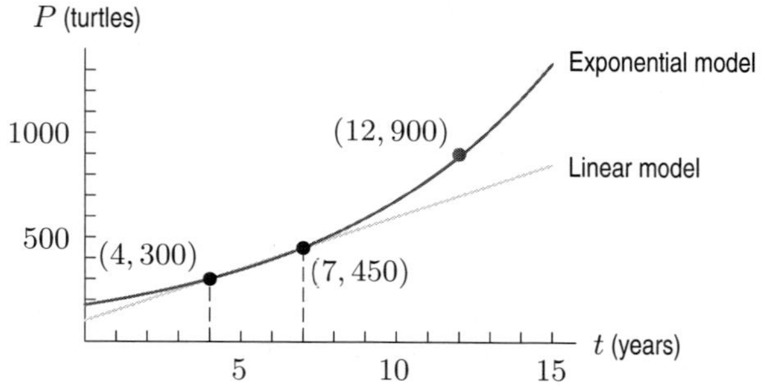

Figure 3.8: Comparison of the linear and exponential models of the turtle population

Similarities and Differences between Linear and Exponential Functions

In some ways the general formulas for linear and exponential functions are similar. If y is a linear function of x and x is a positive integer, we can write $y = b + mx$ as

$$y = b + \underbrace{m + m + m + \ldots + m}_{x \text{ times}}.$$

Similarly, if y is an exponential function of x, so that $y = a \cdot b^x$ and x is a positive integer, we can write

$$y = a \cdot \underbrace{b \cdot b \cdot b \cdot \ \ldots \ \cdot b}_{x \text{ times}}.$$

So linear functions involve repeated sums whereas exponential functions involve repeated products. In both cases, x determines the number of repetitions.

There are other similarities between the formulas for linear and exponential functions. The slope m of a linear function gives the rate of change of a physical quantity and the y-intercept gives the starting value. Similarly, in $y = a \cdot b^x$, the value of b gives the growth factor and a gives the starting value.

Example 2 The following tables contain values from an exponential or linear function. For each table, decide if the function is linear or exponential, and find a possible formula for the function.

(a)

x	$f(x)$
0	65
1	75
2	85
3	95
4	105

(b)

x	$g(x)$
0	400
1	600
2	900
3	1350
4	2025

Solution (a) The function values increase by 10 as x increases by 1, so this is a linear function with slope $m = 10$. Since $f(0) = 65$, the vertical intercept is 65. A possible formula is

$$f(x) = 65 + 10x.$$

(b) The function is not linear, since $g(x)$ increases by different amounts as x increases by 1. To determine whether g might be exponential, we look at ratios of consecutive values:

$$\frac{600}{400} = 1.5, \quad \frac{900}{600} = 1.5, \quad \frac{1350}{900} = 1.5, \quad \frac{2025}{1350} = 1.5.$$

Each time x increases by 1, the value of $g(x)$ increases by a factor of 1.5. This is an exponential function with growth factor 1.5. Since $g(0) = 400$, the vertical intercept is 400. A possible formula is

$$g(x) = 400(1.5)^x.$$

Exponential Growth Will Always Outpace Linear Growth in the Long Run

Figure 3.8 shows the graphs of the linear and exponential models for the turtle population from Example 1. The graphs highlight the fact that, although these two graphs remain fairly close for the first ten or so years, the exponential model predicts explosive growth later on.

It can be shown that an exponentially increasing quantity will, in the long run, always outpace a linearly increasing quantity. This fact led the 19th-century clergyman and economist, Thomas Malthus, to make some rather gloomy predictions, which are illustrated in the next example.

Example 3 The population of a country is initially 2 million people and is increasing at 4% per year. The country's annual food supply is initially adequate for 4 million people and is increasing at a constant rate adequate for an additional 0.5 million people per year.

(a) Based on these assumptions, in approximately what year will this country first experience shortages of food?
(b) If the country doubled its initial food supply, would shortages still occur? If so, when? (Assume the other conditions do not change).
(c) If the country doubled the rate at which its food supply increases, in addition to doubling its initial food supply, would shortages still occur? If so, when? (Again, assume the other conditions do not change.)

Solution Let P represent the country's population (in millions) and N the number of people the country can feed (in millions). The population increases at a constant percent rate, so it can be modeled by an exponential function. The initial population is $a = 2$ million people and the annual growth factor is $b = 1 + 0.04 = 1.04$, so a formula for the population is

$$P = 2(1.04)^t.$$

In contrast, the food supply increases by a constant amount each year and is therefore modeled by a linear function. The initial food supply is adequate for $b = 4$ million people and the growth rate is $m = 0.5$ million per year, so the number of people that can be fed is

$$N = 4 + 0.5t.$$

(a) Figure 3.9(a) gives the graphs of P and N over a 105-year span. For many years, the food supply is far in excess of the country's needs. However, after about 78 years the population has begun to grow so rapidly that it catches up to the food supply and then outstrips it. After that time, the country will suffer from shortages.

(b) If the country can initially feed eight million people rather than four, the formula for N is

$$N = 8 + 0.5t.$$

However, as we see from Figure 3.9(b), this measure only buys the country three or four extra years with an adequate food supply. After 81 years, the population is growing so rapidly that the head start given to the food supply makes little difference.

(c) If the country doubles the rate at which its food supply increases, from 0.5 million per year to 1.0 million per year, the formula for N is

$$N = 8 + 1.0t.$$

Unfortunately the country still runs out of food eventually. Judging from Figure 3.9(c), this happens in about 102 years.

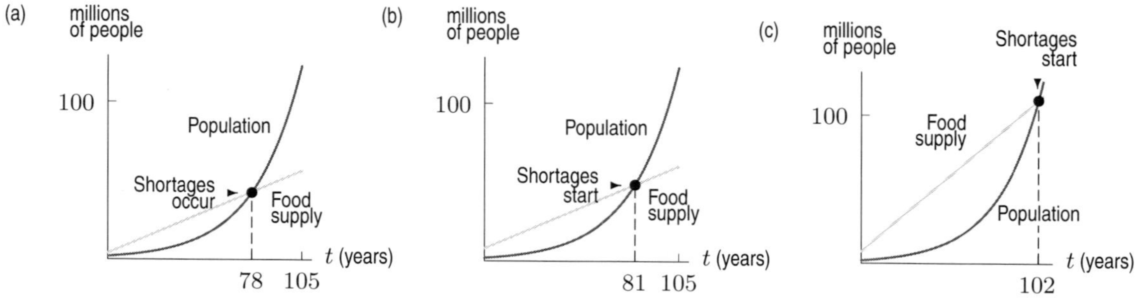

Figure 3.9: These graphs illustrate the fact that an exponentially growing population eventually outstrips a linearly growing food supply

Malthus believed that populations increase exponentially while food production increases linearly. The last example explains his gloomy predictions: Malthus believed that any population eventually outstrips its food supply, leading to famine and war.

Exercises and Problems for Section 3.2

Exercises

1. The following formulas give the populations (in 1000s) of four different cities, A, B, C, and D, where t is in years. Which are changing exponentially? Describe in words how each of these populations is changing over time. Graph those that are exponential.

$$P_A = 200 + 1.3t, \quad P_B = 270(1.021)^t,$$
$$P_C = 150(1.045)^t, \quad P_D = 600(0.978)^t.$$

2. A population has size 5000 at time $t = 0$, with t in years.

(a) If the population decreases by 100 people per year, find a formula for the population, P, at time t.

(b) If the population decreases by 8% per year, find a formula for the population, P, at time t.

3. A population has size 100 at time $t = 0$, with t in years.

(a) If the population grows by 10 people per year, find a formula for the population, P, at time t.

(b) If the population grows by 10% per year, find a formula for the population, P, at time t.

(c) Graph both functions on the same axes.

4. In an environment with unlimited resources and no predators, a population tends to grow by the same percentage each year. Should a linear or exponential function be used to model such a population? Why?

5. Determine whether the function whose values are in Table 3.8 could be exponential.

Table 3.8

x	1	2	4	5	8	9
$f(x)$	4096	1024	64	16	0.25	0.0625

The tables in Problems 6–9 contain values from an exponential or a linear function. In each problem:

(a) Decide if the function is linear or exponential.

(b) Find a possible formula for each function and graph it.

6.

x	$f(x)$
0	12.5
1	13.75
2	15.125
3	16.638
4	18.301

7.

x	$g(x)$
0	0
1	2
2	4
3	6
4	8

8.

x	$h(x)$
0	14
1	12.6
2	11.34
3	10.206
4	9.185

9.

x	$i(x)$
0	18
1	14
2	10
3	6
4	2

Problems

10. Explain the difference between linear and exponential growth. That is, without writing down any formulas, describe how linear and exponential functions progress differently from one value to the next.

In Problems 11–16, find formulas for the exponential functions satisfying the given conditions.

11. $h(0) = 3$ and $h(1) = 15$

12. $f(3) = -3/8$ and $f(-2) = -12$

13. $g(1/2) = 4$ and $g(1/4) = 2\sqrt{2}$

14. $g(0) = 5$ and $g(-2) = 10$

15. $g(1.7) = 6$ and $g(2.5) = 4$

16. $f(1) = 4$ and $f(3) = d$

17. Suppose $f(-3) = 5/8$ and $f(2) = 20$. Find a formula for f assuming it is:

(a) Linear **(b)** Exponential

For Problems 18–23, find formulas for the exponential functions.

18.

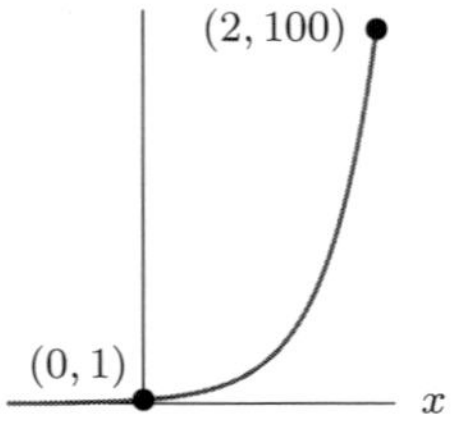

19.

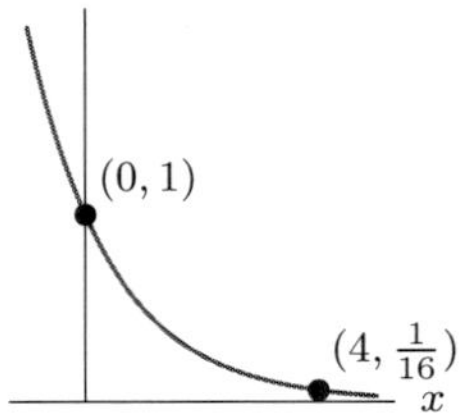

20.

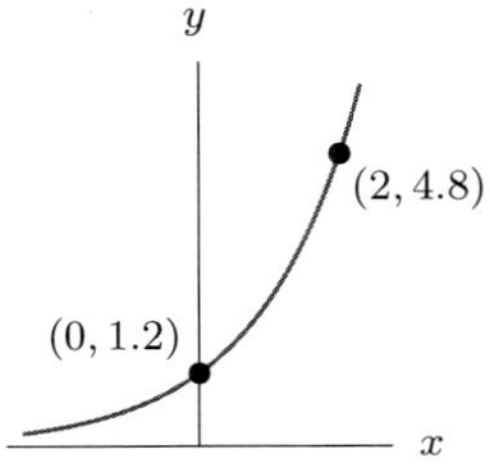

21.

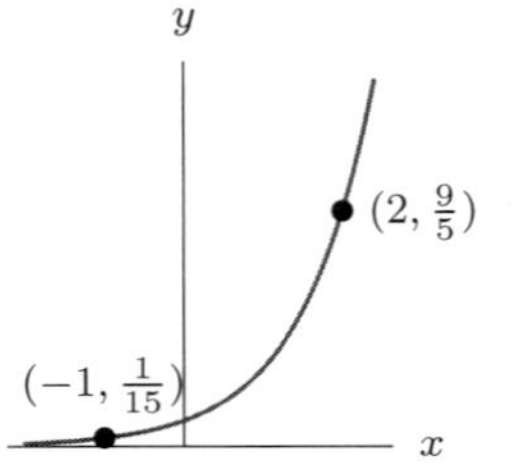

22.

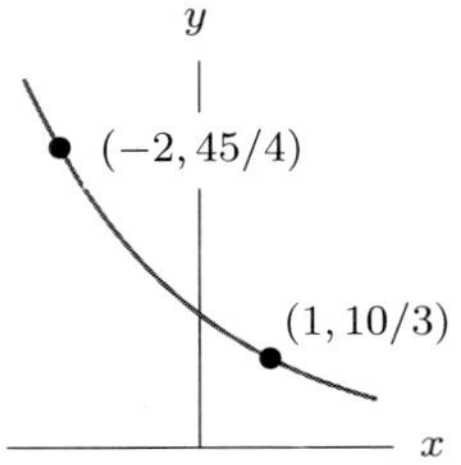

23.

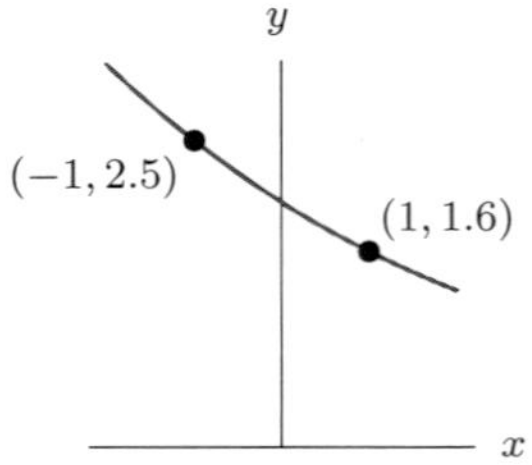

24. Find possible formulas for the functions in Figure 3.10.

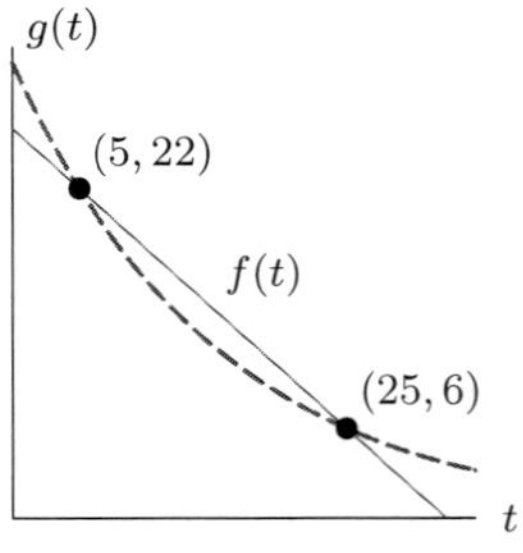

Figure 3.10

25. Find formulas for the exponential functions in Figure 3.11.

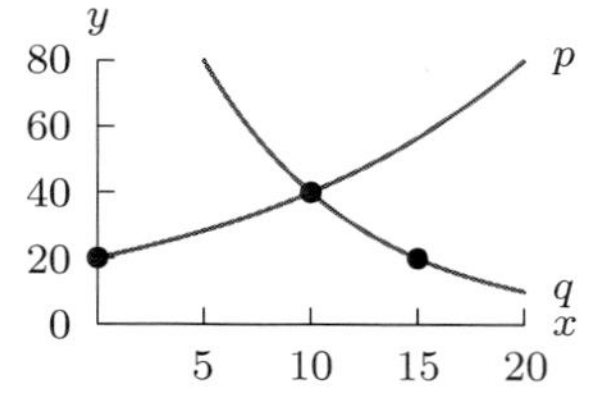

Figure 3.11

In Problems 26–29, could the function be linear or exponential or is it neither? Write possible formulas for the linear or exponential functions.

26.

r	1	3	7	15	31
$p(r)$	13	19	31	55	103

27.

x	6	9	12	18	24
$q(x)$	100	110	121	146.41	177.16

28.

x	10	12	15	16	18
$f(x)$	1	2	4	8	16

29.

t	1	2	3	4	5
$g(t)$	512	256	128	64	32

30. Let $p(x) = 2 + x$ and $q(x) = 2^x$. Estimate the values of x such that $p(x) < q(x)$.

31. Let $P(t)$ be the population of a country, in millions, t years after 1990, with $P(7) = 3.21$ and $P(13) = 3.75$.

(a) Find a formula for $P(t)$ assuming it is linear. Describe in words the country's annual population growth given this assumption.
(b) Find a formula for $P(t)$ assuming it is exponential. Describe in words the country's annual population growth given this assumption.

32. What is the value of the population at the end of 10 years, given each of the following assumptions? Graph each population against time.

(a) A population decreases linearly and the decrease is 10% in the first year.
(b) A population decreases exponentially at the rate of 10% a year.

33. Figure 3.12 shows the balance, P, in a bank account.

(a) Find a possible formula for $P = f(t)$ assuming the balance grows exponentially.
(b) What was the initial balance?
(c) What annual interest rate does the account pay?

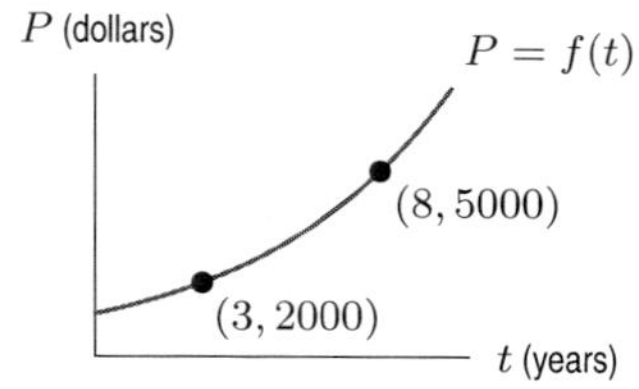

Figure 3.12

34. The number of asthma sufferers in the world was about 84 million in 1990 and 130 million in 2001.[4] Let N represent the number of asthma sufferers (in millions) worldwide t years after 1990.

(a) Write N as a linear function of t. What is the slope? What does it tell you about asthma sufferers?
(b) Write N as an exponential function of t. What is the growth factor? What does it tell you about asthma sufferers?
(c) How many asthma sufferers are predicted worldwide in the year 2010 with the linear model? With the exponential model?

35. Table 3.9 gives the approximate number of cell phone subscribers, S, worldwide.[5]

(a) Explain how you know an exponential function fits the data. Find a formula for S in terms of t, the number of years since 1995.
(b) Interpret the growth rate in terms of cell phone subscribers.
(c) In 2004, there were 1340 million subscribers.[6] Does this fit the pattern?

Table 3.9

Year	1995	1996	1997	1998	1999	2000
Subscribers (m.)	91	138	210	320	485	738

[4] www.who.int/inf-fs/en/fact206.html, August 24, 2002.
[5] The Worldwatch Institute, *Vital Signs* 2002 (New York: W.W. Norton & Company, 2002), p. 84.
[6] *The World Almanac and Book of Facts 2006*, p. 380 (New York).

36. A 1987 treaty to protect the ozone layer produced dramatic declines in global production, P, of chlorofluorocarbons (CFCs). See Figure 3.13.[7] Find a formula for P as an exponential function of the number of years, t, since 1989. What was the annual percent decay rate?

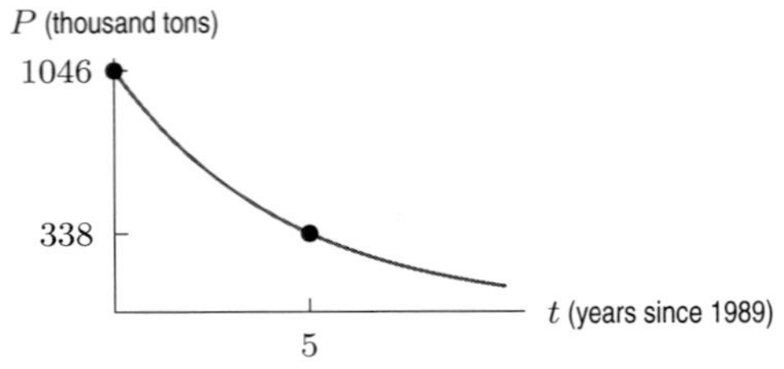

Figure 3.13

37. In 1940, there were about 10 brown tree snakes per square mile on the island of Guam, and in 2002, there were about 20,000 per square mile.[8] Find an exponential formula for the number, N, of brown tree snakes per square mile on Guam t years after 1940. What was, on average, the annual percent increase in the population during this period?

38. Match the stories in (a)–(e) with the formulas in (i)–(v). In each case, state what the variables represent. Assume that the constants P_0, r, B, A are all positive.

(a) The percent of a lake's surface covered by algae, initially at 35%, was halved each year since the passage of anti-pollution laws.
(b) The amount of charge on a capacitor in an electric circuit decreases by 30% every second.
(c) Polluted water is passed through a series of filters. Each filter removes all but 30% of the remaining impurities from the water.
(d) In 1950, the population of a town was 3000 people. Over the course of the next 50 years, the town grew at a rate of 10% per decade.
(e) In 1950, the population of a town was 3000 people. Over the course of the next 50 years, the town grew at a rate of 250 people per year.

(i) $f(x) = P_0 + rx$ (ii) $g(x) = P_0(1+r)^x$
(iii) $h(x) = B(0.7)^x$ (iv) $j(x) = B(0.3)^x$
(v) $k(x) = A(2)^{-x}$

39. A 2006 Lexus costs $61,055 and the car depreciates a total of 46% during its first 7 years.

(a) Suppose the depreciation is exponential. Find a formula for the value of the car at time t.
(b) Suppose instead that the depreciation is linear. Find a formula for the value of the car at time t.
(c) If this were your car and you were trading it in after 4 years, which depreciation model would you prefer (exponential or linear)?

40. On November 27, 1993, the *New York Times* reported that wildlife biologists have found a direct link between the increase in the human population in Florida and the decline of the local black bear population. From 1953 to 1993, the human population increased, on average, at a rate of 8% per year, while the black bear population decreased at a rate of 6% per year. In 1953 the black bear population was 11,000.

(a) The 1993 human population of Florida was 13 million. What was the human population in 1953?
(b) Find the black bear population for 1993.
(c) Had this trend continued,[9] when would the black bear population have numbered less than 100?

41. Suppose the city of Yonkers is offered two alternative fines by the judge. (See Example 8 on page 111.)

Penalty A: $1 million on August 2 and the fine increases by $10 million each day thereafter.

Penalty B: 1¢ on August 2 and the fine doubles each day thereafter.

(a) If the city of Yonkers plans to defy the court order until the end of the month (August 31), compare the fines incurred under Penalty A and Penalty B.
(b) If t represents the number of days after August 2, express the fine incurred as a function of t under

(i) Penalty A (ii) Penalty B

(c) Assume your formulas in part (b) holds for $t \geq 0$, is there a time such that the fines incurred under both penalties are equal? If so, estimate that time.

3.3 GRAPHS OF EXPONENTIAL FUNCTIONS

As with linear functions, an understanding of the significance of the parameters a and b in the formula $Q = ab^t$ helps us analyze and compare exponential functions.

[7] These numbers reflect the volume of the major CFCs multiplied by their respective ozone-depleting potentials (ODPs), as reported by the United Nations Environmental Programme Ozone Secretariat. See hq.uncp.org/ozone.
[8] *Science News*, Vol. 162, August 10, 2002, p. 85.
[9] Since 1993, the black bear population has in fact remained stable: www.myfwc.com/bear, accessed January 5, 2006.

Graphs of the Exponential Family: The Effect of the Parameter a

In the formula $Q = ab^t$, the value of a tells us where the graph crosses the Q-axis, since a is the value of Q when $t = 0$. In Figure 3.14 each graph has the same value of b but different values of a and thus different vertical intercepts.

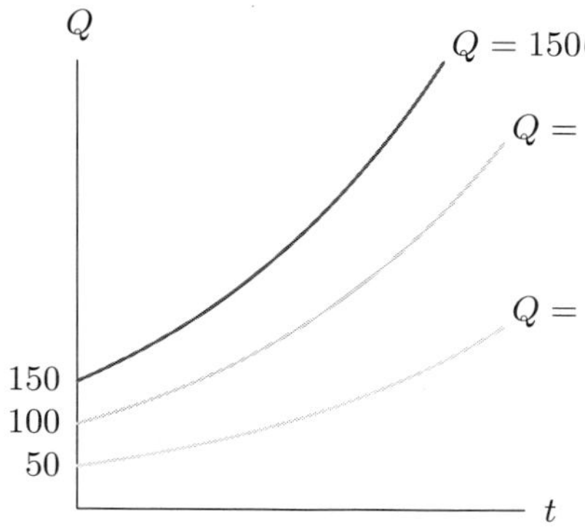

Figure 3.14: Graphs of $Q = a(1.2)^t$ for $a = 50, 100$, and 150

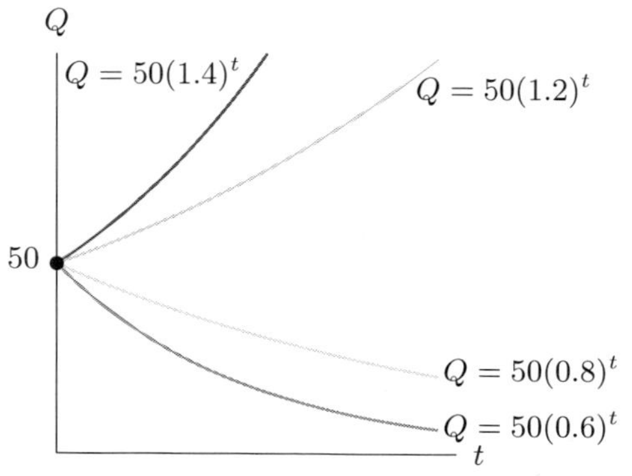

Figure 3.15: Graphs of $Q = 50b^t$ for $b = 0.6, 0.8, 1.2$ and 1.4

Graphs of the Exponential Family: The Effect of the Parameter b

The growth factor, b, is called the *base* of an exponential function. Provided a is positive, if $b > 1$, the graph climbs when read from left to right, and if $0 < b < 1$, the graph falls when read from left to right.

Figure 3.15 shows how the value of b affects the steepness of the graph of $Q = ab^t$. Each graph has a different value of b but the same value of a (and thus the same Q-intercept). For $b > 1$, the greater the value of b, the more rapidly the graph rises. For $0 < b < 1$, the smaller the value of b, the more rapidly the graph falls. In every case, however, the graph is concave up.

Horizontal Asymptotes

The t-axis is a *horizontal asymptote* for the graph of $Q = ab^t$, because Q approaches 0 as t gets large, either positively or negatively. For exponential decay, such as $Q = f(t) = a(0.6)^t$ in Figure 3.15, the value of Q approaches 0 as t gets large and positive. We write

$$Q \to 0 \quad \text{as} \quad t \to \infty.$$

This means that Q is as close to 0 as we like for all sufficiently large values of t. We say that the *limit* of Q as t goes to infinity is 0, and we write

$$\lim_{t\to\infty} f(t) = 0.$$

For exponential growth, the value of Q approaches zero as t grows more negative. See Figure 3.28. In this case, we write

$$Q \to 0 \quad \text{as} \quad t \to -\infty.$$

This means that Q is as close to 0 as we like for all sufficiently large negative values of t. Using limit notation, we write

$$\lim_{t\to-\infty} f(t) = 0.$$

We make the following definition:

> The horizontal line $y = k$ is a **horizontal asymptote** of a function, f, if the function values get arbitrarily close to k as x gets large (either positively or negatively or both). We describe this behavior using the notation
>
> $$f(x) \to k \quad \text{as} \quad x \to \infty$$
>
> or
>
> $$f(x) \to k \quad \text{as} \quad x \to -\infty.$$
>
> Alternatively, using limit notation, we write
>
> $$\lim_{x\to\infty} f(x) = k \quad \text{or} \quad \lim_{x\to-\infty} f(x) = k$$

Example 1 A capacitor is the part of an electrical circuit that stores electric charge. The quantity of charge stored decreases exponentially with time. Stereo amplifiers provide a familiar example: When an amplifier is turned off, the display lights fade slowly because it takes time for the capacitors to discharge. (Thus, it can be unsafe to open a stereo or a computer immediately after it is turned off.)

If t is the number of seconds after the circuit is switched off, suppose that the quantity of stored charge (in micro-coulombs) is given by

$$Q = 200(0.9)^t, \quad t \geq 0,$$

(a) Describe in words how the stored charge changes over time.
(b) What quantity of charge remains after 10 seconds? 20 seconds? 30 seconds? 1 minute? 2 minutes? 3 minutes?
(c) Graph the charge over the first minute. What does the horizontal asymptote of the graph tell you about the charge?

Solution

(a) The charge is initially 200 micro-coulombs. Since $b = 1 + r = 0.9$, we have $r = -0.10$, which means that the charge level decreases by 10% each second.
(b) Table 3.10 gives the value of Q at $t = 0, 10, 20, 30, 60, 120$, and 180. Notice that as t increases, Q gets closer and closer to, but does not quite reach, zero. The charge stored by the capacitor is getting smaller, but never completely vanishes.
(c) Figure 3.16 shows Q over a 60-second interval. The horizontal asymptote at $Q = 0$ corresponds to the fact that the charge gets very small as t increases. After 60 seconds, for all practical purposes, the charge is zero.

Table 3.10 *Charge (in micro-coulombs) stored by a capacitor over time*

t (seconds)	Q, charge level
0	200
10	69.736
20	24.315
30	8.478
60	0.359
120	0.000646
180	0.00000116

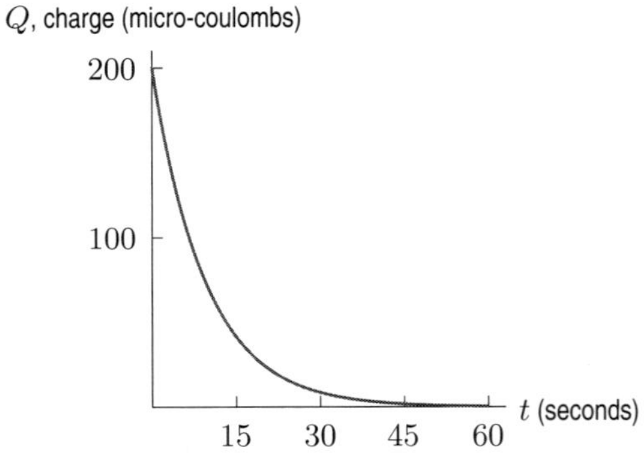

Figure 3.16: The charge stored by a capacitor over one minute

Solving Exponential Equations Graphically

We are often interested in solving equations involving exponential functions. In the following examples, we do this graphically. In Section 4.1, we will see how to solve equations using logarithms.

Example 2 In Example 8 on page 111, the fine, P, imposed on the city of Yonkers is given by $P = 100 \cdot 2^t$ where t is the number of days after August 2. In 1988, the annual budget of the city was \$337 million. If the city chose to disobey the court order, at what point would the fine have wiped out the entire annual budget?

Solution We need to find the day on which the fine reaches \$337 million. That is, we must solve the equation

$$100 \cdot 2^t = 337{,}000{,}000.$$

Using a computer or graphing calculator we can graph $P = 100 \cdot 2^t$ to find the point at which the fine reaches 337 million. From Figure 3.17, we see that this occurs between $t = 21$ and $t = 22$.

At day $t = 21$, August 23, the fine is:

$$P = 100 \cdot 2^{21} = 209{,}715{,}200$$

or just over \$200 million. On day $t = 22$, the fine is

$$P = 100 \cdot 2^{22} = 419{,}430{,}400$$

or almost \$420 million—quite a bit more than the city's entire annual budget!

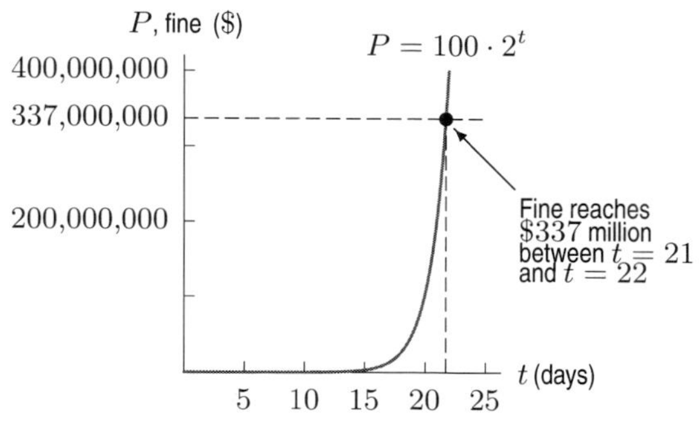

Figure 3.17: The fine imposed on Yonkers exceeds \$337 million after 22 days

Example 3 A 200 μg sample of carbon-14 decays according to the formula

$$Q = 200(0.886)^t$$

where t is in thousands of years. Estimate when there is 25 μg of carbon-14 left.

Solution We must solve the equation

$$200(0.886)^t = 25.$$

At the moment, we cannot find a formula for the solution to this equation. However, we can estimate the solution graphically. Figure 3.18 shows a graph of $Q = 200(0.886)^t$ and the line $Q = 25$. The amount of carbon-14 decays to 25 micrograms at $t \approx 17.180$. Since t is measured in thousands of years, this means in about 17,180 years.

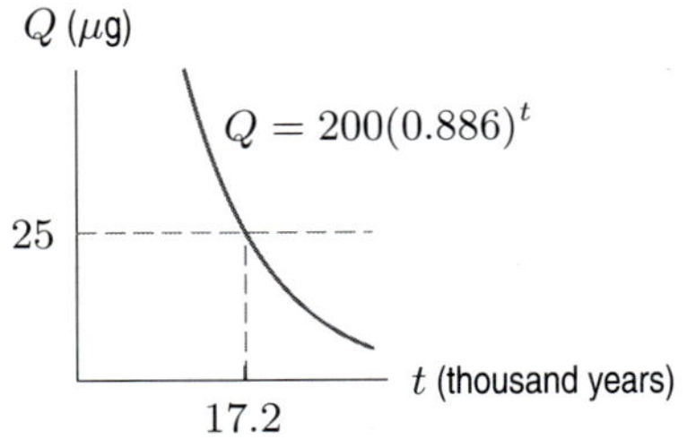

Figure 3.18: Solving the equation $200(0.886)^t = 25$

Fitting Exponential Functions to Data

The data in Table 3.11 gives population data for the Houston Metro Area since 1900. In Section 1.6, we saw how to fit a linear function to data, but Figure 3.19 suggests that it may make more sense to fit an exponential function using *exponential regression*.

Table 3.11 *Population (in thousands) of Houston Metro Area, t years after 1900*

t	N	t	N
0	184	60	1583
10	236	70	2183
20	332	80	3122
30	528	90	3733
40	737	100	4672
50	1070		

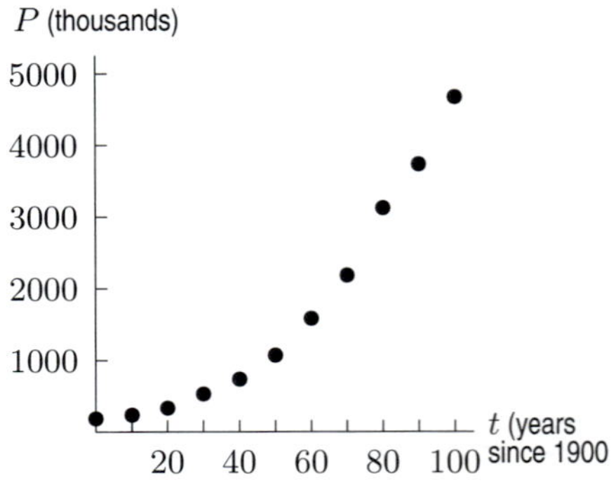

Figure 3.19: Population of Houston Metro Area

One algorithm used by a calculator or computer gives the best fitting exponential function as

$$P = 183.5(1.035)^t.$$

Other algorithms may give different formulas. Figure 3.20 shows this function and the data.

Since the base of this exponential function is 1.035, the population was increasing at a rate of about 3.5% per year between 1900 and 2000.

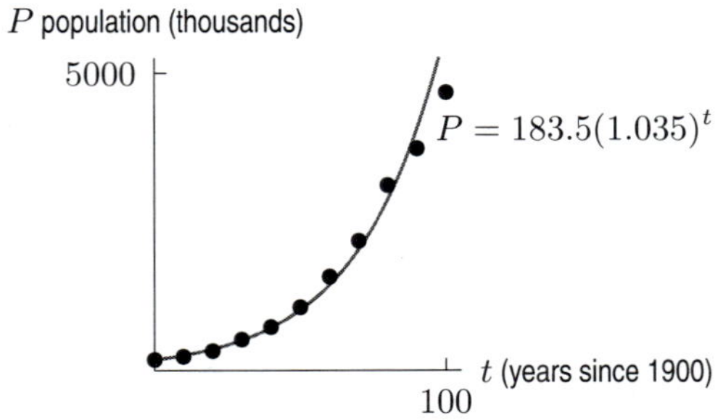

Figure 3.20: The population data since 1900 together with an exponential model

Exercises and Problems for Section 3.3

Exercises

1. **(a)** Make a table of values for $f(x) = 2^x$ for $x = -3, -2, -1, 0, 1, 2, 3$.
(b) Graph $f(x)$. Describe the graph in words.

2. **(a)** Make a table of values for $f(x) = \left(\frac{1}{2}\right)^x$ for $x = -3, -2, -1, 0, 1, 2, 3$.
(b) Graph $f(x)$. Describe the graph in words.

3. The graphs of $f(x) = (1.1)^x$, $g(x) = (1.2)^x$, and $h(x) = (1.25)^x$ are in Figure 3.21. Explain how you can match these formulas and graphs without a calculator.

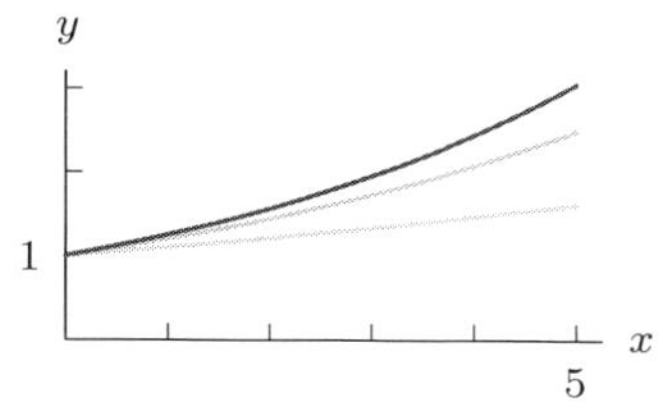

Figure 3.21

4. The graphs of $f(x) = (0.7)^x$, $g(x) = (0.8)^x$, and $h(x) = (0.85)^x$ are in Figure 3.22. Explain how you can match these formulas and graphs without a calculator.

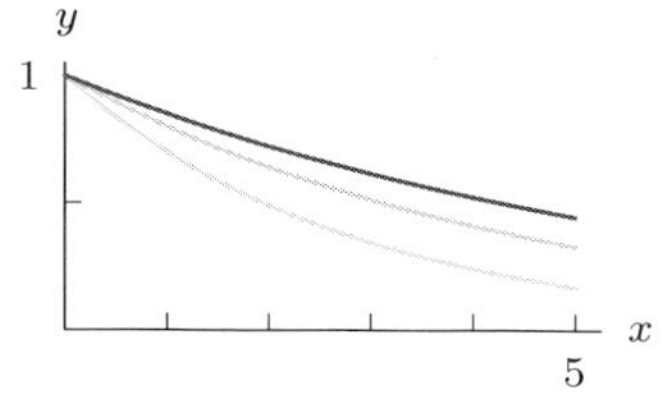

Figure 3.22

For Exercises 5–8, use Figure 3.23. Assume the equations for A, B, C, and D can all be written in the form $y = ab^t$.

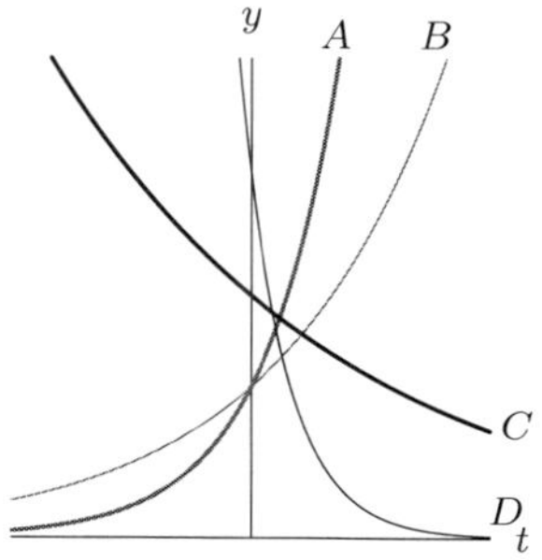

Figure 3.23

5. Which function has the largest value for a?

6. Which two functions have the same value for a?

7. Which function has the smallest value for b?

8. Which function has the largest value for b?

9. Solve $y = 46(1.1)^x$ graphically for x if $y = 91$.

10. Solve $p = 22(0.87)^q$ graphically for q if $p = 10$.

11. Solve $4m = 17(2.3)^w$ graphically for w if $m = 12$.

12. Solve $P/7 = (0.6)^t$ graphically for t if $P = 2$.

13. If $b > 1$, what is the horizontal asymptote of $y = ab^t$ as $t \to -\infty$?

14. If $0 < b < 1$, what is the horizontal asymptote of $y = ab^t$ as $t \to \infty$?

Problems

15. The city of Baltimore has been declining in population for the last fifty years.[10] In the year 2000, the population of Baltimore was 651 thousand and declining at a rate of 0.75% per year. If this trend continues:

(a) Give a formula for the population of Baltimore, P, in thousands, as a function of years, t, since 2000.
(b) What is the predicted population in 2010?
(c) To two decimal places, estimate t when the population is 550 thousand.

16. Let $P = f(t) = 1000(1.04)^t$ be the population of a community in year t.

(a) Evaluate $f(0)$ and $f(10)$. What do these expressions represent in terms of the population?
(b) Using a calculator or a computer, find appropriate viewing windows on which to graph the population for the first 10 years and for the first 50 years. Give the viewing windows you used and sketch the resulting graphs.
(c) If the percentage growth rate remains constant, approximately when will the population reach 2500 people?

[10] *The World Almanac and Book of Facts 2006*, p. 480 (New York).

17. Suppose y, the number of cases of a disease, is reduced by 10% each year.

(a) If there are initially 10,000 cases, express y as a function of t, the number of years elapsed.
(b) How many cases will there be 5 years from now?
(c) How long does it take to reduce the number of cases to 1000?

18. The earth's atmospheric pressure, P, in terms of height above sea level is often modeled by an exponential decay function. The pressure at sea level is 1013 millibars and that the pressure decreases by 14% for every kilometer above sea level.

(a) What is the atmospheric pressure at 50 km?
(b) Estimate the altitude h at which the pressure equals 900 millibars.

19. Consider the exponential functions graphed in Figure 3.24 and the six constants a, b, c, d, p, q.

(a) Which of these constants are definitely positive?
(b) Which of these constants are definitely between 0 and 1?
(c) Which of these constants could be between 0 and 1?
(d) Which two of these constants are definitely equal?
(e) Which one of the following pairs of constants could be equal?

a and p $\quad$ b and d $\quad$ b and q $\quad$ d and q

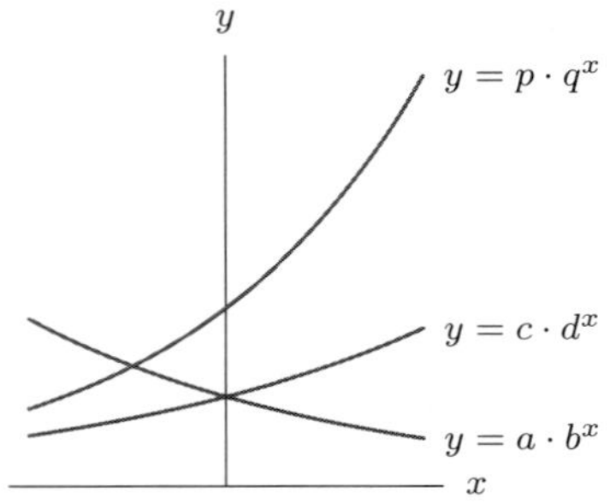

Figure 3.24

Problems 20–21 use Figure 3.25, where y_0 is the y-coordinate of the point of intersection of the graphs. Describe what happens to y_0 if the following changes are made, assuming the other quantities remain the same.

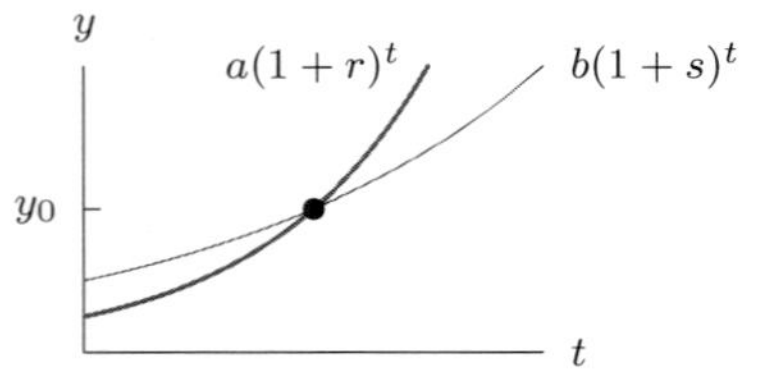

Figure 3.25

20. r is increased $\quad$ **21.** a is increased

Problems 22–23 use Figure 3.26, where t_0 is the t-coordinate of the point of intersection of the graphs. Describe what happens to t_0 if the following changes are made, assuming the other quantities remain the same.

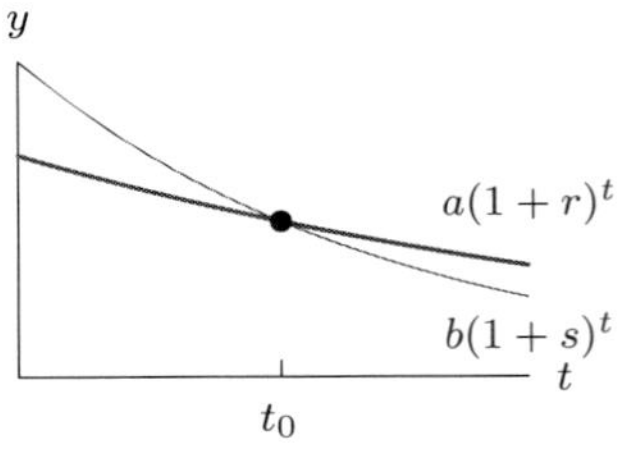

Figure 3.26

22. b is decreased $\quad$ **23.** r is increased

24. For which value(s) of a and b is $y = ab^x$ an increasing function? A decreasing function? Concave up?

25. Write a paragraph that compares the function $f(x) = a^x$, where $a > 1$, and $g(x) = b^x$, where $0 < b < 1$. Include graphs in your answer.

26. Set a window of $-4 \le x \le 4, -1 \le y \le 6$ and graph the following functions using several different values of a for each. Include some values of a with $a < 1$.

(a) $y = a2^x,\ 0 < a < 5$.
(b) $y = 2a^x,\ 0 < a < 5$.

In Problems 27–30, graph $f(x)$, a function defined for all real numbers and satisfying the condition.

27. $f(x) \to 5$ as $x \to \infty$

28. $f(x) \to 2$ as $x \to -\infty$ and $f(x) \to -1$ as $x \to \infty$

29. $\lim_{x\to-\infty} f(x) = 3$

30. $\lim_{x\to-\infty} f(x) = 0$ and $\lim_{x\to\infty} f(x) = -\infty$

In Problems 31–32, assume that all important features are shown in the graph of $y = f(x)$. Estimate

(a) $\lim_{x\to-\infty} f(x)$ $\quad$ **(b)** $\lim_{x\to\infty} f(x)$

31.

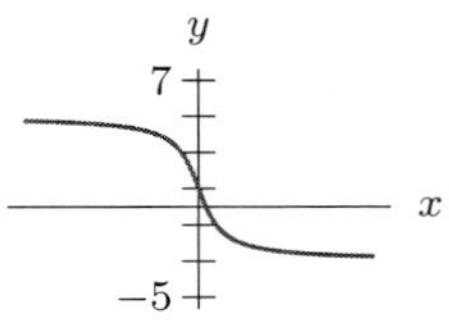

32.

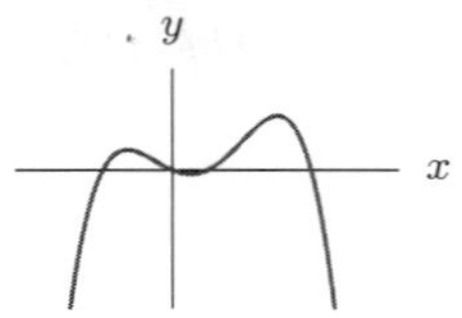

33. Find **(a)** $\lim_{x\to\infty} 7(0.8)^x$ **(b)** $\lim_{t\to-\infty} 5(1.2)^t$

(c) $\lim_{t\to\infty} 0.7(1-(0.2)^t)$

34. If the exponential function ab^x has the property that $\lim_{x\to-\infty} ab^x = 0$, what can you say about the value of b?

In Problems 35–36, graph the function to find horizontal asymptotes.

35. $f(x) = 8 - 2^x$ **36.** $f(x) = 3^{-x^2} + 2$

37. Suppose you use your calculator to graph $y = 1.04^{5x}$. You correctly enter $y = 1.04\hat{}(5x)$ and see the graph in Figure 3.27. A friend graphed the function by entering $y = 1.04\hat{}5x$ and said, "The graph is a straight line, so I must have the wrong window." Explain why changing the window will not correct your friend's error.

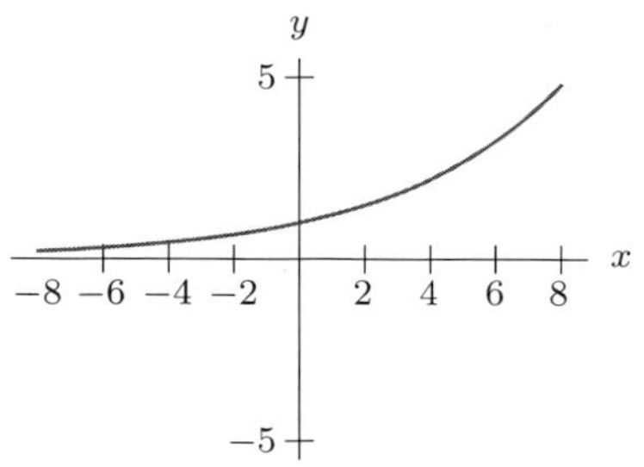

Figure 3.27

38. The population of a colony of rabbits grows exponentially. The colony begins with 10 rabbits; five years later there are 340 rabbits.

(a) Give a formula for the population of the colony of rabbits as a function of the time.
(b) Use a graph to estimate how long it takes for the population of the colony to reach 1000 rabbits.

39. Table 3.12 shows global wind energy generating capacity, W (in megawatts), as a function of the number of years, t, since 1995.[11]

(a) Plot the data and explain why it is reasonable to approximate these data with an exponential function.
(b) Use a calculator or computer to fit an exponential function to these data.
(c) What annual percent growth rate does the exponential model show?

Table 3.12

t	0	1	2	3	4
W	4780	6070	7640	10,150	13,930
t	5	6	7	8	9
W	18,450	24,930	32,037	39,664	47,760

40. Sales of energy-efficient compact fluorescent lamps in China have been growing approximately exponentially. Table 3.13 shows the sales in millions.[12]

(a) Use a calculator or computer to find the exponential regression function for sales, S (in millions), as a function of the number of years, t, since 1994.
(b) Plot the function with the data. Does it appear to fit the data well?
(c) What annual percent growth rate does the exponential model show?
(d) If this growth rate continues, what sales are predicted in the year 2010?

Table 3.13

Year	1994	1996	1998	2000	2002	2003
Sales (millions)	20	30	60	125	295	440

41. What are the domain and range of the exponential function $Q = ab^t$ where a and b are both positive constants?

42. The functions $f(x) = (\frac{1}{2})^x$ and $g(x) = 1/x$ are similar in that they both tend toward zero as x becomes large. Using a calculator, determine which function, f or g, approaches zero faster.

43. Let f be a piecewise-defined function given by

$$f(x) = \begin{cases} 2^x, & x < 0 \\ 0, & x = 0 \\ 1 - \frac{1}{2}x, & x > 0. \end{cases}$$

(a) Graph f for $-3 \le x \le 4$.
(b) The domain of $f(x)$ is all real numbers. What is its range?
(c) What are the intercepts of f?
(d) What happens to $f(x)$ as $x \to \infty$ and $x \to -\infty$?
(e) Over what intervals is f increasing? Decreasing?

[11]The Worldwatch Institute, *Vital Signs* 2005 (New York: W.W. Norton & Company, 2005), p. 35.
[12]Nadel, S. and Hong, "Market Data on Efficient Lighting," Right Light 6 Conference, Session 8, May, 2005.

44. Three scientists, working independently of each other, arrive at the following formulas to model the spread of a species of mussel in a system of fresh water lakes:

$$f_1(x) = 3(1.2)^x,\ f_2(x) = 3(1.21)^x,\ f_3(x) = 3.01(1.2)^x,$$

where $f_n(x)$, $n = 1, 2, 3$, is the number of individual mussels (in 1000s) predicted by model number n to be living in the lake system after x months have elapsed.

(a) Graph these three functions for $0 \leq x \leq 60$, $0 \leq y \leq 40{,}000$.

(b) The graphs of these three models do not seem all that different from each other. But do the three functions make significantly different predictions about the future mussel population? To answer this, graph the difference function, $f_2(x) - f_1(x)$, of the population sizes predicted by models 1 and 2, as well as the difference functions, $f_3(x) - f_1(x)$ and $f_3(x) - f_2(x)$. (Use the same window as in part (a).)

(c) Based on your graphs in part (b), discuss the assertion that all three models are in good agreement as far as long-range predictions of mussel population are concerned. What conclusions can you draw about exponential functions in general?

3.4 CONTINUOUS GROWTH AND THE NUMBER e

If \$1.00 is invested in a bank account that pays 100% interest once a year, then, assuming no other deposits or withdrawals, after one year we have

$$\$1.00(1 + 100\%) = \$2.00.$$

If \$1.00 is invested in a bank account that pays 50% interest twice a year, then, since $50\% = 100\%/2$, we have

$$\text{Balance after first six months} = \$1.00\left(1 + \frac{100\%}{2}\right) = \$1.50$$

$$\text{Balance after second six months} = \$1.50\left(1 + \frac{100\%}{2}\right) = \$1.00\left(1 + \frac{100\%}{2}\right)^2 = \$2.25.$$

The balance is larger because interest earned in the first six months itself earns interest in the second six months.

Similarly, if \$1.00 is invested in a bank account that pays 25% interest 4 times a year, then after one year, since $25\% = 100\%/4$, we have

$$\text{Balance} = \$1.00\left(1 + \frac{100\%}{4}\right)^4 = \$2.441406.$$

Table 3.14 shows the balance after one year as the interest is calculated more and more frequently.

Table 3.14 *Approximate balance for various frequencies*

Frequency	Approximate balance
1 (annually)	\$2.00
2 (semi-annually)	\$2.25
4 (quarterly)	\$2.441406
12 (monthly)	\$2.613035
365 (daily)	\$2.714567

As the frequency increases, the balance increases, because the interest earns more interest. How large can the balance grow? Table 3.15 shows balances for computations made each hour, each minute, and each second. It appears that the balance never gets larger than 2.7182. In fact, as the frequency of computation increases, the balance approaches 2.71828182.... This irrational number, introduced by Euler[13] in 1727, is so important that it is given a special name, e.

Table 3.15 *Approximate balance for greater frequencies*

Frequency	Approximate balance
8760 (hourly)	\$2.718127
525,600 (each minute)	\$2.718279
31,536,000 (each second)	\$2.718282

The Number e

The number $e = 2.71828182\ldots$ is often used for the base, b, of the exponential function. Base e is called the *natural base*. This may seem mysterious, as what could possibly be natural about using an irrational base such as e? The answer is that the formulas of calculus are much simpler if e is used as the base for exponentials. Some of the remarkable properties of the number e are introduced in Problems 25–28 on page 135. Since $2 < e < 3$, the graph of $Q = e^t$ lies between the graphs of $Q = 3^t$ and $Q = 2^t$. See Figure 3.28.

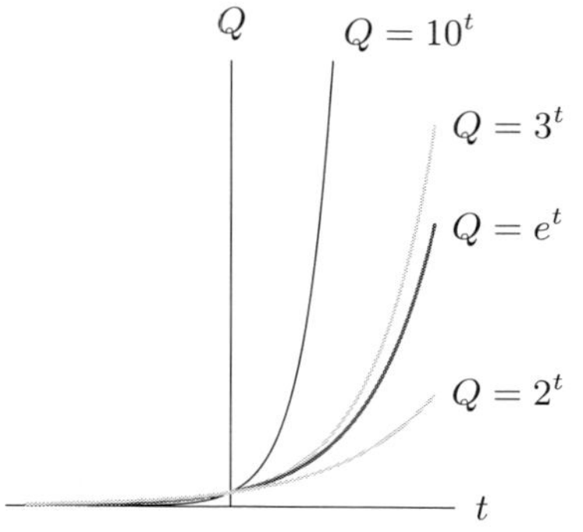

Figure 3.28: Graphs of exponential functions with various bases

Exponential Functions with Base e Represent Continuous Growth

Any positive base b can be written as a power of e:

$$b = e^k.$$

If $b > 1$, then k is positive; if $0 < b < 1$, then k is negative. The function $Q = ab^t$ can be rewritten in terms of e:

$$Q = ab^t = a\left(e^k\right)^t = ae^{kt}.$$

The constant k is called the *continuous growth rate*. In general:

> For the exponential function $Q = ab^t$, the **continuous growth rate**, k, is given by solving $e^k = b$. Then
>
> $$Q = ae^{kt}.$$
>
> If a is positive,
> - If $k > 0$, then Q is increasing.
> - If $k < 0$, then Q is decreasing.

[13]Leonhard Euler (1707-1783), a Swiss mathematician, introduced e, $f(x)$ notation, π, and i (for $\sqrt{-1}$).

The value of the continuous growth rate, k, may be given as a decimal or a percent. If t is in years, for example, then the units of k are given per year; if t is in minutes, then k is given per minute.

Example 1 Give the continuous growth rate of each of the following functions and graph each function:

$$P = 5e^{0.2t}, \qquad Q = 5e^{0.3t}, \qquad \text{and} \quad R = 5e^{-0.2t}.$$

Solution The function $P = 5e^{0.2t}$ has a continuous growth rate of 20%, and $Q = 5e^{0.3t}$ has a continuous 30% growth rate. The function $R = 5e^{-0.2t}$ has a continuous growth rate of -20%. The negative sign in the exponent tells us that R is decreasing instead of increasing.

Because $a = 5$ in all three formulas, all three functions cross the vertical axis at 5. Note that the graphs of these functions in Figure 3.29 have the same shape as the exponential functions in Section 3.3. They are concave up and have horizontal asymptotes of $y = 0$. (Note that $P \to 0$ and $Q \to 0$ as $t \to -\infty$, whereas $R \to 0$ as $t \to \infty$.)

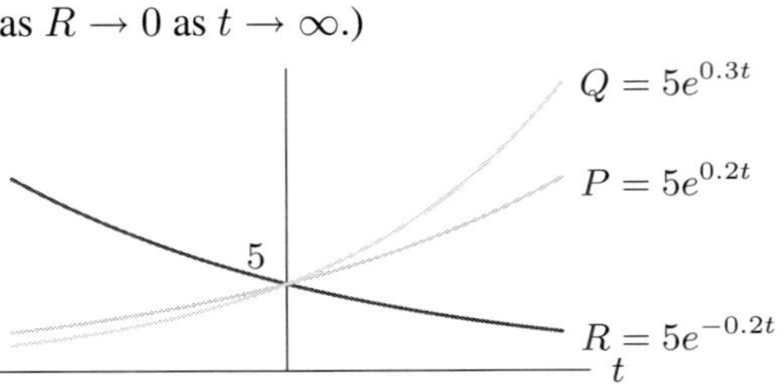

Figure 3.29: Exponential functions with different continuous growth rates

---

Example 2 A population increases from 7.3 million at a continuous rate of 2.2% per year. Write a formula for the population, and estimate graphically when the population reaches 10 million.

Solution We express the formula in base e since the continuous growth rate is given. If P is the population (in millions) in year t, then

$$P = 7.3e^{0.022t}.$$

See Figure 3.30. We see that $P = 10$ when $t \approx 14.3$. Thus, it takes about 14.3 years for the population to reach 10 million.

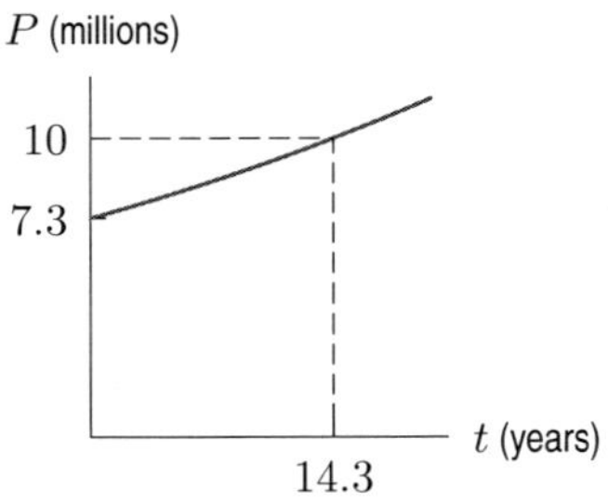

Figure 3.30

Example 3 Caffeine leaves the body at a continuous rate of 17% per hour. How much caffeine is left in the body 8 hours after drinking a cup of coffee containing 100 mg of caffeine?

Solution If A is the amount of caffeine in the body t hours after drinking the coffee, then

$$A = 100e^{-0.17t}.$$

After 8 hours, we have $A = 100e^{-0.17(8)} = 25.67$ mg.

The Difference Between Annual and Continuous Growth Rates

If $P = P_0(1.07)^t$, with t in years, we say that P is growing at an *annual* rate of 7%. If $P = P_0e^{0.07t}$, with t in years, we say that P is growing at a *continuous* rate of 7% per year. Since $e^{0.07} = 1.0725\ldots$, we can rewrite $P_0e^{0.07t} = P_0(1.0725)^t$. In other words, a 7% continuous rate and a 7.25% annual rate generate the same increases in P. We say the two rates are equivalent.

We can check that $e^{0.0677} = 1.07\ldots$, so a 7% annual growth rate is equivalent to a 6.77% continuous growth rate. The continuous growth rate is always smaller than the equivalent annual rate.

The bank account example at the start of this section reminds us why a quantity growing at continuous rate of 7% per year increases faster than a quantity growing at an annual rate of 7%: In the continuous case, the interest earns more interest.

Example 4 In November 2005, the Wells Fargo Bank offered interest at a 2.323% continuous yearly rate.[14] Find the equivalent annual rate.

Solution Since $e^{0.02323} = 1.0235$, the equivalent annual rate is 2.35%. As expected, the equivalent annual rate is larger than the continuous yearly rate.

Exercises and Problems for Section 3.4

Exercises

1. Without a calculator, match the functions $y = e^x$, $y = 2e^x$, and $y = 3e^x$ to the graphs in Figure 3.31.

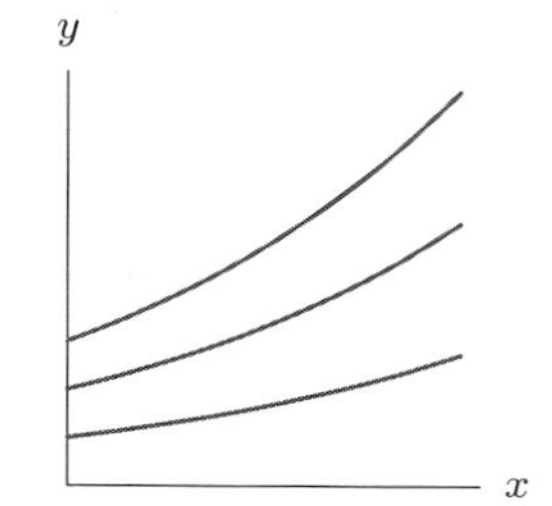

Figure 3.31

2. Without a calculator, match the functions $y = 2^x$, $y = 3^x$, and $y = e^x$ with the graphs in Figure 3.32.

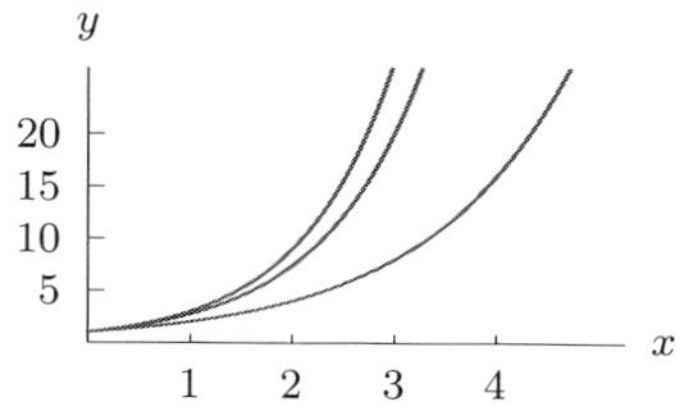

Figure 3.32

3. Without a calculator, match the functions (a)–(d) with the graphs (I)–(IV) in Figure 3.33.

(a) $e^{0.25t}$ **(b)** $(1.25)^t$ **(c)** $(1.2)^t$ **(d)** $e^{0.3t}$

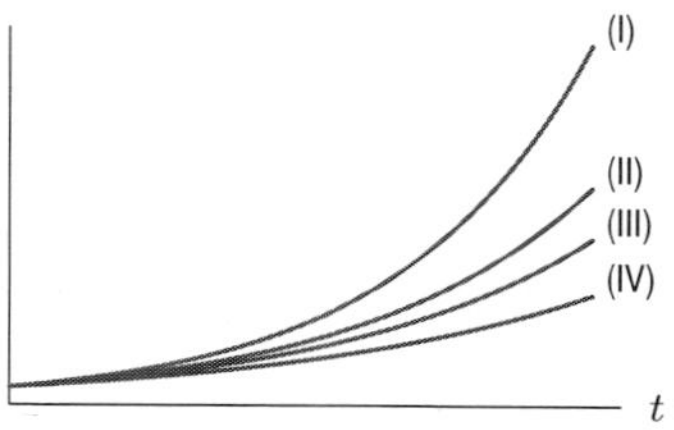

Figure 3.33

4. Without graphing on a calculator, match the functions (a)–(d) with the graphs (I)–(IV) in Figure 3.34.

(a) 1.5^x **(b)** $e^{0.45x}$ **(c)** $e^{0.47x}$ **(d)** $e^{0.5x}$

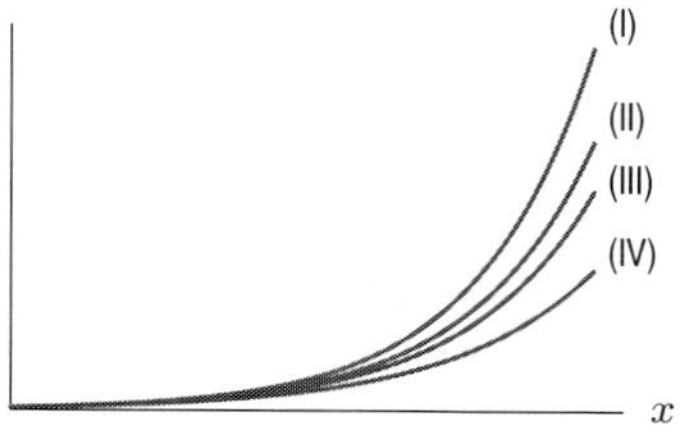

Figure 3.34

[14] http://money.cnn.com/2005/11/30/debt/informa_rate.

5. Without a calculator, match the functions $y = e^x$, $y = e^{-x}$, and $y = -e^x$ to the graphs in Figure 3.35.

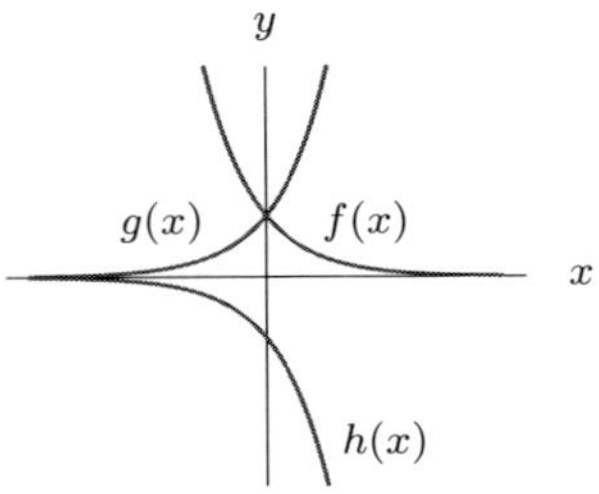

Figure 3.35

6. Without a calculator, match each formula to one of the graphs (I)–(IV) in Figure 3.36.

(a) $e^{-0.01t}$ **(b)** $e^{0.05t}$ **(c)** $e^{-0.10t}$ **(d)** $e^{0.20t}$

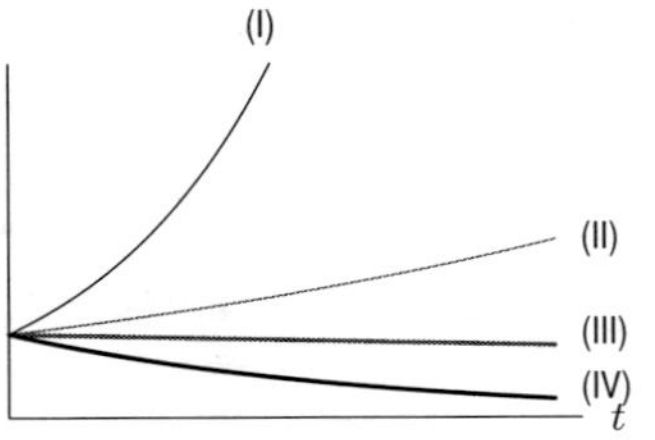

Figure 3.36

7. Without a calculator, match the functions (a)–(d) with the graphs (I)–(IV) in Figure 3.37.

(a) $y = e^x$ **(b)** $y = e^{-x}$

(c) $y = e^{-2x}$ **(d)** $y = e^{-3x}$

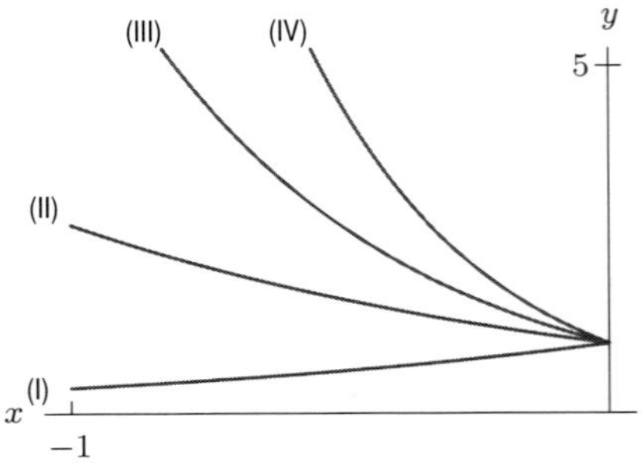

Figure 3.37

8. At time t in years, the value, V, of an investment of \$1000 is given by $V = 1000e^{0.02t}$. When is the investment worth \$3000?

9. How long does it take an investment to double if it grows according to the formula $V = 537e^{0.015t}$? Assume t is in years.

Problems

10. Without a calculator, arrange the following quantities in ascending order:

(a) $3^{2.2}, e^{2.2}, (\sqrt{2})^{2.2}$
(b) $3^{-2.2}, e^{-2.2}$

In Problems 11–14, find the limits.

11. $\lim_{x\to\infty} e^{-3x}$ **12.** $\lim_{t\to-\infty} 5e^{0.07t}$

13. $\lim_{t\to\infty} (2 - 3e^{-0.2t})$ **14.** $\lim_{t\to-\infty} 2e^{-0.1t+6}$

15. If $\lim_{t\to\infty} ae^{kt} = \infty$, what can you say about the values of a and k?

16. From time $t = 0$, with t in years, a \$1200 deposit in a bank account grows according to the formula

$$B = 1200e^{0.03t}.$$

(a) What is the balance in the account at the end of 100 years?

(b) When does the balance first go over \$50,000?

17. Calculate the amount of money in a bank account if \$2000 is deposited for 15 years at an interest rate of

(a) 5% annually
(b) 5% continuously per year

18. A population of 3.2 million grows at a constant percentage rate.

(a) What is the population one century later if there is:
(i) An annual growth rate of 2%
(ii) A continuous growth rate of 2% per year

(b) Explain how you can tell which of the answers would be larger before doing the calculations.

19. A population is 25,000 in year $t = 0$ and grows at a continuous rate of 7.5% per year.

(a) Find a formula for $P(t)$, the population in year t.
(b) By what percent does the population increase each year? Why is this more than 7.5%?

20. A population grows from its initial level of 22,000 at a continuous growth rate of 7.1% per year.

(a) Find a formula for $P(t)$, the population in year t.
(b) By what percent does the population increase each year?

21. A radioactive substance decays at a continuous rate of 14% per year, and 50 mg of the substance is present in the year 2000.

(a) Write a formula for the amount present, A (in mg), t years after 2000.
(b) How much will be present in the year 2010?
(c) Estimate when the quantity drops below 5 mg.

22. In 2004, the gross world product, W, (total output in goods and services) was 54.7 trillion dollars and growing at a continuous rate of 3.8% per year.[15] Write a formula for W, in trillions of dollars, as a function of years, t, since 2004. Estimate the value of t when the gross world product is predicted to reach 80 trillion dollars.

23. World poultry production was 77.2 million tons in the year 2004 and increasing at a continuous rate of 1.6% per year.[16] Assume that this growth rate continues.

(a) Write an exponential formula for world poultry production, P, in million tons, as a function of the number of years, t, since 2004.
(b) Use the formula to estimate world poultry production in the year 2010.
(c) Use a graph to estimate the year in which world poultry production goes over 90 million tons.

24. What can you say about the value of the constants a, k, b, l in Figure 3.38?

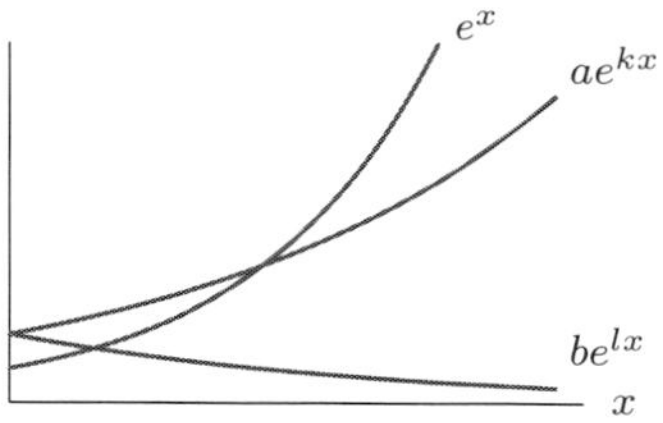

Figure 3.38

25. This problem explores the value of $(1+1/n)^n$ for integer values of n as n gets large.

(a) Use a calculator or computer to evaluate $(1+1/n)^n$, correct to seven decimal places, for $n = 1000$, 10,000, 100,000, and 1,000,000. Does $(1+1/n)^n$ appear to be an increasing or decreasing function of n?
(b) The limit of the sequence of values in part (a) is $e = 2.718281828\ldots$. What power of 10 is needed to give a value of e correct to 6 decimal places?
(c) What happens if you evaluate $(1+1/n)^n$ using much larger values of n? For example, try $n = 10^{16}$ on a calculator.

26. This problem uses a calculator or computer to explore graphically the value of $(1+1/x)^x$ as x gets large.

(a) Graph $y = (1+1/x)^x$ for $1 \le x \le 10$.
(b) Are the values of y in part (a) increasing or decreasing?
(c) Do the values of y in part (a) appear to approach a limiting value?
(d) Graph $y = (1+1/x)^x$, for $1 \le x \le 100$, and then for $1 \le x \le 1000$. Do the y values appear to approach a limiting value? If so, approximately what is it?
(e) Graph $y = (1+1/x)^x$ and $y = e$ on the same axes, for $1 \le x \le 10{,}000$. What does the graph suggest?
(f) By checking $x = 10{,}000$, $x = 20{,}000$, and so on, decide how large (as a multiple of 10,000) x should be to give a value of e correct to 4 decimal places.

27. (a) Using a computer or calculator, graph $f(x) = 2^x$.
(b) Find the slope of the line tangent to f at $x = 0$ to an accuracy of two decimals. [Hint: Zoom in on the graph until it is indistinguishable from a line and estimate the slope using two points on the graph.]
(c) Find the slope of the line tangent to $g(x) = 3^x$ at $x = 0$ to an accuracy of two decimals.
(d) Find b (to two decimals) such that the line tangent to the function $h(x) = b^x$ at $x = 0$ has slope 1.

28. With more terms giving a better approximation, it can be shown that

$$e = 1 + \frac{1}{1} + \frac{1}{1\cdot 2} + \frac{1}{1\cdot 2\cdot 3} + \frac{1}{1\cdot 2\cdot 3\cdot 4} + \cdots.$$

(a) Use a calculator to sum the five terms shown.
(b) Find the sum of the first seven terms.
(c) Compare your sums with the calculator's displayed value for e (which you can find by entering e^1) and state the number of correct digits in the five and seven term sum.
(d) How many terms of the sum are needed in order to give a nine decimal digit approximation equal to the calculator's displayed value for e?

[15] The Worldwatch Institute, *Vital Signs 2005* (New York: W.W. Norton & Company, 2005), p. 45.
[16] The Worldwatch Institute, *Vital Signs* 2005 (New York: W.W. Norton & Company, 2005), p. 24.

3.5 COMPOUND INTEREST

What is the difference between a bank account that pays 12% interest once per year and one that pays 1% interest every month? Imagine we deposit $\$1000$ into the first account. Then, after 1 year, we have (assuming no other deposits or withdrawals)

$$\$1000(1.12) = \$1120.$$

But if we deposit $\$1000$ into the second account, then after 1 year, or 12 months, we have

$$\$1000 \underbrace{(1.01)(1.01)\ldots(1.01)}_{\text{12 months of 1\% monthly interest}} = 1000(1.01)^{12} = \$1126.83.$$

Thus, we earn $\$6.83$ more in the second account than in the first. To see why this happens, notice that the 1% interest we earn in January itself earns interest at a rate of 1% per month. Similarly, the 1% interest we earn in February earns interest, and so does the interest earned in March, April, May, and so on. The extra $\$6.83$ comes from interest earned on interest. This effect is known as *compounding*. We say that the first account earns 12% interest *compounded annually* and the second account earns 12% interest *compounded monthly*.

Nominal Versus Effective Rate

The expression 12% compounded monthly means that interest is added twelve times per year and that $12\%/12 = 1\%$ of the current balance is added each time. We refer to the 12% as the *nominal rate* (nominal means "in name only"). When the interest is compounded more frequently than once a year, the account effectively earns more than the nominal rate. Thus, we distinguish between nominal rate and *effective annual rate*, or *effective rate*. The effective annual rate tells you how much interest the investment actually earns.

Example 1 What are the nominal and effective annual rates of an account paying 12% interest, compounded annually? Compounded monthly?

Solution Since an account paying 12% annual interest, compounded annually, grows by exactly 12% in one year, we see that its nominal rate is the same as its effective rate: both are 12%.

The account paying 12% interest, compounded monthly, also has a nominal rate of 12%. On the other hand, since it pays 1% interest every month, after 12 months, its balance increases by a factor of

$$\underbrace{(1.01)(1.01)\ldots(1.01)}_{\text{12 months of 1\% monthly growth}} = 1.01^{12} \approx 1.1268250.$$

Thus, effectively, the account earns 12.683% interest in a year.

Example 2 What is the effective annual rate of an account that pays interest at the nominal rate of 6% per year, compounded daily? Compounded hourly?

Solution Since there are 365 days in a year, daily compounding pays interest at the rate of

$$\frac{6\%}{365} = 0.0164384\% \text{ per day.}$$

Thus, the daily growth factor is

$$1 + \frac{0.06}{365} = 1.000164384.$$

If at the beginning of the year the account balance is P, after 365 days the balance is

$$\underbrace{P \cdot \left(1 + \frac{0.06}{365}\right)^{365}}_{\substack{365 \text{ days of} \\ 0.0164384\% \text{ daily interest}}} = P \cdot (1.0618313).$$

Thus, this account earns interest at the effective annual rate of 6.18313%.

Notice that daily compounding results in a higher rate than yearly compounding (6.183% versus 6%), because with daily compounding the interest has the opportunity to earn interest.

If interest is compounded hourly, since there are $24 \cdot 365$ hours in a year, the balance at year's end is

$$P \cdot \left(1 + \frac{0.06}{24 \cdot 365}\right)^{24 \cdot 365} = P \cdot (1.0618363).$$

The effective rate is now 6.18363% instead of 6.18313%—that is, just slightly better than the rate of the account that compounds interest daily. The effective rate increases with the frequency of compounding.

To summarize:

> If interest at an annual rate of r is compounded n times a year, then r/n times the current balance is added n times a year. Therefore, with an initial deposit of $\$P$, the balance t years later is
>
> $$B = P \cdot \left(1 + \frac{r}{n}\right)^{nt}.$$
>
> Note that r is the nominal rate; for example, $r = 0.05$ if the annual rate is 5%.

Continuous Compounding and the Number e

In Example 2 we calculated the effective interest rates for two accounts with a 6% per year nominal interest rate, but different compounding periods. We see that the account with more frequent compounding earns a higher effective rate, though the increase is small.

This suggests that compounding more and more frequently—every minute or every second or many times per second—would increase the effective rate still further. As we saw in Section 3.4, there is a limit to how much more an account can earn by increasing the frequency of compounding.

Table 3.16 *Effect of increasing the frequency of compounding*

Compounding frequency	Annual growth factor	Effective annual rate
Annually	1.0600000	6%
Monthly	1.0616778	6.16778%
Daily	1.0618313	6.18313%
Hourly	1.0618363	6.18363%
⋮	⋮	⋮
Continuously	$e^{0.06} \approx 1.0618365$	6.18365%

Table 3.16 shows several compounding periods with their annual growth factors and effective annual rates. As the compounding periods become shorter, the growth factor approaches $e^{0.06}$. Using a calculator, we check that

$$e^{0.06} \approx 1.0618365,$$

which is the final value for the annual growth factors in Table 3.16. If an account with a 6% nominal interest rate delivers this effective yield, we say that the interest has been *compounded continuously*.

In general:

> If interest on an initial deposit of $\$P$ is *compounded continuously* at an annual rate r, the balance t years later can be calculated using the formula
>
> $$B = Pe^{rt}.$$
>
> Again, r is the nominal rate, and, for example, $r = 0.06$ when the annual rate is 6%.

It is important to realize that the functions $B = Pe^{0.06t}$ and $B = P(1.0618365)^t$ both give the balance in a bank account growing at a continuous rate of 6% per year. These formulas both represent the *same* exponential function—they just describe it in different ways.[17]

Example 3 Which is better: An account that pays 8% annual interest compounded quarterly or an account that pays 7.95% annual interest compounded continuously?

Solution The account that pays 8% interest compounded quarterly pays 2% interest 4 times a year. Thus, in one year the balance is

$$P(1.02)^4 \approx P(1.08243),$$

which means the effective annual rate is 8.243%.

The account that pays 7.95% interest compounded continuously has a year-end balance of

$$Pe^{0.0795} \approx P(1.08275),$$

so the effective annual rate is 8.275%. Thus, 7.95% compounded continuously pays more than 8% compounded quarterly.

[17]Actually, this is not precisely true, because we rounded off when we found $b = 1.0618365$. However, we can find b to as many digits as we want, and to this extent the two formulas are the same.

Exercises and Problems for Section 3.5

Exercises

In Exercises 1–4, what is the balance after 1 year if an account containing \$500 earns the stated yearly nominal interest, compounded

(a) Annually **(b)** Weekly (52 weeks per year)
(c) Every minute (525,600 per year)
(d) Continuously?

1. 1% **2.** 3% **3.** 5% **4.** 8%

5. The same amount of money is deposited into two different bank accounts paying the same nominal rate, one compounded annually and the other compounded continuously. Which curve in Figure 3.39 corresponds to which compounding method? What is the initial deposit?

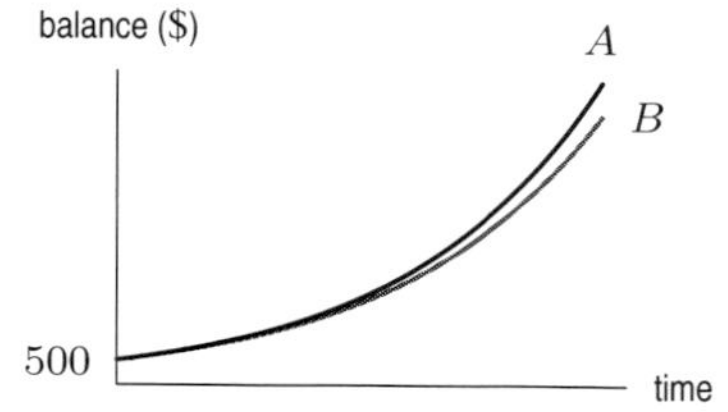

Figure 3.39

6. Find the effective annual yield and the continuous growth rate if $Q = 5500\,e^{0.19\,t}$.

7. If \$5000 is deposited in an account paying a nominal interest rate of 4% per year, how much is in the account 10 years later if interest is compounded

(a) Annually?
(b) Continuously?

8. Suppose \$1000 is deposited into an account paying interest at a nominal rate of 8% per year. Find the balance three years later if the interest is compounded

(a) Monthly **(b)** Weekly
(c) Daily **(d)** Continuously

9. Find the effective annual rate if \$1000 is deposited at 5% annual interest, compounded continuously.

10. A bank pays interest at the nominal rate of 4.2% per year. What is the effective annual yield if compounding is:

(a) Annual **(b)** Monthly **(c)** Continuous

In Problems 11–14, what are the nominal and effective annual rates for an account paying the stated annual interest, compounded

(a) Annually? **(b)** Quarterly?
(c) Daily? **(d)** Continuously?

11. 1% **12.** 100% **13.** 3% **14.** 6%

Problems

15. A bank account pays 6% annual interest. As the number of compounding periods increases, the effective interest rate earned also rises.

(a) Find the annual interest rate earned by the account if the interest is compounded:

(i) Quarterly (ii) Monthly
(iii) Weekly (iv) Daily

(b) Evaluate $e^{0.06}$, where $e = 2.71828\ldots$ Explain what your result tells you about the bank account.

16. If you need \$25,000 six years from now, what is the minimum amount of money you need to deposit into a bank account that pays 5% annual interest, compounded:

(a) Annually **(b)** Monthly **(c)** Daily

(d) Your answers get smaller as the number of times of compounding increases. Why is this so?

17. Suppose \$300 was deposited into one of five bank accounts and t is time in years. For each verbal description (i)–(v), state which formulas (a)–(e) could represent it.

(a) $B = 300(1.2)^t$ **(b)** $B = 300(1.12)^t$
(c) $B = 300(1.06)^{2t}$ **(d)** $B = 300(1.06)^{t/2}$
(e) $B = 300(1.03)^{4t}$

(i) This investment earned 12% annually, compounded annually.
(ii) This investment earned, on average, more than 1% each month.
(iii) This investment earned 12% annually, compounded semi-annually.
(iv) This investment earned, on average, less than 3% each quarter.
(v) This investment earned, on average, more than 6% every 6 months.

18. Three different investments are given.

(a) Find the balance of each of the investments after the two-year period.
(b) Rank them from best to worst in terms of rate of return. Explain your reasoning.

- Investment A: \$875 deposited at 13.5% per year compounded daily for 2 years.
- Investment B: \$1000 deposited at 6.7% per year compounded continuously for 2 years.
- Investment C: \$1050 deposited at 4.5% per year compounded monthly for 2 years.

19. Rank the following three bank deposit options from best to worst.

- Bank A: 7% compounded daily
- Bank B: 7.1% compounded monthly
- Bank C: 7.05% compounded continuously

20. Which is better, an account paying 5.3% interest compounded continuously or an account paying 5.5% interest compounded annually? Justify your answer.

21. If the balance, M, at time t in years, of a bank account that compounds its interest payments monthly is given by

$$M = M_0(1.07763)^t.$$

(a) What is the effective annual rate for this account?
(b) What is the nominal annual rate?

22. A sum of \$850 is invested for 10 years and the interest is compounded quarterly. There is \$1000 in the account at the end of 10 years. What is the nominal annual rate?

23. In the 1980s a northeastern bank experienced an unusual robbery. Each month an armored car delivered cash deposits from local branches to the main office, a trip requiring only one hour. One day, however, the delivery was six hours late. This delay turned out to be a scheme devised by an employee to defraud the bank. The armored car drivers had lent the money, a total of approximately \$200,000,000, to arms merchants who then used it as collateral against the purchase of illegal weapons. The interest charged for this loan was 20% per year compounded continuously. How much was the fee for the six-hour period?

24. An investment grows by 3% per year for 10 years. By what percent does it increase over the 10-year period?

25. An investment grows by 30% over a 5-year period. What is its effective annual percent growth rate?

26. An investment decreases by 60% over a 12-year period. At what effective annual percent rate does it decrease?

CHAPTER SUMMARY

- **Exponential Functions**
  Value of $f(t)$ changes at constant percent rate with respect to t.
- **General Formula for Exponential Functions**
  Exponential function: $f(t) = ab^t, \quad b > 0$.
  f increasing for $b > 1$, decreasing for $0 < b < 1$.
  Growth factor: $b = 1 + r$.
  Growth rate: r, percent change as a decimal.
- **Comparing Linear and Exponential Functions**
  An increasing exponential function eventually overtakes any linear function.
- **Graphs of Exponential Functions**
  Concavity; asymptotes; effect of parameters a and b. Solving exponential equations graphically; fitting exponential functions to data.
- **The Number e**
  Continuous growth: $f(t) = ae^{kt}$.
  f is increasing for $k > 0$, decreasing for $k < 0$.
  Continuous growth rate: k.
- **Compound Interest**
  For compounding n times per year, balance,
  $$B = P\left(1 + \frac{r}{n}\right)^{nt}.$$
  For continuous compounding, $B = Pe^{kt}$.
  Nominal rate, r or k, versus effective rate earned over one year.
- **Horizontal Asymptotes and Limits to Infinity**

REVIEW EXERCISES AND PROBLEMS FOR CHAPTER THREE

Exercises

Are the functions in Exercises 1–9 exponential? If so, write the function in the form $f(t) = ab^t$.

1. $g(w) = 2\left(2^{-w}\right)$

2. $m(t) = (2 \cdot 3^t)^3$

3. $f(x) = \dfrac{3^{2x}}{4}$

4. $G(t) = 3(t)^t$

5. $q(r) = \dfrac{-4}{3^r}$

6. $j(x) = 2^x 3^x$

7. $Q(t) = 8^{t/3}$

8. $K(x) = \dfrac{2^x}{3 \cdot 3^x}$

9. $p(r) = 2^r + 3^r$

10. A town has population 3000 people at year $t = 0$. Write a formula for the population, P, in year t if the town

(a) Grows by 200 people per year.
(b) Grows by 6% per year.
(c) Grows at a continuous rate of 6% per year.
(d) Shrinks by 50 people per year.
(e) Shrinks by 4% per year.
(f) Shrinks at a continuous rate of 4% per year.

11. The following formulas each describe the size of an animal population, P, in t years since the start of the study. Describe the growth of each population in words.

(a) $P = 200(1.028)^t$ **(b)** $P = 50e^{-0.17t}$
(c) $P = 1000(0.89)^t$ **(d)** $P = 600e^{0.20t}$
(e) $P = 2000 - 300t$ **(f)** $P = 600 + 50t$

12. If $f(x) = 12 + 20x$ and $g(x) = \frac{1}{2} \cdot 3^x$, for what values of x is $g(x) < f(x)$?

For Exercises 13–18, find a formula for the exponential function.

13.

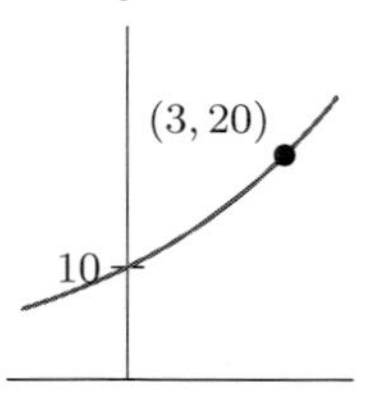

14.

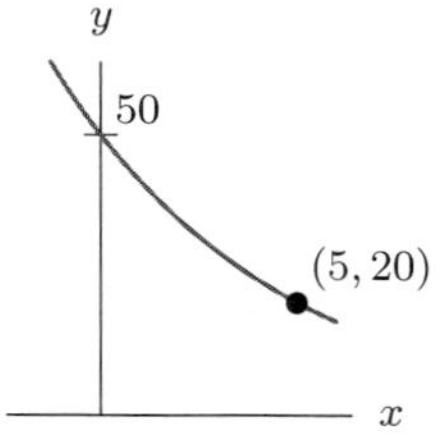

15.

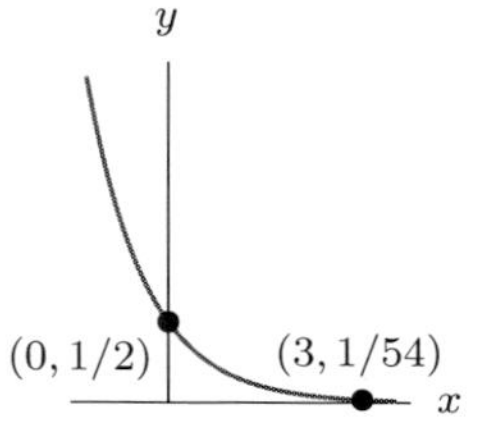

16.

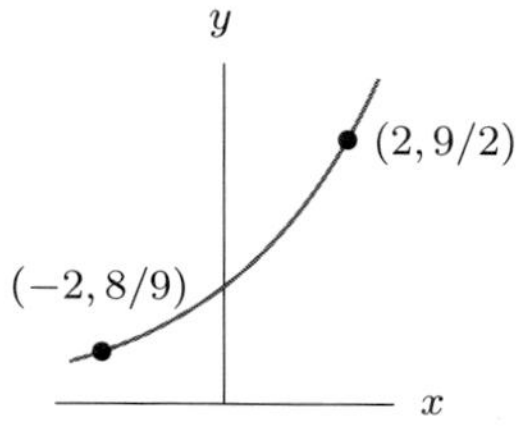

17.

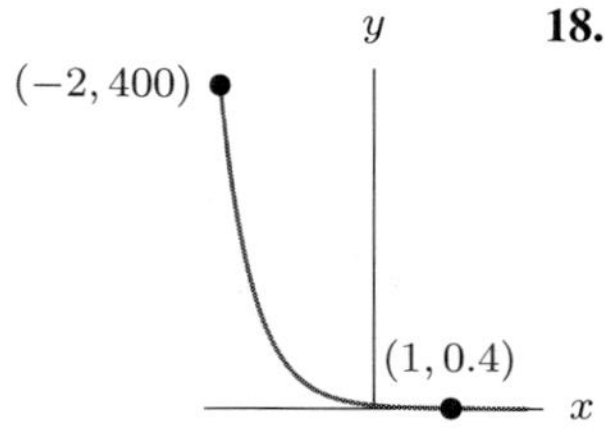

18.

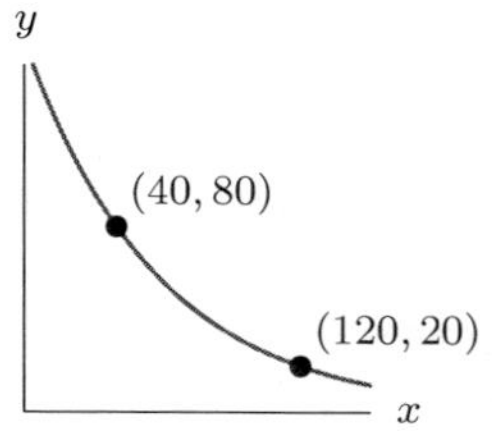

19. Without a calculator, match each of the formulas to one of the graphs in Figure 3.40.

(a) $y = 0.8^t$ **(b)** $y = 5(3)^t$
(c) $y = -6(1.03)^t$ **(d)** $y = 15(3)^{-t}$
(e) $y = -4(0.98)^t$ **(f)** $y = 82(0.8)^{-t}$

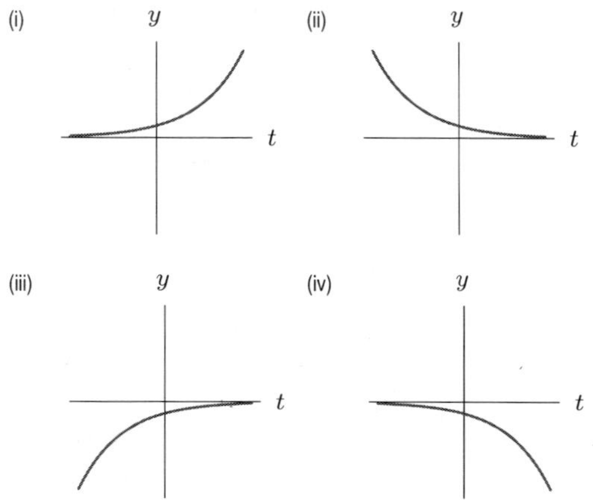

Figure 3.40

20. Without a calculator, match each of the following formulas to one of the graphs in Figure 3.40.

(a) $y = 8.3e^{-t}$ **(b)** $y = 2.5e^t$ **(c)** $y = -4e^{-t}$

Problems

In Problems 21–25, graph $f(x)$, a function defined for all real numbers and satisfying the condition.

21. $f(x) \to 3$ as $x \to -\infty$

22. $\lim_{x\to\infty} f(x) = 5$

23. $\lim_{x\to-\infty} f(x) = 2$ and $\lim_{x\to\infty} f(x) = -1$

24. $f(x) \to 0$ as $x \to -\infty$ and $f(x) \to -\infty$ as $x \to \infty$

25. $f(x)$ has a horizontal asymptote of $y = 5$.

Find the limits in Problems 26–31.

26. $\lim_{x\to\infty} 257(0.93)^x$

27. $\lim_{t\to\infty} 5.3e^{-0.12t}$

28. $\lim_{x\to-\infty} (15 - 5e^{3x})$

29. $\lim_{t\to-\infty} (21(1.2)^t + 5.1)$

30. $\lim_{x\to\infty} (7.2 - 2e^{3x})$

31. $\lim_{x\to-\infty} (5e^{-7x} + 1.5)$

32. An account pays interest at a nominal rate of 8% per year. Find the effective annual yield if interest is compounded

(a) Monthly **(b)** Weekly

(c) Daily **(d)** Continuously

33. Find $g(t) = ab^t$ if $g(10) = 50$ and $g(30) = 25$.

34. Find a formula for $f(x)$, an exponential function such that $f(-8) = 200$ and $f(30) = 580$.

35. Suppose that $f(x)$ is exponential and that $f(-3) = 54$ and $f(2) = \frac{2}{9}$. Find a formula for $f(x)$.

36. Find a formula for $f(x)$, an exponential function such that $f(2) = 1/27$ and $f(-1) = 27$.

37. Find the equation of an exponential curve through the points $(-1, 2)$, $(1, 0.3)$.

Find possible formulas for the functions in Problems 38–40.

38. V gives the value of an account that begins in year $t = 0$ with \$12,000 and earns 4.2% annual interest, compounded continuously.

39. The exponential function $p(t)$ given that $p(20) = 300$ and $p(50) = 40$.

40. The linear function $q(x)$ whose graph intersects the graph of $y = 5000e^{-x/40}$ at $x = 50$ and $x = 150$.

41. An investment worth $V = \$2500$ in year $t = 0$ earns 4.2% annual interest, compounded continuously. Find a formula for V in terms of t.

42. A population is represented by $P = 12{,}000e^{-0.122t}$. Give the values of a, k, b, and r, where $P = ae^{kt} = ab^t$. What do these values tell you about the population?

Decide whether the functions in Problems 43–45 could be approximately linear, approximately exponential, or are neither. For those that could be nearly linear or nearly exponential, find a formula.

43.

t	3	10	14
$Q(t)$	7.51	8.7	9.39

44.

t	5	9	15
$R(t)$	2.32	2.61	3.12

45.

t	5	12	16
$S(t)$	4.35	6.72	10.02

46. In January 2005, the population of California was 36.8 million and growing at an annual rate of 1.3%. Assume that growth continues at the same rate.

(a) By how much will the population increase between 2005 and 2030? Between 2030 and 2055?

(b) Explain how you can tell before doing the calculations which of the two answers in part (a) is larger.

47. There were 178.8 million licensed drivers in the US in 1989 and 187.2 million in 1999.[18] Find a formula for the number, N of licensed drivers in the US as a function of t, the number of years since 1989, assuming growth is

(a) Linear **(b)** Exponential

48. The population of a small town increases by a growth factor of 1.134 over a two-year period.

(a) By what percent does the town increase in size during the two-year period?

(b) If the town grows by the same percent each year, what is its annual percent growth rate?

[18] *The World Almanac* 2002 (New York: World Almanac Education Group, Inc., 2002), p. 228.

49. If t is in years, the formulas for dollar balances of two different bank accounts are:

$$f(t) = 1100(1.05)^t \quad \text{and} \quad g(t) = 1500e^{0.05t}.$$

(a) Describe in words the bank account modeled by f.
(b) Describe the account modeled by g. State the effective annual yield.

50. Accion is a non-profit microlending organization which makes small loans to entrepreneurs who do not qualify for bank loans.[19] A New York woman who sells clothes from a cart has the choice of a \$1000 loan from Accion to be repaid by \$1160 a year later and a \$1000 loan from a loan shark with an annual interest rate of 22%, compounded annually.

(a) What is the annual interest rate charged by Accion?
(b) To pay off the loan shark for a year's loan of \$1000, how much would the woman have to pay?
(c) Which loan is a better deal for the woman? Why?

Problems 51–54 use Figure 3.41, which show $f(x) = ab^x$ and $g(x) = cd^x$ on three different scales. Their point of intersection is marked.

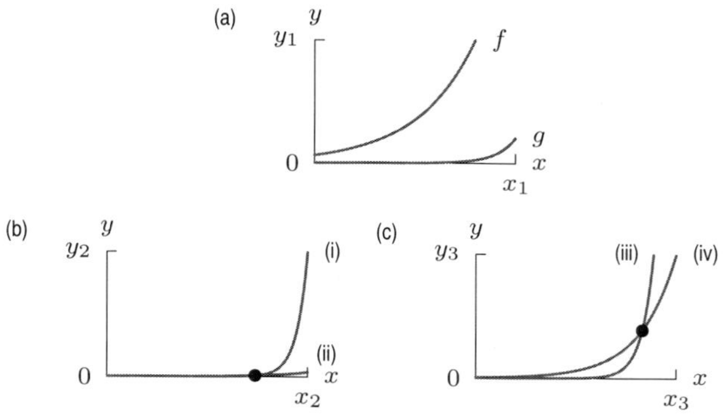

Figure 3.41

51. Which is larger, a or c?

52. Which is larger, b or d?

53. Rank in order from least to greatest: x_1, x_2, x_3.

54. Match f and g to the graphs labeled (i)–(iv) in (b) and (c).

55. Find the annual growth rates of a quantity which:

(a) Doubles in size every 7 years
(b) Triples in size every 11 years
(c) Grows by 3% per month
(d) Grows by 18% every 5 months

56. In 1995, the population of a town was 18,500 and it grew by 250 people by the end of the year. By 2005, its population had reached 22,500.

(a) Can this population be best described by a linear or an exponential model, or neither? Explain.
(b) If possible, find a formula for $P(t)$, the population t years after 1995.

57. In 1995, the population of a town was 20,000, and it grew by 4.14% that year. By 2005, the town's population had reached 30,000.

(a) Can this population be best described by a linear or an exponential model, or neither? Explain.
(b) If possible, find a formula for $P(t)$, this population t years after 1995.

58. Forty percent of a radioactive substance decays in five years. By what percent does the substance decay each year?

59. Table 3.17 shows the concentration of theophylline, a common asthma drug, in the blood stream as a function of time after injection of a 300 mg initial dose.[20] It is claimed that this data set is consistent with an exponential decay model $C = ab^t$ where C is the concentration and t is the time.

(a) Estimate the values of a and b, using ratios to estimate b. How good is this model?
(b) Use a calculator or computer to find the exponential regression function for concentration as a function of time. Compare answers from parts (a) and (b).

Table 3.17

Time (hours)	0	1	3	5	7	9
Concentration (mg/l)	12.0	10.0	7.0	5.0	3.5	2.5

60. Figure 3.42 gives the voltage, $V(t)$, across a circuit element at time t seconds. For $t < 0$, the voltage is a constant 80 volts; for $t \geq 0$, the voltage decays exponentially.

(a) Find a piecewise formula for $V(t)$.
(b) At what value of t will the voltage reach 0.1?

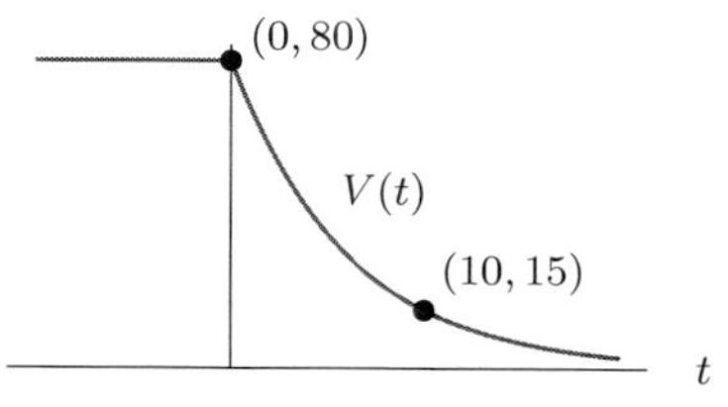

Figure 3.42

[19] www.accion.org.
[20] Based on D. N. Burghes, I. Huntley, and J. McDonald, *Applying Mathematics*. (Ellis Horwood, 1982).

61. Hong Kong shifted from British to Chinese rule in 1997. Figure 3.43 shows[21] the number of people who emigrated from Hong Kong during each of the years from 1980 to 1992.

(a) Find an exponential function that approximates the data.
(b) What does the model predict about the number of emigrants in 1997?
(c) Briefly explain why this model is or is not useful to predict emigration in the year 2010.

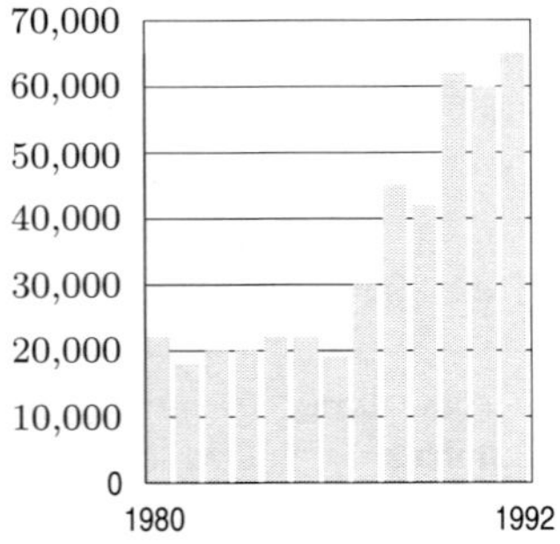

Figure 3.43

62. The annual inflation rate, r, for a five-year period is given in Table 3.18.

(a) By what total percent did prices rise between the start of 2000 and the end of 2004?
(b) What is the average annual inflation rate for this time period?
(c) At the beginning of 2000, a shower curtain costs \$20. Make a prediction for the good's cost at the beginning of 2010, using the average inflation rate found in part (b).

Table 3.18

t	2000	2001	2002	2003	2004
r	3.4%	2.8%	1.6%	2.3%	2.7%

63. Before the AIDS epidemic, Botswana[22] had a rapidly growing population as shown in Table 3.19. In 2005, the population started falling.

(a) Fit an exponential growth model, $P = ab^t$, to this data set, where P is the population in millions and t measures the years since 1975 in 5-year intervals—so $t = 1$ corresponds to 1980. Estimate a and b. Plot the data set and $P = ab^t$ on the same graph.
(b) Starting from 1975, how long does it take for the population of Botswana to double? When is the population of Botswana projected to exceed 214 million, the 1975 population of the US?

Table 3.19

Year	1975	1980	1985	1990
Population (millions)	0.755	0.901	1.078	1.285

64. It is a well-documented fact that the earning power of men is higher than that of women.[23] Table 3.20 gives the median income of year-round full-time workers in the US in dollars.

(a) Plot the data and connect the points.
(b) Let t be the year. Construct two functions of the form $W(t) = ae^{b(t-1950)}$, one each for the men's and women's earning power data.
(c) Graph the two functions from 1950 to 2000 and again from 2000 to 2080.
(d) Do the graphs in part (c) predict women's salaries will catch up with men's? If so, when?
(e) Comment on your predictions in part (d).

Table 3.20

Year	1950	1960	1970	1980	1990	2000
Female	953	1261	2237	4920	10,070	16,063
Male	2570	4080	6670	12,530	20,293	28,343

65. According to a letter to the *New York Times* on April 10, 1993, "... the probability of [a driver's] involvement in a single-car accident increases exponentially with increasing levels of blood alcohol." The letter goes on to state that when a driver's blood-alcohol content (BAC) is 0.15, the risk of such an accident is about 25 times greater than for a nondrinker.

(a) Let p_0 be a nondrinker's probability of being involved in a single-car accident. Let $f(x)$ be the probability of an accident for a driver whose blood alcohol level is x. Find a formula for $f(x)$. (This only makes sense for some values of x.)

[21] Adapted from the *New York Times* July 5, 1995.
[22] *World Population Growth and Aging* by N. Keyfitz, University of Chicago Press, 1990.
[23] *The World Almanac and Book of Facts 2006*, p. 84.

(b) At the time of the letter, the legal definition of intoxication was a BAC of 0.1 or higher. According to your formula for $f(x)$, how many times more likely to be involved in a single-car accident was a driver at the legal limit than a nondrinker?

(c) Suppose that new legislation is proposed to change the definition of legal intoxication. The new definition states that a person is legally intoxicated when their likelihood of involvement in a single-car accident is three times that of a non-drinker. To what BAC would the new definition of legal intoxication correspond?

CHECK YOUR UNDERSTANDING

Are the statements in Problems 1–32 true or false? Give an explanation for your answer.

1. Exponential functions are functions that increase or decrease at a constant percent rate.

2. The independent variable in an exponential function is always found in the exponent.

3. If $y = 40(1.05)^t$ then y is an exponential function of t.

4. The following table shows a function that could be exponential.

x	1	2	4	5	6
y	1	2	4	7	11

5. If your salary, S, grows by 4% each year, then $S = S_0(0.04)^t$ where t is in years.

6. If $f(t) = 4(2)^t$ then $f(2) = 64$.

7. If $f(t) = 3(\frac{2}{5})^t$ then f is a decreasing function.

8. If $Q = f(t) = 1000(0.5)^t$ then when $Q = 125, t = 3$.

9. If $Q = f(t) = ab^t$ then a is the initial value of Q.

10. If we are given two data points, we can find a linear function and an exponential function that go through these points.

11. A population that has 1000 members and decreases at 10% per year can be modeled as $P = 1000(0.10)^t$.

12. A positive increasing exponential function always becomes larger than any increasing linear function in the long run.

13. A possible formula for an exponential function that passes through the point $(0, 1)$ and the point $(2, 10)$ is $y = 4.5t + 1$.

14. If a population increases by 50% each year, then in two years it increases by 100%.

15. In the formula $Q = ab^t$, the value of a tells us where the graph crosses the Q-axis.

16. In the formula $Q = ab^t$, if $a > 1$, the graph always rises as we read from left to right.

17. The symbol e represents a constant whose value is approximately 2.71828.

18. If $f(x) \to k$ as $x \to \infty$ we say that the line $y = k$ is a horizontal asymptote.

19. Exponential graphs are always concave up.

20. If there are 110 grams of a substance initially and its decay rate is 3% per minute, then the amount after t minutes is $Q = 110(0.03)^t$ grams.

21. If a population had 200 members at time zero and was growing a 4% per year, then the population size after t years can be expressed as $P = 200(1.04)^t$.

22. If $P = 5e^{0.2t}$, we say the continuous growth rate of the function is 2%.

23. If $P = 4e^{-0.90t}$, we say the continuous growth rate of the function is 10%.

24. If $Q = 3e^{0.2t}$, then when $t = 5$, $Q = 3$.

25. If $Q = Q_0e^{kt}$, with Q_0 positive and k negative, then Q is decreasing.

26. If an investment earns 5% compounded monthly, its effective rate will be more than 5%.

27. If a \$500 investment earns 6% per year, compounded quarterly, we can find the balance after three years by evaluating the formula $B = 500(1 + \frac{6}{4})^{3 \cdot 4}$.

28. If interest on a \$2000 investment is compounded continuously at 3% per year, the balance after five years is found by evaluating the formula $B = 2000e^{(0.03)(5)}$.

29. Investing \$10,000 for 20 years at 5% earns more if interest is compounded quarterly than if it is compounded annually.

30. Investing \$$P$ for T years always earns more if interest is compounded continuously than if it is compounded annually.

31. There is no limit to the amount a twenty-year \$10,000 investment at 5% interest can earn if the number of times the interest is compounded becomes greater and greater.

32. If you put \$1000 into an account that earns 5.5% compounded continuously, then it takes about 18 years for the investment to grow to \$2000.

TOOLS FOR CHAPTER 3: EXPONENTS

We list the definition and properties that are used to manipulate exponents.

Definition of Zero, Negative, and Fractional Exponents
If m and n are positive integers:[24]
- $a^0 = 1$
- $a^{-n} = \dfrac{1}{a^n}$
- $a^{1/n} = \sqrt[n]{a}$, the nth root of a
- $a^{m/n} = \sqrt[n]{a^m} = (\sqrt[n]{a})^m$

Properties of Exponents
- $a^m \cdot a^n = a^{m+n}$ For example, $2^4 \cdot 2^3 = (2\cdot 2\cdot 2\cdot 2)\cdot(2\cdot 2\cdot 2) = 2^7$.
- $\dfrac{a^m}{a^n} = a^{m-n}$, $a \neq 0$ For example, $\dfrac{2^4}{2^3} = \dfrac{2\cdot 2\cdot 2\cdot 2}{2\cdot 2\cdot 2} = 2^1$.
- $(a^m)^n = a^{mn}$ For example, $(2^3)^2 = 2^3 \cdot 2^3 = 2^6$.
- $(ab)^n = a^n b^n$
- $\left(\dfrac{a}{b}\right)^n = \dfrac{a^n}{b^n}$, $b \neq 0$

Be aware of the following notational conventions:

$$ab^n = a(b^n), \qquad \text{but } ab^n \neq (ab)^n,$$
$$-b^n = -(b^n), \qquad \text{but } -b^n \neq (-b)^n,$$
$$-ab^n = (-a)(b^n).$$

For example, $-2^4 = -(2^4) = -16$, but $(-2)^4 = (-2)(-2)(-2)(-2) = +16$. Also, be sure to realize that for $n \neq 1$,

$$(a+b)^n \neq a^n + b^n \qquad \text{Power of a sum} \neq \text{Sum of powers.}$$

Example 1 Evaluate without a calculator:
(a) $(27)^{2/3}$ (b) $(4)^{-3/2}$ (c) $8^{1/3} - 1^{1/3}$

Solution (a) We have $(27)^{2/3} = \sqrt[3]{27^2} = \sqrt[3]{729} = 9$, or, equivalently, $(27)^{2/3} = \left(27^{1/3}\right)^2 = \left(\sqrt[3]{27}\right)^2 = 3^2 = 9$.
(b) We have $(4)^{-3/2} = (2)^{-3} = \dfrac{1}{2^3} = \dfrac{1}{8}$.
(c) We have $8^{1/3} - 1^{1/3} = 2 - 1 = 1$.

[24]We assume that the base is restricted to the values for which the power is defined.

Example 2 Use the rules of exponents to simplify the following:

(a) $\dfrac{100x^2y^4}{5x^3y^2}$ (b) $\dfrac{y^4(x^3y^{-2})^2}{2x^{-1}}$ (c) $\sqrt[3]{-8x^6}$ (d) $\left(\dfrac{M^{1/5}}{3N^{-1/2}}\right)^2$

Solution (a) We have

$$\frac{100x^2y^4}{5x^3y^2} = 20(x^{2-3})(y^{4-2}) = 20x^{-1}y^2 = \frac{20y^2}{x}.$$

(b) We have

$$\frac{y^4\left(x^3y^{-2}\right)^2}{2x^{-1}} = \frac{y^4x^6y^{-4}}{2x^{-1}} = \frac{y^{(4-4)}x^{(6-(-1))}}{2} = \frac{y^0x^7}{2} = \frac{x^7}{2}.$$

(c) We have

$$\sqrt[3]{-8x^6} = \sqrt[3]{-8}\cdot\sqrt[3]{x^6} = -2x^2.$$

(d) We have

$$\left(\frac{M^{1/5}}{3N^{-1/2}}\right)^2 = \frac{\left(M^{1/5}\right)^2}{\left(3N^{-1/2}\right)^2} = \frac{M^{2/5}}{3^2N^{-1}} = \frac{M^{2/5}N}{9}.$$

Example 3 Solve for x:

(a) $\dfrac{10x^7}{4x^2} = 37$ (b) $\dfrac{x^2}{3x^5} = 10$ (c) $\sqrt{9x^5} = 10$

Solution (a) We have

$$\begin{aligned}\frac{10x^7}{4x^2} &= 37\\ 2.5x^5 &= 37\\ x^5 &= 14.8\\ x &= (14.8)^{1/5} = 1.714.\end{aligned}$$

(b) We have

$$\begin{aligned}\frac{x^2}{3x^5} &= 10\\ \frac{1}{3}x^{-3} &= 10\\ \frac{1}{x^3} &= 30\\ x^3 &= \frac{1}{30}\\ x &= \left(\frac{1}{30}\right)^{1/3} = 0.322.\end{aligned}$$

(c) We have

$$\begin{aligned}\sqrt{9x^5} &= 10\\ 3x^{5/2} &= 10\\ x^{5/2} &= \frac{10}{3}\\ x &= \left(\frac{10}{3}\right)^{2/5} = 1.619.\end{aligned}$$

Exercises to Tools for Chapter 3

For Exercises 1–43, evaluate without a calculator.

1. 4^3 **2.** $(-5)^2$ **3.** 11^2

4. 10^4 **5.** $(-1)^{12}$ **6.** $(-1)^{13}$

7. $\frac{5^3}{5^2}$ **8.** $\frac{5^3}{5}$ **9.** $\frac{10^8}{10^5}$

10. $\frac{6^4}{6^4}$ **11.** 8^0 **12.** $\sqrt{4}$

13. $\sqrt{4^2}$ **14.** $\sqrt{4^3}$ **15.** $\sqrt{4^4}$

16. $\sqrt{(-4)^2}$ **17.** $\frac{1}{7^{-2}}$ **18.** $\frac{2^7}{2^3}$

19. $(-1)^{445}$ **20.** -11^2 **21.** $(-2)3^2$

22. $\left(5^0\right)^3$ **23.** $2.1\left(10^3\right)$ **24.** $32^{1/5}$

25. $16^{1/2}$ **26.** $16^{1/4}$ **27.** $16^{3/4}$

28. $16^{5/4}$ **29.** $16^{5/2}$ **30.** $100^{5/2}$

31. $\sqrt[3]{-125}$ **32.** $\sqrt{(-4)^2}$ **33.** $(-1)^3\sqrt{36}$

34. $(0.04)^{1/2}$ **35.** $(-8)^{2/3}$ **36.** 3^{-1}

37. 3^{-2} **38.** $3^{-3/2}$ **39.** 25^{-1}

40. 25^{-2} **41.** $25^{-3/2}$ **42.** $(1/27)^{-1/3}$

43. $(0.125)^{1/3}$

Simplify the expressions in Exercises 44–77 and leave without radicals if possible. Assume all variables are positive.

44. $\sqrt{x^4}$ **45.** $\sqrt{y^8}$

46. $\sqrt{w^8z^4}$ **47.** $\sqrt{x^5y^4}$

48. $\sqrt{16x^3}$ **49.** $\sqrt{49w^9}$

50. $\sqrt{25x^3z^4}$ **51.** $\sqrt{r^2}$

52. $\sqrt{r^3}$ **53.** $\sqrt{r^4}$

54. $\sqrt{36t^2}$ **55.** $\sqrt{64s^7}$

56. $\sqrt{50x^4y^6}$ **57.** $\sqrt{48u^{10}v^{12}y^5}$

58. $\sqrt{8m}\sqrt{2m^3}$ **59.** $\sqrt{6s^2t^3v^5}\sqrt{6st^5v^3}$

60. $(0.1)^2\left(4xy^2\right)^2$ **61.** $3\left(3^{x/2}\right)^2$

62. $\left(4L^{2/3}P\right)^{3/2}(P)^{-3/2}$ **63.** $7\left(5w^{1/2}\right)\left(2w^{1/3}\right)$

64. $\left(S\sqrt{16xt^2}\right)^2$ **65.** $\sqrt{e^{2x}}$

66. $(3AB)^{-1}\left(A^2B^{-1}\right)^2$ **67.** $e^{kt}\cdot e^3\cdot e$

68. $\sqrt{M+2}(2+M)^{3/2}$ **69.** $\left(3x\sqrt{x^3}\right)^2$

70. $x^e\left(x^e\right)^2$ **71.** $\left(y^{-2}e^y\right)^2$

72. $\frac{4x^{(3\pi+1)}}{x^2}$ **73.** $\frac{4A^{-3}}{(2A)^{-4}}$

74. $\frac{a^{n+1}3^{n+1}}{a^n3^n}$ **75.** $\frac{12u^3}{3\left(uv^2w^4\right)^{-1}}$

76. $\left(a^{-1}+b^{-1}\right)^{-1}$

77. $\left(\dfrac{35(2b+1)^9}{7(2b+1)^{-1}}\right)^2$ (Do not expand $(2b+1)^9$.)

If possible, evaluate the quantities in Exercises 78–86. Check your answers with a calculator.

78. $(-32)^{3/5}$ **79.** $-32^{3/5}$ **80.** $-625^{3/4}$

81. $(-625)^{3/4}$ **82.** $(-1728)^{4/3}$ **83.** $64^{-3/2}$

84. $-64^{3/2}$ **85.** $(-64)^{3/2}$ **86.** $81^{5/4}$

In Exercises 87–92, solve for x.

87. $\dfrac{10x^5}{x^2} = 2$ **88.** $\dfrac{5x^3}{x^5} = 125$

89. $\sqrt{4x^3} = 5$ **90.** $7x^4 = 20x^2$

91. $5x^{-2} = 500$ **92.** $2(x+2)^3 = 100$

In Exercises 93–94, use algebra to find the point of intersection.

93.

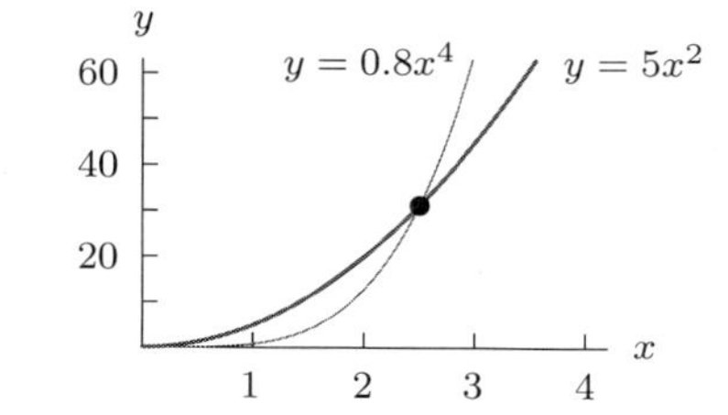

94.

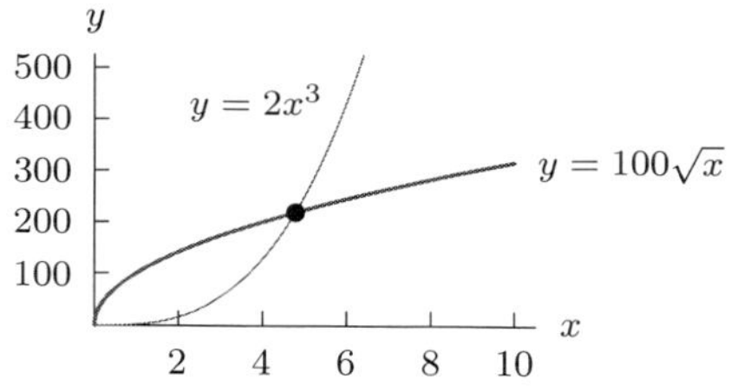

Are the statements in Exercises 95–104 true or false.?

95. $t^3t^4 = t^{12}$ **96.** $x^2y^5 = (xy)^{10}$

97. $(p^3)^8 = p^{11}$ **98.** $5u^2 + 5u^3 = 10u^5$

99. $(3r)^2 9s^2 = 81r^2s^2$ **100.** $\sqrt[3]{-64b^3c^6} = -4bc^2$

101. $\dfrac{m^8}{2m^2} = \dfrac{1}{2}m^4$ **102.** $5z^{-4} = \dfrac{1}{5z^4}$

103. $-4w^2 - 3w^3 = -w^2(4+3w)$

104. $(u+v)^{-1} = \dfrac{1}{u} + \dfrac{1}{v}$

Solve the equations in Problems 105–106 in terms of r and s given that

$$2^r = 5 \quad \text{and} \quad 2^s = 7.$$

105. $2^x = 35.$ **106.** $2^x = 140.$

Let $2^a = 5$ and $2^b = 7$. Using exponent rules, solve the equations in Problems 107–112 in terms of a and b.

107. $5^x = 32$ **108.** $7^x = \dfrac{1}{8}$

109. $25^x = 64$ **110.** $14^x = 16$

111. $5^x = 7$ **112.** $0.4^x = 49$

Chapter Four

LOGARITHMIC FUNCTIONS

In this chapter, we introduce the inverse of exponential functions, logarithmic functions. We study the properties of logarithms and use logarithms to solve exponential equations; logarithms are used to calculate doubling time and half-life. We also look at log scales, and we see how logs are used to linearize data.

The Tools Section on page 189 reviews the properties of logarithms.

4.1 LOGARITHMS AND THEIR PROPERTIES

What is a Logarithm?

Suppose that a population grows according to the formula $P = 10^t$, where P is the colony size at time t, in hours. When will the population be 2500? We want to solve the following equation for t:

$$10^t = 2500.$$

In Section 3.2, we used a graphical method to approximate t. This time, we introduce a function which returns precisely the exponent of 10 we need.

Since $10^3 = 1000$ and $10^4 = 10{,}000$, and $1000 < 2500 < 10{,}000$, the exponent we are looking for is between 3 and 4. But how do we find the exponent exactly?

To answer this question, we define the *common logarithm function*, or simply the *log function*, written $\log_{10} x$, or $\log x$, as follows.

> If x is a positive number,
>
> $$\log x \text{ is the exponent of 10 that gives } x.$$
>
> In other words, if
>
> $$y = \log x \quad \text{then} \quad 10^y = x.$$

For example, $\log 100 = 2$, because 2 is the exponent of 10 that gives 100, or $10^2 = 100$.

To solve the equation $10^t = 2500$, we must find the power of 10 that gives 2500. Using the log button on a calculator, we can approximate this exponent. We find

$$\log 2500 \approx 3.398, \quad \text{which means that} \quad 10^{3.398} \approx 2500.$$

As predicted, this exponent is between 3 and 4. The precise exponent is $\log 2500$; the approximate value is 3.398. Thus, it takes roughly 3.4 hours for the population to reach 2500.

Example 1 Rewrite the following statements using exponents instead of logs.

(a) $\log 100 = 2$ (b) $\log 0.01 = -2$ (c) $\log 30 = 1.477$

Solution For each statement, we use the fact that if $y = \log x$ then $10^y = x$.

(a) $2 = \log 100$ means that $10^2 = 100$.

(b) $-2 = \log 0.01$ means that $10^{-2} = 0.01$.

(c) $1.477 = \log 30$ means that $10^{1.477} = 30$. (Actually, this is only an approximation. Using a calculator, we see that $10^{1.477} = 29.9916\ldots$ and that $\log 30 = 1.47712125\ldots$.)

Example 2 Rewrite the following statements using logs instead of exponents.

(a) $10^5 = 100{,}000$ (b) $10^{-4} = 0.0001$ (c) $10^{0.8} = 6.3096$.

Solution For each statement, we use the fact that if $10^y = x$, then $y = \log x$.

(a) $10^5 = 100{,}000$ means that $\log 100{,}000 = 5$.
(b) $10^{-4} = 0.0001$ means that $\log 0.0001 = -4$.
(c) $10^{0.8} = 6.3096$ means that $\log 6.3096 = 0.8$. (This, too, is only an approximation because $10^{0.8}$ actually equals $6.30957344\ldots$.)

Logarithms Are Exponents

Note that logarithms are just exponents! Thinking in terms of exponents is often a good way to answer a logarithm problem.

Example 3 Without a calculator, evaluate the following, if possible:

(a) $\log 1$ (b) $\log 10$ (c) $\log 1{,}000{,}000$
(d) $\log 0.001$ (e) $\log \dfrac{1}{\sqrt{10}}$ (f) $\log(-100)$

Solution

(a) We have $\log 1 = 0$, since $10^0 = 1$.
(b) We have $\log 10 = 1$, since $10^1 = 10$.
(c) Since $1{,}000{,}000 = 10^6$, the exponent of 10 that gives 1,000,000 is 6. Thus, $\log 1{,}000{,}000 = 6$.
(d) Since $0.001 = 10^{-3}$, the exponent of 10 that gives 0.001 is -3. Thus, $\log 0.001 = -3$.
(e) Since $1/\sqrt{10} = 10^{-1/2}$, the exponent of 10 that gives $1/\sqrt{10}$ is $-\frac{1}{2}$. Thus $\log(1/\sqrt{10}) = -\frac{1}{2}$.
(f) Since 10 to any power is positive, -100 cannot be written as a power of 10. Thus, $\log(-100)$ is undefined.

Logarithmic and Exponential Functions are Inverses

The operation of taking a logarithm "undoes" the exponential function; the logarithm and the exponential are inverse functions. For example, $\log(10^6) = 6$ and $10^{\log 6} = 6$. In particular

> For any N,
> $$\log(10^N) = N$$
> and for $N > 0$,
> $$10^{\log N} = N.$$

Example 4 Evaluate without a calculator: (a) $\log\left(10^{8.5}\right)$ (b) $10^{\log 2.7}$ (c) $10^{\log(x+3)}$

Solution Using $\log(10^N) = N$ and $10^{\log N} = N$, we have:

(a) $\log\left(10^{8.5}\right) = 8.5$ (b) $10^{\log 2.7} = 2.7$ (c) $10^{\log(x+3)} = x + 3$

You can check the first two results on a calculator.

Properties of Logarithms

In Chapter 3, we saw how to solve exponential equations such as $100 \cdot 2^t = 337{,}000{,}000$, graphically. To use logarithms to solve these equations, we use the properties of logarithms, which are justified on page 156.

Properties of the Common Logarithm

- By definition, $y = \log x$ means $10^y = x$.
- In particular,
$$\log 1 = 0 \quad \text{and} \quad \log 10 = 1.$$
- The functions 10^x and $\log x$ are inverses, so they "undo" each other:
$$\log(10^x) = x \qquad \text{for all } x,$$
$$10^{\log x} = x \qquad \text{for } x > 0.$$
- For a and b both positive and any value of t,
$$\log(ab) = \log a + \log b$$
$$\log\left(\frac{a}{b}\right) = \log a - \log b$$
$$\log(b^t) = t \cdot \log b.$$

We can now use logarithms to solve the equation that we solved graphically in Section 3.2.

Example 5 Solve $100 \cdot 2^t = 337{,}000{,}000$ for t.

Solution Dividing both sides of the equation by 100 gives
$$2^t = 3{,}370{,}000.$$
Taking logs of both sides gives
$$\log\left(2^t\right) = \log(3{,}370{,}000).$$
Since $\log(2^t) = t \cdot \log 2$, we have
$$t \log 2 = \log(3{,}370{,}000),$$
so, solving for t, we have
$$t = \frac{\log(3{,}370{,}000)}{\log 2} = 21.684.$$
In Example 2 on page 125, we found the graphical approximation of between 21 and 22 days as the time for the Yonkers fine to exceed the city's annual budget.

The Natural Logarithm

When e is used as the base for exponential functions, computations are easier with the use of another logarithm function, called log base e. The log base e is used so frequently that it has its own notation: $\ln x$, read as the *natural log of* x. We make the following definition:

For $x > 0$,

$$\ln x \text{ is the power of } e \text{ that gives } x$$

or, in symbols,

$$\ln x = y \quad \text{means} \quad e^y = x,$$

and y is called the **natural logarithm** of x.

Just as the functions 10^x and $\log x$ are inverses, so are e^x and $\ln x$. The function $\ln x$ has similar properties to the common log function:

Properties of the Natural Logarithm

- By definition, $y = \ln x$ means $x = e^y$.
- In particular,

$$\ln 1 = 0 \quad \text{and} \quad \ln e = 1.$$

- The functions e^x and $\ln x$ are inverses, so they "undo" each other:

$$\begin{aligned} \ln(e^x) &= x \qquad \text{for all } x \\ e^{\ln x} &= x \qquad \text{for } x > 0. \end{aligned}$$

- For a and b both positive and any value of t,

$$\begin{aligned} \ln(ab) &= \ln a + \ln b \\ \ln\left(\frac{a}{b}\right) &= \ln a - \ln b \\ \ln(b^t) &= t \cdot \ln b. \end{aligned}$$

Example 6 Solve for x:

(a) $5e^{2x} = 50$ (b) $3^x = 100$.

Solution (a) We first divide both sides by 5 to obtain

$$e^{2x} = 10.$$

Taking the natural log of both sides, we have

$$\begin{aligned} \ln(e^{2x}) &= \ln 10 \\ 2x &= \ln 10 \\ x &= \frac{\ln 10}{2} \approx 1.151. \end{aligned}$$

(b) Taking natural logs of both sides,

$$\begin{aligned} \ln(3^x) &= \ln 100 \\ x \ln 3 &= \ln 100 \\ x &= \frac{\ln 100}{\ln 3} \approx 4.192. \end{aligned}$$

Misconceptions and Calculator Errors Involving Logs

It is important to know how to use the properties of logarithms. It is equally important to recognize statements that are *not* true. Beware of the following:

- $\log(a+b)$ is not the same as $\log a + \log b$
- $\log(a-b)$ is not the same as $\log a - \log b$
- $\log(ab)$ is not the same as $(\log a)(\log b)$
- $\log\left(\frac{a}{b}\right)$ is not the same as $\frac{\log a}{\log b}$
- $\log\left(\frac{1}{a}\right)$ is not the same as $\frac{1}{\log a}$.

There are no formulas to simplify either $\log(a+b)$ or $\log(a-b)$. Also the expression $\log 5x^2$ is not the same as $2 \cdot \log 5x$, because the exponent, 2, applies only to the x and not to the 5. However, it is correct to write

$$\log 5x^2 = \log 5 + \log x^2 = \log 5 + 2\log x.$$

Using a calculator to evaluate expressions like $\log(\frac{17}{3})$ requires care. On some calculators, entering $\log 17/3$ gives 0.410, which is incorrect. This is because the calculator assumes that you mean $(\log 17)/3$, which is not the same as $\log(17/3)$. Notice also that

$$\frac{\log 17}{\log 3} \approx \frac{1.230}{0.477} \approx 2.579,$$

which is not the same as either $(\log 17)/3$ or $\log(17/3)$. Thus, the following expressions are all different.

$$\log\frac{17}{3} \approx 0.753, \qquad \frac{\log 17}{3} \approx 0.410, \qquad \text{and} \qquad \frac{\log 17}{\log 3} \approx 2.579.$$

Justification of $\log(a \cdot b) = \log a + \log b$ and $\log(a/b) = \log a - \log b$

If a and b are both positive, we can write $a = 10^m$ and $b = 10^n$, so $\log a = m$ and $\log b = n$. Then, the product $a \cdot b$ can be written

$$a \cdot b = 10^m \cdot 10^n = 10^{m+n}.$$

Therefore $m + n$ is the power of 10 needed to give $a \cdot b$, so

$$\log(a \cdot b) = m + n,$$

which gives

$$\boxed{\log(a \cdot b) = \log a + \log b.}$$

Similarly, the quotient a/b can be written as

$$\frac{a}{b} = \frac{10^m}{10^n} = 10^{m-n}.$$

Therefore $m - n$ is the power of 10 needed to give a/b, so

$$\log\left(\frac{a}{b}\right) = m - n,$$

and thus

$$\boxed{\log\left(\frac{a}{b}\right) = \log a - \log b.}$$

Justification of $\log(b^t) = t \cdot \log b$

Suppose that b is positive, so we can write $b = 10^k$ for some value of k. Then

$$b^t = (10^k)^t.$$

We have rewritten the expression b^t so that the base is a power of 10. Using a property of exponents, we can write $(10^k)^t$ as 10^{kt}, so

$$b^t = (10^k)^t = 10^{kt}.$$

Therefore kt is the power of 10 which gives b^t, so

$$\log(b^t) = kt.$$

But since $b = 10^k$, we know $k = \log b$. This means

$$\log(b^t) = (\log b)t = t \cdot \log b.$$

Thus, for $b > 0$ we have

$$\boxed{\log\left(b^t\right) = t \cdot \log b.}$$

Exercises and Problems for Section 4.1

Exercises

Rewrite the statements in Exercises 1–6 using exponents instead of logs.

1. $\log 19 = 1.279$

2. $\log 4 = 0.602$

3. $\ln 26 = 3.258$

4. $\ln(0.646) = -0.437$

5. $\log P = t$

6. $\ln q = z$

Rewrite the statements in Exercises 7–10 using logs.

7. $10^8 = 100{,}000{,}000$

8. $e^{-4} = 0.0183$

9. $10^v = \alpha$

10. $e^a = b$

Solve the equations in Exercises 11–18 using logs.

11. $2^x = 11$

12. $(1.45)^x = 25$

13. $e^{0.12x} = 100$

14. $10 = 22(0.87)^q$

15. $48 = 17(2.3)^w$

16. $2/7 = (0.6)^{2t}$

17. $0.00012 = 0.001^{m/2}$

18. $500 = 25(1.1)^{3x}$

19. Evaluate without a calculator.

(a) $\log 1$ **(b)** $\log 0.1$ **(c)** $\log(10^0)$
(d) $\log\sqrt{10}$ **(e)** $\log(10^5)$ **(f)** $\log(10^2)$
(g) $\log\left(\frac{1}{\sqrt{10}}\right)$ **(h)** $10^{\log 100}$ **(i)** $10^{\log 1}$
(j) $10^{\log(0.01)}$

20. Evaluate without a calculator.

(a) $\ln 1$ **(b)** $\ln e^0$ **(c)** $\ln e^5$
(d) $\ln\sqrt{e}$ **(e)** $e^{\ln 2}$ **(f)** $\ln\left(\frac{1}{\sqrt{e}}\right)$

Problems

21. Evaluate 10^n for $n = 3, n = 3.5, n = 3.48, n = 3.477$, and $n = 3.47712$. Based on your answer, estimate the value of $\log 3000$.

22. Given that $10^{1.3} \approx 20$, approximate the value of $\log 200$ without using a calculator.

23. Evaluate without a calculator.

(a) $\log(\log 10)$ **(b)** $\sqrt{\log 100} - \log\sqrt{100}$
(c) $\log(\sqrt{10}\sqrt[3]{10}\sqrt[5]{10})$ **(d)** $1000^{\log 3}$
(e) $0.01^{\log 2}$ **(f)** $\dfrac{1}{\log(1/\log\sqrt[10]{10})}$

24. Express the following in terms of x without logs.

(a) $\log 100^x$ **(b)** $1000^{\log x}$ **(c)** $\log 0.001^x$

25. Express the following in terms of x without natural logs.

(a) $\ln e^{2x}$ **(b)** $e^{\ln(3x+2)}$
(c) $\ln\left(\dfrac{1}{e^{5x}}\right)$ **(d)** $\ln\sqrt{e^x}$

26. True or false?

(a) $\log AB = \log A + \log B$
(b) $\dfrac{\log A}{\log B} = \log A - B$
(c) $\log A \log B = \log A + \log B$
(d) $p \cdot \log A = \log A^p$
(e) $\log\sqrt{x} = \frac{1}{2}\log x$
(f) $\sqrt{\log x} = \log(x^{1/2})$

27. True or false?

(a) $\ln(ab^t) = t\ln(ab)$ **(b)** $\ln(1/a) = -\ln a$
(c) $\ln a \cdot \ln b = \ln(a+b)$ **(d)** $\ln a - \ln b = \dfrac{\ln a}{\ln b}$

28. Suppose that $x = \log A$ and that $y = \log B$. Write the following expressions in terms of x and y.

(a) $\log(AB)$ **(b)** $\log(A^3 \cdot \sqrt{B})$
(c) $\log(A - B)$ **(d)** $\dfrac{\log A}{\log B}$
(e) $\log\dfrac{A}{B}$ **(f)** AB

29. Let $p = \log m$ and $q = \log n$. Write the following expressions in terms of p and/or q without using logs.

(a) m **(b)** n^3
(c) $\log(mn^3)$ **(d)** $\log\sqrt{m}$

30. Let $p = \ln m$ and $q = \ln n$. Write the following expressions in terms of p and/or q without using logs.

(a) $\ln(nm^4)$ **(b)** $\ln\left(\dfrac{1}{n}\right)$
(c) $\dfrac{\ln m}{\ln n}$ **(d)** $\ln(n^3)$

31. A graph of $P = 25(1.075)^t$ is given in Figure 4.1.

(a) What is the initial value of P (when $t = 0$)? What is the percent growth rate?
(b) Use the graph to estimate the value of t when $P = 100$.
(c) Use logs to find the exact value of t when $P = 100$.

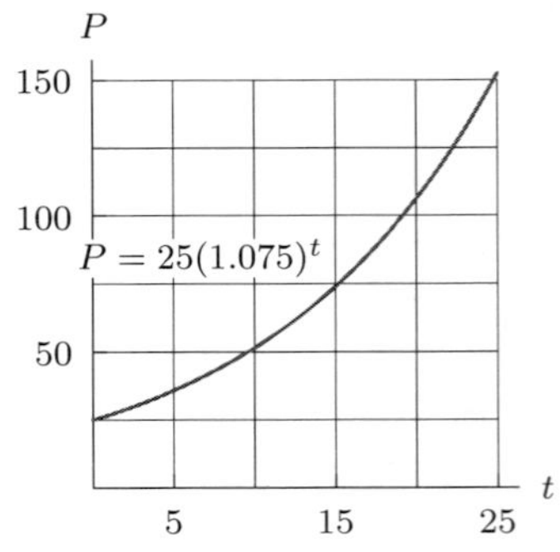

Figure 4.1

32. A graph of $Q = 10e^{-0.15t}$ is given in Figure 4.2.

(a) What is the initial value of Q (when $t = 0$)? What is the continuous percent decay rate?
(b) Use the graph to estimate the value of t when $Q = 2$.
(c) Use logs to find the exact value of t when $Q = 2$.

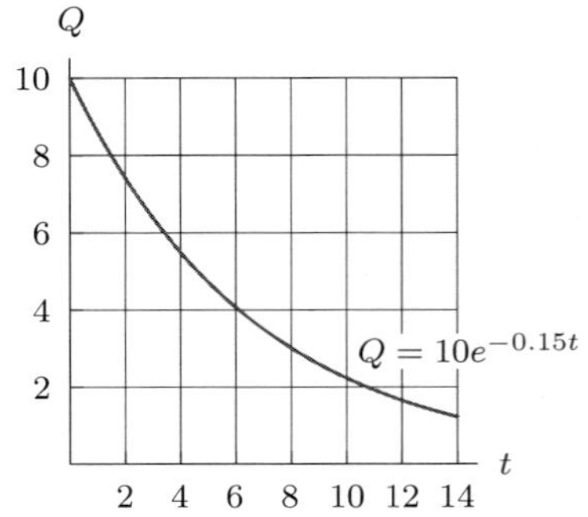

Figure 4.2

33. Find a possible formula for the exponential function S in Figure 4.3, if $R(x) = 5.1403(1.1169)^x$.

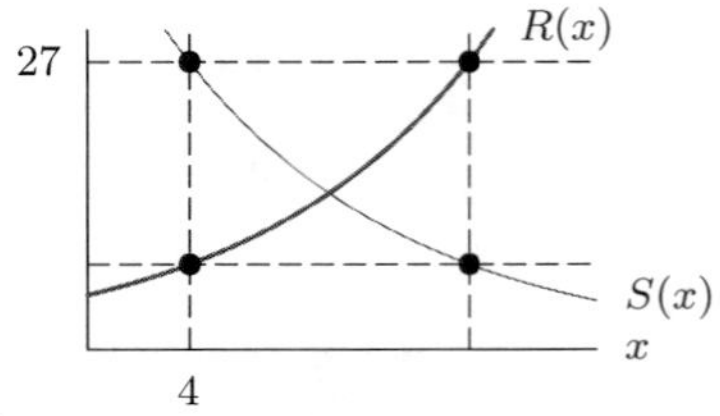

Figure 4.3

In Problems 34–50, solve the equations exactly for x or t.

34. $5(1.031)^x = 8$

35. $4(1.171)^x = 7(1.088)^x$

36. $3\log(2x+6) = 6$

37. $3 \cdot 2^x + 8 = 25$

38. $e^{x+4} = 10$

39. $e^{x+5} = 7 \cdot 2^x$

40. $b^x = c$

41. $ab^x = c$

42. $Pa^x = Qb^x$

43. $Pe^{kx} = Q$

44. $121e^{-0.112t} = 88$

45. $58e^{4t+1} = 30$

46. $17e^{0.02t} = 18e^{0.03t}$

47. $44e^{0.15t} = 50(1.2)^t$

48. $\log(1-x) - \log(1+x) = 2$

49. $\log(2x+5) \cdot \log(9x^2) = 0$

50. $\sqrt{\log \sqrt[3]{x} - \log \sqrt[4]{x}} = \dfrac{1}{2}$

51. Solve each of the following equations exactly for x.

(a) $e^{2x} + e^{2x} = 1$ **(b)** $2e^{3x} + e^{3x} = b$

52. Suppose $0 < A < AB < 1 < B$. Rank the following in order from least to greatest:

$$0, \log A, \log B, \log A + \log B, \log(B^A).$$

53. Suppose $\log A < 0 < \log AB < 1 < \log B$. Rank the following in order from least to greatest:

$$0,\ 1,\ 100,\ A,\ A^2B^2,\ B^2.$$

54. Three students try to solve the equation

$$11 \cdot 3^x = 5 \cdot 7^x.$$

The first student finds that $x = \dfrac{\log(11/5)}{\log(7/3)}$. The second finds that $x = \dfrac{\log(5/11)}{\log(3/7)}$. The third finds that $x = \dfrac{\log 11 - \log 5}{\log 7 - \log 3}$. Which student (or students) is (are) correct? Explain.

4.2 LOGARITHMS AND EXPONENTIAL MODELS

The log function is often useful when answering questions about exponential models. Because logarithms "undo" the exponential functions, we use them to solve many exponential equations.

Example 1 In Example 3 on page 125 we solved the equation $200(0.886^t) = 25$ graphically, where t is in thousands of years. We found that a 200 microgram sample of carbon-14 decays to 25 micrograms in approximately 17,200 years. Now solve $200(0.886)^t = 25$ using logarithms.

Solution First, isolate the power on one side of the equation

$$200(0.886^t) = 25$$
$$0.886^t = 0.125.$$

Take the log of both sides, and use the fact that $\log(0.886^t) = t\log 0.886$. Then

$$\log(0.886^t) = \log 0.125$$
$$t\log 0.886 = \log 0.125$$

so

$$t = \frac{\log 0.125}{\log 0.886} \approx 17.18 \text{ thousand years.}$$

This answer is close to the value we found from the graph, 17,200.

Example 2 The US population, P, in millions, is currently growing according to the formula[1]

$$P = 299e^{0.009t},$$

where t is in years since 2006. When is the population predicted to reach 350 million?

[1] Based on data from www.census.gov and www.cia.gov/cia/publications/factbook, accessed July 31, 2006.

Solution We want to solve the following equation for t:

$$299e^{0.009t} = 350.$$

Dividing by 299 gives

$$e^{0.009t} = \frac{350}{299}$$

So $0.009t$ is the power of e which gives $350/299$. Thus, by the definition of the natural log,

$$0.009t = \ln\left(\frac{350}{299}\right).$$

Solving for t and evaluating $\ln(350/299)$ on a calculator gives

$$t = \frac{\ln(350/299)}{0.009} = 17.5 \text{ years.}$$

The US population is predicted to reach 350 million during the year 2024.

Example 3 The population of City A begins with 50,000 people and grows at 3.5% per year. The population of City B begins with a larger population of 250,000 people but grows at the slower rate of 1.6% per year. Assuming that these growth rates hold constant, will the population of City A ever catch up to the population of City B? If so, when?

Solution If t is time measured in years and P_A and P_B are the populations of these two cities, then

$$P_A = 50{,}000(1.035)^t \quad \text{and} \quad P_B = 250{,}000(1.016)^t.$$

We want to solve the equation

$$50{,}000(1.035)^t = 250{,}000(1.016)^t.$$

We first get the exponential terms together by dividing both sides of the equation by $50{,}000(1.016)^t$:

$$\frac{(1.035)^t}{(1.016)^t} = \frac{250{,}000}{50{,}000} = 5.$$

Since $\frac{a^t}{b^t} = \left(\frac{a}{b}\right)^t$, this gives

$$\left(\frac{1.035}{1.016}\right)^t = 5.$$

Taking logs of both sides and using $\log b^t = t \log b$, we have

$$\log\left(\frac{1.035}{1.016}\right)^t = \log 5$$

$$t \log\left(\frac{1.035}{1.016}\right) = \log 5$$

$$t = \frac{\log 5}{\log(1.035/1.016)} \approx 86.865.$$

Thus, the cities' populations will be equal in just under 87 years. To check this, notice that when $t = 86.865$,

$$P_A = 50{,}000(1.035)^{86.865} = 992{,}575$$

and

$$P_B = 250{,}000(1.016)^{86.865} = 992{,}572.$$

The answers are not exactly equal because we rounded off the value of t. Rounding can introduce significant errors, especially when logs and exponentials are involved. Using $t = 86.86480867$, the computed values of P_A and P_B agree to three decimal places.

Doubling Time

Eventually, any exponentially growing quantity doubles, or increases by 100%. Since its percent growth rate is constant, the time it takes for the quantity to grow by 100% is also a constant. This time period is called the *doubling time*.

Example 4 (a) Find the time needed for the turtle population described by the function $P = 175(1.145)^t$ to double its initial size.

(b) How long does this population take to quadruple its initial size? To increase by a factor of 8?

Solution (a) The initial size is 175 turtles; doubling this gives 350 turtles. We need to solve the following equation for t:

$$\begin{aligned} 175(1.145)^t &= 350 \\ 1.145^t &= 2 \\ \log\left(1.145^t\right) &= \log 2 \\ t \cdot \log 1.145 &= \log 2 \\ t = \frac{\log 2}{\log 1.145} &\approx 5.119 \text{ years.} \end{aligned}$$

We check this by noting that

$$175(1.145)^{5.119} = 350,$$

which is double the initial population. In fact, at any time it takes the turtle population about 5.119 years to double in size.

(b) Since the population function is exponential, it increases by 100% every 5.119 years. Thus it doubles its initial size in the first 5.119 years, quadruples its initial size in two 5.119 year periods, or 10.238 years, and increases by a factor of 8 in three 5.119 year periods, or 15.357 years. We check this by noting that

$$175(1.145)^{10.238} = 700,$$

or 4 times the initial size, and that

$$175(1.145)^{15.357} = 1400,$$

or 8 times the initial size.

Example 5 A population doubles in size every 20 years. What is its continuous growth rate?

Solution We are not given the initial size of the population, but we can solve this problem without that information. Let the symbol P_0 represent the initial size of the population.We have $P = P_0e^{kt}$. After 20 years, $P = 2P_0$, and so

$$\begin{aligned} P_0e^{k\cdot 20} &= 2P_0 \\ e^{20k} &= 2 \\ 20k &= \ln 2 \qquad \text{Taking ln of both sides} \\ k = \frac{\ln 2}{20} &\approx 0.03466. \end{aligned}$$

Thus, the population grows at the continuous rate of 3.466% per year.

Half-Life

Just as an exponentially growing quantity doubles in a fixed amount of time, an exponentially decaying quantity decreases by a factor of 2 in a fixed amount of time, called the *half-life* of the quantity.

Example 6 Carbon-14 decays radioactively at a constant annual rate of 0.0121%. Show that the half-life of carbon-14 is about 5728 years.

Solution We are not given an initial amount of carbon-14, but we can solve this problem without that information. Let the symbol Q_0 represent the initial quantity of carbon-14 present. The growth rate is -0.000121 because carbon-14 is decaying. So the growth factor is $b = 1 - 0.000121 = 0.999879$. Thus, after t years the amount left will be

$$Q = Q_0(0.999879)^t.$$

We want to find how long it takes for the quantity to drop to half its initial level. Thus, we need to solve for t in the equation

$$\frac{1}{2}Q_0 = Q_0(0.999879)^t.$$

Dividing each side by Q_0, we have

$$\frac{1}{2} = 0.999879^t.$$

Taking logs

$$\begin{aligned} \log\frac{1}{2} &= \log\left(0.999879^t\right) \\ \log 0.5 &= t \cdot \log 0.999879 \\ t &= \frac{\log 0.5}{\log 0.999879} \approx 5728.143. \end{aligned}$$

Thus, no matter how much carbon-14 there is initially, after about 5728 years, half will remain.

Similarly, we can determine the growth rate given the half-life or doubling time.

Example 7 The quantity, Q, of a substance decays according to the formula $Q = Q_0e^{-kt}$, where t is in minutes. The half-life of the substance is 11 minutes. What is the value of k?

Solution We know that after 11 minutes, $Q = \frac{1}{2}Q_0$. Thus, solving for k, we get

$$\begin{aligned} Q_0e^{-k\cdot 11} &= \frac{1}{2}Q_0 \\ e^{-11k} &= \frac{1}{2} \\ -11k &= \ln\frac{1}{2} \\ k &= \frac{\ln(1/2)}{-11} \approx 0.06301, \end{aligned}$$

so $k = 0.063$ per minute. This substance decays at the continuous rate of 6.301% per minute.

Converting Between $Q = ab^t$ and $Q = ae^{kt}$

Any exponential function can be written in either of the two forms:

$$Q = ab^t \quad \text{or} \quad Q = ae^{kt}.$$

If $b = e^k$, so $k = \ln b$, the two formulas represent the same function.

Example 8 Convert the exponential function $P = 175(1.145)^t$ to the form $P = ae^{kt}$.

Solution Since the new formula represents the same function, we want $P = 175$ when $t = 0$. Thus, substituting $t = 0$, gives $175 = ae^{k(0)} = a$, so $a = 175$. The parameter a in both functions represents the initial population. For all t,

$$175(1.145)^t = 175(e^k)^t,$$

so we must find k such that

$$e^k = 1.145.$$

Therefore k is the power of e which gives 1.145. By the definition of ln, we have

$$k = \ln 1.145 \approx 0.1354.$$

Therefore,

$$P = 175e^{0.1354t}.$$

Example 9 Convert the formula $Q = 7e^{0.3t}$ to the form $Q = ab^t$.

Solution Using the properties of exponents,

$$Q = 7e^{0.3t} = 7(e^{0.3})^t.$$

Using a calculator, we find $e^{0.3} \approx 1.3499$, so

$$Q = 7(1.3499)^t.$$

Example 10 Assuming t is in years, find the continuous and annual percent growth rates in Examples 8 and 9.

Solution In Example 8, the annual percent growth rate is 14.5% and the continuous percent growth rate per year is 13.54%. In Example 9, the continuous percent growth rate is 30% and the annual percent growth rate is 34.985%.

Example 11 Find the continuous percent growth rate of $Q = 200(0.886)^t$, where t is in thousands of years.

Solution Since this function describes exponential decay, we expect a negative value for k. We want

$$e^k = 0.886.$$

Solving for k gives

$$k = \ln(0.886) = -0.12104.$$

So we have $Q = 200e^{-0.12104t}$ and the continuous growth rate is -12.104% per thousand years.

Exponential Growth Problems That Cannot Be Solved By Logarithms

Some equations with the variable in the exponent cannot be solved using logarithms.

Example 12 With t in years, the population of a country (in millions) is given by $P = 2(1.02)^t$, while the food supply (in millions of people that can be fed) is given by $N = 4 + 0.5t$. Determine the year in which the country first experiences food shortages.

Solution The country starts to experience shortages when the population equals the number of people that can be fed—that is, when $P = N$. We attempt to solve the equation $P = N$ by using logs:

$$\begin{aligned} 2(1.02)^t &= 4 + 0.5t \\ 1.02^t &= 2 + 0.25t \quad \text{Dividing by 2} \\ \log 1.02^t &= \log(2 + 0.25t) \\ t \log 1.02 &= \log(2 + 0.25t). \end{aligned}$$

Unfortunately, we cannot isolate t, so, this equation cannot be solved using logs. However, we can approximate the solution of the original equation numerically or graphically, as shown in Figure 4.4. The two functions, P and N, are equal when $t \approx 199.381$. Thus, it will be almost 200 years before shortages occur.

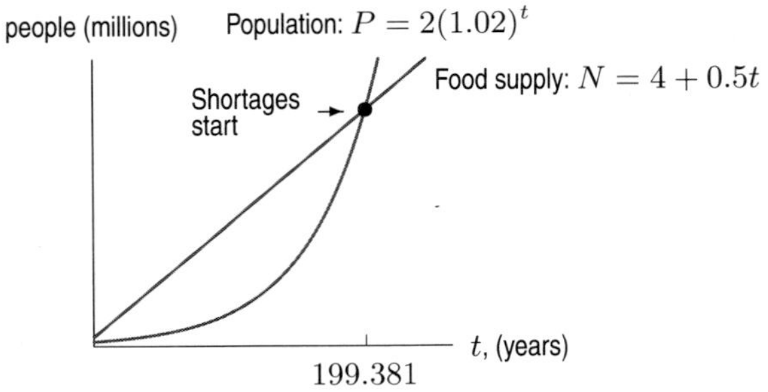

Figure 4.4: Finding the intersection of linear and exponential graphs

Exercises and Problems for Section 4.2

Exercises

In Exercises 1–8, give the starting value a, the growth rate r, and the continuous growth rate k.

1. $Q = 230(1.182)^t$

2. $Q = 0.181\left(e^{0.775}\right)^t$

3. $Q = 0.81(2)^t$

4. $Q = 5 \cdot 2^{t/8}$

5. $Q = 12.1 \cdot 10^{-0.11t}$

6. $Q = 40e^{(t-5)/12}$

7. $Q = 2e^{(1-3t/4)}$

8. $Q = 2^{-(t-5)/3}$

In Exercises 9–12, convert to the form $Q = ae^{kt}$.

9. $Q = 4 \cdot 7^t$

10. $Q = 2 \cdot 3^t$

11. $Q = 4 \cdot 8^{1.3t}$

12. $Q = 973 \cdot 6^{2.1t}$

In Exercises 13–16, convert to the form $Q = ab^t$.

13. $Q = 4e^{7t}$

14. $Q = 0.3e^{0.7t}$

15. $Q = \frac{14}{5}e^{0.03t}$

16. $Q = e^{-0.02t}$

For Exercises 17–18, write the exponential function in the form $y = ab^t$. Find b accurate to four decimal places. If t is measured in years, give the percent annual growth or decay rate and the continuous percent growth or decay rate per year.

17. $y = 25e^{0.053t}$

18. $y = 100e^{-0.07t}$

For Exercises 19–20, write the exponential function in the form $y = ae^{kt}$. Find k accurate to four decimal places. If t is measured in years, give the percent annual growth rate and the continuous percent growth rate per year.

19. $y = 6000(0.85)^t$

20. $y = 5(1.12)^t$

Find the doubling time in Exercises 21–24.

21. A population growing according to $P = P_0e^{0.2t}$.

22. A city is growing by 26% per year.

23. A bank account is growing by 2.7% per year.

24. A company's profits are increasing by an annual growth factor of 1.12.

Find the half-lives of the substances in Exercises 25–27.

25. Tritium, which decays at a rate of 5.471% per year.

26. Einsteinium-253, which decays at a rate of 3.406% per day.

27. A radioactive substance that decays at a continuous rate of 11% per minute.

Problems

In Problems 28–30, find a formula for the exponential function.

28. $V = f(t)$ gives the value after t years of \$2500 invested at 3.25% annual interest, compounded quarterly (four times per year).

29. The graph of g contains $(-50, 20)$ and $(120, 70)$.

30. $V = h(t)$ gives the value of an item initially worth \$10,000 that loses half its value every 5 years.

31. A population grows from 11000 to 13000 in three years. Assuming the growth is exponential, find the:

(a) Annual growth rate **(b)** Continuous growth rate
(c) Why are your answers to parts (a) and (b) different?

32. A population doubles in size every 15 years. Assuming exponential growth, find the

(a) Annual growth rate **(b)** Continuous growth rate

33. A population increases from 5.2 million at an annual rate of 3.1%. Find the continuous growth rate.

34. If 17% of a radioactive substance decays in 5 hours, what is the half-life of the substance?

35. Sketch the exponential function $y = u(t)$ given that it has a starting value of 0.8 and a doubling time of 12 years. Label the axes and indicate the scale.

Solve the equations in Problems 36–41 if possible. Give an exact solution if there is one.

36. $1.7(2.1)^{3x} = 2(4.5)^x$

37. $3^{4\log x} = 5$

38. $5(1.044)^t = t + 10$

39. $12(1.221)^t = t + 3$

40. $10e^{3t} - e = 2e^{3t}$

41. $\log x + \log(x-1) = \log 2$

42. Use algebra to show that the time it takes for a quantity growing exponentially to double is independent of the starting quantity and the time. To do this, let d represent the time it takes for P to double. Show that if P becomes $2P$ at time $t + d$, then d depends only on the growth factor b, but not on the starting quantity a and time t. (Assume $P \neq 0$.)

43. The US census projects the population of the state of Washington using the function $N(t) = 5.4e^{0.013t}$, where $N(t)$ is in millions and t is in years since 1995.

(a) What is the population's continuous growth rate?
(b) What is the population of Washington in year $t = 0$?
(c) How many years is it before the population triples?
(d) In what year does this model indicate a population of only one person? Is this reasonable or unreasonable?

44. The voltage V across a charged capacitor is given by $V(t) = 5e^{-0.3t}$ where t is in seconds.

(a) What is the voltage after 3 seconds?
(b) When will the voltage be 1?
(c) By what percent does the voltage decrease each second?

45. A colony of bacteria grows exponentially. The colony begins with 3 bacteria, but 3 hours after the beginning of the experiment, it has grown to 100 bacteria.

(a) Give a formula for the number of bacteria as a function of time.
(b) How long does it take for the colony to triple in size?

46. In July 2005, the Internet was linked by a global network of about 353.2 million host computers.[2] The number of host computers has been growing approximately exponentially and was about 36.7 million in July 1998.

(a) Find a formula for the number, N, of internet host computers as an exponential function of t, the number of years since July 1998, using the form $N = ae^{kt}$.
(b) At what continuous annual percent rate does N increase?
(c) What is the doubling time?

47. The number of cases of sepsis, an immune response to infection or trauma, has been growing exponentially and has doubled in this country in the last 5 years.[3]

(a) Find the continuous annual percent growth rate in the number of cases.
(b) If this growth rate continues, how many years will it take for the number of cases to triple?

48. The half-life of iodine-123 is about 13 hours. You begin with 50 grams of iodine-123.

(a) Write an equation that gives the amount of iodine-123 remaining after t hours.
(b) Determine the number of hours needed for your sample to decay to 10 grams.

[2] www.isc.org/ds/, accessed January 11, 2006.
[3] "Improvement seen in Recognizing, Treating Sepsis", Watertown Daily Times, September 25, 2002.

49. Scientists observing owl and hawk populations collect the following data. Their initial count for the owl population is 245 owls, and the population grows by 3% per year. They initially observe 63 hawks, and this population doubles every 10 years.

(a) Find a formula for the size of the population of owls in terms of time t.
(b) Find a formula for the size of the population of hawks in terms of time t.
(c) Use a graph to find how long it will take for these populations to be equal in number.

50. Figure 4.5 shows the graphs of the exponential functions f and g, and the linear function, h.

(a) Find formulas for f, g, and h.
(b) Find the exact value(s) of x such that $f(x) = g(x)$.
(c) Estimate the value(s) of x such that $f(x) = h(x)$.

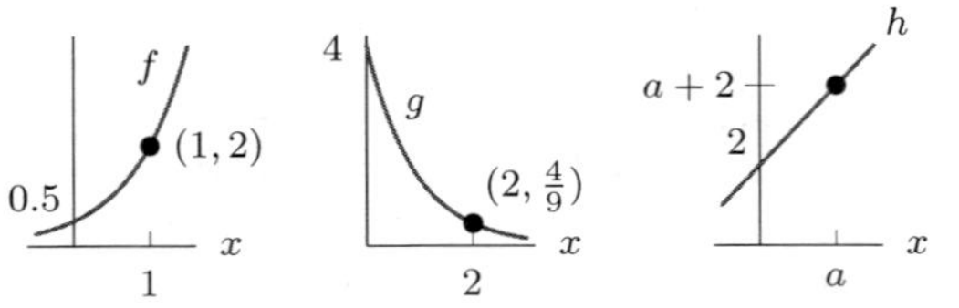

Figure 4.5

51. In 1991, the body of a man was found in melting snow in the Alps of Northern Italy. An examination of the tissue sample revealed that 46% of the carbon-14 present in his body at the time of his death had decayed. The half-life of carbon-14 is approximately 5728 years. How long ago did this man die?

52. In a country where inflation is a concern, prices have risen by 40% over a 5-year period.

(a) By what percent do the prices rise each year?
(b) How long does it take for prices to rise by 5%?
(c) Use a continuous growth rate to model inflation by a function of the form $P = P_0e^{rt}$.

53. A person's blood alcohol content (BAC) is a measure of how much alcohol is in the blood stream. When the person stops drinking, the BAC declines over time as the alcohol is metabolized. If Q is the amount of alcohol and Q_0 the initial amount, then $Q = Q_0e^{-t/\tau}$, where τ is known as the *elimination time*. How long does it take for a person's BAC to drop from 0.10 to 0.04 if the elimination time is 2.5 hours?

54. The probability of a transistor failing within t months is given by $P(t) = 1 - e^{-0.016t}$.

(a) What is the probability of failure within the first 6 months? Within the second six months?
(b) Within how many months will the probability of failure be 99.99%?

55. You deposit \$4000 into an account that earns 6% annual interest, compounded annually. A friend deposits \$3500 into an account that earns 5.95% annual interest, compounded continuously. Will your friend's balance ever equal yours? If so, when?

56. **(a)** Find the time required for an investment to triple in value if it earns 4% annual interest, compounded continuously.
(b) Now find the time required assuming that the interest is compounded annually.

57. The temperature, H, in °F, of a cup of coffee t hours after it is set out to cool is given by the equation:

$$H = 70 + 120(1/4)^t.$$

(a) What is the coffee's temperature initially (that is, at time $t = 0$)? After 1 hour? 2 hours?
(b) How long does it take the coffee to cool down to 90°F? 75°F?

58. The size of a population, P, of toads t years after it is introduced into a wetland is given by

$$P = \frac{1000}{1 + 49(1/2)^t}.$$

(a) How many toads are there in year $t = 0$? $t = 5$? $t = 10$?
(b) How long does it take for the toad population to reach 500? 750?
(c) What is the maximum number of toads that the wetland can support?

Problems 59–60 use Figure 4.6, where t_0 is the t-coordinate of the point of intersection of the graphs. Describe what happens to t_0 if the following changes are made, assuming the other quantities remain the same.

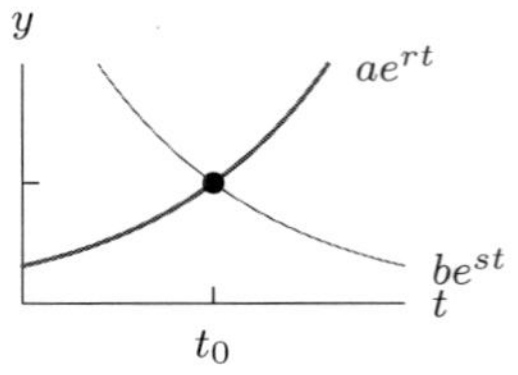

Figure 4.6

59. a is increased

60. s is decreased

61. In Figure 4.6, suppose that a and b are fixed and r increases. Describe how s changes if t_0 remains unchanged.

4.3 THE LOGARITHMIC FUNCTION

The Graph, Domain, and Range of the Common Logarithm

In Section 4.1 we defined the log function (to base 10) for all positive numbers. In other words,

Domain of $\log x$ is all positive numbers.

By considering its graph in Figure 4.7, we determine the range of $y = \log x$. The log graph crosses the x-axis at $x = 1$, because $\log 1 = \log(10^0) = 0$. The graph climbs to $y = 1$ at $x = 10$, because $\log 10 = \log(10^1) = 1$. In order for the log graph to climb to $y = 2$, the value of x must reach 100, or 10^2, and in order for it to climb to $y = 3$, the value of x must be 10^3, or 1000. To reach the modest height of $y = 20$ requires x to equal 10^{20}, or 100 billion billion! The log function increases so slowly that it often serves as a benchmark for other slow-growing functions. Nonetheless, the graph of $y = \log x$ eventually climbs to any value we choose.

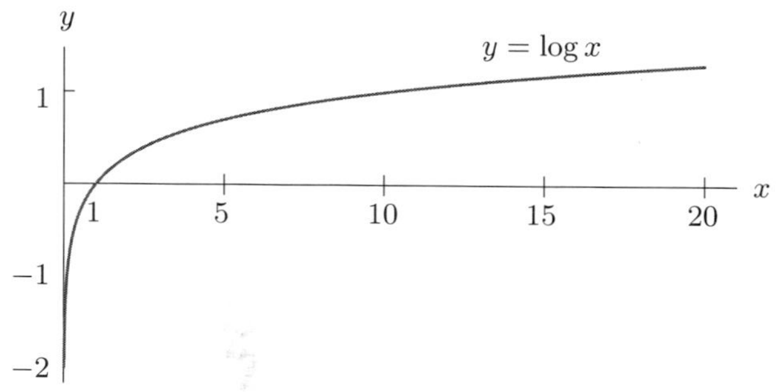

Figure 4.7: The log function grows very rapidly for $0 < x < 1$ and very slowly for $x > 1$. It has a vertical asymptote at $x = 0$ but never touches the y-axis.

Although x cannot equal zero in the log function, we can choose $x > 0$ to be as small as we like. As x decreases toward zero, the values of $\log x$ get large and negative. For example,

$$\begin{aligned}
\log 0.1 &= \log 10^{-1} = -1, \\
\log 0.01 &= \log 10^{-2} = -2, \\
&\vdots \\
\log 0.0000001 &= \log 10^{-7} = -7,
\end{aligned}$$

and so on. So, small positive values of x give exceedingly large negative values of y. Thus,

Range of $\log x$ is all real numbers.

The log function is increasing and its graph is concave down, since its rate of change is decreasing.

Graphs of the Inverse Functions $y = \log x$ and $y = 10^x$

The fact that $y = \log x$ and $y = 10^x$ are inverses means that their graphs are related. Looking at Tables 4.1 and 4.2, we see that the point $(0.01, -2)$ is on the graph of $y = \log x$ and the point

$(-2, 0.01)$ is on the graph of $y = 10^x$. In general, if the point (a, b) is on the graph of $y = \log x$, the point (b, a) is on the graph of $y = 10^x$. Thus, the graph of $y = \log x$ is the graph of $y = 10^x$ with x and y-axes interchanged. If the x- and y-axes have the same scale, this is equivalent to reflecting the graph of $y = 10^x$ across the diagonal line $y = x$. See Figure 4.8.

Table 4.1 *Log function*

x	$y = \log x$
0.01	−2
0.1	−1
1	0
10	1
100	2
1000	3

Table 4.2 *Exponential function*

x	$y = 10^x$
−2	0.01
−1	0.1
0	1
1	10
2	100
3	1000

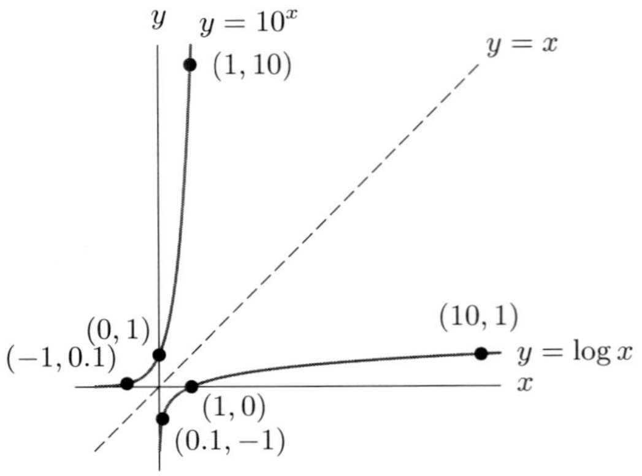

Figure 4.8: The functions $y = \log x$ and $y = 10^x$ are inverses of one another

Asymptotes

In Section 3.3 we saw that the graph of an exponential function has a horizontal asymptote. In Figure 4.8, we see that $y = 10^x$ has horizontal asymptote $y = 0$, because

$$\text{as } x \to -\infty, \quad 10^x \to 0.$$

Correspondingly, as x gets closer to zero, $y = \log x$ takes on larger and larger negative values. We write

$$\text{as } x \to 0^+, \qquad f(x) \to -\infty.$$

The notation $x \to 0^+$ is read "x approaches zero from the right" and means that we are choosing smaller and smaller positive values of x—that is, we are sliding toward $x = 0$ through small positive values. We say the graph of the log function $y = \log x$ has a *vertical asymptote* of $x = 0$.

To describe vertical asymptotes in general, we use the notation

$$x \to a^+$$

to mean that x slides toward a from the right (that is, through values larger than a) and

$$x \to a^-$$

to mean that x slides toward a from the left (that is, through values smaller than a).

If $f(x) \to \infty$ as $x \to a^+$, we say the *limit* of $f(x)$ as x approaches a from the right is infinity,[4] and write

$$\lim_{x \to a^+} f(x) = \infty.$$

If $f(x) \to \infty$ as $x \to a^-$, we write

$$\lim_{x \to a^-} f(x) = \infty.$$

[4]Some authors say that these limits do not exist.

If both $\lim_{x\to a^+} f(x) = \infty$ and $\lim_{x\to a^-} f(x) = \infty$, we say the limit of $f(x)$ as x approaches a is infinity, and write

$$\lim_{x\to a} f(x) = \infty.$$

Similarly, we can write

$$\lim_{x\to a^+} f(x) = -\infty \quad \text{or} \quad \lim_{x\to a^-} f(x) = -\infty \quad \text{or} \quad \lim_{x\to a} f(x) = -\infty.$$

We summarize the information about both horizontal and vertical asymptotes:

> Let $y = f(x)$ be a function and let a be a finite number.
> - The graph of f has a **horizontal asymptote** of $y = a$ if
>
> $$\lim_{x\to\infty} f(x) = a \quad \text{or} \quad \lim_{x\to-\infty} f(x) = a \quad \text{or both.}$$
>
> - The graph of f has a **vertical asymptote** of $x = a$ if
>
> $$\lim_{x\to a^+} f(x) = \infty \text{ or } \lim_{x\to a^+} f(x) = -\infty \text{ or } \lim_{x\to a^-} f(x) = \infty \text{ or } \lim_{x\to a^-} f(x) = -\infty.$$

Notice that the process of finding a vertical asymptote is different from the process for finding a horizontal asymptote. Vertical asymptotes occur where the function values grow larger and larger, either positively or negatively, as x approaches a finite value (i.e. where $f(x) \to \infty$ or $f(x) \to -\infty$ as $x \to a$.) Horizontal asymptotes are determined by whether the function values approach a finite number as x takes on large positive or large negative values (i.e., as $x \to \infty$ or $x \to -\infty$).

Graph of Natural Logarithm

In addition to similar algebraic properties, the natural log and the common log have similar graphs.

Example 1 Graph $y = \ln x$ for $0 < x < 10$.

Solution Values of $\ln x$ are in Table 4.3. Like the common log, the natural log is only defined for $x > 0$ and has a vertical asymptote at $x = 0$. The graph is slowly increasing and concave down.

Table 4.3 *Values of* $\ln x$ *(rounded)*

x	$\ln x$
0	Undefined
1	0
2	0.7
e	1
3	1.1
4	1.4
⋮	⋮

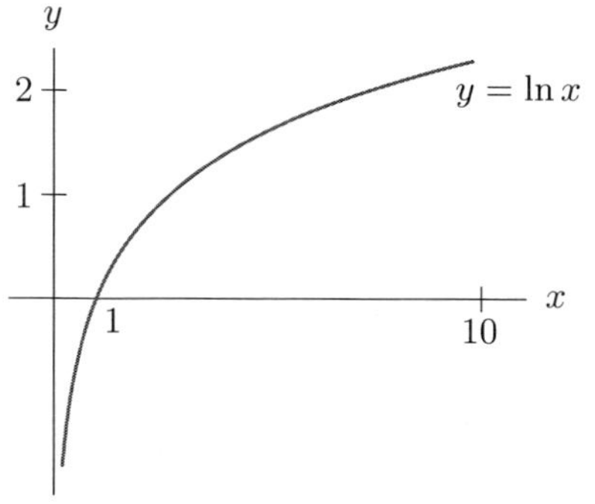

Figure 4.9: Graph of the natural logarithm

The functions $y = \ln x$ and $y = e^x$ are inverses. If the scales on the axes are the same, their graphs are reflections of one another across the line $y = x$. See Figure 4.10.

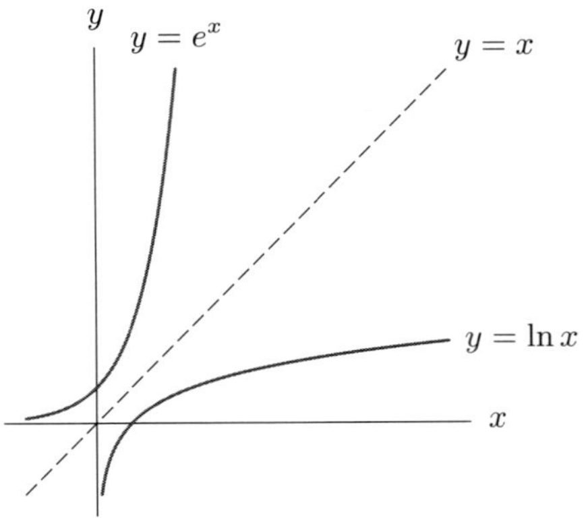

Figure 4.10: The functions $y = \ln x$ and $y = e^x$ are inverses of one another

Chemical Acidity

Logarithms are useful in measuring quantities whose magnitudes vary widely, such as acidity (pH), sound (decibels), and earthquakes (the Richter scale). In chemistry, the acidity of a liquid is expressed using pH. The acidity depends on the hydrogen ion concentration in the liquid (in moles per liter); this concentration is written $[H^+]$. The greater the hydrogen ion concentration, the more acidic the solution. The pH is defined as:

$$\text{pH} = -\log[H^+].$$

Example 2 The hydrogen ion concentration of seawater is $[H^+] = 1.1 \cdot 10^{-8}$. Estimate the pH of seawater. Then check your answer with a calculator.

Solution We want to estimate $\text{pH} = -\log(1.1 \cdot 10^{-8})$. Since $1.1 \cdot 10^{-8} \approx 10^{-8}$ and $\log 10^{-8} = -8$, we know that

$$\text{pH} = -\log(1.1 \cdot 10^{-8}) \approx -(-8) = 8.$$

Using a calculator, we have

$$\text{pH} = -\log(1.1 \cdot 10^{-8}) = 7.959.$$

Example 3 A vinegar solution has a pH of 3. Determine the hydrogen ion concentration.

Solution Since $3 = -\log[H^+]$, we have $-3 = \log[H^+]$. This means that $10^{-3} = [H^+]$. So the hydrogen ion concentration is 10^{-3} moles per liter.

Logarithms and Orders of Magnitude

We often compare sizes or quantities by computing their ratios. If A is twice as tall as B, then

$$\frac{\text{Height of } A}{\text{Height of } B} = 2.$$

If one object is 10 times heavier than another, we say it is an *order of magnitude* heavier. If one quantity is two factors of 10 greater than another, we say it is two orders of magnitude greater, and so on. For example, the value of a dollar is two orders of magnitude greater than the value of a penny, because we have

$$\frac{\$1}{\$0.01} = 100 = 10^2.$$

The order of magnitude is the logarithm of their ratio.

Example 4 The sound intensity of a refrigerator motor is 10^{-11} watts/cm^2. A typical school cafeteria has sound intensity of 10^{-8} watts/cm^2. How many orders of magnitude more intense is the sound of the cafeteria?

Solution To compare the two intensities, we compute their ratio:

$$\frac{\text{Sound intensity of cafeteria}}{\text{Sound intensity of refrigerator}} = \frac{10^{-8}}{10^{-11}} = 10^{-8-(-11)} = 10^3.$$

Thus, the sound intensity of the cafeteria is 1000 times greater than the sound intensity of the refrigerator. The log of this ratio is 3. We say that the sound intensity of the cafeteria is three orders of magnitude greater than the sound intensity of the refrigerator.

Decibels

The intensity of audible sound varies over an enormous range. The range is so enormous that we consider the logarithm of the sound intensity. This is the idea behind the *decibel* (abbreviated dB). To measure a sound in decibels, the sound's intensity, I, is compared to the intensity of a standard benchmark sound, I_0. The intensity of I_0 is defined to be 10^{-16} watts/cm^2, roughly the lowest intensity audible to humans. The comparison between a sound intensity I and the benchmark sound intensity I_0 is made as follows:

$$\text{Noise level in decibels} = 10 \cdot \log\left(\frac{I}{I_0}\right).$$

For instance, let's find the decibel rating of the refrigerator in Example 4. First, we find how many orders of magnitude more intense the refrigerator sound is than the benchmark sound:

$$\frac{I}{I_0} = \frac{\text{Sound intensity of refrigerator}}{\text{Benchmark sound intensity}} = \frac{10^{-11}}{10^{-16}} = 10^5.$$

Thus, the refrigerator's intensity is 5 orders of magnitude more than I_0, the benchmark intensity. We have

$$\text{Decibel rating of refrigerator} = 10 \cdot \underbrace{\text{Number of orders of magnitude}}_{5} = 50 \text{ dB}.$$

Note that 5, the number of orders of magnitude, is the log of the ratio I/I_0. We use the log function because it "counts" the number of powers of 10. Thus if N is the decibel rating, then

$$N = 10 \log\left(\frac{I}{I_0}\right).$$

Example 5 (a) If a sound doubles in intensity, by how many units does its decibel rating increase?

(b) Loud music can measure 110 dB whereas normal conversation measures 50 dB. How many times more intense is loud music than normal conversation?

Solution (a) Let I be the sound's intensity before it doubles. Once doubled, the new intensity is $2I$. The decibel rating of the original sound is $10\log(I/I_0)$, and the decibel rating of the new sound is $10\log(2I/I_0)$. The difference in decibel ratings is given by

$$\begin{aligned}
\text{Difference in decibel ratings} &= 10\log\left(\frac{2I}{I_0}\right) - 10\log\left(\frac{I}{I_0}\right) \\
&= 10\left(\log\left(\frac{2I}{I_0}\right) - \log\left(\frac{I}{I_0}\right)\right) && \text{Factoring out 10} \\
&= 10\cdot\log\left(\frac{2I/I_0}{I/I_0}\right) && \text{Using the property } \log a - \log b = \log(a/b) \\
&= 10\cdot\log 2 && \text{Canceling } I/I_0 \\
&\approx 3.010 \text{ dB} && \text{Because } \log 2 \approx 0.3.
\end{aligned}$$

Thus, if the sound intensity is doubled, the decibel rating goes up by approximately 3 dB.

(b) If I_M is the sound intensity of loud music, then

$$10\log\left(\frac{I_M}{I_0}\right) = 110 \text{ dB}.$$

Similarly, if I_C is the sound intensity of conversation, then

$$10\log\left(\frac{I_C}{I_0}\right) = 50 \text{ dB}.$$

Computing the difference of the decibel ratings gives

$$10\log\left(\frac{I_M}{I_0}\right) - 10\log\left(\frac{I_C}{I_0}\right) = 60.$$

Dividing by 10 gives

$$\begin{aligned}
\log\left(\frac{I_M}{I_0}\right) - \log\left(\frac{I_C}{I_0}\right) &= 6 \\
\log\left(\frac{I_M/I_0}{I_C/I_0}\right) &= 6 && \text{Using the property } \log b - \log a = \log(b/a) \\
\log\left(\frac{I_M}{I_C}\right) &= 6 && \text{Canceling } I_0 \\
\frac{I_M}{I_C} &= 10^6 && \log x = 6 \text{ means that } x = 10^6.
\end{aligned}$$

So $I_M = 10^6 I_C$, which means that loud music is 10^6 times, or one million times, as intense as normal conversation.

Exercises and Problems for Section 4.3

Exercises

1. What is the equation of the asymptote of the graph of $y = 10^x$? Of the graph of $y = 2^x$? Of the graph of $y = \log x$?

2. What is the equation for the asymptote of the graph of $y = e^x$? Of the graph of $y = e^{-x}$? Of the graph of $y = \ln x$?

3. Without a calculator, match the functions $y = 10^x$, $y = e^x$, $y = \log x$, $y = \ln x$ with the graphs in Figure 4.11.

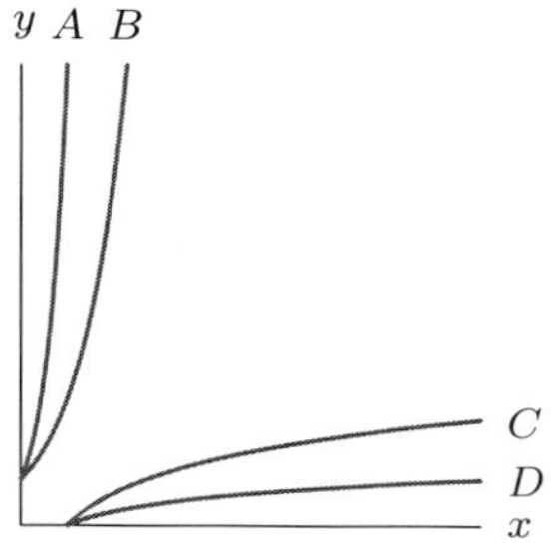

Figure 4.11

4. Without a calculator, match the functions $y = 2^x$, $y = e^{-x}$, $y = 3^x$, $y = \ln x$, $y = \log x$ with the graphs in Figure 4.12.

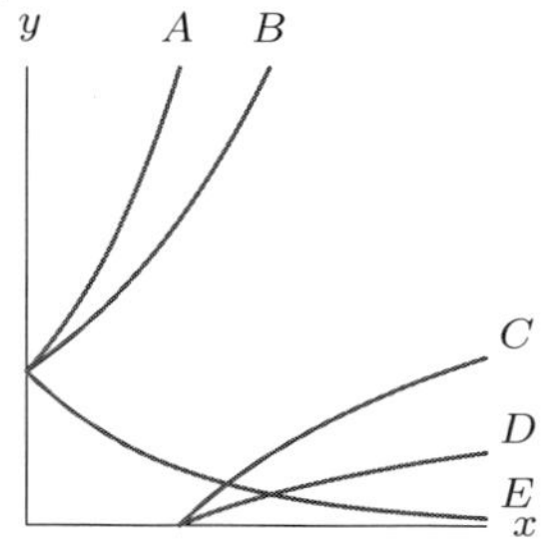

Figure 4.12

5. What is the value (if any) of the following?

(a) 10^{-x} as $x \to \infty$ (b) $\log x$ as $x \to 0^+$

6. What is the value (if any) of the following?

(a) e^x as $x \to -\infty$ (b) $\ln x$ as $x \to 0^+$

Graph the functions in Problems 7–10. Label all asymptotes and intercepts.

7. $y = 2 \cdot 3^x + 1$

8. $y = -e^{-x}$

9. $y = \log(x - 4)$

10. $y = \ln(x + 1)$

In Problems 11–12, graph the function. Identify any vertical asymptotes. State the domain of the function.

11. $y = 2\ln(x - 3)$

12. $y = 1 - \ln(2 - x)$

In Exercises 13–17, find the hydrogen ion concentration, $[H^+]$, for the substances.[5] [Hint: pH $= -\log[H^+]$.]

13. Lye, with a pH of 13.

14. Battery acid, with a pH of 1.

15. Baking soda, with a pH of 8.3.

16. Tomatoes, with a pH of 4.5.

17. Hydrochloric acid, with a pH of 0.

18. Find **(a)** $\lim_{x \to 0^+} \log x$ **(b)** $\lim_{x \to 0^-} \ln(-x)$

Problems

19. Match the statements (a)–(d) with the functions (I)–(IV).

(a) $\lim_{x \to 0^+} f(x) = -\infty$ **(b)** $\lim_{x \to 0^-} f(x) = 0$

(c) $\lim_{x \to \infty} f(x) = \infty$ **(d)** $\lim_{x \to -\infty} f(x) = 0$

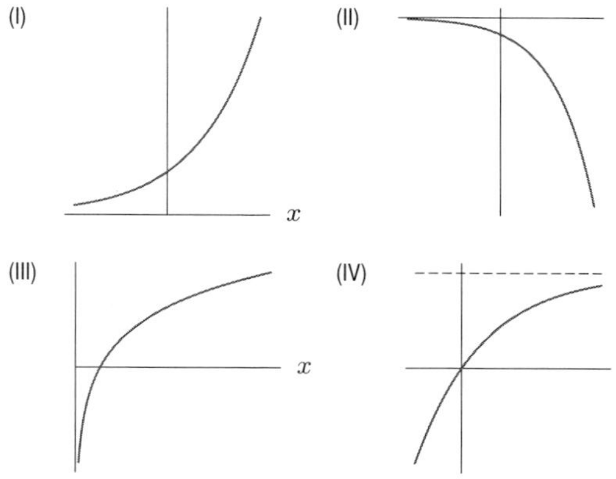

[5] Data from www.miamisci.org/ph/hhoh.html

20. Match the graphs (a)–(c) to one of the functions $r(x)$, $s(x)$, $t(x)$ whose values are in the tables.

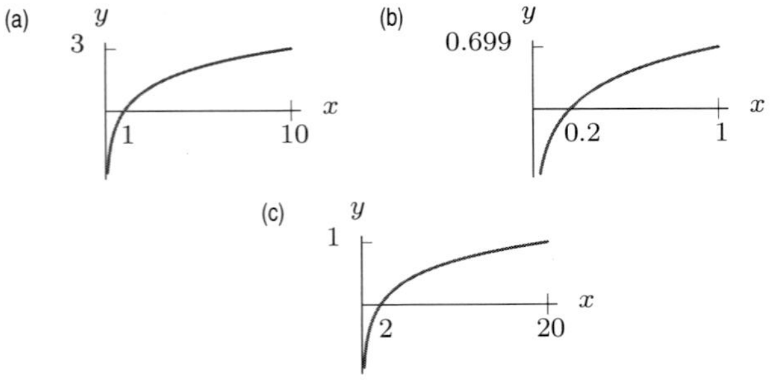

x	2	4	10
$r(x)$	1	1.301	1.699

x	0.5	5	10
$s(x)$	−0.060	0.379	0.699

x	0.1	2	100
$t(x)$	−3	0.903	6

21. Immediately following the gold medal performance of the US women's gymnastic team in the 1996 Olympic Games, an NBC commentator, John Tesh, said of one team member: "Her confidence and performance have grown logarithmically." He clearly thought this was an enormous compliment. Is it a compliment? Is it realistic?

In Problems 22–27, find possible formulas for the functions using logs or exponentials.

22.

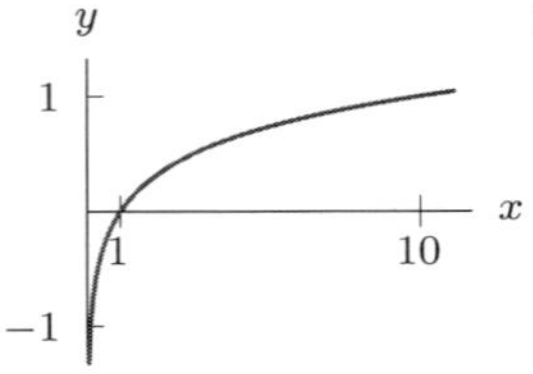

23.

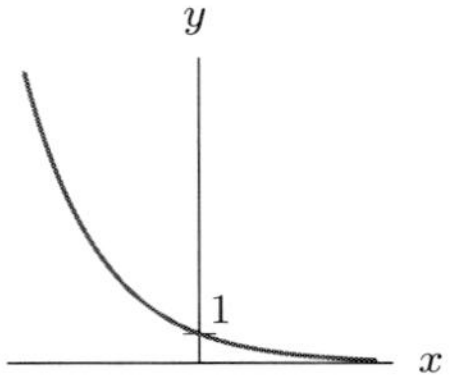

24.

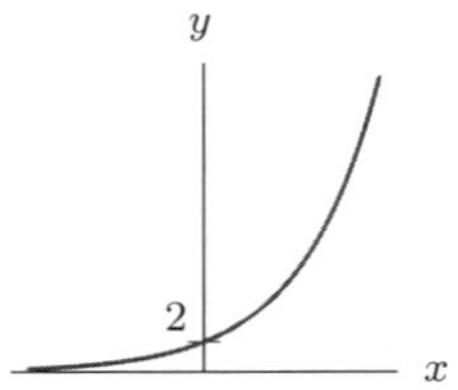

25.

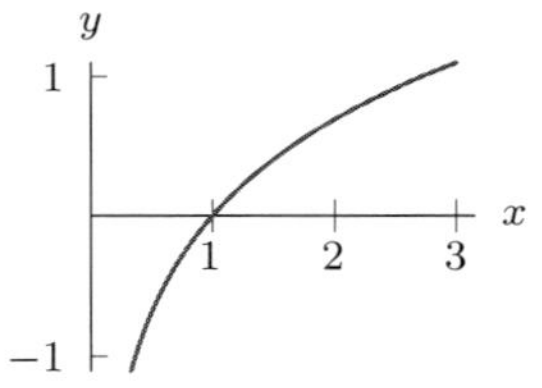

26.

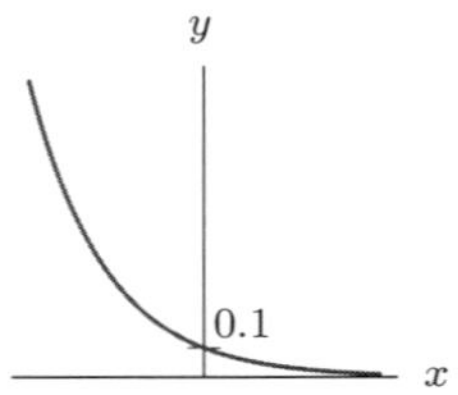

27.

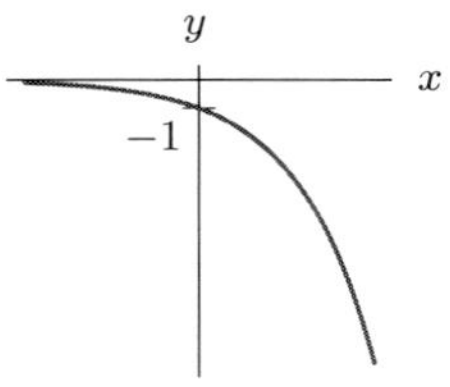

Find the domain of the functions in Problems 28–31.

28. $h(x) = \ln(x^2)$

29. $g(x) = (\ln x)^2$

30. $f(x) = \ln(\ln x)$

31. $k(x) = \ln(x - 3)$

32. **(a)** Using the definition of pH on page 170, find the concentrations of hydrogen ions in solutions with

(i) pH $= 2$ (ii) pH $= 4$ (iii) pH $= 7$

(b) A high concentration of hydrogen ions corresponds to an acidic solution. From your answer to part (a), decide if solutions with high pHs are more or less acidic than solutions with low pHs.

33. **(a)** A 12 oz cup of coffee contains about $2.41 \cdot 10^{18}$ hydrogen ions. What is the concentration (moles/liter) of hydrogen ions in a 12 oz cup of coffee? [Hint: One liter equals 30.3 oz. One mole of hydrogen ions equals $6.02 \cdot 10^{23}$ hydrogen ions.]

(b) Based on your answer to part (a) and the formula for pH, what is the pH of a 12 oz cup of coffee?

34. **(a)** The pH of lemon juice is about 2.3. What is the concentration of hydrogen ions in lemon juice?

(b) A person squeezes 2 oz of lemon juice into a cup. Based on your answer to part (a), how many hydrogen ions does this juice contain?

35. Sound A measures 30 decibels and sound B is 5 times as loud as sound A. What is the decibel rating of sound B to the nearest integer?

36. **(a)** Let D_1 and D_2 represent the decibel ratings of sounds of intensity I_1 and I_2, respectively. Using log properties, find a simplified formula for the difference between the two ratings, $D_2 - D_1$, in terms of the two intensities, I_1 and I_2. (Decibels are introduced on page 171.)

(b) If a sound's intensity doubles, how many decibels louder does the sound become?

37. The magnitude of an earthquake is measured relative to the strength of a "standard" earthquake, whose seismic waves are of size W_0. The magnitude, M, of an earthquake with seismic waves of size W is defined to be

$$M = \log\left(\frac{W}{W_0}\right).$$

The value of M is called the *Richter scale* rating of the strength of an earthquake.

(a) Let M_1 and M_2 represent the magnitude of two earthquakes whose seismic waves are of sizes W_1 and W_2, respectively. Using log properties, find a simplified formula for the difference $M_2 - M_1$ in terms of W_1 and W_2.

(b) The 1989 earthquake in California had a rating of 7.1 on the Richter scale. How many times larger were the seismic waves in the March 2005 earthquake off the coast of Sumatra, which measured 8.7 on the Richter scale? Give your answer to the nearest integer.

38. The average time T, in milliseconds (ms), it takes a person to move a mouse cursor a distance D across a computer screen to a target button of length S is given by *Fitts' Law*, which states that $T = a + b\log(D/S + 1)$, where a and b are constants.[6]

(a) Briefly explain why it does not matter what units are used for S and D, provided the units are the same. Why is it useful for a formula concerning computer screens not to be based on fixed sizes?

(b) The cursor is moved 15 cm to a target of length 3 cm. Letting $a = 50$ and $b = 500$, estimate the time required to complete this operation.

(c) Graph T for $S = 3$ cm and for $0 \leq D \leq 20$.

(d) What is the T-intercept of the graph in part (c)? What does this tell you about cursor movement?

(e) Suppose the distance moved, D, is doubled and the target length S stays the same. Without knowing the values of D and S, but assuming $a = 50$ and $b = 500$, what can you say about the change in T, the time required?

4.4 LOGARITHMIC SCALES

The Solar System and Beyond

Table 4.4 gives the distance from the sun to a number of different astronomical objects. The planet Mercury is 58,000,000 km from the sun, that earth is 149,000,000 km from the sun, and that Pluto is 5,900,000,000 km, or almost 6 billion kilometers from the sun. The table also gives the distance to Proxima Centauri, the star closest to the sun, and to the Andromeda Galaxy, the spiral galaxy closest to our own galaxy, the Milky Way.

Table 4.4 *Distance from the sun to various astronomical objects*

Object	Distance (million km)	Saturn	1426
Mercury	58	Uranus	2869
Venus	108	Neptune	4495
Earth	149	Pluto	5900
Mars	228	Proxima Centauri	$4.1 \cdot 10^7$
Jupiter	778	Andromeda Galaxy	$2.4 \cdot 10^{13}$

Linear Scales

We can represent the information in Table 4.4 graphically in order to get a better feel for the distances involved. Figure 4.13 shows the distance from the sun to the first five planets on a *linear scale*, which means that the evenly spaced units shown in the figure represent equal distances. In this case, each unit represents 100 million kilometers.

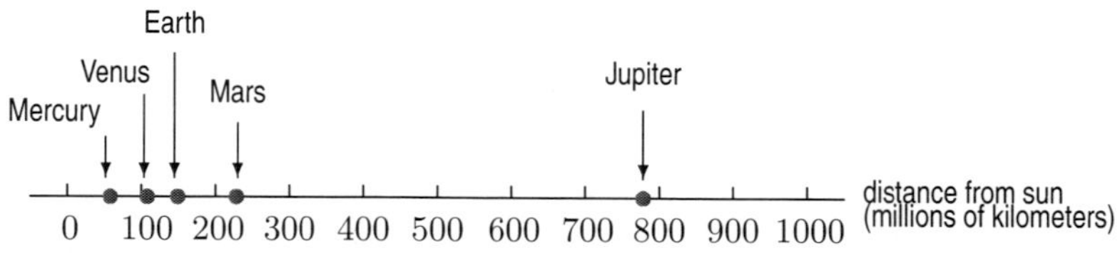

Figure 4.13: The distance from the sun of the first five planets (in millions of kilometers)

[6]Raskin, J, *The Humane Interface* (Reading: Addison Wesley, 2000), p 93. Jef Raskin designed the Apple Macintosh computer using, among other resources, mathematical formulas such as Fitts' Law.

The drawback of Figure 4.13 is that the scale is too small to show all of the astronomical distances described by the table. For example, to show the distance to Pluto on this scale would require over six times as much space on the page. Even worse, assuming that each 100 million km unit on the scale measures half an inch on the printed page, we would need 3 miles of paper to show the distance to Proxima Centauri!

You might conclude that we could fix this problem by choosing a larger scale. In Figure 4.14 each unit on the scale is 1 billion kilometers. Notice that all five planets shown by Figure 4.13 are crowded into the first unit of Figure 4.14; even so, the distance to Pluto barely fits. The distances to the other objects certainly don't fit. For instance, to show the Andromeda Galaxy, Figure 4.14 would have to be almost 200,000 miles long. Choosing an even larger scale will not improve the situation.

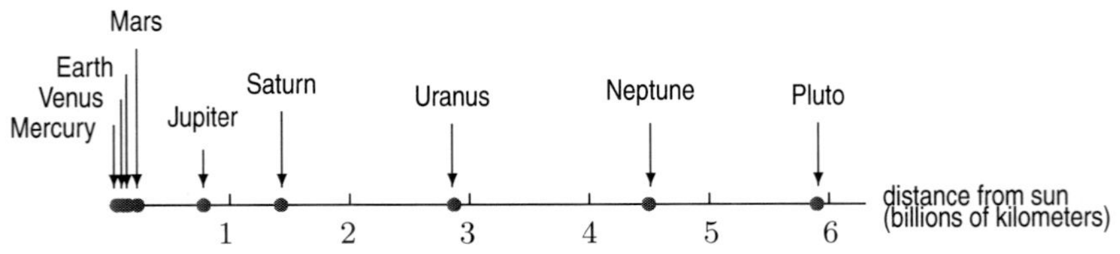

Figure 4.14: The distance to all nine planets (in billions of kilometers)

Logarithmic Scales

We conclude that the data in Table 4.4 cannot easily be represented on a linear scale. If the scale is too small, the more distant objects do not fit; if the scale is too large, the less distant objects are indistinguishable. The problem is not that the numbers are too big or too small; the problem is that the numbers vary too greatly in size.

We consider a different type of scale on which equal distances are not evenly spaced. All the objects from Table 4.4 are represented in Figure 4.15. The nine planets are still cramped, but it is possible to tell them apart. Each tick mark on the scale in Figure 4.15 represents a distance ten times larger than the one before it. This kind of scale is called *logarithmic*.

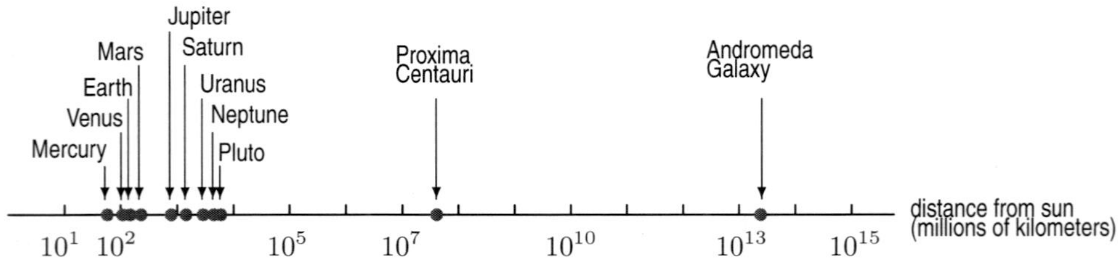

Figure 4.15: The distance from the sun (in millions of kilometers)

How Do We Plot Data on a Logarithmic Scale?

A logarithmic scale is marked with increasing powers of 10: 10^1, 10^2, 10^3, and so on. Notice that even though the distances in Figure 4.15 are not evenly spaced, the exponents are evenly spaced. Therefore the distances in Figure 4.15 are spaced according to their logarithms.

In order to plot the distance to Mercury, 58 million kilometers, we use the fact that

$$10 < 58 < 100,$$

so Mercury's distance is between 10^1 and 10^2, as shown in Figure 4.15. To plot Mercury's distance more precisely, calculate $\log 58 = 1.763$, so $10^{1.763} = 58$, and use 1.763 to represent Mercury's position. See Figure 4.16.

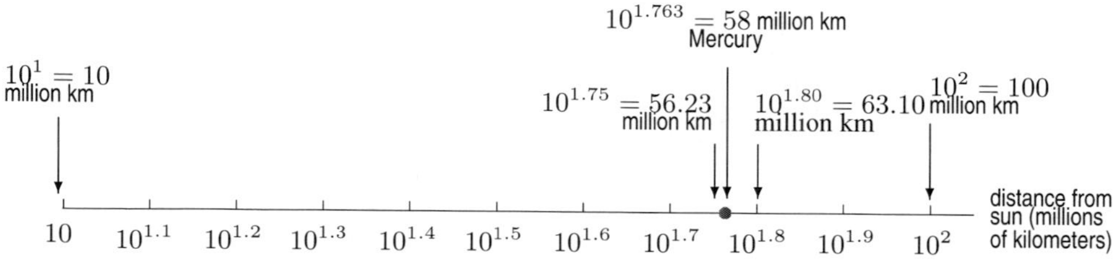

Figure 4.16: Mercury's distance, 58 million kilometers falls between $10^{1.75} = 56.23$ and $10^{1.80} = 63.10$ million kilometers

---

Example 1 Where should Saturn be on the logarithmic scale? What about the Andromeda Galaxy?

Solution Saturn's distance is 1426 million kilometers, so we want the exponent of 10 that gives 1426, which is

$$\log 1426 \approx 3.154119526.$$

Thus $10^{3.154} \approx 1426$, so we use 3.154 to indicate Saturn's distance.

Similarly, the distance to the Andromeda Galaxy is $2.4 \cdot 10^{13}$ million kilometers, and since

$$\log(2.4 \cdot 10^{13}) \approx 13.38,$$

we use 13.38 to represent the galaxy's distance. See Figure 4.17.

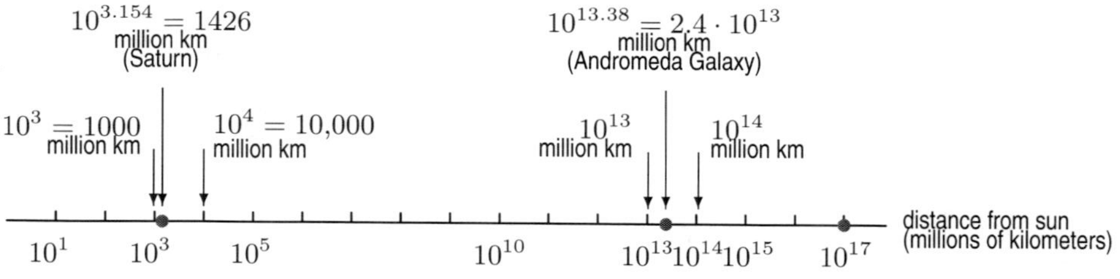

Figure 4.17: Saturn's distance is $10^{3.154}$ and the Andromeda Galaxy's distance is $10^{13.38}$

Logs of Small Numbers

The history of the world, like the distance to the stars and planets, involves numbers of vastly different sizes. Table 4.5 gives the ages of certain events[7] and the logarithms of their ages. The logarithms have been used to plot the events in Figure 4.18.

Table 4.5 *Ages of various events in earth's history and logarithms of the ages*

Event	Age (millions of years)	log (age)	Event	Age (millions of years)	log (age)
Man emerges	1	0	Rise of dinosaurs	245	2.39
Ape man fossils	5	0.70	Vertebrates appear	570	2.76
Rise of cats, dogs, pigs	37	1.57	First plants	2500	3.40
Demise of dinosaurs	67	1.83	Earth forms	4450	3.65

[7] *CRC Handbook*, 75th *ed.* 14-8.

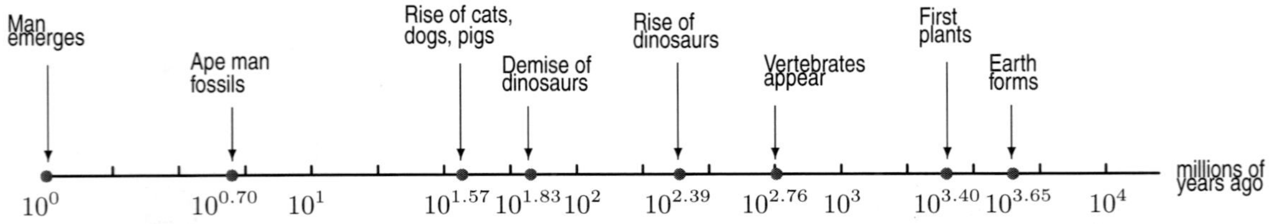

Figure 4.18: Logarithmic scale showing the ages of various events (in millions of years ago)

The events described by Table 4.5 all happened at least 1 million years ago. How do we indicate events which occurred less than 1 million years ago on the log scale?

Example 2 Where should the building of the pyramids be indicated on the log scale?

Solution The pyramids were built about 5000 years ago, or

$$\frac{5000}{1{,}000{,}000} = 0.005 \text{ million years ago.}$$

Notice that 0.005 is between 0.001 and 0.01, that is,

$$10^{-3} < 0.005 < 10^{-2}.$$

Since

$$\log 0.005 \approx -2.30,$$

we use -2.30 for the pyramids. See Figure 4.19.

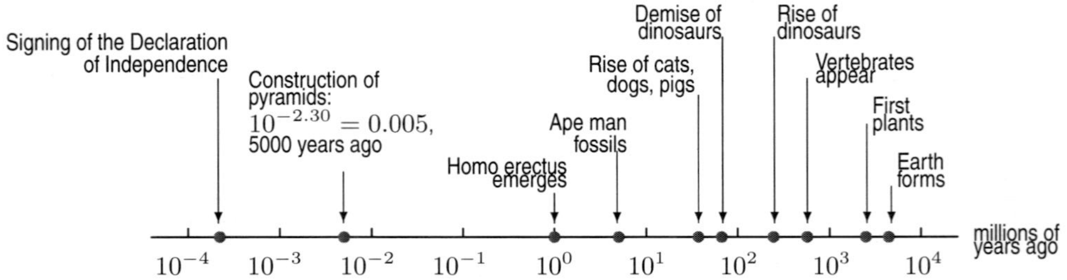

Figure 4.19: Logarithmic scale showing the ages of various events. Note that events that are less than 1 million years old are indicated by negative exponents

---

Another Way to Label a Log Scale

In Figures 4.18 and 4.19, the log scale has been labeled so that exponents are evenly spaced. Another way to label a log scale is with the values themselves instead of the exponents. This has been done in Figure 4.20.

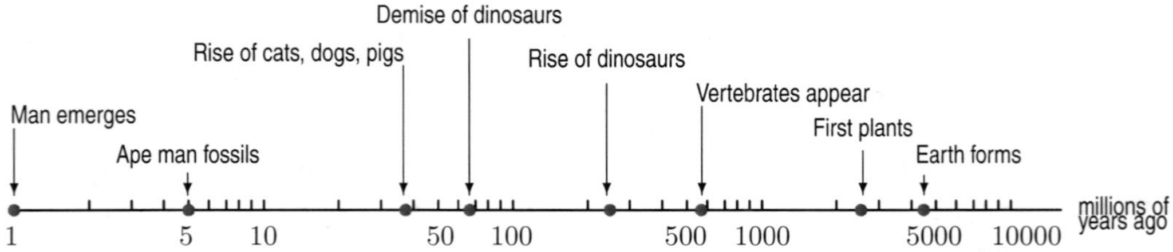

Figure 4.20: Axis labeled using the actual values, not with logs

Notice the characteristic way that the labels and tick marks "pile up" on each interval. The even spacing between exponents on log scales leads to uneven spacing in values. Although the values 10, 20, 30, 40, and 50 are evenly spaced, their corresponding exponents are not: $\log 10 = 1$, $\log 20 = 1.30$, $\log 30 = 1.48$, $\log 40 = 1.60$, and $\log 50 = 1.70$. Therefore, when we label an axis according to values on a scale that is spaced according to exponents, the labels get bunched up.

Log-Log Scales

Table 4.6 shows the average metabolic rate in kilocalories per day (kcal/day) for animals of different weights.[8] (A kilocalorie is the same as a standard nutritional calorie.) For instance, a 1-lb rat consumes about 35 kcal/day, whereas a 1750-lb horse consumes almost 9500 kcal/day.

Table 4.6 *The metabolic rate (in kcal/day) for animals of different weights*

Animal	Weight (lbs)	Rate (kcal/day)
Rat	1	35
Cat	8	166
Human	150	2000
Horse	1750	9470

It is not practical to plot these data on an ordinary set of axes. The values span too broad a range. However, we can plot the data using log scales for both the horizontal (weight) axis and the vertical (rate) axes. See Figure 4.21. Figure 4.22 shows a close-up view of the data point for cats to make it easier to see how the labels work. Once again, notice the characteristic piling up of labels and gridlines. This happens for the same reason that it happened in Figure 4.20.

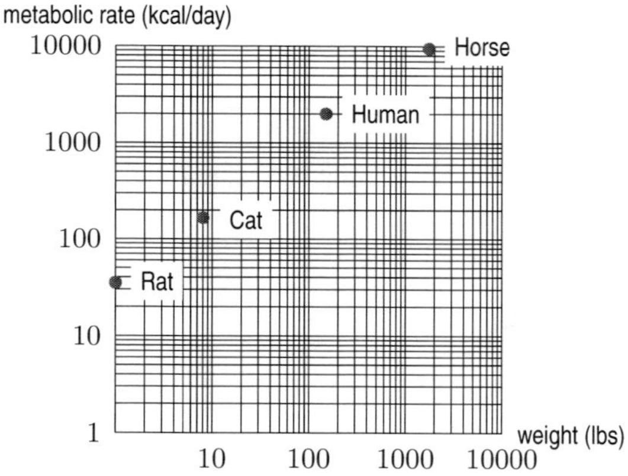

Figure 4.21: Metabolic rate (in kcal/hr) plotted against body weight

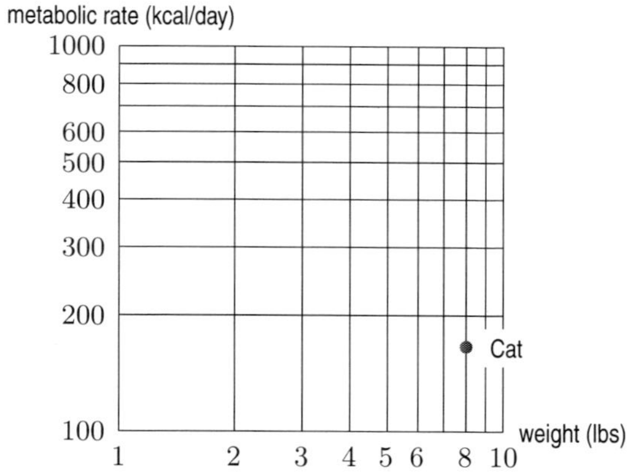

Figure 4.22: A close-up view of the Cat data point

Using Logs to Fit an Exponential Function to Data

In Section 1.6 we used linear regression to find the equation for a line of best fit for a set of data. What if the data do not lie close to a straight line, but instead approximate the graph of some other function? In this section we see how logarithms help us fit data with an exponential function of the form $Q = a \cdot b^t$.

[8] *The New York Times*, January 11, 1999.

Sales of Compact Discs

Table 4.7 shows the fall in the sales of vinyl long-playing records (LPs) and the rise of compact discs (CDs) during for the years 1982 through 1993.[9]

Table 4.7 *CD and LP sales*

t, years since 1982	c, CDs (millions)	l, LPs (millions)
0	0	244
1	0.8	210
2	5.8	205
3	23	167
4	53	125
5	102	107
6	150	72
7	207	35
8	287	12
9	333	4.8
10	408	2.3
11	495	1.2

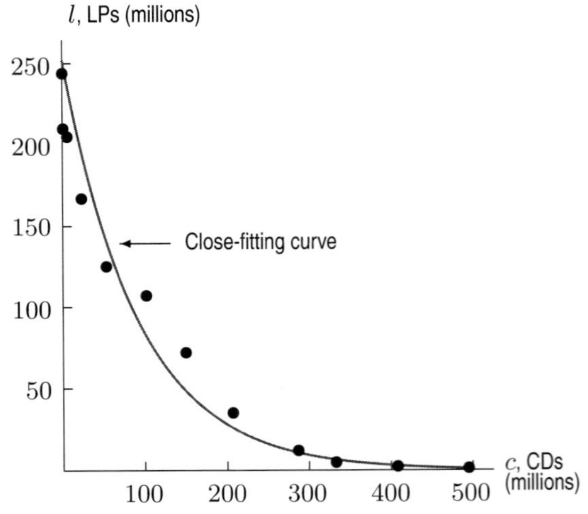

Figure 4.23: The number of LPs sold, l, as a function of number of CDs sold, c

From Table 4.7, we see that as CD sales rose dramatically during the 1980s and early 1990s, LP sales declined equally dramatically. Figure 4.23 shows the number of LPs sold in a given year as a function of the number of CDs sold that year.

Using a Log Scale to Linearize Data

In Section 4.4, we saw that a log scale allows us to compare values that vary over a wide range. Let's see what happens when we use a log scale to plot the data shown in Figure 4.23. Table 4.8 shows values $\log l$, where l is LP sales. These are plotted against c, CD sales, in Figure 4.24. Notice that plotting the data in this way tends to *linearize* the graph—that is, make it look more like a line. A line has been drawn in to emphasize the trend in the data.

Table 4.8 *Values of $y = \ln l$ and c.*

c, CDs	l, LPs	$y = \ln l$
0	244	5.50
0.8	210	5.35
5.8	205	5.32
23	167	5.12
53	125	4.83
102	107	4.67
150	72	4.28
207	35	3.56
287	12	2.48
333	4.8	1.57
408	2.3	0.83
495	1.2	0.18

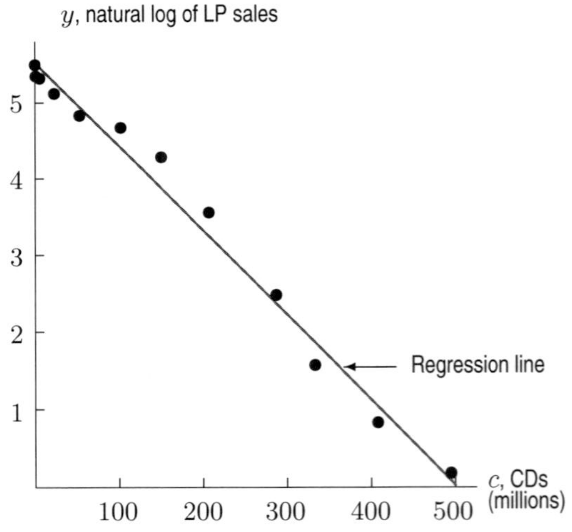

Figure 4.24: The y-axis of this graph gives the natural log of LP sales

[9]Data from Recording Industry Association of America, Inc., 1998

Finding a Formula for the Curve

We say that the data in the third column of Table 4.8 have been *transformed*. A calculator or computer gives a regression line for the transformed data:[10]

$$y = 5.52 - 0.011c.$$

Notice that this equation gives y in terms of c. To transform the equation back to our original variables, l and c, we substitute $\ln l$ for y, giving

$$\ln l = 5.52 - 0.011c.$$

We solve for l by raising e to both sides:

$$\begin{aligned} e^{\ln l} &= e^{5.52-0.011c} \\ &= (e^{5.52})(e^{-0.011c}) \qquad \text{Using an exponent rule.} \end{aligned}$$

Since $e^{\ln l} = l$ and $e^{5.52} \approx 250$, we have

$$l = 250e^{-0.011c}.$$

This is the equation of the curve in Figure 4.23.

Fitting An Exponential Function To Data

In general, to fit an exponential formula, $N = ae^{kt}$, to a set of data of the form (t, N), we use three steps. First, we transform the data by taking the natural log of both sides and making the substitution $y = \ln N$. This leads to the equation

$$\begin{aligned} y = \ln N &= \ln\left(ae^{kt}\right) \\ &= \ln a + \ln e^{kt} \\ &= \ln a + kt. \end{aligned}$$

Setting $b = \ln a$ gives a linear equation with k as the slope and b as the y-intercept.

$$y = b + kt.$$

Secondly we can now use linear regression on the variables t and y. (Remember that $y = \ln N$.) Finally, as step three, we transform the linear regression equation back into our original variables by substituting $\ln N$ for y and solving for N.

Exercises and Problems for Section 4.4

Exercises

In Exercises 1–4, you wish to graph the quantities on a standard piece of paper. On which should you use a logarithmic scale? On which a linear scale? Why?

1. The wealth of 20 different people, one of whom is a multi-billionaire.

2. The number of diamonds owned by 20 people, one of whom is a multi-billionaire.

3. The number of meals per week eaten in restaurants for a random sample of 20 people worldwide.

4. The number of tuberculosis bacteria in 20 different people, some never exposed to the disease, some slightly exposed, some with mild cases, and some dying of it.

[10]The values obtained by a computer or another calculator may vary slightly from the ones given.

5. **(a)** Use a calculator to fill in the following tables (round to 4 decimal digits).

n	1	2	3	4	5	6	7	8	9
$\log n$									

n	10	20	30	40	50	60	70	80	90
$\log n$									

(b) Using the results of part (a), plot the integer points 2 through 9 and the multiples of 10 from 20 to 90 on the log scaled axis shown in Figure 4.25.

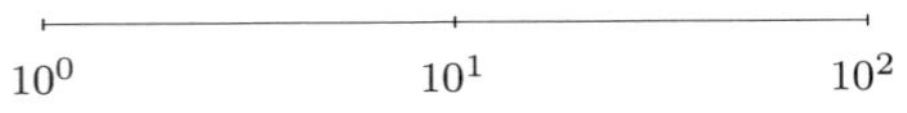

Figure 4.25

For the tables in Problem 6–8,

(a) Use linear regression to find a linear function $y = b + mx$ that fits the data. Record the correlation coefficient.

(b) Use linear regression on the values x and $\ln y$ to fit a function of the form $\ln y = b + mx$. Record the correlation coefficient. Convert to an exponential function $y = ae^{kx}$.

(c) Compare the correlation coefficients. Graph the data and the two functions to assess which function fits best.

6.

x	y
30	70
85	120
122	145
157	175
255	250
312	300

7.

x	y
8	23
17	150
23	496
26	860
32	2720
37	8051

8.

x	y
3.2	35
4.7	100
5.1	100
5.5	150
6.8	200
7.6	300

Problems

9. The signing of the Declaration of Independence is marked on the log scale in Figure 4.19 on page 178. To two decimal places, what is its position?

10. **(a)** Draw a line segment about 5 inches long. On it, choose an appropriate linear scale and mark points that represent the integral powers of two from zero to the sixth power. What is true about the location of the points as the exponents get larger?
(b) Draw a second line segment. Repeat the process in (a) but this time use a logarithmic scale so that the units are now powers of ten. What do you notice about the location of these points?

11. Table 4.9 shows the numbers of deaths in 2002 due to various causes in the US.[11]

Table 4.9

Cause	Deaths
Scarlet fever	2
Whooping cough	18
Asthma	4261
HIV	14,011
Kidney Diseases	40,974
Accidents	106,742
Malignant neoplasms	550,271
Cardiovascular Disease	918,828
All causes	2,443,387

(a) Explain why a log scale is necessary to plot the data from Table 4.9.
(b) Find the log of each value given.
(c) Plot the data using a log scale. Label each point with the related cause.

12. Table 4.10 shows the typical body masses in kilograms for various animals.[12]
(a) Find the log of the body mass of each animal to two decimal places.
(b) Plot the body masses for each animal in Table 4.10 on a linear scale using A to identify the Blue Whale, B to identify the African Elephant, and so on down to L to identify the Hummingbird.

[11] www.cdc/gov/nchs/data/nvsr/nvsr53/nvsr53_05.pdf, accessed December 22, 2005.
[12] R. McNiell Alexander, *Dynamics of Dinosaurs and Other Extinct Giants.* (New York: Columbia University Press, 1989) and H. Tennekes, *The Simple Science of Flight* (Cambridge: MIT Press, 1996).

(c) Plot and label the body masses for each animal in Table 4.10 on a logarithmic scale.
(d) Which scale, (b) or (c), is more useful?

Table 4.10

Animal	Body mass	Animal	Body mass
Blue Whale	91000	Lion	180
African Elephant	5450	Human	70
White Rhinoceros	3000	Albatross	11
Hippopotamus	2520	Hawk	1
Black Rhinoceros	1170	Robin	0.08
Horse	700	Hummingbird	0.003

13. Figure 4.26 shows the populations of eleven different places, with the scale markings representing the logarithm of the population. Give the approximate populations of each place.

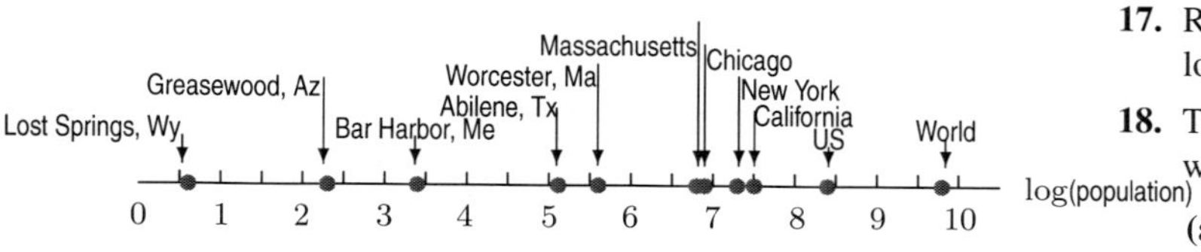

Figure 4.26

14. Table 4.11 shows the dollar value of some items in 2004. Plot and label these values on a log scale.

Table 4.11

Item	Dollar value	Item	Dollar value
Pack of gum	0.50	New house	264,400
Movie ticket	9.00	Lottery winnings	100 million
New computer	1200	Bill Gates' worth	46.6 bn
Year at college	27,500	National debt	7,500 bn
Luxury car	60,400	US GDP	11,700 bn

15. Table 4.12 shows the sizes of various organisms. Plot and label these values on a log scale.

Table 4.12

Animal	Size (cm)	Animal	Size (cm)
Virus	0.0000005	Domestic cat	60
Bacterium	0.0002	Wolf (with tail)	200
Human cell	0.002	Thresher shark	600
Ant	0.8	Giant squid	2200
Hummingbird	12	Sequoia	7500

16. **(a)** Complete the Table 4.13 with values of $y = 3^x$.
(b) Complete Table 4.14 with values for $y = \log(3^x)$. What kind of function is this?
(c) Complete tables for $f(x) = 2 \cdot 5^x$ and $g(x) = \log(2 \cdot 5^x)$. What kinds of functions are these?
(d) What seems to be true about a function which is the logarithm of an exponential function? Is this true in general?

Table 4.13

x	0	1	2	3	4	5
$y = 3^x$						

Table 4.14

x	0	1	2	3	4	5
$y = \log(3^x)$						

17. Repeat part (b) and (c) of Problem 16 using the natural log function. Is your answer to part (d) the same?

18. Table 4.15 shows the value, y, of US imports from China with x in years since 1985.
(a) Find a formula for a linear function $y = b + mx$ that approximates the data.
(b) Find $\ln y$ for each y value, and use the x and $\ln y$ values to find a formula for a linear function $\ln y = b + mx$ that approximates the data.
(c) Use the equation in part (b) to find an exponential function of the form $y = ae^{kx}$ that fits the data.

Table 4.15 *Value of US imports from China in millions of dollars*

	1985	1986	1987	1988	1989	1990	1991
x	0	1	2	3	4	5	6
y	3862	4771	6293	8511	11,990	15,237	18,976

19. Table 4.16 shows newspapers' share of the expenditure of national advertisers. Using the method of Problem 18, fit an exponential function of the form $y = ae^{kx}$ to the data, where y is percent share and x is the number of years since 1950.

Table 4.16 *Newspapers' share of advertising*

	1950	1960	1970	1980	1990	1992
x	0	10	20	30	40	42
y	16.0	10.8	8.0	6.7	5.8	5.0

20. To study how recognition memory decreases with time, the following experiment was conducted. The subject read a list of 20 words slowly aloud, and later, at different time intervals, was shown a list of 40 words containing the 20 words that he or she had read. The percentage, P, of words recognized was recorded as a function of t, the time elapsed in minutes. Table 4.17 shows the averages for 5 different subjects.[13] This is modeled by $P = a\ln t + b$.

(a) Find $\ln t$ for each value of t, and then use regression on a calculator or computer to estimate a and b.
(b) Graph the data points and regression line on a coordinate system of P against $\ln t$.
(c) When does this model predict that the subjects will recognize no words? All words?
(d) Graph the data points and curve $P = a\ln t + b$ on a coordinate system with P against t, with $0 \le t \le 10{,}500$.

Table 4.17 *Percentage of words recognized*

t, min	5	15	30	60	120	240
P%	73.0	61.7	58.3	55.7	50.3	46.7
t, min	480	720	1440	2880	5760	10,080
P%	40.3	38.3	29.0	24.0	18.7	10.3

21. Table 4.18 gives the length ℓ (in cm) and weight w (in gm) of 16 different fish known as threadfin bream (*Nemipterus marginatus*) found in the South China Sea.[14]

(a) Let $W = \ln w$ and $L = \ln \ell$. For these sixteen data points, plot W on the vertical axis and L on the horizontal axis. Describe the resulting scatterplot.
(b) Fitting a line to the scatterplot you drew in part (a), find a possible formula for W in terms of L.
(c) Based on your formula for part (b), find a possible formula for w in terms of ℓ.
(d) Comment on your formula, keeping in mind what you know about units as well the typical relationship between weight, volume, and length.

Table 4.18 *Length and weight of fish*

Type	1	2	3	4	5	6	7	8
ℓ	8.1	9.1	10.2	11.9	12.2	13.8	14.8	15.7
w	6.3	9.6	11.6	18.5	26.2	36.1	40.1	47.3
Type	9	10	11	12	13	14	15	16
ℓ	16.6	17.7	18.7	19.0	20.6	21.9	22.9	23.5
w	65.6	69.4	76.4	82.5	106.6	119.8	169.2	173.3

22. A light, flashing regularly, consists of cycles, each cycle having a dark phase and a light phase. The frequency of this light is measured in cycles per second. As the frequency is increased, the eye initially perceives a series of flashes of light, then a coarse flicker, a fine flicker, and ultimately a steady light. The frequency at which the flickering disappears is called the fusion frequency.[15] Table 4.19 shows the results of an experiment[16] in which the fusion frequency F was measured as a function of the light intensity I. It is modeled by $F = a\ln I + b$.

(a) Find $\ln I$ for each value of I, and then use linear regression on a calculator or computer to estimate a and b in the equation $F = a\ln I + b$.
(b) Plot F against $\ln I$, showing the data points and the line.
(c) Plot F against I, showing the data points and the curve and give its equation.
(d) The units of I are arbitrary, that is, not given. If the units of I were changed, which of the constants a and b would be affected, and in what way?

Table 4.19 *Fusion frequency, F, as a function of the light intensity, I*

I	0.8	1.9	4.4	10.0	21.4	48.4	92.5	218.7	437.3	980.0
F	8.0	12.1	15.2	18.5	21.7	25.3	28.3	31.9	35.2	38.2

23. (a) Plot the data given by Table 4.20. What kind of function might fit this data well?
(b) Using the substitution $z = \ln x$, transform the data in Table 4.20, and compile your results into a new table. Plot the transformed data as $\ln x$ versus y.
(c) What kind of function gives a good fit to the plot you made in part (b)? Find a formula for y in terms of z that fits the data well.
(d) Using the formula from part (c), find a formula for y in terms of x that gives a good fit to the data in Table 4.20.
(e) What does your formula from part (d) tell you about x as a function of y (as opposed to y as a function of x)?

Table 4.20

x	0.21	0.55	1.31	3.22	5.15	12.48
y	−11	−2	6.5	16	20.5	29

[13] Adapted from D. Lewis, *Quantitative Methods in Psychology.* (New York: McGraw-Hill, 1960).
[14] Data taken from *Introduction to Tropical Fish Stock Assessment* by Per Sparre, Danish Institute for Fisheries Research, and Siebren C. Venema, FAO Fisheries Department, available at http://www.fao.org/docrep/W5449E/w5449e00.htm. This source cites the following original reference: Pauly, D., 1983. Some simple methods for the assessment of tropical fish stocks.
[15] R. S. Woodworth, *Experimental Psychology.* (New York: Holt and Company, 1948).
[16] D. Lewis, *Quantitative Methods in Psychology.* (New York: McGraw-Hill, 1960).

CHAPTER SUMMARY

- **Logarithms**
  Common log: $y = \log x$ means $10^y = x$.
  $\log 10 = 1, \log 1 = 0$.
  Natural log: $y = \ln x$ means $e^y = x$.
  $\ln e = 1, \ln 1 = 0$.
- **Properties of Logs**

$$\log(ab) = \log a + \log b.$$
$$\log(a/b) = \log a - \log b.$$
$$\log(b^t) = t \log b.$$
$$\log(10^x) = 10^{\log x} = x.$$

- **Converting between base b and base e**
  If $Q = ab^t$ and $Q = ae^{kt}$, then $k = \ln b$.
- **Solving Equations using Logs**
  Solve equations such as $ab^t = c$ and $ae^{kt} = c$ using logs.
  Not all exponential equations can be solved with logs, e.g. $2^t = 3 + t$.
- **Logarithmic Functions**
  Graph; domain; range; concavity; asymptotes.
- **Applications of Logarithms**
  Doubling time; half life;
  Chemical acidity; orders of magnitude; decibels.
- **Logarithmic Scales**
  Plotting data; log-log scales. Linearizing data and fitting curves to data using logs.
- **Limits and Limits from the Right and from the Left**

REVIEW EXERCISES AND PROBLEMS FOR CHAPTER FOUR

Exercises

In Exercises 1–4, convert to the form $Q = ae^{kt}$.

1. $Q = 12(0.9)^t$
2. $Q = 16(0.487)^t$
3. $Q = 14(0.862)^{1.4t}$
4. $Q = 721(0.98)^{0.7t}$

In Exercises 5–6, convert to the form $Q = ab^t$.

5. $Q = 7e^{-10t}$
6. $Q = 5e^t$

For Exercises 7–10, use logarithms to solve for t.

7. $3^t = 50$
8. $e^{0.15t} = 25$
9. $40e^{-0.2t} = 12$
10. $5 \cdot 2^t = 100$

In Exercises 11–12, give the starting value a, the growth factor b, the growth rate r, and the continuous growth rate k.

11. $Q = \dfrac{(1.13)^t}{20}$
12. $Q = \dfrac{200}{e^{0.177(t-2)}}$

13. The following populations, $P(t)$, are given in millions in year t. Describe the growth of each population in words. Give both the percent annual growth rate and the continuous growth rate per year.

(a) $P(t) = 51(1.03)^t$
(b) $P(t) = 15e^{0.03t}$
(c) $P(t) = 7.5(0.94)^t$
(d) $P(t) = 16e^{-0.051t}$
(e) $P(t) = 25(2)^{t/18}$
(f) $P(t) = 10(\frac{1}{2})^{t/25}$

14. A population, $P(t)$ (in millions) in year t, increases exponentially. Suppose $P(8) = 20$ and $P(15) = 28$.

(a) Find a formula for $P(t)$ without using e.
(b) If $P(t) = ae^{kt}$, find k.

Solve the equations in Exercises 15–26 exactly if possible.

15. $400(1.112)^t = 1328$
16. $0.007e^{-1.22t} = 0.002$
17. $55e^{0.571t} = 28e^{0.794t}$
18. $100^{2x+3} = \sqrt[3]{10{,}000}$
19. $5 \cdot 1.1^x = 55$
20. $7e^{2t} = 2e^{0.9t}$
21. $0.01^x = \dfrac{1}{\sqrt{10}}$
22. $16.3(1.072)^t = 18.5$
23. $13e^{0.081t} = 25e^{0.032t}$
24. $87e^{0.066t} = 3t + 7$
25. $\dfrac{\log x^2 + \log x^3}{\log(100x)} = 3$
26. $220e^{1.323t} = 500e^{1.118t}$

In Exercises 27–29, simplify fully.

27. $\log\left(100^{x+1}\right)$
28. $\ln\left(e \cdot e^{2+M}\right)$
29. $\ln(A + B) - \ln(A^{-1} + B^{-1})$

Problems

In Problems 30–33, find a formula for the exponential function.

30. $P = f(t)$ gives the size of a population that begins with 22,000 members and grows at a continuous annual rate of 1.76%.

31. The graph of g contains $(-30, 200)$ and $(20, 60)$.

32. $V = h(t)$ is the value after t years of \$650 invested at 2.95% annual interest, compounded monthly.

33. $Q = p(t)$ describes a value that begins at 5000 and that doubles every 30 years.

34. **(a)** What is the domain and range of $f(x) = 10^x$? What is the asymptote of $f(x) = 10^x$?
(b) What does your answer to part (a) tell you about the domain, range, and asymptotes of $g(x) = \log x$?

35. What is the domain of $y = \ln(x^2 - x - 6)$?

36. The balance B (in \$) in an account after t years is given by $B = 5000(1.12)^t$.

(a) What is the balance after 5 years? 10 years?
(b) When is the balance \$10,000? \$20,000?

37. The populations (in thousands) of two cities are given by

$$P_1 = 51(1.031)^t \quad \text{and} \quad P_2 = 63(1.052)^t,$$

where t is the number of years since 1980. When does the population of P_1 equal that of P_2?

38. **(a)** Let $B = 5000(1.06)^t$ give the balance of a bank account after t years. If the formula for B is written $B = 5000e^{kt}$, estimate the value of k correct to four decimal places. What is the financial meaning of k?
(b) The balance of a bank account after t years is given by the formula $B = 7500e^{0.072t}$. If the formula for B is written $B = 7500b^t$, find b exactly, and give the value of b correct to four decimal places. What is the financial meaning of b?

39. In 2005, the population of the country Erehwon was 50 million people and increasing by 2.9% every year. The population of the country Ecalpon, on other hand, was 45 million people and increasing by 3.2% every year.

(a) For each country, write a formula expressing the population as a function of time t, where t is the number of years since 2005.
(b) Find the value(s) of t, if any, when the two countries have the same population.
(c) When is the population of Ecalpon double that of Erehwon?

40. The following is excerpted from an article that appeared in the January 8, 1990 *Boston Globe*.

> Men lose roughly 2 percent of their bone mass per year in the same type of loss that can severely affect women after menopause, a study indicates. "There is a problem with osteoporosis in men that hasn't been appreciated. It's a problem that needs to be recognized and addressed," said Dr. Eric Orwoll, who led the study by the Oregon Health Sciences University. The bone loss was detected at all ages and the 2 percent rate did not appear to vary, Orwoll said.

(a) Assume that the average man starts losing bone mass at age 30. Let M_0 be the average man's bone mass at this age. Express the amount of remaining bone mass as a function of the man's age, a.
(b) At what age will the average man have lost half his bone mass?

41. The number of bacteria present in a culture after t hours is given by the formula $N = 1000e^{0.69t}$.

(a) How many bacteria will there be after $1/2$ hour?
(b) How long before there are 1,000,000 bacteria?
(c) What is the doubling time?

42. Oil leaks from a tank. At hour $t = 0$ there are 250 gallons of oil in the tank. Each hour after that, 4% of the oil leaks out.

(a) What percent of the original 250 gallons has leaked out after 10 hours? Why is it less than $10 \cdot 4\% = 40\%$?
(b) If $Q(t) = Q_0 e^{kt}$ is the quantity of oil remaining after t hours, find the value of k. What does k tell you about the leaking oil?

43. A population increases from 30,000 to 34,000 over a 5-year period at a constant annual percent growth rate.

(a) By what percent did the population increase in total?
(b) At what constant percent rate of growth did the population increase each year?
(c) At what continuous annual growth rate did this population grow?

44. Radioactive carbon-14 decays according to the function $Q(t) = Q_0 e^{-0.000121t}$ where t is time in years, $Q(t)$ is the quantity remaining at time t, and Q_0 is the amount of present at time $t = 0$. Estimate the age of a skull if 23% of the original quantity of carbon-14 remains.

45. Suppose 2 mg of a drug is injected into a person's bloodstream. As the drug is metabolized, the quantity diminishes at the continuous rate of 4% per hour.

(a) Find a formula for $Q(t)$, the quantity of the drug remaining in the body after t hours.
(b) By what percent does the drug level decrease during any given hour?
(c) The person must receive an additional 2 mg of the drug whenever its level has diminished to 0.25 mg. When must the person receive the second injection?
(d) When must the person receive the third injection?

46. Suppose that $u = \log 2$ and $v = \log 5$.

(a) Find possible formulas for the following expressions in terms of u and/or v. Your answers should not involve logs.

(i) $\log(0.4)$ (ii) $\log 0.25$
(iii) $\log 40$ (iv) $\log\sqrt{10}$

(b) Justify the statement: $\log(7) \approx \frac{1}{2}(u + 2v)$.

47. Solve the following equations. Give approximate solutions if exact ones can't be found.

(a) $e^{x+3} = 8$ **(b)** $4(1.12^x) = 5$
(c) $e^{-0.13x} = 4$ **(d)** $\log(x - 5) = 2$
(e) $2\ln(3x) + 5 = 8$ **(f)** $\ln x - \ln(x-1) = 1/2$
(g) $e^x = 3x + 5$ **(h)** $3^x = x^3$
(i) $\ln x = -x^2$

48. Solve for x exactly.

(a) $\dfrac{3^x}{5^{x-1}} = 2^{x-1}$
(b) $-3 + e^{x+1} = 2 + e^{x-2}$
(c) $\ln(2x - 2) - \ln(x - 1) = \ln x$
(d) $9^x - 7 \cdot 3^x = -6$
(e) $\ln\left(\dfrac{e^{4x} + 3}{e}\right) = 1$
(f) $\dfrac{\ln(8x) - 2\ln(2x)}{\ln x} = 1$

Problems 49–50 involve the Rule of 70, which gives quick estimates of the doubling time of an exponentially growing quantity. If $r\%$ is the annual growth rate of the quantity, then the Rule of 70 says

$$\text{Doubling time in years} \approx \frac{70}{r}.$$

49. Use the Rule of 70 by estimate how long it takes a $\$1000$ investment to double if it grows at the following annual rates: 1%, 2%, 5%, 7%, 10%. Compare with the actual doubling times.

50. Using natural logs, solve for the doubling time for $Q = ae^{kt}$. Use your result to explain why the Rule of 70 works.

51. You want to borrow $\$25{,}000$ to buy a Ford Explorer XL. The best available annual interest rate is 6.9%, compounded monthly. Determine how long it will take to pay off the loan if you can only afford monthly payments of $\$330$. To do this, use the loan payment formula

$$P = \frac{Lr/12}{1 - (1 + (r/12))^{-m}},$$

where P is the monthly payment, L is the amount borrowed, r is the annual interest rate, and m is the number of months the loan is carried.

52. A rubber ball is dropped onto a hard surface from a height of 6 feet, and it bounces up and down. At each bounce it rises to 90% of the height from which it fell.

(a) Find a formula for $h(n)$, the height reached by the ball on bounce n.
(b) How high will the ball bounce on the 12th bounce?
(c) How many bounces before the ball rises no higher than an inch?

53. A manager at Saks Fifth Avenue wants to estimate the number of customers to expect on the last shopping day before Christmas. She collects data from three previous years, and determines that the crowds follow the same general pattern. When the store opens at 10 am, 500 people enter, and the total number in the store doubles every 40 minutes. When the number of people in the store reaches $10{,}000$, security guards need to be stationed at the entrances to control the crowds. At what time should the guards be commissioned?

54. The Richter scale is a measure of the ground motion that occurs during an earthquake. The intensity, R, of an earthquake as measured on the Richter scale is given by

$$R = \log\left(\frac{a}{T}\right) + B$$

where a is the amplitude (in microns) of vertical ground motion, T is the period (in seconds) of the seismic wave, and B is a constant. Let $B = 4.250$ and $T = 2.5$. Find a if

(a) $R = 6.1$ **(b)** $R = 7.1$

(c) Compare the values of R in parts (a) and (b). How do the corresponding values of a compare?

55. Since $e = 2.718\ldots$ we know that $2 < e < 3$, which means that $2^2 < e^2 < 3^2$. Without using a calculator, explain why

(a) $1 < \ln 3 < 2$ **(b)** $1 < \ln 4 < 2$

CHECK YOUR UNDERSTANDING

Are the statements in Problems 1–29 true or false? Give an explanation for your answer.

1. If x is a positive number, $\log x$ is the exponent of 10 that gives x.

2. If $10^y = x$ then $\log x = y$.

3. The quantity 10^{-k} is a negative number when k is positive.

4. For any n, we have $\log(10^n) = n$.

5. If $n > 0$, then $10^{\log n} = n$.

6. If a and b are positive, $\log\left(\frac{a}{b}\right) = \frac{\log a}{\log b}$.

7. If a and b are positive, $\ln(a + b) = \ln a + \ln b$.

8. For any value a, $\log a = \ln a$.

9. For any value x, $\ln(e^{2x}) = 2x$.

10. The function $y = \log x$ has an asymptote at $y = 0$.

11. The graph of the function $y = \log x$ is concave down.

12. The reflected graph of $y = \log x$ across the line $y = x$ is the graph of $y = 10^x$.

13. If $y = \log \sqrt{x}$ then $y = \frac{1}{2}\log x$.

14. The function $y = \log(b^t)$is always equal to $y = (\log b)^t$.

15. The values of $\ln e$ and $\log 10$ are both 1.

16. If $7.32 = e^t$ then $t = \frac{7.32}{e}$.

17. If $50(0.345)^t = 4$, then $t = \frac{\log(4/50)}{\log 0.345}$.

18. If $ab^t = n$, then $t = \frac{\log(n/a)}{\log b}$.

19. The doubling time of a quantity $Q = Q_0e^{kt}$ is the time it takes for any t-value to double.

20. The half-life of a quantity is the time it takes for the quantity to be reduced by half.

21. If the half-life of a substance is 5 hours then there will be $\frac{1}{4}$ of the substance in 25 hours.

22. If $y = 6(3)^t$, then $y = 6e^{(\ln 3)t}$.

23. If a population doubles in size every 20 years, its annual continuous growth rate is 20%.

24. If $Q = Q_0e^{kt}$, then $t = \frac{\ln(Q/Q_0)}{k}$.

25. Log scales provide a way to graph quantities that have vastly different magnitudes.

26. In a graph made using a log-log scale, consecutive powers of 10 are equally spaced on the horizontal axis and on the vertical axis.

27. One million and one billion differ by one order of magnitude.

28. After fitting a data set with both an exponential function, $y = Ae^{kx}$, and a power function, $y = Bx^n$, we must have $B = A$.

29. Given the points on a cubic curve, $(1, 1)$, $(2, 8)$, $(3, 27)$ and $(4, 64)$ it is not possible to fit an exponential function to this data.

TOOLS FOR CHAPTER 4: LOGARITHMS

We list the definitions and properties of the common and natural logarithms.

Properties of Logarithms If $M, N > 0$:

- Logarithm of a product: $\log MN = \log M + \log N$ $\quad$ $\ln MN = \ln M + \ln N$
- Logarithm of a quotient: $\log M/N = \log M - \log N$ $\quad$ $\ln M/N = \ln M - \ln N$
- Logarithm of a power: $\log M^P = P \log M$ $\quad$ $\ln M^P = P \ln M$
- Logarithm of 1: $\log 1 = 0$ $\quad$ $\ln 1 = 0$
- Logarithm of the base: $\log 10 = 1$ $\quad$ $\ln e = 1$

Be aware of the following two common errors,

$$\log(M + N) \neq (\log M)(\log N)$$

and

$$\log(M - N) \neq \frac{\log M}{\log N}.$$

Relationships Between Logarithms and Exponents

- $\log N = x$ if and only if $10^x = N$ $\quad$ $\ln N = x$ if and only if $e^x = N$
- $\log 10^x = x$ $\quad$ $\ln e^x = x$
- $10^{\log x} = x$, for $x > 0$ $\quad$ $e^{\ln x} = x$, for $x > 0$

Example 1 Evaluate without a calculator:

(a) $\log 10{,}000$ $\qquad$ (b) $\ln 1$

Solution

(a) Common logarithms are powers of 10. The power of 10 needed to get 10,000 is 4, so $\log 10{,}000 = 4$.

(b) Natural logarithms are powers of e. The power of e needed to get 1 is zero, so $\ln 1 = 0$.

Example 2 Write the equation in exponential form

(a) $\log x = -3$ $\qquad$ (b) $\ln x = \sqrt{2}$

Solution

(a) By definition $\log x = -3$ means $10^{-3} = x$.

(b) By definition $\ln x = \sqrt{2}$ means $e^{\sqrt{2}} = x$.

Example 3 Write the equation in logarithmic form

(a) $10^x = 1000$ $\qquad$ (b) $10^{-2} = 0.01$ $\qquad$ (c) $e^{-1} = 0.368$

Solution

(a) By definition $10^x = 1000$ means $\log 1000 = x$.

(b) By definition $10^{-2} = 0.01$ means $\log 0.01 = -2$.

(c) By definition $e^{-1} = 0.368$ means $\ln 0.368 = -1$.

Example 4 Write the expression using sums and/or differences of logarithmic expressions which do not contain the logarithms of products, quotients or powers.

(a) $\log(10x)$ (b) $\ln\left(\frac{e^2}{\sqrt{x}}\right)$

Solution (a)

$$\begin{aligned}\log(10x) &= \log 10 + \log x \quad \text{Logarithm of a product}\\ &= 1 + \log x \quad \text{Logarithm of the base.}\end{aligned}$$

(b)

$$\begin{aligned}\ln\left(\frac{e^2}{\sqrt{x}}\right) &= \ln e^2 - \ln\sqrt{x} \quad \text{Logarithm of a quotient}\\ &= \ln e^2 - \ln x^{1/2}\\ &= 2\ln e - \frac{1}{2}\ln x \quad \text{Logarithm of a power}\\ &= 2\cdot 1 - \frac{1}{2}\ln x \quad \text{Logarithm of the base}\\ &= 2 - \frac{1}{2}\ln x.\end{aligned}$$

Example 5 Write the expression as a single logarithm.

(a) $\ln x - 2\ln y$ (b) $3(\log x + \frac{4}{3}\log y)$

Solution (a)

$$\begin{aligned}\ln x - 2\ln y &= \ln x - \ln y^2 \quad \text{Logarithm of a power}\\ &= \ln\left(\frac{x}{y^2}\right) \quad \text{Logarithm of a quotient.}\end{aligned}$$

(b)

$$\begin{aligned}3\left(\log x + \frac{4}{3}\log y\right) &= 3\log x + 4\log y\\ &= \log x^3 + \log y^4 \quad \text{Logarithm of a power}\\ &= \log(x^3y^4) \quad \text{Logarithm of a product.}\end{aligned}$$

Example 6 Express in terms of x without logarithms.

(a) $e^{3\ln x}$ (b) $\log 10^{2x}$

Solution (a)

$$\begin{aligned}e^{3\ln x} &= e^{\ln x^3} \quad \text{Logarithm of a power}\\ &= x^3 \quad \text{Logarithm of the base.}\end{aligned}$$

(b) $\log 10^{2x} = 2x$ Logarithm of the base.

Example 7 Solve the equation for x.

(a) $12e^x = 5$ (b) $2^{-3x} = 17$

Solution (a)

$$\begin{aligned} 12e^x &= 5 \\ e^x &= \frac{5}{12} && \text{Dividing by 12} \\ \ln e^x &= \ln\left(\frac{5}{12}\right) && \text{Taking ln of both sides} \\ x &= \ln\left(\frac{5}{12}\right) && \text{Logarithm of the base.} \end{aligned}$$

(b)

$$\begin{aligned} 2^{-3x} &= 17 \\ \log 2^{-3x} &= \log 17 && \text{Taking logs of both sides} \\ -3x\log 2 &= \log 17 && \text{Logarithm of a power} \\ -3x &= \frac{\log 17}{\log 2} && \text{Dividing by log 2} \\ x &= -\frac{\log 17}{3\log 2} && \text{Dividing by -3.} \end{aligned}$$

Example 8 Solve the equation for x.

(a) $2(\log(2x+50)) - 4 = 0$ (b) $\ln(x+2) = 3$

Solution (a)

$$\begin{aligned} 2(\log(2x+50)) - 4 &= 0 \\ 2(\log(2x+50)) &= 4 \\ \log(2x+50) &= 2 \\ 10^{\log(2x+50)} &= 10^2 && \text{Converting to exponential form} \\ 2x+50 &= 10^2 \\ 2x+50 &= 100 \\ 2x &= 50 \\ x &= 25. \end{aligned}$$

(b)

$$\begin{aligned} \ln(x+2) &= 3 \\ e^{\ln(x+2)} &= e^3 && \text{Raise } e \text{ to each side} \\ x+2 &= e^3 \\ x &= e^3 - 2. \end{aligned}$$

Exercises to Tools for Chapter 4

For Exercises 1–10, evaluate without a calculator.

1. $\log(\log 10)$
2. $\ln(\ln e)$
3. $\log 0.0001$
4. $2\ln e^4$
5. $\dfrac{\log 100^6}{\log 100^2}$
6. $\ln\left(\dfrac{1}{e^5}\right)$
7. $\dfrac{\log 1}{\log 10^5}$
8. $e^{\ln 3} - \ln e$
9. $\sqrt{\log 10{,}000}$
10. $10^{\log 7}$

For Exercises 11–16, rewrite the exponential equation in equivalent logarithmic form.

11. $10^5 = 100{,}000$
12. $10^{-4} = 0.0001$
13. $10^{0.477} = 3$
14. $e^2 = 7.389$
15. $e^{-2} = 0.135$
16. $e^{2x} = 7$

For Exercises 17–20, rewrite the logarithmic equation in equivalent exponential form.

17. $\log 0.01 = -2$
18. $\log(x+3) = 2$
19. $\ln x = -1$
20. $\ln 4 = x^2$

For Exercises 21–31, if possible, write the expression using sums and/or differences of logarithmic expressions which do not contain the logarithms of products, quotients or powers.

21. $\log 2x$
22. $\ln(x(7-x)^3)$
23. $\dfrac{\ln x}{2}$
24. $\log\left(\dfrac{x}{5}\right)$
25. $\log\left(\dfrac{x^2+1}{x^3}\right)$
26. $\ln\sqrt{\dfrac{x-1}{x+1}}$
27. $\log(x^2+y^2)$
28. $\ln\left(\dfrac{xy^2}{z}\right)$
29. $\log(x^2-y^2)$
30. $(\log x)(\log y)$
31. $\dfrac{\ln x^2}{\ln(x+2)}$

For Exercises 32–40, rewrite the expression as a single logarithm.

32. $\log 12 + \log x$
33. $\ln x^3 + \ln x^2$
34. $\ln x^2 - \ln(x+10)$
35. $\frac{1}{2}\log x + 4\log y$
36. $\log 3 + 2\log\sqrt{x}$
37. $\frac{1}{3}\log 8 - \frac{1}{2}\log 25$
38. $3\left(\log(x+1) + \frac{2}{3}\log(x+4)\right)$
39. $\ln x + \ln\left(\frac{y}{2}(x+4)\right) + \ln z^{-1}$
40. $2\log(9-x^2) - (\log(3+x) + \log(3-x))$

For Exercises 41–52, simplify the expression if possible.

41. $10^{-\log 5x}$
42. $e^{-3\ln t}$
43. $2\ln e^{\sqrt{x}}$
44. $\log(A^2+B^2)$
45. $t\ln e^{t/2}$
46. $10^{2+\log x}$
47. $\log 10x - \log x$
48. $2\ln x^{-2} + \ln x^4$
49. $\ln\sqrt{x^2+16}$
50. $\log 100^{2z}$
51. $\dfrac{\ln e}{\ln e^2}$
52. $\ln\dfrac{1}{e^x+1}$

For Exercises 53–62, solve for x using logarithms.

53. $4^x = 9$
54. $12^x = 7$
55. $3\cdot 5^x = 9$
56. $4\cdot 13^{3x} = 17$
57. $e^x = 8$
58. $2e^x = 13$
59. $e^{-5x} = 9$
60. $e^{7x} = 5e^{3x}$
61. $12^{5x} = 3\cdot 15^{2x}$
62. $19^{6x} = 77\cdot 7^{4x}$

In Exercises 63–68, solve for x.

63. $\log(2x+7) = 2$
64. $3\log(4x+9) - 6 = 2$
65. $4\log(9x+17) - 5 = 1$
66. $\log(2x) = \log(x+10)$
67. $\ln(3x+4) = 5$
68. $2\ln(6x-1) + 5 = 7$

Chapter Five

TRANSFORMATIONS OF FUNCTIONS AND THEIR GRAPHS

We have introduced the families of linear, exponential, and logarithmic functions and we will study several other families in the later chapters. Before going on to the next family, we introduce some tools that are useful for analyzing every family of functions. These tools allow us to transform members of a family into one another by shifting, flipping, and stretching their graphs. In the process, we construct another family, the family of quadratic functions.

Throughout the chapter, we consider the relationship between changes made to the formula of a function and changes made to its graph.

The Tools Section on page 239 reviews completing the square.

5.1 VERTICAL AND HORIZONTAL SHIFTS

Suppose we shift the graph of some function vertically or horizontally, giving the graph of a new function. In this section we investigate the relationship between the formulas for the original function and the new function.

Vertical and Horizontal Shift: The Heating Schedule For an Office Building

We start with an example of a vertical shift in the context of the heating schedule for a building.

Example 1 To save money, an office building is kept warm only during business hours. Figure 5.1 shows the temperature, H, in °F, as a function of time, t, in hours after midnight. At midnight ($t = 0$), the building's temperature is 50°F. This temperature is maintained until 4 am. Then the building begins to warm up so that by 8 am the temperature is 70°F. At 4 pm the building begins to cool. By 8 pm, the temperature is again 50°F.

Suppose that the building's superintendent decides to keep the building 5°F warmer than before. Sketch a graph of the resulting function.

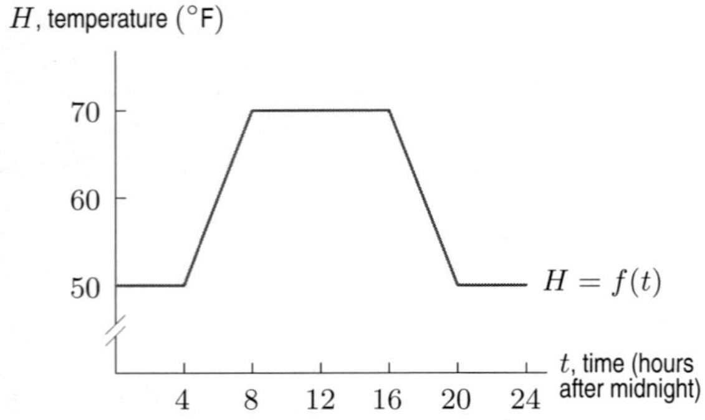

Figure 5.1: The heating schedule at an office building

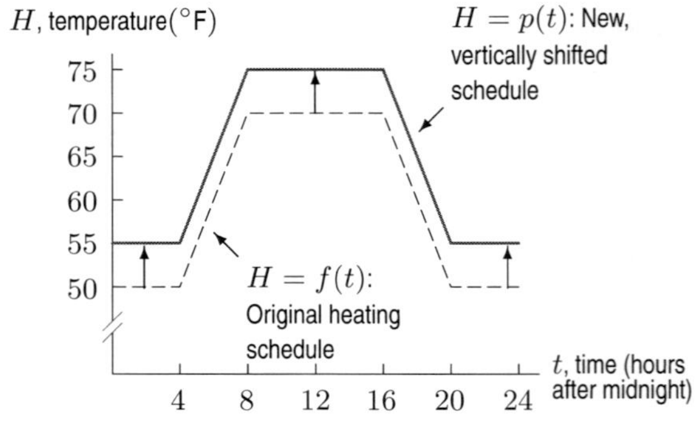

Figure 5.2: Graph of new heating schedule, $H = p(t)$, obtained by shifting original graph, $H = f(t)$, upward by 5 units

Solution The graph of f, the heating schedule function of Figure 5.1, is shifted upward by 5 units. The new heating schedule, $H = p(t)$, is graphed in Figure 5.2. The building's overnight temperature is now 55°F instead of 50°F and its daytime temperature is 75°F instead of 70°F. The 5°F increase in temperature corresponds to the 5-unit vertical shift in the graph.

The next example involves shifting a graph horizontally.

Example 2 The superintendent then changes the original heating schedule to start two hours earlier. The building now begins to warm at 2 am instead of 4 am, reaches 70°F at 6 am instead of 8 am, begins cooling off at 2 pm instead of 4 pm, and returns to 50°F at 6 pm instead of 8 pm. How are these changes reflected in the graph of the heating schedule?

Solution Figure 5.3 gives a graph of $H = q(t)$, the new heating schedule, which is obtained by shifting the graph of the original heating schedule, $H = f(t)$, two units to the left.

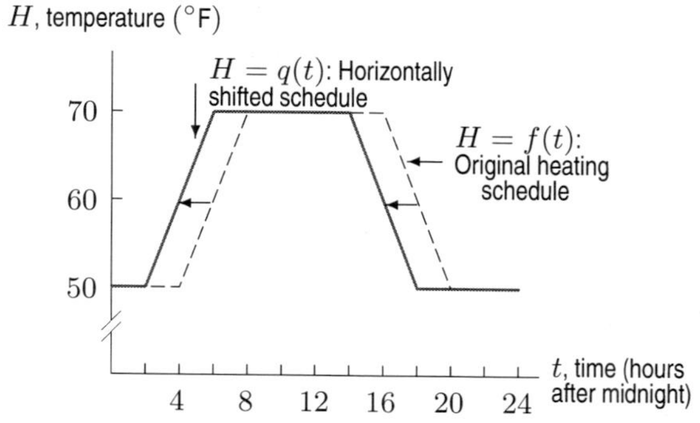

Figure 5.3: Graph of new heating schedule, $H = q(t)$, found by shifting, f, the original graph 2 units to the left

Notice that the upward shift in Example 1 results in a warmer temperature, whereas the leftward shift in Example 2 results in an earlier schedule.

Formulas for a Vertical or Horizontal Shift

How does a horizontal or vertical shift of a function's graph affect its formula?

Example 3 In Example 1, the graph of the original heating schedule, $H = f(t)$, was shifted upward by 5 units; the result was the warmer schedule $H = p(t)$. How are the formulas for $f(t)$ and $p(t)$ related?

Solution The temperature under the new schedule, $p(t)$, is always 5°F warmer than the temperature under the old schedule, $f(t)$. Thus,

$$\text{New temperature at time } t = \text{Old temperature at time } t + 5.$$

Writing this algebraically:

$$\underbrace{p(t)}_{\substack{\text{New temperature}\\ \text{at time } t}} = \underbrace{f(t)}_{\substack{\text{Old temperature}\\ \text{at time } t}} + 5.$$

The relationship between the formulas for p and f is given by the equation $p(t) = f(t) + 5$.

We can get information from the relationship $p(t) = f(t) + 5$, although we do not have an explicit formula for f or p.

Suppose we need to know the temperature at 6 am under the schedule $p(t)$. The graph of $f(t)$ shows that under the old schedule $f(6) = 60$. Substituting $t = 6$ into the equation relating f and p gives $p(6)$:

$$p(6) = f(6) + 5 = 60 + 5 = 65.$$

Thus, at 6 am the temperature under the new schedule is 65°F.

Example 4 In Example 2 the heating schedule was changed to 2 hours earlier, shifting the graph horizontally 2 units to the left. Find a formula for q, this new schedule, in terms of f, the original schedule.

Solution The old schedule always reaches a given temperature 2 hours after the new schedule. For example, at 4 am the temperature under the new schedule reaches 60°. The temperature under the old schedule reaches 60° at 6 am, 2 hours later. The temperature reaches 65° at 5 am under the new schedule, but not until 7 am, under the old schedule. In general, we see that

$$\begin{array}{c} \text{Temperature under new schedule} \\ \text{at time } t \end{array} = \begin{array}{c} \text{Temperature under old schedule} \\ \text{at time } (t+2)\text{, two hours later.} \end{array}$$

Algebraically, we have

$$q(t) = f(t+2).$$

This is a formula for q in terms of f.

Let's check the formula from Example 4 by using it to calculate $q(14)$, the temperature under the new schedule at 2 pm. The formula gives

$$q(14) = f(14+2) = f(16).$$

Figure 5.1 shows that $f(16) = 70$. Thus, $q(14) = 70$. This agrees with Figure 5.3.

Translations of a Function and Its Graph

In the heating schedule example, the function representing a warmer schedule,

$$p(t) = f(t) + 5,$$

has a graph which is a vertically shifted version of the graph of f. On the other hand, the earlier schedule is represented by

$$q(t) = f(t+2)$$

and its graph is a horizontally shifted version of the graph of f. Adding 5 to the temperature, or output value, $f(t)$, shifted its graph *up* five units. Adding 2 to the time, or input value, t, shifted its graph to the *left* two units. Generalizing these observations to any function g:

> If $y = g(x)$ is a function and k is a constant, then the graph of
> - $y = g(x) + k$ is the graph of $y = g(x)$ shifted vertically $|k|$ units. If k is positive, the shift is up; if k is negative, the shift is down.
> - $y = g(x+k)$ is the graph of $y = g(x)$ shifted horizontally $|k|$ units. If k is positive, the shift is to the left; if k is negative, the shift is to the right.

A vertical or horizontal shift of the graph of a function is called a *translation* because it does not change the shape of the graph, but simply translates it to another position in the plane. Shifts or translations are the simplest examples of *transformations* of a function. We will see others in later sections of Chapter 5.

Inside and Outside Changes

Since $y = g(x + k)$ involves a change to the input value, x, it is called an *inside change* to g. Similarly, since $y = g(x) + k$ involves a change to the output value, $g(x)$, it is called an *outside change*. In general, an inside change in a function results in a horizontal change in its graph, whereas an outside change results in a vertical change.

In this section, we consider changes to the input and output of a function. For the function

$$Q = f(t),$$

a change inside the function's parentheses can be called an "inside change" and a change outside the function's parentheses can be called an "outside change."

Example 5 If $n = f(A)$ gives the number of gallons of paint needed to cover a house of area A ft^2, explain the meaning of the expressions $f(A + 10)$ and $f(A) + 10$ in the context of painting.

Solution These two expressions are similar in that they both involve adding 10. However, for $f(A + 10)$, the 10 is added on the inside, so 10 is added to the area, A. Thus,

$$n = f(\underbrace{A + 10}_{\text{Area}}) = \begin{array}{c}\text{Amount of paint needed}\\ \text{to cover an area of } (A + 10) \text{ ft}^2\end{array} = \begin{array}{c}\text{Amount of paint needed to cover}\\ \text{an area 10 ft}^2 \text{ larger than } A.\end{array}$$

The expression $f(A) + 10$ represents an outside change. We are adding 10 to $f(A)$, which represents an amount of paint, not an area. We have

$$n = \underbrace{f(A)}_{\substack{\text{Amount}\\ \text{of paint}}} + 10 = \begin{array}{c}\text{Amount of paint needed}\\ \text{to cover region of area } A\end{array} + 10 \text{ gals} = \begin{array}{c}\text{10 gallons more paint than}\\ \text{amount needed to cover area } A.\end{array}$$

In $f(A + 10)$, we added 10 square feet on the inside of the function, which means that the area to be painted is now 10 ft^2 larger. In $f(A) + 10$, we added 10 gallons to the outside, which means that we have 10 more gallons of paint than we need.

Example 6 Let $s(t)$ be the average weight (in pounds) of a baby at age t months. The weight, V, of a particular baby named Jonah is related to the average weight function $s(t)$ by the equation

$$V = s(t) + 2.$$

Find Jonah's weight at ages $t = 3$ and $t = 6$ months. What can you say about Jonah's weight in general?

Solution At $t = 3$ months, Jonah's weight is

$$V = s(3) + 2.$$

Since $s(3)$ is the average weight of a 3-month old boy, we see that at 3 months, Jonah weighs 2 pounds more than average. Similarly, at $t = 6$ months we have

$$V = s(6) + 2,$$

which means that, at 6 months, Jonah weighs 2 pounds more than average. In general, Jonah weighs 2 pounds more than average for babies of his age.

Example 7 The weight, W, of another baby named Ben is related to $s(t)$ by the equation

$$W = s(t+4).$$

What can you say about Ben's weight at age $t = 3$ months? At $t = 6$ months? Assuming that babies increase in weight over the first year of life, decide if Ben is of average weight for his age, above average, or below average.

Solution Since $W = s(t+4)$, at age $t = 3$ months Ben's weight is given by

$$W = s(3+4) = s(7).$$

We defined $s(7)$ to be the average weight of a 7-month old baby. At age 3 months, Ben's weight is the same as the average weight of 7-month old babies. Since, on average, a baby's weight increases as the baby grows, this means that Ben is heavier than the average for a 3-month old. Similarly, at age $t = 6$, Ben's weight is given by

$$W = s(6+4) = s(10).$$

Thus, at 6 months, Ben's weight is the same as the average weight of 10-month old babies. In both cases, we see that Ben is above average in weight.

Notice that in Example 7, the equation

$$W = s(t+4)$$

involves an inside change, or a change in months. This equation tells us that Ben weighs as much as babies who are 4 months older than he is. However in Example 6, the equation

$$V = s(t) + 2$$

involves an outside change, or a change in weight. This equation tells us that Jonah is 2 pounds heavier than the average weight of babies his age. Although both equations tell us that the babies are heavier than average for their age, they vary from the average in different ways.

Combining Horizontal and Vertical Shifts

We have seen what happens when we shift a function's graph either horizontally or vertically. What happens if we shift it both horizontally and vertically?

Example 8 Let r be the transformation of the heating schedule function, $H = f(t)$, defined by the equation

$$r(t) = f(t-2) - 5.$$

(a) Sketch the graph of $H = r(t)$.
(b) Describe in words the heating schedule determined by r.

Solution

(a) To graph r, we break this transformation into two steps. First, we sketch a graph of $H = f(t-2)$. This is an inside change to the function f and it results in the graph of f being shifted 2 units to the right. Next, we sketch a graph of $H = f(t-2) - 5$. This graph can be found by shifting our sketch of $H = f(t-2)$ down 5 units. The resulting graph is shown in Figure 5.4. The graph of r is the graph of f shifted 2 units to the right and 5 units down.

(b) The function r represents a schedule that is both 2 hours later and 5 degrees cooler than the original schedule.

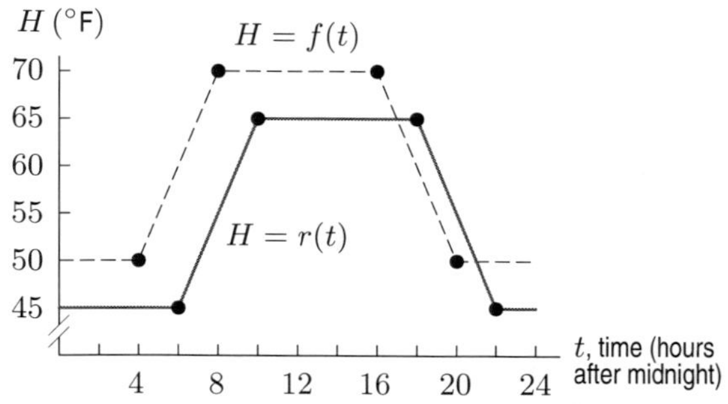

Figure 5.4: Graph of $r(t) = f(t-2) - 5$ is graph of $H = f(t)$ shifted right by 2 and down by 5

We can use transformations to understand an unfamiliar function by relating it to a function we already know.

Example 9 A graph of $f(x) = x^2$ is in Figure 5.5. Define g by shifting the graph of f to the right 2 units and down 1 unit; see Figure 5.6. Find a formula for g in terms of f. Find a formula for g in terms of x.

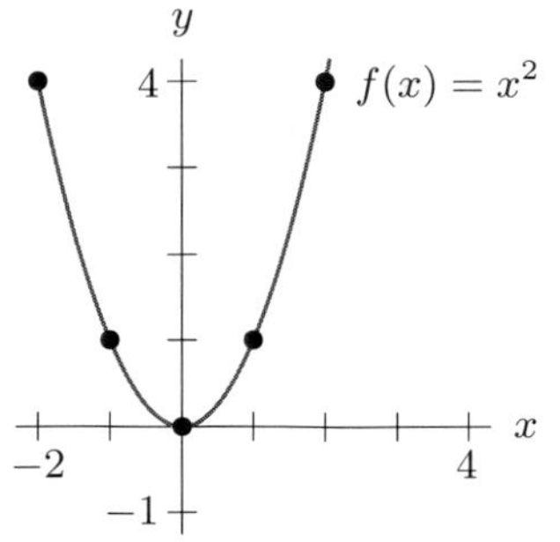

Figure 5.5: The graph of $f(x) = x^2$

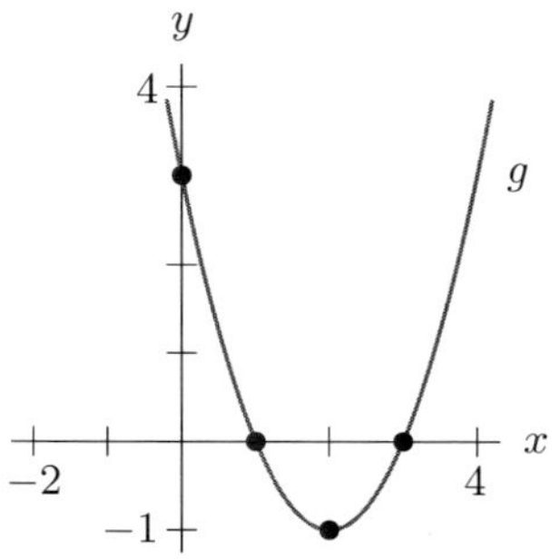

Figure 5.6: The graph of g, a transformation of f

Solution The graph of g is the graph of f shifted to the right 2 units and down 1 unit, so a formula for g is $g(x) = f(x-2) - 1$. Since $f(x) = x^2$, we have $f(x-2) = (x-2)^2$. Therefore,

$$g(x) = (x-2)^2 - 1.$$

It is a good idea to check by graphing $g(x) = (x-2)^2 - 1$ and comparing the graph with Figure 5.6.

Exercises and Problems for Section 5.1

Exercises

1. Using Table 5.1, complete the tables for g, h, k, m, where:

(a) $g(x) = f(x-1)$ **(b)** $h(x) = f(x+1)$
(c) $k(x) = f(x) + 3$ **(d)** $m(x) = f(x-1)+3$

Explain how the graph of each function relates to the graph of $f(x)$.

Table 5.1

x	-2	-1	0	1	2
$f(x)$	-3	0	2	1	-1

x	-1	0	1	2	3
$g(x)$					

x	-3	-2	-1	0	1
$h(x)$					

x	-2	-1	0	1	2
$k(x)$					

x	-1	0	1	2	3
$m(x)$					

In Exercises 2–5, graph the transformations of $f(x)$ in Figure 5.7.

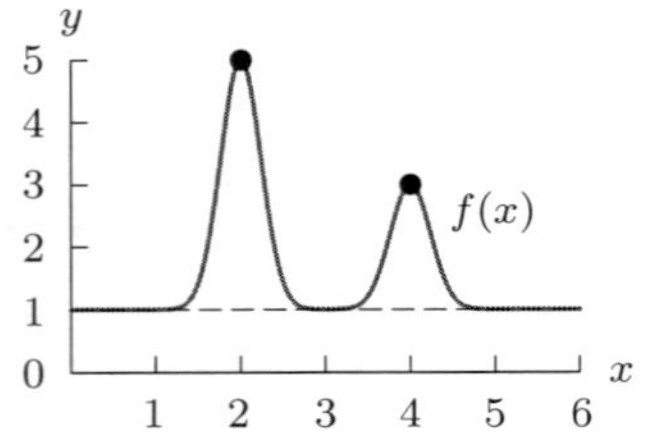

Figure 5.7

2. $y = f(x+2)$ **3.** $y = f(x) + 2$

4. $y = f(x-1) - 5$ **5.** $y = f(x+6) - 4$

6. Let $f(x) = 4^x$, $g(x) = 4^x + 2$, and $h(x) = 4^x - 3$. What is the relationship between the graph of $f(x)$ and the graphs of $h(x)$ and $g(x)$?

7. Let $f(x) = \left(\frac{1}{3}\right)^x$, $g(x) = \left(\frac{1}{3}\right)^{x+4}$, and $h(x) = \left(\frac{1}{3}\right)^{x-2}$. How do the graphs of $g(x)$ and $h(x)$ compare to the graph of $f(x)$?

8. Match the graphs in (a)–(f) with the formulas in (i)–(vi).

(i) $y = |x|$ (ii) $y = |x| - 1.2$
(iii) $y = |x - 1.2|$ (iv) $y = |x| + 2.5$
(v) $y = |x + 3.4|$ (vi) $y = |x - 3| + 2.7$

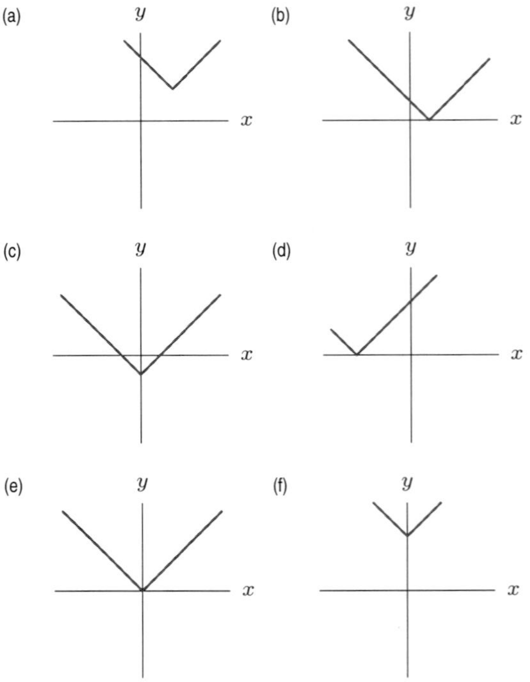

9. The graph of $f(x)$ contains the point $(3, -4)$. What point must be on the graph of

(a) $f(x) + 5$? **(b)** $f(x+5)$?
(c) $f(x-3) - 2$?

10. The domain of the function $g(x)$ is $-2 < x < 7$. What is the domain of $g(x-2)$?

11. The range of the function $R(s)$ is $100 \leq R(s) \leq 200$. What is the range of $R(s) - 150$?

Write a formula and graph the transformations of $m(n) = \frac{1}{2}n^2$ in Exercises 12–19.

12. $y = m(n) + 1$ **13.** $y = m(n+1)$

14. $y = m(n) - 3.7$ **15.** $y = m(n - 3.7)$

16. $y = m(n) + \sqrt{13}$ **17.** $y = m(n + 2\sqrt{2})$

18. $y = m(n+3) + 7$ **19.** $y = m(n - 17) - 159$

Write a formula and graph the transformations of $k(w) = 3^w$ in Exercises 20–25.

20. $y = k(w) - 3$

21. $y = k(w - 3)$

22. $y = k(w) + 1.8$

23. $y = k(w + \sqrt{5})$

24. $y = k(w + 2.1) - 1.3$

25. $y = k(w - 1.5) - 0.9$

Problems

26. **(a)** Using Table 5.2, evaluate

(i) $f(x)$ for $x = 6$.

(ii) $f(5) - 3$.

(iii) $f(5 - 3)$.

(iv) $g(x) + 6$ for $x = 2$.

(v) $g(x + 6)$ for $x = 2$.

(vi) $3g(x)$ for $x = 0$.

(vii) $f(3x)$ for $x = 2$.

(viii) $f(x) - f(2)$ for $x = 8$.

(ix) $g(x + 1) - g(x)$ for $x = 1$.

(b) Solve

(i) $g(x) = 6$.
(ii) $f(x) = 574$.
(iii) $g(x) = 281$.

(c) The values in the table were obtained using the formulas $f(x) = x^3 + x^2 + x - 10$ and $g(x) = 7x^2 - 8x - 6$. Use the table to find two solutions to the equation $x^3 + x^2 + x - 10 = 7x^2 - 8x - 6$.

Table 5.2

x	0	1	2	3	4	5	6	7	8	9
$f(x)$	−10	−7	4	29	74	145	248	389	574	809
$g(x)$	−6	−7	6	33	74	129	198	281	378	489

27. The graph of $g(x)$ contains the point $(-2, 5)$. Write a formula for a translation of g whose graph contains the point

(a) $(-2, 8)$
(b) $(0, 5)$

28. **(a)** Let $f(x) = \left(\frac{x}{2}\right)^3 + 2$. Calculate $f(-6)$.
(b) Solve $f(x) = -6$.
(c) Find points that correspond to parts (a) and (b) on the graph of $f(x)$ in Figure 5.8.
(d) Calculate $f(4) - f(2)$. Draw a vertical line segment on the y-axis that illustrates this calculation.
(e) If $a = -2$, compute $f(a + 4)$ and $f(a) + 4$.
(f) In part (e), what x-value corresponds to $f(a + 4)$? To $f(a) + 4$?

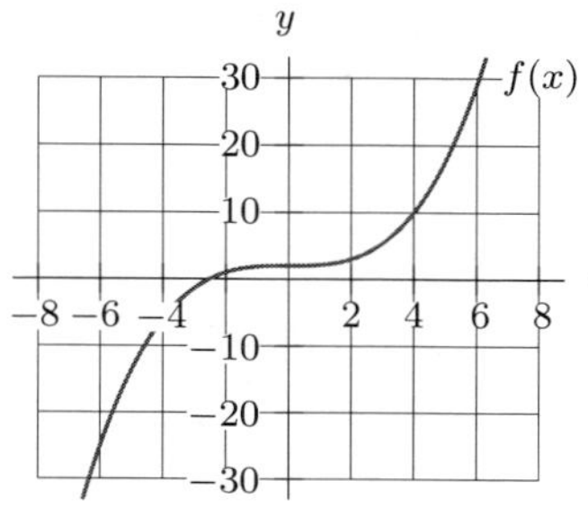

Figure 5.8

29. The function $P(t)$ gives the number of people in a certain population in year t. Interpret in terms of population:

(a) $P(t) + 100$
(b) $P(t + 100)$

30. Describe a series of shifts which translates the graph of $y = (x + 3)^3 - 1$ onto the graph of $y = x^3$.

31. Graph $f(x) = \ln(|x - 3|)$ and $g(x) = \ln(|x|)$. Find the vertical asymptotes of both functions.

32. Graph $y = \log x$, $y = \log(10x)$, and $y = \log(100x)$. How do the graphs compare? Use a property of logs to show that the graphs are vertical shifts of one another.

Explain in words the effects of the transformations in Exercises 33–38 on the graph of $q(z)$. Assume a, b are positive constants.

33. $q(z) + 3$

34. $q(z) - a$

35. $q(z + 4)$

36. $q(z - a)$

37. $q(z + b) - a$

38. $q(z - 2b) + ab$

39. Suppose $S(d)$ gives the height of high tide in Seattle on a specific day, d, of the year. Use shifts of the function $S(d)$ to find formulas for each of the following functions:

(a) $T(d)$, the height of high tide in Tacoma on day d, given that high tide in Tacoma is always one foot higher than high tide in Seattle.
(b) $P(d)$, the height of high tide in Portland on day d, given that high tide in Portland is the same height as the previous day's high tide in Seattle.

40. Table 5.3 contains values of $f(x)$. Each function in parts (a)–(c) is a translation of $f(x)$. Find a possible formula for each of these functions in terms of f. For example, given the data in Table 5.4, you could say that $k(x) = f(x) + 1$.

Table 5.3

x	0	1	2	3	4	5	6	7
$f(x)$	0	0.5	2	4.5	8	12.5	18	24.5

Table 5.4

x	0	1	2	3	4	5	6	7
$k(x)$	1	1.5	3	5.5	9	13.5	19	25.5

(a)

x	0	1	2	3	4	5	6	7
$h(x)$	-2	-1.5	0	2.5	6	10.5	16	22.5

(b)

x	0	1	2	3	4	5	6	7
$g(x)$	0.5	2	4.5	8	12.5	18	24.5	32

(c)

x	0	1	2	3	4	5	6	7
$i(x)$	-1.5	0	2.5	6	10.5	16	22.5	30

41. For $t \geq 0$, let $H(t) = 68 + 93(0.91)^t$ give the temperature of a cup of coffee in degrees Fahrenheit t minutes after it is brought to class.

(a) Find formulas for $H(t + 15)$ and $H(t) + 15$.
(b) Graph $H(t)$, $H(t + 15)$, and $H(t) + 15$.
(c) Describe in practical terms a situation modeled by the function $H(t + 15)$. What about $H(t) + 15$?
(d) Which function, $H(t+15)$ or $H(t)+15$, approaches the same final temperature as the function $H(t)$? What is that temperature?

42. At a jazz club, the cost of an evening is based on a cover charge of \$20 plus a beverage charge of \$7 per drink.

(a) Find a formula for $t(x)$, the total cost for an evening in which x drinks are consumed.
(b) If the price of the cover charge is raised by \$5, express the new total cost function, $n(x)$, as a transformation of $t(x)$.
(c) The management increases the cover charge to \$30, leaves the price of a drink at \$7, but includes the first two drinks for free. For $x \geq 2$, express $p(x)$, the new total cost, as a transformation of $t(x)$.

43. A hot brick is removed from a kiln and set on the floor to cool. Let t be time in minutes after the brick was removed. The difference, $D(t)$, between the brick's temperature, initially 350°F, and room temperature, 70°F, decays exponentially over time at a rate of 3% per minute. The brick's temperature, $H(t)$, is a transformation of $D(t)$. Find a formula for $H(t)$. Compare the graphs of $D(t)$ and $H(t)$, paying attention to the asymptotes.

44. Suppose $T(d)$ gives the average temperature in your hometown on the dth day of last year (where $d = 1$ is January 1st, and so on).

(a) Graph $T(d)$ for $1 \leq d \leq 365$.
(b) Give a possible value for each of the following: $T(6)$; $T(100)$; $T(215)$; $T(371)$.
(c) What is the relationship between $T(d)$ and $T(d + 365)$? Explain.
(d) If you were to graph $w(d) = T(d + 365)$ on the same axes as $T(d)$, how would the two graphs compare?
(e) Do you think the function $T(d) + 365$ has any practical significance? Explain.

45. Let $f(x) = e^x$ and $g(x) = 5e^x$. If $g(x) = f(x - h)$, find h.

5.2 REFLECTIONS AND SYMMETRY

In Section 5.1 we saw that a horizontal shift of the graph of a function results from a change to the input of the function. (Specifically, adding or subtracting a constant inside the function's parentheses.) A vertical shift corresponds to an outside change.

In this section we consider the effect of reflecting a function's graph about the x or y-axis. A reflection about the x-axis corresponds to an outside change to the function's formula; a reflection about the y-axis and corresponds to an inside change.

A Formula for a Reflection

Figure 5.9 shows the graph of a function $y = f(x)$ and Table 5.5 gives a corresponding table of values. Note that we do not need an explicit formula for f.

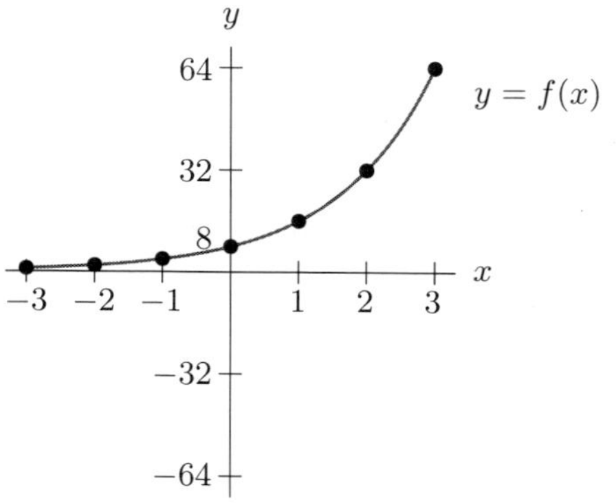

Figure 5.9: A graph of the function $y = f(x)$

Table 5.5 *Values of the function* $y = f(x)$

x	y
-3	1
-2	2
-1	4
0	8
1	16
2	32
3	64

Figure 5.10 shows a graph of a function $y = g(x)$, resulting from a vertical reflection of the graph of f about the x-axis. Figure 5.11 is a graph of a function $y = h(x)$, resulting from a horizontal reflection of the graph of f about the y-axis. Figure 5.12 is a graph of a function $y = k(x)$, resulting from a horizontal reflection of the graph of f about the y-axis followed by a vertical reflection about the x-axis.

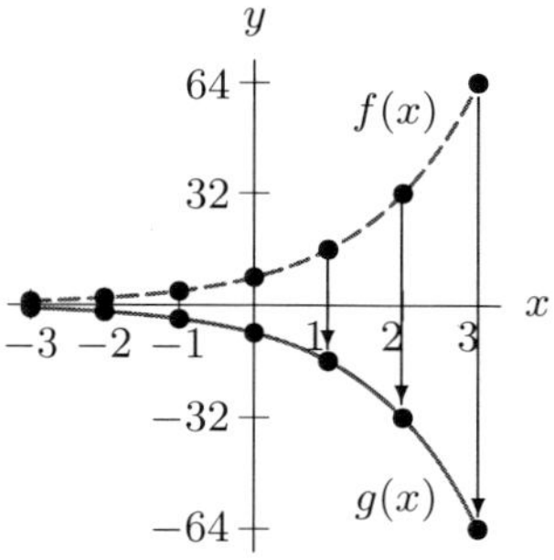

Figure 5.10: Graph reflected about x-axis

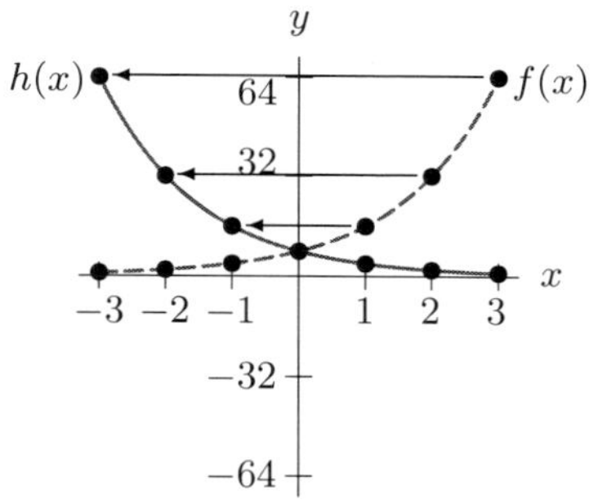

Figure 5.11: Graph reflected about y-axis

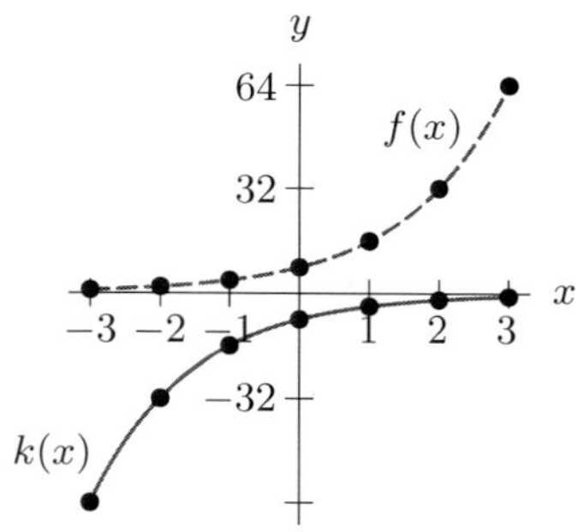

Figure 5.12: Graph reflected about y- and x-axes

Example 1 Find a formula in terms of f for (a) $y = g(x)$ (b) $y = h(x)$ (c) $y = k(x)$

Solution (a) The graph of $y = g(x)$ is obtained by reflecting the graph of f vertically about the x-axis. For example, the point $(3, 64)$ on the graph of f reflects to become the point $(3, -64)$ on the graph of g. The point $(2, 32)$ on the graph of f becomes $(2, -32)$ on the graph of g. See Table 5.6.

Table 5.6 *Values of the functions* $g(x)$ *and* $f(x)$ *graphed in Figure 5.10*

x	-3	-2	-1	0	1	2	3
$g(x)$	-1	-2	-4	-8	-16	-32	-64
$f(x)$	1	2	4	8	16	32	64

Notice that when a point is reflected vertically about the x-axis, the x-value stays fixed, while the y-value changes sign. That is, for a given x-value,

$$y\text{-value of } g \text{ is the negative of } y\text{-value of } f.$$

Algebraically, this means

$$g(x) = -f(x).$$

(b) The graph of $y = h(x)$ is obtained by reflecting the graph of $y = f(x)$ horizontally about the y-axis. In part (a), a vertical reflection corresponded to an outside change in the formula, specifically, multiplying by -1. Thus, you might guess that a horizontal reflection of the graph corresponds to an inside change in the formula. This is correct. To see why, consider Table 5.7.

Table 5.7 *Values of the functions $h(x)$ and $f(x)$ graphed in Figure 5.11*

x	-3	-2	-1	0	1	2	3
$h(x)$	64	32	16	8	4	2	1
$f(x)$	1	2	4	8	16	32	64

Notice that when a point is reflected horizontally about the y-axis, the y-value remains fixed, while the x-value changes sign. For example, since $f(-3) = 1$ and $h(3) = 1$, we have $h(3) = f(-3)$. Since $f(-1) = 4$ and $h(1) = 4$, we have $h(1) = f(-1)$. In general,

$$h(x) = f(-x).$$

(c) The graph of the function $y = k(x)$ results from a horizontal reflection of the graph of f about the y-axis, followed by a vertical reflection about the x-axis. Since a horizontal reflection corresponds to multiplying the inputs by -1 and a vertical reflection corresponds to multiplying the outputs by -1, we have

Vertical reflection across the x-axis

$$k(x) = -f(-x).$$

Horizontal reflection across the y-axis

Let's check a point. If $x = 1$, then the formula $k(x) = -f(-x)$ gives:

$$k(1) = -f(-1) = -4 \qquad \text{since } f(-1) = 4.$$

This result is consistent with the graph, since $(1, -4)$ is on the graph of $k(x)$.

For a function f:

- The graph of $y = -f(x)$ is a reflection of the graph of $y = f(x)$ about the x-axis.
- The graph of $y = f(-x)$ is a reflection of the graph of $y = f(x)$ about the y-axis.

Symmetry About the y-Axis

The graph of $p(x) = x^2$ in Figure 5.13 is *symmetric* about the y-axis. In other words, the part of the graph to the left of the y-axis is the mirror image of the part to the right of the y-axis. Reflecting the graph of $p(x)$ about the y-axis gives the graph of $p(x)$ again.

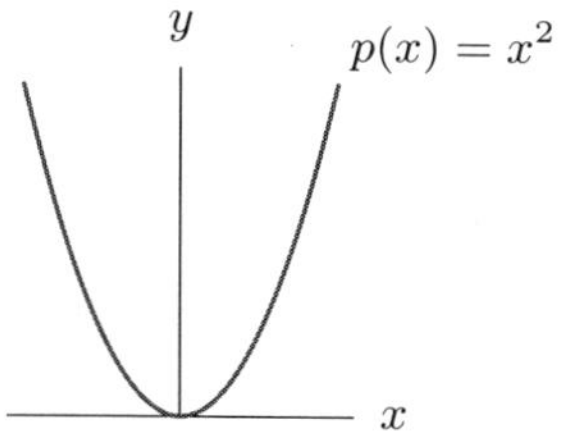

Figure 5.13: Reflecting the graph of $p(x) = x^2$ about the y-axis does not change its appearance

Symmetry about the y-axis is called *even symmetry*, because power functions with even exponents, such as $y = x^2$, $y = x^4$, $y = x^6$, ... have this property. Since $y = p(-x)$ is a reflection of the graph of p about the y-axis and $p(x)$ has even symmetry, we have

$$p(-x) = p(x).$$

To check this relationship, let $x = 2$. Then $p(2) = 2^2 = 4$, and $p(-2) = (-2)^2 = 4$, so $p(-2) = p(2)$. This means that the point $(2, 4)$ and its reflection about the y-axis, $(-2, 4)$, are both on the graph of $p(x)$.

Example 2 For the function $p(x) = x^2$, check algebraically that $p(-x) = p(x)$ for all x.

Solution Substitute $-x$ into the formula for $p(x)$ giving

$$\begin{aligned} p(-x) &= (-x)^2 = (-x)\cdot(-x) \\ &= x^2 \\ &= p(x). \end{aligned}$$

Thus, $p(-x) = p(x)$.

In general,

> If f is a function, then f is called an **even function** if, for all values of x in the domain of f,
>
> $$f(-x) = f(x).$$
>
> The graph of f is symmetric about the y-axis.

Symmetry About the Origin

Figures 5.14 and 5.15 show the graph of $q(x) = x^3$. Reflecting the graph of q first about the y-axis and then about the x-axis (or vice-versa) gives the graph of q again. This kind of symmetry is called symmetry about the origin, or *odd symmetry*.

In Example 1, we saw that $y = -f(-x)$ is a reflection of the graph of $y = f(x)$ about both the y-axis and the x-axis. Since $q(x) = x^3$ is symmetric about the origin, q is the same function as this double reflection. That is,

$$q(x) = -q(-x) \quad \text{which means that} \quad q(-x) = -q(x).$$

To check this relationship, let $x = 2$. Then $q(2) = 2^3 = 8$, and $q(-2) = (-2)^3 = -8$, so $q(-2) = -q(2)$. This means the point $(2, 8)$ and its reflection about the origin, $(-2, -8)$, are both on the graph of q.

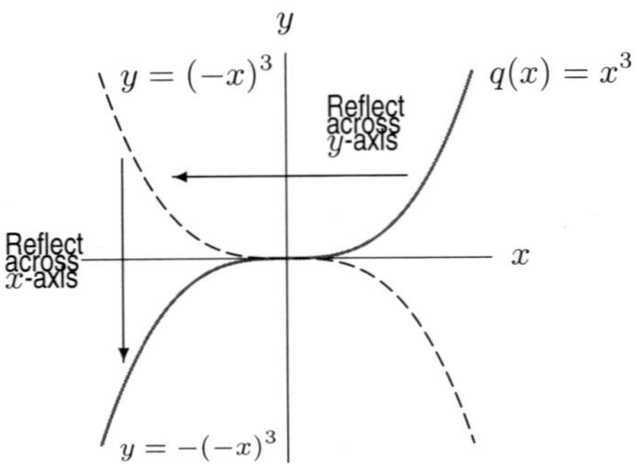

Figure 5.14: If the graph is reflected about the y-axis and then about the x-axis, it does not change

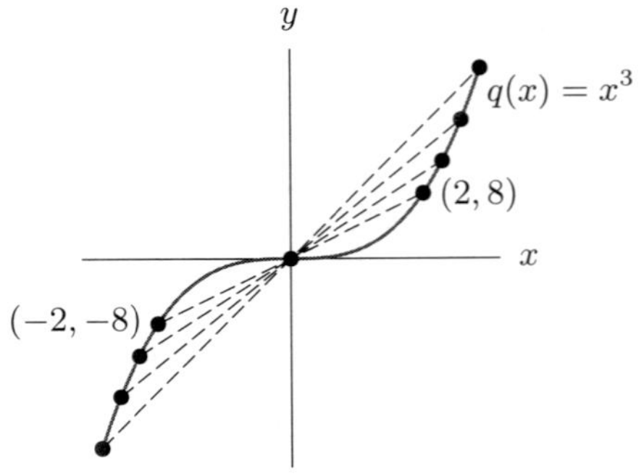

Figure 5.15: If every point on this graph is reflected about the origin, the graph is unchanged

---

Example 3 For the function $q(x) = x^3$, check algebraically that $q(-x) = -q(x)$ for all x.

Solution We evaluate $q(-x)$ giving

$$\begin{aligned} q(-x) &= (-x)^3 = (-x)\cdot(-x)\cdot(-x) \\ &= -x^3 \\ &= -q(x). \end{aligned}$$

Thus, $q(-x) = -q(x)$.

In general,

> If f is a function, then f is called an **odd function** if, for all values of x in the domain of f,
>
> $$f(-x) = -f(x).$$
>
> The graph of f is symmetric about the origin.

Example 4 Determine whether the following functions are symmetric about the y-axis, the origin, or neither.
(a) $f(x) = |x|$ (b) $g(x) = 1/x$ (c) $h(x) = -x^3 - 3x^2 + 2$

Solution The graphs of the functions in Figures 5.16, 5.17, and 5.18 can be helpful in identifying symmetry.

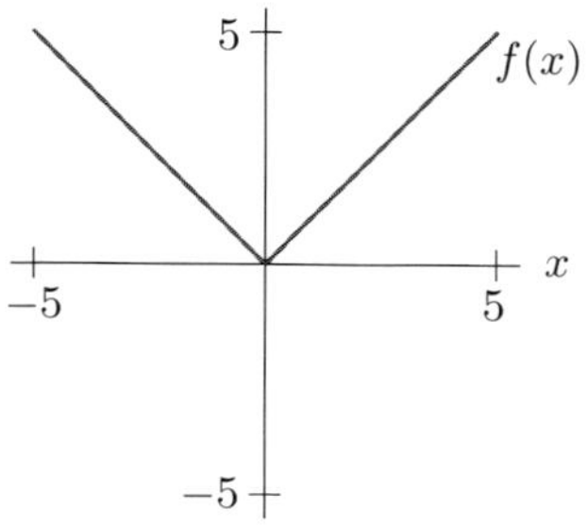

Figure 5.16: The graph of $f(x) = |x|$ appears to be symmetric about the y-axis

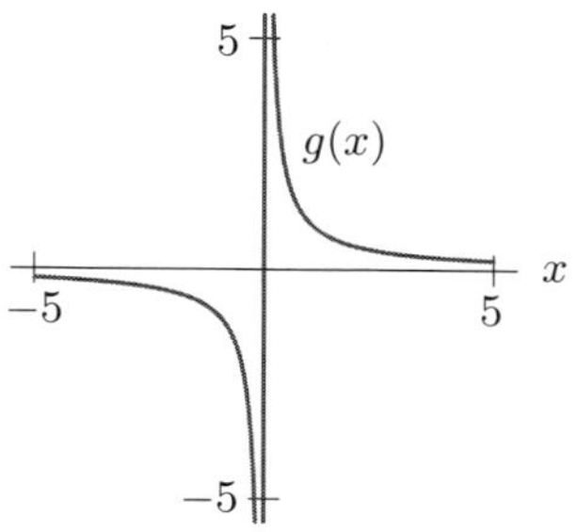

Figure 5.17: The graph of $g(x) = 1/x$ appears to be symmetric about the origin

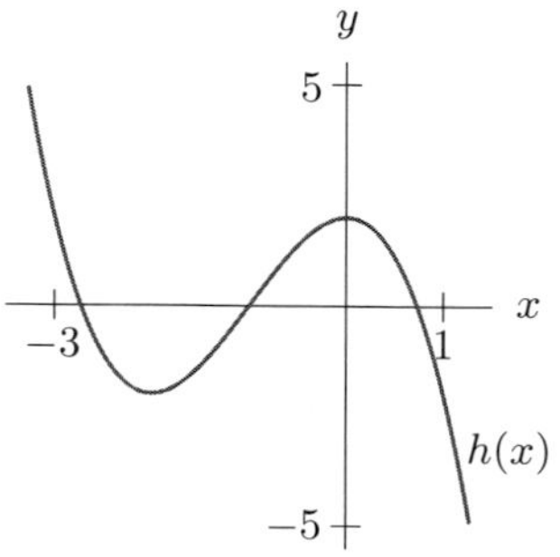

Figure 5.18: The graph of $h(x) = -x^3 - 3x^2 + 2$ is symmetric neither about the y-axis nor about the origin

From the graphs it appears that f is symmetric about the y-axis (even symmetry), g is symmetric about the origin (odd symmetry), and h has neither type of symmetry. However, how can we be sure that $f(x)$ and $g(x)$ are really symmetric? We check algebraically.

If $f(-x) = f(x)$, then f has even symmetry. We check by substituting $-x$ in for x:

$$\begin{aligned} f(-x) &= |-x| \\ &= |x| \\ &= f(x). \end{aligned}$$

Thus, f does have even symmetry.

If $g(-x) = -g(x)$, then g is symmetric about the origin. We check by substituting $-x$ for x:

$$\begin{aligned} g(-x) &= \frac{1}{-x} \\ &= -\frac{1}{x} \\ &= -g(x). \end{aligned}$$

Thus, g is symmetric about the origin.

The graph of h does not exhibit odd or even symmetry. To confirm, look at an example, say $x = 1$:

$$h(1) = -1^3 - 3 \cdot 1^2 + 2 = -2.$$

Now substitute $x = -1$, giving

$$h(-1) = -(-1)^3 - 3 \cdot (-1)^2 + 2 = 0.$$

Thus $h(1) \neq h(-1)$, so the function is not symmetric about the y-axis. Also, $h(-1) \neq -h(1)$, so the function is not symmetric about the origin.

Combining Shifts and Reflections

We can combine the horizontal and vertical shifts from Section 5.1 with the horizontal and vertical reflections of this section to make more complex transformations of functions

Example 5 A cold yam is placed in a hot oven. Newton's Law of Heating tells us that the difference between the oven's temperature and the yam's temperature decays exponentially with time. The yam's temperature is initially 0°F, the oven's temperature is 300°F, and the temperature difference decreases by 3% per minute. Find a formula for $Y(t)$, the yam's temperature at time t.

Solution Let $D(t)$ be the difference between the oven's temperature and the yam's temperature, which is given by an exponential function $D(t) = ab^t$. The initial temperature difference is $300°\text{F} - 0°\text{F} = 300°\text{F}$, so $a = 300$. The temperature difference decreases by 3% per minute, so $b = 1 - 0.03 = 0.97$. Thus,

$$D(t) = 300(0.97)^t.$$

If the yam's temperature is represented by $Y(t)$, then the temperature difference is given by

$$D(t) = 300 - Y(t),$$

so, solving for $Y(t)$, we have

$$Y(t) = 300 - D(t),$$

giving

$$Y(t) = 300 - 300(0.97)^t.$$

Writing $Y(t)$ in the form

$$Y(t) = \underbrace{-D(t)}_{\text{Reflect}} + \underbrace{300}_{\text{Shift}}$$

shows that the graph of Y is obtained by reflecting the graph of D about the t-axis and then shifting it vertically up 300 units. Notice that the horizontal asymptote of D, which is on the t-axis, is also shifted upward, resulting in a horizontal asymptote at 300°F for Y.

Figures 5.19 and 5.20 give the graphs of D and Y. Figure 5.20 shows that the yam heats up rapidly at first and then its temperature levels off toward $300°F$, the oven temperature.

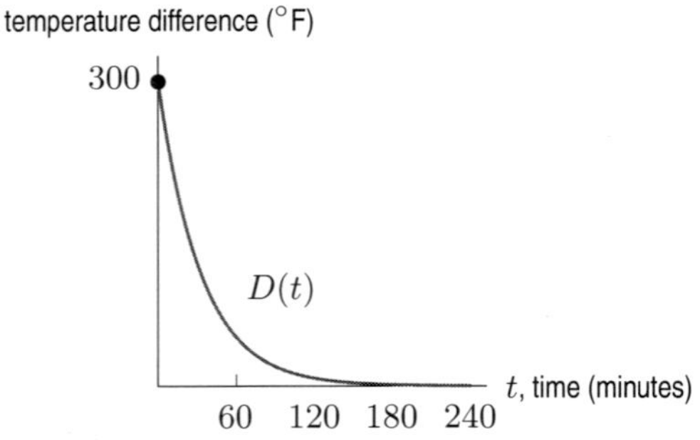

Figure 5.19: Graph of $D(t) = 300(0.97)^t$, the temperature difference between the yam and the oven

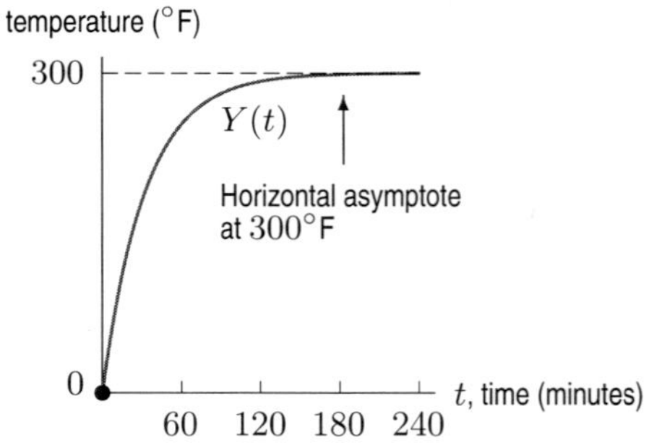

Figure 5.20: The transformation $Y(t) = -D(t) + 300$, where $D(t) = 300(0.97)^t$

Note that the temperature difference, D, is a decreasing function, so its average rate of change is negative. However, Y, the yam's temperature, is an increasing function, so its average rate of change is positive. Reflecting the graph of D about the t-axis to obtain the graph of Y changed the sign of the average rate of change.

Exercises and Problems for Section 5.2

Exercises

1. The graph of $y = f(x)$ contains the point $(2, -3)$. What point must lie on the reflected graph if the graph is reflected

(a) About the y-axis? **(b)** About the x-axis?

2. The graph of $P = g(t)$ contains the point $(-1, -5)$.

(a) If the graph has even symmetry, which other point must lie on the graph?
(b) What point must lie on the graph of $-g(t)$?

3. The graph of $H(x)$ is symmetric about the origin. If $H(-3) = 7$, what is $H(3)$?

4. The range of $Q(x)$ is $-2 \leq Q(x) \leq 12$. What is the range of $-Q(x)$?

5. If the graph of $y = e^x$ is reflected about the x-axis, what is the formula for the resulting graph? Check by graphing both functions together.

6. If the graph of $y = e^x$ is reflected about the y-axis, what is the formula for the resulting graph? Check by graphing both functions together.

7. Complete the following tables using $f(p) = p^2 + 2p - 3$, and $g(p) = f(-p)$, and $h(p) = -f(p)$. Graph the three functions. Explain how the graphs of g and h are related to the graph of f.

p	-3	-2	-1	0	1	2	3
$f(p)$							

p	-3	-2	-1	0	1	2	3
$g(p)$							

p	-3	-2	-1	0	1	2	3
$h(p)$							

8. Graph $y = f(x) = 4^x$ and $y = f(-x)$ on the same set of axes. How are these graphs related? Give an explicit formula for $y = f(-x)$.

9. Graph $y = g(x) = \left(\frac{1}{3}\right)^x$ and $y = -g(x)$ on the same set of axes. How are these graphs related? Give an explicit formula for $y = -g(x)$.

Give a formula and graph for each of the transformations of $m(n) = n^2 - 4n + 5$ in Exercises 10–13.

10. $y = m(-n)$

11. $y = -m(n)$

12. $y = -m(-n)$

13. $y = -m(-n) + 3$

Give a formula and graph for each of the transformations of $k(w) = 3^w$ in Exercises 14–19.

14. $y = k(-w)$

15. $y = -k(w)$

16. $y = -k(-w)$

17. $y = -k(w - 2)$

18. $y = k(-w) + 4$

19. $y = -k(-w) - 1$

In Exercises 20–23, show that the function is even, odd, or neither.

20. $f(x) = 7x^2 - 2x + 1$

21. $f(x) = 4x^7 - 3x^5$

22. $f(x) = 8x^6 + 12x^2$

23. $f(x) = x^5 + 3x^3 - 2$

Problems

24. **(a)** Graph the function obtained from $f(x) = x^3$ by first reflecting about the x-axis, then translating up two units. Write a formula for the resulting function.
(b) Graph the function obtained from f by first translating up two units, then reflecting about the x-axis. Write a formula for the resulting function.
(c) Are the functions in parts (a) and (b) the same?

25. **(a)** Graph the function obtained from $g(x) = 2^x$ by first reflecting about the y-axis, then translating down three units. Write a formula for the resulting function.
(b) Graph the function obtained from g by first translating down three units, then reflecting about the y-axis. Write a formula for the resulting function.
(c) Are the functions in parts (a) and (b) the same?

26. If the graph of a line $y = b + mx$ is reflected about the y-axis, what are the slope and intercepts of the resulting line?

27. Graph $y = \log(1/x)$ and $y = \log x$ on the same axes. How are the two graphs related? Use the properties of logarithms to explain the relationship algebraically.

28. The function $d(t)$ graphed in Figure 5.21 gives the winter temperature in °F at a high school, t hours after midnight.

(a) Describe in words the heating schedule for this building during the winter months.
(b) Graph $c(t) = 142 - d(t)$.
(c) Explain why c might describe the cooling schedule for summer months.

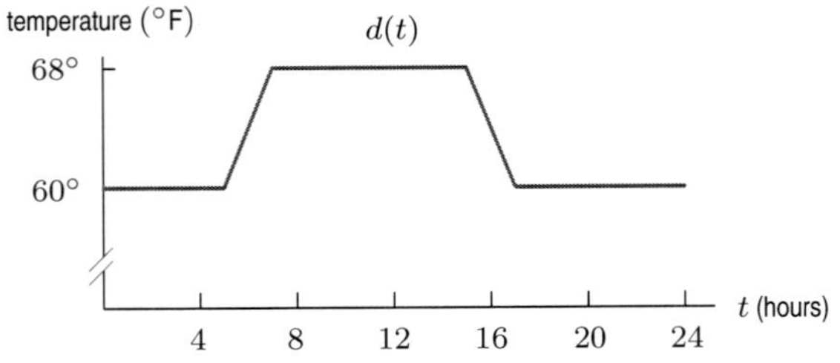

Figure 5.21

29. Using Figure 5.22, match the formulas (i)–(vi) with a graph from (a)–(f).

(i) $y = f(-x)$ (ii) $y = -f(x)$
(iii) $y = f(-x) + 3$ (iv) $y = -f(x-1)$
(v) $y = -f(-x)$ (vi) $y = -2 - f(x)$

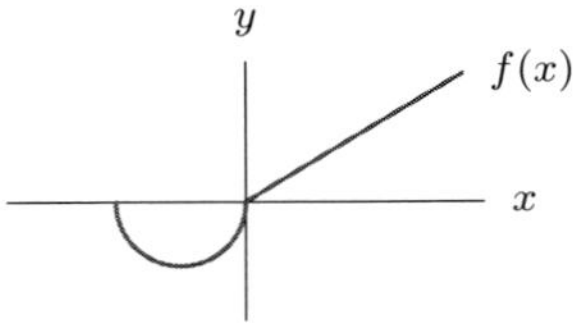

Figure 5.22

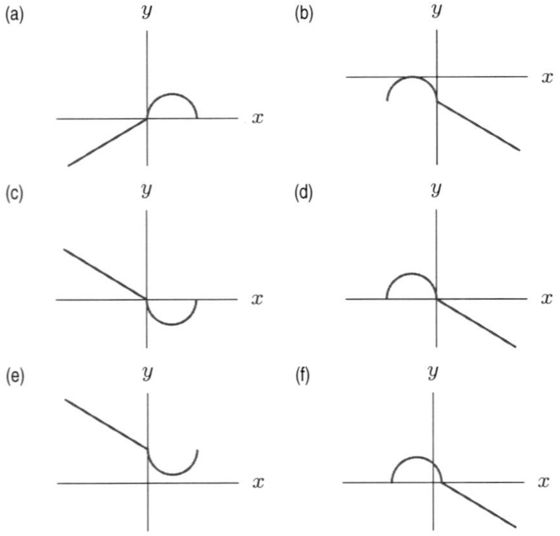

30. In Table 5.8, fill in as many y-values as you can if you know that f is

(a) An even function **(b)** An odd function.

Table 5.8

x	-3	-2	-1	0	1	2	3
y	5		-4			-8	

31. Figure 5.23 shows the graph of a function f in the second quadrant. In each of the following cases, sketch $y = f(x)$, given that f is symmetric about

(a) The y-axis. **(b)** The origin. **(c)** The line $y = x$.

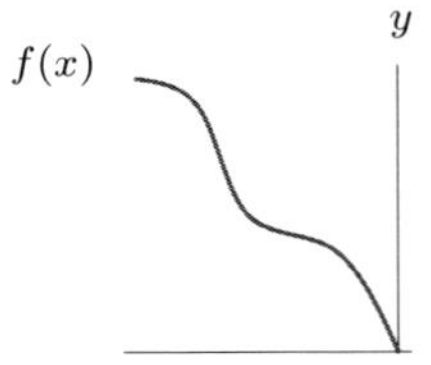

Figure 5.23

32. For each table, decide whether the function could be symmetric about the y-axis, about the origin, or neither.

(a)

x	-3	-2	-1	0	1	2	3
$f(x)$	6	1	-2	-3	-2	1	6

(b)

x	-3	-2	-1	0	1	2	3
$g(x)$	-8.1	-2.4	-0.3	0	0.3	2.4	8.1

(c)

x	-3	-2	-1	0	1	2	3
$f(x)+g(x)$	-2.1	-1.4	-2.3	-3	-1.7	3.4	14.1

(d)

x	-3	-2	-1	0	1	2	3
$f(x+1)$	1	-2	-3	-2	1	6	13

33. A function is called symmetric about the line $y = x$ if interchanging x and y gives the same graph. The simplest example is the function $y = x$. Graph another straight line that is symmetric about the line $y = x$ and give its equation.

34. Show that the graph of the function h is symmetric about the origin, given that

$$h(x) = \frac{1 + x^2}{x - x^3}.$$

35. Comment on the following justification that the function $f(x) = x^3 - x^2 + 1$ is an even function: Because $f(0) = 1 \neq -f(0)$, we know that $f(x)$ is not odd. If a function is not odd, it must be even.

36. Is it possible for an odd function whose domain is all real numbers to be strictly concave up?

37. If f is an odd function and defined at $x = 0$, what is the value of $f(0)$? Explain how you can use this result to show that $c(x) = x + 1$ and $d(x) = 2^x$ are not odd.

38. In the first quadrant an even function is increasing and concave down. What can you say about the function's behavior in the second quadrant?

39. Show that the power function $f(x) = x^{1/3}$ is odd. Give a counterexample to the statement that all power functions of the form $f(x) = x^p$ are odd.

40. Graph $s(x) = 2^x + (\frac{1}{2})^x$, $c(x) = 2^x - (\frac{1}{2})^x$, and $n(x) = 2^x - (\frac{1}{2})^{x-1}$. State whether you think these functions are even, odd or neither. Show that your statements are true using algebra. That is, prove or disprove statements such as $s(-x) = s(x)$.

41. There are functions which are *neither* even nor odd. Is there a function that is *both* even and odd?

42. Some functions are symmetric about the y-axis. Is it possible for a function to be symmetric about the x-axis?

5.3 VERTICAL STRETCHES AND COMPRESSIONS

We have studied translations and reflections of graphs. In this section, we consider vertical stretches and compressions of graphs. As with a vertical translation, a vertical stretch or compression of a function is represented by an outside change to its formula.

Vertical Stretch: A Stereo Amplifier

A stereo amplifier takes a weak signal from a cassette-tape deck, compact disc player, or radio tuner, and transforms it into a stronger signal to power a set of speakers.

Figure 5.24 shows a graph of a typical radio signal (in volts) as a function of time, t, both before and after amplification. In this illustration, the amplifier has boosted the strength of the signal by a factor of 3. (The amount of amplification, or *gain*, of most stereos is considerably greater than this.)

Notice that the wave crests of the amplified signal are 3 times as high as those of the original signal; similarly, the amplified wave troughs are 3 times deeper than the original wave troughs. If f is the original signal function and V is the amplified signal function, then

$$\underbrace{\text{Amplified signal strength at time } t}_{V(t)} = 3 \cdot \underbrace{\text{Original signal strength at time } t}_{f(t)},$$

so we have

$$V(t) = 3 \cdot f(t).$$

This formula tells us that values of the amplified signal function are 3 times the values of the original signal. The graph of V is the graph of f stretched vertically by a factor of 3. As expected, a vertical stretch of the graph of $f(t)$ corresponds to an outside change in the formula.

Notice that the t-intercepts remain fixed under a vertical stretch, because the f-value of these points is 0, which is unchanged when multiplied by 3.

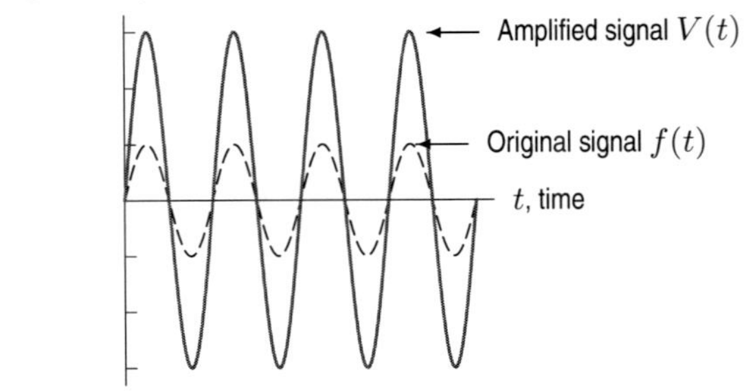

Figure 5.24: A stereo amplifier transforms a weak signal into a signal 3 times as strong

Negative Stretch Factor

What happens if we multiply a function by a negative stretch factor? Figure 5.25 gives a graph of a function $y = f(x)$, together with a graph of $y = -2 \cdot f(x)$. The stretch factor of f is $k = -2$. We think of $y = -2f(x)$ as a combination of two separate transformations of $y = f(x)$. First, the graph is stretched by a factor of 2, then it is reflected across the x-axis.

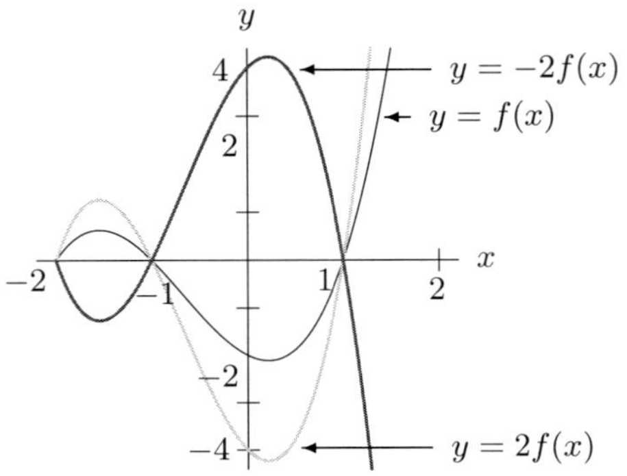

Figure 5.25: The graph of $y = -2f(x)$ is a vertically stretched version of the graph of $y = f(x)$ that has been reflected across the x-axis

Formula for Vertical Stretch or Compression

Generalizing the examples gives the following result:

If f is a function and k is a constant, then the graph of $y = k \cdot f(x)$ is the graph of $y = f(x)$

- Vertically stretched by a factor of k, if $k > 1$.
- Vertically compressed by a factor of k, if $0 < k < 1$.
- Vertically stretched or compressed by a factor $|k|$ and reflected across x-axis, if $k < 0$.

Example 1 A yam is placed in a 300°F oven. Suppose that Table 5.9 gives values of $H = r(t)$, the yam's temperature t minutes after being placed in the oven. Figure 5.26 shows these data points with a curve drawn in to emphasize the trend.

Table 5.9 *Temperature of a yam*

t, time (min)	$r(t)$, temperature (°F)
0	0
10	150
20	225
30	263
40	281
50	291
60	295

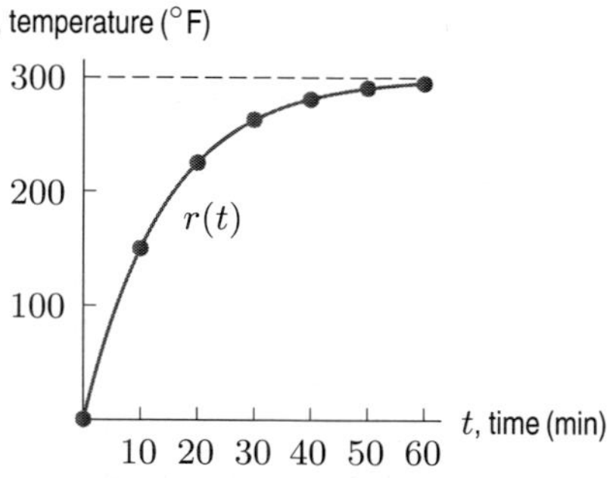

Figure 5.26: The temperature of a yam at time t

(a) Describe the function r in words. What do the data indicate about the yam's temperature?
(b) Make a table of values for $q(t) = 1.5r(t)$. Sketch a graph of this new function. Under what condition might q describe the yam's temperature?

Solution

(a) The function r is increasing and concave down. The yam starts out at 0°F and warms up quickly, reaching 150°F after 10 minutes. It continues to heat up, although more slowly, climbing by 75°F in the next 10 minutes and by 38°F in the 10 minutes after that. The temperature levels off around 300°F, the oven's temperature, represented by a horizontal asymptote.

(b) We calculate values of $q(t)$ from values of r. For example, Table 5.9 gives $r(0) = 0$ and $r(10) = 150$. Thus,

$$\begin{aligned} q(0) &= 1.5r(0) \\ &= 1.5 \cdot 0 \\ &= 0. \end{aligned}$$

Similarly, $q(10) = 1.5(150) = 225$, and so on. The values for $q(t)$ are 1.5 times as large as the corresponding values for $r(t)$.

The data in Table 5.10 are plotted in Figure 5.27. The graph of q is a vertically stretched version of the graph of r, because the stretch factor of $k = 1.5$ is larger than 1. The horizontal asymptote of r was $H = 300$, so the horizontal asymptote of q is $H = 1.5 \cdot 300 = 450$. This suggests that the yam has been placed in a 450°F oven instead of a 300°F oven.

Table 5.10 *Values of* $q(t) = 1.5r(t)$

t, time (min)	$q(t)$, temperature (°F)
0	0
10	225
20	337.5
30	394.5
40	421.5
50	436.5
60	442.5

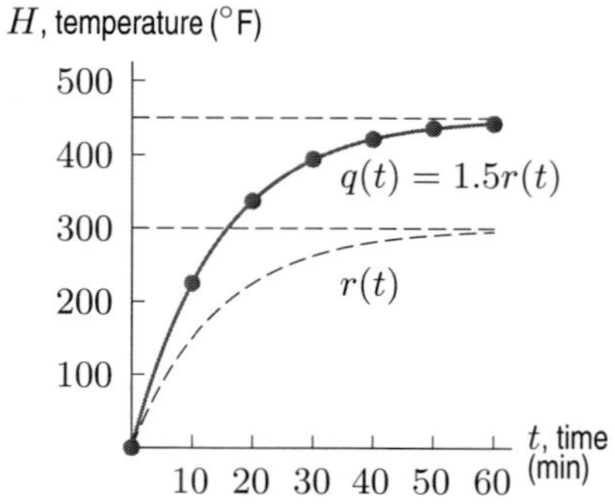

Figure 5.27: Graph of $q(t) = 1.5r(t)$

Stretch Factors and Average Rates of Change

Consider again the graph of the radio signal and its amplification in Figure 5.24. Notice that the amplified signal, V, is increasing on the same intervals as the original signal, f. Similarly, both functions decrease on the same intervals.

Stretching or compressing a function vertically does not change the intervals on which the function increases or decreases. However, the average rate of change of a function, visible in the steepness of the graph, is altered by a vertical stretch or compression.

Example 2 In Example 1, the function $H = r(t)$ gives the temperature (in °F) of a yam placed in a 300°F oven. The function $q(t) = 1.5r(t)$ gives the temperature of the yam placed in a 450°F oven.

(a) Calculate the average rate of change of r over 10-minute intervals. What does this tell you about the yam's temperature?

(b) Now calculate the average rate of change of q over 10-minute intervals. What does this tell you about the yam's temperature?

Solution (a) On the first interval, from $t = 0$ to $t = 10$, we have

$$\begin{aligned}\text{Average rate of change of temperature, } r &= \frac{\Delta H}{\Delta t} = \frac{r(10)-r(0)}{10-0}\\ &= \frac{150-0}{10} \quad \text{(referring to Table 5.9)}\\ &= 15^\circ\text{F/min.}\end{aligned}$$

Thus, during the first 10 minute interval, the yam's temperature increased at an average rate of 15°F per minute.

On the second time interval from $t = 10$ to $t = 20$,

$$\text{Average rate of change of temperature} = \frac{\Delta H}{\Delta t} = \frac{225-150}{10} = 7.5^\circ\text{F/min},$$

and on the third interval from $t = 20$ to $t = 30$,

$$\text{Average rate of change of temperature} = \frac{\Delta H}{\Delta t} = \frac{263-225}{10} = 3.8^\circ\text{F/min}.$$

See Table 5.11.

Table 5.11 *The average rate of change of yam's temperature, $r(t)$*

Time interval (min)	0 – 10	10 – 20	20 – 30	30 – 40	40 – 50	50 – 60
Average rate of change of r (°F/min)	15	7.5	3.8	1.8	1.0	0.4

(b) The data in Table 5.10 was used to calculate the average rate of change of $q(t) = 1.5r(t)$ in Table 5.12.

Table 5.12 *The average rate of change of yam's temperature, $q(t)$*

Time interval (min)	0 – 10	10 – 20	20 – 30	30 – 40	40 – 50	50 – 60
Average rate of change of q (°F/min)	22.5	11.25	5.7	2.7	1.5	0.6

Comparing the average rates of change on each 10-minute interval, we see that q's average rate of change is 1.5 times r's. Thus, q depicts a yam whose temperature increases more quickly than r does.

In the last example, multiplying a function by a stretch factor k has the effect of multiplying the function's average rate of change on each interval by the same factor. We check this statement algebraically for the function $g(x) = k \cdot f(x)$. On the interval from a to b,

$$\begin{array}{c}\text{Average rate of change} \\ \text{of } y = g(x)\end{array} = \frac{\Delta y}{\Delta x} = \frac{g(b) - g(a)}{b - a}.$$

But $g(b) = k \cdot f(b)$ and $g(a) = k \cdot f(a)$. Thus,

$$\begin{aligned}\begin{array}{c}\text{Average rate of change} \\ \text{of } y = g(x)\end{array} &= \frac{\Delta y}{\Delta x} = \frac{k \cdot f(b) - k \cdot f(a)}{b - a} \\ &= k \cdot \frac{f(b) - f(a)}{b - a} \quad \text{(factoring out } k\text{)} \\ &= k \cdot \left(\begin{array}{c}\text{Average rate of change} \\ \text{of } f\end{array} \right).\end{aligned}$$

In general, we have the following result:

> If $g(x) = k \cdot f(x)$, then on any interval,
>
> $$\text{Average rate of change of } g = k \cdot (\text{Average rate of change of } f).$$

Combining Transformations

Any transformations of functions can be combined.

Example 3 The function $y = f(x)$ is graphed in Figure 5.28. Graph the function $g(x) = -\frac{1}{2}f(x + 3) - 1$.

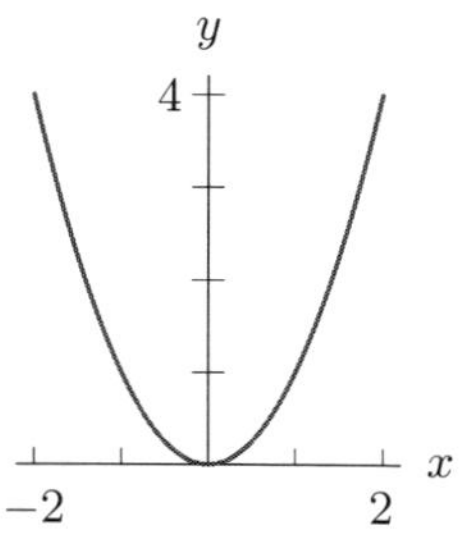

Figure 5.28: Graph of $y = f(x)$

Solution To combine several transformations, always work from inside the parentheses outward as in Figure 5.29. The graphs corresponding to each step are shown in Figure 5.30. Note that we did not need a formula for f to graph g.

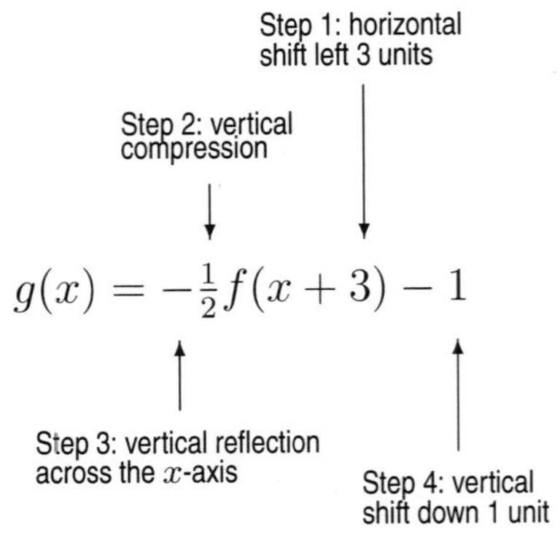

Figure 5.29

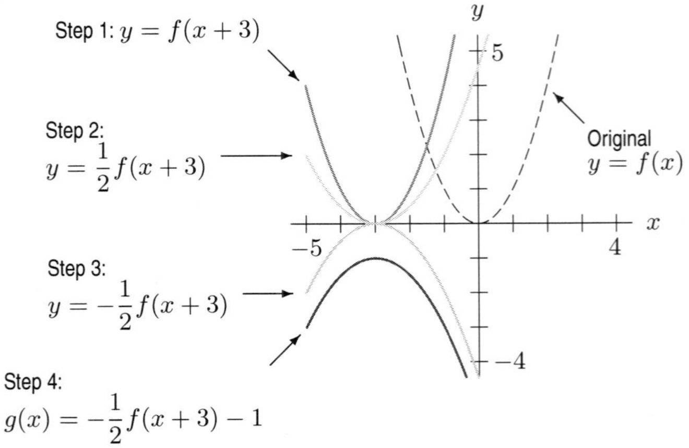

Figure 5.30: The graph of $y = f(x)$ transformed in four steps into $g(x) = -(1/2)f(x+3) - 1$

Exercises and Problems for Section 5.3

Exercises

1. Let $y = f(x)$. Write a formula for the transformation which both increases the y-value by a factor of 10 and shifts the graph to the right by 2 units.

2. The graph of the function $g(x)$ contains the point $(5, \frac{1}{3})$. What point must be on the graph of $y = 3g(x) + 1$?

3. The range of the function $C(x)$ is $-1 \leq C(x) \leq 1$. What is the range of $0.25C(x)$?

In Exercises 4–7, graph and label $f(x)$, $4f(x)$, $-\frac{1}{2}f(x)$, and $-5f(x)$ on the same axes.

4. $f(x) = \sqrt{x}$ **5.** $f(x) = -x^2 + 7x$

6. $f(x) = e^x$ **7.** $f(x) = \ln x$

8. Using Table 5.13, make tables for the following transformations of f on an appropriate domain.

(a) $\frac{1}{2}f(x)$ **(b)** $-2f(x+1)$ **(c)** $f(x) + 5$
(d) $f(x-2)$ **(e)** $f(-x)$ **(f)** $-f(x)$

Table 5.13

x	−3	−2	−1	0	1	2	3
$f(x)$	2	3	7	−1	−3	4	8

9. Using Table 5.14, create a table of values for

(a) $f(-x)$ **(b)** $-f(x)$ **(c)** $3f(x)$

(d) Which of these tables from parts (a), (b), and (c) represents an even function?

Table 5.14

x	−4	−3	−2	−1	0	1	2	3	4
$f(x)$	13	6	1	−2	−3	−2	1	6	13

10. Figure 5.31 is a graph of $y = x^{3/2}$. Match the following functions with the graphs in Figure 5.32.

(a) $y = x^{3/2} - 1$ **(b)** $y = (x-1)^{3/2}$
(c) $y = 1 - x^{3/2}$ **(d)** $y = \frac{3}{2}x^{3/2}$

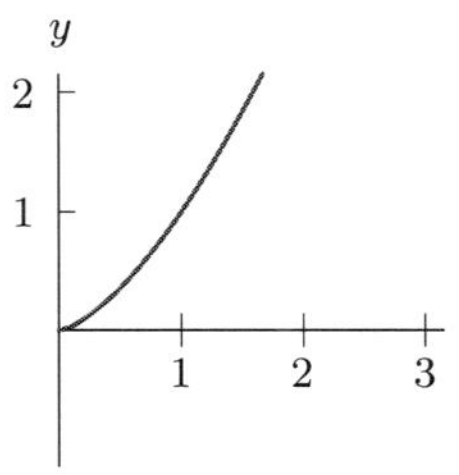

Figure 5.31

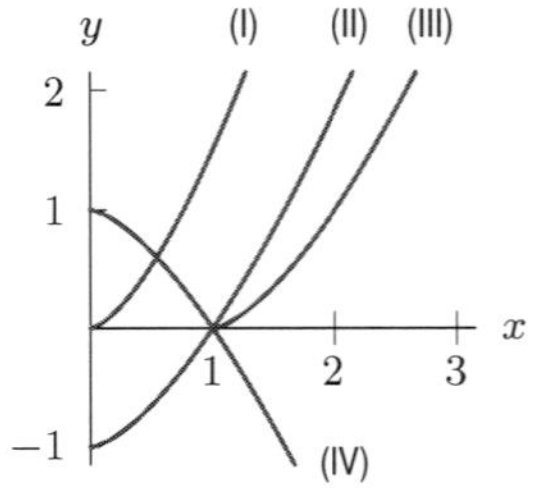

Figure 5.32

Without a calculator, graph the transformations in Exercises 11–16. Label at least three points.

11. $y = f(x+3)$ if $f(x) = |x|$

12. $y = f(x)+3$ if $f(x) = |x|$

13. $y = -g(x)$ if $g(x) = x^2$

14. $y = g(-x)$ if $g(x) = x^2$

15. $y = 3h(x)$ if $h(x) = 2^x$

16. $y = 0.5h(x)$ if $h(x) = 2^x$

17. Using Figure 5.33, match the functions (i)–(v) with a graph (a)–(i).

(i) $y = 2f(x)$ (ii) $y = \frac{1}{3}f(x)$
(iii) $y = -f(x)+1$ (iv) $y = f(x+2)+1$
(v) $y = f(-x)$

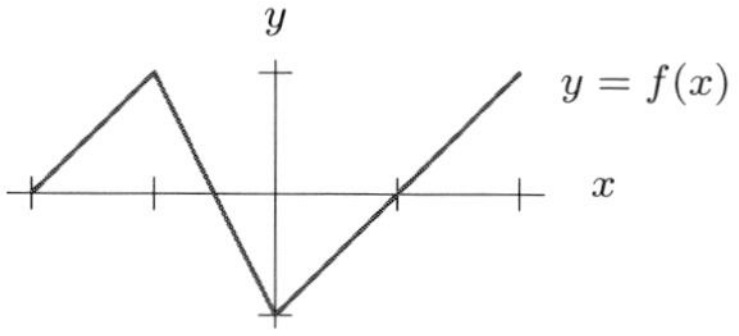

Figure 5.33

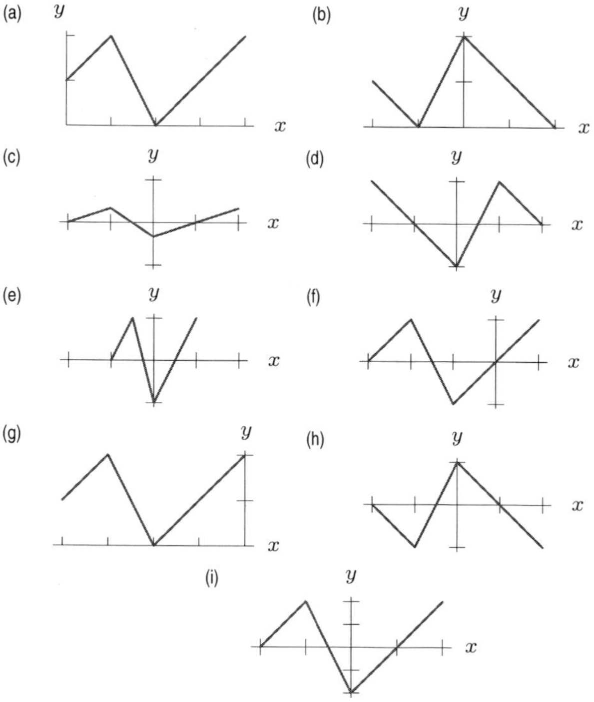

Problems

18. Describe the effect of the transformation $2f(x+1)-3$ on the graph of $y = f(x)$.

19. The function $s(t)$ gives the distance (miles) in terms of time (hours). If the average rate of change of $s(t)$ on $0 \le t \le 4$ is 70 mph, what is the average rate of change of $\frac{1}{2}s(t)$ on this interval?

In Problems 20–24, let $f(t) = 1/(1+x^2)$. Graph the function given, labeling intercepts and asymptotes.

20. $y = f(t)$ **21.** $y = f(t-3)$

22. $y = 0.5f(t)$ **23.** $y = -f(t)$

24. $y = f(t+5)-5$

25. The number of gallons of paint, $n = f(A)$, needed to cover a house is a function of the surface area, in ft^2. Match each story to one expression.

(a) I figured out how many gallons I needed and then bought two extra gallons just in case.
(b) I bought enough paint to cover my house twice.
(c) I bought enough paint to cover my house and my welcome sign, which measures 2 square feet.

(i) $2f(A)$ (ii) $f(A+2)$ (iii) $f(A)+2$

26. The US population in millions is $P(t)$ today and t is in years. Match each statement (I)–(IV) with one of the formulas (a)–(h).

I. The population 10 years before today.

II. Today's population plus 10 million immigrants.

III. Ten percent of the population we have today.

IV. The population after 100,000 people have emigrated.

(a) $P(t)-10$ **(b)** $P(t-10)$ **(c)** $0.1P(t)$
(d) $P(t)+10$ **(e)** $P(t+10)$ **(f)** $P(t)/0.1$
(g) $P(t)+0.1$ **(h)** $P(t)-0.1$

27. Let $R = P(t)$ be the number of rabbits living in the national park in month t. (See Example 5 on page 5.) What do the following expressions represent?

(a) $P(t+1)$ **(b)** $2P(t)$

28. Without a calculator, match each formula (a)–(e) with a graph in Figure 5.34. There may be no answer or several answers.

(a) $y = 3 \cdot 2^x$ **(b)** $y = 5^{-x}$ **(c)** $y = -5^x$
(d) $y = 2-2^{-x}$ **(e)** $y = 1-\left(\frac{1}{2}\right)^x$

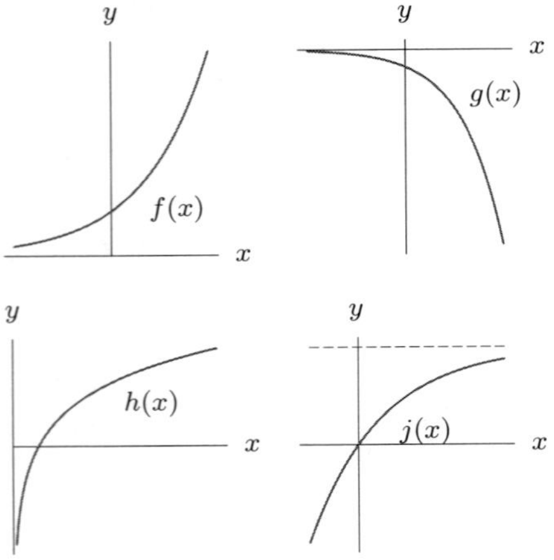

Figure 5.34

Graph the transformations of f in Problems 29–33 using Figure 5.35. Label the points corresponding to A and B.

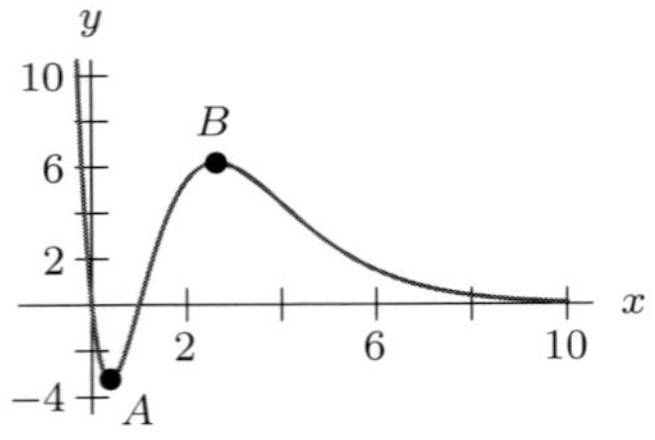

Figure 5.35

29. $y = f(x-3)$

30. $y = f(x) - 3$

31. $y = f(-x)/3$

32. $y = -2f(x)$

33. $y = 5 - f(x+5)$

34. Using Figure 5.36, find formulas, in terms of f, for the horizontal and vertical shifts of the graph of f in parts (a)–(c). What is the equation of each asymptote?

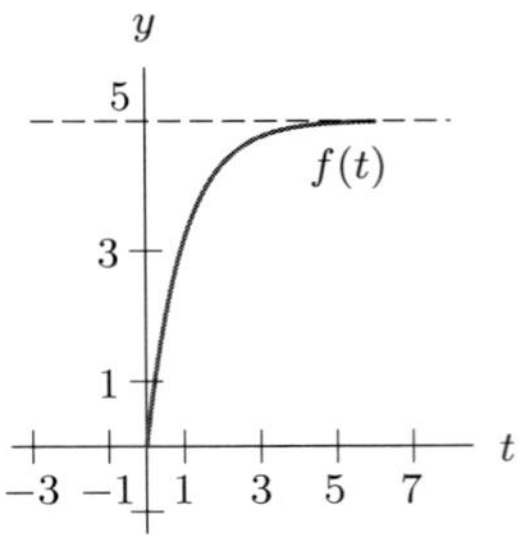

Figure 5.36

(a)

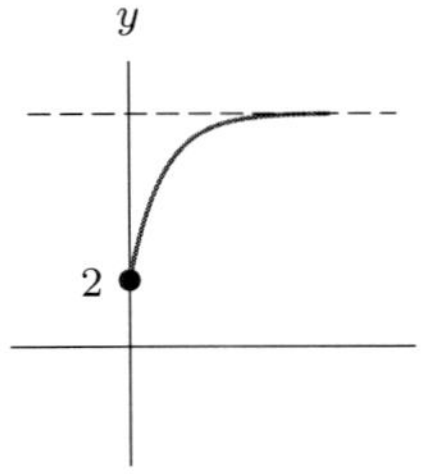

(b)

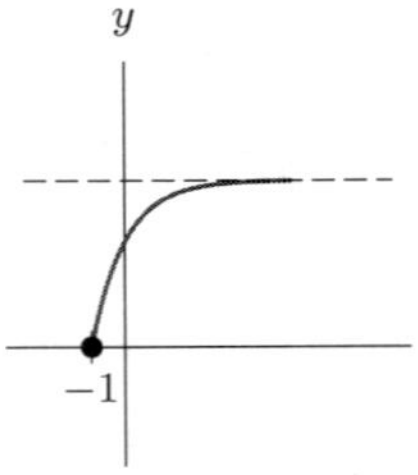

(c)

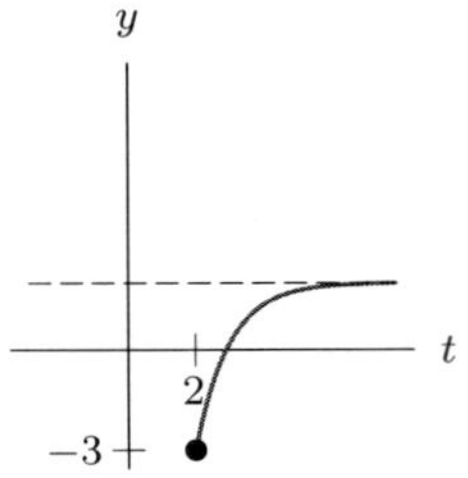

35. Using Figure 5.37, find formulas, in terms of f, for the transformations of f in parts (a)–(c).

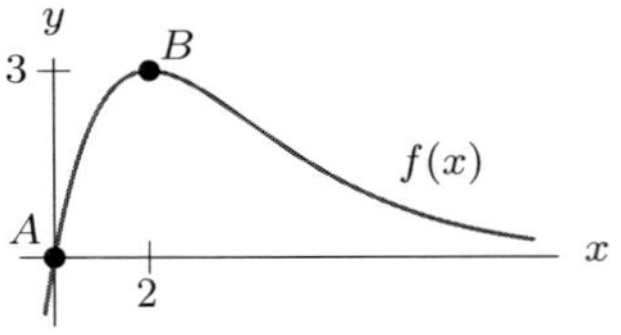

Figure 5.37

(a)

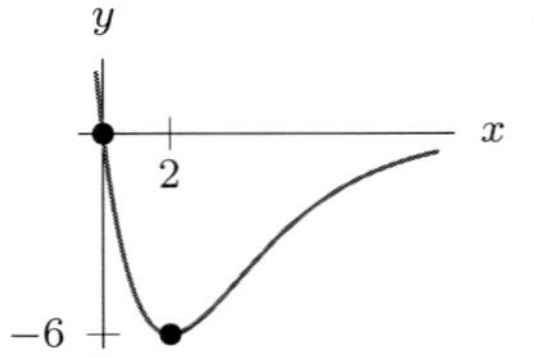

(b)

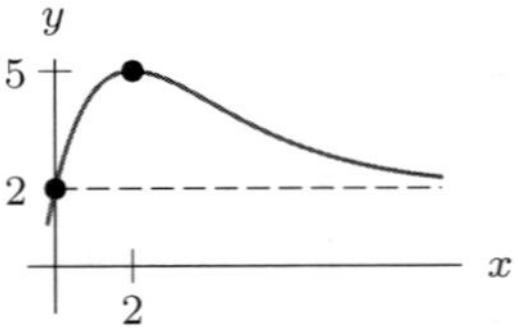

(c)

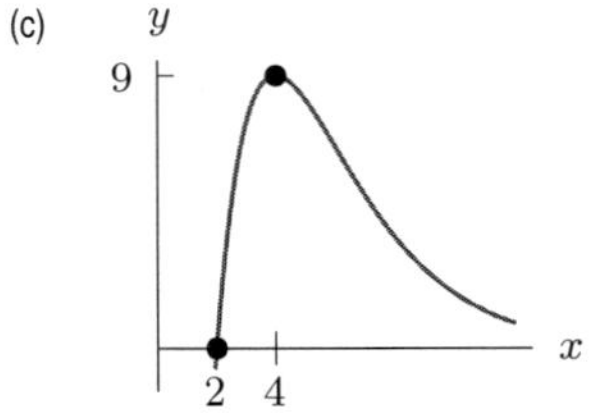

36. In Figure 5.38, the point b is labeled on the x-axis. On the y-axis, locate and label the output values:

(a) $f(b)$ **(b)** $-2f(b)$ **(c)** $-2f(-b)$

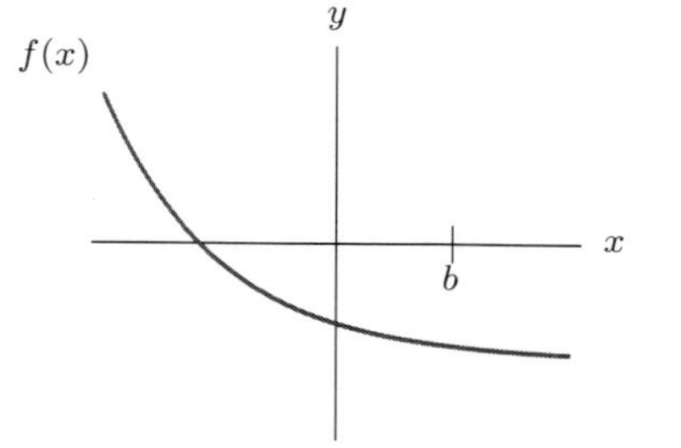

Figure 5.38

37. Figure 5.39 gives a graph of $y = f(x)$. Consider the transformations $y = \frac{1}{2}f(x)$ and $y = 2f(x)$. Which points on the graph of $y = f(x)$ stay fixed under these transformations? Compare the intervals on which all three functions are increasing and decreasing.

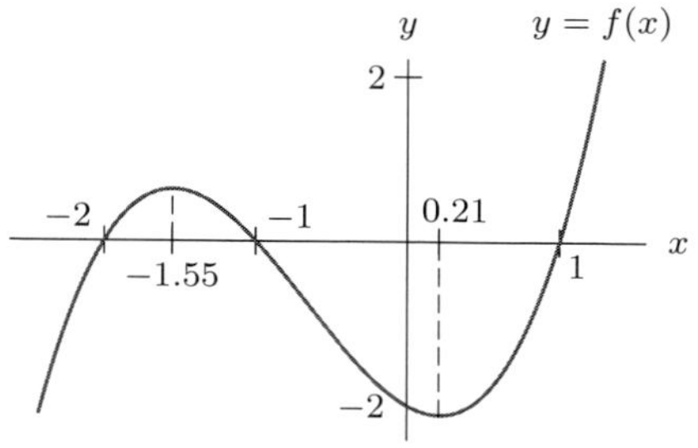

Figure 5.39

38. Let $f(x) = e^x$ and $g(x) = 5e^{x-2}$. If $g(x) = kf(x)$, find k.

5.4 HORIZONTAL STRETCHES AND COMPRESSIONS

In Section 5.3, we observed that a vertical stretch of a function's graph corresponds to an outside change in its formula, specifically, multiplication by a stretch factor. Since horizontal changes generally correspond to inside changes, we expect that a horizontal stretch will correspond to a constant multiple of the inputs. This turns out to be the case.

Horizontal Stretch: A Lighthouse Beacon

The beacon in a lighthouse turns once per minute, and its beam sweeps across a beach house. Figure 5.40 gives a graph of $L(t)$, the intensity, or brightness, of the light striking the beach house as a function of time.

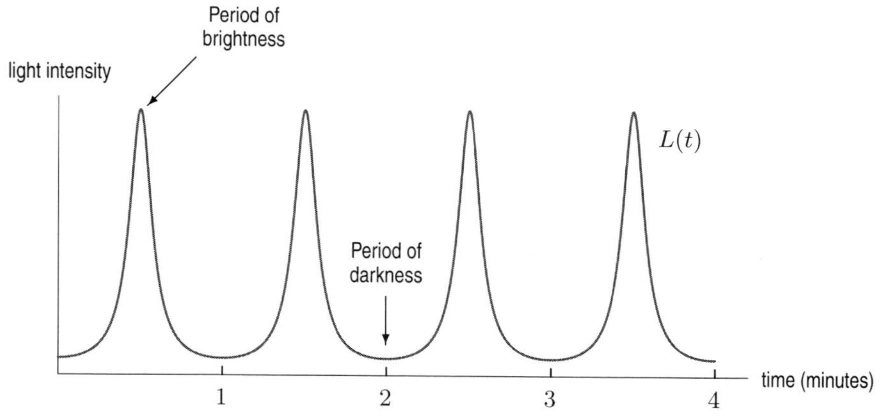

Figure 5.40: Light intensity or brightness, $L(t)$, as a function of time

Now suppose the lighthouse beacon turns twice as fast as before, so that its beam sweeps past the beach house twice instead of once each minute. The periods of brightness now occur twice

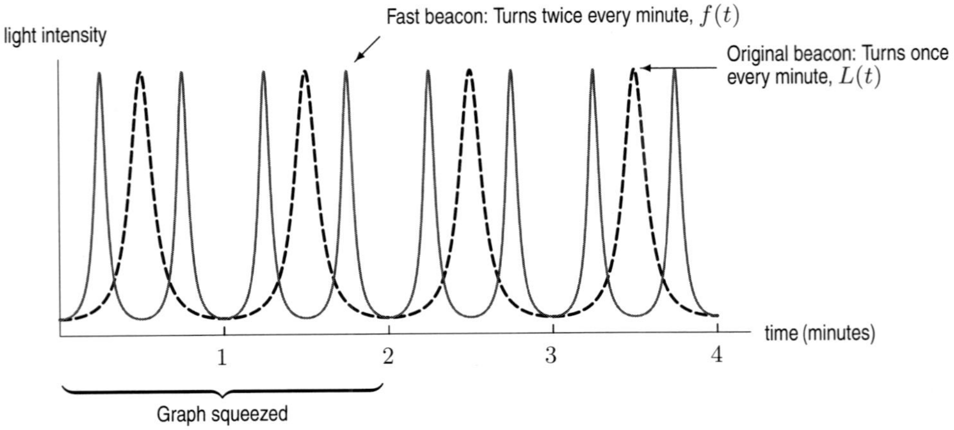

Figure 5.41: Comparing light intensity from the fast beacon, $f(t)$, to light intensity from the original beacon, $L(t)$

as often. See Figure 5.41. The graph of $f(t)$, the intensity of light from this faster beacon, is a horizontal squeezing or compression of the original graph of $L(t)$.

If the lighthouse beacon turns at half its original rate, so that its beam sweeps past the beach house once every two minutes instead of once every minute, the periods of brightness occur half as often as originally. Slowing the beacon's speed results in a horizontal stretch of the original graph, illustrated by the graph of $s(t)$, the light intensity of the slow beacon, in Figure 5.42.

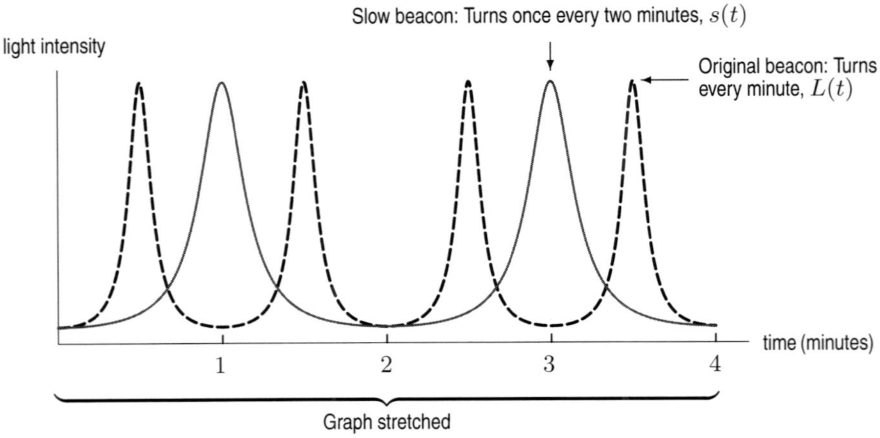

Figure 5.42: Comparing light intensity from the slow beacon, $s(t)$, to light intensity from the original beacon, $L(t)$

Formula for Horizontal Stretch or Compression

How are the formulas for the three light functions related? We expect that multiplying the function's input by a constant will horizontally stretch or compress its graph. The fast beacon corresponds to speeding up by a factor of 2, or multiplying the input times by 2. Thus

$$f(t) = L(2t).$$

Similarly for the slow beacon, the input times are multiplied by $1/2$, so

$$s(t) = L(\tfrac{1}{2}t).$$

Generalizing the lighthouse example gives the following result:

> If f is a function and k a positive constant, then the graph of $y = f(kx)$ is the graph of f
> - Horizontally compressed by a factor of $1/k$ if $k > 1$,
> - Horizontally stretched by a factor of $1/k$ if $k < 1$.
>
> If $k < 0$, then the graph of $y = f(kx)$ also involves a horizontal reflection about the y-axis.

Example 1 Values of the function $f(x)$ are in Table 5.15 and its graph is in Figure 5.43. Make a table and a graph of the function $g(x) = f(\frac{1}{2}x)$.

Table 5.15 *Values of $f(x)$*

x	$f(x)$
−3	0
−2	2
−1	0
0	−1
1	0
2	−1
3	1

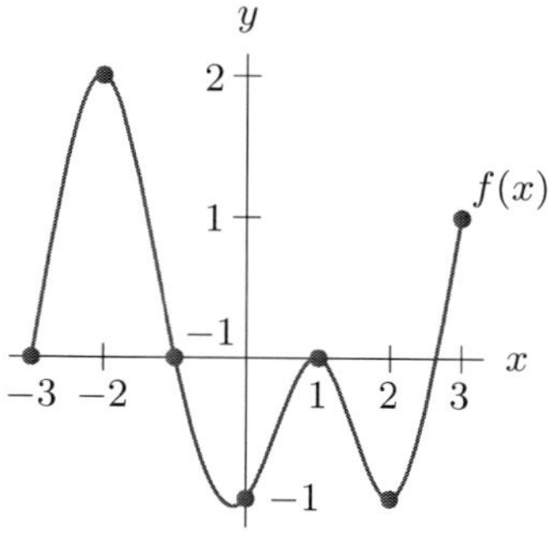

Figure 5.43

Solution To make a table for $g(x) = f(\frac{1}{2}x)$, we substitute values for x. For example, if $x = 4$, then

$$g(4) = f(\tfrac{1}{2} \cdot 4) = f(2).$$

Table 5.15 shows that $f(2) = -1$, so

$$g(4) = f(2) = -1.$$

This result is recorded in Table 5.16. If $x = 6$, since Table 5.15 gives $f(3) = 1$, we have

$$g(6) = f(\tfrac{1}{2} \cdot 6) = f(3) = 1.$$

In Figure 5.44, we see that the graph of g is the graph of f stretched horizontally away from the y-axis. Substituting $x = 0$, gives

$$g(0) = f(\tfrac{1}{2} \cdot 0) = f(0) = -1,$$

so the y-intercept remains fixed (at -1) under a horizontal stretch.

Table 5.16 *Values of $g(x) = f(\frac{1}{2}x)$*

x	$g(x)$
−6	0
−4	2
−2	0
0	−1
2	0
4	−1
6	1

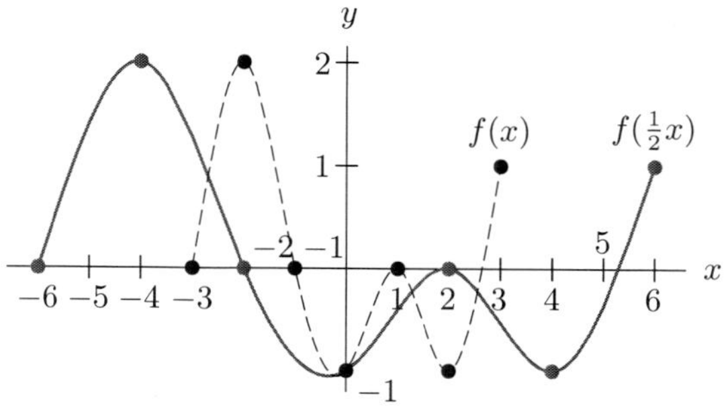

Figure 5.44: The graph of $g(x) = f(\frac{1}{2}x)$ is the graph of $y = f(x)$ stretched away from the y-axis by a factor of 2

Example 1 shows the effect of an inside multiple of $1/2$. The next example shows the effect on the graph of an inside multiple of 2.

Example 2 Let $f(x)$ be the function in Example 1. Make a table and a graph for the function $h(x) = f(2x)$.

Solution We use Table 5.15 and the formula $h(x) = f(2x)$ to evaluate $h(x)$ at several values of x. For example, if $x = 1$, then

$$h(1) = f(2 \cdot 1) = f(2).$$

Table 5.15 shows that $f(2) = -1$, so $h(1) = -1$. These values are recorded in Table 5.17. Similarly, substituting $x = 1.5$, gives

$$h(1.5) = f(2 \cdot 1.5) = f(3) = 1.$$

Since $h(0) = f(2 \cdot 0) = f(0)$, the y-intercept remains fixed (at -1). In Figure 5.45 we see that the graph of h is the graph of f compressed by a factor of 2 horizontally toward the y-axis.

Table 5.17 *Values of $h(x) = f(2x)$*

x	$h(x)$
-1.5	0
-1.0	2
-0.5	0
0.0	-1
0.5	0
1.0	-1
1.5	1

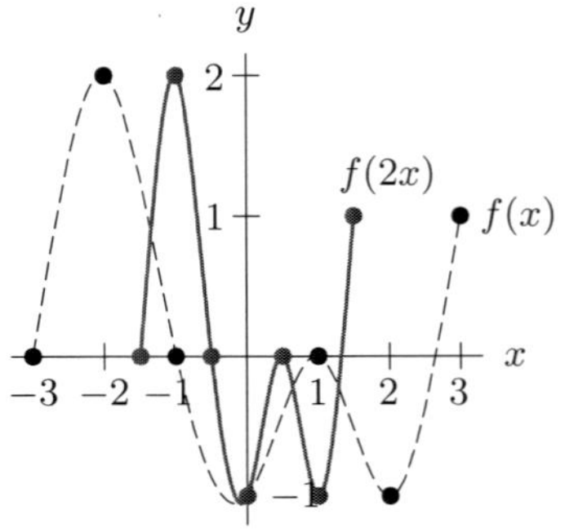

Figure 5.45: The graph of $h(x) = f(2x)$ is the graph of $y = f(x)$ compressed horizontally by a factor of 2

In Chapter 3, we used the function $P = 263e^{0.009t}$ to model the US population in millions. This function is a transformation of the exponential function $f(t) = e^t$, since we can write

$$P = 263e^{0.009t} = 263f(0.009t).$$

The US population is $f(t) = e^t$ stretched vertically by a factor of 263 and stretched horizontally by a factor of $1/0.009 \approx 111$.

Example 3 Match the functions $f(t) = e^t$, $g(t) = e^{0.5t}$, $h(t) = e^{0.8t}$, $j(t) = e^{2t}$ with the graphs in Figure 5.46.

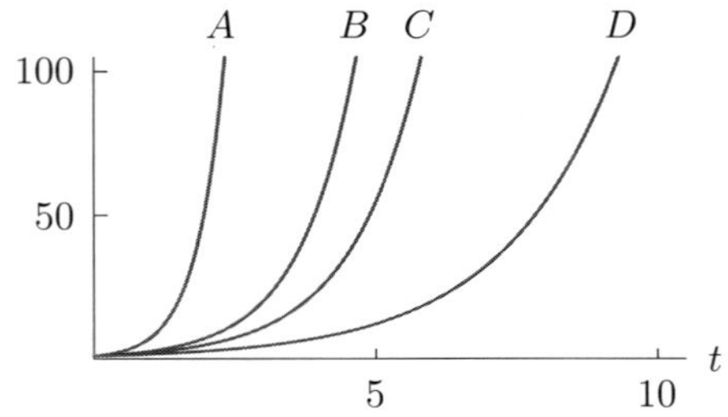

Figure 5.46

Solution Since the function $j(t) = e^{2t}$ climbs fastest of the four and $g(t) = e^{0.5t}$ climbs slowest, graph A must be j and graph D must be g. Similarly, graph B is f and graph C is h.

Exercises and Problems for Section 5.4

Exercises

1. The point $(2, 3)$ lies on the graph of $g(x)$. What point must lie on the graph of $g(2x)$?

2. Describe the effect of the transformation $10f(\frac{1}{10}x)$ on the graph of $f(x)$.

3. Using Table 5.18, make a table of values for $f(\frac{1}{2}x)$ for an appropriate domain.

Table 5.18

x	-3	-2	-1	0	1	2	3
$f(x)$	2	3	7	-1	-3	4	8

4. Fill in all the blanks in Table 5.19 for which you have sufficient information.

Table 5.19

x	-3	-2	-1	0	1	2	3
$f(x)$	-4	-1	2	3	0	-3	-6
$f(\frac{1}{2}x)$							
$f(2x)$							

5. Graph $m(x) = e^x$, $n(x) = e^{2x}$, and $p(x) = 2e^x$ on the same axes and describe how the graphs of $n(x)$ and $p(x)$ compare with that of $m(x)$.

6. Graph $y = h(3x)$ if $h(x) = 2^x$.

In Exercises 7–9, graph and label $f(x)$, $f(\frac{1}{2}x)$, and $f(-3x)$ on the same axes between $x = -2$ and $x = 2$.

7. $f(x) = e^x + x^3 - 4x^2$

8. $f(x) = e^{x+7} + (x-4)^3 - (x+2)^2$

9. $f(x) = \ln(x^4 + 3x^2 + 4)$

10. Using Figure 5.47, match each function to a graph (if any) that represents it:

(i) $y = f(2x)$ (ii) $y = 2f(2x)$ (iii) $y = f(\frac{1}{2}x)$

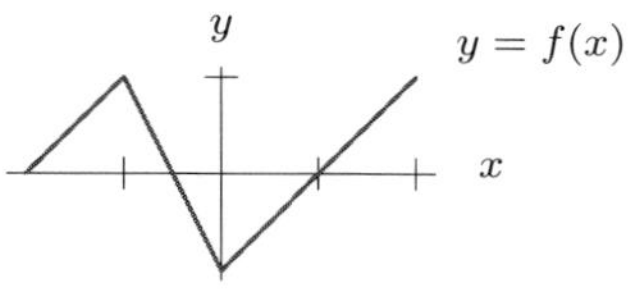

Figure 5.47

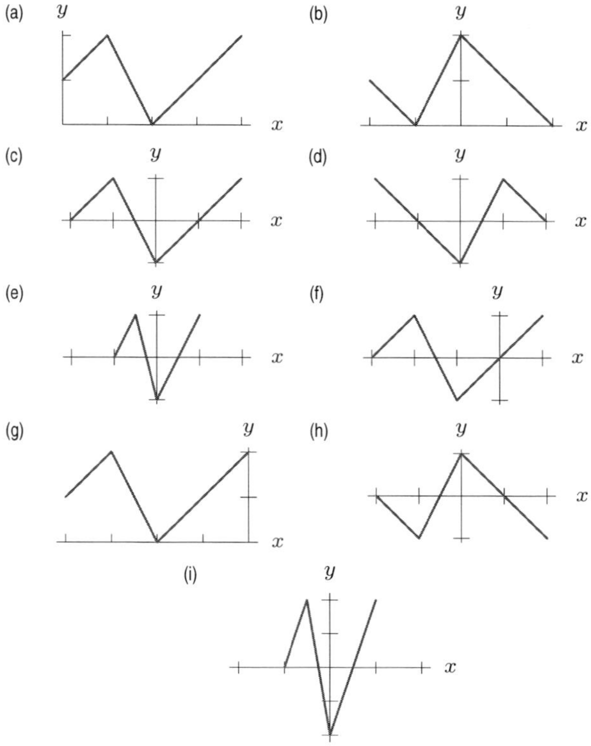

Problems

11. For the function $f(p)$ an input of 2 yields an output value of 4. What value of p would you use to have $f(3p) = 4$?

12. The domain of $l(x)$ is $-12 \le x \le 12$ and its range is $0 \le l(x) \le 3$. What are the domain and range of

(a) $l(2x)$? **(b)** $l(\frac{1}{2}x)$?

13. The point (a, b) lies on the graph of $y = f(x)$. If the graph is stretched away from the y-axis by a factor of d (where $d > 1$), and then translated upward by c units, what are the new coordinates for the point (a, b)?

In Problems 14–15, graph the transformation of f, the function in Figure 5.48.

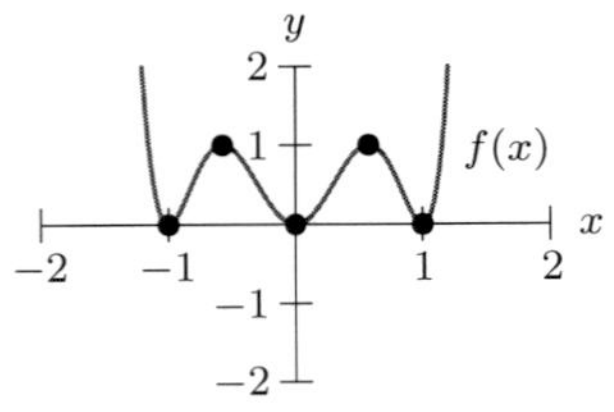

Figure 5.48

14. $y = -2f(x-1)$ **15.** $y = f(x/2) - 1$

16. Every day I take the same taxi over the same route from home to the train station. The trip is x miles, so the cost for the trip is $f(x)$. Match each story in (a)–(d) to a function in (i)–(iv) representing the amount paid to the driver.

(a) I received a raise yesterday, so today I gave my driver a five dollar tip.
(b) I had a new driver today and he got lost. He drove five extra miles and charged me for it.
(c) I haven't paid my driver all week. Today is Friday and I'll pay what I owe for the week.
(d) The meter in the taxi went crazy and showed five times the number of miles I actually traveled.

(i) $5f(x)$ (ii) $f(x)+5$
(iii) $f(5x)$ (iv) $f(x+5)$

17. A company projects a total profit, $P(t)$ dollars, in year t. Explain the economic meaning of $r(t) = 0.5P(t)$ and $s(t) = P(0.5t)$.

18. Let $A = f(r)$ be the area of a circle of radius r.

(a) Write a formula for $f(r)$.
(b) Which expression represents the area of a circle whose radius is increased by 10%? Explain.

(i) $0.10f(r)$ (ii) $f(r+0.10)$ (iii) $f(0.10r)$
(iv) $f(1.1r)$ (v) $f(r)+0.10$

(c) By what percent does the area increase if the radius is increased by 10%?

In Problems 19–20, state which graph represents

(a) $f(x)$ **(b)** $f(-2x)$ **(c)** $f(-\frac{1}{2}x)$ **(d)** $f(2x)$

19.

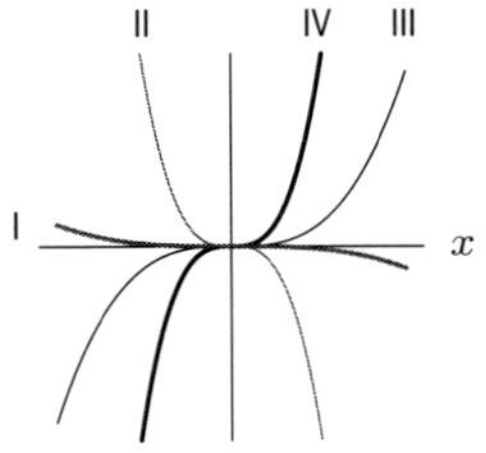

20.

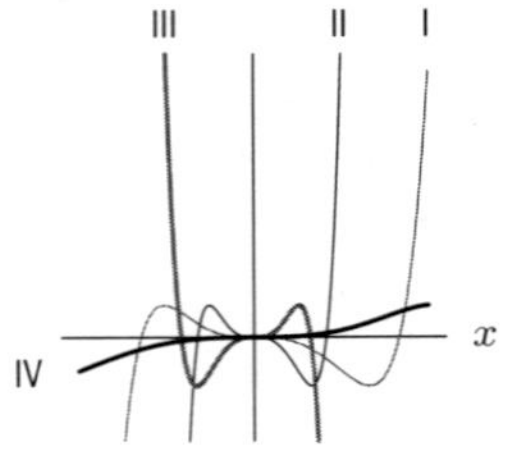

21. Find a formula for the function in Figure 5.50 as a transformation of the function f in Figure 5.49.

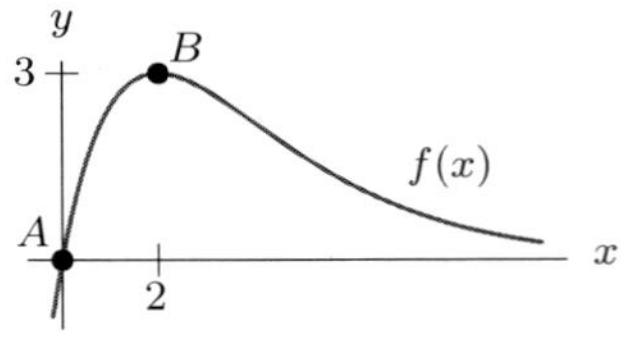

Figure 5.49

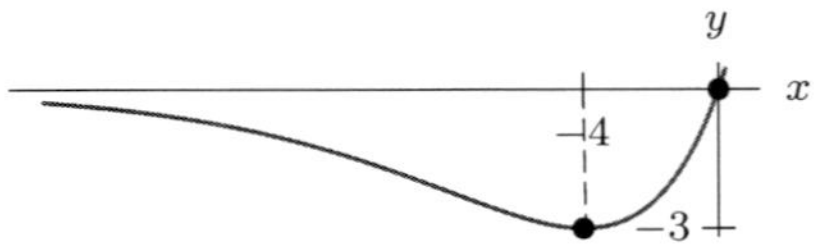

Figure 5.50

22. This problem investigates the effect of a horizontal stretch on the zeros of a function.

(a) Graph $f(x) = 4 - x^2$. Mark the zeros of f on the graph.
(b) Graph and find a formula for $g(x) = f(0.5x)$. What are the zeros of $g(x)$?
(c) Graph and find a formula for $h(x) = f(2x)$. What are the zeros of $h(x)$?
(d) Without graphing, what are the zeros of $f(10x)$?

23. In Figure 5.51, the point c is labeled on the x-axis. On the y-axis, locate and label output values:

(a) $g(c)$ **(b)** $2g(c)$ **(c)** $g(2c)$

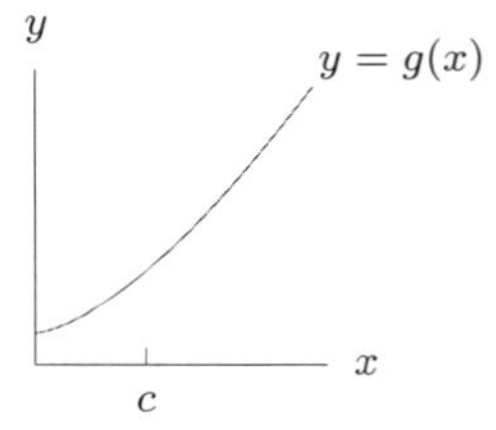

Figure 5.51

Table 5.20 gives values of $T = f(d)$, the average temperature (in °C) at a depth d meters in a borehole in Belleterre, Quebec. The functions in Problems 24–29 describe boreholes near Belleterre. Construct a table of values for each function and describe in words what it tells you about the borehole.[1]

Table 5.20

d, depth (m)	25	50	75	100
T, temp (°C)	5.5	5.2	5.1	5.1
d, depth (m)	125	150	175	200
T, temp (°C)	5.3	5.5	5.75	6

24. $g(d) = f(d) - 3$

25. $h(d) = f(d+5)$

26. $m(d) = f(d-10)$

27. $n(d) = 1.5f(d)$

28. $p(d) = f(0.8d)$

29. $q(d) = 1.5f(d) + 2$

5.5 THE FAMILY OF QUADRATIC FUNCTIONS

In Chapter 2, we looked at the example of a baseball which is popped up by a batter. The height of the ball above the ground was modeled by the quadratic function $y = f(t) = -16t^2 + 64t + 3$, where t is time in seconds after the ball leaves the bat, and y is in feet. The function is graphed in Figure 5.52.

The point on the graph with the largest y value appears to be $(2, 67)$. (We show this in Example 5 on page 229.) This means that the baseball reaches its maximum height of 67 feet 2 seconds after being hit. The maximum point $(2, 67)$ is called the *vertex*.

The graph of a quadratic function is called a *parabola*; its maximum (or minimum, if the parabola opens upward) is the vertex.

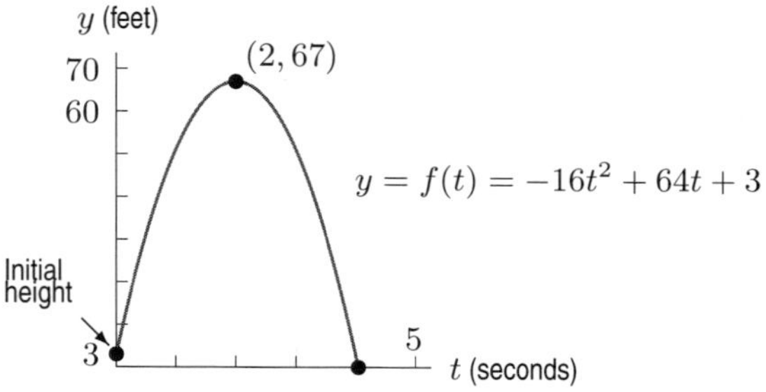

Figure 5.52: Height of baseball at time t

The Vertex of a Parabola

The graph of the function $y = x^2$ is a parabola with vertex at the origin. All other functions in the quadratic family turn out to be transformations of this function. Let's first graph a quadratic function of the form $y = a(x-h)^2 + k$ and locate its vertex.

Example 1 Let $f(x) = x^2$ and $g(x) = -2(x+1)^2 + 3$.

(a) Express the function g in terms of the function f.
(b) Sketch a graph of f. Transform the graph of f into the graph of g.
(c) Multiply out and simplify the formula for g.
(d) Explain how the formula for g can be used to obtain the vertex of the graph of g.

[1] Hugo Beltrami of St. Francis Xavier University and David Chapman of the University of Utah posted this data at http://esrc.stfx.ca/borehole/node3.html, accessed December 20, 2005.

Solution (a) Since $f(x+1) = (x+1)^2$, we have

$$g(x) = -2f(x+1) + 3.$$

(b) The graph of $f(x) = x^2$ is shown at the left in Figure 5.53. The graph of g is obtained from the graph of f in four steps, as shown in Figure 5.53.

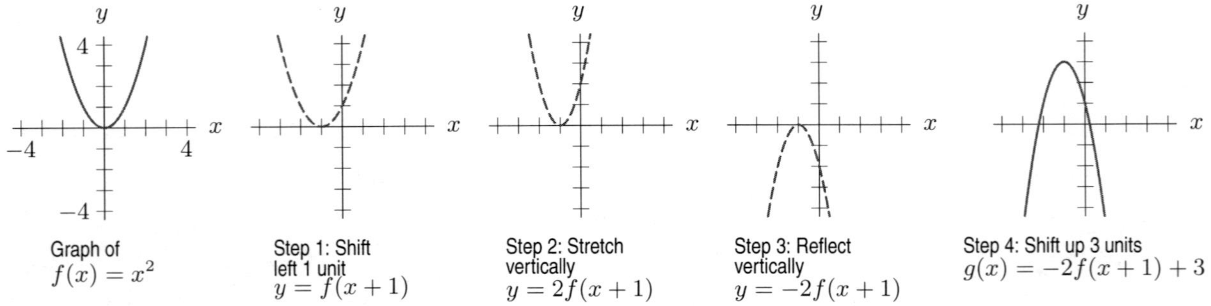

Figure 5.53: The graph of $f(x) = x^2$, on the left, is transformed in four steps into the graph of $g(x) = -2(x+1)^2 + 3$, on the right

(c) Multiplying out gives $g(x) = -2(x^2 + 2x + 1) + 3 = -2x^2 - 4x + 1$

(d) The vertex of the graph of f is $(0, 0)$. In Step 1 the vertex shifts 1 unit to the left (because of the $(x+1)$ in the formula), and in Step 4 the vertex shifts 3 units up (because of the $+3$ in the formula). Thus, the vertex of the graph of g is at $(-1, 3)$.

In general, the graph of $g(x) = a(x-h)^2 + k$ is obtained from the graph of $f(x) = x^2$ by shifting horizontally $|h|$ units, stretching vertically by a factor of a (and reflecting about the x-axis if $a < 0$), and shifting vertically $|k|$ units. In the process, the vertex is shifted from $(0, 0)$ to the point (h, k). The graph of the function is symmetrical about a vertical line through the vertex, called the *axis of symmetry*.

Formulas for Quadratic Functions

The function g in Example 1 can be written in two ways:

$$g(x) = -2(x+1)^2 + 3$$

and

$$g(x) = -2x^2 - 4x + 1.$$

The first version is helpful for understanding the graph of the quadratic function and finding its vertex. In general, we have the following:

The **standard form** for a **quadratic function** is

$$y = ax^2 + bx + c, \quad \text{where } a,\ b,\ c \text{ are constants, } a \neq 0.$$

The **vertex form** is

$$y = a(x-h)^2 + k, \quad \text{where } a,\ h,\ k \text{ are constants, } a \neq 0.$$

The graph of a quadratic function is called a **parabola**. The parabola

- Has vertex (h, k)
- Has axis of symmetry $x = h$
- Opens upward if $a > 0$ or downward if $a < 0$

Thus, any quadratic function can be expressed in both standard form and vertex form. To convert from vertex form to standard form, we multiply out the squared term. To convert from standard form to vertex form, we *complete the square*.

Example 2 Put these quadratic functions into vertex form by completing the square and then graph them.

(a) $s(x) = x^2 - 6x + 8$ (b) $t(x) = -4x^2 - 12x - 8$

Solution (a) To complete the square,[2] find the square of half of the coefficient of the x-term, $(-6/2)^2 = 9$. Add and subtract this number after the x-term:

$$s(x) = \underbrace{x^2 - 6x + 9}_{\text{Perfect square}} - 9 + 8,$$

so

$$s(x) = (x-3)^2 - 1.$$

The vertex of s is $(3, -1)$ and the axis of symmetry is the vertical line $x = 3$. There is no vertical stretch since $a = 1$, and the parabola opens upward. See Figure 5.54.

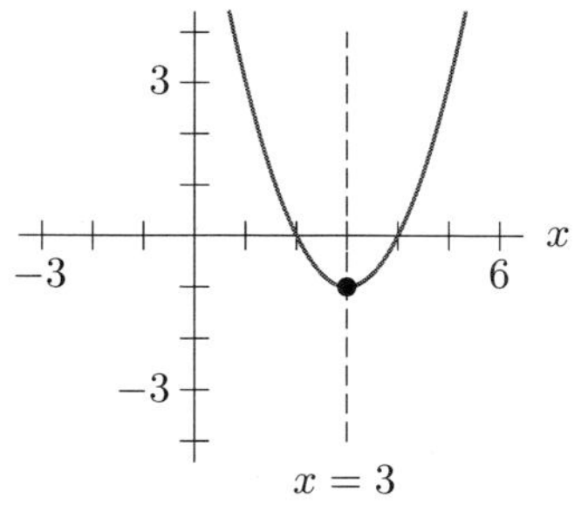

Figure 5.54: $s(x) = x^2 - 6x + 8$

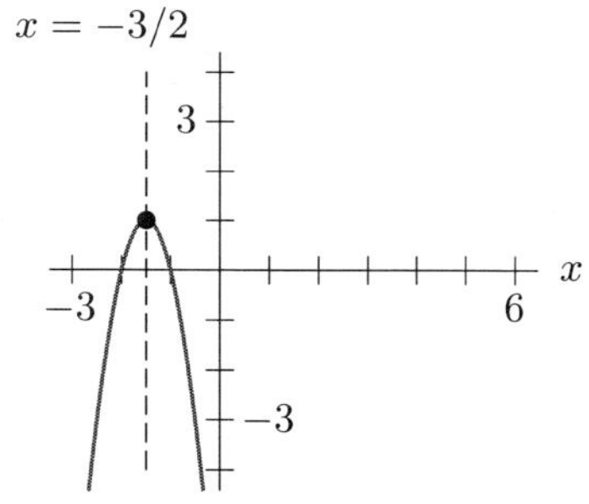

Figure 5.55: $t(x) = -4x^2 - 12x - 8$

[2]A more detailed explanation of this method is in the Tools section for this chapter.

(b) To complete the square, first factor out -4, the coefficient of x^2, giving

$$t(x) = -4(x^2 + 3x + 2).$$

Now add and subtract the square of half the coefficient of the x-term, $(3/2)^2 = 9/4$, inside the parentheses. This gives

$$t(x) = -4\Big(\underbrace{x^2 + 3x + \frac{9}{4}}_{\text{Perfect square}} - \frac{9}{4} + 2\Big)$$

$$t(x) = -4\left(\left(x + \frac{3}{2}\right)^2 - \frac{1}{4}\right)$$

$$t(x) = -4\left(x + \frac{3}{2}\right)^2 + 1.$$

The vertex of t is $(-3/2, 1)$, the axis of symmetry is $x = -3/2$, the vertical stretch factor is 4, and the parabola opens downward. See Figure 5.55.

Finding a Formula From a Graph

If we know the vertex of a quadratic function and one other point, we can use the vertex form to find its formula.

Example 3 Find the formula for the quadratic function graphed in Figure 5.56.

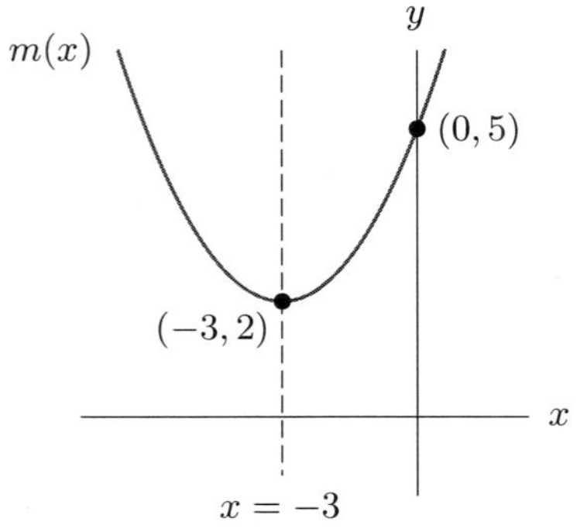

Figure 5.56

Solution Since the vertex is given, we use the form $m(x) = a(x - h)^2 + k$ and find a, h, and k. The vertex is $(-3, 2)$, so $h = -3$ and $k = 2$. Thus,

$$m(x) = a(x - (-3))^2 + 2,$$

so

$$m(x) = a(x + 3)^2 + 2.$$

To find a, use the y-intercept $(0, 5)$. Substitute $x = 0$ and $y = m(0) = 5$ into the formula for $m(x)$ and solve for a:

$$5 = a(0 + 3)^2 + 2$$
$$3 = 9a$$
$$a = \frac{1}{3}.$$

Thus, the formula is

$$m(x) = \frac{1}{3}(x+3)^2 + 2.$$

If we want the formula in standard form, we multiply out:

$$m(x) = \frac{1}{3}x^2 + 2x + 5.$$

Example 4 Find the equation of the parabola in Figure 5.57 using the factored form.

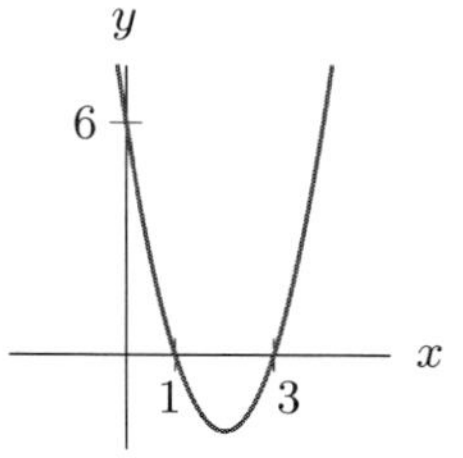

Figure 5.57

Solution Since the parabola has x-intercepts at $x = 1$ and $x = 3$, its formula can be written as

$$y = a(x-1)(x-3).$$

Substituting $x = 0, y = 6$ gives

$$6 = a(3)$$
$$a = 2.$$

Thus, the equation is

$$y = 2(x-1)(x-3).$$

x² -3 7x +3
x² -4x +3
2x² -8x +6

Applications of Quadratic Functions

In applications, it is often useful to find the maximum or minimum value of a quadratic function. First, we return to the baseball example which started this section.

Example 5 For t in seconds, the height of a baseball in feet is given by the formula

$$y = f(t) = -16t^2 + 64t + 3.$$

Using algebra, find the maximum height reached by the baseball and the time at which the ball reaches the ground.

Solution To find the maximum height, complete the square to find the vertex:

$$\begin{aligned} y = f(t) &= -16(t^2 - 4t) + 3 \\ &= -16(t^2 - 4t + 4 - 4) + 3 \\ &= -16(t^2 - 4t + 4) - 16(-4) + 3 \\ &= -16(t-2)^2 + 16 \cdot 4 + 3 \\ &= -16(t-2)^2 + 67. \end{aligned}$$

Thus, the vertex is at the point $(2, 67)$. This means that the ball reaches it maximum height of 67 feet at $t = 2$ seconds.

The time at which the ball hits the ground is found by solving $f(t) = 0$. We have

$$\begin{aligned} -16(t-2)^2 + 67 &= 0 \\ (t-2)^2 &= \frac{67}{16} \\ t - 2 &= \pm\sqrt{\frac{67}{16}} \approx \pm 2.046. \end{aligned}$$

The solutions are $t \approx -0.046$ and $t \approx 4.046$. Since the ball was thrown at $t = 0$, we want $t \geq 0$. Thus, the ball hits the ground approximately 4.046 seconds after being hit.

Example 6 A city decides to make a park by fencing off a section of riverfront property. Funds are allotted to provide 80 meters of fence. The area enclosed will be a rectangle, but only three sides will be enclosed by fence—the other side will be bound by the river. What is the maximum area that can be enclosed in this way?

Solution Two sides are perpendicular to the bank of the river and have equal length, which we call h. The other side is parallel to the bank of the river. Call its length b. See Figure 5.58. Since the fence is 80 meters long,

$$\begin{aligned} 2h + b &= 80 \\ b &= 80 - 2h. \end{aligned}$$

The area of the park, A, is the product of the lengths of two adjacent sides, so

$$\begin{aligned} A = bh &= (80 - 2h)h \\ &= -2h^2 + 80h. \end{aligned}$$

The function $A = -2h^2 + 80h$ is quadratic. Since the coefficient of h^2 is negative, the parabola opens downward and we have a maximum at the vertex. The zeros of this quadratic function are $h = 0$ and $h = 40$, so the axis of symmetry, which is midway between the zeros, is $h = 20$. The vertex of a parabola occurs on its axis of symmetry. Thus, substituting $h = 20$ gives the maximum area:

$$A = (80 - 2 \cdot 20)20 = (80 - 40)20 = 40 \cdot 20 = 800 \text{ meter}^2.$$

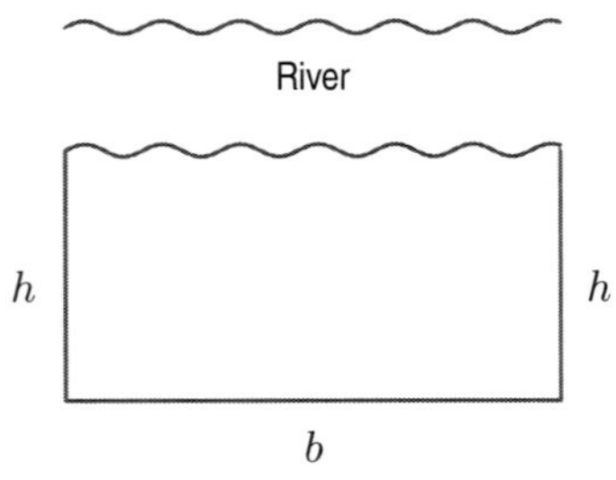

Figure 5.58

Exercises and Problems for Section 5.5

Exercises

For the quadratic functions in Exercises 1–2, state the coordinates of the vertex, the axis of symmetry, and whether the parabola opens upward or downward.

1. $f(x) = 3(x-1)^2 + 2$

2. $g(x) = -(x+3)^2 - 4$

3. Sketch the quadratic functions given in standard form. Identify the values of the parameters a, b, and c. Label the zeros, axis of symmetry, vertex, and y-intercept.

(a) $g(x) = x^2 + 3$ **(b)** $f(x) = -2x^2 + 4x + 16$

4. Find the vertex and axis of symmetry of the graph of $v(t) = t^2 + 11t - 4$.

5. Find the vertex and axis of symmetry of the graph of $w(x) = -3x^2 - 30x + 31$.

6. Show that the function $y = -x^2 + 7x - 13$ has no real zeros.

7. Find the value of k so that the graph of $y = (x-3)^2 + k$ passes through the point $(6, 13)$.

8. The parabola $y = ax^2 + k$ has vertex $(0, -2)$ and passes through the point $(3, 4)$. Find its equation.

In Exercises 9–14, find a formula for the parabola.

9.

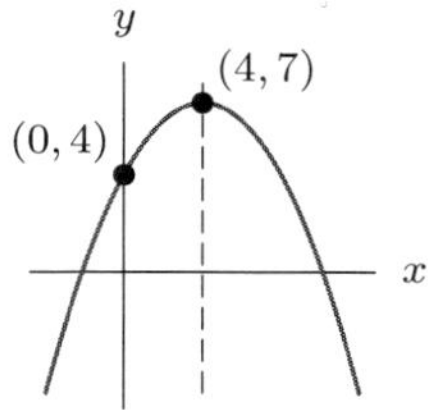

10.

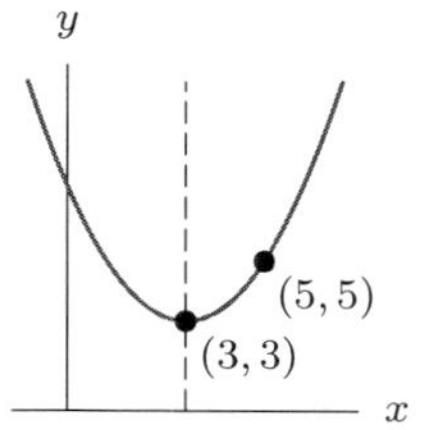

11.

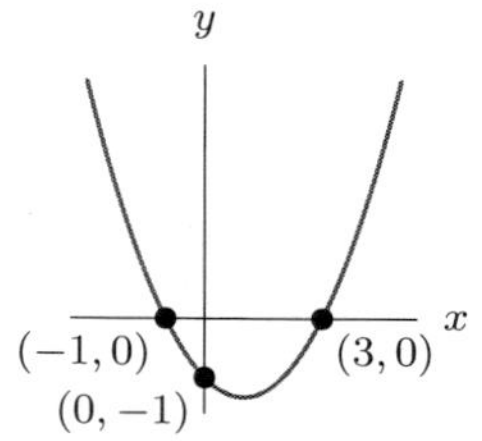

12.

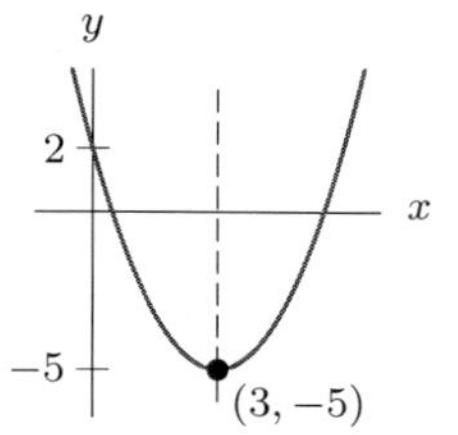

13.

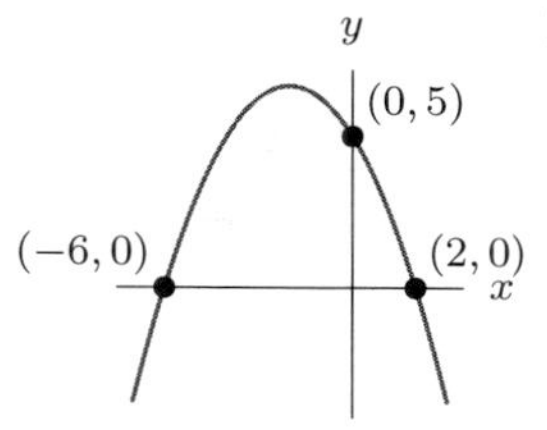

14.

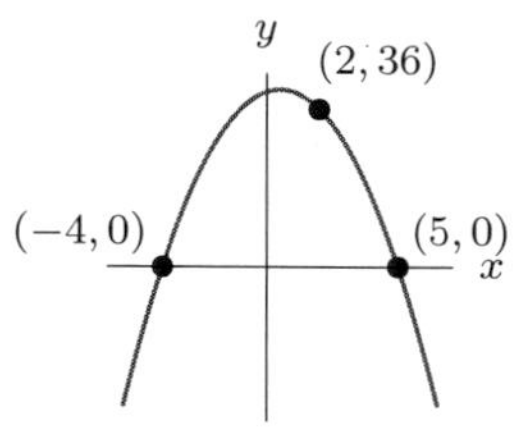

For Exercises 15–18, convert the quadratic functions to vertex form by completing the square. Identify the vertex and the axis of symmetry.

15. $f(x) = x^2 + 8x + 3$

16. $g(x) = -2x^2 + 12x + 4$

17. Using the vertex form, find a formula for the parabola with vertex $(2, 5)$ which passes through the point $(1, 2)$.

18. Using the factored form, find the formula for the parabola whose zeros are $x = -1$ and $x = 5$, and which passes through the point $(-2, 6)$.

Problems

In Problems 19–24, find a formula for the quadratic function whose graph has the given properties.

19. A vertex at $(4, 2)$ and a y-intercept of $y = 6$.

20. A vertex at $(4, 2)$ and a y-intercept of $y = -4$.

21. A vertex at $(4, 2)$ and zeros at $x = -3, 11$.

22. A y-intercept of $y = 7$ and x-intercepts at $x = 1, 4$.

23. A y-intercept of $y = 7$ and one zero at $x = -2$.

24. A vertex at $(-7, -3)$ and contains the point $(-3, -7)$.

25. Graph $y = x^2 - 10x + 25$ and $y = x^2$. Use a shift transformation to explain the relationship between the two graphs.

26. **(a)** Graph $h(x) = -2x^2 - 8x - 8$.
(b) Compare the graphs of $h(x)$ and $f(x) = x^2$. How are these two graphs related? Be specific.

27. Let f be a quadratic function whose graph is a concave up parabola with a vertex at $(1, -1)$, and a zero at the origin.

(a) Graph $y = f(x)$.
(b) Determine a formula for $f(x)$.
(c) Determine the range of f.
(d) Find any other zeros.

28. Let $f(x) = x^2$ and let $g(x) = (x-3)^2 + 2$.

(a) Give the formula for g in terms of f, and describe the relationship between f and g in words.
(b) Is g a quadratic function? If so, find its standard form and the parameters a, b, and c.
(c) Graph g, labeling all important features.

29. If we know a quadratic function f has a zero at $x = -1$ and vertex at $(1, 4)$, do we have enough information to find a formula for this function? If your answer is yes, find it; if not, give your reasons.

30. Gwendolyn, a pleasant parabola, was taking a peaceful nap when her dream turned into a nightmare: she dreamt that a low-flying pterodactyl was swooping toward her. Startled, she flipped over the horizontal axis, darted up (vertically) by three units, and to the left (horizontally) by two units. Finally she woke up and realized that her equation was $y = (x-1)^2 + 3$. What was her equation before she had the bad dream?

31. A tomato is thrown vertically into the air at time $t = 0$. Its height, $d(t)$ (in feet), above the ground at time t (in seconds) is given by

$$d(t) = -16t^2 + 48t.$$

(a) Graph $d(t)$.
(b) Find t when $d(t) = 0$. What is happening to the tomato the first time $d(t) = 0$? The second time?
(c) When does the tomato reach its maximum height?
(d) What is the maximum height that the tomato reaches?

32. An espresso stand finds that its weekly profit is a function of the price, x, it charges per cup. If x is in dollars, the weekly profit is $P(x) = -2900x^2 + 7250x - 2900$ dollars.

(a) Approximate the maximum profit and the price per cup that produces that profit.
(b) Which function, $P(x-2)$ or $P(x)-2$, gives a function that has the same maximum profit? What price per cup produces that maximum profit?
(c) Which function, $P(x+50)$ or $P(x)+50$, gives a function where the price per cup that produces the maximum profit remains unchanged? What is the maximum profit?

33. If you have a string of length 50 cm, what are the dimensions of the rectangle of maximum area that you can enclose with your string? Explain your reasoning. What about a string of length k cm?

34. A football player kicks a ball at an angle of 37° above the ground with an initial speed of 20 meters/second. The height, h, as a function of the horizontal distance traveled, d, is given by:

$$h = 0.75d - 0.0192d^2.$$

(a) Graph the path the ball follows.
(b) When the ball hits the ground, how far is it from the spot where the football player kicked it?
(c) What is the maximum height the ball reaches during its flight?
(d) What is the horizontal distance the ball has traveled when it reaches its maximum height?[3]

35. A ballet dancer jumps in the air. The height, $h(t)$, in feet, of the dancer at time t, in seconds since the start of the jump, is given by[4]

$$h(t) = -16t^2 + 16Tt,$$

where T is the total time in seconds that the ballet dancer is in the air.

(a) Why does this model apply only for $0 \leq t \leq T$?
(b) When, in terms of T, does the maximum height of the jump occur?
(c) Show that the time, T, that the dancer is in the air is related to H, the maximum height of the jump, by the equation

$$H = 4T^2.$$

[3] Adapted from R. Halliday, D. Resnick, and K. Krane, *Physics.* (New York: Wiley, 1992), p.58.
[4] K. Laws, *The Physics of Dance.* (Schirmer, 1984).

CHAPTER SUMMARY

- **Vertical and Horizontal Shifts**
  Vertical: $y = g(x) + k$.
  Upward if $k > 0$; downward if $k < 0$.
  Horizontal: $y = g(x + k)$.
  Left if $k > 0$; right if $k < 0$.
- **Reflections**
  Across x-axis: $y = -f(x)$.
  Across y-axis: $y = f(-x)$.
- **Symmetry**
  About y-axis: $f(-x) = f(x)$; even function.
  About the origin: $f(-x) = -f(x)$; odd function.
- **Stretches and Compressions**
  Vertical: $y = kf(x)$. Stretch if $k > 0$; compress if $0 < k < 1$; reflect across x-axis if $k < 0$.
  Horizontal: $y = f(kx)$. Compress if $k > 0$; stretch if $0 < k < 1$; reflect across y-axis if $k < 0$.
- **Quadratic Functions**
  Standard form: $y = ax^2 + bx + c$.
  Vertex form: $y = a(x - h)^2 + k$.
  Opening upward if $a > 0$; downward if $a < 0$.
  Vertex (h, k), axis of symmetry $x = h$, maximum, minimum.
  Completing the square.

REVIEW EXERCISES AND PROBLEMS FOR CHAPTER FIVE

Exercises

1. Suppose $x = 2$. Determine the value of the input of the function f in each of the following expressions:

(a) $f(2x)$ **(b)** $f(\frac{1}{2}x)$ **(c)** $f(x+3)$ **(d)** $f(-x)$

2. Determine the value of x in each of the following expressions which leads to an input of 2 to the function f:

(a) $f(2x)$ **(b)** $f(\frac{1}{2}x)$ **(c)** $f(x+3)$ **(d)** $f(-x)$

3. The point $(2, 5)$ is on the graph of $y = f(x)$. Give the coordinates of one point on the graph of each of the following functions.

(a) $y = f(x - 4)$ **(b)** $y = f(x) - 4$
(c) $y = f(4x)$ **(d)** $y = 4f(x)$

4. The point $(-3, 4)$ is on the graph of $y = g(x)$. Give the coordinates of one point on the graph of each of the following functions.

(a) $y = g(\frac{1}{3}x)$ **(b)** $y = \frac{1}{3}g(x)$
(c) $y = g(-3x)$ **(d)** $y = -g(3x)$

Are the functions in Exercises 5–10 even, odd, or neither?

5. $a(x) = \dfrac{1}{x}$ **6.** $m(x) = \dfrac{1}{x^2}$

7. $e(x) = x + 3$ **8.** $p(x) = x^2 + 2x$

9. $b(x) = |x|$ **10.** $q(x) = 2^{x+1}$

11. Let $f(x) = 1 - x$. Evaluate and simplify:

(a) $f(2x)$ **(b)** $f(x + 1)$ **(c)** $f(1 - x)$
(d) $f(x^2)$ **(e)** $f(1/x)$ **(f)** $f(\sqrt{x})$

12. Fill in all the blanks in Table 5.21 for which you have sufficient information.

Table 5.21

x	-3	-2	-1	0	1	2	3
$f(x)$	-4	-1	2	3	0	-3	-6
$f(-x)$							
$-f(x)$							
$f(x) - 2$							
$f(x - 2)$							
$f(x) + 2$							
$f(x + 2)$							
$2f(x)$							
$-f(x)/3$							

In Exercises 13–16, find a formula for the parabola.

13.

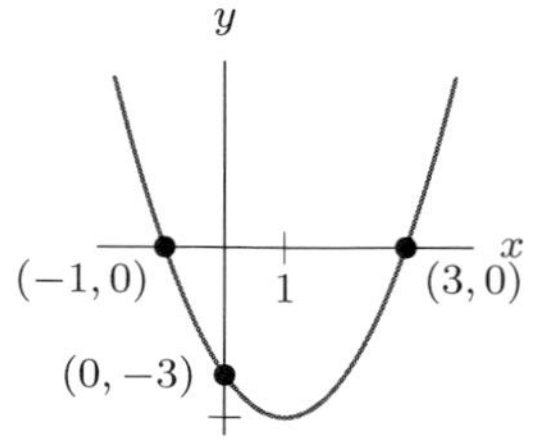

14.

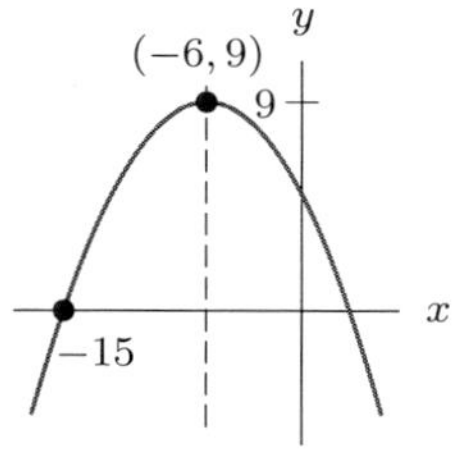

15.

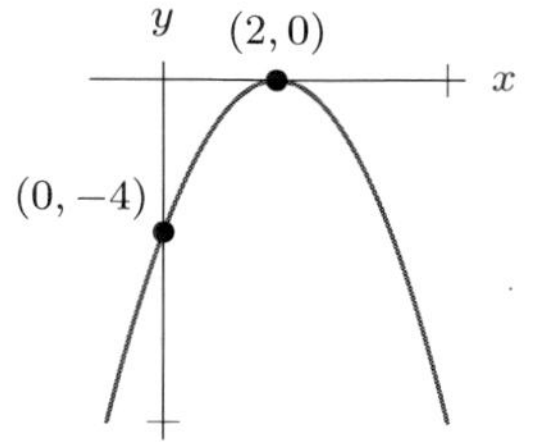

16.

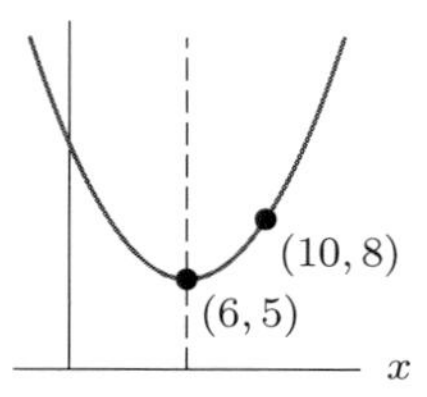

Problems

In Problems 17–18, use Figure 5.59 to sketch the function.

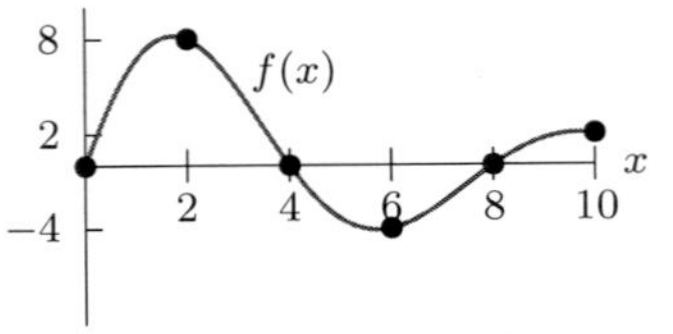

Figure 5.59

17. $y = f(x+2)+2$ **18.** $y = -2f(-x)$

In Problems 19–20, use Figure 5.59 to find a possible formula for the transformation of f shown.

19.

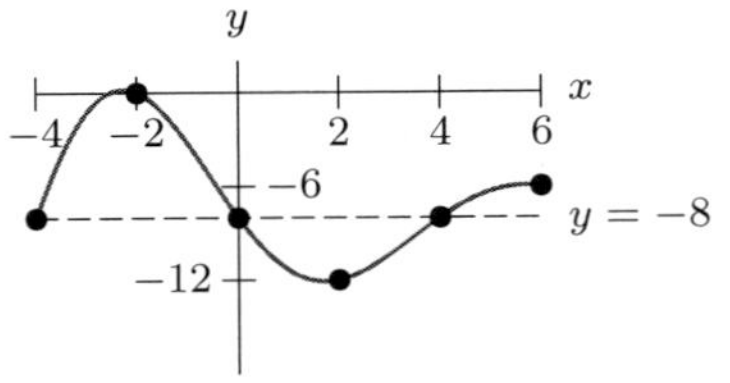

20.

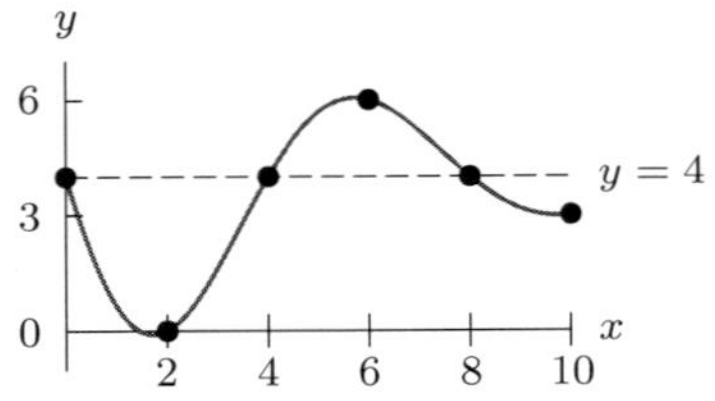

21. Let $D(p)$ be the number of iced cappuccinos sold each week by a coffeehouse when the price is p cents.

(a) What does the expression $D(225)$ represent?
(b) Do you think that $D(p)$ is an increasing function or a decreasing function? Why?
(c) What does the following equation tell you about p? $D(p) = 180$
(d) The coffeehouse sells n iced cappuccinos when they charge the average price in their area, t cents. Thus, $D(t) = n$. What is the meaning of the following expressions: $D(1.5t)$, $1.5D(t)$, $D(t+50)$, $D(t)+50$?

22. Suppose $w = j(x)$ is the average daily quantity of water (in gallons) required by an oak tree of height x feet.

(a) What does the expression $j(25)$ represent? What about $j^{-1}(25)$?
(b) What does the following equation tell you about v: $j(v) = 50$. Rewrite this statement in terms of j^{-1}.
(c) Oak trees are on average z feet high and a tree of average height requires p gallons of water. Represent this fact in terms of j and then in terms of j^{-1}.
(d) Using the definitions of z and p from part (c), what do the following expressions represent?

$$j(2z),\quad 2j(z),\quad j(z+10),\quad j(z)+10,$$

$$j^{-1}(2p),\quad j^{-1}(p+10),\quad j^{-1}(p)+10.$$

23. Without a calculator, match each of the functions (a)–(f) with one of the graphs (I) – (VI).

(a) $y = e^x$ **(b)** $y = e^{5x}$ **(c)** $y = 5e^x$
(d) $y = e^{x+5}$ **(e)** $y = e^{-x}$ **(f)** $y = e^x + 5$

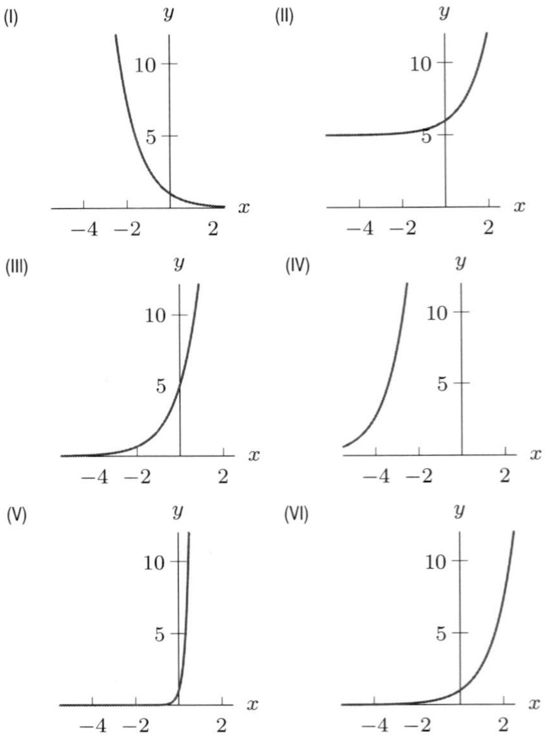

24. The graph in Figure 5.60 gives the number of hours of daylight in Charlotte, North Carolina on day d of the year, where $d = 0$ is January 1. Graph the number of hours of daylight in Buenos Aires, Argentina, which is as far south of the equator as Charlotte is north. [Hint: When it is summer in the Northern Hemisphere, it is winter in the Southern Hemisphere.]

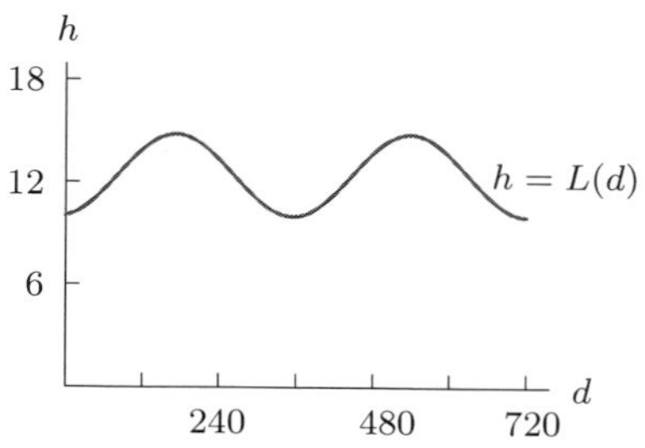

Figure 5.60

25. During a hurricane, a brick breaks loose from the top of a chimney, 38 feet above the ground. As the brick falls, its distance from the ground after t seconds is given by:

$$d(t) = -16t^2 + 38.$$

(a) Find formulas for $d(t) - 15$ and $d(t - 1.5)$.
(b) On the same axes, graph $d(t)$, $d(t) - 15$, $d(t - 1.5)$.
(c) Suppose $d(t)$ represents the height of a brick which began to fall at noon. What might $d(t) - 15$ represent? $d(t - 1.5)$?
(d) Using algebra, determine when the brick hits the ground:
  (i) If $d(t)$ represents the distance of the brick from the ground,
  (ii) If $d(t) - 15$ represents the distance of the brick from the ground.
(e) Use one of your answers in part (d) to determine when the brick hits the ground if $d(t - 1.5)$ represents its distance above the ground at time t.

The functions graphed in Problems 26–27 are transformations of some basic function. Give a possible formula for each one.

26.

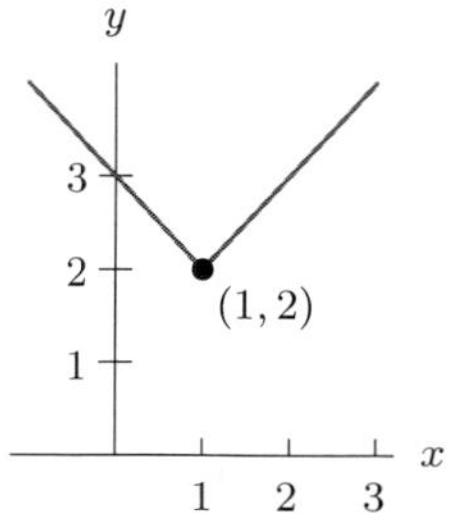

27.

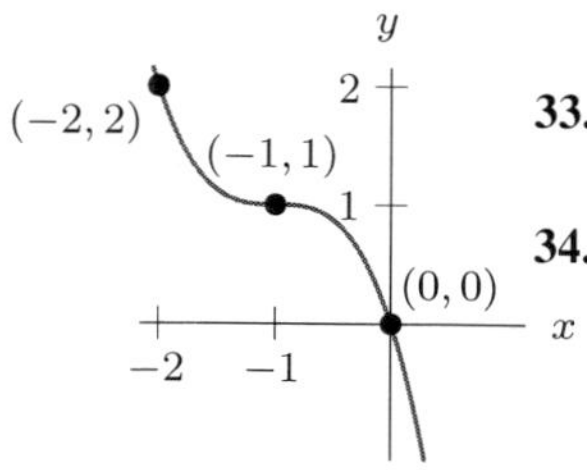

In Problems 28–29, use Figure 5.61 to find a formula for the transformations of $h(x)$.

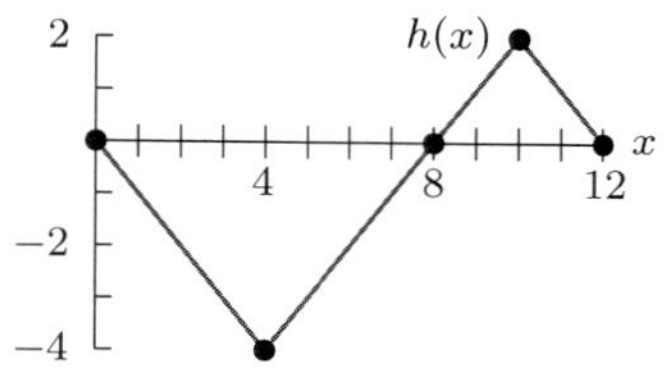

Figure 5.61

28.

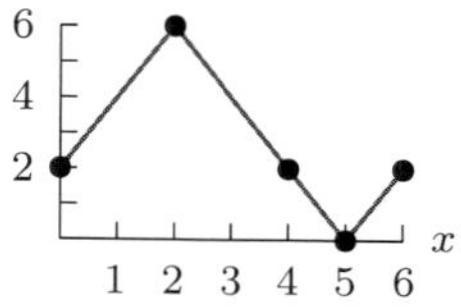

29.

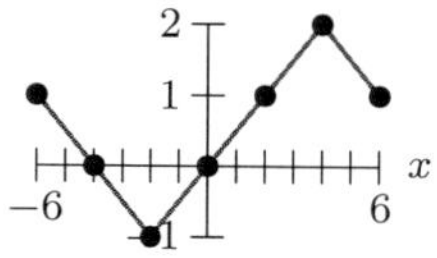

Problems 30–34 use Table 5.22 which gives the total cost, $C = f(n)$, for a carpenter to build n wooden chairs.

Table 5.22

n	0	10	20	30	40	50
$f(n)$	5000	6000	6800	7450	8000	8500

30. Evaluate the following expressions. Explain in everyday terms what they mean.
(a) $f(10)$ **(b)** $f(x)$ if $x = 30$
(c) z if $f(z) = 8000$ **(d)** $f(0)$

31. Find approximate values for p and q if $f(p) = 6400$ and $q = f(26)$.

32. Let $d_1 = f(30) - f(20)$, $d_2 = f(40) - f(30)$, and $d_3 = f(50) - f(40)$.
(a) Evaluate d_1, d_2 and d_3.
(b) What do these numbers tell you about the carpenter's cost of building chairs?

33. Graph $f(n)$. Label the quantities you found in Problems 30–32 on your graph.

34. The carpenter currently builds k chairs per week.
(a) What do the following expressions represent?
  (i) $f(k + 10)$ (ii) $f(k) + 10$
  (iii) $f(2k)$ (iv) $2f(k)$
(b) If the carpenter sells his chairs at 80% above cost, plus an additional 5% sales tax, write an expression for his gross income (including sales tax) each week.

35. In Figure 5.62, the value of d is labeled on the x-axis. Locate the following quantities on the y-axis:

(a) $g(d)$ **(b)** $g(-d)$ **(c)** $-g(-d)$

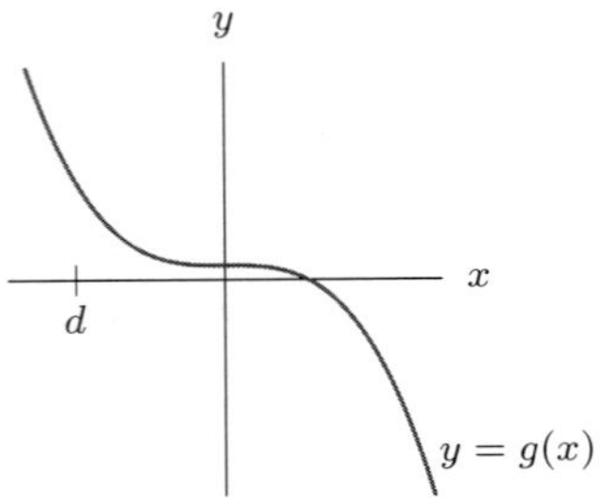

Figure 5.62

36. In Figure 5.63, the values c and d are labeled on the x-axis. On the y-axis, locate the following quantities:

(a) $h(c)$ **(b)** $h(d)$
(c) $h(c+d)$ **(d)** $h(c)+h(d)$

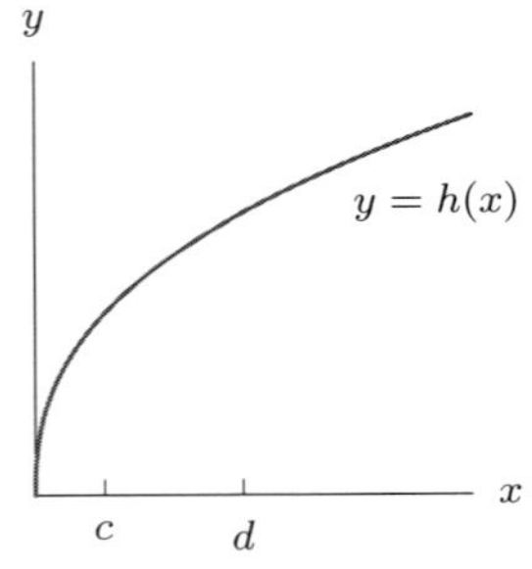

Figure 5.63

In Problems 37–39, use Figure 5.64 to find a formula for the graphs in terms of h.

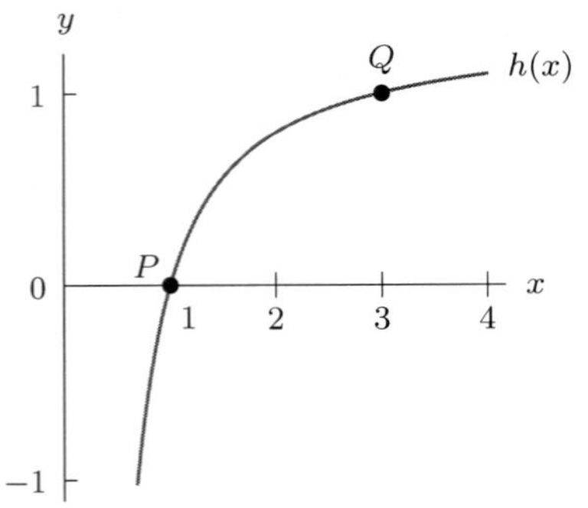

Figure 5.64

37.

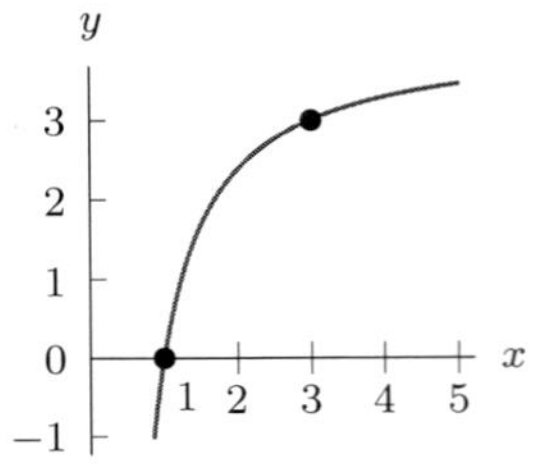

38.

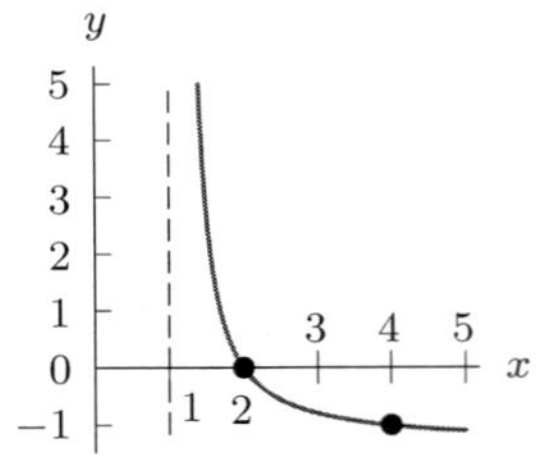

39.

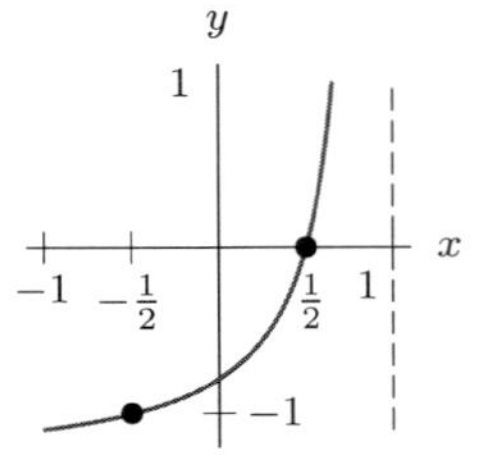

40. Let $T(d)$ be the low temperature in degrees Fahrenheit on the dth day of last year (where $d = 1$ is January 1st, and so on).

(a) Sketch a possible graph of T for your home town for $1 \le d \le 365$.

(b) Suppose n is a new function that reports the low temperature on day, d, in terms of degrees above (or below) freezing. (For example, if the low temperature on the 100th day of the year is 42°F, then $n(100) = 42 - 32 = 10$.) Write an expression for $n(d)$ as a shift of T. Graph n. How does the graph of n relate to the graph of T?

(c) This year, something different has happened. The temperatures for each day are exactly the same as they were last year, except each temperature occurs a week earlier than it did last year. Suppose $p(d)$ gives the low temperature on the dth day of this year. Write an expression for $p(d)$ in terms of $T(d)$. Graph p. How do the graphs of p and T relate?

41. When slam-dunking, a basketball player seems to hang in the air at the height of his jump. The height $h(t)$, in feet above the ground, of a basketball player at time t, in seconds since the start of a jump, is given by

$$h(t) = -16t^2 + 16Tt,$$

where T is the total time in seconds that it takes to complete the jump. For a jump that takes 1 second to complete, how much of this time does the basketball player spend at the top 25% of the trajectory? [Hint: Find the maximum height reached. Then find the times at which the height is 75% of this maximum.]

42. A cube-shaped box has an edge of length x cm.

(a) Write a formula for the function $L(x)$ that gives the length of tape used to run a layer of tape around all the edges.
(b) Explain what would happen if your roll of tape contained $L(x) - 6$ cm of tape. What about $L(x - 6)$ cm of tape?
(c) If the tape must be wrapped beyond each corner for a distance of 1 cm, express the length of tape used in terms of a shift of the function L.
(d) Write a formula in terms of x for the function S that gives the surface area of the box.
(e) Write a formula in terms of x for the function V that gives the volume of the box.

Because the box is fragile, it is packed inside a larger box with room for packing material around it. There must be 5 cm of clearance between the smaller and larger boxes on every side.

(f) Express the surface area of the larger box as a shift of the function S.
(g) Express the volume of the larger box as a shift of V.
(h) The larger box has its edges taped. Express the length of tape used as a shift of the function L.
(i) The outside box needs to be double taped to reinforce it for shipping (every edge receives two layers of tape). Express the length of tape that needs to be used in terms of $L(x)$.
(j) The edge length of the outside box must be 20% longer than the edge length of the original box. Which expression gives the length of tape necessary to tape the edges, $1.2L(x)$ or $L(1.2x)$? Explain.

CHECK YOUR UNDERSTANDING

Are the statements in Problems 1–29 true or false? Give an explanation for your answer.

1. If $g(x) = f(x) + 3$ then the graph of $g(x)$ is a vertical shift of the graph of f.

2. If $g(t) = f(t-2)$ then the graph of $g(t)$ can be obtained by shifting the graph of f two units to the left.

3. If $g(x) = f(x) + k$ and k is negative, the graph of $g(x)$ is the same as the graph of f, but shifted down.

4. Vertical and horizontal shifts are called translations.

5. The reflection of $y = x^2$ across the x-axis is $y = -x^2$.

6. If $f(x)$ is an odd function, then $f(x) = f(-x)$.

7. The graphs of odd functions are symmetric about the y-axis.

8. The graph of $y = -f(x)$ is the reflection of the graph of $y = f(x)$ across the x-axis.

9. The graph of $y = f(-x)$ is the reflection of the graph of $y = f(x)$ across the y-axis.

10. If the graph of a function f is symmetric about the y-axis then $f(x) = f(-x)$.

11. Figure 5.65 suggests that $g(x) = f(x + 2) + 1$.

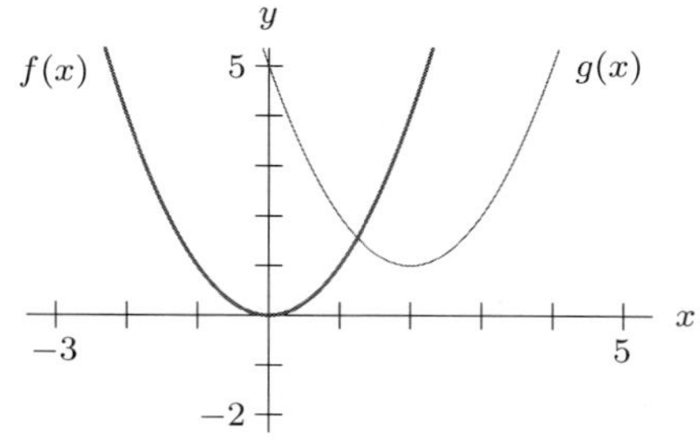

Figure 5.65

12. If $g(x) = x^2 + 4$ then $g(x - 2) = x^2$.

13. For any function f, we have $f(x + k) = f(x) + k$.

14. Figure 5.66 could be the graph of $f(x) = |x - 1| - 2$.

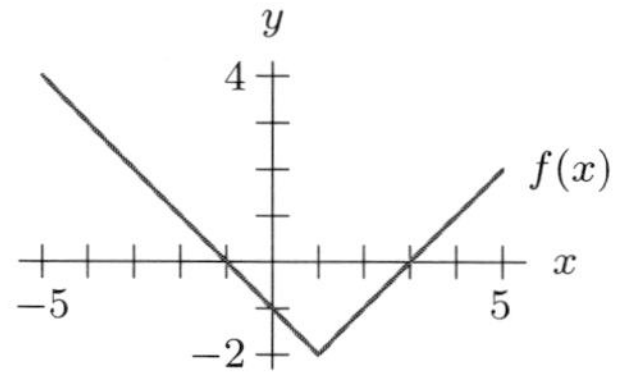

Figure 5.66

15. Let $f(x) = 3^x$. If the graph of $f(x)$ is reflected across the x-axis and then shifted up four units, the new graph has the equation $y = -3^x + 4$.

16. If $q(p) = p^2 + 2p + 4$ then $-q(-p) = p^2 - 2p + 4$.

17. Multiplying a function by a constant k, with $k > 1$, vertically stretches its graph.

18. If $g(x) = kf(x)$, then on any interval the average rate of change of g is k times the average rate of change of f.

19. Figure 5.67 suggests that $g(x) = -2f(x + 1) + 3$.

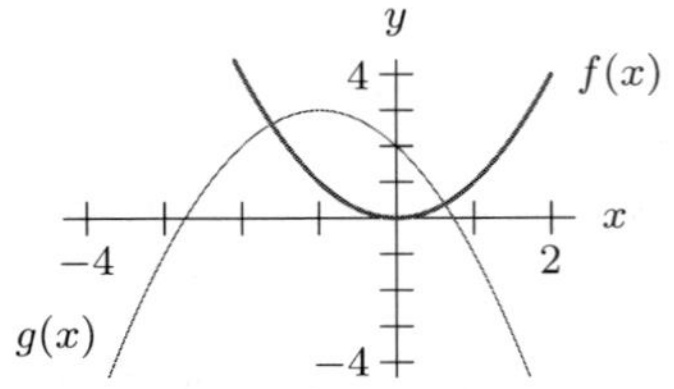

Figure 5.67

20. Using Table 5.23, we can conclude that if $g(x) = -\frac{1}{2}f(x + 1) - 3$, then $g(-2) = -10$.

Table 5.23

x	-3	-2	-1	0	1	2	3
$f(x)$	10	6	4	1	-2	-4	-10

21. Shifting the graph of a function up by one unit and then compressing it vertically by a factor of $\frac{1}{2}$ produces the same result as first compressing the graph by a factor of $\frac{1}{2}$ and then shifting it up by one unit.

22. Figure 5.68 suggests that $g(x) = 3f(\frac{1}{2}x)$.

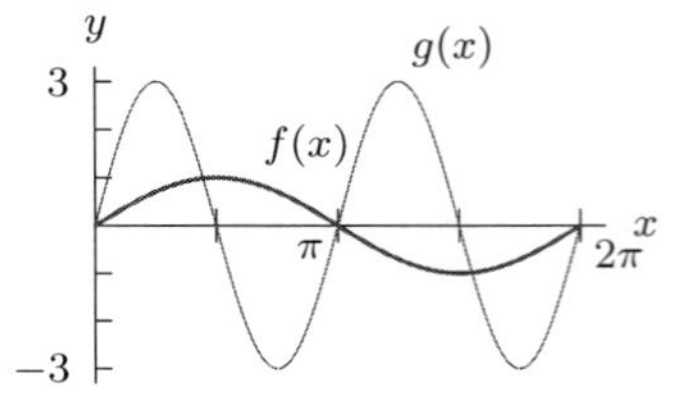

Figure 5.68

23. For the function given in Table 5.24, for $x = -2$, we have $3f(2x) + 1 = -2f(\frac{1}{2}x)$.

Table 5.24

x	-4	-3	-2	-1	0	1	2
$f(x)$	1	4	0	-2	0	0	-2

24. The graph of every quadratic equation is a parabola.

25. The maximum or minimum point of a parabola is called its vertex.

26. If a parabola is concave up its vertex is a maximum point.

27. If the equation of a parabola is written as $y = a(x - h)^2 + k$, then the vertex is located at the point $(-h, k)$.

28. If the equation of a parabola is written as $y = a(x - h)^2 + k$, then the axis of symmetry is found at $x = h$.

29. If the equation of a parabola is $y = ax^2 + bx + c$ and $a < 0$, then the parabola opens downward.

TOOLS FOR CHAPTER 5: COMPLETING THE SQUARE

Completing the Square

Another example of changing the form of an expression is the conversion of $ax^2 + bx + c$ into the form $a(x - h)^2 + k$. We make this conversion by *completing the square*, a method for producing a perfect square within a quadratic expression. A perfect square is an expression of the form:

$$(x + n)^2 = x^2 + 2nx + n^2$$

for some number n. In order to complete the square in a given expression, we must find that number n, which is half the coefficient of x. Before giving a general procedure, let's work through an example.

Example 1 Complete the square to rewrite $x^2 - 10x + 4$ in the form $a(x - h)^2 + k$.

Solution Step 1: We divide the coefficient of x by 2, giving $\frac{1}{2}(-10) = -5$.
Step 2: We squaring the result of step 1, giving $(-5)^2 = 25$.
Step 3: Now add and subtract the 25 after the x-term:

$$\begin{aligned} x^2 - 10x + 4 &= x^2 - 10x + (25 - 25) + 4 \\ &= \underbrace{(x^2 - 10x + 25)}_{\text{Perfect square}} - 25 + 4. \end{aligned}$$

Step 4: Notice that we have created a perfect square, $x^2 - 10x + 25$. The next step is to factor the perfect square and combine the constant terms, $-25 + 4$, giving the final result:

$$x^2 - 10x + 4 = (x - 5)^2 - 21.$$

Thus, $a = +1$, $h = +5$, and $k = -21$.

The procedure we followed can be summarized as follows:

> **To complete the square** in the expression $x^2 + bx + c$, divide the coefficient of x by 2, giving $b/2$. Then add and subtract $(b/2)^2 = b^2/4$ and factor the perfect square:
>
> $$x^2 + bx + c = \left(x + \frac{b}{2}\right)^2 - \frac{b^2}{4} + c.$$
>
> To complete the square in the expression $ax^2 + bx + c$, factor out a first.

The next example has a coefficient a with $a \neq 1$. After factoring out the coefficient, we follow the same steps as in Example 1.

Example 2 Complete the square in the formula $h(x) = 5x^2 + 30x - 10$.

Solution We first factor out 5:

$$h(x) = 5(x^2 + 6x - 2).$$

Now we complete the square in the expression $x^2 + 6x - 2$.
Step 1: Divide the coefficient of x by 2, giving 3.
Step 2: Square the result: $3^2 = 9$.
Step 3: Add the result after the x term, then subtract it:

$$h(x) = 5(\underbrace{x^2 + 6x + 9}_{\text{Perfect square}} - 9 - 2).$$

Step 4: Factor the perfect square and simplify the rest:

$$h(x) = 5\left((x+3)^2 - 11\right).$$

Now that we have completed the square, we can multiply by the 5:

$$h(x) = 5(x+3)^2 - 55.$$

Deriving The Quadratic Formula

We derive a general formula for the zeros of $q(x) = ax^2 + bx + c$, with $a \neq 0$, by completing the square. To find the zeros, set $q(x) = 0$:

$$ax^2 + bx + c = 0.$$

Before we complete the square, we factor out the coefficient of x^2:

$$a\left(x^2 + \frac{b}{a}x + \frac{c}{a}\right) = 0.$$

Since $a \neq 0$, we can divide both sides by a:

$$x^2 + \frac{b}{a}x + \frac{c}{a} = 0.$$

To complete the square, we add and then subtract $((b/a)/2)^2 = b^2/(4a^2)$:

$$\underbrace{x^2 + \frac{b}{a}x + \frac{b^2}{4a^2}}_{\text{Perfect square}} - \frac{b^2}{4a^2} + \frac{c}{a} = 0.$$

We factor the perfect square and simplify the constant term, giving:

$$\left(x + \frac{b}{2a}\right)^2 - \left(\frac{b^2 - 4ac}{4a^2}\right) = 0 \qquad \text{since } \frac{-b^2}{4a^2} + \frac{c}{a} = \frac{-b^2}{4a^2} + \frac{4ac}{4a^2} = -\left(\frac{b^2 - 4ac}{4a^2}\right)$$

$$\left(x+\frac{b}{2a}\right)^2 = \frac{b^2-4ac}{4a^2} \qquad \text{adding } \frac{b^2-4ac}{4a^2} \text{ to both sides}$$

$$x+\frac{b}{2a} = \pm\sqrt{\frac{b^2-4ac}{4a^2}} = \frac{\pm\sqrt{b^2-4ac}}{2a} \qquad \text{taking the square root}$$

$$x = \frac{-b}{2a} \pm \frac{\sqrt{b^2-4ac}}{2a} \qquad \text{subtracting } b/2a$$

$$x = \frac{-b \pm \sqrt{b^2-4ac}}{2a}.$$

Exercises to Tools for Chapter 5

For Exercises 1–12, complete the square for each expression.

1. $x^2 + 8x$
2. $y^2 - 12y$
3. $w^2 + 7w$
4. $2r^2 + 20r$
5. $s^2 + 6s - 8$
6. $3t^2 + 24t - 13$
7. $a^2 - 2a - 4$
8. $n^2 + 4n - 5$
9. $c^2 + 3c - 7$
10. $3r^2 + 9r - 4$
11. $4s^2 + s + 2$
12. $12g^2 + 8g + 5$

In Exercises 13–16, rewrite in the form $a(x-h)^2 + k$.

13. $x^2 - 2x - 3$
14. $10 - 6x + x^2$
15. $-x^2 + 6x - 2$
16. $3x^2 - 12x + 13$

In Exercises 17–26, complete the square to find the vertex of the parabola.

17. $y = x^2 + 6x + 3$
18. $y = x^2 - x + 4$
19. $y = -x^2 - 8x + 2$
20. $y = x^2 - 3x - 3$
21. $y = -x^2 + x - 6$
22. $y = 3x^2 + 12x$
23. $y = -4x^2 + 8x - 6$
24. $y = 5x^2 - 5x + 7$
25. $y = 2x^2 - 7x + 3$
26. $y = -3x^2 - x - 2$

In Exercises 27–36, solve by completing the square.

27. $r^2 - 6r + 8 = 0$
28. $g^2 = 2g + 24$
29. $p^2 - 2p = 6$
30. $n^2 = 3n + 18$
31. $d^2 - d = 2$
32. $2r^2 + 4r - 5 = 0$
33. $2s^2 = 1 - 10s$
34. $5q^2 - 8 = 2q$
35. $7r^2 - 3r - 6 = 0$
36. $5p^2 + 9p = 1$

In Exercises 37–42, solve by using the quadratic formula.

37. $n^2 - 4n - 12 = 0$
38. $2y^2 + 5y = -2$
39. $6k^2 + 11k = -3$
40. $w^2 + w = 4$
41. $z^2 + 4z = 6$
42. $2q^2 + 6q - 3 = 0$

In Exercises 43–56, solve using factoring, completing the square, or the quadratic formula.

43. $r^2 - 2r = 8$
44. $-3t^2 + 4t + 9 = 0$
45. $n^2 + 4n - 3 = 2$

46. $s^2 + 3s = 1$
47. $z^3 + 2z^2 = 3z + 6$
48. $2q^2 + 4q - 5 = 8$
49. $25u^2 + 4 = 30u$
50. $v^2 - 4v - 9 = 0$
51. $3y^2 = 6y + 18$
52. $2p^2 + 23 = 14p$
53. $2w^3 + 24 = 6w^2 + 8w$
54. $4x^2 + 16x - 5 = 0$
55. $49m^2 + 70m + 22 = 0$
56. $8x^2 - 1 = 2x$

Chapter Six

TRIGONOMETRIC FUNCTIONS

Blood pressure in the heart, an alternating electric current, the water level in a reservoir, and the phases of the moon all exhibit repeating behavior. Linear and exponential functions do not adequately model these processes. In this chapter, we introduce the *trigonometric* functions, which model such repetitive behavior.

The Tools Section on page 301 reviews right triangle trigonometry.

6.1 INTRODUCTION TO PERIODIC FUNCTIONS

The World's Largest Ferris Wheel

To celebrate the millennium, British Airways funded construction of the "London Eye," the world's largest ferris wheel.[1] The wheel is located on the south bank of the river Thames, in London, England, measures 450 feet in diameter, and carries up to 800 passengers in 32 capsules. It turns continuously, completing a single rotation once every 30 minutes. This is slow enough for people to hop on and off while it turns.

Ferris Wheel Height As a Function of Time

Suppose you hop on this ferris wheel at time $t = 0$ and ride it for two full turns. Let $f(t)$ be your height above the ground, measured in feet as a function of t, the number of minutes you have been riding. We can figure out some values of $f(t)$.

Let's imagine that the wheel is turning in the counterclockwise direction. At time $t = 0$ you have just boarded the wheel, so your height is 0 ft above the ground (not counting the height of your seat). Thus, $f(0) = 0$. Since the wheel turns all the way around once every 30 minutes, after 7.5 minutes the wheel has turned one-quarter of the way around. Thinking of the wheel as a giant clock, this means you have been carried from the 6 o'clock position to the 3 o'clock position, as shown in Figure 6.1. You are now halfway up the wheel, or 225 feet above the ground, so $f(7.5) = 225$.

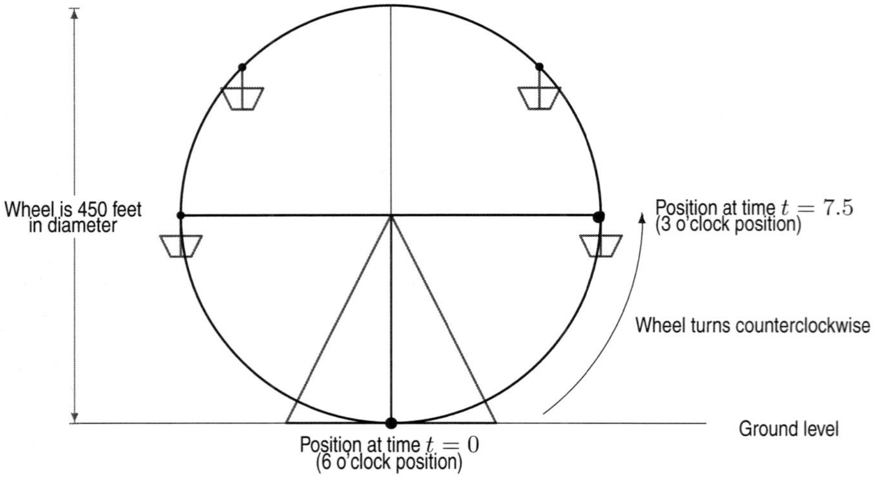

Figure 6.1: The world's largest ferris wheel is 450 ft in diameter and turns around once every 30 minutes. Seats not drawn to scale

After 15 minutes, the wheel has turned halfway around, so you are now at the top, in the 12 o'clock position. Thus, $f(15) = 450$. And after 22.5 minutes, the wheel has turned three quarters of the way around, bringing you into the 9 o'clock position. You have descended from the top of the wheel halfway down to the ground, and you are once again 225 feet above the ground. Thus, $f(22.5) = 225$. (See Figures 6.2 and 6.3.) Finally, after 30 minutes, the wheel has turned all the way around, bringing you back to ground level, so $f(30) = 0$.

[1] http://british-airways-london-eye.visit-london-england.com, accessed May 23, 2002.

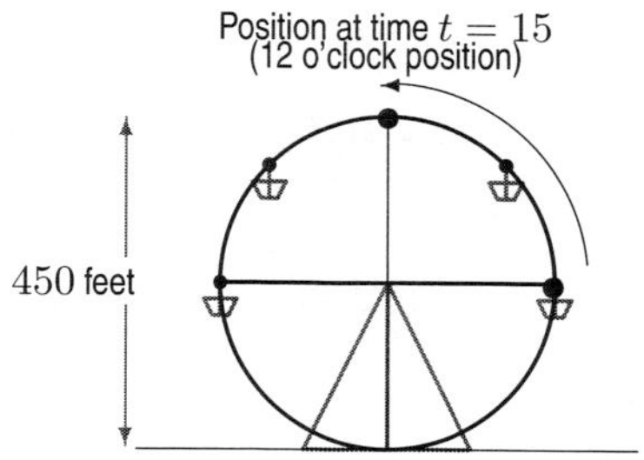

Figure 6.2: At time $t = 15$, the wheel has turned halfway around

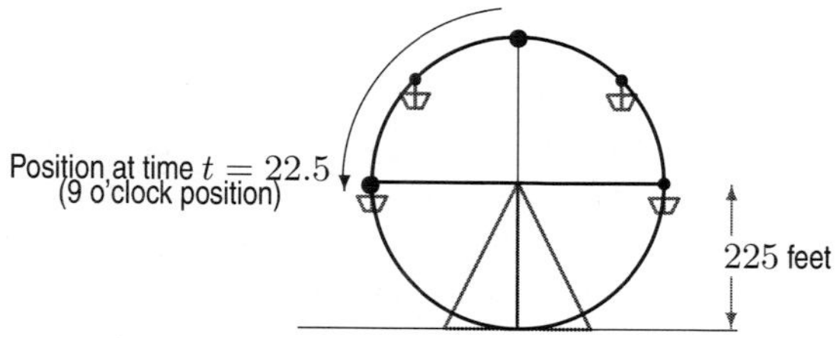

Figure 6.3: At time $t = 22.5$, the wheel has turned three quarters of the way around

The Second Time Around On the Wheel

Since the wheel turns without stopping, at time $t = 30$ it begins its second turn. Thus, at time $t = 37.5$, the wheel has again turned one quarter of the way around, and $f(37.5) = 225$. Similarly, at time $t = 45$ the wheel has again turned halfway around, bringing you back up to the very top. Likewise, at time $t = 52.5$, the wheel has carried you halfway back down to the ground. This means that $f(45) = 450$ and $f(52.5) = 225$. Finally, at time $t = 60$, the wheel has completed its second full turn and you are back at ground level, so $f(60) = 0$.

Table 6.1 *Values of $f(t)$, your height above the ground t minutes after boarding the wheel*

t (minutes)	0	7.5	15	22.5	30	37.5	45	52.5	
$f(t)$ (feet)	0	225	450	225	0	225	450	225	
t (minutes)	60	67.5	75	82.5	90	97.5	105	112.5	120
$f(t)$ (feet)	0	225	450	225	0	225	450	225	0

Repeating Values of the Ferris Wheel Function

Notice that the values of $f(t)$ in Table 6.1 begin repeating after 30 minutes. This is because the second turn is just like the first turn, except that it happens 30 minutes later. If you ride the wheel for more full turns, the values of $f(t)$ continue to repeat at 30-minute intervals.

Graphing the Ferris Wheel Function

The data from Table 6.1 are plotted in Figure 6.4. The graph begins at $y = 0$ (ground level), rises to $y = 225$ (halfway up the wheel) and then to $y = 450$ (the top of the wheel). The graph then falls to $y = 225$ and then down to $y = 0$. This cycle then repeats itself three more times, once for each rotation of the wheel.

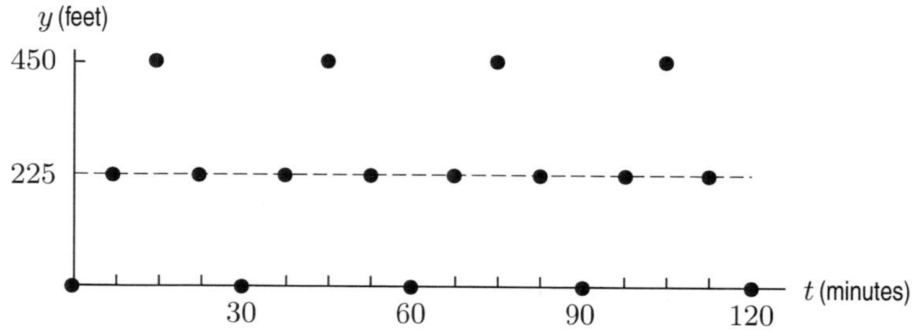

Figure 6.4: Values of $f(t)$, the ferris wheel height function, at 7.5 minute intervals

Filling in the Graph of the Ferris Wheel Function

It is tempting to connect the points in Figure 6.4 with straight lines, but this is not correct. Consider the first 7.5 minutes of your ride, starting at the 6 o'clock position and ending at the 3 o'clock position. (See Figure 6.5). Halfway through this part of the ride, the wheel has turned halfway from the 6 o'clock to the 3 o'clock position. However, as is clear from Figure 6.5, your seat rises less than half the vertical distance from $y = 0$ to $y = 225$. At the same time, the seat glides more than half the horizontal distance. If the points in Figure 6.4 were connected with straight lines, $f(3.75)$ would be halfway between $f(0)$ and $f(7.5)$, which is incorrect.

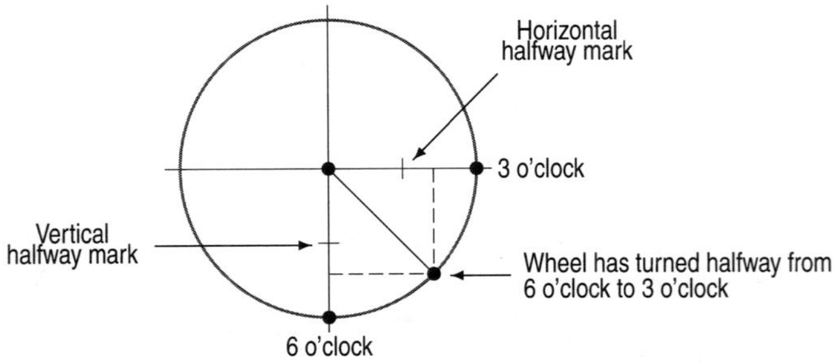

Figure 6.5: As the wheel turns half the way from 6 o'clock to 3 o'clock, the seat rises less than half the vertical distance but glides more than half the horizontal distance

The graph of $f(t)$ in Figure 6.6 is a smooth curve that repeats itself. It looks the same from $t = 0$ to $t = 30$ as from $t = 30$ to $t = 60$, or from $t = 60$ to $t = 90$, or from $t = 90$ to $t = 120$.

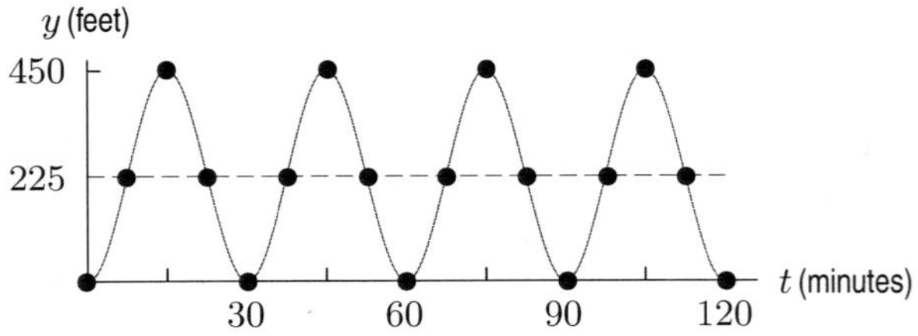

Figure 6.6: The graph of $y = f(t)$ is a smooth wave-shaped curve

Periodic Functions: Period, Midline, and Amplitude

The ferris wheel function, f, is said to be *periodic*. The smallest time interval during which a function completes one full cycle is called its *period* and is represented as a horizontal distance in Figure 6.7.

We can think about the period in terms of horizontal shifts. If the graph of f is shifted to the left by 30 units, the resulting graph looks exactly the same. That is,

$$\underbrace{\text{Graph of } f \text{ shifted left by 30 units}}_{f(t+30)} \quad \text{is the same as} \quad \underbrace{\text{Original graph,}}_{f(t)}$$

so, for all values of t,

$$f(t + 30) = f(t).$$

In general, we make the following definition:

A function f is **periodic** if its values repeat at regular intervals. Graphically, this means that if the graph of f is shifted horizontally by c units, the new graph is identical to the original. In function notation, periodic means that, for all t in the domain of f,

$$f(t+c)=f(t).$$

The smallest positive constant c for which this relationship holds for all values of t is called the **period** of f.

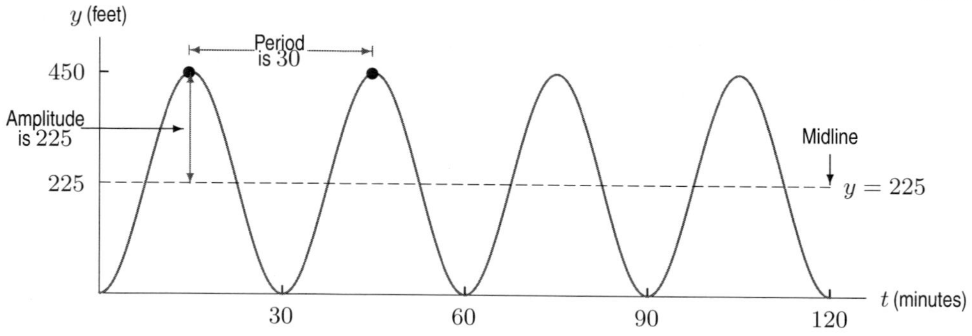

Figure 6.7: The graph of $y=f(t)$ showing the amplitude, period, and midline

In Figure 6.7, the dashed horizontal line is the *midline* of the graph of f. The *amplitude* of a wave-like periodic function is the distance between its maximum and the midline (or the distance between the midline and the minimum). Thus the amplitude of f is 225 because the ferris wheel's maximum height is 450 feet and its midline is at 225 feet. The amplitude is represented graphically as a vertical distance. (See Figure 6.7.) In general:

The **midline** of a periodic function is the horizontal line midway between the function's maximum and minimum values. The **amplitude** is the vertical distance between the function's maximum (or minimum) value and the midline.

Exercises and Problems for Section 6.1

Exercises

In Exercises 1–8, do the functions appear to be periodic with period less than 4?

1.

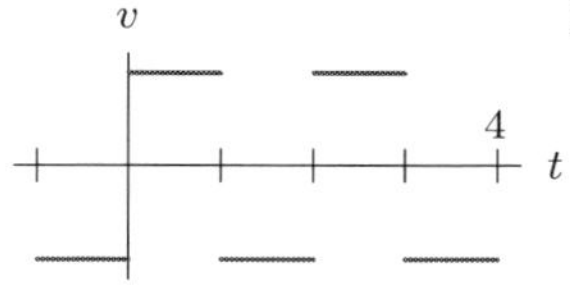

2.

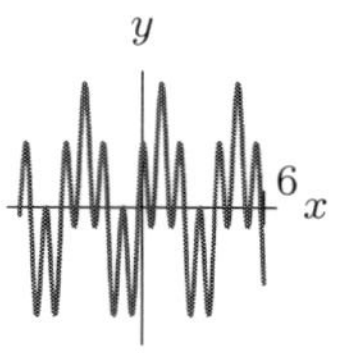

3.

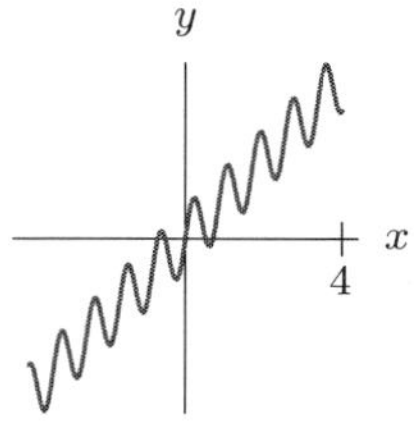

4.

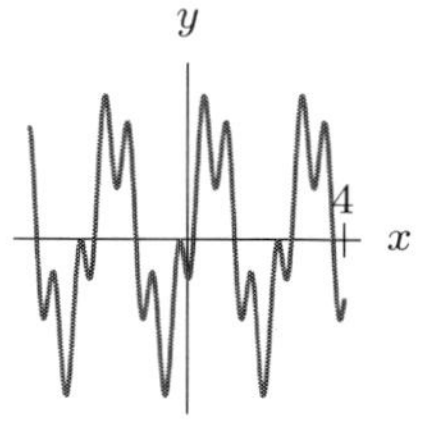

5.

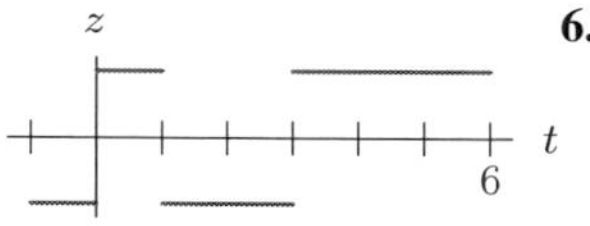

6.

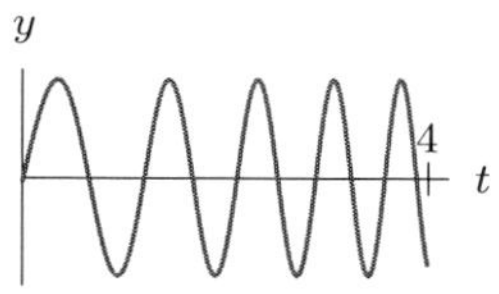

7.

t	0	1	2	3	4	5	6
$f(t)$	1	5	7	1	5	7	1

8.

r	0	π	2π	3π	4π	5π	6π	7π
$q(r)$	0	1	0	-1	1	0	1	0

In Exercises 9–12, estimate the period of the periodic functions.

9.

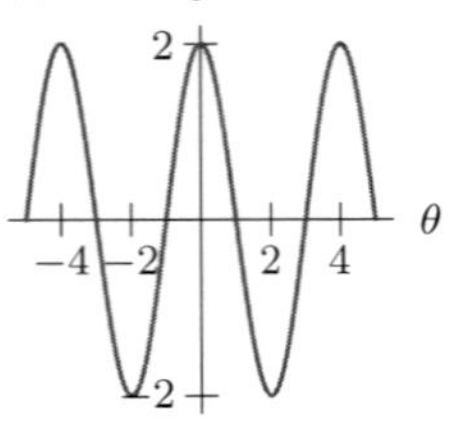

10.

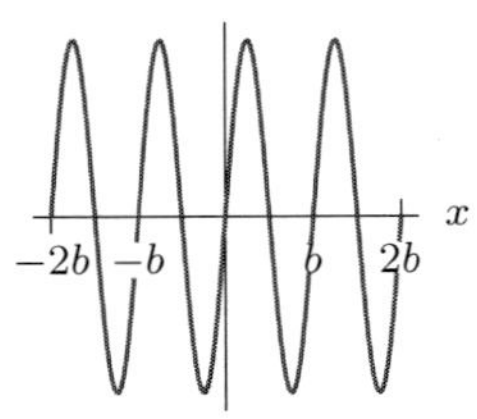

11.

t	0	1	2	3	4	5	6
$f(t)$	12	13	14	12	13	14	12

12.

z	1	11	21	31	41	51	61	71	81
$g(z)$	5	3	2	3	5	3	2	3	5

Problems

You board the London ferris wheel described in this section. In Problems 13–15, graph $h = f(t)$, your height in feet above the ground t minutes after the wheel begins to turn. Label the period, the amplitude, and the midline of each graph, as well as both axes. In each case, first determine an appropriate interval for t, with $t \geq 0$.

13. The London ferris wheel has increased its rotation speed. The wheel completes one full revolution every ten minutes. You get off when you reach the ground after having made two complete revolutions.

14. Everything is the same as Problem 13 (including the rotation speed) except the wheel has a 600 foot diameter.

15. The London ferris wheel is rotating at twice the speed as the wheel in Problem 13.

Problems 16–18 involve different ferris wheels. Graph $h = f(t)$. Label the period, the amplitude, and the midline for each graph. In each case, first determine an appropriate interval for t, with $t \geq 0$.

16. A ferris wheel is 20 meters in diameter and boarded from a platform that is 4 meters above the ground. The six o'clock position on the ferris wheel is level with the loading platform. The wheel completes one full revolution every 2 minutes. At $t = 0$ you are in the twelve o'clock position. You then make two complete revolutions and any additional part of a revolution needed to return to the boarding platform.

17. A ferris wheel is 50 meters in diameter and boarded from a platform that is 5 meters above the ground. The six o'clock position on the ferris wheel is level with the loading platform. The wheel completes one full revolution every 8 minutes. You make two complete revolutions on the wheel, starting at $t = 0$.

18. A ferris wheel is 35 meters in diameter and boarded at ground level. The wheel completes one full revolution every 5 minutes. At $t = 0$ you are in the three o'clock position and ascending. You then make two complete revolutions and return to the boarding platform.

The graphs in Problems 19–22 describe your height, $h = f(t)$, above the ground on different ferris wheels, where h is in meters and t is time in minutes. You boarded the wheel before $t = 0$. For each graph, determine the following: your position and direction at $t = 0$, how long it takes the wheel to complete one full revolution, the diameter of the wheel, at what height above the ground you board the wheel, and the length of time the graph shows you riding the wheel. The boarding platform is level with the bottom of the wheel.

19.

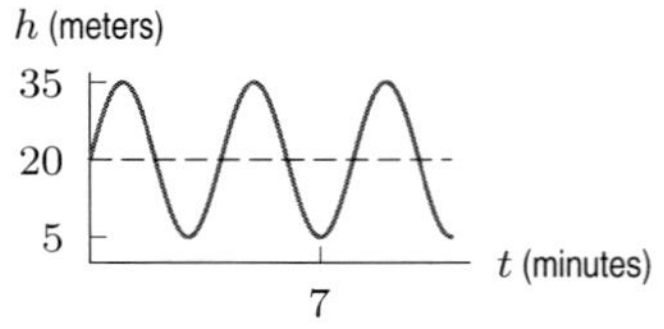

20.

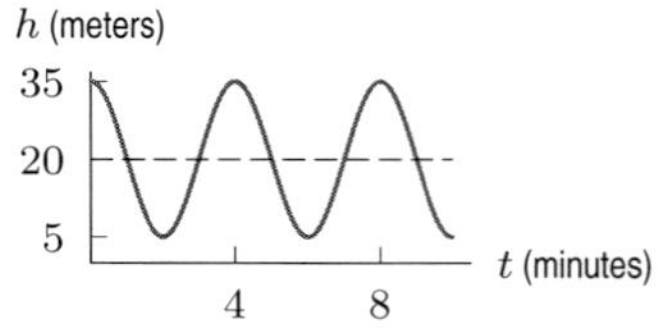

21.

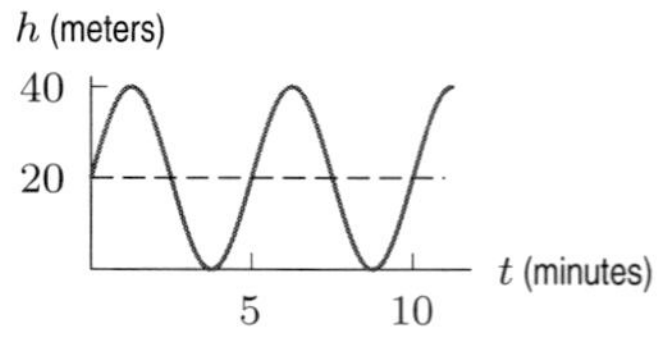

22.

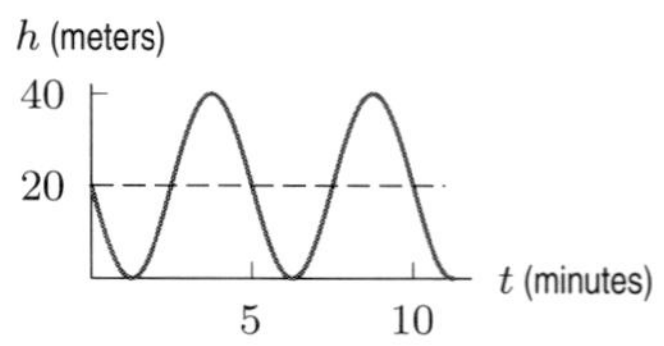

Problems 23–26 concern a weight suspended from the ceiling by a spring. (See Figure 6.8.) Let d be the distance in centimeters from the ceiling to the weight. When the weight is motionless, $d = 10$. If the weight is disturbed, it begins to bob up and down, or *oscillate*. Then d is a periodic function of t, time in seconds, so $d = f(t)$.

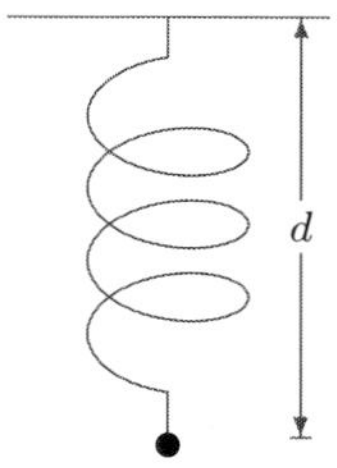

Figure 6.8

23. Determine the midline, period, amplitude, and the minimum and maximum values of f from the graph in Figure 6.9. Interpret these quantities physically; that is, use them to describe the motion of the weight.

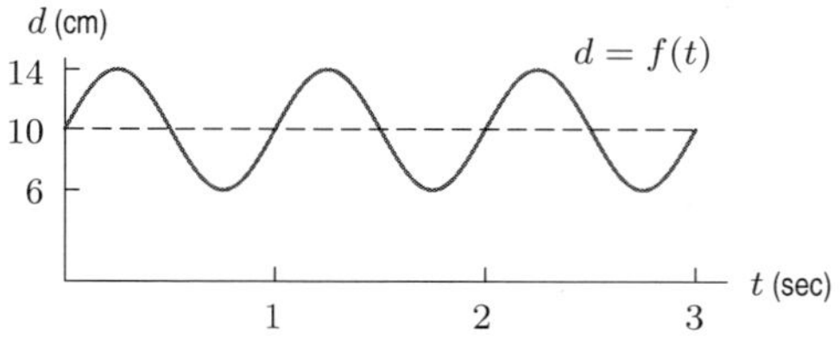

Figure 6.9:

24. A new experiment with the same weight and spring is represented by Figure 6.10. Compare Figure 6.10 to Figure 6.9. How do the oscillations differ? For both figures, the weight was disturbed at time $t = -0.25$ and then left to move naturally; determine the nature of the initial disturbances.

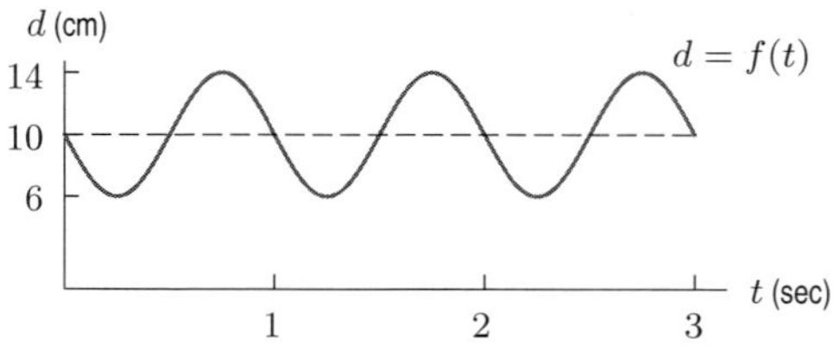

Figure 6.10:

25. The weight in Problem 23 is gently pulled down to a distance of 14 cm from the ceiling and released at time $t = 0$. Sketch its motion for $0 \leq t \leq 3$.

26. Figures 6.11 and 6.12 describe the motion of two different weights, A and B, attached to two different springs. Based on these graphs, which weight:

(a) Is closest to the ceiling when not in motion?
(b) Makes the largest oscillations?
(c) Makes the fastest oscillations?

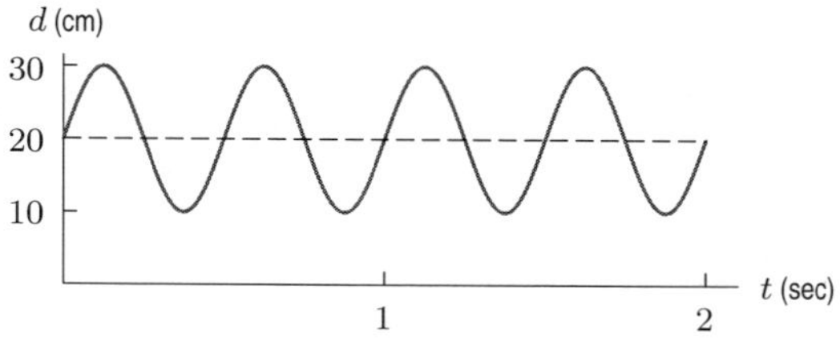

Figure 6.11: Weight A

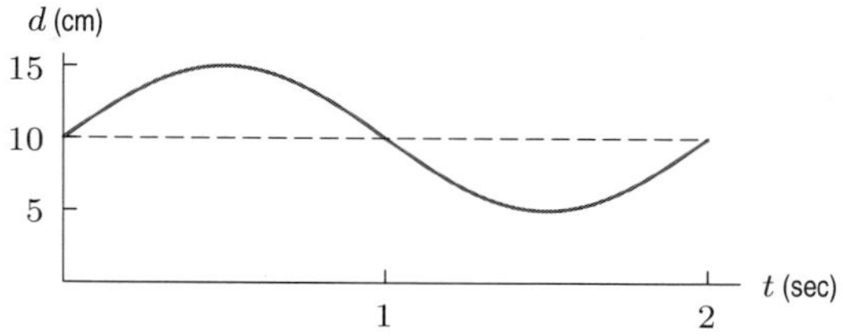

Figure 6.12: Weight B

27. In the US, household electricity is in the form of *alternating current* (AC) at 155.6 volts and 60 hertz. This means that the voltage cycles from -155.6 volts to $+155.6$ volts and back to -155.6 volts, and that 60 cycles occur each second. Suppose that at $t = 0$ the voltage at a given outlet is at 0 volts.

(a) Sketch $V = f(t)$, the voltage as a function of time, for the first 0.1 seconds.
(b) State the period, the amplitude, and the midline of the graph you made in part (a). Describe the physical significance of these quantities.

28. The temperature of a chemical reaction oscillates between a low of 30°C and a high of 110°C. The temperature is at its lowest point when $t = 0$ and completes one cycle over a five-hour period.

(a) Sketch the temperature, T, against the elapsed time, t, over a ten-hour period.
(b) Find the period, the amplitude, and the midline of the graph you drew in part (a).

29. Use a calculator or a computer to decide whether each of the following functions is periodic or not.

(a) $f(x) = \sin(x/\pi)$ **(b)** $f(x) = \sin(\pi/x)$
(c) $f(x) = x\sin x$ **(d)** $f(x) = e^{-x}\sin x$
(e) $f(x) = \pi + \sin x$ **(f)** $f(x) = \sin(e^{-x})$
(g) $f(x) = \sin(x+\pi)$

30. Table 6.2 gives the number of white blood cells (in 10,000s) in a patient with chronic myelogenous leukemia with nearly periodic relapses. Plot these data and estimate the midline, amplitude and period.

Table 6.2

Day	0	10	40	50	60	70	75	80	90
WBC	0.9	1.2	10	9.2	7.0	3.0	0.9	0.8	0.4
Day	100	110	115	120	130	140	145	150	160
WBC	1.5	2.0	5.7	10.7	9.5	5.0	2.7	0.6	1.0
Day	170	175	185	195	210	225	230	240	255
WBC	2.0	6.0	9.5	8.2	4.5	1.8	2.5	6.0	10.0

31. Table 6.3 gives data from a vibrating string experiment, with time, t, in seconds, and height, $h = f(t)$, in centimeters. Find the midline, amplitude and period of f.

Table 6.3

t	0	.1	.2	.3	.4	.5	.6	.7	.8	.9	1
h	2	2.6	3	3	2.6	2	1.4	1	1	1.4	2

32. Table 6.4 gives the height $h = f(t)$ in feet of a weight on a spring where t is time in seconds. Find the midline, amplitude and period of the function f.

Table 6.4

t	0	1	2	3	4	5	6	7
h	4.0	5.2	6.2	6.5	6.2	5.2	4.0	2.8
t	8	9	10	11	12	13	14	15
h	1.8	1.5	1.8	2.8	4.0	5.2	6.2	6.5

6.2 THE SINE AND COSINE FUNCTIONS

In this section we construct a formula for the ferris wheel function. To do this, we consider points on a unit circle and then riders on the ferris wheel.

The Unit Circle

The *unit circle* is the circle of radius one that is centered at the origin. (See Figure 6.13.) Since the distance from the point P with coordinates (x, y) to the origin is 1, we have

$$\sqrt{x^2+y^2} = 1,$$

so squaring both sides gives the equation of the circle:

$$x^2+y^2 = 1.$$

Angles can be used to locate points on the unit circle. Positive angles are measured counterclockwise from the positive x-axis; negative angles are measured clockwise.

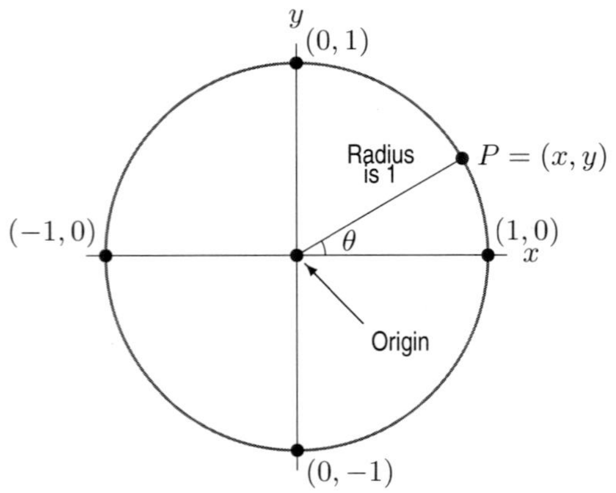

Figure 6.13: The unit circle, with equation $x^2 + y^2 = 1$

Example 1 Figure 6.14 shows the point P determined by $\theta = 90°$. The coordinates of P are $(0, 1)$. Figure 6.15 shows the point Q corresponding to $\alpha = 180°$. It has coordinates $(-1, 0)$. Figure 6.16 shows the point R determined by the angle $\phi = 210°$. Shortly, we see how to find the coordinates of point R.

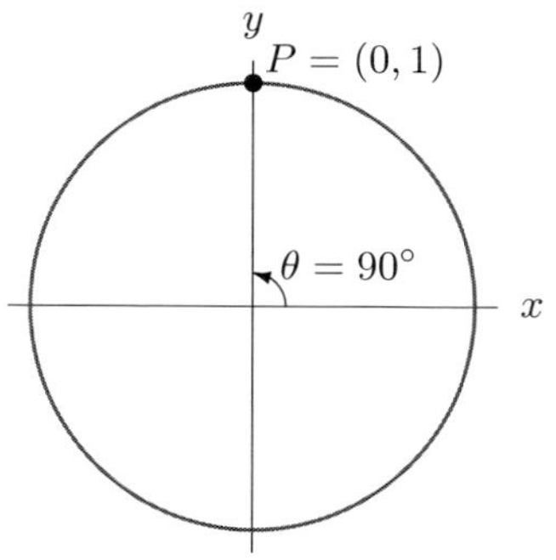

Figure 6.14: The angle $\theta = 90°$ specifies the point $P = (0, 1)$ on the unit circle

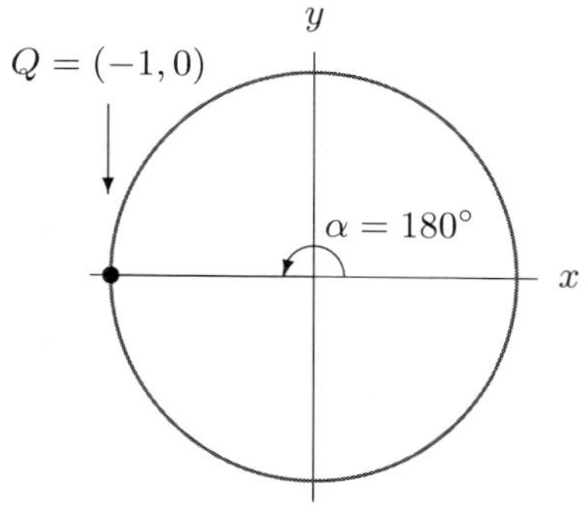

Figure 6.15: The angle $\alpha = 180°$ specifies the point Q on the unit circle

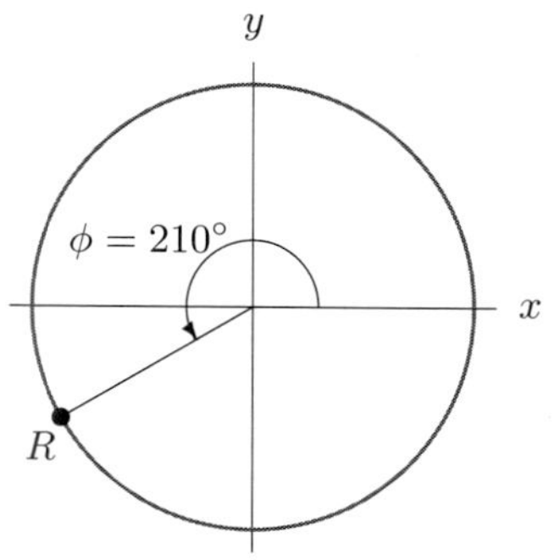

Figure 6.16: The angle $\phi = 210°$ specifies the point R on the unit circle

Definition of Sine and Cosine

In Example 1, angles were used to locate points on the unit circle. The trigonometric functions *sine* and *cosine* give the coordinates of a point in terms of its angle.

> Suppose $P = (x, y)$ is the point on the unit circle specified by the angle θ. We define the functions, **cosine** of θ, or $\cos\theta$, and the **sine** of θ, or $\sin\theta$, by the formulas
>
> $$\cos\theta = x \qquad \text{and} \qquad \sin\theta = y.$$
>
> In other words, $\cos\theta$ is the x-coordinate of the point on the unit circle specified by the angle θ and $\sin\theta$ is the y-coordinate. (See Figure 6.17.)

Notice that we often omit the parentheses around the independent variable, writing $\cos\theta$ instead of $\cos(\theta)$. The independent variable here is θ, not x. In fact, x and y are both functions of θ.

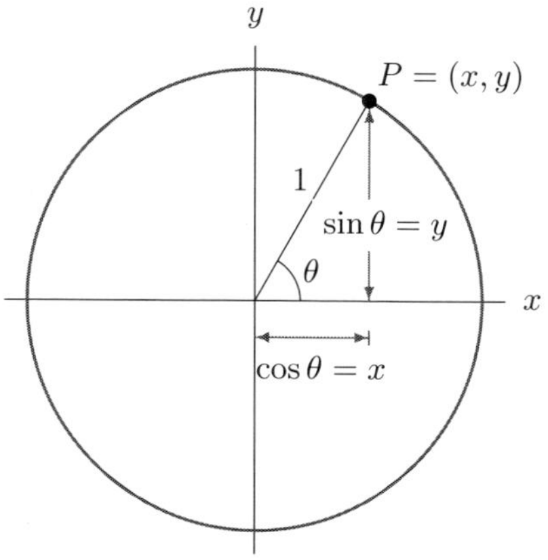

Figure 6.17: For point P, $\cos\theta$ is the x-coordinate and $\sin\theta$ is the y-coordinate

Values of the Sine and Cosine Functions

In principle, we can find values of $\sin\theta$ and $\cos\theta$ for any value of θ. However, there are no easy algebraic formulas for calculating the sine and cosine in terms of θ. So in practice we rely on tables of values, or on calculators and computers. However, there are some values of $\sin\theta$ and $\cos\theta$ we can find on our own.

Example 2 Find the values of $\cos 90^\circ$, $\sin 90^\circ$, $\cos 180^\circ$, $\sin 180^\circ$, $\cos 210^\circ$, $\sin 210^\circ$.

Solution See Figure 6.14 on page 251. The point P has coordinates $(0, 1)$, so

$$\cos 90^\circ = 0 \quad \text{and} \quad \sin 90^\circ = 1.$$

In Figure 6.15, point Q has coordinates $(-1, 0)$, so

$$\cos 180^\circ = -1 \quad \text{and} \quad \sin 180^\circ = 0.$$

In Figure 6.16 on page 251, we do not know the coordinates of point R. To find the values of $\cos 210^\circ$ and $\sin 210^\circ$, we use a calculator, which tells us that, approximately

$$\cos 210^\circ = -0.866 \quad \text{and} \quad \sin 210^\circ = -0.500.$$

Notice that $\cos 210^\circ$ and $\sin 210^\circ$ are both negative because the point R is in the third quadrant.

Example 3
(a) In Figure 6.18, find the coordinates of the point Q on the unit circle.
(b) Find the lengths of the line segments labeled m and n in Figure 6.18.

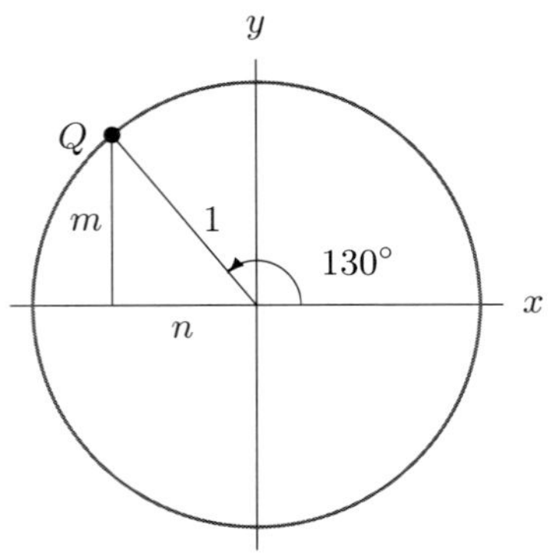

Figure 6.18: The point Q designated by 130° on the unit circle

Solution
(a) The coordinates of the point Q are $(\cos 130^\circ, \sin 130^\circ)$. A calculator gives, approximately,

$$\cos 130^\circ = -0.643 \quad \text{and} \quad \sin 130^\circ = 0.766.$$

(b) The length of line segment m is the same as the y-coordinate of point Q, so $m = \sin 130^\circ = 0.766$. The x-coordinate of Q is negative because Q is in the second quadrant. The length of line segment n has the same magnitude as the x-coordinate of Q but is positive. Thus, the length of $n = -\cos 130^\circ = 0.643$.

Coordinates of a Point on a Circle of Radius r

Using the sine and cosine, we can find the coordinates of points on circles of any size. Figure 6.19 shows two concentric circles: the inner circle has radius 1 and the outer circle has radius r. The angle θ designates point P on the unit circle and point Q on the larger circle. We know that the coordinates of point $P = (\cos\theta, \sin\theta)$. We want to find (x, y), the coordinates of Q.

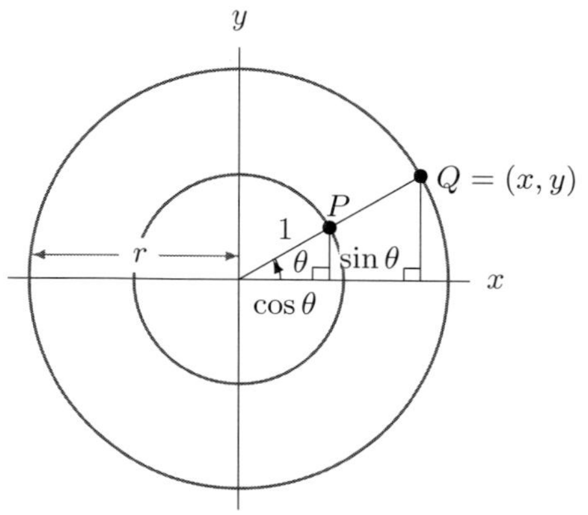

Figure 6.19: Points P and Q on circles of different radii specify the same angle θ

The coordinates of P and Q are the lengths of the sides of the two right triangles in Figure 6.19. Since both right triangles include the angle θ, they are similar and their sides are proportional. This means that the larger triangle is a "magnification" of the smaller triangle. Since the radius of the large circle is r times the radius of the small circle, we have

$$\frac{x}{\cos\theta} = \frac{r}{1} \quad \text{and} \quad \frac{y}{\sin\theta} = \frac{r}{1}.$$

Solving for x and y gives us the following result:

> The coordinates (x, y) of the point Q in Figure 6.19 are given by
>
> $$x = r\cos\theta \quad \text{and} \quad y = r\sin\theta.$$

Example 4 Find the coordinates of the points A, B, and C in Figure 6.20 to three decimal places.

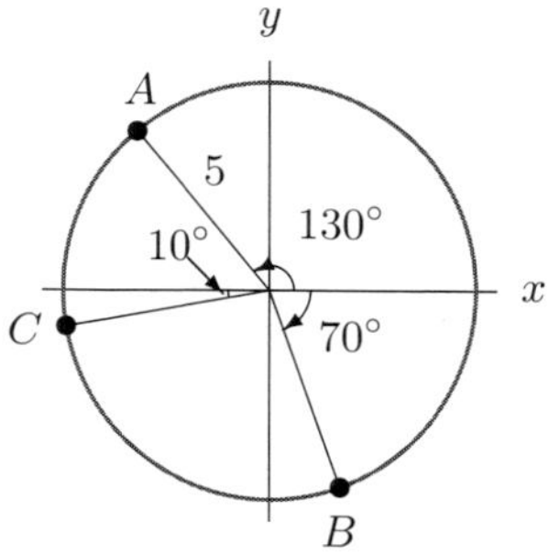

Figure 6.20: Finding coordinates of points on a circle of radius $r = 5$

Solution Since the circle has radius 5, the coordinates of point A are given by

$$x = 5\cos 130^\circ = 5(-0.643) = -3.214,$$
$$y = 5\sin 130^\circ = 5(0.766) = 3.830.$$

Point B corresponds to an angle of -70°, (because the angle is measured clockwise), so B has coordinates

$$x = 5\cos(-70^\circ) = 5(0.342) = 1.710,$$
$$y = 5\sin(-70^\circ) = 5(-0.940) = -4.698.$$

For point C, we must first calculate the corresponding angle, since the 10° is not measured from the positive x-axis. The angle we want is $180^\circ + 10^\circ = 190^\circ$, so

$$x = 5\cos(190^\circ) = 5(-0.985) = -4.924,$$
$$y = 5\sin(190^\circ) = 5(-0.174) = -0.868.$$

Example 5 In Figure 6.21, write the height of the point P above the x-axis as a function of the angle θ.

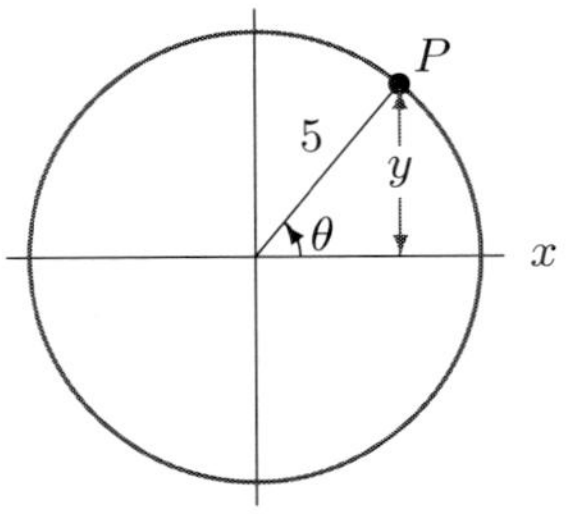

Figure 6.21: Height of point P above x-axis

Solution We want y as a function of the angle θ. Since the radius of the circle is 5, we have

$$y = 5\sin\theta.$$

Height on the Ferris Wheel as a Function of Angle

Imagine the ferris wheel superimposed on a coordinate system with the origin at the center of the ferris wheel and the positive x-axis extending horizontally to the right. Wherever your seat is on the wheel, we can measure the angle that the line from the center of the wheel to your seat makes with this axis. Figure 6.22 shows the seat in two different positions on the wheel. The first position is at 1 o'clock, which corresponds to a 60° angle. The second position is at 10 o'clock, and it corresponds to a 150° angle. A point at the 3 o'clock position corresponds to an angle of 0°.

We now find a formula for the height on the London ferris wheel as a function of the angle θ.

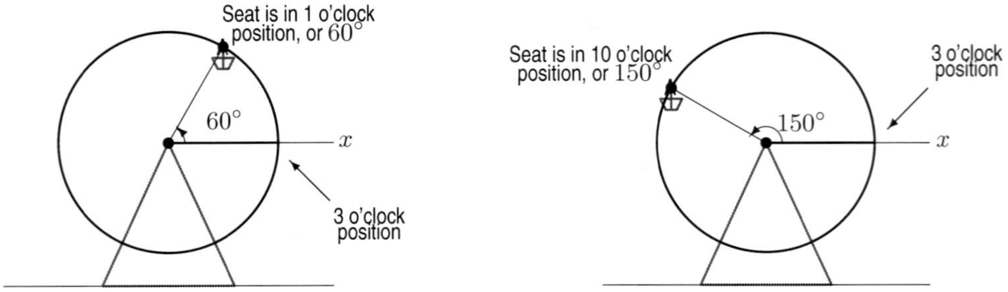

Figure 6.22: The 1 o'clock position forms a 60° angle with the positive x-axis, and the 10 o'clock position forms a 150° angle

Example 6 The ferris wheel described in Section 6.1 has a radius of 225 feet. Find your height above the ground as a function of the angle θ measured from the 3 o'clock position. What is your height when the angle is $60°$?

Solution Think of the ferris wheel as a circle centered at the origin, with your position at point P as shown in Figure 6.23. Since $r = 225$, the y-coordinate of point P is given by

$$y = r\sin\theta = 225\sin\theta.$$

Your height above the ground is given by $225 + y$, so the formula for your height in terms of θ is

$$\text{Height} = 225 + y = 225 + 225\sin\theta.$$

When $\theta = 60°$, we have

$$\begin{aligned}\text{Height} &= 225 + 225\sin 60° \\ &= 225 + 225(0.866) = 419.856 \text{ feet.}\end{aligned}$$

So you are approximately 420 feet above the ground.

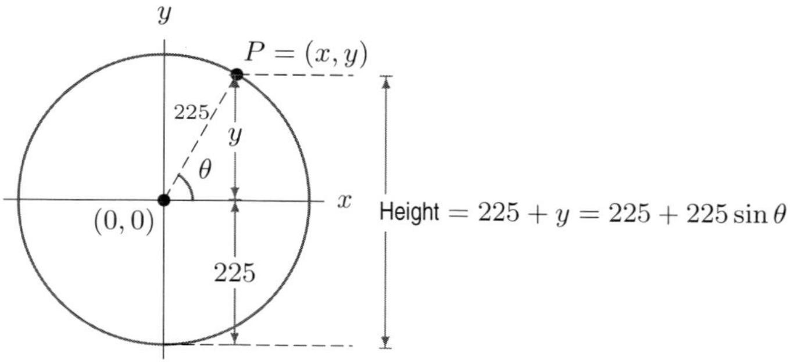

Figure 6.23: Height above the 3 o'clock position is given by the value of y; height above the ground is given by $225 + y$

Exercises and Problems for Section 6.2

Exercises

1. Mark the following angles on a unit circle and give the coordinates of the point determined by each angle.

(a) $100°$ (b) $200°$ (c) $-200°$
(d) $-45°$ (e) $1000°$ (f) $-720°$

In Exercises 2–4, what angle (in degrees) corresponds to the given number of rotations around the unit circle?

2. 4 **3.** -2 **4.** 16.4

For Exercises 5–8, sketch and find the coordinates of the point corresponding to each angle on the unit circle.

5. S is at $225°$, T is at $270°$, and U is at $330°$

6. D is at $-90°$, E is at $-135°$, and F is at $-225°$

7. A is at $390°$, B is at $495°$, and C is at $690°$

8. P is at $540°$, Q is at $-180°$, and R is at $450°$

9. Suppose the angles in Exercise 5 are on a circle of radius 5, evaluate the coordinates of S, T and U.

10. Suppose the angles in Exercise 7 are on a circle of radius 3, evaluate the coordinates of A, B and C.

In Exercises 11–24, find the coordinates of the point at the given angle on a circle of radius 3.8 centered at the origin.

11. $90°$ **12.** $180°$ **13.** $-180°$
14. $-90°$ **15.** $-270°$ **16.** $-540°$
17. $1426°$ **18.** $1786°$ **19.** $45°$
20. $135°$ **21.** $225°$ **22.** $315°$
23. $-10°$ **24.** $-20°$

Problems

25. Sketch the angles $\phi = 420°$ and $\theta = -150°$ as a displacement on a ferris wheel, starting from the 3 o'clock position. What position on the wheel do these angles indicate?

26. Find an angle θ, with $0° < \theta < 360°$, that has the same

(a) Cosine as $240°$ **(b)** Sine as $240°$

27. Find an angle ϕ, with $0° < \phi < 360°$, that has the same

(a) Cosine as $53°$ **(b)** Sine as $53°$

28. **(a)** Given that $P \approx (0.707, 0.707)$ is a point on the unit circle with angle $45°$, estimate $\sin 135°$ and $\cos 135°$ without a calculator.

(b) Given that $Q \approx (0.259, 0.966)$ is a point on the unit circle with angle $75°$, estimate $\sin 285°$ and $\cos 285°$ without a calculator.

29. For the angle ϕ shown in Figure 6.24, sketch each of the following angles.

(a) $180 + \phi$ **(b)** $180 - \phi$ **(c)** $90 - \phi$ **(d)** $360 - \phi$

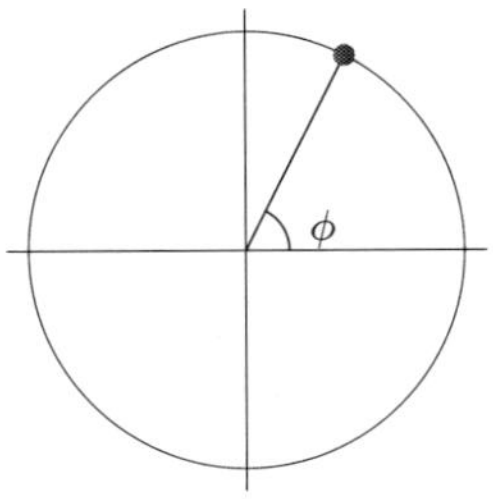

Figure 6.24

30. Let θ be an angle in the first quadrant, and suppose $\sin\theta = a$. Evaluate the following expressions in terms of a. (See Figure 6.25.)

(a) $\sin(\theta + 360°)$ **(b)** $\sin(\theta + 180°)$
(c) $\cos(90° - \theta)$ **(d)** $\sin(180° - \theta)$
(e) $\sin(360° - \theta)$ **(f)** $\cos(270° - \theta)$

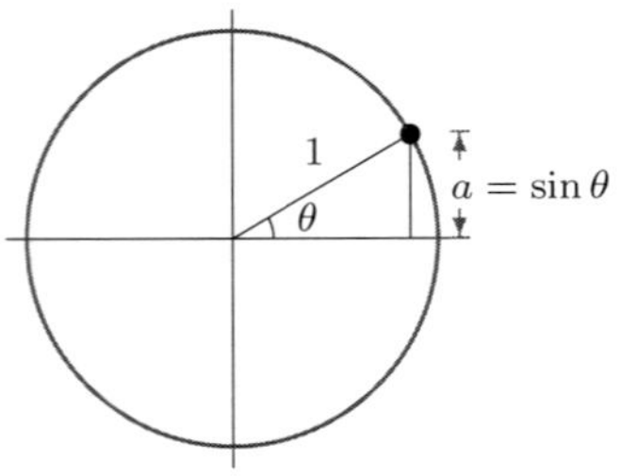

Figure 6.25

31. Explain in your own words the definition of $\sin\theta$ on the unit circle (θ in degrees).

32. You have been riding on the London ferris wheel (see Section 6.1) for 17 minutes and 30 seconds. What is your height above the ground?

33. A ferris wheel is 20 meters in diameter and makes one revolution every 4 minutes. For how many minutes of any revolution will your seat be above 15 meters?

34. A compact disc is 120 millimeters across with a center hole of diameter 15 millimeters. The center of the disc is at the origin. What are the coordinates of the points at which the inner and outer edge intersect the positive x-axis? What are the coordinates of the points at which the inner and outer edges cut a line making an angle θ with the positive x-axis?

35. The revolving door in Figure 6.26 rotates counterclockwise and has four equally spaced panels.

(a) What is the angle between two adjacent panels?
(b) What is the angle created by a panel rotating from B to A?
(c) When the door is as shown in Figure 6.26, a person going outside rotates the door from D to B. What is this angle of rotation?
(d) If the door is initially as shown in Figure 6.26, a person coming inside rotates the door from B to D. What is this angle of rotation?
(e) The door starts in the position shown in Figure 6.26. Where is the panel at A after three people enter and five people exit? Assume that people going in and going out do so in the manner described in parts (d) and (c), and that each person goes completely through the door before the next enters.

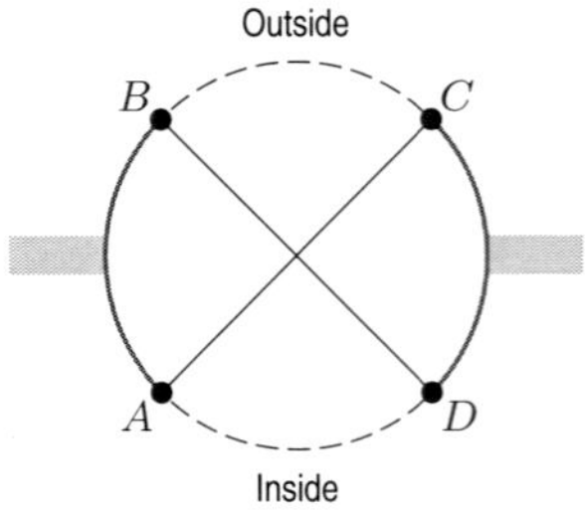

Figure 6.26

36. A revolving door (which rotates counterclockwise in Figure 6.27) was designed with five equally spaced panels for the entrance to the Pentagon. The arcs BC and AD have equal length.

(a) What is the angle between two adjacent panels?
(b) A four-star general enters by pushing on the panel at point B, and leaves the panel at point D. What is the angle of rotation?
(c) With the door in the position shown in Figure 6.27, an admiral leaves the Pentagon by pushing the panel between A and D to point B. What is the angle of rotation?

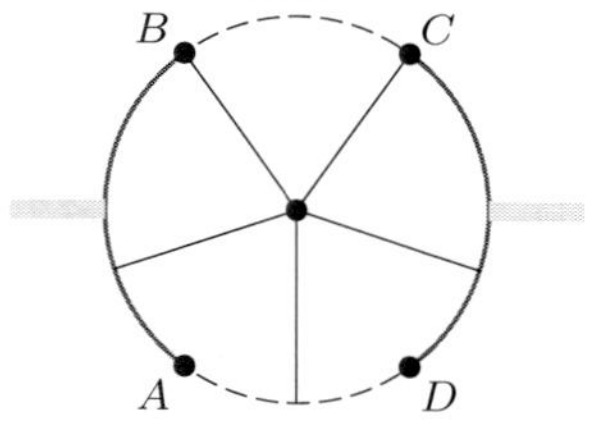

Figure 6.27

6.3 RADIANS

So far we have measured angles in degrees. There is another way to measure an angle, which involves arclength. This is the idea behind radians; it turns out to be very helpful in calculus.

Definition of a Radian

The arc length *spanned*, or cut off, by an angle is shown in Figure 6.28.

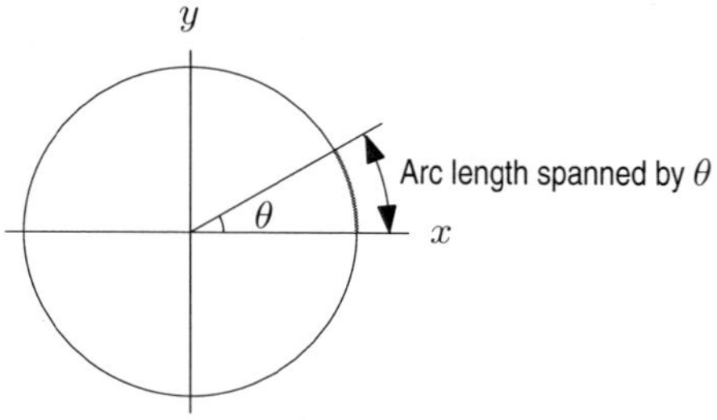

Figure 6.28: Arc length spanned by the angle θ

If the radius of a circle is fixed, (say the radius is 1), the arc length is completely determined by the angle θ. Following Figure 6.29, we make the following definition:

> An angle of **1 radian** is defined to be the angle, in the counterclockwise direction, at the center of a unit circle which spans an arc of length 1.

The radius and arc length must be measured in the same units.

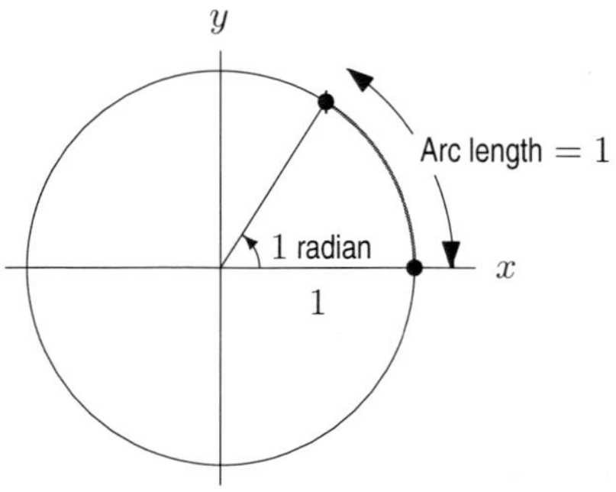

Figure 6.29: One radian cuts off an arc length of one in a unit circle

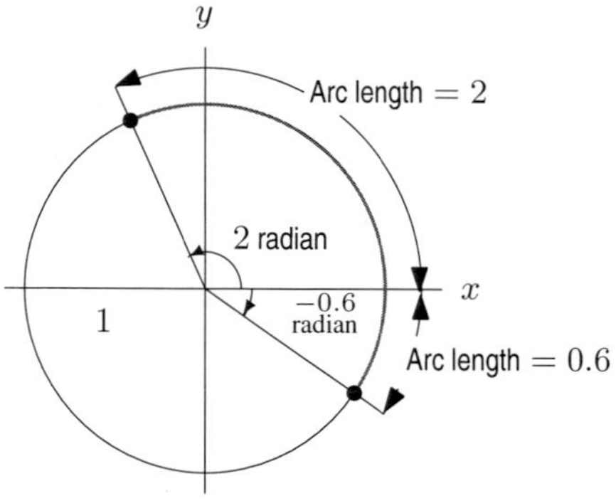

Figure 6.30: Angles of 2 radian and -0.6 radian

An angle of 2 radians cuts off an arc of length 2 in a unit circle; an angle of -0.6 radian is measured clockwise and cuts off an arc of length 0.6. See Figure 6.30. In general:

> The radian measure of a positive angle is the length of the arc spanned by the angle in a unit circle. For a negative angle, the radian measure is the negative of the arc length.

Radians are dimensionless units of measurement for angles (they do not have units of length).

Relationship Between Radians and Degrees

The circumference, C, of a circle of radius r is given by

$$C = 2\pi r.$$

In a unit circle, $r = 1$, so $C = 2\pi$. This means that the arc length spanned by a complete revolution of 360° is 2π, so

$$360^\circ = 2\pi \text{ radians}.$$

Dividing by 2π gives

$$1 \text{ radian} = \frac{360^\circ}{2\pi} \approx 57.296^\circ.$$

Thus, one radian is approximately 57.296°. One-quarter revolution, or 90°, is equal to $\frac{1}{4}(2\pi)$ or $\pi/2$ radians. Figure 6.31 shows several positions on the unit circle described in radian measure. Since $\pi \approx 3.142$, one complete revolution is about 6.283 radians and one-quarter revolution is about 1.571 radians.

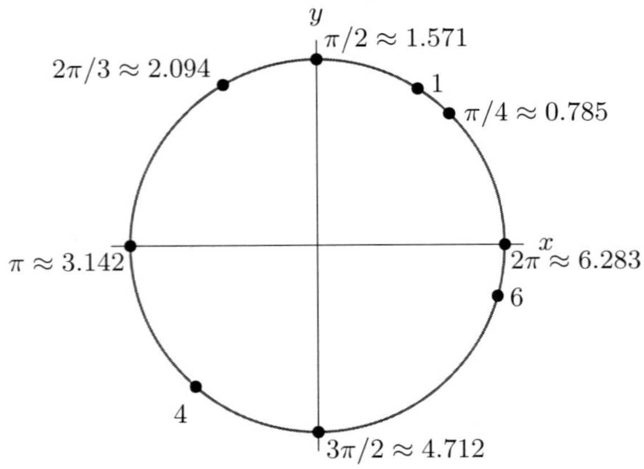

Figure 6.31: Angles in radian measure marked on circle

Example 1 In which quadrant is an angle of 2 radians? An angle of 5 radians?

Solution Refer to Figure 6.31. The second quadrant includes angles between $\pi/2$ and π, (that is, between 1.571 and 3.142 radians), so 2 radians lies in the second quadrant. An angle of 5 radians is between 4.71 and 6.283, that is, between $3\pi/2$ and 2π radians, so 5 radians lies in the fourth quadrant.

Example 2 The ferris wheel described in Section 6.1 makes one rotation every 30 minutes.

(a) If you start at the 3 c'clock position, find the angle in radians that specifies your position after 10 minutes.

(b) Find the angle that specifies your position after t minutes.

Solution (a) After 10 minutes, you have completed $1/3$ of a revolution, so

$$\text{Angle} = \frac{1}{3}(2\pi) = \frac{2\pi}{3} \text{ radians.}$$

(b) After t minutes, you have completed $t/30$ of a revolution, so

$$\text{Angle} = \frac{t}{30}(2\pi) = \frac{\pi t}{15} \text{ radians.}$$

Converting Between Degrees and Radians

To convert degrees to radians, or vice versa, we use the fact that 2π radians $= 360°$. So

$$1 \text{ radian} = \frac{180°}{\pi} \approx 57.296°.$$

Similarly

$$1° = \frac{\pi}{180} \approx 0.01745 \text{ radians.}$$

Thus, to convert from radians to degrees, multiply the radian measure by $180°/\pi$ radians. To convert from degrees to radians, multiply the degree measure by π radians$/180°$.

Example 3 (a) Convert 3 radians to degrees. (b) Convert 3 degrees to radians.

Solution (a) $3 \text{ radians} \cdot \dfrac{180°}{\pi \text{ radians}} = \dfrac{540°}{\pi \text{ radians}} \approx 171.887°.$

(b) $3° \cdot \dfrac{\pi \text{ radians}}{180°} = \dfrac{\pi \text{ radians}}{60} \approx 0.052 \text{ radians.}$

The word radians is often dropped, so if an angle or rotation is referred to without units, it is understood to be in radians. We can write, for instance, $90° = \pi/2$ and $\pi = 180°$.

Example 4 Find the arc length spanned by an angle of $30°$ in a circle of radius 1 meter.

Solution Convert $30°$ to radians, giving

$$30° = 30° \cdot \frac{\pi}{180°} = \frac{\pi}{6}.$$

The definition of a radian tells us that the arc length spanned by an angle of $\pi/6$ in a circle of radius 1 meter is $\pi/6 \approx 0.524$ meter. See Figure 6.32.

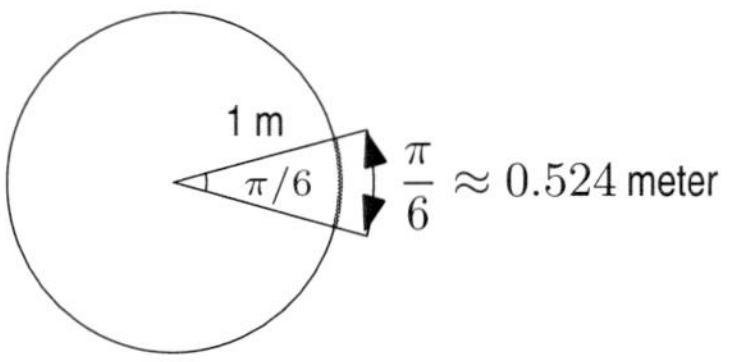

Figure 6.32: Arc length spanned by $30° = \pi/6$ radian in a unit circle

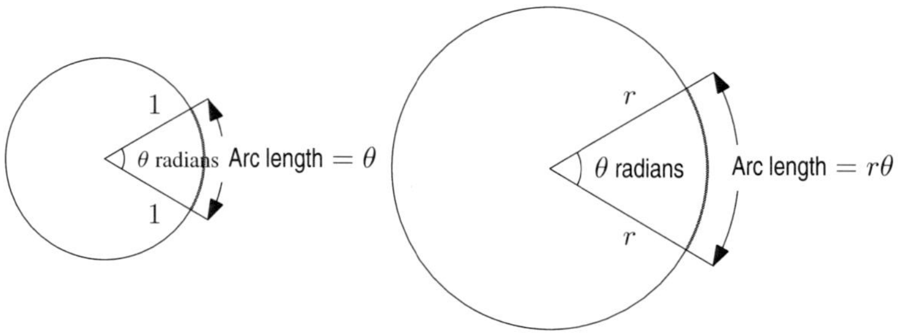

Figure 6.33: Arc length spanned by angle θ is proportional to the radius of the circle

Arc Length

We defined a radian using arc length in a unit circle. However, radians can be used to calculate arc length in a circle of any size. An angle of θ radians spans an arc of length θ in a unit circle. An angle of θ radians spans an arc of length $r\theta$ in a circle of radius r. Figure 6.33 reflects the following result:

> The **arc length**, s, spanned in a circle of radius r by an angle of θ radians, $0 \leq \theta \leq 2\pi$, is given by
> $$s = r\theta.$$

Thus if the size of an angle is fixed, the arc length it spans is proportional to the radius of the circle. (See Figure 6.34.) Note that θ must be in radians in this arc length formula.

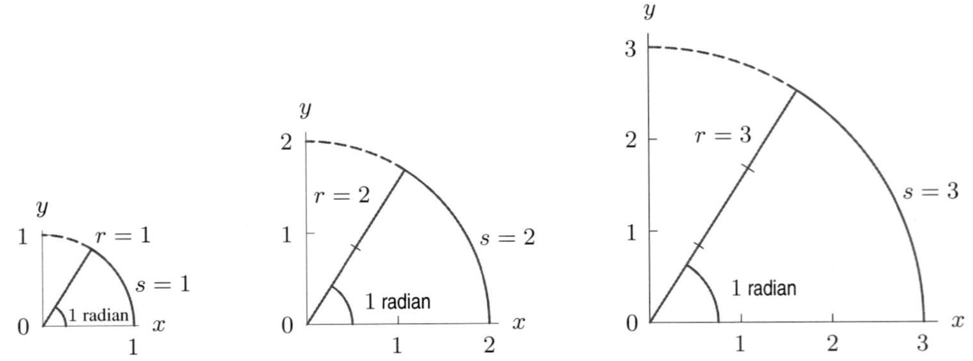

Figure 6.34: On a circle of radius 1, one radian spans an arc of length 1. On a circle of radius 2, it spans an arc of length 2; on a circle of radius 3, it spans an arc of length 3

Example 5 What length of arc is cut off by an angle of 120° on a circle of radius 12 cm?

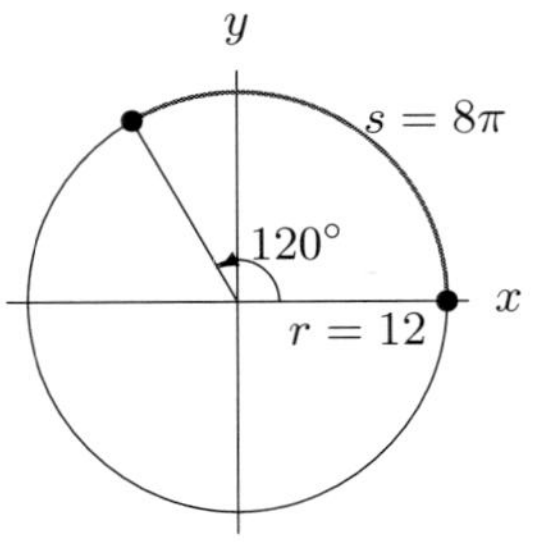

Figure 6.35: An arc cut off by 120°

Solution Converting 120° to radians gives

$$\text{Angle} = 120^\circ \cdot \frac{\pi}{180^\circ} = \frac{2}{3}\pi \text{ radians},$$

so

$$\text{Arc length} = 12 \cdot \frac{2}{3}\pi = 8\pi \text{ cm}.$$

Example 6 You walk 4 miles around a circular lake. Give an angle in radians which represents your final position relative to your starting point if the radius of the lake is: (a) 1 mile (b) 3 miles

Solution (a) The path around the lake is a unit circle, so you have traversed an angle in radians equal to the arc length traveled, 4 miles. An angle of 4 radians is in the middle of the third quadrant relative to your starting point. (See Figure 6.36.)

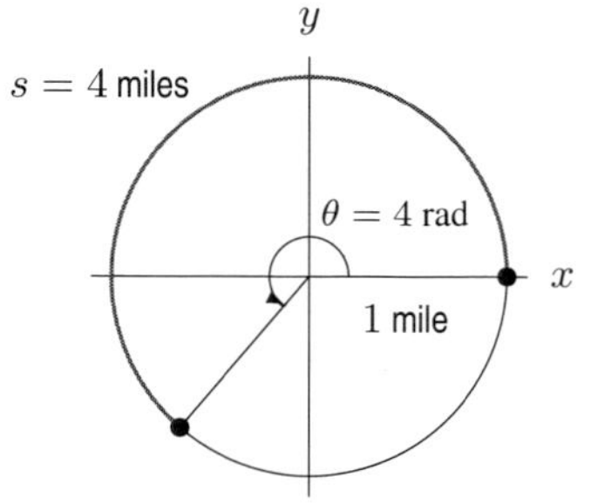

Figure 6.36: Arc length 4 and radius 1, so angle is 4 radians

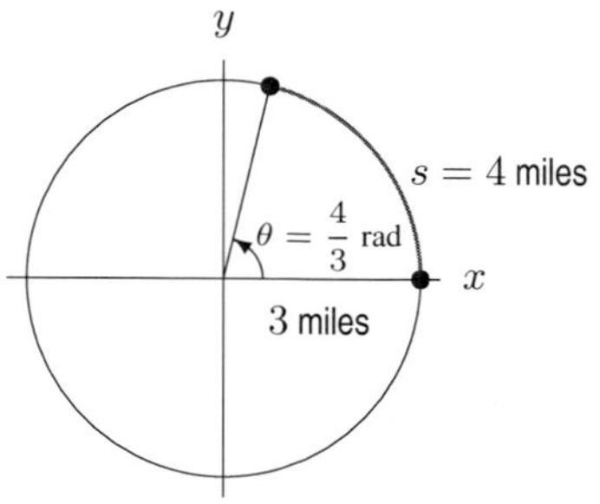

Figure 6.37: Arc length 4 and radius 3, so angle is $4/3$ radians

(b) See Figure 6.37. To find θ, use $s = r\theta$ giving

$$\theta = \frac{s}{r}.$$

The angle you have traversed is

$$\frac{s}{r} = \frac{4}{3} \approx 1.33 \text{ radians}.$$

One-quarter revolution is $\pi/2 \approx 1.571$ radians, so you are nearly one-quarter of the way around the lake.

Sine and Cosine of a Number

We have defined the sine and cosine of an angle. For any real number t, we define $\cos t$ and $\sin t$ by interpreting t as an angle of t radians.

Exercises and Problems for Section 6.3

Exercises

Determine the radian measure of the angles in Exercises 1–4.

1. 60° **2.** 45° **3.** 100° **4.** 17°

In Exercises 5–8, convert the angle to radians.

5. 150° **6.** 120° **7.** -270° **8.** π°

In Exercises 9–13, convert the angle given in radians to degrees.

9. $\frac{7}{2}\pi$ **10.** 5π **11.** 90 **12.** 2 **13.** 45

14. Convert each of the following angles to radians in two forms: as a multiple of π and as a decimal approximation rounded to two decimal places.

(a) 30° **(b)** 120° **(c)** 200° **(d)** 315°

15. An angle with radian measure $\pi/4$ corresponds to a point on the unit circle in quadrant I. Give the quadrants corresponding to angles with the following radian measures:

(a) 1 radian **(b)** 2 radians **(c)** 3 radians
(d) 4 radians **(e)** 5 radians **(f)** 6 radians
(g) 7 radians **(h)** 8 radians **(i)** 9 radians
(j) 10 radians

In Exercises 16–19, what angle in radians corresponds to the given number of rotations around the unit circle?

16. 1 **17.** -2 **18.** 0.75 **19.** 4.27

In Exercises 20–23, find the arc length corresponding to the given angle on a circle of radius 6.2.

20. $-180°$ **21.** $45°$ **22.** $\dfrac{180°}{\pi}$ **23.** $a°$

Problems

24. What is the radius of a circle in which an angle of 3 radians cuts off an arc of 30 cm?

25. What is the length of an arc which is cut off by an angle of 225° in a circle of radius 4 feet?

26. What is the length of an arc cut off by an angle of 2 radians on a circle of radius 8 inches?

27. What is the angle determined by an arc of length 2π meters on a circle of radius 18 meters?

In Problems 28–33, where possible give the radius r, the measure of θ in both degrees and radians, the arc length s, and the coordinates of point P.

28.

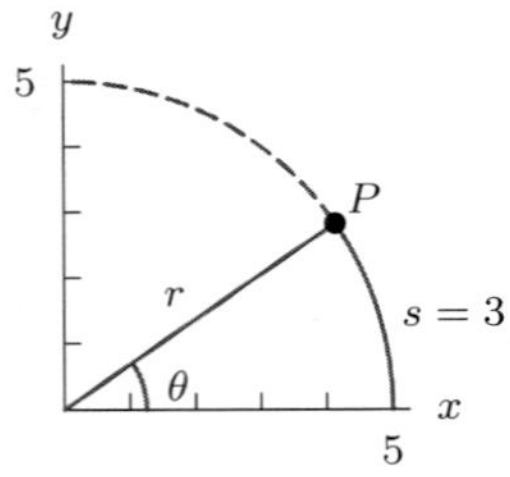

29.

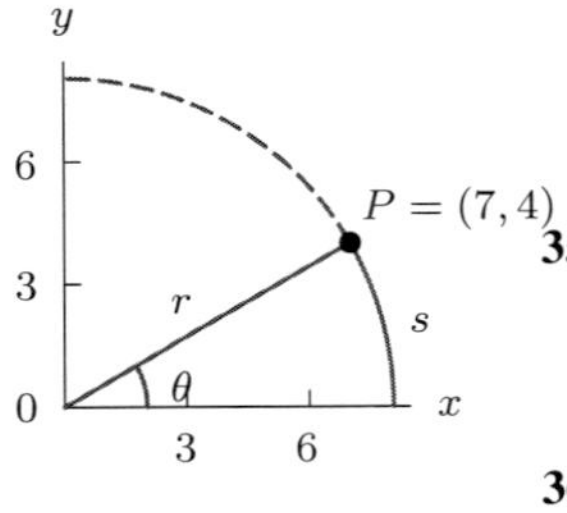

30.

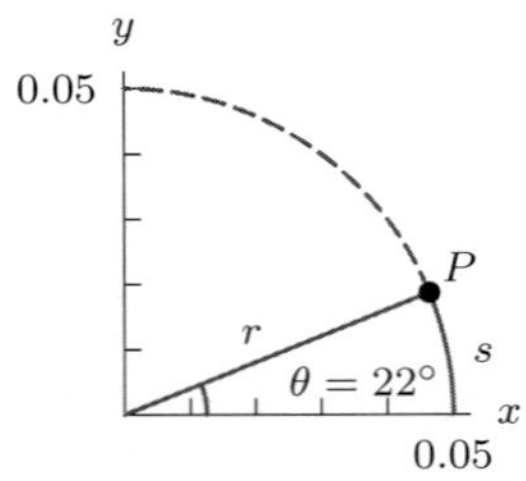

31.

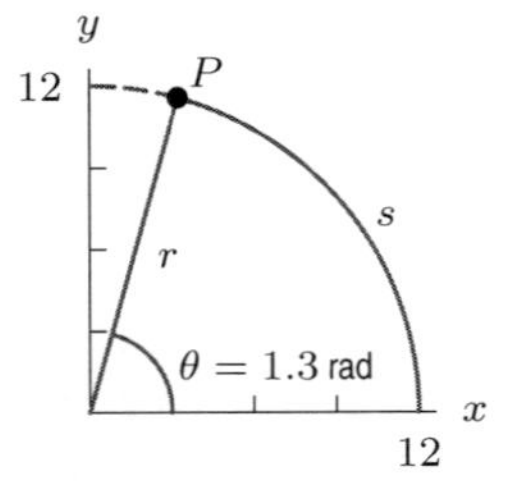

32.

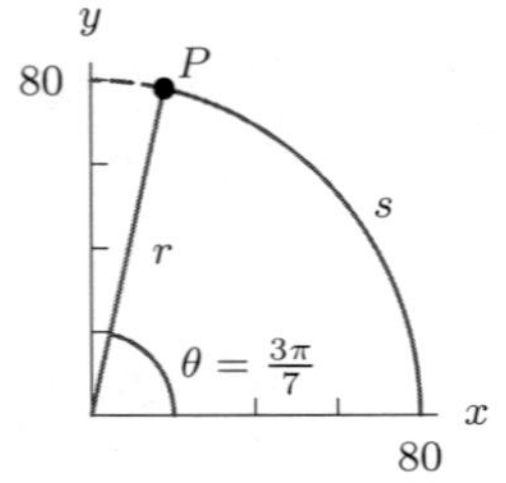

33.

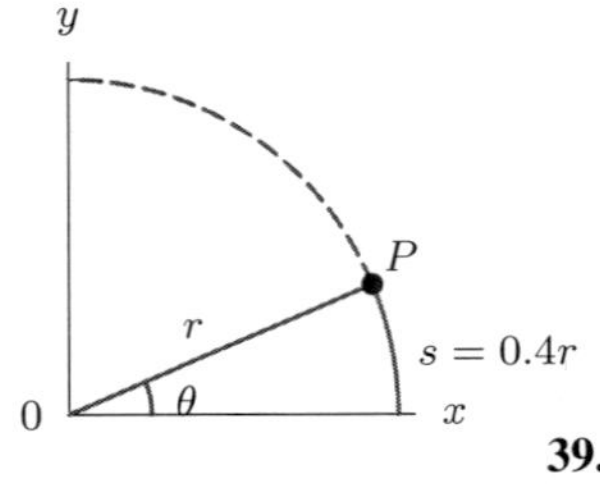

34. A radian can be defined to be the angle at the center of a circle which cuts off an arc of length equal to the radius of the circle. This is because the arc length formula in radians, $s = r \cdot \theta$, becomes just $s = r$ when $\theta = 1$. Figure 6.38 shows circles of radii 2 cm and 3 cm.

(a) Does the angle 1 radian appear to be the same in both circles?

(b) Estimate the number of arcs of length 2 cm that fit in the circumference of the circle of radius 2 cm.

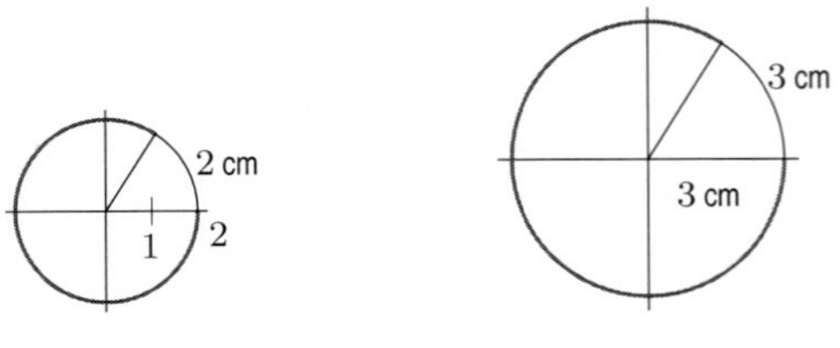

Figure 6.38

35. Without using a calculator, give the sign of each of the following numbers:

(a) $\cos 3$ **(b)** $\sin 4$ **(c)** $\sin(-4)$ **(d)** $\cos 7$

36. Without using a calculator, rank the following in order from smallest to largest.

(a) The angles: $\dfrac{2\pi}{3}, 2.3, \dfrac{2}{3}, -\dfrac{2\pi}{3}$

(b) The numbers:

$$\cos\left(\frac{2\pi}{3}\right), \quad \cos(2.3), \quad \cos\left(\frac{2}{3}\right), \quad \cos\left(-\frac{2\pi}{3}\right)$$

Evaluate $\sin\theta$ and $\cos\theta$ for the angle θ on the unit circle in Problems 37–38.

37.

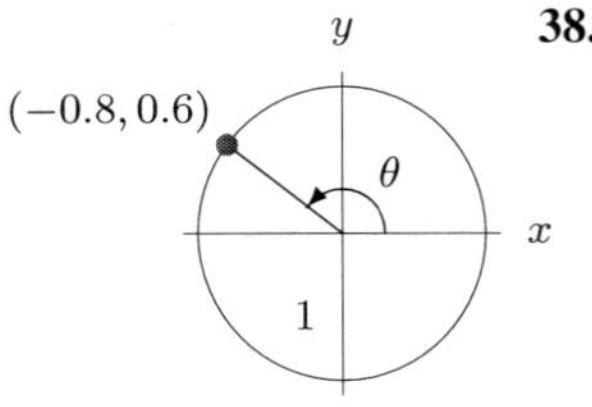

38.

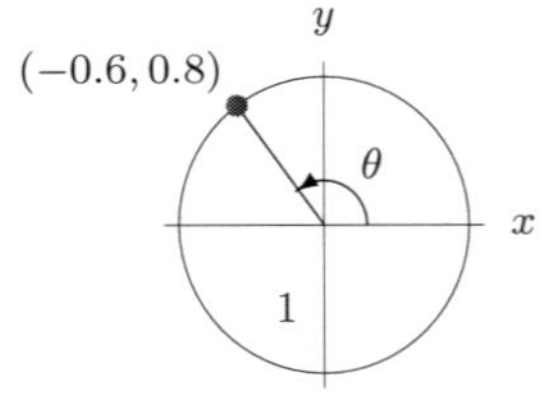

39. An ant starts at the point $(1, 0)$ on the unit circle and walks counterclockwise a distance of 3 units around the circle. Find the x and y coordinates (accurate to 2 decimal places) of the final location of the ant.

40. For ϕ in Figure 6.39, sketch the following angles.

(a) $\pi + \phi$ **(b)** $\pi - \phi$

(c) $\pi/2 - \phi$ **(d)** $2\pi - \phi$

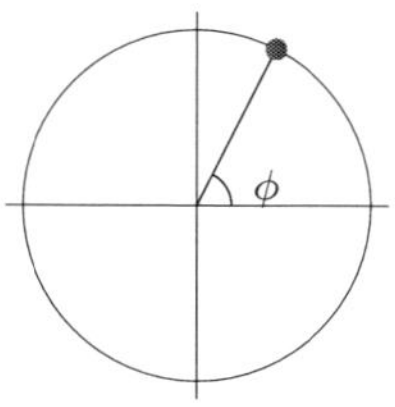

Figure 6.39

41. If a weight hanging on a string of length 3 feet swings through 5° on either side of the vertical, how long is the arc through which the weight moves from one high point to the next high point?

42. How far does the tip of the minute hand of a clock move in 35 minutes if the hand is 6 inches long?

43. Using a weight on a string called a plumb bob, it is possible to erect a pole that is exactly vertical, which means that the pole points directly toward the center of the earth. Two such poles are erected one hundred miles apart. If the poles were extended they would meet at the center of the earth at an angle of 1.4333°. Compute the radius of the earth.

44. Without a calculator, decide which is bigger, t or $\sin t$, for $0 < t < \pi/2$. Illustrate your answer with a sketch.

45. Do you think there is a value of t for which $\cos t = t$? If so, estimate the value of t. If not, explain why not.

6.4 GRAPHS OF THE SINE AND COSINE

Before graphing, we calculate values of the sine and cosine. Some values can be found exactly.

Exact Values of the Sine and Cosine

In Example 2 on page 252, we found sine and cosine of 90° and 180°. Because $90^\circ = \pi/2$ and $180^\circ = \pi$ we have

$$\cos\frac{\pi}{2} = 0, \quad \sin\frac{\pi}{2} = 1, \quad \cos\pi = -1, \quad \sin\pi = 0.$$

Example 1 Evaluate $\sin\theta$ and $\cos\theta$ for $\theta = 0, 3\pi/2$, and 2π.

Solution Figure 6.40 gives the coordinates of the points on the unit circle specified by $0, 3\pi/2$, and 2π. We use these coordinates to evaluate the sines and cosines of these angles.

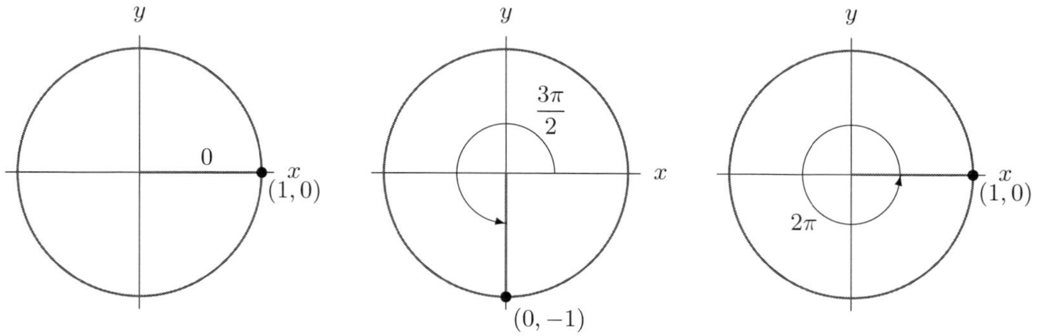

Figure 6.40: The coordinates of the points on the unit circle specified by the angles $0, 3\pi/2$, and 2π

The definition of the cosine function tells us that $\cos 0$ is the x-coordinate of the point on the unit circle specified by the angle 0. Since the x-coordinate of this point is $x = 1$, we have

$$\cos 0 = 1.$$

Similarly, the y-coordinate of the point $(1, 0)$ is $y = 0$ which gives

$$\sin 0 = 0.$$

From Figure 6.40 we see that the coordinates of the remaining points are $(0, -1)$, and $(1, 0)$, so

$$\frac{3\pi}{2} = 0, \qquad \sin\frac{3\pi}{2} = -1, \qquad \text{and} \qquad \cos 2\pi = 1, \qquad \sin 2\pi = 0.$$

In the Tools section on page 301, triangles are used to compute the exact values of the sine and cosine of $30° = \pi/6$, $45° = \pi/4$, $60° = \pi/3$. See also Problem 24 at the end of this section. The results are:

$$\cos 30° = \cos\tfrac{\pi}{6} = \frac{\sqrt{3}}{2} \qquad \cos 45° = \cos\tfrac{\pi}{4} = \frac{1}{\sqrt{2}} \qquad \cos 60° = \cos\tfrac{\pi}{3} = \frac{1}{2}$$

$$\sin 30° = \sin\tfrac{\pi}{6} = \frac{1}{2} \qquad \sin 45° = \sin\tfrac{\pi}{4} = \frac{1}{\sqrt{2}} \qquad \sin 60° = \sin\tfrac{\pi}{3} = \frac{\sqrt{3}}{2}$$

Graphs of the Sine and Cosine Functions

Table 6.5 records values of the sine and cosine functions, using $1/\sqrt{2} \approx 0.7$. The graphs are plotted in Figure 6.41. To model periodic phenomena, we usually work in radians. Thus, the axes are labeled in radians and degrees. Notice that both the sine and cosine functions are periodic with a period of $2\pi = 360°$ and an amplitude of 1.

Table 6.5 *Table of values (rounded) for* $\sin\theta$ *and* $\cos\theta$*, with* θ *in degrees and radians*

θ (degrees)	0°	45°	90°	135°	180°	225°	270°	315°	360°	450°	540°	630°	720°	...
θ (radians)	0	$\pi/4$	$\pi/2$	$3\pi/4$	π	$5\pi/4$	$3\pi/2$	$7\pi/4$	2π	$5\pi/2$	3π	$7\pi/2$	4π	...
$\sin\theta$	0	0.7	1	0.7	0	−0.7	−1	−0.7	0	1	0	−1	0	...
$\cos\theta$	1	0.7	0	−0.7	−1	−0.7	0	0.7	1	0	−1	0	1	...

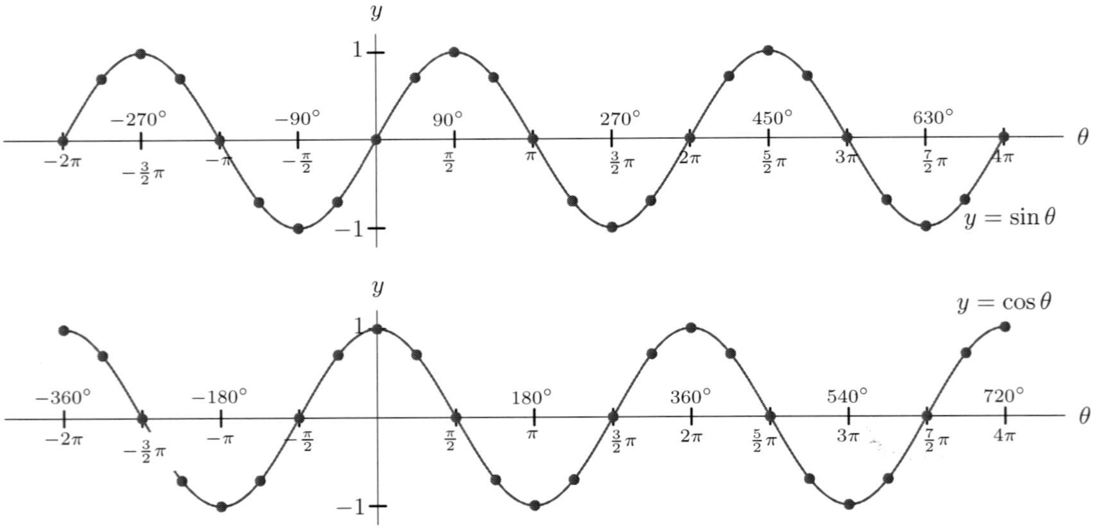

Figure 6.41: The graphs of $\sin\theta$ and $\cos\theta$

Several properties of the sine and cosine functions are illustrated by their graphs. The sine is an odd function and the cosine is an even function. Since the outputs of the sine and cosine functions are the coordinates of points on the unit circle, they lie between -1 and 1. So the range of the sine and cosine are

$$-1 \leq \sin\theta \leq 1 \quad \text{and} \quad -1 \leq \cos\theta \leq 1.$$

The domain of both of these functions is all real numbers, since any angle, positive or negative, specifies a point on the unit circle. Both functions are periodic with period 2π, because adding a multiple of 2π (or 360°) to an angle does not change the position of the point that it designates on the unit circle. The midline of both graphs is $y = 0$ and their amplitude is 1.

Amplitude

We can determine the amplitude of a trigonometric function from its formula. Recall that the amplitude is the distance between the midline and the highest or lowest point on the graph.

Example 2 Compare the graph of $y = \sin t$ to the graphs of $y = 2\sin t$ and $y = -0.5\sin t$, for $0 \leq t \leq 2\pi$. How are these graphs similar? How are they different? What are their amplitudes?

Solution The graphs are in Figure 6.42. The amplitude of $y = \sin t$ is 1, the amplitude of $y = 2\sin t$ is 2 and the amplitude of $y = -0.5\sin t$ is 0.5. The graph of $y = -0.5\sin t$ is "upside-down" relative to $y = \sin t$. These observations are consistent with the fact that the constant A in the equation

$$y = A\sin t$$

stretches or shrinks the graph vertically, and reflects it about the t-axis if A is negative.

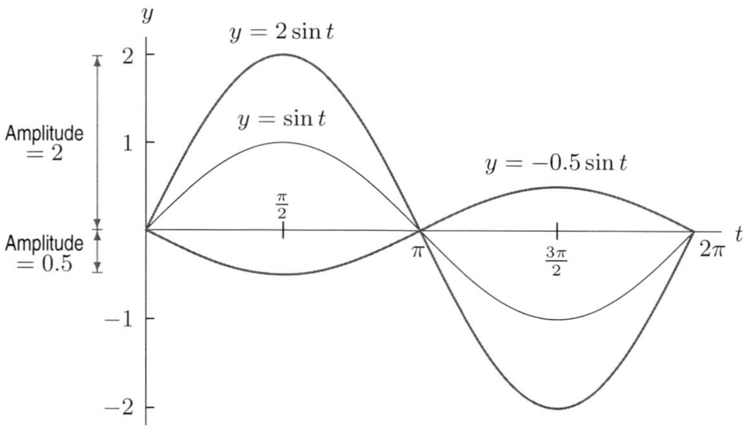

Figure 6.42: The graphs of $y = \sin t$, $y = 2\sin t$, and $y = -0.5\sin t$ all have different amplitudes

Midline

Recall that the graph of $y = f(t) + k$ is the graph of $y = f(t)$ shifted vertically by k units. For example, the graph of $y = \cos t + 2$ is the graph of $y = \cos t$ shifted up by 2 units, as shown in Figure 6.43. Notice that the midline of the new graph is the line $y = 2$; it has been shifted up 2 units from its old position at $y = 0$.

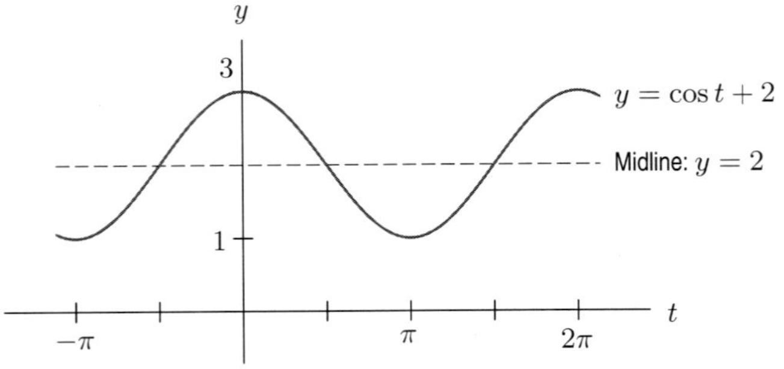

Figure 6.43: The graph of $y = \cos t + 2$ and its midline $y = 2$

Generalizing, we conclude that the graphs of $y = \sin t + k$ and $y = \cos t + k$ have midlines $y = k$. Notice that the expression $\sin t + k$ could also be written as $k + \sin t$; it is *not* the same as $\sin(t + k)$.

Example 3 Graph the ferris wheel function giving your height, $h = f(\theta)$, in feet, above ground as a function of the angle θ:

$$f(\theta) = 225 + 225 \sin \theta.$$

(See Example 6 on page 255.) What are the period, midline, and amplitude?

Solution Using a calculator, we get the graph in Figure 6.44. The period of this function is $360°$, because $360°$ is one full rotation, so the function repeats every $360°$. The midline is $h = 225$ feet, since the values of h oscillate about this value. The amplitude is also 225 feet, since the maximum value of h is 450 feet.

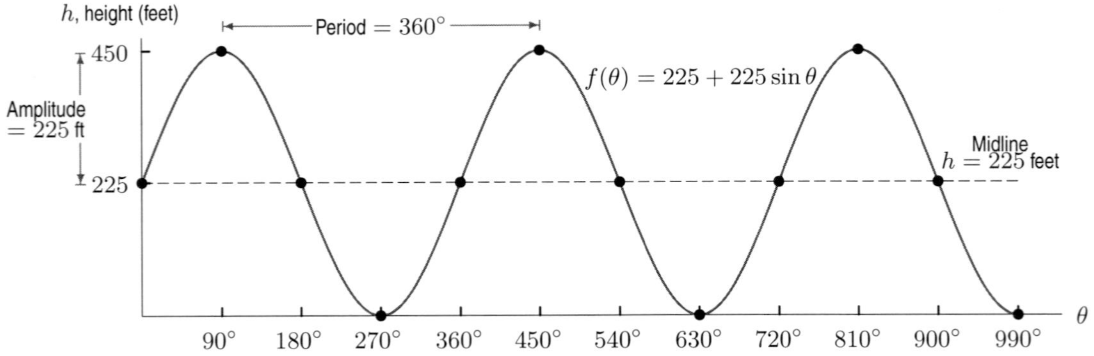

Figure 6.44: On the ferris wheel: Height, h, above ground as a function of the angle, θ

Exercises and Problems for Section 6.4

Exercises

In Exercises 1–8, find the midline and amplitude of the periodic function.

1. $y = \sin(4x)$

2. $y = 6\sin(4x) + 5$

3. $y = -7\cos(2x) - 4$

4.

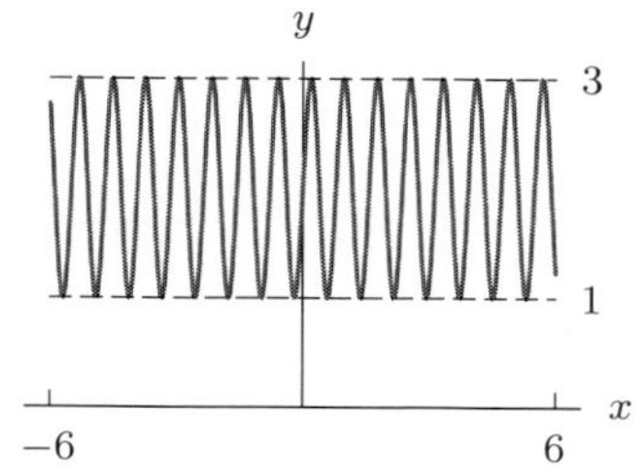

5.

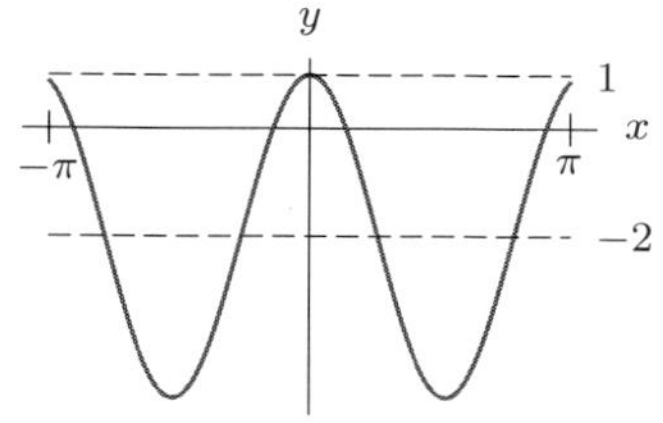

6.

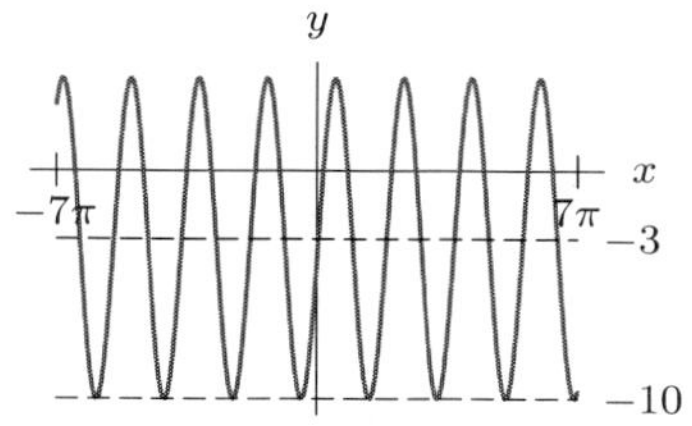

7. The height in cm of the tip of the hour hand on a vertical clock face is a function, $h(t)$, of the time, t, in hours. The hour hand is 15 cm long, and the middle of the clock face is 185 cm above the ground.

8. The height in cm of the tip of the minute hand on a vertical clock face is a function, $i(t)$, of the time, t, in minutes. The minute hand is 20 cm long, and the middle of the clock face is 223 cm above the ground.

9. Graph $y = \sin t$ for $-\pi \leq t \leq 3\pi$.

(a) Indicate the interval(s) on which the function is

(i) Positive (ii) Increasing (iii) Concave up

(b) Estimate the t values at which the function is increasing most rapidly.

10. Compare the values of $\sqrt{1/2}$ and $\sqrt{2}/2$ using a calculator. Explain your observations.

11. Compare the values of $\sqrt{3/4}$ and $\sqrt{3}/2$ using your calculator. Explain your observations.

In Exercises 12–15, find an exact value without a calculator.

12. $\cos \dfrac{7\pi}{6}$

13. $\sin \dfrac{2\pi}{3}$

14. $\sin\left(-\dfrac{\pi}{3}\right)$

15. $\cos\left(\dfrac{-\pi}{6}\right)$

Problems

16. Figure 6.45 shows the graphs of $y = (\sin x) - 1$ and $y = (\sin x) + 1$. Identify which is which.

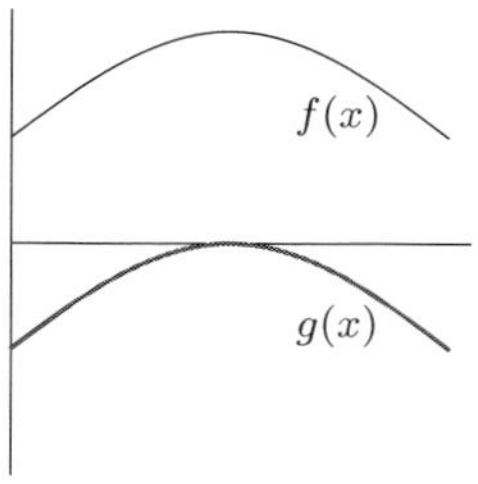

Figure 6.45

17. Figure 6.46 shows $y = \sin x$ and $y = \cos x$ starting at $x = 0$. Which is $y = \cos x$? Find values for a and b.

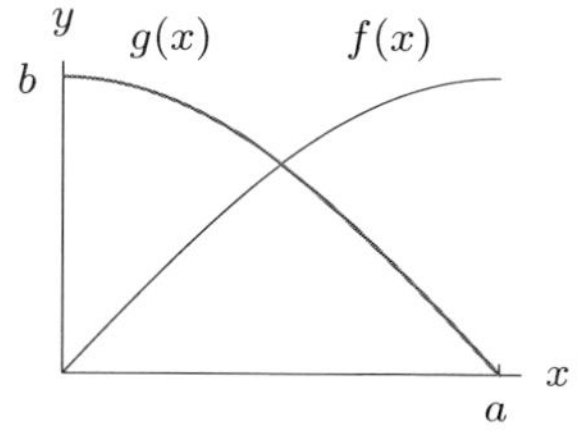

Figure 6.46

18. In Figure 6.47, assume that $\sin A = \sin B$.

(a) If $A = \pi/8$, find B. **(b)** If $A = 1$, find B.

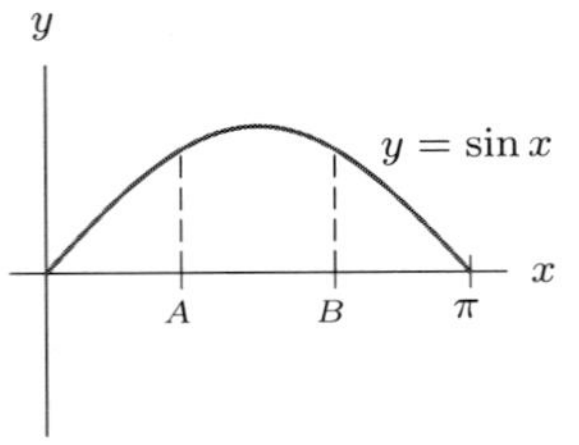

Figure 6.47

19. Figure 6.48 shows $y = \sin(x - \frac{\pi}{2})$ and $y = \sin(x + \frac{\pi}{2})$ starting at $x = 0$. Identify which is which.

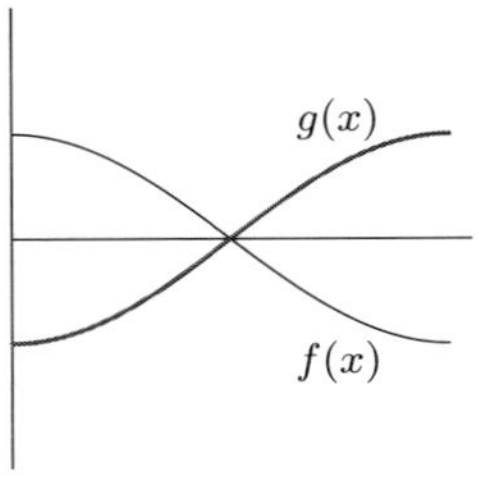

Figure 6.48

20. Match each of the letters A–G in Figure 6.49 to one of the following values of x (in radians): $1, 2, 4, 5, \pi/2, \pi$, and $3\pi/2$.

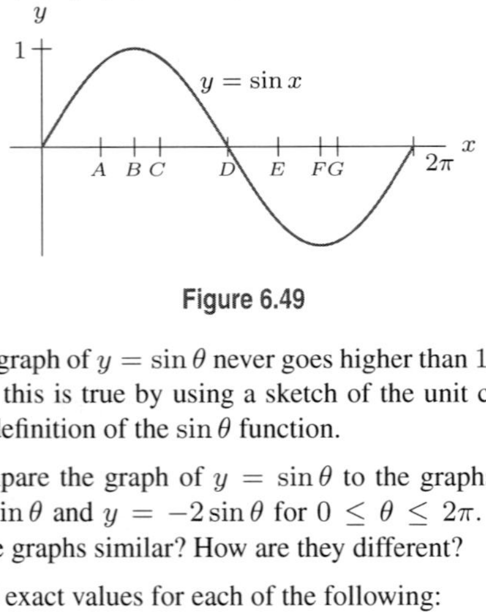

Figure 6.49

21. The graph of $y = \sin\theta$ never goes higher than 1. Explain why this is true by using a sketch of the unit circle and the definition of the $\sin\theta$ function.

22. Compare the graph of $y = \sin\theta$ to the graphs of $y = 0.5\sin\theta$ and $y = -2\sin\theta$ for $0 \leq \theta \leq 2\pi$. How are these graphs similar? How are they different?

23. Find exact values for each of the following:

(a) $\sin\left(\frac{3\pi}{4}\right)$ **(b)** $\cos\left(\frac{5\pi}{3}\right)$ **(c)** $\cos\left(\frac{7\pi}{6}\right)$

(d) $\sin\left(\frac{11\pi}{6}\right)$ **(e)** $\sin\left(\frac{9\pi}{4}\right)$

24. Calculate $\sin 45^\circ$ and $\cos 45^\circ$ exactly. Use the fact that the point P corresponding to 45° on the unit circle, $x^2 + y^2 = 1$, lies on the line $y = x$.

25. Find exact values for the coordinates of point W in Figure 6.50.

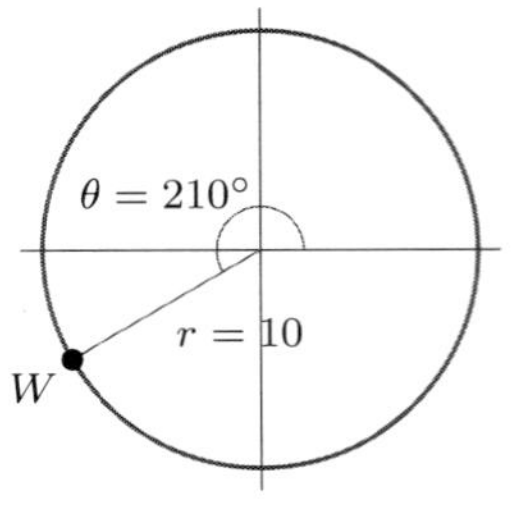

Figure 6.50

26. **(a)** For each of the following angles, sketch an arc of length θ on a unit circle and a line segment of length $\cos\theta$: (i) $\theta = 1.1$ (ii) $\theta = 5.2$

(b) Now sketch $y = \cos t$. Mark on your graph of $y = \cos t$ line segments of length θ and $\cos\theta$ for the same values of θ that you used in part (a).

27. **(a)** Match the lengths p, q, r, s marked on the unit circle in Figure 6.51 with the following values:

(i) $t = 0.8$ (ii) $t = \pi - 2.9$
(iii) $\cos(0.8)$ (iv) $-\cos(2.9)$

(b) On a graph of $y = \cos t$, sketch segments corresponding to each of the values in part (a).

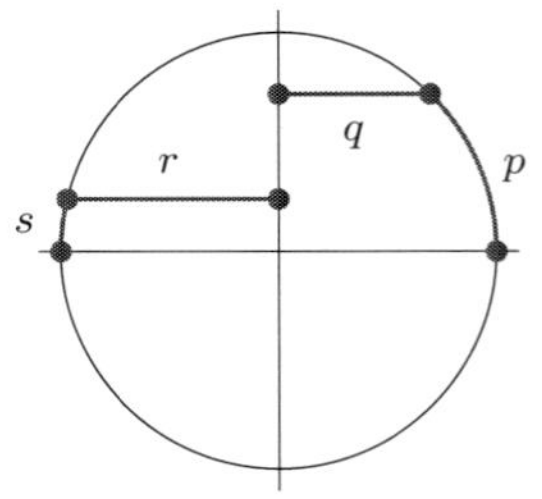

Figure 6.51

28. A circle of radius 5 is centered at the point $(-6, 7)$. Find a formula for $f(\theta)$, the x-coordinate of the point P in Figure 6.52.

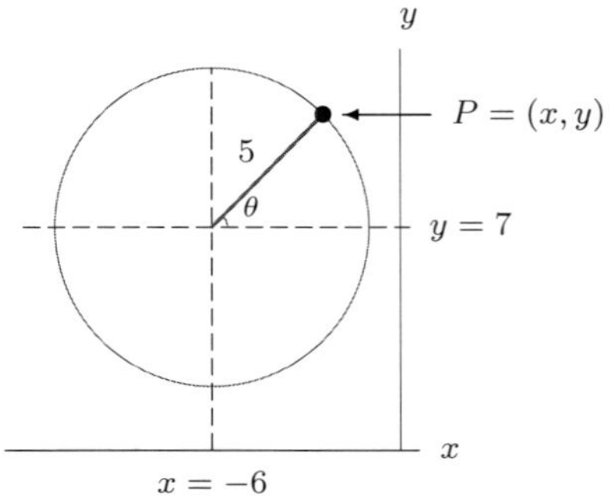

Figure 6.52

29. **(a)** Write an expression for the slope of the line segment joining P and Q in Figure 6.53.
(b) Evaluate your expression for $a = \pi/4$, $b = 4\pi/3$. Give an exact value for your answer.

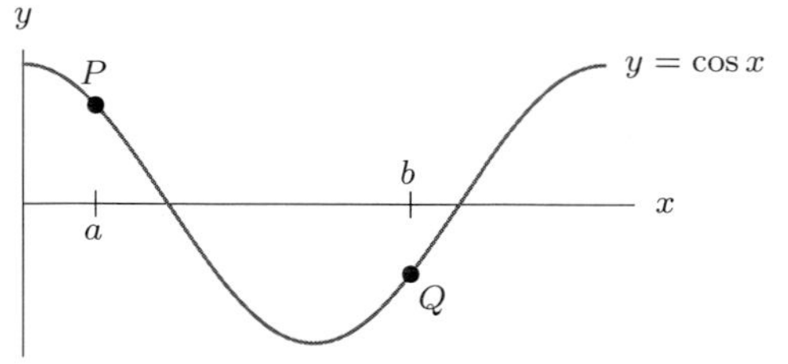

Figure 6.53

30. **(a)** Write an expression for the slope of the line segment joining S and T in Figure 6.54.
(b) Evaluate your expression for $a = 1.7$, $h = 0.05$. Round your answer to two decimal places.

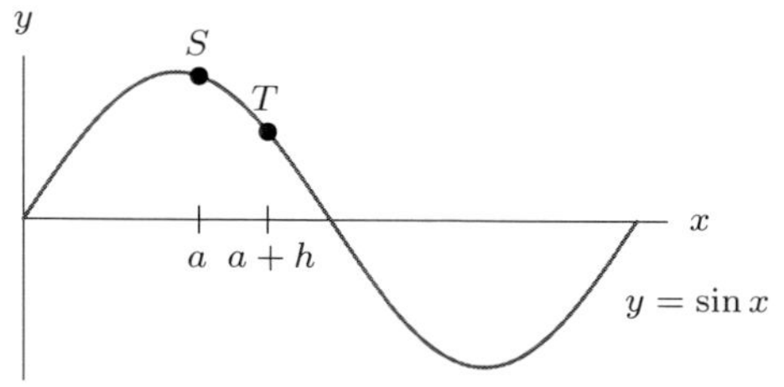

Figure 6.54

6.5 SINUSOIDAL FUNCTIONS

Section 6.1 introduced $f(t)$, your height above the ground while riding a ferris wheel. The graph of f looks like a transformation of $y = \sin t$. Transformations of the sine and cosine are called *sinusoidal* functions, and can be expressed in the form

$$y = A\sin(B(t - h)) + k \qquad \text{and} \qquad y = A\cos(B(t - h)) + k,$$

where A, B, h, and k are constants. Their graphs resemble the graphs of sine and cosine, but may also be shifted, flipped, or stretched. These transformations may change the period, amplitude, and midline of the function as well as its value at $t = 0$.

From Section 6.4 we know the following:

> The functions of $y = A\sin t$ and $y = A\cos t$ have **amplitude** $|A|$.
>
> The **midline** of the functions $y = \sin t + k$ and $y = \cos t + k$ is the horizontal line $y = k$.

Period

Next, we consider the effect of the constant B. We usually have $B > 0$.

Example 1 Graph $y = \sin t$ and $y = \sin 2t$ for $0 \le t \le 2\pi$. Describe any similarities and differences. What are their periods?

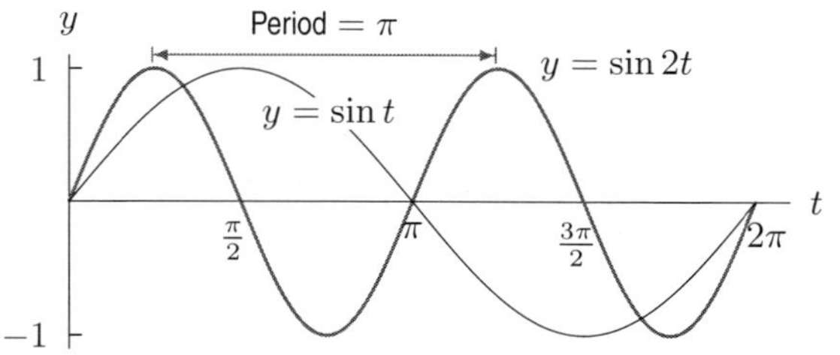

Figure 6.55: The functions $y = \sin t$ and $y = \sin 2t$ have different periods

Solution The graphs are in Figure 6.55. The two functions have the same amplitude and midline, but their periods are different. The period of $y = \sin t$ is 2π, but the period of $y = \sin 2t$ is π. This is because the factor of 2 causes a horizontal compression, squeezing the graph twice as close to the y-axis.

If $B > 0$ the function $y = \sin(Bt)$ resembles the function $y = \sin t$ except that it is stretched or compressed horizontally. The constant B determines how many cycles the function completes on an interval of length 2π. For example, we see from Figure 6.55 that the function $y = \sin 2t$ completes two cycles on the interval $0 \le t \le 2\pi$.

Since, for $B > 0$, the graph of $y = \sin(Bt)$ completes B cycles on the interval $0 \le t \le 2\pi$, each cycle has length $2\pi/B$. The period is thus $2\pi/B$. In general, for B of any sign, we have:

> The functions $y = \sin(Bt)$ and $y = \cos(Bt)$ have **period** $P = 2\pi/|B|$.

The number of cycles in one unit of time is $|B|/2\pi$, the *frequency*.

Example 2 Find possible formulas for the functions f and g shown in Figures 6.56 and 6.57.

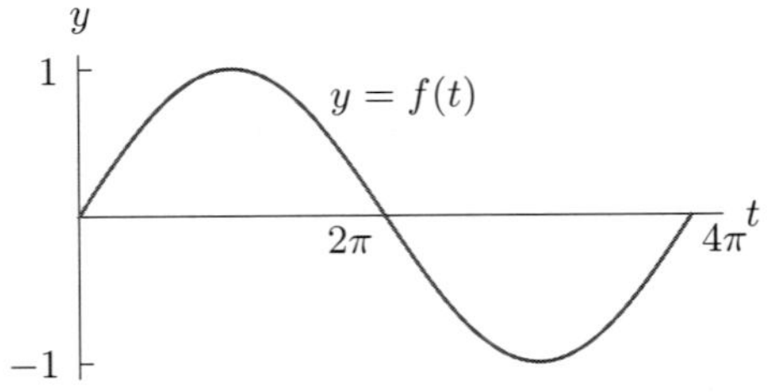

Figure 6.56: This function has period 4π

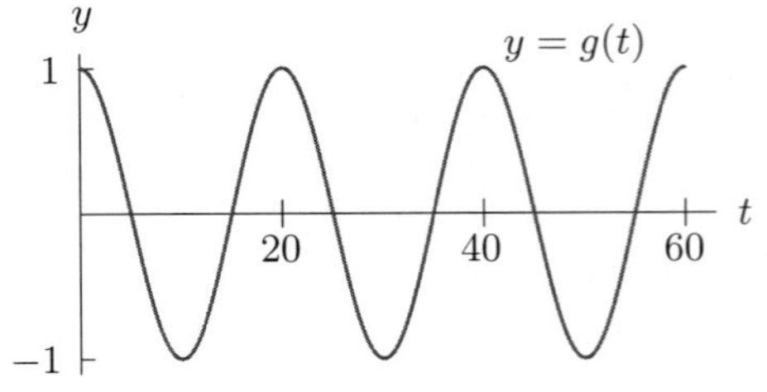

Figure 6.57: This function has period 20

Solution The graph of f resembles the graph of $y = \sin t$ except that its period is $P = 4\pi$. Using $P = 2\pi/B$ gives

$$4\pi = \frac{2\pi}{B} \quad \text{so} \quad B = \frac{1}{2}.$$

Thus, $f(t) = \sin\left(\frac{1}{2}t\right)$.

The function g resembles the function $y = \cos t$ except that its period is $P = 20$. This gives

$$20 = \frac{2\pi}{B} \quad \text{so} \quad B = \frac{\pi}{10}.$$

Thus, $g(t) = \cos\left(\frac{\pi}{10}t\right)$.

Example 3 Household electrical power in the US is provided in the form of alternating current. Typically the voltage cycles smoothly between $+155.6$ volts and -155.6 volts 60 times per second.[2] Use a cosine function to model the alternating voltage.

Solution If V is the voltage at time, t, in seconds, then V begins at $+155.6$ volts, drops to -155.6 volts, and then climbs back to $+155.6$ volts, repeating this process 60 times per second. We use a cosine with amplitude $A = 155.6$. Since the function alternates 60 times in one second, the period is $1/60$ of a second. We know that $P = 2\pi/B = 1/60$, so $B = 120\pi$. We have $V = 155.6\cos(120\pi t)$.

[2] A voltage cycling between $+155.6$ volts and -155.6 volts has an average magnitude, over time, of 110 volts.

Example 4 Describe in words the function $y = 300\cos(0.2\pi t) + 600$ and sketch its graph.

Solution This function resembles $y = \cos t$ except that it has an amplitude of 300, a midline of $y = 600$, and a period of $P = 2\pi/(0.2\pi) = 10$. See Figure 6.58.

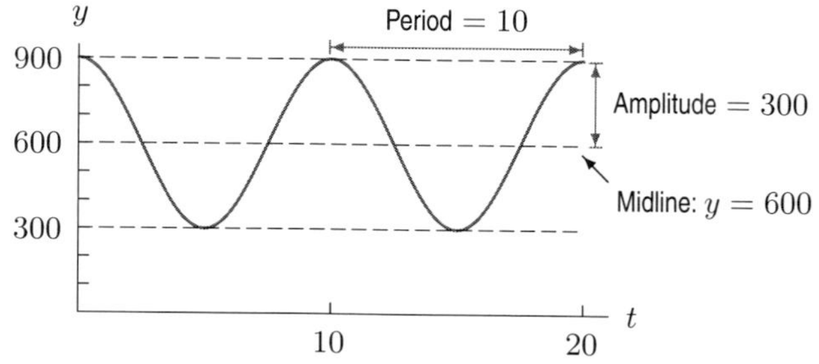

Figure 6.58: The function $y = 300\cos(0.2\pi t) + 600$

Horizontal Shift

Figure 6.59 shows the graphs of two trigonometric functions, f and g, with period $P = 12$. The graph of f resembles a sine function, so a possible formula for f is $f(t) = \sin Bt$. Since the period of f is 12, we have $12 = 2\pi/B$, so $B = 2\pi/12$, so $f(t) = \sin(\pi t/6)$.

The graph of g looks like the graph of f shifted to the right by 2 units. Thus a possible formula for g is

$$g(t) = f(t-2),$$

or

$$g(t) = \sin\left(\frac{\pi}{6}(t-2)\right).$$

Notice that we can also write the formula for $g(t)$ as

$$g(t) = \sin\left(\frac{\pi}{6}t - \frac{\pi}{3}\right),$$

but $\pi/3$ is *not* the horizontal shift in the graph! To pick out the horizontal shift from the formula, we must write the formula in factored form, that is, as $\sin(B(t-h))$.

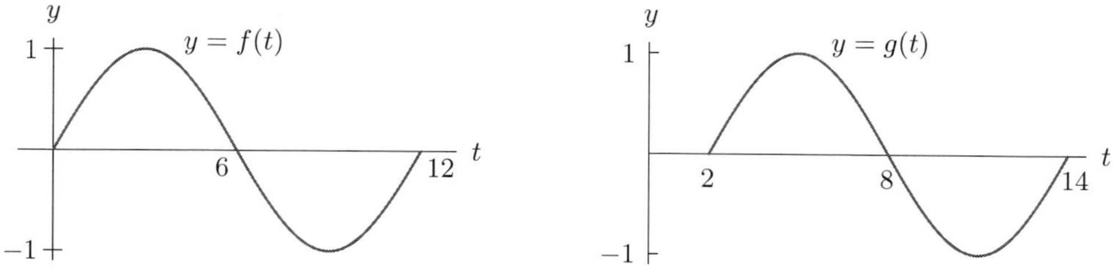

Figure 6.59: The graphs of two trigonometric functions f and g, related by a horizontal shift

> The graphs of $y = \sin(B(t-h))$ and $y = \cos(B(t-h))$ are the graphs of $y = \sin Bt$ and $y = \cos Bt$ **shifted horizontally** by h units.

Example 5 Describe in words the graph of the function $g(t) = \cos(3t - \pi/4)$.

Solution Write the formula for g in the form $\cos(B(t-h))$ by factoring 3 out from the expression $3t - \pi/4$ to get $g(t) = \cos(3(t - \pi/12))$. The period of g is $2\pi/3$ and the graph is the graph of $f = \cos 3t$ shifted $\pi/12$ units to the right, as shown in Figure 6.60.

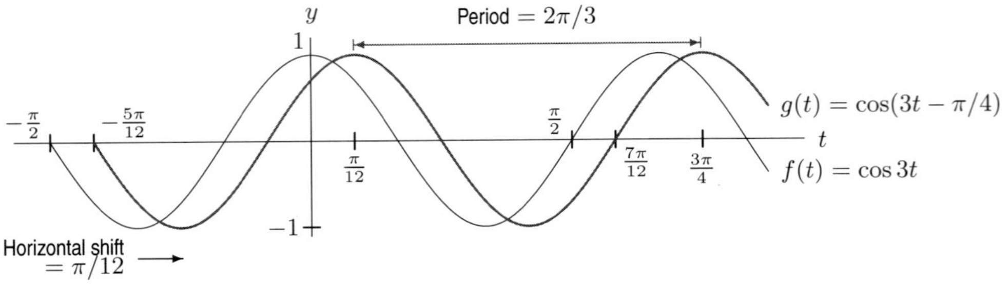

Figure 6.60: The graphs of $g(t) = \cos\left(3t - \frac{\pi}{4}\right)$ and $f(t) = \cos 3t$

Summary of Transformations

The parameters A, B, h, and k determine the graph of a transformed sine or cosine function.

For the **sinusoidal** functions

$$y = A\sin(B(t-h)) + k \quad \text{and} \quad y = A\cos(B(t-h)) + k,$$

- $|A|$ is the amplitude
- $2\pi/|B|$ is the period
- h is the horizontal shift
- $y = k$ is the midline
- $|B|/2\pi$ is the frequency; that is, the number of cycles completed in unit time.

Phase Shift

In Example 5, we factored $(3t - \pi/4)$ to write the function as $g(t) = \cos(3(t - \pi/12))$. This allowed us to recognize the horizontal shift, $\pi/12$. However, in most physical applications, the quantity $\pi/4$, known as the *phase shift*, is more important than the horizontal shift. We define

$$\text{Phase shift} = \text{Fraction of period} \times 2\pi.$$

The phase shift enables us to calculate the fraction of a full period that the curve has been shifted. For instance, in Example 5, the wave has been shifted

$$\frac{\text{Phase shift}}{2\pi} = \frac{\pi/4}{2\pi} = \frac{1}{8} \text{ of a full period,}$$

and the graph of g in Figure 6.61 is the graph of $f(t) = \cos 3t$ shifted $1/8$ of its period to the right.

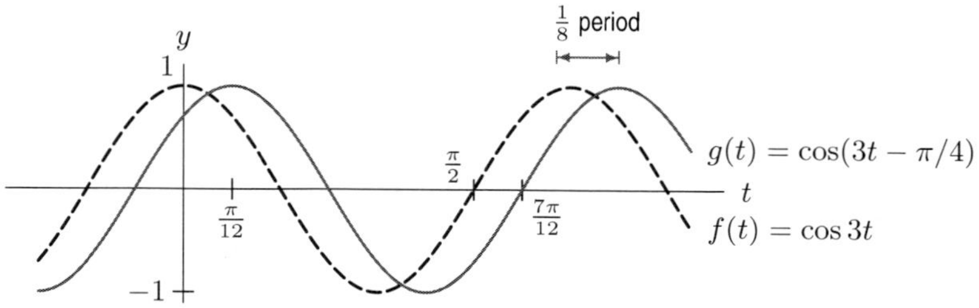

Figure 6.61: The graph of $g(t) = \cos(3t - \frac{\pi}{4})$ has phase shift $\frac{\pi}{4}$ relative to $f(t) = \cos 3t$

Phase shift is significant because in many applications, such as optical interference, we want to know if two waves reinforce or cancel each other. For two waves of the same period, a phase shift of 0 or 2π tells us that the two waves reinforce each other; a phase shift of π tells us that the two waves cancel. Thus, the phase shift tells us the relative positions of two waves of the same period.[3]

For the sinusoidal functions written in the form

$$y = A\sin(Bt + \phi) \quad \text{and} \quad y = A\cos(Bt + \phi),$$

ϕ is the **phase shift.**

Example 6 (a) In Figure 6.62, by what fraction of a period is the graph of $g(t)$ shifted from the graph of $f(t)$?
(b) What is the phase shift?

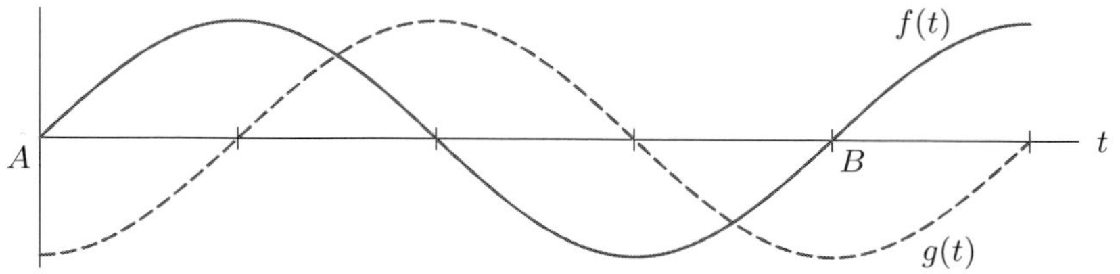

Figure 6.62

Solution (a) The period of $f(t)$ is the length of the interval from A to B. The graph of $g(t)$ is shifted $1/4$ period to the right.
(b) The phase shift is $\frac{1}{4}(2\pi) = \pi/2$.

Using the Transformed Sine and Cosine Functions

Sinusoidal functions are used to model oscillating quantities. Starting with $y = A\sin(B(t-h))+k$ or $y = A\cos(B(t-h)) + k$, we calculate values of the parameters A, B, h, k.

[3]The phase shift which tells us that two waves cancel is independent of their period. The horizontal shift that gives the same information is not independent of period.

Example 7 The temperature, T, in °C, of the surface water in a pond varies according to the graph in Figure 6.63. If t is the number of hours since sunrise at 6 am, find a possible formula for $T = f(t)$.

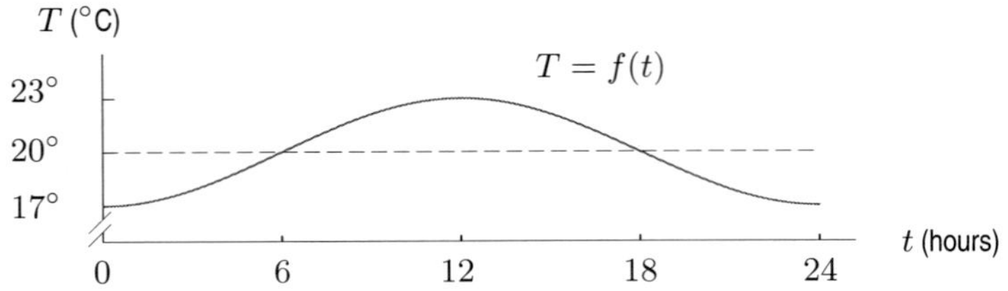

Figure 6.63: Surface water temperature of a pond since sunrise

Solution The graph of $T = f(t)$ resembles a cosine function with amplitude $|A| = 3$, period $= 24$ and midline $T = 20$. Compared to $y = \cos t$, there is no horizontal shift, but the graph has been reflected across the midline, so A is negative, $A = -3$. We have

$$24 = \frac{2\pi}{B} \quad \text{so} \quad B = \frac{2\pi}{24} = \frac{\pi}{12}.$$

So a possible formula for f is

$$T = f(t) = -3\cos\left(\frac{\pi}{12}t\right) + 20.$$

Example 8 A rabbit population in a national park rises and falls each year. It is at its minimum of 5000 rabbits in January. By July, as the weather warms up and food grows more abundant, the population triples in size. By the following January, the population again falls to 5000 rabbits, completing the annual cycle. Use a trigonometric function to find a possible formula for $R = f(t)$, where R is the size of the rabbit population as a function of t, the number of months since January.

Solution Notice that January is month 0, so July is month 6. The five points in Table 6.6 have been plotted in Figure 6.64 and a curve drawn in. This curve has midline $k = 10{,}000$, amplitude $|A| = 5000$, and period $= 12$ so $B = 2\pi/12 = \pi/6$. It resembles a cosine function reflected across its midline. Thus, a possible formula for this curve is

$$R = f(t) = -5000\cos\left(\frac{\pi}{6}t\right) + 10{,}000.$$

There are other possible formulas. For example, we could use a sine function and write

$$R = 5000\sin\left(\frac{\pi}{6}(t-3)\right) + 10{,}000.$$

Table 6.6 *Rabbit population over time*

t (month)	R
0	5000
6	15,000
12	5000
18	15,000
24	5000

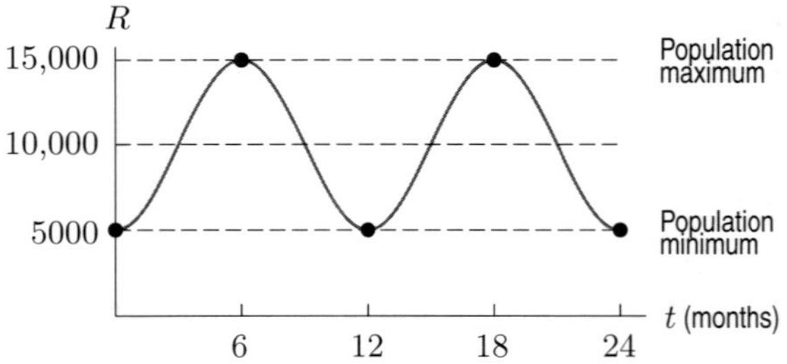

Figure 6.64: The number of rabbits, R, as a function of time in months, t, for $0 \le t \le 24$

Now let's return to the ferris wheel example in Section 6.1.

Example 9 Use the sinusoidal function $f(t) = A\sin(B(t-h)) + k$ to represent your height above ground at time t while riding the ferris wheel.

Solution The diameter of the ferris wheel is 450 feet, so the midline is $k = 225$ and the amplitude, A, is also 225. The period of the ferris wheel is 30 minutes, so

$$B = \frac{2\pi}{30} = \frac{\pi}{15}.$$

Figure 6.65 shows a sine graph shifted 7.5 minutes to the right because we reach $y = 225$ (the 3 o'clock position) when $t = 7.5$. Thus, the horizontal shift is $h = 7.5$, so

$$f(t) = 225\sin\left(\frac{\pi}{15}(t-7.5)\right) + 225.$$

Since we can expand and write

$$f(t) = 225\sin\left(\frac{\pi}{15}t - \frac{\pi}{2}\right) + 225,$$

the phase shift is $\pi/2$, corresponding to the fact that we reach $y = 225$ after a quarter of a cycle ($\pi/4$ is a quarter of 2π).

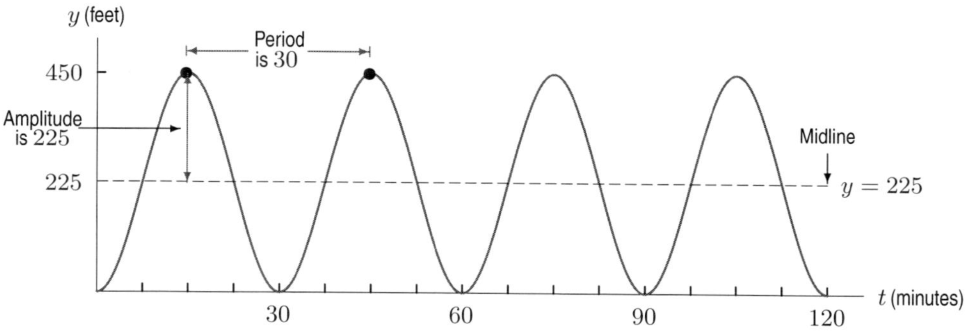

Figure 6.65: Graph of the ferris wheel height function $f(t) = 225\sin(\frac{\pi}{15}(t-7.5)) + 225$

There are other possible formulas for the function graphed in Figure 6.65. For example, we could have used a cosine reflected about its midline, as in Example 8. (See Problem 32.)

Exercises and Problems for Section 6.5

Exercises

In Exercises 1–4, state the period, amplitude, and midline.

1. $y = 6\sin(t+4)$

2. $y = 7\sin(4(t+7)) - 8$

3. $2y = \cos(8(t-6)) + 2$

4. $y = \pi\cos(2t+4) - 1$

In Exercises 5–6, what are the horizontal and phase shifts?

5. $y = 2\cos(3t+4) - 5$

6. $y = -4\cos(7t+13) - 5$

7. Let $f(x) = \sin(2\pi x)$ and $g(x) = \cos(2\pi x)$. State the periods, amplitudes, and midlines of f and g.

8. Which of the following functions are periodic? Justify your answers. State the periods of those that are periodic.

(a) $y = \sin(-t)$ **(b)** $y = 4\cos(\pi t)$
(c) $y = \sin(t) + t$ **(d)** $y = \sin(t/2) + 1$

In Exercises 9–10, estimate the period, amplitude, and midline.

9.

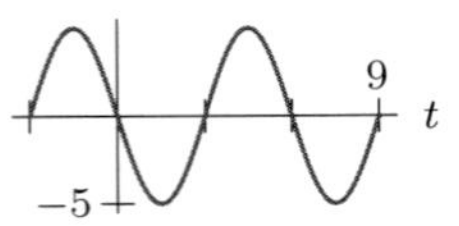

10.

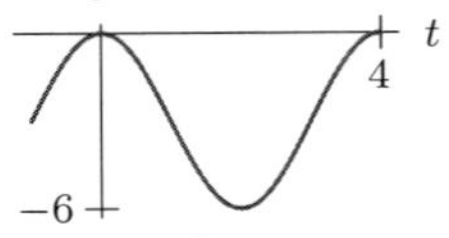

In Exercises 11–18, find formulas for the trigonometric functions.

11.

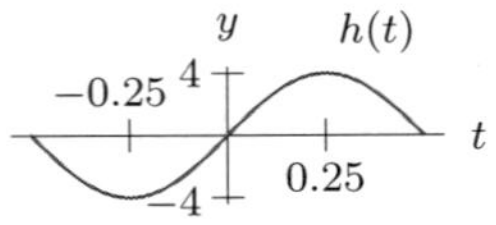

12.

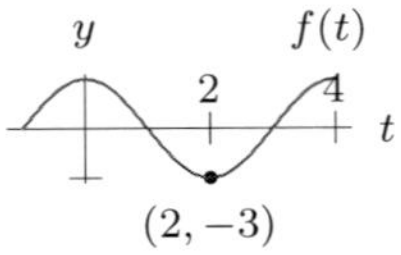

13.

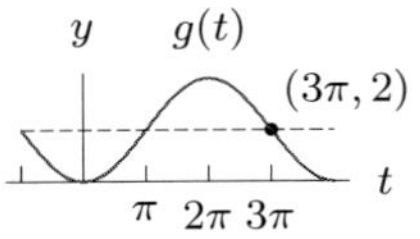

14.

15.

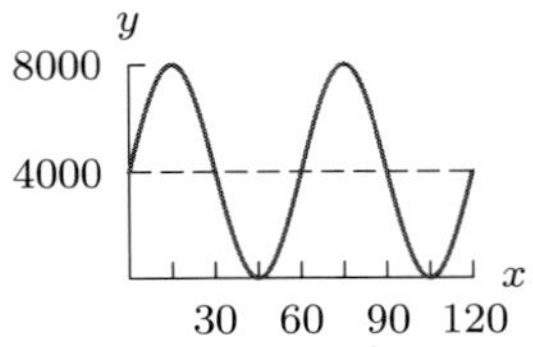

16.

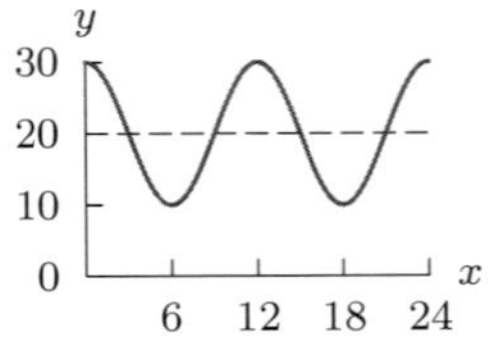

17.

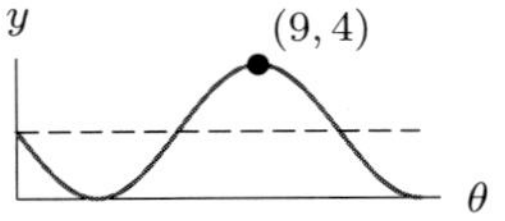

18.

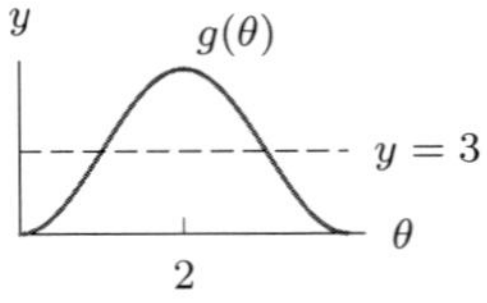

Problems

Without a calculator, graph two periods of the functions in Exercises 19–22.

19. $y = 40\sin\left(\frac{2\pi t}{7}\right) + 60$

20. $y = 3 - 2\cos(50\pi t)$

21. $y = 0.001 - 0.003\sin(0.002\pi t)$

22. $y = 12\cos\left(\frac{2\pi}{50}t\right) - 24$

For Problems 23–24, you are given a formula for a sinusoidal function f, and graphs of f and a second sinusoidal function g. First, decide by what fraction of a period of f (phase shift) has been shifted to obtain g. Using this, find a formula for g.

23. $f(x) = 3\sin((\pi/4)x)$

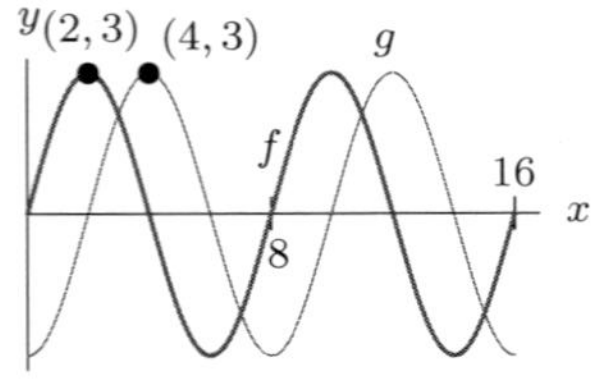

24. $f(x) = 10\sin((\pi/5)x)$

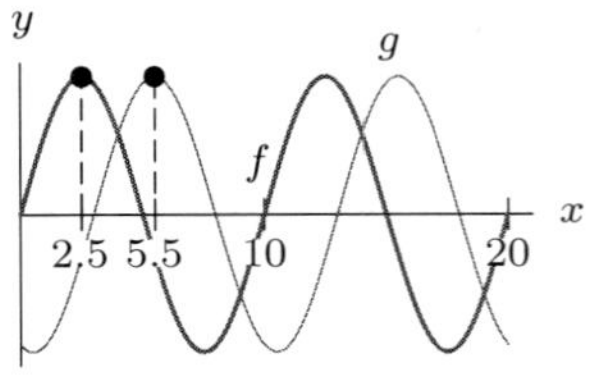

25. Figure 6.66 shows $y = \sin x$ and $y = \sin 2x$. Which graph is $y = \sin x$? Identify the points a to e.

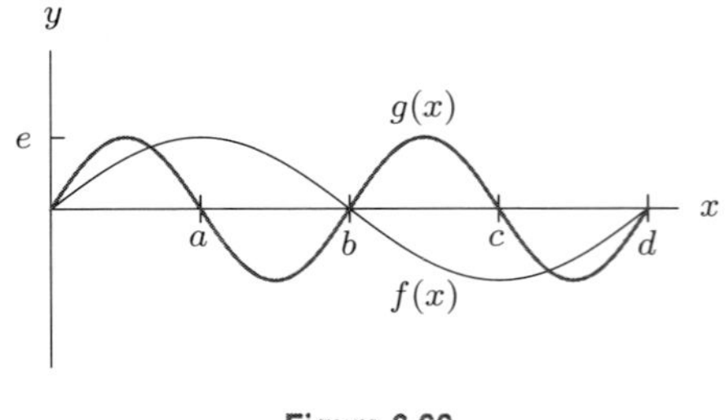

Figure 6.66

26. Describe in words how you can obtain the graph of $y = \cos\left(5t + \frac{\pi}{4}\right)$ from the graph of $y = \cos(5t)$.

27. A person's blood pressure, P, (in millimeters of mercury, abbreviated mm Hg) is given, for time, t, in seconds, by

$$P = 100 - 20\cos\left(\frac{8\pi}{3}t\right),$$

Graph this function. State the period and amplitude and explain the practical significance of these quantities.

In Problems 28–29, find a formula, using the sine function, for your height above ground after t minutes on the ferris wheel. Graph the function to check that it is correct.

28. A ferris wheel is 35 meters in diameter and boarded at ground level. The wheel completes one full revolution every 5 minutes. At $t = 0$ you are in the three o'clock position and ascending.

29. A ferris wheel is 20 meters in diameter and boarded in the six o'clock position from a platform that is 4 meters above the ground. The wheel completes one full revolution every 2 minutes. At $t = 0$ you are in the twelve o'clock position.

The graphs in Problems 30–31 show your height h meters above ground after t minutes on various ferris wheels. Using the sine function, find a formula for h as a function of t.

30.

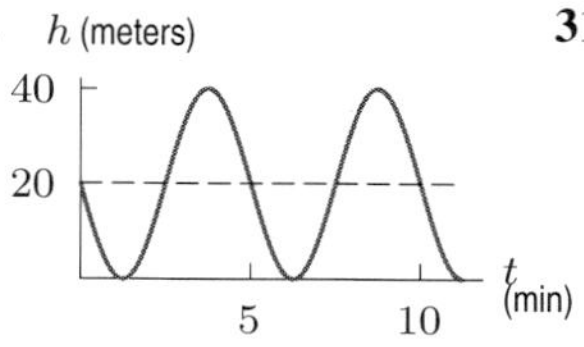

31.

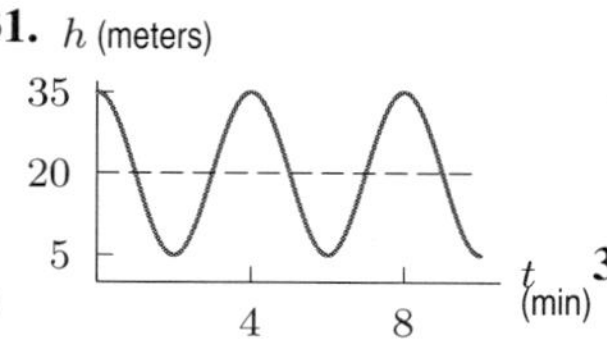

32. Find formula of the form $y = A\cos(B(t - h)) + k$ for the graph in Figure 6.65 on page 275.

33. The London ferris wheel has diameter 450 feet and one complete revolution takes 30 minutes.

(a) Find the rate (in degrees per minute) that the London ferris wheel is rotating.
(b) Let t, the time in minutes, be 0 when you are in the 6 o'clock position. Write θ, measured from the 3 o'clock position, as a function of t.
(c) Find a formula for the ferris wheel function, $h = f(t)$, giving your height in feet above the ground.
(d) Graph $h = f(t)$. What are the period, midline, and amplitude?

34. The following formulas give animal populations as functions of time, t, in years. Describe the growth of each population in words.

(a) $P = 1500 + 200t$ **(b)** $P = 2700 - 80t$
(c) $P = 1800(1.03)^t$ **(d)** $P = 800e^{-0.04t}$
(e) $P = 230\sin\left(\frac{2\pi}{7}t\right) + 3800$

35. A population of animals oscillates between a low of 1300 on January 1 ($t = 0$) and a high of 2200 on July 1 ($t = 6$).

(a) Find a formula for the population, P, in terms of the time, t, in months.
(b) Interpret the amplitude, period, and midline of the function $P = f(t)$.
(c) Use a graph to estimate when $P = 1500$.

36. Find a possible formula for the trigonometric function whose values are in the following table.

x	0	.1	.2	.3	.4	.5	.6	.7	.8	.9	1
$g(x)$	2	2.6	3	3	2.6	2	1.4	1	1	1.4	2

For Problems 37–40, let $f(x) = \sin(2\pi x)$ and $g(x) = \cos(2\pi x)$. Find a possible formula in terms of f or g for the graph.

37.

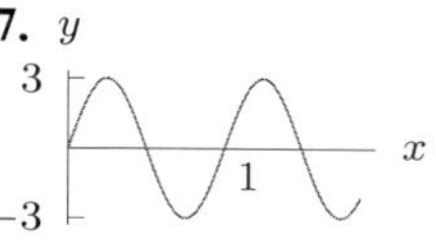

38.

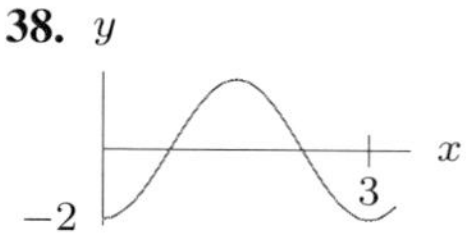

39.

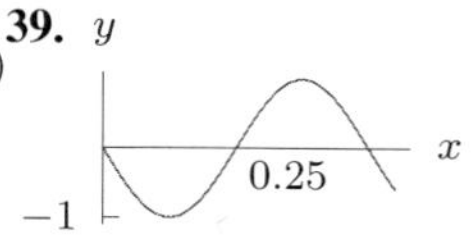

40.

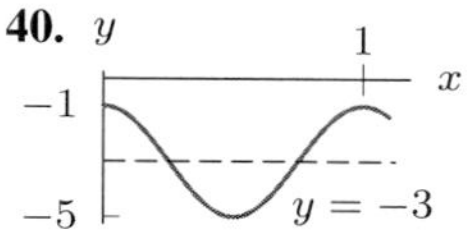

41. A company sells $S(t)$ thousand electric blankets in month t (with $t = 0$ being January), where

$$S(t) \approx 72.25 + 41.5\sin\left(\frac{\pi t}{6} + \frac{\pi}{2}\right).$$

Graph this function over one year. Find its period and amplitude and explain their practical significance.

42. The pressure, P (in lbs/ft^2), in a pipe varies over time. Five times an hour, the pressure oscillates from a low of 90 to a high of 230 and then back to a low 90. The pressure at $t = 0$ is 90.

(a) Graph $P = f(t)$, where t is time in minutes. Label your axes.
(b) Find a possible formula for $P = f(t)$.
(c) By graphing $P = f(t)$ for $0 \le t \le 2$, estimate when the pressure first equals 115 lbs/ft^2.

43. A flight from La Guardia Airport in New York City to Logan Airport in Boston has to circle Boston several times before landing. Figure 6.67 shows the graph of the distance, d, of the plane from La Guardia as a function of time, t. Construct a function $f(t)$ whose graph approximates this one. Do this by using different formulas on the intervals $0 \leq t \leq 1$ and $1 \leq t \leq 2$.

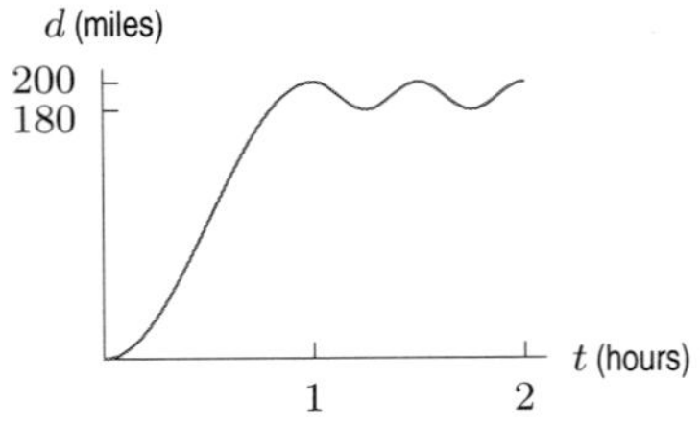

Figure 6.67

44. **(a)** Graph the data in Table 6.7, which gives the population, P, in thousands, of Somerville, MA, with t in years since 1920.[4]

(b) Based on the data, a researcher decides the population varies in an approximately periodic way with time. Do you agree?

(c) On your graph, sketch in a sine curve that fits your data as closely as possible. Your sketch should capture the overall trend of the data but need not pass through all the data points. [Hint: Start by choosing a midline.]

(d) Find a formula for the curve you drew in part (c).

(e) According to the US census, the population of Somerville in 1910 was 77,236. How well does this agree with the value given by your formula?

Table 6.7

t	0	10	20	30	40	50	60	70	80
P	93	104	102	102	95	89	77	76	77

45. Table 6.8 shows the average daily maximum temperature in degrees Fahrenheit in Boston each month.[5]

(a) Plot the average daily maximum temperature as a function of the number of months past January.

(b) What are the amplitude and period of the function?

(c) Find a trigonometric approximation of this function.

(d) Use your formula to estimate the daily maximum temperature for October. How well does this estimate agree with the data?

Table 6.8

Month	Jan	Feb	Mar	Apr	May	Jun
Temperature	36.4	37.7	45.0	56.6	67.0	76.6
Month	Jul	Aug	Sep	Oct	Nov	Dec
Temperature	81.8	79.8	72.3	62.5	47.6	35.4

46. The website arXiv.org is used by scientists to share research papers. Figure 6.68 shows usage[6] of this site, $n = f(t)$, the number of new connections on day t, where $t = 0$ is Monday, August 5, 2002.

(a) The graph suggests that f is approximately periodic with a period of 7 days. However, f is not exactly periodic. How can you tell?

(b) Why might the usage of the arXiv.org website lend itself to being modeled by a periodic function?

(c) Find a trigonometric function that gives a reasonable approximation for f. Explain what the period, amplitude, and midline of your function tell you about the usage of the arXiv.org website.

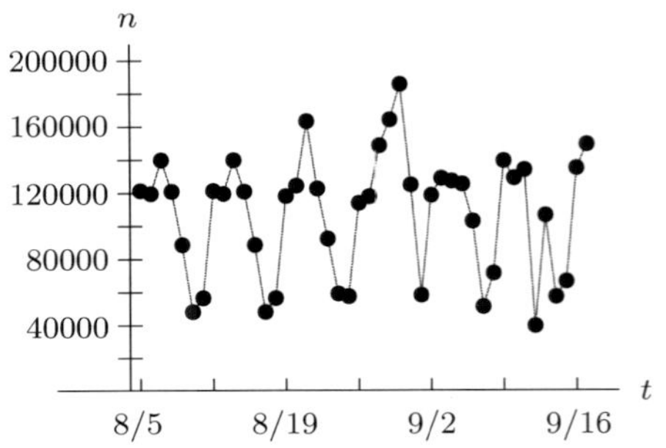

Figure 6.68

47. Table 6.9 gives US petroleum imports, P, in quadrillion BTUs, for t in years since 1900. Plot the data on $73 \leq t \leq 92$ and find a trigonometric function $f(t)$ that approximates US petroleum imports as a function of t.

Table 6.9

t	73	74	75	76	77	78	79	80	81	82
P	14.2	14.5	14	16.5	18	17	17	16	15	14
t	83	84	85	86	87	88	89	90	91	92
P	13	15	12	14	16	17	18	17	16	16

[4] From the US census.

[5] Statistical Abstract of the United States.

[6] Data obtained from usage statistics made available at the arXive.org site. Usage statistics for August 28, 2002, are unavailable and have been estimated.

6.6 OTHER TRIGONOMETRIC FUNCTIONS

The Tangent Function

Another useful trigonometric function is called the tangent. Suppose $P = (x, y)$ is the point on the unit circle corresponding to the angle θ. We define the *tangent* of θ, or $\tan\theta$, by

$$\tan\theta = \frac{y}{x}.$$

The graphical interpretation of $\tan\theta$ is as a slope. In Figure 6.69, the slope of the line passing from the origin through P is given by

$$m = \frac{\Delta y}{\Delta x} = \frac{y-0}{x-0} = \frac{y}{x},$$

so

$$m = \tan\theta.$$

In words, $\tan\theta$ is the slope of the line passing through the origin and point P.

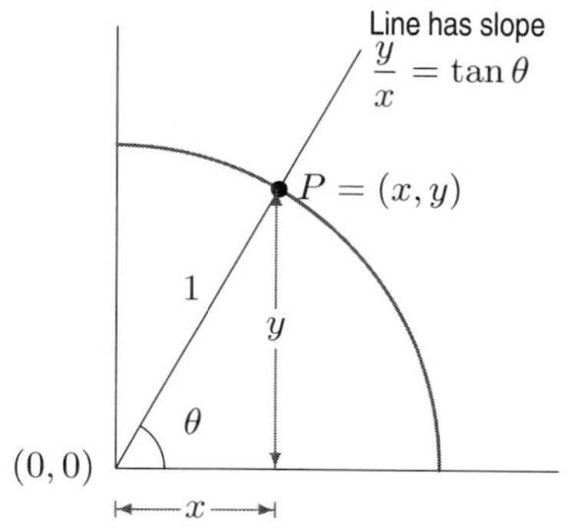

Figure 6.69: The slope of the line passing through the origin and point P is $\tan\theta$

Example 1 Find the slope of the line passing through the origin at an angle of (a) $30°$ (b) $\pi/4$

Solution (a) A calculator set in degree mode gives

$$\tan 30° \approx 0.577.$$

Thus, the slope of the line is about 0.577.

(b) A calculator set in radian mode gives $\tan(\pi/4) = 1$. This makes sense because an angle of $\pi/4$ describes a ray halfway between the x- and y-axes, or the line $y = x$, which has slope of 1.

Graph of the Tangent Function

Table 6.10 contains values of the tangent function. Figure 6.70 illustrates the connection between an angle θ on the unit circle and the graph of the tangent function. As the angle θ opens from 0 to $\pi/2$, the slope of the line, and therefore the tangent of θ, increases from 0 to $+\infty$. At $\pi/2$, this line

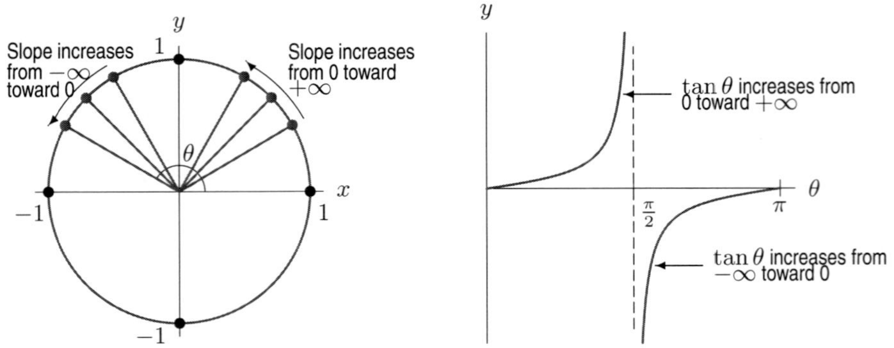

Figure 6.70: The connection between the unit circle and the graph of the tangent function

is vertical, and its slope is undefined. Thus, $\tan(\pi/2)$ is undefined and the graph of $y = \tan\theta$ has a vertical asymptote at $\theta = \pi/2$.

Table 6.10 *Values of tangent function (rounded), with θ in degrees and radians*

θ	$0°$	$30°$	$60°$	$90°$	$120°$	$150°$	$180°$
θ	0	$\pi/6$	$\pi/3$	$\pi/2$	$2\pi/3$	$5\pi/6$	π
$\tan\theta$	0	0.6	1.7	Undefined	-1.7	-0.6	0

For θ between $\pi/2$ and π, the slopes are negative. The line is very steep near $\pi/2$, but becomes less steep as θ approaches π, where it is horizontal. Thus, $\tan\theta$ becomes less negative as θ increases in the second quadrant and reaches 0 at $\theta = \pi$.

For values of θ between π and 2π, observe that

$$\tan(\theta + \pi) = \tan\theta,$$

because the angles θ and $\theta + \pi$ determine the same line through the origin, and hence the same slope. Thus, $y = \tan\theta$ has period π. The completed graph of $y = \tan\theta$ is shown in Figure 6.71.

Notice the differences between the tangent function and the sinusoidal functions. The tangent function has a period of π, whereas the sine and cosine both have periods of 2π. The tangent function has vertical asymptotes at odd multiples of $\pi/2$; the sine and cosine have no asymptotes. Even though the tangent function is periodic, it does not have an amplitude or midline because it does not have a maximum or minimum value.

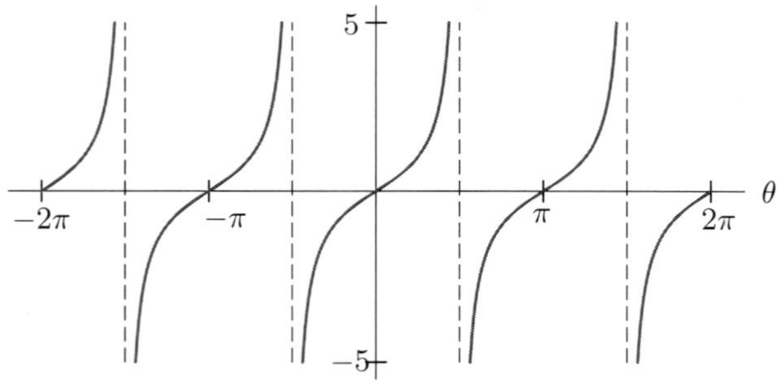

Figure 6.71: The graph of $y = \tan\theta$ in radians

Relationships Between the Trigonometric Functions

There is a relationship among the three trigonometric functions we have defined. If (x, y) are the coordinates of the point P determined by the angle θ on the unit circle, we defined

$$\cos\theta = x \qquad \text{and} \qquad \sin\theta = y.$$

Since $\tan\theta = y/x$, if $x \neq 0$, we have

$$\boxed{\tan\theta = \frac{\sin\theta}{\cos\theta}.}$$

In Problem 26, we see that $\tan\theta$ is an odd function.

Example 2 Calculate the exact value of $\tan(\pi/4)$.

Solution Since we know $\cos(\pi/4) = \sin(\pi/4) = 1/\sqrt{2}$, we have

$$\tan\left(\frac{\pi}{4}\right) = \frac{\sin(\pi/4)}{\cos(\pi/4)} = \frac{1/\sqrt{2}}{1/\sqrt{2}} = 1.$$

Note that this is the same result as in Example 1.

Because of their definitions in terms of the unit circle, the sine and cosine functions are also related. We have seen that any point on the unit circle has coordinates $(\cos\theta, \sin\theta)$. But we also know that points on the unit circle must satisfy its equation

$$x^2 + y^2 = 1.$$

Substituting $x = \cos\theta$ and $y = \sin\theta$ gives

$$(\cos\theta)^2 + (\sin\theta)^2 = 1,$$

or, writing $\cos^2\theta$ instead of $(\cos\theta)^2$ and $\sin^2\theta$ instead of $(\sin\theta)^2$, we have

$$\boxed{\cos^2\theta + \sin^2\theta = 1.}$$

If we know which quadrant a given angle is in and we know any one of its three trigonometric values, we can calculate the other two by using these relationships.

Example 3 Check that the relationships between $\sin\theta$, $\cos\theta$, and $\tan\theta$ are satisfied for $\theta = 40°$.

Solution To three decimal places, a calculator (in degree mode) gives $\sin 40° = 0.643$, $\cos 40° = 0.766$, and $\tan 40° = 0.839$. The fact that

$$\frac{0.643}{0.766} = 0.839,$$

confirms that, to three decimal places,

$$\frac{\sin 40°}{\cos 40°} = \tan 40°.$$

Similarly, to three decimal places, we find that

$$(0.643)^2 + (0.766)^2 = 1,$$

confirming, to the accuracy with which we are working, that

$$\sin^2 40^\circ + \cos^2 40^\circ = 1.$$

Example 4 Suppose that $\cos\theta = 2/3$ and $3\pi/2 \le \theta \le 2\pi$. Find $\sin\theta$ and $\tan\theta$.

Solution Use the relationship $\cos^2\theta + \sin^2\theta = 1$ to find $\sin\theta$. Substitute $\cos\theta = 2/3$:

$$\begin{aligned}\left(\frac{2}{3}\right)^2 + \sin^2\theta &= 1\\ \frac{4}{9} + \sin^2\theta &= 1\\ \sin^2\theta &= 1 - \frac{4}{9} = \frac{5}{9}\\ \sin\theta &= \pm\sqrt{\frac{5}{9}} = \pm\frac{\sqrt{5}}{3}.\end{aligned}$$

Because θ is in the fourth quadrant, $\sin\theta$ is negative, so $\sin\theta = -\sqrt{5}/3$. To find $\tan\theta$, use the relationship

$$\tan\theta = \frac{\sin\theta}{\cos\theta} = \frac{-\sqrt{5}/3}{2/3} = -\frac{\sqrt{5}}{2}.$$

The Reciprocals of the Trigonometric Functions: Secant, Cosecant, Cotangent

The reciprocals of the trigonometric functions are given special names. Where the denominators are not equal to zero, we have

$$\begin{aligned}\text{secant}\,\theta = \sec\theta &= \frac{1}{\cos\theta}.\\ \text{cosecant}\,\theta = \csc\theta &= \frac{1}{\sin\theta}.\\ \text{cotangent}\,\theta = \cot\theta &= \frac{1}{\tan\theta} = \frac{\cos\theta}{\sin\theta}.\end{aligned}$$

The Pythagorean identity, $\cos^2\theta + \sin^2\theta = 1$, can be rewritten in terms of other trigonometric functions. Dividing through by $\cos^2\theta$ gives, provided $\cos\theta \ne 0$,

$$\begin{aligned}\frac{\cos^2\theta}{\cos^2\theta} + \frac{\sin^2\theta}{\cos^2\theta} &= \frac{1}{\cos^2\theta}\\ 1 + \left(\frac{\sin\theta}{\cos\theta}\right)^2 &= \left(\frac{1}{\cos\theta}\right)^2\end{aligned}$$

so

$$\boxed{1 + \tan^2\theta = \sec^2\theta.}$$

A similar identity relates $\cot\theta$ and $\csc\theta$. See Problem 27.

Example 5 Use a graph of $g(\theta) = \cos\theta$ to explain the shape of the graph of $f(\theta) = \sec\theta$.

Solution Figure 6.72 shows the graphs of $\cos\theta$ and $\sec\theta$. In the first quadrant $\cos\theta$ decreases from 1 to 0, so the reciprocal of $\cos\theta$ increases from 1 toward $+\infty$. The values of $\cos\theta$ are negative in the second quadrant and decrease from 0 to -1, so the values of $\sec\theta$ increase from $-\infty$ to -1. The graph of $y = \cos\theta$ is symmetric about the vertical line $\theta = \pi$, so the graph of $f(\theta) = \sec\theta$ is symmetric about the same line. Thus, the graph of $f(\theta) = \sec\theta$ on the interval $\pi \le \theta \le 2\pi$ is the mirror image of the graph on $0 \le \theta \le \pi$. Note that $\sec\theta$ is undefined wherever $\cos\theta = 0$, namely, at $\theta = \pi/2$ and $\theta = 3\pi/2$. The graph of $f(\theta) = \sec\theta$ has vertical asymptotes at those values.

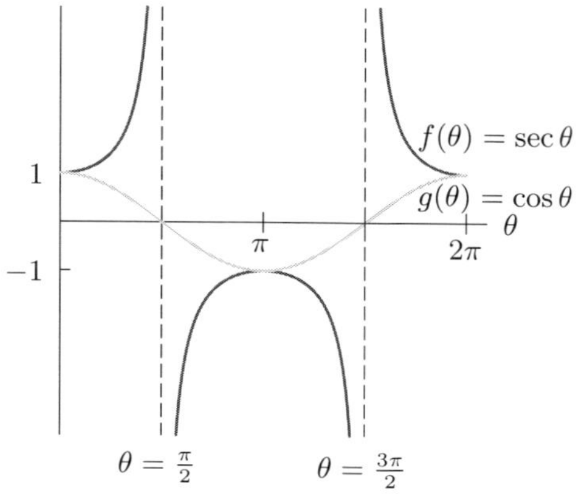

Figure 6.72: Graph of secant

The graphs of $y = \csc\theta$ and $y = \cot\theta$ are obtained in a similar fashion from the graphs of $y = \sin\theta$ and $t = \tan\theta$, respectively. See Figure 6.73 and 6.74.

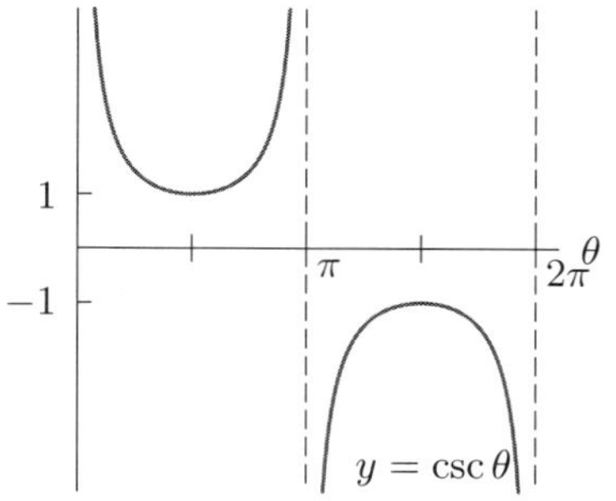

Figure 6.73: Graph of cosecant

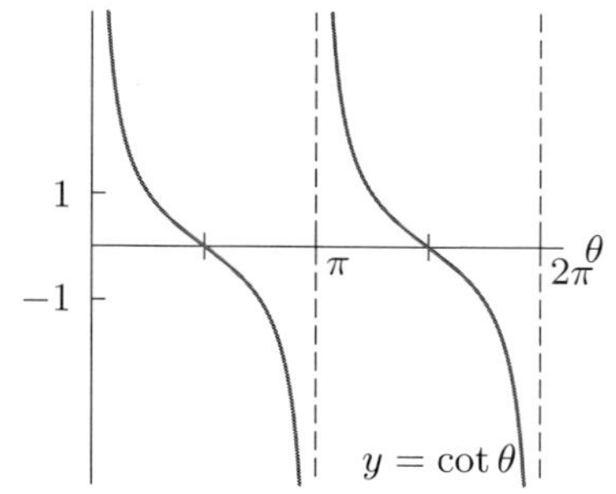

Figure 6.74: Graph of cotangent

Exercises and Problems for Section 6.6

Exercises

1. Find exact values for $\sin 0^\circ$, $\cos 0^\circ$ and $\tan 0^\circ$.

Without a calculator, find exact values for the quantities in Exercises 2–16.

2. $\cos 90^\circ$ **3.** $\sin 90^\circ$ **4.** $\tan 90^\circ$

5. $\sin 270^\circ$ **6.** $\tan 225^\circ$ **7.** $\tan 135^\circ$

8. $\tan 540^\circ$ **9.** $\tan \frac{5\pi}{4}$ **10.** $\tan \frac{\pi}{3}$

11. $\tan \frac{2\pi}{3}$ **12.** $\tan \frac{11\pi}{6}$ **13.** $\csc \frac{5\pi}{4}$

14. $\cot \frac{5\pi}{3}$ **15.** $\sec\left(-\frac{\pi}{6}\right)$ **16.** $\sec \frac{11\pi}{6}$

Problems

In Exercises 17–20, give exact answers for $0 \leq \theta \leq \pi/2$.

17. If $\cos\theta = \frac{1}{2}$, what is $\sec\theta$? $\tan\theta$?

18. If $\cos\theta = \frac{1}{2}$, what is $\csc\theta$? $\cot\theta$?

19. If $\sin\theta = \frac{1}{3}$, what is $\sec\theta$? $\tan\theta$?

20. If $\sec\theta = 17$, what is $\sin\theta$? $\tan\theta$?

In Exercises 21–22, give a possible formula for the function.

21.

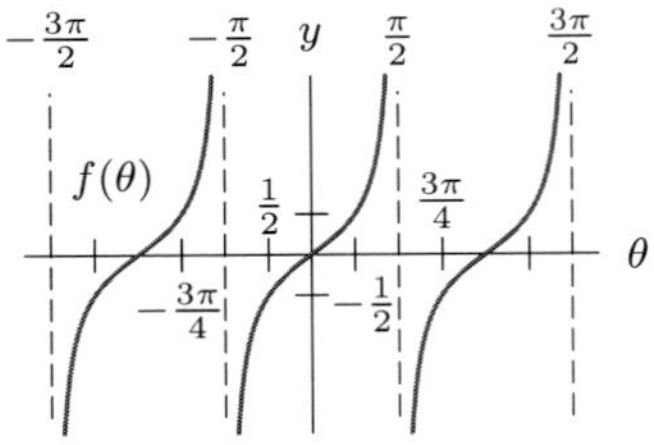

22.

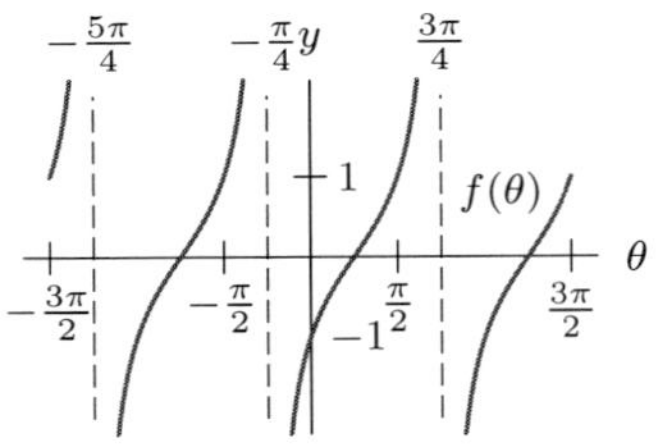

23. **(a)** $\cos\alpha = -\sqrt{3}/5$ and α is in the third quadrant. Find exact values for $\sin\alpha$ and $\tan\alpha$.
(b) $\tan\beta = 4/3$ and β is in the third quadrant. Find exact values for $\sin\beta$ and $\cos\beta$.

24. **(a)** $\cos\phi = 0.4626$ and $3\pi/2 < \phi < 2\pi$. Find decimal approximations for $\sin\phi$ and $\tan\phi$.
(b) $\sin\theta = -0.5917$ and $\pi < \theta < 3\pi/2$. Find decimal approximations for $\cos\theta$ and $\tan\theta$.

25. Suppose that $y = \sin\theta$ for $0° < \theta < 90°$. Evaluate $\cos\theta$ in terms of y.

26. Use the fact that sine is an odd function and cosine is an even function to show that tangent is an odd function.

27. Show how to obtain the identity $\cot^2\theta + 1 = \csc^2\theta$ from the Pythagorean identity.

Problems 28–31 give an expression for one of the three functions $\sin\theta$, $\cos\theta$, or $\tan\theta$, with θ in the first quadrant. Find expressions for the other two functions. Your answers will be algebraic expressions in terms of x.

28. $\sin\theta = x/3$

29. $\cos\theta = 4/x$

30. $x = 2\cos\theta$

31. $x = 9\tan\theta$

32. **(a)** Find an equation for the line l in Figure 6.75.
(b) Find the x-intercept of the line.

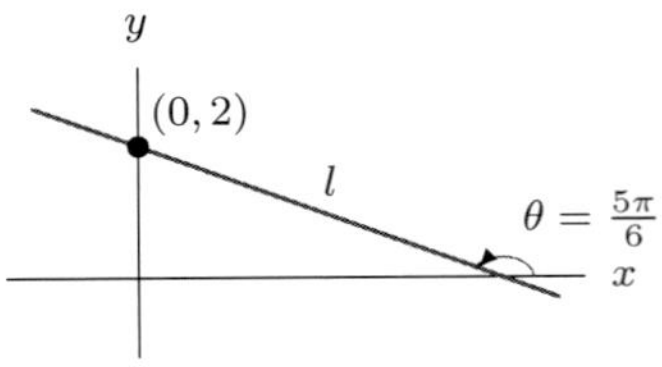

Figure 6.75

33. Use Figure 6.76 to find an equation for the line l in terms of x_0, y_0, and θ.

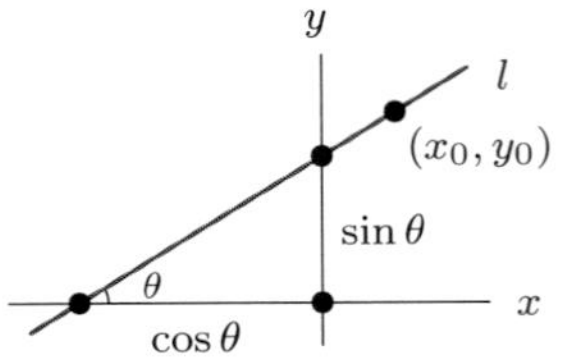

Figure 6.76

34. **(a)** At what values of t does the graph of $y = \tan t$ have vertical asymptotes? What can you say about the graph of $y = \cos t$ at the same values of t?
(b) At what values of t does the graph of $y = \tan t$ have t-intercepts? What can you say about the graph of $y = \sin t$ at the same values of t?

35. Graph $y = \cos x \cdot \tan x$. Is this function exactly the same as $y = \sin x$? Why or why not?

Find exact values for the lengths of the labeled segments in Problems 36–37.

36.

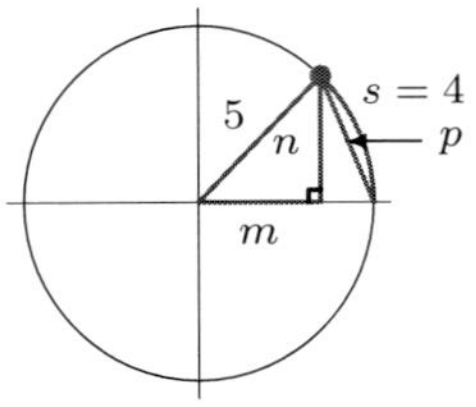

37.

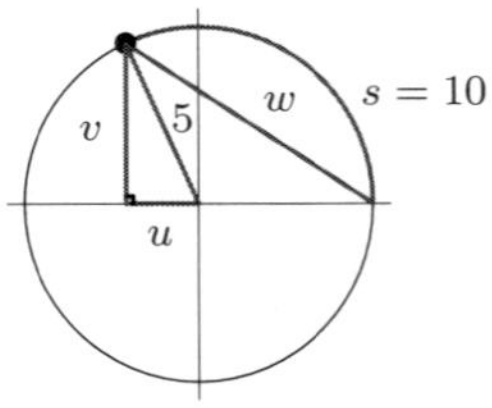

38. Graph the functions $y = \sec x$ and $y = \csc x$, and $y = \cot x$. Describe the behavior of each one.

6.7 INVERSE TRIGONOMETRIC FUNCTIONS

Solving Trigonometric Equations Graphically

A trigonometric equation is an equation that involves trigonometric functions. Consider, for example, the rabbit population of Example 8 on page 274:

$$R = -5000 \cos\left(\frac{\pi}{6}t\right) + 10000.$$

To find when the population reaches 12,000, we need to solve the trigonometric equation

$$-5000 \cos\left(\frac{\pi}{6}t\right) + 10000 = 12{,}000.$$

We can find approximate solutions to trigonometric equations by using a graph. We start with a simpler example.

Example 1 Use a graph to approximate solutions to the equation

$$\cos t = 0.4$$

Solution We draw a graph of $y = \cos t$ and trace along it on a calculator to find points at which $y = 0.4$ and read off the t-values at these points. In Figure 6.77, the points t_0, t_1, t_2, t_3 represent values of t satisfying $\cos t = 0.4$. If t is in radians, we find $t_0 = -1.159$, $t_1 = 1.159$, $t_2 = 5.12$, $t_3 = 7.44$. We can check these values by evaluating:

$$\cos(-1.159) = 0.40, \quad \cos(1.159) = 0.40, \quad \cos(5.12) = 0.40, \quad \cos(7.44) = 0.40.$$

Notice that because the cosine function is periodic, the equation $\cos t = 0.4$ has infinitely many solutions. The symmetry of the graph suggests that the solutions are related.

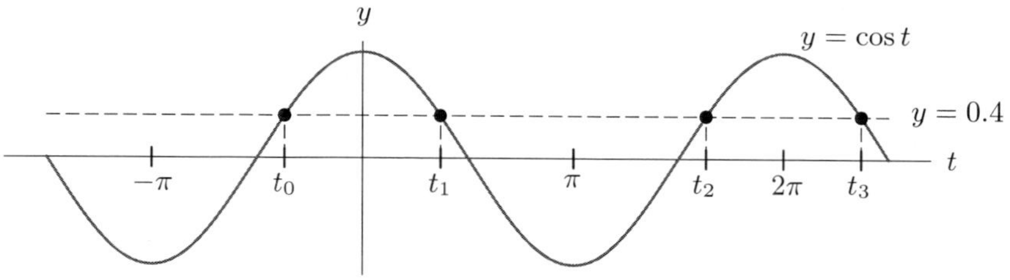

Figure 6.77: The points t_0, t_1, t_2, t_3 are solutions to the equation $\cos t = 0.4$

Solving Trigonometric Equations Using the Inverse Cosine

While trigonometric equations can be solved graphically, there is an another method. Finding a solution to

$$\cos t = 0.4$$

means finding an angle whose cosine is 0.4. In other words, given an output for the cosine function, we want to find the corresponding input. This is what the *inverse cosine*, labeled $\boxed{\cos^{-1}}$ on a calculator, gives us. Evaluating an inverse cosine produces an angle whose cosine is the given value.

A calculator in radian mode gives one of the t-values we found in Example 1:

$$\cos^{-1}(0.4) = 1.159, \qquad \text{where} \qquad \cos(1.159) = 0.4.$$

Notice that the $\boxed{\cos^{-1}}$ key gives only one solution to a trigonometric equation. There are other solutions, which we can find using the symmetry of the cosine graph. Since $t_1 = 1.159$, we see in Figure 6.78 that $t_0 = -1.159$ because the cosine function is symmetric about the y-axis. In addition, the arch of the cosine graph from $-\pi/2$ to $\pi/2$ is exactly the same shape as the arch from $3\pi/2$ to $5\pi/2$, so

$$t_2 = 2\pi - 1.159 = 5.12 \qquad \text{and} \qquad t_3 = 2\pi + 1.159 = 7.44.$$

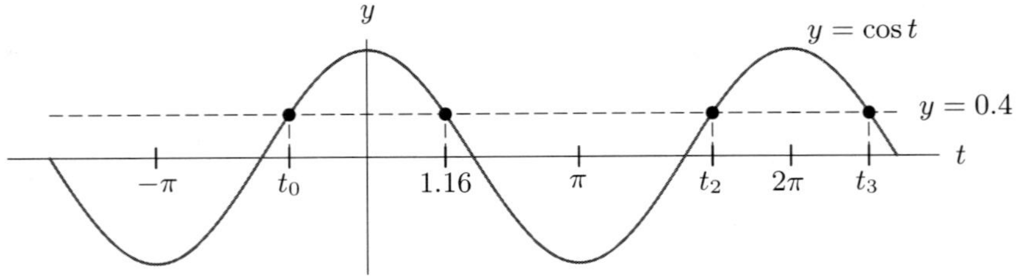

Figure 6.78: Symmetry of cosine graph shows relationship between solutions of $\cos t = 0.4$

Example 2 Use the inverse cosine to estimate when the rabbit population, R, in Example 8 on page 274 reaches 12,000.

Solution The population is given by

$$R = -5000 \cos\left(\frac{\pi}{6}t\right) + 10000,$$

so we solve the equation

$$-5000 \cos\left(\frac{\pi}{6}t\right) + 10000 = 12000.$$

We first isolate the trigonometric expression $\cos(\pi t/6)$:

$$-5000 \cos\left(\frac{\pi}{6}t\right) = 2000$$

$$\cos\left(\frac{\pi}{6}t\right) = -0.4$$

To solve for t, we need an angle whose cosine is -0.4. This is what $\cos^{-1}(-0.4)$ gives us. So

$$\frac{\pi}{6}t = \cos^{-1}(-0.4),$$

giving

$$t = \frac{6}{\pi}\cos^{-1}(-0.4).$$

Using a calculator to evaluate $\cos^{-1}(-0.4)$, we see that the solution t_1 in Figure 6.79 is

$$t_1 = \frac{6}{\pi}(1.98) = 3.786.$$

But how do we find the second solution, t_2? Since the graph of R is symmetric about the line $t = 6$, we have

$$t_2 = 12 - t_1 \approx 8.214.$$

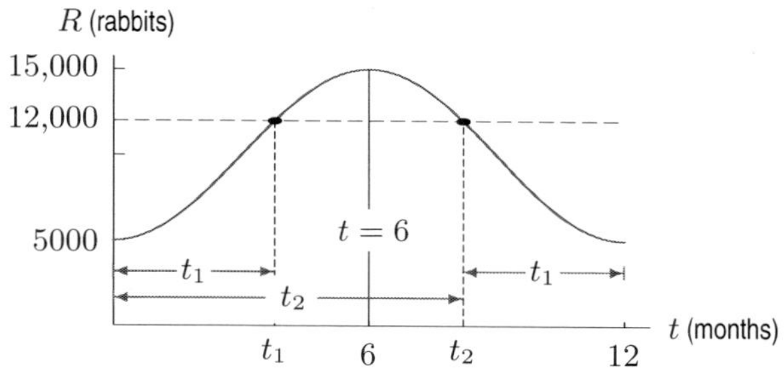

Figure 6.79: By symmetry about the line $t = 6$, we see that $t_2 = 12 - t_1$

The Inverse Cosine Function

As we saw in Figure 6.78 on page 286, the equation

$$\cos t = 0.4$$

has infinitely many solutions. However, the inverse cosine key on a calculator gives only one of them, namely

$$\cos^{-1}(0.4) \approx 1.159,$$

Why does the calculator select this particular solution?

From the graph in Figure 6.80, we see that the angles in the interval $0 \leq t \leq \pi$ produce all values in the range of $\cos t$ once and once only. By choosing output values for the inverse cosine in this interval, we define a new *function* that assigns just one angle to each possible value of the cosine.[7] (Recall that a function can have only one output for each input value.) This is the function that your calculator evaluates when you press the $\boxed{\cos^{-1}}$ key.

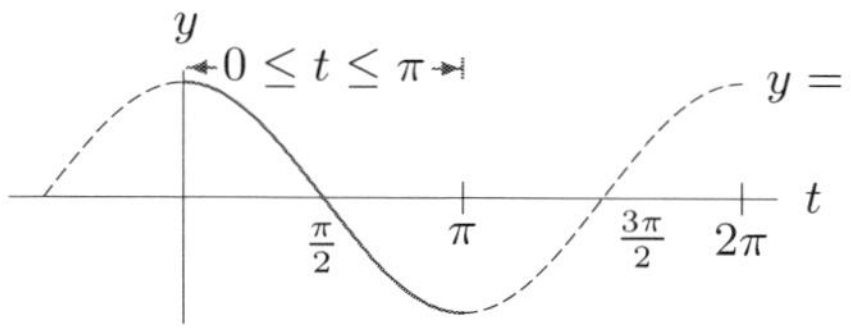

Figure 6.80: The solid portion of this graph represents a function that has only one input value for each output value

The inverse cosine function inputs values of y between -1 and 1 (all possible values of $\cos t$) and outputs angles between 0 and π. We interpret the value of $\cos^{-1}(y)$ as "the angle between 0 and π whose cosine is y." Because an angle in radians determines an arc of the same measure on a unit circle, the inverse cosine of y is sometimes called the *arccosine* of y. We summarize:

> The **inverse cosine** function, also called the **arccosine** function, is denoted by $\cos^{-1} y$ or $\arccos y$. We define
>
> $$t = \cos^{-1} y \quad \text{provided that} \quad y = \cos t \quad \text{and} \quad 0 \leq t \leq \pi.$$
>
> In other words, if $t = \arccos y$, then t is the angle between 0 and π whose cosine is y.
> The domain of the inverse cosine is $-1 \leq y \leq 1$ and its range is $0 \leq t \leq \pi$.

[7] Other intervals, such as $-\pi \leq t \leq 0$, could also be used. The interval $0 \leq t \leq \pi$ has become the agreed upon choice.

Example 3 Evaluate (a) $\cos^{-1}(0)$ (b) $\arccos(1)$ (c) $\cos^{-1}(-1)$

Solution

(a) $\cos^{-1}(0)$ means the angle between 0 and π whose cosine is 0. Since $\cos(\pi/2) = 0$, we have $\cos^{-1}(0) = \pi/2$.
(b) $\arccos(1)$ means the angle between 0 and π whose cosine is 1. Since $\cos(0) = 1$, we have $\arccos(1) = 0$.
(c) $\cos^{-1}(-1)$ means the angle between 0 and π whose cosine is -1. Since $\cos(\pi) = -1$, we have $\cos^{-1}(-1) = \pi$.

Warning! It is important to realize that the notation $\cos^{-1} y$ does *not* indicate the reciprocal of $\cos y$. In other words, $\cos^{-1} y$ is not the same as $(\cos y)^{-1}$. For example,

$$\cos^{-1}(0) = \frac{\pi}{2} \quad \text{because} \quad \cos\left(\frac{\pi}{2}\right) = 0,$$

but

$$(\cos 0)^{-1} = \frac{1}{\cos 0} = \frac{1}{1} = 1.$$

The Inverse Sine and Inverse Tangent Functions

To solve equations involving sine or tangent, we need an *inverse sine* and an *inverse tangent* function. For example, suppose we want to solve

$$\sin t = 0.8, \qquad 0 \le t \le 2\pi.$$

The inverse sine of 0.8, or $\sin^{-1}(0.8)$, gives us an angle whose sine is 0.8. But there are many angles with a given sine and the inverse sine function specifies only one of them. Just as we did with the inverse cosine, we choose an interval on the t-axis that produces all values in the range of the sine function once and once only.

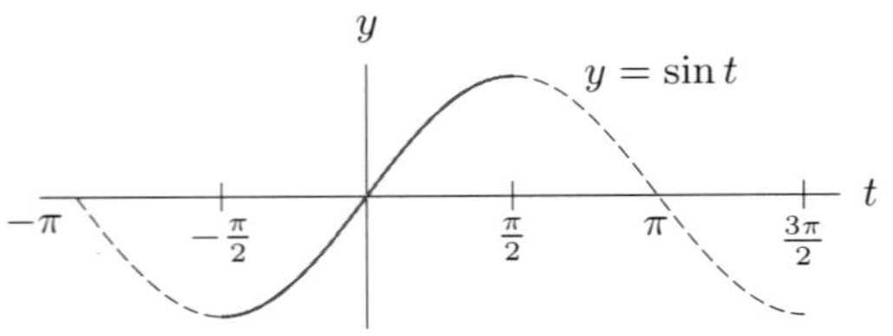

Figure 6.81: The graph of $y = \sin t$, $-\frac{\pi}{2} \le t \le \frac{\pi}{2}$

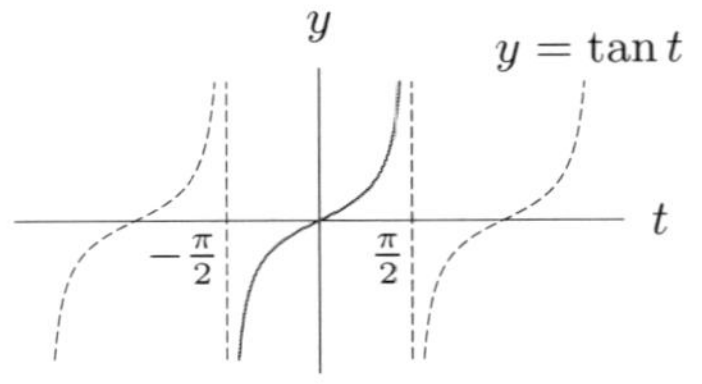

Figure 6.82: The graph of $y = \tan t$, $-\frac{\pi}{2} < t < \frac{\pi}{2}$

Figure 6.81 shows that we cannot choose the same interval that we chose for $\cos t$, which was $0 \le t \le \pi$. However, the interval $-\pi/2 \le t \le \pi/2$ includes a unique angle for each value of $\sin t$. This interval is chosen because it is the smallest interval around $t = 0$ that includes all values of $\sin t$. This same interval, except for the endpoints, is also used to define the inverse tangent function, as illustrated in Figure 6.82.

The **inverse sine** function, also called the **arcsine** function, is denoted by $\sin^{-1} y$ or $\arcsin y$. We define

$$t = \sin^{-1} y \quad \text{provided that} \quad y = \sin t \text{ and } -\frac{\pi}{2} \le t \le \frac{\pi}{2}.$$

The domain of the inverse sine is $-1 \le y \le 1$ and the range is $-\pi/2 \le t \le \pi/2$.
The **inverse tangent** function, also called the **arctangent** function, is denoted by $\tan^{-1} y$ or $\arctan y$. We define

$$t = \tan^{-1} y \quad \text{provided that} \quad y = \tan t \text{ and } -\frac{\pi}{2} < t < \frac{\pi}{2}.$$

The domain of the inverse tangent is $-\infty < y < \infty$ and the range is $-\pi/2 < t < \pi/2$.

Example 4 Evaluate (a) $\sin^{-1}(1)$ (b) $\arcsin(-1)$ (c) $\tan^{-1}(0)$ (d) $\arctan(1)$

Solution

(a) $\sin^{-1}(1)$ means the angle between $-\pi/2$ and $\pi/2$ whose sine is 1. Since $\sin(\pi/2) = 1$, we have $\sin^{-1}(1) = \pi/2$.
(b) $\arcsin(-1) = -\pi/2$ since $\sin(-\pi/2) = -1$.
(c) $\tan^{-1}(0)$ since $\tan 0 = 0$.
(d) $\arctan(1) = \pi/4$ since $\tan(\pi/4) = 1$.

Example 5 Evaluate (a) $\sin^{-1}(-0.5)$ (b) $\arctan(-1)$

Solution

(a) $\sin^{-1}(-0.5)$ is the angle between $-\pi/2 \le t \le \pi/2$ whose sine is -0.5. From page 264 we have $\sin(\pi/6) = 0.5$. The sine is an odd function, so $\sin(-\pi/6) = -0.5$, and $\sin^{-1}(-0.5) = -\pi/6$.
(b) The tangent is also an odd function, and $\tan \pi/4 = 1$, so $\tan(-\pi/4) = -1$, and therefore $\arctan(-1) = -\pi/4$.

We can use the inverse sine and inverse tangent to solve equations.

Example 6 While riding the London Eye ferris wheel, how much time during the first turn do you spend above 400 feet?

Solution

Since your first turn takes 30 minutes and your height is given by $f(t) = 225 \sin\left(\frac{\pi}{15}(t - 7.5)\right) + 225$, we must solve the inequality

$$225 \sin\left(\frac{\pi}{15}(t - 7.5)\right) + 225 \ge 400 \qquad \text{for } 0 \le t \le 30.$$

This can be simplified as follows:

$$225 \sin\left(\frac{\pi}{15}(t - 7.5)\right) \ge 175$$
$$\sin\left(\frac{\pi}{15}(t - 7.5)\right) \ge \frac{175}{225}.$$

From Figure 6.83, we see that the equation $\sin\left(\frac{\pi}{15}(t-7.5)\right) = \frac{175}{225}$ has two solutions on the interval $0 \le t \le 30$. One of them can be found by using the arcsine function:

$$\frac{\pi}{15}(t-7.5) = \arcsin\frac{175}{225}$$
$$t-7.5 = \frac{15}{\pi}\arcsin\frac{175}{225}$$
$$t = \underbrace{7.5 + \frac{15}{\pi}\arcsin\frac{175}{225}}_{\text{Exact solution}} \approx \underbrace{11.75.}_{\text{Approximate solution}}$$

Since the period of this function is 30, by symmetry the other solution is $30 - 11.75 = 18.25$ minutes. Thus, you spend $18.25 - 11.75 = 6.5$ minutes at a height of 400 feet or more.

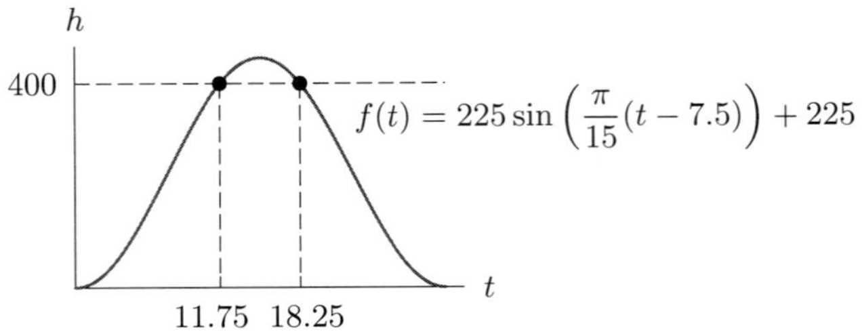

Figure 6.83: On the ferris wheel between $t = 11.75$ minutes and $t = 18.25$ minutes, you are above 400 feet

Solving Equations Using Reference Angles

Because of the symmetry of the unit circle, the values of sine, cosine, and tangent of angles in the first quadrant can be used to find values of these functions for angles in the other three quadrants.

Example 7 Use the values of the sine and cosine of $65°$ to find the sine and cosine of $-65°, 245°$, and $785°$.

Solution Let $P = (\cos 65°, \sin 65°) = (0.422, 0.906)$ be the point on the unit circle given by the angle $65°$. In Figure 6.84, we see that $-65°$ gives a point labeled Q that is the reflection of P across the x-axis. Thus, the y-coordinate of Q is the negative of the y-coordinate of P, so $Q = (0.422, -0.906)$. This means that

$$\sin(-65°) = -0.906 \quad \text{and} \quad \cos(-65°) = 0.422.$$

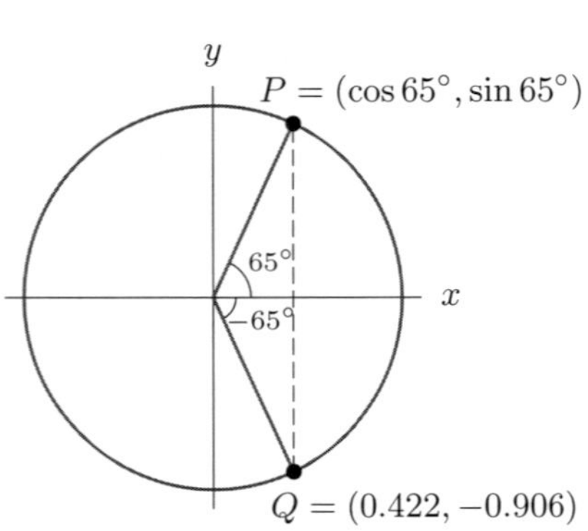

Figure 6.84: The angles $65°$ and $-65°$

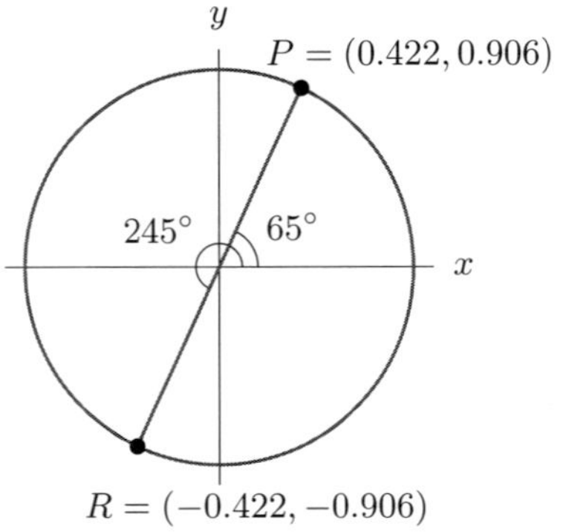

Figure 6.85: The angles $65°$ and $245°$

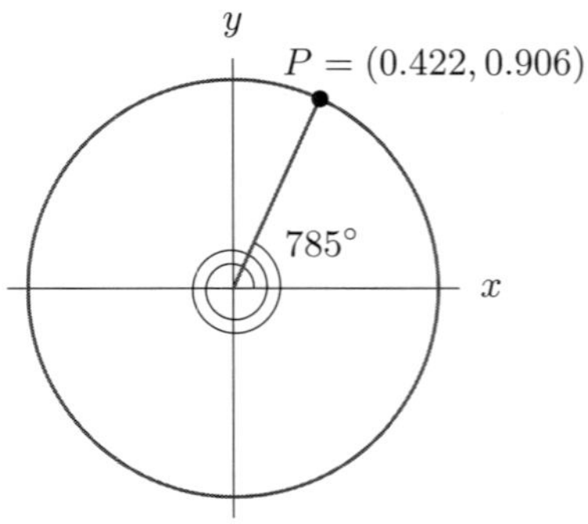

Figure 6.86: The angle $785°$

In Figure 6.85, we see that $245^\circ = 180^\circ + 65^\circ$ gives point R that is diametrically opposite the point P. The coordinates of R are the negatives of the coordinates of P, so $R = (-0.422, -0.906)$. Thus,

$$\sin 245^\circ = -0.906 \quad \text{and} \quad \cos 245^\circ = -0.422.$$

Finally in Figure 6.86, we see that $785^\circ = 720^\circ + 65^\circ$, so this angle specifies the same point as 65°. This means that

$$\sin 785^\circ = 0.906 \quad \text{and} \quad \cos 785^\circ = 0.422.$$

Because we have used the sine and cosine values of 65° to find the sine and cosine values of -65°, 245°, and 785°, we call 65° the *reference angle* for the angles -65°, 245°, and 785°.

> For an angle θ corresponding to the point P on the unit circle, the **reference angle** of θ is the angle between the line joining P to the origin and the nearest part of the x-axis. See Figure 6.87. A reference angle is always between 0° and 90°; that is, between 0 and $\pi/2$.

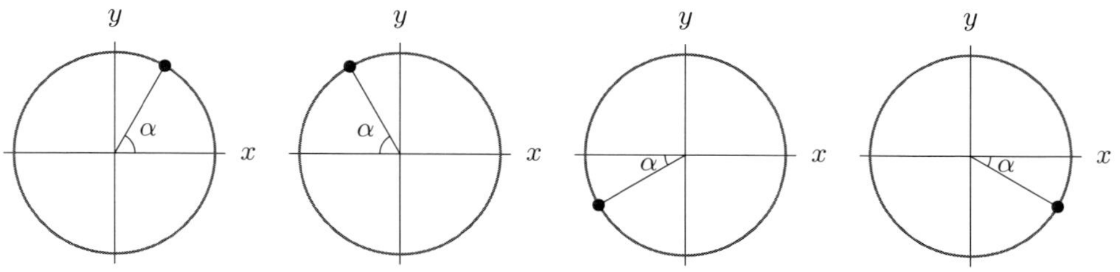

Figure 6.87: In each figure, α is the reference angle for the angle corresponding to P

Example 8 Use reference angles to solve the equations

(a) $\cos\theta = 0.422$ for $0^\circ \le \theta \le 360^\circ$. (b) $\tan\theta = 2.145$ for $0^\circ \le \theta \le 720^\circ$.

Solution

(a) Since $\cos(65^\circ) = 0.422$, a calculator set in degrees gives $\cos^{-1}(0.422) = 65^\circ$. We see in Figure 6.84 that all the angles with a cosine of 0.422 correspond either to the point P or to the point Q. We want solutions between 0° and 360°, so Q is represented by $360^\circ - 65^\circ = 295^\circ$. Thus, the solutions are

$$\theta = 65^\circ \quad \text{and} \quad \theta = 295^\circ.$$

(b) A calculator gives $\tan^{-1}(2.145) = 65^\circ$. Since $\tan\theta$ is positive in the first and third quadrants, the angles with a tangent of 2.145 correspond either to the point P or the point R in Figure 6.85. Since we are interested in solutions between 0° and 720°, the solutions are

$$\theta = 65^\circ, \quad 245^\circ, \quad 65^\circ + 360^\circ, \quad 245^\circ + 360^\circ.$$

That is

$$\theta = 65^\circ, \quad 245^\circ, \quad 425^\circ, \quad 605^\circ.$$

Example 9 Find all solutions of $\cos\theta = -0.422$ for $0° \leq \theta \leq 720°$.

Solution Since $\cos^{-1}(-0.422) = 115°$, the angles we want correspond to the points P or Q in Figure 6.88. Both these points have reference angles of $65°$, so the solutions are

$$\theta = 115° \text{ and } \theta = 180° + 65° = 245°.$$
$$\theta = 360° + 115° = 475° \text{ and } \theta = 360° + 245° = 605°.$$

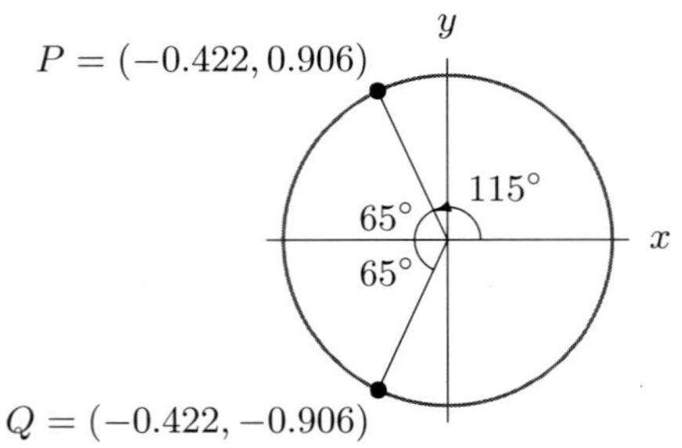

Figure 6.88: Points corresponding to angles with $\cos\theta = -0.422$

Example 10 Use reference angles to solve the equation from Example 6:

$$\sin\left(\frac{\pi}{15}(t-7.5)\right) = \frac{175}{225} \quad \text{for } 0 \leq t \leq 30.$$

Solution As in Example 6, the first solution is given by the arcsine. Solving for t in the equation

$$\frac{\pi}{15}(t-7.5) = \arcsin\frac{175}{225} = 0.891$$

gives $t = 11.75$ minutes. The second solution corresponds to another point on the circle with the same reference angle, 0.891, and a positive value of the sine. This is in the second quadrant, so we have

$$\frac{\pi}{15}(t-7.5) = \pi - 0.891.$$

Solving for t in this equation gives $t = 18.25$ minutes, as in Example 6.

Exercises and Problems for Section 6.7

Exercises

In Exercises 1–6, find a solution with θ in radians (if possible).

1. $\tan\theta = 947$

2. $\sin\theta = 4/7$

3. $3\cos\theta = 0.714$

4. $\cos(\theta - 2) = -50$

5. $\tan(5\theta + 7) = -0.241$

6. $2\sin(4\theta) = 0.667$

In Exercises 7–14, solve the equation exactly for $0 \leq t \leq 2\pi$.

7. $\sin t = -1$

8. $\sin t = 1/2$

9. $\cos t = -1$

10. $\cos t = 1/2$

11. $\tan t = 1$

12. $\tan t = -1$

13. $\tan t = \sqrt{3}$

14. $\tan t = 0$

15. **(a)** Use a graph of $y = \cos t$ to estimate two solutions to the equation $\cos t = -0.3$ for $0 \leq t \leq 2\pi$.
(b) Solve the same equation using the inverse cosine.

16. **(a)** Find exact solutions to the equation $\cos t = 1/2$ with $-2\pi \leq t \leq 2\pi$. Plot them on a graph of $y = \cos t$.
(b) Using a unit circle, explain how many solutions to the equation $\cos t = 1/2$ you would expect in the interval $0 \leq t \leq 2\pi$.

In Exercises 17–24, find the reference angle.

17. $13\pi/6$ **18.** $-7\pi/3$ **19.** $10\pi/3$ **20.** $11\pi/4$

21. $73\pi/3$ **22.** $-46\pi/7$ **23.** 18 **24.** -22

25. Without a calculator, find exact values for

(a) $\cos 120°$ **(b)** $\sin 135°$
(c) $\cos 225°$ **(d)** $\sin 300°$

26. Without a calculator, find exact values for:

(a) $\sin\left(\frac{2\pi}{3}\right)$ **(b)** $\cos\left(\frac{3\pi}{4}\right)$
(c) $\tan\left(-\frac{3\pi}{4}\right)$ **(d)** $\cos\left(\frac{11\pi}{6}\right)$

In Exercises 27–29, use a graph to estimate all the solutions of the equations between 0 and 2π.

27. $\sin\theta = 0.65$ **28.** $\tan x = 2.8$ **29.** $\cos t = -0.24$

Problems

In Problems 30–35, find all solutions to the equation for $0 \leq x \leq 2\pi$.

30. $\cos x = 0.6$ **31.** $2\sin x = 1 - \sin x$
32. $5\cos x = 1/\cos x$ **33.** $\sin 2x = 0.3$
34. $\sin(x - 1) = 0.25$ **35.** $5\cos(x + 3) = 1$

In Problems 36–39, use inverse trigonometric functions to find a solution to the equation in the given interval. Then, use a graph to find all other solutions to the equation on this interval.

36. $\cos x = 0.6, \quad 0 \leq x \leq 4\pi$
37. $\sin x = 0.3, \quad 0 \leq x \leq 2\pi$
38. $\cos x = -0.7, \quad 0 \leq x \leq 2\pi$
39. $\sin x = -0.8, \quad 0 \leq x \leq 4\pi$

Find exact solutions to the equations in Problems 40–42.

40. $\sin\theta = -\sqrt{2}/2$ **41.** $\cos\theta = \sqrt{3}/2$
42. $\tan\theta = -\sqrt{3}/3$

43. Find the angle θ, in radians, in the second quadrant whose tangent is -3.

44. Solve for α exactly: $\sec^2\alpha + 3\tan\alpha = \tan\alpha$ with $0 \leq \alpha < 2\pi$.

Solve the equations in Problems 45–48 for $0 \leq t \leq 2\pi$. First estimate answers from a graph; then find exact answers.

45. $\cos(2t) = \frac{1}{2}$ **46.** $\tan t = \frac{1}{\tan t}$

47. $2\sin t\cos t - \cos t = 0$ **48.** $3\cos^2 t = \sin^2 t$

49. A tree 50 feet tall casts a shadow 60 feet long. Find the angle of elevation θ of the sun.

50. A staircase is to rise 17.3 feet over a horizontal distance of 10 feet. At approximately what angle with respect to the floor should it be built?

51. A company's sales are seasonal with the peak in mid-December and the lowest point in mid-June. The company makes \$100,000 in sales in December, and only \$20,000 in June.

(a) Find a trigonometric function, $s = f(t)$, representing sales at time t months after mid-January.
(b) What would you expect the sales to be for mid-April?
(c) Find the t-values for which $s = 60{,}000$. Interpret your answer.

52. In a tidal river, the time between high tide and low tide is 6.2 hours. At high tide the depth of the water is 17.2 feet, while at low tide the depth is 5.6 feet. Assume the water depth is a trigonometric function of time.

(a) Graph the depth of the water over time if there is a high tide at 12:00 noon. Label your graph, indicating the high and low tide.
(b) Write an equation for the curve you drew in part (a).
(c) A boat requires a depth of 8 feet to sail, and is docked at 12:00 noon. What is the latest time in the afternoon it can set sail? Your answer should be accurate to the nearest minute.

53. Approximate the x-coordinates of points P and Q shown in Figure 6.89, assuming that the curve is a sine curve. [Hint: Find a formula for the curve.]

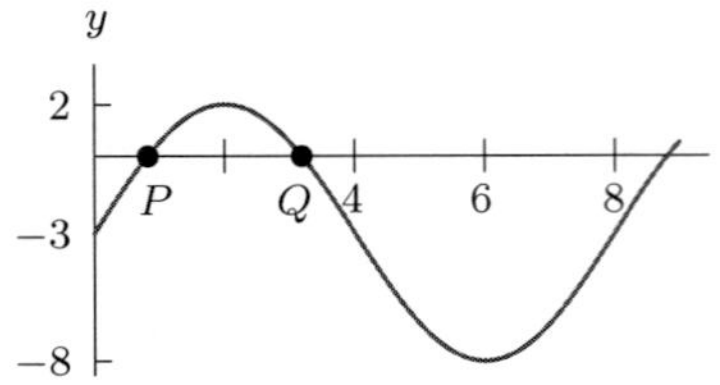

Figure 6.89

54. In your own words, explain what each of the following expressions means. Evaluate each expression for $x = 0.5$. Give an exact answer if possible.

(a) $\sin^{-1} x$ **(b)** $\sin(x^{-1})$ **(c)** $(\sin x)^{-1}$.

55. Evaluate the following expressions in radians. Give an exact answer if possible.

(a) $\arccos(0.5)$ **(b)** $\arccos(-1)$ **(c)** $\arcsin(0.1)$

56. Use a graph to find all the solutions to the equation $12 - 4\cos 3t = 14$ between 0 and $2\pi/3$ (one cycle). How many solutions are there between 0 and 2π?

57. Approximate the zero(s) of $f(t) = 3 - 5\sin(4t)$ for $0 \leq t < \pi/2$.

(a) Graphically. **(b)** Using the arcsine function.

58. State the domain and range of the following functions and explain what your answers mean in terms of evaluating the functions.

(a) $f(x) = \sin^{-1} x$ **(b)** $g(x) = \cos^{-1} x$
(c) $h(x) = \tan^{-1} x$

59. One of the following statements is always true; the other is true for some values of x and not for others. Which is which? Justify your answer with an example.

I. $\arcsin(\sin x) = x$ II. $\sin(\arcsin x) = x$

60. Let k be a positive constant and t be an angle measured in radians. Consider the equation

$$k \sin t = t^2.$$

(a) Explain why any solution to the equation must be between $-\sqrt{k}$ and $\sqrt{k}$, inclusive.
(b) Approximate every solution to the equation when $k = 2$.
(c) Explain why the equation has more solutions for larger values of k than it does for small values.
(d) Approximate the least value of k, if any, for which the equation has a negative solution.

61. You are perched in the crow's nest, C, on top of the mast of a ship, S. See Figure 6.90. You will calculate how far you can see when you are x meters above the surface of the ocean.

(a) Find formulas for d, the distance you can see to the horizon, H, and l, the distance to the horizon along the earth's surface, in terms of x, the height of the ship's mast, and r, the radius of the earth.
(b) How far is the horizon from the top of a 50-meter mast? How far, measured along the earth's surface, is the horizon from the ship's position on the ocean? Use $r = 6{,}370{,}000$ meters.

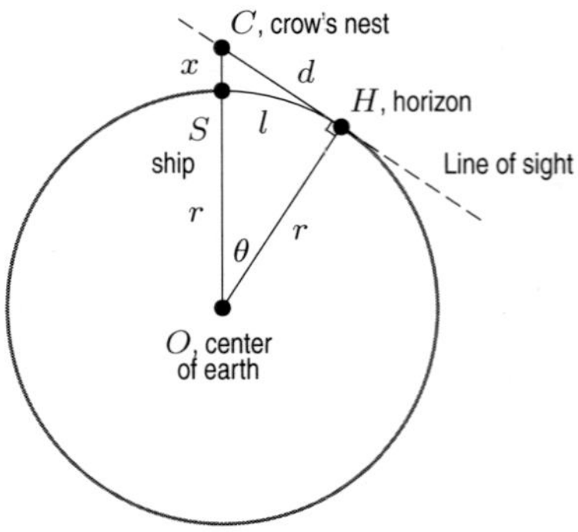

Figure 6.90

62. Let ℓ_1 and ℓ_2 be non-parallel lines of slope m_1 and m_2, respectively, where $m_1 > m_2 > 0$. Find a formula for θ, the angle formed by the intersection of these two lines, in terms of m_1 and m_2. See Figure 6.91.

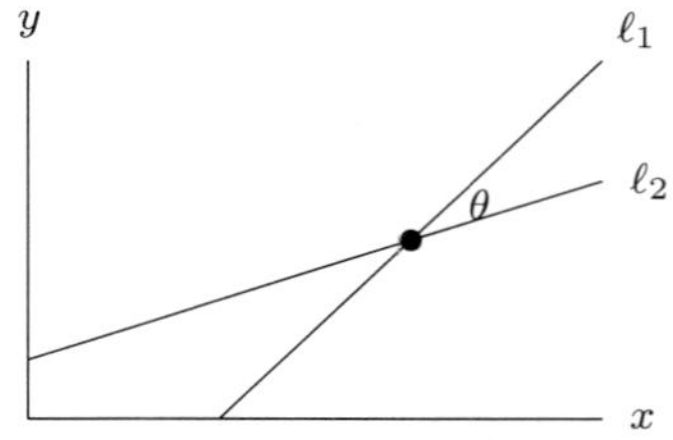

Figure 6.91

CHAPTER SUMMARY

- **Periodic Functions**
  Period: smallest c such that $f(x+c) = f(x)$.

  $\sin(t+2\pi) = \sin t; \cos(t+2\pi) = \cos t; \tan(t+\pi) = \tan t.$

- **Sine and Cosine**
  Angles as rotations correspond to points on circle.
  Coordinates on unit circle: $x = \cos\theta$; $y = \sin\theta$.

- **Radians**
  Definition. Relationship to degrees: 2π radians $= 360°$.
  Arc length: $s = r\theta$.

- **Sinusoidal Functions**
  General formulas

$$y = A\sin(B(t-h)) + k,$$
$$y = A\cos(B(t-h)) + k.$$

  Amplitude $= |A|$; period $= 2\pi/|B|$; midline $y = k$.
  Phase shift $= Bh$; horizontal shift $= h$.

- **Other Trigonometric Functions**
  Tangent, secant, cosecant, and cotangent.

- **Identities:**

$$\tan\theta = \frac{\sin\theta}{\cos\theta};$$
$$\cos^2\theta + \sin^2\theta = 1.$$

- **Inverse Trig Functions**
  Inverse cosine:

$$\cos^{-1} y = t \text{ means } y = \cos t \text{ for } 0 \le t \le \pi.$$

  Inverse sine:

$$\sin^{-1} y = t \text{ means } y = \sin t \text{ for } -\pi/2 \le t \le \pi/2.$$

  Inverse tangent:

$$\tan^{-1} y = t \text{ means } y = \tan t \text{ for } -\pi/2 \le t \le \pi/2.$$

  Reference angles; solving equations.

REVIEW EXERCISES AND PROBLEMS FOR CHAPTER SIX

Exercises

1. The sinusoidal curves in Figure 6.92 model three different traffic patterns: (i) moderately heavy traffic with some slow-downs; (ii) stop-and-go rush hour traffic; and (iii) light, fast-moving traffic. Which is which?

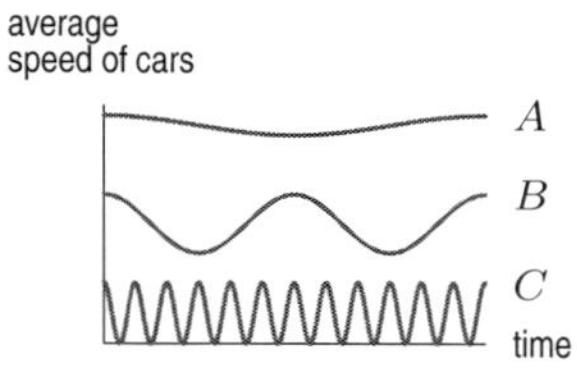

Figure 6.92

2. Find approximations to two decimal places for the coordinates of point Z in Figure 6.93.

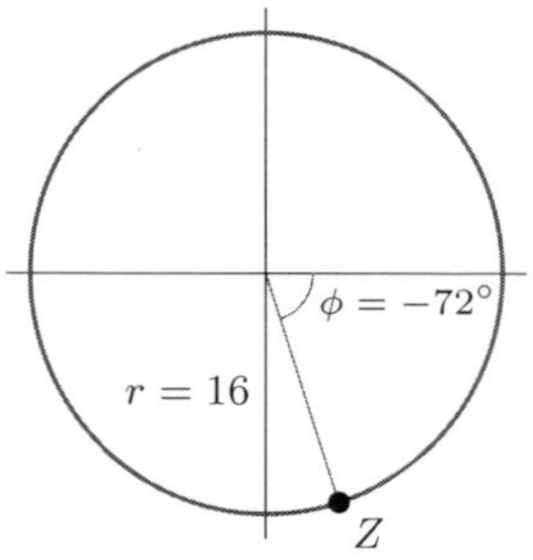

Figure 6.93

3. In which quadrants do the following statements hold?

(a) $\sin\theta > 0$ and $\cos\theta > 0$
(b) $\tan\theta > 0$
(c) $\tan\theta < 0$
(d) $\sin\theta < 0$ and $\cos\theta > 0$
(e) $\cos\theta < 0$ and $\tan\theta > 0$

In Exercises 4–7, convert the angle to radians.

4. $330°$ **5.** $315°$ **6.** $-225°$ **7.** $6\pi°$

In Exercises 8–10, convert the angle from radians to degrees.

8. $\frac{3}{2}\pi$ **9.** 180 **10.** $5\pi/\pi$

In Exercises 11–13, what angle in radians corresponds to the given number of rotations around the unit circle?

11. 4 **12.** -6 **13.** 16.4

14. If you start at the point $(1, 0)$ on the unit circle and travel counterclockwise through the given angle (in radians), in which quadrant will you be?

(a) 2 **(b)** 4 **(c)** 6 **(d)** 1.5 **(e)** 3.2

In Exercises 15–17, find the arc length corresponding to the given angle on a circle of radius 6.2.

15. 17° **16.** -585° **17.** $-\dfrac{360^\circ}{\pi}$

18. Without a calculator, match the graphs in Figure 6.94 to the following functions:

(a) $y = \sin(2t)$ **(b)** $y = (\sin t) + 2$
(c) $y = 2\sin t$ **(d)** $y = \sin(t + 2)$

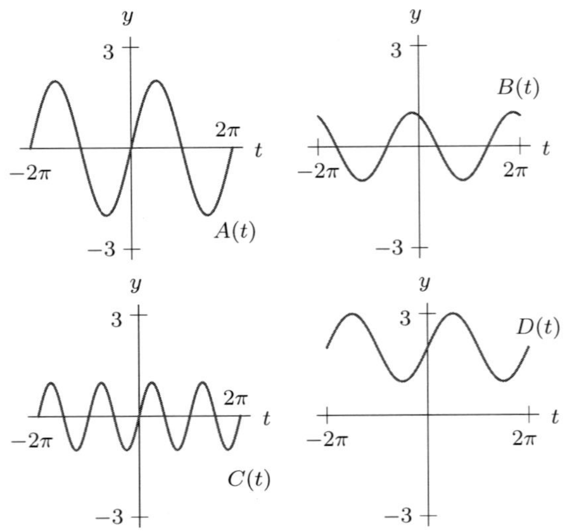

Figure 6.94

In Exercises 19–21, state the period, amplitude, and midline.

19. $y = \cos t + 3$
20. $y = \sin(t + 3) + 7$
21. $6y = 12\sin(\pi t - 7) + 42$

State the amplitude, period, phase shift, and horizontal shifts for the function in Exercises 22–25. Without a calculator, graph the function on the given interval.

22. $y = -4\sin t, \quad -2\pi \le t \le 2\pi$
23. $y = -20\cos(4\pi t), \quad -\frac{3}{4} \le t \le 1$
24. $y = \cos\left(2t + \dfrac{\pi}{2}\right), \quad -\pi \le t \le 2\pi$
25. $y = 3\sin(4\pi t + 6\pi), \quad -\frac{3}{2} \le t \le \frac{1}{2}$

In Exercises 26–29, estimate the amplitude, midline, and period of the sinusoidal functions.

26.

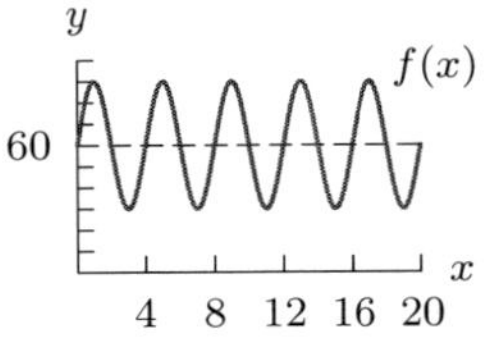

27.

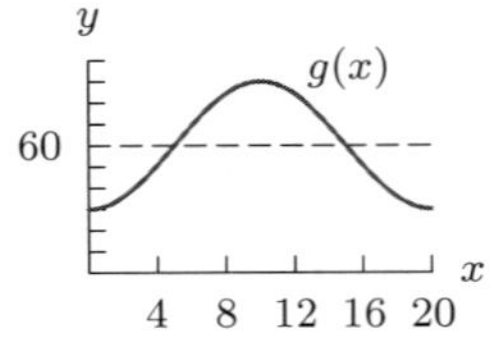

28.

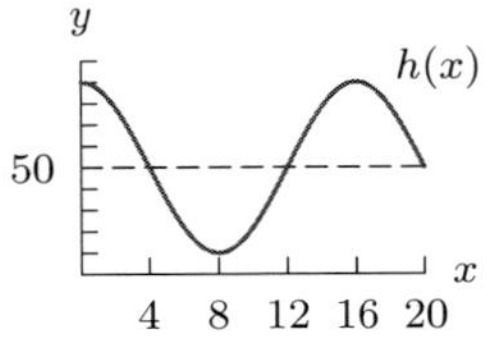

29.

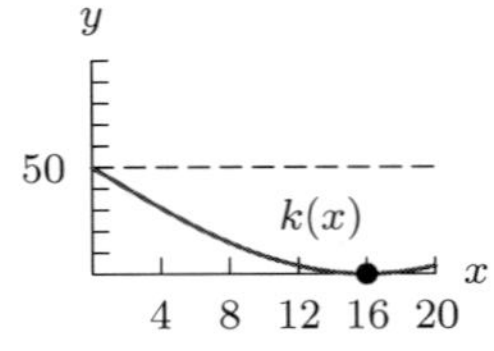

30. Using only vertical shifts, stretches, and flips, and horizontal stretches (but no horizontal shifts or flips), would we start with $y = \sin x$ or $y = \cos x$ if we wanted to find a formula for the functions shown in Problems 26–29?

Without a calculator, graph two periods of the functions in Exercises 31–34.

31. $y = \sin(\frac{1}{2}t)$ **32.** $y = 4\cos(t + \frac{\pi}{4})$
33. $y = 5 - \sin t$ **34.** $y = \cos(2t) + 4$

Problems

In Exercises 35–37, find exact values without a calculator.

35. $\cos 540^\circ$ **36.** $\sin \dfrac{7\pi}{6}$ **37.** $\tan\left(-\dfrac{2\pi}{3}\right)$

38. Find $\tan\theta$ exactly if $\sin\theta = -3/5$, and θ is in the fourth quadrant.

In Exercises 39–40, find a solution with θ in radians (if possible).

39. $\sin\theta = 2/5$ **40.** $\tan(\theta - 1) = 0.17$

41. Find the radian value of x in Figure 6.95.

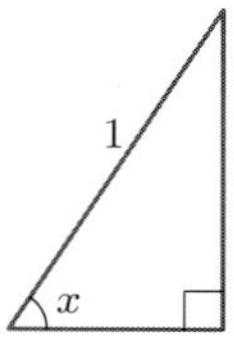

Figure 6.95

42. **(a)** Find a sinusoidal formula for the graph in Figure 6.96.
(b) Find x_1 and x_2.

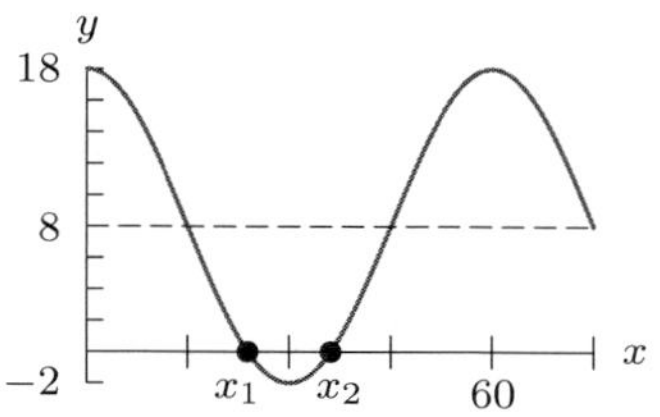

Figure 6.96

43. **(a)** Find a sinusoidal formula for f in Figure 6.97.
(b) Find the x-coordinates of the three indicated points.

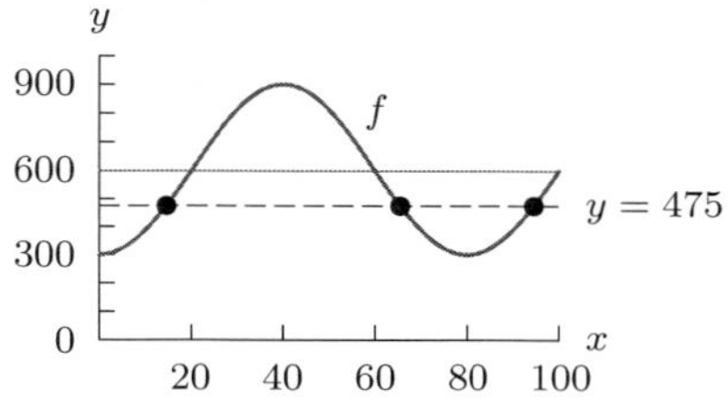

Figure 6.97

For the functions graphed in Problems 44–47, find four possible formulas for the function: $f_1(x) = A\cos(B(x-h))+k$, where $A > 0$, $f_2(x) = A\cos(B(x-h))+k$, where $A < 0$, $f_3(x) = A\sin(B(x-h))+k$, where $A > 0$, and $f_4(x) = A\sin(B(x-h))+k$, where $A < 0$.

44.

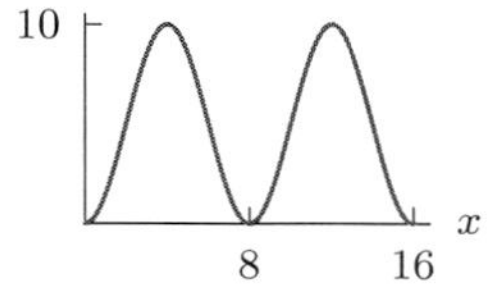

45.

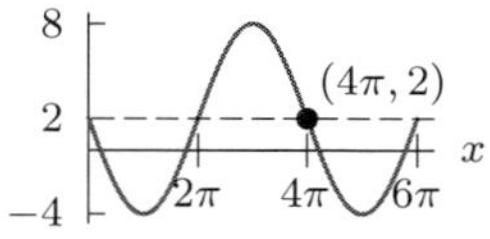

46.

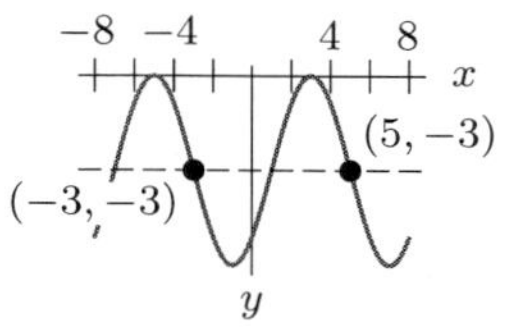

47.

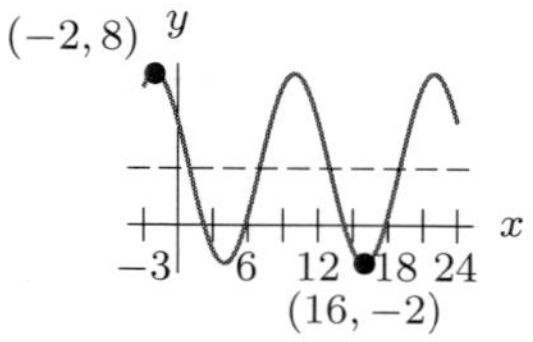

In Problems 48–53, solve the equations with $0 \le \alpha < 2\pi$. Give exact answers if possible.

48. $2\cos\alpha = 1$

49. $\tan\alpha = \sqrt{3} - 2\tan\alpha$

50. $\sin(2\alpha) + 3 = 4$

51. $4\tan\alpha + 3 = 2$

52. $3\sin^2\alpha + 4 = 5$

53. $\tan^2\alpha = 2\tan\alpha$

54. How far does the tip of the minute hand of a clock move in 1 hour and 27 minutes if the hand is 2 inches long?

55. How many miles on the surface of the earth correspond to one degree of latitude? (The earth's radius is 3960 miles.)

56. A person on earth is observing the moon, which is 238,860 miles away. The moon has a diameter of 2160 miles. What is the angle in degrees spanned by the moon in the eye of the beholder?

57. A compact disk is 12 cm in diameter and rotates at 100 rpm (revolutions per minute) when being played. The hole in the center is 1.5 cm in diameter. Find the speed in cm/min of a point on the outer edge of the disk and the speed of a point on the inner edge.

58. Explain in words the difference between the expressions $\sin^2\theta$ and $\sin\theta^2$. Give a specific numerical example illustrating the difference between these expressions.

59. A weather satellite orbits the earth in a circular orbit 500 miles above the earth's surface. What is the radian measure of the angle (measured at the center of the earth) through which the satellite moves in traveling 600 miles along its orbit? (The radius of the earth is 3960 miles.)

60. A weight is suspended from the ceiling by a spring. Figure 6.98 shows a graph of the distance from the ceiling to the weight, $d = f(t)$, as a function of time.

(a) Find a possible formula for $f(t)$.
(b) Solve $f(t) = 12$ exactly. Interpret your results.

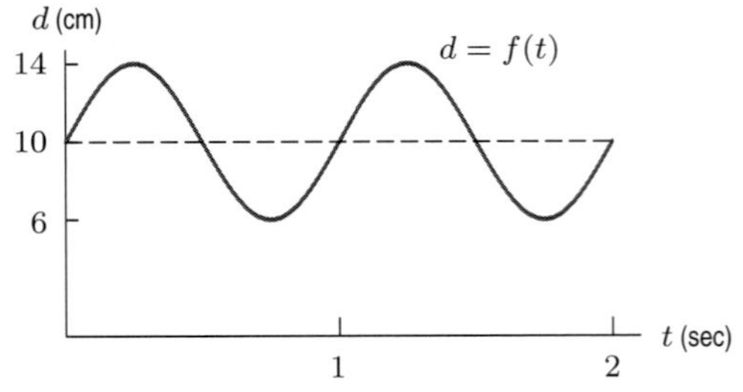

Figure 6.98

61. An animal population increases from a low of 1200 in year $t = 0$, to a high of 3000 four years later and then decreases back to 1200 over the next four years. Model this behavior by a sinusoidal function.

62. Table 6.11 gives the average monthly temperature, y, in degrees Fahrenheit for the city of Fairbanks, Alaska, as a function of t, the month, where $t = 0$ indicates January.

(a) Plot the data points. On the same graph, draw a curve which fits the data.
(b) What kind of function best fits the data? Be specific.
(c) Find a formula for a function, $f(t)$, that models the temperature. [Note: There are many correct answers.]
(d) Use your answer to part (c) to solve $f(t) = 32$. Interpret your results.
(e) Check your results from part (d) graphically.
(f) In the southern hemisphere, the times at which summer and winter occur are reversed, relative to the northern hemisphere. Modify your formula from part (c) so that it represents the average monthly temperature for a southern-hemisphere city whose summer and winter temperatures are similar to Fairbanks.

Table 6.11

t	0	1	2	3	4	5
y	−11.5	−9.5	0.5	18.0	36.9	53.1
t	6	7	8	9	10	11
y	61.3	59.9	48.8	31.7	12.2	−3.3

63. The data in Table 6.12 gives the height above the floor of a weight bobbing on a spring attached to the ceiling. Fit a sine function to this data.

Table 6.12

t, sec	0.0	0.1	0.2	0.3	0.4	0.5
y, cm	120	136	165	180	166	133
t, sec	0.6	0.7	0.8	0.9	1.0	1.1
y, cm	120	135	164	179	165	133

64. In climates that require central heating, the setting of a household thermostat gives the temperature at which the furnace turns on. Water is boiled, and steam is forced into radiators that warm the house. Once the household temperature reaches the thermostat setting, the furnace shuts off. However, the radiators remain hot for some time, and so the household temperature continues rising for a while before it starts to drop. Once it drops back to the thermostat setting, the furnace relights, and the cycle begins anew. As it takes some time for the radiators to reheat, the house continues to cool even after the furnace has turned back on. The household temperature is represented by the graph in Figure 6.99.

(a) The household temperature, T, can be modeled by a trigonometric function. Assume the furnace spends as much time on as it does off. What is the setting on the thermostat?
(b) According to the graph, describe what is happening at $t = 0, 0.25, 0.5, 0.75, 1$.
(c) Give a formula for $T = f(t)$, the temperature in terms of time, t, in hours.
(d) Describe the physical significance of the period, the amplitude, and the midline.
(e) The house described in this problem takes as much time to heat up as it does to cool down. Suppose another house takes 15 minutes to heat up but 45 minutes to cool. Sketch (roughly) its temperature over one hour. Label any significant points (such as times the furnace turns on or off). Is this the graph of a trigonometric function?

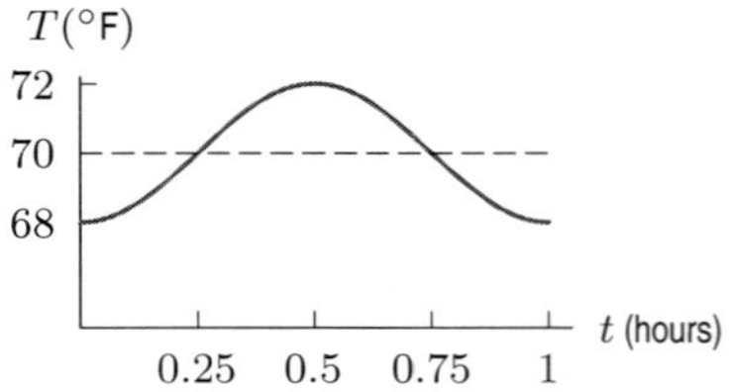

Figure 6.99

CHECK YOUR UNDERSTANDING

Are the statements in Problems 1–11 true or false for all values of x? Give an explanation for your answer.

1. $\sin(-x) = -\sin x$

2. $\cos(-x) = -\cos x$

3. $\sin(-x) = \sin x$

4. $\cos(-x) = \cos x$

5. $\sin(x + \pi) = -\sin x$

6. $\cos(x + \pi) = \sin x$

7. $\cos(x + 4\pi) = \cos x$

8. $2\cos x = \cos(2x)$

9. $\cos\left(\frac{1}{x}\right) = \frac{\cos 1}{\cos x}$

10. $\sec^2 x + 1 = \tan^2 x$.

11. $\cos(x + 1) = \cos x + \cos 1$

Are the statements in Problems 12–87 true or false? Give an explanation for your answer.

12. If $f(t)$ and $g(t)$ are periodic functions with period A and $f(t) = g(t)$ for $0 \le t < A$, then $f(t) = g(t)$ for all t.

13. The function $\sin x$ has period 2π.

14. The function $\sin(\pi x)$ has period π.

15. A parabola is a periodic function.

16. Iff is a periodic function, then there exists a constant c such that $f(x + c) = f(x)$ for all x in the domain of f.

17. The smallest positive constant c for which $f(x + c) = f(x)$ is called the period of f.

18. The amplitude of a periodic function is the difference between its maximum and minimum values.

19. The midline of a periodic function is the horizontal line $y = \dfrac{\text{Maximum} + \text{Minimum}}{2}$.

20. A unit circle may have a radius of 3.

21. The angle $\theta = 180^\circ$ specifies the point $(0, -1)$ on the unit circle.

22. The point $(1, 0)$ on the unit circle corresponds to $\theta = 0^\circ$.

23. If $P = (x, y)$ is a point on the unit circle and θ is the corresponding angle then $\sin\theta = y$.

24. The coordinates of the point of intersection of the terminal ray of a 240° angle with a circle of radius 2 are $(-\sqrt{3}, -1)$.

25. If a point (x, y) is on the circumference of a circle of radius r and the corresponding angle is θ, then $x = r\cos\theta$.

26. In Figure 6.100, the points P and Q on the unit circle correspond to angles with the same cosine values.

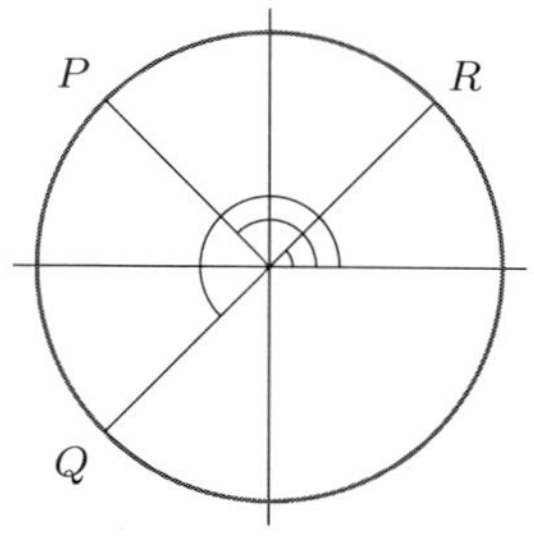

Figure 6.100

27. In Figure 6.100, the points P and R on the unit circle correspond to angles with the same sine values.

28. The point on the unit circle whose coordinates are $(\cos 200^\circ, \sin 200^\circ)$ is in the third quadrant.

29. The point on the unit circle whose coordinates are $(\cos(-200^\circ), \sin(-200^\circ))$ is in the third quadrant.

30. An angle of one radian is about equal to an angle of one degree.

31. The radian measure of an angle is the length of the arc spanned by the angle in a unit circle.

32. An angle of three radians corresponds to a point in the third quadrant.

33. To convert an angle from degrees to radians you multiply the angle by $\dfrac{180^\circ}{\pi}$.

34. The length of an arc s spanned in a circle of radius 3 by an angle of $\dfrac{\pi}{3}$ is 180.

35. An angle of $2\pi/3$ radians corresponds to a point in the second quadrant.

36. In a unit circle, one complete revolution about the circumference is about 6.28 radians.

37. The cosine of 30° is the same as $\sin(\pi/3)$.

38. $\sin(\pi/6) = \sqrt{3}/2$.

39. $\sin(\pi/4) = \cos(\pi/4)$.

40. $-\sin(\pi/3) = \sin(-\pi/3)$.

41. The value of $\cos 315^\circ$ is $-\sqrt{2}/2$.

42. The function in Figure 6.101 appears to be periodic with period less than 5.

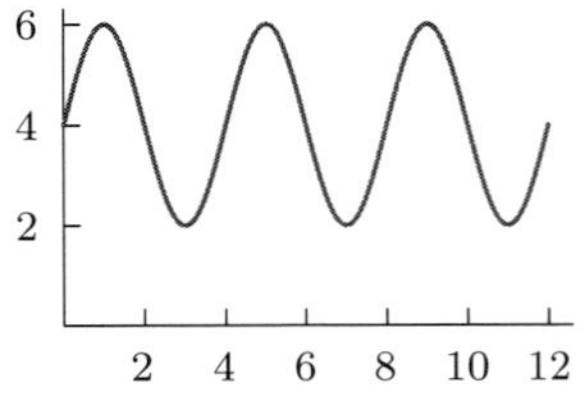

Figure 6.101

43. In Figure 6.101, the amplitude of the function is 6.

44. In Figure 6.101, the period of the function is 4.

45. In Figure 6.101, the midline has equation $y = 4$.

46. In Figure 6.101, the function $g(x) = f(2x)$ has the same period as f.

47. The amplitude of $y = -3\sin(2x) + 4$ is -3.

48. The amplitude of $y = 25 + 10\cos x$ is 25.

49. The period of $y = 25 + 10\cos x$ is 2π.

50. The maximum y-value of $y = 25 + 10\cos x$ is 10.

51. The minimum y-value of $y = 25 + 10\cos x$ is 15.

52. The midline equation for $y = 25 + 10\cos x$ is $y = 35$.

53. The function $\cos x$ is a sinusoidal function.

54. The function $y = -2\sin x + k$ has amplitude -2.

55. The graph of the function $y = 3\cos x - 4$ is the graph of the function $y = \cos x$ reflected across the x-axis.

56. The function $f(t) = \sin(2t)$ has period π.

57. The function $f(x) = \cos(3x)$ has a period three times as large as the function $g(x) = \cos x$.

58. Changing the value of B in the function $y = A\sin(Bx) + k$ changes the period of the function.

59. The graph of $y = A\sin(2x + h) + k$ is the graph of $y = A\sin(2x) + k$ shifted to the left by h units.

60. A sinusoidal function that has a midline of $y = 5$, an amplitude of 3, and completes 4 cycles in the interval $0 \leq x \leq 2\pi$ could have the equation $y = -3\cos(4x) + 5$.

61. The function graphed in Figure 6.102 could have the equation $y = \frac{1}{2}\sin(2x) + 1$.

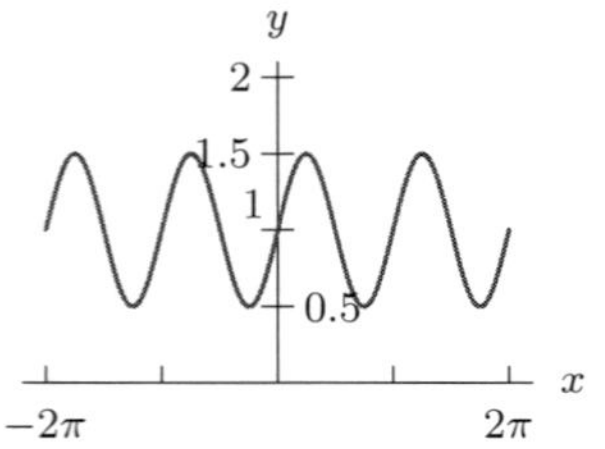

Figure 6.102

62. The function graphed in Figure 6.102 could have the equation $y = -0.5\cos(2x + \frac{\pi}{3}) + 1$.

63. For all values of θ where $\cos\theta \neq 0$, we have $\tan\theta = \dfrac{\sin\theta}{\cos\theta}$.

64. The tangent function is defined everywhere on the interval $0 \leq x \leq 2\pi$.

65. The tangent function has a period of π.

66. The tangent of $\pi/2$ is infinite.

67. For any value x, we have $\sin^2 5x + \cos^2 5x = 1$.

68. If θ is in the second quadrant, then $\tan\theta$ could equal $\frac{3}{4}$.

69. The value of $\sec\pi = -1$.

70. Since the value of $\sin\pi = 0$, the value of $\csc\pi$ is undefined.

71. The reciprocal of the sine function is the cosine function.

72. If $y = \arccos 0.5$, then $y = \frac{\pi}{3}$.

73. If $y = \arctan(-1)$, then $\sin y = -\sqrt{2}/2$.

74. $\sin^{-1}(\sqrt{3}/2) = \pi/3$.

75. $\cos(\cos^{-1}(2/3)) = 2/3$.

76. If $y = \sin^{-1} x$ then $y = \dfrac{1}{\sin x}$.

77. The domain of the inverse cosine is all real numbers.

78. If $\cos t = 1$, then $\tan t = 0$.

79. The reference angle for $120°$ is $30°$.

80. The reference angle for $300°$ is $60°$.

81. If $\arcsin x = 0.5$ then $x = \pi/6$.

82. If $\cos\theta = \dfrac{\sqrt{2}}{2}$ then θ must be $\dfrac{\pi}{4}$.

83. For all angles θ in radians, $\arccos(\cos\theta) = \theta$.

84. For all values of x between -1 and 1, $\cos(\arccos x) = x$.

85. If $\tan A = \tan B$, then $\dfrac{A - B}{\pi}$ is an integer.

86. If $\cos A = \cos B$, then $\sin A = \sin B$.

87. If $\cos A = \cos B$, then $B = A + 2n\pi$ for some integer n.

TOOLS FOR CHAPTER 6: RIGHT TRIANGLES

For most of recorded history, people have used the properties of triangles to make indirect measurements of the world around them. In the fourth century BC, the Greek mathematician Eudoxus used trigonometry (which means, literally, *triangle measurement*) to calculate the radius of the earth. Today scientists and engineers, surveyors, and contractors still use trigonometry.

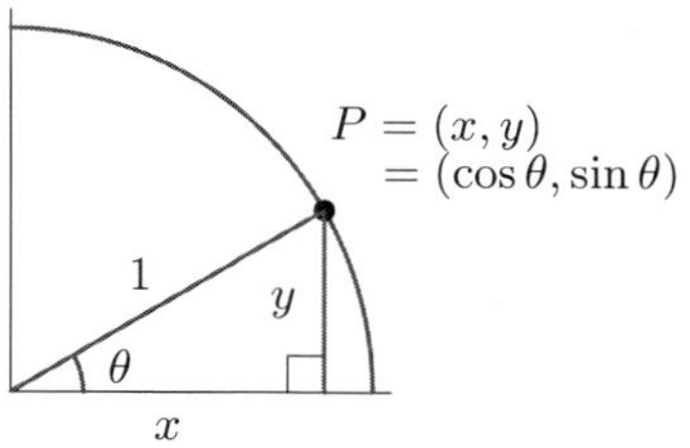

Figure 6.103: A right triangle shown with the unit circle

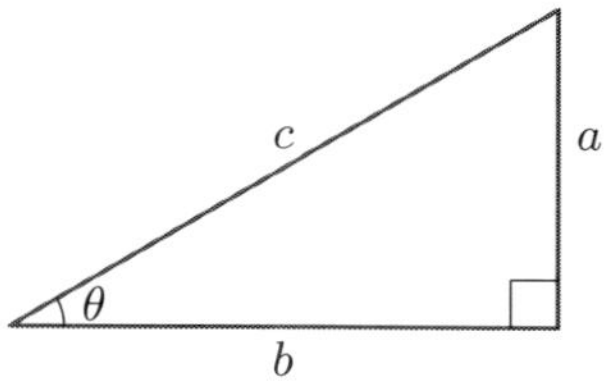

Figure 6.104: A triangle which is similar to the triangle in Figure 6.103

There is a relationship between the trigonometric functions and right triangles. An angle θ determines a point P on the unit circle in Figure 6.103. The angle θ also determines the right triangle with hypotenuse of length 1; the other sides of this triangle have lengths $x = \cos\theta$ and $y = \sin\theta$. The right triangle in Figure 6.104 also has an angle θ, so these two triangles are similar. Therefore, the ratios of the lengths of their corresponding sides are equal:

$$\frac{a}{c} = \frac{\sin\theta}{1} = \sin\theta \qquad \text{and} \qquad \frac{b}{c} = \frac{\cos\theta}{1} = \cos\theta.$$

In addition,

$$\frac{a}{b} = \frac{\sin\theta}{\cos\theta} = \tan\theta.$$

The side directly across from the angle θ is referred to as the *opposite* side, and the other side, which forms one side of the angle θ, is called the *adjacent* side. Using this terminology we have:

> If θ is an angle in a right triangle (other than the right angle),
>
> $$\sin\theta = \frac{\text{Opposite}}{\text{Hypotenuse}}, \qquad \cos\theta = \frac{\text{Adjacent}}{\text{Hypotenuse}}, \qquad \tan\theta = \frac{\text{Opposite}}{\text{Adjacent}}.$$

Special Angles: 30°, 45°, 60°

We use right triangles to calculate exact values of the sine, cosine, and tangent of 30°, 45°, and 60°.

Example 1 Figure 6.105 shows the point $P = (x, y)$ corresponding to the angle 45° on the unit circle. A right triangle has been drawn in. The triangle is isosceles; it has two equal angles (both 45°) and two equal sides, so $x = y$. Therefore, the Pythagorean theorem $x^2 + y^2 = 1$ gives

$$\begin{aligned} x^2 + x^2 &= 1 \\ 2x^2 &= 1 \\ x &= \sqrt{\frac{1}{2}} = \frac{1}{\sqrt{2}}. \end{aligned}$$

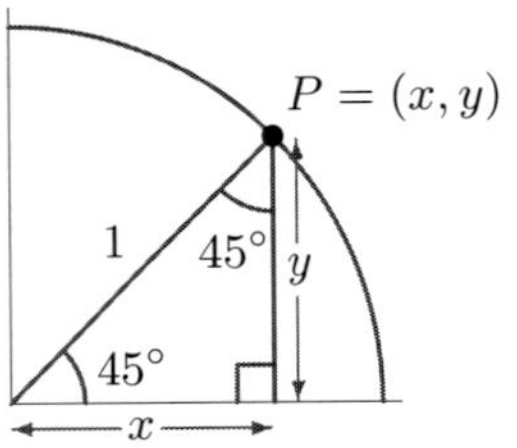

Figure 6.105: This triangle has two equal angles and two equal sides, so $x = y$

We know that x is positive because P is in the first quadrant. Since $x = y$, we see that $y = 1/\sqrt{2}$ as well. Thus, since x and y are the coordinates of P, we have $\cos 45° = 1/\sqrt{2}$, $\sin 45° = 1/\sqrt{2}$, and $\tan 45° = 1$.

Example 2 Figure 6.106 shows the point $Q = (x, y)$ corresponding to the angle 30° on the unit circle. A right triangle has been drawn in, and a mirror image of this triangle is shown below the x-axis. Together these two triangles form the triangle ΔOQA. This triangle has three equal 60° angles and so has three equal sides, each side of length 1. The length of side $\overline{QA}$ can also be written as $2y$, and so we have $2y = 1$, or $y = 1/2$. By the Pythagorean theorem,

$$
\begin{aligned}
x^2 + y^2 &= 1 \\
x^2 + \left(\frac{1}{2}\right)^2 &= 1 \\
x^2 &= \frac{3}{4} \\
x &= \sqrt{\frac{3}{4}} = \frac{\sqrt{3}}{2}.
\end{aligned}
$$

Note that x is positive because Q is in the first quadrant. Since x and y are the coordinates of Q, this means that $\cos 30° = \sqrt{3}/2$, $\sin 30° = 1/2$, and $\tan 30° = 1/\sqrt{3}$.

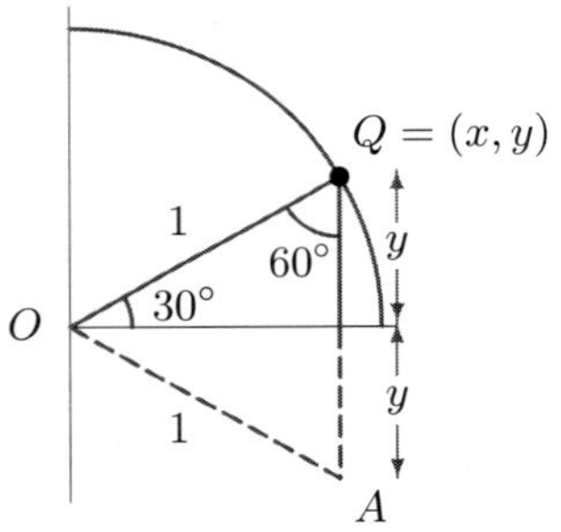

Figure 6.106: The triangle ΔOQA has three equal angles and three equal sides, so $2y = 1$

A similar argument shows that $\cos 60° = 1/2$, $\sin 60° = \sqrt{3}/2$, and $\tan 60° = \sqrt{3}$.

It is worth memorizing the values of sine and cosine for these special angles. See Table 6.13. As the following example shows, special angles and reference angles[8] can be used to calculate exact values of the trigonometric functions of some angles greater than 90° or less than 0°.

Table 6.13 *Trigonometric functions of special angles*

θ	$\cos\theta$	$\sin\theta$	$\tan\theta$
30°	$\sqrt{3}/2$	$1/2$	$1/\sqrt{3}$
45°	$1/\sqrt{2}$	$1/\sqrt{2}$	1
60°	$1/2$	$\sqrt{3}/2$	$\sqrt{3}$

Example 3 Find the exact coordinates of a point B designated by 315° on a circle of radius 6.

Solution Point B is in the fourth quadrant, where $\cos 315°$ is positive and $\sin 315°$ is negative. See Figure 6.107. The reference angle for 315° is 45°, so $\cos 315° = \cos 45°$ and $\sin 315° = -\sin 45°$.

The coordinates of point B are given by

$$x = r\cos\theta = 6\cos 315° = 6\left(\frac{\sqrt{2}}{2}\right) = 3\sqrt{2} \quad \text{and} \quad y = r\sin\theta = 6\sin 315° = 6\left(\frac{-\sqrt{2}}{2}\right) = -3\sqrt{2}$$

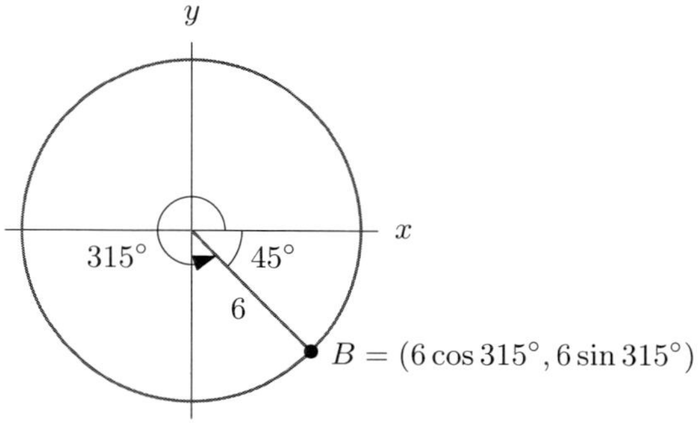

Figure 6.107: Exact coordinates of B found using reference angle of 45°

Thus, the coordinates of B are $(3\sqrt{2}, -3\sqrt{2})$.

[8]See page 291.

Exercises on Tools for Chapter 6

1. Use Figure 6.108 to find the following exactly:

(a) $\tan\theta$ **(b)** $\sin\theta$ **(c)** $\cos\theta$

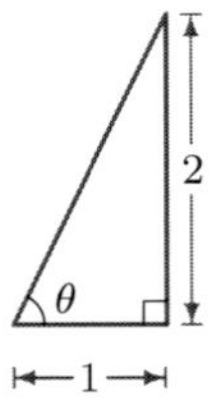

Figure 6.108

2. Find exact values of the sine, cosine and tangent for the angles θ and ϕ in Figure 6.109.

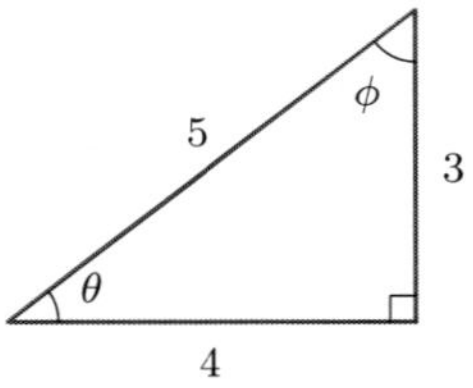

Figure 6.109

3. Using Figure 6.110, find exactly:

(a) $\sin\theta$ **(b)** $\sin\phi$ **(c)** $\cos\theta$
(d) $\cos\phi$ **(e)** $\tan\theta$ **(f)** $\tan\phi$

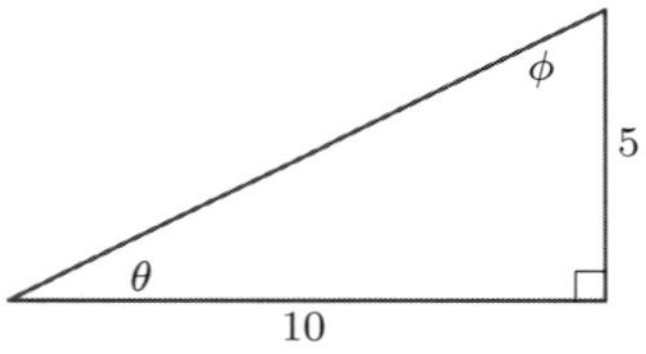

Figure 6.110

4. Find the lengths h and x in Figure 6.111.

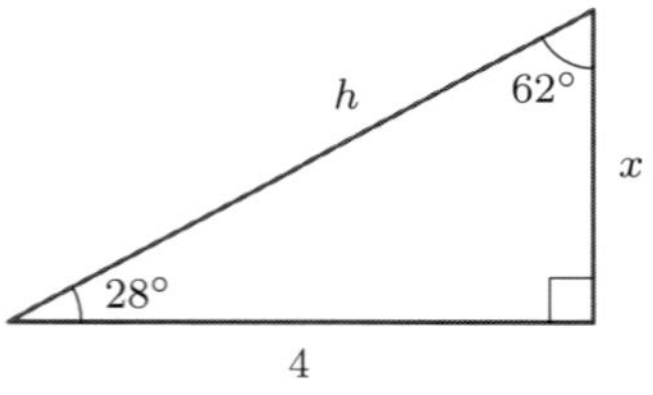

Figure 6.111

In Exercises 5–10, use Figure 6.112 to find exactly

(a) $\sin\theta$ **(b)** $\cos\theta$ **(c)** $\tan\theta$

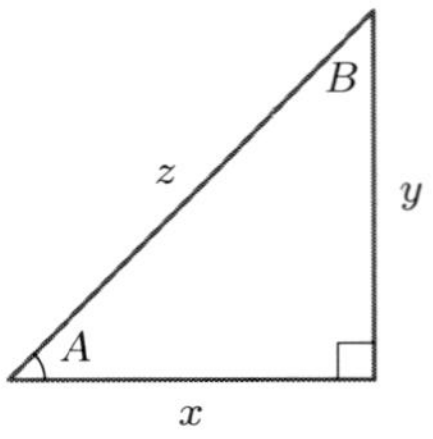

Figure 6.112

5. $x = 2,\ z = 7,\ A = \theta$

6. $x = 9,\ y = 5,\ A = \theta$

7. $y = 8,\ z = 12,\ A = \theta$

8. $z = 17,\ A = \theta,\ B = \theta$

9. $y = 2,\ z = 11,\ B = \theta$

10. $x = a,\ y = b,\ B = \theta$

In Exercises 11–16, use Figure 6.113 to find exact values of q and r.

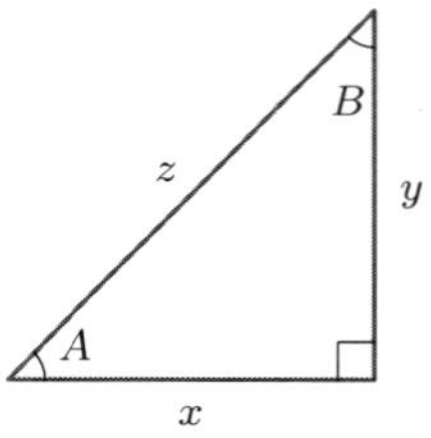

Figure 6.113

11. $A = 17^\circ,\ B = 73^\circ,\ x = q,\ y = r,\ z = 7$

12. $A = 12^\circ,\ B = 78^\circ,\ x = q,\ y = 4,\ z = r$

13. $A = 37^\circ,\ B = 53^\circ,\ x = 6,\ y = q,\ z = r$

14. $A = 40°,\ x = q,\ y = r,\ z = 15$

15. $B = 77°,\ x = 9,\ y = r,\ z = q$

16. $B = 22°,\ x = \lambda,\ y = q,\ z = r$

Find the missing sides and angles in the right triangles in Problems 17–20, where a is the side across from angle A, b across from B, and c across from the right angle.

17. $a = 20, b = 28$

18. $a = 20, c = 28$

19. $c = 20, A = 28°$

20. $a = 20, B = 28°$

21. You have been asked to build a ramp for Dan's Daredevil Motorcycle Jump. The dimensions you are given are indicated in Figure 6.114. Find all the other dimensions.

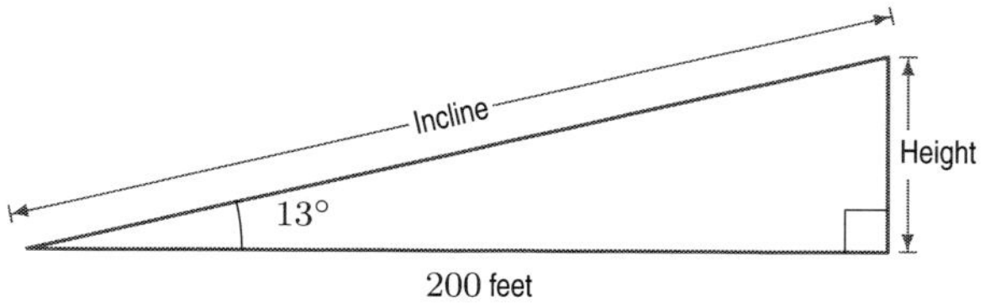

Figure 6.114

22. A kite flier wondered how high her kite was flying. She used a protractor to measure an angle of 38° from level ground to the kite string. If she used a full 100 yard spool of string, how high, in feet, was the kite? (Disregard the string sag and the height of the string reel above the ground.)

23. The top of a 200-foot vertical tower is to be anchored by cables that make an angle of 30° with the ground. How long must the cables be? How far from the base of the tower should anchors be placed?

24. Find approximately the acute angle formed by the line $y = -2x + 5$ and the x-axis.

25. The front door to the student union is 20 feet above the ground, and it is reached by a flight of steps. The school wants to build a wheel-chair ramp, with an incline of 15 degrees, from the ground to the door. How much horizontal distance is needed for the ramp?

26. A ladder 3 meters long leans against a house, making an angle α with the ground. How far is the base of the ladder from the base of the wall, in terms of α? Include a sketch.

27. A plane is flying at an elevation of 35,000 feet when the Gateway Arch in St. Louis, Missouri comes into view. The pilot wants to estimate her horizontal distance from the arch, so she notes the angle of depression, θ, between the horizontal and a line joining her eye to a point on the ground directly below the arch. Make a sketch. Express her horizontal distance to that point as a function of θ.

28. You see a friend, whose height you know is 5 feet 10 inches, some distance away. Using a surveying device called a transit, you determine the angle between the top of your friend's head and ground level to be 8°. You want to find the distance d between you and your friend.

(a) First assume you cannot find your calculator to evaluate trigonometric functions. Find d by approximating the arc length of an 8° angle with your friend's height. (See Figure 6.115.)

(b) Now assume that you have a calculator. Use it to find the distance d. (See Figure 6.116.)

(c) Would the difference between these two values for d increase or decrease if the angle were smaller?

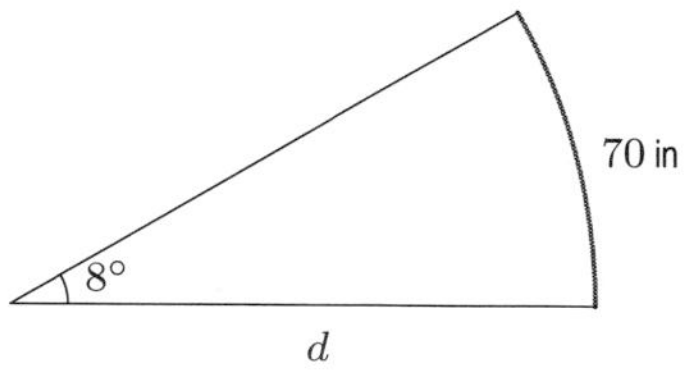

Figure 6.115: Arc of 70 in with 8° angle

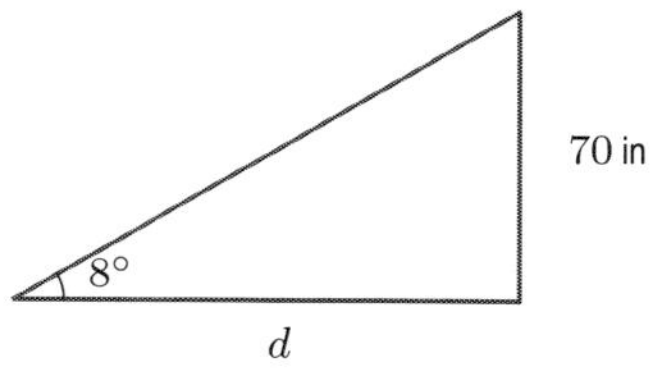

Figure 6.116: Triangle with side 70 in and angle 8°

29. Hampton is a small town on a straight stretch of coast line running north and south. A lighthouse is located 3 miles off-shore directly east of Hampton. The light house has a revolving search light that makes two revolutions per minute. The angle that the beam makes with the east-west line through Hampton is called ϕ. Find the distance from Hampton to the point where the beam strikes the shore, as a function of ϕ. Include a sketch.

30. You are parasailing on a rope that is 125 feet long behind a boat. See Figure 6.117.

(a) At first, you stabilize at a height that forms a 45° angle with the water. What is that height?

(b) After enjoying the scenery, you encounter a strong wind that blows you down to a height that forms a 30° angle with the water. At what height are you now?

(c) Find c and d, i.e. the horizontal distances between you and the boat, in parts (a) and (b).

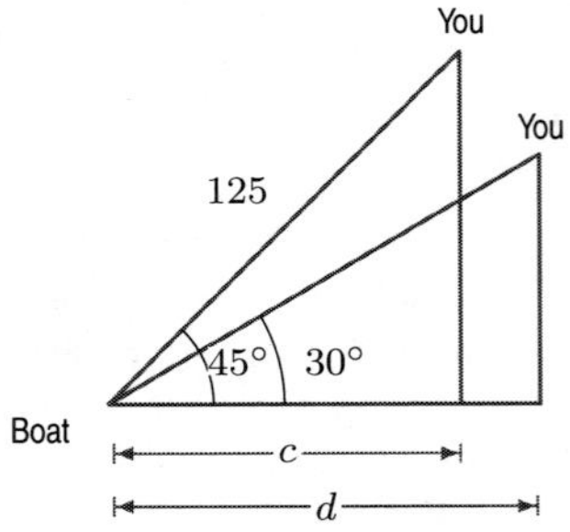

Figure 6.117

31. A bridge over a river was damaged in an earthquake and you are called in to determine the length, d, of the steel beam needed to fill the gap. (See Figure 6.118.) You cannot be on the bridge, but you are able to drop a line from T, the beginning of the bridge, and measure a distance of 50 ft to the point P. From P you find the angles of elevation to the two ends of the gap to be 42° and 35°. How wide is the gap?

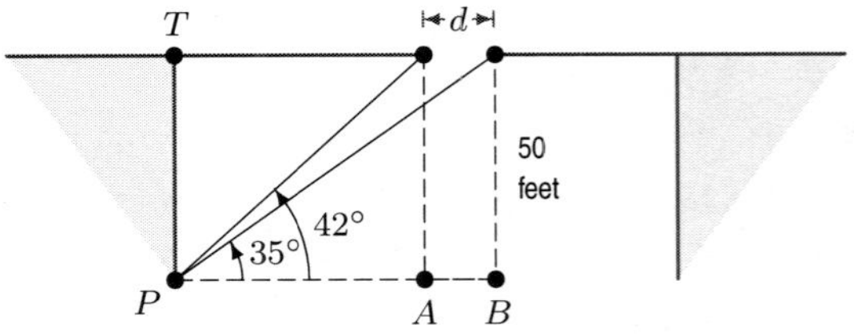

Figure 6.118: Gap in damaged bridge

Chapter Seven

TRIGONOMETRY

Chapter 6 concentrated on the use of trigonometric functions to represent periodic phenomena. This chapter uses trigonometry to analyze triangles.

From the definitions of the trigonometric functions, we derive identities which relate the trigonometric functions to one another. These identities are used to simplify expressions, solve equations, and understand trigonometric graphs. The chapter concludes with applications of trigonometry to mathematical models and to polar coordinates.

7.1 GENERAL TRIANGLES: LAWS OF SINES AND COSINES

Sines and cosines relate the angles of a right triangle to its sides. Similar, although more complicated relationships exist for all triangles, not just right triangles.

The Law of Cosines

The Pythagorean theorem relates the three sides of a right triangle. The *Law of Cosines* relates the three sides of any triangle.

Law of Cosines: For a triangle with sides a, b, c, and angle C opposite side c, we have

$$c^2 = a^2 + b^2 - 2ab\cos C$$

We use Figure 7.1 to derive the Law of Cosines. The dashed line of length h is at right angles to side a and it divides this side into two pieces, one of length x and one of length $a - x$.

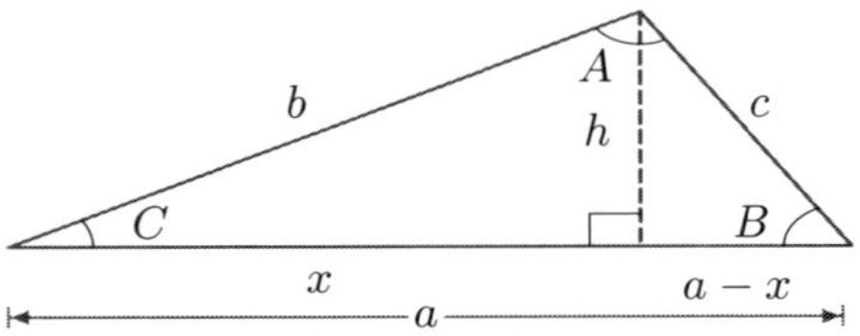

Figure 7.1: Triangle used to derive the Law of Cosines

Applying the Pythagorean theorem to the right-hand right triangle, we get

$$(a - x)^2 + h^2 = c^2$$
$$a^2 - 2ax + x^2 + h^2 = c^2.$$

Applying the Pythagorean theorem to the left-hand triangle, we get $x^2 + h^2 = b^2$. Substituting into the previous equation gives

$$a^2 - 2ax + \underbrace{x^2 + h^2}_{b^2} = c^2$$
$$a^2 + b^2 - 2ax = c^2.$$

But $\cos C = x/b$, so $x = b\cos C$. This gives the Law of Cosines:

$$a^2 + b^2 - 2ab\cos C = c^2.$$

Notice that if C happens to be a right angle, that is, if $C = 90°$, then $\cos C = 0$. In this case, the Law of Cosines reduces to the Pythagorean theorem:

$$a^2 + b^2 - 2ab \cdot 0 = c^2$$
$$a^2 + b^2 = c^2.$$

Therefore, the Law of Cosines is a generalization of the Pythagorean theorem that works for any triangle. Notice also that in Figure 7.1 we assumed that angle C is acute, that is, less than $90°$. Problem 35 concerns the derivation for the case where $90° < C < 180°$.

Example 1 A person leaves her home and walks 5 miles due east and then 3 miles northeast. How far has she walked? How far away from home is she?

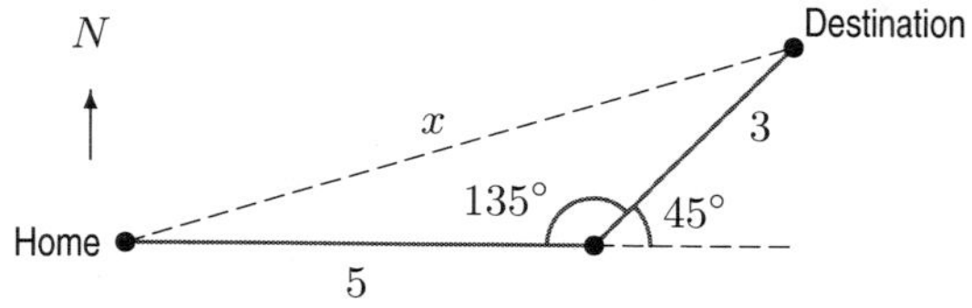

Figure 7.2: A person walks 5 miles east and then 3 miles northeast

Solution She has walked $5 + 3 = 8$ miles. One side of the triangle in Figure 7.2 is 5 miles long, while the second side is 3 miles long and forms an angle of 135° with the first. This is because when the person turns northeast, she turns through an angle of 45°. Thus, we know two sides of this triangle, 5 and 3, and the angle between them, which is 135°. To find the length of the third side, x, we use the Law of Cosines:

$$\begin{aligned} x^2 &= 5^2 + 3^2 - 2 \cdot 5 \cdot 3 \cos 135^\circ \\ &= 34 - 30\left(-\frac{\sqrt{2}}{2}\right) \\ &= 55.213. \end{aligned}$$

This gives $x = \sqrt{55.213} = 7.431$ miles. Notice that this is less than 8 miles, the total distance walked.

In the previous example, two sides of a triangle and the angle between them were known. The Law of Cosines is also useful if all three sides of a triangle are known.

Example 2 At what angle must the person from Example 1 walk to go directly home?

Solution According to Figure 7.3, if the person faces due west and then turns south through an angle of θ, she heads directly home. This same angle θ is opposite the side of length 3 in the triangle. The Law of Cosines tells us that

$$\begin{aligned} 5^2 + 7.431^2 - 2 \cdot 5 \cdot 7.431 \cos\theta &= 3^2 \\ -74.31 \cos\theta &= -71.220 \\ \cos\theta &= 0.958 \\ \theta &= \cos^{-1}(0.958) = 16.582^\circ. \end{aligned}$$

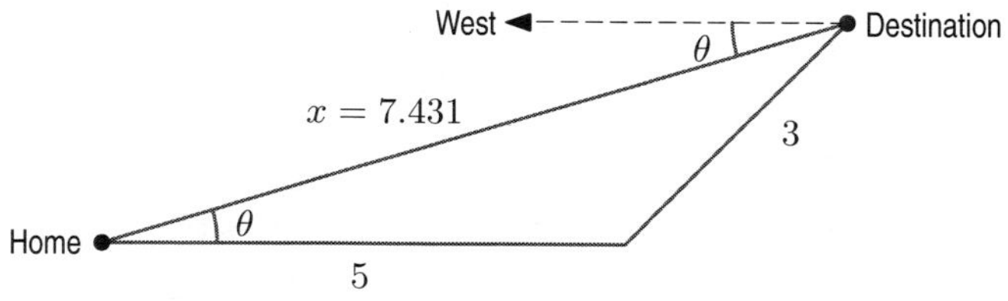

Figure 7.3: The person faces at an angle θ south of west to head home

In Examples 1 and 2, notice that we used the Law of Cosines in two different ways for the same triangle.

The Law of Sines

Figure 7.4 shows the same triangle as in Figure 7.1. Since $\sin C = h/b$ and $\sin B = h/c$, we have $h = b\sin C$ and $h = c\sin B$. This means that

$$b\sin C = c\sin B$$

so

$$\frac{\sin B}{b} = \frac{\sin C}{c}.$$

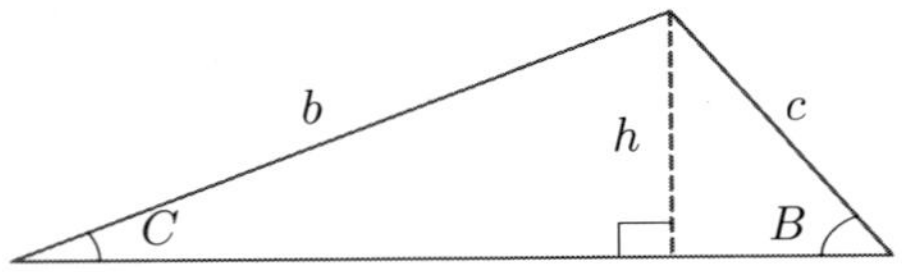

Figure 7.4: Triangle used to derive the Law of Sines

A similar type of argument (see Problem 36 on page 312) shows that

$$a\sin B = b\sin A,$$

which leads to the Law of Sines:

> **Law of Sines:** For a triangle with sides $a,\ b,\ c$ opposite angles $A,\ B,\ C$ respectively:
>
> $$\frac{\sin A}{a} = \frac{\sin B}{b} = \frac{\sin C}{c}.$$

The Law of Sines is useful when we know a side and the angle opposite it.

Example 3 An aerial tram starts at a point one half mile from the base of a mountain whose face has a 60° angle of elevation. (See Figure 7.5.) The tram ascends at an angle of 20°. What is the length of the cable from T to A?

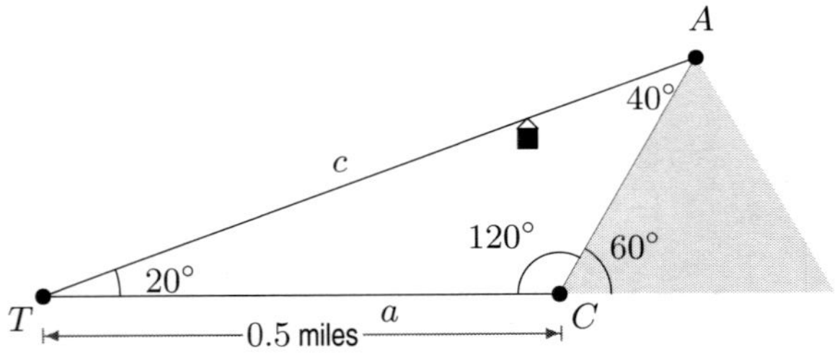

Figure 7.5

Solution The Law of Cosines does not help us here because we only know the length of one side of the triangle. We do however know two angles in this diagram. Thus, we can use the Law of Sines:

$$\frac{\sin A}{a} = \frac{\sin C}{c}$$

$$\frac{\sin 40^\circ}{0.5} = \frac{\sin 120^\circ}{c}$$

so $c = 0.5\left(\dfrac{\sin 120^\circ}{\sin 40^\circ}\right) = 0.674$. Therefore, the cable from T to A is 0.674 miles.

The Ambiguous Case

There is a drawback to using the Law of Sines for finding angles. The problem is that the Law of Sines does not tell us the angle, but only its sine, and there are two angles between $0°$ and $180°$ with a given sine. For example, if the sine of an angle is $1/2$, the angle may be either $30°$ or $150°$.

Example 4 Solve the following triangles for θ and ϕ.

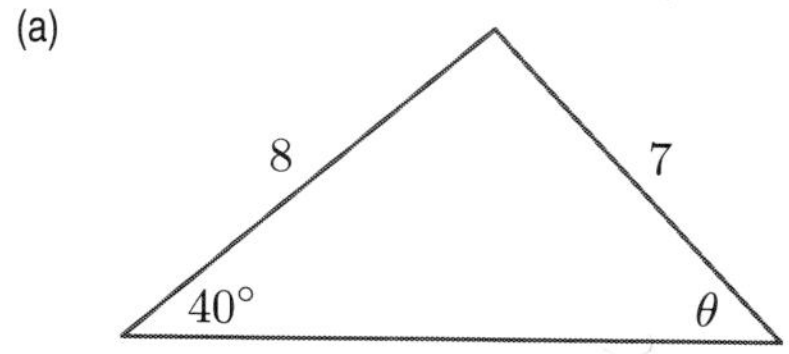

Figure 7.6

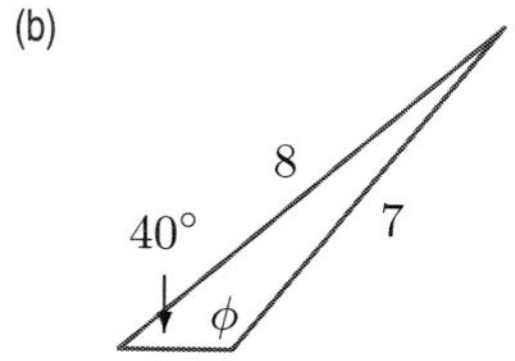

Figure 7.7

Solution (a) Using the Law of Sines in Figure 7.6, we have

$$\frac{\sin\theta}{8} = \frac{\sin 40°}{7}$$
$$\sin\theta = \frac{8}{7}\sin 40° = 0.735$$
$$\theta = \sin^{-1}(0.735) \approx 47.275°.$$

(b) From Figure 7.7, we get

$$\frac{\sin\phi}{8} = \frac{\sin 40°}{7}$$
$$\sin\phi = \frac{8}{7}\sin 40° = 0.735$$

This is the same equation we had for θ in part (a). However, judging from the figures, ϕ is not equal to θ. Knowing the sine of an angle is not enough to tell us the angle. In fact, there are *two* angles between $0°$ and $180°$ whose sine is 0.735. One of them is $\theta = \sin^{-1}(0.735) = 47.275°$, and the other is $\phi = 180° - \theta = 132.725°$.

Exercises and Problems for Section 7.1

Exercises

In Exercises 1–19, use Figure 7.8 to find the missing sides, a, b, c, and angles, A, B, C (if possible). If there are two solutions, find both.

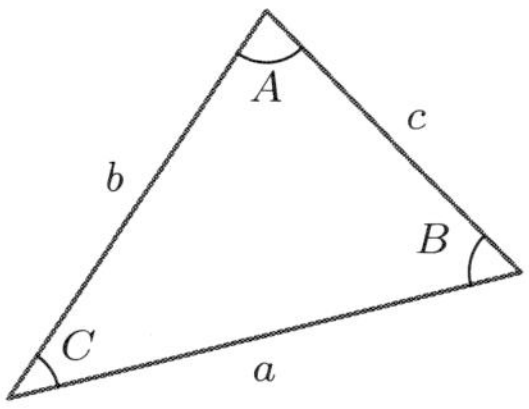

Figure 7.8

1. $a = 20, b = 28, c = 41$
2. $a = 14, b = 12, C = 23°$
3. $a = 20, B = 81°, c = 28$
4. $a = 20, b = 28, C = 12°$
5. $a = 9,\ b = 8,\ C = 80°$
6. $a = 8, b = 11, C = 114°$.
7. $a = 5, b = 11$, and $C = 32°$.
8. $A = 13°,\ B = 25°, c = 4$
9. $A = 105°,\ B = 9°,\ c = 15$
10. $A = 95°,\ B = 22°,\ c = 7$
11. $A = 77°,\ B = 42°,\ c = 9$
12. $a = 8$, and $C = 98°, c = 17$
13. $A = 12°,\ C = 150°,\ c = 5$
14. $A = 92°,\ C = 35°,\ c = 9$
15. $A = 5°,\ C = 9°,\ c = 3$
16. $B = 95°,\ b = 5,\ c = 10$
17. $B = 72°,\ b = 13,\ c = 4$
18. $B = 75°,\ b = 7,\ c = 2$
19. $B = 17°,\ b = 5,\ c = 8$

Find all sides and angles of the triangles in Exercises 20–23. Sketch each triangle. If there is more than one possible triangle, solve and sketch both. Note that α is the angle opposite side a, and β is the angle opposite side b, and γ is the angle opposite side c.

20. $a = 18.7$ cm, $c = 21.0$ cm, $\beta = 22^\circ$

21. $a = 2.00$ m, $\alpha = 25.80^\circ$, $\beta = 10.50^\circ$

22. $b = 510.0$ ft, $c = 259.0$ ft, $\gamma = 30.0^\circ$

23. $a = 16.0$ m, $b = 24.0$ m, $c = 20.0$ m

Problems

24. **(a)** Find an expression for $\sin\theta$ in Figure 7.9 and $\sin\phi$ in Figure 7.10.
(b) Explain how you can find θ and ϕ by using the inverse sine function. In what way does the method used for θ differ from the method used for ϕ?

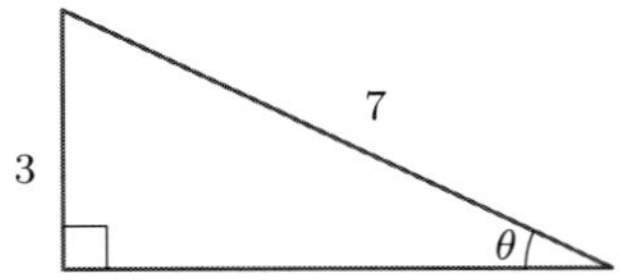

Figure 7.9

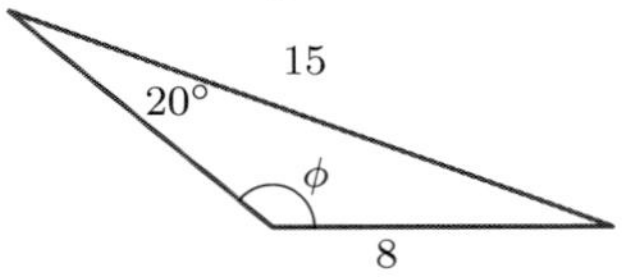

Figure 7.10

25. In Figure 7.11: **(a)** Find $\sin\theta$ **(b)** Solve for θ
(c) Find the area of the triangle.

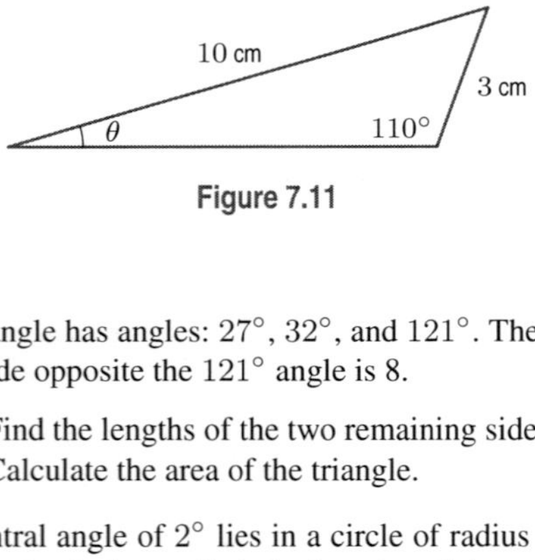

Figure 7.11

26. A triangle has angles: 27°, 32°, and 121°. The length of the side opposite the 121° angle is 8.

(a) Find the lengths of the two remaining sides.
(b) Calculate the area of the triangle.

27. A central angle of 2° lies in a circle of radius 5 feet. To six decimal places, find the lengths of the arc and the chord determined by this angle.

28. Repeat Problem 27 for an angle of 30°.

29. Two fire stations are located 56.7 miles apart, at points A and B. There is a forest fire at point C. If $\angle CAB = 54^\circ$ and $\angle CBA = 58^\circ$, which fire station is closer? How much closer?

30. To measure the height of the Eiffel Tower in Paris, a person stands away from the base and measures the angle of elevation to the top of the tower to be 60°. Moving 210 feet closer, the angle of elevation to the top of the tower is 70°. How tall is the Eiffel Tower?[1]

31. Two airplanes leave Kennedy airport in New York at 11 am. The air traffic controller reports that they are traveling away from each other at an angle of 103°. The DC-10 travels 509 mph and the L-1011 travels at 503 mph.[2] How far apart are they at 11:30 am?

32. A parcel of land is in the shape of an isosceles triangle. The base has length 425 feet; the other sides, which are of equal length, meet at an angle of 39°. How long are they?

33. In video games, images are drawn on the screen using xy-coordinates. The origin, $(0, 0)$, is the lower-left corner of the screen. An image of an animated character moves from its position at $(8, 5)$ through a distance of 12 units along a line at an angle of $25^\circ$to the horizontal. What are its new coordinates?

34. A computer-generated image begins at screen coordinates $(5, 3)$. The image is rotated counterclockwise about the origin through an angle of 42°. Find the new coordinates of the image.

35. Derive the Law of Cosines assuming the angle C is obtuse, as in Figure 7.12.

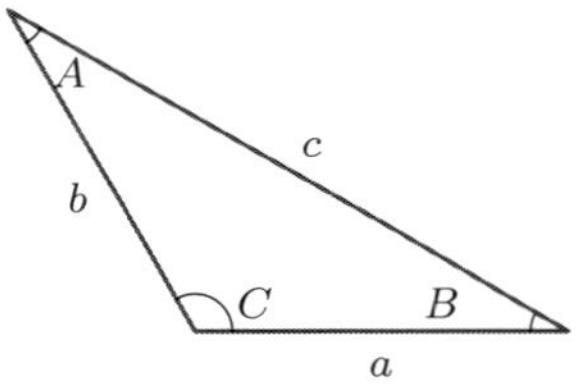

Figure 7.12

36. Use Figure 7.13 to show that $\dfrac{\sin A}{a} = \dfrac{\sin B}{b}$.

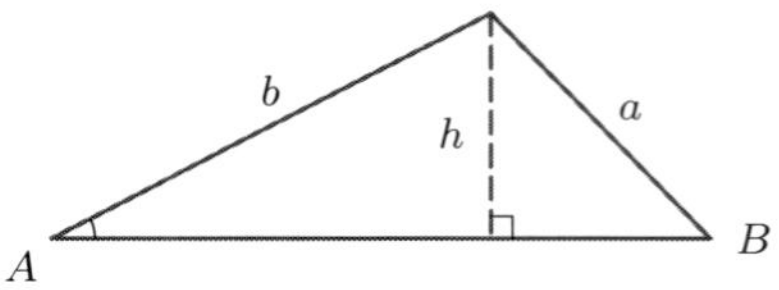

Figure 7.13

[1]The World Almanac Book of Facts, 1999 p. 629
[2]The World Almanac Book of Facts, 2002 p. 217

37. In baseball, the four bases form a square. See Figure 7.14.[3]

(a) The pitcher is standing on the pitching rubber and runners are coming to both first and second bases. Which is the shorter throw? How much shorter?

(b) A ball is hit from home plate to a point 30 feet past second base. An outfielder comes in to catch the ball. How far is the throw to home plate? To third base?

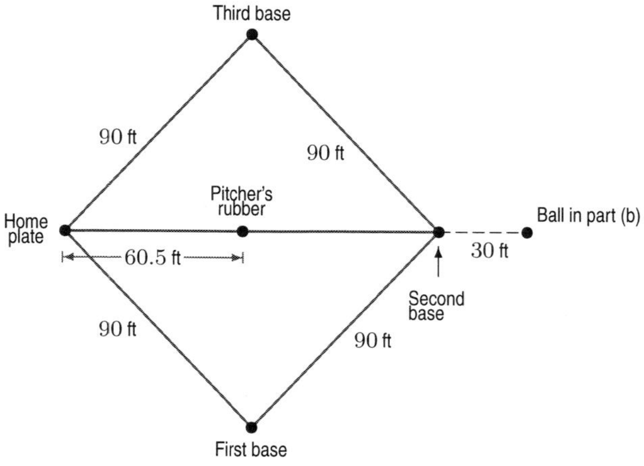

Figure 7.14

38. A park director wants to build a bridge across a river to a bird sanctuary. He hires a surveyor to determine the length of the bridge, represented by AB in Figure 7.15. The surveyor places a transit (an instrument for measuring vertical and horizontal angles) at point A and measures angle BAC to be $93°$. The surveyor then moves the transit 102 feet to point C and measures angle BCA to be $49°$. How long should the bridge be?

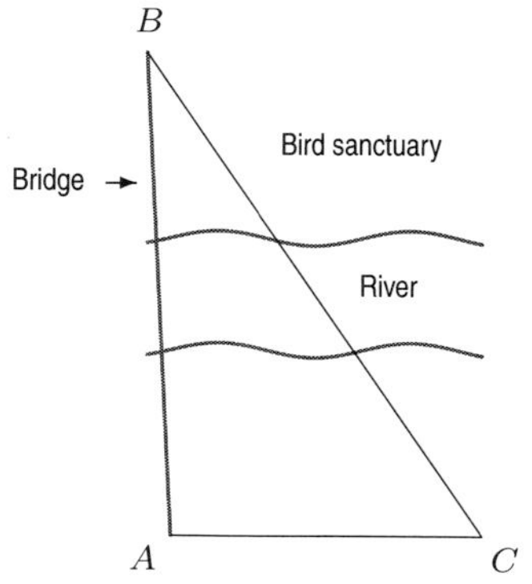

Figure 7.15

39. To estimate the width of an archaeological mound, archaeologists place two stakes on opposite ends of the widest point. See Figure 7.16. They set a third stake 82 feet from one stake and 97 feet from the other stake. The angle formed is $125°$. Find the width of the mound.

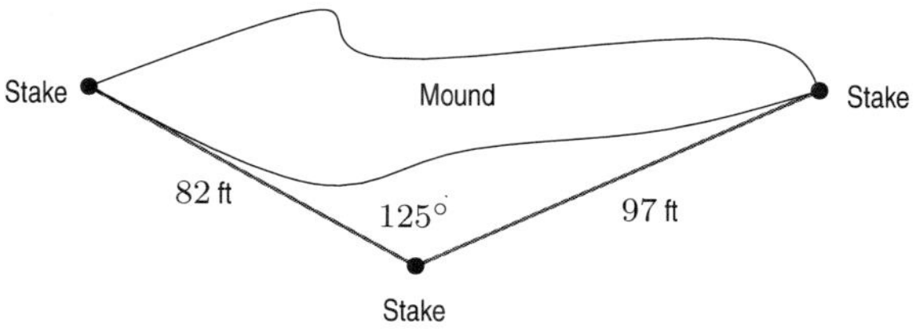

Figure 7.16

40. Every triangle has three sides and three angles. Make a chart showing the set of all possible triangle configurations where three of these six measures are known, and the other three measures can be deduced (or partially deduced) from the three known measures.

7.2 TRIGONOMETRIC IDENTITIES

Equations Versus Identities

When we solve an equation, we find a value of the variable that makes the equation true. For example, the equation

$$2(x-1)=x$$

is true for $x=2$, but is not true for any other value of x.

Now consider the equation

$$2(x-1)=2x-2.$$

This equation is true for all values of x: the left and right sides always have the same value. An equation that is true for all values of x is called an *identity*.

[3]www.majorleaguebaseball.com

We can get an idea whether an equation is an identity by graphing the functions defined by each side of the equation. The equation $2(x-1) = x$ is not an identity because the graphs of $f(x) = 2(x-1)$ and $g(x) = x$ are not identical. On the other hand, the graphs of $f(x) = 2(x-1)$ and $g(X) = 2x - 2$ look identical. This suggests that the functions produce the same output for any input x and that $2(x-1) = 2x - 2$ is an identity. However graphs may look identical even if they are not. To be sure that an equation is an identity, we use algebra.

Example 1 Which of the following is an identity?

(a) $\cos t = 1/2$ (b) $\cos(t + 2\pi) = \cos t$ (c) $\sin(t + 4\pi) = \sin t$

Solution

(a) This is not an identity because it is not true for all values of t. For example, $\cos 0 \neq 1/2$.
(b) This is an identity because the cosine is periodic with period 2π.
(c) This is an identity because the sine is periodic with period 2π.

The Tangent and Pythagorean Identities

There are many identities involving trigonometric functions. In Section 6.6 we saw that

$$\boxed{\tan\theta = \frac{\sin\theta}{\cos\theta}.}$$

This identity holds whenever $\cos\theta$ is not zero. We also derived the *Pythagorean identity*,

$$\boxed{\sin^2\theta + \cos^2\theta = 1.}$$

(Try graphing the function $f(\theta) = \sin^2\theta + \cos^2\theta$ to visualize this identity.) The Pythagorean identity is often used in one of two alternate forms,

$$\sin^2\theta = 1 - \cos^2\theta \quad \text{or} \quad \cos^2\theta = 1 - \sin^2\theta.$$

Why We Need Identities

Identities are useful because they allow us to replace one expression by a more convenient one. For example, a useful strategy in solving trigonometric equations is to rewrite the equation in terms of a single trigonometric function. The next two examples illustrate this technique.

Example 2 Solve $2\sin\theta = \sqrt{2}\cos\theta$, for $0 \le \theta \le 2\pi$.

Solution We divide both sides by $\cos\theta$, giving

$$2\frac{\sin\theta}{\cos\theta} = \sqrt{2}.$$

Replacing $\dfrac{\sin\theta}{\cos\theta}$ by $\tan\theta$, we get

$$2\tan\theta = \sqrt{2}$$
$$\tan\theta = \frac{\sqrt{2}}{2}.$$

One solution is $\theta = \tan^{-1}(\sqrt{2}/2) = 0.615$. The tangent function has period π, so a second solution is $\theta = \pi + 0.615 = 3.757$. We cannot divide both sides of an equation by 0, so dividing by $\cos\theta$ is not valid if $\cos\theta = 0$. We must check separately whether values of θ that make $\cos\theta = 0$, namely $\theta = \pi/2$ and $\theta = 3\pi/2$, are solutions. Neither of these values satisfies the original equation.

Example 3 Solve $3\sin^2 t = 5 - 5\cos t$, for $0 \le t \le \pi$.

Solution We can get approximate solutions to the equation by graphing $y = 3(\sin t)^2$ and $y = 5 - 5\cos t$ and finding the points of intersection for $0 \le t \le \pi$. To solve this equation algebraically, we want to write the equation entirely in terms of $\sin t$ or entirely in terms of $\cos t$. Since $\sin^2 t = 1 - \cos^2 t$, we convert to $\cos t$, giving

$$\begin{aligned} 3(1 - \cos^2 t) &= 5 - 5\cos t \\ 3 - 3\cos^2 t &= 5 - 5\cos t \\ 3\cos^2 t - 5\cos t + 2 &= 0. \end{aligned}$$

The left side factors:

$$(3\cos t - 2)(\cos t - 1) = 0,$$

so

$$3\cos t - 2 = 0 \quad \text{or} \quad \cos t - 1 = 0.$$

This gives $\cos t = 1$, with solution $t = 0$, and $\cos t = 2/3$, with solution $t = \cos^{-1}(2/3) = 0.841$. Note that both solutions fall in the interval $0 \le t \le \pi$.

Double-Angle Formula for Sine

We now find a formula for $\sin 2\theta$ in terms of $\sin\theta$ and $\cos\theta$. First, note that $\sin 2\theta$ is not the same as $2\sin\theta$! For one thing, the graph of $y = \sin 2\theta$ has a different period and amplitude from the graph of $y = 2\sin\theta$. A graph suggests that $\sin 2\theta$ and $2\sin\theta\cos\theta$ could be the same function. In fact, they are, as we now show.

We derive our formula by using Figure 7.17. The right triangles OAB and OCB each have an angle θ at 0. The lengths of OA and OC are 1; the length of AC is $2\sin\theta$. Writing α for the angle at A and applying the Law of Sines to triangle OAC gives

$$\frac{\sin 2\theta}{2\sin\theta} = \frac{\sin\alpha}{1}.$$

In triangle OAB, the length of side OB is $\cos\theta$, and the hypotenuse is 1, so

$$\sin\alpha = \frac{\text{Opposite}}{\text{Hypotenuse}} = \frac{\cos\theta}{1} = \cos\theta.$$

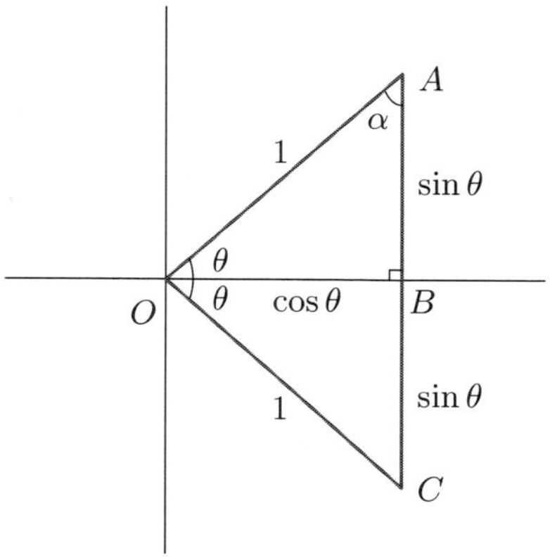

Figure 7.17: Triangle used to derive the double angle formula for sine

Thus, substituting $\cos\theta$ for $\sin\alpha$, we have

$$\frac{\sin 2\theta}{2\sin\theta} = \cos\theta$$

so

$$\boxed{\sin 2\theta = 2\sin\theta\cos\theta.}$$

This identity is known as the *double-angle formula for sine*. We have shown the double-angle formula is true for $0 \le \theta \le \pi/2$. Problems 50 and 51 extend the result to all θ.

Example 4 Find all solutions to $\sin 2t = \sin t$, for $0 \le t \le 2\pi$.

Solution Using the double-angle formula $\sin 2t = 2\sin t\cos t$, we have

$$\begin{aligned} 2\sin t\cos t &= \sin t \\ 2\sin t\cos t - \sin t &= 0 \\ \sin t(2\cos t - 1) &= 0 \quad \text{Factoring out } \sin t. \end{aligned}$$

Thus,

$$\sin t = 0 \quad \text{or} \quad 2\cos t - 1 = 0.$$

Now, $\sin t = 0$ for $t = 0$, π, and 2π. We solve the second equation by writing

$$\begin{aligned} 2\cos t - 1 &= 0 \\ \cos t &= \frac{1}{2}. \end{aligned}$$

We know that $\cos t = 1/2$ for $t = \pi/3$ and $t = 5\pi/3$. Thus there are five solutions to the original equation on the interval $0 \le t \le 2\pi$: $t = 0, \pi/3, \pi, 5\pi/3$, and 2π. Figure 7.18 illustrates these solutions graphically as the points where the graphs of $\sin 2t$ and $\sin t$ intersect.

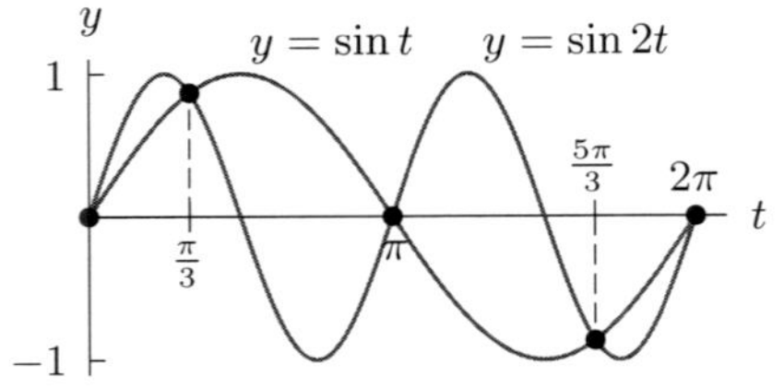

Figure 7.18: There are five solutions to the equation $\sin 2t = \sin t$, for $0 \le t \le 2\pi$

Double-Angle Formulas for Cosine and Tangent

There is also a double-angle formula for cosine, which we derive by applying the Law of Cosines to triangle OAC in Figure 7.17. Side AC has length $2\sin\theta$, so

$$\begin{aligned} (2\sin\theta)^2 &= 1^2 + 1^2 - 2\cdot 1\cdot 1\cos 2\theta \\ 4\sin^2\theta &= 2 - 2\cos 2\theta. \end{aligned}$$

Solving for $\cos 2\theta$ gives

$$\boxed{\cos 2\theta = 1 - 2\sin^2\theta.}$$

Problems 52 and 53 extend the result to all θ.

This is the double-angle formula for cosine. In Problem 12, we use the Pythagorean identity to write the formula for $\cos 2t$ in two other ways:

$$\boxed{\cos 2t = 2\cos^2 t - 1} \quad \text{and} \quad \boxed{\cos 2t = \cos^2 t - \sin^2 t.}$$

Finally, Problem 13 asks you to derive the double-angle formula for tangent:

$$\boxed{\tan 2t = \frac{2\tan t}{1 - \tan^2 t}.}$$

Other Identities

Because the cosine is an even function and the sine and tangent are odd functions, we have

$$\cos(-t) = \cos t.$$
$$\sin(-t) = -\sin t \quad \text{and} \quad \tan(-t) = -\tan t.$$

Several other identities are suggested by graphing. In Figure 7.19 we see that shifting the graph of $y = \cos t$ to the right by $\pi/2$ units gives the graph of $y = \sin t$. This suggests, as is shown in Section 7.3, that

$$\sin t = \cos\left(t - \frac{\pi}{2}\right) \qquad \text{for all } t.$$

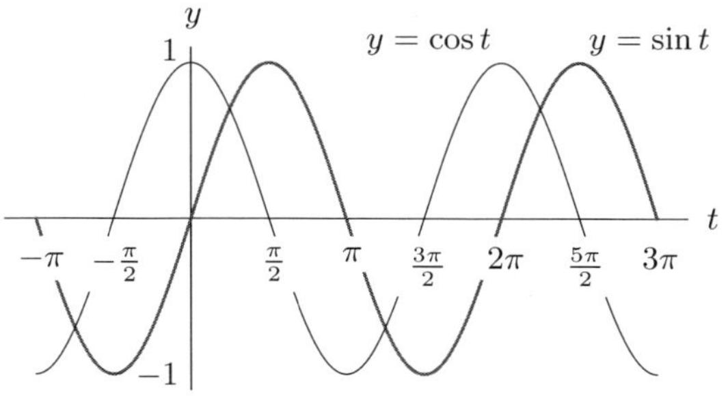

Figure 7.19: The graph of $y = \sin t$ can be obtained by shifting the graph of $y = \cos t$ to the right by $\pi/2$

Similarly, shifting the graph of $y = \sin t$ to the left by $\pi/2$ gives the graph of $y = \cos t$, so

$$\cos t = \sin\left(t + \frac{\pi}{2}\right) \qquad \text{for all } t.$$

There is a summary of the identities in this section inside the cover of the book.

Solving Trigonometric Equations

In Examples 2 and 3 on page 314, we solved equations involving more than one trigonometric function. If the equation involves more than one angle, we can use identities to write all the functions in terms of a single angle.

Example 5 Solve $\sin 2t = \sqrt{3}\sin(t + \pi/2)$ for $0 \le t < 2\pi$.

Solution Use the double-angle formula for sine to replace $\sin 2t$ by $2\sin t\cos t$. We also replace $\sin(t+\pi/2)$ by $\cos t$, giving

$$2\sin t\cos t = \sqrt{3}\cos t.$$

Now all of the trigonometric functions are functions of the same argument, t. Next we write the equation as

$$2\sin t\cos t - \sqrt{3}\cos t = 0$$

and factor the left side to get

$$\cos t(2\sin t - \sqrt{3}) = 0,$$

so

$$\cos t = 0 \quad \text{or} \quad 2\sin t - \sqrt{3} = 0$$

Now each equation involves only one trigonometric function. The solutions of $\cos t = 0$ are $t = \pi/2$ and $t = 3\pi/2$. The solutions of the second equation are $t = \sin^{-1}(\sqrt{3}/2) = \pi/3$ and $t = 2\pi/3$.

Exercises and Problems for Section 7.2

Exercises

Simplify the expressions in Exercises 1–8.

1. $\tan x\cos x$

2. $\cos^2(2\theta) + \sin^2(2\theta)$

3. $\dfrac{\sin 2\alpha}{\cos\alpha}$

4. $\dfrac{\cos^2\theta - 1}{\sin\theta}$

5. $\dfrac{\cos 2t}{\cos t + \sin t}$

6. $\dfrac{1}{1-\sin\theta} + \dfrac{1}{1+\sin\theta}$

7. $\dfrac{\cos\phi - 1}{\sin\phi} + \dfrac{\sin\phi}{\cos\phi + 1}$

8. $\dfrac{1}{\sin t\cos t} - \dfrac{1}{\tan t}$

9. Complete the following table, using exact values where possible. State the trigonometric identities that relate the quantities in the table.

θ (radians)	$\sin^2\theta$	$\cos^2\theta$	$\sin 2\theta$	$\cos 2\theta$
1				
$\pi/2$				
2				
$5\pi/6$				

10. Using the Pythagorean identity, give an expression for $\sin\theta$ in terms of $\cos\theta$.

11. Use graphs to check that $\cos 2t$ and $1 - 2\sin^2 t$ have the same sign for all values of t.

12. Use the Pythagorean identity to write the double angle formula for cosine in these two alternate forms:

(a) $\cos 2t = 2(\cos t)^2 - 1$
(b) $\cos 2t = (\cos t)^2 - (\sin t)^2$

13. Use $\tan 2t = \dfrac{\sin 2t}{\cos 2t}$ to derive a double angle formula for tangent. [Hint: Use Problem 12 (b).]

For Problems 14–16, use algebra to prove the identity.

14. $\dfrac{\sin t}{1-\cos t} = \dfrac{1+\cos t}{\sin t}$

15. $\dfrac{\cos x}{1-\sin x} - \tan x = \dfrac{1}{\cos x}$

16. $\dfrac{\sin x\cos y + \cos x\sin y}{\cos x\cos y - \sin x\sin y} = \dfrac{\tan x + \tan y}{1 - \tan x\tan y}$

Use trigonometric identities to solve each of the trigonometric equations in Exercises 17–20 exactly for $0 \le \theta \le 2\pi$.

17. $\sin^2\theta - \cos^2\theta = \sin\theta$

18. $\sin(2\theta) - \cos\theta = 0$

19. $\sec^2\theta = 1 - \tan\theta$

20. $\tan(2\theta) + \tan\theta = 0$

Problems

Problems 21–23 refer to the unit circle in Figure 7.20. The point P corresponds to an angle of t radians.

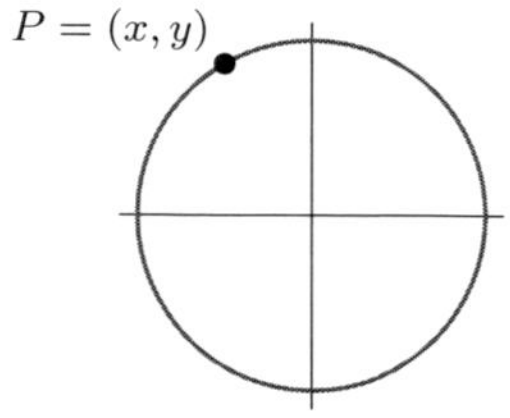

Figure 7.20

21. **(a)** What are the coordinates of the point Q corresponding to angle $t + 2\pi$?
(b) Show that $\cos(t + 2\pi) = \cos t$.
(c) Show that $\sin(t + 2\pi) = \sin t$.

22. **(a)** What are the coordinates of the point Q corresponding to angle $t + \pi$?
(b) Show that $\cos(t + \pi) = -\cos t$.
(c) Show that $\sin(t + \pi) = -\sin t$.
(d) Show that $\tan(t + \pi) = \tan t$.

23. **(a)** What are the coordinates of the point Q corresponding to angle $-t$?
(b) Show that $\cos(-t) = \cos t$.
(c) Show that $\sin(-t) = -\sin t$
(d) Show that $\tan(-t) = -\tan t$.

24. Use graphs to find five pairs of expressions which appear to be identically equal.

(a) $2\cos^2 t + \sin t + 1$ **(b)** $\cos^2 t$
(c) $\cos(2t)$ **(d)** $1 - 2\sin^2 t$
(e) $2\sin t\cos t$ **(f)** $\cos^2 t - \sin^2 t$
(g) $\sin(2t)$ **(h)** $\sin(3t)$
(i) $-2\sin^2 t + \sin t + 3$ **(j)** $\sin(2t)\cos t + \cos(2t)\sin t$
(k) $\dfrac{1 - \sin t}{\sin t}$ **(l)** $\dfrac{1 + \cos(2t)}{2}$
(m) $\dfrac{\left(\dfrac{1}{\cos t} - 1\right)}{1 - \cos t}$

For Problems 25–40, use a graph to decide whether the equation is an identity. If the equation is an identity, prove it algebraically. If it is not an identity, find a value of x for which the equation is false.

25. $\dfrac{1}{2 + x} = \dfrac{1}{2} + \dfrac{1}{x}$

26. $\sin\left(\dfrac{1}{x}\right) = \dfrac{1}{\sin x}$

27. $\sqrt{64 - x^2} = 8 - x$

28. $\sin(2x) = 2\sin x$

29. $\cos(x^2) = (\cos x)^2$

30. $\sin(x^2) = 2\sin x$

31. $\tan x = \dfrac{\sin(2x)}{1 + \cos(2x)}$

32. $\dfrac{\sin(2A)}{\cos(2A)} = 2\tan A$

33. $\dfrac{\sin^2\theta - 1}{\cos\theta} = -\cos\theta$

34. $\cos\left(x + \dfrac{\pi}{3}\right) = \cos x + \cos\dfrac{\pi}{3}$

35. $\sin x \tan x = \dfrac{1 - (\cos x)^2}{\cos x}$

36. $\tan t + \dfrac{1}{\tan t} = \dfrac{1}{\sin t\cos t}$

37. $\sin\left(\dfrac{1}{x}\right) = \sin 1 - \sin x$

38. $\sin(2x) = \dfrac{2\tan x}{1 + (\tan x)^2}$

39. $\dfrac{\sin\theta}{\cos\theta} - \dfrac{\cos\theta}{\sin\theta} = \dfrac{\cos(2\theta)}{\sin(2\theta)}$

40. $\cos(2x) = \dfrac{1 - (\tan x)^2}{1 + (\tan x)^2}$

41. Find exactly all solutions to the equations.

(a) $\cos 2\theta + \cos\theta = 0, \quad 0 \le \theta < 360°$
(b) $2\cos^2\theta = 3\sin\theta + 3, \quad 0 \le \theta \le 2\pi$

42. Use Figure 7.21 to express the following in terms of θ.

(a) y **(b)** $\cos\varphi$
(c) $1 + y^2$ **(d)** The triangle's area

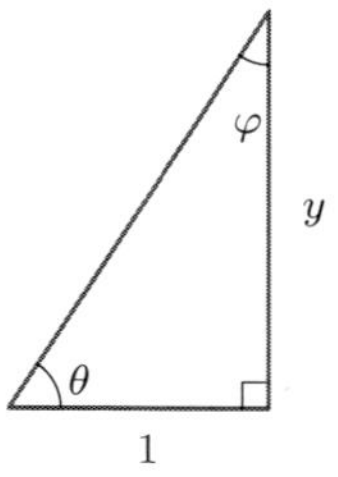

Figure 7.21

43. Let y be the side opposite to the angle θ in a right triangle whose hypotenuse is 1. (See Figure 7.22.) Without trigonometric functions, express the following expressions in terms of y.

(a) $\cos\theta$ **(b)** $\tan\theta$ **(c)** $\cos(2\theta)$
(d) $\sin(\pi-\theta)$ **(e)** $\sin^2(\cos^{-1}(y))$

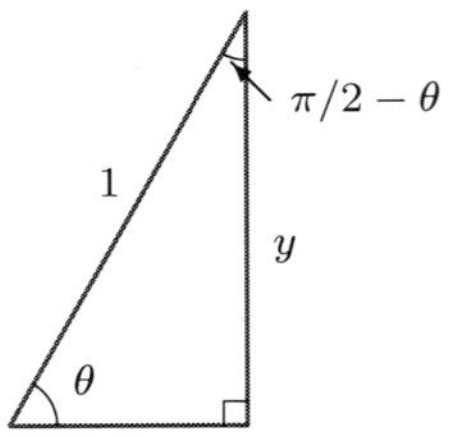

Figure 7.22

44. Suppose that $\sin\theta = 3/5$ and θ is in the second quadrant. Find $\sin(2\theta)$, $\cos(2\theta)$, and $\tan(2\theta)$ exactly.

45. If $x = 3\cos\theta$, $0 < \theta < \pi/2$, express $\sin(2\theta)$ in terms of x.

46. If $x+1 = 5\sin\theta$, $0 < \theta < \pi/2$, express $\cos(2\theta)$ in terms of x.

47. Express in terms of x without trigonometric functions. [Hint: Let $\theta = \cos^{-1} x$ in part (a).]

(a) $\tan(2\cos^{-1} x)$ **(b)** $\sin(2\tan^{-1} x)$

48. Find an identity for $\cos(4\theta)$ in terms of $\cos\theta$. (You need not simplify your answer.)

49. Use trigonometric identities to find an identity for $\sin(4\theta)$ in terms of $\sin\theta$ and $\cos\theta$.

50. In the text we showed that $\sin 2t = 2\sin t\cos t$ for $0 < t < \pi/2$ and substitution for t shows that it is also true for $t = 0$ and $t = \pi/2$. In this problem assume that $\pi/2 < t \le \pi$.

(a) Show that $\sin 2(\pi - t) = 2\sin(\pi-t)\cos(\pi-t)$.
(b) Use the periodicity and the oddness of the sine function to show that $\sin 2(\pi-t) = -\sin 2t$.
(c) Use Problem 22 to show that $\cos(\pi-t) = -\cos t$ and $\sin(\pi-t) = \sin t$.
(d) Show that $\sin 2t = 2\sin t\cos t$.

51. In the text and Problem 50 we showed that $\sin 2t = 2\sin t\cos t$ for $0 \le t \le \pi$. In this problem assume that $-\pi \le t < 0$. Show that

(a) $\sin(-2t) = 2\sin(-t)\cos(-t)$.
(b) $\sin 2t = 2\sin t\cos t$.

52. In the text we showed that $\cos 2t = 1 - 2\sin^2 t$ for $0 < t < \pi/2$ and substitution for t shows that it is also true for $t = 0$ and $t = \pi/2$. In this problem assume that $\pi/2 < t \le \pi$.

(a) Show that $\cos 2(\pi-t) = 1 - 2\sin^2(\pi-t)$.
(b) Use the periodicity and the evenness of the cosine function to show that $\cos 2(\pi-t) = \cos 2t$.
(c) Use Problem 22 to show that $1 - 2\sin^2(\pi-t) = 1 - 2\sin^2 t$.
(d) Show that $\cos 2t = 1 - 2\sin^2 t$.

53. In the text and Problem 52 we showed that $\cos 2t = 1 - 2\sin^2 t$ for $0 \le t \le \pi$. In this problem assume that $-\pi \le t < 0$. Show that

(a) $\cos(-2t) = 1 - 2\sin^2(-t)$.
(b) $\cos 2t = 1 - 2\sin^2 t$.

7.3 SUM AND DIFFERENCE FORMULAS FOR SINE AND COSINE

The double-angle formulas introduced in Section 7.2 are special cases of a group of more general trigonometric identities, the sum and difference formulas. If we know the sine and cosine values for two angles, say, θ and ϕ, can we use those values to find the sine and cosine of the angle $\theta+\phi$? For example, can we use the exact values for the sine and cosine of 30° and 45° to find an exact value for the sine of 75°? You can use your calculator to check that the sine of 75° is *not* equal to $\sin 30^\circ + \sin 45^\circ$, and, in general, $\sin(\theta+\phi) \ne \sin\theta + \sin\phi$. At the end of this section, we show that the following formula holds for any angles θ and ϕ:

$$\sin(\theta+\phi) = \sin\theta\cos\phi + \cos\theta\sin\phi.$$

Before proving this result on page 323–325, we see how it is used.

Example 1 Find an exact value for $\sin 75^\circ$.

Solution Use the sum formula with $\theta = 30^\circ$ and $\phi = 45^\circ$. Then

$$\begin{aligned}\sin 75^\circ &= \sin(30^\circ + 45^\circ)\\ &= \sin 30^\circ \cos 45^\circ + \cos 30^\circ \sin 45^\circ\\ &= \frac{1}{2}\cdot\frac{\sqrt{2}}{2} + \frac{\sqrt{3}}{2}\cdot\frac{\sqrt{2}}{2} = \frac{\sqrt{2}+\sqrt{6}}{4}.\end{aligned}$$

Rewriting $\sin t + \cos t$

The sum formula can be used to simplify sums of sines and cosines. Consider the graph of $f(t) = \sin t + \cos t$ in Figure 7.23. This graph looks remarkably like the graph of the sine function itself, having the same period but a larger amplitude (somewhere between 1 and 2). It appears to have a phase shift of about $\pi/4$.

We would like to determine whether f is actually sinusoidal, that is, whether f can be written in the form $f(t) = A\sin(Bt + \phi) + k$. If so, Figure 7.23 suggests that $B = 1$, that the phase shift $\phi = \pi/4$, and that $k = 0$. What about the value of A?

We use the identity $\sin(\theta + \phi) = \sin\theta\cos\phi + \sin\phi\cos\theta$ to study $\sin(t + \pi/4)$. Substituting $\theta = t$ and $\phi = \pi/4$, the identity becomes

$$\begin{aligned}\sin\left(t + \frac{\pi}{4}\right) &= \sin t \cos\frac{\pi}{4} + \sin\frac{\pi}{4}\cos t\\ &= \frac{1}{\sqrt{2}}\sin t + \frac{1}{\sqrt{2}}\cos t.\end{aligned}$$

Multiplying both sides of this equation by $\sqrt{2}$ gives

$$\sqrt{2}\sin\left(t + \frac{\pi}{4}\right) = \sin t + \cos t.$$

Thus,

$$f(t) = \sqrt{2}\sin\left(t + \frac{\pi}{4}\right).$$

We see that f is indeed a sine function with period 2π, amplitude $A = \sqrt{2} \approx 1.414$, and phase shift $\phi = \pi/4$. These values can be seen in the graph in Figure 7.23.

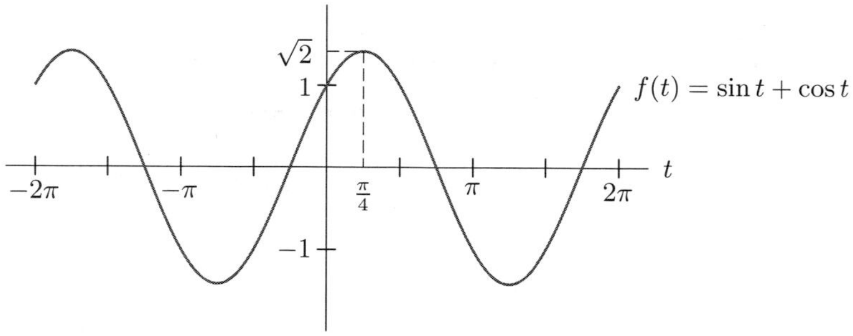

Figure 7.23: A graph of $f(t)$ reflecting sinusoidal behavior, with amplitude $\sqrt{2}$ and phase shift $\pi/4$

Rewriting a_1 sin Bt + a_2 cos Bt

We have rewritten $f(t) = \sin t + \cos t$ as a single sine function. We now show that this result can be extended: The sum of any two cosine and sine functions having the same periods can be written as a single sine function.[4] We start with

$$a_1 \sin(Bt) + a_2 \cos(Bt),$$

where $a_1,\ a_2$, and B are constants. We want to write this expression in the form $A\sin(Bt+\phi)$. To do this, we show that we can find the constants A and ϕ for any a_1 and a_2. We use the identity $\sin(\theta+\phi) = \sin\theta\cos\phi + \sin\phi\cos\theta$ with $\theta = Bt$. Multiplying by A, we have

$$\begin{aligned} A\sin(Bt+\phi) &= A\sin(Bt)\cos\phi + A\sin\phi\cos(Bt) \\ &= A\cos\phi\sin(Bt) + A\sin\phi\cos(Bt). \end{aligned}$$

This looks very much like $a_1\sin(Bt) + a_2\cos(Bt)$ provided $a_1 = A\cos\phi$ and $a_2 = A\sin\phi$. Can we solve for A and ϕ? We have

$$a_1^2 + a_2^2 = A^2\cos^2\phi + A^2\sin^2\phi = A^2,$$

so

$$A = \sqrt{a_1^2 + a_2^2}.$$

From $a_1 = A\cos\phi$ and $a_2 = A\sin\phi$, we get $\cos\phi = a_1/A$ and $\sin\phi = a_2/A$. If $a_1 \neq 0$, then

$$\frac{a_2}{a_1} = \frac{A\sin\phi}{A\cos\phi} = \frac{\sin\phi}{\cos\phi} = \tan\phi.$$

These formulas allow A and ϕ to be determined from a_1 and a_2. Applying them to the example $\sin t + \cos t$, where $a_1 = 1$ and $a_2 = 1$, gives, as before, $A = \sqrt{2}$ and $\cos\phi = \sin\phi = 1/\sqrt{2}$, so $\tan\phi = 1$, and $\phi = \pi/4$. In summary:

> Provided their periods are equal, the sum of a sine function and a cosine function can be written as a single sine function. We have
>
> $$a_1\sin(Bt) + a_2\cos(Bt) = A\sin(Bt+\phi)$$
>
> where
>
> $$A = \sqrt{a_1^2 + a_2^2} \qquad \text{and} \qquad \tan\phi = \frac{a_2}{a_1}.$$
>
> The angle ϕ is determined by the equations $\cos\phi = a_1/A$ and $\sin\phi = a_2/A$.

To find the angle ϕ, find its quadrant from the signs of $\cos\phi = a_1/A$ and $\sin\phi = a_2/A$. Then find the angle ϕ in that quadrant with $\tan\phi = a_2/a_1$.

Example 2 If $g(t) = 2\sin 3t + 5\cos 3t$, write $g(t)$ as a single sine function.

[4] The sum of a sine and a cosine with the same periods can also be written as a single cosine function.

Solution We have $a_1 = 2$, $a_2 = 5$, and $B = 3$. Thus,

$$A = \sqrt{2^2 + 5^2} = \sqrt{29} \qquad \text{and} \qquad \tan\phi = \frac{5}{2}.$$

Since $\cos\phi = 2/\sqrt{29}$ and $\sin\phi = 5/\sqrt{29}$ are both positive, ϕ is in the first quadrant. Since $\arctan(5/2) = 1.190$ is in the first quadrant, we take $\phi = 1.190$. Therefore,

$$g(t) = \sqrt{29}\sin(3t + 1.190).$$

The four identities concerning the sums and differences of angles are listed inside the cover of the book. The derivation follows.

Sums and Differences of Sines and Cosines

The four formulas for the sines and cosines of the sums and differences of angles lead to four other formulas for the sums and differences of the trigonometric functions themselves.

We begin with the fact that

$$\cos(\theta + \phi) = \cos\theta\cos\phi - \sin\theta\sin\phi$$
$$\cos(\theta - \phi) = \cos\theta\cos\phi + \sin\theta\sin\phi.$$

If we add these two equations, the terms involving the sine function cancel out, giving

$$\cos(\theta + \phi) + \cos(\theta - \phi) = 2\cos\theta\cos\phi.$$

The left-hand side of this equation can be rewritten by substituting $u = \theta + \phi$ and $v = \theta - \phi$:

$$\cos\underbrace{(\theta + \phi)}_{u} + \cos\underbrace{(\theta - \phi)}_{v} = 2\cos\theta\cos\phi.$$

The right-hand side of the equation can also be rewritten in terms of u and v. To do this, solve for θ and ϕ by adding and subtracting the equations $u = \theta + \phi$ and $v = \theta - \phi$, giving $u + v = 2\theta$ and $u - v = 2\phi$, so that

$$\theta = \frac{u+v}{2} \qquad \text{and} \qquad \phi = \frac{u-v}{2}.$$

With these substitutions, we have

$$\cos u + \cos v = 2\cos\frac{u+v}{2}\cos\frac{u-v}{2}.$$

This identity relates the sum of two cosine functions to the product of two new cosine functions.

There are similar formulas for the sum of two sine functions, and the difference between sine functions and cosine functions. See Problems 14, 15, 16. A summary is inside the cover of the book.

Justification of $\cos(\theta - \phi) = \cos\theta\cos\phi + \sin\theta\sin\phi$

To show why this identity for $\cos(\theta - \phi)$ is true, we find the distance between points A and B in Figure 7.24 in two ways. Points A and B correspond to the angles θ and ϕ on the unit circle, so their coordinates are $A = (\cos\theta, \sin\theta)$ and $B = (\cos\phi, \sin\phi)$. The angle AOB is $(\theta - \phi)$, so by the Law of Cosines, the distance AB is given by

$$AB^2 = 1^2 + 1^2 - 2 \cdot 1 \cdot 1\cos(\theta - \phi) = 2 - 2\cos(\theta - \phi).$$

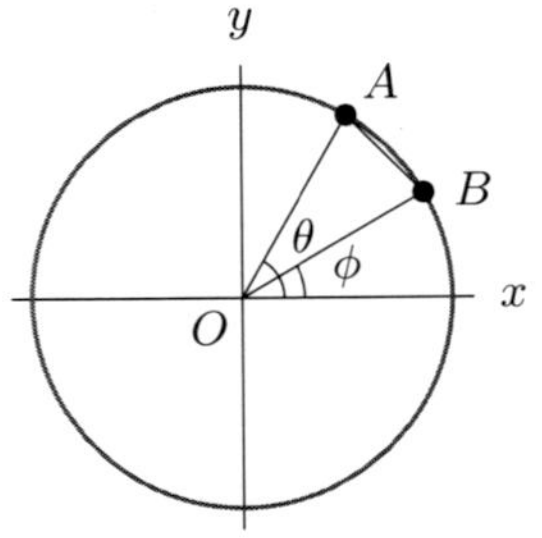

Figure 7.24: To justify the identity for $\cos(\theta - \phi)$, find the distance between A and B two ways

Finding the distance AB by the distance formula and multiplying out, we get

$$\begin{aligned} AB^2 &= (\cos\theta - \cos\phi)^2 + (\sin\theta - \sin\phi)^2 \\ &= \cos^2\theta - 2\cos\theta\cos\phi + \cos^2\phi + \sin^2\theta - 2\sin\theta\sin\phi + \sin^2\phi \\ &= \cos^2\theta + \sin^2\theta + \cos^2\phi + \sin^2\phi - 2\cos\theta\cos\phi - 2\sin\theta\sin\phi \\ &= 2 - 2(\cos\theta\cos\phi + \sin\theta\sin\phi). \end{aligned}$$

Setting the two distances equal gives

$$2 - 2\cos(\theta - \phi) = 2 - 2(\cos\theta\cos\phi + \sin\theta\sin\phi).$$

So

$$\boxed{\cos(\theta - \phi) = \cos\theta\cos\phi + \sin\theta\sin\phi.}$$

Obtaining Other Identities from the Identity for cos $(\theta - \phi)$

Letting $\phi = \pi/2$, in the identity for $\cos(\theta - \phi)$, we obtain

$$\cos\left(\theta - \frac{\pi}{2}\right) = \cos\theta\cos\frac{\pi}{2} + \sin\theta\sin\frac{\pi}{2} = \cos\theta \cdot 0 + \sin\theta \cdot 1 = \sin\theta.$$

So

$$\boxed{\sin\theta = \cos\left(\theta - \frac{\pi}{2}\right).}$$

Replacing θ by $(\theta - \pi/2)$ in this identity gives

$$\sin\left(\theta - \frac{\pi}{2}\right) = \cos\left(\theta - \frac{\pi}{2} - \frac{\pi}{2}\right) = \cos(\theta - \pi).$$

Substituting $\phi = \pi$ in the identity for $\cos(\theta - \phi)$ gives

$$\cos(\theta - \pi) = \cos\theta\cos\pi + \sin\theta\sin\pi = \cos\theta(-1) + \sin\theta \cdot 0 = -\cos\theta.$$

So

$$\boxed{\sin\left(\theta - \frac{\pi}{2}\right) = -\cos\theta.}$$

Justification of Other Sum and Difference Formulas

Using $\sin\theta = \cos(\theta - \pi/2)$ with θ replaced by $(\theta+\phi)$ gives

$$\sin(\theta+\phi) = \cos\left(\theta+\phi-\frac{\pi}{2}\right).$$

Rewriting $(\theta+\phi-\pi/2)$ as $(\theta-\pi/2-(-\phi))$, we use the identity for $\cos(\theta-\phi)$ with θ replaced by $(\theta-\pi/2)$ and ϕ replaced by $-\phi$:

$$\begin{aligned}\cos\left(\theta+\phi-\frac{\pi}{2}\right) &= \cos\left(\theta-\frac{\pi}{2}-(-\phi)\right)\\ &= \cos\left(\theta-\frac{\pi}{2}\right)\cos(-\phi)+\sin\left(\theta-\frac{\pi}{2}\right)\sin(-\phi).\end{aligned}$$

Since $\cos(-\phi)=\cos\phi$ and $\sin(-\phi)=-\sin\phi$, and since $\cos(\theta-\pi/2)=\sin\theta$ and $\sin(\theta-\pi/2)=-\cos\theta$, we have

$$\cos\left(\theta+\phi-\frac{\pi}{2}\right) = \sin\theta\cos\phi+(-\cos\theta)(-\sin\phi).$$

So

$$\boxed{\sin(\theta+\phi)=\sin\theta\cos\phi+\cos\theta\sin\phi.}$$

Replacing ϕ by $-\phi$ in the formula for $\sin(\theta+\phi)$ gives

$$\sin(\theta-\phi)=\sin(\theta+(-\phi))=\sin\theta\cos(-\phi)+\cos\theta\sin(-\phi),$$

and since $\cos(-\phi)=\cos\phi$ and $\sin(-\phi)=-\sin\phi$,

$$\boxed{\sin(\theta-\phi)=\sin\theta\cos\phi-\cos\theta\sin\phi.}$$

Replacing ϕ by $-\phi$ in the formula for $\cos(\theta-\phi)$ gives

$$\cos(\theta+\phi)=\cos(\theta-(-\phi))=\cos\theta\cos(-\phi)+\sin\theta\sin(-\phi).$$

$$\boxed{\cos(\theta+\phi)=\cos\theta\cos\phi-\sin\theta\sin\phi.}$$

Exercises and Problems for Section 7.3

Exercises

In Exercises 1–4, write in the form $A\sin(Bt+\phi)$ using sum or difference formulas.

1. $8\sin t - 6\cos t$

2. $8\sin t + 6\cos t$

3. $-\sin t + \cos t$

4. $-2\sin 3t + 5\cos 3t$

5. Use identities and the exact values of $\sin\theta$ and $\cos\theta$ for $\theta = 30°, 45°, 60°$ to find exact values of $\sin\theta$ and $\cos\theta$ for $\theta = 15°$ and $\theta = 75°$.

In Exercises 6–10, find exact values.

6. $\cos 165° - \cos 75°$

7. $\cos 75° + \cos 15°$

8. $\sin 345°$

9. $\sin 105°$

10. $\cos 285°$

11. Test each of the following identities by first evaluating both sides to see that they are numerically equal when $u = 35°$ and $v = 40°$. Then let $u = x$ and $v = 25°$ and compare the graphs of each side of the identity as functions of x (in degrees). For example, in part (a) graph $y = \cos x + \cos 25$ and $y = 2\cos(\frac{x+25}{2})\cos(\frac{x-25}{2})$.

(a) $\cos u + \cos v = 2\cos\left(\frac{u+v}{2}\right)\cos\left(\frac{u-v}{2}\right)$
(b) $\cos u - \cos v = -2\sin\left(\frac{u+v}{2}\right)\sin\left(\frac{u-v}{2}\right)$
(c) $\sin u + \sin v = 2\sin\left(\frac{u+v}{2}\right)\cos\left(\frac{u-v}{2}\right)$
(d) $\sin u - \sin v = 2\cos\left(\frac{u+v}{2}\right)\sin\left(\frac{u-v}{2}\right)$

12. Test each of the following identities first by evaluating both sides to see that they are numerically equal when $u = 15°$ and $v = 42°$. Then let $u = x$ and $v = 20°$ and compare the graphs of each side of the identity as functions of x (in degrees). For example, in part (a) graph $y = \sin(x + 20)$ and $y = \sin x\ \cos 20 + \sin 20\ \cos x$.

(a) $\sin(u + v) = \sin u \cos v + \sin v \cos u$
(b) $\sin(u - v) = \sin u \cos v - \sin v \cos u$
(c) $\cos(u + v) = \cos u \cos v - \sin v \sin u$
(d) $\cos(u - v) = \cos u \cos v + \sin v \sin u$

13. Starting from the addition formulas for sine and cosine, derive the following identities:

(a) $\sin t = \cos(t - \pi/2)$ **(b)** $\cos t = \sin(t + \pi/2)$

14. Prove: $\cos u - \cos v = -2\sin\left(\frac{u+v}{2}\right)\sin\left(\frac{u-v}{2}\right)$.

15. Prove: $\sin u + \sin v = 2\sin\left(\frac{u+v}{2}\right)\cos\left(\frac{u-v}{2}\right)$.

16. Prove: $\sin u - \sin v = 2\cos\left(\frac{u+v}{2}\right)\sin\left(\frac{u-v}{2}\right)$.

Problems

17. Show that $\cos 3t = 4\cos^3 t - 3\cos t$.

18. Solve the equation $\sin 4x + \sin x = 0$ for values of x in the interval $0 < x < 2\pi$ by applying an identity to the left side of the equation.

19. Use the cosine addition formula and other identities to find a formula for $\cos 3\theta$ in terms of $\cos\theta$. Your formula should contain no trigonometric functions other than $\cos\theta$ and its powers.

20. We have calculated exact, rational values for $\sin r\pi$ when $r = 0, \pm 1, \pm\frac{1}{2}$. Surprisingly, these are the only rational values of $\sin r\pi$ when r is rational. Prove the following identity, which is used to obtain the result:[5]

$$2\cos((n+1)\theta) = (2\cos\theta)(2\cos(n\theta)) - 2\cos((n-1)\theta).$$

21. Use the addition formula $\cos(u + v) = \cos u \cos v - \sin u \sin v$ to derive the following identity for the average rate of change of the cosine function:

$$\frac{\cos(x+h) - \cos x}{h} = \cos x\left(\frac{\cos h - 1}{h}\right) - \sin x\left(\frac{\sin h}{h}\right).$$

22. Use the addition formula $\sin(u + v) = \sin u \cos v + \sin v \cos u$ to derive the following identity for the average rate of change of the sine function:

$$\frac{\sin(x+h) - \sin x}{h} = \sin x\left(\frac{\cos h - 1}{h}\right) + \cos x\left(\frac{\sin h}{h}\right).$$

23. Use the addition formula $\tan(u+v) = \dfrac{\tan u + \tan v}{1 - \tan u \tan v}$ to derive the following identity for the average rate of change of the tangent function:

$$\frac{\tan(x+h) - \tan x}{h} = \frac{1}{\cos^2 x}\left(\frac{\sin h}{h}\right)\frac{1}{\cos h - \sin h \tan x}.$$

24. We can obtain formulas involving half-angles from the double angle formulas. To find an expression for $\sin\frac{1}{2}v$, we solve $\cos 2u = 1 - 2\sin^2 u$ for $\sin^2 u$ to obtain

$$\sin^2 u = \frac{1 - \cos 2u}{2}.$$

Taking the square root gives:

$$\sin u = \pm\sqrt{\frac{1 - \cos 2u}{2}},$$

and finally we make the substitution $u = \frac{1}{2}v$ to get:

$$\sin\frac{1}{2}v = \pm\sqrt{\frac{1 - \cos v}{2}}.$$

(a) Show that $\cos\frac{1}{2}v = \pm\sqrt{\frac{1 + \cos v}{2}}$.

(b) Show that $\tan\frac{1}{2}v = \pm\sqrt{\frac{1 - \cos v}{1 + \cos v}}$.

[5] I. Niven and H. Zuckerman, *An Introduction to Number Theory*, 3rd ed., p. 143 (Wiley & Sons: New York, 1972).

The sign in these formulas depends on the value of $v/2$. If $v/2$ is in quadrant I then the sign of $\sin v/2$ is $+$, the sign of $\cos v/2$ is $+$, and the sign of $\tan v/2$ is $+$.

(c) If $v/2$ is in quadrant II find the signs for $\sin v/2$, for $\cos v/2$ and for $\tan v/2$.

(d) If $v/2$ is in quadrant III find the signs for $\sin v/2$, for $\cos v/2$ and for $\tan v/2$.

(e) If $v/2$ is in quadrant IV find the signs for $\sin v/2$, for $\cos v/2$ and for $\tan v/2$.

25. We can find the area of triangle ABC in Figure 7.25 by adding the areas of the two right triangles formed by the perpendicular from vertex C to the opposite side.

(a) Find values from $\sin\theta$, $\cos\theta$, $\sin\phi$, and $\cos\phi$ in terms of a, b, c_1, c_2, and h.

(b) Show that

$$\text{Area } \Delta CAD = (1/2)ab\sin\theta\cos\phi$$

$$\text{Area } \Delta CDB = (1/2)ab\cos\theta\sin\phi.$$

(c) Use the sum formula to explain why

$$\text{Area } \Delta ABC = (1/2)ab\sin C.$$

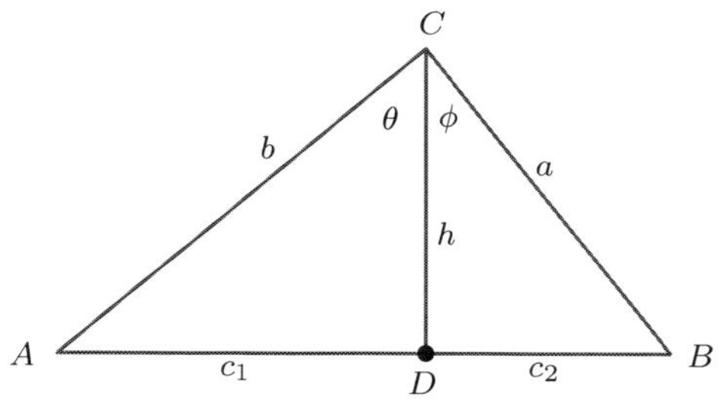

Figure 7.25

26. **(a)** Show that for any triangle ABC, we have $a = b\cos C + c\cos B$.

(b) Use part (a) and the Law of Sines to obtain the sum formula for sine.
[Hint: Why is $\sin A = \sin(B + C)$?]

27. In this problem, you will derive the formula $\cos(\theta+\phi) = \cos\theta\cos\phi - \sin\theta\sin\phi$ directly.

(a) Find the coordinates of points P_1, P_2, P_3, and P_4 in the unit circle in Figure 7.26.

(b) Draw the line segments P_1P_2 and P_3P_4 and explain why $P_1P_2 = P_3P_4$, using facts from geometry.

(c) Use the distance formula to calculate the lengths P_1P_2 and P_3P_4.

(d) Write an equation and simplify to obtain the sum formula for cosine.

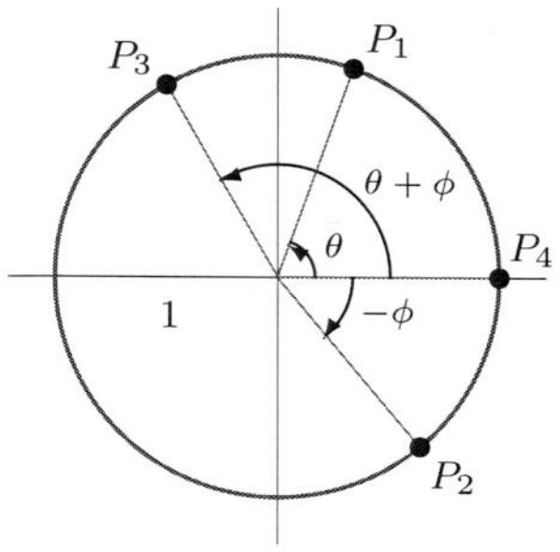

Figure 7.26

7.4 TRIGONOMETRIC MODELS

This section gives examples of mathematical models using trigonometric functions.

Sums of Trigonometric Functions

The first example shows mathematical models involving the sums of trigonometric functions. We see how the identities enable us to predict and explain the observed phenomena.

Example 1 A utility company serves two different cities. Let P_1 be the power requirement in megawatts (mw) for City 1 and P_2 be the requirement for City 2. Both P_1 and P_2 are functions of t, the number of hours elapsed since midnight. Suppose P_1 and P_2 are given by the following formulas:

$$P_1 = 40 - 15\cos\left(\frac{\pi}{12}t\right) \quad \text{and} \quad P_2 = 50 + 10\sin\left(\frac{\pi}{12}t\right).$$

(a) Describe the power requirements of each city in words.
(b) What is the maximum total power the utility company must be prepared to provide?

Solution (a) The power requirement of City 1 is at a minimum of $40 - 15 = 25$ mw at $t = 0$, or midnight. It rises to a maximum of $40 + 15 = 55$ mw at noon and falls back to 25 mw by the following midnight. The power requirement of City 2 is at a maximum of 60 mw at 6 am. It falls to a minimum of 40 mw by 6 pm but by the following morning has climbed back to 60 mw, again at 6 am. Figure 7.27 shows P_1 and P_2 over a two-day period.

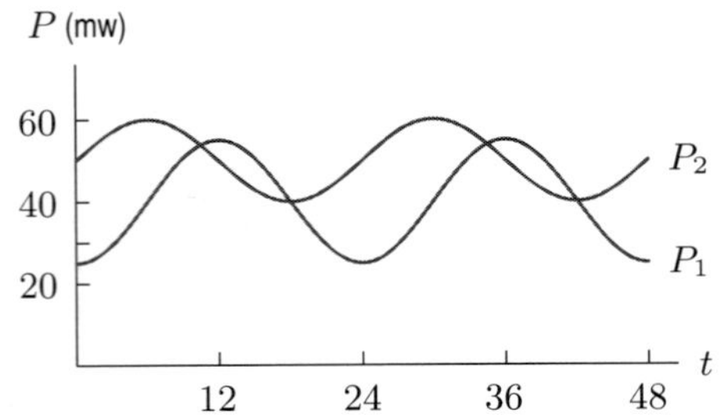

Figure 7.27: Power requirements for cities 1 and 2

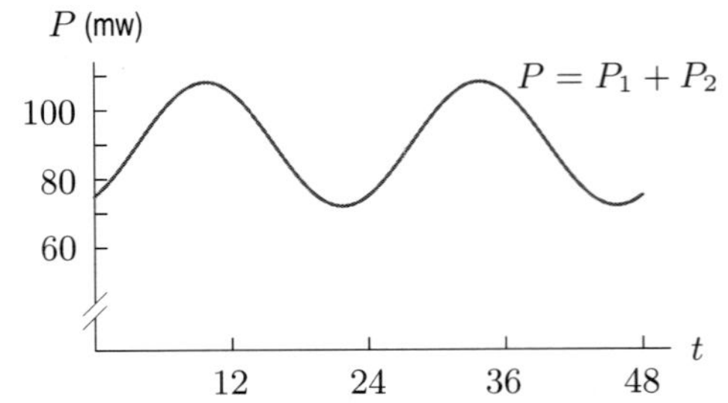

Figure 7.28: Total power demand for both cities combined

(b) The utility company must provide enough power to satisfy the needs of both cities. The total power required is given by

$$P = P_1 + P_2 = 90 + 10\sin\left(\frac{\pi}{12}t\right) - 15\cos\left(\frac{\pi}{12}t\right).$$

The graph of total power in Figure 7.28 looks like a sinusoidal function. It varies between about 108 mw and 72 mw, giving it an amplitude of roughly 18 mw. Since the maximum value of P is about 108 mw, the utility company must be prepared to provide at least this much power at all times. Since the maximum value of P_1 is 55 and the maximum value of P_2 is 60, you might have expected that the maximum value of P would be $55 + 60 = 115$. The reason that this is not true is that the maximum values of P_1 and P_2 occur at different times. However, the midline of P is a horizontal line at 90, which does equal the midline of P_1 plus the midline of P_2. The period of P is 24 hours because the values of both P_1 and P_2 begin repeating after 24 hours, so their sum repeats that frequently as well.

The result on page 322 allows us to find a sinusoidal function for the total power:

$$P = P_1 + P_2 = 90 + 10\sin\left(\frac{\pi}{12}t\right) - 15\cos\left(\frac{\pi}{12}t\right) = 90 + A\sin(Bt + \phi),$$

where

$$A = \sqrt{10^2 + (-15)^2} = 18.028 \text{ mw}.$$

Since $\cos\phi = 10/18.028 = 0.555$ and $\sin\phi = -15/18.028 = -0.832$, we know ϕ must be in the fourth quadrant. Also,

$$\tan\phi = \frac{-15}{10} = -1.5 \quad \text{and} \quad \tan^{-1}(-1.5) = -0.983,$$

and since -0.983 is in the fourth quadrant, we take $\phi = -0.983$. Thus,

$$P = P_1 + P_2 = 90 + 18\sin\left(\frac{\pi}{12}t - 0.983\right).$$

The next example shows that the sum of sine or cosine functions is not always sinusoidal.

Example 2 Sketch and describe the graph of $y = \sin 2x + \sin 3x$.

Solution Figure 7.29 shows that the function $y = \sin 2x + \sin 3x$ is not sinusoidal. It is, however, periodic. Its period seems to be 2π, since it repeats twice on the interval of length 4π shown in the figure.

We can see that the function $y = \sin 2x + \sin 3x$ has period 2π by looking at the periods of $\sin 2x$ and $\cos 3x$. Since the period of $\sin 2x$ is π and the period of $\sin 3x$ is $2\pi/3$, on any interval of length 2π, the function $y = \sin 2x$ completes two cycles and the function $y = \sin 3x$ completes three cycles. Both functions are at the beginning of a new cycle after an interval of 2π, so their sum begins to repeat at this point. (See Figure 7.30.) Notice that even though the maximum value of each of the functions $\sin 2x$ and $\sin 3x$ is 1, the maximum value of their sum is not 2; it is a little less than 2. This is because $\sin 2x$ and $\sin 3x$ achieve their maximum values for different x values.

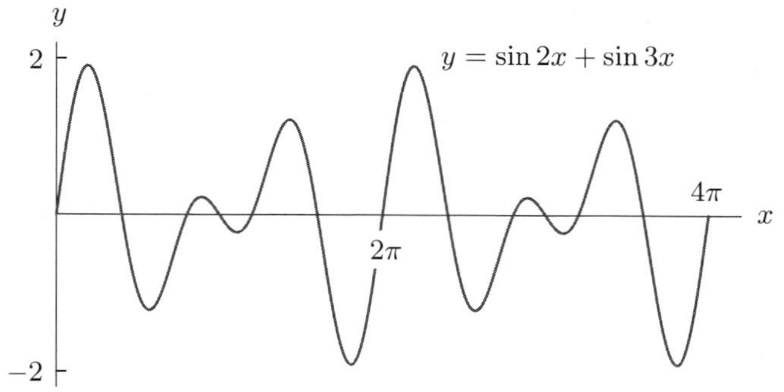

Figure 7.29: A graph of the sum $y = \sin 2x + \sin 3x$

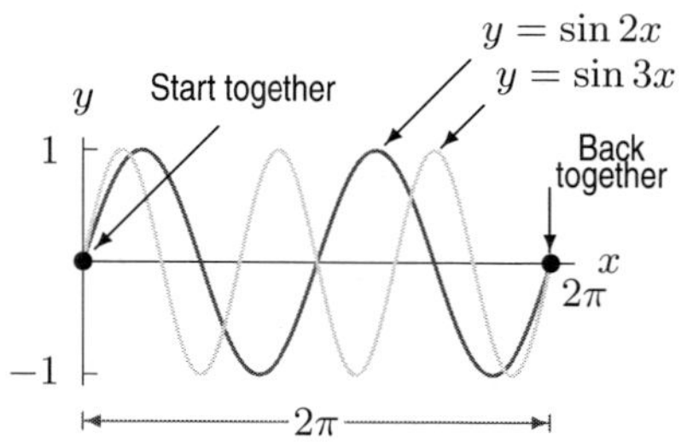

Figure 7.30: Graphs of $y = \sin 2x$ and $y = \sin 3x$ on an interval of 2π

Damped Oscillation

In Problems 23–26 in Section 6.1, we considered a weight attached to the ceiling by a spring. If the weight is disturbed, it begins bobbing up and down; we modeled the weight's motion using a trigonometric function. Figure 7.31 shows such a weight at rest. Suppose d is the displacement in centimeters from the weight's position at rest. For instance, if $d = 5$ then the weight is 5 cm above its rest position; if $d = -5$, then the weight is 5 cm below its rest position.

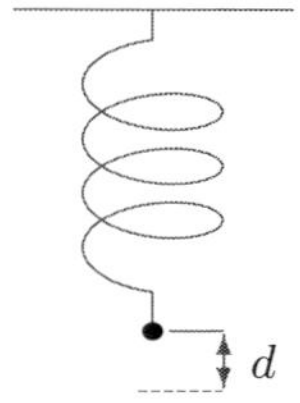

Figure 7.31: The value of d represents the weight's displacement from its at-rest position, $d = 0$

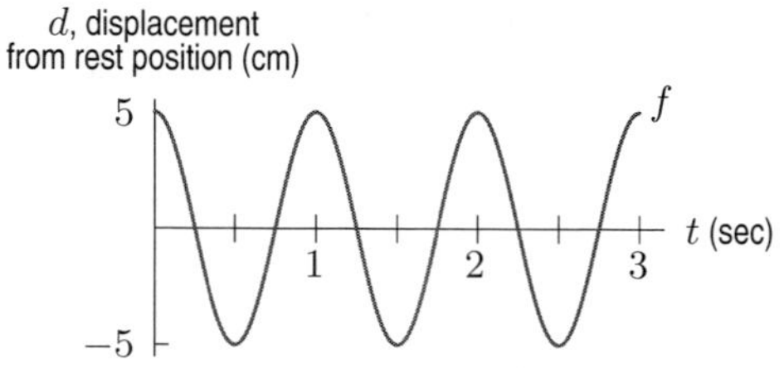

Figure 7.32: The predicted motion of the weight for the first 3 seconds

Imagine that we raise the weight 5 cm above its rest position and release it at time $t = 0$, where t is in seconds. Suppose the weight bobbed up and down once every second for the first few seconds. We could model this behavior by the function

$$d = f(t) = 5\cos(2\pi t).$$

One full cycle is completed each second, so the period is 1, and the amplitude is 5. Figure 7.32 gives a graph of d for the first three seconds of the weight's motion.

This trigonometric model of the spring's motion is flawed, however, because it predicts that the weight will bob up and down forever. In fact, we know that as time passes, the amplitude of the bobbing diminishes and eventually the weight comes to rest. How can we alter our formula to model this kind of behavior? We need an amplitude which decreases over time.

For example, the amplitude of the spring's motion might decrease at a constant rate, so that after 5 seconds, the weight stops moving. In other words, we could imagine that the amplitude is a decreasing linear function of time. Using A to represent the amplitude, this means that $A = 5$ at $t = 0$, and that $A = 0$ at $t = 5$. Thus, $A(t) = 5 - t$. Then, instead of writing

$$d = f(t) = \underbrace{\text{Constant amplitude}}_{5} \cdot \cos 2\pi t,$$

we write

$$d = f(t) = \underbrace{\text{Decreasing amplitude}}_{(5-t)} \cdot \cos 2\pi t$$

so that the formula becomes

$$d = f(t) = (5 - t) \cdot \cos 2\pi t.$$

Figure 7.33 shows a graph of this function.

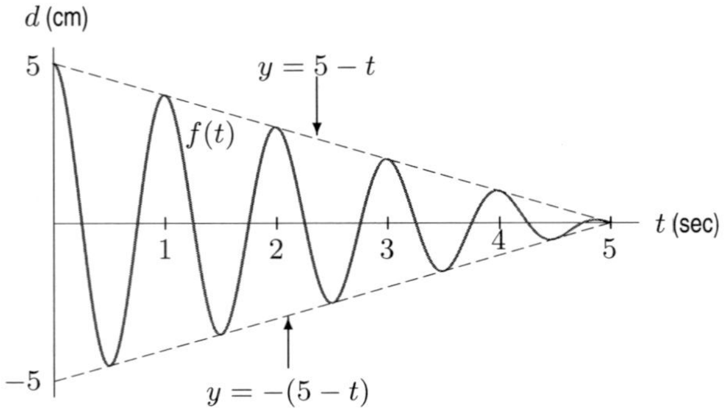

Figure 7.33: Graph of the weight's displacement assuming amplitude is a decreasing linear function of time

While the function $d = f(t) = (5 - t)\cos 2\pi t$ is a better model for the behavior of the spring for $t < 5$, Figure 7.34 shows that this model does not work for $t > 5$. The breakdown in the model occurs because the magnitude of $A(t) = 5 - t$ starts to increase when t grows larger than 5.

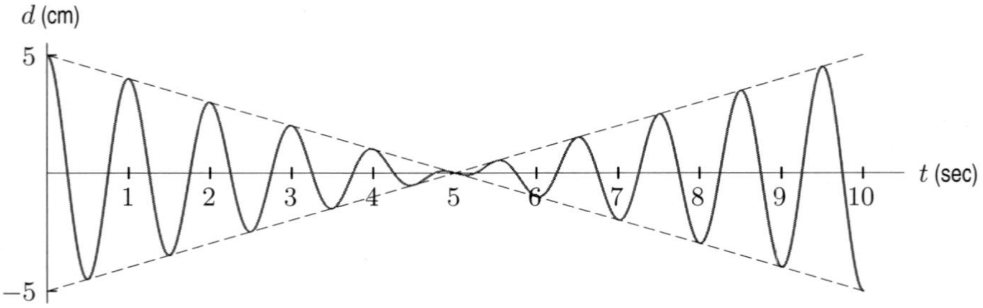

Figure 7.34: The formula $d = f(t) = (5 - t)\cos 2\pi t$ makes inaccurate predictions for values of t larger than 5

To improve the model, we keep the idea of representing the amplitude as a decreasing function of time, but we pick a different decreasing function. The problem with the linear function $A(t) = 5 - t$ is two-fold. It approaches zero too abruptly, and, after attaining zero, it becomes negative. We want a function that approaches zero gradually and does not become negative.

Let's try a decreasing exponential function. Suppose that the amplitude of the spring's motion is halved each second. Then at $t = 0$ the amplitude is 5, and the amplitude at time t is given by

$$A(t) = 5\left(\frac{1}{2}\right)^t.$$

Thus, a formula for the motion is

$$d = f(t) = \underbrace{5\left(\frac{1}{2}\right)^t}_{\text{Decreasing amplitude}} \cdot \cos 2\pi t.$$

Figure 7.35 shows a graph of this function. The dashed curves in Figure 7.35 show the decreasing exponential function. (There are two curves, $y = 5(\frac{1}{2})^t$ and $y = -5(\frac{1}{2})^t$, because the amplitude measures distance on both sides of the midline.)

Figure 7.35 predicts that the weight's oscillations diminish gradually, so that at time $t = 5$ the weight is still oscillating slightly. Figure 7.36 shows that the weight continues to make small oscillations long after 5 seconds have elapsed.

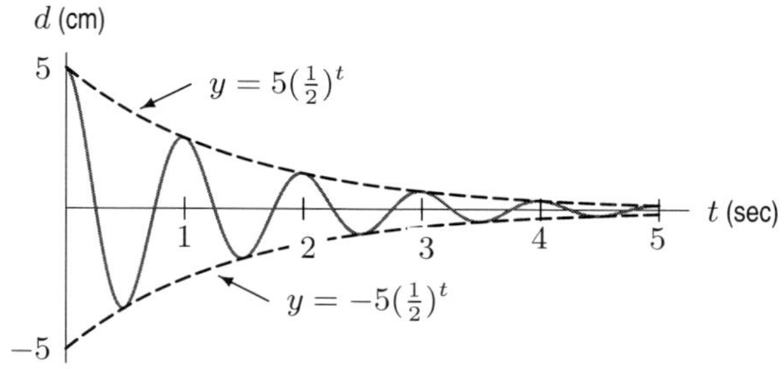

Figure 7.35: A graph of the weight's displacement assuming that the amplitude is a decreasing exponential function of time

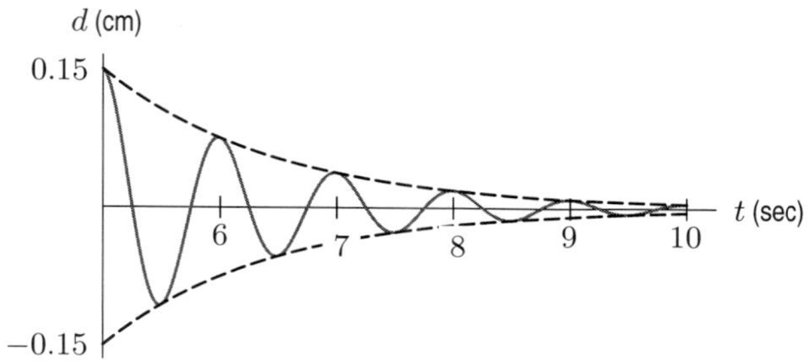

Figure 7.36: Graph showing that the weight is still making small oscillations for $t \geq 5$. Note reduced scale on d-axis; t-axis starts at $t = 5$

If A_0, B, and C, and k are constants, $k > 0$, a function of the form

$$y = A_0 e^{-kt}\cos(Bt) + C \qquad \text{or} \qquad y = A_0 e^{-kt}\sin(Bt) + C$$

can be used to model an oscillating quantity whose amplitude decreases exponentially according to $A(t) = A_0 e^{-kt}$ where A_0 is the initial amplitude. Our model for the displacement of a weight is in this form with $k = \ln 2$.

Oscillation With a Rising Midline

In the next example, we consider an oscillating quantity which does not have a horizontal midline, but whose amplitude of oscillation is in some sense constant.

Example 3 In Section 6.5, Example 8, we represented a rabbit population undergoing seasonal fluctuations by the function

$$R = f(t) = 10000 - 5000\cos\left(\frac{\pi}{6}t\right),$$

where R is the size of the rabbit population t months after January. See Figure 7.37. The rabbit population varies periodically about the midline, $y = 10000$. The average number of rabbits is 10,000, but, depending on the time of year, the actual number may be above or below the average.

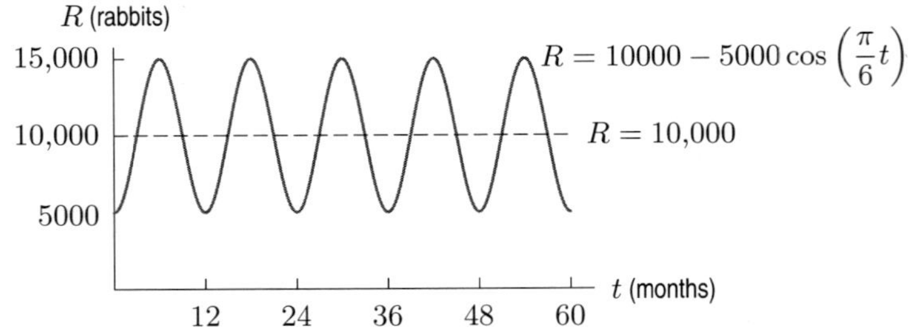

Figure 7.37: The rabbit population over a 5-year (60 month) period

This suggests the following way of thinking about the formula for R:

$$R = \underbrace{10{,}000}_{\text{Average value}} \underbrace{- 5000\cos\left(\frac{\pi}{6}t\right)}_{\text{Seasonal variation}}.$$

Notice that we can't say the average value of the rabbit population is 10,000 unless we look at the population over year-long units. For example, if we looked at the population over the first two months, an interval on which it is always below 10,000, then the average would be less than 10,000.

Now let us imagine a different situation. What if the average, even over long periods of time, does not remain constant? For example, suppose that, due to conservation efforts, there is a steady increase of 50 rabbits per month in the average rabbit population. Thus, instead of writing

$$P = f(t) = \underbrace{10{,}000}_{\text{Constant midline}} \underbrace{- 5000\cos\left(\frac{\pi}{6}t\right)}_{\text{Seasonal variation}},$$

we could write

$$P = f(t) = \underbrace{10{,}000 + 50t}_{\substack{\text{Midline population increasing}\\ \text{by 50 every month}}} \underbrace{- 5000\cos(\frac{\pi}{6}t)}_{\text{Seasonal variation}}.$$

Figure 7.38 gives a graph of this new function over a five-year period.

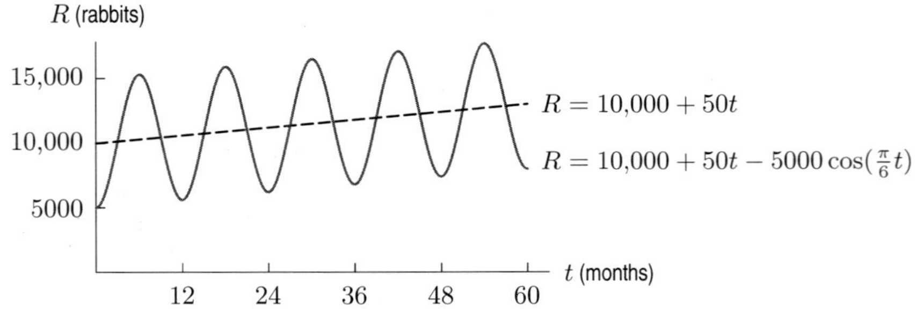

Figure 7.38: A graph of the gradually increasing rabbit population

Acoustic Beats

By international agreement, on a perfectly tuned piano, the A above middle C has a frequency of 440 cycles per second, also written 440 hertz (hz). The lowest-pitched note on the piano (the key at the left-most end) has frequency 55 hertz. Suppose a frequency of 55 hertz is struck on a tuning fork together with a note on an out-of-tune piano, whose frequency is 61 hertz. The intensities, I_1 and I_2, of these two tones are represented by the functions

$$I_1 = \cos(2\pi f_1 t) \qquad \text{and} \qquad I_2 = \cos(2\pi f_2 t),$$

where $f_1 = 55$, and $f_2 = 61$, and t is in seconds. If both tones are sounded at the same time, then their combined intensity is the sum of their separate intensities:

$$I = I_1 + I_2 = \cos(2\pi f_1 t) + \cos(2\pi f_2 t).$$

The graph of this function in Figure 7.39 resembles a rapidly varying sinusoidal function except that its amplitude increases and decreases. The ear perceives this variation in amplitude as a variation in loudness, so the tone appears to waver (or *beat*) in a regular way. This is an example of *acoustic beats*. A piano can be tuned by adjusting it until the beats fade.

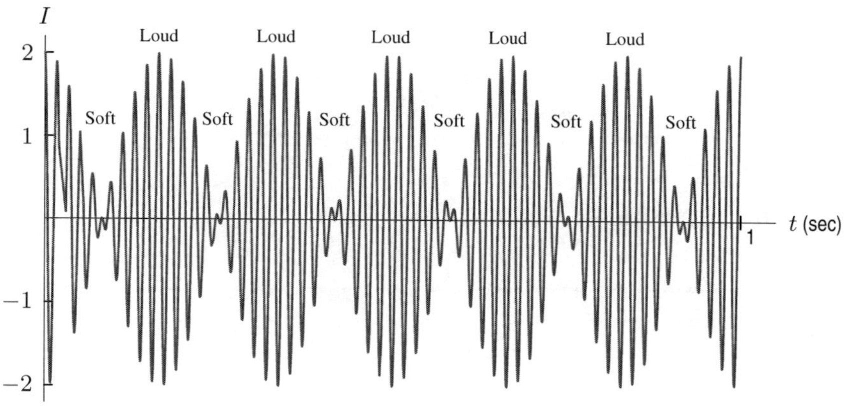

Figure 7.39: A graph of $I = \cos(2\pi f_1 t) + \cos(2\pi f_2 t)$

How can we explain the graph in Figure 7.39? Using the identity for a sum of cosines on page 323, we rewrite the intensity as

$$\begin{aligned} I = \cos(2\pi f_2 t) + \cos(2\pi f_1 t) &= 2\cos\frac{2\pi f_2 t + 2\pi f_1 t}{2} \cdot \cos\frac{2\pi f_2 t - 2\pi f_1 t}{2} \\ &= 2\cos\frac{2\pi \cdot (61+55)t}{2} \cdot \cos\frac{2\pi \cdot (61-55)t}{2} \\ &= 2\cos(2\pi \cdot 58t)\cos(2\pi \cdot 3t). \end{aligned}$$

We can think of this formula in the following way:

$$I = 2\cos(2\pi \cdot 3t) \cdot \cos(2\pi \cdot 58t) = A(t)p(t),$$

where $A(t) = 2\cos(2\pi \cdot 3t)$ gives a (slowly) changing amplitude and $p(t) = \cos(2\pi \cdot 58t)$ gives a pure tone of 58 hz. Thus, we can think of the tone described by I as having a pitch of 58 hz, which is midway between the tones sounded by the tuning fork and the out-of-tune piano. As the amplitude rises and falls, the tone grows louder and softer, but its pitch remains a constant 58 hz. (See Figure 7.40.)

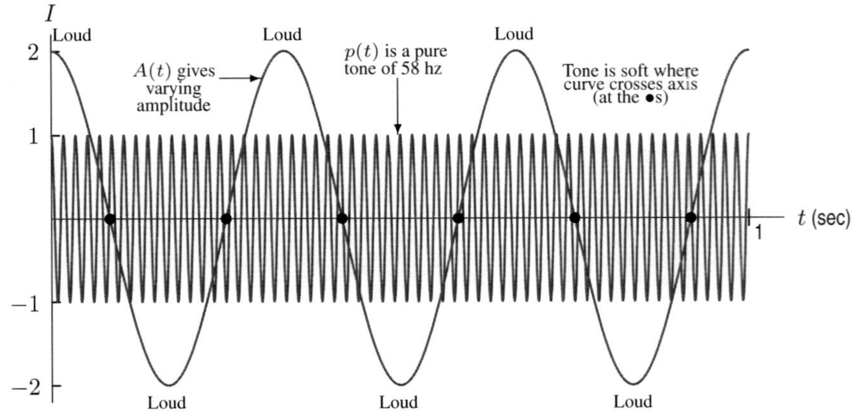

Figure 7.40: The function $A(t) = 2\cos(2\pi \cdot 3t)$ gives a varying amplitude, while the function $p(t) = \cos(2\pi \cdot 58t)$ gives a pure tone of 58 hz

Notice from Figure 7.40 that the function $A(t)$ completes three full cycles on the interval $0 \leq t \leq 1$. The tone is loudest when $A(t) = 1$ or $A(t) = -1$. Since both of these values occur once per cycle, the tone grows loud six times every second.

Exercises and Problems for Section 7.4

Exercises

1. Graph $y = t + 5\sin t$ and $y = t$ on $0 \leq t \leq 2\pi$. Where do the two graphs intersect if t is not restricted to $0 \leq t \leq 2\pi$?

2. Graph the function $f(x) = e^{-x}(2\cos x + \sin x)$ on the interval $0 \leq x \leq 12$. Use a calculator or computer to find the maximum and minimum values of the function.

Problems

3. From its past behavior, John knows that the value of a stock has a cyclical component which increases for the first three months of each year, falls for the next six, and rises again for the last three. In addition, inflation adds a linear component to the stock's price. John is seeking a model of the form

$$f(t) = mt + b + A\sin\frac{\pi t}{6}.$$

with t in months since Jan 1. He has the following data:

Date	Jan 1	Apr 1	Jul 1	Oct 1	Jan 1
Price	\$20.00	\$37.50	\$35.00	\$32.50	\$50.00

(a) Find values of m, b, and A so that f fits the data.

(b) During which month(s) does this stock appreciate the most?

(c) During what period each year is this stock actually losing value?

4. A power company serves two different cities, City A and City B. The power requirements of both cities vary in a predictable fashion over the course of a typical day.

(a) At midnight, the power requirement of City A is at a minimum of 40 megawatts. (A megawatt is a unit of power.) By noon the city has reached its maximum power consumption of 90 megawatts and by midnight it once again requires only 40 megawatts. This pattern repeats every day. Find a possible formula for $f(t)$, the power, in megawatts, required by City A as a function of t, in hours since midnight.

(b) The power requirements, $g(t)$ megawatts, of City B differ from those of City A. For t, in hours since midnight,

$$g(t) = 80 - 30\sin\left(\frac{\pi}{12}t\right).$$

Give the amplitude and the period of $g(t)$, and a physical interpretation of these quantities.

(c) Graph and find all t such that

$$f(t) = g(t), \qquad 0 \le t < 24.$$

Interpret your solution(s) in terms of power usage.

(d) Why should the power company be interested in the maximum value of the function

$$h(t) = f(t) + g(t), \qquad 0 \le t < 24?$$

What is the approximate maximum of this function, and approximately when is it attained?

(e) Find a formula for $h(t)$ as a single sine function. What is the exact maximum of this function?

5. Let $f(t) = \cos(e^t)$, where t is measured in radians.

(a) Note that $f(t)$ has a horizontal asymptote as $t \to -\infty$. Find its equation, and explain why f has this asymptote.
(b) Describe the behavior of f as $t \to \infty$. Explain why f behaves this way.
(c) Find the vertical intercept of f.
(d) Let t_1 be the least zero of f. Find t_1 exactly. [Hint: What is the smallest positive zero of the cosine function?]
(e) Find an expression for t_2, the least zero of f greater than t_1.

6. Let $f(x) = \sin\left(\frac{1}{x}\right)$ for $x > 0$, x in radians.

(a) $f(x)$ has a horizontal asymptote as $x \to \infty$. Find the equation for the asymptote and explain carefully why $f(x)$ has this asymptote.
(b) Describe the behavior of $f(x)$ as $x \to 0$. Explain why $f(x)$ behaves in this way.
(c) Is $f(x)$ a periodic function?
(d) Let z_1 be the greatest zero of $f(x)$. Find the exact value of z_1.
(e) How many zeros do you think the function $f(x)$ has?
(f) Suppose a is a zero of f. Find a formula for b, the largest zero of f less than a.

7. Derive the following identity used in an electrical engineering text[6] to represent the received AM signal function. Note that A, ω_c, ω_d are constants and $M(t)$ is a function of time, t.

$$\begin{aligned} r(t) &= (A + M(t))\cos\omega_c t + I\cos(\omega_c + \omega_d)t \\ &= ((A + M(t)) + I\cos\omega_d t)\cos\omega_c t - I\sin\omega_d t\sin\omega_c t \end{aligned}$$

8. In the July 1993 issue of *Standard and Poor's Industry Surveys* the editors stated:[7]

> The strength (of sales) of video games, seven years after the current fad began, is amazing .... What will happen next year is anything but clear. While video sales ended on a strong note last year, the toy industry is nothing if not cyclical.

(a) Why might sales of video games be cyclical?
(b) Does $s(t) = a\sin(bt)$, where t is time, serve as a reasonable model for the sales graph in Figure 7.41? What about $s(t) = a\cos(bt)$?
(c) Modify your choice in part (b) to provide for the higher amplitude in the years 1985–1992 as compared to 1979–1982.
(d) Graph the function created in part (c) and compare your results to Figure 7.41. Modify your function to improve your approximation.
(e) Use your function to predict sales for 1993.

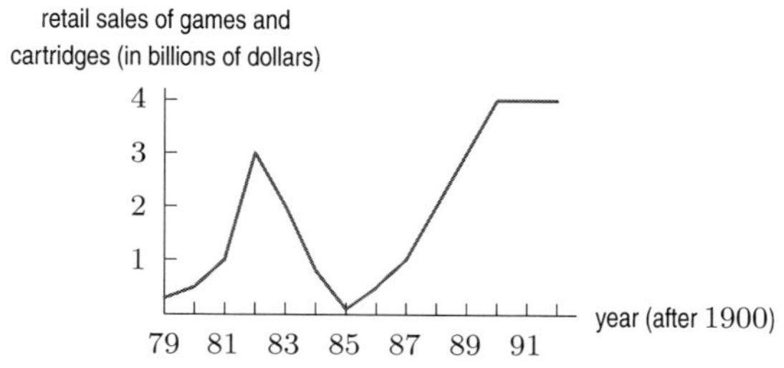

Figure 7.41

9. An amusement park has a giant double ferris wheel as in Figure 7.42. The double ferris wheel has a 30-meter rotating arm attached at its center to a 25-meter main support. At each end of the rotating arm is attached a ferris wheel measuring 20 meters in diameter, rotating in the direction shown in Figure 7.42. The rotating arm takes 6 minutes to complete one full revolution, and each wheel takes 4 minutes to complete a revolution about that wheel's hub. At time $t = 0$ the rotating arm is parallel to

[6] *Modern Digital and Analog Communication Systems* by B. P. Lathi, 2nd ed., page 269.
[7] Source: Nintendo of America

the ground and your seat is at the 3 o'clock position of the rightmost wheel.

(a) Find a formula for $h = f(t)$, your height above the ground in meters, as a function of time in minutes. [Hint: Your height above ground equals the height of your wheel hub above ground plus your height above that hub.]
(b) Graph $f(t)$. Is $f(t)$ periodic? If so, what is its period?
(c) Approximate the least value of t such that h is at a maximum value. What is this maximum value?

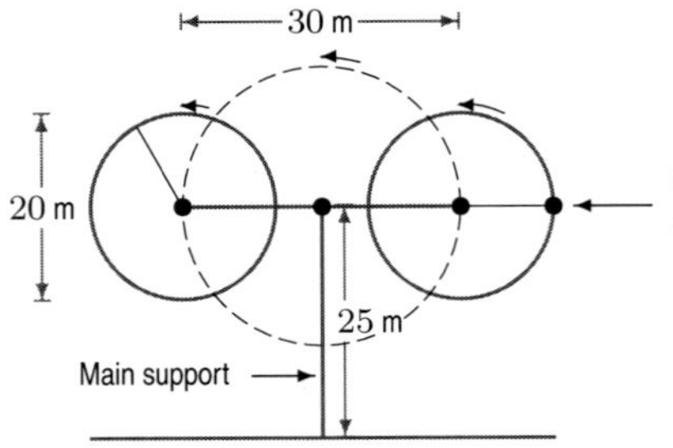

Figure 7.42

10. A rope has one free end. If we give the free end a small upward shake, a wiggle travels down the length of the rope. If we repeatedly shake the free end, a periodic series of wiggles travels down the rope. This situation can be described by a wave function:

$$y(x,t) = A\sin(kx - \omega t).$$

Here, x is the distance along the rope in meters; y is the displacement distance perpendicular to the rope; t is time in seconds; A is the amplitude; $2\pi/k$ is the distance from peak to peak, called the wavelength, λ, measured in meters; and $2\pi/\omega$ is the time in seconds for one wavelength to pass by. Suppose $A = 0.06$, $k = 2\pi$, and $\omega = 4\pi$.

(a) What is the wavelength of the motion?
(b) How many peaks of the wave pass by a given point each second?
(c) Construct the graph of this wave from $x = 0$ to $x = 1.5$ m when t is fixed at 0.
(d) What other values of t would give the same graph as the one found in part (d)?

7.5 POLAR COORDINATES

A point, P, in the plane is often identified by its *Cartesian coordinates* (x, y), where x is the horizontal distance to the point from the origin and y is the vertical distance.[8] Alternatively, we can identify the point, P, by specifying its distance, r, from the origin and the angle, θ, shown in Figure 7.43. The angle θ is measured counterclockwise from the positive x-axis to the line joining P to the origin. The labels r and θ are called the *polar coordinates* of point P.

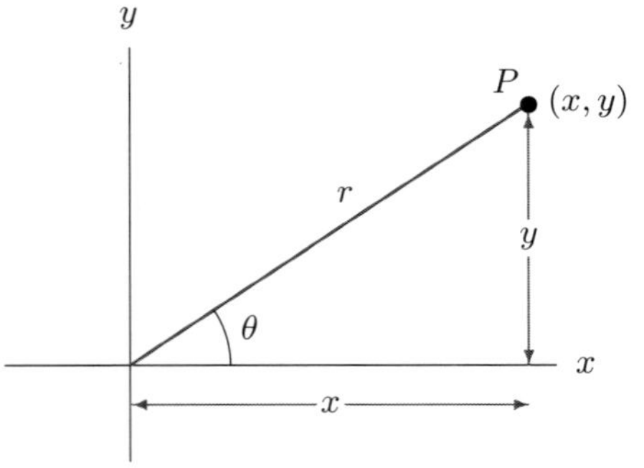

Figure 7.43: Cartesian and polar coordinates for the point P

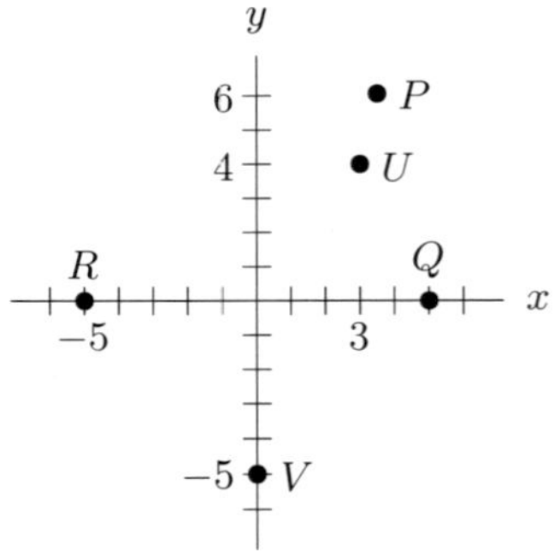

Figure 7.44: Points in the Cartesian plane

Relation Between Cartesian and Polar Coordinates

From the right triangle in Figure 7.43, we see that

- $x = r\cos\theta$ and $y = r\sin\theta$
- $r = \sqrt{x^2 + y^2}$ and $\tan\theta = \dfrac{y}{x}$, $x \neq 0$

The angle θ is determined by the equations $\cos\theta = x/\sqrt{x^2+y^2}$ and $\sin\theta = y/\sqrt{x^2+y^2}$.

[8]Cartesian coordinates can also be called rectangular coordinates.

Warning: In general $\theta \neq \tan^{-1}(y/x)$. It is not possible to determine which quadrant θ is in from the value of $\tan\theta$ alone.

Example 1 (a) Give Cartesian coordinates for the points with polar coordinates (r, θ) given by $P = (7, \pi/3)$, $Q = (5, 0)$, $R = (5, \pi)$.
(b) Give polar coordinates for the points with Cartesian coordinates (x, y) given by $U = (3, 4)$ and $V = (0, -5)$.

Solution (a) See Figure 7.44. Point P is a distance of 7 from the origin. The angle $\theta = \pi/3$ radians (60°). The Cartesian coordinates of P are

$$x = r\cos\theta = 7\cos\frac{\pi}{3} = \frac{7}{2} \quad \text{and} \quad y = r\sin\theta = 7\sin\frac{\pi}{3} = \frac{7\sqrt{3}}{2}.$$

Point Q is located a distance of 5 units along the positive x-axis with Cartesian coordinates

$$x = r\cos\theta = 5\cos 0 = 5 \quad \text{and} \quad y = r\sin\theta = 5\sin 0 = 0.$$

For point R, which is on the negative x-axis,

$$x = r\cos\theta = 5\cos\pi = -5 \quad \text{and} \quad y = r\sin\theta = 5\sin\pi = 0.$$

(b) For $U = (3, 4)$, we have $r = \sqrt{3^2 + 4^2} = 5$ and $\tan\theta = 4/3$. A possible value for θ is $\theta = \arctan 4/3 = 0.927$ radians, or about $53°$. The polar coordinates of U are $(5, 0.927)$. The point V falls on the y-axis, so we can choose $r = 5$, $\theta = 3\pi/2$ for its polar coordinates. In this case, we cannot use $\tan\theta = y/x$ to find θ, because $\tan\theta = y/x = -5/0$ is undefined.

Because the angle θ can be allowed to wrap around the origin more than once, there are many possibilities for the polar coordinates of a point. For the point V in Example 1, we can also choose $\theta = -\pi/2$ or $\theta = 7\pi/2$, so that $(5, -\pi/2)$, $(5, 7\pi/2)$, and $(5, 3\pi/2)$ are all polar coordinates for V. However, we usually choose θ between 0 and 2π. In some situations, negative values of r are used, but we shall not do so.

Example 2 Give three possible sets of polar coordinates for the point P in Figure 7.45.

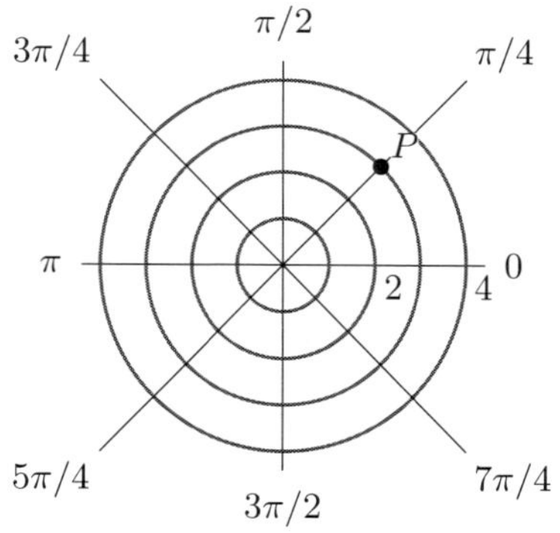

Figure 7.45

Solution Because $r = 3$ and $\theta = \pi/4$, one set of polar coordinates for P is $(3, \pi/4)$. We can also use $\theta = \pi/4 + 2\pi = 9\pi/4$ and $\theta = \pi/4 - 2\pi = -7\pi/4$, to get $(3, 9\pi/4)$ and $(3, -7\pi/4)$.

Graphing Equations in Polar Coordinates

The equations for certain graphs are much simpler when expressed in polar coordinates than in Cartesian coordinates. On the other hand, some graphs that have simple equations in Cartesian coordinates have complicated equations in polar coordinates.

Example 3

(a) Describe in words the graphs of the equation $y = 1$ (in Cartesian coordinates) and the equation $r = 1$ (in polar coordinates).

(b) Write the equation $r = 1$ using Cartesian coordinates. Write the equation $y = 1$ using polar coordinates.

Solution

(a) The equation $y = 1$ describes a horizontal line. Since the equation $y = 1$ places no restrictions on the value of x, it describes every point having a y-value of 1, no matter what the value of its x-coordinate. Similarly, the equation $r = 1$ places no restrictions on the value of θ. Thus, it describes every point having an r-value of 1, that is, having a distance of 1 from the origin. This set of points is the unit circle. See Figure 7.46.

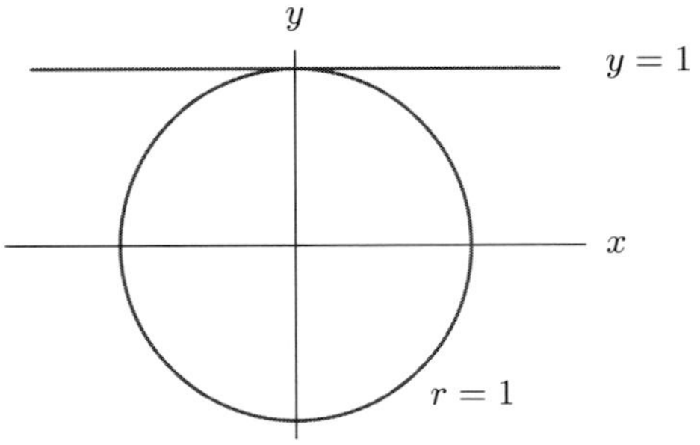

Figure 7.46: The graph of the equation $r = 1$ is the unit circle because $r = 1$ for every point regardless of the value of θ. The graph of $y = 1$ is a horizontal line since $y = 1$ for any x

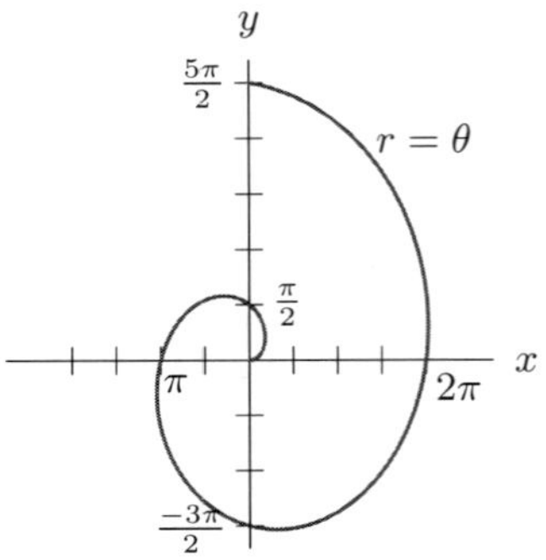

Figure 7.47: A graph of the Archimedean spiral $r = \theta$

(b) Since $r = \sqrt{x^2 + y^2}$, we rewrite the equation $r = 1$ using Cartesian coordinates as $\sqrt{x^2 + y^2} = 1$, or, squaring both sides, as $x^2 + y^2 = 1$. We see that the equation for the unit circle is simpler in polar coordinates than it is in Cartesian coordinates.

On the other hand, since $y = r \sin\theta$, we can rewrite the equation $y = 1$ in polar coordinates as $r \sin\theta = 1$, or, dividing both sides by $\sin\theta$, as $r = 1/\sin\theta$. We see that the equation for this horizontal line is simpler in Cartesian coordinates than in polar coordinates.

Example 4

Graph the equation $r = \theta$. The graph is called an *Archimedean spiral* after the Greek mathematician Archimedes who described its properties (although not by using polar coordinates).

Solution

To construct this graph, use the values in Table 7.1. To help us visualize the shape of the spiral, we convert the angles in Table 7.1 to degrees and the r-values to decimals. See Table 7.2.

Table 7.1 *Points on the Archimedean spiral $r = \theta$, with θ in radians*

θ	0	$\frac{\pi}{6}$	$\frac{\pi}{3}$	$\frac{\pi}{2}$	$\frac{2\pi}{3}$	$\frac{5\pi}{6}$	π	$\frac{7\pi}{6}$	$\frac{4\pi}{3}$	$\frac{3\pi}{2}$
r	0	$\frac{\pi}{6}$	$\frac{\pi}{3}$	$\frac{\pi}{2}$	$\frac{2\pi}{3}$	$\frac{5\pi}{6}$	π	$\frac{7\pi}{6}$	$\frac{4\pi}{3}$	$\frac{3\pi}{2}$

Table 7.2 *Points on the Archimedean spiral $r = \theta$, with θ in degrees*

θ	0	30°	60°	90°	120°	150°	180°	210°	240°	270°
r	0.00	0.52	1.05	1.57	2.09	2.62	3.14	3.67	4.19	4.71

Notice that as the angle θ increases, points on the curve move farther from the origin. At 0°, the point is at the origin. At 30°, it is 0.52 units away from the origin, at 60° it is 1.05 units away, and at 90° it is 1.57 units away. As the angle winds around, the point traces out a curve that moves away from the origin, giving a spiral. (See Figure 7.47.)

Example 5 Curves of the form $r = a \sin n\theta$ or $r = a \cos n\theta$ are called *roses*. Graph the roses

(a) $r = 3 \sin 2\theta$ (b) $r = 4 \cos 3\theta$

Solution

(a) Using a calculator or making a table of values, we see that the graph is a rose with four petals, each of length 3. See Figure 7.48.

(b) The graph is a rose with three petals, each of length 4. See Figure 7.49.

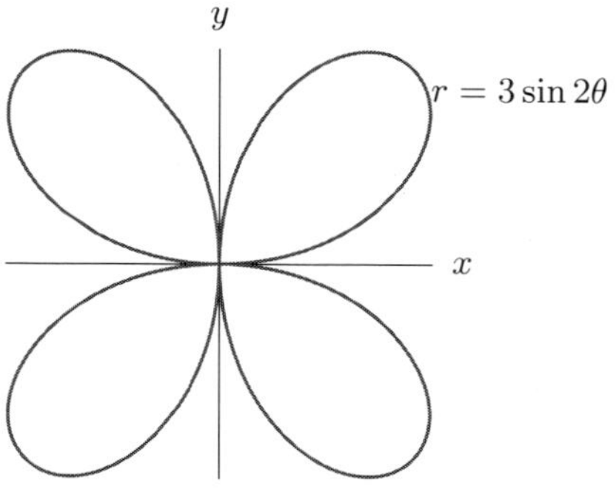

Figure 7.48: Graph of $r = 3 \sin 2\theta$

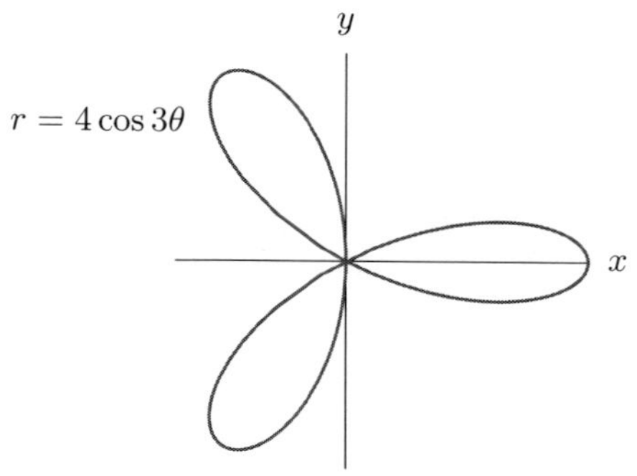

Figure 7.49: Graph of $r = 4 \cos 3\theta$

Example 6 Curves of the form $r = a + n \sin\theta$ or $r = a + b \cos\theta$ are called limaçons. Graph $r = 1 + 2\cos\theta$ and $r = 3 + 2\cos\theta$.

Solution See Figures 7.50 and 7.51.

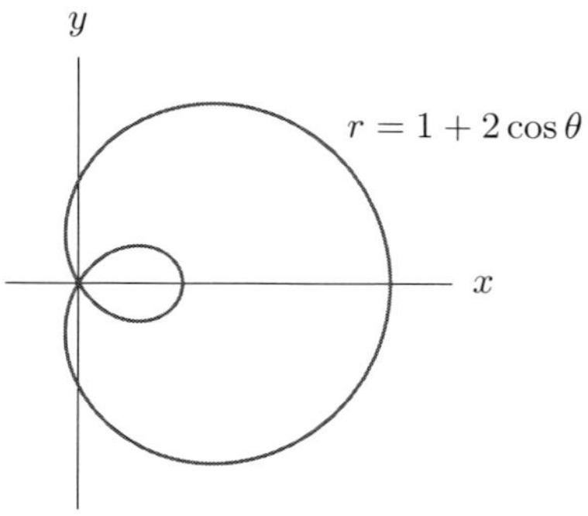

Figure 7.50: Graph of $r = 1 + 2\cos\theta$

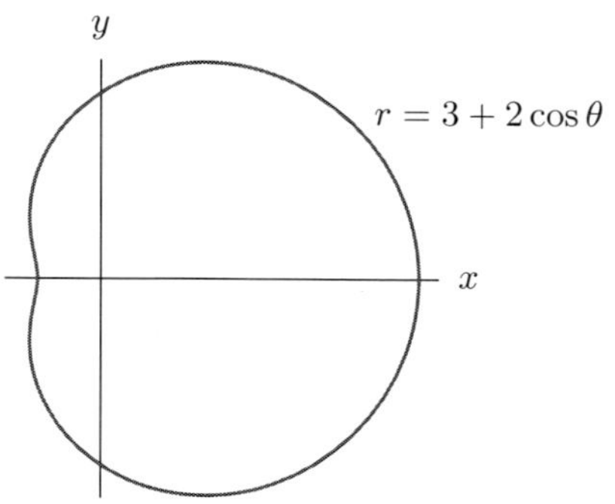

Figure 7.51: Graph of $r = 3 + 2\cos\theta$

Polar coordinates can be used with inequalities to describe regions that are obtained from circles. Such regions are often much harder to represent in Cartesian coordinates.

Example 7 Using inequalities, describe a compact disk with outer diameter 120 mm and inner diameter 15 mm.

Solution The compact disk lies between two circles of radius 7.5 mm and 60 mm. See Figure 7.52. Thus, if the origin is at the center, the disk is represented by

$$7.5 \le r \le 60 \quad \text{and} \quad 0 \le \theta \le 2\pi.$$

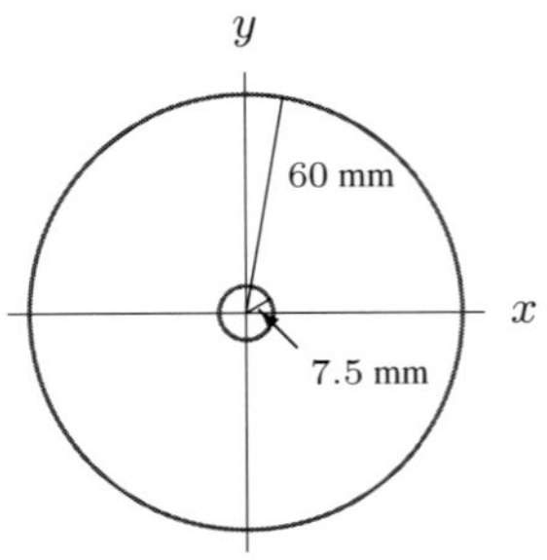

Figure 7.52: Compact disk

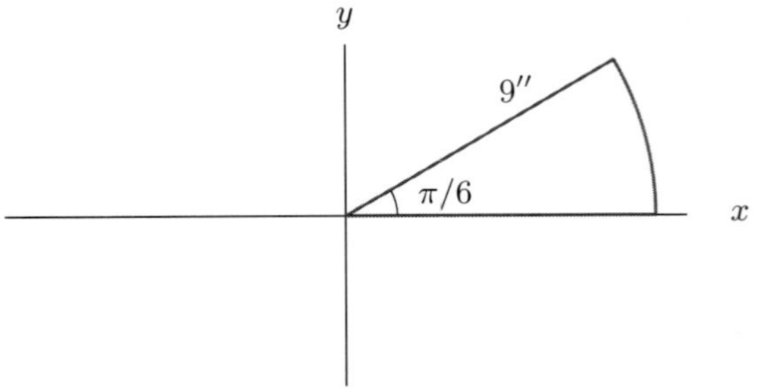

Figure 7.53: Pizza slice

Example 8 An 18 inch pizza is cut into 12 slices. Use inequalities to describe one of the slices.

Solution The pizza has radius 9 inches; the angle at the center is $2\pi/12 = \pi/6$. See Figure 7.53. Thus, if the origin is at center of the original pizza, the slice is represented by

$$0 \le r \le 9 \quad \text{and} \quad 0 \le \theta \le \frac{\pi}{6}.$$

Exercises and Problems for Section 7.5

Exercises

In Exercises 1–9, in which quadrant is a point with the polar coordinate θ?

1. $\theta = 315^\circ$ **2.** $\theta = -290^\circ$ **3.** $\theta = 470^\circ$

4. $\theta = 2.4\pi$ **5.** $\theta = 3.2\pi$ **6.** $\theta = -2.9\pi$

7. $\theta = 10.3\pi$ **8.** $\theta = -13.4\pi$ **9.** $\theta = -7$

In Exercises 10–13 mark the point on the xy-plane. In which range, 0° to 90°, 90° to 180°, 180° to 270°, or 270° to 360°, is its polar coordinate θ?

10. $(2, 3)$ **11.** $(-1, 7)$

12. $(2, -3)$ **13.** $(-1.1, -3.2)$

Convert the Cartesian coordinates in Problems 14–17 to polar coordinates.

14. $(1, 1)$ **15.** $(-1, 0)$

16. $(\sqrt{6}, -\sqrt{2})$ **17.** $(-\sqrt{3}, 1)$

Convert the polar coordinates in Exercises 18–21 to Cartesian coordinates. Give exact answers.

18. $(1, 2\pi/3)$ **19.** $(\sqrt{3}, -3\pi/4)$

20. $(2\sqrt{3}, -\pi/6)$ **21.** $(2, 5\pi/6)$

Problems

Convert the equations in Problems 22–25 to rectangular coordinates.

22. $r = 2$ **23.** $r = 6\cos\theta$

24. $\theta = \pi/4$ **25.** $\tan\theta = r\cos\theta - 2$

Convert the equations in Problems 26–29 to polar coordinates. Express your answer as $r = f(\theta)$.

26. $3x - 4y = 2$ **27.** $x^2 + y^2 = 5$

28. $y = x^2$ **29.** $2xy = 1$

For Problems 30–37, the origin is at the center of a clock, with the positive x-axis going through 3 and the positive y-axis going through 12. The hour hand is 3 cm long and the minute hand is 4 cm long. What are the Cartesian coordinates and polar coordinates of the tips of the hour hand and minute hand, H and M, respectively, at the following times?

30. 12 noon **31.** 3 pm **32.** 9 am **33.** 1 pm

34. 1:30 pm **35.** 7 am **36.** 3:30 pm **37.** 9:15 am

In Problems 38–40, give inequalities for r and θ which describe the following regions in polar coordinates.

38.

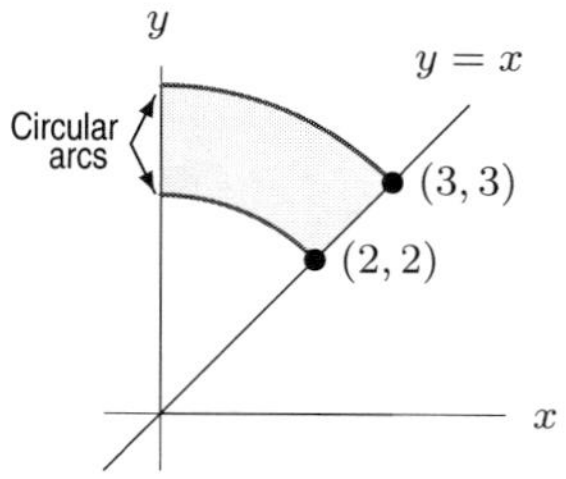

39.

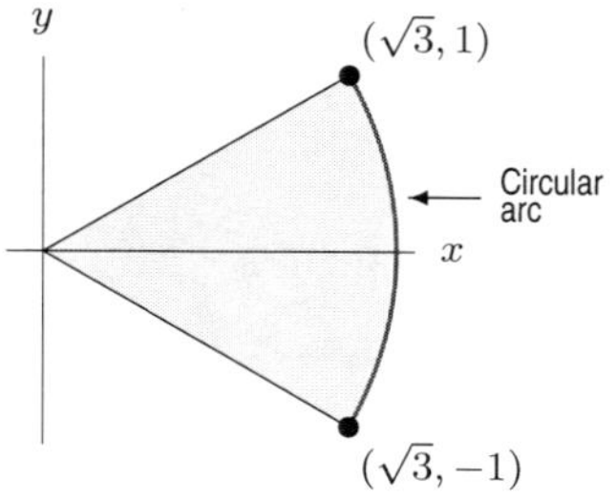

40.

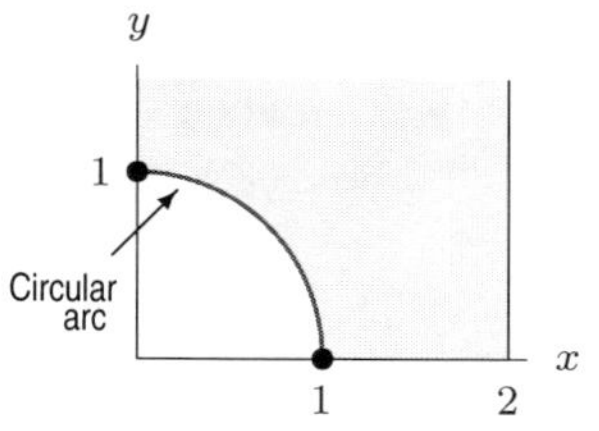

Note: Region extends indefinitely in the y-direction.

41. **(a)** Make a table of values for the equation $r = 1 - \sin\theta$. Include $\theta = 0, \pi/3, \pi/2, 2\pi/3, \pi, \cdots$.
(b) Use the table to graph the equation $r = 1 - \sin\theta$ in the xy-plane. This curve is called a *cardioid*.
(c) At what point(s) does the cardioid $r = 1 - \sin\theta$ intersect a circle of radius 1/2 centered at the origin?
(d) Graph the curve $r = 1 - \sin 2\theta$ in the xy-plane. Compare this graph to the cardioid $r = 1 - \sin\theta$.

42. Graph the equation $r = 1 - \sin(n\theta)$, for $n = 1, 2, 3, 4$. What is the relationship between the value of n and the shape of the graph?

43. Graph the equation $r = 1 - \sin\theta$, with $0 \le \theta \le n\pi$, for $n = 2, 3, 4$. What is the relationship between the value of n and the shape of the graph?

44. Graph the equation $r = 1 - n\sin\theta$, for $n = 2, 3, 4$. What is the relationship between the value of n and the shape of the graph?

45. Graph the equation $r = 1 - \cos\theta$. Describe its relationship to $r = 1 - \sin\theta$.

46. Give inequalities that describe the flat surface of a washer that is one inch in diameter and has an inner hole with a diameter of 3/8 inch.

47. Graph the equation $r = 1 - \sin(2\theta)$ for $0 \le \theta \le 2\pi$. There are two loops. For each loop, give a restriction on θ that shows all of that loop and none of the other loop.

48. A slice of pizza is one eighth of a circle of radius 1 foot. The slice is in the first quadrant, with one edge along the x-axis, and the center of the pizza at the origin. Give inequalities describing this region using:
(a) Polar coordinates (b) Rectangular coordinates

7.6 COMPLEX NUMBERS AND POLAR COORDINATES

Extending the Definition of Functions

Some of the functions we have studied, such as x^2, e^x, and $\sin x$, are defined for all real numbers, whereas other functions, such as $\sqrt{x}$, are defined for some real numbers but not for others. The function $\sqrt{x}$ is undefined for $x < 0$.

In this section, we expand our idea of number to include *complex numbers*. These numbers make it possible to interpret expressions like $\sqrt{-4}$ in a meaningful way.

Using Complex Numbers to Solve Equations

The quadratic equation

$$x^2 - 2x + 2 = 0$$

is not satisfied by any real number x. Applying the quadratic formula gives

$$x = \frac{2 \pm \sqrt{4-8}}{2} = 1 \pm \frac{\sqrt{-4}}{2}.$$

But -4 does not have a square root which is a real number. To overcome this problem, we define the imaginary number i such that:

$$i^2 = -1.$$

Using i, we see that $(2i)^2 = -4$. Now we can write our solution of the quadratic as:

$$x = 1 \pm \frac{\sqrt{-4}}{2} = 1 \pm \frac{2i}{2} = 1 \pm i.$$

The numbers $1+i$ and $1-i$ are examples of complex numbers.

A **complex number** is defined as any number that can be written in the form

$$z = a + bi,$$

where a and b are real numbers and $i = \sqrt{-1}$.
The *real part* of z is the number a; the *imaginary part* is the number b.

Calling the number i imaginary makes it sound like i does not exist in the same way as real numbers. In some cases, it is useful to make such a distinction between real and imaginary numbers. If we measure mass or position, we want our answers to be real. But imaginary numbers are just as legitimate mathematically as real numbers. For example, complex numbers are used in studying wave motion in electric circuits.

Algebra of Complex Numbers

Numbers such as 0, 1, $\frac{1}{2}$, π, e and $\sqrt{2}$ are called *purely real*, because their imaginary part is zero. Numbers such as i, $2i$, and $\sqrt{2}i$ are called *purely imaginary*, because they contain only the number i multiplied by a nonzero real coefficient.

Two complex numbers are called *conjugates* if their real parts are equal and if their imaginary parts differ only in sign. The complex conjugate of the complex number $z = a + bi$ is denoted $\overline{z}$, so

$$\overline{z} = a - bi.$$

(Note that z is real if and only if $z = \overline{z}$.)

- Two complex numbers, $z = a + bi$ and $w = c + di$, are equal only if $a = c$ and $b = d$.
- Adding two complex numbers is done by adding real and imaginary parts separately:
  $$(a+bi) + (c+di) = (a+c) + (b+d)i.$$
- Subtracting is similar: $(a+bi) - (c+di) = (a-c) + (b-d)i.$

- Multiplication works just like for polynomials, using $i^2 = -1$ to simplify:

$$\begin{aligned}(a+bi)(c+di) &= a(c+di)+bi(c+di) = ac+adi+bci+bdi^2\\ &= ac+adi+bci-bd = (ac-bd)+(ad+bc)i.\end{aligned}$$

- Powers of i: We know that $i^2 = -1$; then, $i^3 = i \cdot i^2 = -i$, and $i^4 = (i^2)^2 = (-1)^2 = 1$. Then $i^5 = i \cdot i^4 = i$, and so on. Thus we have

$$i^n = \begin{cases} i & \text{for } n = 1, 5, 9, 13, \ldots \\ -1 & \text{for } n = 2, 6, 10, 14, \ldots \\ -i & \text{for } n = 3, 7, 11, 15, \ldots \\ 1 & \text{for } n = 4, 8, 12, 16, \ldots \end{cases}$$

- The product of a number and its conjugate is always real and nonnegative:

$$z \cdot \overline{z} = (a+bi)(a-bi) = a^2 - abi + abi - b^2i^2 = a^2 + b^2.$$

- Dividing is done by multiplying the numerator and denominator by the conjugate of the denominator, thereby making the denominator real:

$$\frac{a+bi}{c+di} = \frac{a+bi}{c+di}\cdot\frac{c-di}{c-di} = \frac{ac-adi+bci-bdi^2}{c^2+d^2} = \frac{ac+bd}{c^2+d^2} + \frac{bc-ad}{c^2+d^2}i.$$

Example 1 Compute the sum $(3+6i)+(5-2i)$

Solution Adding real and imaginary parts gives $(3+6i)+(5-2i) = (3+5)+(6i-2i) = 8+4i$.

Example 2 Simplify $(2+7i)(4-6i)-i$.

Solution Multiplying out gives $(2+7i)(4-6i)-i = 8+28i-12i-42i^2-i = 8+15i+42 = 50+15i$.

Example 3 Compute $\dfrac{2+7i}{4-6i}$.

Solution The conjugate of the denominator is $4+6i$, so we multiply by $(4+6i)/(4+6i)$.

$$\frac{2+7i}{4-6i} = \frac{2+7i}{4-6i}\cdot\frac{4+6i}{4+6i} = \frac{8+12i+28i+42i^2}{4^2+6^2} = \frac{-34+40i}{52} = -\frac{17}{26}+\frac{10}{13}i.$$

The Complex Plane and Polar Coordinates

It is often useful to picture a complex number $z = x + iy$ in the plane, with x along the horizontal axis and y along the vertical. The xy-plane is then called the *complex plane*. Figure 7.54 shows the complex numbers $3i$, $-2 + 3i$, -3, $-2i$, 2, and $1 + i$.

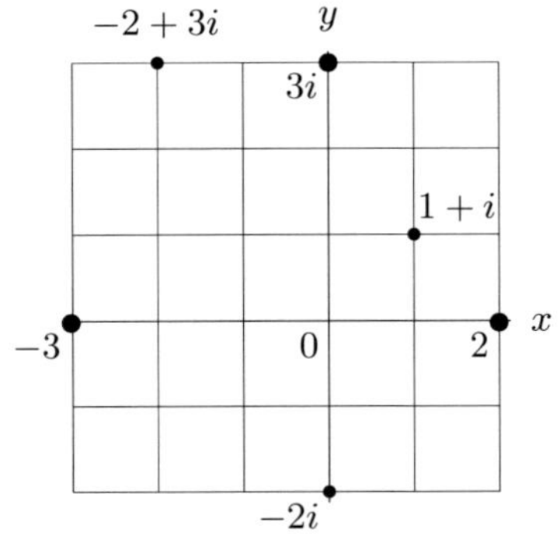

Figure 7.54: Points in the complex plane

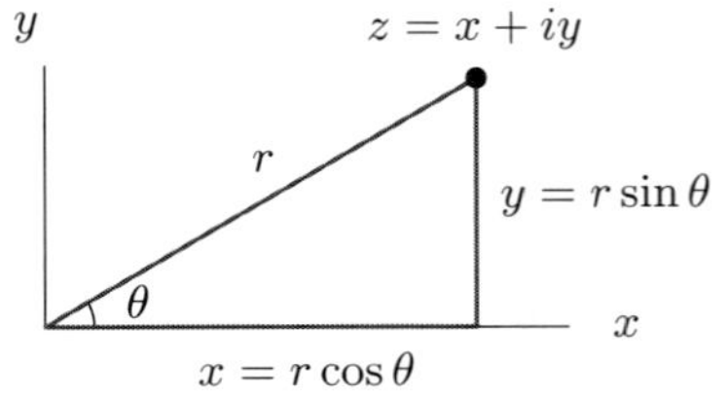

Figure 7.55: The point $z = x + iy$ in the complex plane, showing polar coordinates

Figure 7.55 shows that a complex number can be written using polar coordinates as follows:

$$z = x + iy = r\cos\theta + ir\sin\theta.$$

Example 4 Express $z = -2i$ and $z = -2 + 3i$ using polar coordinates. (See Figure 7.54.)

Solution For $z = -2i$, the distance of z from the origin is 2, so $r = 2$. Also, one value for θ is $\theta = 3\pi/2$. Thus, using polar coordinates,

$$z = -2i = 2\cos(3\pi/2) + i\,2(\sin 3\pi/2).$$

For $z = -2+3i$, we have $x = -2$, $y = 3$, so $r = \sqrt{(-2)^2 + 3^2} = 3.606$ and $\tan\theta = 3/(-2)$, so, for example, $\theta = 2.159$. Thus, using polar coordinates,

$$z = -2 + 3i = 3.606\cos(2.159) + i\,3.606\sin(2.159).$$

Example 5 What complex number is represented by the point with polar coordinates $r = 5$ and $\theta = 3\pi/4$?

Solution Since $x = r\cos\theta$ and $y = r\sin\theta$ we see that $z = x + iy = 5\cos(3\pi/4) + i(5\sin(3\pi/4)) = -5/\sqrt{2} + i\,5/\sqrt{2}$.

Euler's Formula

Based on what we have seen so far, there is no reason to expect a connection between the exponential function e^x and the trigonometric functions $\sin x$ and $\cos x$. However, in the eighteenth century, the Swiss mathematician Leonhard Euler discovered a surprising connection between these functions that involves complex numbers. This result, called *Euler's formula*, states that, for real θ in radians,

$$e^{i\theta} = \cos\theta + i\sin\theta$$

Example 6 Evaluate $e^{i\pi}$.

Solution Using Euler's formula,

$$e^{i\pi} = \cos\pi + i\sin\pi = -1 + i\cdot 0 = -1.$$

This statement, known as *Euler's identity*, is sometimes written $e^{i\pi} + 1 = 0$. It is famous because it relates five of the most fundamental constants in mathematics: 0, 1, e, π, and i.

You may be wondering what it means to raise a number to an imaginary power. Euler's formula does not tell us what complex powers mean, but it does provide a consistent way to evaluate them. We use Euler's formula without proof; we also assume that complex exponentials, such as 4^i and $e^{i\theta}$, obey the usual rules of algebraic manipulation.

Example 7 Convert 4^i to the form $a + bi$.

Solution Using Euler's formula and the exponent rules, we can evaluate the expression 4^i. Since $4 = e^{\ln 4}$,

$$\begin{aligned}
4^i &= \left(e^{\ln 4}\right)^i \\
&= e^{i(\ln 4)} && \text{Using an exponent rule} \\
&= \cos(\ln 4) + i\sin(\ln 4) && \text{Using Euler's formula} \\
&= 0.183 + 0.983i && \text{Using a calculator.}
\end{aligned}$$

Polar Form of a Complex Number

Euler's formula allows us to write the complex number z in *polar form*:

> If the point representing z has polar coordinates (r, θ), then
>
> $$z = r(\cos\theta + i\sin\theta) = re^{i\theta}.$$
>
> The expression $re^{i\theta}$ is called the **polar form** of the complex number z.

Similarly, since $\cos(-\theta) = \cos\theta$ and $\sin(-\theta) = -\sin\theta$, we have

$$re^{-i\theta} = r\left(\cos(-\theta) + i\,\sin(-\theta)\right) = r(\cos\theta - i\,\sin\theta).$$

Thus, $re^{-i\theta}$ is the conjugate of $re^{i\theta}$.

Example 8 Express the complex number represented by the point with polar coordinates $r = 8$ and $\theta = 3\pi/4$, in Cartesian form, $a + bi$, and in polar form, $z = re^{i\theta}$.

Solution Using Cartesian coordinates, the complex number is

$$z = 8\left(\cos\left(\frac{3\pi}{4}\right) + i\sin\left(\frac{3\pi}{4}\right)\right) = \frac{-8}{\sqrt{2}} + i\frac{8}{\sqrt{2}}.$$

The polar form is

$$z = 8e^{i\,3\pi/4}.$$

The polar form of complex numbers makes finding powers and roots of complex numbers easier. Using the polar form, we find any power of z as follows:

$$z^p = (re^{i\theta})^p = r^p e^{ip\theta}.$$

To find roots, let p be a fraction, as in the following example.

Example 9 Find a cube root of the complex number represented by the point with polar coordinates $(8, 3\pi/4)$.

Solution From Example 8, we have $z = 8e^{i3\pi/4}$. Thus,

$$\begin{aligned}\sqrt[3]{z} = \left(8e^{i\,3\pi/4}\right)^{1/3} = 8^{1/3}e^{i(3\pi/4)\cdot(1/3)} = 2e^{\pi i/4} &= 2\left(\cos\left(\frac{\pi}{4}\right) + i\sin\left(\frac{\pi}{4}\right)\right) \\ &= 2\left(\frac{1}{\sqrt{2}} + \frac{i}{\sqrt{2}}\right) = \frac{2}{\sqrt{2}} + \frac{2i}{\sqrt{2}}.\end{aligned}$$

Euler's Formula and Trigonometric Identities

We can use Euler's formula to derive trigonometric identities.

Example 10 Derive the Pythagorean identity, $\cos^2\theta + \sin^2\theta = 1$, using Euler's formula.

Solution From the exponent rules, we know

$$e^{i\theta} \cdot e^{-i\theta} = e^{i\theta - i\theta} = e^0 = 1.$$

Using Euler's formula, we rewrite this as

$$\underbrace{(\cos\theta + i\sin\theta)}_{e^{i\theta}}\underbrace{(\cos\theta - i\sin\theta)}_{e^{-i\theta}} = 1.$$

Multiplying out the left-hand side gives

$$(\cos\theta + i\sin\theta)(\cos\theta - i\sin\theta) = \cos^2\theta - i\sin\theta\cos\theta + i\sin\theta\cos\theta - i^2\sin^2\theta = 1.$$

Since $-i^2 = +1$, simplifying the left side gives the Pythagorean identity

$$\cos^2\theta + \sin^2\theta = 1.$$

Example 11 Derive the identities for $\cos(\theta + \phi)$ and $\sin(\theta + \phi)$.

Solution Using the exponent rules, we see from Euler's formula that

$$\begin{aligned} e^{i(\theta+\phi)} &= e^{i\theta} \cdot e^{i\phi} \\ &= (\cos\theta + i\sin\theta)(\cos\phi + i\sin\phi) \\ &= \cos\theta\cos\phi + \underbrace{i\cos\theta\sin\phi + i\sin\theta\cos\phi}_{i(\cos\theta\sin\phi + \sin\theta\cos\phi)} + \underbrace{i^2\sin\theta\sin\phi}_{-\sin\theta\sin\phi} \\ &= \underbrace{\cos\theta\cos\phi - \sin\theta\sin\phi}_{\text{Real part}} + i\underbrace{(\sin\theta\cos\phi + \cos\theta\sin\phi)}_{\text{Imaginary part}}. \end{aligned}$$

But Euler's formula also gives

$$e^{i(\theta+\phi)} = \underbrace{\cos(\theta+\phi)}_{\text{Real part}} + i\underbrace{\sin(\theta+\phi)}_{\text{Imaginary part}}.$$

Two complex numbers are equal only if their real and imaginary parts are equal. Setting real parts equal gives

$$\cos(\theta + \phi) = \cos\theta\cos\phi - \sin\theta\sin\phi.$$

Setting imaginary parts equal gives

$$\sin(\theta + \phi) = \sin\theta\cos\phi + \sin\phi\cos\theta.$$

Example 12 Show that $\cos x = \dfrac{e^{ix} + e^{-ix}}{2}$.

Solution We know from Euler's formula that

$$e^{ix} = \cos x + i\sin x.$$

Since $\cos(-x) = \cos x$ and $\sin(-x) = -\sin x$, we have

$$e^{-ix} = e^{i(-x)} = \cos(-x) + i\sin(-x) = \cos x - i\sin x.$$

Then

$$\begin{aligned} e^{ix} + e^{-ix} &= (\cos x + i\sin x) + (\cos x - i\sin x) \\ &= 2\cos x, \end{aligned}$$

so

$$\cos x = \frac{e^{ix} + e^{-ix}}{2}.$$

Exercises and Problems for Section 7.6

Exercises

For Exercises 1–6, express the complex number in polar form, $z = re^{i\theta}$.

1. -5 **2.** $-i$ **3.** 0

4. $2i$ **5.** $-3 - 4i$ **6.** $-1 + 3i$

Perform the calculations in Exercises 7–14. Give your answer in Cartesian form, $z = x + iy$.

7. $(2 + 3i)^2$

8. $(2 + 3i)(5 + 7i)$

9. $(2 + 3i) + (-5 - 7i)$

10. $(0.5 - i)(1 - i/4)$

11. $(e^{i\pi/3})^2$

12. $(2i)^3 - (2i)^2 + 2i - 1$

13. $\sqrt{e^{i\pi/3}}$

14. $\sqrt[4]{10e^{i\pi/2}}$

Problems

15. **(a)** Find the polar form of the complex number i.

(b) If $z = re^{i\theta}$ is any point in the complex plane, show that the point iz can by found by rotating the segment from 0 to z by 90° counterclockwise, as in Figure 7.56.

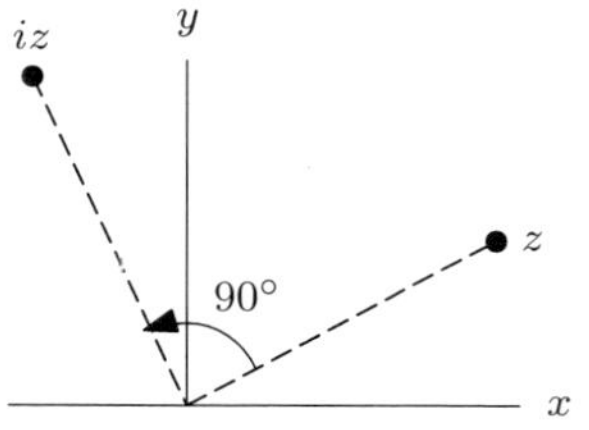

Figure 7.56

16. **(a)** Let $z = 3 + 2i$. Plot z and iz on the complex plane.

(b) Show that the line from the origin to z is perpendicular to the line from the origin to iz.

By writing the complex numbers in polar form, $z = re^{i\theta}$, find a value for the quantities in Problems 17–25. Give your answer in Cartesian form, $z = x + iy$.

17. $\sqrt{4i}$ **18.** $\sqrt{-i}$ **19.** $\sqrt[3]{i}$

20. $\sqrt{7i}$ **21.** $\sqrt[4]{-1}$ **22.** $(1 + i)^{2/3}$

23. $(\sqrt{3} + i)^{1/2}$ **24.** $(\sqrt{3}+i)^{-1/2}$ **25.** $(\sqrt{5} + 2i)^{\sqrt{2}}$

Solve the simultaneous equations in Problems 26–27 for the complex numbers A_1 and A_2.

26. $A_1 + A_2 = 2$
$(1-i)A_1 + (1+i)A_2 = 0$

27. $A_1 + A_2 = i$
$iA_1 - A_2 = 3$

28. Let $z_1 = -3 - i\sqrt{3}$ and $z_2 = -1 + i\sqrt{3}$.

(a) Find $z_1 z_2$ and z_1/z_2. Give your answer in Cartesian form, $z = x + iy$.

(b) Put z_1 and z_2 into polar form, $z = re^{i\theta}$. Find $z_1 z_2$ and z_1/z_2 using the polar form, and check that you get the same answer as in part (a).

29. If the roots of the equation $x^2 + 2bx + c = 0$ are the complex numbers $p \pm iq$, find expressions for p and q in terms of a and b.

For Problems 30–33, use Euler's formula to derive the identity. (Note that if a, b, c, d are real numbers, $a + bi = c + di$ means that $a = c$ and $b = d$.)

30. $\sin 2\theta = 2\sin\theta\cos\theta$

31. $\cos 2\theta = \cos^2\theta - \sin^2\theta$

32. $\cos(-\theta) = \cos\theta$

33. $\sin(-\theta) = -\sin\theta$

CHAPTER SUMMARY

- **Identities** (See inside cover)
  Pythagorean: $\cos^2\theta + \sin^2\theta = 1$
  Tangent: $\tan\theta = \sin\theta/\cos\theta$.
  Negative angles: $\cos(-\theta) = \cos\theta, \sin(-\theta) = -\sin\theta$.
  Relating sine and cosine: $\sin\theta = \cos(\theta - \pi/2)$;
  $\cos\theta = \sin(\theta - \pi/2)$.
- **Double Angle Formulas** (See inside cover)
$$\begin{aligned}\sin 2t &= 2\sin t\cos t,\\ \cos 2t &= 1 - 2\sin^2 t\\ &= 2\cos^2 t - 1,\\ \tan 2t &= \frac{2\tan t}{1-\tan^2 t}.\end{aligned}$$
- **Sum and Difference Formulas** (See inside cover)
  For example, $\sin(\theta + \phi)$, $\cos u + \cos v$, $A\sin(Bt + \phi)$.
- **Modeling with Trig Functions**
  Damped oscillations; rising midline; acoustic beats.
- **Polar Coordinates**
$$x = r\cos\theta; \quad y = r\sin\theta;$$
$$r = \sqrt{x^2 + y^2}; \quad \tan\theta = \frac{y}{x}.$$
  Graphing with polar coordinates.
- **Complex Numbers**
  Operations and graphing with $z = a + bi$.
  Finding roots of complex numbers.
- **Euler's Formula**
  $e^{i\theta} = \cos\theta + i\sin\theta$.
  Polar form: $z = r(\cos\theta + i\sin\theta) = re^{i\theta}$.

REVIEW EXERCISES AND PROBLEMS FOR CHAPTER SEVEN

Exercises

1. **(a)** In the right triangle in Figure 7.57, angle $A = 30°$ and $b = 2\sqrt{3}$. Find the lengths of the other sides and the other angles.

(b) Repeat part (a), this time assuming only that $a = 25$ and $c = 24$.

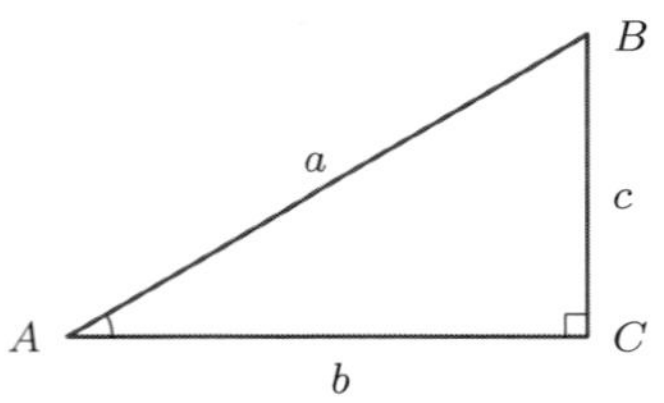

Figure 7.57

Find all sides and angles of the triangles in Exercises 2–5. (Not necessarily drawn to scale.)

2.

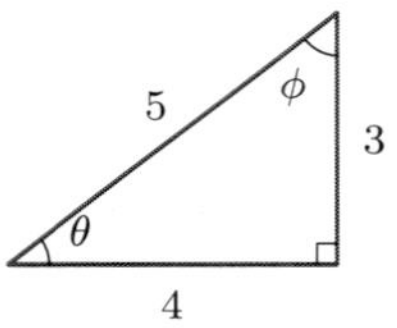

3.

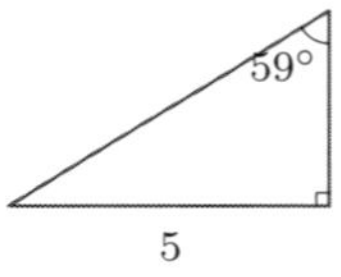

4.

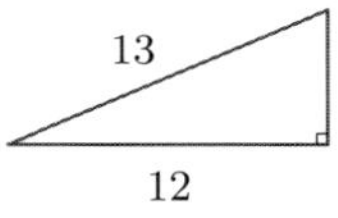

5.

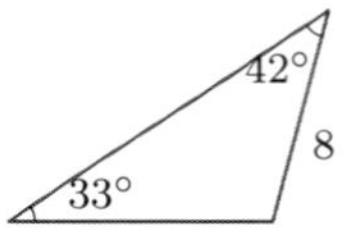

6. Find the value of the angle θ in Figure 7.58.

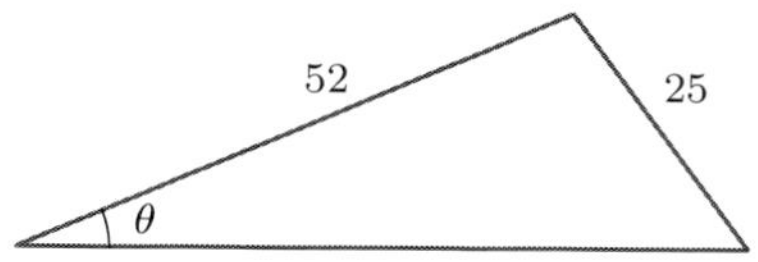

Figure 7.58

Simplify the expressions in Exercises 7–10.

7. $\dfrac{\sin\theta}{\tan\theta}$ **8.** $\dfrac{\cos^2 A}{1+\sin A}$ **9.** $\dfrac{\cos 2\phi+1}{\cos\phi}$

10. $\cos^2\theta(1+\tan\theta)(1-\tan\theta)$

In Exercises 11–14, find exact values using sum and difference formulas.

11. $\sin 165^\circ - \sin 75^\circ$ **12.** $\sin 285^\circ$

13. $\cos 105^\circ$ **14.** $\cos 345^\circ$

In Exercises 15–18, in which quadrant is a point with the polar coordinate θ?

15. $\theta = -299^\circ$ **16.** $\theta = 730^\circ$

17. $\theta = -7.7\pi$ **18.** $\theta = 14.4\pi$

Convert each of the polar coordinates in Exercises 19–22 to Cartesian coordinates. Use decimal approximations where appropriate.

19. $(\pi/2, 0)$ **20.** $(2, 2)$ **21.** $(0, \pi/2)$ **22.** $(3, 40^\circ)$

Problems

23. **(a)** Find expressions in terms of a, b, and c for the sine, cosine, and tangent of the angle ϕ in Figure 7.59.
(b) Using your answers in part (a), show that $\sin\phi = \cos\theta$ and $\cos\phi = \sin\theta$.

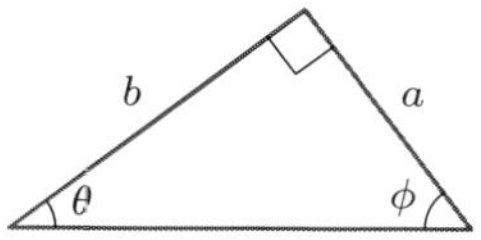

Figure 7.59

24. One student said that $\sin 2\theta \neq 2\sin\theta$, but another student said let $\theta = \pi$ and then $\sin 2\theta = 2\sin\theta$. Who is right?

25. Suppose that $\cos(2\theta) = 2/7$ and θ is in the first quadrant. Find $\cos\theta$ exactly.

26. Use a trigonometric identity to find exactly all solutions to:
$\cos 2\theta = \sin\theta, \quad 0 \le \theta < 2\pi.$

27. Suppose $\sin\theta = 1/7$ for $\pi/2 < \theta < \pi$.
(a) Use the Pythagorean identity to find $\cos\theta$.
(b) What is θ?

28. Solve exactly: $\cos(2\alpha) = -\sin\alpha$ with $0 \le \alpha < 2\pi$.

29. Suppose that $\sin(\ln x) = \frac{1}{3}$, and that $\sin(\ln y) = \frac{1}{5}$. If $0 < \ln x < \frac{\pi}{2}$ and $0 < \ln y < \frac{\pi}{2}$, use the sum-of-angle sine formula to evaluate

$$\sin(\ln(xy)).$$

30. Simplify the expression $\sin\left(2\cos^{-1}\left(\frac{5}{13}\right)\right)$ to a rational number.

Prove the identities in Problems 31–34.

31. $\tan\theta = \dfrac{1-\cos 2\theta}{2\cos\theta\sin\theta}$

32. $(\sin^2 2\pi t + \cos^2 2\pi t)^3 = 1$

33. $\sin^4 x - \cos^4 x = \sin^2 x - \cos^2 x$

34. $\dfrac{1+\sin\theta}{\cos\theta} = \dfrac{\cos\theta}{1-\sin\theta}$

35. Graph $\tan^2 x+\sin^2 x$, $(\tan^2 x)(\sin^2 x)$, $\tan^2 x-\sin^2 x$, $\tan^2 x/\sin^2 x$ to see if any two expressions appear to be identical. Prove any identities you find.

36. Graph $f(t) = \cos t$ and $g(t) = \sin(t+\pi/2)$. Explain what you see.

37. **(a)** Graph $g(\theta) = \sin\theta - \cos\theta$.
(b) Write $g(\theta)$ as a sine function without using cosine. Write $g(\theta)$ as a cosine function without using sine.

38. Let x be the side adjacent to the angle θ, with $0 < \theta < \pi/4$, in a right triangle whose hypotenuse is 1. (See Figure 7.60.) Without trigonometric functions, express the following in terms of x.

(a) $\cos\theta$ **(b)** $\cos(\frac{\pi}{2} - \theta)$ **(c)** $\tan^2\theta$
(d) $\sin(2\theta)$ **(e)** $\cos(4\theta)$ **(f)** $\sin(\cos^{-1} x)$

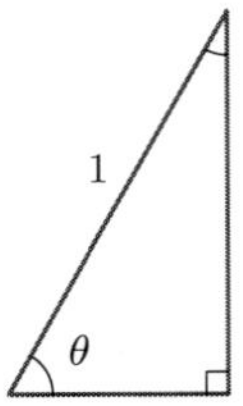

Figure 7.60

39. The planet Betagorp travels around its sun, Trigosol, in an elliptical orbit. The distance, r, of Betagorp from Trigosol at angle θ is given by

$$r = \frac{5.35 \cdot 10^7}{1 - 0.234\cos\theta}.$$

Find the least positive θ (in decimal degrees to 3 significant digits) so that the distance is $5.12 \cdot 10^7$ miles.

40. To check the calibration of their transit (an instrument to measure angles), two student surveyors used the set-up in Figure 7.61. What angles in degrees for α and β should they get if their transit is accurate?

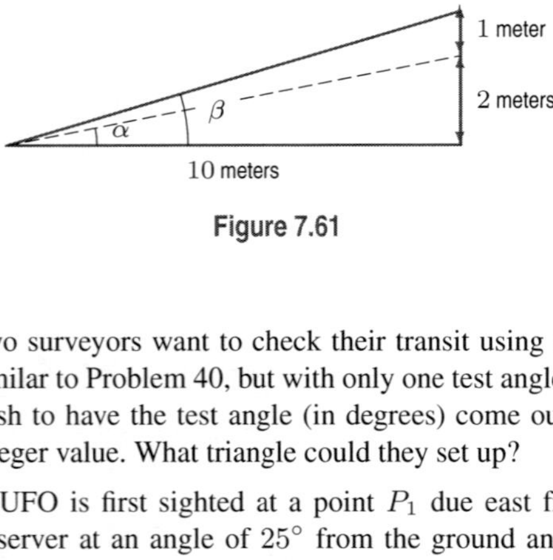

Figure 7.61

41. Two surveyors want to check their transit using a setup similar to Problem 40, but with only one test angle. They wish to have the test angle (in degrees) come out to an integer value. What triangle could they set up?

42. A UFO is first sighted at a point P_1 due east from an observer at an angle of 25° from the ground and at an altitude (vertical distance above ground) of 200 m. (See Figure 7.62.) The UFO is next sighted at a point P_2 due east at an angle of 50° and an altitude of 400 m. What is the distance from P_1 to P_2?

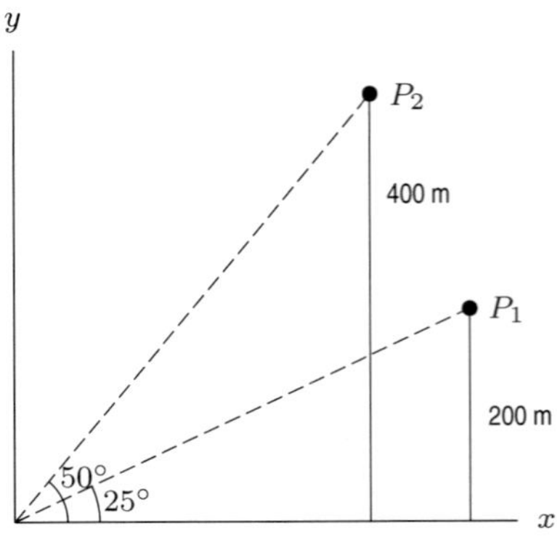

Figure 7.62

43. Knowing the height of the Columbia Tower in Seattle, determine the height of the Seafirst Tower and the distance between the towers. See Figure 7.63.

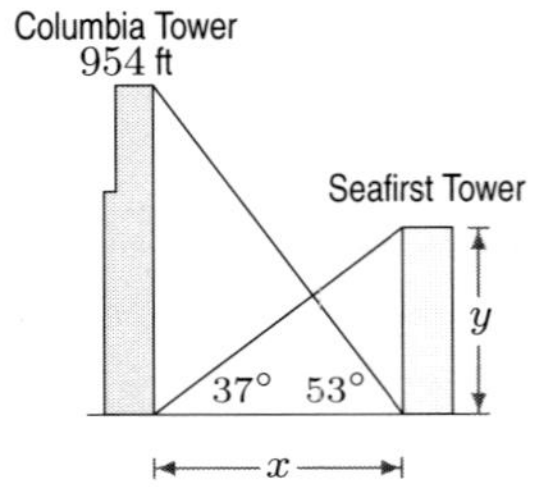

Figure 7.63

44. The telephone company needs to run a wire from the telephone pole across a street to a new house. See Figure 7.64. Since they cannot measure across the busy street, they hire a surveyor to determine the amount of wire needed. The surveyor measures a distance of 23.5 feet from the telephone pole to a stake, D, which she sets in the ground. She measures 145.3 feet from the house to a second stake, E. She sets a third stake, F, at a distance of 105.2 feet from the second stake. The surveyor measures angle DEF to be 83° and angle EFD to be 68°. A total of 20 feet of wire is needed to make connections at the house and pole, and wire is sold in 100 feet rolls. How many rolls of wire are needed?

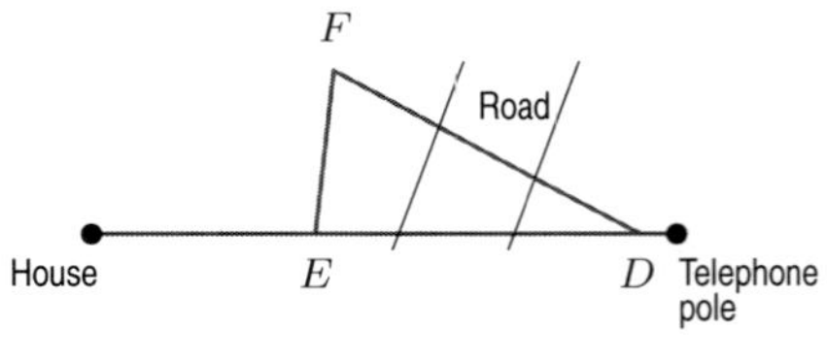

Figure 7.64

45. **(a)** Explain why $r = 2\cos\theta$ and $(x-1)^2 + y^2 = 1$ are equations for the same circle.
(b) Give Cartesian and polar coordinates for the 12, 3, 6, and 9 o'clock positions on the circle in part (a).

46. For each of the following expressions, find a line segment in Figure 7.65 with length equal to the value of the expression.

(a) $\sin\theta$ **(b)** $\cos\theta$ **(c)** $\tan\theta$

(d) $\dfrac{1}{\sin\theta}$ **(e)** $\dfrac{1}{\cos\theta}$ **(f)** $\dfrac{1}{\tan\theta}$

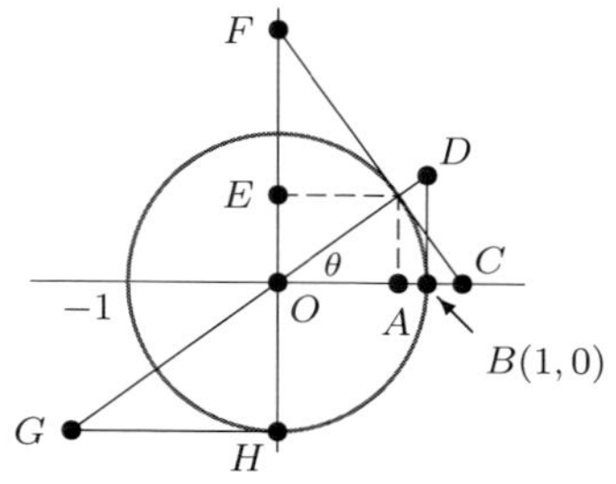

Figure 7.65

CHECK YOUR UNDERSTANDING

Are the statements in Problems 1–58 true or false? Give an explanation for your answer.

1. The function $f(t) = e^{-t}\cos t$ is a damped oscillation.

2. If ΔABC is an isosceles right triangle with angle $BAC = 90°$, then $\sin B = \frac{\sqrt{3}}{2}$.

3. A 16 foot ladder leans against a wall. The angle formed by the ladder and the ground is $60°$. Then, the top of the ladder will meet the wall at a height of 8 feet above the ground.

4. The Law of Cosines may be applied only to right triangles.

5. If the lengths of three sides of a triangle are known, the Law of Cosines may be used to determine any of the angles.

6. In ΔNPR, the lengths of the sides opposite angles N, P and R are n, p and r respectively. Angle P is given by $\cos^{-1}\left(\dfrac{n^2 + r^2 - p^2}{2rn}\right)$.

7. When used to find the length of the side of a triangle opposite a right angle, the Law of Cosines reduces to the Pythagorean Theorem.

8. If two angles and the length of one side of a triangle are known, the Law of Cosines may be used to find the length of another side.

9. In ΔQXR, the lengths of the sides opposite angles Q, X and R are q, x and r respectively. Then $r\sin Q = q\sin R$.

10. In ΔLAB, the ratio of the length of side LA to the length of side BA is $\dfrac{\sin B}{\sin L}$.

11. If the lengths of two sides of a triangle and the angle included between them are known, the Law of Cosines may be used to determine the length of the third side of the triangle.

12. If two angles and the length of one side of a triangle are known, the Law of Sines may be used to determine the length of another side.

13. It is possible for triangle ΔMAT to have the following dimensions: $MA = 8$, $MT = 12$ and angle $T = 60°$.

14. For all values of ϕ for which the expression $\tan\phi$ is defined, $\tan\phi \cdot \cos\phi = \sin\phi$.

15. If the left side of an equation and the right side of the equation are equal for all values of the variable for which the equation is defined, then the equation is an identity.

16. The function $f(\theta) = \sin\theta\tan\theta$ is an odd function.

17. For all values of θ for which the expressions are defined, $1 - 2\tan^2\theta\cos^2\theta = 2\cos^2\theta - 1$

18. The equation $\sin\beta^2 + \cos\beta^2 = 1$ holds for all values of β.

19. The identity $\sin t = \cos(t - \frac{\pi}{2})$ signifies that the graph of $y = \sin t$ is the graph of $y = \cos t$, but shifted $\frac{\pi}{2}$ units to the right.

20. If $\cos\alpha = -\frac{2}{3}$ and the point corresponding to α is in the third quadrant, then $\sin(2\alpha) = \dfrac{4\sqrt{5}}{9}$.

21. Since $\sin 45° = \cos 45°$, we see that $\sin\theta = \cos\theta$ is an identity.

22. For all values of θ, we have $\sin^2\theta = \frac{1}{2}(1 - \cos 2\theta)$.

23. For all values of θ, we have $\cos(4\theta) = -\cos(-4\theta)$.

24. $\sin(\theta + \phi) = \sin\theta + \sin\phi$ for all θ and ϕ.

25. $\sin 2\theta = 2\sin\theta$ for all θ.

26. $\sin(\theta + \frac{\pi}{2}) = \cos\theta$ for all θ.

27. The function $f(t) = \sin t\cos t$ is periodic with period 2π.

28. For all values of t, we have $\sin(t-\pi) = \sin(t+\pi)$.

29. For all values of t, we have $\cos(t-\pi/2) = \cos(t+\pi/2)$.

30. The graph of $A\cos(Bt)$ is a horizontal shift of the graph of $A\sin(Bt)$.

31. The function $f(x) = \sin 2x + \sin 3x$ is a periodic function.

32. The function $f(x) = \sin 2x + \sin 3x$ is a sinusoidal function.

33. The sum of two sinusoidal functions is sinusoidal.

34. When writing $a_1 \sin(Bt) + a_2 \cos(Bt)$ as a single sine function, the amplitude of the resulting single sine function is larger than that of either function in the sum.

35. The function $f(t) = 3\sin(2\pi t) + 4\cos(2\pi t)$ is sinusoidal.

36. The function $g(t) = (1/2)e^t \sin(t/2)$ has a decreasing amplitude.

37. If a company has a power requirement of $g(t) = 55 - 10\sin(\pi t/12)$, then the peak requirement is 65 units.

38. A CRT screen that has a refresh rating of 60 hertz will refresh every 1/60th of a second.

39. The function $f(t) = \cos(\pi t) + 1$ has all of its zeros at odd integer values.

40. The parameter B in $g(x) = A\sin(Bx) + C$ affects the amplitude of the function.

41. The graph of $r = 1$ is a straight line.

42. The graph of $\theta = 1$ is the unit circle.

43. The polar coordinates $(3, \pi)$ correspond to $(-3, \pi)$ in Cartesian coordinates.

44. In both polar and Cartesian coordinate systems, each point in the xy-plane has a set of unique coordinate values.

45. The point $(0, -3)$ in Cartesian coordinates is represented by $(3, 3\pi/2)$ in polar coordinates.

46. The region defined in polar coordinates by $0 \le r \le 2$ and $0 \le \theta \le 2$ looks like a two-by-two square.

47. Every nonnegative real number has a real square root.

48. For any complex number z, the product $z \cdot \bar{z}$ is a real number.

49. The square of any complex number is a real number.

50. If f is a polynomial, and $f(z) = i$, then $f(\bar{z}) = i$.

51. Every nonzero complex number z can be written in the form $z = e^w$, where w is a complex number.

52. If $z = x+iy$, where x and y are positive, then $z^2 = a+ib$ has a and b positive.

53. One solution of $3e^t \sin t - 2e^t \cos t = 0$ is $t = \arctan(\frac{2}{3})$.

54. $0 \le \cos^{-1}(\sin(\cos^{-1} x)) \le \frac{\pi}{2}$ for all x with $-1 \le x \le 1$.

55. If $a > b > 0$, then the domain of $f(t) = \ln(a + b\sin t)$ is all real numbers t.

56. For all values of θ we have $e^{i\theta} = \cos\theta + i\sin\theta$.

57. $(1+i)^2 = 2i$.

58. $i^{101} = i$.

Chapter Eight

COMPOSITIONS, INVERSES, AND COMBINATIONS OF FUNCTIONS

In Chapter 5, we studied transformations of functions. The composite functions in this chapter are generalizations of these transformations. In Chapter 2, we introduced the inverse function and its notation, f^{-1} . In Chapter 3, we defined the inverse of the exponential function (that is, the logarithm) and in Chapter 6, we defined the inverses of the trigonometric functions. In this chapter, we consider inverse functions in more detail.

8.1 COMPOSITION OF FUNCTIONS

The Effect of a Drug on Heart Rates

A therapeutic drug has the side effect of raising a patient's heart rate. Table 8.1 gives the relationship between Q, the amount of drug in the patient's body (in milligrams), and r, the patient's heart rate (in beats per minute). We see that the higher the drug level, the faster the heart rate.

Table 8.1 *Heart rate, $r = f(Q)$, as a function of drug level, Q*

Q, drug level (mg)	0	50	100	150	200	250
r, heart rate (beats per minute)	60	70	80	90	100	110

A patient is given a 250 mg injection of the drug. Over time, the level of drug in the patient's bloodstream falls. Table 8.2 gives the drug level, Q, as a function of time, t.

Table 8.2 *Drug level, $Q = g(t)$, as a function of time, t, since the medication was given*

t, time (hours)	0	1	2	3	4	5	6	7	8
Q, drug level (mg)	250	200	160	128	102	82	66	52	42

Since heart rate depends on the drug level and drug level depends on time, the heart rate also depends on time. Tables 8.1 and 8.2 can be combined to give the patient's heart rate, r, as a function of t. For example, according to Table 8.2, at time $t = 0$ the drug level is 250 mg. According to Table 8.1, at this drug level, the patient's heart rate is 110 beats per minute. So $r = 110$ when $t = 0$. The results of similar calculations have been compiled in Table 8.3. Note that many of the entries, such as $r = 92$ when $t = 2$, are estimates.

Table 8.3 *Heart rate, $r = h(t)$, as a function of time, t*

t, time (hours)	0	1	2	3	4	5	6	7	8
r, heart rate (beats per minute)	110	100	92	86	80	76	73	70	68

Now, since

$$r = f(Q) \quad \text{or} \quad \underbrace{\text{Heart rate}}_{r} = f(\underbrace{\text{drug level}}_{Q}),$$

and

$$Q = g(t) \quad \text{or} \quad \underbrace{\text{Drug level}}_{Q} = g(\underbrace{\text{time}}_{t}),$$

we can substitute $Q = g(t)$ into $r = f(Q)$, giving

$$r = f(\underbrace{\ Q\ }_{g(t)}) = f(g(t)).$$

The function h in Table 8.3 is said to be the *composition* of the function f and g, written

$$h(t) = f(g(t)).$$

This formula represents the process that we used to find the values of $r = h(t)$ in Table 8.3.

Example 1 Use Tables 8.1 and 8.2 to estimate the values of: (a) $h(0)$ (b) $h(4)$

Solution

(a) If $t = 0$, then
$$r = h(0) = f(g(0)).$$
Table 8.2 shows that $g(0) = 250$, so
$$r = h(0) = f(\underbrace{250}_{g(0)}).$$
We see from Table 8.1 that $f(250) = 110$. Thus,
$$r = h(0) = \underbrace{110}_{f(250)}.$$
As before, this tells us that the patient's heart rate at time $t = 0$ is 110 beats per minute.

(b) If $t = 4$, then
$$h(4) = f(g(4)).$$
Working from the inner set of parentheses outward, we start by evaluating $g(4)$. Table 8.2 shows that $g(4) = 102$. Thus,
$$h(4) = f(\underbrace{102}_{g(4)}).$$
Table 8.1 does not have a value for $f(102)$. But since $f(100) = 80$ and $f(150) = 90$, we estimate that $f(102)$ is close to 80. Thus, we let
$$h(4) \approx \underbrace{80}_{f(102)}.$$
This indicates that four hours after the injection, the patient's heart rate is approximately 80 beats per minute.

> The function $f(g(t))$ is said to be a **composition** of f with g. The function $f(g(t))$ is defined by using the output of the function g as the input to f.

The composite function $f(g(t))$ is only defined for values in the domain of g whose $g(t)$ values are in the domain of f.

Formulas for Composite Functions

A possible formula for $r = f(Q)$, the heart rate as a function of drug level is
$$r = f(Q) = 60 + 0.2Q.$$
A possible formula for $Q = g(t)$, the drug level as a function of time is
$$Q = g(t) = 250(0.8)^t.$$
To find a formula for $r = h(t) = f(g(t))$, the heart rate as a function of time, we use the function $g(t)$ as the input to f. Thus,
$$r = f(\underbrace{\text{input}}_{g(t)}) = 60 + 0.2(\underbrace{\text{input}}_{g(t)}),$$
so
$$r = f(g(t)) = 60 + 0.2g(t).$$
Now, substitute the formula for $g(t)$. This gives
$$r = h(t) = f(g(t)) = 60 + 0.2 \cdot \underbrace{250(0.8)^t}_{g(t)}$$

so

$$r = h(t) = 60 + 50(0.8)^t.$$

We can check the formula against Table 8.3. For example, if $t = 4$

$$h(4) = 60 + 50(0.8)^4 = 80.48.$$

This result is in agreement with the value $h(4) \approx 80$ that we estimated in Table 8.3.

Example 2 Let $p(x) = 2x + 1$ and $q(x) = x^2 - 3$. Suppose $u(x) = p(q(x))$ and $v(x) = q(p(x))$.
(a) Calculate $u(3)$ and $v(3)$.
(b) Find formulas for $u(x)$ and $v(x)$.

Solution (a) We want

$$u(3) = p(q(3)).$$

We start by evaluating $q(3)$. The formula for q gives $q(3) = 3^2 - 3 = 6$, so

$$u(3) = p(6).$$

The formula for p gives $p(6) = 2 \cdot 6 + 1 = 13$, so

$$u(3) = 13.$$

To calculate $v(3)$, we have

$$\begin{aligned} v(3) &= q(p(3)) \\ &= q(7) && \text{Because } p(3) = 2 \cdot 3 + 1 = 7 \\ &= 46 && \text{Because } q(7) = 7^2 - 3 \end{aligned}$$

Notice that, $v(3) \neq u(3)$. The functions $v(x) = q(p(x))$ and $u(x) = p(q(x))$ are different.

(b) In the formula for u,

$$\begin{aligned} u(x) &= p(\underbrace{\ q(x)\ }_{\text{Input for } p}) \\ &= 2q(x) + 1 && \text{Because } p(\text{Input}) = 2 \cdot \text{Input} + 1 \\ &= 2(x^2 - 3) + 1 && \text{Substituting } q(x) = x^2 - 3 \\ &= 2x^2 - 5. \end{aligned}$$

Check this formula by evaluating $u(3)$, which we know to be 13:

$$u(3) = 2 \cdot 3^2 - 5 = 13.$$

In the formula for v,

$$\begin{aligned} v(x) &= q(\underbrace{\ p(x)\ }_{\text{Input for } q}) \\ &= q(2x + 1) && \text{Because } p(x) = 2x + 1 \\ &= (2x + 1)^2 - 3 && \text{Because } q(\text{Input}) = \text{Input}^2 - 3 \\ &= 4x^2 + 4x - 2. \end{aligned}$$

Check this formula by evaluating $v(3)$, which we know to be 46:

$$v(3) = 4 \cdot 3^2 + 4 \cdot 3 - 2 = 46.$$

So far we have considered examples of two functions composed together, but there is no limit on the number of functions that can be composed. Functions can even be composed with themselves.

Example 3 Let $p(x) = \sin x + 1$ and $q(x) = x^2 - 3$. Find a formula in terms of x for $w(x) = p(p(q(x)))$.

Solution We work from inside the parentheses outward. First we find $p(q(x))$, and then input the result to p.

$$\begin{aligned} w(x) &= p(p(q(x))) \\ &= p(p(x^2 - 3)) \\ &= p(\underbrace{\sin(x^2 - 3) + 1}_{\text{Input for } p}) \\ &= \sin(\sin(x^2 - 3) + 1) + 1. \qquad \text{Because } p(\text{Input}) = \sin(\text{Input}) + 1 \end{aligned}$$

Composition of Functions Defined by Graphs

So far we have composed functions defined by tables and formulas. In the next example, we compose functions defined by graphs.

Example 4 Let u and v be two functions defined by the graphs in Figure 8.1. Evaluate:

(a) $v(u(-1))$ (b) $u(v(5))$ (c) $v(u(0)) + u(v(4))$

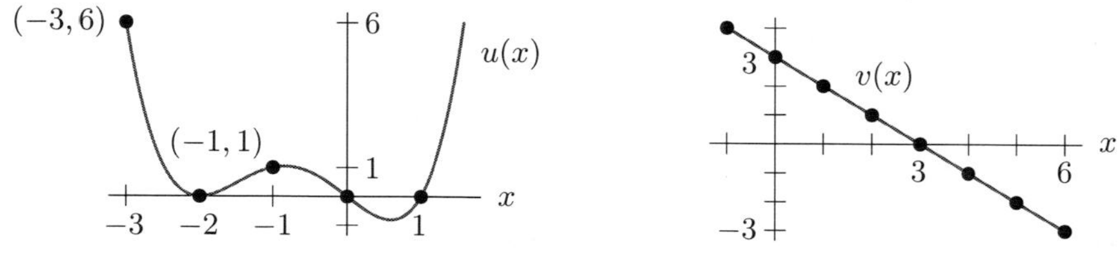

Figure 8.1: Evaluate the composition of functions u and v defined by their graphs

Solution (a) To evaluate $v(u(-1))$, start with $u(-1)$. From Figure 8.1, we see that $u(-1) = 1$. Thus,

$$v(u(-1)) = v(1).$$

From the graph we see that $v(1) = 2$, so

$$v(u(-1)) = 2.$$

(b) Since $v(5) = -2$, we have $u(v(5)) = u(-2) = 0$.

(c) Since $u(0) = 0$, we have $v(u(0)) = v(0) = 3$.
Since $v(4) = -1$, we have $u(v(4)) = u(-1) = 1$.
Thus $v(u(0)) + u(v(4)) = 3 + 1 = 4$.

Decomposition of Functions

Sometimes we reason backward to find the functions which went into a composition. This process is called *decomposition*.

Example 5 Let $h(x) = f(g(x)) = e^{x^2+1}$. Find possible formulas for $f(x)$ and $g(x)$.

Solution In the formula $h(x) = e^{x^2+1}$, the expression $x^2 + 1$ is in the exponent. Thus, we can take the inside function to be $g(x) = x^2 + 1$. This means that we can write

$$h(x) = e^{\overbrace{x^2+1}} = e^{g(x)}.$$

$$\qquad\qquad g(x)$$

Then the outside function is $f(x) = e^x$. We check that composing f and g gives h:

$$f(g(x)) = f(x^2 + 1) = e^{x^2+1} = h(x).$$

There are many possible solutions to Example 5. For example, we might choose $f(x) = e^{x+1}$ and $g(x) = x^2$. Then

$$f(g(x)) = e^{g(x)+1} = e^{x^2+1} = h(x).$$

Alternatively, we might choose $f(x) = e^{x^2+1}$ and $g(x) = x$. Although this satisfies the condition that $h(x) = f(g(x))$, it is not very useful, because f is the same as h. This kind of decomposition is referred to as *trivial*. Another example of a trivial decomposition of $h(x)$ is $f(x) = x$ and $g(x) = e^{x^2+1}$.

Example 6 The vertex formula for the family of quadratic functions is

$$p(x) = a(x - h)^2 + k.$$

Decompose the formula into three simple functions. That is, find formulas for u, v, and w where

$$p(x) = u(v(w(x))),$$

Solution We work from inside the parentheses outward. In the formula $p(x) = a(x - h)^2 + k$, we have the expression $x - h$ inside the parentheses. In the formula $p(x) = u(v(w(x)))$, the innermost function is $w(x)$. Thus, we let

$$w(x) = x - h.$$

In the formula $p(x) = a(x - h)^2 + k$, the first operation done to $x - h$ is squaring. Thus, we let

$$v(\text{Input}) = \text{Input}^2$$
$$v(x) = x^2.$$

So we have

$$v(w(x)) = v(x - h) = (x - h)^2.$$

Finally, to obtain $p(x) = a(x - h)^2 + k$, we multiply $(x - h)^2$ by a and add k. Thus, we let

$$u(\text{Input}) = a \cdot \text{Input} + k$$
$$u(x) = ax + k.$$

To check, we compute

$$\begin{aligned} u(v(w(x))) &= u(v(x - h)) && \text{Since } w(x) = x - h \\ &= u(\underbrace{(x - h)^2}_{\text{Input for } u}) && \text{Since } v(x - h) = (x - h)^2 \\ &= a(x - h)^2 + k && \text{Since } u((x - h)^2) = a \cdot (x - h)^2 + k. \end{aligned}$$

Exercises and Problems for Section 8.1

Exercises

1. Use Table 8.4 to construct a table of values for $r(x) = p(q(x))$.

Table 8.4

x	0	1	2	3	4	5
$p(x)$	1	0	5	2	3	4
$q(x)$	5	2	3	1	4	8

2. Let p and q be the functions in Exercise 1. Construct a table of values for $s(x) = q(p(x))$.

3. Using Tables 8.5 and 8.6, complete Table 8.7:

Table 8.5

x	$f(x)$
0	0
$\pi/6$	$1/2$
$\pi/4$	$\sqrt{2}/2$
$\pi/3$	$\sqrt{3}/2$
$\pi/2$	1

Table 8.6

y	$g(y)$
0	$\pi/2$
$1/4$	π
$\sqrt{2}/4$	0
$1/2$	$\pi/3$
$\sqrt{2}/2$	$\pi/4$
$3/4$	0
$\sqrt{3}/2$	$\pi/6$
1	0

Table 8.7

x	$g(f(x))$
0	
$\pi/6$	
$\pi/4$	
$\pi/3$	
$\pi/2$	

4. Let $f(x) = \sin 4x$ and $g(x) = \sqrt{x}$. Find formulas for $f(g(x))$ and $g(f(x))$.

5. Let $h(x) = 2^x$ and $k(x) = x^2$. Find formulas for $h(k(x))$ and $k(h(x))$.

6. Let $m(x) = 3 + x^2$ and $n(x) = \tan x$. Find formulas for $m(n(x))$ and $n(m(x))$.

Find formulas for the functions in Exercises 7–12 and simplify. Let $f(x) = x^2 + 1$, $g(x) = \dfrac{1}{x-3}$, and $h(x) = \sqrt{x}$.

7. $f(g(x))$

8. $g(f(x))$

9. $f(h(x))$

10. $h(f(x))$

11. $g(g(x))$

12. $g(f(h(x)))$

In Exercises 13–17, identify the function $f(x)$.

13. $h(x) = e^{f(x)} = e^{\sin x}$

14. $j(x) = \sqrt{f(x)} = \sqrt{\ln(x^2+4)}$

15. $k(x) = \sin(f(x)) = \sin(x^3 + 3x + 1)$

16. $l(x) = (f(x))^2 = \cos^2 2x$

17. $m(x) = \ln f(x) = \ln(5 + 1/x)$

Problems

In Problems 18–21, give a practical interpretation in words of the function.

18. $f(h(t))$, where $A = f(r)$ is the area of a circle of radius r and $r = h(t)$ is the radius of the circle at time t.

19. $k(g(t))$, where $L = k(H)$ is the length of a steel bar at temperature H and $H = g(t)$ is temperature at time t.

20. $R(Y(q))$, where R gives a farmer's revenue as a function of corn yield per acre, and Y gives the corn yield as a function of the quantity, q, of fertilizer.

21. $t(f(H))$, where $t(v)$ is the time of a trip at velocity v, and $v = f(H)$ is velocity at temperature H.

22. Complete Table 8.8 if $r(t) = q(p(t))$.

Table 8.8

t	$p(t)$	$q(t)$	$r(t)$
0	4	??	??
1	??	2	1
2	??	??	0
3	2	0	4
4	1	5	??
5	0	1	3

23. Complete Table 8.9 given that $h(x) = f(g(x))$.

Table 8.9

x	$f(x)$	$g(x)$	$h(x)$
0	1	2	5
1	9	0	
2		1	

24. Using Figure 8.2, estimate the following:

(a) $f(g(2))$ **(b)** $g(f(2))$ **(c)** $f(f(3))$ **(d)** $g(g(3))$

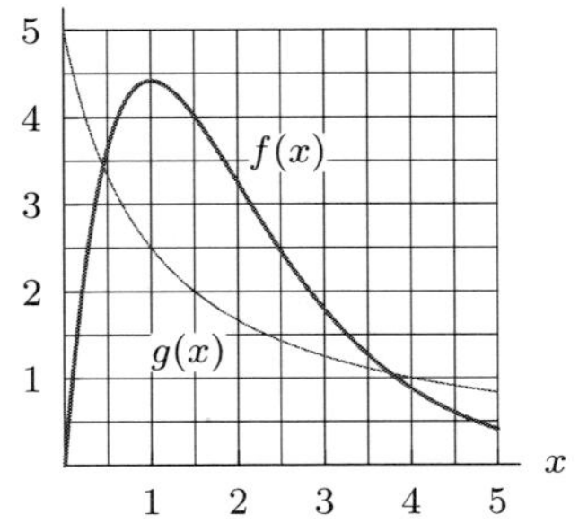

Figure 8.2

25. Use Figure 8.3 to calculate the following:

(a) $f(f(1))$ **(b)** $g(g(1))$
(c) $f(g(2))$ **(d)** $g(f(2))$

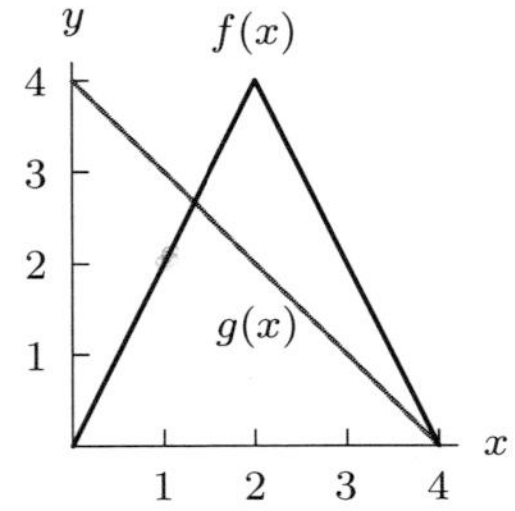

Figure 8.3

26. Use Figure 8.3 to find all solutions to the equations:

(a) $f(g(x)) = 0$ **(b)** $g(f(x)) = 0$

In Problems 27–30, use the information from Figures 8.4, and 8.5 to graph the functions.

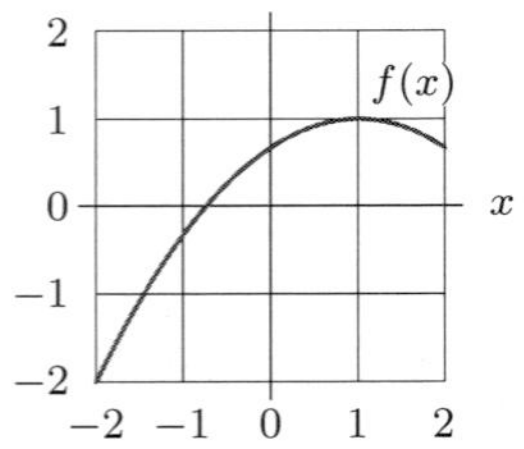

Figure 8.4

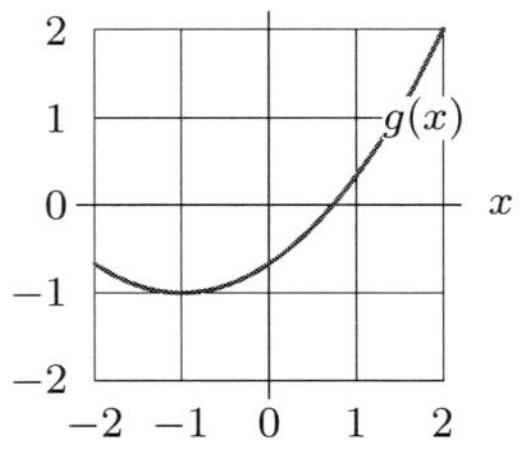

Figure 8.5

27. $f(g(x))$ **28.** $g(f(x))$ **29.** $f(f(x))$ **30.** $g(g(x))$

31. Complete Table 8.10, Table 8.11, and Table 8.12 given that $h(x) = g(f(x))$. Assume that different values of x lead to different values of $f(x)$.

Table 8.10

x	$f(x)$
−2	4
−1	
0	
1	5
2	1

Table 8.11

x	$g(x)$
1	
2	1
3	2
4	0
5	−1

Table 8.12

x	$h(x)$
−2	
−1	1
0	2
1	
2	−2

32. **(a)** Use Table 8.13 and Figure 8.6 to calculate:

(i) $f(g(4))$ (ii) $g(f(4))$
(iii) $f(f(0))$ (iv) $g(g(0))$

(b) Solve $g(g(x)) = 1$ for x.

Table 8.13

x	0	1	2	3	4	5
$f(x)$	2	5	3	4	1	0

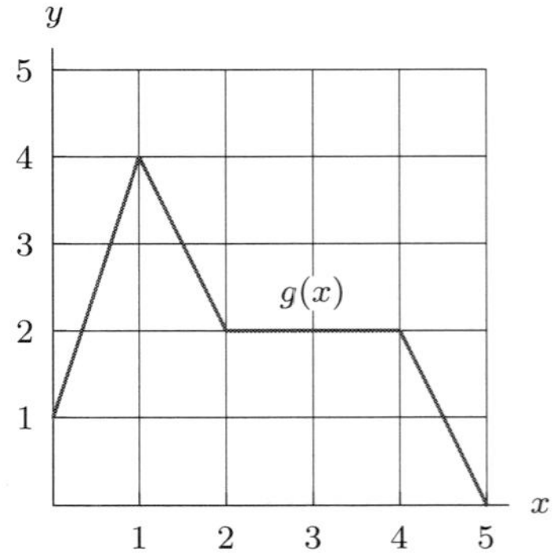

Figure 8.6

33. Let $f(x)$ and $g(x)$ be the functions in Figure 8.3.

(a) Graph the functions $f(g(x))$ and $g(f(x))$.
(b) On what interval(s) is $f(g(x))$ increasing?
(c) On what interval(s) is $g(f(x))$ increasing?

34. Find $f(f(1))$ for

$$f(x) = \begin{cases} 2 & \text{if } x \le 0 \\ 3x+1 & \text{if } 0 < x < 2 \\ x^2 - 3 & \text{if } x \ge 2 \end{cases}$$

35. Using your knowledge of the absolute value function, explain in a few sentences the relationship between the graph of $y = |\sin x|$ and the graph of $y = \sin x$.

36. Graph the following functions for $-2\pi \le x \le 2\pi$.

(a) $f(x) = \sin x$ **(b)** $g(x) = |\sin x|$

(c) $h(x) = \sin|x|$ **(d)** $i(x) = |\sin|x||$

(e) Do any two of these functions have identical graphs? If so, explain why this makes sense.

37. Let $p(t) = 10(0.01)^t$ and $q(t) = \log t^2$. Solve the equation $q(p(t)) = 0$ for t.

38. Let $f(t) = \sin t$ and $g(t) = 3t - \pi/4$. Solve the equation $f(g(t)) = 1$ for t in the interval $0 \le t \le 2\pi/3$.

In Problems 39–42, find a simplified formula for the difference quotient

$$\frac{f(x+h) - f(x)}{h}.$$

39. $f(x) = x^2 + x$

40. $f(x) = \sqrt{x}$

41. $f(x) = \dfrac{1}{x}$

42. $f(x) = 2^x$

In Problems 43–45, let $x > 0$ and $k(x) = e^x$. Find a possible formula for $f(x)$.

43. $k(f(x)) = e^{2x}$

44. $f(k(x)) = e^{2x}$

45. $k(f(x)) = x$

Decompose the functions in Problems 46–51 into $u(v(x))$ for given u or v.

46. $y = \dfrac{1+x^2}{2+x^2}$ given that

(a) $v(x) = x^2$ **(b)** $v(x) = x^2 + 1$

47. $y = e^{-\sqrt{x}}$ given that

(a) $u(x) = e^x$ **(b)** $v(x) = \sqrt{x}$

48. $y = \sqrt{1 - x^3}$ given that

(a) $u(x) = \sqrt{1 + x^3}$ **(b)** $v(x) = x^3$

49. $y = 2^{x+1}$ given that

(a) $u(x) = 2x$ **(b)** $v(x) = -x$

50. $y = \sin^2 x$ given that

(a) $u(x) = x^2$ **(b)** $v(x) = x^2$

51. $y = e^{2\cos x}$ given that

(a) $u(x) = e^x$ **(b)** $v(x) = \cos x$

Decompose the functions in Problems 52–59 into two new functions, u and v, where v is the inside function, $u(x) \ne x$, and $v(x) \ne x$.

52. $f(x) = \sqrt{3 - 5x}$

53. $g(x) = \sin(x^2)$

54. $h(x) = \sin^2 x$

55. $k(x) = e^{\sin x} + \sin x$

56. $F(x) = (2x+5)^3$

57. $G(x) = \dfrac{2}{1+\sqrt{x}}$

58. $H(x) = 3^{2x-1}$

59. $J(x) = 8 - 2|x|$

60. You have two money machines, both of which increase any money inserted into them. The first machine doubles your money. The second adds five dollars. The money that comes out is described by $d(x) = 2x$, in the first case, and $a(x) = x + 5$, in the second, where x is the number of dollars inserted. The machines can be hooked up so that the money coming out of one machine goes into the other. Find formulas for each of the two possible composition machines. Is one machine more profitable than the other?

61. Currency traders often move investments from one country to another in order to make a profit. Table 8.14 gives exchange rates for US dollars, Japanese yen, and the European Union's euro.[1] In January, 2006, for example, 1 US dollar purchases 114.64 Japanese yen or 0.829 European euro. Similarly, 1 European euro purchases 138.29 Japanese yen or 1.2063 US dollars. Suppose

$f(x)$ = Number of yen one can buy with x dollars
$g(x)$ = Number of euros one can buy with x dollars
$h(x)$ = Number of euros one can buy with x yen

(a) Find formulas for f, g, and h.

(b) Evaluate and interpret in terms of currency: $h(f(1000))$.

Table 8.14 *Exchange rate for US dollars, Japanese yen and euros, January 10, 2006*

Amount invested	Dollars purchased	Yen purchased	Euros purchased
1 dollar	1.0000	114.64	0.829
1 yen	0.00872	1.0000	0.00723
1 euro	1.2063	138.29	1.0000

[1] www.x-rates.com, January 10, 2006. Currency exchange rates fluctuate constantly.

8.2 INVERSE FUNCTIONS

Inverse functions were introduced in Section 2.4. In Section 4.1, we defined the logarithm as the inverse function of the exponential function. In Section 6.7, we defined the arccosine as the inverse of cosine. We now study inverse functions in general.

Definition of Inverse Function

Recall that the statement $f^{-1}(50) = 20$ means that $f(20) = 50$. In fact, the values of f^{-1} are determined in just this way. In general,

> Suppose $Q = f(t)$ is a function with the property that each value of Q determines exactly one value of t. Then f has an **inverse function**, f^{-1} and
>
> $$f^{-1}(Q) = t \quad \text{if and only if} \quad Q = f(t).$$
>
> If a function has an inverse, it is said to be **invertible**.

The definitions of the logarithm and of the inverse cosine have the same form as the definition of f^{-1}. Since $y = \log x$ is the inverse function of $y = 10^x$, we have

$$x = \log y \quad \text{if and only if} \quad y = 10^x,$$

and since $y = \cos^{-1} t$ is the inverse function of $y = \cos t$,

$$t = \cos^{-1} y \quad \text{if and only if} \quad y = \cos t.$$

Example 1 Solve the equation $\sin x = 0.8$ using an inverse function.

Solution The solution is $x = \sin^{-1}(0.8)$. A calculator (set in radians) gives $x = \sin^{-1}(0.8) \approx 0.927$.

Example 2 Suppose that g is an invertible function, with $g(10) = -26$ and $g^{-1}(0) = 7$. What other values of g and g^{-1} do you know?

Solution Because $g(10) = -26$, we know that $g^{-1}(-26) = 10$; because $g^{-1}(0) = 7$, we know that $g(7) = 0$.

Example 3 A population is given by the formula $P = f(t) = 20 + 0.4t$ where P is the number of people (in thousands) and t is the number of years since 1980. Evaluate the following quantities. Explain in words what each tells you about the population.

(a) $f(25)$ (b) $f^{-1}(25)$

(c) Show how to estimate $f^{-1}(25)$ from a graph of f.

Solution (a) Substituting $t = 25$, we have

$$f(25) = 20 + 0.4 \cdot 25 = 30.$$

Thus, in 2005 (year $t = 25$), we have $P = 30$, so the population was 30,000 people.

(b) We have $t = f^{-1}(P)$. Thus, in $f^{-1}(25)$, the 25 is a population. So $f^{-1}(25)$ is the year in which the population reaches 25 thousand. We find t by solving the equation

$$20 + 0.4t = 25$$
$$0.4t = 5$$
$$t = 12.5.$$

Therefore, $f^{-1}(25) = 12.5$, which means that the population reached 25,000 people 12.5 years after 1980, or midway into 1992.

(c) We can estimate $f^{-1}(25)$ by reading the graph of $P = f(t)$ backward as shown in Figure 8.7.

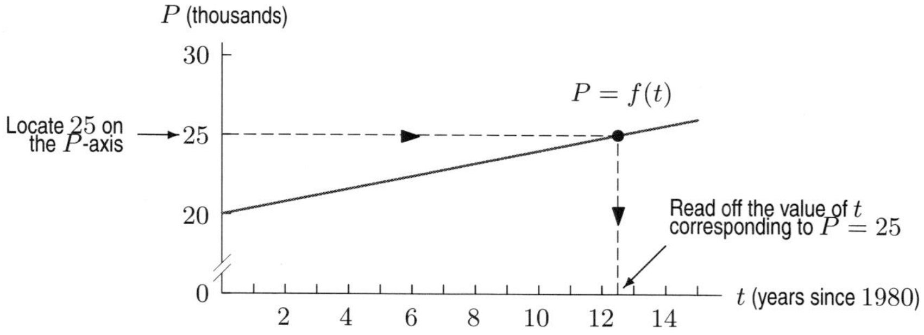

Figure 8.7: Using a graph of the function $P = f(t)$ to read off values of the inverse function $f^{-1}(P)$

Finding a Formula for an Inverse Function

It is sometimes possible to find a formula for an inverse function, f^{-1} from a formula for f. If the function $P = f(t)$ gives the population (in thousands) of a town in year t, then $f^{-1}(P)$ is the year in which the population reaches the value P. In Example 3 we found $f^{-1}(P)$ for $P = 25$. We now perform the same calculations for a general P. Since

$$P = 20 + 0.4t,$$

solving for t gives

$$0.4t = P - 20$$
$$t = \frac{P - 20}{0.4},$$

so

$$f^{-1}(P) = 2.5P - 50.$$

The values in Table 8.15 were calculated using the formula $P = f(t) = 20 + 0.4t$; the values in Table 8.16 were calculated using the formula for $t = f^{-1}(P) = 2.5P - 50$. The table for f^{-1} can be obtained from the table for f by interchanging its columns, because the inverse function reverses the roles of inputs and outputs.

Table 8.15

t	$P = f(t)$
0	20
5	22
10	24
15	26
20	28

Table 8.16

P	$t = f^{-1}(P)$
20	0
22	5
24	10
26	15
28	20

Example 4 Suppose you deposit \$500 into a savings account that pays 4% interest compounded annually. The balance, in dollars, in the account after t years is given by $B = f(t) = 500(1.04)^t$.

(a) Find a formula for $t = f^{-1}(B)$.
(b) What does the inverse function represent in terms of the account?

Solution (a) To find a formula for f^{-1}, we solve for t in terms of B:

$$\begin{aligned} B &= 500(1.04)^t \\ \frac{B}{500} &= (1.04)^t \\ \log\left(\frac{B}{500}\right) &= t\log 1.04 \qquad \text{Taking logs of both sides} \\ t &= \frac{\log(B/500)}{\log 1.04}. \end{aligned}$$

Thus, a formula for the inverse function is

$$t = f^{-1}(B) = \frac{\log(B/500)}{\log 1.04}.$$

(b) The function $t = f^{-1}(B)$ gives the number of years for the balance to grow to $\$B$.

In the previous example the variables of the function $B = f(t)$ had contextual meaning, so the inverse function was written as $t = f^{-1}(B)$. In abstract mathematical examples, a function $y = f(x)$ will often have its inverse function written with x as the independent variable. This situation is shown in the next example.

Example 5 Find the inverse of the function

$$f(x) = \frac{3x}{2x+1}.$$

Solution First, we solve the equation $y = f(x)$ for x:

$$\begin{aligned} y &= \frac{3x}{2x+1} \\ 2xy + y &= 3x \\ 2xy - 3x &= -y \\ x(2y-3) &= -y \\ x &= \frac{-y}{2y-3} = \frac{y}{3-2y} \end{aligned}$$

As before, we write $x = f^{-1}(y) = \dfrac{y}{3-2y}$.

Since y is now the independent variable, by convention we rewrite the inverse function with x as the independent variable. We have

$$f^{-1}(x) = \frac{x}{3-2x}.$$

Noninvertible Functions: Horizontal Line Test

Not every function has an inverse function. A function $Q = f(t)$ has no inverse if it returns the same Q-value for two different t-values. When that happens, the value of t cannot be uniquely determined from the value of Q.

For example, if $q(x) = x^2$ then $q(-3) = 9$ and $q(+3) = 9$. This means that we cannot say what the value $q^{-1}(9)$ would be. (Is it $+3$ or -3?). Thus, q is not invertible. In Figure 8.8, notice that the horizontal line $y = 9$ intersects the graph of $q(x) = x^2$ at two different points: $(-3, 9)$ and $(3, 9)$. This corresponds to the fact that the function q returns $y = 9$ for two different x-values, $x = +3$ and $x = -3$.

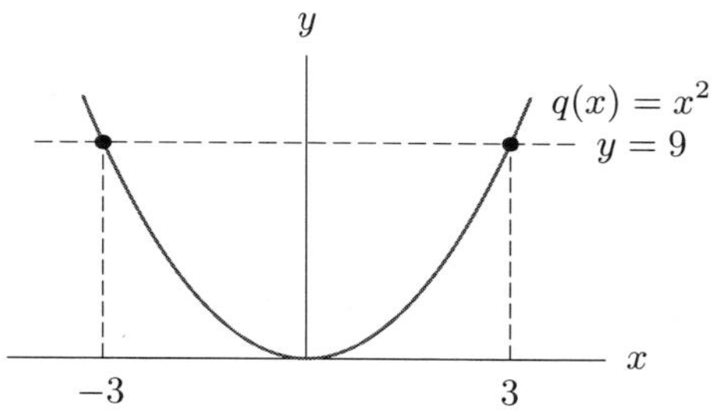

Figure 8.8: The graph of $q(x) = x^2$ fails the horizontal line test

We have the following general result:

> **The Horizontal Line Test** If there is a horizontal line which intersects a function's graph in more than one point, then the function does not have an inverse. If every horizontal line intersects a function's graph at most once, then the function has an inverse.

Evaluating an Inverse Function Graphically

Finding a formula for an inverse function can be difficult. However, this does not mean that the inverse function does not exist. Even without a formula, it may be possible to find values of the inverse function.

Example 6 Let $u(x) = x^3 + x + 1$. Explain why a graph suggests the function u is invertible. Assuming u has an inverse, estimate $u^{-1}(4)$.

Solution To show that u is invertible, we could try to find a formula for u^{-1}. To do this, we would solve the equation $y = x^3 + x + 1$ for x. Unfortunately, this is difficult. However, the graph in Figure 8.9 suggests that u passes the horizontal line test and therefore that u is invertible. To estimate $u^{-1}(4)$, we find an x-value such that

$$x^3 + x + 1 = 4.$$

In Figure 8.9, the graph of $y = u(x)$ and the horizontal line $y = 4$ intersect at the point $x \approx 1.213$. Thus, tracing along the graph, we estimate $u^{-1}(4) \approx 1.213$.

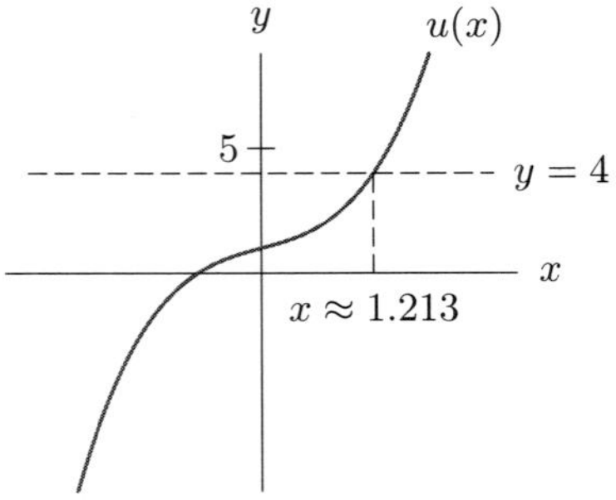

Figure 8.9: The graph of $u(x)$ passes the horizontal line test. Since $u(1.213) \approx 4$, we have $u^{-1}(4) \approx 1.213$

In Example 6, even without a formula for u^{-1}, we can approximate $u^{-1}(a)$ for any value of a.

Example 7 Let $P(x) = 2^x$.

(a) Show that P is invertible.
(b) Find a formula for $P^{-1}(x)$.
(c) Sketch the graphs of P and P^{-1} on the same axes.
(d) What are the domain and range of P and P^{-1}?

Solution

(a) Since P is an exponential function with base 2, it is always increasing, and therefore passes the horizontal line test. (See the graph of P in Figure 8.10.) Thus, P has an inverse function.

(b) To find a formula for $P^{-1}(x)$, we solve for x in the equation

$$2^x = y.$$

We can take the log of both sides to get

$$\begin{aligned} \log 2^x &= \log y \\ x \log 2 &= \log y \\ x = P^{-1}(y) &= \frac{\log y}{\log 2}. \end{aligned}$$

Thus, we have a formula for P^{-1} with y as the input. To graph P and P^{-1} on the same axes, we write P^{-1} as a function of x:

$$P^{-1}(x) = \frac{\log x}{\log 2} = \frac{1}{\log 2} \cdot \log x = 3.322 \log x.$$

(c) Table 8.17 gives values of $P(x)$ for $x = -3, -2, \ldots, 3$. Interchanging the columns of Table 8.17 gives Table 8.18 for $P^{-1}(x)$. We use these tables to sketch Figure 8.10.

Table 8.17 *Values of* $P(x) = 2^x$

x	$P(x) = 2^x$
-3	0.125
-2	0.25
-1	0.5
0	1
1	2
2	4
3	8

Table 8.18 *Values of* $P^{-1}(x)$

x	$P^{-1}(x)$
0.125	-3
0.25	-2
0.5	-1
1	0
2	1
4	2
8	3

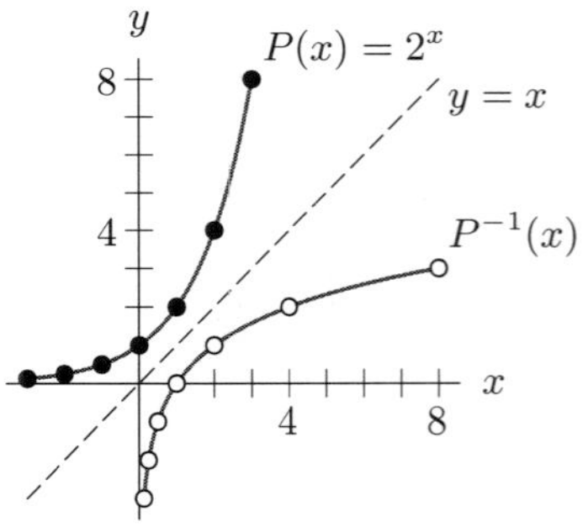

Figure 8.10: The graphs of $P(x) = 2^x$ and its inverse are symmetrical across the line $y = x$

(d) The domain of P, an exponential function, is all real numbers, and its range is all positive numbers. The domain of P^{-1}, a logarithmic function, is all positive numbers and its range is all real numbers.

The Graph, Domain, and Range of an Inverse Function

In Figure 8.10, we see that the graph of P^{-1} is the mirror-image of the graph of P across the line $y = x$. In general, this is true if the x- and y-axes have the same scale. To understand why this occurs, consider how a function is related to its inverse.

If f is an invertible function with, for example, $f(2) = 5$, then $f^{-1}(5) = 2$. Thus, the point $(2, 5)$ is on the graph of f and the point $(5, 2)$ is on the graph of f^{-1}. Generalizing, if (a, b) is any point on the graph of f, then (b, a) is a point on the graph of f^{-1}. Figure 8.11 shows how reflecting the point (a, b) across the line $y = x$ gives the point (b, a). Consequently, the graph of f^{-1} is the reflection of the graph of f across the line $y = x$.

Notice that outputs from a function are inputs to its inverse function. Similarly, outputs from the inverse function are inputs to the original function. This is expressed in the statement $f^{-1}(b) = a$ if and only if $f(a) = b$, and also in the fact that we can obtain a table for f^{-1} by interchanging the columns of a table for f. Consequently, the domain and range for f^{-1} are obtained by interchanging the domain and range of f. In other words,

Domain of f^{-1} = Range of f and Range of f^{-1} = Domain of f.

In Example 7, the function P has all real numbers as its domain and all positive numbers as its range; the function P^{-1} has all positive numbers as its domain and all real numbers as its range.

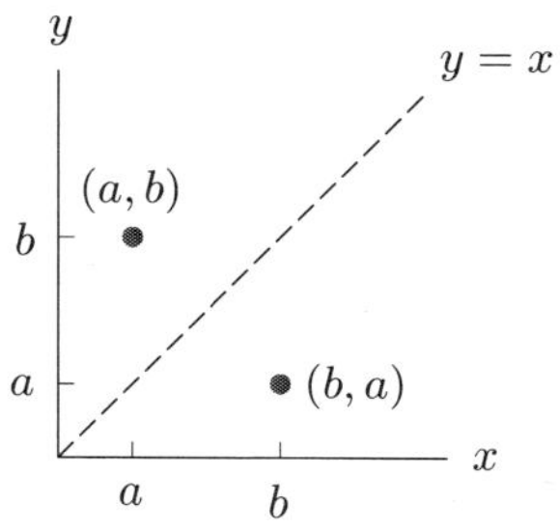

Figure 8.11: The reflection of the point (a, b) across the line $y = x$ is the point (b, a)

A Property of Inverse Functions

The fact that Tables 8.17 and 8.18 contain the same values, but with the columns switched, reflects the special relationship between the values of $P(x)$ and $P^{-1}(x)$. For the population function in Example 7

$$P^{-1}(2) = 1 \quad \text{and} \quad P(1) = 2 \quad \text{so} \quad P^{-1}(P(1)) = 1,$$

and

$$P^{-1}(0.25) = -2 \quad \text{and} \quad P(-2) = 0.25 \quad \text{so} \quad P^{-1}(P(-2)) = -2.$$

This result holds for any input x, so in general,

$$P^{-1}(P(x)) = x.$$

In addition, $P(P^{-1}(2)) = 2$ and $P(P^{-1}(0.25)) = 0.25$, and for any x

$$P(P^{-1}(x)) = x.$$

Similar reasoning holds for any other invertible function, suggesting the general result:

> If $y = f(x)$ is an invertible function and $y = f^{-1}(x)$ is its inverse, then
> - $f^{-1}(f(x)) = x$ for all values of x for which $f(x)$ is defined,
> - $f(f^{-1}(x)) = x$ for all values of x for which $f^{-1}(x)$ is defined.

This property tell us that composing a function and its inverse function returns the original value as the end result. We can use this property to decide whether two functions are inverses.

Example 8 (a) Check that $f(x) = \dfrac{x}{2x+1}$ and $f^{-1}(x) = \dfrac{x}{1-2x}$ are inverse functions of each other.
(b) Graph f and f^{-1} on axes with the same scale. What are the domains and ranges of f and f^{-1}?

Solution (a) To check that these functions are inverses, we compose

$$f^{-1}(f(x)) = \frac{f(x)}{1-2f(x)} = \frac{\dfrac{x}{2x+1}}{1-2\left(\dfrac{x}{2x+1}\right)}$$

$$= \frac{\dfrac{x}{2x+1}}{\dfrac{2x+1}{2x+1} - \dfrac{2x}{2x+1}}$$

$$= \frac{\dfrac{x}{2x+1}}{\dfrac{1}{2x+1}}$$

$$= x.$$

Similarly, you can check that $f(f^{-1}(x)) = x$.

(b) The graphs of f and f^{-1} in Figure 8.12 are symmetric about the line $y = x$.

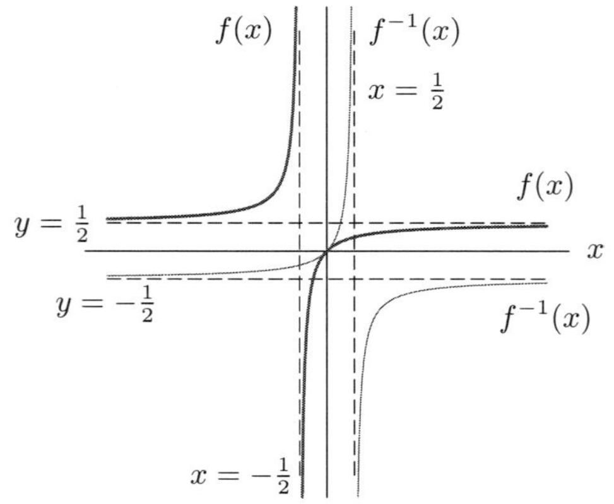

Figure 8.12: The graph of $f(x) = x/(2x+1)$ and the inverse $f^{-1}(x) = x/(1-2x)$

The function $f(x) = x/(2x+1)$ is undefined at $x = -1/2$, so its domain consists of all real numbers except $-1/2$. Figure 8.12 suggests that f has a horizontal asymptote at $y = 1/2$ which it does not cross and that its range is all real numbers except $1/2$.

Because the inverse function $f^{-1}(x) = x/(1-2x)$ is undefined at $x = 1/2$, its domain is all real numbers except $1/2$. Note that this is the same as the range of f. The graph of f^{-1} appears to have a horizontal asymptote which it does not cross at $y = -1/2$ suggesting that its range is all real numbers except $-1/2$. Note that this is the same as the domain of f.

The ranges of the functions f and f^{-1} can be confirmed algebraically.

Restricting the Domain

A function that fails the horizontal line test is not invertible. For this reason, the function $f(x) = x^2$ does not have an inverse function. However, by considering only part of the graph of f, we can eliminate the duplication of y-values. Suppose we consider the half of the parabola with $x \geq 0$. See Figure 8.13. This part of the graph does pass the horizontal line test because there is only one (positive) x-value for each y-value in the range of f.

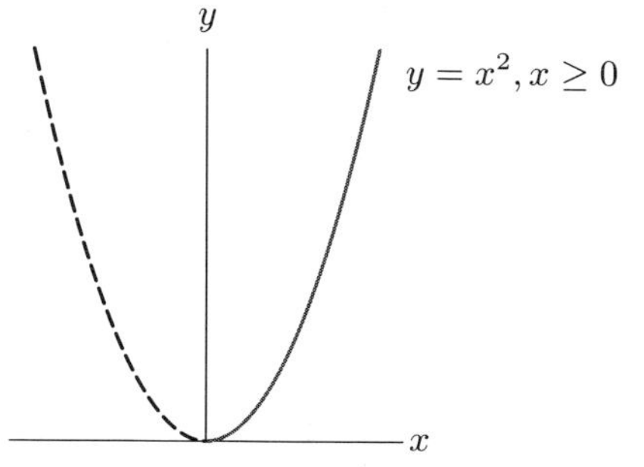

Figure 8.13

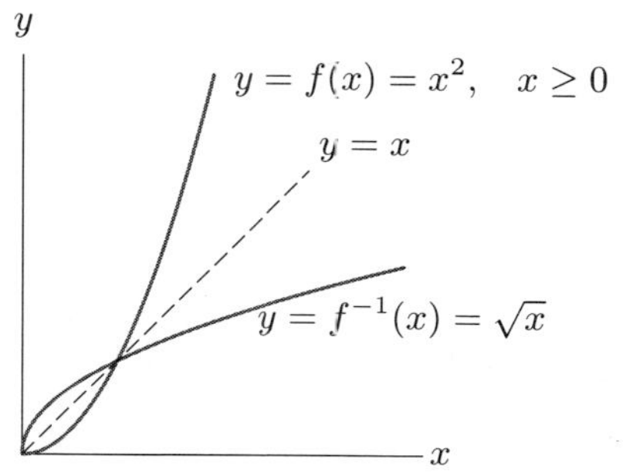

Figure 8.14

We can find an inverse for $f(x) = x^2$ on its restricted domain,[2] $x \geq 0$. Using the fact that $x \geq 0$ and solving $y = x^2$ for x gives

$$x = \sqrt{y}.$$

Thus a formula for the inverse function is

$$x = f^{-1}(y) = \sqrt{y}.$$

Rewriting the formula for f^{-1} with x as the input, we have

$$f^{-1}(x) = \sqrt{x}.$$

The graphs of f and f^{-1} are shown in Figure 8.14. Note that the domain of f is the the range of f^{-1}, and the domain of f^{-1} ($x \geq 0$) is the range of f.

In Section 6.7 we restricted the domains of the sine, cosine, and tangent functions in order to define their inverse functions:

$$\begin{array}{llll} y = \sin^{-1} x & \text{if and only if} & x = \sin y & \text{and} \quad -\dfrac{\pi}{2} \leq y \leq \dfrac{\pi}{2} \\ y = \cos^{-1} x & \text{if and only if} & x = \cos y & \text{and} \quad 0 \leq y \leq \pi \\ y = \tan^{-1} x & \text{if and only if} & x = \tan y & \text{and} \quad -\dfrac{\pi}{2} < y < \dfrac{\pi}{2}. \end{array}$$

[2]Technically, changing the domain results in a new function, but we will continue to call it $f(x)$.

The graphs of each of the inverse trigonometric functions are shown in Figures 8.15-8.17. Note the symmetry about the line $y = x$ for each trigonometric function and its inverse.

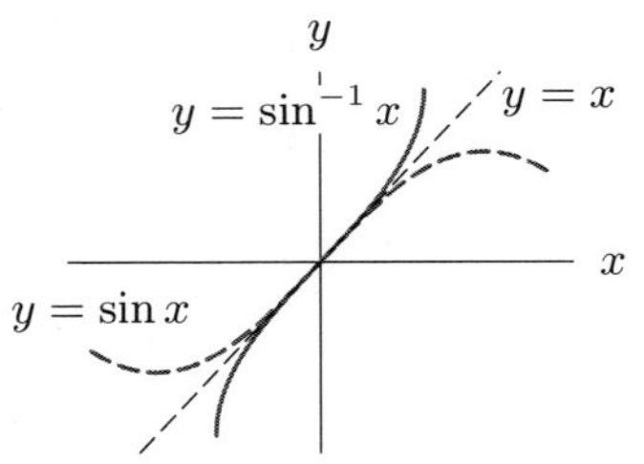

Figure 8.15

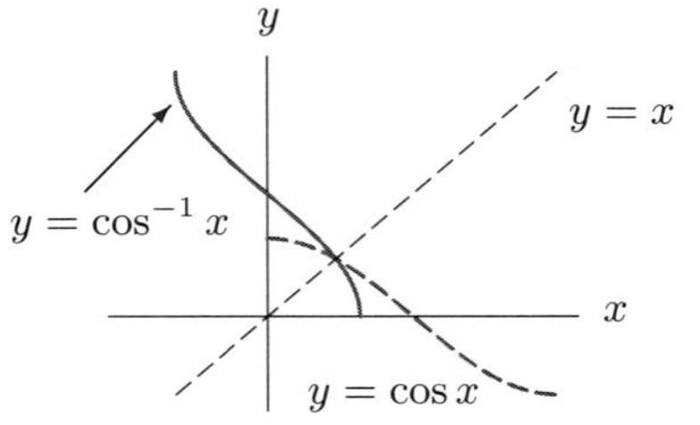

Figure 8.16

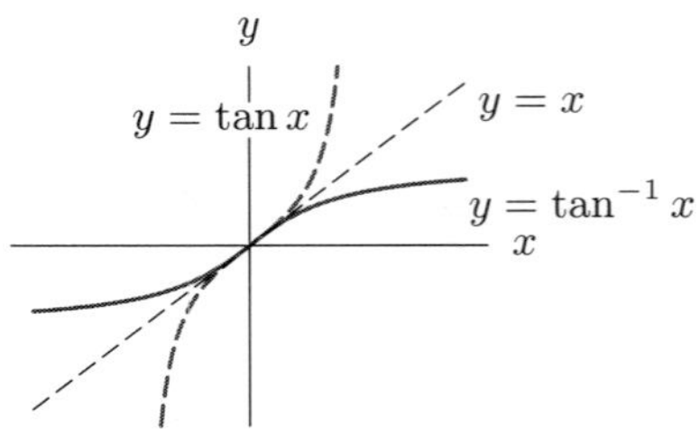

Figure 8.17

In each case, the restricted domain of the function is the range of the inverse function. In addition, the domain of the inverse is the range of the original function. For example

$y = \sin x$	has restricted domain	$-\frac{\pi}{2} \le x \le \frac{\pi}{2}$	and	range	$-1 \le y \le 1$
$y = \sin^{-1} x$	has domain	$-1 \le x \le 1$	and	range	$-\frac{\pi}{2} \le y \le \frac{\pi}{2}.$

Exercises and Problems for Section 8.2

Exercises

In Exercises 1–6, use a graph to decide whether or not the function is invertible.

1. $y = x^6 + 2x^2 - 10$

2. $y = x^4 - 6x + 12$

3. $y = e^{x^2}$

4. $y = \cos(x^3)$

5. $y = |x|$

6. $y = x + \ln x$

In Exercises 7–14, check that the functions are inverses.

7. $f(x) = \frac{x}{4} - \frac{3}{2}$ and $g(t) = 4\left(t + \frac{3}{2}\right)$

8. $f(x) = 1 + 7x^3$ and $f^{-1}(x) = \sqrt[3]{\frac{x-1}{7}}$

9. $g(x) = 1 - \frac{1}{x-1}$ and $g^{-1}(x) = 1 + \frac{1}{1-x}$

10. $h(x) = \sqrt{2x}$ and $k(t) = \frac{t^2}{2}$, for $x, t \ge 0$

11. $f(x) = e^{x+1}$ and $f^{-1}(x) = \ln x - 1$

12. $f(x) = e^{2x}$ and $f^{-1}(x) = \frac{\ln x}{2}$

13. $f(x) = e^{x/2}$ and $f^{-1}(x) = 2\ln x$

14. $f(x) = \ln(x/2)$ and $f^{-1}(x) = 2e^x$

Find the inverses of the functions in Exercises 15–27.

15. $h(x) = 12x^3$

16. $h(x) = \frac{x}{2x+1}$

17. $k(x) = 3 \cdot e^{2x}$

18. $g(x) = e^{3x+1}$

19. $n(x) = \log(x - 3)$

20. $h(x) = \ln(1 - 2x)$

21. $h(x) = \frac{\sqrt{x}}{\sqrt{x}+1}$

22. $g(x) = \frac{x-2}{2x+3}$

23. $f(x) = \sqrt{\frac{4-7x}{4-x}}$

24. $f(x) = \frac{\sqrt{x}+3}{11-\sqrt{x}}$

25. $f(x) = \ln\left(1 + \frac{1}{x}\right)$

26. $s(x) = \frac{3}{2 + \log x}$

27. $q(x) = \ln(x + 3) - \ln(x - 5)$

Problems

28. Let $f(x) = e^x$ and $g(x) = \ln x$.

(a) Find $f(g(x))$ and $g(f(x))$. What can you conclude about the relationship between these two functions?

(b) Graph $f(x)$ and $g(x)$ together on axes with the same scales. What is the line of symmetry of the graph?

29. Let $p(t) = 10^t$ and $q(t) = \log t$.

(a) Find $p(q(t))$ and $q(p(t))$. What can you conclude about the relationship between these two functions?

(b) Graph $p(t)$ and $q(t)$ together on ty-axes with the same scales. What is the line of symmetry of the graph?

30. Let $C = f(q) = 200 + 0.1q$ give the cost in dollars to manufacture q kg of a chemical. Find and interpret $f^{-1}(C)$.

31. Let $P = f(t) = 10e^{0.02t}$ give the population in millions at time t in years. Find and interpret $f^{-1}(P)$.

32. If $t = g(v)$ represents the time in hours it takes to drive to the next town at velocity v mph, what does $g^{-1}(t)$ represent? What are its units?

33. The noise level, N, of a sound in decibels is given by

$$N = f(I) = 10 \log\left(\frac{I}{I_0}\right),$$

where I is the intensity of the sound and I_0 is a constant. Find and interpret $f^{-1}(N)$.

Solve the equations in Problems 34–39 exactly. Use an inverse function when appropriate.

34. $7\sin(3x) = 2$

35. $2^{x+5} = 3$

36. $x^{1.05} = 1.09$

37. $\ln(x + 3) = 1.8$

38. $\dfrac{2x + 3}{x + 3} = 8$

39. $\sqrt{x + \sqrt{x}} = 3$

40. Values of f and g are in Table 8.19. Based on this table:

(a) Is $f(x)$ invertible? If not, explain why; if so, construct a table of values of $f^{-1}(x)$ for all values of x for which $f^{-1}(x)$ is defined.

(b) Answer the same question as in part (a) for $g(x)$.

(c) Make a table of values for $h(x) = f(g(x))$, with $x = -3, -2, -1, 0, 1, 2, 3$.

(d) Explain why you cannot define a function $j(x)$ by the formula $j(x) = g(f(x))$.

Table 8.19

x	-3	-2	-1	0	1	2	3
$f(x)$	9	7	6	-4	-5	-8	-9
$g(x)$	3	1	3	2	-3	-1	3

41. Figure 8.18 defines the function f. Rank the following quantities in order from least to greatest: $0, f(0), f^{-1}(0), 3, f(3), f^{-1}(3)$.

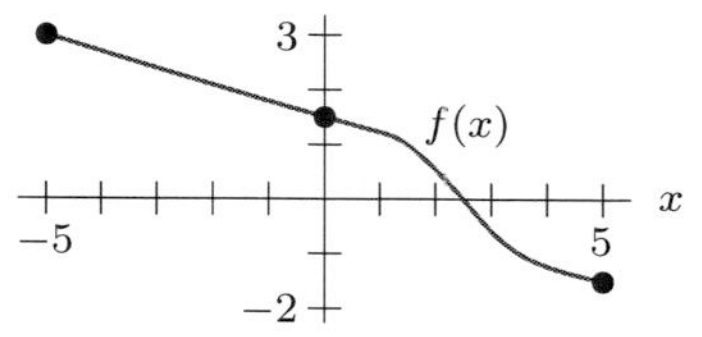

Figure 8.18

42. Let $f(x) = e^x$. Solve each of the following equations exactly for x.

(a) $(f(x))^{-1} = 2$

(b) $f^{-1}(x) = 2$

(c) $f(x^{-1}) = 2$

43. Simplify the expression $\cos^2(\arcsin t)$, using the property that inverses "undo" each other.

44. Let $Q = f(t) = 20(0.96)^{t/3}$ be the number of grams of a radioactive substance remaining after t years.

(a) Describe the behavior of the radioactive substance as a function of time.

(b) Evaluate $f(8)$. Explain the meaning of this quantity in practical terms.

(c) Find a formula for $f^{-1}(Q)$ in terms of Q.

(d) Evaluate $f^{-1}(8)$. Explain the meaning of this quantity in practical terms.

45. **(a)** What is the formula for the area of a circle in terms of its radius?

(b) Graph this function for the domain all real numbers.

(c) What domain actually applies in this situation? On separate axes, graph the function for this domain.

(d) Find the inverse of the function in part (c).

(e) Graph the inverse function on the domain you gave in part (c) on the same axes used in part (c).

(f) If area is a function of the radius, is radius a function of area? Explain carefully.

46. A company believes there is a linear relationship between the consumer demand for its products and the price charged. When the price was \$3 per unit, the quantity demanded was 500 units per week. When the unit price was raised to \$4, the quantity demanded dropped to 300 units per week. Let $D(p)$ be the quantity per week demanded by consumers at a unit price of \$$p$.

(a) Estimate and interpret $D(5)$.
(b) Find a formula for $D(p)$ in terms of p.
(c) Calculate and interpret $D^{-1}(5)$.
(d) Give an interpretation of the slope of $D(p)$ in terms of demand.
(e) Currently, the company can produce 400 units every week. What should the price of the product be if the company wants to sell all 400 units?
(f) If the company produced 500 units per week instead of 400 units per week, would its weekly revenues increase, and if so, by how much?

47. Table 8.20 gives the number of cows in a herd.

(a) Find an exponential function that approximates the data.
(b) Find the inverse function of the function in part (a).
(c) When do you predict that the herd will contain 400 cows?

Table 8.20

t (years)	0	1	2
$P(t)$ (cows)	150	165	182

48. Suppose $P = f(t)$ is the population (in thousands) in year t, and that $f(7) = 13$ and $f(12) = 20$,

(a) Find a formula for $f(t)$ assuming f is exponential.
(b) Find a formula for $f^{-1}(P)$.
(c) Evaluate $f(25)$ and $f^{-1}(25)$. Explain what these expressions mean in terms of population.

49. A gymnast at Ringling Brothers, Barnum, & Bailey Circus is fired straight up in the air from a cannon. While she is in the air, a trampoline is moved into the spot where the cannon was. Figure 8.19 is a graph of the gymnast's height, h, as a function of time, t.

(a) Approximately what is her maximum height?
(b) Approximately when does she land on the trampoline?
(c) Restrict the domain of $h(t)$ so that $h(t)$ has an inverse. That is, pick a piece of the graph on which $h(t)$ does have an inverse. Graph this new restricted function.
(d) Change the story to go with your graph in part (c).
(e) Graph the inverse of the function in part (c). Explain in your story why it makes sense that the inverse is a function.

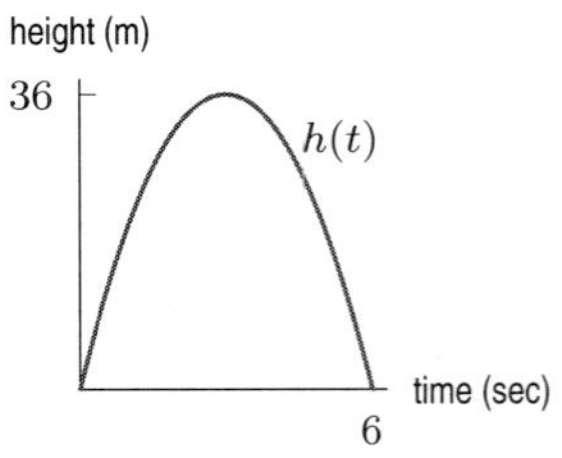

Figure 8.19

50. A 100 ml solution contains 99% alcohol and 1% water. Let $y = C(x)$ be the concentration of alcohol in the solution after x ml of alcohol are removed, so

$$C(x) = \frac{\text{Amount of alcohol}}{\text{Amount of solution}}.$$

(a) What is $C(0)$?
(b) Find a formula in terms of x for $C(x)$.
(c) Find a formula in terms of y for $C^{-1}(y)$.
(d) Explain the physical significance of $C^{-1}(y)$.

51. **(a)** How much alcohol do you think should be removed from the 99% solution in Problem 50 in order to obtain a 98% solution? (Make a guess.)
(b) Express the exact answer to part (a) using the function C^{-1} you found in Problem 62.
(c) Determine the exact answer to part (a). Are you surprised by your result?

The predicted pulse in beats per minute (bpm) of a healthy person fifteen minutes after consuming q milligrams of caffeine is given by $r = f(q)$. The amount of caffeine in a serving of coffee is q_c, and $r_c = f(q_c)$. Assume that f is an increasing function for non-toxic levels of caffeine. What do the statements in Problems 52–57 tell you about caffeine and a person's pulse?

52. $f(2q_c)$

53. $f^{-1}(r_c + 20)$

54. $2f^{-1}(r_c) + 20$

55. $f(q_c) - f(0)$

56. $f^{-1}(r_c + 20) - q_c$

57. $f^{-1}(1.1f(q_c))$

58. Suppose that f, g, and h are invertible and that

$$f(x) = g(h(x)).$$

Find a formula for $f^{-1}(x)$ in terms of g^{-1} and h^{-1}.

59. The von Bertalanffy growth model predicts the mean length L of fish of age t (in years):[3]

$$L = f(t) = L_\infty\left(1 - e^{-k(t+t_0)}\right), \text{ for constant } L_\infty, k, t_0.$$

Find a formula for f^{-1}. Describe in words what f^{-1} tells you about fish. What is the domain of f^{-1}?

8.3 COMBINATIONS OF FUNCTIONS

Like numbers, functions can be combined using addition, subtraction, multiplication, and division.

The Difference of Two Functions Defined by Formulas: A Measure of Prosperity

We can define new functions as the sum or difference of two functions. In Chapter 3, we discussed Thomas Malthus, who predicted widespread food shortages because he believed that human populations increase exponentially, whereas the food supply increases linearly. We considered a country with population $P(t)$ million in year t. The population is initially 2 million and grows at the rate of 4% per year, so

$$P(t) = 2(1.04)^t.$$

Let $N(t)$ be the number of people (in millions) that the country can feed in year t. The annual food supply is initially adequate for 4 million people and it increases by enough for an additional 0.5 million people every year. Thus

$$N(t) = 4 + 0.5t.$$

This country first experiences shortages in about 78 years. (See Figure 8.20.) When is it most prosperous?

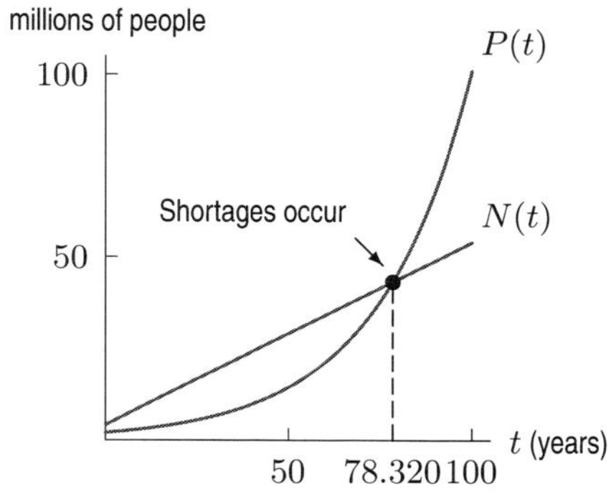

Figure 8.20: Predicted population, $P(t)$, and number of people who can be fed, $N(t)$, over a 100-year period

The answer depends on how we decide to measure prosperity. We could measure prosperity in one of the following ways:

- By the food surplus—that is, the amount of food the country has over and above its needs. This surplus food could be warehoused or exported in trade.
- By the per capita food supply—that is, how much food there is per person. (The term *per capita* means per person, or literally, "per head.") This indicates the portion of the country's wealth each person might enjoy.

[3] *Introduction to Tropical Fish Stock Assessment* by Per Sparre, Danish Institute for Fisheries Research, and Siebren C. Venema, FAO Fisheries Department, available at http://www.fao.org/docrep/W5449E/w5449e00.htm.

First, we choose to measure prosperity in terms of food surplus, $S(t)$, in year t, where

$$S(t) = \underbrace{\text{Number of people that can be fed}}_{N(t)} - \underbrace{\text{Number of people living in the country}}_{P(t)}$$

so

$$S(t) = N(t) - P(t).$$

For example, to determine the surplus in year $t = 25$, we evaluate

$$S(25) = N(25) - P(25).$$

Since $N(25) = 4 + 0.5(25) = 16.5$ and $P(25) = 2(1.04)^{25} \approx 5.332$, we have

$$S(25) \approx 16.5 - 5.332 = 11.168.$$

Thus, in year 25 the food surplus could feed 11.168 million additional people.

We use the formulas for N and P to find a formula for S:

$$S(t) = \underbrace{N(t)}_{4+0.5t} - \underbrace{P(t)}_{2(1.04)^t},$$

so

$$S(t) = 4 + 0.5t - 2(1.04)^t.$$

A graph of S is shown in Figure 8.21. The maximum surplus occurs sometime during the 48th year. In that year, there is surplus food sufficient for an additional 14.865 million people.

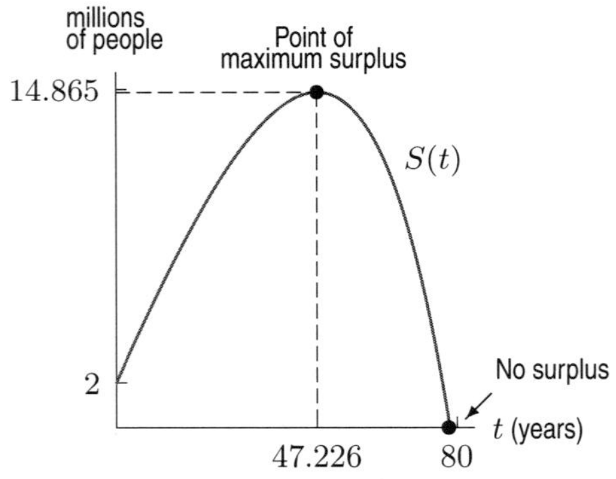

Figure 8.21: Surplus graphed using the formula $S(t) = 4 + 0.5t - 2(1.04)^t$

The Sum and Difference of Two Functions Defined by Graphs

How does the graph of the surplus function S, shown in Figure 8.21, relate to the graphs of N and P in Figure 8.20? Since

$$S(t) = N(t) - P(t),$$

the value of $S(t)$ is represented graphically as the vertical distance between the graphs of $N(t)$ and $P(t)$. See Figure 8.22. Figure 8.23 shows the surplus plotted against time.

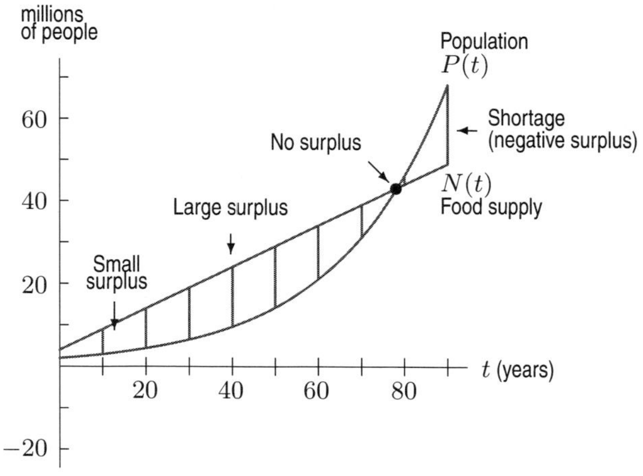

Figure 8.22: Surplus, $S(t) = N(t) - P(t)$, as vertical distance between $N(t)$ and $P(t)$ graphs

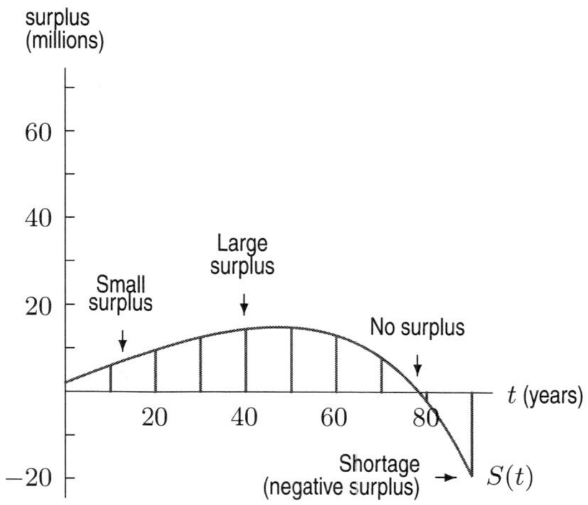

Figure 8.23: Surplus as a function of time

From year $t = 0$ to $t \approx 78.320$, the food supply is more than the population needs. Therefore the surplus, $S(t)$, is positive on this time interval. At time $t = 78.320$, the food supply is exactly sufficient for the population, so $S(t) = 0$, resulting in the horizontal intercept $t = 78.320$ on the graph of $S(t)$ in Figure 8.23. For times $t > 78.320$, the food supply is less than the population needs. Therefore the surplus is negative, representing a food shortage.

In the next example we consider a sum of two functions.

Example 1 Let $f(x) = x$ and $g(x) = \dfrac{1}{x}$. By adding vertical distances on the graphs of f and g, sketch

$$h(x) = f(x) + g(x) \quad \text{for } x > 0.$$

Solution The graphs of f and g are shown in Figure 8.24. For each value of x, we add the vertical distances that represent $f(x)$ and $g(x)$ to get a point on the graph of $h(x)$. Compare the graph of $h(x)$ to the values shown in Table 8.21.

Table 8.21 *Adding function values*

x	$\frac{1}{4}$	$\frac{1}{2}$	1	2	4
$f(x) = x$	$\frac{1}{4}$	$\frac{1}{2}$	1	2	4
$g(x) = 1/x$	4	2	1	$\frac{1}{2}$	$\frac{1}{4}$
$h(x) = f(x) + g(x)$	$4\frac{1}{4}$	$2\frac{1}{2}$	2	$2\frac{1}{2}$	$4\frac{1}{4}$

Note that as x increases, $g(x)$ decreases toward zero, so the values of $h(x)$ get closer to the values of $f(x)$. On the other hand, as x approaches zero, $h(x)$ gets closer to $g(x)$.

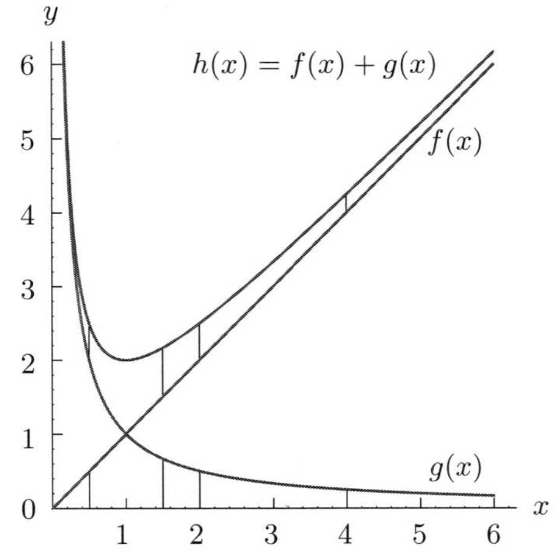

Figure 8.24: Graph of $h(x) = f(x) + g(x)$ constructed by adding vertical distances under f and g

Factoring a Function's Formula into a Product

It is often useful to be able to express a given function as a product of functions.

Example 2 Find exactly all the zeros of the function

$$p(x) = 2^x \cdot 6x^2 - 2^x \cdot x - 2^{x+1}.$$

Solution We could approximate the zeros by finding the points where the graph of the function p crosses the x-axis. Unfortunately, these solutions are not exact. Alternatively, we can express p as a product. Using the fact that

$$2^{x+1} = 2^x \cdot 2^1 = 2 \cdot 2^x,$$

we rewrite the formula for p as

$$\begin{aligned} p(x) &= 2^x \cdot 6x^2 - 2^x x - 2 \cdot 2^x \\ &= 2^x(6x^2 - x - 2) \quad \text{Factoring out } 2^x \\ &= 2^x(2x+1)(3x-2) \quad \text{Factoring the quadratic.} \end{aligned}$$

Thus p is the product of the exponential function 2^x and two linear functions. Since p is a product, it equals zero if one or more of its factors equals zero. But 2^x is never equal to 0, so $p(x)$ equals zero if and only if one of the linear factors is zero:

$$(2x+1) = 0 \quad \text{or} \quad (3x-2) = 0$$

$$x = -\frac{1}{2} \qquad x = \frac{2}{3}.$$

The Quotient of Functions Defined by Formulas and Graphs: Prosperity

Now let's think about our second proposed measure of prosperity, the per capita food supply, $R(t)$. With this definition of prosperity

$$R(t) = \frac{\text{Number of people that can be fed}}{\text{Number of people living in the country}} = \frac{N(t)}{P(t)}.$$

For example,

$$R(25) = \frac{N(25)}{P(25)} = \frac{16.5}{5.332} \approx 3.1.$$

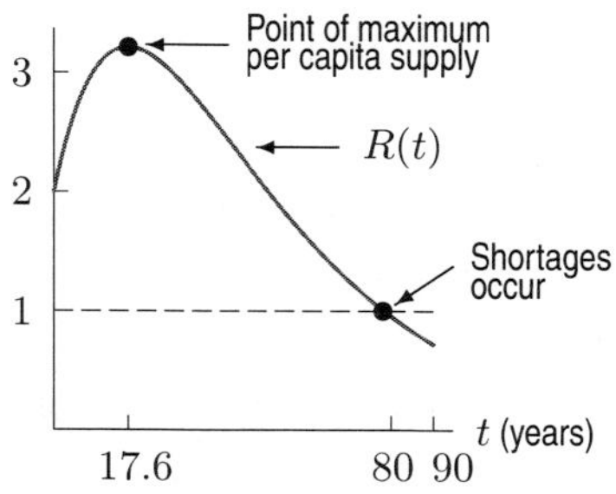

Figure 8.25: Per capita food supply, $R(t) = \frac{N(t)}{P(t)}$

This means that in year 25, everybody in the country could, on average, have more than three times as much food as he or she needs. The formula for $R(t)$ is

$$R(t) = \frac{N(t)}{P(t)} = \frac{4 + 0.5t}{2(1.04)^t}.$$

From the graph of R in Figure 8.25, we see that the maximum per capita food supply occurs during the 18th year. Notice this maximum prosperity prediction is different from the one made using the surplus function $S(t)$.

However, both prosperity models predict that shortages begin after time $t = 78.320$. This is not a coincidence. The food surplus model predicts shortages when $S(t) = N(t) - P(t) < 0$, or $N(t) < P(t)$. The per capita food supply model predicts shortages when $R(t) < 1$, meaning that the amount of food available per person is less than the amount necessary to feed 1 person. Since $R(t) = \frac{N(t)}{P(t)} < 1$ is true only when $N(t) < P(t)$, the same condition leads to shortages.

The Quotient of Functions Defined by Tables: Per Capita Crime Rate

Table 8.22 gives the number of violent crimes committed in two cities between 2000 and 2005. It appears that crime in both cities is on the rise and that there is less crime in City B than in City A.

Table 8.22 *Number of violent crimes committed each year in two cities*

Year	2000	2001	2002	2003	2004	2005
t, years since 2000	0	1	2	3	4	5
Crimes in City A	793	795	807	818	825	831
Crimes in City B	448	500	525	566	593	652

Table 8.23 gives the population for these two cities from 2000 to 2005. The population of City A is larger than that of City B and both cities are growing.

Table 8.23 *Population of the two cities*

Year	2000	2001	2002	2003	2004	2005
t, years since 2000	0	1	2	3	4	5
Population of City A	61,000	62,100	63,220	64,350	65,510	66,690
Population of City B	28,000	28,588	29,188	29,801	30,427	31,066

Can we attribute the growth in crime in both cities to the population growth? Can we attribute the larger number of crimes in City A to its larger population? To answer these questions, we consider the per capita crime rate in each city.

Let's define $N_A(t)$ to be the number of crimes in City A during year t (where $t = 0$ means 2000). Similarly, let's define $P_A(t)$ to be the population of City A in year t. Then the per capita crime rate in City A, $r_A(t)$, is given by

$$r_A(t) = \frac{\text{Number of crimes in year } t}{\text{Number of people in year } t} = \frac{N_A(t)}{P_A(t)}.$$

We have defined a new function, $r_A(t)$, as the quotient of $N_A(t)$ and $P_A(t)$. For example, the data in Tables 8.22 and 8.23 shows that the per capita crime rate for City A in year $t = 0$ is

$$r_A(0) = \frac{N_A(0)}{P_A(0)} = \frac{793}{61{,}000} = 0.0130 \text{ crimes per person.}$$

Similarly, the per capita crime rate for the year $t = 1$ is

$$r_A(1) = \frac{N_A(1)}{P_A(1)} = \frac{795}{62{,}100} = 0.0128 \text{ crimes per person.}$$

Thus, the per capita crime rate in City A actually decreased from 0.0130 crimes per person in 2000 to 0.0128 crimes per person in 2001.

Example 3

(a) Make a table of values for $r_A(t)$ and $r_B(t)$, the per capita crime rates of Cities A and B.
(b) Use the table to decide which city is more dangerous.

Solution

(a) Table 8.24 gives values of $r_A(t)$ for $t = 0, 1, \ldots, 5$. The per capita crime rate in City A declined between 2000 and 2005 despite the fact that the total number of crimes rose during this period. Table 8.24 also gives values of $r_B(t)$, the per capita crime rate of City B, defined by

$$r_B(t) = \frac{N_B(t)}{P_B(t)},$$

where $N_B(t)$ is the number of crimes in City B in year t and $P_B(t)$ is the population of City B in year t. For example, the per capita crime rate in City B in year $t = 0$ is

$$r_B(0) = \frac{N_B(0)}{P_B(0)} = \frac{448}{28{,}000} = 0.016 \text{ crimes per person.}$$

Table 8.24 *Values of $r_A(t)$ and $r_B(t)$, the per capita violent crime rates of Cities A and B*

Year	2000	2001	2002	2003	2004	2005
t, years since 2000	0	1	2	3	4	5
$r_A(t) = N_A(t)/P_A(t)$	0.0130	0.0128	0.01276	0.01271	0.01259	0.01246
$r_B(t) = N_B(t)/P_B(t)$	0.01600	0.01749	0.01799	0.01899	0.01949	0.02099

(b) From Table 8.24, we see that between 2000 and 2005, City A has a lower per capita crime rate than City B. The crime rate of City A is decreasing, whereas the crime rate of City B is increasing. Thus, even though Table 8.22 indicates that there are more crimes committed in City A, Table 8.24 tells us that City B is, in some sense, more dangerous. Table 8.24 also tells us that, even though the number of crimes is rising in both cities, City A is getting safer, while City B is getting more dangerous.

Exercises and Problems for Section 8.3

Exercises

In Exercises 1–6, find the following functions.

(a) $f(x) + g(x)$ **(b)** $f(x) - g(x)$
(c) $f(x)g(x)$ **(d)** $f(x)/g(x)$

1. $f(x) = x + 1 \quad g(x) = 3x^2$

2. $f(x) = x^2 + 4 \quad g(x) = x + 2$

3. $f(x) = x + 5 \quad g(x) = x - 5$

4. $f(x) = x^2 + 4 \quad g(x) = x^2 + 2$

5. $f(x) = x^3 \quad g(x) = x^2$

6. $f(x) = \sqrt{x} \quad g(x) = x^2 + 2$

In Exercises 7–12, find a simplified formula for the function. Let $m(x) = 3x^2 - x$, $n(x) = 2x$, and $o(x) = \sqrt{x+2}$.

7. $f(x) = m(x) + n(x)$

8. $g(x) = (o(x))^2$

9. $h(x) = n(x)o(x)$

10. $i(x) = m(o(x))n(x)$

11. $j(x) = (m(x))/n(x)$

12. $k(x) = m(x) - n(x) - o(x)$

In Exercises 13–16, let $u(x) = e^x$ and $v(x) = 2x + 1$. Find a simplified formula for the function.

13. $f(x) = u(x)v(x)$

14. $g(x) = u(x)^2 + v(x)^2$

15. $h(x) = (v(u(x)))^2$

16. $k(x) = v(u(x)^2)$

Find formulas for the functions in Exercises 17–22. Let $f(x) = \sin x$ and $g(x) = x^2$.

17. $f(x) + g(x)$

18. $g(x)f(x)$

19. $f(x)/g(x)$

20. $f(g(x))$

21. $g(f(x))$

22. $1 - (f(x))^2$

Problems

23. **(a)** On the same set of axes, graph $f(x) = (x-4)^2 - 2$ and $g(x) = -(x-2)^2 + 8$.
(b) Make a table of values for f and g for $x = 0, 1, 2, ..., 6$.
(c) Make a table of values for $y = f(x) - g(x)$ for $x = 0, 1, 2, ..., 6$.
(d) On your graph, sketch the vertical line segment of length $f(x) - g(x)$ for each integer value of x from 0 to 6. Check that the segment lengths agree with the values from part (c).
(e) Plot the values from your table for the function $y = f(x) - g(x)$ on your graph.
(f) Simplify the formulas for $f(x)$ and $g(x)$ in part (a). Find a formula for $y = f(x) - g(x)$.
(g) Use part (f) to graph $y = f(x) - g(x)$ on the same axes as f and g. Does the graph pass through the points you plotted in part (e)?

24. Figure 8.26 shows a weight attached to the end of a spring which is hanging from the ceiling. The weight is pulled down from the ceiling and then released. The weight oscillates up and down, but over time, friction decreases the magnitude of the vertical oscillations. Which of the following functions could describe the distance of the weight from the ceiling, d, as a function of time, t?

(i) $d = 2 + \cos t$ (ii) $d = 2 + e^{-t}\cos t$
(iii) $d = 2 + \cos(e^t)$ (iv) $d = 2 + e^{\cos t}$

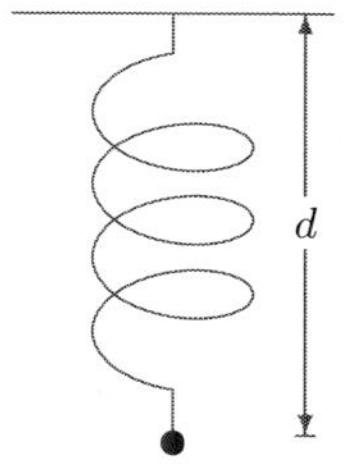

Figure 8.26

25. Table 8.25 gives the upper household income limits for the tenth and ninety-fifth percentiles t years after 1993.[4] For instance, $P_{10}(5) = 9{,}700$ tells us that in 1998 the maximum income for a household in the poorest 10% of all households was $9,700. Let $f(t) = P_{95}(t) - P_{10}(t)$ and $g(t) = P_{95}(t)/P_{10}(t)$.

(a) Make tables of values for f and g.

[4]US Census Bureau, *The Changing Shape of the Nation's Income Distribution*, accessed December 29, 2005, at www.census.gov/prod/2000pubs/p60-204.pdf.

(b) Describe in words what f and g tell you about household income.

Table 8.25

t (yrs)	0	1	2	3	4	5
$P_{10}(t)$ ($)	8670	8830	9279	9256	9359	9700
$P_{95}(t)$ ($)	118,036	120,788	120,860	124,187	128,521	132,199

26. Use Table 8.26 to make tables of values for $x = -1, 0, 1, 2, 3, 4$ for the following functions.

(a) $h(x) = f(x) + g(x)$ **(b)** $j(x) = 2f(x)$
(c) $k(x) = (g(x))^2$ **(d)** $m(x) = g(x)/f(x)$

Table 8.26

x	-1	0	1	2	3	4
$f(x)$	-4	-1	2	5	8	11
$g(x)$	4	1	0	1	4	9

27. Use Figure 8.27 to graph the following functions.

(a) $y = g(x) - 3$ **(b)** $y = g(x) + x$

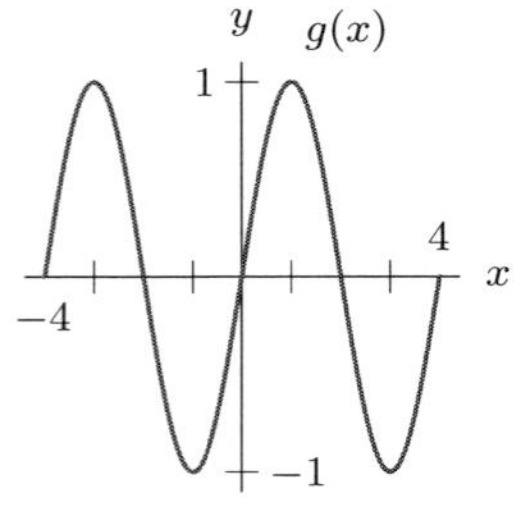

Figure 8.27

28. Graph $h(x) = f(x) + g(x)$ using Figure 8.28.

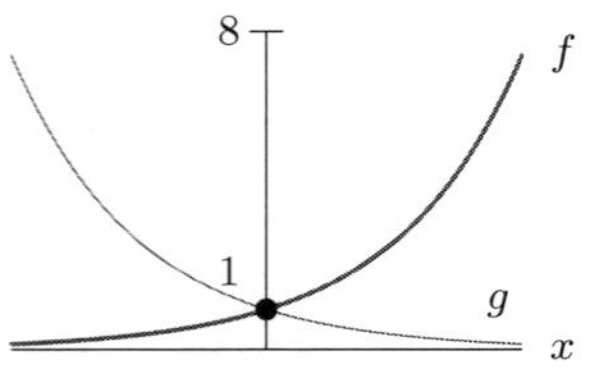

Figure 8.28

29. Use Figure 8.29 to graph $h(x) = g(x) - f(x)$. On the graph of $h(x)$, label the points whose x-coordinates are $x = a$, $x = b$, and $x = c$. Label the y-intercept.

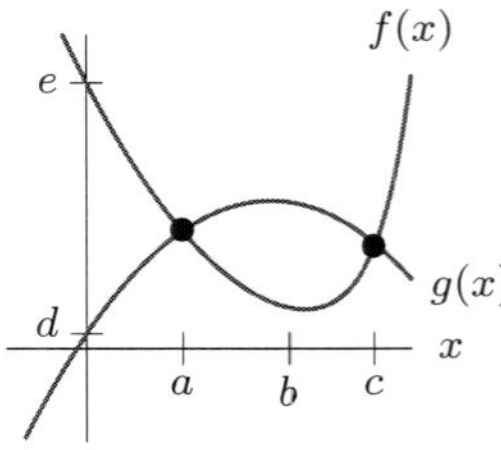

Figure 8.29

30. **(a)** Find possible formulas for the functions in Figure 8.30.

(b) Let $h(x) = f(x) \cdot g(x)$. Graph $f(x)$, $g(x)$ and $h(x)$ on the same set of axes.

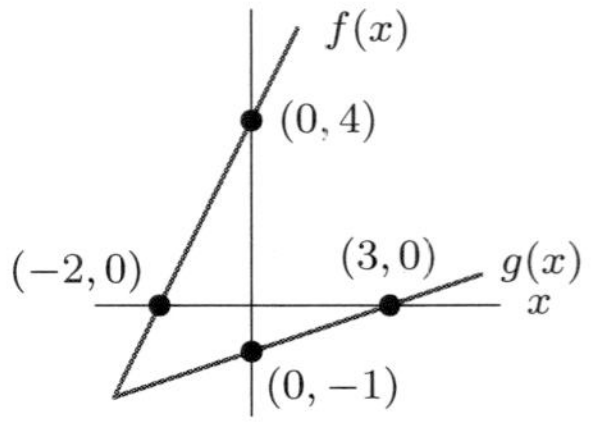

Figure 8.30

31. Use Figure 8.31 to graph $c(x) = a(x) \cdot b(x)$. [Hint: There is not enough information to determine formulas for a and b but you can use the method of Problem 30.]

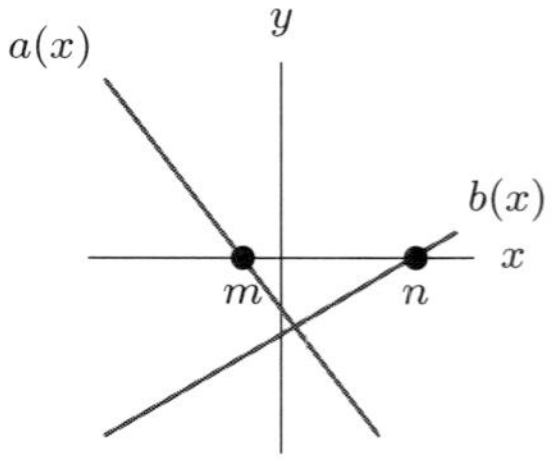

Figure 8.31

32. Sketch two linear functions whose product is the function f graphed in Figure 8.32(a). Explain why this is not possible for the function q graphed in Figure 8.32(b).

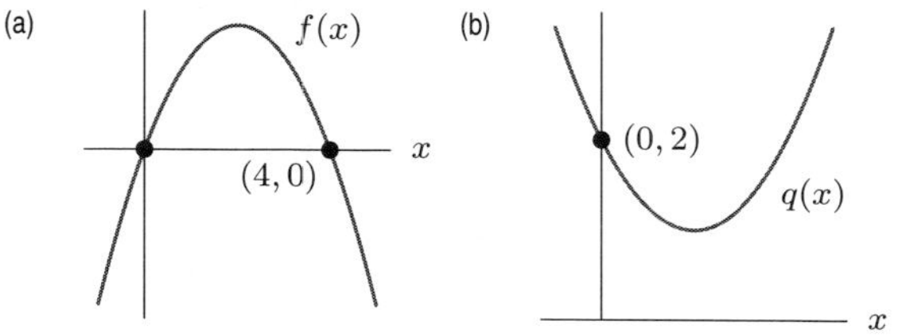

Figure 8.32

33. Let $f(x) = x + 1$ and $g(x) = x^2 - 1$. In parts (a)–(e), write a formula in terms of $f(x)$ and $g(x)$ for the function. Then evaluate the formula for $x = 3$. Write a formula in terms of x for each function. Check your formulas for $x = 3$.

(a) $h(x)$ is the sum of $f(x)$ and $g(x)$.
(b) $j(x)$ is the difference between $g(x)$ and twice $f(x)$.
(c) $k(x)$ is the product of $f(x)$ and $g(x)$.
(d) $m(x)$ is the ratio of $g(x)$ to $f(x)$.
(e) $n(x)$ is defined by $n(x) = (f(x))^2 - g(x)$.

34. An average of 50,000 people visit Riverside Park each day in the summer. The park charges \$15.00 for admission. Consultants predict that for each \$1.00 increase in the entrance price, the park would lose an average of 2500 customers per day. Express the daily revenue from ticket sales as a function of the number of \$1.00 price increases. What ticket price maximizes the revenue from ticket sales?

35. In Figure 8.33, the line l_2 is fixed and the point P moves along l_2. Define $f(\theta)$ as the y-coordinate of P.

(a) Find a formula for $f(\theta)$ if $0 < \theta < \pi/2$. [Hint: Use the equation for l_2.]
(b) Graph $y = f(\theta)$ on the interval $-\pi \le \theta \le \pi$.
(c) How does the y-coordinate of P change as θ changes? Is $y = f(\theta)$ periodic?

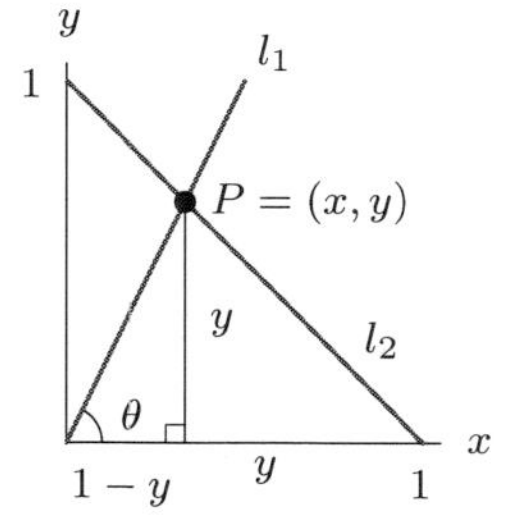

Figure 8.33

36. Describe the similarities and differences between the graphs of $y = \sin(1/x)$ and $y = 1/\sin x$.

37. Let $f(x) = kx^2 + B$ and $g(x) = C^{2x}$ and

$$h(x) = kx^2C^{2x} + BC^{2x} + C^{2x}.$$

Suppose $f(3) = 7$ and $g(3) = 5$. Evaluate $h(3)$.

38. Is the following statement true or false? If $f(x) \cdot g(x)$ is an odd function, then both $f(x)$ and $g(x)$ are odd functions. Explain your answer.

39. **(a)** Is the sum of two even functions even, odd, or neither? Justify your answer.
(b) Is the sum of two odd functions even, odd, or neither? Justify your answer.
(c) Is the sum of an even and an odd function even, odd, or neither? Justify your answer.

40. Let $f(t)$ be the number of men and $g(t)$ be the number of women in Canada in year t. Let $h(t)$ be the average income, in Canadian dollars, of women in Canada in year t.

(a) Find the function $p(t)$ which gives the number of people in Canada in year t.
(b) Find the total amount of money $m(t)$ earned by Canadian women in year t.

41. At the Mauna Loa Observatory, measurements of CO_2 levels (ppm) in the atmosphere reveal a slow exponential increase due to deforestation and burning of fossil fuels, and a periodic seasonal variation.

(a) In 1960, the average CO_2 level was 316.75 parts per million and was rising by 0.4% per year. Write the average annual CO_2 level as an exponential function of time, t, in years since 1960.
(b) Each year, the CO_2 level oscillates once between 3.25 ppm above and 3.25 ppm below the average level. Write a sinusoidal function for the seasonal variation in CO_2 levels in terms of time, t.
(c) Graph the sum of your functions in parts (a) and (b).

42. In 1961 and 1962, large amounts of the radioactive isotope carbon-14 were produced by tests of nuclear bombs.[5] If t is years since 1963, the amount of carbon-14, as a percent in excess of the normal level, is given by

$$C(t) = 108(e^{-0.1t} - e^{-0.7t}).$$

(a) Graph the function.
(b) Approximately when was the level of carbon-14 the highest?
(c) What happens to the level of carbon-14 in the long-run?

43. Table 8.27 gives data on strawberry production from 2000 through 2004,[6] where t is in years since 2000. Let

[5] Adapted from Bolton, *Patterns in Physics*.
[6] www.usda.gov/nass/pubs/agstats.htm, accessed January 15, 2006.

f_{CA}(t), f_{FL}(t), and f_{US}(t) be the harvested area in year t for strawberries grown in California, Florida, and the US overall, respectively. Likewise, let g_{CA}(t), g_{FL}(t), and g_{US}(t) give the yield in thousands of pounds per acre for these three regions.

(a) Let $h_{CA}(t) = f_{CA}(t) \cdot g_{CA}(t)$. Create a table of values for h_{CA}(t) for $0 \le t \le 4$. Describe in words what h_{CA}(t) tells you about strawberry production.
(b) Let $p(t)$ be the fraction of all US strawberries (by weight) grown in Florida and California in year t. Find a formula for $p(t)$ in terms of f_{CA}, f_{FL}, f_{US}. Use Table 8.27 to make a table of values for $p(t)$.

Table 8.27

	Harvested area (acres)			Yield (1000 lbs per acre)		
t	CA	FL	US total	CA	FL	US total
0	27,600	6,300	47,650	59.0	35.0	42.0
1	26,400	6,500	46,100	52.5	26.0	37.8
2	28,500	6,900	47,900	56.5	25.5	39.4
3	29,600	7,100	48,700	62.0	22.0	42.8
3	33,200	7,100	51,600	59.0	23.0	42.9

CHAPTER SUMMARY

- **Composition of Functions**
  Notation: $h(t) = f(g(t))$.
  Domain and range; Decomposition.
- **Inverse Functions**
  Definition: $f^{-1}(Q) = t$ if and only if $Q = f(t)$.
  Invertibility; horizontal line test.
  Domain and range of an inverse function.
  Restricting domain of a function to construct an inverse.
- **Combinations of Functions**
  Sums, differences, products, quotients.

REVIEW EXERCISES AND PROBLEMS FOR CHAPTER EIGHT

Exercises

In Exercises 1–6, let $f(x) = 3x^2$, $g(x) = 9x - 2$, $m(x) = 4x$, and $r(x) = \sqrt{3x}$. Find and simplify the composite function.

1. $r(g(x))$ **2.** $f(r(x))$
3. $r(f(x))$ **4.** $g(f(x))$
5. $g(m(f(x)))$ **6.** $f(m(g(x)))$

In Exercises 7–8, find and simplify for $f(x) = 2^x$ and $g(x) = \dfrac{x}{x+1}$.

7. $f(g(x))$ **8.** $g(f(x))$

In Exercises 9–14, find simplified formulas if $f(x) = e^x$, $g(x) = 2x - 1$, and $h(x) = \sqrt{x}$.

9. $g(f(x))$ **10.** $g(x)f(x)$ **11.** $g(g(x))$
12. $f(g(h(x)))$ **13.** $f(g(x))h(x)$ **14.** $(f(h(x)))^2$

15. Find formulas for the following functions, given that

$$f(x) = x^2 + x, \qquad g(x) = 2x - 3, \qquad h(x) = \frac{x}{1-x}.$$

(a) $f(2x)$ **(b)** $g(x^2)$ **(c)** $h(1-x)$
(d) $(f(x))^2$ **(e)** $g^{-1}(x)$ **(f)** $(h(x))^{-1}$
(g) $f(x) \cdot g(x)$ **(h)** $h(f(x))$

In Exercises 16–18, find simplified formulas if $u(x) = \dfrac{1}{1+x^2}$, $v(x) = e^x$, and $w(x) = \ln x$.

16. $v(x)/u(x)$ **17.** $u(v(x)) \cdot w(v(x))$

18. $\dfrac{w(2+h) - w(2)}{h}$

In Exercises 19–22 find simplified formulas if $f(x) = x^{3/2}$, $g(x) = \dfrac{(3x-1)^2}{4}$, and $h(x) = \tan 2x$.

19. $f(x)h(x)$ **20.** $\dfrac{h(x)}{f(g(x))}$

21. $h(g(x)) - f(9x)$ **22.** $h(x/2)\cos x$

In Exercises 23–30, find a formula for the inverse function. Assume these functions are defined on domains on which they are invertible.

23. $f(x) = 3x - 7$

24. $j(x) = \sqrt{1+\sqrt{x}}$

25. $h(x) = \dfrac{2x+1}{3x-2}$

26. $k(x) = \dfrac{3-\sqrt{x}}{\sqrt{x}+2}$

27. $g(x) = \dfrac{\ln x - 5}{2\ln x + 7}$

28. $h(x) = \log\left(\dfrac{x+5}{x-4}\right)$

29. $f(x) = \cos\sqrt{x}$

30. $g(x) = 2^{\sin x}$

Problems

31. Complete Table 8.28 given that $w(t) = v(u(t))$.

Table 8.28

t	u	v	w
0	2	3	—
1	—	—	2
2	1	1	4
3	—	2	0
4	0	0	—

32. Let $f(x) = \dfrac{1}{x+1}$. Find and simplify $f\left(\dfrac{1}{x}\right) + \dfrac{1}{f(x)}$.

In Problems 33–36, suppose that $f(x) = g(h(x))$. Find possible formulas for $g(x)$ and $h(x)$ (There may be more than one possible answer. Assume $g(x) \neq x$ and $h(x) \neq x$.)

33. $f(x) = (x+3)^2$

34. $f(x) = \sqrt{1+\sqrt{x}}$

35. $f(x) = 9x^2 + 3x$

36. $f(x) = \dfrac{1}{x^2+8x+16}$

Evaluate the expressions in Problems 37–38 using Figures 8.34 and 8.35, giving estimates if necessary.

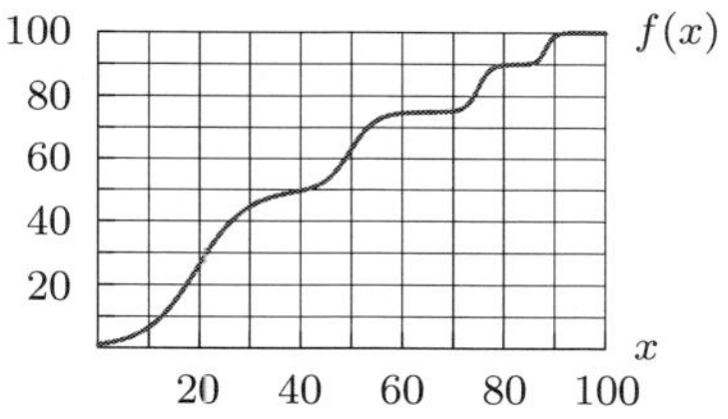

Figure 8.34

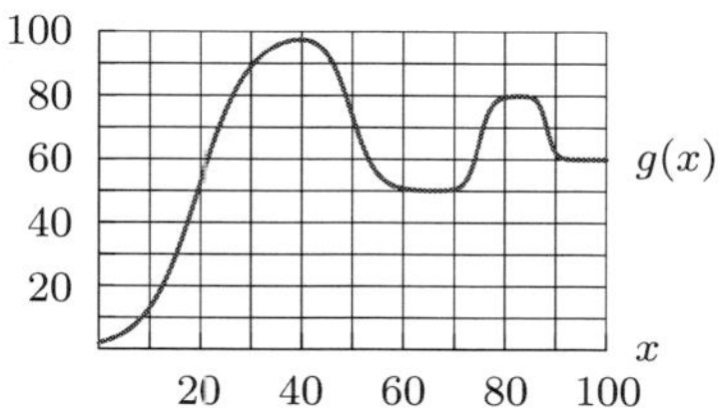

Figure 8.35

37. $f(g(65))$

38. $v(50)$ where $v(x) = g(x)f(x)$

Using Figures 8.36 and 8.37, graph the functions in Problems 39–42.

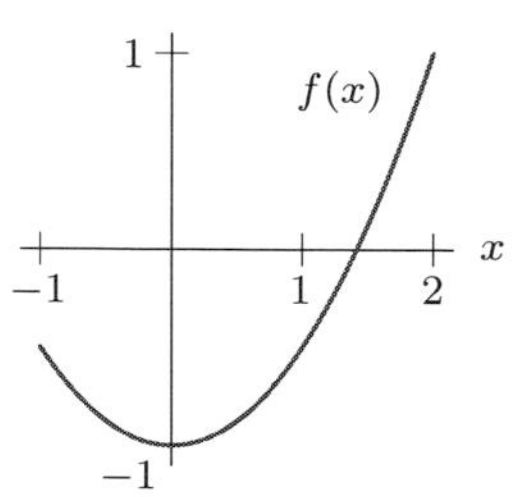

Figure 8.36

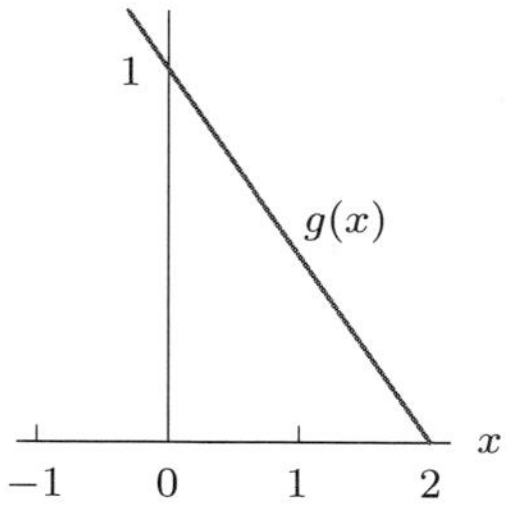

Figure 8.37

39. $f(x) - g(x)$

40. $f(g(x))$

41. $g(f(x))$

42. $g(f(x-2))$

43. Using Figure 8.38, match the functions (a)–(g) and graphs (I)–(IV). There may be some functions whose graphs are not shown.

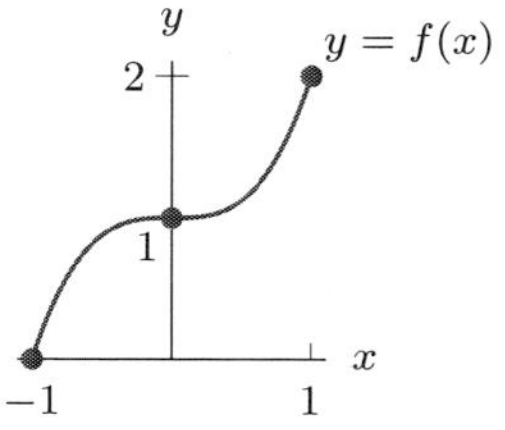

Figure 8.38

(a) $y = -f(x)$ **(b)** $y = f(-x)$
(c) $y = f(-x) - 2$ **(d)** $y = f^{-1}(x)$
(e) $y = -f^{-1}(x)$ **(f)** $y = f(x+1)$
(g) $y = -(f(x) - 2)$

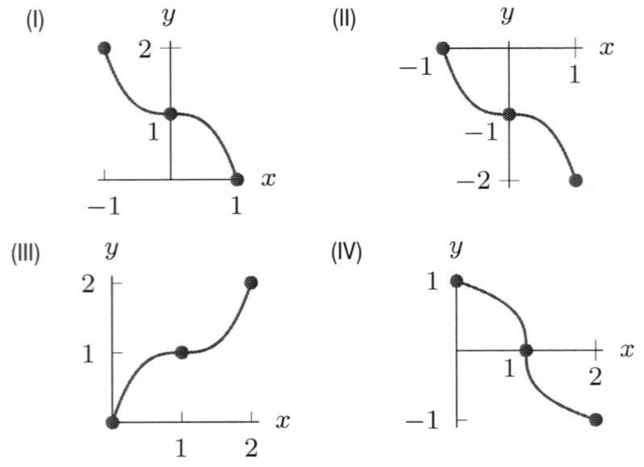

44. Use Figure 8.39.

(a) Evaluate $f(g(a))$.
(b) Evaluate $g(f(c))$.
(c) Evaluate $f^{-1}(b) - g^{-1}(b)$.
(d) For what positive value(s) of x is $f(x) \leq g(x)$?

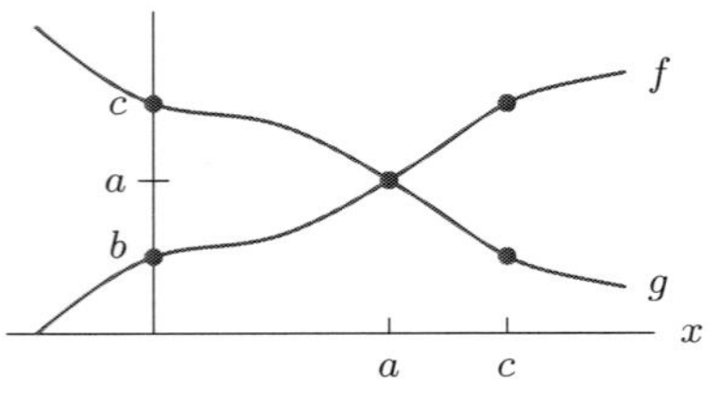

Figure 8.39

45. Let $f(x) = 12 - 4x$, $g(x) = 1/x$, and $h(x) = \sqrt{x-4}$. Find the domain of the functions:

(a) $g(f(x))$ **(b)** $h(f(x))$

For Problems 46–50, let

$$p(x) = 2x - 3 \qquad q(x) = \sqrt{x} - 3$$
$$r(x) = \frac{2x-1}{2x+1} \qquad s(x) = (x-1)^2$$

46. Find and simplify $p(q(x))$.

47. Find and simplify $r^{-1}(x)$.

48. Solve for x exactly: $p(x) = r(x)$.

49. Find $u(x)$ given that $q(x) = p(u(x))$.

50. Graph and label $p(s(x))$.

51. A research facility on the Isle of Shoals has 800 gallons of fresh water for a two-month period.

(a) There are 7 members of the research team and each is allotted 2 gallons of water per day for cooking and drinking. Find a formula for $f(t)$, the amount of fresh water left on the island after t days has elapsed.
(b) Evaluate and interpret the following expressions

(i) $f(0)$ (ii) $f^{-1}(0)$
(iii) t if $f(t) = \frac{1}{2}f(0)$ (iv) $800 - f(t)$

52. The rating, r, of an earthquake of intensity I is given by $r = f(I) = \log(I/I_0)$, where I_0 is a constant. Find and interpret $f^{-1}(r)$.

53. There is a linear relationship between the number of units, $N(x)$, of a product that a company sells and the amount of money, x, spent on advertising. If the company spends \$25,000 on advertising, it sells 400 units, and for each additional \$5,000 spent, it sells 20 units more.

(a) Calculate and interpret $N(20{,}000)$.
(b) Find a formula for $N(x)$ in terms of x.
(c) Give interpretations of the slope and the x- and y-intercepts of $N(x)$ if possible.
(d) Calculate and interpret $N^{-1}(500)$.
(e) An internal audit reveals that the profit made by the company on the sale of 10 units of its product, before advertising costs have been accounted for, is \$2,000. What are the implications regarding the company's advertising campaign? Discuss.

54. A hot brick is removed from a kiln at 200°C above room temperature. Over time, the brick cools off. After 2 hours have elapsed, the brick is 20°C above room temperature. Let t be the time in hours since the brick was removed from the kiln. Let $y = H(t)$ be the difference between the brick's and the room's temperature at time t. Assume that $H(t)$ is an exponential function.

(a) Find a formula for $H(t)$.
(b) How many degrees does the brick's temperature drop during the first quarter hour? During the next quarter hour?
(c) Find and interpret $H^{-1}(y)$.
(d) How much time elapses before the brick's temperature is 5°C above room temperature?
(e) Interpret the physical meaning of the horizontal asymptote of $H(t)$.

In Problems 55–61, you hire either Ace Construction or Space Contractors to build office space. Let $f(x)$ be the average total cost in dollars of building x square feet of office space, as estimated by Ace. Let $h(x)$ be the total number of square feet of office space you can build with x dollars, as estimated by Space.

55. Describe in words what the following statement tells you: $f(2000) = 200{,}000$

56. Let $g(x) = f(x)/x$. Using the information from Problem 55, evaluate $g(2000)$, and describe in words what $g(2000)$ represents. [Hint: Think about the units.]

57. Ace tells you that, due to the economies of scale, "Building twice as much office space always costs less than twice as much." Express this statement symbolically, in terms of f and x. [Hint: If you are building x square feet, how do you represent the cost? How would you represent twice the cost? How do you represent the cost of building twice as many square feet?]

58. Suppose that $q > p$ and $p > 1$. Assuming that the contractor's statement in Problem 57 is correct, rank the following in increasing order, using inequality signs: $f(p)$, $g(p)$, $f(q)$, $g(q)$.

59. What does the statement $h(200{,}000) = 1500$ tell you?

60. Let $j(x)$ be the average cost in dollars per square foot of office space as estimated by Space Contractors. Give a formula for $j(x)$. (Your formula will have $h(x)$ in it.)

61. Research reveals that $h(f(x)) < x$ for every value of x you check. Explain the implications of this statement. [Hint: Which company seems more economical?]

In Problems 62–65, let $f(x)$ be an increasing function and let $g(x)$ be a decreasing function. Are the following functions increasing, decreasing, or is it impossible to tell? Explain.

62. $f(f(x))$

63. $g(f(x))$

64. $f(x) + g(x)$

65. $f(x) - g(x)$

66. For a positive integer x, let $f(x)$ be the remainder obtained by dividing x by 3. For example, $f(6) = 0$, because 6 divided by 3 equals 2 with a remainder of 0. Likewise, $f(7) = 1$, because 7 divided by 3 equals 2 with a remainder of 1.

(a) Evaluate $f(8)$, $f(17)$, $f(29)$, $f(99)$.
(b) Find a formula for $f(3x)$.
(c) Is $f(x)$ invertible?
(d) Find a formula in terms of $f(x)$ for $f(f(x))$.
(e) Does $f(x + y)$ necessarily equal $f(x) + f(y)$?

CHECK YOUR UNDERSTANDING

Let $f(x) = \dfrac{1}{x}$, $g(x) = \sqrt{x}$, and $h(x) = x - 5$. Are the statements in Problems 1–10 true or false? Give an explanation for your answer.

1. $f(4) + g(4) = (f + g)(8)$.

2. $\dfrac{h(x)}{f(x)} = \dfrac{x-5}{x}$.

3. $f(4) + g(4) = 2\frac{1}{4}$.

4. $f(g(x))$ is defined for all x.

5. $g(f(x)) = \sqrt{\dfrac{1}{x}}$.

6. $f(x)g(x) = f(g(x))$.

7. $2f(2) = g(1)$.

8. $f(1)g(1)h(1) = -4$.

9. $\dfrac{f(3) + g(3)}{h(3)} = \dfrac{\frac{1}{3} + \sqrt{3}}{-2}$.

10. $4h(2) = h(8)$.

Are the statements in Problems 11–37 true or false? Give an explanation for your answer.

11. If $f(x) = x^2$ and $g(x) = \sqrt{x+3}$. Then $f(g(x))$ is defined for all x.

12. If $f(x) = x^2$ and $g(x) = \sqrt{x+3}$. Then $f(g(6)) = 9$.

13. In general $f(g(x)) = g(f(x))$.

14. The formula for the area of a circle is $A = \pi r^2$ and the formula for the circumference of a circle is $C = 2\pi r$. Then the area of a circle as a function of the circumference is $A = \dfrac{C^2}{2\pi}$.

15. If $f(x) = x^2 + 2$ then $f(f(1)) = 11$.

16. If $h(x) = f(g(x))$, $h(2) = 1\frac{1}{4}$ and $g(2) = \frac{1}{2}$ then $f(x)$ might be equal to $x^2 + 1$.

17. If $f(x) = \dfrac{1}{x}$ then $f(x + h) = \dfrac{1}{x} + \dfrac{1}{h}$.

18. If $f(x) = x^2 + x$, then $\dfrac{f(x+h) - f(x)}{h} = 2x + h$.

19. If $f(x) = x^2$ and $g(x) = \sin x$ then $f(g(x)) = x^2 \sin x$.

20. If $f(x)$ and $g(x)$ are linear, then $f(g(x))$ is linear.

21. If $f(x)$ and $g(x)$ are quadratic, then $f(g(x))$ is quadratic.

22. There is more than one way to write $h(x) = 3(x^2 + 1)^3$ as a composition $h(x) = f(g(x))$.

23. The composition $f(g(x))$ is never the same as the composite $g(f(x))$.

24. If f and g are both increasing, then $h(x) = f(g(x))$ is also increasing.

25. The functions f and g given in Tables 8.29 and 8.30 satisfy $g(f(2)) = f(g(3))$.

Table 8.29

t	1	2	3	4
$f(t)$	3	1	2	4

Table 8.30

x	1	2	3	4
$g(x)$	3	4	1	2

26. If f is increasing and invertible, then f^{-1} is decreasing.

27. If there is a vertical line that intersects a graph in more than one point, then the graph does not represent a function.

28. Every function has an inverse.

29. If $g(3) = g(5)$, then g is not invertible.

30. If no horizontal line intersects the graph of a function in more than one point, then the function has an inverse.

31. Most quadratic functions have an inverse.

32. All linear functions of the form $f(x) = mx + b, m \neq 0$ have inverses.

33. The following table describes y as a function of x.

Table 8.31

x	1	2	3	4	5	6
y	3	4	7	8	5	3

34. The table in Problem 33 describes a function from x to y that is invertible.

35. The graph in Figure 8.40 is the graph of a function.

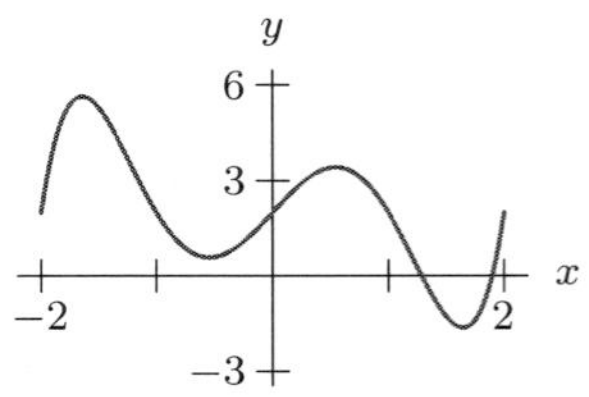

Figure 8.40

36. The function graphed in Figure 8.40 is invertible.

37. For an invertible function g, it is always true that $g^{-1}(g(x)) = x$.

Chapter Nine

POLYNOMIAL AND RATIONAL FUNCTIONS

This chapter begins with power functions. Sums and differences of power functions lead to the family of polynomial functions. Ratios of polynomials lead to the family of rational functions.

The chapter ends by comparing power, exponential, and logarithmic functions and by fitting functions to data.

The Tools Section on page 441 reviews algebraic fractions.

9.1 POWER FUNCTIONS

Proportionality and Power Functions

The following two examples introduce proportionality and power functions.

Example 1 The area, A, of a circle is proportional to the square of its radius, r:

$$A = \pi r^2.$$

Example 2 The weight, w, of an object is inversely proportional to the square of the object's distance, d, from the earth's center:[1]

$$w = \frac{k}{d^2} = kd^{-2}.$$

For an object with weight 44 pounds on the surface of the earth, which is about 3959 miles from the earth's center, we get the data listed in Table 9.1 and graphed in Figure 9.1.

Table 9.1 *Weight of an object, w, inversely proportional to the square of the objects's distance, d, from the earth's center*

d, miles	$w = f(d)$, lbs
4000	43.3
5000	27.8
6000	19.2
7000	14.1
8000	10.8

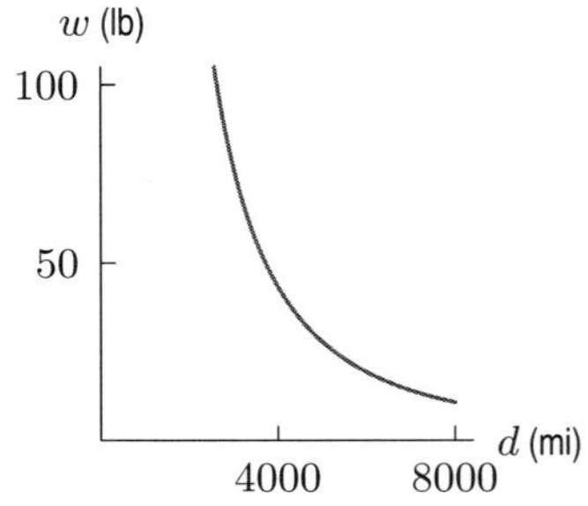

Figure 9.1: Weight, w, inversely proportional to the square of the object's distance, d, from the earth's center

A quantity y is **(directly) proportional to a power** of x if

$$y = kx^n, \qquad k \text{ and } n \text{ are constants.}$$

A quantity y is **inversely proportional** to x^n if

$$y = \frac{k}{x^n}, \qquad k \text{ and } n \text{ are constants, with } n > 0.$$

[1]There is a distinction between mass and weight. For example, an astronaut in orbit may be weightless, but he still has mass.

The functions in Examples 1 and 2 are power functions. Generalizing, we define:

> A **power function** is a function of the form
>
> $$f(x) = kx^p, \qquad \text{where } k \text{ and } p \text{ are constants.}$$

Example 3 Which of the following functions are power functions? For each power function, state the value of the constants k and p in the formula $y = kx^p$.

(a) $f(x) = 13\sqrt[3]{x}$ (b) $g(x) = 2(x+5)^3$ (c) $u(x) = \sqrt{\dfrac{25}{x^3}}$ (d) $v(x) = 6 \cdot 3^x$

Solution The functions f and u are power functions; the functions g and v are not.

(a) The function $f(x) = 13\sqrt[3]{x}$ is a power function because we can write its formula as
$$f(x) = 13x^{1/3}.$$
Here, $k = 13$ and $p = 1/3$.

(b) Although the value of $g(x) = 2(x+5)^3$ is proportional to the cube of $x+5$, it is *not* proportional to a power of x. We cannot write $g(x)$ in the form $g(x) = kx^p$; thus, g is not a power function.

(c) We can rewrite the formula for $u(x) = \sqrt{25/x^3}$ as
$$u(x) = \frac{\sqrt{25}}{\sqrt{x^3}} = \frac{5}{(x^3)^{1/2}} = \frac{5}{x^{3/2}} = 5x^{-3/2}.$$
Thus, u is a power function. Here, $k = 5$ and $p = -3/2$.

(d) Although the value of $v(x) = 6 \cdot 3^x$ is proportional to a power of 3, the power is not a constant—it is the variable x. In fact, $v(x) = 6 \cdot 3^x$ is an exponential function, not a power function. Notice that $y = 6 \cdot x^3$ is a power function. However, $6 \cdot x^3$ and $6 \cdot 3^x$ are quite different.

The Effect of the Power p

We now study functions whose constant of proportionality is $k = 1$ so that we can focus on the effect of the power p.

Graphs of the Special Cases $y = x^0$ and $y = x^1$

The power functions corresponding to $p = 0$ and $p = 1$ are both linear. The graph of $y = x^0 = 1$ is a horizontal line through the point $(1, 1)$. The graph of $y = x^1 = x$ is a line through the origin with slope $+1$.

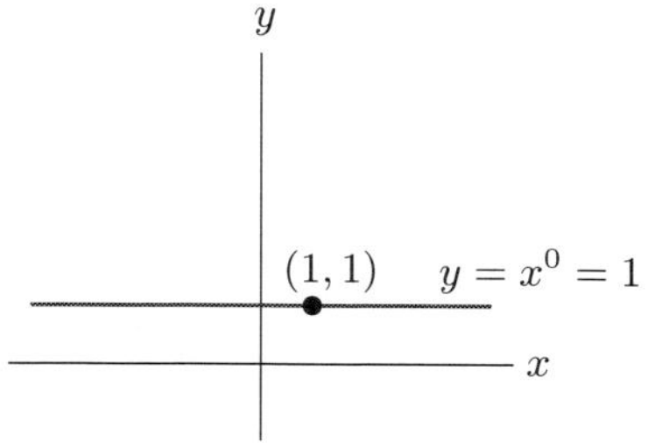

Figure 9.2: Graph of $y = x^0 = 1$

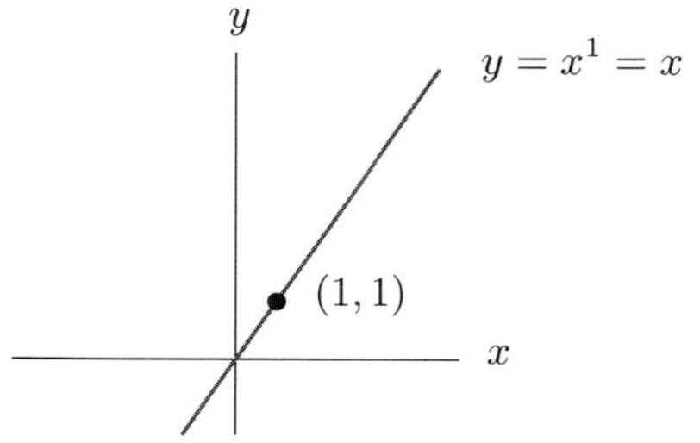

Figure 9.3: Graph of $y = x^1 = x$

Positive Integer Powers: $y = x^3,\ x^5,\ x^7 \ldots$, and $y = x^2,\ x^4, x^6 \ldots$

The graphs of all power functions with p a positive even integer have the same characteristic $\bigcup$-shape and are symmetric about the y-axis. For instance, the graphs of $y = x^2$ and $y = x^4$ in Figure 9.4 are similar in shape, although the graph of $y = x^4$ is flatter near the origin and steeper away from the origin than the graph of $y = x^2$.

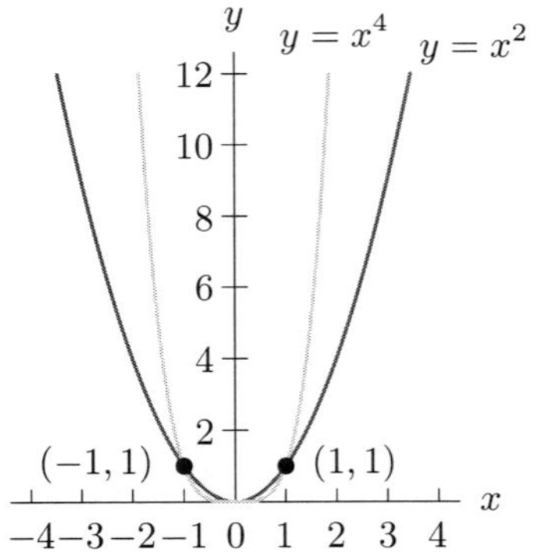

Figure 9.4: Graphs of positive even powers of x are $\bigcup$-shaped

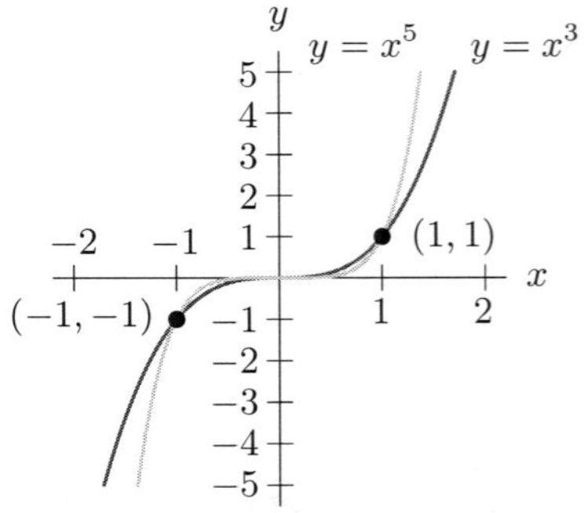

Figure 9.5: Graphs of positive odd powers of x are "chair"-shaped

The graphs of power functions with p a positive odd integer resemble the side view of a chair and are symmetric about the origin. Figure 9.5 shows the graphs of $y = x^3$ and $y = x^5$. The graph of $y = x^5$ is flatter near the origin and steeper far from the origin than the graph of $y = x^3$.

Negative Integer Powers: $y = x^{-1}, x^{-3}, x^{-5}$, ... and $y = x^{-2}, x^{-4}, x^{-6}$, ...

For negative powers, if we rewrite

$$y = x^{-1} = \frac{1}{x}$$

and

$$y = x^{-2} = \frac{1}{x^2},$$

then it is clear that as $x > 0$ increases, the denominators increase and the functions decrease. The graphs of power functions with odd negative powers, $y = x^{-3}$, x^{-5}, ... resemble the graph of $y = x^{-1} = 1/x$. The graphs of even integer powers, $y = x^{-4}$, x^{-6}, ... are similar in shape to the graph of $y = x^{-2} = 1/x^2$. See Figures 9.6 and 9.7.

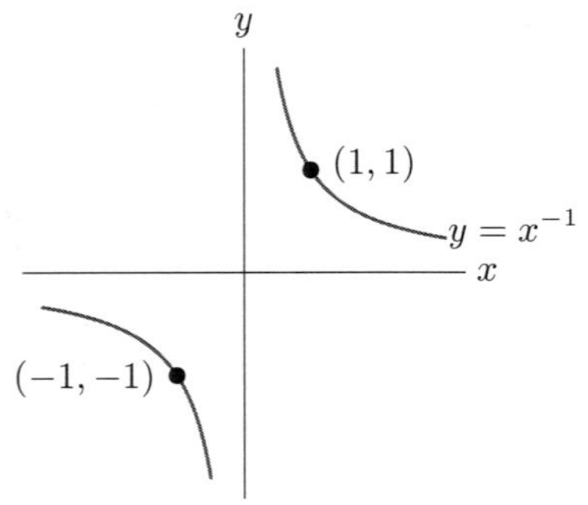

Figure 9.6: Graph of $y = x^{-1} = 1/x$

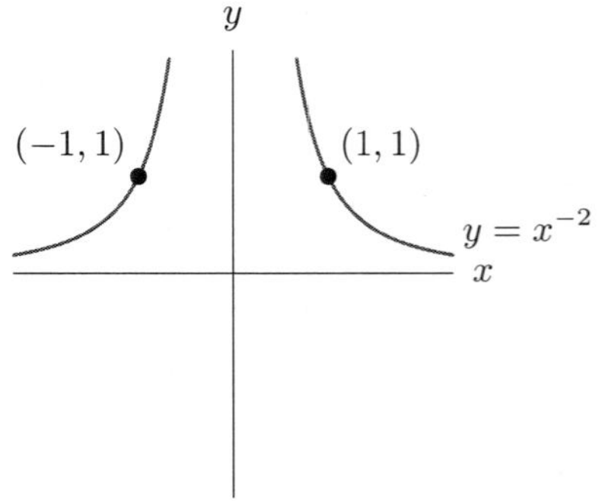

Figure 9.7: Graph of $y = x^{-2} = 1/x^2$

We see in Figures 9.6 and 9.7 that $y = 0$ is a horizontal asymptote and $x = 0$ is a vertical asymptote for the graphs of $y = 1/x$ and $y = 1/x^2$.

Numerically, the values of $1/x$ and $1/x^2$ can be made as close to zero as we like by choosing a sufficiently large x. See Table 9.2. Graphically, this means that the curves $y = 1/x$ and $y = 1/x^2$ get closer and closer to the x-axis for large values of x. We write $y \to 0$ as $x \to \infty$. Using limit notation, we see

$$\lim_{x\to\infty}\left(\frac{1}{x}\right) = 0 \quad \text{and} \quad \lim_{x\to\infty}\left(\frac{1}{x^2}\right) = 0$$

Table 9.2 *Values of x^{-1} and x^{-2} approach zero as x grows large*

x	0	10	20	30	40	50
$y = 1/x$	Undefined	0.1	0.05	0.033	0.025	0.02
$y = 1/x^2$	Undefined	0.01	0.0025	0.0011	0.0006	0.0004

On the other hand, as x gets close to zero, the values of $1/x$ and $1/x^2$ get very large. See Table 9.3. Graphically, this means that the curves $y = 1/x$ and $y = 1/x^2$ get very close to the y-axis as x gets close to zero. From Figure 9.6, we see that[2]

$$\lim_{x\to 0+}\left(\frac{1}{x}\right) = \infty \quad \text{and} \quad \lim_{x\to 0-}\left(\frac{1}{x}\right) = -\infty$$

From Figure 9.7, we see that

$$\lim_{x\to 0}\left(\frac{1}{x^2}\right) = \infty.$$

Table 9.3 *Values of x^{-1} and x^{-2} grow large as x approaches zero from the positive side*

x	0.1	0.05	0.01	0.001	0.0001	0
$y = 1/x$	10	20	100	1000	10,000	Undefined
$y = 1/x^2$	100	400	10,000	1,000,000	100,000,000	Undefined

Graphs of Positive Fractional Powers: $y = x^{1/2}, x^{1/3}, x^{1/4}, \ldots$

Figure 9.8 shows the graphs of $y = x^{1/2}$ and $y = x^{1/4}$. These graphs have the same shape, although $y = x^{1/4}$ is steeper near the origin and flatter away from the origin than $y = x^{1/2}$. The same can be said about the graphs of $y = x^{1/3}$ and $y = x^{1/5}$ in Figure 9.9. In general, if n is a positive integer, then the graph of $y = x^{1/n}$ resembles the graph of $y = x^{1/2}$ if n is even; if n is odd, the graph resembles the graph of $y = x^{1/3}$.

Notice that the graphs of $y = x^{1/2}$ and $y = x^{1/3}$ bend in a direction opposite to that of the graphs of $y = x^2$ and x^3. For example, the graph of $y = x^2$ is concave up, but the graph of $y = x^{1/2}$ is concave down. However, all these functions become infinitely large as x increases.

[2]Some authors say that these limits do not exist.

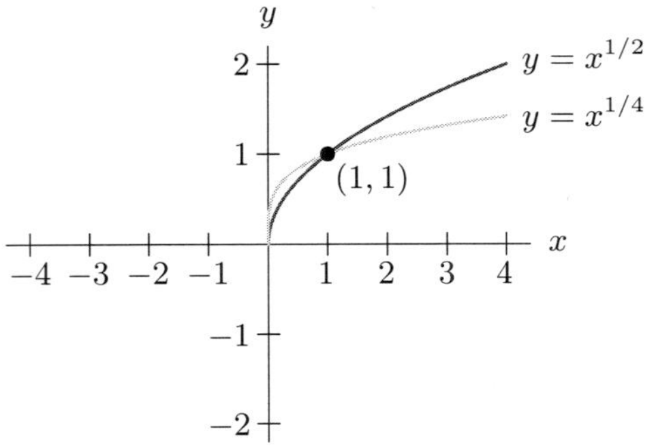

Figure 9.8: The graphs of $y = x^{1/2}$ and $y = x^{1/4}$

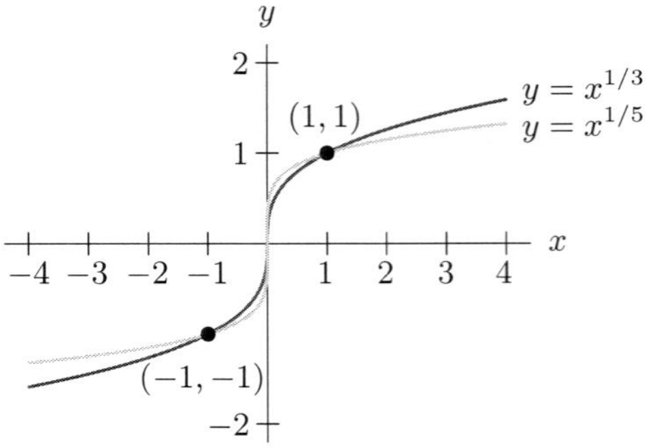

Figure 9.9: The graphs of $y = x^{1/3}$ and $y = x^{1/5}$

Example 4 The radius of a sphere is directly proportional to the cube root of its volume. If a sphere of radius 18.2 cm has a volume of 25,252.4 cm^3, what is the radius of a sphere whose volume is 30,000 cm^3?

Solution Since the radius of the sphere is proportional to the cube root of its volume, we know that

$$r = kV^{1/3}, \qquad \text{for } k \text{ constant.}$$

We also know that $r = 18.2$ cm when $V = 25{,}252.4$ cm^3, therefore

$$18.2 = k(25{,}252.4)^{1/3}$$

giving

$$k = \frac{18.2}{(25{,}252.4)^{1/3}} \approx 0.620.$$

Thus, when $V = 30{,}000$, we get $r = 0.620(30{,}000)^{1/3} \approx 19.3$, so the radius of the sphere is approximately 19.3 cm.

Finding the Formula for a Power Function

As is the case for linear and exponential functions, the formula of a power function can be found from two points on its graph.

Example 5 Water is leaking out of a container with a hole in the bottom. Torricelli's Law states that at any instant, the velocity v with which water escapes from the container is a power function of d, the depth of the water at that moment. When $d = 1$ foot, then $v = 8$ ft/sec; when $d = 1/4$ foot, then $v = 4$ ft/sec. Express v as a function of d.

Solution Torricelli's Law tells us that $v = kd^p$, where k and p are constants. The fact that $v = 8$ when $d = 1$, gives $8 = k(1)^p$, so $k = 8$, and therefore $v = 8d^p$. Also $v = 4$ when $d = 1/4$, so

$$4 = 8\left(\frac{1}{4}\right)^p.$$

Rewriting $(1/4)^p = 1/4^p$, we can solve for 4^p:

$$4 = 8 \cdot \frac{1}{4^p}$$
$$4^p = \frac{8}{4} = 2.$$

Since $4^{1/2} = 2$, we must have $p = 1/2$. Therefore we have $v = 8d^{1/2}$. *Note*: Torricelli's Law is often written in the form $v = \sqrt{2gd}$, where $g = 32$ ft/sec^2 is the acceleration due to gravity.

Exercises and Problems for Section 9.1

Exercises

Are the functions in Exercises 1–6 power functions? If so, write the function in the form $f(x) = kx^p$.

1. $g(x) = \dfrac{(-x^3)^3}{6}$

2. $R(t) = \dfrac{4}{\sqrt{16t}}$

3. $Q(t) = \left(\dfrac{1}{2\sqrt{t}}\right)^3$

4. $K(w) = \dfrac{w^4}{4\sqrt{w^3}}$

5. $T(s) = (6s^{-2})(es^{-3})$

6. $h(x) = 22(7^x)^2$

Do the power functions in Exercises 7–10 appear to have odd, even, or fractional powers?

7.

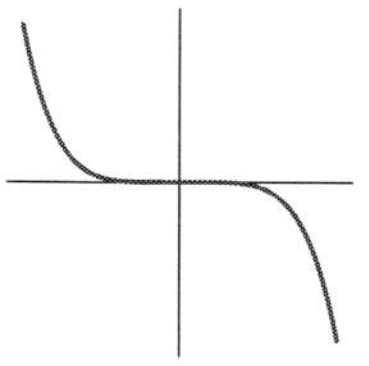

8.

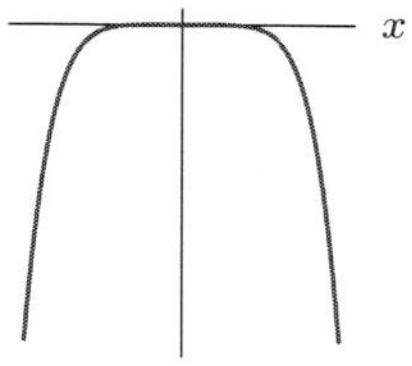

9.

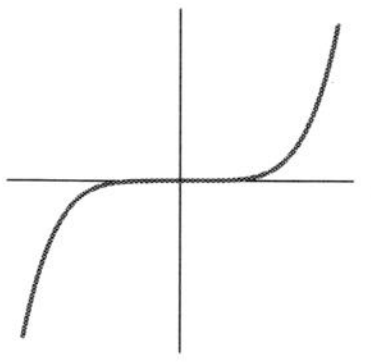

10.

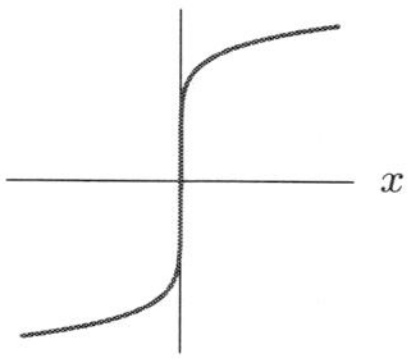

In Exercises 11–13, find a power function through the two points.

11. $(7, 8)$ $(1, 0.7)$ **12.** $(1, 5)$ $(3, 27)$ **13.** $(6, 17)$ $(1, 2)$

14. Find a possible formula for the power function $f(t)$ given that $f(3) = 5$ and $f(5) = 3$.

15. Suppose y is directly proportional to x. If $y = 6$ when $x = 4$, find the constant of proportionality and write the formula for y as a function of x. Use your formula to find x when $y = 8$.

16. Suppose y is inversely proportional to x. If $y = 6$ when $x = 4$, find the constant of proportionality and write the formula for y as a function of x. Use your formula to find x when $y = 8$.

17. Suppose c is directly proportional to the square of d. If $c = 45$ when $d = 3$, find the constant of proportionality and write the formula for c as a function of d. Use your formula to find c when $d = 5$.

18. Suppose c is inversely proportional to the square of d. If $c = 45$ when $d = 3$, find the constant of proportionality and write the formula for c as a function of d. Use your formula to find c when $d = 5$.

In Exercises 19–22, find possible formulas for the power functions.

19.

x	0	1	2	3
$j(x)$	0	2	16	54

20.

x	2	3	4	5
$f(x)$	12	27	48	75

21.

x	-6	-2	3	4
$g(x)$	36	$4/3$	$-9/2$	$-32/3$

22.

x	-2	$-1/2$	$1/4$	4
$h(x)$	$-1/2$	-8	-32	$-1/8$

23. Find **(a)** $\lim\limits_{x\to\infty} x^{-4}$ **(b)** $\lim\limits_{x\to-\infty} 2x^{-1}$

24. Find **(a)** $\lim\limits_{t\to\infty} (t^{-3}+2)$ **(b)** $\lim\limits_{y\to-\infty} (5-7y^{-2})$

Problems

25. Compare the graphs of $y = x^{-2}$, $y = x^{-4}$, and $y = x^{-6}$. Describe the similarities and differences.

26. Describe the behavior of $y = x^{-10}$ and $y = -x^{10}$ as

(a) $x \to 0$ **(b)** $x \to \infty$ **(c)** $x \to -\infty$

27. Describe the behavior of $y = x^{-3}$ and $y = x^{1/3}$ as

(a) $x \to 0$ from the right **(b)** $x \to \infty$

28. If $f(x) = kx^p$, p an integer, show that f is an even function if p is even, and an odd function if p is odd.

29. **(a)** Figure 9.10 shows $g(x)$, a mystery power function. If you learn that the point $(-1, 3)$ lies on its graph, do you have enough information to write a formula for $g(x)$?

(b) If you are told that the point $(1, -3)$ also lies on the graph, what new deductions can you make?

(c) If the point $(2, -96)$ lies on the graph g, in addition to the points already given, state three other points which also lie on it.

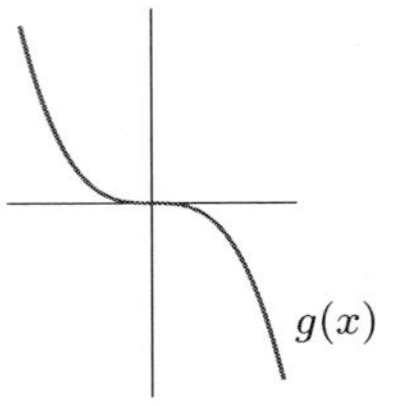

Figure 9.10

30. Figure 9.11 shows the power function $y = c(t)$. Is $c(t) = 1/t$ the only possible formula for c? Could there be others?

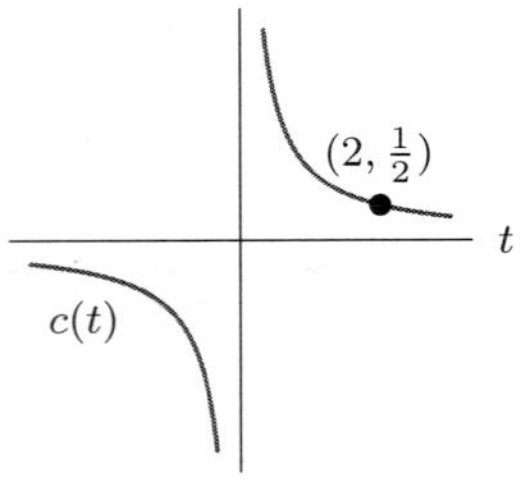

Figure 9.11

31. **(a)** One of the graphs in Figure 9.12 is $y = x^n$ and the other is $y = x^{1/n}$, where n is a positive integer. Which is which? How do you know?

(b) What are the coordinates of point A?

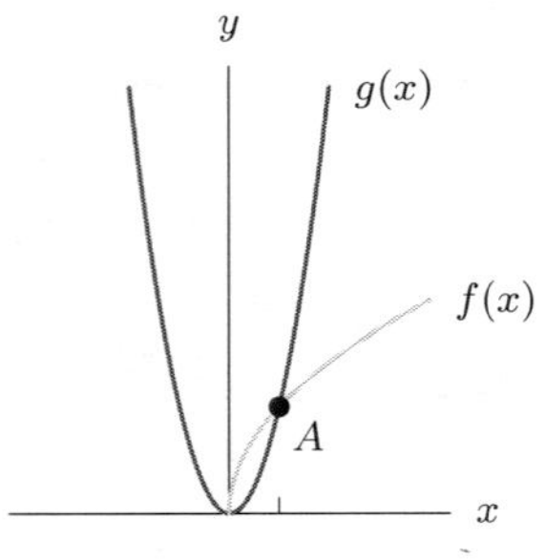

Figure 9.12

32. The circulation time of a mammal—that is, the average time it takes for all the blood in the body to circulate once and return to the heart—is governed by the equation

$$t = 17.4m^{1/4},$$

where m is the body mass of the mammal in kilograms, and t is the circulation time in seconds.[3]

(a) Complete Table 9.4 which shows typical body masses in kilograms for various mammals.[4]

(b) If the circulation time of one mammal is twice that of another, what is the relationship between their body masses?

Table 9.4

Animal	Body mass (kg)	Circulation time (sec)
Blue whale	91000	
African elephant	5450	
White rhinoceros	3000	
Hippopotamus	2520	
Black rhinoceros	1170	
Horse	700	
Lion	180	
Human	70	

33. Three ounces of broiled ground beef contains 245 calories.[5] Is the number of calories directly or inversely proportional to the number of ounces? Explain your reasoning and write a formula for the proportion. How many calories are there in 4 ounces of broiled hamburger?

34. A 30-second commercial during Super Bowl XL in 2006 cost advertisers \$2.5 million. For the first Super Bowl in 1967, an advertiser could have purchased approximately 28.699 minutes of advertising time for the same amount of money.[6]

(a) Assuming that cost is proportional to time, find the cost of advertising, in dollars/second, during the 1967 and 2006 Super Bowls.

(b) How many times more expensive was Super Bowl advertising in 2006 than in 1967?

35. A group of friends rent a house at the beach for spring break. If nine of them share the house, it costs \$150 each. Is the cost to each person directly or inversely proportional to the number of people sharing the house? Explain your reasoning and write a formula for the proportion. How many people are needed to share the house if each student wants to pay a maximum of \$100 each?

---

[3]K. Schmidt-Nielsen, *Scaling, Why is Animal Size so Important?* (Cambridge: CUP, 1984).
[4]R. McNeill Alexander, *Dynamics of Dinosaurs and Other Extinct Giants.* (New York: Columbia University Press, 1989).
[5]The World Almanac Book of Facts, 1999 p. 718
[6]money.cnn.com/2006/01/03/news/companies/superbowlads, accessed January 15, 2006.

36. Driving at 55 mph, it takes approximately 3.5 hours to drive from Long Island to Albany, NY. Is the time the drive takes directly or inversely proportional to the speed? Explain your reasoning and write a formula for the proportion. To get to Albany in 3 hours, how fast would you have to drive?

37. On a map, $1/2$ inch represents 5 miles. Is the map distance between two locations directly or inversely proportional to the actual distance which separates the two locations? Explain your reasoning and write a formula for the proportion. How far apart are two towns if the distance between these two towns on the map is 3.25 inches?

38. A volcano erupts in a powerful explosion. The sound from the explosion is heard in all directions for many hundreds of kilometers. The speed of sound is about 340 meters per second.

(a) Fill in Table 9.5 showing the distance, d, that the sound of the explosion has traveled at time t. Write a formula for d as a function t.
(b) How long after the explosion will a person living 200 km away hear the explosion?
(c) Fill Table 9.5 showing the land area, A, over which the explosion can be heard as a function of time. Write a formula for A as a function of t.
(d) The average population density around the volcano is 31 people per square kilometer. Write a formula for P as function of t, where P is the number of people who have heard the explosion at time t.
(e) Graph the function $P = f(t)$. How long will it take until 1 million people have heard the explosion?

Table 9.5

Time, t	5 sec	10 sec	1 min	5 min
Distance, d (km)				
Area, A (km^2)				

39. The thrust, T, delivered by a ship's propeller is proportional[7] to the square of the propeller rotation speed, R, times the fourth power of the propeller diameter, D.

(a) Write a formula for T in terms of R and D.
(b) What happens to the thrust if the propeller speed is doubled?
(c) What happens to the thrust if the propeller diameter is doubled?
(d) If the propeller diameter is increased by 50%, by how much can the propeller speed be reduced to deliver the same thrust?

40. Two oil tankers crash in the Pacific ocean. The spreading oil slick has a circular shape, and the radius of the circle is increasing at 200 meters per hour.

(a) Express the radius of the spill, r, as a power function of time, t, in hours since the crash.
(b) Express the area of the spill, A, as a power function of time, t.
(c) Clean-up efforts begin 7 hours after the spill. How large an area is covered by oil at that time?

41. When an aircraft flies horizontally, its *stall velocity* (the minimum speed required to keep the aircraft aloft) is directly proportional to the square root of the quotient of its weight by its wing area. If a breakthrough in materials science allowed the construction of an aircraft with the same weight but twice the wing area, would the stall velocity increase or decrease? By what percent?

42. One of Kepler's three laws of planetary motion states that the square of the period, P, of a body orbiting the sun is proportional to the cube of its average distance, d, from the sun. The earth has a period of 365 days and its distance from the sun is approximately 93,000,000 miles.

(a) Find P as a function of d.
(b) The planet Jupiter has an average distance from the sun of 483,000,000 miles. How long in earth days is a Jupiter year?

43. A person's weight, w, on a planet of radius d is given by

$$w = kd^{-2}, \quad k > 0,$$

where the constant k depends on the masses of the person and the planet.

(a) A man weighs 180 lb on the surface of the earth. How much does he weigh on the surface of a planet whose mass is the same the earth's, but whose radius is three times as large? One-third as large?
(b) What fraction of the earth's radius must an equally massive planet have if, on this planet, the weight of the man in part (a) is one ton?

44. The following questions involve the behavior of the power function $y = x^{-p}$, for p a positive integer. If a distinction between even and odd values of p is significant, the significance should be indicated.

(a) What is the domain of $y = x^{-p}$? What is the range?
(b) What symmetries does the graph of $y = x^{-p}$ have?
(c) What is the behavior of $y = x^{-p}$ as $x \to 0$?
(d) What is the behavior of $y = x^{-p}$ for large positive values of x? For large negative values of x?

[7]Gillner, Thomas C., *Modern Ship Design*, (US Naval Institute Press, 1972).

45. Let $f(x) = 16x^4$ and $g(x) = 4x^2$.

(a) If $f(x) = g(h(x))$, find a possible formula for $h(x)$, assuming $h(x) \leq 0$ for all x.
(b) If $f(x) = j(2g(x))$, find a possible formula for $j(x)$, assuming $j(x)$ is a power function.

46. Consider the power function $y = t(x) = k \cdot x^{p/3}$ where p is any integer, $p \neq 0$.

(a) For what values of p does $t(x)$ have domain restrictions? What are those restrictions?
(b) What is the range of $t(x)$ if p is even? If p is odd?
(c) What symmetry does the graph of $t(x)$ exhibit if p is even? If p is odd?

9.2 POLYNOMIAL FUNCTIONS

A *polynomial function* is a sum of power functions whose exponents are nonnegative integers. We use what we learned about power functions to study polynomials.

Example 1 You make five separate deposits of \$1000 each into a savings account, one deposit per year, beginning today. What annual interest rate gives a balance in the account of \$6000 five years from today? (Assume the interest rate is constant over these five years.)

Solution Let r be the annual interest rate. Our goal is to determine what value of r gives you \$6000 in five years. In year $t = 0$, you make a \$1000 deposit. One year later, you have \$1000 plus the interest earned on that amount. At that time, you add another \$1000.

To picture how this works, imagine the account pays 5% annual interest, compounded annually. Then, after one year, your balance would be

$$\begin{aligned}\text{Balance} &= (100\% \text{ of Initial deposit}) + (5\% \text{ of Initial deposit}) + \text{Second deposit} \\ &= 105\% \text{ of } \underbrace{\text{Initial deposit}}_{\$1000} + \underbrace{\text{Second deposit}}_{\$1000} \\ &= 1.05(1000) + 1000.\end{aligned}$$

Let x represent the annual growth factor, $1 + r$. For example, if the account paid 5% interest, then $x = 1 + 0.05 = 1.05$. We write the balance after one year in terms of x:

$$\text{Balance after one year} = 1000x + 1000.$$

After two years, you would have earned interest on the first-year balance. This gives

$$\text{Balance after earning interest} = \underbrace{(1000x + 1000)}_{\text{First-year balance}}x = 1000x^2 + 1000x.$$

The third \$1000 deposit brings your balance to

$$\text{Balance after two years} = 1000x^2 + 1000x + \underbrace{1000.}_{\text{Third deposit}}$$

A year's worth of interest on this amount, plus the fourth \$1000 deposit, brings your balance to

$$\begin{aligned}\text{Balance after three years} &= \underbrace{(1000x^2 + 1000x + 1000)}_{\text{Second-year balance}}x + \underbrace{1000}_{\text{Fourth deposit}} \\ &= 1000x^3 + 1000x^2 + 1000x + 1000.\end{aligned}$$

The pattern is this: Each of the \$1000 deposits grows to $\$1000x^n$ by the end of its nth year in the bank. Thus,

$$\text{Balance after five years} = 1000x^5 + 1000x^4 + 1000x^3 + 1000x^2 + 1000x.$$

If the interest rate is chosen correctly, then the balance will be \$6000 in five years. This gives us

$$1000x^5 + 1000x^4 + 1000x^3 + 1000x^2 + 1000x = 6000.$$

Dividing by 1000 and moving the 6 to the left side, we have the equation

$$x^5 + x^4 + x^3 + x^2 + x - 6 = 0.$$

Solving this equation for x determines how much interest we must earn. Using a computer or calculator, we find where the graph of $Q(x) = x^5 + x^4 + x^3 + x^2 + x - 6$ crosses the x-axis. Figure 9.13 shows that this occurs at $x \approx 1.0614$. Since $x = 1 + r$, this means $r = 0.0614$. So the account must earn 6.14% annual interest[8] for the balance to be \$6000 at the end of five years.

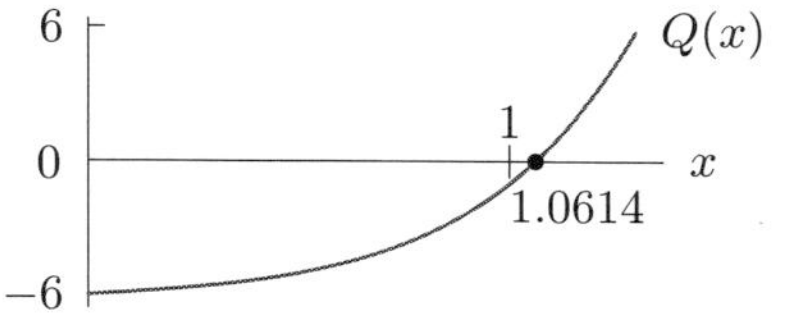

Figure 9.13: Finding where $Q(x)$ crosses the x-axis, for $x \geq 0$

You may wonder if Q crosses the x-axis more than once. For $x \geq 0$, graphing Q on a larger scale suggests that Q increases for all values of x and crosses the x-axis only once. For $x > 1$, we expect Q to be an increasing function, because larger values of x indicate higher interest rates and therefore larger values of $Q(x)$. Having crossed the axis once, the graph of Q does not "turn around" to cross it again.

The function $Q(x) = x^5 + x^4 + x^3 + x^2 + x - 6$ is the sum of power functions; Q is called a *polynomial*. (Note that the expression -6 can be written as $-6x^0$, so it, too, is a power function.)

A General Formula for the Family of Polynomial Functions

The general formula for a polynomial function can be written as

$$p(x) = a_n x^n + a_{n-1}x^{n-1} + ... + a_1 x + a_0,$$

where n is called the *degree* of the polynomial and a_n is the *leading coefficient*. For example, the function

$$g(x) = 3x^2 + 4x^5 + x - x^3 + 1,$$

is a polynomial of degree 5 because the term with the highest power is $4x^5$. It is customary to write a polynomial with the powers in decreasing order from left to right:

$$g(x) = 4x^5 - x^3 + 3x^2 + x + 1.$$

The function g has one other term, $0 \cdot x^4$, which we don't bother to write down. The values of g's coefficients are $a_5 = 4, a_4 = 0, a_3 = -1, a_2 = 3, a_1 = 1$, and $a_0 = 1$. In summary:

[8]This is 6.14% interest per year, compounded annually.

The general formula for the family of polynomial functions can be written as

$$p(x) = a_n x^n + a_{n-1}x^{n-1} + \ldots + a_1 x + a_0,$$

where n is a positive integer called the **degree** of p and where $a_n \neq 0$.

- Each power function $a_n x^n$ in this sum is called a **term**.
- The constants $a_n, a_{n-1}, \ldots, a_0$ are called **coefficients**.
- The term a_0 is called the **constant term**. The highest-powered term, $a_n x^n$, is called the **leading term**.
- To write a polynomial in **standard form**, we arrange its terms from highest power to lowest power, going from left to right.

Like the power functions from which they are built, polynomials are defined for all values of x. Except for polynomials of degree zero (whose graphs are horizontal lines), the graphs of polynomials do not have horizontal or vertical asymptotes. The shape of the graph depends on its degree; typical graphs are shown in Figure 9.14.

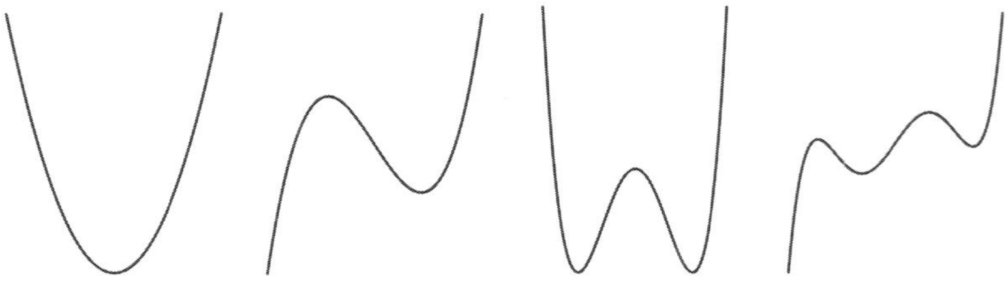

Figure 9.14: Graphs of typical polynomials of degree n

The Long-Run Behavior of Polynomial Functions

We have seen that, as x grows large, $y = x^2$ increases fast, $y = x^3$ increases faster, and $y = x^4$ increases faster still. In general, power functions with larger positive powers eventually grow much faster than those with smaller powers. This tells us about the behavior of polynomials for large x. For instance, consider the polynomial $g(x) = 4x^5 - x^3 + 3x^2 + x + 1$. Provided x is large enough, the absolute value of the term $4x^5$ is much larger than the absolute value of the other terms combined. For example, if $x = 100$,

$$4x^5 = 4(100)^5 = 40{,}000{,}000{,}000,$$

and the other terms in $g(x)$ are

$$\begin{aligned} -x^3 + 3x^2 + x + 1 &= -(100)^3 + 3(100)^2 + 100 + 1 \\ &= -1{,}000{,}000 + 30{,}000 + 100 + 1 = -969{,}899. \end{aligned}$$

Therefore $p(100) = 39{,}999{,}030{,}101$, which is approximately equal to the value of the $4x^5$ term. In general, if x is large enough, the most important contribution to the value of a polynomial p is made by the leading term; we can ignore the lower-powered terms.

When viewed on a large enough scale, the graph of the polynomial $p(x) = a_n x^n + a_{n-1}x^{n-1} + \cdots + a_1 x + a_0$ looks like the graph of the power function $y = a_n x^n$. This behavior is called the **long-run behavior** of the polynomial. Using limit notation, we write

$$\lim_{x\to\infty} p(x) = \lim_{x\to\infty} a_n x^n \quad \text{and} \quad \lim_{x\to-\infty} p(x) = \lim_{x\to-\infty} a_n x^n.$$

Example 2 Find a window in which the graph of $f(x) = x^3 + x^2$ resembles the power function $y = x^3$.

Solution Figure 9.15 gives the graphs of $f(x) = x^3 + x^2$ and $y = x^3$. On this scale, f does not look like a power function. On the larger scale in Figure 9.16, the graph of f resembles the graph of $y = x^3$. On this larger scale, the "bumps" in the graph of f are too small to be seen. On an even larger scale, as in Figure 9.17, the graph of f is indistinguishable from the graph of $y = x^3$.

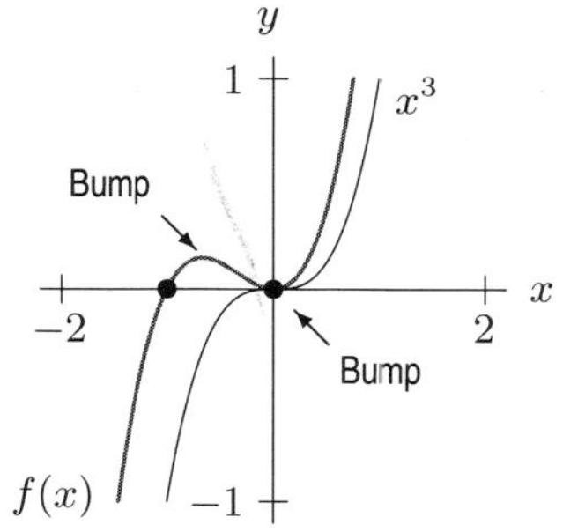

Figure 9.15: On this scale, $f(x) = x^3 + x^2$ does not look like a power function

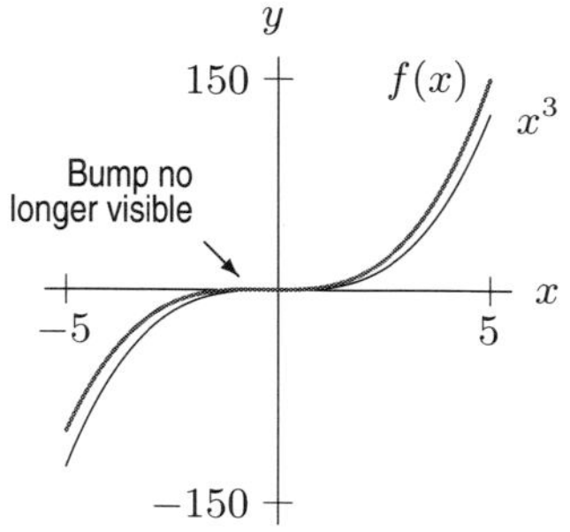

Figure 9.16: On this scale, $f(x) = x^3 + x^2$ resembles the power function $y = x^3$

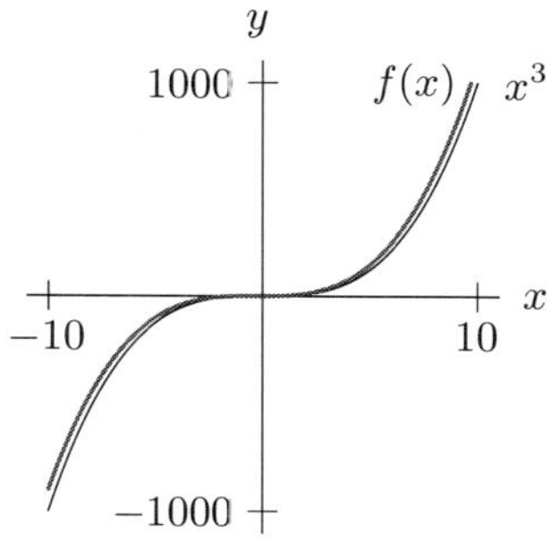

Figure 9.17: On this scale, $f(x) = x^3 + x^2$ is nearly indistinguishable from $y = x^3$

Zeros of Polynomials

The *zeros* of a polynomial p are values of x for which $p(x) = 0$. The zeros are also the x-intercepts, because they tell us where the graph of p crosses the x-axis. Factoring can sometimes be used to find the zeros of a polynomial; however, the numerical and graphical method of Example 1 can always be used. In addition, the long-run behavior of the polynomial can give us clues as to how many zeros (if any) there may be.

Example 3 Given the polynomial

$$q(x) = 3x^6 - 2x^5 + 4x^2 - 1,$$

where $q(0) = -1$, is there a reason to expect a solution to the equation $q(x) = 0$? If not, explain why not. If so, how do you know?

Solution The equation $q(x) = 0$ must have at least two solutions. We know this because on a large scale, q looks like the power function $y = 3x^6$. (See Figure 9.18.) The function $y = 3x^6$ takes on large positive values as x grows large (either positive or negative). Since the graph of q is smooth and unbroken, it must cross the x-axis at least twice to get from $q(0) = -1$ to the positive values it attains as $x \to \infty$ and $x \to -\infty$.

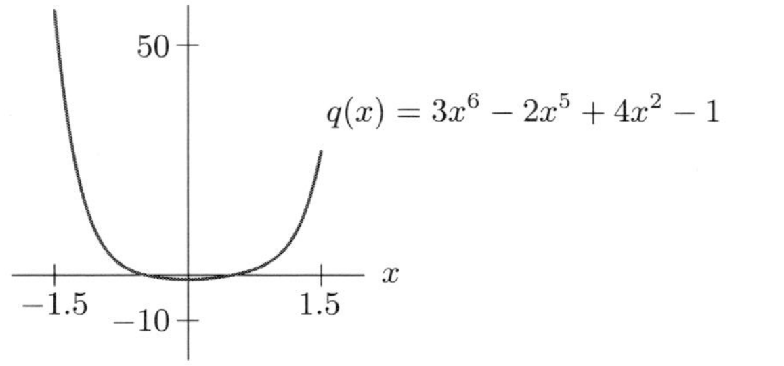

Figure 9.18: Graph must cross x-axis at least twice since $q(0) = -1$ and $q(x)$ looks like $3x^6$ for large x

A sixth degree polynomial such as q in Example 3 can have as many as six real zeros. We consider the zeros of a polynomial in more detail in Section 9.3.

Exercises and Problems for Section 9.2

Exercises

Are the functions in Exercises 1–6 polynomials? If so, of what degree?

1. $y = 5 + x$

2. $y = 5^x - 2$

3. $y = 4x^2 + 2$

4. $y = 4x^4 - 3x^3 + 2e^x$

5. $y = 4x^2 - 7\sqrt{x^9} + 10$

6. $y = 7t^6 - 8t + 7.2$

Describe in words the long-run behavior as $x \to \infty$ of the functions in Exercises 7–10. What power function does each resemble?

7. $y = 4x^4 - 2x^2 + 3$

8. $y = 16x^3 - 4023x^2 - 2$

9. $y = 3x^3 + 2x^2/x^{-7} - 7x^5 + 2$

10. $y = 5x^2/x^{3/2} + 2$

11. Find

(a) $\lim_{x\to\infty}(3x^2 - 5x + 7)$ **(b)** $\lim_{x\to-\infty}(7x^2 - 9x^3)$

Problems

12. Estimate the zeros of $f(x) = x^4 - 3x^2 - x + 2$.

13. Estimate the minimum value of $g(x) = x^4 - 3x^3 - 8$.

14. Compare the graphs of $f(x) = x^3 + 5x^2 - x - 5$ and $g(x) = -2x^3 - 10x^2 + 2x + 10$ on a window that shows all intercepts. How are the graphs similar? Different? Discuss.

15. Find a possible formula for a polynomial with zeros at (and only at) $x = -2, 2, 5$, a y-intercept at $y = 5$, and long-run behavior of $y \to -\infty$ as $x \to \pm\infty$.

16. Let $u(x) = -\frac{1}{5}(x-3)(x+1)(x+5)$ and $v(x) = -\frac{1}{5}x^2(x-5)$.

(a) Graph u and v for $-10 \le x \le 10$, $-10 \le y \le 10$. How are the graphs similar? How are they different?

(b) Compare the graphs of u and v on the window $-20 \le x \le 20$, $-1600 \le y \le 1600$, the window $-50 \le x \le 50$, $-25{,}000 \le y \le 25{,}000$, and the window $-500 \le x \le 500$, $-25{,}000{,}000 \le y \le 25{,}000{,}000$. Discuss.

17. Find the equation of the line through the y-intercept of $y = x^4 - 3x^5 - 1 + x^2$ and the x-intercept of $y = 2x - 4$.

18. Let $f(x) = \left(\frac{1}{50{,}000}\right)x^3 + \left(\frac{1}{2}\right)x$.

(a) For small values of x, which term of f is more important? Explain your answer.

(b) Graph $y = f(x)$ for $-10 \le x \le 10$, $-10 \le y \le 10$. Is this graph linear? How does the appearance of this graph agree with your answer to part (a)?

(c) How large a value of x is required for the cubic term of f to be equal to the linear term?

19. The polynomial function $f(x) = x^3 + x + 1$ is invertible—that is, this function has an inverse.

(a) Graph $y = f(x)$. Explain how you can tell from the graph that f is invertible.

(b) Find $f(0.5)$ and an approximate value for $f^{-1}(0.5)$.

20. If $f(x) = x^2$ and $g(x) = (x+2)(x-1)(x-3)$, find all x for which $f(x) < g(x)$.

21. Let V represent the volume in liters of air in the lungs during a 5-second respiratory cycle. If t is time in seconds, V is given by

$$V = 0.1729t + 0.1522t^2 - 0.0374t^3.$$

(a) Graph this function for $0 \le t \le 5$.
(b) What is the maximum value of V on this interval? What is the practical significance of the maximum value?
(c) Explain the practical significance of the t- and V-intercepts on the interval $0 \le t \le 5$.

22. Let $C(x)$ be a firm's total cost, in millions of dollars, for producing a quantity x thousand units of an item.

(a) Graph $C(x) = (x-1)^3 + 1$.
(b) Let $R(x)$ be the revenue to the firm (in millions of dollars) for selling a quantity x thousand units of the good. Suppose $R(x) = x$. What does this tell you about the price of each unit?
(c) Profit equals revenue minus cost. For what values of x does the firm make a profit? Break even? Lose money?

23. The town of Smallsville was founded in 1900. Its population y (in hundreds) is given by the equation

$$y = -0.1x^4 + 1.7x^3 - 9x^2 + 14.4x + 5,$$

where x is the number of years since 1900. Use a the graph in the window $0 \le x \le 10$, $-2 \le y \le 13$.

(a) What was the population of Smallsville when it was founded?
(b) When did Smallsville become a ghost town (nobody lived there anymore)? Give the year and the month.
(c) What was the largest population of Smallsville after 1905? When did Smallsville reach that population? Again, include the month and year. Explain your method.

24. The volume, V, in milliliters, of 1 kg of water as a function of temperature T is given, for $0 \le T \le 30$°C, by:

$$V = 999.87 - 0.06426T + 0.0085143T^2 - 0.0000679T^3.$$

(a) Graph V.
(b) Describe the shape of your graph. Does V increase or decrease as T increases? Does the graph curve upward or downward? What does the graph tell us about how the volume varies with temperature?
(c) At what temperature does water have the maximum density? How does that appear on your graph? (Density = Mass/Volume. In this problem, the mass of the water is 1 kg.)

25. Let f and g be polynomial functions. Are the compositions

$$f(g(x)) \quad \text{and} \quad g(f(x))$$

also polynomial functions? Explain your answer.

26. Let $f(x) = x - \dfrac{x^3}{6} + \dfrac{x^5}{120}$.

(a) Graph $y = f(x)$ and $y = \sin x$ for $-2\pi \le x \le 2\pi$, $-3 \le y \le 3$.
(b) The graph of f resembles the graph of $\sin x$ on a small interval. Based on your graphs from part (a), give the approximate interval.
(c) Your calculator uses a function similar to f in order to evaluate the sine function. How reasonable an approximation does f give for $\sin(\pi/8)$?
(d) Explain how you could use the function f to approximate the value of $\sin\theta$, where $\theta = 18$ radians. [Hint: Use the fact that the sine function is periodic.]

27. Suppose f is a polynomial function of degree n, where n is a positive even integer. For each of the following statements, write *true* if the statement is always true, *false* otherwise. If the statement is false, give an example that illustrates why it is false.

(a) f is an even function.
(b) f has an inverse.
(c) f cannot be an odd function.
(d) If $f(x) \to +\infty$ as $x \to +\infty$, then $f(x) \to -\infty$ as $x \to -\infty$.

28. Let g be a polynomial function of degree n, where n is a positive odd integer. For each of the following statements, write *true* if the statement is always true, *false* otherwise. If the statement is false, give an example that illustrates why it is false.

(a) g is an odd function.
(b) g has an inverse.
(c) $\lim_{x\to\infty} g(x) = \infty$.
(d) If $\lim_{x\to-\infty} g(x) = -\infty$, then $\lim_{x\to\infty} g(x) = \infty$.

29. A function that is not a polynomial can often be approximated by a polynomial. For example, for certain x-values, the function $f(x) = e^x$ can be approximated by the fifth-degree polynomial

$$p(x) = 1 + x + \frac{x^2}{2} + \frac{x^3}{6} + \frac{x^4}{24} + \frac{x^5}{120}.$$

(a) Show that $p(1) \approx f(1) = e$. How good is the estimate?
(b) Calculate $p(5)$. How well does $p(5)$ approximate $f(5)$?
(c) Graph $p(x)$ and $f(x)$ together on the same set of axes. Based on your graph, for what range of values of x does $p(x)$ give a good estimate for $f(x)$?

30. Table 9.6 gives values of v, the speed of sound (in m/sec) in water as a function of the temperature T (in °C).[9]

(a) An approximate linear formula for v is given by $v = 1402.385 + 5.038813T$. Over what temperature range does this formula agree with the values in Table 9.6 to within 1°C?

(b) The formula in part (a) can be improved by adding the quadratic term $-5.799136 \cdot 10^{-2}T^2$. Repeat part (a) using this adjusted formula.

(c) The formula in part (b) can be further improved by adding the cubic term $3.287156 \cdot 10^{-4}T^3$. Repeat part (a) using this adjusted formula.

(d) The speed of sound in water at 50°C is 1542.6 m/s. If we want to improve our formula still further by adding a quartic (fourth-degree) term, should this term be positive or negative?

Table 9.6

T	0	5	10	15	20	25	30
v	1402.4	1426.2	1447.3	1466.0	1482.4	1496.7	1509.2

9.3 THE SHORT-RUN BEHAVIOR OF POLYNOMIALS

The long-run behavior of a polynomial is determined by its leading term. However, polynomials with the same leading term may have very different short-run behaviors.

Example 1 Compare the graphs of the polynomials f, g, and h given by

$$f(x) = x^4 - 4x^3 + 16x - 16, \quad g(x) = x^4 - 4x^3 - 4x^2 + 16x, \quad h(x) = x^4 + x^3 - 8x^2 - 12x.$$

Solution Each of these functions is a fourth-degree polynomial, and each has x^4 as its leading term. Thus, all their graphs resemble the graph of x^4 on a large scale. See Figure 9.19.

However, on a smaller scale, the functions look different. See Figure 9.20. Two of the graphs go through the origin while the third does not. The graphs also differ from one another in the number of bumps each one has and in the number of times each one crosses the x-axis. Thus, polynomials with the same leading term look similar on a large scale, but may look dissimilar on a small scale.

Figure 9.19: On a large scale, the polynomials f, g, and h resemble the power function $y = x^4$

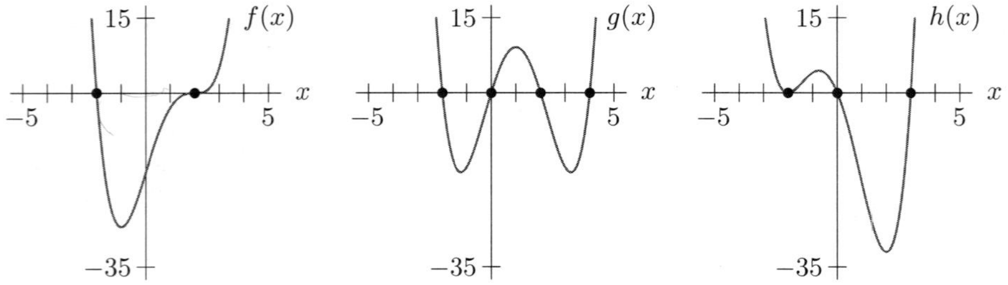

Figure 9.20: On a smaller scale, the polynomials f, g, and h look quite different from each other

[9]Data determined using the Marczak formula described at the UK National Physical Laboratory website, http://www.npl.co.uk.

Factored Form, Zeros, and the Short-Run Behavior of a Polynomial

To predict the long-run behavior of a polynomial, we write it in standard form. However, to determine the zeros and the short-run behavior of a polynomial, we write it in factored form, as a product of other polynomials. Some, but not all, polynomials can be factored.

Example 2 Investigate the short-run behavior of the third-degree polynomial $u(x) = x^3 - x^2 - 6x$.

(a) Rewrite $u(x)$ as a product of linear factors.
(b) Find the zeros of $u(x)$.
(c) Describe the graph of $u(x)$. Where does it cross the x-axis? the y-axis? Where is $u(x)$ positive? Negative?

Solution (a) By factoring out an x and then factoring the quadratic, $x^2 - x - 6$, we rewrite $u(x)$ as

$$u(x) = x^3 - x^2 - 6x = x(x^2 - x - 6) = x(x-3)(x+2).$$

Thus, we have expressed $u(x)$ as the product of three linear factors, x, $x-3$, and $x+2$.

(b) The polynomial equals zero if and only if at least one of its factors is zero. We solve the equation:

$$x(x-3)(x+2) = 0,$$

giving

$$x = 0, \quad \text{or} \quad x - 3 = 0, \quad \text{or} \quad x + 2 = 0,$$

so

$$x = 0, \quad \text{or} \quad x = 3, \quad \text{or} \quad x = -2.$$

These are the zeros, or x-intercepts, of u. To check, evaluate $u(x)$ for these x-values; you should get 0. There are no other zeros.

(c) To describe the graph of u, we give the x- and y-intercepts, and the long-run behavior.

The factored form, $u(x) = x(x-3)(x+2)$, shows that the graph crosses the x-axis at $x = 0, 3, -2$. The graph of u crosses the y-axis at $u(0) = 0^3 - 0^2 - 6 \cdot 0 = 0$; that is, at $y = 0$. For large values of x, the graph of $y = u(x)$ resembles the graph of its leading term, $y = x^3$. Figure 9.21 shows where u is positive and where u is negative.

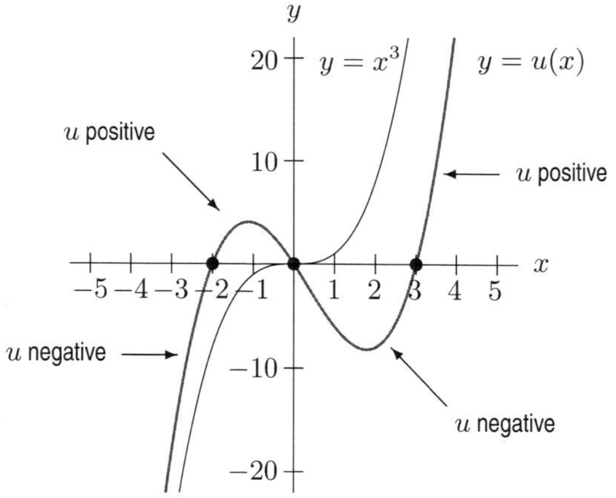

Figure 9.21: The graph of $u(x) = x^3 - x^2 - 6x$ has zeros at $x = -2$, 0, and 3. Its long-run behavior resembles $y = x^3$

In Example 2, each linear factor produced a zero of the polynomial. Now suppose that we do not know the polynomial p, but we do know that it has zeros at $x = 0, -12, 31$. Then we know that the factored form of the polynomial must include the factors $(x - 0)$ or x, and $(x - (-12))$ or $(x + 12)$, and $(x - 31)$. It may include other factors too. In summary:

> Suppose p is a polynomial. If the formula for p has a **linear factor**, that is, a factor of the form $(x - k)$, then p has a zero at $x = k$.
> Conversely, if p has a **zero** at $x = k$, then p has a linear factor of the form $(x - k)$.

The Number of Factors, Zeros, and Bumps

The number of linear factors is always less than or equal to the degree of a polynomial. For example, a fourth degree polynomial can have no more than four linear factors. This makes sense because if we had another factor in the product and multiplied out, the highest power of x would be greater than four. Since each zero corresponds to a linear factor, the number of zeros is less than or equal to the degree of the polynomial.

We can now say that there is a maximum number of bumps in the graph of a polynomial of degree n. Between any two consecutive zeros, there is a bump because the graph changes direction. In Figure 9.21, the graph, which decreases at $x = 1$, must come back up in order to cross the x-axis at $x = 3$. In summary:

> The graph of an nth degree polynomial has at most n zeros and turns at most $(n - 1)$ times.

Multiple Zeros

The functions $s(x) = (x - 4)^2$ and $t(x) = (x + 1)^3$ are both polynomials in factored form. Each is a horizontal shift of a power function. We refer to the zeros of s and t as *multiple zeros*, because in each case the factor contributing the value of $y = 0$ is repeated more than once. For instance, we say that $x = 4$ is a *double zero* of s, since

$$s(x) = (x - 4)^2 = \underbrace{(x - 4)(x - 4)}_{\text{Repeated twice}}.$$

Likewise, we say that $x = -1$ is a *triple zero* of t, since

$$t(x) = (x + 1)^3 = \underbrace{(x + 1)(x + 1)(x + 1)}_{\text{Repeated three times}}.$$

The graphs of s and t in Figures 9.22 and 9.23 show typical behavior near multiple zeros.

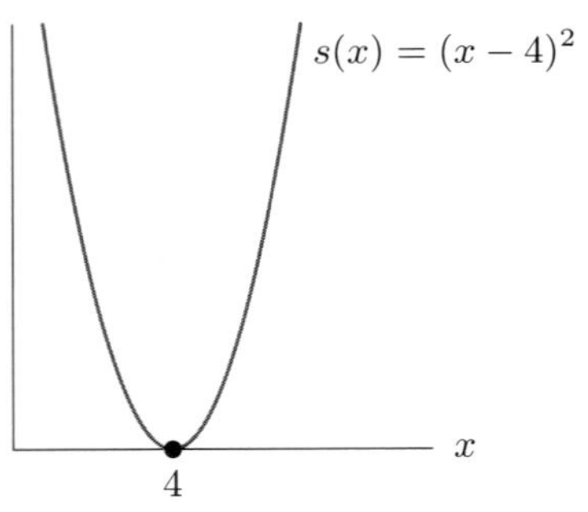

Figure 9.22: Double zero at $x = 4$

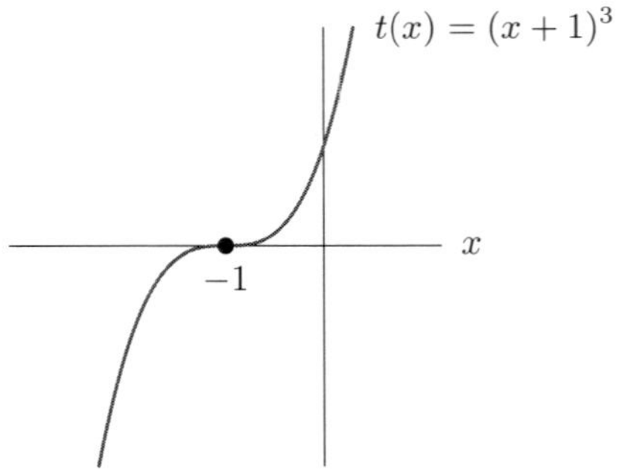

Figure 9.23: Triple zero at $x = -1$

In general:

> If p is a polynomial with a repeated linear factor, then p has a **multiple zero**.
> - If the factor $(x-k)$ is repeated an even number of times, the graph of $y=p(x)$ does not cross the x-axis at $x=k$, but "bounces" off the x-axis at $x=k$. (See Figure 9.22.)
> - If the factor $(x-k)$ is repeated an odd number of times, the graph of $y=p(x)$ crosses the x-axis at $x=k$, but it looks flattened there. (See Figure 9.23.)

Example 3 Describe in words the zeros of the 4th-degree polynomials $f(x)$, $g(x)$, and $h(x)$, in Figure 9.24.

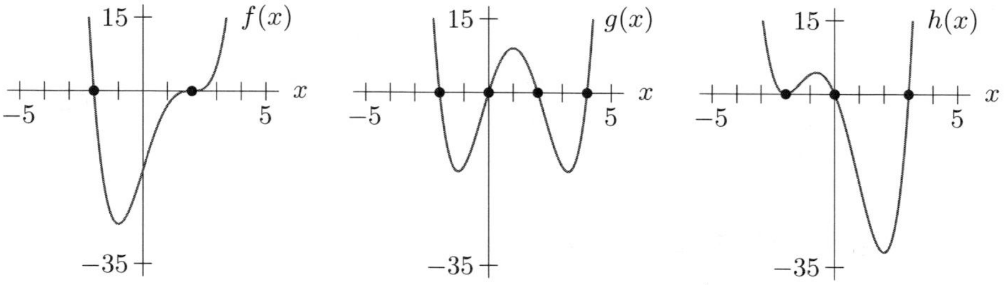

Figure 9.24

Solution The graph suggests that f has a single zero at $x=-2$. The flattened appearance near $x=2$ suggests that f has a multiple zero there. Since the graph crosses the x-axis at $x=2$ (instead of bouncing off it), this zero must be repeated an odd number of times. Since f is 4th degree, f has at most 4 factors, so there must be a triple zero at $x=2$.

The graph of g has four single zeros. The graph of h has two single zeros (at $x=0$ and $x=3$) and a double zero at $x=-2$. The multiplicity of the zero at $x=-2$ is not higher than two because h is of degree $n=4$.

Finding the Formula for a Polynomial from its Graph

The graph of a polynomial often enables us to find a possible formula for the polynomial.

Example 4 Find a possible formula for the polynomial function f graphed in Figure 9.25.

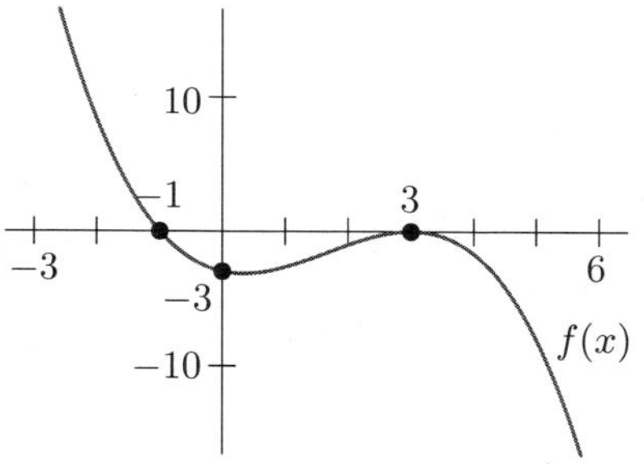

Figure 9.25: Features of the graph lead to a possible formula for this polynomial

Solution Based on its long-run behavior, f is of odd degree greater than or equal to 3. The polynomial has zeros at $x=-1$ and $x=3$. We see that $x=3$ is a multiple zero of even power, because the graph bounces off the x-axis here instead of crossing it. Therefore, we try the formula

$$f(x) = k(x+1)(x-3)^2$$

where k represents a stretch factor. The shape of the graph shows that k must be negative.

To find k, we use the fact that $f(0) = -3$, so

$$f(0) = k(0+1)(0-3)^2 = -3$$

which gives

$$9k = -3 \qquad \text{so} \qquad k = -\frac{1}{3}.$$

Thus, $f(x) = -\frac{1}{3}(x+1)(x-3)^2$ is a possible formula for this polynomial.

The formula for f we found in Example 4 is the polynomial of least degree we could have chosen. However, there are other polynomials, such as $y = -\frac{1}{27}(x+1)(x-3)^4$, with the same overall behavior as the function shown in Figure 9.25.

Exercises and Problems for Section 9.3

Exercises

In Exercises 1–4, find the zeros of the functions.

1. $y = 7(x+3)(x-2)(x+7)$
2. $y = a(x+2)(x-b)$, where a, b are nonzero constants
3. $y = x^3 + 7x^2 + 12x$
4. $y = (x^2 + 2x - 7)(x^3 + 4x^2 - 21x)$
5. Use the graph of $g(x)$ in Figure 9.20 on page 402 to determine the factored form of
$$g(x) = x^4 - 4x^3 - 4x^2 + 16x.$$
6. Use the graph of $f(x)$ in Figure 9.20 on page 402 to determine the factored form of
$$f(x) = x^4 - 4x^3 + 16x - 16.$$
7. Use the graph of $h(x)$ in Figure 9.20 on page 402 to determine the factored form of
$$h(x) = x^4 + x^3 - 8x^2 - 12x.$$
8. Factor $f(x) = 8x^3 - 4x^2 - 60x$ completely, and determine the zeros of f.

Without a calculator, graph the polynomials in Exercises 9–10. Label all the x-intercepts and y-intercepts.

9. $f(x) = -5(x^2 - 4)(25 - x^2)$
10. $g(x) = 5(x-4)(x^2 - 25)$

Problems

11. **(a)** Let $f(x) = (2x-1)(3x-1)(x-7)(x-9)$. What are the zeros of this polynomial?
    (b) Is it possible to find a viewing window that shows all of the zeros and all of the the turning points of f?
    (c) Find two separate viewing windows which together show all the zeros and all the turning points of f.
12. **(a)** Experiment with various viewing windows to determine the zeros of $f(x) = 2x^4 + 9x^3 - 7x^2 - 9x + 5$. Then write f in factored form.
    (b) Find a single viewing window that clearly shows all of the turning points of f.
13. Let $p(x) = x^4 + 10x^3 - 68x^2 + 102x - 45$. By experimenting with various viewing windows, determine the zeros of p and use this information to write $p(x)$ in factored form.
14. Let $u(x) = \frac{1}{8}x^3$ and $v(x) = \frac{1}{8}x(x - 0.01)^2$. Do v and u have the exact same graph? Sketch u and v in the window $-10 \le x \le 10$, $-10 \le y \le 10$. Now do you think that v and u have the same graph? If so, explain why their formulas are different; if not, find a viewing window on which their graphs' differences are prominent.

15. Without using a calculator, decide which of the equations A–E best describes the polynomial in Figure 9.26.

A $y = (x+2)(x+1)(x-2)(x-3)$
B $y = x(x+2)(x+1)(x-2)(x-3)$
C $y = -\frac{1}{2}(x+2)(x+1)(x-2)(x-3)$
D $y = \frac{1}{2}(x+2)(x+1)(x-2)(x-3)$
E $y = -(x+2)(x+1)(x-2)(x-3)$

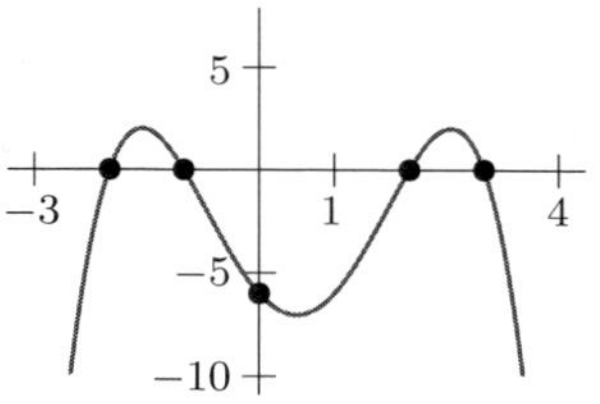

Figure 9.26

In Problems 16–21, find possible formulas for polynomials with the given properties.

16. f has degree ≤ 2, $f(0) = f(1) = f(2) = 1$.
17. f has degree ≤ 2, $f(0) = f(2) = 0$ and $f(3) = 3$.
18. f has degree ≤ 2, $f(0) = 0$ and $f(1) = 1$.
19. f is third degree with $f(-3) = 0$, $f(1) = 0$, $f(4) = 0$, and $f(2) = 5$.
20. g is fourth degree, g has a double zero at $x = 3$, $g(5) = 0$, $g(-1) = 0$, and $g(0) = 3$.
21. Least possible degree through the points $(-3, 0)$, $(1, 0)$, and $(0, -3)$.

Give a possible formula for the polynomials in Problems 22–35.

22.

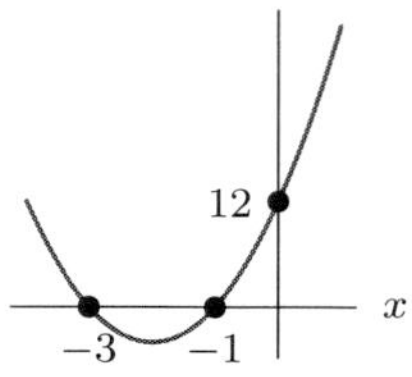

23.

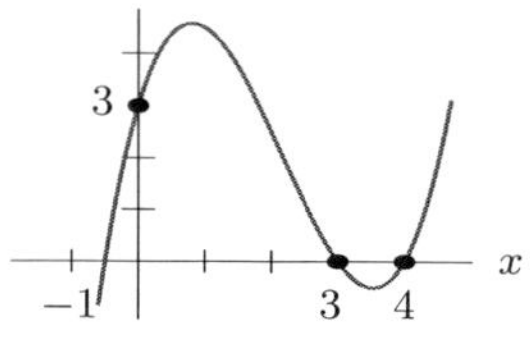

24.

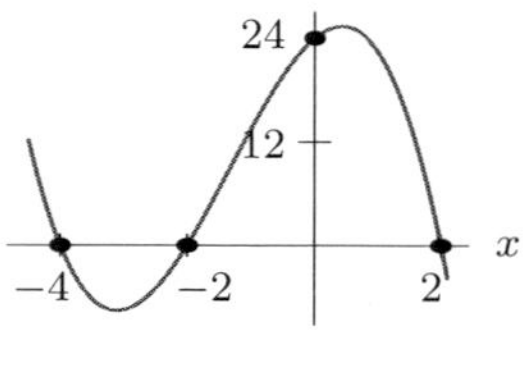

25.

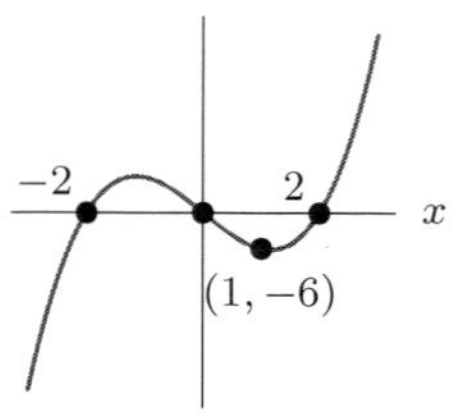

26.

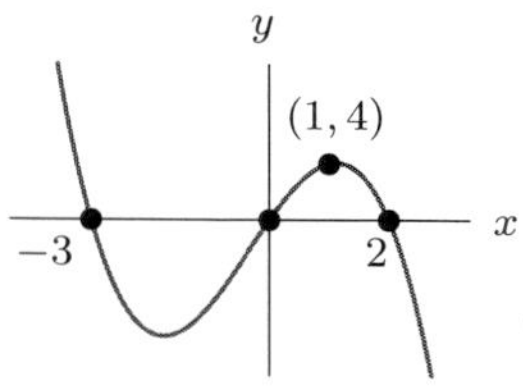

27.

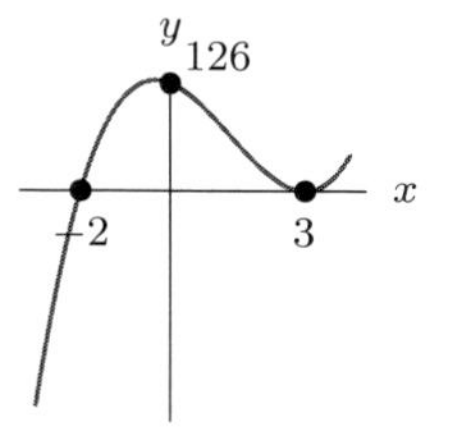

28.

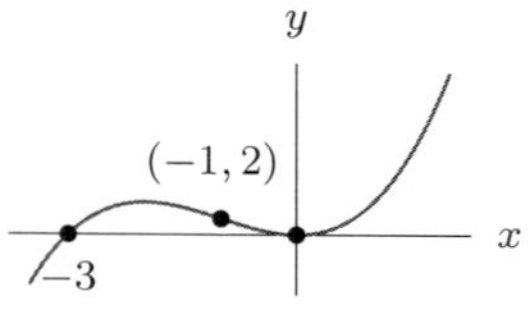

29.

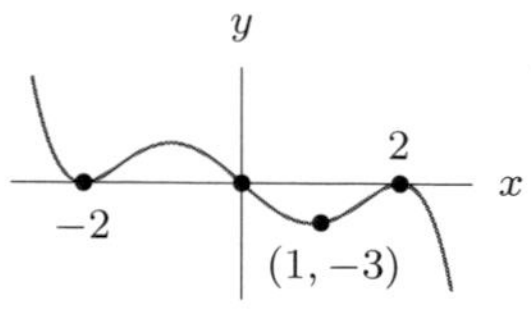

30.

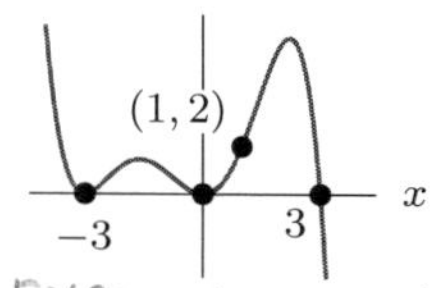

31.

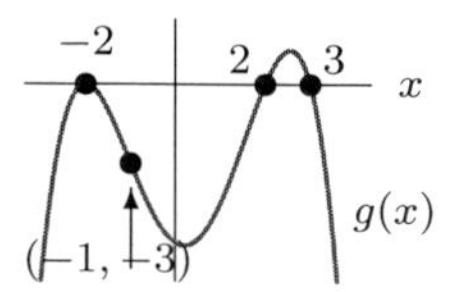

32. Example in notes on Oct. 21st

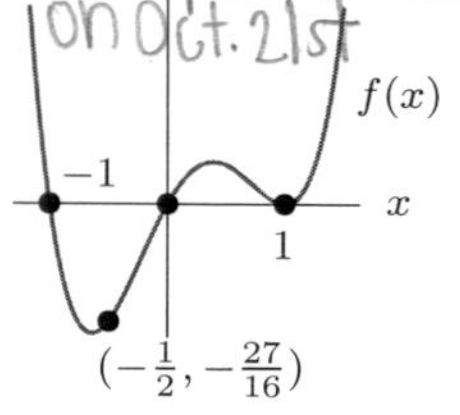

33.

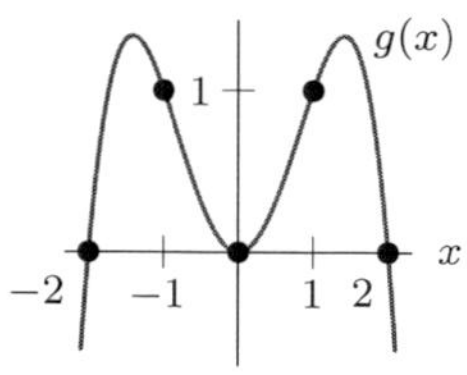

34.

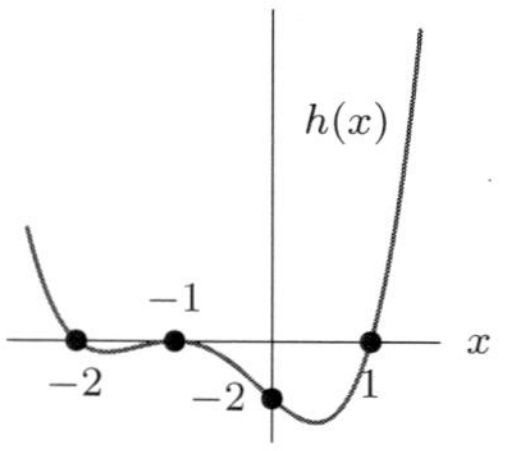

35.

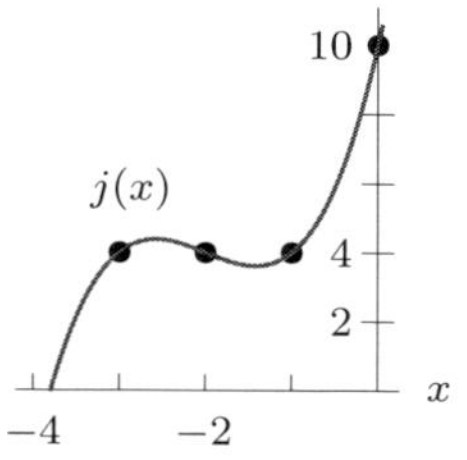

36. Which of these functions have inverses that are functions? Discuss.

(a) $f(x) = (x-2)^3 + 4$.
(b) $g(x) = x^3 - 4x^2 + 2$.

For Problems 37–42, find the real zeros (if any) of the polynomials.

37. $y = x^2 + 5x + 6$

38. $y = x^4 + 6x^2 + 9$

39. $y = 4x^2 - 1$

40. $y = 4x^2 + 1$

41. $y = 2x^2 - 3x - 3$

42. $y = 3x^5 + 7x + 1$

43. An open-top box is to be constructed from a 6 in by 8 in rectangular sheet of tin by cutting out squares of equal size at each corner, then folding up the resulting flaps. Let x denote the length of the side of each cut-out square. Assume negligible thickness.

(a) Find a formula for the volume of the box as a function of x.
(b) For what values of x does the formula from part (a) make sense in the context of the problem?
(c) Sketch a graph of the volume function.
(d) What, approximately, is the maximum volume of the box?

44. You wish to pack a cardboard box inside a wooden crate. In order to have room for the packing materials, you need to leave a 0.5-ft space around the front, back, and sides of the box, and a 1-ft space around the top and bottom of the box. If the cardboard box is x feet long, $(x+2)$ feet wide, and $(x-1)$ feet deep, find a formula in terms of x for the amount of packing material needed.

45. Take an 8.5 by 11-inch piece of paper and cut out four equal squares from the corners. Fold up the sides to create an open box. Find the dimensions of the box that has maximum volume.

46. Consider the function $a(x) = x^5 + 2x^3 - 4x$.

(a) Without using a calculator or computer, what can you say about the graph of a?
(b) Use a calculator or a computer to determine the zeros of this function to three decimal places.
(c) Explain why you think that you have all the possible zeros.
(d) What are the zeros of $b(x) = 2x^5 + 4x^3 - 8x$? Does your answer surprise you?

47. **(a)** Sketch a graph of $f(x) = x^4 - 17x^2 + 36x - 20$ for $-10 \le x \le 10$, $-10 \le y \le 10$.
(b) Your graph should appear to have a vertical asymptote at $x = -5$. Does f actually have a vertical asymptote here? Explain.
(c) How many zeros does f have? Can you find a window in which all of the zeros of f are clearly visible?
(d) Write the formula of f in factored form.
(e) How many turning points does the graph of f have? Can you find a window in which all the turning points of f are clearly visible? Explain.

48. In each of the following cases, find a possible formula for the polynomial f.

(a) Suppose f has zeros at $x = -2$, $x = 3$, $x = 5$ and a y-intercept of 4.
(b) In addition to the properties in part (a), suppose f has the following long-run behavior: As $x \to \pm\infty$, $y \to -\infty$. [Hint: Assume f has a double zero.]
(c) In addition to the properties in part (a), suppose f has the following long-run behavior: As $x \to \pm\infty$, $y \to +\infty$.

49. The following statements about $f(x)$ are true:

- $f(x)$ is a polynomial function
- $f(x) = 0$ at exactly four different values of x
- $f(x) \to -\infty$ as $x \to \pm\infty$

For each of the following statements, write *true* if the statement must be true, *never true* if the statement is never true, or *sometimes true* if it is sometimes true and sometimes not true.

(a) $f(x)$ is an odd function
(b) $f(x)$ is an even function
(c) $f(x)$ is a fourth degree polynomial
(d) $f(x)$ is a fifth degree polynomial
(e) $f(-x) \to -\infty$ as $x \to \pm\infty$
(f) $f(x)$ is invertible

9.4 RATIONAL FUNCTIONS

The Average Cost of Producing a Therapeutic Drug

A pharmaceutical company wants to begin production of a new drug. The total cost C, in dollars, of making q grams of the drug is given by the linear function

$$C(q) = 2{,}500{,}000 + 2000q.$$

The fact that $C(0) = 2{,}500{,}000$ tells us that the company spends $\$2{,}500{,}000$ before it starts making the drug. This quantity is known as the *fixed cost* because it does not depend on how much of the drug is made. It represents the cost for research, testing, and equipment. In addition, the slope of C tells us that each gram of the drug costs an extra $2000 to make. This quantity is known as the *variable cost* per unit. It represents the additional cost, in labor and materials, to make an additional gram of the drug.

The fixed cost of $2.5 million is large compared to the variable cost of $2000 per gram. This means that it is impractical for the company to make a small amount of the drug. For instance, the total cost for 10 grams is

$$C(10) = 2{,}500{,}000 + 2000 \cdot 10 = 2{,}520{,}000,$$

which works out to an average cost of $\$252{,}000$ per gram. The company would probably never sell such an expensive drug.

However, as larger quantities of the drug are manufactured, the initial expenditure of $2.5 million seems less significant. The fixed cost averages out over a large quantity. For example, if the company makes 10,000 grams of the drug,

$$\text{Average cost} = \frac{\text{Cost of producing } 10{,}000 \text{ grams}}{10{,}000} = \frac{2{,}500{,}000 + 2000 \cdot 10{,}000}{10{,}000} = 2250,$$

or $\$2250$ per gram of drug produced.

We define the average cost, $a(q)$, as the cost per gram to produce q grams of the drug:

$$a(q) = \frac{\text{Average cost of}}{\text{producing } q \text{ grams}} = \frac{\text{Total cost}}{\text{Number of grams}} = \frac{C(q)}{q} = \frac{2{,}500{,}000 + 2000q}{q}.$$

Figure 9.27 gives a graph of $y = a(q)$ for $q > 0$. The horizontal asymptote reflects the fact that for large values of q, the value of $a(q)$ is close to 2000. This is because, as more of the drug is produced, the average cost gets closer to $2000 per gram. See Table 9.7.

The vertical asymptote of $y = a(q)$ is the y-axis, which tells us that the average cost per gram is very large if a small amount of the drug is made. This is because the initial $2.5 million expenditure is averaged over very few units. We saw that producing only 10 grams costs a staggering $\$252{,}000$ per gram.

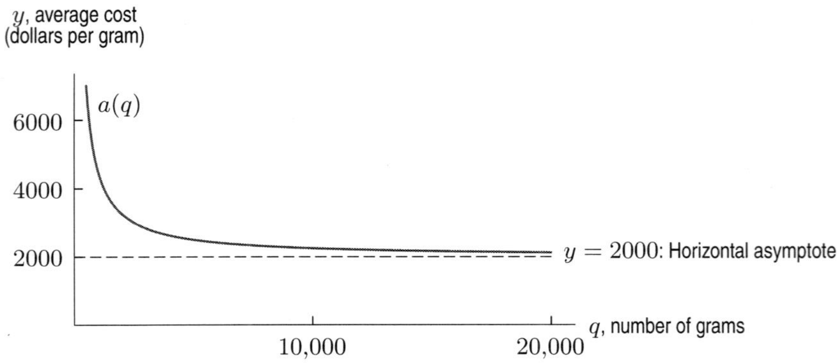

Figure 9.27: The graph of $y = a(q)$, a rational function, has a horizontal asymptote at $y = 2000$ and a vertical asymptote at $q = 0$

Table 9.7 *As quantity q increases, the average cost $a(q)$ draws closer to \$2000 per gram*

Quantity, q	Total cost, $C(q) = 2{,}500{,}000 + 2000q$	Average cost, $a(q) = C(q)/q$
10,000	$2{,}500{,}000 + 20{,}000{,}000 = 22{,}500{,}000$	2250
50,000	$2{,}500{,}000 + 100{,}000{,}000 = 102{,}500{,}000$	2050
100,000	$2{,}500{,}000 + 200{,}000{,}000 = 202{,}500{,}000$	2025
500,000	$2{,}500{,}000 + 1{,}000{,}000{,}000 = 1{,}002{,}500{,}000$	2005

What is a Rational Function?

The formula for $a(q)$ is the ratio of the polynomial $2{,}500{,}000 + 2000q$ and the polynomial q. Since $a(q)$ is given by the ratio of two polynomials, $a(q)$ is an example of a *rational function*. In general:

> If r can be written as the ratio of polynomial functions $p(x)$ and $q(x)$, that is, if
>
> $$r(x) = \frac{p(x)}{q(x)},$$
>
> then r is called a **rational function**. (We assume that $q(x)$ is not the constant polynomial $q(x) = 0$.)

The Long-Run Behavior of Rational Functions

In the long-run, every rational function behaves like a power function. For example, consider

$$f(x) = \frac{6x^4 + x^3 + 1}{-5x + 2x^2}.$$

Since the long-run behavior of a polynomial is determined by its highest power term, for large x the numerator behaves like $6x^4$ and the denominator behaves like $2x^2$. The long-run behavior of f is

$$f(x) = \frac{6x^4 + x^3 + 1}{-5x + 2x^2} \approx \frac{6x^4}{2x^2} = 3x^2,$$

so

$$\lim_{x\to\pm\infty} f(x) = \lim_{x\to\pm\infty} (3x^2) = \infty.$$

See Figure 9.28.

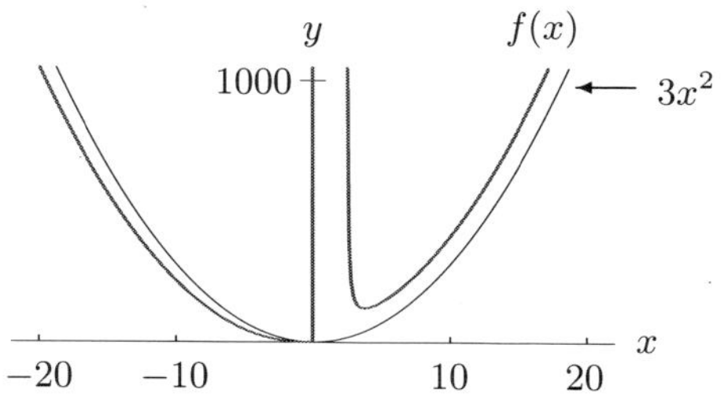

Figure 9.28: In the long-run, the graph of $f(x)$ looks like the graph of $3x^2$

In general, if r is any rational function, then for large enough values of x,

$$r(x) = \frac{a_n x^n + a_{n-1}x^{n-1} + \cdots + a_0}{b_m x^m + b_{m-1}x^{m-1} + \cdots + b_0} \approx \frac{a_n x^n}{b_m x^m} = \frac{a_n}{b_m} x^{n-m}.$$

This means that on a large scale r resembles the function $y = \left(\frac{a_n}{b_m}\right) x^{n-m}$, which is a power function of the form $y = kx^p$, where $k = a_n/b_m$ and $p = n - m$. In summary:

> For large enough values of x (either positive or negative), the graph of the rational function r looks like the graph of a power function. If $r(x) = p(x)/q(x)$, then the **long-run behavior** of $y = r(x)$ is given by
>
> $$y = \frac{\text{Leading term of } p}{\text{Leading term of } q}.$$
>
> Using limits, we write
>
> $$\lim_{x\to\pm\infty} \frac{p(x)}{q(x)} = \lim_{x\to\pm\infty} \frac{\text{Leading term of } p}{\text{Leading term of } q}.$$

Example 1 For positive x, describe the long-run behavior of the rational function

$$r(x) = \frac{x+3}{x+2}.$$

Solution If x is a large positive number, then

$$r(x) = \frac{\text{Big number} + 3}{\text{Same big number} + 2} \approx \frac{\text{Big number}}{\text{Same big number}} = 1.$$

For example, if $x = 100$, we have

$$r(x) = \frac{103}{102} = 1.0098\ldots \approx 1.$$

If $x = 10{,}000$, we have

$$r(x) = \frac{10,003}{10,002} = 1.00009998\ldots \approx 1,$$

For large positive x-values, $r(x) \approx 1$. Thus, for large enough values of x, the graph of $y = r(x)$ looks like the line $y = 1$, its horizontal asymptote. We write $\lim_{x\to\infty} r(x) = 1$. See Figure 9.29. However, for $x > 0$, the graph of r is above the line since the numerator is larger than the denominator.

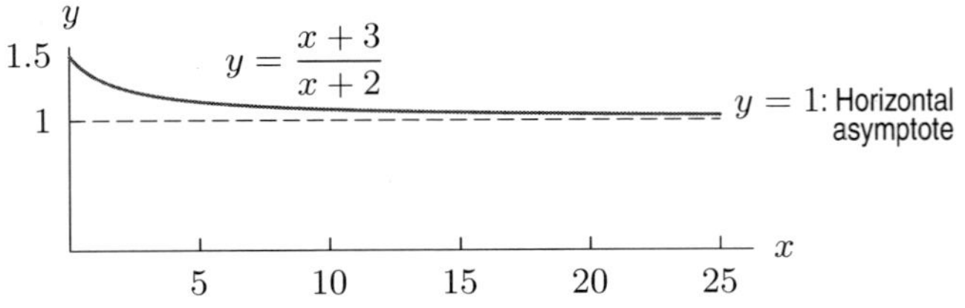

Figure 9.29: For large positive values of x, the graph of $r(x) = (x+3)/(x+2)$ looks like the horizontal line $y = 1$

Example 2 For positive x, describe the positive long-run behavior of the rational function

$$g(x) = \frac{3x+1}{x^2+x-2}.$$

Solution The leading term in the numerator is $3x$ and the leading term in the denominator is x^2. Thus for large enough values of x,

$$g(x) \approx \frac{3x}{x^2} = \frac{3}{x},$$

so

$$\lim_{x\to\infty} g(x) = \lim_{x\to\infty}\left(\frac{3}{x}\right) = 0.$$

Figure 9.30 shows the graphs of $y = g(x)$ and $y = 3/x$. For large values of x, the two graphs are nearly indistinguishable. Both graphs have a horizontal asymptote at $y = 0$.

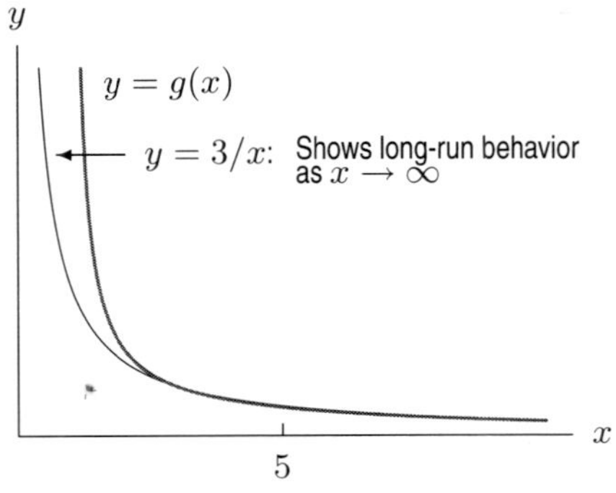

Figure 9.30: For large enough values of x, the function g looks like the function $y = 3x^{-1}$

What Causes Asymptotes?

The graphs of rational functions often behave differently from the graphs of polynomials. Polynomial graphs (except constant functions) cannot level off to a horizontal line like the graphs of rational functions can. In Example 1, the numerator and denominator are approximately equal for large x, producing the horizontal asymptote $y = 1$. In Example 2, the denominator grows faster than the numerator, driving the quotient toward zero.

The rapid rise (or fall) of the graph of a rational function near its vertical asymptote is due to the denominator becoming small (close to zero). It is tempting to assume that any function which has a denominator has a vertical asymptote. However, this is not true. To have a vertical asymptote, the denominator must become close to zero. For example, suppose that

$$r(x) = \frac{1}{x^2+3}.$$

The denominator is always greater than 3; it is never 0. We see from Figure 9.31 that r does not have a vertical asymptote.

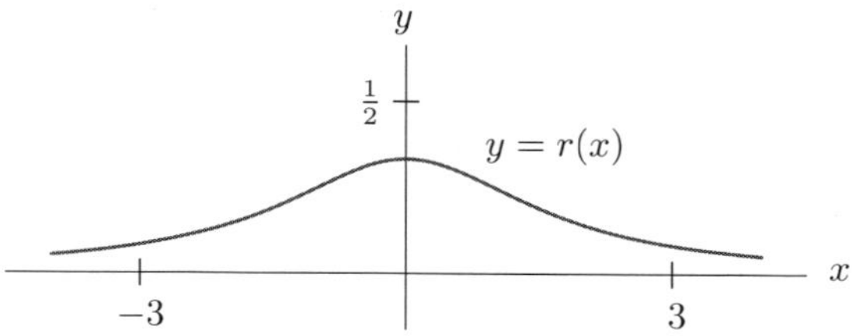

Figure 9.31: The rational function $r(x) = 1/(x^2+3)$ has no vertical asymptote

Exercises and Problems for Section 9.4

Exercises

Are the functions in Exercises 1–6 rational functions? If so, write them in the form $p(x)/q(x)$, the ratio of polynomials.

1. $f(x) = \frac{x^2}{2} + \frac{1}{x}$

2. $f(x) = \frac{\sqrt{x}+1}{x+1}$

3. $f(x) = \frac{4^x+3}{3^x-1}$

4. $f(x) = \frac{x^2+4}{e^x}$

5. $f(x) = \frac{x^2}{x-3} - \frac{5}{x-3}$

6. $f(x) = \frac{9x-1}{4\sqrt{x}+7} + \frac{5x^3}{x^2-1}$

7. Find

(a) $\lim_{x\to\infty} \frac{2x+1}{x-5}$ **(b)** $\lim_{x\to-\infty} \frac{2+5x}{6x+3}$

8. Find

(a) $\lim_{x\to\infty} \frac{x(x^2-4)}{5+5x^3}$ **(b)** $\lim_{x\to-\infty} \frac{3x(x-1)(x-2)}{5-6x^4}$

Find the horizontal asymptote, if it exists, of the functions in Problems 9–11.

9. $f(x) = \frac{1}{1+\frac{1}{x}}$

10. $g(x) = \frac{(1-x)(2+3x)}{2x^2+1}$

11. $h(x) = 3 - \frac{1}{x} + \frac{x}{x+1}$

12. Compare and discuss the long-run behaviors of the following functions:

$$f(x) = \frac{x^2+1}{x^2+5}, \quad g(x) = \frac{x^3+1}{x^2+5}, \quad h(x) = \frac{x+1}{x^2+5}.$$

Problems

13. Let $r(x) = p(x)/q(x)$, where p and q are polynomial of degrees m and n, respectively. What conditions on m and n ensure that the following statements are true?

(a) $\lim_{x\to\infty} r(x) = 0$
(b) $\lim_{x\to\infty} r(x) = k$, with $k \neq 0$.

14. Give examples of rational functions with even symmetry, odd symmetry, and neither. How does the symmetry of $f(x) = p(x)/q(x)$ depend on the symmetry of $p(x)$ and $q(x)$?

15. Find a formula for $f^{-1}(x)$ given that

$$f(x) = \frac{4-3x}{5x-4}.$$

16. Let t be the time in weeks. At time $t = 0$, organic waste is dumped into a pond. The oxygen level in the pond at time t is given by

$$f(t) = \frac{t^2 - t + 1}{t^2 + 1}.$$

Assume $f(0) = 1$ is the normal level of oxygen.

(a) Graph this function.
(b) Describe the shape of the graph. What is the significance of the minimum for the pond?
(c) What eventually happens to the oxygen level?
(d) Approximately how many weeks must pass before the oxygen level returns to 75% of its normal level?

17. The following procedure approximates the cube root of a number. If x is a guess for $\sqrt[3]{2}$, for example, then x^3 equals 2 only if the guess is correct. If $x^3 = 2$ we can also write $x = 2/x^2$. If our guess, x, is less than $\sqrt[3]{2}$, then $2/x^2$ is greater than $\sqrt[3]{2}$. If x is greater than $\sqrt[3]{2}$, then $2/x^2$ is less than $\sqrt[3]{2}$. In either case, if x is an estimate for $\sqrt[3]{2}$, then the average of x and $2/x^2$ provides a better estimate. Define $g(x)$ to be this improved estimate.

(a) Find a possible formula for $g(x)$, expressed as one reduced fraction.
(b) Use $1.26 \approx \sqrt[3]{2}$ as a first guess. Use the function $g(x)$ to estimate the value of $\sqrt[3]{2}$, accurate to five decimal places. Construct a table showing any intermediate results. Explain how you know you have reached the required accuracy.

18. Problem 17 outlines a method of approximating $\sqrt[3]{2}$. An initial guess, x, is averaged with $2/x^2$ to obtain a better guess, denoted by $g(x)$. A better method involves taking a weighted average of x and $2/x^2$.

(a) Let x be a guess for $\sqrt[3]{2}$. Define $h(x)$ by

$$h(x) = \frac{1}{3}\left(x + x + \frac{2}{x^2}\right).$$

Express $h(x)$ as one reduced fraction. Explain why $h(x)$ is referred to as a weighted average.
(b) Explain why $h(x)$ is a better function to use for estimating $\sqrt[3]{2}$ than is $g(x)$. Include specific, numerical examples in your answer.

19. Bronze is an alloy, or mixture, of copper and tin. The alloy initially contains 3 kg copper and 9 kg tin. You add x kg of copper to this 12 kg of alloy. The concentration of copper in the alloy is a function of x:

$$f(x) = \text{Concentration of copper} = \frac{\text{Total amount of copper}}{\text{Total amount of alloy}}.$$

(a) Find a formula for f in terms of x, the amount of copper added.
(b) Evaluate the following expressions and explain their significance for the alloy:

(i) $f(\frac{1}{2})$ (ii) $f(0)$ (iii) $f(-1)$
(iv) $f^{-1}(\frac{1}{2})$ (v) $f^{-1}(0)$

(c) Graph $f(x)$ for $-5 \leq x \leq 5$, $-0.25 \leq y \leq 0.5$. Interpret the intercepts in the context of the alloy.
(d) Graph $f(x)$ for $-3 \leq x \leq 100$, $0 \leq y \leq 1$. Describe the appearance of your graph for large x-values. Does the appearance agree with what you expect to happen when large amounts of copper are added to the alloy?

20. A chemist is studying the properties of a bronze alloy (mixture) of copper and tin. She begins with 2 kg of an alloy that is one-half tin. Keeping the amount of copper constant, she adds small amounts of tin to the alloy. Letting x be the total amount of tin added, define

$$C(x) = \text{Concentration of tin} = \frac{\text{Total amount of tin}}{\text{Total amount of alloy}}.$$

(a) Find a formula for $C(x)$.
(b) Evaluate $C(0.5)$ and $C(-0.5)$. Explain the physical significance of these quantities.
(c) Graph $y = C(x)$, labeling all interesting features. Describe the physical significance of the features you have labeled.

21. The total cost $C(n)$ for a producer to manufacture n units of a good is given by

$$C(n) = 5000 + 50n.$$

The average cost of producing n units is $a(n) = C(n)/n$.

(a) Evaluate and interpret the economic significance of:

(i) $C(1)$ (ii) $C(100)$
(iii) $C(1000)$ (iv) $C(10000)$

(b) Evaluate and interpret the economic significance of:

(i) $a(1)$ (ii) $a(100)$
(iii) $a(1000)$ (iv) $a(10000)$

(c) Based on part (b), what trend do you notice in the values of $a(n)$ as n gets large? Explain this trend in economic terms.

22. Figure 9.32 shows the cost function, $C(n)$, from Problem 21, together with a line, l, that passes through the origin.

(a) What is the slope of line l?
(b) How does line l relate to $a(n_0)$, the average cost of producing n_0 units (as defined in Problem 21)?

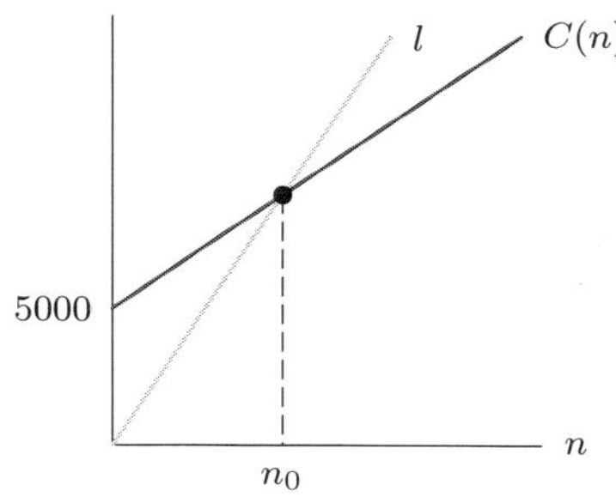

Figure 9.32

23. Typically, the average cost of production (as defined in Problem 21) decreases as the level of production increases. Is this always the case for the goods whose total cost function is graphed in Figure 9.33? Use the result of Problem 22 and explain your reasoning.

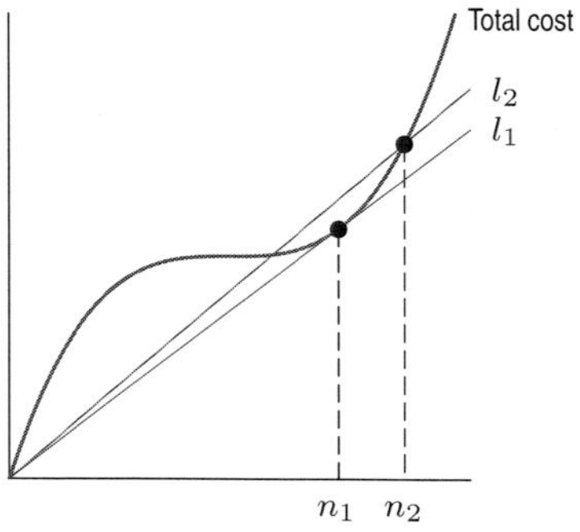

Figure 9.33

24. It costs a company \$30,000 to begin production of a good, plus \$3 for every unit of the good produced. Let x be the number of units produced by the company.

(a) Find a formula for $C(x)$, the total cost for the production of x units of the good.
(b) Find a formula for the company's average cost per unit, $a(x)$.
(c) Graph $y = a(x)$ for $0 < x \leq 50{,}000$, $0 \leq y \leq 10$. Label the horizontal asymptote.
(d) Explain in economic terms why the graph of a has the long-run behavior that it does.
(e) Explain in economic terms why the graph of a has the vertical asymptote that it does.
(f) Find a formula for $a^{-1}(y)$. Give an economic interpretation of $a^{-1}(y)$.
(g) The company makes a profit if the average cost of its good is less than \$5 per unit. Find the minimum number of units the company can produce and make a profit.

9.5 THE SHORT-RUN BEHAVIOR OF RATIONAL FUNCTIONS

The short-run behavior of a polynomial can often be determined from its factored form. The same is true of rational functions. If r is a rational function given by

$$r(x) = \frac{p(x)}{q(x)}, \qquad p, q \text{ polynomials},$$

then the short-run behavior of p and q tell us about the short-run behavior of r.

The Zeros and Vertical Asymptotes of a Rational Function

A fraction is equal to zero if and only if its numerator equals zero (and its denominator does not equal zero). Thus, the rational function $r(x) = p(x)/q(x)$ has a zero wherever p has a zero, provided q does not have a zero there.

Just as we can find the zeros of a rational function by looking at its numerator, we can find the vertical asymptotes by looking at its denominator. A rational function is large wherever its denominator is small. This means that r has a vertical asymptote wherever its denominator has a zero, provided its numerator does not also have a zero there.

Example 1 Find the zeros and vertical asymptotes of the rational function $r(x) = \frac{x+3}{x+2}$.

Solution We see that $r(x) = 0$ if

$$\frac{x+3}{x+2} = 0.$$

This ratio equals zero only if the numerator is zero (and the denominator is not zero), so

$$\begin{aligned} x + 3 &= 0 \\ x &= -3. \end{aligned}$$

The only zero of r is $x = -3$. To check, note that $r(-3) = 0/(-1) = 0$. The denominator has a zero at $x = -2$, so the graph of $r(x)$ has a vertical asymptote there.

Example 2 Graph $r(x) = \frac{25}{(x+2)(x-3)^2}$, showing all the important features.

Solution Since the numerator of this function is never zero, r has no zeros, meaning that the graph of r never crosses the x-axis. The graph of r has vertical asymptotes at $x = -2$ and $x = 3$ because this is where the denominator is zero. What does the graph of r look like near its asymptote at $x = -2$? At $x = -2$, the numerator is 25 and the value of the factor $(x-3)^2$ is $(-2-3)^2 = 25$. Thus, near $x = -2$,

$$r(x) = \frac{25}{(x+2)(x-3)^2} \approx \frac{25}{(x+2)(25)} = \frac{1}{x+2}.$$

So, near $x = -2$, the graph of r looks like the graph of $y = 1/(x+2)$. Note that the graph of $y = 1/(x+2)$ is the graph of $y = 1/x$ shifted to the left by 2 units. We see that

$$\lim_{x\to -2^-} r(x) = -\infty \quad \text{and} \quad \lim_{x\to -2^+} r(x) = \infty.$$

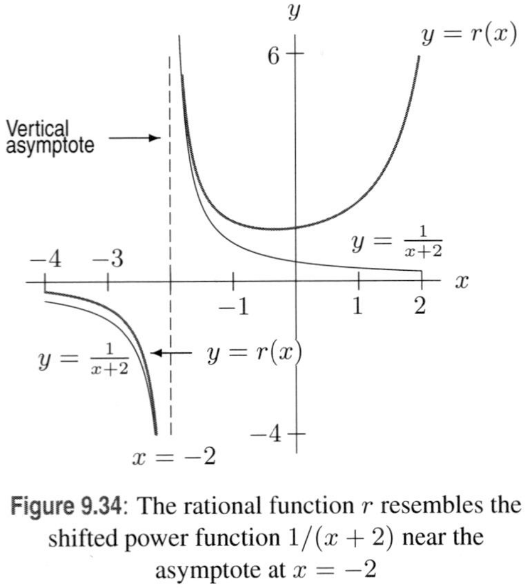

Figure 9.34: The rational function r resembles the shifted power function $1/(x+2)$ near the asymptote at $x = -2$

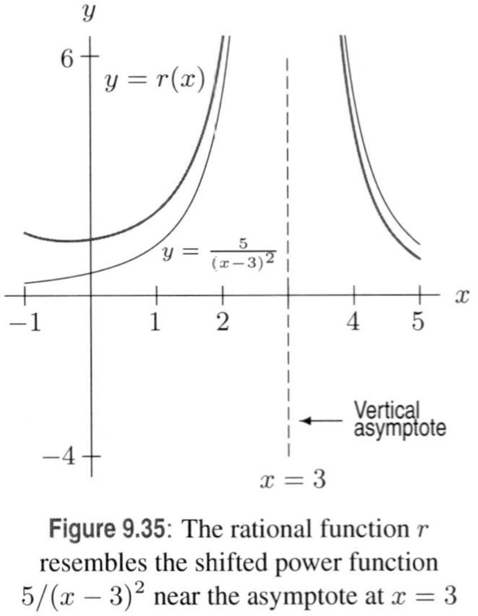

Figure 9.35: The rational function r resembles the shifted power function $5/(x-3)^2$ near the asymptote at $x = 3$

What does the graph of r look like near its vertical asymptote at $x = 3$? Near $x = 3$, the numerator is 25 and value of the factor $(x + 2)$ is approximately $(3 + 2) = 5$. Thus, near $x = 3$,

$$r(x) \approx \frac{25}{(5)(x-3)^2} = \frac{5}{(x-3)^2}.$$

Near $x = 3$, the graph of r looks like the the graph of $y = 5/(x-3)^2$. We see that

$$\lim_{x\to 3} r(x) = \infty.$$

The graph of $y = 5/(x-3)^2$ is the graph of $y = 5/x^2$ shifted to the right 3 units. Since

$$r(0) = \frac{25}{(0+2)(0-3)^2} = \frac{25}{18} \approx 1.4,$$

the graph of r crosses the y-axis at $25/18$. The long-run behavior of r is given by the ratio of the leading term in the numerator to the leading term in the denominator. The numerator is 25 and if we multiply out the denominator, we see that its leading term is x^3. Thus, the long-run behavior of r is given by $y = 25/x^3$, which has a horizontal asymptote at $y = 0$. See Figure 9.36. We see that

$$\lim_{x\to\pm\infty} r(x) = 0.$$

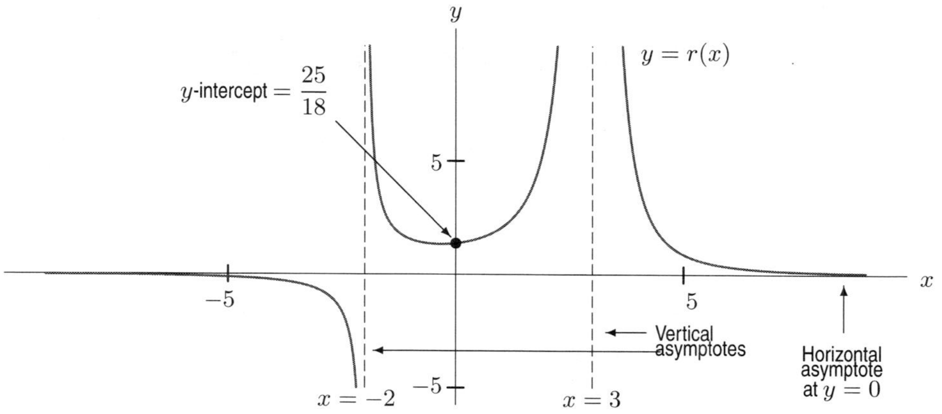

Figure 9.36: A graph of the rational function $r(x) = \dfrac{25}{(x+2)(x-3)^2}$, showing intercepts and asymptotes

The Graph of a Rational Function

We can now summarize what we have learned about the graphs of rational functions.

> If r is a rational function given by $r(x) = \dfrac{p(x)}{q(x)}$, where p and q are polynomials with different zeros, then:
>
> - The **long-run behavior** of r is given by the ratio of the leading terms of p and q.
> - The **zeros** of r are the same as the zeros of the numerator, p.
> - The graph of r has a **vertical asymptote** at each of the zeros of the denominator, q.

If p and q have zeros at the same x-values, the rational function may behave differently. See page 419.

Can a Graph Cross an Asymptote?

The graph of a rational function never crosses a vertical asymptote. However, the graphs of some rational functions cross their horizontal asymptotes. The difference is that a vertical asymptote occurs where the function is undefined, so there can be no y-value there, whereas a horizontal asymptote represents the limiting value of the function as $x \to \pm\infty$. There is no reason that the function cannot take on this limiting y-value for some finite x-value. For example, the graph of $r(x) = \dfrac{x^2 + 2x - 3}{x^2}$ crosses the line $y = 1$, its horizontal asymptote; the graph does not cross the vertical asymptote, the y-axis. See Figure 9.37.

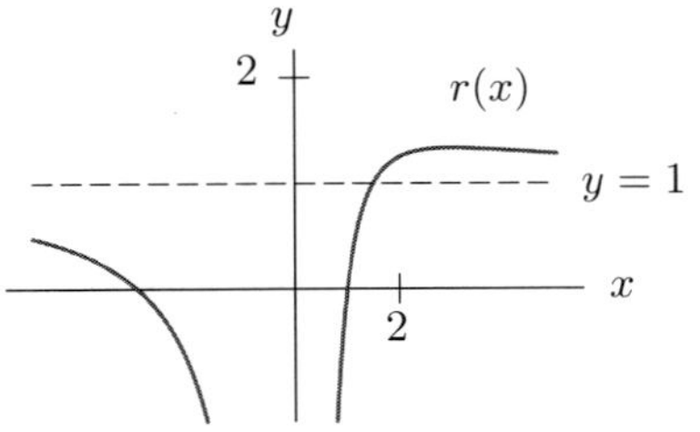

Figure 9.37: A rational function can cross its horizontal asymptote

Rational Functions as Transformations of Power Functions

The average cost function on page 409 can be written as

$$a(q) = \frac{2{,}500{,}000 + 2000q}{q} = 2{,}500{,}000q^{-1} + 2000.$$

Thus, the graph of a is the graph of the power function $y = 2{,}500{,}000q^{-1}$ shifted up 2000 units. Many rational functions can be viewed as translations of power functions.

Finding a Formula for a Rational Function from its Graph

The graph of a rational function can give a good idea of its formula. Zeros of the function correspond to factors in the numerator and vertical asymptotes correspond to factors in the denominator.

Example 3 Find a possible formula for the rational function, $g(x)$, graphed in Figure 9.38.

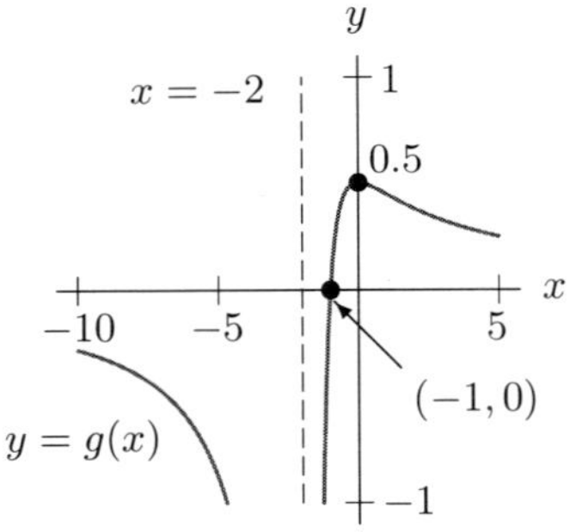

Figure 9.38: The graph of $y = g(x)$ a rational function

Solution From the graph, we see that g has a zero at $x = -1$ and a vertical asymptote at $x = -2$. This means that the numerator of g has a zero at $x = -1$ and the denominator of g has a zero at $x = -2$. The zero of g does not seem to be a multiple zero because the graph crosses the x-axis instead of bouncing and does not have a flattened appearance. Thus, we conclude that the numerator of g has one factor of $(x + 1)$.

The values of $g(x)$ have the same sign on both sides of the vertical asymptote. Thus, the behavior of g near its vertical asymptote is more like the behavior of $y = 1/(x+2)^2$ than like $y = 1/(x+2)$. We conclude that the denominator of g has a factor of $(x+2)^2$. This suggests

$$g(x) = k \cdot \frac{x+1}{(x+2)^2},$$

where k is a stretch factor. To find the value of k, use the fact that $g(0) = 0.5$. So

$$0.5 = k \cdot \frac{0+1}{(0+2)^2}$$
$$0.5 = k \cdot \frac{1}{4}$$
$$k = 2.$$

Thus, a possible formula for g is $g(x) = \dfrac{2(x+1)}{(x+2)^2}$.

When Numerator and Denominator Have the Same Zeros: Holes

The rational function $h(x) = \dfrac{x^2+x-2}{x-1}$ is undefined at $x = 1$ because the denominator equals zero at $x = 1$. However, the graph of h does not have a vertical asymptote at $x = 1$ because the numerator of h also equals zero at $x = 1$. At $x = 1$,

$$h(1) = \frac{x^2+x-2}{x-1} = \frac{1^2+1-2}{1-1} = \frac{0}{0},$$

and this ratio is undefined. What does the graph of h look like? Factoring the numerator of h gives

$$h(x) = \frac{(x-1)(x+2)}{x-1} = \frac{x-1}{x-1}(x+2).$$

For any $x \neq 1$, we can cancel $(x-1)$ top and bottom and rewrite the formula for h as

$$h(x) = x+2, \qquad \text{provided } x \neq 1.$$

Thus, the graph of h is the line $y = x+2$ except at $x = 1$, where h is undefined. The line $y = x+2$ contains the point $(1,3)$, but the graph of h does not. Therefore, we say that the graph of h has a *hole* in it at the point $(1,3)$. See Figure 9.39.

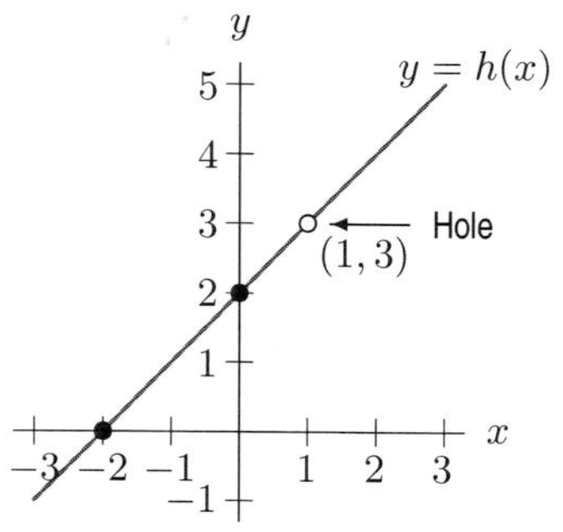

Figure 9.39: The graph of $y = h(x)$ is the line $y = x+2$, except at the point $(1,3)$, where it has a hole

Exercises and Problems for Section 9.5

Exercises

For the rational functions in Exercises 1–4, find all zeros and vertical asymptotes and describe the long-run behavior. Then graph the function without a calculator.

1. $y = \dfrac{x+3}{x+5}$

2. $y = \dfrac{x+3}{(x+5)^2}$

3. $y = \dfrac{x-4}{x^2-9}$

4. $y = \dfrac{x^2-4}{x-9}$

In Exercises 5–8, what are the x-intercepts, y-intercepts, and horizontal and vertical asymptotes (if any)?

5. $f(x) = \dfrac{x-2}{x-4}$

6. $g(x) = \dfrac{x^2-9}{x^2+9}$

7. $h(x) = \dfrac{x^2-4}{x^3+4x^2}$

8. $k(x) = \dfrac{x(4-x)}{x^2-6x+5}$

9. Let $f(x) = \dfrac{1}{x-3}$.

(a) Complete Table 9.8 for x-values close to 3. What happens to the values of $f(x)$ as x approaches 3 from the left? From the right?

Table 9.8

x	2	2.9	2.99	3	3.01	3.1	4
$f(x)$							

(b) Complete Tables 9.9 and 9.10. What happens to the values of $f(x)$ as x takes very large positive values? As x takes very large negative values?

Table 9.9

x	5	10	100	1000
$f(x)$				

Table 9.10

x	-5	-10	-100	-1000
$f(x)$				

(c) Without a calculator, graph $y = f(x)$. Give equations for the horizontal and vertical asymptotes.

10. Let $g(x) = \dfrac{1}{(x+2)^2}$.

(a) Complete Table 9.11 for x-values close to -2. What happens to the values of $g(x)$ as x approaches -2 from the left? From the right?

Table 9.11

x	-3	-2.1	-2.01	-2	-1.99	-1.9	-1
$g(x)$							

(b) Complete Tables 9.12 and 9.13. What happens to the values of $g(x)$ as x takes very large positive values? As x takes very large negative values?

Table 9.12

x	5	10	100	1000
$g(x)$				

Table 9.13

x	-5	-10	-100	-1000
$g(x)$				

(c) Without a calculator, graph $y = g(x)$. Give equations for the horizontal and vertical asymptotes.

Problems

Graph the functions in Exercises 11–12 without a calculator.

11. $y = 2 + \dfrac{1}{x}$

12. $y = \dfrac{2x^2-10x+12}{x^2-16}$

In Problems 13–14, estimate the one-sided limits

(a) $\lim\limits_{x\to a^+} f(x)$ **(b)** $\lim\limits_{x\to a^-} f(x)$

13. $f(x) = \dfrac{x}{5-x}$ with $a = 5$

14. $f(x) = \dfrac{5-x}{(x-2)^2}$ with $a = 2$

In Problems 15–16,

(a) Estimate $\lim\limits_{x\to\infty} f(x)$ and $\lim\limits_{x\to-\infty} f(x)$.

(b) What does the vertical asymptote tell you about limits?

15.

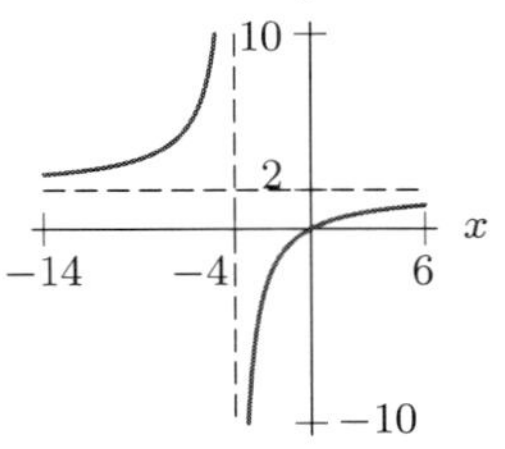

16.

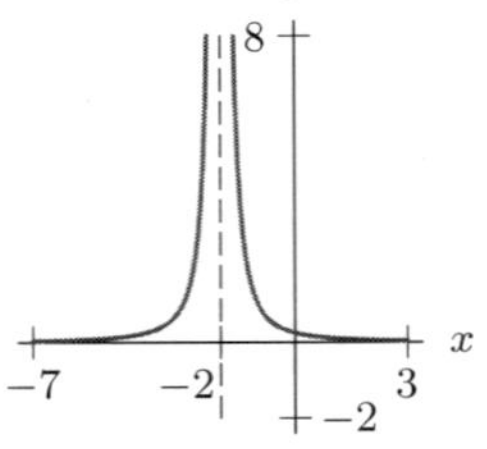

17. Without a calculator, match the functions (a)–(f) with their graphs in (i)–(vi) by finding the zeros, asymptotes, and end behavior for each function.

(a) $y = \dfrac{-1}{(x-5)^2} - 1$ **(b)** $y = \dfrac{x-2}{(x+1)(x-3)}$

(c) $y = \dfrac{2x+4}{x-1}$ **(d)** $y = \dfrac{1}{x+1} + \dfrac{1}{x-3}$

(e) $y = \dfrac{1-x^2}{x-2}$ **(f)** $y = \dfrac{1-4x}{2x+2}$

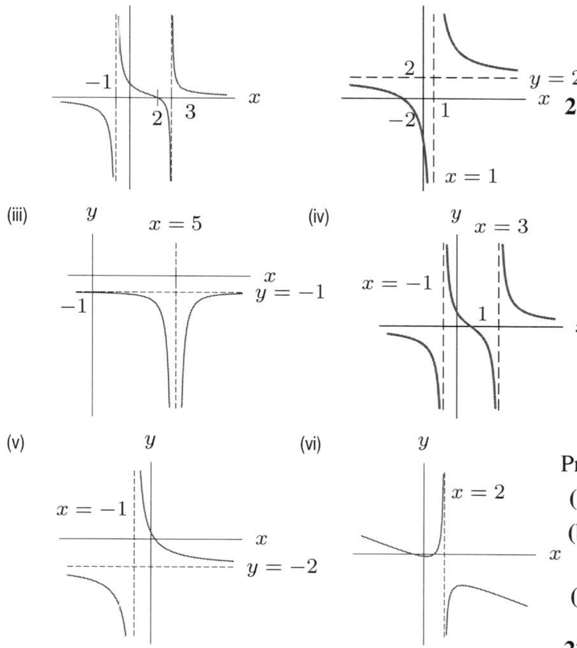

18. Let $f(x) = x^2 + 5x + 6$ and $g(x) = x^2 + 1$.

(a) What are the zeros of f and g?

(b) Let $r(x) = f(x)/g(x)$. Graph r. Does r have zeros? Vertical asymptotes? What is its long-run behavior as $x \to \pm\infty$?

(c) Let $s(x) = g(x)/f(x)$. If you graph s in the window $-10 \le x \le 10$, $-10 \le y \le 10$, it appears to have a zero near the origin. Does it? Does s have a vertical asymptote? What is its long-run behavior?

19. Suppose that n is a constant and that $f(x)$ is a function defined when $x = n$. Complete the following sentences.

(a) If $f(n)$ is large, then $\dfrac{1}{f(n)}$ is ...

(b) If $f(n)$ is small, then $\dfrac{1}{f(n)}$ is ...

(c) If $f(n) = 0$, then $\dfrac{1}{f(n)}$ is ...

(d) If $f(n)$ is positive, then $\dfrac{1}{f(n)}$ is ...

(e) If $f(n)$ is negative, then $\dfrac{1}{f(n)}$ is ...

20. **(a)** Use the results of Problem 19 to graph $y = 1/f(x)$ given the graph of $y = f(x)$ in Figure 9.40.

(b) Find a possible formula for the function in Figure 9.40. Use this formula to check your graph for part (a).

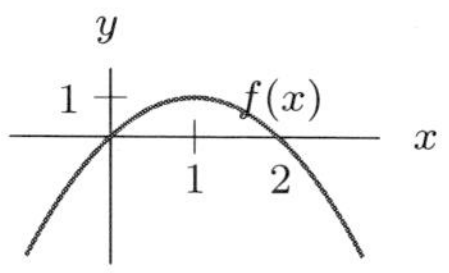

Figure 9.40

21. Use the graph of f in Figure 9.41 to graph

(a) $y = -f(-x) + 2$ (b) $y = \dfrac{1}{f(x)}$

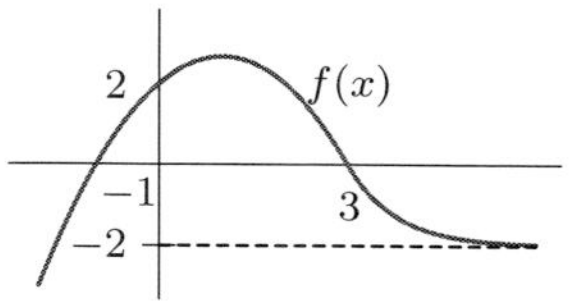

Figure 9.41

Problems 22–24 show a translation of $y = 1/x$.

(a) Find a possible formula for the graph.

(b) Write the formula from part (a) as the ratio of two linear polynomials.

(c) Find the coordinates of the intercepts of the graph.

22.

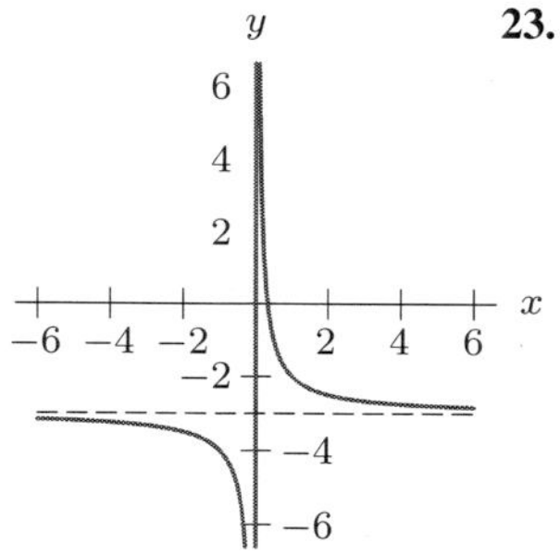

23.

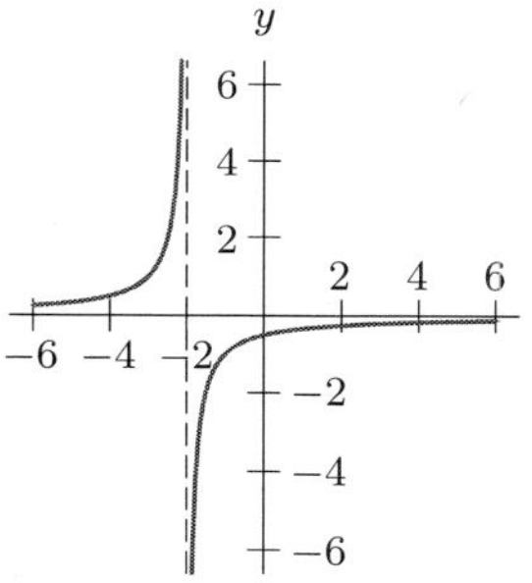

24.

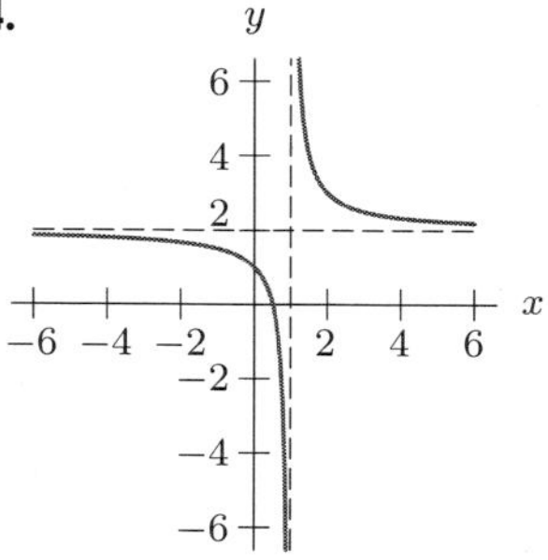

Problems 25–27 show a translation of $y = 1/x^2$.

(a) Find a formula for the graph.

(b) Write the formula from part (a) as the ratio of two polynomials.

(c) Find the coordinates of any intercepts of the graph.

25.

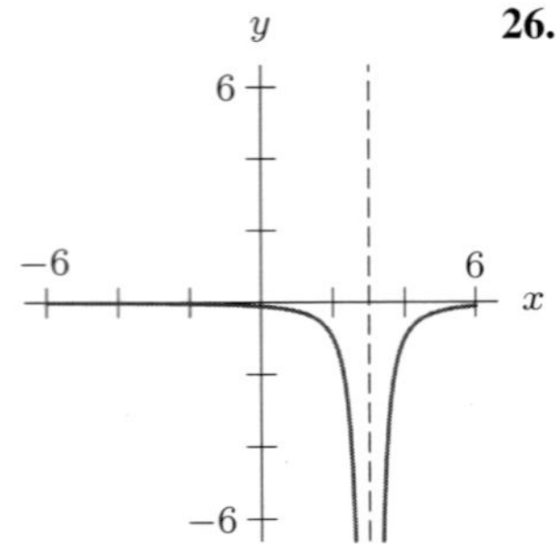

26.

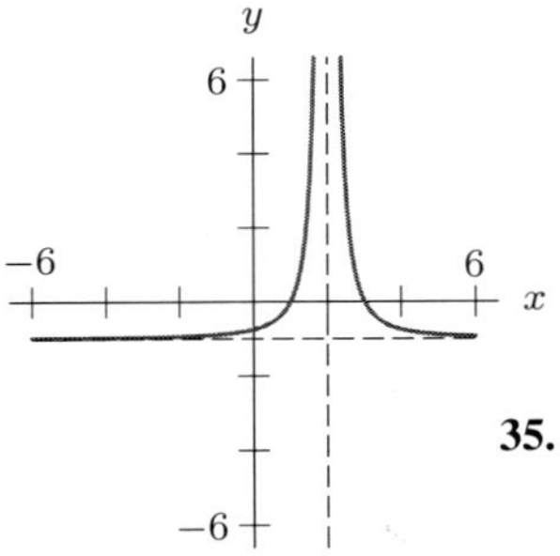

27.

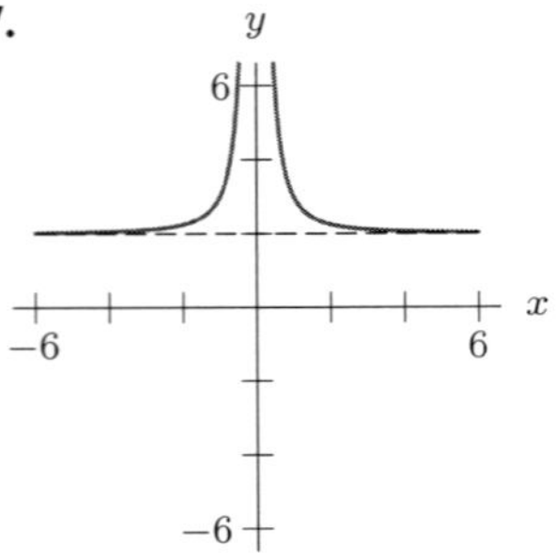

Problems 28–31 give values of translations of either $y = 1/x$ or $y = 1/x^2$. In each case

(a) Determine if the values are from a translation of $y = 1/x$ or $y = 1/x^2$. Explain your reasoning.

(b) Find a possible formula for the function.

28.

x	y
2.7	12.1
2.9	101
2.95	401
3	Undefined
3.05	401
3.1	101
3.3	12.1

29.

x	y
−1000	0.499
−100	0.490
−10	0.400
10	0.600
100	0.510
1000	0.501

30.

x	y
−1000	1.000001
−100	1.00001
−10	1.01
10	1.01
100	1.0001
1000	1.000001

31.

x	y
1.5	−1.5
1.9	−9.5
1.95	−19.5
2	Undefined
2.05	20.5
2.1	10.5
2.5	2.5

32. Cut four equal squares from the corners of a $8.5'' \times 11''$ piece of paper. Fold up the sides to create an open box. Find the dimensions of the box with the maximum volume per surface area.

Find possible formulas for the functions in Problems 33–39.

33.

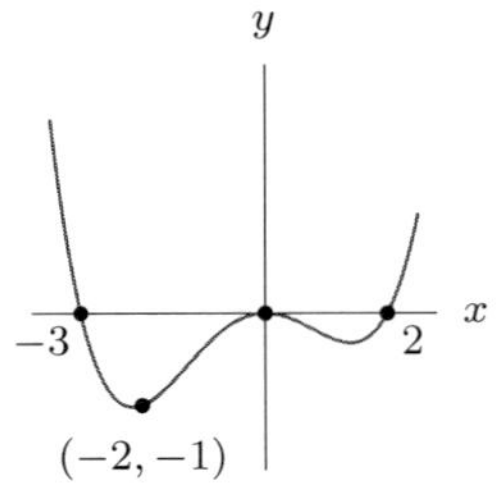

34.

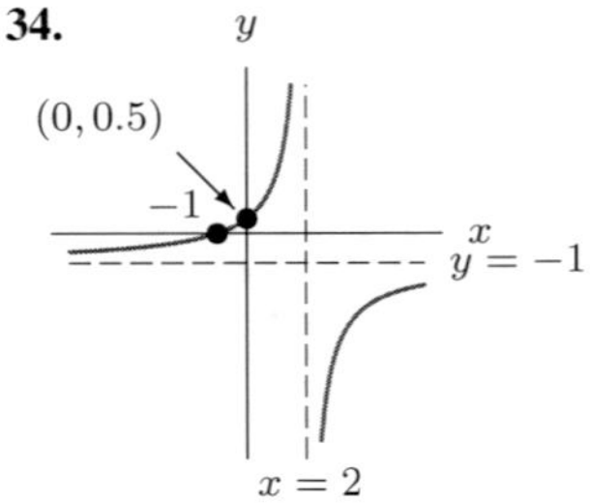

35.

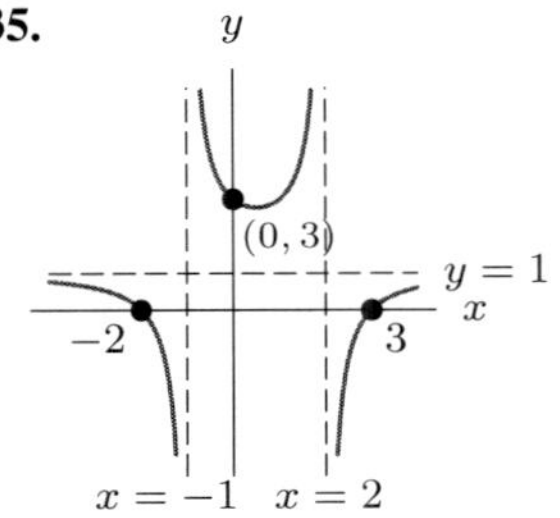

36.

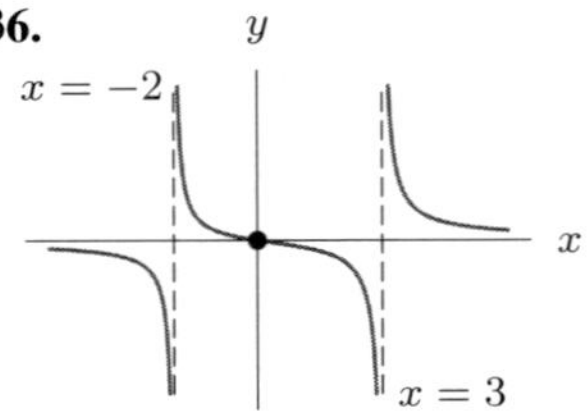

37.

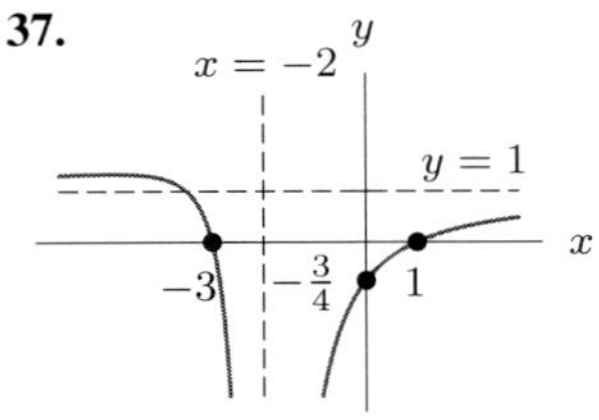

38.

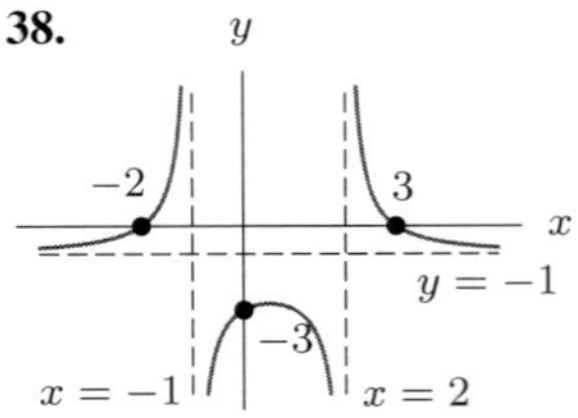

39.

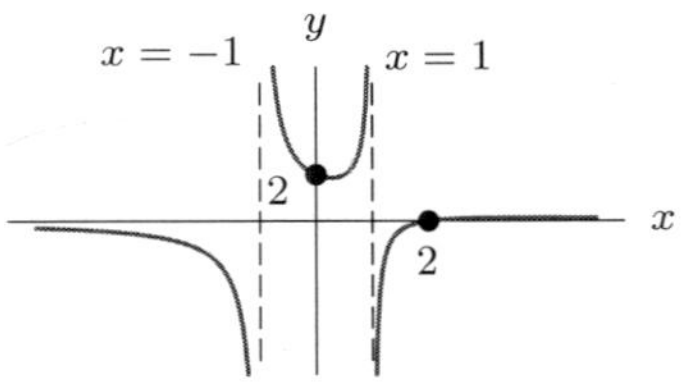

In Problems 40–42, find a possible formula for the rational functions.

40. The graph of $y = f(x)$ has one vertical asymptote, at $x = -1$, and a horizontal asymptote at $y = 1$. The graph of f crosses the y-axis at $y = 3$ and crosses the x-axis once, at $x = -3$.

41. The graph of $y = g(x)$ has two vertical asymptotes: one at $x = -2$ and one at $x = 3$. It has a horizontal asymptote of $y = 0$. The graph of g crosses the x-axis once, at $x = 5$.

42. The graph of $y = h(x)$ has two vertical asymptotes: one at $x = -2$ and one at $x = 3$. It has a horizontal asymptote of $y = 1$. The graph of h touches the x-axis once, at $x = 5$.

9.6 COMPARING POWER, EXPONENTIAL, AND LOG FUNCTIONS

In preceding chapters, we encountered exponential, power, and logarithmic functions. In this section, we compare the long and short-run behaviors of these functions.

Comparing Power Functions

For power functions $y = kx^p$ for large x, the higher the power of x, the faster the function climbs. See Figure 9.42. Not only are the higher powers larger, but they are *much* larger. This is because if $x = 100$, for example, 100^5 is one hundred times as big as 100^4, which is one hundred times as big as 100^3. As x gets larger (written as $x \to \infty$), any positive power of x grows much faster than all lower powers of x. We say that, as $x \to \infty$, higher powers of x *dominate* lower powers.

As x approaches zero (written $x \to 0$), the story is entirely different. Figure 9.43 is a close-up view near the origin. For x between 0 and 1, x^3 is bigger than x^4, which is bigger than x^5. (Try $x = 0.1$ to confirm this.) For values of x near zero, smaller powers dominate.

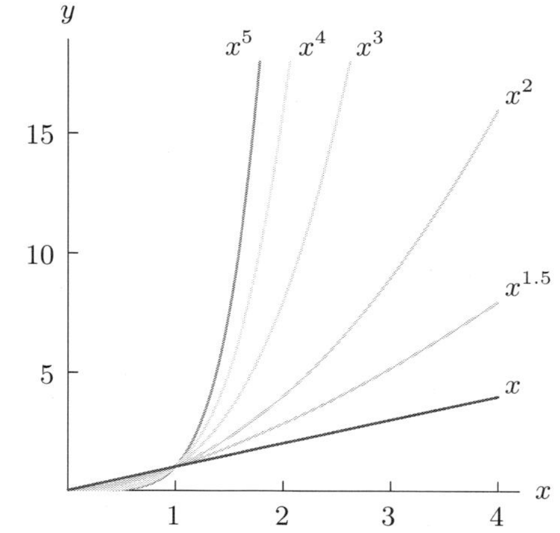

Figure 9.42: For large x: Large powers of x dominate

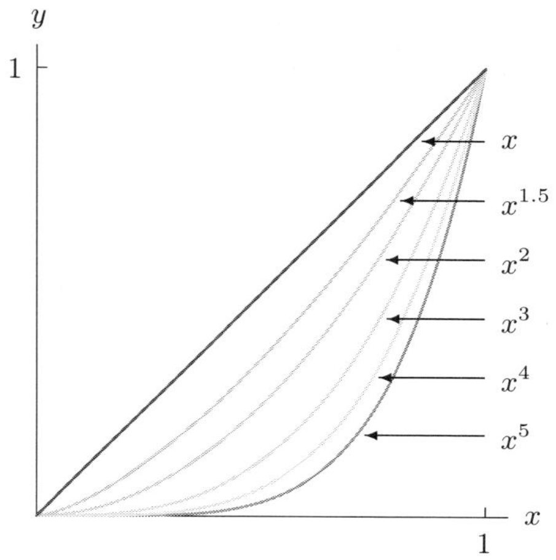

Figure 9.43: For $0 \leq x \leq 1$: Small powers of x dominate

In Chapter 5 we saw the effect of k on the graph of $f(x) = kx^p$. The coefficient k stretches or compresses the graph vertically; if k is negative, the graph is reflected across the x-axis. How does the value of k affect the long-term growth rate of $f(x) = kx^p$? Is the growth of a power function affected more by the size of the coefficient or by the size of the power?

Example 1 Let $f(x) = 100x^3$ and $g(x) = x^4$ for $x > 0$. Compare the long-term behavior of these two functions using graphs.

Solution For $x < 10$, Figure 9.44 suggests that f is growing faster than g and that f dominates g. Eventually, however, the fact that g has a higher power than f asserts itself. In Figure 9.45, we see that $g(x)$ has caught up to $f(x)$ at $x = 100$. In Figure 9.46, we see that for $x > 100$, values of g are larger than values of f.

Could the graphs of f and g intersect again for some value of $x > 100$? To show that this cannot be the case, solve the equation $g(x) = f(x)$:

$$\begin{aligned} x^4 &= 100x^3 \\ x^4 - 100x^3 &= 0 \\ x^3(x - 100) &= 0. \end{aligned}$$

Since the only solutions to this equation are $x = 0$ and $x = 100$, the graphs of f and g do not cross for $x > 100$.

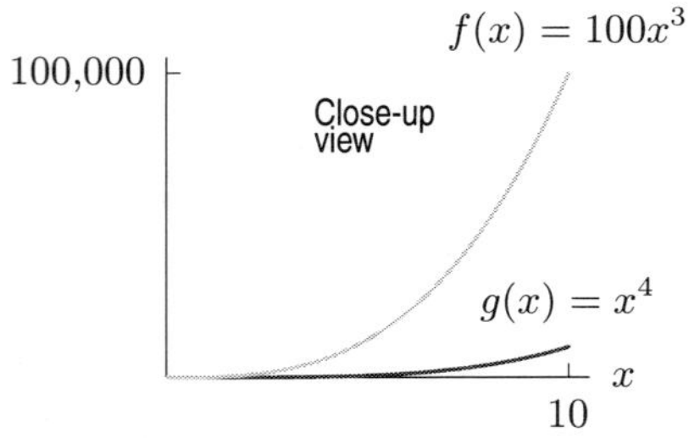

Figure 9.44: On this interval, f climbs faster than g

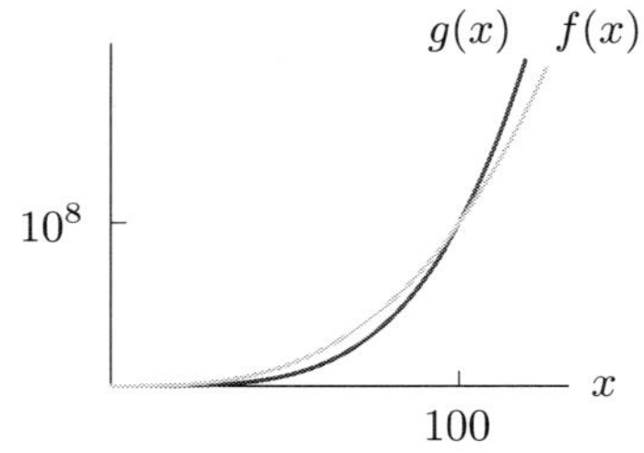

Figure 9.45: On this interval, g catches up to f

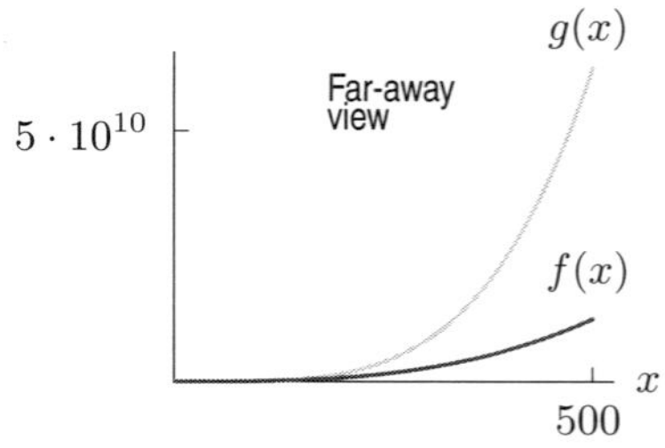

Figure 9.46: On this interval, g ends up far ahead of f

Comparing Exponential Functions and Power Functions

Both power functions and exponential functions can increase at phenomenal rates. For example, Table 9.14 shows values of $f(x) = x^4$ and $g(x) = 2^x$.

Table 9.14 *The exponential function $g(x) = 2^x$ eventually grows faster than the power function $f(x) = x^4$*

x	0	5	10	15	20
$f(x) = x^4$	0	625	10,000	50,625	160,000
$g(x) = 2^x$	1	32	1024	32,768	1,048,576

Despite the impressive growth in the value of the power function $f(x) = x^4$, in the long run $g(x) = 2^x$ grows faster. By the time $x = 20$, the value of $g(20) = 2^{20}$ is over six times as large as $f(20) = 20^4$. Figure 9.47 shows the exponential function $g(x) = 2^x$ catching up to $f(x) = x^4$.

But what about a more slowly growing exponential function? After all, $y = 2^x$ increases at a 100% growth rate. Figure 9.48 compares $y = x^4$ to the exponential function $y = 1.005^x$. Despite the fact that this exponential function creeps along at a 0.5% growth rate, at around $x = 7000$, it overtakes the power function. In summary,

> *Any* positive increasing exponential function eventually grows faster than *any* power function.

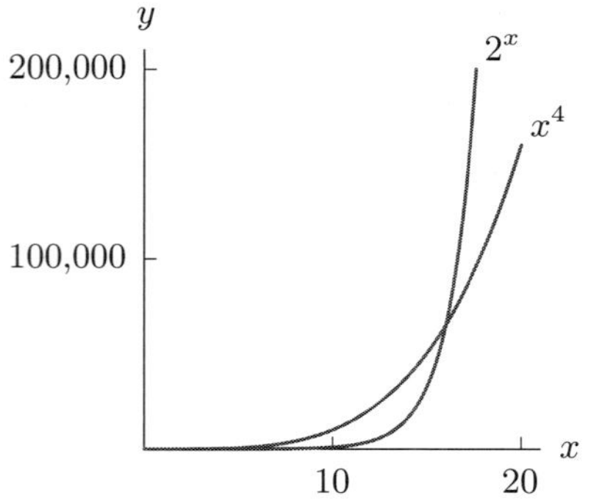

Figure 9.47: The exponential function $y = 2^x$ dominates the power function $y = x^4$

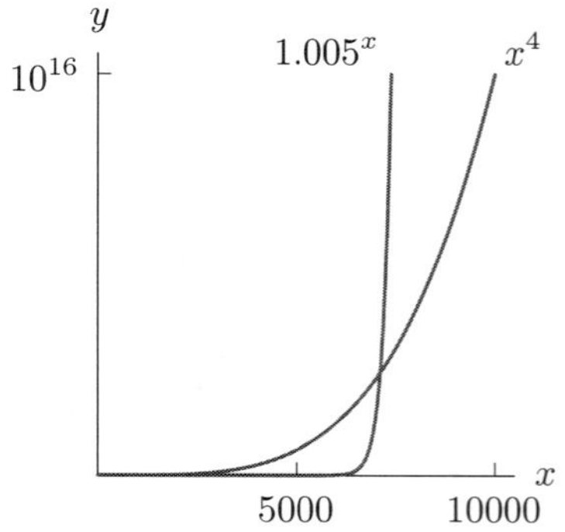

Figure 9.48: The exponential function $y = 1.005^x$ dominates the power function $y = x^4$

Decreasing Exponential Functions and Decreasing Power Functions

Just as an increasing exponential function eventually outpaces any increasing power function, an exponential decay function wins the race toward the x-axis. In general:

> *Any* positive decreasing exponential function eventually approaches the horizontal axis faster than any positive decreasing power function.

For example, let's compare the long term behavior of the decreasing exponential function $y = 0.5^x$ with the decreasing power function $y = x^{-2}$. By rewriting

$$y = 0.5^x = \left(\frac{1}{2}\right)^x = \frac{1}{2^x} \quad \text{and} \quad y = x^{-2} = \frac{1}{x^2}$$

we can see the comparison more easily. In the long run, the smallest of these two fractions is the one with the largest denominator. The fact that 2^x is eventually larger than x^2 means that $1/2^x$ is eventually smaller than $1/x^2$.

Figure 9.49 shows $y = 0.5^x$ and $y = x^{-2}$. Both graphs have the x-axis as a horizontal asymptote. As x increases, the exponential function $y = 0.5^x$ approaches the x-axis faster than the power function $y = x^{-2}$. Figure 9.50 shows what happens for large values of x. The exponential function approaches the x-axis so rapidly that it becomes invisible compared to $y = x^{-2}$.

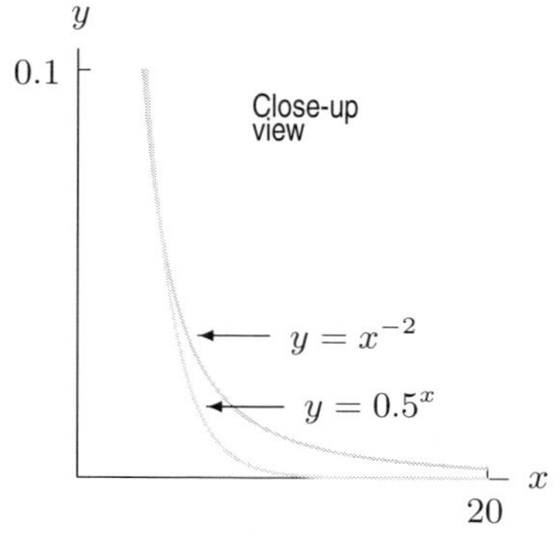

Figure 9.49: Graphs of $y = x^{-2}$ and $y = 0.5^x$

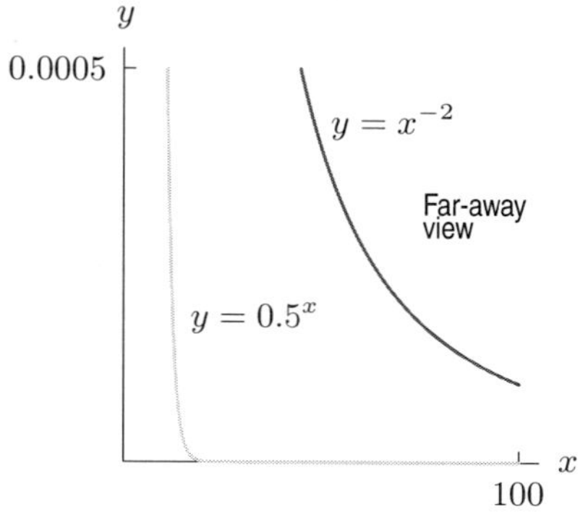

Figure 9.50: Graphs of $y = x^{-2}$ and $y = 0.5^x$

Comparing Log and Power Functions

Power functions like $y = x^{1/2}$ and $y = x^{1/3}$ grow quite slowly. However, they grow rapidly in comparison to log functions. In fact:

> *Any* positive increasing power function eventually grows more rapidly than $y = \log x$ and $y = \ln x$.

For example, Figure 9.51 shows the graphs of $y = x^{1/2}$ and $y = \log x$. The fact that exponential functions grow so fast should alert you to the fact that their inverses, the logarithms, grow very slowly. See Figure 9.52.

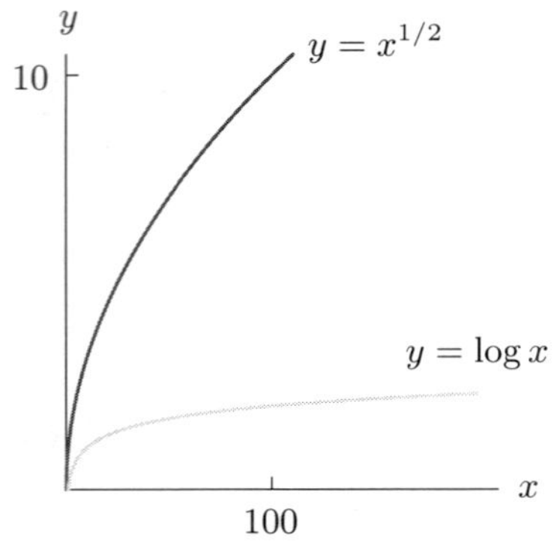

Figure 9.51: Graphs of $y = x^{1/2}$ and $y = \log x$

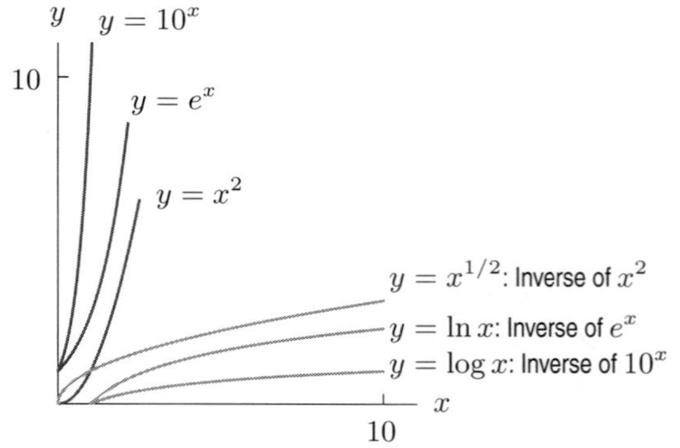

Figure 9.52: Graphs of $y = 10^x$, $y = e^x$, $y = x^2$, $y = x^{1/2}$, $y = \ln x$, and $y = \log x$

Exercises and Problems for Section 9.6

Exercises

Can the formulas in Exercises 1–6 be written in the form of an exponential function or a power function? If not, explain why the function does not fit either form.

1. $m(x) = 3(3x+1)^2$

2. $n(x) = 3 \cdot 2^{3x+1}$

3. $p(x) = (5^x)^2$

4. $q(x) = 5^{(x^2)}$

5. $r(x) = 2 \cdot 3^{-2x}$

6. $s(x) = \dfrac{4}{5x^{-3}}$

7. Without a calculator, match the following functions with the graphs in Figure 9.53.

(i) $y = x^5$ (ii) $y = x^2$ (iii) $y = x$ (iv) $y = x^3$

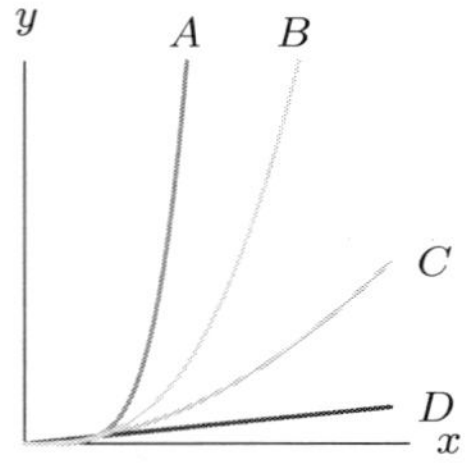

Figure 9.53

8. Without a calculator, match the following functions with the graphs in Figure 9.54.

(i) $y = x^5$ (ii) $y = x^2$ (iii) $y = x$ (iv) $y = x^3$

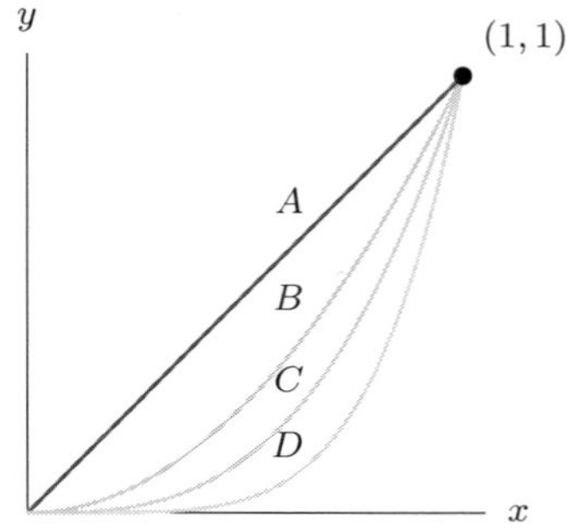

Figure 9.54

9. Let $f(x) = 3^x$ and $g(x) = x^3$.

(a) Complete the following table of values:

x	-3	-2	-1	0	1	2	3
$f(x)$							
$g(x)$							

(b) Describe the long-run behaviors of f and g as $x \to -\infty$ and as $x \to +\infty$.

In Exercises 10–13, which function dominates as $x \to \infty$?

10. $y = 50x^{1.1}$, $y = 1000x^{1.08}$

11. $y = 4e^x$, $y = 2x^{50}$

12. $y = 7(0.99)^x$, $y = 6x^{35}$

13. $y = ax^3$, $y = bx^2$, $a, b > 0$

Problems

14. The functions $y = x^{-3}$ and $y = 3^{-x}$ both approach zero as $x \to \infty$. Which function approaches zero faster? Support your conclusion numerically.

15. The functions $y = x^{-3}$ and $y = e^{-x}$ both approach zero as $x \to \infty$. Which function approaches zero faster? Support your conclusion numerically.

16. Let $f(x) = x^x$. Is f a power function, an exponential function, both, or neither? Discuss.

In Problems 17–19, find a possible formula for f if f is

(a) Linear **(b)** Exponential **(c)** Power function.

17. $f(1) = 18$ and $f(3) = 1458$

18. $f(1) = 16$ and $f(2) = 128$

19. $f(-1) = \frac{3}{4}$ and $f(2) = 48$

20. **(a)** Match the functions $f(x) = x^2, g(x) = 2x^2$, and $h(x) = x^3$ to their graphs in Figure 9.55.
(b) Do graphs A and B intersect for $x > 0$? If so, for what value(s) of x? If not, explain how you know.
(c) Do graphs C and A intersect for $x > 0$? If so, for what value(s) of x? If not, explain how you know.

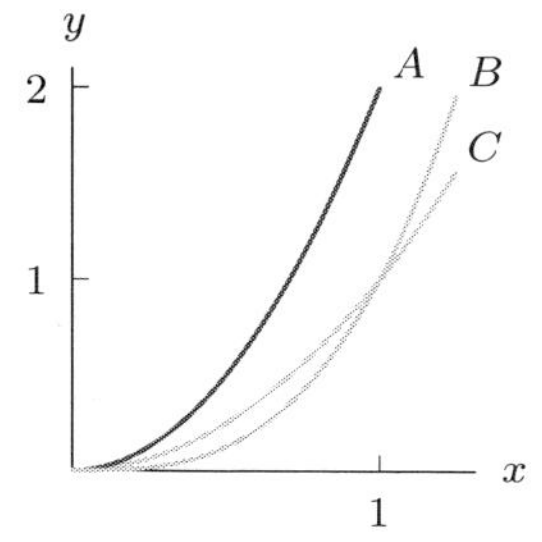

Figure 9.55

21. Match the graphs in Figure 9.56 with the functions $y = kx^{9/16}, y = kx^{3/8}, y = kx^{5/7}, y = kx^{3/11}$.

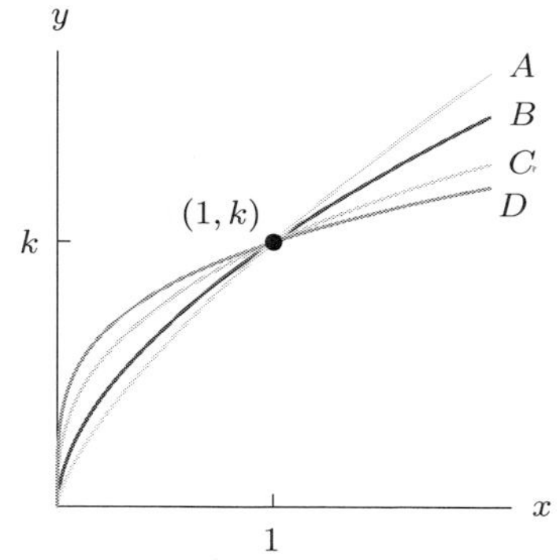

Figure 9.56

22. In Figure 9.57, find the values of m, t, and k.

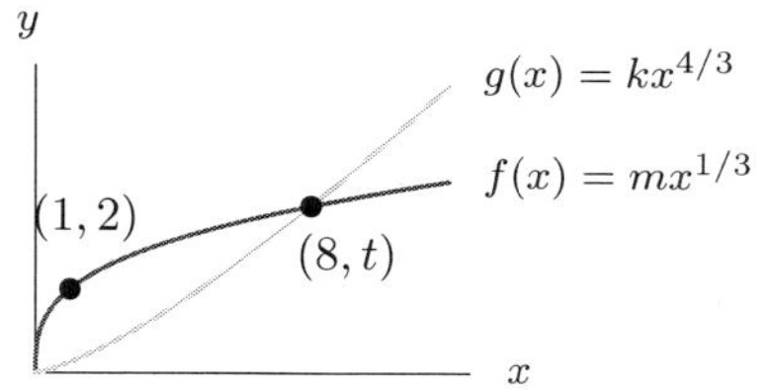

Figure 9.57

23. **(a)** Given $t(x) = x^{-2}$ and $r(x) = 40x^{-3}$, find v such that $t(v) = r(v)$.
(b) For $0 < x < v$, which is greater, $t(x)$ or $r(x)$?
(c) For $x > v$, which is greater, $t(x)$ or $r(x)$?

What is the long-run behavior of the functions in Problems 24–35?

24. $y = \dfrac{x^2+5}{x^8}$

25. $y = \dfrac{5-t^2}{(7+t+\sqrt{t})t^5}$

26. $y = \dfrac{x(x+5)(x-7)}{4+x^2}$

27. $y = \dfrac{2^x+3}{x^2+5}$

28. $y = \dfrac{2^t+7}{5^t+9}$

29. $y = \dfrac{3^{-t}}{4^t+7}$

30. $y = \dfrac{e^x+5}{x^{100}+50}$

31. $y = \dfrac{e^{2t}}{e^{3t}+5}$

32. $y = \dfrac{\ln x}{\sqrt{x}+5}$

33. $y = \dfrac{e^t+t^2}{\ln|t|}$

34. $y = \dfrac{e^x-e^{-x}}{2}$

35. $y = \dfrac{e^x-e^{-x}}{e^x+e^{-x}}$

36. Table 9.15 gives approximate values for three functions, f, g, and h. One is exponential, one is trigonometric, and one is a power function. Determine which is which and find possible formulas for each.

Table 9.15

x	-2	-1	0	1	2
$f(x)$	4	2	4	6	4
$g(x)$	20.0	2.5	0.0	-2.5	-20.0
$h(x)$	1.33	0.67	0.33	0.17	0.08

37. **(a)** The functions in Table 9.16 are of the form $y = a \cdot r^{3/4}$ and $y = b \cdot r^{5/4}$. Explain how you can tell which is which from the values in the table.
(b) Determine the constants a and b.

Table 9.16

r	2.5	3.2	3.9	4.6
$y = g(r)$	15.9	19.1	22.2	25.1
$y = h(r)$	9.4	12.8	16.4	20.2

38. A woman opens a bank account with an initial deposit of \$1000. At the end of each year thereafter, she deposits an additional \$1000.

(a) The account earns 6% annual interest, compounded annually. Complete Table 9.17.
(b) Does the balance of this account grow linearly, exponentially, or neither? Justify your answer.

Table 9.17

Years elapsed	Start-of-year balance	End-of-year deposit	End-of-year interest
0	\$1000.00	\$1000	\$60.00
1	\$2060.00	\$1000	\$123.60
2	\$3183.60	\$1000	
3		\$1000	
4		\$1000	
5		\$1000	

39. Suppose the annual percentage rate (APR) paid by the account in Problem 38 is r, where r does not necessarily equal 6%. Define $p_n(r)$ as the balance of the account after n years have elapsed. (For example, $p_2(0.06) =$ \$3183.60, because, according to Table 9.17, the balance after 2 years is \$3183.60 if the APR is 6%.)

(a) Find formulas for $p_5(r)$ and $p_{10}(r)$.
(b) What is APR if the woman in Problem 38 has \$10,000 in 5 years?

40. Values of f and g are in Table 9.18 and 9.19. One function is of the form $y = a \cdot d^{p/q}$ with $p > q$; the other is of the form $y = b \cdot d^{p/q}$ with $p < q$. Which is which? How can you tell?

Table 9.18

d	2	2.2	2.4	2.6	2.8
$f(d)$	151.6	160.5	169.1	177.4	185.5

Table 9.19

d	10	10.2	10.4	10.6	10.8
$g(d)$	7.924	8.115	8.306	8.498	8.691

9.7 FITTING EXPONENTIALS AND POLYNOMIALS TO DATA

In Section 1.6 we used linear regression to find the equation for a line of best fit for a set of data. In this section, we fit an exponential or a power function to a set of data.

The Spread of AIDS

The data in Table 9.20 give the total number of deaths in the US from AIDS from 1981 to 1996. Figure 9.58 suggests that a linear function may not give the best possible fit for these data.

Table 9.20 *Domestic deaths from AIDS, 1981–96*

t	N	t	N
1	159	9	90039
2	622	10	121577
3	2130	11	158193
4	5635	12	199287
5	12607	13	243923
6	24717	14	292586
7	41129	15	340957
8	62248	16	375904

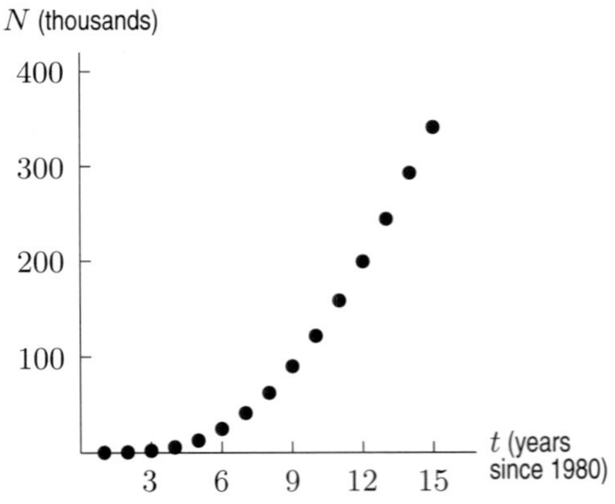

Figure 9.58: Domestic deaths from AIDS, 1981–96

Fitting an Exponential

We first fit an exponential function to the data[10] in Table 9.20

$$N = ae^{kt},$$

where N is the total number of deaths t years after 1980.

Using exponential regression on a calculator or computer, we obtain

$$N \approx 630e^{0.47t}.$$

Figure 9.59 on page 429 shows how the graph of this formula fits the data points.

Fitting a Power Function

Now we fit the AIDS data with a power function of the form

$$N = at^p,$$

where a and p are constants. Some scientists have suggested that a power function may be a better model for the growth of AIDS than an exponential function.[11] Using power function regression on a calculator or a computer, we obtain

$$N \approx 107t^{3.005}.$$

Figure 9.59 shows the graph of this power function with the data.

Which Function Best Fits the Data?

Both the exponential function

$$N = 630e^{0.47t}$$

and the power function

$$N = 107t^{3.005}$$

fit the AIDS data reasonably well. By visual inspection alone, the power function arguably provides the better fit. If we fit a linear function to the original data we get

$$N = -97311 + 25946t.$$

Even this linear function gives a possible fit for $t \geq 4$, that is, for 1984 to 1996. (See Figure 9.59.)

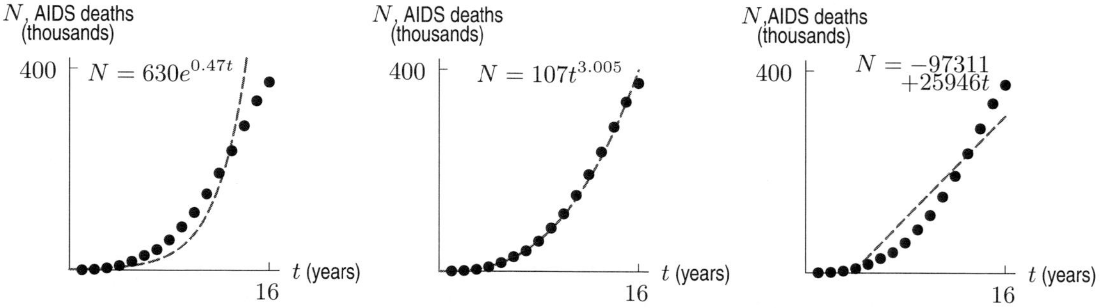

Figure 9.59: The AIDS data since 1980 together with an exponential model, a power-function model, and a linear model

[10] *HIV/AIDS Surveillance Report*, Year-end Edition, Vol 9, No 2, Table 13, US Department of Health and Human Services, Centers for Disease Control and Prevention, Atlanta. 2000–2004 data from *HIV/AIDS Surveillance Report*, Vol 16, at www.cdc.gov/hiv/stats/hastlink.htm, accessed January 15, 2006. Data does not include 450 people whose dates of death are unknown.

[11] " Risk behavior-based model of the cubic growth of acquired immunodeficiency syndrome in the United States", by Stirling A. Colgate, E. Ann Stanley, James M. Hyman, Scott P. Layne, and Alifford Qualls, in *Proc. Natl. Acad. Sci. USA*, Vol 86, June 1989, Population Biology.

Despite the fact that all three functions fit the data reasonably well up to 1996, it's important to realize that they give wildly different predictions for the future. If we use each model to estimate the total number of AIDS deaths by the year 2010 (when $t = 30$), the exponential model gives

$$N = 630e^{(0.47)30} \approx 837{,}322{,}467, \quad \text{about triple the current US population;}$$

the power model gives

$$N = 107(30)^{3.005} \approx 2{,}938{,}550, \quad \text{or about } 1\% \text{ of the current population;}$$

and the linear model gives

$$N = -97311 + 25946 \cdot 30 \approx 681{,}069, \quad \text{or about } 0.25\% \text{ of the current population.}$$

Which function is the best predictor of the future? To explore this question, let us add some more recent data to our previous data on AIDS deaths. See Table 9.21.

Table 9.21
Domestic deaths from AIDS, 1997–2004

t	N
17	406,444
18	424,841
19	442,013
20	457,258
21	462,653
22	501,669
23	524,060
24	529,113

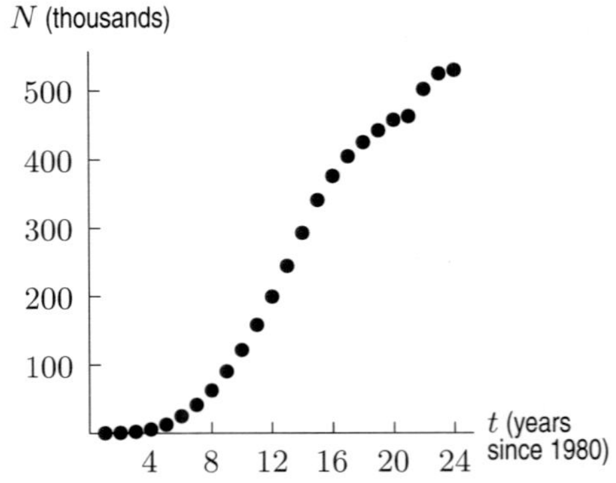

Figure 9.60: Domestic deaths from AIDS, 1981–2004

When data from the entire period from 1981 to 2004 are plotted together (see Figure 9.60), we see that the rate of increase of AIDS deaths reaches a peak sometime around 1995 and then begins to taper off. Since none of the three types of functions we have used to model AIDS deaths exhibit this type of behavior, some other type of function is needed to describe the number of AIDS deaths accurately over the entire 23-year period.

This example illustrates that while a certain type of function may fit a set of data over a short period of time, care must be taken when using a mathematical model to make predictions about the future. An understanding of the processes leading to the data is crucial in answering any long-term question.

Exercises and Problems for Section 9.7

Exercises

1. Find a formula for the power function $f(x)$ such that $f(1) = 1$ and $f(2) = c$.

2. Find a formula for the power function $g(x)$.

x	2	3	4	5
$g(x)$	4.5948	7.4744	10.5561	13.7973

3. Find a formula for an exponential function $h(x)$.

x	2	3	4	5
$h(x)$	4.5948	7.4744	10.5561	13.7973

4. Anthropologists suggest that the relationship between the body weight and brain weight of primates can be modeled with a power function. Table 9.22 lists various body weights and the corresponding brain weights of different primates.[12]

(a) Using Table 9.22, find a power function that gives the brain weight, Q (in mg), as a function of the body weight, b (in gm).

(b) The Erythrocebus (Patas monkey) has a body weight of 7800 gm. Estimate its brain weight.

Table 9.22

b	6667	960	6800	9500	1088
Q	56,567	18,200	110,525	120,100	20,700
b	2733	3000	6300	1500	665
Q	78,250	58,200	96,400	31,700	25,050

5. Students in the School of Forestry & Environmental Studies at Yale University collected data measuring Sassafras trees. Table 9.23 lists the diameter at breast height (dbh, in cm) and the total dry weight (w, in gm) of different trees.[13]

(a) Find a power function that fits the data.

(b) Predict the total weight of a tree with a dbh of 20 cm.

(c) If a tree has a total dry weight of 100,000 gm, what is its expected dbh?

Table 9.23

dbh	5	23.4	11.8	16.7	4.2	5.6
w	5,353	169,290	30,696	76,730	3,436	5,636
dbh	3.8	4.3	6.5	21.9	17.7	25.5
w	14,983	2,098	7,364	177,596	100,848	171,598

6. According to the National Marine Fisheries Service, the Maine lobster catch (in millions of pounds) has greatly increased in the past 30 years.[14] See Table 9.24. With t in years since 1965, use a calculator or computer to fit the data with

(a) A power function of the form $y = at^b$.

(b) A quadratic function of the form $y = at^2 + bt + c$.

(c) Discuss which is a better fit and why.

Table 9.24

Year	1970	1975	1980	1985	1990	1995	2000
t	5	10	15	20	25	30	35
Lobster	17	19	22	20	27	36	56

In Exercises 7–12, find an equation for y in terms of x.

7.

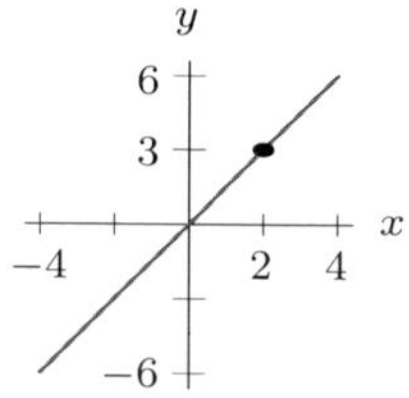

8.

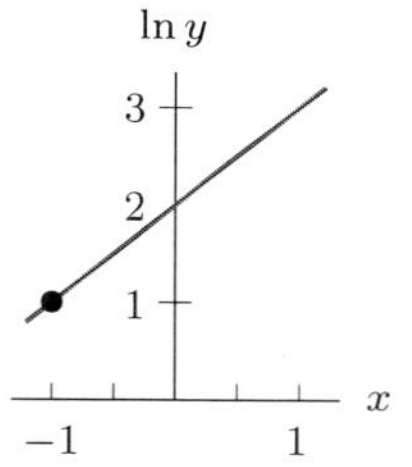

9.

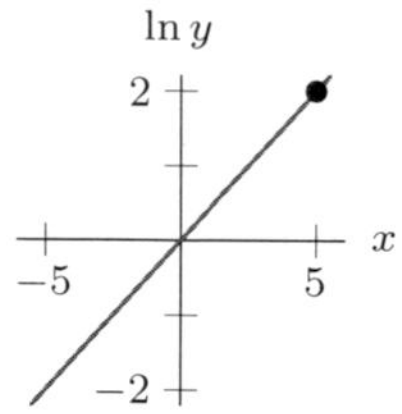

10.

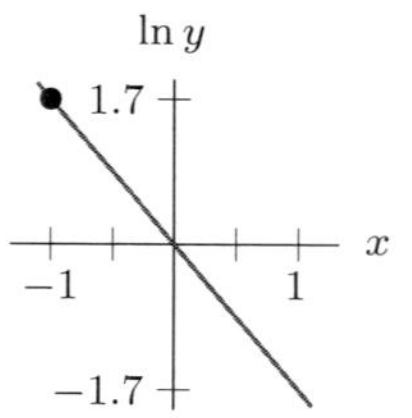

11.

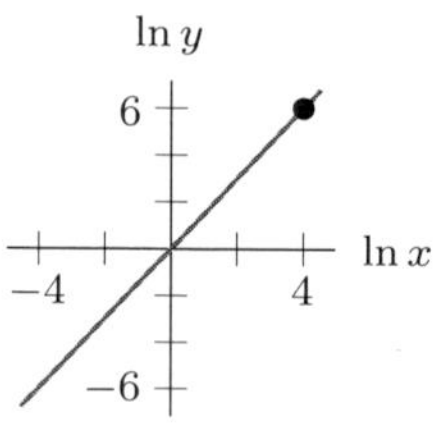

12.

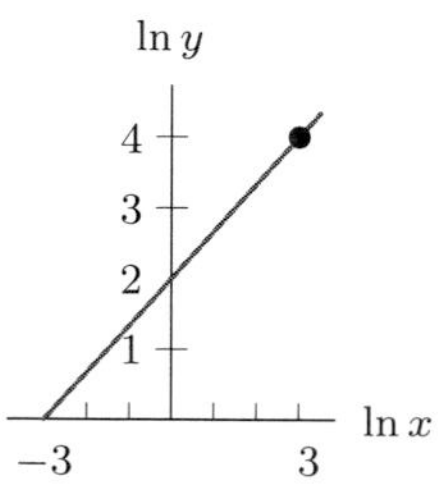

[12]http://mac-huwis.lut.ac.uk/ wis/lectures/primate-adaptation/10PrimateBrains.pdf (December 15, 2002).
[13]www.yale.edu/fes519b/totoket/allom/allom.htm (December 15, 2002).
[14]Adapted from *The New York Times*, p.16, May 31, 2001.

Problems

13. An analog radio dial can be measured in millimeters from left to right. Although the scale of the dial can be different from radio to radio, Table 9.25 gives typical measurements.

(a) Which radio band data appear linear? Graph and connect the data points for each band.
(b) Which radio band data appear exponential?
(c) Find a possible formula for the FM station number in terms of x.
(d) Find a possible formula for the AM station number in terms of x.

Table 9.25

x, millimeters	5	15	25	35	45	55
FM (mhz)	88	92	96	100	104	108
AM (khz/10)	53	65	80	100	130	160

14. (a) Find a linear function that fits the data in Table 9.26. How good is the fit?
(b) The data in the table was generated using the power function $y = 5x^3$. Explain why (in this case) a linear function gives such a good fit to a power function. Does the fit remain good for other values of x?

Table 9.26

x	2.00	2.01	2.02	2.03	2.04	2.05
y	40.000	40.603	41.212	41.827	42.448	43.076

15. In this problem you will fit a quartic polynomial to the AIDS data.

(a) With N as the total number of AIDS deaths t years after 1980, use a calculator or computer to fit the data in Table 9.20 on page 428 with a polynomial of the form

$$N = at^4 + bt^3 + ct^2 + dt + e.$$

(b) Graph the data and your quartic for $0 \leq t \leq 16$. Comment on the fit.
(c) Graph the data and your quartic for $0 \leq t \leq 30$. Comment on the predictions made by this model.

16. The managers of a furniture store have compiled data showing the daily demand for recliners at various prices.

(a) In Table 9.27, fill in the revenue generated by selling the number of recliners at the corresponding price.
(b) Find the quadratic function that best fits the data.
(c) According to the function you found, what price should the store charge for their recliners to maximize revenue? What is the maximum revenue?

Table 9.27

Recliner price (\$)	399	499	599	699	799
Demand (recliners)	62	55	47	40	34
Revenue (\$)					

17. Cellular telephone use has increased over the past two decades. Table 9.28 gives the number of cellular telephone subscriptions, in thousands, from 1985 to 2004.[15]

(a) Fit an exponential function to this data with time in years since 1985.
(b) Based on your model, by what percent was the number of cell subscribers increasing each year?
(c) In the long-run, what do you expect of the rate of growth? What does this mean in terms of the shape of the graph?

Table 9.28

Year	1985	1990	1995	2000	2004
Subscriptions	340	5283	33786	109478	182140

18. The use of one-way pagers declined as cell phones became more popular.[16] The number of users is given in Table 9.29 and plotted in Figure 9.61, along with a quadratic regression function.

(a) How well does the graph of the quadratic function fit the data?
(b) Find a cubic regression function. Does it fit better?

Table 9.29

Year	1990	1991	1992	1993	1994	1995
Users, millions	10	12	15	19	25	32
Year	1996	1997	1998	1999	2000	
Users, millions	38	43	44	43	37	

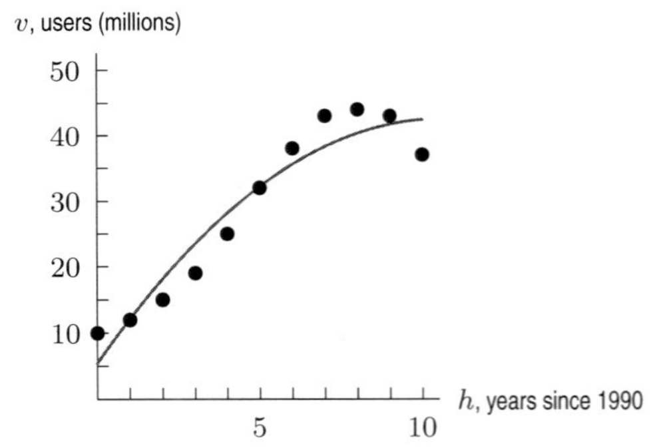

Figure 9.61

[15]World Almanac and Book of Fact, 2006. pg.380
[16]*The New York Times*, p.16, April 11, 2002.

19. Table 9.30 gives the estimated population, in thousands, of the American Colonies from 1650 to 1770.[17]

(a) Make a scatterplot of the data using $t = 0$ to represent the year 1650.
(b) Fit an exponential function to the data.
(c) Explain the meaning of the parameters in your model.
(d) Use your function to predict the population in 1750. Is it high or low?
(e) According to the US Census Bureau, the US population in 1800 was 5,308,483.[18] Use your function to make a prediction for 1800. Is it high or low?

Table 9.30

Year	1650	1670	1690	1700
Population	50.4	111.9	210.4	250.9
Year	1720	1740	1750	1770
Population	466.2	905.6	1170.8	2148.1

20. The US Census Bureau began recording census data in 1790. Table 9.31 gives the population of the US in millions from 1790 to 1860.[19]

(a) With t = 0 representing the number of years since 1790, fit an exponential function to the data.
(b) The 1800 census value is 5.3 million. Find the population predicted by your function for 1800. Problem 19 gave a prediction of 5.5 million using the data for 1650 to 1750. Using the two data sets, explain the difference in predicted values.
(c) Use your function to project the population of the US in 2000. Is this prediction reasonable?

Table 9.31

Year	1790	1800	1810	1820
Population	3.929	5.308	7.240	9.638
Year	1830	1840	1850	1860
Population	12.861	17.063	23.192	31.443

21. Table 9.32 gives N, the number of transistors per integrated circuit chip, t years after 1970.[20]

(a) Plot N versus t and fit an exponential curve to the data.
(b) According to the formula of your curve of best fit, approximately how often does the number of transistors double?

Table 9.32

Chip name	t	N
4004	1	2,300
8008	2	2,500
8080	4	4,500
8086	8	29,000
Intel286	12	134,000
Intel386	15	275,000
Intel486	19	1,200,000
Pentium	23	3,100,000
Pentium II	27	7,500,000
Pentium III	29	9,500,000
Pentium 4	30	42,000,000
Itanium	31	25,000,000
Itanium 2	33	220,000,000
Itanium 2 (9MB cache)	34	592,000,000
Dual Core Itanium	36	1,720,000,000

22. The US export of edible fishery produce, in thousands of metric tons, is shown in Table 9.33.[21] With t in years since 1935, fit the data with a function of the form

(a) $y = at^b$ (b) $y = ab^t$ (c) $y = at^2+bt+c$

(d) Discuss the reliability for estimating 2010 exports with each function.

Table 9.33

Year	1940	1945	1950	1955	1960	1965	1970
Fish export	66	62	55	50	31	50	73
Year	1975	1980	1985	1990	1995	2000	2005
Fish export	109	275	305	883	929	982	1329

23. The data in Table 9.33 show a big jump in fish exports between 1985 and 1990. This suggests fitting a piecewise defined function. With t in years since 1935, fit a quadratic function to the data from

(a) 1940 to 1985 (b) 1990 to 2005

(c) Write a piecewise defined function using parts (a) and (b). Graph the function and the data.

[17] *The World Almanac and Book of Facts, 2002*, New York, NY, p. 376.
[18] http://www.census.gov/, January 15, 2003.
[19] http://www.census.gov/, January 15, 2003.
[20] The Intel Corporation, http://www.intel.com/museum/archives/history_docs/mooreslaw.htm.
[21] www.st.nmfs.gov/st1/trade/trade2001.pdf, accessed December 15, 2002 and www.st.nmfs.gov/st1/trade/documents/TRADE2005.pdf, accessed July 25, 2006.

24. **(a)** Using the data in Table 9.20 on page 428, plot $\ln N$ against t. If the original data were exponential, the points would lie on a line.
(b) Fit a line to the graph from part (a).
(c) From the equation of the line, obtain the formula for N as an exponential function of t.

25. **(a)** Let $N = at^p$, with a, p constant. Explain why if you plot $\ln N$ against $\ln t$, you get a line.
(b) To decide if a function of the form $N = at^p$ fits some data, you plot $\ln N$ against $\ln t$. Explain why this plot is useful.

26. **(a)** Using the data in Table 9.20 on page 428, plot $\ln N$ against $\ln t$. If a power function fitted the original data, the points would lie on a line
(b) Fit a line to the graph from part (a).
(c) From the equation of the line, obtain the formula for N as a power function of t.

27. According to the US Census Bureau, the 2004 mean income by age is given in Table 9.34.[22]

(a) Choose the best type of function to fit the data: linear, exponential, power, or quadratic.
(b) Using a mid-range age value for each interval, find an equation to fit the data.
(c) Interpolation estimates incomes for ages within the range of the data. Predict the income of a 37-year old.
(d) Extrapolation estimates incomes outside the range of data. Use your function to predict the income of a 10-year old. Is it reasonable?

Table 9.34

Age	Mean income, dollars
15 to 24	12,789
25 to 34	31,987
35 to 44	42,582
45 to 54	45,179
55 to 64	40,788
65 to 74	27,579
75+ years	19,955

28. German physicist Arnd Leike of the University of Munich won the 2002 Ig Nobel prize in Physics for experiments with beer foam conducted with his students.[23]

The data in Table 9.35 give the height (in cm) of beer foam after t seconds for three different types of beer, Erdinger Weissbier, Augustinerbräu München, and Budweiser Budvar. The heights are denoted h_e, h_a, and h_b, respectively.

(a) Plot these points and fit exponential functions to them. Give the equations in the form $h = h_0 e^{-t/\tau}$.
(b) What does the value of h_0 tell you for each type of beer? What does the value of τ tell you for each type of beer?

Table 9.35

t	h_e	h_a	h_b	t	h_e	h_a	h_b
0	17.0	14.0	14.0	120	10.7	6.0	7.0
15	16.1	11.8	12.1	150	9.7	5.3	6.2
30	14.9	10.5	10.9	180	8.9	4.4	5.5
45	14.0	9.3	10.0	210	8.3	3.5	4.5
60	13.2	8.5	9.3	240	7.5	2.9	3.5
75	12.5	7.7	8.6	300	6.3	1.3	2.0
90	11.9	7.1	8.0	360	5.2	0.7	0.9
105	11.2	6.5	7.5				

29. Table 9.36 gives the development time t (in days) for eggs of the pea weevil (*Bruchus pisorum*) at temperature H (°C).[24]

(a) Plot these data and fit a power function.
(b) Ecologists define the development rate $r = 1/t$ where t is the development time. Plot r against H, and fit a linear function.
(c) At a certain temperature, the value of r drops to 0 and pea weevil eggs will not develop. What is this temperature according to the model from part (a)? part (b)? Which model's prediction do you think is more reasonable?

Table 9.36

H, °C	10.7	14.4	16.2	18.1	21.4	23.7	24.7	26.9
t, days	38.0	19.5	15.6	9.6	9.5	7.3	4.5	4.5

[22]www.census.gov/hhes/www/income/histinc/p10ar.html, accessed July 25, 2006.
[23]http://ignobel.com/ig/ig-pastwinners.html. The Ig Nobel prize is a spoof of the Nobel prize and honors researchers whose achievements "cannot or should not be reproduced." The data here is taken from *Demonstration of the Exponential Decay Law Using Beer Froth*, Arnd Leike, European Journal of Physics, vol. 23, January 2002, pp. 21-26.
[24]From website created by A. Sharov, http://www.ento.vt.edu/ sharov/PopEcol/lec8/quest8.html. The site attributes the data to Smith, A. M., 1992, Environ. Entomol. 21:314-321.

30. In this problem, we will determine whether or not the compact disc data from Table 4.7 on page 180 can be well modeled using a power function of the form $l = kc^p$, where l and c give the number of LPs and CDs (in millions) respectively, and where k and p are constant.

(a) Based on the plot of the data in Figure 4.23 on page 180, what do you expect to be true about the sign of the power p?

(b) Fit a power function to the data. One data point may have to be omitted. Which point and why?

(c) Let $y = \ln l$ and $x = \ln c$. Find a linear formula for y in terms of x by making substitutions in the equation $l = kc^p$.

(d) Transform the data in Table 4.7 to create a table comparing $x = \ln c$ and $y = \ln l$. What data point must be omitted?

(e) Plot your transformed data from part (d). Based on your plot, do you think a power function gives a good fit to the data? Explain.

CHAPTER SUMMARY

- **Proportionality**
  Direct and indirect.
- **Power Functions**
  $y = kx^p$.
- **Polynomials**
  General formula: $p(x) = a_n x^n + a_{n-1}x^{n-1} + \cdots + a_1 x + a_0$.
  All terms have non-negative, integer exponents. Leading term $a_n x^n$; coefficients $a_0, \ldots, a_n$.
  Long-run behavior: Like $y = a_n x^n$.
  Short-run behavior: Zeros corresponding to each factor; multiple zeros.
- **Rational Functions**
  Ratio of polynomials: $r(x) = \dfrac{p(x)}{q(x)}$.
  Long-run behavior: Horizontal asymptote of $r(x)$:
  Given by ratio of highest-degree terms.
  Short-run behavior: Vertical asymptote of $r(x)$:
  At zeros of $q(x)$ (if $p(x) \neq 0$).
  Short-run behavior: Zeros of $r(x)$:
  At zeros of $p(x)$ (if $q(x) \neq 0$).
  Using limits to understand short- and long-run behavior
- **Comparing Functions**
  Exponential functions eventually dominate power functions. Power functions eventually dominate logs.
- **Fitting Exponentials and Polynomials to Data**

REVIEW EXERCISES AND PROBLEMS FOR CHAPTER NINE

Exercises

In Exercises 1–6, is y a power function of x? If so, write it in the form $y = kx^p$.

1. $y = 6x^3 + 2$

2. $3y = 9x^2$

3. $y - 9 = (x+3)(x-3)$

4. $y = 4(x-2)(x+2)+16$

5. $y = 4(x+7)^2$

6. $y - 1 = 2x^2 - 1$

Do the power functions in Exercises 7–8 appear to have odd, even, or fractional powers?

7.

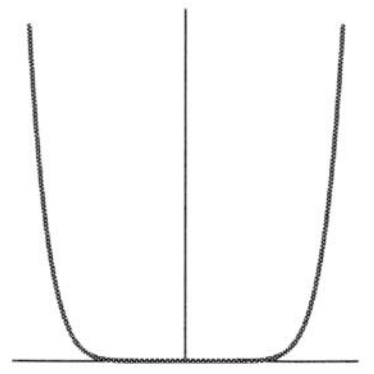

8.

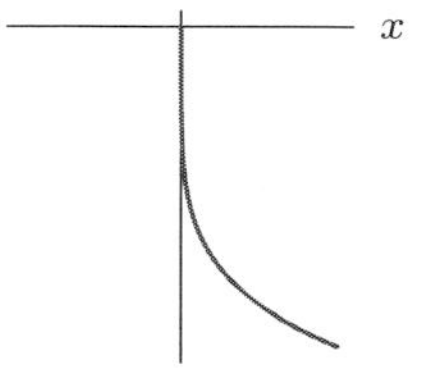

In Exercises 9–10, find possible formulas for the power functions with the properties given.

9. $f(1) = \frac{3}{2}$ and $f(2) = \frac{3}{8}$

10. $g\left(-\frac{1}{5}\right) = 25$ and $g(2) = -\frac{1}{40}$

11. Show that the function $u(x) = x(x-3)(x+2)$ is a polynomial. What is its degree?

For the polynomials in Exercises 12–14, state the degree, the number of terms, and describe the long-run behavior.

12. $y = 2x^3 - 3x + 7$

13. $y = 1 - 2x^4 + x^3$

14. $y = (x+4)(2x-3)(5-x)$

In Exercises 15–16, find the zeros of the functions.

15. $y = (x^2 - 8x + 12)(x - 3)$

16. $y = ax^2(x^2+4)(x+3)$, where a is a nonzero constant.

Are the functions in Exercises 17–18 rational functions? If so, write them in the form $p(x)/q(x)$, the ratio of polynomials.

17. $f(x) = \dfrac{x^3}{2x^2} + \dfrac{1}{6}$

18. $f(x) = \dfrac{x^4 + 3^x - x^2}{x^3 - 2}$

In Exercises 19–20, which function dominates as $x \to \infty$?

19. $y = 12x^3, \quad y = 7/x^{-4}$

20. $y = 4/e^{-x}, \quad y = 17x^{43}$

Evaluate the limits in Exercises 21–24.

21. $\lim\limits_{x\to\infty} (2x^{-3} + 4)$

22. $\lim\limits_{x\to\infty} (3x^{-2} + 5x + 7)$

23. $\lim\limits_{x\to\infty} \dfrac{4x + 3x^2}{4x^2 + 3x}$

24. $\lim\limits_{x\to-\infty} \dfrac{3x^2 + x}{2x^2 + 5x^3}$

Problems

In Problems 25–28, find a viewing window on which the graph of $f(x) = x^3 + x^2$ resembles the plot.

25.

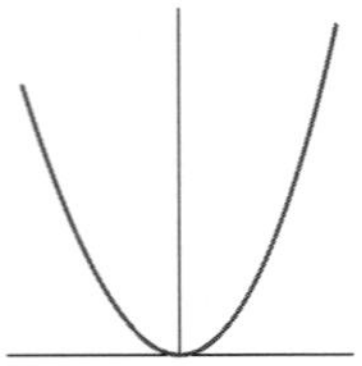

26.

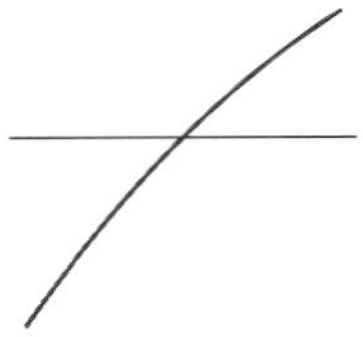

27.

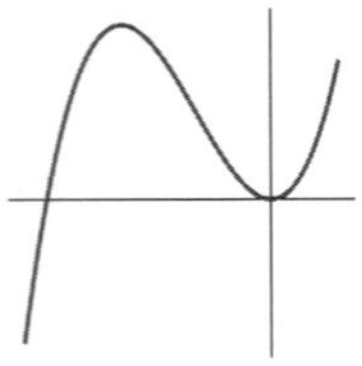

28.

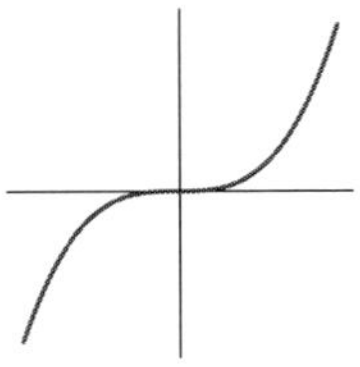

29. Without a calculator, match graphs (i)–(iv) with the functions in Table 9.37.

(i)

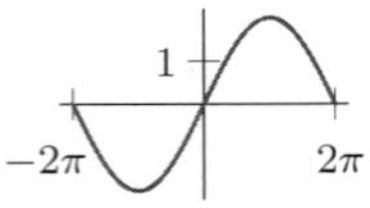

(ii)

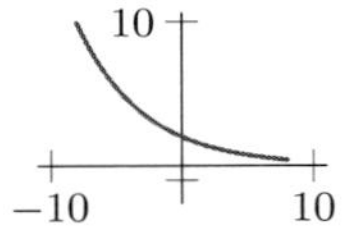

(iii)

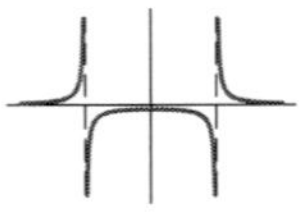

(iv)

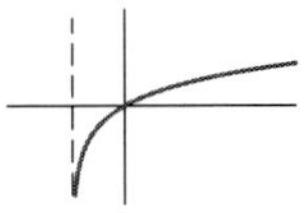

Table 9.37

(A) $y = 0.5\sin(2x)$	(J) $y = 2\sin(0.5x)$
(B) $y = -\ln x$	(K) $y = \ln(x-1)$
(C) $y = 10(0.6)^x$	(L) $y = 2e^{-0.2x}$
(D) $y = 2\sin(2x)$	(M) $y = 1/(x-6)$
(E) $y = \ln(-x)$	(N) $y = (x-2)/(x^2-9)$
(F) $y = -15(3.1)^x$	(O) $y = 1/(x^2-4)$
(G) $y = 0.5\sin(0.5x)$	(P) $y = x/(x-3)$
(H) $y = \ln(x+1)$	(Q) $y = (x-1)/(x+3)$
(I) $y = 7(2.5)^x$	(R) $y = 1/(x^2+4)$

30. Without a calculator, match each graph (i)–(viii) with a functions in Table 9.38.

(i)

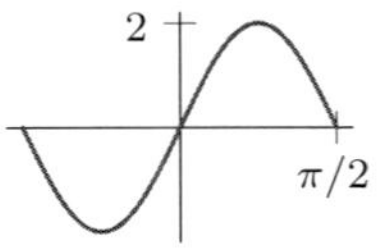

(ii)

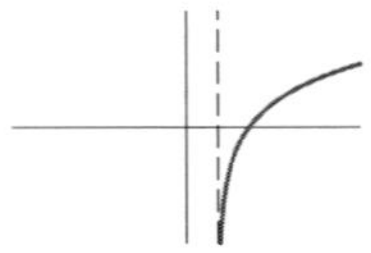

(iii)

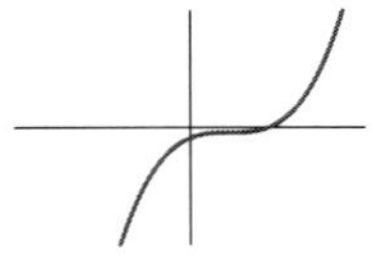

(iv)

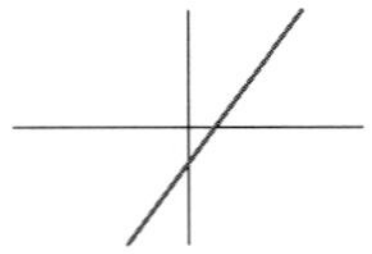

(v)

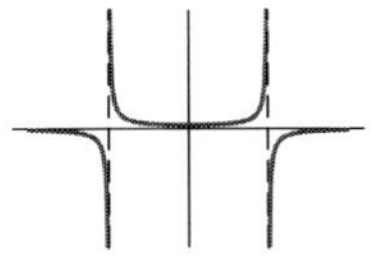

(vi)

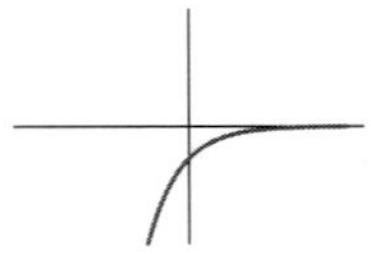

(vii)

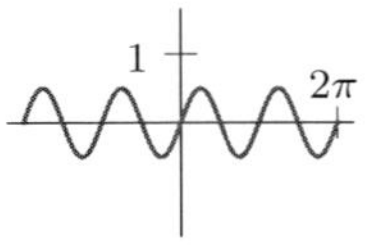

(viii)

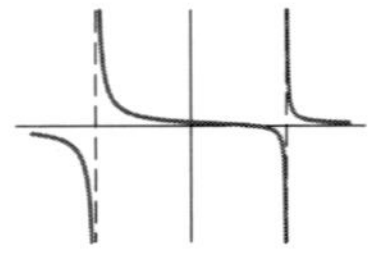

Table 9.38

(A) $y = 0.5\sin(2x)$	(M) $y = (x+3)/(x^2-4)$
(B) $y = 2\sin(2x)$	(N) $y = (x^2-4)/(x^2-1)$
(C) $y = 0.5\sin(0.5x)$	(O) $y = (x+1)^3 - 1$
(D) $y = 2\sin(0.5x)$	(P) $y = -2x - 4$
(E) $y = (x-2)/(x^2-9)$	(Q) $y = 3e^{-x}$
(F) $y = (x-3)/(x^2-1)$	(R) $y = -3e^{x}$
(G) $y = (x-1)^3 - 1$	(S) $y = -3e^{-x}$
(H) $y = 2x - 4$	(T) $y = 3e^{-x^2}$
(I) $y = -\ln x$	(U) $y = 1/(4-x^2)$
(J) $y = \ln(-x)$	(V) $y = 1/(x^2+4)$
(K) $y = \ln(x+1)$	(W) $y = (x+1)^3 + 1$
(L) $y = \ln(x-1)$	(X) $y = 2(x+2)$

Find possible polynomial formulas in Problems 31–36.

31.

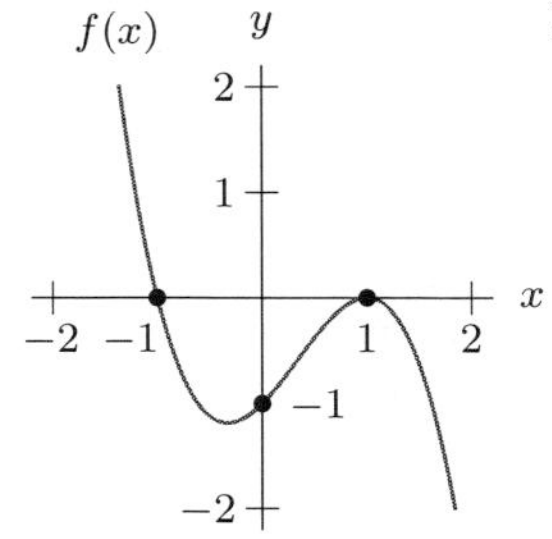

32.

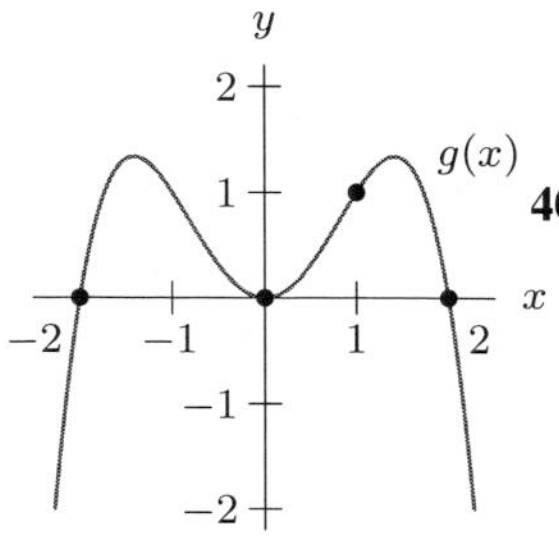

33.

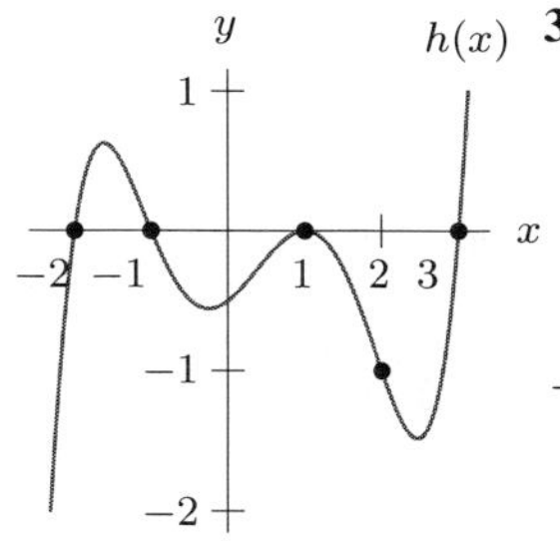

34.

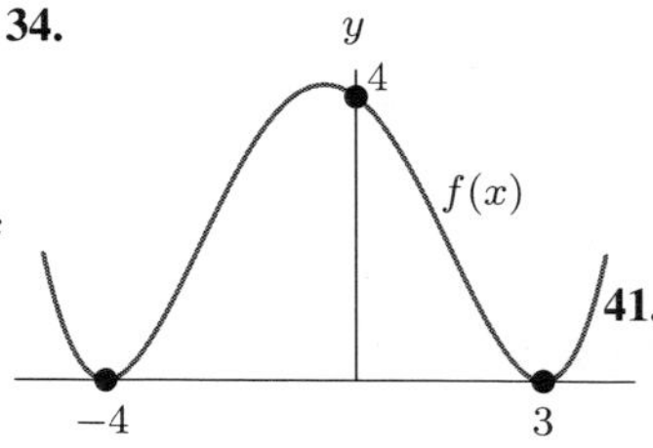

35.

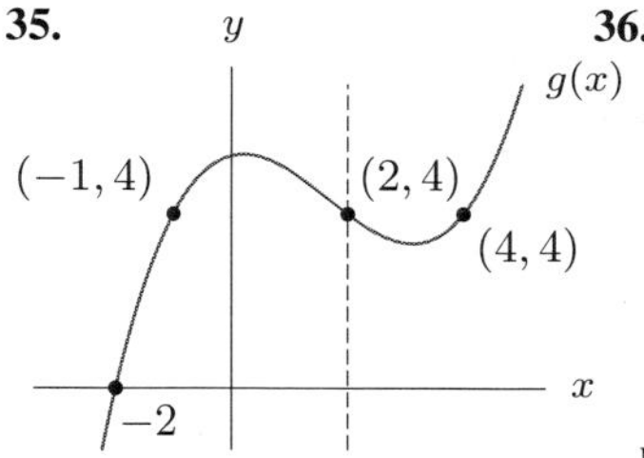

36.

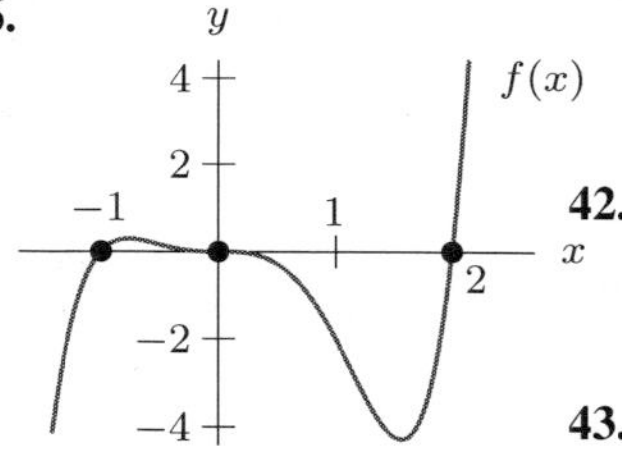

Note appearance near origin

37. For each of the following functions, state whether it is even, odd, or neither.

(a) $f(x) = x^2 + 3$ **(b)** $g(x) = x^3 + 3$
(c) $h(x) = 5/x$ **(d)** $j(x) = |x - 4|$
(e) $k(x) = \log x$ **(f)** $l(x) = \log(x^2)$
(g) $m(x) = 2^x + 2$ **(h)** $n(x) = \cos x + 2$

38. Assume that $x = a$ and $x = b$ are zeros of the second degree polynomial, $y = q(x)$.

(a) Explain what you know and don't know about the graph of q. (Intercepts, vertex, end behavior.)
(b) Explain why a possible formula for $q(x)$ is $q(x) = k(x - a)(x - b)$, with k unknown.

39. **(a)** Suppose $f(x) = ax^2 + bx + c$. What must be true about the coefficients if f is an even function?
(b) Suppose $g(x) = ax^3 + bx^2 + cx + d$. What must be true about the coefficients if g is an odd function?

40. The gravitational force exerted by a planet is inversely proportional to the square of the distance to the center of the planet. Thus, the weight, w, of an object at a distance, r, from a planet's center is given by

$$w = \frac{k}{r^2}, \quad \text{with } k \text{ constant.}$$

A gravitational force of one ton (2000 lbs) will kill a 150-pound person. Suppose the earth's radius were to shrink with its mass remaining the same. What is the smallest radius at which the 150-pound person could survive? Give your answer as a percentage of the earth's radius.

41. The period, p, of the orbit of a planet whose average distance (in millions of miles) from the sun is d, is given by $p = kd^{3/2}$, where k is a constant. The average distance from the earth to the sun is 93 million miles.

(a) If the period of the earth's orbit were twice the current 365 days, what would be the average distance from the sun to earth?
(b) Is there a planet in our solar system whose period is approximately twice the earth's?

42. Refer to Problem 41. If the distance between the sun and the earth were halved, how long (in earth days) would a "year" be?

43. Ship designers usually construct scale models before building a real ship. The formula that relates the speed u to the hull length l of a ship is

$$u = k\sqrt{l},$$

where k is a positive constant. This constant k varies depending on the ship's design, but scale models of a real ship have the same k as the real ship after which they are modeled.[25]

(a) How fast should a scale model with hull length 4 meters travel to simulate a real ship with hull length 225 meters traveling 9 meters/sec?
(b) A new ship is to be built whose speed is to be 10% greater than the speed of an existing ship with the same design. What is the relationship between the hull lengths of the new ship and the existing ship?

[25] R. McNeill Alexander, *Dynamics of Dinosaurs and Other Extinct Giants.* (New York: Columbia University Press, 1989).

44. An alcohol solution consists of 5 gallons of pure water and x gallons of alcohol, $x > 0$. Let $f(x)$ be the ratio of the volume of alcohol to the total volume of liquid. [Note that $f(x)$ is the concentration of the alcohol in the solution.]

(a) Find a possible formula for $f(x)$.
(b) Evaluate and interpret $f(7)$ in the context of the mixture.
(c) What is the zero of f? Interpret your result in the context of the mixture.
(d) Find an equation for the horizontal asymptote of f. Explain its significance in the context of the mixture.

45. The function $f(x)$ defined in Problem 44 gives the concentration of a solution of x gallons of alcohol and 5 gallons of water.

(a) Find a formula for $f^{-1}(x)$.
(b) Evaluate and interpret $f^{-1}(0.2)$ in the context of the mixture.
(c) What is the zero of f^{-1}? Interpret your result in the context of the mixture.
(d) Find an equation for the horizontal asymptote of f^{-1}. Explain its significance in the context of the mixture.

Find possible formulas for the polynomials and rational functions in Problems 46–50.

46. The zeros of f are $x = -3$, $x = 2$, and $x = 5$, and the y-intercept is $y = -6$.

47. This function has zeros at $x = -3$, $x = 2$, $x = 5$, and a double-zero at $x = 6$. It has a y-intercept of 7.

48. This function has zeros at $x = 2$ and $x = 3$. It has a vertical asymptote at $x = 5$. It has a horizontal asymptote of $y = -3$.

49. The polynomial $h(x) = 7$ at $x = -5, -1, 4$, and the y-intercept is 3. [Hint: Visualize h as a vertically shifted version of another polynomial.]

50. The graph of w intercepts the graph of $v(x) = 2x + 5$ at $x = -4, 1, 3$ and has a y-intercept of 2. [Hint: Let $w(x) = p(x) + v(x)$ where p is another polynomial.]

51. The ancient Greeks placed great importance on a number known as the *golden ratio*, ϕ, which is defined geometrically as follows. Starting with a square, add a rectangle to one side of the square, so that the resulting rectangle has the same proportions as the rectangle that was added. The golden ratio is defined as the ratio of either rectangle's length to its width. (See Figure 9.62.) Given this information, show that $\phi = (1 + \sqrt{5})/2$.

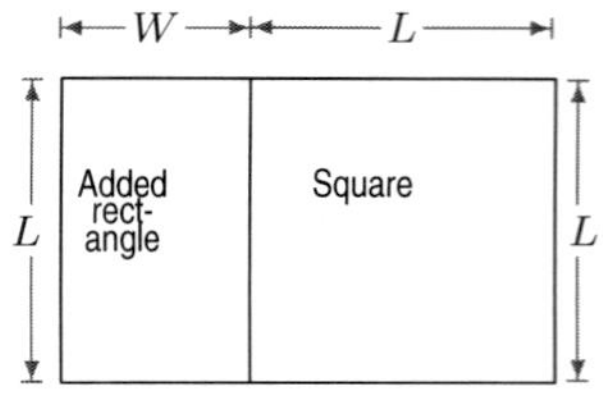

Figure 9.62: Large rectangle has the same proportions as small rectangle

52. The number ϕ, defined in Problem 51, has the property:

$$\phi^k + \phi^{k+1} = \phi^{k+2}.$$

(a) Check that the property holds true for $k = 3$ and $k = 10$.
(b) Show that this property follows from the definition of ϕ.

53. The resolution, r %, of a gamma ray telescope depends on the energy v (in millions of electron volts, or MeVs) of the detected gamma rays.[26] The smaller the value of r, the better the telescope is at distinguishing two gamma ray photons of slightly different energies, and the more detailed observations that can be made. Table 9.39 gives values of r for gamma rays at different energies.

(a) Plot the data in Table 9.39, with r on the vertical axis.
(b) Based on this data, is the telescope better able to distinguish between high-energy photons or low-energy photons?
(c) Fit both power and exponential functions to the data, and give their formulas. Which appears to give the better fit?
(d) The telescope is predicted to grow rapidly worse and worse at distinguishing photons as the energy level drops toward 0 MeV. Which curve, power or exponential, is most consistent with this prediction?

Table 9.39

v, MeV	0.5	0.7	0.9	1.3	1.8	4.0	4.4
r, %	16.0	13.5	12.0	8.5	7.0	4.5	4.0

[26] *The LXeGRIT Compton Status and Future Prospects*, E. Aprile, et al., posted at http://arxiv.org as arXiv:astro-ph/0212005v2, December 4, 2002.

54. We launch an unpowered spacecraft, initially in free space, so that it hits a planet of radius R. When viewed from a distance, the planet looks like a disk of area πR^2. In the absence of gravity, we would need to aim the spacecraft directly at this disk. However, because of gravity, the planet draws the spacecraft toward it, so that even if we are somewhat off the mark, the spacecraft might still hit its target. Thus, the area we must aim for is actually larger than the apparent area of the planet. This area is called the planet's *capture cross-section*. The more massive the planet, the larger its capture cross-section, because it exerts a stronger pull on passing objects.

A spacecraft with a large initial velocity has a greater chance of slipping past the planet even if its aim is only slightly off, whereas a spacecraft with a low initial velocity has a good chance of drifting into the planet even if its aim is poor. Thus, the planet's capture cross-section is a function of the initial velocity of the spacecraft, v. If the planet's capture cross-section is denoted by A and M is the planet's mass, then it can be shown that, for a positive constant G,

$$A(v) = \pi R^2 \left(1 + \frac{2MG/R}{v^2}\right).$$

(a) Show from the formula for $A(v)$ that, for any initial velocity v,
$$A(v) > \pi R^2.$$
Explain why this makes sense physically.

(b) Consider two planets: The first is twice as massive as the second, and the radius of the second is twice the radius of the first. Which has the larger capture cross-section?

(c) The graph of $A(v)$ has both horizontal and vertical asymptotes. Find their equations, and explain their physical significance.

55. A group of x people is tested for the presence of a certain virus. Unfortunately, the test is imperfect and incorrectly identifies some healthy people as being infected and some sick people as being noninfected. There is no way to know when the test is right and when it is wrong. Since the disease is so rare, only 1% of those tested are actually infected.

(a) Write expressions in terms of x for the number of people tested that are actually infected and the number who are not.

(b) The test correctly identifies 98% of all infected people as being infected. (It incorrectly identifies the remaining 2% as being healthy.) Write, in terms of x, an expression representing the number of infected people who are correctly identified as being infected. (This group is known as the *true-positive* group.)

(c) The test incorrectly identifies 3% of all healthy people as being infected. (It correctly identifies the remaining 97% as being healthy.) Write, in terms of x, an expression representing the number of noninfected people who are incorrectly identified as being infected. (This group is known as the *false-positive* group.)

(d) Write, in terms of x, an expression representing the total number of people the test identifies as being infected, including both true- and false-positives.

(e) Write, in terms of x, an expression representing the fraction of those testing positive who are actually infected. Can this expression be evaluated without knowing the value of x?

(f) Suppose you are among the group of people who test positive for the presence of the virus. Based on your test result, do you think it is likely that you are actually infected?

CHECK YOUR UNDERSTANDING

Are the statements in Problems 1–47 true or false? Give an explanation for your answer.

1. All quadratic functions are power functions.

2. The function $y = 3 \cdot 2^x$ is a power function.

3. Let $g(x) = x^p$. If p is a positive, even integer, then the graph of g passes through the point $(-1, 1)$.

4. Let $g(x) = x^p$. If p is a positive, even integer, then the graph of g is symmetric about the y-axis.

5. Let $g(x) = x^p$. If p is a positive, even integer, then the graph of g is concave up.

6. The graph of $f(x) = x^{-1}$ passes through the origin.

7. The graph of $f(x) = x^{-2}$ has the x-axis as its only asymptote.

8. If $f(x) = x^{-1}$ then $f(x)$ approaches $+\infty$ as x approaches zero.

9. As x grows very large, the values of $f(x) = x^{-1}$ approach zero.

10. The function 2^x eventually grows faster than x^b for any b.

11. The function $f(x) = x^{0.5}$ eventually grows faster than $g(x) = \ln x$.

12. We have $2^x \geq x^2$ on the interval $0 \leq x \leq 4$.

13. The function $f(x) = x^{-3}$ approaches the x-axis faster than $g(x) = e^{-x}$ as x grows very large.

14. The function $f(x) = 3^x$ is an example of a power function.

15. The function $y = 3x$ is an example of a power function.

16. Every quadratic function is a polynomial function.

17. The power of the first term of a polynomial is its degree.

18. Far from the origin, the graph of a polynomial looks like the graph of its highest degree term.

19. A zero of a polynomial p is the value $p(0)$.

20. The zeros of a polynomial are the x-coordinates where its graph intersects the x-axis.

21. The y-intercept of a polynomial $y = p(x)$ can be found by evaluating $p(0)$.

22. For very large x-values $f(x) = 1000x^3 + 345x^2 + 17x + 394$ is less than $g(x) = 0.01x^4$.

23. If $y = f(x)$ is a polynomial of degree n, where n is a positive even number, then f has an inverse.

24. If $y = f(x)$ is a polynomial of degree n, where n is a positive odd number, then f has an inverse.

25. If $p(x)$ is a polynomial and $x - a$ is a factor of p, then $x = a$ is a zero of p.

26. A polynomial of degree n cannot have more than n zeros.

27. The polynomial in Figure 9.63 has a multiple zero at $x = -2$.

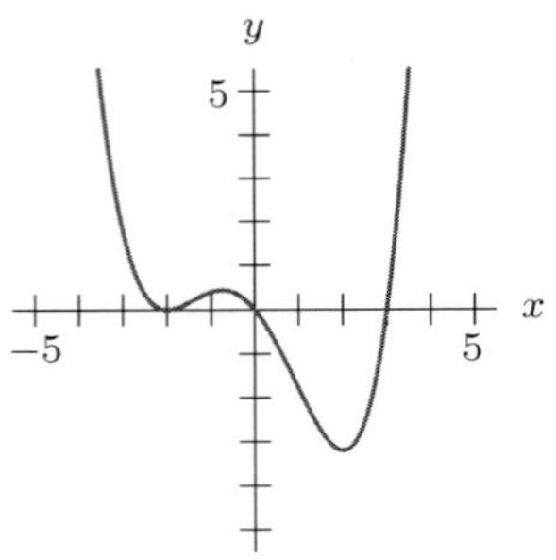

Figure 9.63

28. The polynomial in Figure 9.63 has a multiple zero at $x = 0$.

29. A rational function is the quotient of two polynomials. We assume the denominator is not equal to zero.

30. The function $f(x) = \dfrac{1}{x}$ is a rational function.

31. In order to determine the long-run behavior of a rational function, it is sufficient to consider only the ratio of the highest degree term in the numerator to the highest degree term in the denominator.

32. As x grows through large positive values, $y = \dfrac{x+18}{x+9}$ approaches $y = 2$.

33. As x grows through large positive values, $y = \dfrac{2x+125}{x^2-1}$ approaches $y = 0$.

34. As x grows through large positive values, $y = \dfrac{x^3+4x^2-16x+12}{4x^3-16x+1}$ has an asymptote at $y = 4$.

35. As x grows through large positive values, $y = \dfrac{1-4x^2}{x^2+1}$ approaches $y = 0$.

36. As x grows through large positive values, $y = \dfrac{5x}{x+1}$ approaches $y = -5$.

37. As x grows through large positive values, $y = \dfrac{3x^4-6x^3+10x^2-16x+7}{-3x+x^2}$ behaves like $y = -x^3$.

38. As x decreases through large negative values, $f(x) = \dfrac{x^3-7x^2+28x+76}{-x^2-101x+72}$ approaches positive infinity.

39. A fraction is equal to zero if and only if its numerator equals zero and its denominator does not.

40. The zeros of a function $y = f(x)$ are the values of x that make $y = 0$.

41. The function $f(x) = \dfrac{x+4}{x-3}$ has a zero at $x = -4$.

42. The rational function $y = \dfrac{x+2}{x^2-4}$ has a zero at $x = -2$.

43. The rational function $g(w) = \dfrac{12}{(w-2)(w+3)}$ has exactly two zeros.

44. If $p(x)$ and $q(x)$ have no zeros in common, then the rational function $r(x) = \dfrac{p(x)}{q(x)}$ has an asymptote at each of the zeros of $p(x)$.

45. In general, the rational function $r(x) = \dfrac{p(x)}{q(x)}$ must have at least one zero.

46. Rational functions can never cross an asymptote.

47. The rational function $g(w) = \dfrac{3w-3}{(w-12)(w+4)}$ has a vertical asymptote at $w = 1$.

TOOLS FOR CHAPTER 9: ALGEBRAIC FRACTIONS

Algebraic fractions are combined in the same way as numeric fractions according to the following rules:

> Add numerators when denominators are equal: $\dfrac{a}{c} + \dfrac{b}{c} = \dfrac{a+b}{c}$
>
> Find a common denominator: $\dfrac{a}{b} + \dfrac{c}{d} = \dfrac{a \cdot d}{b \cdot d} + \dfrac{b \cdot c}{b \cdot d} = \dfrac{ad+bc}{bd}$
>
> Multiply numerators and denominators for a product: $\dfrac{a}{b} \cdot \dfrac{c}{d} = \dfrac{ac}{bd}$
>
> To divide by a fraction, multiply by its reciprocal: $\dfrac{a/b}{c/d} = \dfrac{a}{b} \cdot \dfrac{d}{c} = \dfrac{ad}{bc}$
>
> The sign of a fraction is changed by changing the sign of the numerator or the denominator (but not both):
>
> $$-\frac{a}{b} = \frac{-a}{b} = \frac{a}{-b}$$
>
> We assume that no denominators are zero, since we cannot divide by zero; that is, $a/0$ is not defined.

We can simplify a fraction in which either the numerator or denominator is itself a fraction as follows:

$$\frac{a/b}{c} = \frac{a/b}{c/1} = \frac{a}{b} \cdot \frac{1}{c} = \frac{a}{bc} \quad \text{and} \quad \frac{a}{b/c} = \frac{a/1}{b/c} = \frac{a}{1} \cdot \frac{c}{b} = \frac{ac}{b}.$$

Example 1 Perform the indicated operations and express the answers as a single fraction.

(a) $\dfrac{4}{x^2+1} - \dfrac{1-x}{x^2+1}$ (b) $\dfrac{M}{M^2-2M-3} + \dfrac{1}{M^2-2M-3}$

(c) $\dfrac{-H^2P}{17} \cdot \dfrac{\left(PH^{1/3}\right)^2}{K^{-1}}$ (d) $\dfrac{2z/w}{w(w-3z)}$

Solution

(a) $\dfrac{4}{x^2+1} - \dfrac{1-x}{x^2+1} = \dfrac{4-(1-x)}{x^2+1} = \dfrac{3+x}{x^2+1}$

(b) $\dfrac{M}{M^2-2M-3} + \dfrac{1}{M^2-2M-3} = \dfrac{M+1}{(M^2-2M-3)} = \dfrac{M+1}{(M+1)(M-3)} = \dfrac{1}{M-3}$

(c) $\dfrac{-H^2P}{17} \cdot \dfrac{\left(PH^{1/3}\right)^2}{K^{-1}} = \dfrac{-H^2P\left(P^2H^{2/3}\right)}{17K^{-1}} = -\dfrac{H^{8/3}P^3K}{17}$

(d) $\dfrac{2z/w}{w(w-3z)} = \dfrac{2z}{w} \cdot \dfrac{1}{w(w-3z)} = \dfrac{2z}{w^2(w-3z)}$

Example 2 Simplify the following expressions, giving your answer as a single fraction.

(a) $2x^{-1/2} + \dfrac{\sqrt{x}}{3}$ (b) $2\sqrt{t+3} + \dfrac{1-2t}{\sqrt{t+3}}$

Solution

(a) $$2x^{-1/2}+\frac{\sqrt{x}}{3}=\frac{2}{\sqrt{x}}+\frac{\sqrt{x}}{3}=\frac{2\cdot 3+\sqrt{x}\sqrt{x}}{3\sqrt{x}}=\frac{6+x}{3\sqrt{x}}=\frac{6+x}{3x^{1/2}}$$

(b) $$\begin{aligned}2\sqrt{t+3}+\frac{1-2t}{\sqrt{t+3}}&=\frac{2\sqrt{t+3}}{1}+\frac{1-2t}{\sqrt{t+3}}\\&=\frac{2\sqrt{t+3}\sqrt{t+3}+1-2t}{\sqrt{t+3}}\\&=\frac{2(t+3)+1-2t}{\sqrt{t+3}}\\&=\frac{7}{\sqrt{t+3}}=\frac{7}{(t+3)^{1/2}}\end{aligned}$$

Finding a Common Denominator

We can multiply (or divide) both the numerator and denominator of a fraction by the same nonzero number without changing the fraction's value. This is equivalent to multiplying by a factor of $+1$. We are using this rule when we add or subtract fractions with different denominators. For example, to add $\frac{x}{3a}+\frac{1}{a}$, we multiply $\frac{1}{a}\cdot\frac{3}{3}=\frac{3}{3a}$. Then

$$\frac{x}{3a}+\frac{1}{a}=\frac{x}{3a}+\frac{3}{3a}=\frac{x+3}{3a}.$$

Example 3 Perform the indicated operations:

(a) $3-\dfrac{1}{x-1}$

(b) $\dfrac{2}{x^2+x}+\dfrac{x}{x+1}$

Solution

(a) $$3-\frac{1}{x-1}=3\frac{(x-1)}{(x-1)}-\frac{1}{x-1}=\frac{3(x-1)-1}{x-1}=\frac{3x-3-1}{x-1}=\frac{3x-4}{x-1}$$

(b) $$\frac{2}{x^2+x}+\frac{x}{x+1}=\frac{2}{x(x+1)}+\frac{x}{x+1}=\frac{2}{x(x+1)}+\frac{x(x)}{(x+1)(x)}=\frac{2+x^2}{x(x+1)}$$

Note: We can multiply (or divide) the numerator and denominator by the same nonzero number because this is the same as multiplying by a factor of $+1$, and multiplying by a factor of 1 does not change the value of the expression. However, we cannot perform any other operation that would change the value of the expression. For example, we cannot add the same number to the numerator and denominator of a fraction nor can we square both, take the logarithm of both, etc., without changing the fraction.

Reducing Fractions: Canceling

We can reduce a fraction when we have the same (nonzero) factor in both the numerator and the denominator. For example,

$$\frac{ac}{bc}=\frac{a}{b}\cdot\frac{c}{c}=\frac{a}{b}\cdot 1=\frac{a}{b}.$$

Example 4 Reduce the following fractions (if possible).

(a) $\dfrac{2x}{4y}$ (b) $\dfrac{2+x}{2+y}$

(c) $\dfrac{5n-5}{1-n}$ (d) $\dfrac{x^2(4-2x)-(4x-x^2)2x}{x^4}$

Solution

(a) $\dfrac{2x}{4y}=\dfrac{2}{2}\cdot\dfrac{x}{2y}=\dfrac{x}{2y}$

(b) $\dfrac{2+x}{2+y}$ cannot be reduced further.

(c) $\dfrac{5n-5}{1-n}=\dfrac{5(n-1)}{(-1)(n-1)}=-5$

(d)

$$\begin{aligned}\frac{x^2(4-2x)-\left(4x-x^2\right)2x}{x^4} &= \frac{x^2(4-2x)-(4-x)2x^2}{x^4}\\ &= \frac{(4-2x)-2(4-x)}{x^2}\left(\frac{x^2}{x^2}\right)\\ &= \frac{4-2x-8+2x}{x^2}=\frac{-4}{x^2}.\end{aligned}$$

Complex Fractions

A *complex fraction* is a fraction whose numerator or denominator (or both) contains one or more fractions. To simplify a complex fraction, we change the numerator and denominator to single fractions and then divide.

Example 5 Write the following as simple fractions in reduced form.

(a) $\dfrac{\dfrac{1}{x+h}-\dfrac{1}{x}}{h}$ (b) $\dfrac{a+b}{a^{-2}-b^{-2}}$

Solution

(a) $$\frac{\dfrac{1}{x+h}-\dfrac{1}{x}}{h}=\frac{\dfrac{x-(x+h)}{x(x+h)}}{h}=\frac{\dfrac{-h}{x(x+h)}}{\dfrac{h}{1}}=\frac{-h}{x(x+h)}\cdot\frac{1}{h}=\frac{-1}{x(x+h)}\frac{(h)}{(h)}=\frac{-1}{x(x+h)}$$

(b) $$\frac{a+b}{a^{-2}-b^{-2}}=\frac{a+b}{\dfrac{1}{a^2}-\dfrac{1}{b^2}}=\frac{a+b}{\dfrac{b^2-a^2}{a^2b^2}}=\frac{a+b}{1}\cdot\frac{a^2b^2}{b^2-a^2}=\frac{(a+b)(a^2b^2)}{(b+a)(b-a)}=\frac{a^2b^2}{b-a}$$

Splitting Expressions

We can reverse the rule for adding fractions to split up an expression into two fractions,

$$\frac{a+b}{c}=\frac{a}{c}+\frac{b}{c}.$$

Example 6 Split $\dfrac{3x^2+2}{x^3}$ into two reduced fractions.

Solution $$\frac{3x^2+2}{x^3}=\frac{3x^2}{x^3}+\frac{2}{x^3}=\frac{3}{x}+\frac{2}{x^3}$$

Sometimes we can alter the form of the fraction even further if we can create a duplicate of the denominator within the numerator. This technique is useful when graphing some rational functions. For example, we may rewrite the fraction $\frac{x+3}{x-1}$ by creating a factor of $(x-1)$ within the numerator. To do this, we write

$$\frac{x+3}{x-1} = \frac{x-1+1+3}{x-1}$$

which can be written as

$$\frac{(x-1)+4}{x-1}.$$

Then, splitting this fraction, we have

$$\frac{x+3}{x-1} = \frac{x-1}{x-1} + \frac{4}{x-1} = 1 + \frac{4}{x-1}.$$

Note: It is not possible to split a sum that occurs in the denominator of a fraction. For example,

$$\frac{a}{b+c} \text{ does not equal } \frac{a}{b} + \frac{a}{c}.$$

Exercises to Tools for Chapter 9

For Exercises 1–35, perform the operations. Express answers in reduced form.

1. $\frac{3}{5} + \frac{4}{7}$

2. $\frac{7}{10} - \frac{2}{15}$

3. $\frac{1}{2x} - \frac{2}{3}$

4. $\frac{6}{7y} + \frac{9}{y}$

5. $\frac{-2}{yz} + \frac{4}{z}$

6. $\frac{-2z}{y} + \frac{4}{y}$

7. $\frac{2}{x^2} - \frac{3}{x}$

8. $\frac{6}{y} + \frac{7}{y^3}$

9. $\frac{3/4}{7/20}$

10. $\frac{5/6}{15}$

11. $\frac{3/x}{x^2/6}$

12. $\frac{3/x}{6/x^2}$

13. $\frac{13}{x-1} + \frac{14}{2x-2}$

14. $\frac{14}{x-1} + \frac{13}{2x-2}$

15. $\frac{4z}{x^2y} - \frac{3w}{xy^4}$

16. $\frac{10}{y-2} + \frac{3}{2-y}$

17. $\frac{8y}{y-4} + \frac{32}{y-4}$

18. $\frac{8y}{y-4} + \frac{32}{4-y}$

19. $\dfrac{\frac{1}{x} - \frac{2}{x^2}}{\frac{2x-4}{x^5}}$

20. $\frac{9}{x^2+5x+6} + \frac{12}{x+3}$

21. $\frac{8}{3x^2-x-4} - \frac{9}{x+1}$

22. $\frac{5}{(x-2)^2(x+1)} - \frac{18}{(x-2)}$

23. $\frac{15}{(x-3)^2(x+5)} + \frac{7}{(x-3)(x+5)^2}$

24. $\frac{3}{x-4} - \frac{2}{x+4}$

25. $\frac{x^2}{x-1} - \frac{1}{1-x}$

26. $\frac{1}{2r+3} + \frac{3}{4r^2+6r}$

27. $u + a + \frac{u}{u+a}$

28. $\frac{1}{\sqrt{x}} - \frac{1}{(\sqrt{x})^3}$

29. $\frac{1}{e^{2x}} + \frac{1}{e^x}$

30. $\frac{a+b}{2} \cdot \frac{8x+2}{b^2-a^2}$

31. $\frac{0.07}{M} + \frac{3}{4}M^2$

32. $\frac{1}{r_1} + \frac{1}{r_2} + \frac{1}{r_3}$

33. $\frac{8y}{y-4} - \frac{32}{y-4}$

34. $\frac{a}{a^2-9} + \frac{1}{a-3}$

35. $\frac{x^3}{x-4} \Big/ \frac{x^2}{x^2-2x-8}$

In Exercises 36–49, simplify, if possible.

36. $\dfrac{1/(x+y)}{x+y}$

37. $\dfrac{(w+2)/2}{w+2}$

38. $\dfrac{\dfrac{1}{(x+h)^2}-\dfrac{1}{x^2}}{h}$

39. $\dfrac{a^{-2}+b^{-2}}{a^2+b^2}$

40. $\dfrac{a^2-b^2}{a^2+b^2}$

41. $\dfrac{4-(x+h)^2-(4-x^2)}{h}$

42. $\dfrac{b^{-1}(b-b^{-1})}{b+1}$.

43. $\dfrac{1-a^{-2}}{1+a^{-1}}$.

44. $\dfrac{x^{-1}+x^{-2}}{1-x^{-2}}$.

45. $p-\dfrac{q}{\dfrac{p}{q}+\dfrac{q}{p}}$

46. $\dfrac{\dfrac{3}{xy}-\dfrac{5}{x^2y}}{\dfrac{6x^2-7x-5}{x^4y^2}}$

47. $\dfrac{\dfrac{1}{x}\left(3x^2\right)-(\ln x)(6x)}{\left(3x^2\right)^2}$

48. $\dfrac{2x(x^3+1)^2-x^2(2)(x^3+1)(3x^2)}{[(x^3+1)^2]^2}$

49. $\dfrac{\frac{1}{2}(2x-1)^{-1/2}(2)-(2x-1)^{1/2}(2x)}{(x^2)^2}$

In Exercises 50–55, split into a sum or difference of reduced fractions.

50. $\dfrac{26x+1}{2x^3}$

51. $\dfrac{\sqrt{x}+3}{3\sqrt{x}}$

52. $\dfrac{6l^2+3l-4}{3l^4}$

53. $\dfrac{7+p}{p^2+11}$

54. $\dfrac{\frac{1}{3}x-\frac{1}{2}}{2x}$

55. $\dfrac{t^{-1/2}+t^{1/2}}{t^2}$

In Exercises 56–61, rewrite in the form $1+(A/B)$.

56. $\dfrac{x-2}{x+5}$

57. $\dfrac{q-1}{q-4}$

58. $\dfrac{R+1}{R}$

59. $\dfrac{3+2u}{2u+1}$

60. $\dfrac{\cos x+\sin x}{\cos x}$

61. $\dfrac{1+e^x}{e^x}$

Are the statements in Exercises 62–67 true or false?

62. $\dfrac{a+c}{a}=1+c$

63. $\dfrac{rs-s}{s}=r-1$

64. $\dfrac{y}{y+z}=1+\dfrac{y}{z}$

65. $\dfrac{2u^2-w}{u^2-w}=2$

66. $\dfrac{x^2yz}{2x^2y}=\dfrac{z}{2}$

67. $x^{5/3}-3x^{2/3}=\dfrac{x^2-3x}{x^{1/3}}$

Chapter Ten

VECTORS AND MATRICES

Vectors are used to represent quantities, such as displacement and velocity, which have both magnitude and direction. Matrices are arrays of numbers. This chapter shows how vectors and matrices are added, subtracted, and multiplied.

10.1 VECTORS

Distance Versus Displacement

If you start at home and walk 3 miles east and then 4 miles north, you have walked a total of 7 miles. However, your distance from home is not 7 miles, but $5 = \sqrt{3^2 + 4^2}$ miles, by the Pythagorean Theorem. See Figure 10.1. As this example illustrates, there is a distinction between *distance* and *displacement*. Distance is a number that measures separation, while displacement consists of separation and direction. For example, walking 3 miles north and walking 3 miles east give different displacements, but both correspond to a distance of 3 miles.

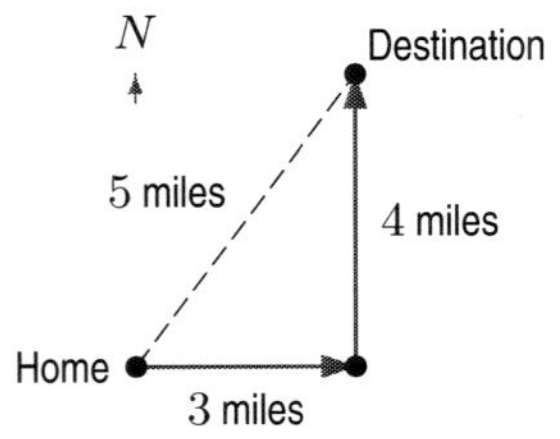

Figure 10.1: After walking 3 miles east then 4 miles north, your distance from home is 5 miles

Adding Displacements Using Triangles

Displacements are added by treating them as arrows called *vectors*. The length of each arrow indicates the magnitude of the displacement, while the orientation of the arrow (which way the arrow points) indicates the direction. To add two displacements, join the tail of the second arrow to the head of the first arrow. Together, the arrows form two sides of a triangle whose third side represents their sum. The sum is the *net displacement* resulting from the two displacements.

Example 1 A person leaves her home and walks 5 miles due east and then 3 miles northeast. How far has she walked? How far away from home is she? What is her net displacement?

Solution In Figure 10.2, the first arrow is 5 units long because she walks 5 miles, while the second arrow is 3 units long and at an angle of 45° to the first. The length of the third side, x, of the triangle can be found by the Law of Cosines:

$$\begin{aligned} x^2 &= 5^2 + 3^2 - 2 \cdot 5 \cdot 3 \cos 135^\circ \\ &= 34 - 30\left(-\frac{\sqrt{2}}{2}\right) \\ &= 55.213. \end{aligned}$$

This gives $x = \sqrt{55.213} = 7.431$ miles for the distance from home to destination, which is less than the total distance walked ($5 + 3 = 8$ miles). To specify her net displacement we use the angle θ in Figure 10.2. We find θ using the Law of Sines:

$$\frac{\sin\theta}{3} = \frac{\sin 135^\circ}{7.431}$$

giving

$$\sin\theta = 3 \cdot \frac{\sqrt{2}/2}{7.431} = 0.285.$$

Taking the arcsin of both sides, we obtain $\theta = \sin^{-1}(0.285) = 16.588^\circ$. Thus, the net displacement is 7.431 miles in a direction 16.588° north of east.

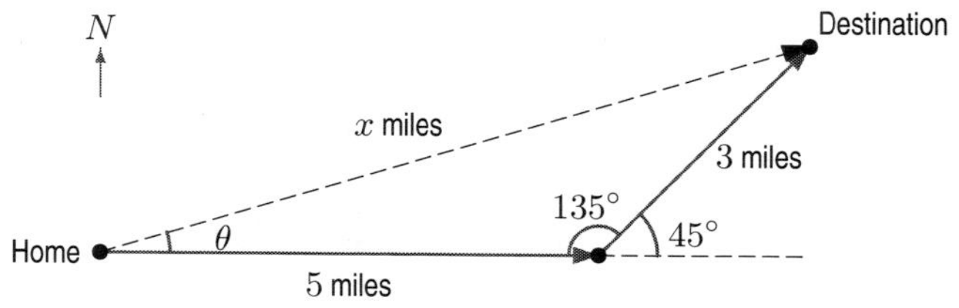

Figure 10.2: A person walks 5 miles east and then 3 miles northeast

The Order of Addition Does Not Affect the Sum

When two displacements are added, the sum is independent of the order in which the two arrows are joined. For example, in Figure 10.3 the 5-mile displacement followed by the 3-mile displacement leads to the same destination as the 3-mile displacement followed by the 5-mile displacement.

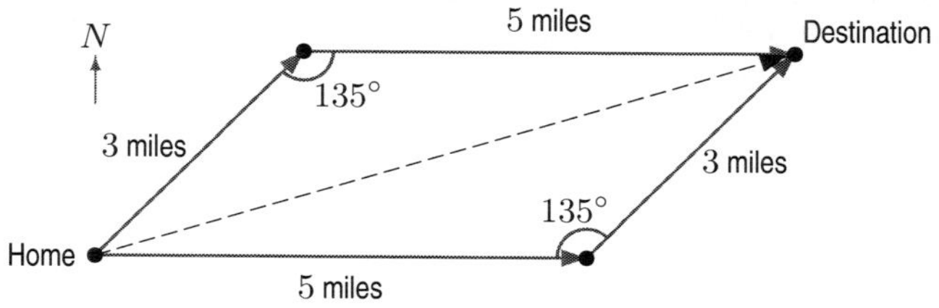

Figure 10.3: No matter which order the arrows are joined, the resulting net displacement is the same

Vectors

Many physical quantities add in the same way as displacements. Such quantities are known as *vectors*. Like displacements, vectors are often represented as arrows. A vector has *magnitude* (the arrow's length) and *direction* (which way the arrow points). Examples of vectors include:

- **Velocity** This is the speed and direction of travel.
- **Force** This is, loosely speaking, the strength and direction of a push or a pull.
- **Magnetic fields** A vector gives the direction and intensity of a magnetic field at a point.
- **Vectors in economics** Vectors are used to keep track of prices and quantities.
- **Vectors in computer animation** Computers generate animations by performing enormous numbers of vector-based calculations.
- **Population vectors** In biology, populations of different animals or age groups can be represented using vectors.

Vector Notation

In this book we write vectors as variables with arrows over them: $\vec{v}$. The notation is intended to ensure that a vector $\vec{v}$ is not mistaken for a *scalar*, which is another name for an ordinary number.

Notation for the Magnitude of a Vector

The length or magnitude of a vector $\vec{v}$ is written $||\vec{v}||$. This notation looks like the absolute value notation, $|x|$, used for scalars. The absolute value of a number is its size without regard to sign, so $|-10| = |+10| = 10$. Similarly, the vectors $\vec{u}$ and $\vec{v}$ in Figure 10.4 are both of length 5, even though they point in different directions. Therefore, $||\vec{u}|| = ||\vec{v}|| = 5$, (although $\vec{u} \neq \vec{v}$).

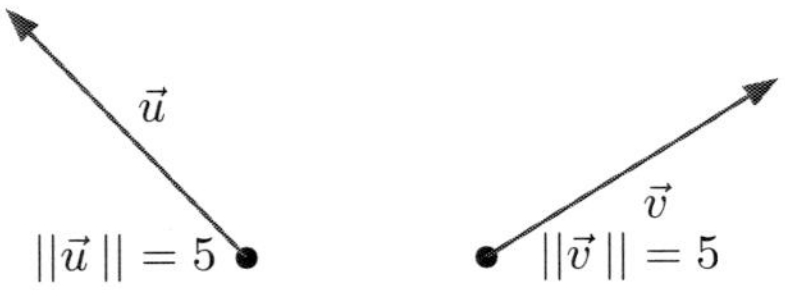

Figure 10.4: The vectors $\vec{u}$ and $\vec{v}$ are both of magnitude 5, but they point in different directions

Addition of Vectors

As with displacements, the sum of two vectors $\vec{u}$ and $\vec{v}$ represented by arrows is found by joining the tail of $\vec{v}$ to the head of $\vec{u}$. Then $\vec{w} = \vec{u} + \vec{v}$ is represented by the arrow drawn from the tail of of $\vec{u}$ to the head of $\vec{v}$. See Figure 10.5.[1]

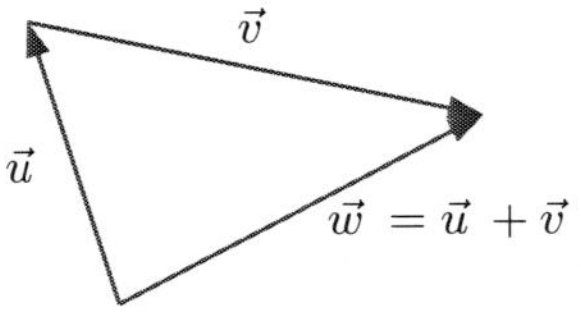

Figure 10.5: Constructing the vector sum $\vec{w} = \vec{u} + \vec{v}$

Example 2 Spacecraft such as *Voyager* and *Galileo* experience gravitational forces exerted by the sun, earth, and other planets. Vectors are used to represent the strength and direction of the gravitational forces.

Suppose a Jupiter-bound spacecraft is coasting between the orbits of Mars and Jupiter. Assume that Jupiter exerts a gravitational force of 8 units on the spacecraft, directed toward Jupiter, and that Mars exerts a force of 3 units on the spacecraft, directed toward Mars. The force $\vec{F}_{\text{M}}$ exerted by Mars is at an angle of 70° to the force $\vec{F}_{\text{J}}$ exerted by Jupiter, as shown in Figure 10.6. (For simplicity, we ignore the forces due to the sun and other planets.)

(a) Find $||\vec{F}_{\text{MJ}}||$, the magnitude of the net force on the spacecraft due to Mars and Jupiter.
(b) What is the direction of $\vec{F}_{\text{MJ}}$?
(c) Without engine power, will the spacecraft stay on course?

Figure 10.6: Forces $\vec{F}_{\text{M}}$ and $\vec{F}_{\text{J}}$ exerted by Mars and Jupiter, respectively

[1] If $\vec{u}$ and $\vec{v}$ are parallel, the vectors are added by the same method. However, in this case the figure is not a triangle.

Solution

(a) The net force, $\vec{F}_{\rm MJ}$, is the sum of $\vec{F}_{\rm M}$ and $\vec{F}_{\rm J}$, the forces due to Mars and Jupiter, so

$$\vec{F}_{\rm MJ} = \vec{F}_{\rm M} + \vec{F}_{\rm J}.$$

Figure 10.7 shows $\vec{F}_{\rm MJ}$, obtained by joining the tail of $\vec{F}_{\rm J}$ to the head of $\vec{F}_{\rm M}$. The Law of Cosines tells us that $||\vec{F}_{\rm MJ}||$, the length of the resulting vector, is given by

$$\begin{aligned} ||\vec{F}_{\rm MJ}||^2 &= 8^2 + 3^2 - 2\cdot 8 \cdot 3 \cos 110^\circ \\ &= 73 - 48\cos 110^\circ \\ &= 89.417 \end{aligned}$$

so $||\vec{F}_{\rm MJ}|| = \sqrt{89.417} = 9.456$ units. Notice that 9.456 is less than the sum of the magnitudes of the individual forces $3 + 8 = 11$ units. The magnitude of the net force is less than 11 units because the forces exerted by Mars and Jupiter are not perfectly aligned.

(b) The direction of $\vec{F}_{\rm MJ}$, denoted by θ in Figure 10.7, can be found using the Law of Sines:

$$\begin{aligned} \frac{\sin\theta}{3} &= \frac{\sin 110^\circ}{||\vec{F}_{\rm MJ}||} \\ \sin\theta &= 3\left(\frac{\sin 110^\circ}{||\vec{F}_{\rm MJ}||}\right) = 3\left(\frac{0.939}{9.456}\right) = 0.298, \end{aligned}$$

so

$$\theta = \arcsin(0.298) = 17.345^\circ.$$

(c) The gravity of Mars is pulling the spacecraft off its Jupiter-bound course by about 17.345°. The spacecraft will veer off course if it does not use its engines.

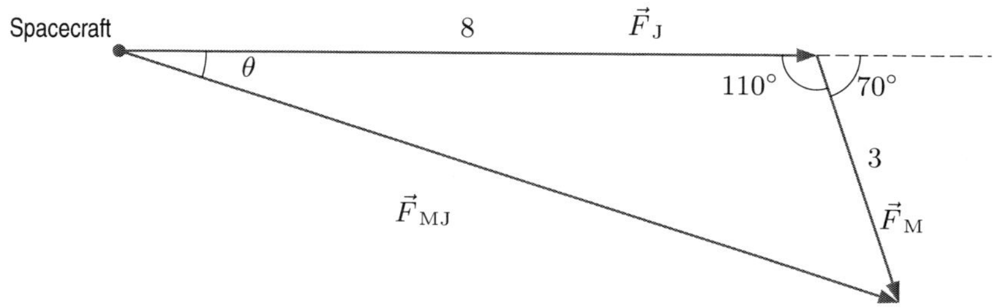

Figure 10.7: The net force, $\vec{F}_{\rm MJ}$, on the spacecraft

Subtraction of Vectors

We have seen how vectors are added. Now we see how to subtract one vector from another. If

$$\vec{w} = \vec{u} - \vec{v},$$

then it is reasonable to assume that by adding $\vec{v}$ to both sides we obtain

$$\vec{w} + \vec{v} = \vec{u}.$$

In other words, $\vec{w}$ is the vector that when added to $\vec{v}$ gives $\vec{u}$. This is illustrated in Figure 10.8.

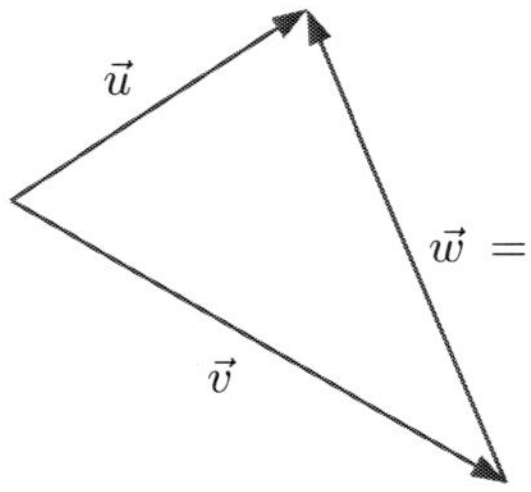

Figure 10.8: If $\vec{w} = \vec{u} - \vec{v}$, then $\vec{w}$ is the vector that when added to $\vec{v}$ gives $\vec{u}$

Figure 10.8 suggests the following rule: To find $\vec{w} = \vec{u} - \vec{v}$, join the tails of $\vec{v}$ and $\vec{u}$, and then $\vec{w}$ is the vector drawn from the head of $\vec{v}$ to the head of $\vec{u}$.

Example 3 Suppose the spacecraft in Example 2 can fire its engine thrusters in any direction, but that the strength of the thrust is always 4 units. In what direction should the spacecraft aim its engine thrusters in order to stay on course toward Jupiter?

Solution The engine-thrust vector $\vec{F}_{\text{engines}}$ must be chosen so as to push the spacecraft back on course. The combined force on the craft due to Mars and Jupiter is given by $\vec{F}_{\text{MJ}}$. When the engines are on, the net force due to Mars, Jupiter, and the engines is given by

$$\vec{F}_{\text{net}} = \vec{F}_{\text{MJ}} + \vec{F}_{\text{engines}}.$$

We choose $\vec{F}_{\text{engines}}$ so that $\vec{F}_{\text{net}}$ points directly toward Jupiter. Solving for $\vec{F}_{\text{engines}}$ gives

$$\vec{F}_{\text{engines}} = \vec{F}_{\text{net}} - \vec{F}_{\text{MJ}}.$$

We do not know the length of $\vec{F}_{\text{net}}$, but we do know that it points toward Jupiter. From Example 2, the length of $\vec{F}_{\text{MJ}}$ is 9.456 units and the vector is directed at a 17.345° angle clockwise from $\vec{F}_{\text{net}}$. Figure 10.9 shows $\vec{F}_{\text{engines}} = \vec{F}_{\text{net}} - \vec{F}_{\text{MJ}}$. (The force $\vec{F}_{\text{engines}}$ is actually applied to the spacecraft, as shown in Figure 10.9. However, we can move a vector from one point to another when adding.)

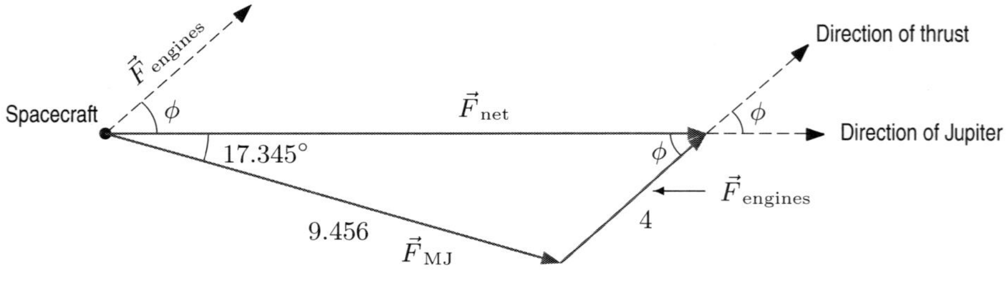

Figure 10.9: The engine-thrust vector is given by $\vec{F}_{\text{engines}} = \vec{F}_{\text{net}} - \vec{F}_{\text{MJ}}$, where $\vec{F}_{\text{net}}$ points directly toward Jupiter and $\vec{F}_{\text{MJ}}$ is as shown in Figure 10.7

We want to find the direction of thrust, labeled ϕ in Figure 10.9. By the Law of Sines, we have

$$\frac{\sin\phi}{9.456} = \frac{\sin 17.345^\circ}{4}$$
$$\sin\phi = 9.456\left(\frac{\sin 17.345^\circ}{4}\right).$$

One solution to this equation is

$$\phi = \arcsin\left(\frac{9.456}{4}\sin 17.345^\circ\right) = 44.811^\circ.$$

The engine thrusters should be aimed 44.811° counterclockwise from the direction of Jupiter.

There is another possible value for ϕ, the direction of the engine thrust in Example 3. This is because there are two solutions to the equation

$$\sin\phi = \frac{9.456}{4}\sin 17.345^\circ.$$

The second solution, $180^\circ - 44.811^\circ = 135.189^\circ$, is shown in Figure 10.10. Both values of ϕ serve to counteract the gravity of Mars, but the 44.811° direction helps push the craft toward Jupiter, whereas the 135.189° direction helps push it away from Jupiter. Thus, the first solution is probably preferable.

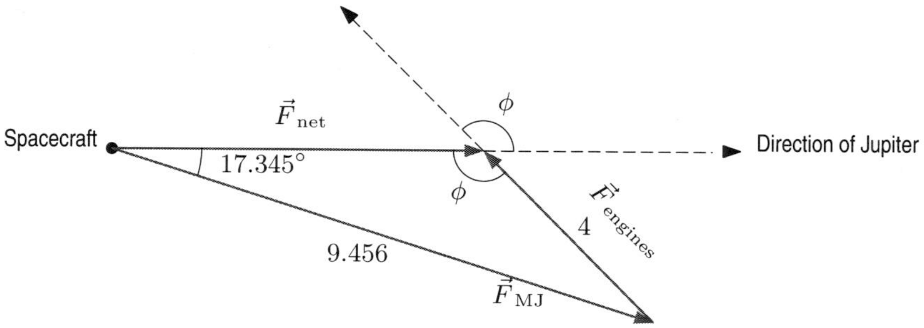

Figure 10.10: There are two possible choices for the direction of thrust. One has the effect of pushing the spacecraft toward Jupiter, the other (shown here) pushes the spacecraft away from Jupiter

Scalar Multiplication

A vector $\vec{u}$ can be added to itself. In Figure 10.11. The resulting vector, $\vec{u} + \vec{u}$, points in the same direction as $\vec{u}$ but is twice as long. The vector $\vec{u} + \vec{u} + \vec{u}$ also points in the same direction as $\vec{u}$, but is three times as long. We write

$$2\vec{u} \quad \text{for} \quad \vec{u} + \vec{u} \qquad \text{and} \qquad 3\vec{u} \quad \text{for} \quad \vec{u} + \vec{u} + \vec{u}.$$

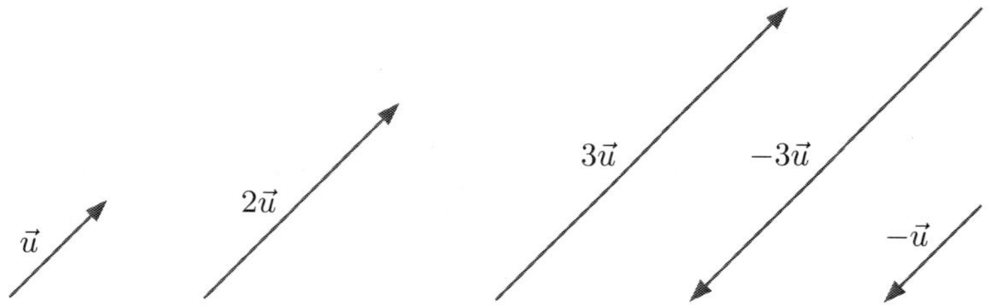

Figure 10.11: The vector $2\vec{u} = \vec{u} + \vec{u}$, points in the same direction as $\vec{u}$ but is twice as long as $\vec{u}$. Similarly, $3\vec{u} = \vec{u} + \vec{u} + \vec{u}$ points in the same direction and is 3 times as long as $\vec{u}$, and $-\vec{u}$ is in the opposite direction as $\vec{u}$

Similarly, $0.5\vec{u}$ points in the same direction as $\vec{u}$ but is only half as long. In general, if $\vec{v}$ is a vector and k a positive number (a positive scalar), then $k\vec{v}$ is a vector that is k times as long as $\vec{v}$ and pointing in the same direction as $\vec{v}$. The vector $k\vec{v}$ is called a *scalar multiple* of $\vec{v}$.

If k is a negative number, the vector $k\vec{v}$ has length $|k|$ times the length of $\vec{v}$, but points in the opposite direction to $\vec{v}$. For example, $-3\vec{u}$ is the vector that is three times as long as $\vec{u}$ but pointing in the opposite direction. See Figure 10.11.

The Zero Vector

The *zero vector*, denoted by the symbol $\vec{0}$, represents no displacement, which is the displacement that leaves you where you started. The zero vector has no direction.

The following rules summarize scalar multiplication.

- If $k > 0$, then $k\vec{v}$ points in the same direction as $\vec{v}$ and is k times as long.
- If $k < 0$, then $k\vec{v}$ points in the opposite direction as $\vec{v}$ and is $|k|$ times as long.
- If $k = 0$, then $k\vec{v} = \vec{0}$, the zero vector.

Alternate view of Subtraction

Since the vector $-\vec{v}$ points in the opposite direction to $\vec{v}$, the difference $\vec{u} - \vec{v}$ is the same as the sum $\vec{u} + (-\vec{v})$. See Figure 10.12.

Figure 10.12: Illustration of the fact that $\vec{u} - \vec{v} = \vec{u} + (-\vec{v})$

Properties of Vector Addition and Scalar Multiplication

The following properties hold true for any three vectors $\vec{u}$, $\vec{v}$, $\vec{w}$ and any two scalars a and b:

1. *Commutativity of addition:* $\vec{u} + \vec{v} = \vec{v} + \vec{u}$
2. *Associativity of addition:* $(\vec{u} + \vec{v}) + \vec{w} = \vec{u} + (\vec{v} + \vec{w})$
3. *Associativity of scalar multiplication:* $a(b\vec{v}) = (ab)\vec{v}$
4. *Distributivity of scalar multiplication:* $(a + b)\vec{v} = a\vec{v} + b\vec{v}$ and $a(\vec{u} + \vec{v}) = a\vec{u} + a\vec{v}$
5. *Identities:* $\vec{v} + \vec{0} = \vec{v}$ and $1 \cdot \vec{v} = \vec{v}$

These properties are analogous to the corresponding properties for addition and multiplication of numbers. In other words, vector addition and scalar multiplication of vectors behave as expected.

Exercises and Problems for Section 10.1

Exercises

In Exercises 1–4, is the given quantity a vector or a scalar?

1. The temperature at a point on the earth's surface.

2. The wind velocity at a point on the earth's surface.

3. The distance from Seattle to St. Louis.

4. The population of the US.

5. The vectors $\vec{w}$ and $\vec{u}$ are in Figure 10.13. Match the vectors $\vec{p}, \vec{q}, \vec{r}, \vec{s}, \vec{t}$ with five of the following vectors: $\vec{u}+\vec{w}, \vec{u}-\vec{w}, \vec{w}-\vec{u}, 2\vec{w}-\vec{u}, \vec{u}-2\vec{w}, 2\vec{w}, -2\vec{w}, 2\vec{u}, -2\vec{u}, -\vec{w}, -\vec{u}$.

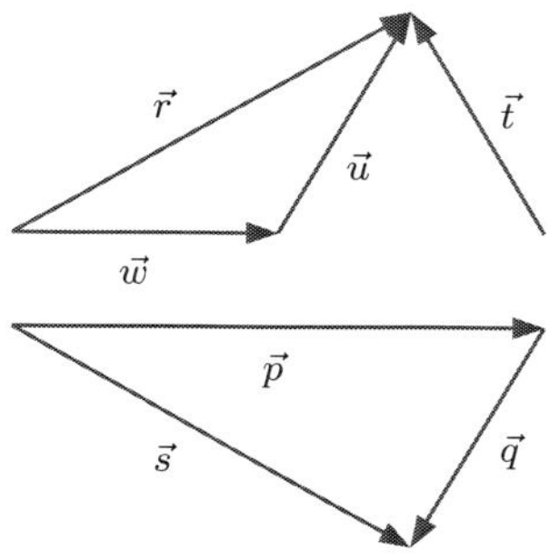

Figure 10.13

Given the displacement vectors $\vec{v}$ and $\vec{w}$ in Figure 10.14, draw the vectors in Exercises 6–11.

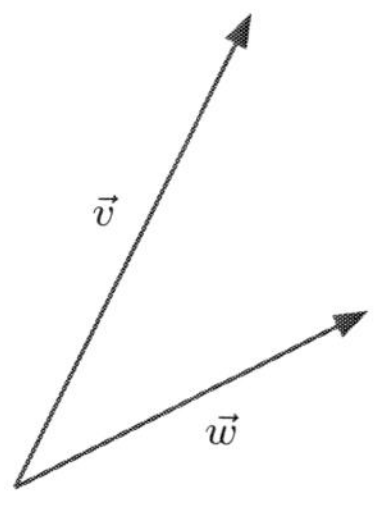

Figure 10.14

6. $\vec{v}+\vec{w}$
7. $\vec{v}-\vec{w}$
8. $2\vec{v}$
9. $\vec{w}+\vec{v}/2$
10. $2\vec{v}+\vec{w}$
11. $\vec{v}-2\vec{w}$

Problems

12. A person leaves home and walks 2 miles due west. She then walks 3 miles southwest. How far away from home is she? In what direction must she walk to head directly home? (Your answer should be an angle in degrees and should include a sketch.)

13. The person from Problem 12 next walks 4 miles southeast. How far away from home is she? In what direction must she walk to head directly home?

14. A hockey puck starts on the edge of the rink and slides with a constant velocity $\vec{v}$, at a speed of 7 ft/sec and an angle of 35° with the edge. After 2 seconds, how far has the puck traveled? How far is it from the edge?

15. Oracle Road heads due north from its intersection with Route 10, which heads 20° west of north.

    (a) If you travel 5 miles up Route 10 from the intersection, how far are you from Oracle Road?
    (b) How far do you have to travel up Route 10 from the intersection to be 2 miles from Oracle Road?

16. A helicopter is hovering at 3000 meters directly over the eastern perimeter of a secret Air Force installation. The eastern perimeter is 5000 meters from the installation headquarters. The helicopter receives a transmission from headquarters that a UFO has been sighted at 7500 meters directly over the installation's western perimeter, which is 9000 meters from the headquarters. How far must the helicopter travel to intercept the UFO? In what direction must it head? Be specific.

17. Suppose instead the UFO in Problem 16 is sighted directly over the installation's northern perimeter, which is 12,000 meters from headquarters. How far must the helicopter travel to intercept the UFO? In what direction must it head? Be specific.

18. **(a)** A kite on a 50 foot string is flying at an angle of 20° with flat ground. What is the magnitude and the direction of the vector from the kite to the ground?
    (b) A wind blows the kite upward so that its angle with the ground is 40°. How far is the new position from the original position?

19. A ball is thrown horizontally at 5 feet per second relative to still air. At the same time a wind blows at 3 feet per second at an angle of 45° to the ball's path. What is the velocity of the ball relative to the ground? (There are two answers)

Use the definitions of addition and scalar multiplication to explain each of the properties in Problems 20–26.

20. $\vec{w}+\vec{v}=\vec{v}+\vec{w}$
21. $(a+b)\vec{v}=a\vec{v}+b\vec{v}$
22. $a(\vec{v}+\vec{w})=a\vec{v}+a\vec{w}$
23. $\vec{v}+\vec{0}=\vec{v}$
24. $1\vec{v}=\vec{v}$
25. $\vec{v}+(-1)\vec{w}=\vec{v}-\vec{w}$
26. $(\vec{u}+\vec{v})+\vec{w}=\vec{u}+(\vec{v}+\vec{w})$

10.2 THE COMPONENTS OF A VECTOR

So far, vectors have been described in terms of length and direction. There is another way to think about a vector, and that is in terms of *components*. We begin with an example illustrating the use of components; we define this term precisely later.

Example 1 A ship travels 200 miles in a direction that its compass says is due east. The captain then discovers that the compass is faulty and that the ship has actually been heading in a direction that is 17.4° to the north of due east. How far north is the ship from its intended course?

Solution In Figure 10.15, a vector $\vec{v}$ has been drawn for the ship's actual displacement. We know that $||\vec{v}\,|| =$ 200 miles and the direction of $\vec{v}$ is 17.4° to the north of due east. We see that $\vec{v}$ can be considered the sum of two vectors, one pointing due east and the other pointing due north, so

$$\vec{v} = \vec{v}_{\text{north}} + \vec{v}_{\text{east}}.$$

From Figure 10.15, we have

$$\sin 17.4^\circ = \frac{||\vec{v}_{\text{north}}||}{200},$$

so

$$||\vec{v}_{\text{north}}|| = 200 \sin 17.4^\circ = 59.808 \text{ miles}.$$

Therefore, the ship is 59.808 miles due north of its intended course.

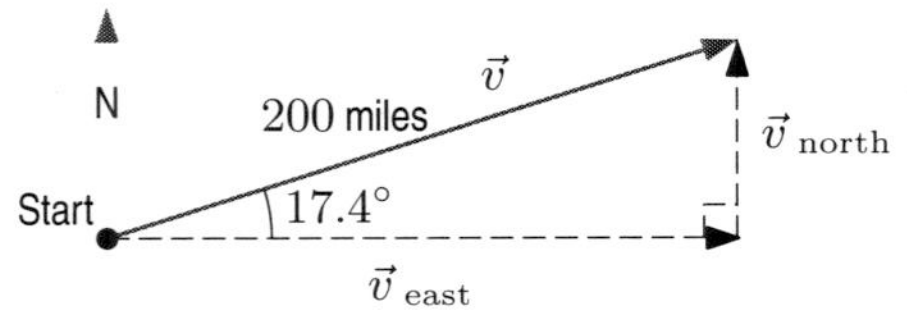

Figure 10.15: How far north is the ship from its intended eastward course?

Unit Vectors

A *unit vector* is a vector of length 1 unit. The unit vectors in the directions of the positive x- and y-axes are called $\vec{i}$ and $\vec{j}$, respectively. See Figure 10.16. These two vectors are important because any vector in the plane can be expressed in terms of $\vec{i}$ and $\vec{j}$.

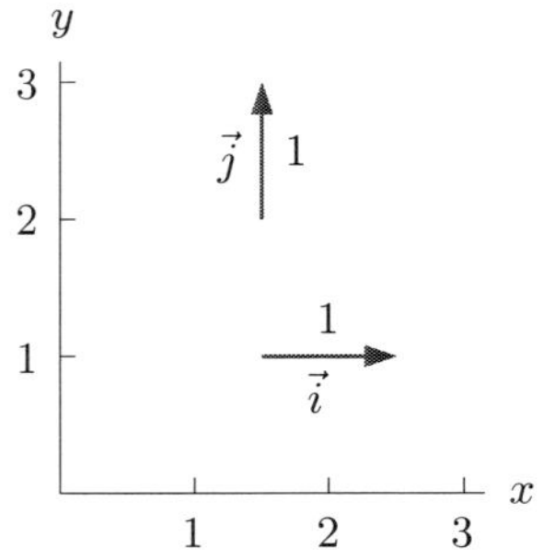

Figure 10.16: The unit vectors $\vec{i}$ and $\vec{j}$

Example 2 Let the x-axis point east and the y-axis point north. A person walks 3 miles east and then 4 miles north. Express her displacement in terms of the unit vectors $\vec{i}$ and $\vec{j}$.

Solution The unit of distance is miles, so the unit vector $\vec{i}$ represents a displacement of 1 mile east, and the unit vector $\vec{j}$ represents a displacement of 1 mile north. The person's displacement can be written

$$\text{Displacement} = 3 \text{ miles east} + 4 \text{ miles north} = 3\vec{i} + 4\vec{j}.$$

Example 3 Let the x-axis points east and the y-axis points north. Describe the displacement $5\vec{i} - 7\vec{j}$ in words.

Solution Since $\vec{i}$ represents a 1 mile displacement to the east, $5\vec{i}$ represents a 5 mile displacement to the east. Since $\vec{j}$ represents a 1 mile displacement to the north, $-7\vec{j}$ represents a 7 mile displacement to the south. Thus, $5\vec{i} - 7\vec{j}$ represents a 5 mile displacement to the east followed by a 7 mile displacement to the south.

Resolving a Vector in the Plane into Components

In this book, we focus chiefly on vectors that lie flat in the plane. Such vectors can always be broken (or *resolved*) into sum of two vectors,[2] one parallel to the x-axis and the other to the y-axis. These two vectors are called the *components* parallel to the axes.

For the vector $\vec{v}$ in Figure 10.17, we can write

$$\vec{v} = \vec{v}_1 + \vec{v}_2,$$

where $\vec{v}_1$ is called the *x-component* of $\vec{v}$ and $\vec{v}_2$ is the *y-component* of $\vec{v}$. What can we say about the components $\vec{v}_1$ and $\vec{v}_2$?

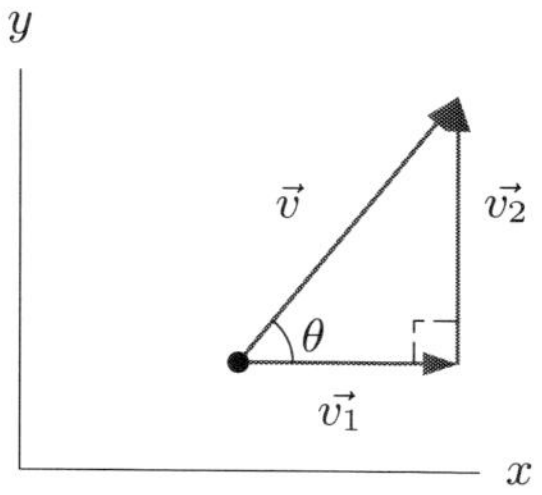

Figure 10.17: A vector $\vec{v}$ can be written as the sum of components, $\vec{v}_1$ and $\vec{v}_2$, parallel to the x- and y-axes

In Figure 10.17, the vector $\vec{v}$ has length $||\vec{v}||$ and makes an angle θ with the positive x-axis, with $0 \le \theta < 2\pi$. Thus, we see that

$$\cos\theta = \frac{||\vec{v}_1||}{||\vec{v}||} \quad \text{and} \quad \sin\theta = \frac{||\vec{v}_2||}{||\vec{v}||}.$$

Therefore

$$||\vec{v}_1|| = ||\vec{v}||\cos\theta \quad \text{and} \quad ||\vec{v}_2|| = ||\vec{v}||\sin\theta.$$

We know that $\vec{v}_1$ is parallel to the x-axis, so it is a scalar multiple of the unit vector $\vec{i}$. Similarly, $\vec{v}_2$ is a scalar multiple of $\vec{j}$. After checking that the signs of $\cos\theta$ and $\sin\theta$ give the correct directions for $\vec{v}_1$ and $\vec{v}_2$, we have

$$\vec{v}_1 = (||\vec{v}||\cos\theta)\vec{i} \quad \text{and} \quad \vec{v}_2 = (||\vec{v}||\sin\theta)\vec{j}.$$

These are the x- and y- components of $\vec{v}$. We have the following result:

> In the plane, a vector $\vec{v}$ of length $||\vec{v}||$ which makes an angle θ with the positive x-axis can be written in terms of its components:
>
> $$\vec{v} = (||\vec{v}||\cos\theta)\vec{i} + (||\vec{v}||\sin\theta)\vec{j}.$$

[2]However, one or both of the vectors may be of zero length.

Example 4 In Example 1, suppose the x-axis points east and the y-axis points north. Resolve the displacement vector $\vec{v}$ into components parallel to the axes.

Solution The vector $\vec{v}_{\text{north}}$ is the y-component of $\vec{v}$ and $\vec{v}_{\text{east}}$ is the x-component. We already know that $||\vec{v}_{\text{north}}|| = 59.808$ miles. From Figure 10.15, we see that $||\vec{v}_{\text{east}}|| = 200\cos 17.4^\circ = 190.848$ miles. (This last piece of information tells us that although the ship traveled a total of 200 miles, its eastward progress amounted to only 190.848 miles.) The easterly and northerly components are

$$\vec{v}_{\text{east}} = 190.848\vec{i} \qquad \text{and} \qquad \vec{v}_{\text{north}} = 59.808\vec{j}.$$

Thus, the course taken by the ship can be written as follows:

$$\vec{v} = \vec{v}_{\text{east}} + \vec{v}_{\text{north}} = 190.848\vec{i} + 59.808\vec{j}.$$

To check the calculation, notice that

$$||\vec{v}_{\text{east}}||^2 + ||\vec{v}_{\text{north}}||^2 = (190.848)^2 + (59.808)^2 \approx 40,000 = ||\vec{v}||^2.$$

Sums and scalar multiples are easy to compute if vectors are written in terms of components. For a sum, add corresponding components; for a multiple, each component is multiplied by the scalar.

Example 5 Let $\vec{u} = 5\vec{i} + 7\vec{j}$, $\vec{v} = \vec{i} + \vec{j}$, and $\vec{w} = 6\vec{j} - 3\vec{i} + 2\vec{j} - 2\vec{i} - 4\vec{j} + 7\vec{i}$. Describe in words the similarities and differences between the following displacements:

$$2\vec{i} + 4\vec{j}, \quad \vec{u} - 3\vec{v}, \quad \vec{w}.$$

Solution We show that the displacements $2\vec{i} + 4\vec{j}$, $\vec{u} - 3\vec{v}$, and $\vec{w}$ are all the same. Using the properties on page 454, we have

$$\begin{aligned}\vec{u} - 3\vec{v} &= (5\vec{i} + 7\vec{j}) - 3(\vec{i} + \vec{j})\\ &= (5\vec{i} + 7\vec{j}) - (3\vec{i} + 3\vec{j})\\ &= (5\vec{i} - 3\vec{i}) + (7\vec{j} - 3\vec{j}) = 2\vec{i} + 4\vec{j}.\end{aligned}$$

The vector $\vec{w}$ can be simplified as

$$\vec{w} = (-3\vec{i} - 2\vec{i} + 7\vec{i}) + (6\vec{j} + 2\vec{j} - 4\vec{j}) = 2\vec{i} + 4\vec{j}.$$

Alternatively, we can see that all three displacements are the same by doing head-to-tail addition for each case, with the same starting point. Figure 10.18 shows that all three end points are the same, although the path followed is different in each case.

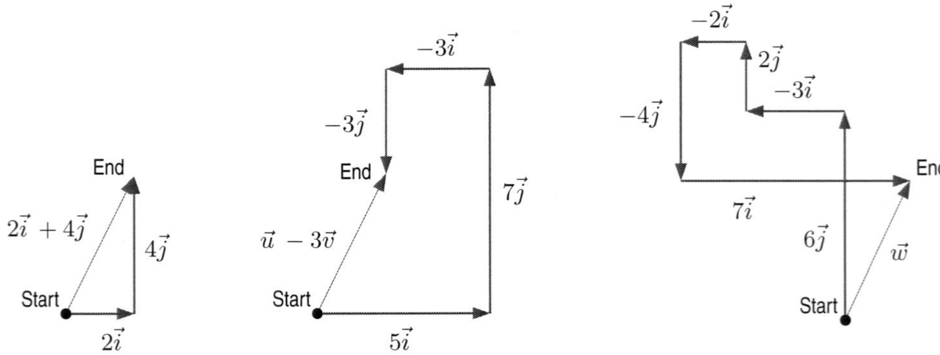

Figure 10.18: These three displacements, $2\vec{i} + 4\vec{j}$, $\vec{u} - 3\vec{v}$, $\vec{w}$, are all the same

Displacement Vectors

Vectors that indicate displacement are often designated by giving two points that specify the initial and final positions. For example, the vector $\vec{v} = \overrightarrow{PQ}$ describes a displacement or motion from point P to point Q. If we know the coordinates of P and Q, we can easily find the components of $\vec{v}$.

Example 6 If P is the point $(-3, 1)$ and Q is the point $(2, 4)$, find the components of the vector $\overrightarrow{PQ}$.

Solution Figure 10.19 shows the vector $\overrightarrow{PQ}$ and its components $\vec{v}_1$ and $\vec{v}_2$. The length of $\vec{v}_1$ is the horizontal distance from P to Q, so

$$||\vec{v}_1|| = 2 - (-3) = 5.$$

The length of $\vec{v}_2$ is the vertical distance from P to Q, so

$$||\vec{v}_2|| = 4 - 1 = 3.$$

Thus, the components are $\vec{v}_1 = 5\vec{i}$ and $\vec{v}_2 = 3\vec{j}$. We write $\vec{v} = \overrightarrow{PQ} = 5\vec{i} + 3\vec{j}$.

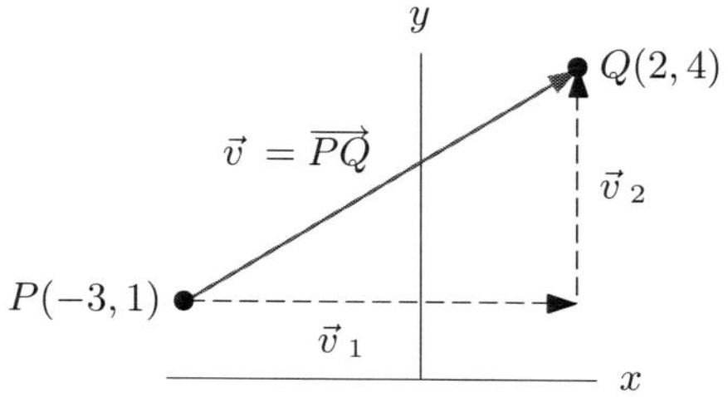

Figure 10.19: Finding components $\vec{v}_1$ and $\vec{v}_2$ from the coordinates of the points P and Q

In general, we have the following result:

> If $P = (x_1, y_1)$ and $Q = (x_2, y_2)$, then the vector $\vec{v} = \overrightarrow{PQ}$ from P to Q is given by
>
> $$\vec{v} = (x_2 - x_1)\vec{i} + (y_2 - y_1)\vec{j}.$$

Vectors in n Dimensions

So far we have considered 2-dimensional vectors which have two components. Vectors in space need three components and they are also very useful. In addition, there are *n-dimensional* vectors, which have n components, where n is a positive integer.

Example 7 A balloon rises vertically a distance of 2 miles, floats west a distance of 3 miles, and floats north a distance of 4 miles. The x-axis points eastward, the y-axis points northward, and the z-axis points upward. The balloon's displacement vector has three components: a vertical component, an eastward component, and a northward component. Therefore, this displacement is a 3-dimensional vector. We write this vector as

$$\text{Displacement} = -3\vec{i} + 4\vec{j} + 2\vec{k}.$$

Exercises and Problems for Section 10.2

Exercises

For Exercises 1–4, perform the indicated computations.

1. $5(2\vec{i} - \vec{j}) + \vec{j}$
2. $(4\vec{i} + 2\vec{j}) - (3\vec{i} - \vec{j})$
3. $-(\vec{i} + \vec{j}) + 2(2\vec{i} - 3\vec{j})$
4. $(\vec{i} + 2\vec{j}) + (-3)(2\vec{i} + \vec{j})$

Resolve the vectors in Exercises 5–7 into components.

5. A vector starting at the point $Q = (4, 6)$ and ending at the point $P = (1, 2)$.
6. A vector starting at the point $P = (1, 2)$ and ending at the point $Q = (4, 6)$.
7. The vector shown in Figure 10.20, with components expressed in inches.

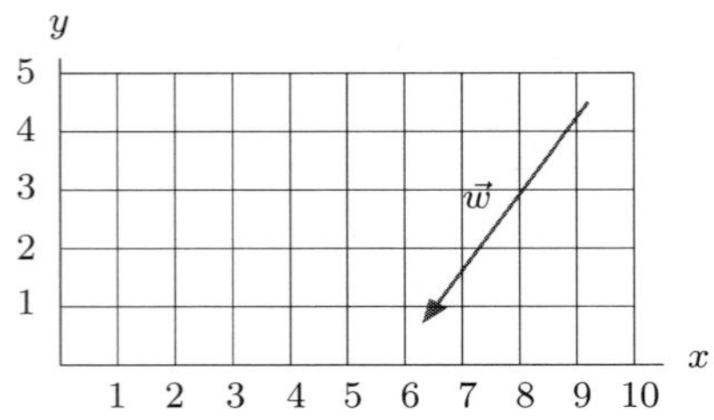

Figure 10.20: Scale: 1 unit on the x- and y-axes is 0.25 inches

8. On the graph in Figure 10.21, draw the vector $\vec{v} = 4\vec{i} + \vec{j}$ twice, once with its tail at the origin and once with its tail at the point $(3, 2)$.

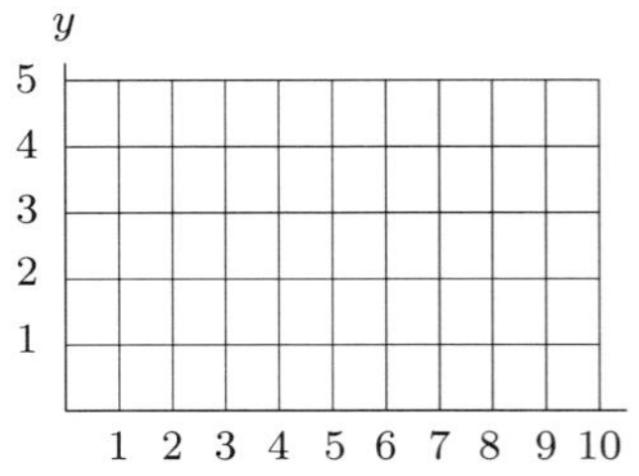

Figure 10.21

Find the length of the vectors in Exercises 9–12.

9. $\vec{v} = \vec{i} - \vec{j} + 2\vec{k}$
10. $\vec{v} = \vec{i} - \vec{j} + 3\vec{k}$
11. $\vec{v} = 7.2\vec{i} - 1.5\vec{j} + 2.1\vec{k}$
12. $\vec{v} = 1.2\vec{i} - 3.6\vec{j} + 4.1\vec{k}$

Problems

13. A car is traveling at a speed of 50 km/hr. The positive y-axis is north and the positive x-axis is east. Resolve the car's velocity vector into two components if the car is traveling in each of the following directions:

    (a) East **(b)** South
    (c) Southeast **(d)** Northwest

In Problems 14–16, use the information in Figure 10.22. Each grid square is 1 unit along each side.

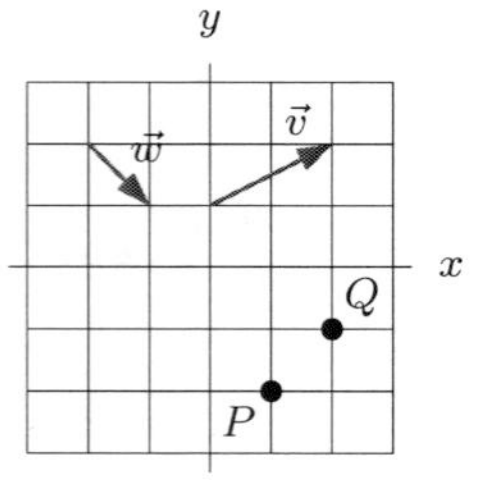

Figure 10.22

14. Write the following vectors in component form.

    (a) $\vec{v}$ **(b)** $2\vec{w}$ **(c)** $\vec{v} + \vec{w}$ **(d)** $\vec{w} - \vec{v}$
    (e) The displacement vector $\overrightarrow{PQ}$.
    (f) The vector from the point P to the point $(2, 0)$.
    (g) A vector perpendicular to the y-axis.
    (h) A vector perpendicular to the x-axis.

15. What is the angle between the vector $\vec{w}$ and the negative y-axis?
16. What is the angle between the vector $\vec{w}$ and the displacement vector $\overrightarrow{PQ}$?
17. Shortly after takeoff, a plane is climbing northwest through still air at an airspeed of 200 km/hr and rising at a rate of 300 m/min. Resolve into components its velocity vector in a coordinate system in which the x-axis points east, the y-axis points north, and the z-axis points up.

18. **(a)** Find a unit vector from the point $P = (1, 2)$ toward the point $Q = (4, 6)$.
(b) Find a vector of length 10 pointing in the same direction.

19. Which is traveling faster, a car whose velocity vector is $21\vec{i} + 35\vec{j}$, or a car whose velocity vector is $40\vec{i}$, assuming that the units are the same for both directions?

20. A truck is traveling due north at 30 km/hr toward a crossroad. On a perpendicular road a police car is traveling west toward the intersection at 40 km/hr. Both vehicles will reach the crossroad in exactly one hour. Find the vector currently representing the displacement of the truck with respect to the police car.

21. Which of the following vectors are parallel?

$$\vec{u} = 2\vec{i} + 4\vec{j} - 2\vec{k}, \quad \vec{v} = \vec{i} - \vec{j} + 3\vec{k},$$
$$\vec{w} = -\vec{i} - 2\vec{j} + \vec{k}, \quad \vec{p} = \vec{i} + \vec{j} + \vec{k},$$
$$\vec{q} = 4\vec{i} - 4\vec{j} + 12\vec{k}, \vec{r} = \vec{i} - \vec{j} + \vec{k}.$$

22. **(a)** A man swims northeast at 5 mph. Give a vector representing his velocity.
(b) He now swims in a river at the same velocity relative to still water; the river's current flows north at 1.2 mph. What is his velocity relative to the riverbed?

23. The coastline is along the x-axis and the sea is in the region $y > 0$. A sailing boat has velocity $\vec{v} = 2.5\vec{i} + 3.1\vec{j}$ mph. What is the speed of the boat? What angle does the boat's path make with the coastline?

24. The hour hand and the minute hand of a clock are represented by the vectors $\vec{h}$ and $\vec{m}$, respectively, with $\|\vec{h}\| = 2$ and $\|\vec{m}\| = 3$. The origin is at the center of the clock and the positive x-axis goes through the three o'clock position.

(a) What are the components of $\vec{h}$ and $\vec{m}$ at the following times? Illustrate with a sketch.

(i) 12 noon (ii) 3 pm
(iii) 1 pm (iv) 1:30 pm

(b) Sketch the displacement vector from the tip of the hour hand to the tip of the minute hand at 3 pm. What are the components of this displacement vector?
(c) Sketch the vector representing the sum $\vec{h} + \vec{m}$ at 1:30 pm. What are the components of this sum?

A cat on the ground at the point $(1, 4, 0)$ watches a squirrel at the top of a tree. The tree is one unit high with base at the point $(2, 4, 0)$. Find the displacement vectors in Problems 25–28.

25. From the origin to the cat.

26. From the bottom of the tree to the squirrel.

27. From the bottom of the tree to the cat.

28. From the cat to the squirrel.

10.3 APPLICATION OF VECTORS

Alternate Notation for the Components of a Vector

The vector $\vec{v} = 3\vec{i} + 4\vec{j}$ is sometimes written $\vec{v} = (3, 4)$. This notation can be confused with the coordinate notation used for points. For instance, $(3, 4)$ might mean the point $x = 3$, $y = 4$ or the vector $3\vec{i} + 4\vec{j}$. Nevertheless, this notation is useful for vectors in n dimensions.

Population Vectors

Vectors in n dimensions are useful for keeping track of n quantities. We add them by adding the corresponding components; scalar multiples are obtained by multiplying each component by the scalar.

Example 1 The 2005 population of the six New England states (Connecticut, Maine, Massachusetts, New Hampshire, Rhode Island, and Vermont) can be thought of as a 6-dimensional vector $\vec{P}$. The components of $\vec{P}$ are the populations of the six states. Using the alternate notation, we write

$$\vec{P} = (P_{\mathrm{CT}}, P_{\mathrm{ME}}, P_{\mathrm{MA}}, P_{\mathrm{NH}}, P_{\mathrm{RI}}, P_{\mathrm{VT}}),$$

where P_{CT} is the population of Connecticut, P_{ME} is the population of Maine, and so on.

The population vector $\vec{P}$ from the previous example does not have a geometrical interpretation. We cannot draw a picture of $\vec{P}$, or interpret its length and direction as we do for displacement vectors. However, the following example shows how it is used.

Example 2 The vectors $\vec{P}$ and $\vec{Q}$ give the populations of the six New England states in 1995 and 2005, respectively. According to the Census Bureau,[3] these vectors are given, in millions of people, by

$$\begin{aligned}\vec{P} &= (3.28, 1.24, 6.07, 1.15, 0.99, 0.59)\\ \vec{Q} &= (3.51, 1.32, 6.40, 1.31, 1.08, 0.62).\end{aligned}$$

For instance, the population of Connecticut was 3.28 million in 1995 and 3.51 million in 2005.

(a) Find $\vec{R} = \vec{Q} - \vec{P}$. Explain its significance in terms of the population of New England.
(b) Let $\vec{S}$ be the estimated population of New England in the year 2015. One expert predicts that $\vec{S} = \vec{Q} + 2\vec{R}$. Find $\vec{S}$ and explain the assumption this expert is making about the New England population.
(c) Another expert makes a different prediction, claiming that $\vec{T} = 1.08\vec{Q}$ will give the population of New England in 2015. Find $\vec{T}$ and explain the assumption that this expert is making.

Solution (a) We have

$$\begin{aligned}\vec{R} &= \vec{Q} - \vec{P}\\ &= (3.51, 1.32, 6.40, 1.31, 1.08, 0.62) - (3.28, 1.24, 6.07, 1.15, 0.99, 0.59)\\ &= (3.51 - 3.28, 1.32 - 1.24, 6.40 - 6.07, 1.31 - 1.15, 1.08 - 0.99, 0.62 - 0.59)\\ &= (0.23, 0.08, 0.33, 0.16, 0.09, 0.03).\end{aligned}$$

The components of $\vec{R}$ give the change in population for each New England state. For instance, the population of Connecticut rose from 3.28 million in 1995 to 3.51 million in 2005, a change of $3.51 - 3.28 = 0.23$ million people, so $R_{\text{CT}} = 0.23$.

(b) The formula $\vec{S} = \vec{Q} + 2\vec{R}$ means that

$$\text{Population in 2015} = \text{Population in 2005} + 2 \cdot \text{Change between 1995 and 2005}.$$

According to this expert, states will see their populations climb twice as much between 2005 and 2015 as they did between 1995 and 2005. Algebraically, we have

$$\begin{aligned}\vec{S} &= \vec{Q} + 2\vec{R}\\ &= (3.51, 1.32, 6.40, 1.31, 1.08, 0.62) + 2 \cdot (0.23, 0.08, 0.33, 0.16, 0.09, 0.03)\\ &= (3.51, 1.32, 6.40, 1.31, 1.08, 0.62) + (0.46, 0.16, 0.66, 0.32, 0.18, 0.06)\\ &= (3.97, 1.48, 7.06, 1.63, 1.26, 0.68).\end{aligned}$$

For instance, Connecticut's population climbed by 0.23 million between 1995 and 2005. The predicted population of Connecticut in 2015 is given by

$$S_{\text{CT}} = Q_{\text{CT}} + 2 \cdot R_{\text{CT}} = 3.51 + 2 \cdot (0.23) = 3.97.$$

The 0.46 million jump in Connecticut's population between 2005 and 2015 is twice as large as the 0.23 million jump between 1995 and 2005.

[3]From: http://www.census.gov, January 10 2006.

(c) The formula $\vec{T} = 1.08\vec{Q}$ means that each component of $\vec{T}$ is 1.08 times as large as the corresponding component of $\vec{Q}$. In other words, the population of each state is predicted to grow by 8%. We have

$$\begin{aligned}\vec{T} = 1.08\vec{Q} &= 1.08 \cdot (3.51, 1.32, 6.40, 1.31, 1.08, 0.62)\\ &= (1.08 \cdot 3.51, 1.08 \cdot 1.32, 1.08 \cdot 6.40, 1.08 \cdot 1.31, 1.08 \cdot 1.08, 1.08 \cdot 0.62)\\ &= (3.79, 1.43, 6.91, 1.41, 1.17, 0.67).\end{aligned}$$

For instance, the population of New Hampshire is predicted to grow from $Q_{\text{NH}} = 1.31$ million in 2005 to $T_{\text{NH}} = 1.08Q_{\text{NH}} = 1.08(1.31) = 1.41$ million in the year 2015.

Notice how vector addition and subtraction was used in parts (a) and (b) of the previous example, and scalar multiplication was used in parts (b) and (c).

Economics

The blockbuster movie *Harry Potter and the Philosopher's Stone* grossed \$292 million at domestic (US) box offices, more than any other domestic release in 2001.[4] It made even more money at international box offices, where it grossed \$459 million. Rounding out its wild success is the \$224 million for video and DVD sales rentals as of midyear 2002, for a grand total of just under one billion dollars. The *revenue vector* $\vec{r} = (292, 459, 224)$ summarizes the domestic, international, and video/DVD gross revenues in millions of dollars.

Example 3 After *Harry Potter*, the three next most financially successful US movies in 2001 were, in order, *Shrek*, *Pearl Harbor*, and *The Mummy Returns*. If $\vec{s}$, $\vec{t}$, and $\vec{u}$ are the revenue vectors for these three movies, then

$$\vec{s} = (268, 198, 436) \qquad \vec{t} = (199, 252, 301) \qquad \vec{u} = (202, 227, 192).$$

Find $\vec{R}$, the total revenue vector for all three of these movies.

Solution The total revenue, $\vec{R}$, is the sum of $\vec{s}$, $\vec{t}$, and $\vec{u}$:

$$\vec{R} = \vec{s} + \vec{t} + \vec{u} = (268 + 199 + 202, 198 + 252 + 227, 436 + 301 + 192) = (669, 677, 929).$$

Price and Consumption Vectors

A car dealership has several different models of cars in its inventory, with different prices for each model. A *price vector*, $\vec{P}$, gives the price of each model:

$$\vec{P} = (P_1, P_2, \ldots, P_n).$$

Here P_1 is the price of car model 1 and P_2 is the price of model 2, and so on. A *consumption vector* $\vec{C}$ gives the number of each model of car purchased (or consumed) during a given month:

$$\vec{C} = (C_1, C_2, \ldots, C_n).$$

[4] Data compiled from www.variety.com and www.videobusiness.com. Video and DVD sales and rental data as of midyear 2002.

Example 4 Suppose $\vec{I}$ gives the number of cars of each model a car dealer currently has in inventory. Explain the meaning of the following in terms of the car dealership.

(a) $\vec{I} - \vec{C}$ (b) $\vec{I} - 2\vec{C}$ (c) $\vec{C} = 0.3\vec{I}$

Solution

(a) This expression represents the difference between the number of each model currently in inventory and the number of that model purchased each month. So $\vec{I} - \vec{C}$ represents the numbers of each model that remain on the lot after one month.
(b) This expression represents the number of each model the dealer has left after two months, assuming no new cars are added to inventory and that $\vec{C}$ does not change.
(c) This equation tells us that 30% of the inventory of each model is purchased by consumers each month.

Example 5 Let $\vec{E}$ represent expenses incurred by the dealer for acquisition, insurance, and overhead, for each model. What does the vector $\vec{P} - \vec{E}$ represent?

Solution This represents the dealer's profit for each model. For instance, a model that the dealer sells for \$29,000 may cost \$25,000 to acquire from the factory, insure, and maintain. The difference of \$4,000 represents profit for the dealer. The vector $\vec{P} - \vec{E}$ keeps track of this profit for each model.

Physics

Newton's law of gravitation states that the magnitude of force exerted on an object of mass m by the earth is given by

$$\|\vec{F}_E\| = G\frac{mM_E}{r_E^2},$$

where M_E is the mass of the earth, r_E is the distance from the object to earth's center, and G is a constant. Similarly, the force exerted on the object by the moon has magnitude

$$\|\vec{F}_L\| = G\frac{mM_L}{r_L^2},$$

where M_L is the mass of the moon and r_L is the distance to the moon's center. The force exerted by the earth is directed toward the earth; the force exerted by the moon is exerted toward the moon.

Example 6 Figure 10.23 shows the position of a 100,000 kg spacecraft relative to the earth and moon. The vector $\vec{r}$ is the position of the moon relative to the earth, $\vec{r}_E$ is the position of the spacecraft relative to the earth, and $\vec{r}_L$ is the position of the spacecraft relative to the moon. With distances in thousands of kilometers,

$$\vec{r} = 384\vec{i} \quad \text{and} \quad \vec{r}_E = 280\vec{i} + 90\vec{j}.$$

Which force is stronger, the pull of the moon on the spacecraft or the pull of the earth? [Use $M_E = 5.98 \cdot 10^{24}$ kg, $M_L = 7.34 \cdot 10^{22}$ kg, and $G = 6.67 \cdot 10^{-23}$ if distance is in thousands of kilometers.]

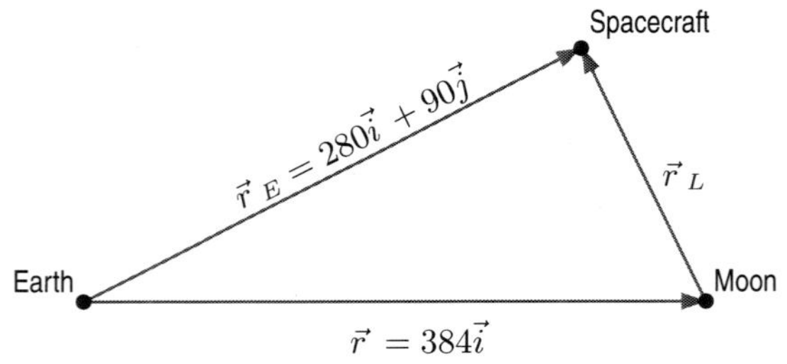

Figure 10.23: Gravitational forces between the earth, the moon, and a 100,000 kg spacecraft

Solution To use Newton's law of gravitation, we first find $||\vec{r}_E||$ and $||\vec{r}_L||$. Since $\vec{r}_E = 280\vec{i} + 90\vec{j}$,

$$||\vec{r}_E|| = \sqrt{(280)^2 + (90)^2} = \sqrt{86500} = 294.109.$$

Since $\vec{r}_L = \vec{r}_E - \vec{r}$, we have

$$\vec{r}_L = (280\vec{i} + 90\vec{j}) - 384\vec{i} = -104\vec{i} + 90\vec{j}.$$

Thus,

$$||\vec{r}_L|| = \sqrt{(-104)^2 + (90)^2} = \sqrt{18916} = 137.535.$$

Using the given values of m, M_E, M_L, and G, we calculate

$$F_E = G\frac{mM_E}{||\vec{r}_E||^2} = 461.116 \quad \text{and} \quad F_L = G\frac{mM_L}{||\vec{r}_L||^2} = 25.882.$$

Thus, the pull of the earth on this spacecraft is much stronger than the pull of the moon; in fact, it is over ten times as strong. The earth's pull is stronger, even though the spaceship is closer to the moon, because the earth is much more massive than the moon.

Computer Graphics: Position Vectors

Video games usually incorporate computer generated graphics, as do the flight simulators used by commercial and military flight training schools. Hollywood makes increasing use of computer graphics in its films. Enormous amounts of computation are involved in creating such effects and much of this computation involves vectors.

A computer screen can be thought of as an xy-grid with the origin at the lower left corner. The position of a point on the screen is specified by a vector pointing from the origin to the point. Such a vector is called a *position vector*. The tail of a position vector is always fixed at the origin.

Example 7 A video game shows two airplanes on the screen at the points $(3, 5)$ and $(7, 2)$. See Figure 10.24. Both airplanes move a distance of 3 units at an angle of 70° counterclockwise from the x-axis. What are the new positions of the airplanes?

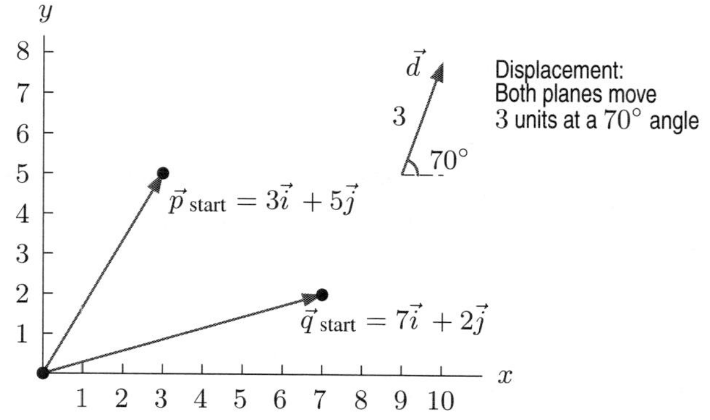

Figure 10.24: Computer screen showing two airplanes moving by a displacement vector $\vec{v}$

Solution The initial position of the first airplane is given by the position vector $\vec{p}_{\text{start}} = 3\vec{i} + 5\vec{j}$. The plane's displacement can also be thought of as a vector, $\vec{d}$. Resolving $\vec{d}$ into components:

$$\vec{d} = (3\cos 70°)\vec{i} + (3\sin 70°)\vec{j} = 1.026\vec{i} + 2.819\vec{j}.$$

The airplane's final position, $\vec{p}_{\text{end}}$, is given by

$$\vec{p}_{\text{end}} = \vec{p}_{\text{start}} + \vec{d} = 3\vec{i} + 5\vec{j} + 1.026\vec{i} + 2.819\vec{j}$$
$$= 4.026\vec{i} + 7.819\vec{j}.$$

The second airplane's initial position is given by the position vector $\vec{q}_{\text{start}} = 7\vec{i} + 2\vec{j}$. The second plane's displacement, however, is exactly the same as the first airplane's, $\vec{d} = 1.026\vec{i} + 2.819\vec{j}$. Therefore, the final position of the second airplane, $\vec{q}_{\text{end}}$, is given by

$$\vec{q}_{\text{end}} = \vec{q}_{\text{start}} + \vec{d} = 7\vec{i} + 2\vec{j} + 1.026\vec{i} + 2.819\vec{j}$$
$$= 8.026\vec{i} + 4.819\vec{j}.$$

Example 7 illustrates the difference between position vectors and other vectors: The tail of a position vector is fixed to the origin, but the tail of the displacement vector $\vec{v}$ can be anywhere, so long as its length and orientation do not change.

Exercises and Problems for Section 10.3

Exercises

In Exercises 1–6, find the vector using $\vec{N} = (5, 6, 7, 8, 9, 10)$ and $\vec{M} = (1, 1, 2, 3, 5, 8)$.

1. $\vec{G} = \vec{N} + \vec{M}$

2. $\vec{A} = \vec{M} - \vec{N}$

3. $\vec{\epsilon} = 2\vec{N} - 7\vec{M}$

4. $\vec{K} = \dfrac{\vec{N}}{3} + \dfrac{2\vec{N}}{3}$

5. $\vec{p} = 1.067\vec{M} + 2.361\vec{N}$

6. $\vec{Z} = \dfrac{\vec{N}}{3} + \dfrac{\vec{M}}{2}$

In Exercises 7–10, use $\vec{Q}$ from Example 2 on page 462.

7. If the population of each New England state increases by 120,000 people from 2005 to 2010, what is $\vec{S}$, the population vector for 2010?

8. If the population of each New England state increases by 2% from 2005 to 2010, what is $\vec{R}$, the population vector for 2010?

9. If the population of each New England state decreases by 43,000 people from 2005 to 2010, what is $\vec{U}$, the population vector for 2010?

10. If the population of each New England state decreases by 22% from 2005 to 2010, what is $\vec{T}$, the population vector for 2010?

Problems

11. There are five students in a class. Their scores on the midterm (out of 100) are given by the vector $\vec{v} = (73, 80, 91, 65, 84)$. Their scores on the final (out of 100) are given by $\vec{w} = (82, 79, 88, 70, 92)$. The final counts twice as much as the midterm. Find a vector giving the total scores (out of 100) of the students.

12. An airplane is heading northeast at an airspeed of 700 km/hr, but there is a wind blowing from the west at 60 km/hr. In what direction does the plane end up flying? What is its speed relative to the ground? [Hint: Resolve the velocity vectors for the airplane and the wind into components.]

13. An airplane is flying at an airspeed of 600 km/hr in a cross-wind that is blowing from the northeast at a speed of 50 km/hr. In what direction should the plane head to end up going due east?

14. Two children are throwing a ball back-and-forth straight across the back seat of a car. Suppose the ball is being thrown at 10 mph relative to the car and the car is going 25 mph down the road.

(a) Make a sketch showing the relevant velocity vectors.
(b) If one child does not catch the ball and it goes out an open window, what angle does the ball's horizontal motion make with the road?

15. A man walks 5 miles in a direction 30° north of east. He then walks a distance x miles due east. He turns around to look back at his starting point, which is at an angle of 10° south of west.

(a) Make a sketch. Give vectors in $\vec{i}$ and $\vec{j}$ components, for each part of the man's walk.
(b) What is x?
(c) How far is the man from his starting point?

16. Three different electric charges q_1, q_2, and q_3 exert forces on a test charge Q. The forces are, respectively, $\vec{F_1} = (3, 6)$, $\vec{F_2} = (-2, 5)$, and $\vec{F_3} = (7, -4)$. The net force, $\vec{F}_{\text{net}}$ is given by $\vec{F}_{\text{net}} = \vec{F_1} + \vec{F_2} + \vec{F_3}$.

(a) Calculate $\vec{F}_{\text{net}}$.

(b) If a fourth charge q_4 is added, what force $\vec{F_4}$ must it exert on Q so that Q feels no net force at all, that is, so that $\vec{F}_{\text{net}} = \vec{0}$?

17. Let $q = 20$ be an electric charge at the origin $(0, 0)$, and let $Q = 30$ be another charge at position $\vec{r} = (r_x, r_y)$. Then q exerts a force $\vec{F}$ on Q given by *Coulomb's Law*[5]

$$||\vec{F}\,|| = \frac{qQ}{||\vec{r}\,||^2}.$$

The direction of $\vec{F}$ is the same as the direction of $\vec{r}$, so that Q tends to be pushed away[6] from q.

(a) Let the position of Q be $\vec{r} = (3, 5)$. Find $||\vec{r}\,||$.

(b) Using Coulomb's Law, find $||\vec{F}\,||$.

(c) Find the components of $\vec{F} = (F_x, F_y)$. [Hint: $\vec{F}$ points in the same direction as $\vec{r}$, so both vectors make the same angle with the x-axis.]

18. Figure 10.25 shows a rectangle whose four corners are the points

$$a = (2, 1), \quad b = (4, 1), \quad c = (4, 2), \quad \text{and} \quad d = (2, 2).$$

As part of a video game, this rectangle is rotated counterclockwise through an angle of 35° about the origin. See Figure 10.26. What are the new coordinates of the corners of the rectangle?

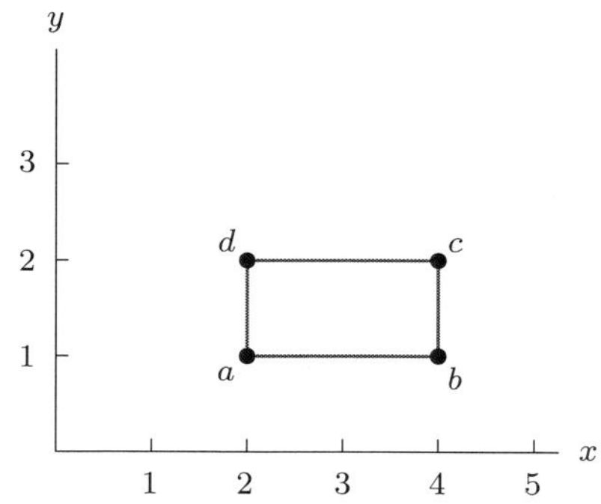

Figure 10.25

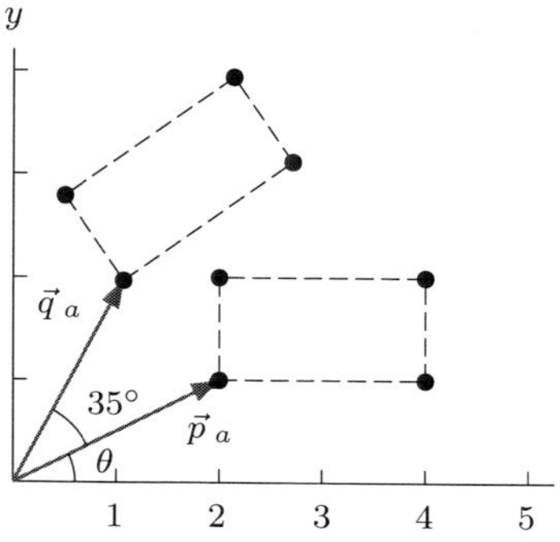

Figure 10.26: Rectangle rotated about the origin through an angle of 35°

19. Provided that $x_1 \neq x_2$ and $y_1 \neq y_2$, the points $p_1 = (x_1, y_1)$ and $p_2 = (x_2, y_2)$ define a rectangle with p_1 in one corner and p_2 in the opposite corner. We represent this rectangle as a 4-vector $\vec{r} = (x_1, y_1, x_2, y_2)$. For instance, Figure 10.27 shows the rectangle described by $\vec{r} = (2, 2, 7, 6)$, which has $p_1 = (2, 2)$ and $p_2 = (7, 6)$.

(a) Sketch the rectangle described by $\vec{s} = (2, 4, 5, 8)$.

(b) Let $\vec{t}$ be a rectangle and let $\vec{u} = (1, 0, 1, 0)$, $\vec{v} = (0, 1, 0, 1)$, and $\vec{w} = \vec{u} + \vec{v}$. Describe in words how the following rectangles are related to $\vec{t}$.

(i) $\vec{t} + \vec{u}$	(ii) $\vec{t} + \vec{v}$
(iii) $\vec{t} - \vec{u}$	(iv) $\vec{t} + 2\vec{u} + 3\vec{v}$
(v) $\vec{t} + \vec{w}$	(vi) $\vec{t} + k\vec{w}$, k a constant.

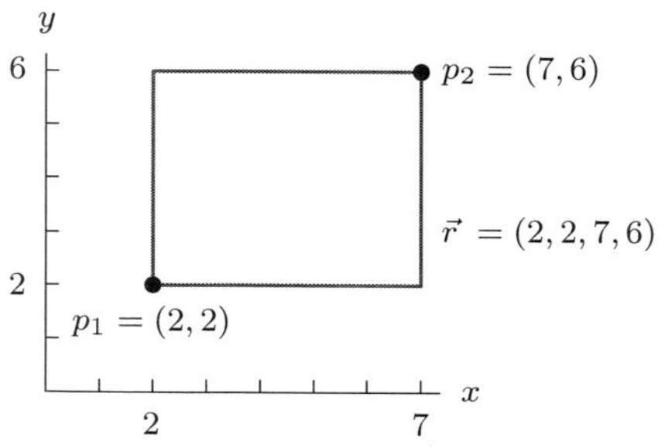

Figure 10.27

[5]Units in the cgs system are centimeters for distance, dynes for force, and electrostatic units (esu) for charge.
[6]This is true because q and Q have the same sign. The force would be opposite to $\vec{r}$ if Q and q had opposite signs.

10.4 THE DOT PRODUCT

We have added and subtracted vectors and we have multiplied vectors by scalars. We now see how to multiply one vector by another vector.

To compute the product of $\vec{u}$ and $\vec{v}$, we multiply each coordinate of $\vec{u}$ by the corresponding coordinate of $\vec{v}$ and add these products. The result is called the *dot product*, written $\vec{u} \cdot \vec{v}$. Notice that the dot product of two vectors is a scalar, not a vector.

Example 1 A car dealer sells five different models of car. The number of each model sold each week is given by the consumption vector $\vec{C} = (22, 14, 8, 12, 19)$. For instance, we see that in one week the dealer sells 22 of the first model, 14 of the second, and so forth.

The price of each model is given by the vector $\vec{P} = (19, 23, 40, 47, 32)$, where the units are \$1000s. Thus, the price of the first model is \$19,000, the price of the second is \$23,000, and so forth. Find the dealer's weekly revenue.

Solution The total revenue earned by the dealer each week (in 1000s of dollars) is given by

$$\begin{aligned}\text{Revenue (in \$1000s)} &= 22\,\text{cars} \cdot 19\,\text{per car} + 14\,\text{cars} \cdot 23\,\text{per car} + 8\,\text{cars} \cdot 40\,\text{per car}\\ &\quad +12\,\text{cars} \cdot 47\,\text{per car} + 19\,\text{cars} \cdot 32\,\text{per car}\\ &= 2232.\end{aligned}$$

The dealer brings in \$2,232,000 each week. Since the revenue is obtained by multiplying the corresponding coordinates of $\vec{C}$ and $\vec{P}$ and adding, we can write revenue as a dot product:

$$\text{Revenue} = \vec{C} \cdot \vec{P}.$$

Note the revenue is a scalar, not a vector, as the dot product always gives a number.

In general, we define the dot product of two vectors as follows:

> if $\vec{u} = (u_1, u_2, \ldots, u_n)$ and $\vec{v} = (v_1, v_2, \ldots, v_n)$ are two n-dimensional vectors, then the **dot product**, $\vec{u} \cdot \vec{v}$, is the scalar given by
>
> $$\vec{u} \cdot \vec{v} = u_1v_1 + u_2v_2 + \cdots + u_nv_n.$$

Example 2 The population vector $\vec{P}_{\text{NewEngland}} = (3.28, 1.24, 6.07, 1.15, 0.99, 0.59)$ gives the populations (in millions) of the six New England states (CT, ME, MA, NH, RI, VT) in 1995. The vector $\vec{r} = (7.0\%, 6.5\%, 5.4\%, 13.9\%, 9.1\%, 5.1\%)$ gives the percent change in population between 1995 and 2005 for each state. We see that

$$\text{Change in Connecticut population} = 0.07 \cdot 3.28 = 0.2296, \qquad \text{and so on.}$$

If $\Delta P_{\text{New England}}$ represents the overall change in the population of New England, then

$$\begin{aligned}\Delta P_{\text{New England}} &= 0.07 \cdot 3.28 + 0.065 \cdot 1.24 + 0.054 \cdot 6.07\\ &\quad +0.139 \cdot 1.15 + 0.091 \cdot 0.99 + 0.054 \cdot 0.56\\ &= 0.9180,\end{aligned}$$

so the population of New England increased by 0.9180 million people (918,000) between 1995 and 2005. Notice that the total change in the New England population is given by the dot product:

$$\Delta P_{\text{New England}} = \vec{r} \cdot \vec{P}_{\text{NewEngland}}.$$

Properties of the Dot Product

It can be shown that the following properties hold:

- $\vec{u} \cdot \vec{v} = \|\vec{u}\| \cdot \|\vec{v}\| \cos\theta$, where θ is the angle between $\vec{u}$ and $\vec{v}$.
- $\vec{u} \cdot \vec{v} = \vec{v} \cdot \vec{u}$ (*Commutative Law*)
- $\vec{u} \cdot (\vec{v} + \vec{w}) = \vec{u} \cdot \vec{v} + \vec{u} \cdot \vec{w}$ (*Distributive Law*)
- $\vec{v} \cdot \vec{v} = \|\vec{v}\|^2$

From the formula $\vec{u} \cdot \vec{v} = \|\vec{u}\| \cdot \|\vec{v}\| \cos\theta$, we see that if $0^\circ < \theta < 90^\circ$, then $\vec{u} \cdot \vec{v}$ is positive, and if $90^\circ < \theta < 180^\circ$, then $\vec{u} \cdot \vec{v}$ is negative. This property provides a useful way to calculate the dot product without using coordinates. We use the Law of Cosines to show why it works for 2-dimensional vectors.

The commutative law holds because multiplication of coordinates is commutative. A proof of the distributive law using coordinates is outlined in the Problem 20. For the fourth property, note that

$$\vec{v} \cdot \vec{v} = (v_1, v_2) \cdot (v_1, v_2) = v_1^2 + v_2^2.$$

By the Pythagorean theorem, $\|\vec{v}\|^2 = {v_1}^2 + {v_2}^2$, so $\vec{v} \cdot \vec{v} = \|v\|^2$. A similar argument applies to vectors of higher dimension.

Justification of $\vec{u} \cdot \vec{v} = \|\vec{u}\| \cdot \|\vec{v}\| \cos\theta$

Figure 10.28 shows two vectors, $\vec{u}$ and $\vec{v}$, with an angle θ between them. The vectors $\vec{u}$, $\vec{v}$, and $\vec{w} = \vec{u} - \vec{v}$ form a triangle. Now, we know (from the fourth property of dot products) that $\vec{w} \cdot \vec{w} = \|\vec{w}\|^2$. We can also calculate $\vec{w} \cdot \vec{w}$ using the distributive and commutative laws:

$$\begin{aligned} \vec{w} \cdot \vec{w} &= (\vec{v} - \vec{u}) \cdot (\vec{v} - \vec{u}) \\ &= \vec{v} \cdot \vec{v} - \vec{u} \cdot \vec{v} - \vec{v} \cdot \vec{u} + \vec{u} \cdot \vec{u} \\ &= \|\vec{v}\|^2 + \|\vec{u}\|^2 - 2\vec{u} \cdot \vec{v}. \end{aligned}$$

Since $\vec{w} \cdot \vec{w} = \|\vec{w}\|^2$, we have shown that

$$\|\vec{w}\|^2 = \|\vec{v}\|^2 + \|\vec{u}\|^2 - 2\vec{u} \cdot \vec{v}.$$

But applying the Law of Cosines to the triangle in Figure 10.28 gives

$$||\vec{w}||^2 = ||\vec{u}||^2 + ||\vec{v}||^2 - 2||\vec{u}|| \cdot ||\vec{v}|| \cos\theta.$$

Thus setting these two expressions for $||\vec{w}||^2$ equal gives

$$||\vec{u}||^2 + ||\vec{v}||^2 - 2\vec{u} \cdot \vec{v} = ||\vec{u}||^2 + ||\vec{v}||^2 - 2||\vec{u}|| \cdot ||\vec{v}|| \cos\theta,$$

which simplifies to the formula we wanted:

$$\vec{u} \cdot \vec{v} = ||\vec{u}|| \cdot ||\vec{v}|| \cos\theta.$$

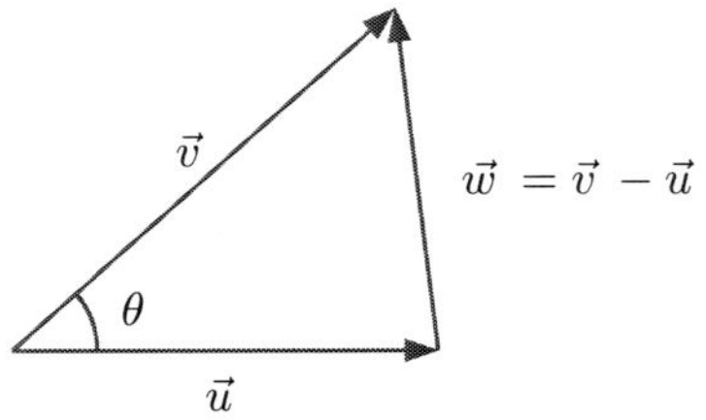

Figure 10.28: Triangle used to justify $\vec{u} \cdot \vec{v} = ||\vec{u}|| \cdot ||\vec{v}|| \cos\theta$

Example 3

(a) Find $\vec{v} \cdot \vec{w}$ where $\vec{v} = 3\vec{i} + 4\vec{j}$ and $\vec{w} = 2\vec{i} + 5\vec{j}$.

(b) One person walks 3 miles east and then 4 miles north to point A. Another person walks 2 miles east and then 5 miles north to point B. Both started from the same spot, O. What is the angle of separation of these two people? (See Figure 10.29.)

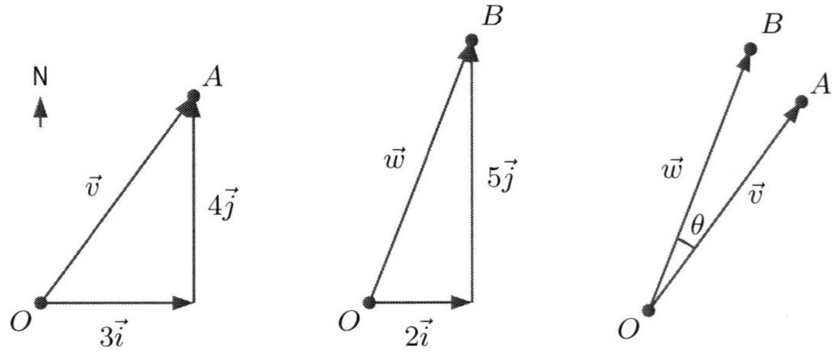

Figure 10.29: What is the angle of separation θ between the people at A and B?

Solution

(a) We have $\vec{v} \cdot \vec{w} = (3\vec{i} + 4\vec{j}) \cdot (2\vec{i} + 5\vec{j}) = 3 \cdot 2 + 4 \cdot 5 = 26$.

(b) Assuming $\vec{i}$ points east and $\vec{j}$ points north, we see that $\vec{v}$ gives the first person's position and $\vec{w}$ gives the second person's position. The angle of separation between these two people is labeled θ in Figure 10.29. We use the formula

$$\vec{v} \cdot \vec{w} = ||\vec{v}|| \cdot ||\vec{w}|| \cos\theta.$$

Since $||\vec{v}|| = \sqrt{3^2 + 4^2} = 5$ and $||\vec{w}|| = \sqrt{2^2 + 5^2} = \sqrt{29}$ and, from part (a), $\vec{v} \cdot \vec{w} = 26$, we have

$$\begin{aligned} 26 &= 5\sqrt{29}\cos\theta, \\ \cos\theta &= \frac{26}{5\sqrt{29}}, \\ \theta &= \arccos\frac{26}{5\sqrt{29}} = 15.068^\circ. \end{aligned}$$

What Does the Dot Product Mean?

The dot product can be interpreted as a measure of the alignment of two vectors. If two vectors $\vec{u}$ and $\vec{v}$ are perpendicular, then the angle between them is $\theta = 90°$. (See Figure 10.30.) In this case, $\cos\theta = \cos 90° = 0$, so

$$\vec{u} \cdot \vec{v} = ||\vec{u}|| \cdot ||\vec{v}|| \cos 90° = 0.$$

A dot product of zero tells us that the two vectors are perpendicular.

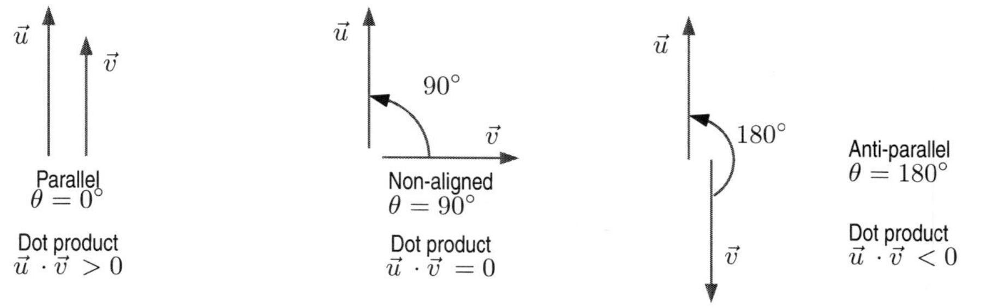

Figure 10.30: Dot product is positive, zero, or negative depending on alignment of vectors

If the two vectors are perfectly aligned, they are parallel, so $\theta = 0°$ and $\cos\theta = \cos 0° = 1$. In this case,

$$\vec{u} \cdot \vec{v} = ||\vec{u}|| \cdot ||\vec{v}|| \cos 0° = ||\vec{u}|| \cdot ||\vec{v}||,$$

so the dot product is positive.

If θ is between $0°$ and $90°$, the vectors are partially aligned and $\cos\theta$ is between 0 and 1. In this case $\vec{u} \cdot \vec{v}$ is positive, but smaller than $||\vec{u}|| \cdot ||\vec{v}||$.

Two vectors are also perfectly aligned if they are pointing in opposite directions. In this case, $\theta = 180°$ and $\cos\theta = -1$, and we have

$$\vec{u} \cdot \vec{v} = ||\vec{u}|| \cdot ||\vec{v}|| \cos 180° = -||\vec{u}|| \cdot ||\vec{v}||,$$

so the dot product is negative.

To summarize:

- Perfect alignment results in the largest possible value for $\vec{u} \cdot \vec{v}$. It occurs if $\vec{u}$ and $\vec{v}$ are parallel, with $\theta = 0°$.
- Perpendicularity results in $\vec{u} \cdot \vec{v} = 0$. It occurs if $\vec{u}$ and $\vec{v}$ are at angle of $\theta = 90°$.
- Perfect alignment in opposite directions results in the most negative value for $\vec{u} \cdot \vec{v}$. It occurs if $\theta = 180°$.

Work

In physics, the concept of *work* is represented by the dot product. In everyday language, work means effort expended. In physics, the term has a similar, but more precise, meaning.

Suppose you load a heavy refrigerator onto a truck. The refrigerator is on casters and glides with little effort along the floor. However, to lift the refrigerator takes a lot of work. The *work* done, in moving the refrigerator against the force of gravity is defined by

$$\text{Work} = \vec{F} \cdot \vec{d},$$

where $\vec{F}$ is the force exerted (assumed constant) and $\vec{d}$ is the displacement. If we measure distance in feet and force in pounds, work is measured in foot-pounds, where 1 foot-pound is the amount of work required to raise 1 pound a distance of 1 foot.

Suppose we push the refrigerator up a ramp that makes a $10°$ angle with the floor. If the ramp is 12 ft long and the force exerted on the refrigerator is 350 lbs vertically upward, then the displacement $\vec{d}$ has a magnitude of 12, and the angle θ between $\vec{F}$ and $\vec{d}$ is $90° - 10° = 80°$. (See Figure 10.31.) The work done by the force $\vec{F}$ is

$$\text{Work} = ||\vec{F}\,|| \cdot ||\vec{d}\,|| \cos 80° = 350 \cdot 12 \cos 80° = 729.322 \text{ ft-lbs.}$$

To push the refrigerator up the ramp, we do 729.322 ft-lbs of work.

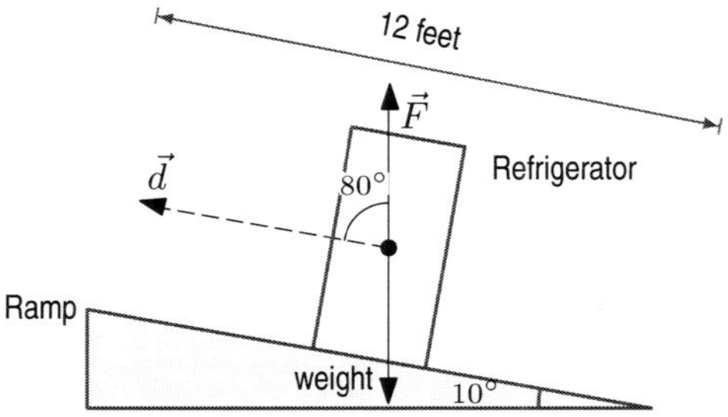

Figure 10.31: Refrigerator being pushed up ramp; the angle between $\vec{F}$ and $\vec{r}$ is $\theta = 100°$

It is informative to consider the two extreme cases: horizontal and vertical ramps. A horizontal ramp leads to a $90°$ angle between $\vec{F}$ and $\vec{d}$, so

$$\text{Work} = ||\vec{F}\,|| \cdot ||\vec{d}\,|| \cos 90° = 350 \cdot 12 \cdot 0 = 0 \text{ ft-lbs.}$$

Since we push the refrigerator in a direction perpendicular to its weight, we don't have to fight the weight at all (assuming frictionless casters). For a vertical ramp, we have

$$\text{Work} = ||\vec{F}\,|| \cdot ||\vec{d}\,|| \cos 0° = 350 \cdot 12 \cdot 1 = 4200 \text{ ft-lbs,}$$

In this case, we are pushing (or hoisting) the refrigerator in a direction opposite to that of its weight, so we feel its full force.

Exercises and Problems for Section 10.4

Exercises

For Exercises 1–10, perform the following operations on the given 3-dimensional vectors.

$$\vec{a} = 2\vec{j} + \vec{k} \quad \vec{b} = -3\vec{i} + 5\vec{j} + 4\vec{k} \quad \vec{c} = \vec{i} + 6\vec{j}$$
$$\vec{y} = 4\vec{i} - 7\vec{j} \quad \vec{z} = \vec{i} - 3\vec{j} - \vec{k}$$

1. $\vec{a} \cdot \vec{z}$

2. $\vec{c} \cdot \vec{y}$

3. $\vec{a} \cdot \vec{y}$

4. $\vec{a} \cdot \vec{b}$

5. $\vec{b} \cdot \vec{z}$

6. $\vec{c} \cdot \vec{a} + \vec{a} \cdot \vec{y}$

7. $\vec{a} \cdot (\vec{c} + \vec{y})$

8. $(\vec{a} \cdot \vec{b})\vec{a}$

9. $((\vec{c} \cdot \vec{c})\vec{a}) \cdot \vec{a}$

10. $(\vec{a} \cdot \vec{y})(\vec{c} \cdot \vec{z})$

Problems

11. The force of gravity acting on a ball is 2 lb downward. How much work is done by gravity if the ball

(a) Falls 3 feet?
(b) Moves upward 5 feet?

12. How much work is done in pushing a 350 lb refrigerator up a 12 ft ramp which makes a 30° angle with the floor?

13. Which pairs of the vectors $\sqrt{3}\vec{i} + \vec{j}$, $3\vec{i} + \sqrt{3}\vec{j}$, $\vec{i} - \sqrt{3}\vec{j}$ are parallel and which are perpendicular?

14. Compute the angle between the vectors $\vec{i} + \vec{j} + \vec{k}$ and $\vec{i} - \vec{j} - \vec{k}$.

15. For what values of t are $\vec{u} = t\vec{i} - \vec{j} + \vec{k}$ and $\vec{v} = t\vec{i} + t\vec{j} - 2\vec{k}$ perpendicular? Are there values of t for which $\vec{u}$ and $\vec{v}$ are parallel?

16. Suppose that $||\vec{a}|| = 7$ and $||\vec{b}|| = 4$, but that $\vec{a}$ and $\vec{b}$ can point in any direction. What are the maximum and minimum possible lengths for the vectors $\vec{a} + \vec{b}$ and $\vec{a} - \vec{b}$? Illustrate your answers with sketches.

17. A 100-meter dash is run on a track in the direction of the vector $\vec{v} = 2\vec{i} + 6\vec{j}$. The wind velocity $\vec{w}$ is $5\vec{i} + \vec{j}$ km/hr. The rules say that a legal wind speed measured in the direction of the dash must not exceed 5 km/hr. Will the race results be disqualified due to an illegal wind? Justify your answer.

18. Let A, B, C be the points $A = (1, 2); B = (4, 1); C = (2, 4)$. Is triangle ΔABC a right triangle?

19. Show that the vectors $(\vec{b} \cdot \vec{c})\vec{a} - (\vec{a} \cdot \vec{c})\vec{b}$ and $\vec{c}$ are perpendicular.

20. In this problem, you will check the distributive law for 2-vectors. Show that if $\vec{u} = (u_1, u_2)$, $\vec{v} = (v_1, v_2)$, and $\vec{w} = (w_1, w_2)$, then

$$\vec{u} \cdot (\vec{v} + \vec{w}) = \vec{u} \cdot \vec{v} + \vec{u} \cdot \vec{w}.$$

21. Let $\vec{u} = u_1\vec{i} + u_2\vec{j}$ and $\vec{v} = v_1\vec{i} + v_2\vec{j}$. Using $\vec{u} \cdot \vec{v} = ||\vec{u}|| \cdot ||\vec{v}|| \cos\theta$ and the distributive law, show that $\vec{u} \cdot \vec{v} = u_1v_1 + u_2v_2$.

22. **(a)** Bread, eggs, and milk cost \$3.00 per loaf, \$2.00 per dozen, and \$4.00 per gallon, respectively, at Acme Store. Use a price vector $\vec{a}$ and a consumption vector $\vec{c}$ to write a vector equation that describes what may be bought for \$40.

(b) At Beta Mart, where the food is fresher, the price vector is $\vec{b} = (3.20, 1.80, 4.50)$. Explain the meaning of $(\vec{b} - \vec{a}) \cdot \vec{c}$ in practical terms. Is $\vec{b} - \vec{a}$ ever perpendicular to $\vec{c}$?

(c) Some people think Beta Mart's freshness makes each grocery item at Beta equivalent to 110% of the corresponding Acme item. What does it mean for a consumption vector to satisfy the inequality $(1/1.1)\vec{b} \cdot \vec{c} < \vec{a} \cdot \vec{c}$?

23. Recall that in 2 or 3 dimensions, if θ is the angle between $\vec{v}$ and $\vec{w}$, the dot product is given by

$$\vec{v} \cdot \vec{w} = ||\vec{v}|| ||\vec{w}|| \cos\theta.$$

We use this relationship to define the angle between two vectors in n-dimensions. If $\vec{v}, \vec{w}$ are n-vectors, then the dot product, $\vec{v} \cdot \vec{w} = v_1w_1 + v_2w_2 + \cdots + v_nw_n$, is used to define the angle θ by

$$\cos\theta = \frac{\vec{v} \cdot \vec{w}}{||\vec{v}|| ||\vec{w}||} \qquad \text{provided } ||\vec{v}||, ||\vec{w}|| \neq 0.$$

We now use this idea of angle to measure how close two populations are to one another genetically. Table 10.1 shows the relative frequencies of four alleles (variants of a gene) in four populations.

Table 10.1

Allele	Eskimo	Bantu	English	Korean
A_1	0.29	0.10	0.20	0.22
A_2	0.00	0.08	0.06	0.00
B	0.03	0.12	0.06	0.20
O	0.67	0.69	0.66	0.57

Let $\vec{a}_1, \vec{a}_2, \vec{a}_3, \vec{a}_4$ be the 4-vectors showing the relative frequencies in the Eskimo, Bantu, English, Korean populations, respectively. The genetic distance between two populations is defined as the angle between the corresponding vectors. Using this definition, is the English population closer genetically to the Bantus or to the Koreans? Explain.[7]

24. A basketball gymnasium is 25 meters high, 80 meters wide and 200 meters long. For a half time stunt, the cheerleaders want to run two strings, one from each of the two corners of the gym above one basket to the diagonally opposite corners of the gym floor. What is the angle made by the strings as they cross?

[7]Adapted from Cavalli-Sforza and Edwards, "Models and Estimation Procedures," Am J. Hum. Genet., Vol. 19 (1967), pp. 223-57.

25. Let S be the triangle with vertices $A = (2, 2, 2)$, $B = (4, 2, 1)$, and $C = (2, 3, 1)$.

(a) Find the length of the shortest side of S.
(b) Find the cosine of the angle BAC at vertex A.

26. We can represent a rectangle as a 4-vector $\vec{r} = (x_1, y_1, x_2, y_2)$, where (x_1, y_1) and (x_2, y_2) are opposite corners. (See Problem 19 on page 467.) Assume that $x_2 > x_1$ and $y_2 > y_1$. Let $\vec{w} = (-1, 0, 1, 0)$ and $\vec{h} = (0, -1, 0, 1)$. What do the following quantities tell you about the rectangle?

(a) $\vec{r} \cdot \vec{w}$
(b) $\vec{r} \cdot \vec{h}$
(c) $2\vec{r} \cdot (\vec{w} + \vec{h})$

10.5 MATRICES

Table 10.2 shows the latest census data (in 1000s) by age group for the six New England states.[8]

Table 10.2 *Year 2000 Population (1000s) by age group for the six New England states*

	CT	ME	MA	NH	RI	VT
under 15	710	244	244	249	207	119
15–24	364	145	145	142	125	67
25–34	434	147	147	165	144	73
35–44	572	214	214	210	166	98
45–54	473	194	194	185	140	94
55–64	312	118	118	115	89	57
65–74	218	95	95	75	68	44
over 74	216	83	83	60	71	35

We can treat the array of numbers in the table as a mathematical object in its own right, independent of the row and column headers. This rectangular grid of numbers is called a *matrix*,[9] usually written inside parentheses. Since the numbers in the table are populations, we use $\mathbf{P}$ to denote this matrix:

$$\mathbf{P} = \begin{pmatrix} 710 & 244 & 244 & 249 & 207 & 119 \\ 364 & 145 & 145 & 142 & 125 & 67 \\ 434 & 147 & 147 & 165 & 144 & 73 \\ 572 & 214 & 214 & 210 & 166 & 98 \\ 473 & 194 & 194 & 185 & 140 & 94 \\ 312 & 118 & 118 & 115 & 89 & 57 \\ 218 & 95 & 95 & 75 & 68 & 44 \\ 216 & 83 & 83 & 60 & 71 & 35 \end{pmatrix}.$$

The individual entries are called *entries* in the matrix; we write p_{ij} for the entry in the ith row of the jth column. Thus for example, $p_{21} = 364$ and $p_{16} = 119$.

[8]The US Census Bureau, www.census.gov.

[9]In general, matrices can include objects other than numbers, such as complex numbers, functions, or even operators from calculus like $\partial/\partial x$.

Addition, Subtraction, and Scalar Multiplication

Like vectors, matrices provide a convenient way of organizing information. And, as with vectors, we can manipulate matrices algebraically by adding them, subtracting them, or multiplying them by scalars.

As for vectors, to multiply a matrix by a scalar, we multiply each entry by the scalar. To add or subtract two matrices, we add or subtract the corresponding entries.[10]

Example 1 Evaluate the following matrices. What do they tell you about the population of New England?

(a) Evaluate $1.1\mathbf{P}$.
(b) Evaluate $1.1\mathbf{P} - \mathbf{P}$.
(c) Show that $1.1\mathbf{P} - \mathbf{P} = 0.1\mathbf{P}$.

Solution (a) To multiply a matrix by a scalar, we multiply each entry by the scalar. For instance, the first (upper-left) entry of $\mathbf{P}$ is 710, so the first entry of $1.1\mathbf{P}$ is $1.1(710) = 781$. This matrix tells us what the population for each age group would be for each state after a 10% increase:

$$1.1\mathbf{P} = \begin{pmatrix} 781.0 & 268.4 & 268.4 & 273.9 & 227.7 & 130.9 \\ 400.4 & 159.5 & 159.5 & 156.2 & 137.5 & 73.7 \\ 477.4 & 161.7 & 161.7 & 181.5 & 158.4 & 80.3 \\ 629.2 & 235.4 & 235.4 & 231.0 & 182.6 & 107.8 \\ 520.3 & 213.4 & 213.4 & 203.5 & 154.0 & 103.4 \\ 343.2 & 129.8 & 129.8 & 126.5 & 97.9 & 62.7 \\ 239.8 & 104.5 & 104.5 & 82.5 & 74.8 & 48.4 \\ 237.6 & 91.3 & 91.3 & 66.0 & 78.1 & 38.5 \end{pmatrix}.$$

(b) To subtract one matrix from another, we subtract the corresponding entries. For instance, the first entry of $1.1\mathbf{P}$ is 781, and the first entry of $\mathbf{P}$ is 710, and so the first entry of $1.1\mathbf{P} - \mathbf{P} = 781 - 710 = 71$. This matrix tells us how much the population in part (a) increased for each age group for each state:

$$1.1\mathbf{P} - \mathbf{P} = \begin{pmatrix} 71.0 & 24.4 & 24.4 & 24.9 & 20.7 & 11.9 \\ 36.4 & 14.5 & 14.5 & 14.2 & 12.5 & 6.7 \\ 43.4 & 14.7 & 14.7 & 16.5 & 14.4 & 7.3 \\ 57.2 & 21.4 & 21.4 & 21.0 & 16.6 & 9.8 \\ 47.3 & 19.4 & 19.4 & 18.5 & 14.0 & 9.4 \\ 31.2 & 11.8 & 11.8 & 11.5 & 8.9 & 5.7 \\ 21.8 & 9.5 & 9.5 & 7.5 & 6.8 & 4.4 \\ 21.6 & 8.3 & 8.3 & 6.0 & 7.1 & 3.5 \end{pmatrix}.$$

(c) To find each entry of $1.1\mathbf{P} - \mathbf{P}$, we multiply the original entry by 1.1 and then subtract the original entry. For instance, the first entry is $1.1(710) - 710 = 0.1(710)$. You can check for yourself that multiplying every entry of $\mathbf{P}$ by 0.1 gives the same result as obtained in part (b).

---

[10]We can add or subtract two matrices provided that they have the same number of rows and columns.

If $\mathbf{A}$ is a matrix having m rows and n columns, we say that $\mathbf{A}$ is an $m \times n$ matrix, and we let a_{ij} stand for the entry at row i, column j. If $m = 1$ or $n = 1$, the matrix has only one row or one column, and is a vector. A $1 \times n$ matrix is often called a *row* vector and a $m \times 1$ matrix is often called a *column* vector.

If $\mathbf{A}$ and $\mathbf{B}$ are $m \times n$ matrices and k is a constant:

- $\mathbf{C} = k\mathbf{A}$ is an $m \times n$ matrix such that $c_{ij} = ka_{ij}$. This is called *scalar multiplication* of a matrix.
- $\mathbf{C} = \mathbf{A} + \mathbf{B}$ is an $m \times n$ matrix such that $c_{ij} = a_{ij} + b_{ij}$. This is called *matrix addition*.
- $\mathbf{C} = \mathbf{A} - \mathbf{B}$ is an $m \times n$ matrix such that $c_{ij} = a_{ij} - b_{ij}$. This is called *matrix subtraction*.

As with vector subtraction, matrix subtraction can be defined in terms of matrix addition by rewriting $\mathbf{A} + (-1)\mathbf{B}$ as $\mathbf{A} - \mathbf{B}$.

Properties Of Scalar Multiplication and Matrix Addition

The properties of scalar multiplication and matrix addition are similar to the properties for vectors.

- *Commutativity of addition*: $\mathbf{A} + \mathbf{B} = \mathbf{B} + \mathbf{A}$.
- *Associativity of addition*: $(\mathbf{A} + \mathbf{B}) + \mathbf{C} = \mathbf{A} + (\mathbf{B} + \mathbf{C})$
- *Associativity of scalar multiplication*: $k_1(k_2\mathbf{A}) = (k_1k_2)\mathbf{A}$.
- *Distributivity of scalar multiplication*: $(k_1 + k_2)\mathbf{A} = k_1\mathbf{A} + k_2\mathbf{A}$ and $k(\mathbf{A} + \mathbf{B}) = k\mathbf{A} + k\mathbf{B}$.

Multiplication of a Matrix and a Vector

Although there are more general cases of matrix multiplication, in this text we focus on multiplying a $n \times n$ square matrix by an n-dimensional column vector. We introduce this topic by considering two dependent populations.

Matrix Multiplication Of 2-Dimensional Vectors

A country begins with $x_0 = 4$ million people and another country begins with $y_0 = 2$ million people. Every year, 20% of the people in country X move to country Y, and 30% of the people in country Y move to country X. Suppose also that no one is born and no one dies. What happens to the populations of these two countries over time? After one year, x_1, the number of people (in millions) in country X, is given by

$$\begin{aligned} x_1 &= x_0 - \underbrace{0.2x_0}_{20\% \text{ leave } X} + \underbrace{0.3y_0}_{30\% \text{ leave } Y} \\ &= 0.8x_0 + 0.3y_0. \end{aligned}$$

Likewise, y_1, the number of people (in millions) in country Y, is given by

$$\begin{aligned} y_1 &= y_0 + \underbrace{0.2x_0}_{20\% \text{ leave } X} - \underbrace{0.3y_0}_{30\% \text{ leave } Y} \\ &= 0.2x_0 + 0.7y_0. \end{aligned}$$

Thus, we have

$$\begin{aligned} x_1 &= 0.8(4) + 0.3(2) = 3.2 + 0.6 = 3.8 \\ y_1 &= 0.2(4) + 0.7(2) = 0.8 + 1.4 = 2.2. \end{aligned}$$

The population of country X drops by 0.2 million, while the population of country Y goes up by 0.2 million. The same reasoning shows that after two years,

$$\begin{aligned} x_2 &= 0.8x_1 + 0.3y_1 \\ y_2 &= 0.2x_1 + 0.7y_1, \end{aligned}$$

so

$$\begin{aligned} x_2 &= 0.8(3.8) + 0.3(2.2) = 3.04 + 0.66 = 3.7 \\ y_2 &= 0.2(3.8) + 0.7(2.2) = 0.76 + 1.54 = 2.3. \end{aligned}$$

In general, we see that

$$\begin{aligned} x_{\text{new}} &= 0.8x_{\text{old}} + 0.3y_{\text{old}} \\ y_{\text{new}} &= 0.2x_{\text{old}} + 0.7y_{\text{old}}. \end{aligned}$$

Thinking in terms of vectors, we can write $\vec{P}_{\text{new}} = (x_{\text{new}}, y_{\text{new}})$ and $\vec{P}_{\text{old}} = (x_{\text{old}}, y_{\text{old}})$. Then, using the dot product, we have

$$\begin{aligned} x_{\text{new}} &= (0.8, 0.3) \cdot \vec{P}_{\text{old}} \\ y_{\text{new}} &= (0.2, 0.7) \cdot \vec{P}_{\text{old}}. \end{aligned}$$

This pair of equations tells us how the components of $\vec{P}_{\text{new}}$ are related to the components of $\vec{P}_{\text{old}}$. Notice that each component in $\vec{P}_{\text{new}}$ is a combination of both components of $\vec{P}_{\text{old}}$. Using matrices, we write the pair of equations as a single equation:

$$\vec{P}_{\text{new}} = \begin{pmatrix} x_{\text{new}} \\ y_{\text{new}} \end{pmatrix} = \underbrace{\begin{pmatrix} 0.8 & 0.3 \\ 0.2 & 0.7 \end{pmatrix} \vec{P}_{\text{old}}}_{\text{Matrix multiplication}}.$$

We place the two coefficients from the first equation, $x_{\text{new}} = 0.8x_{\text{old}} + 0.3y_{\text{old}}$, in the first row of the matrix. We place the two coefficients from the second equation, $y_{\text{new}} = 0.2x_{\text{old}} + 0.7y_{\text{old}}$, in the second row of the matrix. In general, we use the following notation:

If $\vec{P}_{\text{new}} = (x_{\text{new}}, y_{\text{new}})$ and $\vec{P}_{\text{old}} = (x_{\text{old}}, y_{\text{old}})$, then we can write the equations

$$\begin{aligned} x_{\text{new}} &= ax_{\text{old}} + by_{\text{old}} \\ y_{\text{new}} &= cx_{\text{old}} + dy_{\text{old}} \end{aligned}$$

in the compact form

$$\vec{P}_{\text{new}} = \begin{pmatrix} a & b \\ c & d \end{pmatrix} \vec{P}_{\text{old}}.$$

The coefficients in the first row of the matrix tell us how x_{new} is related to $\vec{P}_{\text{old}}$ and the coefficients in the second row tell us how y_{new} is related to $\vec{P}_{\text{old}}$. If we let $\mathbf{A} = \begin{pmatrix} a & b \\ c & d \end{pmatrix}$, then we can write

$$\vec{P}_{\text{new}} = \mathbf{A}\vec{P}_{\text{old}}.$$

Example 2 What does the following matrix equation tell you about the relationship between $\vec{P}_{\text{new}}$ and $\vec{P}_{\text{old}}$?

$$\vec{P}_{\text{new}} = \begin{pmatrix} 0.9 & 0.4 \\ 0.1 & 0.6 \end{pmatrix} \vec{P}_{\text{old}}.$$

Solution We have

$$\begin{aligned} x_{\text{new}} &= (0.9, 0.4) \cdot \vec{P}_{\text{old}} &&= 0.9x_{\text{old}} + 0.4y_{\text{old}} \\ y_{\text{new}} &= (0.1, 0.6) \cdot \vec{P}_{\text{old}} &&= 0.1x_{\text{old}} + 0.6y_{\text{old}}. \end{aligned}$$

This tells us that each year, 10% of the population of country X moves to country Y, and 40% of the population of country Y moves to country X.

Matrix Multiplication Of n-Dimensional Vectors

Let $\vec{u} = (u_1, u_2, \ldots, u_n)$ be an n-dimensional vector and let $\mathbf{A}$ by an $n \times n$ square matrix with entries $a_{11}, a_{12}, \ldots, a_{nn}$. Let $\vec{v}$ be the n-dimensional vector with coordinates given by equations:

$$\begin{aligned} v_1 &= (a_{11}, a_{12}, \ldots, a_{1n}) \cdot (u_1, u_2, \ldots, u_n) &&= a_{11}u_1 + \cdots + a_{1n}u_n \\ v_2 &= (a_{21}, a_{22}, \ldots, a_{2n}) \cdot (u_1, u_2, \ldots, u_n) &&= a_{21}u_1 + \cdots + a_{2n}u_n \\ &\vdots && \vdots \\ v_n &= (a_{n1}, a_{n2}, \ldots, a_{nn}) \cdot (u_1, u_2, \ldots, u_n) &&= a_{n1}u_1 + \cdots + a_{nn}u_n. \end{aligned}$$

Then, if

$$\mathbf{A} = \begin{pmatrix} a_{11} & \cdots & a_{1n} \\ \vdots & & \vdots \\ a_{n1} & \cdots & a_{nn} \end{pmatrix},$$

we see that $\vec{v}$ is the product of $\mathbf{A}$ and $\vec{u}$:

$$\vec{v} = \mathbf{A}\vec{u}.$$

Example 3 Let e be the number of employed people in a certain city, and u be the number of unemployed people. We define $\vec{E} = (e, u)$ as the *employment vector* of this city. Suppose that each year, 10% of employed peopled become unemployed, and 20% of unemployed people become employed. Find a matrix $\mathbf{A}$ such that $\vec{E}_{\text{new}} = \mathbf{A}\vec{E}_{\text{old}}$.

Solution We have

$$\begin{aligned} e_{\text{new}} &= e_{\text{old}} - \underbrace{0.1e_{\text{old}}}_{10\% \text{ become unemployed}} + \underbrace{0.2u_{\text{old}}}_{20\% \text{ become employed}} \\ &= 0.9e_{\text{old}} + 0.2u_{\text{old}} \\ u_{\text{new}} &= u_{\text{old}} - \underbrace{0.2u_{\text{old}}}_{20\% \text{ become employed}} + \underbrace{0.1e_{\text{old}}}_{10\% \text{ become unemployed}} \\ &= 0.1e_{\text{old}} + 0.8u_{\text{old}}. \end{aligned}$$

We can rewrite this pair of equations using matrix notation:

$$\vec{E}_{\text{new}} = \mathbf{A}\vec{E}_{\text{old}} = \begin{pmatrix} 0.9 & 0.2 \\ 0.1 & 0.8 \end{pmatrix} \vec{E}_{\text{old}}.$$

Example 4 A country produces agricultural products valued at \$70 billion and industrial products valued at \$50 billion. Some of these agricultural products are used up in order to produce other products (for instance, cotton is needed to produce cloth, or soybeans are needed to feed cattle), and some of these industrial products are used up in order to produce other products (for instance, tractors are needed to harvest cotton, or ball bearings are needed to build engines). The production vector of this country is $\vec{P} = (70, 50)$.

(a) Let $\mathbf{C}$ be the consumption matrix defined in the following equation. Calculate $\vec{U}$, the amount of agricultural and industrial products used during production:

$$\vec{U} = \mathbf{C}\vec{P} = \begin{pmatrix} 0.2 & 0.1 \\ 0.3 & 0.2 \end{pmatrix} \vec{P}.$$

Find $\vec{U}$ and describe what it tells you about the country's economy.

(b) Let the surplus vector $\vec{S} = \vec{P} - \mathbf{C}\vec{P}$. Find $\vec{S}$ and describe what it tells you about the country's economy.

Solution (a) We have

$$\vec{U} = \begin{pmatrix} 0.2 & 0.1 \\ 0.3 & 0.2 \end{pmatrix} \begin{pmatrix} 70 \\ 50 \end{pmatrix},$$

so

$$\begin{aligned} U_1 &= 0.2(70) + 0.1(50) = 19 \\ U_2 &= 0.3(70) + 0.2(50) = 31. \end{aligned}$$

Thus, $\vec{U} = (19, 31)$. This tells us that \$19 billion in agricultural products are used up during production, and \$31 billion in industrial products are used up during production.

(b) We have

$$\begin{aligned}\vec{S} &= \vec{P} - \mathbf{C}\vec{P} \\ &= \vec{P} - \vec{U} \qquad \text{because } \vec{U} = \mathbf{C}\vec{P} \\ &= (70, 50) - (19, 31) \qquad \text{from part (a),}\end{aligned}$$

so $\vec{S} = (51, 19)$. This tells us that, after internal consumption is accounted for, \$51 billion in agricultural products are available for general consumption, as are \$19 billion in industrial products.

Example 5 Let $\vec{P_0} = (x_0, y_0)$ be a position vector. This vector is rotated about the origin through an angle ϕ without changing its length. What is its new position, $\vec{P_1}$? Find a matrix $\mathbf{R}$ such that $\vec{P_1} = \mathbf{R}\vec{P_0}$.

Solution Let $r = \|\vec{P_0}\| = \|\vec{P_1}\|$. We see from Figure 10.32 that $x_0 = r\cos\theta$, $y_0 = r\sin\theta$, $x_1 = r\cos(\theta + \phi)$, and $y_1 = r\sin(\theta+\phi)$. Notice that we can rewrite the coordinates of $\vec{P_1}$ as follows:

$$\begin{aligned}x_1 &= r\cos(\theta+\phi) \\ &= r(\cos\theta\cos\phi - \sin\theta\sin\phi) \\ &= (r\cos\theta)\cos\phi - (r\sin\theta)\sin\phi \\ &= x_0\cos\phi - y_0\sin\phi \\ y_1 &= r\sin(\theta+\phi) \\ &= r(\sin\theta\cos\phi + \cos\theta\sin\phi) \\ &= (r\sin\theta)\cos\phi + (r\cos\theta)\sin\phi \\ &= y_0\cos\phi + x_0\sin\phi.\end{aligned}$$

We can rewrite this pair of equations using a matrix $\mathbf{R}$:

$$\vec{P_1} = \mathbf{R}\vec{P_0} = \begin{pmatrix}\cos\phi & -\sin\phi \\ \sin\phi & \cos\phi\end{pmatrix}\vec{P_0}.$$

Notice that this equation holds for any original vector, $\vec{P_0}$, and any angle of rotation, ϕ.

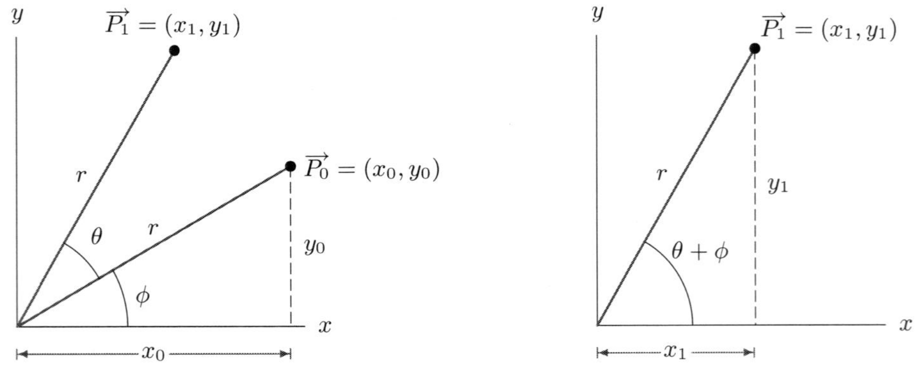

Figure 10.32: The vector $\vec{P_0}$ is rotated through an angle ϕ

Exercises and Problems for Section 10.5

Exercises

1. Evaluate the following expressions given that:

$$\mathbf{R} = \begin{pmatrix} 3 & 7 \\ 2 & -1 \end{pmatrix} \quad \text{and} \quad \mathbf{S} = \begin{pmatrix} 1 & -5 \\ 0 & 8 \end{pmatrix}.$$

(a) $5\mathbf{R}$ **(b)** $-2\mathbf{S}$
(c) $\mathbf{R}+\mathbf{S}$ **(d)** $\mathbf{S}-3\mathbf{R}$
(e) $\mathbf{R}+2\mathbf{R}+2(\mathbf{R}-\mathbf{S})$ **(f)** $k\mathbf{S}$, k constant

2. Evaluate the following expressions given that:

$$\mathbf{A} = \begin{pmatrix} 2 & 5 & 7 \\ 4 & -6 & 3 \\ 16 & -5 & 0 \end{pmatrix} \quad \text{and} \quad \mathbf{B} = \begin{pmatrix} 8 & -6 & 0 \\ 5 & 3 & -2 \\ 3 & 7 & 12 \end{pmatrix}.$$

(a) $2\mathbf{A}$ **(b)** $-3\mathbf{B}$ **(c)** $\mathbf{A}+\mathbf{B}$
(d) $2\mathbf{A}-3\mathbf{B}$ **(e)** $5(\mathbf{A}+\mathbf{B})$ **(f)** $\mathbf{A}+(\mathbf{A}+(\mathbf{A}+\mathbf{B}))$

3. Evaluate the following expressions given that

$$\mathbf{U} = \begin{pmatrix} 3 & 2 & 5 & 1 \\ 4 & 6 & 7 & 3 \\ 1 & 9 & 5 & 8 \\ 0 & -2 & 4 & 6 \end{pmatrix} \text{and } \mathbf{V} = \begin{pmatrix} 1 & 6 & 4 & 2 \\ 3 & 5 & -1 & 7 \\ 9 & 4 & 7 & 3 \\ 2 & 8 & 4 & 5 \end{pmatrix}.$$

(a) $4\mathbf{U}$ **(b)** $-2\mathbf{V}$
(c) $\mathbf{U}-\mathbf{V}$ **(d)** $3\mathbf{U}-3\mathbf{V}$
(e) $\mathbf{U}+\mathbf{U}-(\mathbf{U}-\mathbf{V})$ **(f)** $2(2\mathbf{U}-\mathbf{V})$

4. Let $\mathbf{R}$ and $\mathbf{S}$ be the matrices from Problem 1, and let $\vec{p} = (3,1)$ and $\vec{q} = (-1,5)$ be vectors. Evaluate the following expressions.

(a) $\mathbf{R}\vec{p}$ **(b)** $\mathbf{S}\vec{q}$ **(c)** $\mathbf{S}(\vec{q}+\vec{p})$
(d) $(\mathbf{R}+\mathbf{S})\vec{p}$ **(e)** $\mathbf{R}\vec{p}\cdot\mathbf{S}\vec{q}$ **(f)** $(\vec{p}\cdot\vec{q})\mathbf{S}$

5. Let $\mathbf{A}$ and $\mathbf{B}$ be the matrices defined in Problem 2, and let $\vec{u} = (3,2,5)$ and $\vec{v} = (-1,0,3)$ be vectors. Evaluate the following expressions.

(a) $\mathbf{A}\vec{u}$ **(b)** $\mathbf{B}\vec{v}$ **(c)** $\mathbf{A}(\vec{u}+\vec{v})$
(d) $(\mathbf{A}+\mathbf{B})\vec{v}$ **(e)** $\mathbf{A}\vec{u}\cdot\mathbf{B}\vec{v}$ **(f)** $(\vec{u}\cdot\vec{v})\mathbf{A}$

6. Let $\mathbf{U}$ and $\mathbf{V}$ be the matrices defined in Problem 3, and let $\vec{s} = (2,0,-1,7)$ and $\vec{t} = (4,5,1,-1)$ be vectors. Evaluate the following expressions.

(a) $\mathbf{U}\vec{t}$ **(b)** $\mathbf{V}\vec{s}$
(c) $\mathbf{U}\vec{t}\cdot\mathbf{U}\vec{s}$ **(d)** $\mathbf{U}(\vec{s}-\vec{t})$
(e) $(\mathbf{U}+\mathbf{V})(\vec{s}+\vec{t})$ **(f)** $(\vec{s}\cdot\vec{t})(\mathbf{U}+\mathbf{V})$

Problems

7. A new virus emerges in a population. Each day, 10% of the susceptible population becomes infected, and 50% of the infected population recovers. Additionally, 2% of the recovered population becomes reinfected. Let $\vec{p} = (s,i,r)$ be a population vector where s is the number of susceptible (never-infected) people, i is the number of infected people, and r is the number of recovered people, all numbers in millions. Assume that initially $\vec{p_0} = (2,0,0)$.

(a) Find a matrix $\mathbf{T}$ such that $\vec{p}_{\text{new}} = \mathbf{T}\vec{p}_{\text{old}}$ where $\vec{p}_{\text{old}}$ is the starting population and $\vec{p}_{\text{new}}$ is the population a year later.

(b) Find $\vec{p_1}$, $\vec{p_2}$, and $\vec{p_3}$, the populations on days 1, 2, and 3, respectively.

8. A certain species of insect has a three-year life cycle. The vector $\vec{P} = (f,s,t)$ gives the number of insects in their first, second, and third years of life, respectively. Each year, insects of all 3 ages lay eggs, and so the value of f_{new} depends on the values of f_{old}, s_{old}, and t_{old}. Only some first-year insects survive until their second year, and so s_{new} depends only on f_{old}; likewise t_{new} depends only on s_{old}.

(a) Suppose $\vec{P}_{\text{new}} = \mathbf{T}\vec{P}_{\text{old}} = \begin{pmatrix} 0.3 & 0.6 & 0.5 \\ 0.7 & 0 & 0 \\ 0 & 0.4 & 0 \end{pmatrix}\vec{P}_{\text{old}}$.
Describe in words what this tells you about the life cycle of this insect. Be specific.

(b) Let $\vec{P_0} = (2000,0,0)$ be the initial population vector in year $t=0$. Evaluate $\vec{P_1}$, $\vec{P_2}$, and $\vec{P_3}$.

9. Let $\vec{p}_{2005} = (A_{2005}, B_{2005}) = (200,400)$ describe the populations (in 1000s) of two cities in the year 2005. Each year, 3% of the population of city A moves to the city B, and 5% of the population of the city B moves to city A.

(a) Find a matrix $\mathbf{T}$ such that $\vec{p}_{\text{new}} = \mathbf{T}\vec{p}_{\text{old}}$, where $\vec{p}_{\text{old}}$ is the starting population and $\vec{p}_{\text{new}}$ is the population 1 year later.

(b) Find $\vec{p}_{2006}$ and $\vec{p}_{2007}$, the populations in 2006 and 2007, respectively.

10. Let $\mathbf{I_2} = \begin{pmatrix} 1 & 0 \\ 0 & 1 \end{pmatrix}$ and $\mathbf{I_3} = \begin{pmatrix} 1 & 0 & 0 \\ 0 & 1 & 0 \\ 0 & 0 & 1 \end{pmatrix}$.

(a) Evaluate $\mathbf{I_2}\vec{u}$ for (i) $\vec{u} = (3, 2)$, (ii) $\vec{u} = (0, 7)$, and (iii) $\vec{u} = (a, b)$.
(b) Evaluate $\mathbf{I_3}\vec{v}$ for (i) $\vec{v} = (-1, 5, 7)$, (ii) $\vec{v} = (3, 8, 1)$, and (iii) $\vec{v} = (a, b, c)$.
(c) Find a matrix $\mathbf{I_4}$ such that $\mathbf{I_4}\vec{w} = \vec{w}$ where $\vec{w}$ is any 4-vector.

11. In an algebraic equation such as $y = kx$, $k \neq 0$, we can multiply through by $k^{-1} = 1/k$ in order to solve for x, giving $x = k^{-1}y$. We can do a similar thing with matrices: if $\vec{v} = \mathbf{A}\vec{u}$, it is sometimes possible to find an *inverse* of $\mathbf{A}$, written $\mathbf{A}^{-1}$, such that $\vec{u} = \mathbf{A}^{-1}\vec{v}$.

(a) Let $\mathbf{A} = \begin{pmatrix} 2 & 1 \\ 3 & 2 \end{pmatrix}$. It can be shown that $\mathbf{A}^{-1} = \begin{pmatrix} 2 & -1 \\ -3 & 2 \end{pmatrix}$. Check this fact using the vector $\vec{u} = (3, 5)$, first by multiplying $\vec{u}$ by $\mathbf{A}$ to get $\vec{v}$, and then by multiplying $\vec{v}$ by $\mathbf{A}^{-1}$ to show that we get $\vec{u}$ back.
(b) Rework part (a), this time for $\vec{u} = (-1, 7)$.
(c) Rework part (a), this time for $\vec{u} = (a, b)$.

12. In Problem 11, we introduced the notion of the inverse of a matrix $\mathbf{A}$, written $\mathbf{A}^{-1}$. If $\mathbf{A} = \begin{pmatrix} a & b \\ c & d \end{pmatrix}$, then $\mathbf{A}^{-1} = \frac{1}{D}\begin{pmatrix} d & -b \\ -c & a \end{pmatrix}$ where $D = ad - bc$. Notice that $\mathbf{A}^{-1}$ is undefined if $D = 0$.

(a) Use this formula to show that $\mathbf{A}^{-1}$ from Problem 11 is $\begin{pmatrix} 2 & -1 \\ -3 & 2 \end{pmatrix}$.
(b) Let $\mathbf{B} = \begin{pmatrix} 3 & 11 \\ 1 & 7 \end{pmatrix}$. Find $\mathbf{B}^{-1}$. Verify that if $\vec{v} = \mathbf{B}\vec{u}$ then $\vec{u} = \mathbf{B}^{-1}\vec{v}$ for $\vec{u} = (a, b)$.
(c) Show that $\mathbf{C} = \begin{pmatrix} 2 & 8 \\ 3 & 12 \end{pmatrix}$ does not have an inverse.

13. Multiplying a vector by a matrix is a much more complicated process than multiplying a vector by a scalar. However, in certain special cases, matrix and scalar multiplication can lead to the same result. If $\mathbf{A}\vec{v} = \lambda\vec{v}$ where $\mathbf{A}$ is a matrix, $\vec{v}$ a nonzero vector, and λ a scalar, then $\vec{v}$ is said to be an *eigenvector* of $\mathbf{A}$, and λ is said to be an *eigenvalue* of $\mathbf{A}$. For instance, let $\mathbf{A} = \begin{pmatrix} -2 & -1 \\ 8 & 7 \end{pmatrix}$, $\vec{v_1} = (1, -8)$, and $\lambda_1 = 6$. We have:

$$\mathbf{A}\vec{v_1} = \begin{pmatrix} -2 & -1 \\ 8 & 7 \end{pmatrix} \cdot \begin{pmatrix} 1 \\ -8 \end{pmatrix} = \begin{pmatrix} 6 \\ -48 \end{pmatrix}$$
$$\lambda_1\vec{v_1} = 6 \cdot \begin{pmatrix} 1 \\ -8 \end{pmatrix} = \begin{pmatrix} 6 \\ -48 \end{pmatrix}.$$

We see that multiplying $\vec{v_1}$ by $\lambda_1 = 6$ works out the same as multiplying $\vec{v_1}$ by $\mathbf{A}$; we say that $\vec{v_1}$ is an *eigenvector* of $\mathbf{A}$ with an *eigenvalue* of $\lambda_1 = 6$.

(a) Show that $\vec{v_2} = (1, -1)$ is an eigenvector of $\mathbf{A}$. What is the eigenvalue?
(b) Show that $\vec{v_3} = (-3, 3)$ is an eigenvector of $\mathbf{A}$. What is the eigenvalue?
(c) If $\vec{v}$ is an eigenvector of $\mathbf{A}$, explain why the vectors $\vec{v}$ and $\mathbf{A}\vec{v}$ are parallel.

14. In this problem, we will consider a new way to think about matrix multiplication of vectors. A vector such as $\vec{v} = (9, 8)$ can be written as a combination of unit vectors, like this:

$$\vec{v} = (9, 8) = 9\vec{i} + 8\vec{j} = 9\underbrace{(1, 0)}_{\vec{i}} + 8\underbrace{(0, 1)}_{\vec{j}}.$$

But we can also write $\vec{v}$ as a combination[11] of other vectors besides $\vec{i}$ and $\vec{j}$. For instance, if $\vec{c_1} = (1, 2)$, and $\vec{c_2} = (3, 1)$, notice that

$$\vec{v} = (9, 8) = 3\vec{c_1} + 2\vec{c_2} = 3\underbrace{(1, 2)}_{\vec{c_1}} + 2\underbrace{(3, 1)}_{\vec{c_2}}.$$

This is where matrix multiplication comes in: if we think of $\vec{c_1}$ and $\vec{c_2}$ as the columns of a matrix $\mathbf{C}$, we see that

$$\begin{pmatrix} 9 \\ 8 \end{pmatrix} = \Bigg(\underbrace{\boxed{\begin{matrix} 1 \\ 2 \end{matrix}}}_{\vec{c_1}} \; \underbrace{\boxed{\begin{matrix} 3 \\ 1 \end{matrix}}}_{\vec{c_2}} \Bigg) \begin{pmatrix} 3 \\ 2 \end{pmatrix}$$

is the same as

$$\begin{pmatrix} 9 \\ 8 \end{pmatrix} = 3\underbrace{\begin{pmatrix} 1 \\ 2 \end{pmatrix}}_{\vec{c_1}} + 2\underbrace{\begin{pmatrix} 3 \\ 1 \end{pmatrix}}_{\vec{c_2}}.$$

In general, if $\vec{v} = \mathbf{C}\vec{u}$, then the coordinates of $\vec{u}$ tell us how to combine the columns of $\mathbf{C}$ in order to get $\vec{v}$.

(a) Let $\vec{v} = 3\vec{r_1} + 5\vec{r_2}$ where $r_1 = (3, 2)$ and $r_2 = (0, 1)$. Find $\vec{v}$, $\vec{u}$, and $\mathbf{R}$ such that $\vec{v} = \mathbf{R}\vec{u}$.

[11] This sort of combination is called a *linear combination*.

(b) Let $\vec{q} = \mathbf{S}\vec{p} = \begin{pmatrix} 3 & 4 \\ 2 & 3 \end{pmatrix}\begin{pmatrix} 3 \\ -2 \end{pmatrix}$. Show that $\vec{q}$ can be written as a combination of $\vec{s_1} = (3,2)$ and $\vec{s_2} = (4,3)$, the two columns of $\mathbf{S}$.

15. Let $\vec{v} = (2,5)$, $\vec{c_1} = (3,2)$, and $\vec{c_2} = (5,4)$. In this problem we will use the results of Problem 14 to write $\vec{v}$ as a combination of $\vec{c_1}$ and $\vec{c_2}$ given by $\vec{v} = a\vec{c_1} + b\vec{c_2}$.

(a) Let $\mathbf{C}$ be a matrix whose columns are given by $\vec{c_1}$ and $\vec{c_2}$, and let $\vec{u} = (a,b)$. Show that $\mathbf{C}\vec{u} = a\vec{c_1} + b\vec{c_2}$.

(b) Let $\vec{v} = \mathbf{C}\vec{u}$ where $\vec{v} = (2,5)$. Referring to Problem 11 and Problem 12, solve for $\vec{u}$.

(c) Show that $\vec{v}$ is a combination of $\vec{c_1}$ and $\vec{c_2}$ given by $\vec{v} = a\vec{c_1} + b\vec{c_2}$.

16. Following the procedure outlined in Problem 15, write $\vec{v} = (5,8)$ as a combination of $\vec{c_1} = (6,1)$ and $\vec{c_2} = (7,2)$ given by $\vec{v} = a\vec{c_1} + b\vec{c_2}$.

CHAPTER SUMMARY

- **Vectors**
  Displacement; vector notation; components.
  Length: $\|\vec{v}\| = \sqrt{v_1^2 + v_2^2 + v_3^2}$; unit vectors.
- **Addition, Subtraction, and Scalar Multiplication of Vectors**
- **Dot product**
  $\vec{v} \cdot \vec{w} = v_1w_1 + v_2w_2 + v_3w_3 = \|\vec{v}\|\|\vec{w}\|\cos\theta$.
- **Applications**
  Economics; computer graphics; work.
- **Matrices**
  Addition; subtraction; multiplication by scalars, by vectors.

REVIEW EXERCISES AND PROBLEMS FOR CHAPTER TEN

Exercises

Calculate the vectors in Exercises 1–8, using $\vec{a} = (5,1,0)$, $\vec{b} = (2,-1,9)$, $\vec{c} = (1,1,2)$.

1. $3\vec{c}$

2. $\vec{a} + \vec{b}$

3. $\vec{b} - \vec{a}$

4. $\vec{a} + 2(\vec{b} + \vec{c})$

5. $2\vec{a} - 3(\vec{b} - \vec{c})$

6. $2(\vec{b} + 4(\vec{a} + \vec{c}))$

7. $\vec{a} + \vec{b} - (\vec{a} - \vec{b})$

8. $4(\vec{c} + 2(\vec{a} - \vec{c}) - 2\vec{a})$

For Exercises 9–16, perform the indicated computations.

9. $-4(\vec{i} - 2\vec{j}) - 0.5(\vec{i} - \vec{k})$

10. $2(0.45\vec{i} - 0.9\vec{j} - 0.01\vec{k}) - 0.5(1.2\vec{i} - 0.1\vec{k})$

11. $(3\vec{i} + \vec{j}) \cdot (5\vec{i} - 2\vec{j})$

12. $(3\vec{j} - 2\vec{k} + \vec{i}) \cdot (4\vec{k} - 2\vec{i} + 3\vec{j})$

13. $(5\vec{i} - \vec{j} - 3\vec{k}) \cdot (2\vec{i} + \vec{j} + \vec{k})$

14. $(2\vec{i} \cdot 5\vec{j})(\vec{i} + \vec{j} + \vec{k})$

15. $(2\vec{i} + 5\vec{j}) \cdot 3\vec{i}(\vec{i} + \vec{j} + \vec{k})$

16. $(\vec{i} + \vec{j} + \vec{k})(2\vec{i} + 3\vec{j} + \vec{k}) \cdot (3\vec{i} + \vec{j} + 4\vec{k})$

Resolve the vectors in Exercises 17–18 into components.

17.

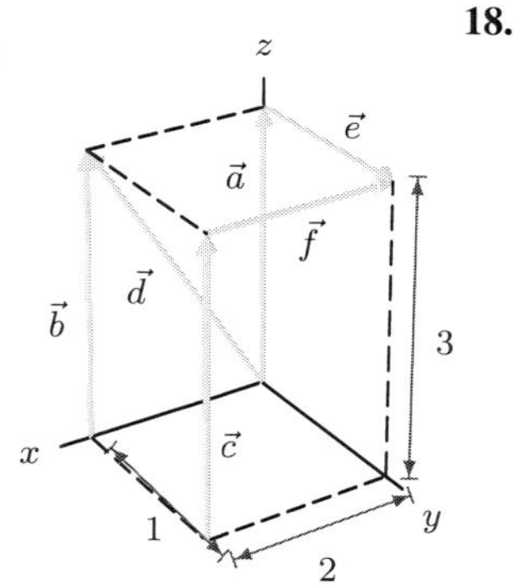

18.

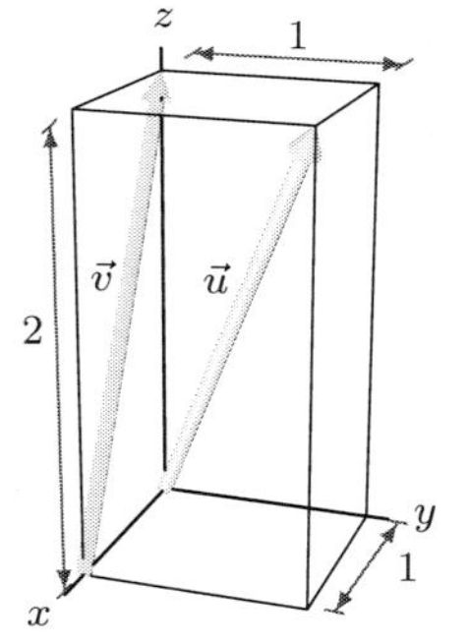

19. Find the length of the vectors $\vec{u}$ and $\vec{v}$ in Problem 18.

20. Find a vector of length 2 that points in the same direction as $\vec{i} - \vec{j} + 2\vec{k}$.

Problems

21. **(a)** Are the vectors $4\vec{i}+a\vec{j}+6\vec{k}$ and $a\vec{i}+(a-1)\vec{j}+3\vec{k}$ parallel for any values of the constant a?
(b) Are these vectors ever perpendicular?

22. A point P is on the rim of a moving bicycle wheel of radius 1 ft. The bicycle is moving forward at 6π ft/sec.

(a) Sketch the velocity of P relative to the wheel's axle at the 3, 6, 9, 12 o'clock positions.
(b) Sketch the velocity of the axle relative to the ground.
(c) Sketch the velocity of P relative to the ground at the positions in part (a).
(d) Does P ever stop, relative to the ground? What is the fastest speed that P moves, relative to the ground?

23. A gymnastics academy offers classes to children of different ages. Software used by the academy tracks enrollment data using vectors. Let $\vec{E} = (e_1, e_2, e_3, e_4, e_5)$ represent the number of children enrolled at each of five different experience levels, from beginner (1) to advanced (5), and let $\vec{E}_{\text{max}} = (40, 40, 30, 15, 10)$ be the maximum allowed enrollment at each level.

(a) Let $\vec{E} = (51, 47, 41, 22, 23)$ give the number of applicants at each level for the next session. After full enrollment is reached, the remaining applicants are placed on a wait list, $\vec{L}$. Find $\vec{L}$.
(b) At the start of the next session, $\vec{E} = \vec{E}_{\text{max}}$. Let $\vec{D} = (8, 4, 9, 7, 6)$ be the number of dropouts after the first week of the next session. Find $\vec{F} = \vec{E} - \vec{D}$ and $\vec{G} = \vec{L} - \vec{D}$. What do these vectors tell you about enrollment?

24. Let $\vec{E}$ be the enrollment vector for the gymnastics academy described in Problem 23 on page 484, where $\vec{E}_{\text{max}} = (40, 40, 30, 15, 10)$. Also, let $\vec{T} = (30, 30, 40, 80, 120)$ be the weekly tuition (in \$).

(a) Evaluate $\vec{E}_{\text{max}} \cdot \vec{T}$. What does this tell you about the gymnastic academy?
(b) Let $\vec{R} = (5, 5, 10, 20, 30)$ describe a rate hike in weekly tuition, and let $\vec{T}_{\text{new}} = \vec{T} + \vec{R}$. Evaluate $\vec{E}_{\text{max}} \cdot \vec{R}$ and $\vec{E}_{\text{max}} \cdot \vec{T}_{\text{new}}$. What do these two quantities tell you about tuition?

In Problems 25–27, a 5-pound block sits on a plank of wood. If one end of the plank is raised, the block slides down the plank. However, friction between the block and the plank prevents it from sliding until the plank has been raised a certain height. It turns out that the *sliding force* exerted by gravity on the block is proportional to the sine of the angle made by the plank with the ground.

25. Find a formula for $F = f(\theta)$, the sliding force (in lbs) exerted on a block if the plank makes an angle of θ with the ground. [Hint: What is sliding force if the plank is horizontal? Vertical?]

26. One end of the plank is lifted at a constant rate of 2 ft per second, while the other end rests on the ground.

(a) Find a formula for $F = h(t)$, the sliding force exerted on the block as a function of time.
(b) Suppose the block begins to slide once the sliding force equals 3 lbs. At what time will the block begin to slide?

27. The 5-lb force exerted on the block by gravity can be resolved into two components, the sliding force $\vec{F_S}$ parallel to the ramp and the normal force $\vec{F_N}$ perpendicular to the ramp. Use this information to show that your formula in Problem 25 is correct.

28. A plane is heading due east and climbing at the rate of 80 km/hr. If its airspeed is 480 km/hr and there is a wind blowing 100 km/hr to the northeast, what is the ground speed of the plane?

29. A particle moving with speed v hits a barrier at an angle of 60° and bounces off at an angle of 60° in the opposite direction with speed reduced by 20 percent, as shown in Figure 10.33. Find the velocity vector of the object after impact.

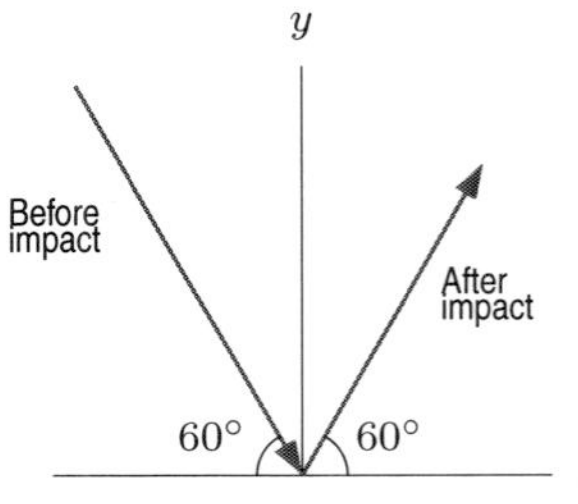

Figure 10.33

30. Figure 10.34 shows a molecule with four atoms at O, A, B and C. Show that every atom in the molecule is 2 units away from every other atom.

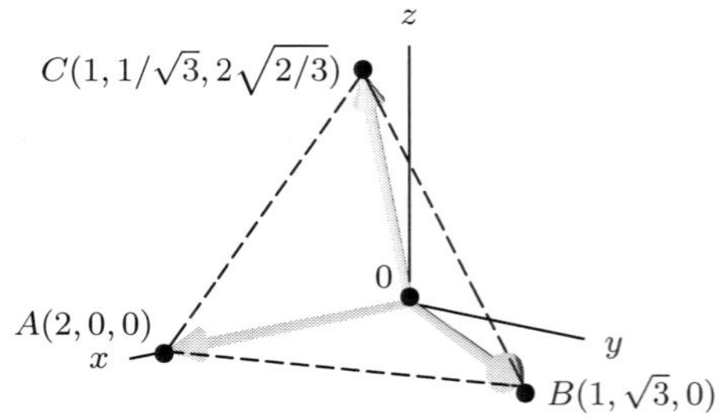

Figure 10.34

31. Two cylindrical cans of radius 2 and height 7 are shown in Figure 10.35. The cans touch down the side. Let A be the point at the top rims where they touch. Let B be the front point on the bottom rim of the left can, and C be the back point on the bottom rim of the right can. The origin is at A and the z-axis points upward; the x-axis points forward (out of the paper) and the y-axis points to the right.

(a) Write vectors in component form for $\overrightarrow{AB}$ and $\overrightarrow{AC}$.
(b) What is the angle between $\overrightarrow{AB}$ and $\overrightarrow{AC}$?

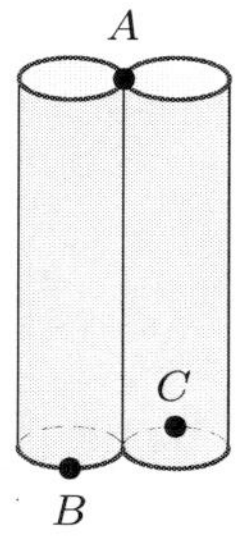

Figure 10.35

32. **(a)** Using the fact that $\vec{u}\cdot\vec{v} = ||\vec{u}||\cdot||\vec{v}||\cos\theta$, show that
$$\vec{u}\cdot(-\vec{v}) = -(\vec{u}\cdot\vec{v}).$$
[Hint: What happens to the angle when you multiply $\vec{v}$ by -1?]

(b) Using the fact that $\vec{u}\cdot\vec{v} = ||\vec{u}||\cdot||\vec{v}||\cos\theta$, show that for any negative scalar λ
$$\vec{u}\cdot(\lambda\vec{v}) = \lambda(\vec{u}\cdot\vec{v})$$
$$(\lambda\vec{u})\cdot\vec{v} = \lambda(\vec{u}\cdot\vec{v}).$$

33. A consumption vector of three goods is given by $\vec{x} = (x_1, x_2, x_3)$, where x_1, x_2 and x_3 are the quantities consumed of the three goods. Consider a budget constraint represented by the equation $\vec{p}\cdot\vec{x} = k$, where $\vec{p}$ is the price vector of the three goods and k is a constant. Show that the difference between two consumption vectors corresponding to points satisfying the same budget constraint is perpendicular to the price vector $\vec{p}$.

34. Consider the grid in Figure 10.36. Write expressions for $\overrightarrow{AB}$ and $\overrightarrow{CD}$ in terms of $\vec{i}$ and $\vec{j}$.

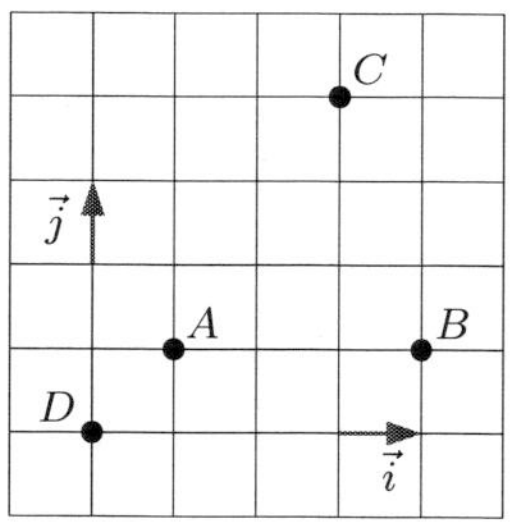

Figure 10.36

35. Consider the grid of equilateral triangles in Figure 10.37. Find expressions for $\overrightarrow{AB}$, $\overrightarrow{BC}$, $\overrightarrow{AC}$, and $\overrightarrow{AD}$ in terms of $\vec{u}$ and $\vec{v}$.

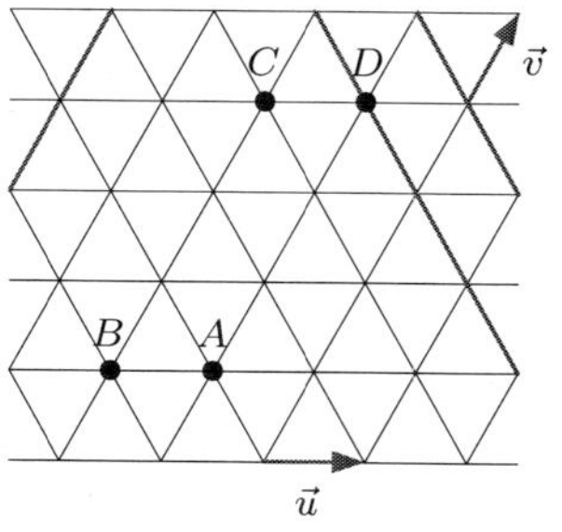

Figure 10.37

36. Consider the regular hexagon in Figure 10.38. Express the six sides and all three diameters in terms of $\vec{m}$ and $\vec{n}$.

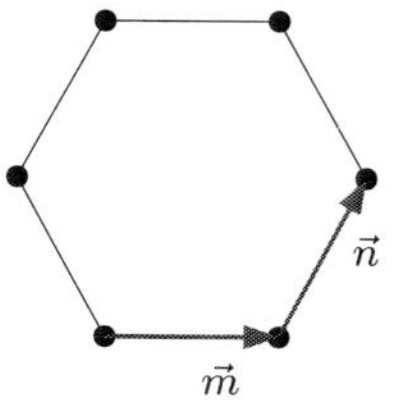

Figure 10.38

37. Consider the grid of regular hexagons in Figure 10.39. Express $\overrightarrow{AC}$, $\overrightarrow{AB}$, $\overrightarrow{AD}$ and $\overrightarrow{BD}$ in terms of $\vec{m}$ and $\vec{n}$.

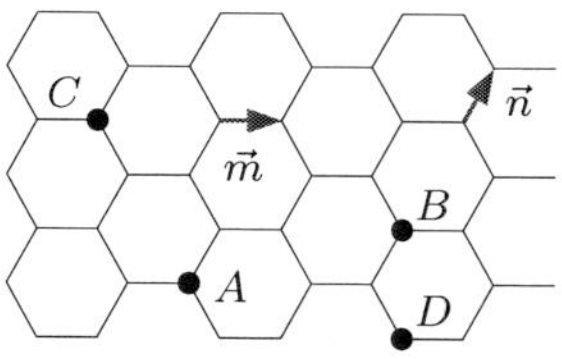

Figure 10.39

CHECK YOUR UNDERSTANDING

Are the statements in Problems 1–25 true or false? Give an explanation for your answer.

1. The vector $0.5\vec{i} + 0.5\vec{j}$ has length 1.
2. If a vector $\vec{v}$ is multiplied by a scalar, k, the resultant vector $k\vec{v}$ is longer than the original vector.
3. If $\|\vec{u}\| = 5$, then $\|-\vec{u}\| = -5$.
4. The vectors $3\vec{i} - \vec{j} + \vec{k}$ and $6\vec{i} - 2\vec{j} + 2\vec{k}$ are parallel.
5. The vectors $2\vec{i} + \vec{j}$ and $2\vec{i} - \vec{j}$ are perpendicular.
6. The two vectors $2\vec{i} + 3\vec{j}$ and $3\vec{i} + 2\vec{j}$ point in opposite directions
7. If $\vec{u} \cdot \vec{v} < 0$ then $\vec{u}$ and $\vec{v}$ form an acute angle (between $0°$ and $90°$).
8. If $\vec{u}$ and $\vec{v}$ are vectors, then their dot product is also a vector.
9. For any two vectors $\vec{u}$ and $\vec{v}$ we have $\vec{u} + \vec{v} = \vec{v} + \vec{u}$.
10. For any two vectors $\vec{u}$ and $\vec{v}$ we have $\|\vec{u} + \vec{v}\| = \|\vec{u}\| + \|\vec{v}\|$.
11. For any two vectors $\vec{u}$ and $\vec{v}$ we have $\vec{u} \cdot \vec{v} = \vec{v} \cdot \vec{u}$.
12. If the dot product of two nonzero vectors is zero, then the vectors are perpendicular.
13. The distance between the points $P = (2,3)$ and $Q = (3,4)$ is the vector $\vec{i} + \vec{j}$.
14. The vector starting at the point $P = (1,2)$ and ending at the point $Q = (-3,7)$ can be represented by $-2\vec{i} + 9\vec{j}$.
15. To add any two vectors, we add their corresponding components.
16. If $\vec{v}$ is any vector, $\vec{v} \cdot \vec{v}$ is the length of $\vec{v}$.
17. The two vectors $2\vec{i} + 3\vec{j}$ and $3\vec{i} + 2\vec{j}$ have the same length.
18. If A is a $m \times n$ matrix, the matrix A has m rows and n columns.
19. If $\mathbf{A} = \begin{pmatrix} 1\,2\,3 \\ 4\,5\,6 \\ 7\,8\,9 \end{pmatrix}$ then $a_{23} = 8$.
20. The matrix $\mathbf{A} = \begin{pmatrix} 1\,2\,3 \\ 4\,5\,6 \end{pmatrix}$ is a 3×2 matrix.
21. If A is a 5×5 matrix, then a_{35} refers to the element of A that is in the third row and fifth column.
22. To rotate a vector $\vec{v}$ through an angle ϕ, multiply $\vec{v}$ by the matrix $R = \begin{pmatrix} -\cos\phi & \sin\phi \\ \sin\phi & \cos\phi \end{pmatrix}$.
23. If A, B are 2×2 matrices, then $AB = BA$.
24. You cannot add a 2×2 matrix $\mathbf{A}$ and a 3×3 matrix $\mathbf{B}$.
25. If $A = \begin{pmatrix} 2 & 3 \\ -1 & 5 \end{pmatrix}$ and $\vec{v} = \begin{pmatrix} 10 \\ 5 \end{pmatrix}$, then $A\vec{v} = \begin{pmatrix} 35 \\ 15 \end{pmatrix}$

Chapter 10

ADDITIONAL EXAMPLES AND EXERCISES

10-A MULTIPLICATION OF MATRICES

Objectives
1. Find the product of two matrices
2. Work with properties of matrices

While addition and subtraction of matrices and the product of a scalar and a matrix are fairly straightforward, defining the *product* of two matrices requires a bit more detail.

We explain first what we mean by the product of a row matrix (a matrix with one row) with a column matrix (a matrix with one column).

Let's look at a simple example. In a given month suppose 23 units of material and 7 units of labor were required to manufacture 4-door sedans. We can represent this by the column matrix

$$\begin{bmatrix} 23 \\ 7 \end{bmatrix}$$

Also, suppose the cost per unit of material is \$450 and the cost per unit of labor is \$600. We represent these costs by the row matrix

$$\begin{bmatrix} 450 & 600 \end{bmatrix}$$

The total cost of producing the sedans is calculated as follows:

$$\begin{aligned} \text{Total cost} &= (\text{Cost per unit of material}) \times (\text{Units of material}) \\ &\quad + (\text{Cost per unit of labor}) \times (\text{Units of labor}) \\ &= 450 \times 23 + 600 \times 7 = 14{,}550 \end{aligned}$$

In terms of the matrix representations,

$$\text{Total cost} = \begin{bmatrix} 450 & 600 \end{bmatrix} \begin{bmatrix} 23 \\ 7 \end{bmatrix} = 450 \times 23 + 600 \times 7 = 14{,}550$$

This leads us to formulate the following definition for multiplying a row matrix times a column matrix:

If $R = [r_1 r_2 \dots r_n]$ is a row matrix of dimension $1 \times n$ and $C = \begin{bmatrix} c_1 \\ c_2 \\ \vdots \\ c_n \end{bmatrix}$ is a column matrix of dimension $n \times 1$, then by the **product of R and C** we mean the number

$$\text{RC} = r_1c_1 + r_2c_2 + r_3c_3 + \dots + r_nc_n$$

Example 1 Finding the Product of a 1×3 Row Matrix and a 3×1 Column Matrix

If $$R = \begin{bmatrix} 1 & 5 & 3 \end{bmatrix} \quad \text{and} \quad C = \begin{bmatrix} 2 \\ -1 \\ 4 \end{bmatrix}$$

then the product of R and C is

$$RC = \begin{bmatrix} 1 & 5 & 3 \end{bmatrix} \begin{bmatrix} 2 \\ -1 \\ 4 \end{bmatrix} = 1 \cdot 2 + 5 \cdot (-1) + 3 \cdot 4 = 9$$

Notice that for the product of a row matrix R and a column matrix C to be defined, if R is a $1 \times n$ row matrix, then C must have dimension $n \times 1$.

Example 2 Finding the Product of a 1×4 Row Matrix and a 4×1 Column Matrix

Let
$$R = \begin{bmatrix} 1 & 0 & 1 & 5 \end{bmatrix} \quad \text{and} \quad C = \begin{bmatrix} 0 \\ -11 \\ 0 \\ 8 \end{bmatrix}$$

Then the product of R and C is

$$RC = \begin{bmatrix} 1 & 0 & 1 & 5 \end{bmatrix} \begin{bmatrix} 0 \\ -11 \\ 0 \\ 8 \end{bmatrix} = 1 \cdot 0 + 0 \cdot (-11) + 1 \cdot 0 + 5 \cdot 8 = 40$$

Now Work Problem 1.

Given two matrices A and B, the rows of A can be thought of as row matrices, while the columns of B can be thought of as column matrices. This observation will be used in the following definition.

Multiplication of Matrices

Let A denote an $m \times r$ matrix, and let B denote an $r \times n$ matrix. The **product** AB is defined as the $m \times n$ matrix whose entry in row i, column j is the product of the ith row of A and the jth column of B.

An example will help to clarify the definition.

1 **Example 3** Finding the Product of Two Matrices

Find the product AB if

$$A = \begin{bmatrix} 2 & 4 & -1 \\ 5 & 8 & 0 \end{bmatrix} \quad \text{and} \quad B = \begin{bmatrix} 2 & 5 & 1 & 4 \\ 4 & 8 & 0 & 6 \\ -3 & 1 & -2 & -1 \end{bmatrix}$$

Solution First, we note that A is 2×3 and B is 3×4, so the product AB is defined and will be a 2×4 matrix.

Suppose that we want the entry in row 2, column 3 of AB. To find it, we find the product of the row vector from row 2 of A and the column vector from column 3 of B.

$$\underset{\text{Row 2 of } A}{\begin{bmatrix} 5 & 8 & 0 \end{bmatrix}} \overset{\text{Column 3 of } B}{\begin{bmatrix} 1 \\ 0 \\ -2 \end{bmatrix}} = 5 \cdot 1 + 8 \cdot 0 + 0(-2) = 5$$

So far we have

$$AB = \begin{bmatrix} \_\_ & \_\_ & \_\_ & \_\_ \\ \_\_ & \_\_ & 5 & \_\_ \end{bmatrix} \begin{matrix} \\ \text{Row 2} \end{matrix}$$

(Column 3)

Now, to find the entry in row 1, column 4 of AB, we find the product of row 1 of A and column 4 of B.

$$\underset{\text{Row 1 of } A}{\begin{bmatrix} 2 & 4 & -1 \end{bmatrix}} \overset{\text{Column 4 of } B}{\begin{bmatrix} 4 \\ 6 \\ -1 \end{bmatrix}} = 2 \cdot 4 + 4 \cdot 6 + (-1)(-1) = 33$$

Continuing in this fashion, we find AB.

$$AB = \begin{bmatrix} 2 & 4 & -1 \\ 5 & 8 & 0 \end{bmatrix} \begin{bmatrix} 2 & 5 & 1 & 4 \\ 4 & 8 & 0 & 6 \\ -3 & 1 & -2 & -1 \end{bmatrix}$$

$$= \begin{bmatrix} \text{Row 1 of } A & \text{Row 1 of } A & \text{Row 1 of } A & \text{Row 1 of } A \\ \text{times} & \text{times} & \text{times} & \text{times} \\ \text{column 1 of } B & \text{column 2 of } B & \text{column 3 of } B & \text{column 4 of } B \\ \text{Row 2 of } A & \text{Row 2 of } A & \text{Row 2 of } A & \text{Row 2 of } A \\ \text{times} & \text{times} & \text{times} & \text{times} \\ \text{column 1 of } B & \text{column 2 of } B & \text{column 3 of } B & \text{column 4 of } B \end{bmatrix}$$

$$= \begin{bmatrix} 2 \cdot 2 + 4 \cdot 4 + (-1)(-3) & 2 \cdot 5 + 4 \cdot 8 + (-1)1 & 2 \cdot 1 + 4 \cdot 0 + (-1)(-2) & 33 \text{ (from earlier)} \\ 5 \cdot 2 + 8 \cdot 4 + 0(-3) & 5 \cdot 5 + 8 \cdot 8 + 0 \cdot 1 & 5 \text{ (from earlier)} & 5 \cdot 4 + 8 \cdot 6 + 0(-1) \end{bmatrix}$$

$$= \begin{bmatrix} 23 & 41 & 4 & 33 \\ 42 & 89 & 5 & 68 \end{bmatrix}$$

A graphing utility can be used to multiply two matrices. Use a graphing utility to do Example 3.

Graphing Utility Solution

Enter the matrices A and B into a graphing utility. Figure 11 shows the product AB.

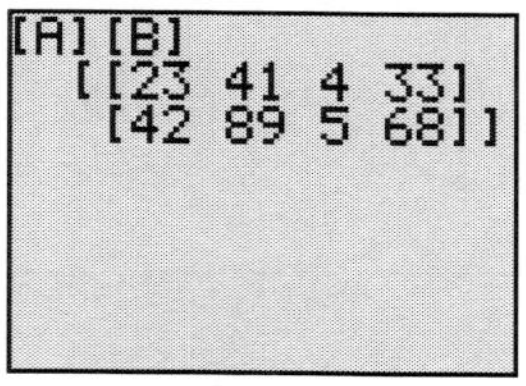

Figure 11

Now Work Problem 27.

Example 4 Manufacturing Cost

One month's production at Motors, Inc., may be given in matrix form as

$$A = \begin{matrix} \text{Sedan} & \text{Convertible} & \text{SUV} \\ \end{matrix}$$
$$A = \begin{bmatrix} 23 & 16 & 10 \\ 7 & 9 & 11 \end{bmatrix} \begin{matrix} \text{Units of material} \\ \text{Units of labor} \end{matrix}$$

(columns: Sedan, Convertible, SUV)

Suppose that in this month's production, the cost for each unit of material is \$450 and the cost for each unit of labor is \$600. What is the total cost to manufacture the sedans, the convertibles, and the SUVs?

Solution

For sedans, the cost is 23 units of material at \$450 each, plus 7 units of labor at \$600 each, for a total cost of

$$23 \cdot \$450 + 7 \cdot \$600 = 10{,}350 + 4200 = \$14{,}550$$

Similarly, for convertibles, the total cost is

$$16 \cdot \$450 + 9 \cdot \$600 = 7200 + 5400 = \$12{,}600$$

Finally, for SUVs, the total cost is

$$10 \cdot \$450 + 11 \cdot \$600 = 4500 + 6600 = \$11{,}100$$

If we represent the cost of units of material and units of labor by the 1×2 matrix

$$U = \begin{bmatrix} 450 & 600 \end{bmatrix}$$

(Unit cost of material; Unit cost of labor)

then the total cost is the product UA.

$$\begin{aligned} UA &= \begin{bmatrix} 450 & 600 \end{bmatrix} \begin{bmatrix} 23 & 16 & 10 \\ 7 & 9 & 11 \end{bmatrix} \\ &= \begin{bmatrix} 450 \cdot 23 + 600 \cdot 7 & 450 \cdot 16 + 600 \cdot 9 & 450 \cdot 10 + 600 \cdot 11 \end{bmatrix} \\ &= \begin{bmatrix} 14{,}550 & 12{,}600 & 11{,}100 \end{bmatrix} \end{aligned}$$

We conclude that the total cost to manufacture the sedans is \$14,550, to manufacture the convertibles is \$12,600, and to manufacture the SUVs is \$11,100.

Now Work Problem 59.

Properties of Matrix Multiplication

Work with properties of 2 matrices

If A is a matrix of dimension $m \times r$ (which has r columns) and B is a matrix of dimension $p \times n$ (which has p rows) and if $r \neq p$, the product AB is not defined.

> Multiplication of matrices is possible only if the number of columns of the matrix on the left equals the number of rows of the matrix on the right.
>
> If A is of dimension $m \times r$ and B is of dimension $r \times n$, then the product AB is of dimension $m \times n$.

Figure 12 illustrates the result stated above.

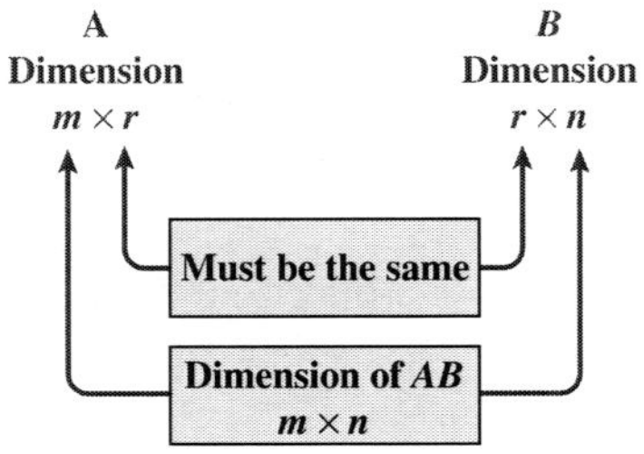

Figure 12

For example, for the matrices given in Example 3 the product BA is not defined, because B is 3×4 and A is 2×3.

Another result of matrix multiplication is illustrated in the next example.

Example 5 Multiplying Two Matrices

If

$$A = \begin{bmatrix} 2 & 1 & 3 \\ 1 & -1 & 0 \end{bmatrix} \quad \text{and} \quad B = \begin{bmatrix} 1 & 0 \\ 2 & 1 \\ 3 & 2 \end{bmatrix}$$

find: **(a)** AB **(b)** BA

Solution

$$\textbf{(a)}\ AB = \underset{2 \times 3}{\begin{bmatrix} 2 & 1 & 3 \\ 1 & -1 & 0 \end{bmatrix}} \underset{3 \times 2}{\begin{bmatrix} 1 & 0 \\ 2 & 1 \\ 3 & 2 \end{bmatrix}} = \underset{2 \times 2}{\begin{bmatrix} 13 & 7 \\ -1 & -1 \end{bmatrix}}$$

$$\textbf{(b)}\ BA = \underset{3 \times 2}{\begin{bmatrix} 1 & 0 \\ 2 & 1 \\ 3 & 2 \end{bmatrix}} \underset{2 \times 3}{\begin{bmatrix} 2 & 1 & 3 \\ 1 & -1 & 0 \end{bmatrix}} = \underset{3 \times 3}{\begin{bmatrix} 2 & 1 & 3 \\ 5 & 1 & 6 \\ 8 & 1 & 9 \end{bmatrix}}$$

Notice in Example 5 that AB is 2×2 and BA is 3×3. It is possible for both AB and BA to be defined yet be unequal. In fact, even if A and B are both $n \times n$ matrices, so that AB and BA are each defined and $n \times n$, AB and BA will usually be unequal.

Example 6 Multiplying Two Square Matrices

If

$$A = \begin{bmatrix} 2 & 1 \\ 0 & 4 \end{bmatrix} \quad \text{and} \quad B = \begin{bmatrix} -3 & 1 \\ 1 & 2 \end{bmatrix}$$

find: **(a)** AB **(b)** BA

Solution

(a) $AB = \begin{bmatrix} 2 & 1 \\ 0 & 4 \end{bmatrix}\begin{bmatrix} -3 & 1 \\ 1 & 2 \end{bmatrix} = \begin{bmatrix} -5 & 4 \\ 4 & 8 \end{bmatrix}$

(b) $BA = \begin{bmatrix} -3 & 1 \\ 1 & 2 \end{bmatrix}\begin{bmatrix} 2 & 1 \\ 0 & 4 \end{bmatrix} = \begin{bmatrix} -6 & 1 \\ 2 & 9 \end{bmatrix}$

The preceding examples demonstrate that an important property of real numbers, the commutative property of multiplication, is not shared by matrices.

Matrix multiplication is not commutative. That is, in general,

AB* is not equal to *BA

Matrix multiplication is associative.

Associative Property of Multiplication

Let A be a matrix of dimension $m \times r$, let B be a matrix of dimension $r \times p$, and let C be a matrix of dimension $p \times n$. Then matrix multiplication is **associative.** That is,

$$A(BC) = (AB)C$$

The resulting matrix ABC is of dimension $m \times n$.

Notice the limitations that are placed on the dimensions of the matrices in order for multiplication to be associative.

Distributive Property

Let A be a matrix of dimension $m \times r$. Let B and C be matrices of dimension $r \times n$. Then the **distributive property** states that

$$A(B + C) = AB + AC$$

The resulting matrix $AB + AC$ is of dimension $m \times n$.

Now Work Problem 41.

The Identity Matrix

For an $n \times n$ square matrix, the entries located in row i, column i, $1 \le i \le n$, are called the **diagonal entries.** An $n \times n$ square matrix whose diagonal entries are 1s, while all other entries are 0s, is called the **identity matrix I_n.** For example,

$$I_2 = \begin{bmatrix} 1 & 0 \\ 0 & 1 \end{bmatrix}, \quad I_3 = \begin{bmatrix} 1 & 0 & 0 \\ 0 & 1 & 0 \\ 0 & 0 & 1 \end{bmatrix}$$

and so on.

Example 7 Finding the Product of a Matrix and the Identity Matrix I_2

For

$$A = \begin{bmatrix} 3 & 2 \\ -4 & 5 \end{bmatrix}$$

compute **(a)** AI_2 **(b)** I_2A

Solution

(a) $AI_2 = \begin{bmatrix} 3 & 2 \\ -4 & 5 \end{bmatrix}\begin{bmatrix} 1 & 0 \\ 0 & 1 \end{bmatrix} = \begin{bmatrix} 3 & 2 \\ -4 & 5 \end{bmatrix} = A$

(b) $I_2A = \begin{bmatrix} 1 & 0 \\ 0 & 1 \end{bmatrix}\begin{bmatrix} 3 & 2 \\ -4 & 5 \end{bmatrix} = \begin{bmatrix} 3 & 2 \\ -4 & 5 \end{bmatrix} = A$

Example 7 demonstrates a more general result.

> If A is a square matrix of dimension $n \times n$, then $AI_n = I_nA = A$.

For square matrices the identity matrix plays the role that the number 1 plays for multiplication in the set of real numbers.

When the matrix A is not square, care must be taken when forming the products AI and IA. For example, if

$$A = \begin{bmatrix} 1 & 2 \\ 2 & 3 \\ 1 & 1 \end{bmatrix}$$

then A is of dimension 3×2 and

$$AI_2 = A\begin{bmatrix} 1 & 0 \\ 0 & 1 \end{bmatrix} = \begin{bmatrix} 1 & 2 \\ 3 & 2 \\ 1 & 1 \end{bmatrix}\begin{bmatrix} 1 & 0 \\ 0 & 1 \end{bmatrix} = \begin{bmatrix} 1 & 2 \\ 3 & 2 \\ 1 & 1 \end{bmatrix} = A$$

Although the product I_2A is not defined, we can calculate the product I_3A as follows:

$$I_3A = \begin{bmatrix} 1 & 0 & 0 \\ 0 & 1 & 0 \\ 0 & 0 & 1 \end{bmatrix}\begin{bmatrix} 1 & 2 \\ 3 & 2 \\ 1 & 1 \end{bmatrix} = \begin{bmatrix} 1 & 2 \\ 3 & 2 \\ 1 & 1 \end{bmatrix} = A$$

The above observations can be generalized as follows:

Identity Property

If A is a matrix of dimension $m \times n$ and if I_n denotes the identity matrix of dimension $n \times n$, and I_m denotes the identity matrix of dimension $m \times m$, then

$$I_mA = A \quad \text{and} \quad AI_n = A$$

EXERCISE 10-A

In Problems 1–16, find the product.

1. $\begin{bmatrix} 1 & 3 \end{bmatrix}\begin{bmatrix} 2 \\ 4 \end{bmatrix}$

2. $\begin{bmatrix} -1 & 4 \end{bmatrix}\begin{bmatrix} 5 \\ 2 \end{bmatrix}$

3. $\begin{bmatrix} 1 & -2 & 3 \end{bmatrix}\begin{bmatrix} 0 \\ 1 \\ 2 \end{bmatrix}$

4. $\begin{bmatrix} -1 & 1 & 0 \end{bmatrix}\begin{bmatrix} 1 \\ -1 \\ 1 \end{bmatrix}$

5. $\begin{bmatrix} 1 & 4 \end{bmatrix}\begin{bmatrix} 2 & 0 \\ 4 & -2 \end{bmatrix}$

6. $\begin{bmatrix} 2 & 0 \\ 4 & -2 \end{bmatrix}\begin{bmatrix} 2 \\ 4 \end{bmatrix}$

7. $\begin{bmatrix} 2 & 0 \\ 4 & -2 \end{bmatrix}\begin{bmatrix} 2 & 1 \\ 3 & -2 \end{bmatrix}$

8. $\begin{bmatrix} 1 & 4 \\ -1 & 2 \end{bmatrix}\begin{bmatrix} 3 & 0 \\ -2 & 2 \end{bmatrix}$

9. $\begin{bmatrix} 1 & -2 & 3 \end{bmatrix}\begin{bmatrix} 0 & 1 \\ 1 & 2 \\ 2 & 3 \end{bmatrix}$

10. $\begin{bmatrix} 1 & -2 & 3 \\ 4 & 0 & 6 \end{bmatrix}\begin{bmatrix} 0 \\ 1 \\ 2 \end{bmatrix}$

11. $\begin{bmatrix} 1 & -2 & 3 \\ 4 & 0 & 6 \end{bmatrix}\begin{bmatrix} 0 & -2 \\ 1 & 0 \\ 2 & -4 \end{bmatrix}$

12. $\begin{bmatrix} 1 & -2 & 3 \\ 4 & 0 & 6 \end{bmatrix}\begin{bmatrix} -1 & 2 & 1 \\ 1 & 3 & 0 \\ 0 & 4 & -2 \end{bmatrix}$

13. $\begin{bmatrix} 2 & 0 \\ 4 & -2 \\ 6 & -1 \end{bmatrix}\begin{bmatrix} 2 & 1 \\ 3 & -2 \end{bmatrix}$

14. $\begin{bmatrix} 1 & 4 \\ -1 & 2 \end{bmatrix}\begin{bmatrix} 2 & 0 & 6 \\ -1 & 4 & 1 \end{bmatrix}$

15. $\begin{bmatrix} 1 & -1 & 6 \\ 2 & 0 & -1 \\ 3 & 1 & 2 \end{bmatrix}\begin{bmatrix} 3 & 2 \\ 0 & 1 \\ 1 & 0 \end{bmatrix}$

16. $\begin{bmatrix} 2 & 0 & 1 \\ -1 & 1 & 1 \end{bmatrix}\begin{bmatrix} 1 & 0 & 0 \\ 2 & 1 & 2 \\ 3 & 2 & 4 \end{bmatrix}$

In Problems 17–26, let A be of dimension 3×4, B be of dimension 3×3, C be of dimension 2×3, and D be of dimension 3×2. Determine which of the following expressions are defined and, for those that are, give the dimension.

17. BA
18. CD
19. AB
20. DC
21. $(BA)C$
22. $A(CD)$
23. $BA + A$
24. $CD + BA$
25. $DC + B$
26. $CB - A$

In Problems 27–42, use the matrices given below. For Problems 27–40 perform the indicated operation(s); for Problems 41 and 42 verify the indicated property.

$$A = \begin{bmatrix} 1 & 2 \\ 0 & 4 \end{bmatrix} \quad B = \begin{bmatrix} 1 & 2 & 3 \\ -1 & 4 & -2 \end{bmatrix} \quad C = \begin{bmatrix} 3 & 1 \\ 4 & -1 \\ 0 & 2 \end{bmatrix} \quad D = \begin{bmatrix} 1 & 0 & 4 \\ 0 & 1 & 2 \\ 0 & -1 & 1 \end{bmatrix} \quad E = \begin{bmatrix} 3 & -1 \\ 4 & 2 \end{bmatrix}$$

27. AB
28. DC
29. BC
30. AA
31. $(D + I_3)C$
32. $DC + C$
33. EI_2
34. I_3D
35. $(2E)B$
36. $E(2B)$
37. $-5E + A$
38. $3A + 2E$
39. $3CB + 4D$
40. $2EA - 3BC$
41. Verify the associative property of matrix multiplication by finding $D(CB)$ and $(DC)B$.
42. Verify the distributive property by finding $(A + E)B$ and $AB + EB$.

In Problems 43–50, use a graphing utility or EXCEL and the matrices given below to perform the indicated operations.

$$A = \begin{bmatrix} -1 & -1 & 3 & 0 \\ 2 & 6 & 2 & 2 \\ -4 & 2 & 3 & 2 \\ 7 & 0 & 5 & -1 \end{bmatrix} \qquad B = \begin{bmatrix} -1 & 2 & 4 & 5 \\ 2 & 0 & 5 & 3 \\ 0.5 & 6 & -7 & 11 \\ 5 & -1 & 2 & 7 \end{bmatrix} \qquad C = \begin{bmatrix} 13 & -8 & 7 & 0 \\ 0 & 5 & 0 & -2 \\ 5 & 0 & 7 & 0 \\ 7 & 7 & 7 & 7 \end{bmatrix}$$

43. AB

44. BA

45. $(AB)C$

46. $A(BC)$

47. $B(A + C)$

48. $(A + C)B$

49. $A(2B - 3C)$

50. $(A + B)(A - B)$

51. For

$$A = \begin{bmatrix} 1 & -1 \\ 2 & 0 \end{bmatrix} \quad \text{and} \quad B = \begin{bmatrix} 3 & 2 \\ -2 & 4 \end{bmatrix}$$

find AB and BA. Notice that $AB \neq BA$.

52. Show that, for all values a, b, c, and d, the matrices

$$A = \begin{bmatrix} a & b \\ -b & a \end{bmatrix} \quad \text{and} \quad B = \begin{bmatrix} c & d \\ -d & c \end{bmatrix}$$

are commutative; that is, $AB = BA$.

53. For what numbers x will the following be true?

$$\begin{bmatrix} x & 4 & 1 \end{bmatrix} \begin{bmatrix} 2 & 1 & 0 \\ 1 & 0 & 2 \\ 0 & 2 & 4 \end{bmatrix} \begin{bmatrix} x \\ -7 \\ \frac{5}{4} \end{bmatrix} = 0$$

54. Let

$$A = \begin{bmatrix} 1 & 2 & 5 \\ 2 & 4 & 10 \\ -1 & -2 & -5 \end{bmatrix}$$

Show that $A \cdot A = A^2 = 0$. Thus the rule in the real number system that if $a^2 = 0$, then $a = 0$ does not hold for matrices.

55. What must be true about a, b, c, and d if we demand that $AB = BA$ for the following matrices?

$$A = \begin{bmatrix} a & b \\ c & d \end{bmatrix} \qquad B = \begin{bmatrix} 1 & 1 \\ -1 & 1 \end{bmatrix}$$

Assume that

$$\begin{bmatrix} a & b \\ c & d \end{bmatrix} \neq \begin{bmatrix} 1 & 0 \\ 0 & 1 \end{bmatrix}$$

56. Let

$$A = \begin{bmatrix} a & b \\ b & a \end{bmatrix}$$

Find a and b such that $A^2 + A = 0$, where $A^2 = A \cdot A$.

57. For the matrix

$$A = \begin{bmatrix} a & 1 - a \\ 1 + a & -a \end{bmatrix}$$

show that $A^2 = A \cdot A = I_2$.

58. Find the row vector $[x_1 \quad x_2]$ for which

$$\begin{bmatrix} x_1 & x_2 \end{bmatrix} \begin{bmatrix} \frac{1}{2} & \frac{1}{2} \\ \frac{1}{4} & \frac{3}{4} \end{bmatrix} = \begin{bmatrix} x_1 & x_2 \end{bmatrix}$$

under the condition that $x_1 + x_2 = 1$. Here, the row vector $[x_1 \quad x_2]$ is called a **fixed vector** of the matrix

$$\begin{bmatrix} \frac{1}{2} & \frac{1}{2} \\ \frac{1}{4} & \frac{3}{4} \end{bmatrix}$$

59. Department Store Purchases Lee went to a department store and purchased 6 pairs of pants, 8 shirts, and 2 jackets. Chan purchased 2 pairs of pants, 5 shirts, and 3 jackets. If pants are \$25 each, shirts are \$18 each, and jackets are \$39 each, use matrix multiplication to find the amounts spent by Lee and Chan.

60. Factory Production Suppose a factory is asked to produce three types of products, which we will call P_1, P_2, P_3. Suppose the following purchase order was received: $P_1 = 7$, $P_2 = 12$, $P_3 = 5$. Represent this order by a row vector and call it P:

$$P = \begin{bmatrix} 7 & 12 & 5 \end{bmatrix}$$

To produce each of the products, raw material of four kinds is needed. Call the raw material M_1, M_2, M_3, and M_4. The matrix below gives the amount of material needed for each product:

$$Q = \begin{matrix} \\ P_1 \\ P_2 \\ P_3 \end{matrix} \begin{matrix} \begin{matrix} M_1 & M_2 & M_3 & M_4 \end{matrix} \\ \begin{bmatrix} 2 & 3 & 1 & 12 \\ 7 & 9 & 5 & 20 \\ 8 & 12 & 6 & 15 \end{bmatrix} \end{matrix}$$

Suppose the cost for each of the materials M_1, M_2, M_3, and M_4 is \$10, \$12, \$15, and \$20, respectively. The cost vector is

$$C = \begin{bmatrix} 10 \\ 12 \\ 15 \\ 20 \end{bmatrix}$$

Compute each of the following and interpret each one:

(a) PQ (b) QC (c) PQC

Problems 61–69 require the following definition:

Powers of Matrices For a square matrix A it is possible to find $A \cdot A = A^2$. We can also compute

$$A^n = \underbrace{A \cdot A \cdot \ldots \cdot A}_{n \text{ factors}}$$

In Problems 61–64, find A^2, A^3, and A^4.

61. $A = \begin{bmatrix} 1 & 0 \\ 3 & 2 \end{bmatrix}$

62. $A = \begin{bmatrix} 3 & 1 \\ -2 & -1 \end{bmatrix}$

63. $A = \begin{bmatrix} 1 & 0 \\ 0 & 1 \end{bmatrix}$

64. $A = \begin{bmatrix} \frac{1}{2} & \frac{1}{2} \\ \frac{1}{4} & \frac{3}{4} \end{bmatrix}$

65. Can you guess what A^n looks like for the matrix given in Problem 63?

66. Can you guess what A^n looks like for the matrix given in Problem 64?

In Problems, 67–69 use a graphing utility to compute A^2, A^{10}, and A^{15}.

67. $A = \begin{bmatrix} -0.5 & -1 & 0.3 & 0 & 0.3 \\ 2 & 1.6 & 1 & -1 & 0.4 \\ -4 & 2 & 0.7 & 2 & 0.2 \\ 1 & 0 & 0 & -1 & 0 \\ 0 & 0 & -0.9 & 0 & 0 \end{bmatrix}$

68. $A = \begin{bmatrix} -1 & 0.02 & 0.24 & 0 \\ 2 & 0 & 0 & 1.3 \\ 0.5 & 6 & -0.7 & 1.1 \\ 2.5 & -1 & 0.02 & 0.7 \end{bmatrix}$

69. $A = \begin{bmatrix} 0 & 1 & 0 \\ 1 & 0 & 1 \\ 1 & 1 & 1 \end{bmatrix}$

70. Make up two matrices A and B for which $AB = BA$.

71. Make up two square matrices A and B for which AB and BA are both defined, but $AB \neq BA$.

72. Make up two matrices A and B for which AB is defined but BA is not.

10-B THE INVERSE OF A MATRIX

Objectives

1. Find the inverse of a matrix
2. Use the inverse of a matrix to solve a system of equations

The *inverse* of a matrix, if it exists, plays the role in matrix algebra that the reciprocal of a number plays in the set of real numbers. For example, the product of 2 and its reciprocal, $\frac{1}{2}$, equals 1, the multiplicative identity. The product of a matrix and its inverse equals the identity matrix.

Inverse of a Matrix

Let A be a matrix of dimension $n \times n$. A matrix B of dimension $n \times n$ is called the inverse of A if $AB = BA = I_n$. We denote the inverse of a matrix A, if it exists, by A^{-1}.

Example 1 Verifying That One Matrix Is the Inverse of Another Matrix

Show that $\begin{bmatrix} \frac{1}{2} & -\frac{1}{2} \\ 0 & 1 \end{bmatrix}$ is the inverse of $\begin{bmatrix} 2 & 1 \\ 0 & 1 \end{bmatrix}$.

Solution Since

$$\begin{bmatrix} 2 & 1 \\ 0 & 1 \end{bmatrix} \begin{bmatrix} \frac{1}{2} & -\frac{1}{2} \\ 0 & 1 \end{bmatrix} = \begin{bmatrix} 1 & 0 \\ 0 & 1 \end{bmatrix}$$

and

$$\begin{bmatrix} \frac{1}{2} & -\frac{1}{2} \\ 0 & 1 \end{bmatrix}\begin{bmatrix} 2 & 1 \\ 0 & 1 \end{bmatrix} = \begin{bmatrix} 1 & 0 \\ 0 & 1 \end{bmatrix}$$

the required condition is met.

Now Work Problem 1.

The next example provides a technique for finding the inverse of a matrix. Although this technique is not the one we shall ultimately use, it is illustrative.

Example 2 Finding the Inverse of a Matrix

Find the inverse of the matrix $A = \begin{bmatrix} 2 & 1 \\ 0 & 1 \end{bmatrix}$.

Solution We begin by assuming that this matrix has an inverse of the form

$$A^{-1} = \begin{bmatrix} a & b \\ c & d \end{bmatrix}$$

Then the product of A and A^{-1} must be the identity matrix:

$$\begin{bmatrix} 2 & 1 \\ 0 & 1 \end{bmatrix}\begin{bmatrix} a & b \\ c & d \end{bmatrix} = \begin{bmatrix} 1 & 0 \\ 0 & 1 \end{bmatrix}$$

Multiplying the matrices on the left side, we get

$$\begin{bmatrix} 2a + c & 2b + d \\ c & d \end{bmatrix} = \begin{bmatrix} 1 & 0 \\ 0 & 1 \end{bmatrix}$$

The condition for equality requires that

$$2a + c = 1 \qquad 2b + d = 0 \qquad c = 0 \qquad d = 1$$

Using $c = 0$ and $d = 1$ in the first two equations, we find

$$a = \tfrac{1}{2} \qquad b = -\tfrac{1}{2} \qquad c = 0 \qquad d = 1$$

The inverse of

$$A = \begin{bmatrix} 2 & 1 \\ 0 & 1 \end{bmatrix} \quad \text{is} \quad A^{-1} = \begin{bmatrix} a & b \\ c & d \end{bmatrix} = \begin{bmatrix} \frac{1}{2} & -\frac{1}{2} \\ 0 & 1 \end{bmatrix}$$

Sometimes a square matrix does not have an inverse.

Example 3 Showing a Matrix Has No Inverse

Show that the matrix below does not have an inverse.

$$A = \begin{bmatrix} 0 & 1 \\ 0 & 0 \end{bmatrix}$$

Solution

We proceed as in Example 2 by assuming that A does have an inverse. It will be of the form

$$A^{-1} = \begin{bmatrix} a & b \\ c & d \end{bmatrix}$$

The product of A and A^{-1} must be the identity matrix.

$$\begin{bmatrix} 0 & 1 \\ 0 & 0 \end{bmatrix}\begin{bmatrix} a & b \\ c & d \end{bmatrix} = \begin{bmatrix} 1 & 0 \\ 0 & 1 \end{bmatrix}$$

Performing the multiplication on the left side, we have

$$\begin{bmatrix} c & d \\ 0 & 0 \end{bmatrix} = \begin{bmatrix} 1 & 0 \\ 0 & 1 \end{bmatrix}$$

But these two matrices can never be equal (look at row 2, column 2: $0 \neq 1$). We conclude that our assumption that A has an inverse is false. That is, A does not have an inverse.

Now Work Problem 2.1.

So far, we have shown that a square matrix may or may not have an inverse. What about nonsquare matrices? Can they have inverses? The answer is "No." By definition, whenever a matrix has an inverse, it will commute with its inverse under multiplication. So if the nonsquare matrix A had the alleged inverse B, then AB would have to be equal to BA. But the fact that A is not square causes AB and BA to have different dimensions and prevents them from being equal. So such a B could not exist.

A nonsquare matrix has no inverse.

The procedure used in Example 2 to find the inverse, if it exists, of a square matrix becomes quite involved as the dimension of the matrix gets larger. A more efficient method that uses the reduced row-echelon form of a matrix is provided next.

Reduced Row-Echelon Technique for Finding Inverses

We will introduce this technique by looking at an example.

Example 4 Finding the Inverse of a Matrix

Find the inverse of the matrix

$$A = \begin{bmatrix} 4 & 2 \\ 3 & 1 \end{bmatrix}$$

Solution

Assuming A has an inverse, we will denote it by

$$X = \begin{bmatrix} x_1 & x_2 \\ x_3 & x_4 \end{bmatrix}$$

Then the product of A and X is the identity matrix of dimension 2×2. That is,

$$AX = I_2$$

$$\begin{bmatrix} 4 & 2 \\ 3 & 1 \end{bmatrix}\begin{bmatrix} x_1 & x_2 \\ x_3 & x_4 \end{bmatrix} = \begin{bmatrix} 1 & 0 \\ 0 & 1 \end{bmatrix}$$

Performing the multiplication on the left yields

$$\begin{bmatrix} 4x_1 + 2x_3 & 4x_2 + 2x_4 \\ 3x_1 + x_3 & 3x_2 + x_4 \end{bmatrix} = \begin{bmatrix} 1 & 0 \\ 0 & 1 \end{bmatrix}$$

This matrix equation can be written as the following system of four equations containing four variables:

$$\begin{cases} 4x_1 + 2x_3 = 1 & 4x_2 + 2x_4 = 0 \\ 3x_1 + x_3 = 0 & 3x_2 + x_4 = 1 \end{cases}$$

We find the solution to be

$$x_1 = -\tfrac{1}{2} \qquad x_2 = 1 \qquad x_3 = \tfrac{3}{2} \qquad x_4 = -2$$

The inverse of A is

$$A^{-1} = \begin{bmatrix} -\frac{1}{2} & 1 \\ \frac{3}{2} & -2 \end{bmatrix}$$

2 Solving a System of *n* Linear Equations Containing *n* Variables Using Inverses

Use the inverse of a matrix to solve a system of equations

The inverse of a matrix can be used to solve a system of n linear equations containing n variables. We begin with a system of three linear equations containing three variables:

$$\begin{cases} x + y + 2z = 1 \\ 2x + y = 2 \\ x + 2y + 2z = 3 \end{cases}$$

If we let

$$A = \begin{bmatrix} 1 & 1 & 2 \\ 2 & 1 & 0 \\ 1 & 2 & 2 \end{bmatrix} \qquad X = \begin{bmatrix} x \\ y \\ z \end{bmatrix} \qquad B = \begin{bmatrix} 1 \\ 2 \\ 3 \end{bmatrix}$$

the above system of equations can be written compactly as the matrix equation

$$AX = B$$

In general, any system of n linear equations containing n variables $x_1, x_2, \dots, x_n$, can be written in the form

$$AX = B$$

where A is the $n \times n$ matrix of the coefficients of the unknowns, B is an $n \times 1$ column matrix whose entries are the numbers appearing to the right of each equal sign in the system, and X is an $n \times 1$ column matrix containing the n variables.

To find X, we start with the matrix equation $AX = B$ and use properties of matrices. Our assumption is that the $n \times n$ matrix A has an inverse A^{-1}.

$$\begin{aligned} AX &= B && \text{A has an inverse } A^{-1} \\ A^{-1}(AX) &= A^{-1}B && \text{Multiply both sides by } A^{-1} \\ (A^{-1}A)X &= A^{-1}B && \text{Apply the Associative Property on the left side} \\ I_nX &= A^{-1}B && \text{Apply the Inverse Property: } A^{-1}A = I_n \\ X &= A^{-1}B && \text{Apply the Identity Property: } I_nX = X \end{aligned}$$

This leads us to formulate the following result:

> A system of n linear equations containing n variables
>
> $$\mathbf{AX = B}$$
>
> for which A is a square matrix and A^{-1} exists, always has a unique solution that is given by
>
> $$\mathbf{X = A^{-1}B}$$

We use the above result in the next example.

Example 7 Solving a System of Equations Using Inverses

Solve the system of equations: $\begin{cases} x + y + 2z = 1 \\ 2x + y = 2 \\ x + 2y + 2z = 3 \end{cases}$

Solution Here

$$A = \begin{bmatrix} 1 & 1 & 2 \\ 2 & 1 & 0 \\ 1 & 2 & 2 \end{bmatrix} \quad X = \begin{bmatrix} x \\ y \\ z \end{bmatrix} \quad B = \begin{bmatrix} 1 \\ 2 \\ 3 \end{bmatrix}$$

From Example 5 we know A has the inverse.

$$A^{-1} = \begin{bmatrix} \frac{1}{2} & \frac{1}{2} & -\frac{1}{2} \\ -1 & 0 & 1 \\ \frac{3}{4} & -\frac{1}{4} & -\frac{1}{4} \end{bmatrix}$$

Based on the result just stated, the solution X of the system is

$$X = A^{-1}B$$

$$\begin{bmatrix} x \\ y \\ z \end{bmatrix} = \begin{bmatrix} \frac{1}{2} & \frac{1}{2} & -\frac{1}{2} \\ -1 & 0 & 1 \\ \frac{3}{4} & -\frac{1}{4} & -\frac{1}{4} \end{bmatrix} \begin{bmatrix} 1 \\ 2 \\ 3 \end{bmatrix} = \begin{bmatrix} 0 \\ 0 \\ -\frac{1}{2} \end{bmatrix}$$

The solution is $x = 0$, $y = 2$, $z = -\frac{1}{2}$.

EXERCISE 10-B

In Problems 1–6, show that the given matrices are inverses of each other.

1. $\begin{bmatrix} 1 & 2 \\ 2 & 3 \end{bmatrix}\begin{bmatrix} -3 & 2 \\ 2 & -1 \end{bmatrix}$

2. $\begin{bmatrix} 1 & 5 \\ 2 & 0 \end{bmatrix}\begin{bmatrix} 0 & \frac{1}{2} \\ \frac{1}{5} & -\frac{1}{10} \end{bmatrix}$

3. $\begin{bmatrix} -1 & -2 \\ 3 & 4 \end{bmatrix}\begin{bmatrix} 2 & 1 \\ -\frac{3}{2} & -\frac{1}{2} \end{bmatrix}$

4. $\begin{bmatrix} 1 & 3 \\ 2 & -1 \end{bmatrix}\begin{bmatrix} \frac{1}{7} & \frac{3}{7} \\ \frac{2}{7} & -\frac{1}{7} \end{bmatrix}$

5. $\begin{bmatrix} 1 & 2 & 3 \\ 2 & 3 & 4 \\ 1 & 2 & 1 \end{bmatrix}\begin{bmatrix} -\frac{5}{2} & 2 & -\frac{1}{2} \\ 1 & -1 & 1 \\ \frac{1}{2} & 0 & -\frac{1}{2} \end{bmatrix}$

6. $\begin{bmatrix} 1 & 3 & 3 \\ 1 & 4 & 3 \\ 1 & 3 & 4 \end{bmatrix}\begin{bmatrix} 7 & -3 & -3 \\ -1 & 1 & 0 \\ -1 & 0 & 1 \end{bmatrix}$

In Problems 7–20, find the inverse of each matrix using the reduced row-echelon technique.

7. $\begin{bmatrix} 3 & 7 \\ 2 & 5 \end{bmatrix}$

8. $\begin{bmatrix} 4 & 1 \\ 3 & 1 \end{bmatrix}$

9. $\begin{bmatrix} 1 & -1 \\ 3 & -4 \end{bmatrix}$

10. $\begin{bmatrix} 5 & 3 \\ 3 & 2 \end{bmatrix}$

11. $\begin{bmatrix} 2 & 1 \\ 4 & 3 \end{bmatrix}$

12. $\begin{bmatrix} 2 & 3 \\ 2 & -2 \end{bmatrix}$

13. $\begin{bmatrix} 0 & 0 & 1 \\ 0 & 1 & 0 \\ 1 & 0 & 0 \end{bmatrix}$

14. $\begin{bmatrix} -1 & 1 & 0 \\ 1 & 0 & 2 \\ 3 & 1 & 0 \end{bmatrix}$

15. $\begin{bmatrix} 1 & 1 & -1 \\ 3 & -1 & 0 \\ 2 & -3 & 4 \end{bmatrix}$

16. $\begin{bmatrix} 1 & 1 & 1 \\ 2 & 1 & 1 \\ 1 & 1 & 2 \end{bmatrix}$

17. $\begin{bmatrix} 1 & 1 & -1 \\ 2 & 1 & 1 \\ 1 & 0 & 1 \end{bmatrix}$

18. $\begin{bmatrix} 2 & 3 & -1 \\ 1 & 1 & 1 \\ 0 & 2 & -1 \end{bmatrix}$

19. $\begin{bmatrix} 1 & 1 & 0 & 0 \\ 0 & 1 & -1 & 1 \\ 1 & -1 & 1 & 1 \\ 0 & 1 & 0 & -1 \end{bmatrix}$

20. $\begin{bmatrix} 1 & 2 & -3 & -2 \\ 0 & 1 & 4 & -2 \\ 3 & -1 & 4 & 0 \\ 2 & 1 & 0 & 3 \end{bmatrix}$

In Problems 21–26, show that each matrix has no inverse.

21. $\begin{bmatrix} 4 & 6 \\ 2 & 3 \end{bmatrix}$

22. $\begin{bmatrix} -1 & 2 \\ 3 & -6 \end{bmatrix}$

23. $\begin{bmatrix} -8 & 4 \\ -4 & 2 \end{bmatrix}$

24. $\begin{bmatrix} 2 & 10 \\ 1 & 5 \end{bmatrix}$

25. $\begin{bmatrix} 1 & 1 & 1 \\ 3 & -4 & 2 \\ 0 & 0 & 0 \end{bmatrix}$

26. $\begin{bmatrix} -1 & 2 & 3 \\ 5 & 2 & 0 \\ 2 & -4 & -6 \end{bmatrix}$

In Problems 27–34, find the inverse, if it exists, of each matrix.

27. $\begin{bmatrix} 1 & 1 \\ 1 & 2 \end{bmatrix}$

28. $\begin{bmatrix} 2 & 1 \\ 1 & 1 \end{bmatrix}$

29. $\begin{bmatrix} 3 & -2 \\ 0 & 4 \end{bmatrix}$

30. $\begin{bmatrix} 4 & -1 \\ -2 & 0 \end{bmatrix}$

31. $\begin{bmatrix} 3 & 2 \\ 6 & 4 \end{bmatrix}$

32. $\begin{bmatrix} 4 & 2 \\ 2 & 1 \end{bmatrix}$

33. $\begin{bmatrix} 1 & -2 & -1 \\ -2 & 5 & 4 \\ 3 & -8 & -5 \end{bmatrix}$

34. $\begin{bmatrix} 1 & 1 & -1 \\ -2 & -1 & 4 \\ 3 & 2 & -8 \end{bmatrix}$

35. Find the inverse of both

$$A = \begin{bmatrix} 1 & 2 \\ 2 & -1 \end{bmatrix} \quad \text{and} \quad B = \begin{bmatrix} 1 & 3 \\ 2 & 1 \end{bmatrix}$$

to determine $A^{-1} - B^{-1}$.

36. Find the inverse of

$$A = \begin{bmatrix} 1 & -4 \\ 2 & -3 \end{bmatrix} \quad \text{and} \quad B = \begin{bmatrix} 2 & -2 \\ 3 & 2 \end{bmatrix}$$

Determine $A^{-1} - B^{-1}$.

37. Write the matrix product $A^{-1}B$ used to find the solution to the system $AX = B$.

$$\begin{cases} x + 3y + 2z = 2 \\ 2x + 7y + 3z = 1 \\ x \qquad\quad + 6z = 3 \end{cases}$$

38. Write the matrix product $A^{-1}B$ used to find the solution to the system $AX = B$.

$$\begin{cases} x + 2y + 2z = 3 \\ 2x + 5y + 7z = 2 \\ 2x + y - 4z = 4 \end{cases}$$

In Problems 39–50, solve each system of equations by the method of Example 7. For Problems 39–44, use the inverse found in Problem 7. For Problems 45–50, use the inverse found in Problem 15.

39. $\begin{cases} 3x + 7y = 10 \\ 2x + 5y = 2 \end{cases}$

40. $\begin{cases} 3x + 7y = -4 \\ 2x + 5y = -3 \end{cases}$

41. $\begin{cases} 3x + 7y = 13 \\ 2x + 5y = 9 \end{cases}$

42. $\begin{cases} 3x + 7y = 0 \\ 2x + 5y = 14 \end{cases}$

43. $\begin{cases} 3x + 7y = 12 \\ 2x + 5y = -4 \end{cases}$

44. $\begin{cases} 3x + 7y = -2 \\ 2x + 5y = 10 \end{cases}$

45. $\begin{cases} x + y - z = 3 \\ 3x - y = -4 \\ 2x - 3y + 4z = 6 \end{cases}$

46. $\begin{cases} x + y - z = 6 \\ 3x - y = 8 \\ 2x - 3y + 4z = -3 \end{cases}$

47. $\begin{cases} x + y - z = 12 \\ 3x - y = -4 \\ 2x - 3y + 4z = 16 \end{cases}$

48. $\begin{cases} x + y - z = -8 \\ 3x - y = 12 \\ 2x - 3y + 4z = -2 \end{cases}$

49. $\begin{cases} x + y - z = 0 \\ 3x - y = -8 \\ 2x - 3y + 4z = -6 \end{cases}$

50. $\begin{cases} x + y - z = 21 \\ 3x - y = 12 \\ 2x - 3y + 4z = 14 \end{cases}$

In Problems 51–56, use a graphing utility or EXCEL to find the inverse, if it exists, of each matrix.

51. $\begin{bmatrix} 25 & 61 & -12 \\ 18 & -2 & 4 \\ 8 & 35 & 21 \end{bmatrix}$

52. $\begin{bmatrix} 18 & -3 & 4 \\ 6 & -20 & 14 \\ 10 & 25 & -15 \end{bmatrix}$

53. $\begin{bmatrix} 44 & 21 & 18 & 6 \\ -2 & 10 & 15 & 5 \\ 21 & 12 & -12 & 4 \\ -8 & -16 & 4 & 9 \end{bmatrix}$

54. $\begin{bmatrix} 16 & 22 & -3 & 5 \\ 21 & -17 & 4 & 8 \\ 2 & 8 & 27 & 20 \\ 5 & 15 & -3 & -10 \end{bmatrix}$

55. $A = \begin{bmatrix} 3 & 0 & 2 & -1 & 3 \\ -2 & 1 & 2 & 3 & 0 \\ 2 & 2 & 1 & 1 & -1 \\ 1 & 2 & 0 & 2 & -3 \\ 4 & 0 & -1 & 1 & -1 \end{bmatrix}$

56. $A = \begin{bmatrix} 0 & 0 & 2 & -1 & 3 \\ -2 & 0 & 2 & 3 & 0 \\ 2 & 2 & 0 & 0 & -1 \\ 1 & 2 & 0 & 2 & -3 \\ 4 & 4 & 0 & 0 & -2 \end{bmatrix}$

In Problems 57–60, use the idea behind Example 7 and either a graphing utility or EXCEL to solve each system of equations.

57. $\begin{cases} 25x + 61y - 12z = 10 \\ 18x - 12y + 7z = -9 \\ 3x + 4y - z = 12 \end{cases}$

58. $\begin{cases} 25x + 61y - 12z = 5 \\ 18x - 12y + 7z = -3 \\ 3x + 4y - z = 12 \end{cases}$

59. $\begin{cases} 25x + 61y - 12z = 21 \\ 18x - 12y + 7z = 7 \\ 3x + 4y - z = -2 \end{cases}$

60. $\begin{cases} 25x + 61y - 12z = 25 \\ 18x - 12y + 7z = 10 \\ 3x + 4y - z = -4 \end{cases}$

61. Show that the inverse of $A = \begin{bmatrix} a & b \\ c & d \end{bmatrix}$

is given by the formula $A^{-1} = \begin{bmatrix} \frac{d}{\Delta} & \frac{-b}{\Delta} \\ \frac{-c}{\Delta} & \frac{a}{\Delta} \end{bmatrix}$

where $\Delta = ad - bc \neq 0$. The number Δ is called the **determinant** of A.

In Problems 62–65, use the result of Problem 61 to find the inverse of each matrix.

62. $\begin{bmatrix} 1 & 2 \\ 2 & 3 \end{bmatrix}$

63. $\begin{bmatrix} 1 & 5 \\ 2 & 0 \end{bmatrix}$

64. $\begin{bmatrix} -1 & -2 \\ 3 & 4 \end{bmatrix}$

65. $\begin{bmatrix} 1 & 2 \\ 8 & 15 \end{bmatrix}$

10-C

Definitions

Previously, we studied the set of equations

$$a_1x + b_1y = c_1$$
$$a_2x + b_2y = c_2$$

Now let us refer to the values

$\begin{vmatrix} a_1 \\ a_2 \end{vmatrix}$ as the column of x coefficients,

$\begin{vmatrix} b_1 \\ b_2 \end{vmatrix}$ as the column of y coefficients, and

$\begin{vmatrix} c_1 \\ c_2 \end{vmatrix}$ as the column of constants.

When we solved this set of equations, we got the solution

$$x = \frac{b_2c_1 - b_1c_2}{a_1b_2 - a_2b_1} \quad \text{and} \quad y = \frac{a_1c_2 - a_2c_1}{a_1b_2 - a_2b_1}$$

> It is common practice to use the same word *determinant* to refer to the *symbol*
>
> $$\begin{vmatrix} a_1 & b_1 \\ a_2 & b_2 \end{vmatrix}$$
>
> as well as to the value $(a_1b_2 - a_2b_1)$. We'll follow this practice.

The denominator $(a_1b_2 - a_2b_1)$ of each of these fractions is called the *determinant* of the coefficients a_1, b_1, a_2, and b_2. It is commonly expressed by the symbol

$$\begin{vmatrix} a_1 & b_1 \\ a_2 & b_2 \end{vmatrix}$$

with it being understood that this signifies the product of the upper-left and lower-right numbers, minus the product of the upper-right and lower-left numbers. Thus

$$\begin{vmatrix} a_1 & b_1 \\ a_2 & b_2 \end{vmatrix} = a_1b_2 - a_2b_1$$

Example 1 In the determinant

$$\begin{vmatrix} 2 & 5 \\ 6 & 1 \end{vmatrix}$$

each of the numbers 2, 5, 6, and 1 is called an *element.* There are two *rows,* the first row containing the elements 2 and 5, and the second row having the elements 6 and 1. There are also two *columns,* the first with elements 2 and 6, and the second with elements 5 and 1. The *value* of this determinant is

$$2(1) - 5(6) = 2 - 30 = -28$$

Thus a determinant is written as a *square array,* enclosed between vertical bars. The number of rows equals the number of columns. The *order* of a determinant is equal to the number of rows or columns that it contains. Thus the determinants above are of second order.

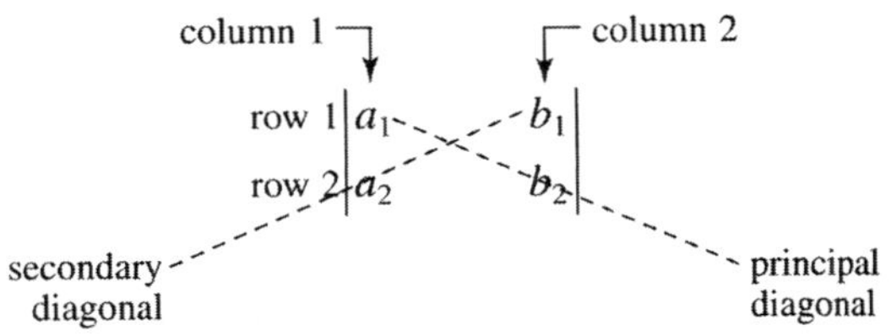

The diagonal running from upper left to lower right is called the *principal diagonal.* The other is called the *secondary diagonal.*

Second-Order Determinant

$$\begin{vmatrix} a_1 & b_1 \\ a_2 & b_2 \end{vmatrix} = a_1b_2 - a_2b_1 \qquad \textbf{65}$$

The value of a second-order determinant is equal to the product of the elements on the principal diagonal minus the product of the elements on the secondary diagonal.

Example 2

$$\begin{vmatrix} 2 & -1 \\ 3 & 5 \end{vmatrix} = 2(5) - 3(-1) = 13$$

Common Errors	Students sometimes add the numbers on a diagonal instead of *multiplying.* Be careful not to do that.
	Don't just ignore a zero element. It causes the product along its diagonal to be zero.

Example 3

$$\begin{vmatrix} 4 & a \\ x & 3b \end{vmatrix} = 4(3b) - x(a) = 12b - ax$$

Example 4

$$\begin{vmatrix} 0 & 3 \\ 5 & 2 \end{vmatrix} = 0(2) - 3(5) = -15$$

Evaluating Determinants by Calculator

Most graphics calculators can evaluate determinants if all of the elements are numbers.

Example 5 Evaluate the following determinant using a graphics calculator:

$$\begin{vmatrix} 3 & 2 \\ 5 & 1 \end{vmatrix}$$

Solution Locate the matrix editor on your calculator. Then follow the instructions for entering each element in the proper order. Finally, choose the command to evaluate the determinant, which is probably labeled DET. The value of the determinant will then be displayed, in this case, a value of -7.

> Unfortunately, calculators and computer algebra systems vary so widely that we cannot give more specific instructions than these. As before, you must consult your user's manual.

Evaluating Determinants on a Computer Algebra System (CAS)

A computer algebra system (CAS) can evaluate determinants, even when the elements contain *algebraic symbols.*

Example 6 Evaluate the following determinant using a CAS:

$$\begin{vmatrix} x & 3 \\ 2y & w \end{vmatrix}$$

Solution We locate the matrix editor on our CAS and enter 2 for the number of rows and of columns where requested. We then enter the four elements, x, 3, $2y$, and w, as directed. Following the instructions for evaluating the determinant of this matrix gives

$$wx - 6y$$

Solving a System of Equations by Determinants

We now return to the set of equations

$$\begin{aligned} a_1x + b_1y &= c_1 \\ a_2x + b_2y &= c_2 \end{aligned} \quad (1)$$

If, in the solution for x,

$$x = \frac{b_2c_1 - b_1c_2}{a_1b_2 - a_2b_1} \quad (2)$$

> This denominator is called the *determinant of the coefficients,* or sometimes the *determinant of the system.* It is usually given the Greek capital letter delta (Δ). If this determinant equals zero, there is no solution to the set of equations.

we replace the denominator by the square array

$$\begin{vmatrix} a_1 & b_1 \\ a_2 & b_2 \end{vmatrix} = a_1b_2 - a_2b_1$$

we get

$$x = \frac{b_2c_1 - b_1c_2}{\begin{vmatrix} a_1 & b_1 \\ a_2 & b_2 \end{vmatrix}} \quad (3)$$

Now look at the numerator of Equation (3), which is $b_2c_1 - b_1c_2$. If this expression were the value of some determinant, it is clear that the elements on the principal diagonal must be b_2 and c_1 and that the elements on the secondary diagonal must be b_1 and c_2. One such determinant is

column of constants

$$\begin{vmatrix} c_1 & b_1 \\ c_2 & b_2 \end{vmatrix} = b_2c_1 - b_1c_2$$

Our solution for x can then be expressed as

$$x = \frac{\begin{vmatrix} c_1 & b_1 \\ c_2 & b_2 \end{vmatrix}}{\begin{vmatrix} a_1 & b_1 \\ a_2 & b_2 \end{vmatrix}}$$

An expression for y can be developed in a similar way. Thus the solution to the system of Equations (1), in terms of determinants, is given by the following rule:

Cramer's Rule

$$x = \frac{\begin{vmatrix} c_1 & b_1 \\ c_2 & b_2 \end{vmatrix}}{\begin{vmatrix} a_1 & b_1 \\ a_2 & b_2 \end{vmatrix}} \quad \text{and} \quad y = \frac{\begin{vmatrix} a_1 & c_1 \\ a_2 & c_2 \end{vmatrix}}{\begin{vmatrix} a_1 & b_1 \\ a_2 & b_2 \end{vmatrix}} \quad \textbf{70}$$

Named after the Swiss mathematician Gabriel Cramer (1704–52), Cramer's rule, as we will see, is valid for systems with any number of linear equations.

The solution for any variable is a fraction whose denominator is the determinant of the coefficients and whose numerator is that same determinant, except that the column of coefficients for the variable for which we are solving is replaced by the column of constants.

Of course, the denominator Δ cannot be zero, or we get division by zero, indicating that the set of equations has no unique solution.

Example 7 Solve by determinants:

$$\begin{aligned} 2x - 3y &= 1 \\ x + 4y &= -5 \end{aligned}$$

Solution We first evaluate the determinant of the coefficients, because if $\Delta = 0$, there is no unique solution, and we have no need to proceed further.

$$\Delta = \begin{vmatrix} 2 & -3 \\ 1 & 4 \end{vmatrix} = 2(4) - 1(-3) = 8 + 3 = 11$$

Solving for x, we have

column of constants

$$x = \frac{\begin{vmatrix} 1 & -3 \\ -5 & 4 \end{vmatrix}}{\Delta} = \frac{1(4) - (-5)(-3)}{11} = \frac{4 - 15}{11} = -1$$

Solving for y yields

column of constants

$$y = \frac{\begin{vmatrix} 2 & 1 \\ 1 & -5 \end{vmatrix}}{\Delta} = \frac{2(-5) - 1(1)}{11} = \frac{-10 - 1}{11} = -1$$

> It's easier to find y by substituting x back into one of the previous equations, but we use determinants instead, to show how it is done.
>
> We would, of course, check our answers by substituting back into the original equations, as we did in Chapter 10.

> **Common Error** Don't forget to first arrange the given equations into the form
>
> $$\begin{aligned} a_1x + b_1y &= c_1 \\ a_2x + b_2y &= c_2 \end{aligned}$$
>
> before writing the determinants.

Example 8 Solve for x and y by determinants:

$$\begin{aligned} 2ax - by + 3a &= 0 \\ 4y &= 5x - a \end{aligned}$$

Solution We first rearrange our equations.

$$\begin{aligned} 2ax - by &= -3a \\ 5x - 4y &= a \end{aligned}$$

The determinant of the coefficients is

$$\Delta = \begin{vmatrix} 2a & -b \\ 5 & -4 \end{vmatrix} = 2a(-4) - 5(-b) = -8a + 5b$$

So

$$x = \frac{\begin{vmatrix} -3a & -b \\ a & -4 \end{vmatrix}}{\Delta} = \frac{(-3a)(-4) - a(-b)}{5b - 8a} = \frac{12a + ab}{5b - 8a}$$

and

$$y = \frac{\begin{vmatrix} 2a & -3a \\ 5 & a \end{vmatrix}}{\Delta} = \frac{2a(a) - 5(-3a)}{5b - 8a} = \frac{2a^2 + 15a}{5b - 8a}$$

THIRD-ORDER DETERMINANTS

The technique we used for finding the value of a second-order determinant *cannot* be used for higher-order determinants. We will now show two other methods—first, what we'll call the *diagonal* method, which works *only for third-order determinants;* and then *development by minors,* which works for *any* determinant.

The Diagonal Method

The value of a third-order determinant can be found as follows. To the right of the determinant, rewrite the first two columns.

$$\left|\begin{array}{ccc|cc} a_1 & b_1 & c_1 & a_1 & b_1 \\ a_2 & b_2 & c_2 & a_2 & b_2 \\ a_3 & b_3 & c_3 & a_3 & b_3 \end{array}\right|$$

Then multiply the elements along each of the six diagonals. The value of the determinant is the sum of the products along those diagonals running downward to the right, *minus* the products along those diagonals running downward to the left.

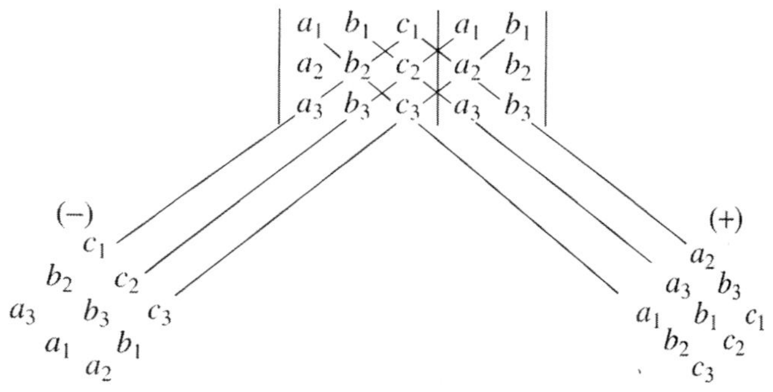

> We will not emphasize this method, because it works only for third-order determinants and also because it is not suitable for computer solution.

The value of the determinant is then given by the following equation:

Third-Order Determinant

$$\begin{vmatrix} a_1 & b_1 & c_1 \\ a_2 & b_2 & c_2 \\ a_3 & b_3 & c_3 \end{vmatrix} = \begin{array}{l} a_1b_2c_3 + a_3b_1c_2 + a_2b_3c_1 \\ -a_3b_2c_1 - a_1b_3c_2 - a_2b_1c_3 \end{array} \qquad \mathbf{66}$$

Example 9 Evaluate the following third-order determinant:

$$\begin{vmatrix} 2 & 1 & 0 \\ 3 & -2 & 1 \\ 5 & 1 & 0 \end{vmatrix}$$

Solution Rewriting the first two columns to the right of the determinant, we have

$$\left|\begin{array}{ccc|cc} 2 & 1 & 0 & 2 & 1 \\ 3 & -2 & 1 & 3 & -2 \\ 5 & 1 & 0 & 5 & 1 \end{array}\right|$$

Multiplying along the diagonals, we obtain

$$\begin{aligned} & 2(-2)(0) + 1(1)(5) + 0(3)(1) - [0(-2)(5) + 2(1)(1) + 1(3)(0)] \\ &= 0 + 5 + 0 - [0 + 2 + 0] \\ &= 5 - 2 = 3 \end{aligned}$$

Minors

The *minor of an element in a determinant* is a determinant of next lower order, obtained by deleting the row and the column in which that element lies.

Example 10 We find the minor of element c in the determinant

$$\begin{vmatrix} a & b & c \\ d & e & f \\ g & h & i \end{vmatrix}$$

by striking out the first row and the third column.

$$\begin{vmatrix} \cancel{a} & \cancel{b} & \cancel{\textcircled{c}} \\ d & e & \cancel{f} \\ g & h & \cancel{i} \end{vmatrix} \quad \text{or} \quad \begin{vmatrix} d & e \\ g & h \end{vmatrix}$$

Sign Factor

The *sign factor* for an element depends upon the position of the element in the determinant. To find the sign factor, add the row number and the column number of the element. If the sum is even, the sign factor is $+1$; if the sum is odd, the sign factor is -1.

> A minor with the sign factor attached is called a *signed minor.* It is also called a *cofactor* of the given element.

Example 11 (a) For the determinant in Example 10, element c is in the first row and the third column. The sum of the row and column numbers

$$1 + 3 = 4$$

is even, so the sign factor for the element c is $+1$.

(b) The sign factor for the element b in Example 10 is -1.

The sign factor for the element in the upper left-hand corner of the determinant is $+1$, and the signs alternate according to the pattern

$$\begin{vmatrix} + & - & + & - & + & - & \cdot \\ - & + & - & + & - & \cdot & \cdot \\ + & - & + & - & \cdot & \cdot & \cdot \\ - & + & - & \cdot & \cdot & \cdot & \cdot \\ + & - & \cdot & \cdot & \cdot & \cdot & \cdot \\ - & \cdot & \cdot & \cdot & \cdot & \cdot & \cdot \\ \cdot & \cdot & \cdot & \cdot & \cdot & \cdot & \cdot \end{vmatrix}$$

so that the sign factor may be found simply by *counting off* from the upper left corner.

Thus for any element in a determinant, we can write three quantities:

1. The element itself.
2. The sign factor of the element.
3. The minor of the element.

Example 12 For the element in the second row and the first column in the following determinant:

$$\begin{vmatrix} -2 & 5 & 1 \\ \mathbf{3} & -2 & 4 \\ 1 & -4 & 2 \end{vmatrix}$$

1. the element itself is 3;
2. the sign factor of that element is -1; and
3. the minor of that element is

$$\begin{vmatrix} 5 & 1 \\ -4 & 2 \end{vmatrix}$$

Evaluating a Determinant by Minors

We now define the *value of a determinant* as follows:

Value of a Determinant

To find the value of a determinant of order n:

1. Multiply each of the n elements in any row (or column) by the sign factor of that element and the minor of that element. **68**
2. Take the algebraic sum of those n products.

Example 13 Evaluate by minors:

$$\begin{vmatrix} 3 & 2 & 4 \\ 1 & 6 & -2 \\ -1 & 5 & -3 \end{vmatrix}$$

Solution We first choose a row or a column for development. The work of expansion is greatly reduced if that row or column contains zeros. Our given determinant has no zeros, so let us choose column 1, which at least contains some 1's.

the elements of column 1

$$\begin{vmatrix} \mathbf{3} & 2 & 4 \\ \mathbf{1} & 6 & -2 \\ \mathbf{-1} & 5 & -3 \end{vmatrix} = 3(1)\begin{vmatrix} 6 & -2 \\ 5 & -3 \end{vmatrix} + 1(-1)\begin{vmatrix} 2 & 4 \\ 5 & -3 \end{vmatrix} + (-1)(1)\begin{vmatrix} 2 & 4 \\ 6 & -2 \end{vmatrix}$$

the sign factor of each element

the minor of each element

$$\begin{aligned} &= 3[(6)(-3) - (-2)(5)] - 1[(2)(-3) - (4)(5)] \\ &\quad - 1[(2)(-2) - 4(6)] \\ &= -24 + 26 + 28 \\ &= 30 \end{aligned}$$

EXERCISE 10-C

Evaluate.

1. $\begin{vmatrix} 6 & 1/2 & -2 \\ 3 & 1/4 & 4 \\ 2 & -1/2 & 3 \end{vmatrix}$

2. $\begin{vmatrix} 2 & 7 & -2 & 8 \\ 4 & 1 & 1 & -3 \\ 0 & 3 & -1 & 4 \\ 6 & 4 & 2 & -8 \end{vmatrix}$

3. $\begin{vmatrix} 0 & n & m \\ -n & 0 & l \\ -m & -l & 0 \end{vmatrix}$

4. $\begin{vmatrix} 8 & 2 & 0 & 1 & 4 \\ 0 & 1 & 4 & 2 & 7 \\ 2 & 6 & 3 & 8 & 0 \\ 1 & 4 & 2 & 6 & 5 \\ 4 & 6 & 8 & 3 & 5 \end{vmatrix}$

5. $\begin{vmatrix} 25 & 23 & 19 \\ 14 & 11 & 9 \\ 21 & 17 & 14 \end{vmatrix}$

6. $\begin{vmatrix} x+y & x-y \\ x-y & x+y \end{vmatrix}$

7. $\begin{vmatrix} 9 & 13 & 17 \\ 11 & 15 & 19 \\ 17 & 21 & 25 \end{vmatrix}$

8. $\begin{vmatrix} 5 & -3 & -2 & 0 \\ 4 & 1 & -6 & 2 \\ -1 & 4 & 3 & -5 \\ 0 & 6 & -4 & 2 \end{vmatrix}$

9. $\begin{vmatrix} 1 & 2 & 3 \\ 3 & 1 & 2 \\ 2 & 3 & 1 \end{vmatrix}$

10. $\begin{vmatrix} 2 & -3 & 1 \\ -2 & 4 & 5 \\ 3 & -1 & -4 \end{vmatrix}$

11. $\begin{vmatrix} 1 & 2 & 2 & 4 \\ 1 & 4 & 4 & 1 \\ 1 & 1 & 2 & 2 \\ 4 & 8 & 11 & 13 \end{vmatrix}$

12. $\begin{vmatrix} a-b & -2a \\ 2b & a-b \end{vmatrix}$

13. $\begin{vmatrix} 3 & 1 & 5 & 2 \\ 4 & 10 & 14 & 6 \\ 8 & 9 & 1 & 4 \\ 6 & 15 & 21 & 9 \end{vmatrix}$

14. $\begin{vmatrix} 7 & 8 & 9 \\ 28 & 35 & 40 \\ 21 & 26 & 30 \end{vmatrix}$

15. $\begin{vmatrix} 1 & 5 & 2 \\ 4 & 7 & 3 \\ 9 & 8 & 6 \end{vmatrix}$

16. $\begin{vmatrix} 3 & -6 \\ -5 & 4 \end{vmatrix}$

17. $\begin{vmatrix} 6 & 4 & 7 \\ 9 & 0 & 8 \\ 5 & 3 & 2 \end{vmatrix}$

18. $\begin{vmatrix} 3 & 2 & 1 & 3 \\ 4 & 2 & 2 & 4 \\ 2 & 3 & 1 & 6 \\ 10 & 4 & 5 & 8 \end{vmatrix}$

Solve by determinants.

19. $\begin{aligned} x + y + z + w &= -4 \\ x + 2y + 3z + 4w &= 0 \\ x + 3y + 6z + 10w &= 9 \\ x + 4y + 10z + 20w &= 24 \end{aligned}$

20. $\begin{aligned} x + 2y + 3z &= 4 \\ 3x + 5y + z &= 18 \\ 4x + y + 2z &= 12 \end{aligned}$

21. $\begin{aligned} 4x + 3y &= 27 \\ 2x - 5y &= -19 \end{aligned}$

22. $\begin{aligned} 8x - 3y - 7z &= 85 \\ x + 6y - 4z &= -12 \\ 2x - 5y + z &= 33 \end{aligned}$

23. $\begin{aligned} x + 2y &= 10 \\ 2x - 3y &= -1 \end{aligned}$

24. $\begin{aligned} x + y + z + u &= 1 \\ 2x + 3y - 4z + 5u &= -31 \\ 3x - 4y + 5z + 6u &= -22 \\ 4x + 5y - 6z - u &= -13 \end{aligned}$

25. $\begin{aligned} 2x - 3y &= 7 \\ 5x + 2y &= 27 \end{aligned}$

26. $\begin{aligned} 6x + y &= 60 \\ 3x + 2y &= 39 \end{aligned}$

27. $\begin{aligned} 2x + 5y &= 29 \\ 2x - 5y &= -21 \end{aligned}$

28. $\begin{aligned} 4x + 3y &= 7 \\ 2x - 3y &= -1 \end{aligned}$

29. $\begin{aligned} 4x - 5y &= 3 \\ 3x + 5y &= 11 \end{aligned}$

30. $\begin{aligned} x + 5y &= 41 \\ 3x - 2y &= 21 \end{aligned}$

31. $\begin{aligned} x + 2y &= 7 \\ x + y &= 5 \end{aligned}$

32. $\begin{aligned} 8x - 3y &= 22 \\ 4x + 5y &= 18 \end{aligned}$

33. $\begin{aligned} 3x + 4y &= 25 \\ 4x + 3y &= 21 \end{aligned}$

34. $\begin{aligned} 37.7x &= 59.2 + 24.6y \\ 28.3x - 39.8y - 62.5 &= 0 \end{aligned}$

35. $\begin{aligned} 5.03a &= 8.16 + 5.11b \\ 3.63b + 7.26a &= 28.8 \end{aligned}$

36. $\begin{aligned} 541x + 216y + 412z &= 866 \\ 211x + 483y - 793z &= 315 \\ 215x + 495y + 378z &= 253 \end{aligned}$

37. $\begin{aligned} 72.3x + 54.2y + 83.3z &= 52.5 \\ 52.2x - 26.6y + 83.7z &= 75.2 \\ 33.4x + 61.6y + 30.2z &= 58.5 \end{aligned}$

38. A link in a certain mechanism starts at a speed v_0 and travels 27.5 inches in 8.25 s, at constant acceleration a. Its motion is described by Eq. A18,

$$s = v_0 t + \tfrac{1}{2}at^2$$

or

$$27.5 = 8.25v_0 + \tfrac{1}{2}a(8.25)^2$$

In another trial, the link is found to travel 34.6 inches in 11.2 s, with the same initial speed and acceleration. Find v_0 and a.

Chapter Eleven

SEQUENCES AND SERIES

This chapter first introduces *sequences*, consisting of a string of terms, each one of them a power function. Then we move to *series*, in which the terms of a sequence are summed to define a new function. Surprisingly enough, even when we have an infinite number of terms, the sum can sometimes be finite. For both sequences and series, we consider two main types, arithmetic and geometric. In an arithmetic sequence or series, the terms are linear. In a geometric sequence or series, the terms are given by power functions with integer powers greater than one.

11.1 SEQUENCES

Elections for president of the United States have been held every four years since 1792, that is, in 1792, 1796, 1800, 1804, and so on up to the year 2004. Any ordered list of numbers, such as

$$1792, 1796, 1800, 1804, \ldots, 2004$$

is called a *sequence*, and the individual numbers are the *terms* of the sequence. A sequence can be a finite list, such as the sequence of past presidential election years, or it can be an infinite list, such as the sequence of positive integers

$$1, 2, 3, 4, \ldots.$$

Example 1

(a) $0, 1, 4, 9, 16, 25, \ldots$ is the sequence of squares of integers.
(b) $2, 4, 8, 16, 32, \ldots$ is the sequence of positive integer powers of 2.
(c) $3, 1, 4, 1, 5, 9, \ldots$ is the sequence of digits in the decimal expansion of π.
(d) $3.9, 5.3, 7.2, 9.6, 12.9, 17.1, 23.1, 38.6, 50.2$ is the sequence of U.S. population figures, in millions, for the first 10 census reports (1790 to 1880).
(e) $3.5, 4.2, 5.1, 5.9, 6.7, 8.1, 9.4, 10.6, 10.1, 7.1, 3.8, 2.1, 1.4, 1.1$ is the sequence of pager subscribers in Japan, in millions, from the years 1989 to 2002.

Notation for Sequences

We denote the terms of a sequence by

$$a_1, a_2, a_3, \ldots, a_n, \ldots$$

so that a_1 is the first term, a_2 is the second term, and so on. We use a_n to denote the nth or *general term* of the sequence. If there is a pattern in the sequence, we may be able to find a formula for a_n.

Example 2

Find the first three terms and the 98th term of the sequence.

(a) $a_n = 1 + \sqrt{n}$ (b) $b_n = (-1)^n \dfrac{n}{n+1}$

Solution

(a) $a_1 = 1 + \sqrt{1}, a_2 = 1 + \sqrt{2} \approx 2.414, a_3 = 1 + \sqrt{3} \approx 2.732$, and $a_{98} = 1 + \sqrt{98} \approx 10.899$.
(b) $b_1 = (-1)^1 \dfrac{1}{1+1} = -\dfrac{1}{2}, b_2 = (-1)^2 \dfrac{2}{2+1} = \dfrac{2}{3}, b_3 = (-1)^3 \dfrac{3}{3+1} = -\dfrac{3}{4}$, and $b_{98} = (-1)^{98} \dfrac{98}{98+1} = \dfrac{98}{99}$. This sequence is called *alternating* because the terms alternate in sign.

A sequence can be thought of as a function whose domain is a set of integers. Each term of the sequence is an output value for the function, so $a_n = f(n)$.

Example 3

List the first 5 terms of the sequence $a_n = f(n)$, where $f(x) = 500 - 10x$.

Solution

Evaluate $f(x)$ for $x = 1, 2, 3, 4, 5$:

$$a_1 = f(1) = 500 - 10 \cdot 1 = 490 \quad \text{and} \quad a_2 = f(2) = 500 - 10 \cdot 2 = 480.$$

Similarly, $a_3 = 470$, $a_4 = 460$, and $a_5 = 450$.

Arithmetic Sequences

You buy a used car that has already been driven 15,000 miles and drive it 8000 miles per year. The odometer registers 23,000 miles 1 year after your purchase, 31,000 miles after 2 years, and so on. The yearly odometer readings form a sequence a_n whose terms are

$$15{,}000, \quad 23{,}000, \quad 31{,}000, \quad 39{,}000, \quad \ldots.$$

Each term of the sequence is obtained from the previous term by adding 8000; that is, the difference between successive terms is 8000. A sequence in which the difference between pairs of successive terms is a fixed quantity is called an *arithmetic sequence*.

Example 4 Which of the following sequences are arithmetic?

(a) $9, 5, 1, -3, -7$
(b) $3, 6, 12, 24, 48$
(c) $2, 2+p, 2+2p, 2+3p$
(d) $10, 5, 0, 5, 10$

Solution

(a) Each term is obtained from the previous term by subtracting 4. This sequence is arithmetic.
(b) This sequence is not arithmetic: each terms is twice the previous term. The differences are 3, 6, 12, 24.
(c) This sequence is arithmetic: p is added to each term to obtain the next term.
(d) This is not arithmetic. The difference between the second and first terms is -5, but the difference between the fifth and fourth terms is 5.

We can write a formula for the general term of an arithmetic sequence. Look at the sequence $2, 6, 10, 14, 18, \ldots$ in which the terms increase by 4, and observe that

$$\begin{aligned} a_1 &= 2 \\ a_2 &= 6 = 2 + 1 \cdot 4 \\ a_3 &= 10 = 2 + 2 \cdot 4 \\ a_4 &= 14 = 2 + 3 \cdot 4. \end{aligned}$$

When we get to the nth term, we have added $(n-1)$ copies of 4, so that $a_n = 2 + (n-1)4$. In general:

> For $n \geq 1$, the nth term of an arithmetic sequence is
>
> $$a_n = a_1 + (n-1)d,$$
>
> where a_1 is the first term, and d is the difference between consecutive terms.

Example 5

(a) Write a formula for the general term of the odometer sequence of the car that is driven 8000 miles per year and had gone 15,000 miles when it was bought.
(b) What is the car's mileage seven years after its purchase?

Solution

(a) The odometer reads 15,000 miles initially, so $a_1 = 15{,}000$. Each year the odometer reading increases by 8000, so $d = 8000$. Thus, $a_n = 15{,}000 + (n-1)8000$.
(b) Seven years from the date of purchase is the start of the 8th year, so $n = 8$. The mileage is

$$a_8 = 15{,}000 + (8-1) \cdot 8000 = 71{,}000.$$

Arithmetic Sequences and Linear Functions

You may have noticed that the arithmetic sequence for the car's odometer reading looks like a linear function. The formula for the nth term, $a_n = 15{,}000 + (n - 1)8000$, can be simplified to $a_n = 7000 + 8000n$, a linear function with slope $m = 8000$ and initial value $b = 7000$. However, for a sequence we consider only positive integer inputs, whereas a linear function is defined for all values of n. We can think of an arithmetic sequence as a linear function whose domain has been restricted to the positive integers.

Geometric Sequences

You are offered a job at a salary of \$40,000 for the first year with a 5% pay raise every year. Under this plan, your annual salaries form a sequence with terms

$$\begin{aligned}
a_1 &= 40{,}000\\
a_2 &= 40{,}000(1.05) = 42{,}000\\
a_3 &= 42{,}000(1.05) = 40{,}000(1.05)^2 = 44{,}100\\
a_4 &= 44{,}100(1.05) = 40{,}000(1.05)^3 = 46{,}305,
\end{aligned}$$

and so on, where each term is obtained from the previous one by multiplying by 1.05. A sequence in which each term is a constant multiple of the preceding term is called a *geometric sequence*. In a geometric sequence, the ratio of successive terms is constant.

Example 6 Which of the following sequences are geometric?

(a) $5, 25, 125, 625, \ldots$ (b) $-8, 4, -2, 1, -\frac{1}{2}, \ldots$ (c) $12, 6, 4, 3, \ldots$

Solution

(a) This sequence is geometric. Each term is 5 times the previous term. Note that the ratio of any term to its predecessor is 5.

(b) This sequence is geometric. The ratio of any term to the previous term is $-\frac{1}{2}$.

(c) This sequence is not geometric. The ratios of successive terms are not constant: $\dfrac{a_2}{a_1} = \dfrac{6}{12} = \dfrac{1}{2}$, but $\dfrac{a_3}{a_2} = \dfrac{4}{6} = \dfrac{2}{3}$.

As for arithmetic sequences, there is a formula for the general term of a geometric sequence. Consider the sequence $8, 2, \frac{1}{2}, \frac{1}{8}, \ldots$ in which each term is $\frac{1}{4}$ times the previous term. We have

$$\begin{aligned}
a_1 &= 8\\
a_2 &= 2 = 8\left(\frac{1}{4}\right)\\
a_3 &= \frac{1}{2} = 8\left(\frac{1}{4}\right)^2\\
a_4 &= \frac{1}{8} = 8\left(\frac{1}{4}\right)^3.
\end{aligned}$$

When we get to the nth term, we have multiplied 8 by $(n-1)$ factors of $\frac{1}{4}$, so that $a_n = 8\left(\frac{1}{4}\right)^{n-1}$. In general,

For $n \geq 1$, the nth term of a geometric sequence is

$$a_n = a_1 r^{n-1},$$

where a_1 is the first term, and r is the ratio of consecutive terms.

Example 7

(a) Write a formula for the general term of the salary sequence which starts at \$40,000 and increases by 5% each year.

(b) What is your salary after 10 years on the job?

Solution

(a) Your starting salary is \$40,000, so $a_1 = 40{,}000$. Each year your salary increases by 5%, so $r = 1.05$. Thus, $a_n = 40{,}000(1.05)^{n-1}$.

(b) After 10 years on the job, you are at the start of your 11th year, so $n = 11$. Your salary is

$$a_{11} = 40{,}000(1.05)^{11-1} \approx 65{,}156 \text{ dollars.}$$

Geometric Sequences and Exponential Functions

The formula for the salary sequence, $a_n = 40{,}000(1.05)^{n-1}$, looks like a formula for an exponential function, $f(t) = a \cdot b^t$. A geometric sequence is an exponential function whose domain is restricted to the positive integers. For many applications, this restricted domain is more realistic than an interval of real numbers. For example, salaries are usually increased once a year, rather than continuously.

Exercises and Problems for Section 11.1

Exercises

Are the sequences in Exercises 1–4 arithmetic?

1. $2, 7, 12, 17, \ldots$

2. $2, -5, -11, -16, \ldots$

3. $2, 7, 11, 14, \ldots$

4. $2, -5, -12, -19, \ldots$

Are the sequences in Exercises 5–7 arithmetic? For those that are, give a formula for the nth term.

5. $1, -1, 2, -2, \ldots$

6. $6, 9, 12, 15, \ldots$

7. $-1, -1.1, -1.2, -1.3, \ldots$

Are the sequences in Exercises 8–11 geometric?

8. $2, 4, 8, 12, \ldots$

9. $2, 6, 18, 54, \ldots$

10. $2, 0.2, 0.02, 0.002, \ldots$

11. $2, -1, \frac{1}{2}, -\frac{1}{4}, \frac{1}{8}, \ldots$

Are the sequences in Exercises 12–17 geometric? For those that are, give a formula for the nth term.

12. $4, 2, 1, \frac{1}{2}, \frac{1}{4}, \ldots$

13. $4, 12, 36, 108, \ldots$

14. $4, 1, \frac{1}{4}, \frac{1}{8}, \ldots$

15. $2, -4, 8, -16, \ldots$

16. $4, 0.4, 0.04, 0.004, \ldots$

17. $1, \frac{1}{1.2}, \frac{1}{(1.2)^2}, \frac{1}{(1.2)^3}, \ldots$

Problems

In Problems 18–21, find the 5th, 50th, n^{th} term of the arithmetic sequences.

18. $3, 5, 7, \ldots$

19. $6, 7.2, \ldots$

20. $a_1 = 2.1, a_3 = 4.7$

21. $a_3 = 5.7, a_6 = 9$

In Problems 22–25, find the 6th and n^{th} of the geometric sequences.

22. $1, 2, 4, \ldots$

23. $7, 5.25, \ldots$

24. $a_1 = 3, a_3 = 48$

25. $a_2 = 6, a_4 = 54$

26. During 2004, about 81 million barrels of oil a day were consumed worldwide.[1] Over the previous decade, consumption had been rising at 1.2% a year; assume that it continues to increase at this rate.

(a) Write the first four terms of the sequence a_n giving daily oil consumption n years after 2003; give a formula for the general term a_n.
(b) In what year is consumption expected to exceed 100 million barrels a day?

27. In 2004, US natural gas consumption was 646.7 billion cubic meters. Asian consumption was 367.7 billion cubic meters.[2] During the previous decade, US consumption increased by 0.6% a year, while Asian consumption grew by 7.9% a year. Assume these rates continue into the future.

(a) Give the first four terms of the sequence, a_n, giving US consumption of natural gas n years after 2003.
(b) Give the first four terms of a similar sequence b_n showing Asian gas consumption.
(c) According to this model, when will Asian yearly gas consumption exceed US consumption?

28. The population[3] of Nevada grew from 2.2 million in 2002 to 2.4 million in 2005. Assuming a constant percent growth rate:

(a) Find a formula for a_n, the population in millions n years after 2005.
(b) When is the population predicted to reach 10 million?

29. Florida's population[4] was 17.960 million in 2005 and 17.613 million in 2004. Assuming the population continues to increase at the same percentage rate:

(a) Find the first three terms of the sequence a_n giving the population, in millions, n years after 2005.
(b) Write a formula for a_n.
(c) What is the doubling time of the population?

The graphs in Problems 30–33 represent either an arithmetic or a geometric sequence; decide which. For an arithmetic sequence, say if the common difference, d, is positive or negative. For a geometric sequence, say if the common ratio, r, is greater or smaller than 1.

30.
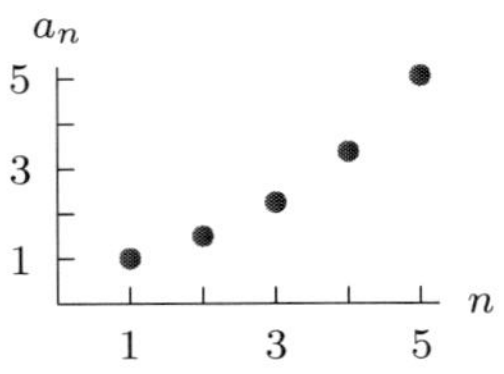

31.
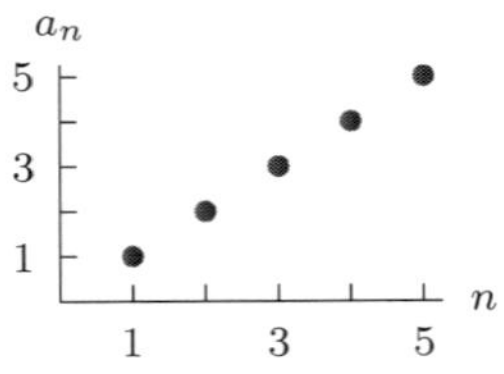

32.
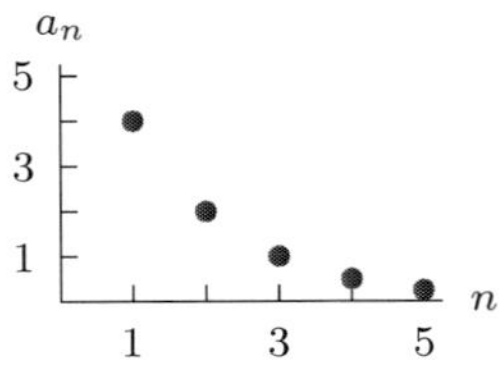

33.
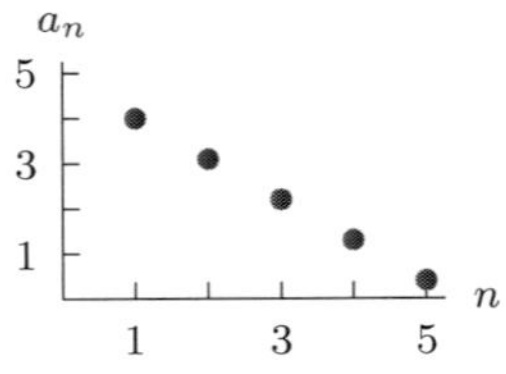

A sequence a_n can be defined by a *recurrence relation*, which gives a_n in terms of the previous term, a_{n-1}, and the first term a_1. In Problems 34–37, find the first four terms of the sequence and a formula for the general term.

34. $a_n = 2a_{n-1};\ a_1 = 3$

35. $a_n = a_{n-1} + 5;\ a_1 = 2$

36. $a_n = -a_{n-1};\ a_1 = 1$

37. $a_n = 2a_{n-1} + 1;\ a_1 = 3$

38. The Fibonacci sequence starts with $1, 1, 2, 3, 5, \ldots$, and each term is the sum of the previous two terms.

(a) Write the next three terms in the sequence.
(b) Write an expression for a_n in terms of a_{n-1} and a_{n-2}.
(c) Suppose r_n is the ratio a_n/a_{n-1} and r_{n-1} is the ratio a_{n-1}/a_{n-2}. Using your answer to part (b), find a formula for r_n in terms of r_{n-1} and without any of the as.
(d) The terms r_n form another sequence. Suppose r_n tends to a fixed value, r as n increases without bound; that is, $r_n \to r$ as $n \to \infty$. Use your answer to part (c) to find an equation for r. Solve this equation. The number that you find is called the *golden ratio*.

[1] www.bp.com/downloads, Statistical Review of World Energy 2005, accessed January 15, 2006.
[2] www.bp.com/downloads, Statistical Review of World Energy 2005, accessed January 15, 2006.
[3] health2k.state.nv.us and www.city-data.com, accessed December 26, 2005.
[4] www.floridacharts.com, accessed December 26, 2005.

39. Some people believe they can make money from a chain letter (they are usually disappointed). A chain letter works roughly like this: A letter arrives with a list of four names attached and instructions to mail a copy to four more friends and to send \$1 to the top name on the list. When you mail the four letters, you remove the top name (to whom the money was sent) and add your own name to the bottom of the list.

(a) If no one breaks the chain, how much money do you receive?

(b) Let d_n be the number of dollars you receive if there are n names on the list instead of 4, but you still mailed to four friends. Find a formula for d_n.

40. For a positive integer n, let a_n be the fraction of the US population with income less than or equal to \$$n$ thousand dollars.

(a) Which is larger, a_{40} or a_{50}? Why?

(b) What does the quantity $a_{50} - a_{40}$ represent in terms of US population?

(c) Is there any value of n with $a_n = 0$? Explain.

(d) What happens to the value of a_n as n increases?

11.2 DEFINING FUNCTIONS USING SUMS: ARITHMETIC SERIES

Domestic Death from AIDS n years after 1980

Table 11.1 shows the number of US AIDS deaths[5] that occurred each year since 1980, where a_n is the number of deaths in year n, and $n = 1$ corresponds to 1981. For each year, we can calculate the total number of death from AIDS since 1980. For example, in 1985

Total number of AIDS deaths (1981-1985) $= 159 + 463 + 1508 + 3505 + 6972 = 12{,}607$.

We write S_n to denote the sum of the first n terms of a sequence. In this example, 1985 corresponds to $n = 5$, so we have $S_5 = 12{,}607$.

Table 11.1 *US deaths from AIDS each year from 1981–2003*

n	a_n	n	a_n	n	a_n	n	a_n	n	a_n
1	159	6	12,110	11	36,616	16	38,025	21	18,524
2	463	7	16,412	12	41,094	17	21,999	22	17,557
3	1508	8	21,119	13	45,598	18	18,397	23	18,017
4	3505	9	27,791	14	50,418	19	17,172		
5	6972	10	31,538	15	51,117	20	15,245		

Example 1 Find and interpret S_8 for the AIDS sequence.

Solution Since S_8 is the sum of the first 8 terms of the sequence, we have.

$$\begin{aligned} S_8 &= \underbrace{a_1 + a_2 + a_3 + a_4 + a_5}_{S_5 = 12{,}607} + a_6 + a_7 + a_8 \\ &= 12{,}607 + 12{,}110 + 16{,}412 + 21{,}119 \\ &= 62{,}248. \end{aligned}$$

Here, S_8 is the number of deaths from AIDS from 1980 to 1988.

The sum of the terms of a sequence is called a *series*. We write S_n for the sum of the first n terms of the sequence, called the nth *partial sum*. We see that S_n is a function of n, the number of terms in the partial sum. In this section we see how to evaluate functions defined by sums.

[5] www.cdc.gov/hiv/stats/hasr1301/table28.htm, from the HIV/AIDS Surveillance Report, 2001, Vol 13, No. 1, p. 34, US Department of Health and Human Services, Centers for Disease Control and Prevention, Atlanta, and www.cdc.gove/hiv/PUBS/Facts/At-A-Glance.htm, accessed January 14, 2006.

Arithmetic Series

Landscape timbers are large beams of wood used to landscape gardens. To make the terrace in Figure 11.1, one timber is set into the slope, followed by a stack of two, then a stack of three, then a stack of four. The stacks are separated by earth.

The total number of timbers in 4 stacks is S_4, the sum of the number of timbers in each stack:

$$\text{Total number of timbers} = S_4 = 1 + 2 + 3 + 4 = 10.$$

For a larger terrace using 5 stacks of timbers, the total number is given by

$$S_5 = 1 + 2 + 3 + 4 + 5 = 15.$$

For an even larger terrace using 6 stacks, the total number is given by

$$S_6 = 1 + 2 + 3 + 4 + 5 + 6 = 21.$$

For a terrace made from n stacks of landscape timbers, S_n, the total number of timbers needed is a function of n, so

$$S_n = f(n) = 1 + 2 + \cdots + n.$$

The symbol $\cdots$ means that all the integers from 1 to n are included in the sum.

Notice that each stack of landscape timbers contains one more timber than the previous one. Thus, the number of timbers in each stack,

$$1, 2, 3, 4, 5, \ldots$$

is an arithmetic sequence, so S_n is the sum of the terms of an arithmetic sequence. Such a sum is called an *arithmetic series*.

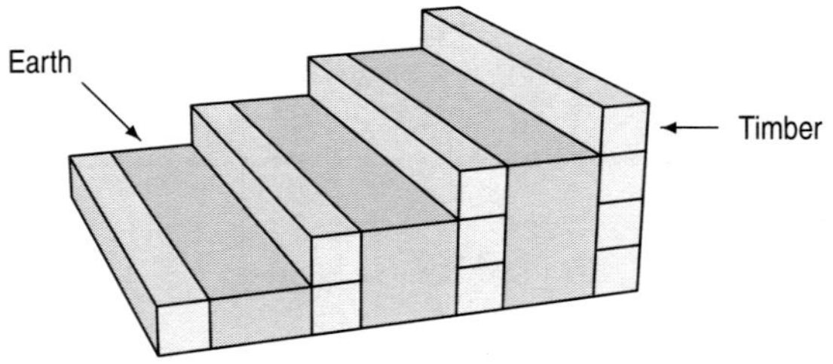

Figure 11.1: A slope terraced for planting using landscape timbers

The Sum of an Arithmetic Series

We now find a formula for the sum of an arithmetic series. A famous story concerning this series[6] is told about the great mathematician Carl Friedrich Gauss (1777–1855), who as a young boy was asked by his teacher to add the numbers from 1 to 100. He did so almost immediately:

$$S_{100} = 1 + 2 + 3 + \cdots + 100 = 5050.$$

[6]As told by E.T. Bell in *The Men of Mathematics*, p.221 (New York: Simon and Schuster, 1937), the series involved was arithmetic, but more complicated than this one.

Of course, no one really knows how Gauss accomplished this, but he probably did not perform the calculation directly, by adding 100 terms. He might have noticed that the terms in the sum can be regrouped into pairs, as follows:

$$\begin{aligned} S_{100} = 1 + 2 + \cdots + 99 + 100 &= \underbrace{(1 + 100) + (2 + 99) + \cdots + (50 + 51)}_{50 \text{ pairs}} \\ &= \underbrace{101 + 101 + \cdots + 101}_{50 \text{ terms}} \quad \text{Each pair adds to 101} \\ &= 50 \cdot 101 = 5050. \end{aligned}$$

The approach of pairing numbers works for the sum from 1 to n, no matter how large n is. Provided n is an even number, we can write

$$\begin{aligned} S_n = 1 + 2 + \cdots + (n-1) + n &= \underbrace{(1 + n) + (2 + (n-1)) + (3 + (n-2)) + \cdots}_{\frac{1}{2}n \text{ pairs}} \\ &= \underbrace{(1 + n) + (1 + n) + \cdots}_{\frac{1}{2}n \text{ pairs}} \quad \text{Each pair adds to } 1 + n \end{aligned}$$

so we have the formula:

$$\boxed{S_n = 1 + 2 + \cdots + n = \frac{1}{2}n(n+1).}$$

Example 2 Check this formula for S_n with $n = 100$.

Solution Using the formula, we get the same answer, 5050, as before:

$$S_{100} = 1 + 2 + \cdots + 100 = \frac{1}{2} \cdot 100(100 + 1) = 50 \cdot 101 = 5050.$$

A similar derivation shows that this formula for S_n also holds for odd values of n.

To find a formula for the sum of a general arithmetic series, we first assume that n is even and write

$$\begin{aligned} S_n &= a_1 + a_2 + \cdots + a_n \\ &= \underbrace{(a_1 + a_n) + (a_2 + a_{n-1}) + (a_3 + a_{n-2}) + \cdots}_{\frac{1}{2}n \text{ pairs}}. \end{aligned}$$

We pair the first term with the last term, the second term with the next to last term, and so on, just as Gauss may have done. Each pair of terms adds up to the same value, just as each of Gauss's pairs added to 101. Using the formula for the terms of an arithmetic sequence, $a_n = a_1 + (n-1)d$, the first pair, $a_1 + a_n$, can be written as

$$a_1 + a_n = a_1 + \underbrace{a_1 + (n-1)d}_{a_n} = 2a_1 + (n-1)d,$$

and the second pair, $a_2 + a_{n-1}$, can be written as

$$a_2 + a_{n-1} = \underbrace{a_1 + d}_{a_2} + \underbrace{a_1 + (n-2)d}_{a_{n-1}} = 2a_1 + (n-1)d.$$

Both the first two pairs have the same sum: $2a_1 + (n-1)d$. The remaining pairs also all have the same sum, so

$$S_n = \underbrace{(a_1 + a_n) + (a_2 + a_{n-1}) + (a_3 + a_{n-2}) + \cdots}_{\frac{1}{2}n \text{ pairs}} = \frac{1}{2}n\left(2a_1 + (n-1)d\right).$$

The same formula gives the sum of the series when n is odd. See Problem 43 on page 500.

> The **sum**, S_n, of the first n terms of the **arithmetic series** with $a_n = a_1 + (n-1)d$ is
>
> $$S_n = \frac{1}{2}n(a_1 + a_n) = \frac{1}{2}n(2a_1 + (n-1)d).$$

Example 3 Calculate the sum $1 + 2 + \cdots + 100$ using the formula for S_n.

Solution Here, $a_1 = 1$, $n = 100$, and $d = 1$. We get the same answer as before:

$$S_{100} = \frac{1}{2}n\left(2a_1 + (n-1)d\right) = \frac{1}{2} \cdot 100\left(2 \cdot 1 + (100-1) \cdot 1\right) = 50 \cdot 101 = 5050.$$

Summation Notation

The symbol Σ is used to indicate addition. This symbol, pronounced *sigma*, is the Greek capital letter for *S*, which stands for sum. Using this notation, we write

$$\sum_{i=1}^{n} a_i \quad \text{to stand for the sum } a_1 + a_2 + \cdots + a_n.$$

The Σ tells us we are adding some numbers. The a_i tells us that the numbers we are adding are called a_1, a_2, and so on. The sum begins with a_1 and ends with a_n because the subscript i starts at $i = 1$ (at the bottom of the Σ sign) and ends at $i = n$ (at the top of the Σ sign):

This tells us that the sum ends at a_n
↓

$$\sum_{i=1}^{n} a_i \longleftarrow$$ This tells us that the numbers we are adding are called $a_1, a_2, a_3, \ldots$

↑
This tells us that the sum starts at a_1

Example 4 Write $S_n = f(n)$, the total number of landscape timbers in n stacks using sigma notation, and give a formula for S_n.

Solution Since $S_n = f(n) = 1 + 2 + \cdots + n$, we have, $a_i = i$. We start at $i = 1$ and end at $i = n$. Thus, $S_n = \sum_{i=1}^{n} i$. The formula for the sum of an arithmetic series with $a_1 = d = 1$ gives

$$S_n = f(n) = \sum_{i=1}^{n} i = \frac{1}{2}n(n+1).$$

Example 5 Use sigma notation to write the sum of the first 20 positive odd numbers. Evaluate this sum.

Solution The odd numbers form an arithmetic sequence: 1, 3, 5, 7, ... with $a_1 = 1$ and $d = 2$. The ith odd number is

$$a_i = 1 + (i-1)2 = 2i - 1.$$

(As a check: $a_1 = 1 + (1-1)2 = 1$ and $a_2 = 1 + (2-1)2 = 3$.) Thus, we have

$$\text{Sum of the first 20 odd numbers} = \sum_{i=1}^{20} a_i = \sum_{i=1}^{20} (2i-1).$$

We evaluate the sum using $n = 20$, $a_1 = 1$, and $d = 2$:

$$\text{Sum} = \frac{1}{2}n\,(2a_1 + (n-1)d) = \frac{1}{2} \cdot 20\,(2 \cdot 1 + (20-1) \cdot 2) = 400.$$

Example 6 If air resistance is neglected, a falling object travels 16 ft during the first second, 48 ft during the next, 80 ft during the next, and so on. These distances form the arithmetic sequence 16, 48, 80, ... In this sequence, $a_1 = 16$ and $d = 32$.

(a) Find a formula for the nth term in the sequence of distances. Calculate the fifth and tenth terms.
(b) Calculate S_1, S_2, and S_3, the total distance an object falls in 1, 2, and 3 seconds, respectively.
(c) Give a formula for S_n, the distance fallen in n seconds.

Solution

(a) The nth term is $a_n = a_1 + (n-1)d = 16 + 32(n-1) = 32n - 16$. Thus, the fifth term is $a_5 = 32(5) - 16 = 144$. The value of $a_{10} = 32(10) - 16 = 304$.
(b) Since $a_1 = 16$ and $d = 32$,

$$\begin{aligned} S_1 &= a_1 = 16 \text{ feet} \\ S_2 &= a_1 + a_2 = a_1 + (a_1 + d) = 16 + 48 = 64 \text{ feet} \\ S_3 &= a_1 + a_2 + a_3 = 16 + 48 + 80 = 144 \text{ feet.} \end{aligned}$$

(c) The formula for S_n is $S_n = \frac{1}{2}n\,(2a_1 + (n-1)d)$. We can check our answer to part (b) using this formula:

$$S_1 = \frac{1}{2}{\cdot}1(2{\cdot}16{+}0{\cdot}32) = 16, \quad S_2 = \frac{1}{2}{\cdot}2(2{\cdot}16{+}1{\cdot}32) = 64, \quad S_3 = \frac{1}{2}{\cdot}3(2{\cdot}16{+}2{\cdot}32) = 144.$$

Example 7 An object falls from 1000 feet starting at time $t = 0$ seconds. What is its height, h, in feet above the ground at $t = 1, 2, 3$ seconds? Show these values on a graph of height against time.

Solution At time $t = 0$, the height $h = 1000$ feet. At time $t = 1$, the object has fallen $S_1 = 16$ feet, so

$$h = 1000 - 16 = 984 \text{ feet.}$$

At time $t = 2$, the object has fallen a total distance of $S_2 = 64$ feet, so

$$h = 1000 - 64 = 936 \text{ feet.}$$

At time $t = 3$, the object has fallen a total distance of $S_3 = 144$ feet, so

$$h = 1000 - 144 = 856 \text{ feet.}$$

These heights are marked on the graph in Figure 11.2.

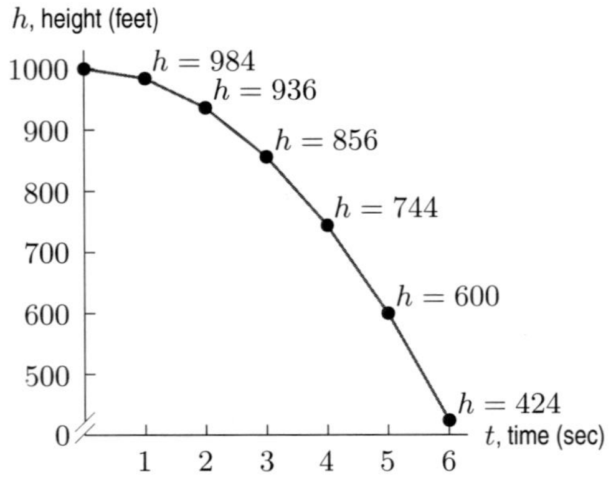

Figure 11.2: Height of a falling object

Exercises and Problems for Section 11.2

Exercises

Are the series in Exercises 1–4 arithmetic?

1. $1 + 2 + 4 + 5 + 7 + 8 + \cdots$

2. $-\frac{1}{3} + \frac{2}{3} + \frac{5}{3} + \frac{8}{3} + \cdots$

3. $2 + 4 + 8 + 16 + \cdots$

4. $10 + 8 + 6 + 4 + 2 + \cdots$

In Exercises 5–9, complete the tables with the terms of the arithmetic series $a_1, a_2, \ldots, a_n$, and the sequence of partial sums, $S_1, S_2, \ldots, S_n$. State the values of a_1 and d where $a_n = a_1 + (n-1)d$.

5.

n	1	2	3	4	5	6	7	8
a_n	2	7	12	17				
S_n	2	9	21	38				

6.

n	1	2	3	4	5	6	7	8
a_n	3	7	11					
S_n	3	10						

7.

n	1	2	3	4	5	6	7	8
a_n	7			16				
S_n								

8.

n	1	2	3	4	5	6	7	8
a_n	2							
S_n	2	13	33	62				

9.

n	1	2	3	4	5	6	7	8
a_n								
S_n						201	273	356

Expand the sums in Exercises 10–15. (Do not evaluate.)

10. $\sum_{i=-1}^{5} i^2$

11. $\sum_{i=10}^{20} (i+1)^2$

12. $\sum_{k=0}^{5} 2k+1$

13. $\sum_{j=1}^{6} 3(j-3)$

14. $\sum_{j=2}^{10} (-1)^j$

15. $\sum_{n=1}^{7} (-1)^{n-1} 2^n$

In Exercises 16–19, write the sum using sigma notation.

16. $3+6+9+12+15+18+21$

17. $10+13+16+19+22$

18. $1/2+1+3/2+2+5/2+3+7/2+4$

19. $30+25+20+15+10+5$

20. **(a)** Use sigma notation to write the sum $2+4+6+\cdots+20$ of the first 10 even numbers.

(b) Without a calculator, find the sum in part (a).

21. Find the sum of the first 1000 integers: $1+2+3+\cdots+1000$.

Without using a calculator, find the sum of the series in Problems 22–29.

22. $\sum_{i=1}^{50} 3i$

23. $\sum_{i=1}^{30} (5i+10)$

24. $\sqrt{2}+2\sqrt{2}+3\sqrt{2}+\cdots+25\sqrt{2}$

25. $-101-91-81-71-61-\cdots-11-1$

26. $26.5+24.5+22.5+\cdots+2.5+0.5$

27. $-3.01-3.02-3.03-\cdots-3.35$

28. $\sum_{n=0}^{15} \left(2+\frac{1}{2}n\right)$

29. $\sum_{n=0}^{10} (8-4n)$

Problems

30. Find the thirtieth positive multiple of 5 and the sum of the first thirty positive multiples of 5.

31. For the AIDS data in Table 11.1 on page 493,

(a) Find and interpret in terms of AIDS

(i) The partial sums S_5, S_6, S_7, S_8.

(ii) $S_6-S_5, S_7-S_6, S_8-S_7$.

(b) Use your answer to part (a) (ii) to explain the value of $S_{n+1}-S_n$ for any positive integer n.

32. Table 11.2 shows US Census figures, in millions. Interpret these figures as partial sums, S_n of a sequence, a_n, where n is the number of decades since 1940, so $S_1=150.7$, $S_2=179.3$, $S_3=203.3$, and so on.

(a) Find and interpret in terms of population

(i) S_4, S_5, S_6

(ii) a_2, a_5, a_6

(iii) $a_6/10$

(b) For any positive integer n, what are the meanings of S_n, a_n, and $a_n/10$ in terms of population?

Table 11.2

Year	1950	1960	1970	1980	1990	2000
Population	150.7	179.3	203.3	226.6	248.7	281.4

Simplify the expressions in Problems 33–36.

33. $\sum_{i=1}^{5} i^2 - \sum_{j=0}^{4} (j+1)^2$

34. $\sum_{i=4}^{20} i - \sum_{j=4}^{20} (-2j)$

35. $\sum_{i=1}^{20} 1$

36. $\sum_{i=1}^{15} i^3 - \sum_{j=3}^{15} j^3$

Problems 37–40 refer to the falling object of Example 6 on page 497.

37. Calculate the distance, S_7, the object falls in 7 seconds.

38. **(a)** Find the total distance that the object falls in 4, 5, 6 seconds.

(b) The object falls from 1000 feet at time $t=0$. Calculate its height at $t=4$, $t=5$, $t=6$ seconds. Explain how you can check your answer using Figure 11.2.

39. Find a formula for $f(n)$, the distance fallen by the object in n seconds.

40. If the object falls from 1000 feet, how long does it take to hit the ground?

41. A boy is dividing M&Ms between himself and his sister. He gives one to his sister and takes one for himself. He gives another to his sister and takes two for himself. He gives a third one to his sister and takes three for himself, and so on.

(a) On the nth round, how many M&Ms does the boy give his sister? How many does he take himself?

(b) After n rounds, how many M&Ms does his sister have? How many does the boy have?

42. An auditorium has 30 seats in the first row, 34 seats in the second row, 38 seats in the third row, and so on. If there are twenty rows in the auditorium, how many seats are there in the last row? How many seats are there in the auditorium?

43. In the text we showed how to calculate the sum of an arithmetic series with an even number of terms. Consider the arithmetic series

$$5 + 12 + 19 + 26 + 33 + 40 + 47 + 54 + 61.$$

Here, there are $n = 9$ terms, and the difference between each term is $d = 7$. Adding these terms directly, we find that their sum is 297. In this problem we find the sum of this arithmetic sequence in two different ways. We then use our results to obtain a general formula for the sum of an arithmetic series with an odd number of terms.

(a) The sum of the first and last terms is $5 + 61 = 66$, the sum of the second and next-to-last terms is $12 + 54 = 66$, and so on. Find the sum of this arithmetic series by pairing off terms in this way. Notice that since the number of terms is odd, one of them will be unpaired.

(b) This arithmetic series can be thought of as a series of eight terms $(5 + \cdots + 54)$ plus an additional term (61). Use the formula we found for the sum of an arithmetic series containing an even number of terms to find the sum of the given arithmetic series.

(c) Find a formula for the sum of an arithmetic series with n terms where n is odd. Let a_1 be the first term in the series, and let d be the difference between consecutive terms. Show that the two approaches used in parts (a) and (b) give the same result, and show that your formula is the same as the formula given for even values of n.

11.3 FINITE GEOMETRIC SERIES

In the previous section, we studied *arithmetic series*. An arithmetic series is the sum of terms in a sequence in which each term is obtained by adding a constant to the preceding term. In this section, we study another type of sum, a *geometric series*. In a geometric series, each term is a constant multiple of the preceding term.

Bank Balance

A person deposits \$2000 every year in an IRA that pays 6% interest per year, compounded annually. After the first deposit (but before any interest has been earned), the balance in the account in dollars is

$$B_1 = 2000.$$

After 1 year has passed, the first deposit has earned interest, so the balance becomes 2000(1.06) dollars. Then the second deposit is made and the balance becomes

$$\begin{aligned} B_2 &= \underbrace{2^{\text{nd}}\text{ deposit}}_{2000} + \underbrace{1^{\text{st}}\text{ deposit with interest}}_{2000(1.06)} \\ &= 2000 + 2000(1.06) \text{ dollars.} \end{aligned}$$

After 2 years have passed, the third deposit is made, and the balance is

$$\begin{aligned} B_3 &= \underbrace{3^{\text{rd}}\text{ deposit}}_{2000} + \underbrace{2^{\text{nd}}\text{ dep. with 1 year interest}}_{2000(1.06)} + \underbrace{1^{\text{st}}\text{ deposit with 2 years interest}}_{2000(1.06)^2} \\ &= 2000 + 2000(1.06) + 2000(1.06)^2. \end{aligned}$$

Let B_n be the balance in dollars after n deposits. Then we see that

$$\begin{aligned}
&\text{After 4 deposits} \quad && B_4 = 2000 + 2000(1.06) + 2000(1.06)^2 + 2000(1.06)^3 \\
&\text{After 5 deposits} \quad && B_5 = 2000 + 2000(1.06) + 2000(1.06)^2 + 2000(1.06)^3 + 2000(1.06)^4 \\
& && \vdots \\
&\text{After } n \text{ deposits} \quad && B_n = 2000 + 2000(1.06) + 2000(1.06)^2 + \cdots + 2000(1.06)^{n-1}.
\end{aligned}$$

Example 1 How much money is in this IRA in 5 years, right after a deposit is made? In 25 years, right after a deposit is made?

Solution After 5 years, we have made 6 deposits. A calculator gives

$$\begin{aligned}
B_6 &= 2000 + 2000(1.06) + 2000(1.06)^2 + \cdots + 2000(1.06)^5 \\
&= \$13{,}950.64.
\end{aligned}$$

In 25 years, we have made 26 deposits. Even using a calculator, it would be tedious to evaluate B_{26} by adding 26 terms. Fortunately, there is a shortcut. Start with the formula for B_{26}:

$$B_{26} = 2000 + 2000(1.06) + 2000(1.06)^2 + \cdots + 2000(1.06)^{25}.$$

Multiply both sides of this equation by 1.06 and add 2000 giving

$$1.06B_{26} + 2000 = 1.06\left(2000 + 2000(1.06) + 2000(1.06)^2 + \cdots + 2000(1.06)^{25}\right) + 2000.$$

We simplify the right-hand side to get

$$1.06B_{26} + 2000 = 2000(1.06) + 2000(1.06)^2 + \cdots + 2000(1.06)^{25} + 2000(1.06)^{26} + 2000.$$

Notice that the right-hand side of this equation and the formula for B_{26} have almost every term in common. We can rewrite this equation as

$$\begin{aligned}
1.06B_{26} + 2000 &= \underbrace{2000 + 2000(1.06) + 2000(1.06)^2 + \cdots + 2000(1.06)^{25}}_{B_{26}} + 2000(1.06)^{26} \\
&= B_{26} + 2000(1.06)^{26}.
\end{aligned}$$

Solving for B_{26} gives

$$\begin{aligned}
1.06B_{26} - B_{26} &= 2000(1.06)^{26} - 2000 \\
0.06B_{26} &= 2000(1.06)^{26} - 2000 \\
B_{26} &= \frac{2000(1.06)^{26} - 2000}{0.06}.
\end{aligned}$$

Using a calculator to evaluate this expression for B_{26}, we find that $B_{26} = 118{,}312.77$ dollars.

Geometric Series

The formula for the IRA balance,

$$B_{26} = 2000 + 2000(1.06) + 2000(1.06)^2 + \cdots + 2000(1.06)^{25},$$

is an example of a *geometric series*. This is a finite geometric series, because there are a finite number of terms (in this case, 26). In general, a geometric series is the sum of the terms of a geometric sequence—that is, in which each term is a constant multiple of the preceding term.

A **finite geometric series** is a sum of the form

$$S_n = a + ar + ar^2 + \cdots + ar^{n-1} = \sum_{i=0}^{n-1} ar^i.$$

Notice that S_n is defined to contain exactly n terms. Since the first term is $a = ar^0$, we stop at ar^{n-1}. For instance, the series

$$50 + 50(1.06) + 50(1.06)^2 + \cdots + 50(1.06)^{25}$$

contains 26 terms, so $n = 26$. For this series, $r = 1.06$ and $a = 50$.

The Sum of a Geometric Series

The shortcut from Example 1 can be used to find the sum of a general geometric series. Let S_n be the sum of a geometric series of n terms, so that

$$S_n = a + ar + ar^2 + \cdots + ar^{n-1}.$$

Multiply both sides of this equation by r and add a, giving

$$\begin{aligned} rS_n + a &= r\left(a + ar + ar^2 + \cdots + ar^{n-1}\right) + a \\ &= \left(ar + ar^2 + ar^3 + \cdots + ar^{n-1} + ar^n\right) + a. \end{aligned}$$

The right-hand side can be rewritten as

$$\begin{aligned} rS_n + a &= \underbrace{a + ar + ar^2 + \cdots + ar^{n-1}}_{S_n} + ar^n \\ &= S_n + ar^n. \end{aligned}$$

Solving the equation $rS_n + a = S_n + ar^n$ for S_n gives

$$\begin{aligned} rS_n - S_n &= ar^n - a \\ S_n(r-1) &= ar^n - a \qquad \text{factoring out } S_n \\ S_n &= \frac{ar^n - a}{r-1} \\ &= \frac{a(r^n - 1)}{r-1}. \end{aligned}$$

By multiplying the numerator and denominator by -1, this formula can be rewritten as follows:

The sum of a finite **geometric series of n terms** is given by

$$S_n = a + ar + ar^2 + \cdots + ar^{n-1} = \frac{a(1-r^n)}{1-r}, \qquad \text{for } r \neq 1.$$

This formula is called a *closed form* of the sum.

Example 2 Use the formula for the sum of a geometric series to solve Example 1.

Solution We need to find B_6 and B_{26} where

$$B_n = 2000 + 2000(1.06) + 2000(1.06)^2 + \cdots + 2000(1.06)^{n-1}.$$

Using the formula for S_n with $a = 2000$ and $r = 1.06$, we get the same answers as before:

$$B_6 = \frac{2000(1-(1.06)^6)}{1-1.06} = 13{,}950.64,$$
$$B_{26} = \frac{2000(1-(1.06)^{26})}{1-1.06} = 118{,}312.77.$$

Drug Levels in The Body

Geometric series arise naturally in many different contexts. The following example illustrates a geometric series with decreasing terms.

Example 3 A patient is given a 20 mg injection of a therapeutic drug. Each day, the patient's body metabolizes 50% of the drug present, so that after 1 day only one-half of the original amount remains, after 2 days only one-fourth remains, and so on. The patient is given a 20 mg injection of the drug every day at the same time. Write a geometric series that gives the drug level in this patient's body right after the nth injection.

Solution Immediately after the 1st injection, the drug level in the body is given by

$$Q_1 = 20.$$

One day later, the original 20 mg has fallen to $20 \cdot \frac{1}{2} = 10$ mg and the second 20 mg injection is given. Right after the second injection, the drug level is given by

$$\begin{aligned} Q_2 &= \underbrace{2^{\text{nd}}\text{ injection}}_{20} + \underbrace{\text{Residue of }1^{\text{st}}\text{ injection}}_{\frac{1}{2}\cdot 20} \\ &= 20 + 20\left(\frac{1}{2}\right). \end{aligned}$$

Two days later, the original 20 mg has fallen to $(20 \cdot \frac{1}{2}) \cdot \frac{1}{2} = 20(\frac{1}{2})^2 = 5$ mg, the second 20 mg injection has fallen to $20 \cdot \frac{1}{2} = 10$ mg, and the third 20 mg injection is given. Right after the third injection, the drug level is given by

$$\begin{aligned} Q_3 &= \underbrace{3^{\text{rd}}\text{ injection}}_{20} + \underbrace{\text{Residue of }2^{\text{nd}}\text{ injection}}_{20\cdot\frac{1}{2}} + \underbrace{\text{Residue of }1^{\text{st}}\text{ injection}}_{20\cdot\frac{1}{2}\cdot\frac{1}{2}} \\ &= 20 + 20\left(\frac{1}{2}\right) + 20\left(\frac{1}{2}\right)^2. \end{aligned}$$

Continuing, we see that

$$\begin{aligned} &\text{After } 4^{\text{th}} \text{ injection} \quad Q_4 = 20 + 20\left(\frac{1}{2}\right) + 20\left(\frac{1}{2}\right)^2 + 20\left(\frac{1}{2}\right)^3 \\ &\text{After } 5^{\text{th}} \text{ injection} \quad Q_5 = 20 + 20\left(\frac{1}{2}\right) + 20\left(\frac{1}{2}\right)^2 + \cdots + 20\left(\frac{1}{2}\right)^4 \\ &\qquad\vdots \\ &\text{After } n^{\text{th}} \text{ injection} \quad Q_n = 20 + 20\left(\frac{1}{2}\right) + 20\left(\frac{1}{2}\right)^2 + \cdots + 20\left(\frac{1}{2}\right)^{n-1}. \end{aligned}$$

This is another example of a geometric series. Here, $a = 20$ and $r = 1/2$ in the geometric series formula

$$Q_n = a + ar + ar^2 + \cdots + ar^{n-1}.$$

To calculate the drug level for a specific value of n, we use the formula for the sum of a geometric series.

Example 4 What quantity of the drug remains in the patient's body after the 10th injection?

Solution After the 10th injection, the drug level in the patient's body is given by

$$Q_{10} = 20 + 20\left(\frac{1}{2}\right) + 20\left(\frac{1}{2}\right)^2 + \cdots + 20\left(\frac{1}{2}\right)^9.$$

Using the formula for S_n with $n = 10$, $a = 20$, and $r = 1/2$, we get

$$Q_{10} = \frac{20(1 - (\frac{1}{2})^{10})}{1 - \frac{1}{2}} = 39.961 \text{ mg}.$$

Exercises and Problems for Section 11.3

Exercises

In Exercises 1–4, decide which of the following are geometric series. For those which are, give the first term and the ratio between successive terms. For those which are not, explain why not.

1. $2 + 1 + \frac{1}{2} + \frac{1}{4} + \frac{1}{8} + \cdots + \frac{1}{128}$

2. $1 - \frac{1}{2} + \frac{1}{4} - \frac{1}{8} + \frac{1}{16} + \cdots + \frac{1}{256}$

3. $1 + \frac{1}{2} + \frac{1}{3} + \frac{1}{4} + \frac{1}{5} + \cdots + \frac{1}{100}$

4. $5 - 10 + 20 - 40 + 80 - \cdots - 2560$

How many terms are there in the series in Exercises 5–6? Find the sum.

5. $\sum_{j=5}^{18} 3 \cdot 2^j$

6. $\sum_{k=2}^{20} (-1)^k 5(0.9)^k$

Find the sum of the series in Exercises 7–11.

7. $3 + \frac{3}{2} + \frac{3}{4} + \frac{3}{8} + \cdots + \frac{3}{2^{10}}$

8. $5 + 15 + 45 + 135 + \cdots + 5(3^{12})$

9. $1/125 + 1/25 + 1/5 + \cdots + 625$

10. $\sum_{n=1}^{10} 4(2^n)$

11. $\sum_{k=0}^{7} 2\left(\frac{3}{4}\right)^k$

Write each of the sums in Exercises 12–15 in sigma notation.

12. $1 + 4 + 16 + 64 + 256$

13. $3 - 9 + 27 - 81 + 243 - 729$

14. $2 + 10 + 50 + 250 + 1250 + 6250 + 31250$

15. $32 - 16 + 8 - 4 + 2 - 1$

Problems

16. Figure 11.3 shows the quantity of the drug atenolol in the blood as a function of time, with the first dose at time $t = 0$. Atenolol is taken in 50 mg doses once a day to lower blood pressure.

(a) If the half-life of atenolol in the blood is 6.3 hours, what percentage of the atenolol present at the start of a 24-hour period is still there at the end?
(b) Find expressions for the quantities $Q_0, Q_1, Q_2, Q_3, \ldots$, and Q_n shown in Figure 11.3. Write the expression for Q_n in closed-form.
(c) Find expressions for the quantities $P_1, P_2, P_3, \ldots$, and P_n shown in Figure 11.3. Write the expression for P_n in closed-form.

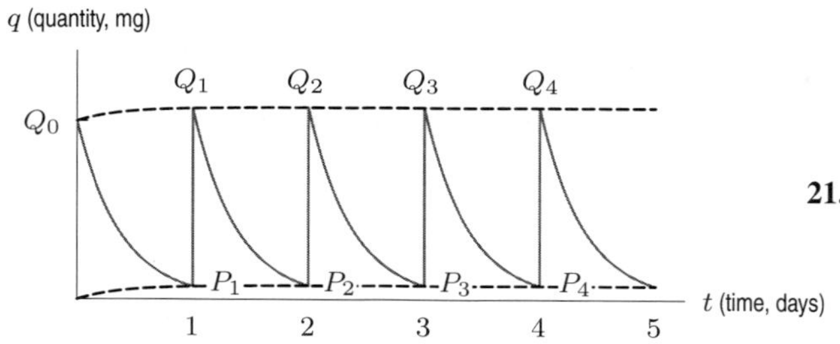

Figure 11.3

17. Annual deposits of \$3000 are made into a bank account earning 5% interest per year. What is the balance in the account right after the 15th deposit if interest is calculated

(a) Annually **(b)** Continuously

18. A bank account with a \$75,000 initial deposit is used to make annual payments of \$1000, starting one year after the initial \$75,000 deposit. Interest is earned at 2% a year, compounded annually, and paid into the account right before the payment is made.

(a) What is the balance in the account right after the 24th payment?
(b) Answer the same question for yearly payments of \$3000.

19. A deposit of \$1000 is made once a year, starting today, into a bank account earning 3% interest per year, compounded annually. If 20 deposits are made, what is the balance in the account on the day of the last deposit?

20. What effect does doubling each of the following quantities (leaving other quantities the same) have on the answer to Problem 19? Is the answer doubled, more than doubled, or less than doubled?

(a) The deposit
(b) The interest rate
(c) The number of deposits made

21. To save for a new car, you put \$500 a month into an account earning interest at 3% per year, compounded continuously.

(a) How much money do you have 2 years after the first deposit, right before you make a deposit?
(b) When does the balance first go over \$10,000?

22. A bank account in which interest is earned at 4% per year, compounded annually, starts with a balance of \$50,000. Payments of \$1000 are made out of the account once a year for ten years, starting today. Interest is earned right before each payment is made.

(a) What is the balance in the account right after the tenth payment is made?
(b) Assume that the tenth payment exhausts the account. What is the largest yearly payment that can be made from this account?

11.4 INFINITE GEOMETRIC SERIES

In this section, we look at geometric series with an infinite number of terms and see under what circumstances they have a finite sum.

Long-term Drug Level in the Body

Suppose the patient from Example 3 on page 503 receives injections over a long period of time. What happens to the drug level in the patient's body? To find out, we calculate the drug level after 10, 15, 20, and 25 injections:

$$Q_{10} = \frac{20(1-(\frac{1}{2})^{10})}{1-\frac{1}{2}} = 39.960938 \text{ mg},$$

$$Q_{15} = \frac{20(1 - (\frac{1}{2})^{15})}{1 - \frac{1}{2}} = 39.998779 \text{ mg},$$

$$Q_{20} = \frac{20(1 - (\frac{1}{2})^{20})}{1 - \frac{1}{2}} = 39.999962 \text{ mg},$$

$$Q_{25} = \frac{20(1 - (\frac{1}{2})^{25})}{1 - \frac{1}{2}} = 39.999999 \text{ mg}.$$

The drug level appears to approach 40 mg. To see why this happens, notice that if there is exactly 40 mg of drug in the body, then half of this amount is metabolized in one day, leaving 20 mg. The next 20 mg injection replaces what was lost, and the level returns to 40 mg. We say that the *equilibrium* drug level is 40 mg.

Initially the patient has less than 40 mg of the drug in the body. Then the amount metabolized in one day is less than 20 mg. Thus, after the next 20 mg injection, the drug level is higher than it was before. For instance, if there are currently 30 mg, then after one day, half of this has been metabolized, leaving 15 mg. At the next injection the level rises to 35 mg, or 5 mg higher than where it started. Eventually, the quantity levels off to 40 mg.

Another way to think about the patient's drug level over time is to consider an *infinite geometric series*. After n injections, the drug level is given by the sum of the finite geometric series

$$Q_n = 20 + 20\left(\frac{1}{2}\right) + 20\left(\frac{1}{2}\right)^2 + \cdots + 20\left(\frac{1}{2}\right)^{n-1} = \frac{20(1 - (\frac{1}{2})^n)}{1 - \frac{1}{2}}.$$

What happens to the value of this sum as the number of terms approaches infinity? It does not seem possible to add up an infinite number of terms. However, we can look at the *partial sums*, Q_n, to see what happens for large values of n. For large values of n, we see that $(1/2)^n$ is very small, so that

$$Q_n = \frac{20\,(1 - \text{Small number})}{1 - \frac{1}{2}} = \frac{20\,(1 - \text{Small number})}{1/2}.$$

We write $\to$ to mean "tends toward." Thus, as $n \to \infty$, we know that $(1/2)^n \to 0$, so

$$Q_n \to \frac{20(1-0)}{1/2} = \frac{20}{1/2} = 40 \text{ mg}.$$

The Sum of an Infinite Geometric Series

Consider the geometric series $S_n = a + ar + ar^2 + \cdots + ar^{n-1}$. In general, if $|r| < 1$, then $r^n \to 0$ as $n \to \infty$, so

$$S_n = \frac{a(1 - r^n)}{1 - r} \to \frac{a(1-0)}{1-r} = \frac{a}{1-r} \quad \text{as } n \to \infty.$$

Thus, if $|r| < 1$, the partial sums S_n approach a finite value, S, as $n \to \infty$. In this case, we say that the series *converges* to S.

For $|r| < 1$, the **sum of the infinite geometric series** is given by

$$S = a + ar + ar^2 + \cdots + ar^n + \cdots = \sum_{i=0}^{\infty} ar^i = \frac{a}{1-r}.$$

On the other hand, if $|r| > 1$, then we say that the series does not converge. The terms in the series get larger and larger as $n \to \infty$, so adding infinitely many of them does not give a finite sum.

What happens when $r = \pm 1$? For $|r| \geq 1$, the value of S_n does not approach a fixed number as $n \to \infty$, so we say that the infinite series does not converge. (To see why, consider what happens to 1^n and $(-1)^n$ as n increases.)

Example 1 Find the sum of the geometric series $\sum_{i=0}^{17} 7(-z)^i$ and $\sum_{i=0}^{\infty}(-z)^i$ provided $|z| < 1$.

Solution We have

$$\sum_{i=0}^{17} 7(-z)^i = 7(-z)^0 + 7(-z)^1 + 7(-z)^2 + \cdots + 7(-z)^{17}.$$

Since $(-z)^0 = 1$, this is a geometric series with $a = 7$ and $r = -z$. There are 18 terms, so

$$\sum_{i=0}^{17} 7(-z)^i = \frac{7(1-(-z)^{18})}{1-(-z)} = \frac{7(1-z^{18})}{1+z}.$$

As for $\sum_{i=0}^{\infty}(-z)^i$, this is an infinite geometric series. Since $|z| < 1$, we have

$$\sum_{i=0}^{\infty}(-z)^i = \frac{1}{1-(-z)} = \frac{1}{1+z}.$$

Present Value of a Series of Payments

When baseball player Alex Rodriguez was signed by the Texas Rangers, before being traded to the New York Yankees, he was given a signing bonus for \$10 million: \$2 million a year for five years. Of course, since much of the money was to be paid in the future, the team's owners did not have to have all \$10 million available on the day of the signing. How much money would the owners have to deposit in a bank account on the day of the signing in order to cover all the future payments? Assuming the account was earning interest, the owners would have to deposit less than \$10 million. This smaller amount is called the *present value* of \$10 million. We calculate the present value of Rodriguez's bonus on the day he signed.

Definition of Present Value

Let's consider a simplified version of this problem, with only one future payment: How much money would we need to deposit in a bank account today in order to have \$2 million in one year? At an annual interest rate of 5%, compounded annually, the deposit would grow by a factor of 1.05. Thus,

$$\text{Deposit} \cdot 1.05 = \$2 \text{ million},$$
$$\text{Deposit} = \frac{\$2{,}000{,}000}{1.05} = \$1{,}904{,}761.91.$$

We need to deposit \$1,904,761.91. This is the *present value* of the \$2 million. Similarly, if we need \$2 million in 2 years, the amount we need to deposit is given by

$$\text{Deposit} \cdot (1.05)^2 = \$2 \text{ million},$$
$$\text{Deposit} = \frac{\$2{,}000{,}000}{(1.05)^2} = \$1{,}814{,}058.96.$$

The \$1,814,058.96 is the present value of \$2 million payable two years from today. In general,

The **present value**, $\$P$, of a future payment, $\$B$, is the amount which would have to be deposited (at some interest rate, r) in a bank account today to have exactly $\$B$ in the account at the relevant time in the future.

If r is the yearly interest rate (compounded annually) and if n is the number of years, then

$$B = P(1+r)^n, \quad \text{or equivalently,} \quad P = \frac{B}{(1+r)^n}.$$

Calculating the Present Value of Rodriguez's Bonus

The present value of Rodriguez's bonus represents what it was worth on the day it was signed. Suppose that he receives his money in 5 payments of \$2 million each, the first payment to be made on the day the contract was signed. We calculate the present values of all 5 payments assuming that interest is compounded annually at a rate of 5% per year. Since the first payment is made the day the contract is signed, we have, in millions of dollars,

$$\text{Present value of first payment} = 2.$$

Since the second payment is made a year in the future, in millions of dollars we have:

$$\text{Present value of second payment} = \frac{2}{(1.05)^1} = \frac{2}{1.05}.$$

The third payment is made two years in the future, so in millions of dollars:

$$\text{Present value of third payment} = \frac{2}{(1.05)^2},$$

and so on. Similarly, in millions of dollars:

$$\text{Present value of fifth payment} = \frac{2}{(1.05)^4}.$$

Thus, in millions of dollars,

$$\text{Total present value} = 2 + \frac{2}{1.05} + \frac{2}{(1.05)^2} + \cdots + \frac{2}{(1.05)^4}.$$

Rewriting this expression, we see that it is a finite geometric series with $a = 2$ and $r = 1/1.05$:

$$\text{Total present value} = 2 + 2\left(\frac{1}{1.05}\right) + 2\left(\frac{1}{1.05}\right)^2 + \cdots + 2\left(\frac{1}{1.05}\right)^4.$$

The formula for the sum of a finite geometric series gives

$$\text{Total present value of contract in millions of dollars} = \frac{2\left(1 - \left(\frac{1}{1.05}\right)^5\right)}{1 - \frac{1}{1.05}} \approx 9.09.$$

Thus, the total present value of the bonus is about \$9.09 million dollars.

Example 2 Suppose Alex Rodriguez's bonus with the Rangers guaranteed him and his heirs an annual payment of \$2 million *forever*. How much would the owners need to deposit in an account today in order to provide these payments?

Solution At 5%, the total present value of an infinite series of payments is given by the infinite series

$$\text{Total present value} = 2 + \frac{2}{1.05} + \frac{2}{(1.05)^2} + \cdots$$
$$= 2 + 2\left(\frac{1}{1.05}\right) + 2\left(\frac{1}{1.05}\right)^2 + \cdots.$$

The sum of this infinite geometric series can be found using the formula:

$$\text{Total present value} = \frac{2}{1 - \frac{1}{1.05}} = 42 \text{ million dollars.}$$

To see that this answer is reasonable, suppose that \$42 million is deposited in an account today, and that a \$2 million payment is immediately made to Alex Rodriguez. Over the course of a year, the remaining \$40 million earns 5% interest, which works out to \$2 million, so the next year the account again has \$42 million. Thus, it would have cost the Texas Rangers only about \$33 million more (than the \$9.09 million) to pay Rodriguez and his heirs \$2 million a year forever.

Example 3 Using summation notation, write the present value of Alex Rodriguez's bonus, first for the 5 payment case and then for the infinite sequence of payments.

Solution For the 5 payments, we have

$$\text{Total present value} = 2 + \frac{2}{1.05} + \frac{2}{(1.05)^2} + \frac{2}{(1.05)^3} + \frac{2}{(1.05)^4} = \sum_{i=0}^{4} 2\left(\frac{1}{1.05}\right)^i.$$

For the infinite sequence of payments, the series goes to ∞ instead of stopping at 4, so we have

$$\text{Total present value} = 2 + \frac{2}{1.05} + \frac{2}{(1.05)^2} + \cdots = \sum_{i=0}^{\infty} 2\left(\frac{1}{1.05}\right)^i.$$

Exercises and Problems for Section 11.4

Exercises

In Exercises 1–4, decide which of the following are geometric series. For those which are, give the first term and the ratio between successive terms. For those which are not, explain why not.

1. $y^2 + y^3 + y^4 + y^5 + \cdots$
2. $1 + x + 2x^2 + 3x^3 + 4x^4 + \cdots$
3. $1 - x + x^2 - x^3 + x^4 - \cdots$
4. $3 + 3z + 6z^2 + 9z^3 + 12z^4 + \cdots$
5. Find the sum of the series in Problem 1.
6. Find the sum of the series in Problem 3.

Find the sum of the series in Exercises 7–12.

7. $-2 + 1 - \frac{1}{2} + \frac{1}{4} - \frac{1}{8} + \frac{1}{16} - \cdots$
8. $3 + 3(0.9) + 3(0.9)^2 + \cdots$
9. $\sum_{i=2}^{\infty} (0.1)^i$
10. $\sum_{i=4}^{\infty} \left(\frac{1}{3}\right)^i$
11. $\sum_{i=0}^{\infty} \frac{3^i + 5}{4^i}$
12. $\sum_{j=1}^{\infty} 7((0.1)^j + (0.2)^{j+2})$

Problems

13. A repeating decimal can always be expressed as a fraction. This problem shows how writing a repeating decimal as a geometric series enables you to find the fraction. Consider the decimal $0.232323\ldots.$

(a) Use the fact that $0.232323\ldots = 0.23 + 0.0023 + 0.000023 + \cdots$ to write $0.232323\ldots$ as a geometric series.

(b) Use the formula for the sum of a geometric series to show that $0.232323\ldots = 23/99$.

In Problems 14–18, use the method of Problem 13 to write each of the decimals as fractions.

14. $0.235235235\ldots$

15. $6.19191919\ldots$

16. $0.12222222\ldots$

17. $0.4788888\ldots$

18. $0.7638383838\ldots$

19. You have an ear infection and are told to take a 250 mg tablet of ampicillin (a common antibiotic) four times a day (every six hours). It is known that at the end of six hours, about 4% of the drug is still in the body. What quantity of the drug is in the body right after the third tablet? The fortieth? Assuming you continue taking tablets, what happens to the drug level in the long run?

20. In Problem 19 we found the quantity Q_n, the amount (in mg) of ampicillin left in the body right after the nth tablet is taken.

(a) Make a similar calculation for P_n, the quantity of ampicillin (in mg) in the body right *before* the nth tablet is taken.

(b) Find a simplified formula for P_n.

(c) What happens to P_n in the long run? Is this the same as what happens to Q_n? Explain in practical terms why your answer makes sense.

21. Draw a graph like that in Figure 11.3 for 250 mg of ampicillin taken every 6 hours, starting at time $t = 0$. Put on the graph the values of $Q_1, Q_2, Q_3, \ldots$ calculated in Problem 19 and the values of $P_1, P_2, P_3, \ldots$ calculated in Problem 20.

22. Basketball player Patrick Ewing received a contract from the New York Knicks in which he was offered \$3 million a year for ten years. Determine the present value of the contract on the date of the first payment if the interest rate is 7% per year, compounded continuously.

Problems 23–25 are about *bonds*, which are issued by a government to raise money. An individual who buys a \$1000 bond gives the government \$1000 and in return receives a fixed sum of money, called the *coupon*, every six months or every year for the life of the bond. At the time of the last coupon, the individual also gets the \$1000, or *principal*, back.

23. What is the present value of a \$1000 bond which pays \$50 a year for 10 years, starting one year from now? Assume interest rate is 6% per year, compounded annually.

24. What is the present value of a \$1000 bond which pays \$50 a year for 10 years, starting one year from now? Assume the interest rate is 4% per year, compounded annually.

25. **(a)** What is the present value of a \$1000 bond which pays \$50 a year for 10 years, starting one year from now? Assume the interest rate is 5% per year, compounded annually.

(b) Since \$50 is 5% of \$1000, this bond is often called a 5% bond. What does your answer to part (a) tell you about the relationship between the principal and the present value of this bond when the interest rate is 5%?

(c) If the interest rate is more than 5% per year, compounded annually, which one is larger: the principal or the value of the bond? Why do you think the bond is then described as *trading at discount*?

(d) If the interest rate is less than 5% per year, compounded annually, why is the bond described as *trading at a premium*?

CHAPTER SUMMARY

- **Sequences**
  Arithmetic: $a_n = a_1 + (n-1)d$.
  Geometric: $a_n = a_1 r^{n-1}$
- **Sigma Notation**
  $\sum_{i=i}^{n} a_i$
- **Arithmetic Series**
  Partial sum:
  $$S_n = a_1 + a_2 + \cdots + a_n = \frac{1}{2}n(2a_1 + (n-1)d).$$
- **Geometric Series**
  Finite:
  $$S_n = a + ar + ar^2 + \cdots + ar^{n-1} = \frac{a(1-r^n)}{1-r}, \quad r \neq 1.$$
  Infinite:
  $$S = a + ar + ar^2 + \cdots = \frac{a}{1-r},$$
  converges for $|r| < 1$, does not converge for $|r| > 1$.
- **Applications**
  Bank balance; drug levels.
  Present value: $P = \dfrac{B}{(1+r)^n}$ or $P = \dfrac{B}{e^{kt}}$.

REVIEW EXERCISES AND PROBLEMS FOR CHAPTER ELEVEN

Exercises

1. **(a)** Write the sum $\sum_{n=1}^{5}(4n-3)$ in expanded form.
(b) Compute the sum.

2. Find the sum of the first eighteen terms of the series: $8 + 11 + 14 + \cdots$. What is the eighteenth term?

3. Write the following using sigma notation: $100 + 90 + 80 + 70 + \cdots + 0$.

In Exercises 4–5, decide which of the following are geometric series. For those which are, give the first term and the ratio between successive terms. For those which are not, explain why not.

4. $1 + 2z + (2z)^2 + (2z)^3 + \cdots$

5. $1 - y^2 + y^4 - y^6 + \cdots$

6. Find the sum of the series in Problem 4.

7. Find the sum of the series in Problem 5.

8. Find the sum of the first nine terms of the series: $7 + 14 + 21 + \cdots$.

Problems

9. Figure 11.4 shows the first four members in a sequence of square-shaped grids. In each successive grid, new dots are shown in black (•). For instance, the second grid has 3 more dots than the first grid, the third has 5 more dots than the second, and the fourth has 7 more dots than the third.

(a) Write down the sequence of the total number of dots in each grid (black and white). Using the fact that each grid is a square, find a formula in terms of n for the number of dots in the nth grid.
(b) Write down the sequence of the number of black dots in each grid. Find a formula in terms of n for the number of black dots in the nth grid.
(c) State the relationship between the two sequences in parts (a) and (b). Use the formulas you found in parts (a) and (b) to confirm this relationship.

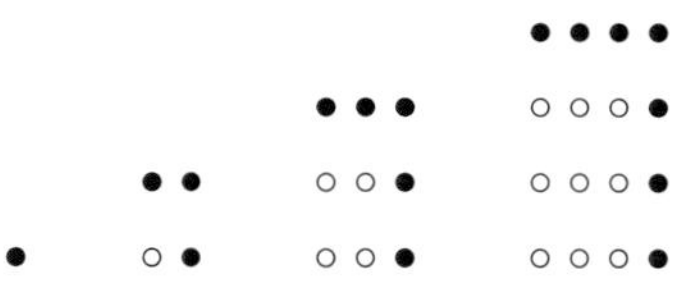

Figure 11.4

10. The numbers in the sequence 1, 4, 9, 16, ... from Problem 9 are known as *square numbers* because they describe the number of dots in successive square grids. Analogously, the numbers in the sequence 1, 3, 6, 10, ... are known as *triangular numbers* because they describe the number of dots in successive triangular patterns, as shown in Figure 11.5: Find a formula for the nth triangular number.

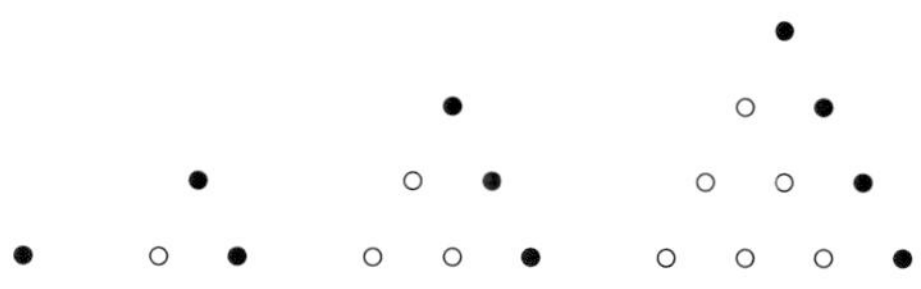

Figure 11.5

11. In a workshop, it costs \$300 to make one piece of furniture. The second piece costs a bit less, \$280. The third costs even less, \$263, and the fourth costs only \$249. The cost for each additional piece of furniture is called the *marginal cost* of production. Table 11.3 gives the marginal cost, c, and the change in marginal cost, Δc, in terms of the number of pierces of furniture, n. As the quantity produced increases, the marginal cost generally decreases and then increases again.

Table 11.3

n	1		2		3		4
c (\$)	300		280		263		249
Δc (\$)		-20		-17		-14	

(a) Assume that the arithmetic sequence -20, -17, $-14, \ldots$, continues. Complete the table for $n = 5$, $6, \ldots, 12$.
(b) Find a formula for c_n, the marginal cost for producing the nth piece of furniture. Use the fact that c_n is found by adding the terms in an arithmetic sequence. Using your formula, find the cost for producing the 12th piece and the 50th piece of furniture.
(c) A piece of furniture can be sold at a profit if it costs less than \$400 to make. How many pieces of furniture should the workshop make each day? Discuss.

12. A store clerk has 108 cans to stack. He can fit 24 cans on the bottom row and can stack the cans 8 rows high. Use arithmetic series to determine how he can stack the cans so that each row contains fewer cans than the row beneath it and that the number of cans in each row decreases at a constant rate.

13. A university with an enrollment of 8000 students in 2007 is projected to grow by 2% in each of the next three years and by 3.5% each of the following seven years. Find the sequence of the university's student enrollment for the next 10 years.

14. Each person in a group of 30 shakes hands with each other person exactly once. How many total handshakes take place?

15. Worldwide consumption of oil was about 81 billion barrels in 2004.[7] Assume that consumption continues to increase at 1.2% per year, the rate for the previous decade.

(a) Write a sum representing the total oil consumption for 25 years, starting with 2004.
(b) Evaluate this sum.

16. One way of valuing a company is to calculate the present value of all its future earnings. A farm expects to sell \$1000 worth of Christmas trees once a year forever, with the first sale in the immediate future. What is the present value of this Christmas tree business? Assume that the interest rate is 4% per year, compounded continuously.

17. You inherit \$100,000 and put the money in a bank account earning 3% per year, compounded annually. You withdraw \$2000 from the account each year, right after the interest is earned. Your first \$2000 is withdrawn before any interest is earned.

(a) Compare the balance in the account right after the first withdrawal and right after the second withdrawal. Which do you expect to be higher?
(b) Calculate the balance in the account right after the 20th withdrawal is made.
(c) What is the largest yearly withdrawal you can take from this account without the balance decreasing over time?

18. After breaking his leg, a patient retrains his muscles by going for walks. The first day, he manages to walk 300 yards. Each day after that he walks 50 yards farther than the day before.

(a) Write a sequence that represents the distances walked each day during the first week.
(b) How far is he walking after two weeks?
(c) How long until he is walking at least one mile?

19. Before email made it easy to contact many people quickly, groups used telephone trees to pass news to their members. In one group, each person is in charge of calling 4 people. One person starts the tree by calling 4 people. At the second stage, each of these 4 people call another 4 people. In the third stage, each of the people in stage two calls 4 people, and so on.

(a) How many people have the news by the end of the 5th stage?
(b) Write a formula for the total number of members in a tree of 10 stages.
(c) How many stages are required to cover a group with 5000 members?

20. A ball is dropped from a height of 10 feet and bounces. Each bounce is $3/4$ of the height of the bounce before. Thus after the ball hits the floor for the first time, the ball rises to a height of $10(3/4) = 7.5$ feet, and after it hits the floor for the second time, it rises to a height of $7.5(3/4) = 10(3/4)^2 = 5.625$ feet.

(a) Find an expression for the height to which the ball rises after it hits the floor for the nth time.

[7] www.bp.com/downloads, Statistical Review of World Energy 2005, accessed January 15, 2006.

(b) Find an expression for the total vertical distance the ball has traveled when it hits the floor for the first, second, third, and fourth times.

(c) Find an expression for the total vertical distance the ball has traveled when it hits the floor for the nth time. Express your answer in closed-form.

21. You might think that the ball in Problem 20 keeps bouncing forever since it takes infinitely many bounces. This is not true! It can be shown that a ball dropped from a height of h feet reaches the ground in $\frac{1}{4}\sqrt{h}$ seconds. It is also true that it takes a bouncing ball the same amount of time to rise h feet. Use these facts to show that the ball in Problem 20 stops bouncing after

$$\frac{1}{4}\sqrt{10} + \frac{1}{2}\sqrt{10}\sqrt{\frac{3}{4}}\left(\frac{1}{1-\sqrt{3/4}}\right)$$

seconds, or approximately 11 seconds.

22. This problem illustrates how banks create credit and can thereby lend out more money than has been deposited. Suppose that initially \$100 is deposited in a bank. Experience has shown bankers that on average only 8% of the money deposited is withdrawn by the owner at any time. Consequently, bankers feel free to lend out 92% of their deposits. Thus \$92 of the original \$100 is loaned out to other customers (to start a business, for example). This \$92 will become someone else's income and, sooner or later, will be redeposited in the bank. Then 92% of \$92, or $\$92(0.92) = \$84.64$, is loaned out again and eventually redeposited. Of the \$84.64, the bank again loans out 92%, and so on.

(a) Find the total amount of money deposited in the bank.

(b) The total amount of money deposited divided by the original deposit is called the *credit multiplier*. Calculate the credit multiplier for this example and explain what this number tells us.

23. Take a rectangle whose sides have length 1 and 2, divide it into two equal pieces, and shade one of them. Now, divide the unshaded half into two equal pieces and shade one of them. Take the unshaded area and divide it into two equal pieces and shade one of them. Continue doing this.

(a) Draw a picture that illustrates this process.

(b) Find a series that describes the shaded area after n divisions.

(c) If you continue this process indefinitely, what would the total shaded area be? Make two arguments, one using your picture, another using the series you found in part (b).

CHECK YOUR UNDERSTANDING

Are the statements in Problems 1–34 true or false? Give an explanation for your answer.

1. If $a_n = n^2 + 1$, then $a_1 = 2$.

2. The first term of a sequence must be a positive integer.

3. The sequence $1, 1, 1, 1, \ldots$ is both arithmetic and geometric.

4. The sequence $a, 2a, 3a, 4a$ is arithmetic.

5. The sequence $a, a+1, a+2, a+3$ is arithmetic.

6. If the difference between successive terms of an arithmetic sequence is d, then $a_n = a_1 + dn - d$.

7. The sequence $3x^2, 3x^4, 3x^6, \ldots$ is geometric.

8. The sequence $2, 4, 2, 4, 2, 4, \ldots$ is alternating.

9. For all series $S_1 = a_1$.

10. The accumulation of college credit is an example of a finite series.

11. The sum formula for a finite series with n terms is $S_n = \frac{1}{2}n(n+1)$.

12. There are four terms in $\sum_{i=0}^{4} a_i$.

13. $\sum_{i=1}^{n} 3 = 3n$.

14. The sum of a finite arithmetic series is the average of the first and last terms, times the number of terms.

15. If C_n denotes your college credits earned by the end of year n, then C_4 is your credit earned during your fourth year of college.

16. If C_n denotes your college credits earned by the end of year n, then $C_3 - C_2$ is your credit earned in year 2.

17. For any series S_n, we must have $S_3 \leq S_4$.

18. If $a_n = (-1)^{n+1}$ then the sequence is alternating.

19. The sum formula for a finite geometric series, $S_n = \dfrac{a(1-r^n)}{1-r}$, can be simplified by dividing by $(1-r)$ to get $S_n = a(1-r^{n-1})$.

20. With the same interest rate, a bank balance after 5 years of \$2000 annual payments is the same as the balance after by 10 years of \$1000 annual payments.

21. $1 - \frac{1}{2} + \frac{1}{4} - \frac{1}{8} + \frac{1}{16} - \frac{1}{32}$ is a geometric series.

22. $2000(1.06)^3 + 2000(1.06)^4 + \cdots + 2000(1.06)^{10}$ is a geometric series.

23. $5000 + 2000(1.06) + 2000(1.06)^2 \cdots + 2000(1.06)^{10}$ is a geometric series.

24. $5000+2000(1.06)+2000(1.06)^2 \cdots +2000(1.06)^{10} = 3000 + \dfrac{2000(1-1.06^{11})}{-0.06}$.

25. Payments of \$2000 are made yearly into an account earning 5% interest compounded annually. If the interest rate is doubled, the balance at the end of 20 years is doubled

26. The present value of a sequence of payments is the amount that you would need to invest at this moment in order to exactly make the future payments.

27. If the present value of a sequence of payments is 1 million dollars, then the present value of a sequence of payments which are doubled in size, would be 2 million.

28. The series $\displaystyle\sum_{i=0}^{\infty} \frac{1}{i^2}$ is geometric.

29. For $Q_n = \displaystyle\sum_{i=1}^{n} (-1)^i$, we see $Q_2 = 0$, $Q_4 = 0$, $Q_6 = 0$, etc. Thus, the series converges to a sum of $Q = \displaystyle\sum_{i=1}^{\infty} (-1)^i = 0$.

30. A patient takes 100 mg of a drug each day; 5% of the amount of the drug in the body decays each day. Then, the long-run level of the drug in the body is 2000 mg.

31. If $S = a + ar + ar^2 + \cdots$ and $|a| < 1$ then the infinite series converges.

32. It is possible to add an infinite number of positive terms and get a finite sum.

33. The sum of an infinite arithmetic series with difference $d = 0.01$ converges.

34. All infinite geometric series converge.

Chapter Twelve

PARAMETRIC EQUATIONS AND CONIC SECTIONS

In this chapter, we focus on conic sections—circles, ellipses, and hyperbolas. These famous geometric shapes are described by quadratic equations. They cannot be represented as functions, but can be represented by parametric equations, which are introduced in this chapter. We discuss the geometric definitions of circles, ellipses, hyperbolas, and parabolas, and investigate the reflective properties of these shapes. We conclude with an introduction to hyperbolic functions.

12.1 PARAMETRIC EQUATIONS

The Mars Pathfinder

On July 4, 1997, the Mars Pathfinder bounced down onto the surface of the red planet, its impact cushioned by airbags. The next day, the Sojourner—a small, six-wheeled robot—rolled out of the spacecraft and began a rambling exploration of the surrounding terrain. Figure 12.1 shows a photograph of the Sojourner and a diagram of the path it took before radio contact was lost.

In Figure 12.1, Sojourner's path is labeled according to the elapsed number of Martian days or *sols* (short for solar periods). The robot's progress was slow because instructions for every movement had to be calculated by NASA engineers and sent from Earth.

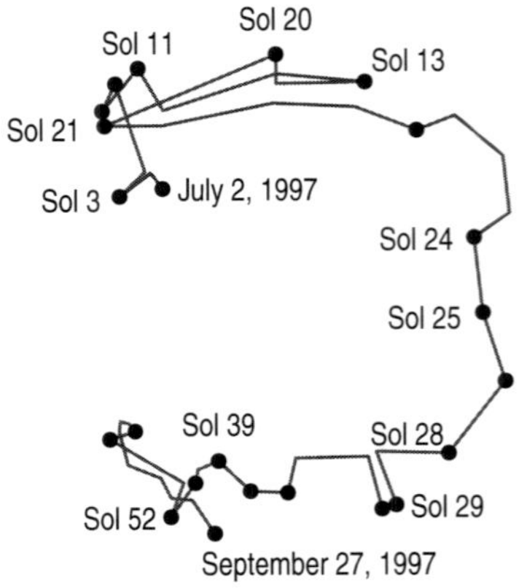

Figure 12.1: The Pathfinder Sojourner robot and the path it followed on the surface of the planet Mars. Image from NASA/Jet Propulsion Laboratory. Diagram adapted from *The New York Times*, July 21, 1998

The Path of the Sojourner Robot

How can Sojourner's path be represented? How can we tell a robot where it should go, how fast, and when? The path in Figure 12.1 is not the graph of a function in the ordinary sense—for instance, it crosses over itself in several different places. However, positions on the path can be determined by knowing t, the elapsed time. In this section, we describe a path by giving the x- and y-coordinates of points on the path as functions of a *parameter* such as t.

Programming a Robot Using Coordinates

One way to program a robot's motion is to send it coordinates to tell it where to go. Imagine that a robot like the Sojourner is moving around the xy-plane.[1] If we choose the origin $(0, 0)$ to be the spot where the robot's spacecraft lands, we can direct its motion by giving it (x, y) coordinates of the points to which it should move.

We select the positive y-axis so that it points north (that is, toward the northern pole of Mars) and the positive x-axis so that it points east. Our units of measurement are meters, so that, the coordinates $(2, 4)$ indicate a point that is 2 meters to the east and 4 meters to the north of the landing site. We program the robot to move to the following points:

$$(0,0) \quad \to \quad (1,1) \quad \to \quad (2,2) \quad \to \quad (2,3) \quad \to \quad (2,4).$$

[1]The surface of Mars is not as flat as the xy-plane, but this is a useful first approximation.

These points have been plotted in Figure 12.2.

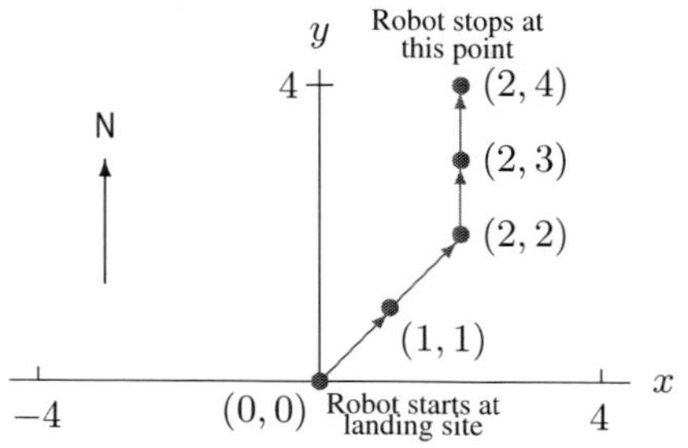

Figure 12.2: The robot is programmed to move along this path

Programming the Robot Using a Parameter

In order to represent the robot's path using a parameter, we need two functions: one for the robot's x-coordinate, $x = f(t)$, and one for its y-coordinate, $y = g(t)$. The function for x describes the robot's east-west motion, and the function for y describes its north-south motion. Together, they are called *parametric equations* for the robot's path.

Example 1 If t is time in minutes, describe the path followed by a robot given by

$$x = 2t, \qquad y = t \qquad \text{for} \quad 0 \le t \le 5.$$

Solution At time $t = 0$, the robot's position is given by

$$x = 2 \cdot 0 = 0, \qquad y = 0, \qquad \text{so it starts at the point } (0,0).$$

One minute later, at $t = 1$, its position is given by

$$x = 2 \cdot 1 = 2, \qquad y = 1, \qquad \text{so it has moved to the point } (2,1).$$

At time $t = 2$, its position is given by

$$x = 2 \cdot 2 = 4, \qquad y = 2, \qquad \text{so it has moved to the point } (4,2).$$

The path followed by the robot is given by

$$(0,0) \quad \to \quad (2,1) \quad \to \quad (4,2) \quad \to \quad (6,3) \quad \to \quad (8,4) \quad \to \quad (10,5).$$

At time $t = 5$, the robot stops at the point $(10, 5)$ because we have restricted the values of t to the interval $0 \le t \le 5$. In Figure 12.3 we see the path followed by the robot; it is a straight line.

In the previous example, we can use substitution to rewrite the formula for x in terms of y. Since $x = 2t$ and $t = y$, we have $x = 2y$. Thus, the path followed by the robot has the equation

$$y = \frac{1}{2}x.$$

Since the parameter t can be easily eliminated from our equations, you may wonder why we use it. One reason is that it is useful to know when the robot gets to each point. The values of x and y tell us where the robot is, while the parameter t tells us when it gets there. In addition, for some pairs of parametric equations, the parameter t cannot be so easily eliminated.

So far, we have assumed that a robot moves in a straight line between two points. We now imagine that a robot can be continuously redirected along a curved path.

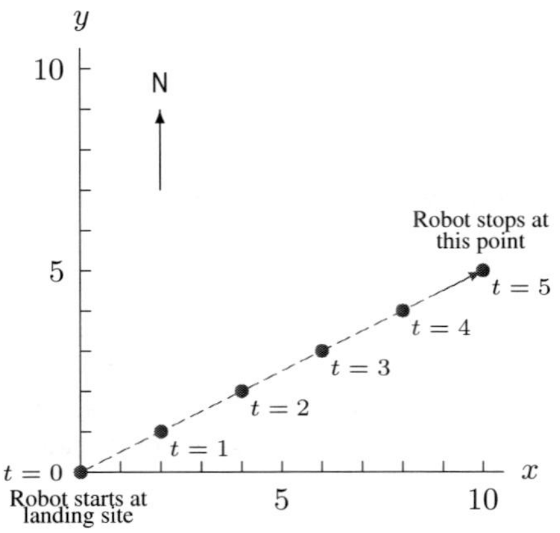

Figure 12.3: Path followed by the robot if $x = 2t, y = t$ for $0 \le t \le 5$

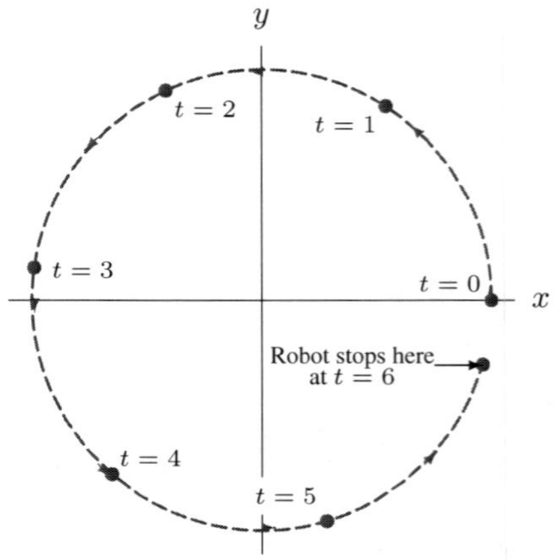

Figure 12.4: Path followed by the robot if $x = \cos t, y = \sin t$ for $0 \le t \le 6$

Example 2 A robot begins at the point $(1, 0)$ and follows the path given by the equations

$$x = \cos t, \qquad y = \sin t \qquad \text{where } t \text{ is in minutes,} \quad 0 \le t \le 6.$$

(a) Describe the path followed by the robot.
(b) What happens when you try to eliminate the parameter t from these equations?

Solution (a) At time $t = 0$, the robot's position is given by

$$x = \cos 0 = 1, \qquad y = \sin 0 = 0$$

so it starts at the point $(1, 0)$, as required. At time $t = 1$, its position is given by

$$x = \cos 1 = 0.54, \qquad y = \sin 1 = 0.84.$$

Thus, the robot has moved west and north. At time $t = 2$, its position is given by

$$x = \cos 2 = -0.42, \qquad y = \sin 2 = 0.91.$$

Now it is farther west and slightly farther north. Continuing, we see that the path followed by the robot is given by

$$(1,0) \to (0.54, 0.84) \quad \to \quad (-0.42, 0.91) \quad \to \quad (-0.99, 0.14)$$
$$\to (-0.65, -0.76) \quad \to \quad (0.28, -0.96) \quad \to \quad (0.96, -0.28).$$

In Figure 12.4 we see the path followed by the robot; it is circular with a radius of one meter. At the end of 6 minutes the robot has not quite returned to its starting point at $(1, 0)$.

(b) One way to eliminate t from this pair of equations is to use the Pythagorean identity,

$$\cos^2 t + \sin^2 t = 1.$$

Since $x = \cos t$ and $y = \sin t$, we can substitute x and y into this equation:

$$\underbrace{(\cos t)}_{x}^2 + \underbrace{(\sin t)}_{y}^2 = 1,$$

giving

$$x^2 + y^2 = 1.$$

This is the equation of a circle of radius 1 centered at $(0,0)$. Attempting to solve for y in terms of x, we have

$$\begin{aligned} x^2 + y^2 &= 1 \\ y^2 &= 1 - x^2 \\ y = +\sqrt{1-x^2} \quad &\text{or} \quad y = -\sqrt{1-x^2}. \end{aligned}$$

Thus, we obtain two different equations for y in terms of x. The first, $y = +\sqrt{1-x^2}$, returns positive values for y (as well as 0), while the second returns negative values for y (as well as 0). The first equation gives the top half of the circle, while the second gives the bottom half.

In the previous example, y is not a function of x, because for all x-values (except 1 and -1) there are two possible y values. This confirms what we already knew, because the graph in Figure 12.4 fails the vertical line test.

Different Motions Along the Same Path

It is possible to parameterize the same curve in more than one way, as the following example illustrates.

Example 3 Describe the motion of the robot that follows the path given by

$$x = \cos\frac{1}{2}t, \qquad y = \sin\frac{1}{2}t \qquad \text{for} \quad 0 \le t \le 6.$$

Solution At time $t = 0$, the robot's position is given by

$$x = \cos\left(\frac{1}{2}\cdot 0\right) = 1, \qquad y = \sin\left(\frac{1}{2}\cdot 0\right) = 0$$

so it starts at the point $(1,0)$, as before. After one minute, that is, at time $t = 1$, its position (rounded to thousandths) is given by

$$x = \cos\left(\frac{1}{2}\cdot 1\right) = 0.878, \qquad y = \sin\left(\frac{1}{2}\cdot 1\right) = 0.479.$$

At time $t = 2$ its position is given by

$$x = \cos\left(\frac{1}{2}\cdot 2\right) = 0.540, \qquad y = \sin\left(\frac{1}{2}\cdot 2\right) = 0.841.$$

Continuing, we see that the path followed by the robot is given by

$$\begin{aligned}(1,0) \to (0.878, 0.479) &\to (0.540, 0.841) \to (0.071, 0.997)\\ \to (-0.416, 0.909) &\to (-0.801, 0.598) \to (-0.990, 0.141).\end{aligned}$$

Figure 12.5 shows the path; it is again circular with a radius of one meter. However, at the end of 6 minutes the robot has not even made it half way around the circle. Because we have multiplied the parameter t by a factor of $1/2$, the robot moves at half its original rate.

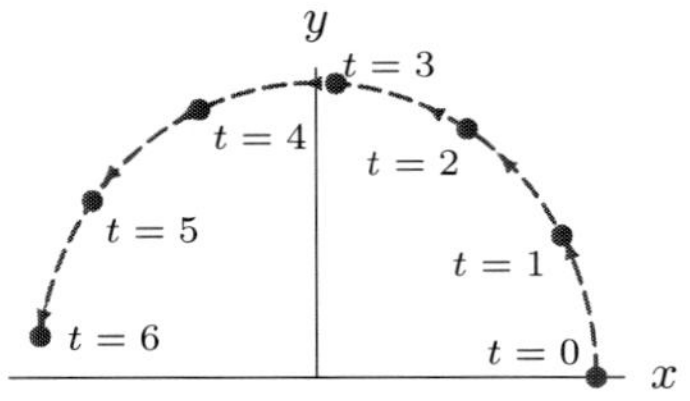

Figure 12.5: Path followed by the robot if $x = \cos\frac{1}{2}t$, $y = \sin\frac{1}{2}t$ for $0 \le t \le 6$

Example 4 Describe the motion of the robot that follows the path given by:
(a) $x = \cos(-t)$, $y = \sin(-t)$ for $0 \le t \le 6$ (b) $x = \cos t$, $y = \sin t$ for $0 \le t \le 10$

Solution
(a) Figure 12.6 shows the robot's path. The robot travels around the circle in the clockwise direction, opposite to that in Examples 2 and 3.
(b) See Figure 12.7. The robot travels around the circle more than once but less than twice, coming to a stop roughly southwest of the landing site.

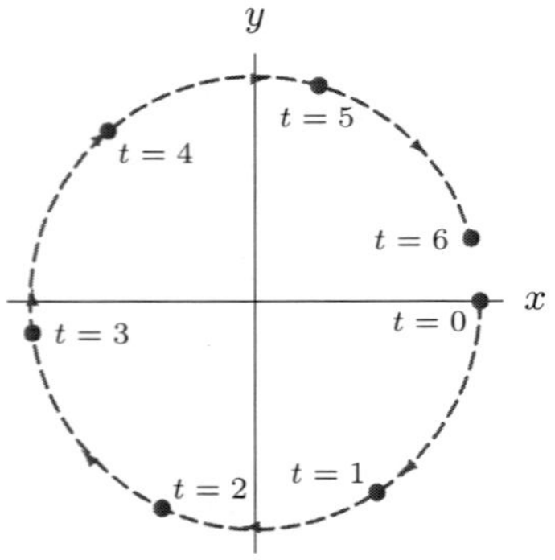

Figure 12.6: Path given by $x = \cos(-t)$, $y = \sin(-t)$ for $0 \le t \le 6$

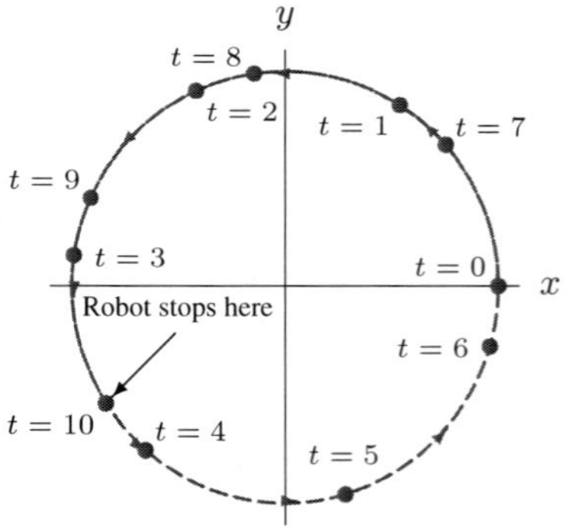

Figure 12.7: Path given by $x = \cos t$, $y = \sin t$ for $0 \le t \le 10$

Other Parametric Curves

Parametric equations can be used to describe extremely complicated motions. For instance, suppose the robot follows the rambling path described by the parametric equations

$$x = 20\cos t + 4\cos(4\sqrt{5}t) \qquad y = 20\sin t + 4\sin(4\sqrt{8}t).$$

See Figure 12.8. The path is roughly circular, and the robot moves in a more or less counterclockwise direction. The dashed circle in Figure 12.8 is given by the parametric equations

$$x = 20\cos t, \qquad y = 20\sin t \qquad \text{for } 0 \le t \le 2\pi.$$

The other terms, $4\cos(4\sqrt{5}t)$ and $4\sin(4\sqrt{8}t)$, are responsible for the robot's deviations from the circle.

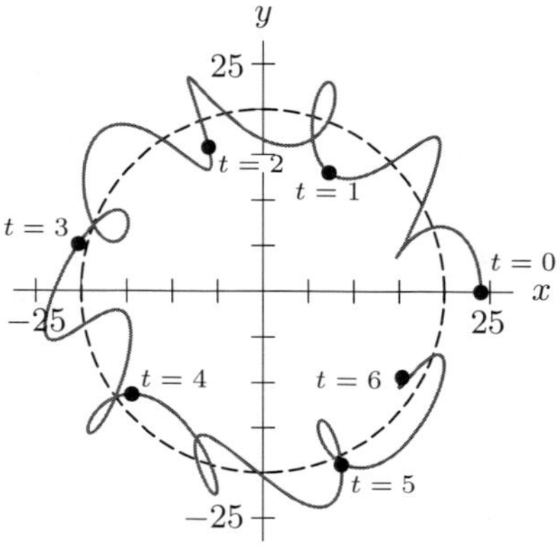

Figure 12.8: Rambling path, parametrically defined

The Archimedean Spiral

In polar coordinates, the Archimedean spiral is the graph of the equation introduced on page 338:

$$r = \theta.$$

Since the relationship between polar coordinates and Cartesian coordinates is

$$x = r\cos\theta \qquad \text{and} \qquad y = r\sin\theta,$$

we can write the Archimedean spiral $r = \theta$ as

$$x = \theta\cos\theta \qquad \text{and} \qquad y = \theta\sin\theta.$$

Replacing θ (an angle) by t (a time), we obtain the parametric equations for the spiral in Figure 12.9:

$$x = t\cos t \qquad y = t\sin t.$$

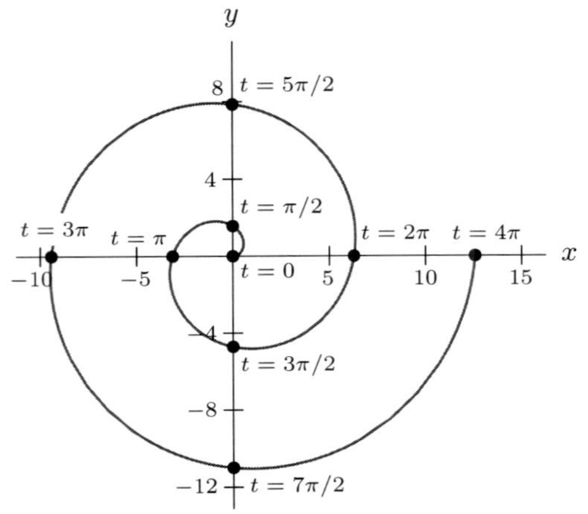

Figure 12.9: Archimedean spiral given by $x = t\cos t, y = t\sin t, 0 \le t \le 4\pi$

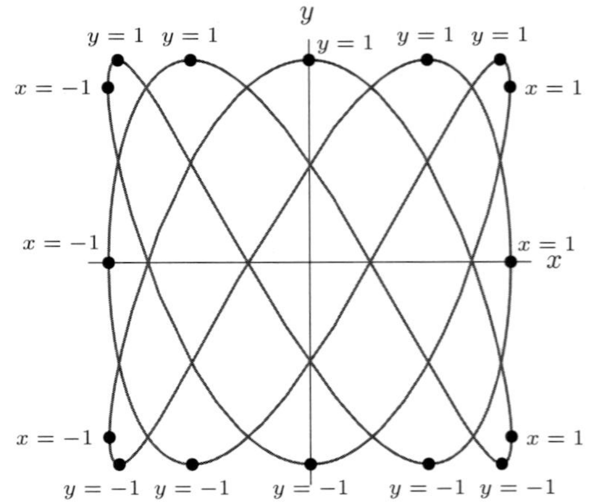

Figure 12.10: Lissajous figure: $x = \cos 3t$, $y = \sin 5t, 0 \le t \le 2\pi$

Lissajous Figures

The beautiful curve in Figure 12.10 is called a *Lissajous figure*. Its equations are

$$x = \cos 3t, \qquad y = \sin 5t \qquad \text{for } 0 \le t \le 2\pi.$$

To explain the shape of this Lissajous figure, notice that $y = \sin 5t$ completes five full oscillations on the interval $0 \le t \le 2\pi$. Thus, since the amplitude of $y = \sin 5t$ is 1, the value of y reaches a maximum of 1 at five different values of t. Similarly, the value of y is -1 at another five different values of t. Figure 12.10 shows the curve climbs to a high point of $y = 1$ five times and falls to a low of $y = -1$ five times.

Meanwhile, since $x = \cos 3t$, the value of x oscillates between 1 and -1 a total of 3 times on the interval $0 \le t \le 2\pi$. Figure 12.10 shows that the curve moves to its right boundary, $x = 1$, three times and to its left boundary, $x = -1$, three times.

Foxes and Rabbits

In a wildlife park, foxes prey on rabbits. Suppose F is the number of foxes, R is the number of rabbits, t is the number of months since January 1, and that

$$R = 1000 - 500\sin\left(\frac{\pi}{6}t\right) \qquad \text{and} \qquad F = 150 + 50\cos\left(\frac{\pi}{6}t\right).$$

These equations are a parameterization of the curve in Figure 12.11. Since this curve fails the vertical line test, F is not a function of R.

We can use the curve to analyze the relationship between the rabbit population and the fox population. In January, there are 1000 rabbits but they are dying off; in April only 500 rabbits remain. By July the population has rebounded to 1000 rabbits, and by October it has soared to 1500 rabbits. Then the rabbit population begins to fall again; the following April there are once more only 500 rabbits.

The fox population also rises and falls, though not at the same time of year as the rabbits. In January, the fox population numbers 200. By April, it has fallen to 150 foxes. The fox population continues to fall, perhaps because there are so few rabbits to eat at that time. By July, only 100 foxes remain. The fox population then begins to increase again. By October, when the rabbit population is largest, the fox population is growing rapidly, so that by January it has returned to its maximum size of 200 foxes. By this point, though, the rabbit population is already dropping, perhaps because there are so many hungry foxes around. Then the cycle repeats.

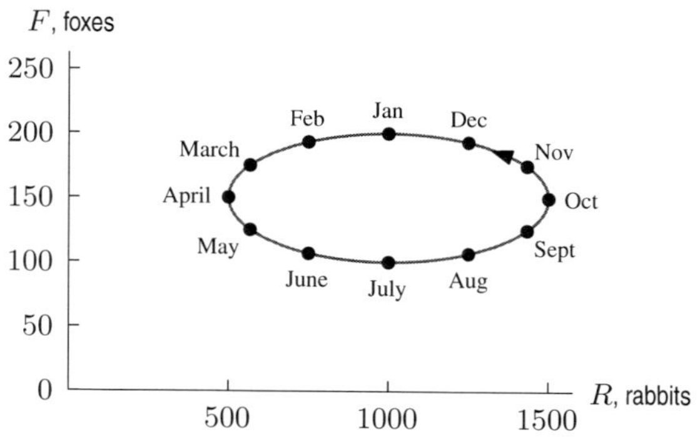

Figure 12.11: The relationship between F, the number of foxes, and R, the number of rabbits

Using Graphs to Parameterize a Curve

Example 5 Figure 12.12 shows the graphs of two functions, $f(t)$ and $g(t)$. Describe the motion of the particle whose coordinates at time t are given by $x = f(t), y = g(t)$.

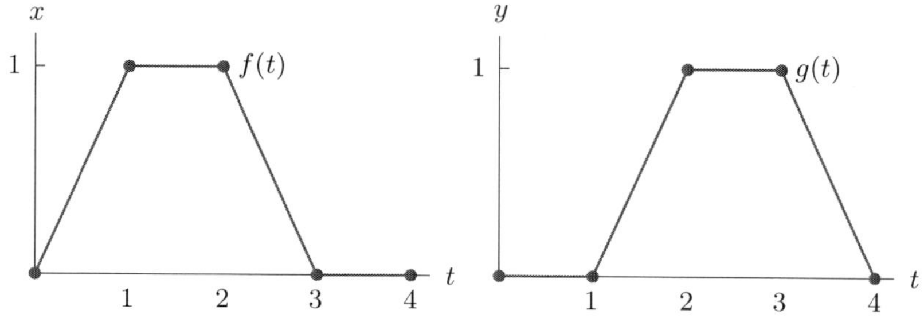

Figure 12.12: Graphs of $x = f(t)$ and $y = g(t)$ used to trace out the path in Figure 12.13

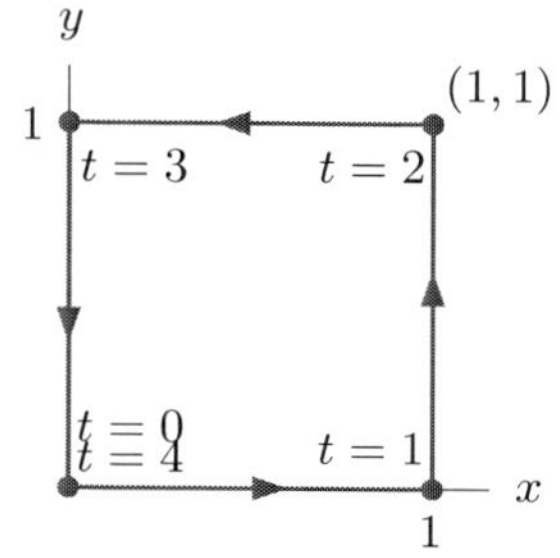

Figure 12.13: Square parameterized by $x = f(t), y = g(t)$ from Figure 12.12

Solution As t increases from 0 to 1, the x-coordinate increases from 0 to 1, while the y-coordinate stays fixed at 0. The particle moves along the x-axis from $(0, 0)$ to $(1, 0)$. As t increases from 1 to 2, the x-coordinate stays fixed at $x = 1$, while the y-coordinate increases from 0 to 1. Thus, the particle moves along the vertical line from $(1, 0)$ to $(1, 1)$. Between times $t = 2$ and $t = 3$, it moves horizontally backward to $(0, 1)$, and between times $t = 3$ and $t = 4$ it moves down the y-axis to $(0, 0)$. Thus, it traces out the square in Figure 12.13.

Exercises and Problems for Section 12.1

Exercises

Graph the parametric equations in Exercises 1–12. Assume that the parameter is restricted to values for which the functions are defined. Indicate the direction of the curve. Eliminate the parameter, t, to obtain an equation for y as a function of x. Check your sketches by graphing y as a function of x.

1. $x = 5 - 2t, \quad y = 1 + 4t$

2. $x = t + 1, \quad y = 3t - 2$

3. $x = \sqrt{t}, \quad y = 2t + 1$

4. $x = \sqrt{t}, \quad y = t^2 + 4$

5. $x = t - 3, \quad y = t^2 + 2t + 1$

6. $x = e^t, \quad y = e^{2t}$

7. $x = e^{2t}, \quad y = e^{3t}$

8. $x = \ln t, \quad y = t^2$

9. $x = t^3, \quad y = 2\ln t$

10. $x = 2\cos t, \quad y = 2\sin t$

11. $x = 2\cos t, \quad y = 3\sin t$

12. $x = 3 + \sin t, \quad y = 2 + \cos t$

For Exercises 13–16, describe the motion of a particle whose position at time t is given by $x = f(t)$, $y = g(t)$.

13.

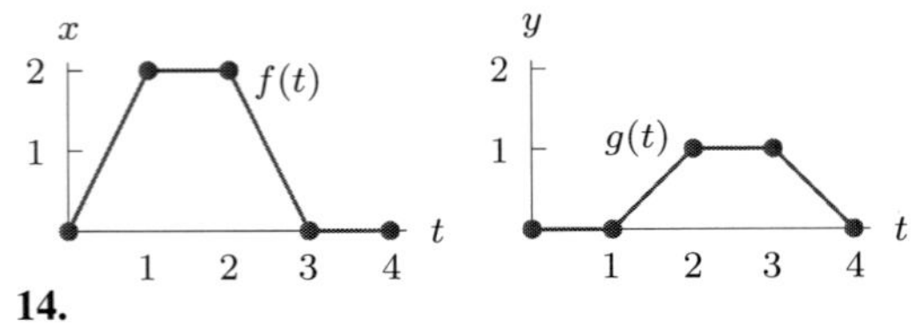

14.

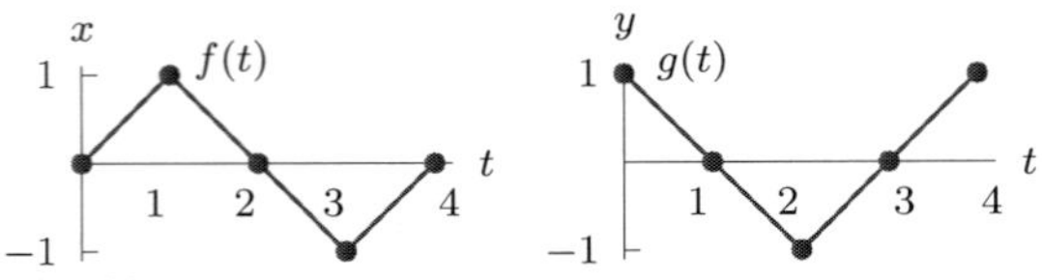

15.

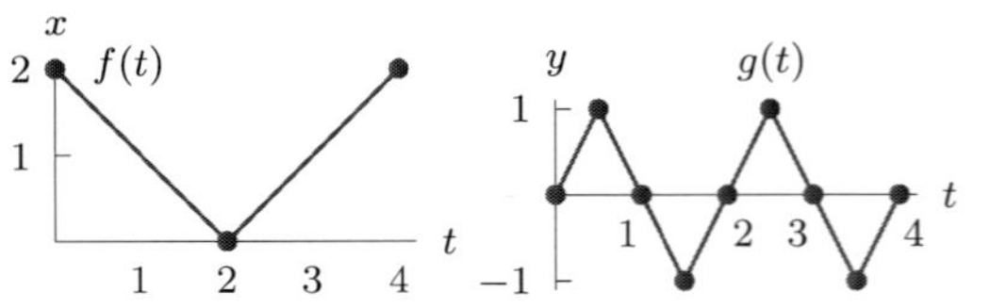

16.

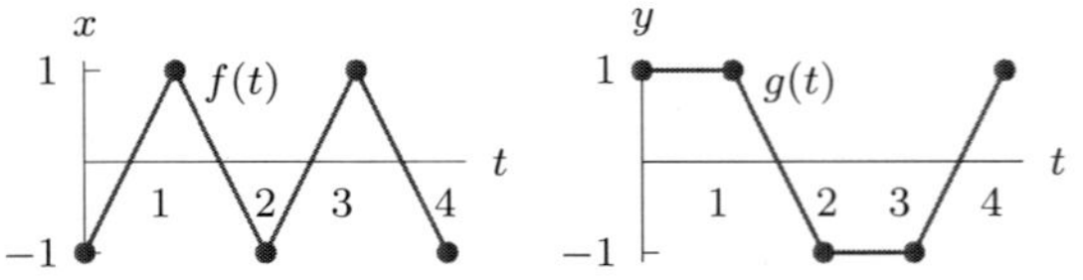

Problems

Problems 17–20 give parameterizations of the unit circle or a part of it. In each case, describe in words how the circle is traced out, including when and where the particle is moving clockwise and when and where it is moving counterclockwise.

17. $x = \cos t, \quad y = -\sin t$

18. $x = \sin t, \quad y = \cos t$

19. $x = \cos(t^2), \quad y = \sin(t^2)$

20. $x = \cos(\ln t), \quad y = \sin(\ln t)$

21. Describe the similarities and differences among the motions in the plane given by the following three pairs of parametric equations:
(a) $x = t, \quad y = t^2$ **(b)** $x = t^2, \quad y = t^4$
(c) $x = t^3, \quad y = t^6$.

22. As t varies, the following parametric equations trace out a line in the plane

$$x = 2 + 3t, \quad y = 4 + 7t.$$

(a) What part of the line is obtained by restricting t to nonnegative numbers?
(b) What part of the line is obtained if t is restricted to $-1 \le t \le 0$?
(c) How should t be restricted to give the part of the line to the left of the y-axis?

23. Write two different parameterizations for each of the curves in the xy-plane.

(a) A parabola whose equation is $y = x^2$.
(b) The parabola from part (a) shifted to the left 2 units and up 1 unit.

Graph the Lissajous figures in Problems 24–27 using a calculator or computer.

24. $x = \cos 2t, \quad y = \sin 5t$

25. $x = \cos 3t, \quad y = \sin 7t$

26. $x = \cos 2t, \quad y = \sin 4t$

27. $x = \cos 2t, \quad y = \sin \sqrt{3}t$

28. Suppose $a, b, c, d, m, n, p, q > 0$. Match each of the following pairs of parametric equations with one of the lines l_1, l_2, l_3, l_4 in Figure 12.14.

I. $\begin{cases} x = a + ct, \\ y = -b + dt. \end{cases}$ II. $\begin{cases} x = m + pt, \\ y = n - qt. \end{cases}$

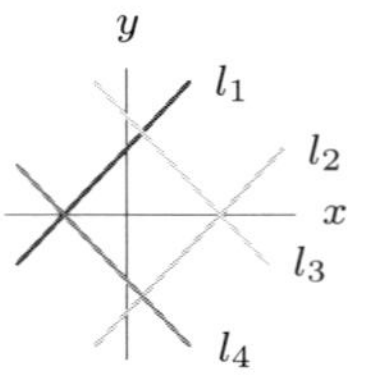

Figure 12.14

29. A bug is crawling around the Cartesian plane. Let $x = f(t)$ be the function denoting the x-coordinate of the bug's position as a function of time, t, and let $y = g(t)$ be the y-coordinate of the bug's position.

(a) Let $f(t) = t$ and $g(t) = t$. What path does the bug follow?
(b) Now let $f(t) = \cos t$ and $g(t) = \sin t$. What path does the bug follow? What is its starting point? (That is, where is the bug when $t = 0$?) When does the bug get back to its starting point?
(c) Now let $f(t) = \cos t$ and $g(t) = 2\sin t$. What path does the bug follow? What is its starting point? When does the bug get back to its starting point?

Write a parameterization for the lines in the xy-plane in Problems 30–31.

30. A vertical line through the point $(-2, -3)$.

31. The line through the points $(2, -1)$ and $(1, 3)$.

32. Motion along a straight line is given by a single equation, say, $x = t^3 - t$ where x is distance along the line. It is difficult to see the motion from a plot; it just traces out the x-line, as in Figure 12.15. To visualize the motion, we introduce a y-coordinate and let it slowly increase, giving Figure 12.16. Try the following on a calculator or computer. Let $y = t$. Now plot the parametric equations $x = t^3 - t$, $y = t$ for, say, $-3 \le t \le 3$. What does the plot in Figure 12.16 tell you about the particle's motion?

Figure 12.15

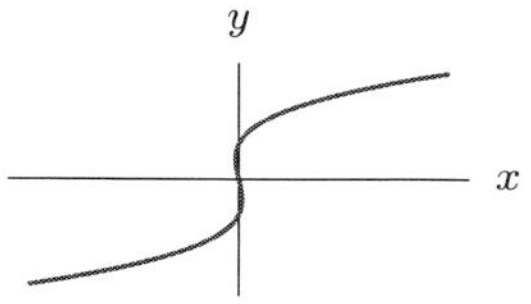

Figure 12.16

For Problems 33–34, plot the motion along the x-line by introducing a y-coordinate, as in Problem 32. What does the plot tell you about the particle's motion?

33. $x = \cos t, \quad -10 \le t \le 10$

34. $x = t^4 - 2t^2 + 3t - 7, \quad -3 \le t \le 2$

35. A ball is thrown vertically into the air at time $t = 0$. Its height $d(t)$, in feet, above the ground at time t, in seconds, is given by:

$$d(t) = -16t^2 + 48t + 6.$$

(a) Write a parameterization for the curve.
(b) Sketch a graph of the curve.
(c) What is the height of the ball at $t = 0$? Explain how this makes sense.
(d) Does the ball ever reach this height again? If so when? Why?
(e) When does the ball reach its maximum height? What is the maximum height?

12.2 IMPLICITLY DEFINED CURVES AND CIRCLES

In the previous section, we saw that the parametrically defined curve $x = \cos t$, $y = \sin t$ is the circle of radius 1 centered at the origin. Eliminating the parameter t gives the equation

$$x^2 + y^2 = 1.$$

The circle is described by the *implicit function* $x^2 + y^2 = 1$. To be explicit is to state a fact outright; to be implicit is to make a statement in a round-about way. If a curve has an equation of the form $y = f(x)$, then we say that y is an *explicit* function of x. However, the equation $x^2 + y^2 = 1$ does not explicitly state that y depends on x. Instead, this dependence is implied by the equation. If we try to solve for y in terms of x, we do not obtain a function. Rather, we obtain *two* functions, one for the top half of the unit circle and one for the bottom half:

$$y = \sqrt{1 - x^2} \quad \text{and} \quad y = -\sqrt{1 - x^2}.$$

Example 1 Graph the equation $y^2 = x^2$. Can you find an explicit formula for y in terms of x?

Solution The equation $y^2 = x^2$ is implicit because it does not tell us what y equals, instead, it tells us what y^2 equals. Suppose $x = 2$. What values can y equal? Since

$$y^2 = 2^2 = 4, \qquad \text{we have} \qquad y = 2 \text{ or } -2,$$

because for either of these y-values $y^2 = 4$. What if x is negative? If $x = -3$, we have

$$y^2 = (-3)^2 = 9, \qquad \text{then} \qquad y = 3 \text{ or } -3,$$

because these are the solutions to $y^2 = 9$.

For almost all x-values, there are two y-values, one given by $y = x$ and one by $y = -x$. The only exception is at $x = 0$, because the only solution to the equation $y^2 = 0$ is $y = 0$. Thus, the graph of $y^2 = x^2$ has two parts, given by

$$y = x \quad \text{and} \quad -x.$$

When both parts of the graph of $y^2 = x^2$ are plotted together, the resulting graph looks like an X. (See Figure 12.17.)

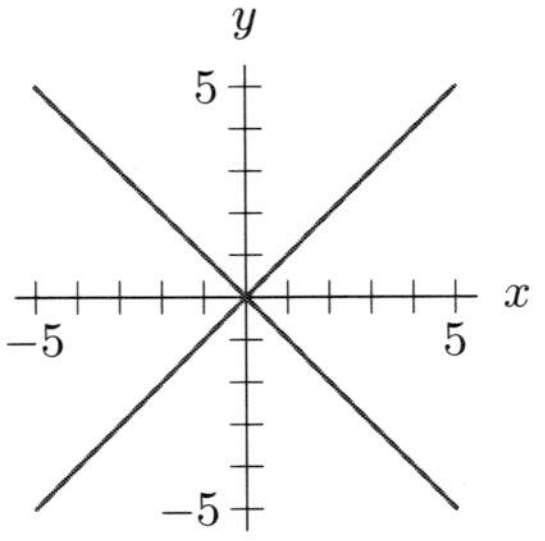

Figure 12.17: Graph of $y^2 = x^2$

Conic Sections

The curves known as *conic sections* include two we are already familiar with, circles and parabolas. They also include *ellipses* and *hyperbolas*. An ellipse is a "squashed" circle; an example is given by Figure 12.11 on page 523. An example of a hyperbola is the graph of the function $y = 1/x$.

Conic sections are so-called because, as was demonstrated by the Greeks, they can be constructed by slicing, or sectioning, a cone. Conic sections arise naturally in physics, since the path of a body orbiting the sun is a conic section. We will study them in terms of parametric and implicit equations. As we have already studied parabolas, we now focus on circles, ellipses and hyperbolas.

Circles

Example 2 Graph the parametric equations

$$x = 5 + 2\cos t \qquad \text{and} \qquad y = 3 + 2\sin t.$$

Solution We see that x varies between 3 and 7 with a midline of 5 while y varies between 1 and 5 with a midline of 3. Figure 12.18 gives a graph of this function. It appears to be a circle of radius 2 centered at the point $(5, 3)$.

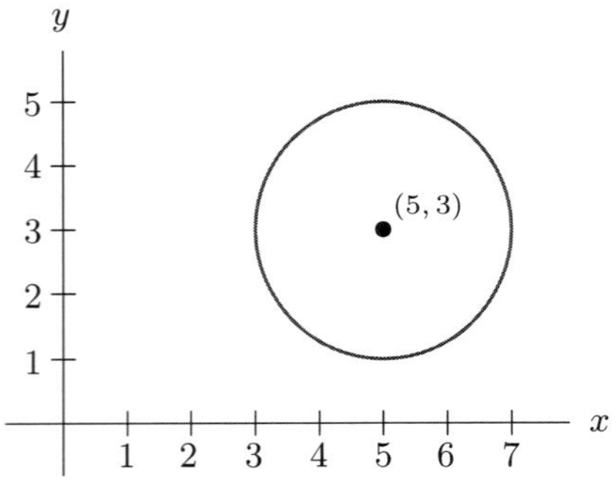

Figure 12.18: The circle defined by $x = 5 + 2\cos t$, $y = 3 + 2\sin t$, $0 \le t \le 2\pi$

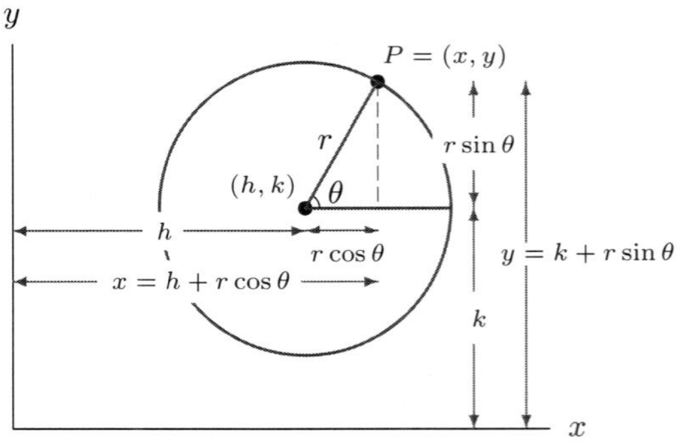

Figure 12.19: A circle of radius r centered at the point (h, k)

Compare Figure 12.18 with Figure 12.19 which shows a circle of radius r centered at the point (h, k). The coordinates of a point $P = (x, y)$, corresponding to an angle θ, are given by

$$x = h + r\cos\theta \qquad \text{and} \qquad y = k + r\sin\theta.$$

The equations in Example 2 are of this form, with t instead of θ and with $h = 5$, $k = 3$, and $r = 2$. This curve is a circle because the point P is always a fixed distance, 2, from the point $(5, 3)$.

> For $r > 0$, the parametric equations of a **circle** of radius r centered at the point (h, k) are:
>
> $$x = h + r\cos t \qquad y = k + r\sin t \qquad 0 \le t \le 2\pi.$$

Eliminating the Parameter t in the Equations for a Circle

To eliminate the parameter t from the parametric equations for a circle, we rewrite the equations as

$$x - h = r\cos t \qquad \text{so} \qquad (x-h)^2 = r^2\cos^2 t$$

and

$$y - k = r\sin t \qquad \text{so} \qquad (y-k)^2 = r^2\sin^2 t.$$

Adding these two equations and applying the Pythagorean identity, $\cos^2 t + \sin^2 t = 1$, gives

$$\begin{aligned}(x-h)^2 + (y-k)^2 &= r^2\cos^2 t + r^2\sin^2 t\\ &= r^2(\cos^2 t + \sin^2 t)\\ &= r^2.\end{aligned}$$

> An implicit equation for the **circle** of radius r centered at the point (h, k) is:
>
> $$(x-h)^2 + (y-k)^2 = r^2.$$
>
> This is called the **standard form** of the equation of a circle.

For the unit circle, we have $h = 0$, $k = 0$, and $r = 1$. This gives

$$x^2 + y^2 = 1.$$

Example 3 Write the circle in Example 2 in standard form.

Solution We have $h = 5$, $k = 3$, and $r = 2$. This gives

$$(x-5)^2 + (y-3)^2 = 2^2 = 4.$$

If we expand the equation for the circle in Example 3, we get a quadratic equation in two variables:

$$x^2 - 10x + 25 + y^2 - 6y + 9 = 4,$$

or

$$x^2 + y^2 - 10x - 6y + 30 = 0.$$

Note that the coefficients of x^2 and y^2 (in this case, 1) are equal. Such equations often describe circles; we put them into standard form, $(x-h)^2 + (y-k)^2 = r^2$, by completing the square.

Example 4 Describe in words the curve defined by the equation

$$x^2 + 10x + 20 = 4y - y^2.$$

Solution Rearranging terms gives

$$x^2 + 10x + y^2 - 4y = -20.$$

We complete the square for the terms involving x and (separately) for the terms involving y:

$$\underbrace{(x^2 + 10x + 25)}_{(x+5)^2} + \underbrace{(y^2 - 4y + 4)}_{(y-2)^2} \quad \underbrace{-25 - 4}_{\text{Compensating terms}} = -20$$

$$(x+5)^2 + (y-2)^2 - 29 = -20$$

$$(x+5)^2 + (y-2)^2 = 9.$$

This equation is a circle of radius $r = 3$ with center $(h, k) = (-5, 2)$.

Exercises and Problems for Section 12.2

Exercises

What are the center and radius of the circles in Exercises 1–4?

1. $4x^2 + 4y^2 - 9 = 0$

2. $10 - 3(x^2 + y^2) = 0$

3. $(x+1)^2 + 2(y+3)^2 = 32 - (x+1)^2$

4. $\dfrac{4 - (x-4)^2}{(y-4)^2} - 1 = 0$

In Exercises 5–12, parameterize the circles.

5. Radius 3, centered at the origin, traversed clockwise starting at $(0, 3)$.

6. Radius 4, centered at the origin, traversed clockwise starting at $(4, 0)$.

7. Radius 7, centered at the origin, traversed counterclockwise starting at $(-7, 0)$.

8. Radius 5, centered at the origin, traversed counterclockwise starting at $(0, -5)$.

9. Radius 4, centered at $(3, 1)$, traversed clockwise starting at $(3, 5)$.

10. Radius 5, centered at $(3, 4)$, traversed counterclockwise starting at $(8, 4)$.

11. Radius 3, centered at $(-1, -2)$, traversed counterclockwise starting at $(-4, -2)$.

12. Radius $\sqrt{5}$, centered at $(-2, 1)$, traversed counterclockwise starting at $(-2, 1 + \sqrt{5})$.

Problems

13. Identify the center and radius for each of the following circles.

(a) $(x-2)^2 + (y+4)^2 = 20$
(b) $2x^2 + 2y^2 + 4x - 8y = 12$

What curves do the parametric equations in Problems 14–17 trace out? Find an implicit or explicit equation for each curve.

14. $x = 2 + \cos t,\ y = 2 - \sin t$

15. $x = 2 + \cos t,\ y = \cos^2 t$

16. $x = 2 + \cos t,\ y = 2 - \cos t$

17. $x = 4\sin^2 t, y = 3 + \sin t$

18. An ant, starting at the origin, moves at 2 units/sec along the x-axis to the point $(1, 0)$. The ant then moves counterclockwise along the unit circle to $(0, 1)$ at a speed of $3\pi/2$ units/sec, then straight down to the origin at a speed of 2 units/sec along the y-axis.

(a) Express the ant's coordinates as a function of time, t, in secs.
(b) Express the reverse path as a function of time.

State whether the equations in Problems 19–22 represent a curve parametrically, implicitly, or explicitly. Give the two other types of representations for the same curve.

19. $xy = 1$ for $x > 0$

20. $x = e^t,\quad y = e^{2t}$ for all t

21. $y = \sqrt{4 - x^2}$

22. $x^2 - 2x + y^2 = 0$ for $y < 0$

23. In Figure 12.20 a wheel of radius 1 meter rests on the x-axis with its center on the y-axis. There is a spot on the rim at the point $(1, 1)$. At time $t = 0$ the wheel starts rolling on the x-axis in the direction shown at a rate of 1 radian per second. Find parametric equations describing the motion of

(a) The center of the wheel.
(b) The motion of the spot on the rim. Plot its path.

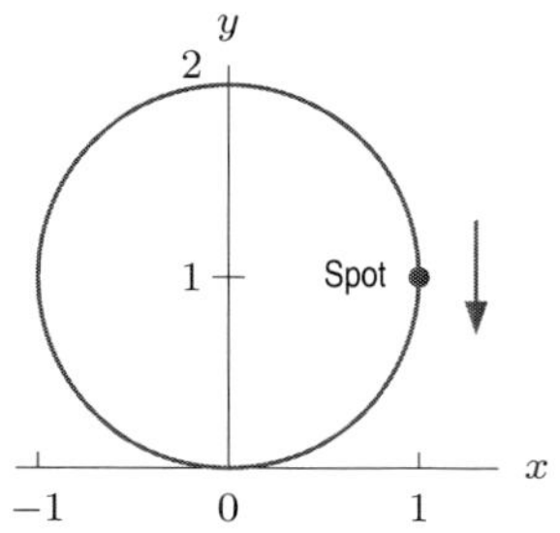

Figure 12.20

24. What can you say about the values of a, b and k if the equations

$$x = a + k\cos t, \quad y = b + k\sin t, \qquad 0 \le t \le 2\pi,$$

trace out each of the circles in Figure 12.21?

(a) C_1 **(b)** C_2 **(c)** C_3

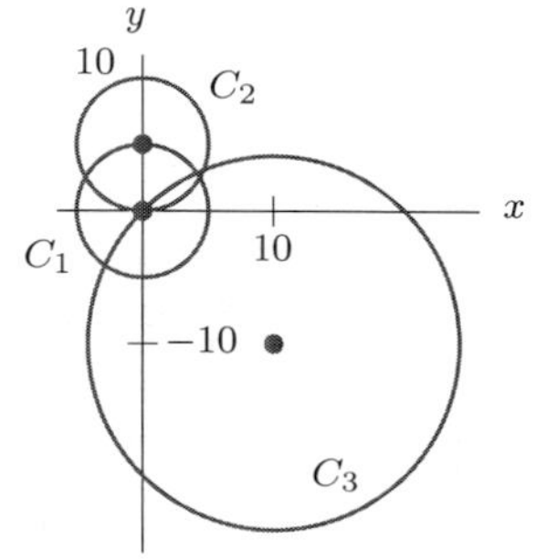

Figure 12.21

12.3 ELLIPSES

In this section we consider the graph of an ellipse, which is a "squashed" circle.

> The parametric equations of an **ellipse** centered at (h, k) are:
>
> $$x = h + a\cos t \qquad y = k + b\sin t \qquad 0 \le t \le 2\pi.$$
>
> We usually take $a > 0$ and $b > 0$.

In the special case where $a = b$, these equations give a circle of radius $r = a$. Thus, a circle is a special kind of ellipse. The parametric equations of an ellipse are transformations of the parametric equations of the unit circle, $x = \cos t$, $y = \sin t$. These transformations have the effect of shifting and stretching the unit circle into an ellipse.

Example 1 Graph the ellipse given by

$$x = 7 + 5\cos t, \qquad y = 4 + 2\sin t, \qquad 0 \le t \le 2\pi.$$

Solution See Figure 12.22. The center of the ellipse is $(h, k) = (7, 4)$. The value $a = 5$ determines the horizontal "radius" of the ellipse, while $b = 2$ determines the vertical "radius". The value of x varies from a maximum of 12 to a minimum of 2 about the vertical midline $x = 7$. Similarly, the value of y varies from a maximum of 6 to a minimum of 2 about the horizontal midline $y = 4$. The ellipse is symmetric about the midlines $x = 7$ and $y = 4$.

In general, the horizontal axis of an ellipse has length $2a$, and the vertical axis has length $2b$. This is similar to a circle, which has diameter $2r$. The difference here is that the diameter of a circle is the same in every direction, whereas the "diameter" of an ellipse depends on the direction.

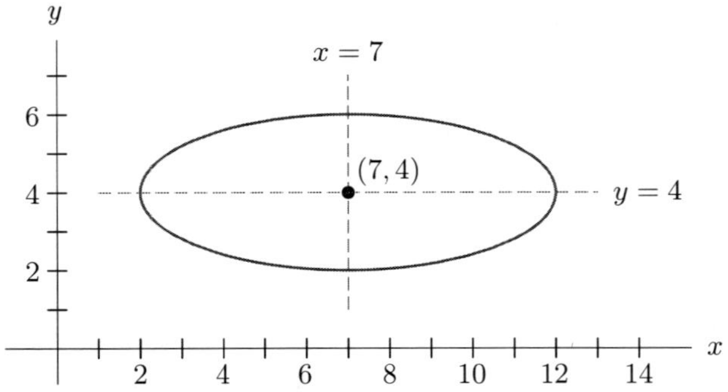

Figure 12.22: The ellipse $x = 7 + 5\cos t$, $y = 4 + 2\sin t, 0 \le t \le 2\pi$

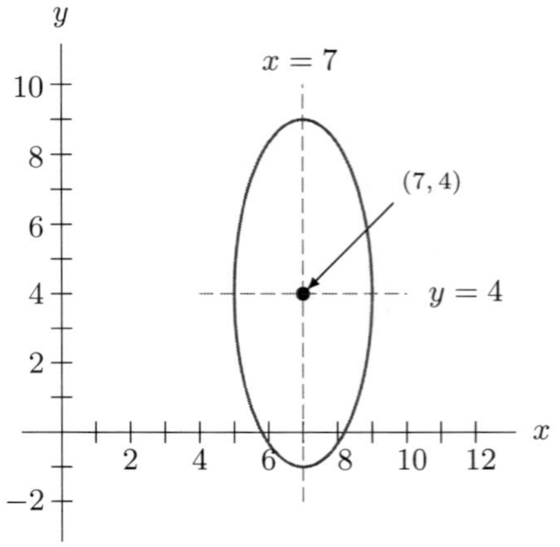

Figure 12.23: $x = 7 + 2\cos t$, $y = 4 + 5\sin t, 0 \le t \le 2\pi$

The longer axis of an ellipse is called the *major* axis and the shorter axis is called the *minor* axis. Either the horizontal or the vertical axis can be the longer axis.

Example 2 Graph the ellipse given by

$$x = 7 + 2\cos t, \qquad y = 4 + 5\sin t, \qquad 0 \le t \le 2\pi.$$

How is this ellipse similar to the one in Example 1? How is it different?

Solution This ellipse has the same center $(h, k) = (7, 4)$ as the ellipse in Example 1. However, as we see in Figure 12.23, the vertical axis of this ellipse is the longer one and the horizontal axis is the shorter one. This is the opposite of the situation in Example 1.

Eliminating the Parameter t in the Equations for an Ellipse

As with circles, we can eliminate t from the parametric equations for an ellipse. We first rewrite the equation for x as follows:

$$\begin{aligned} x &= h + a\cos t \\ x - h &= a\cos t \\ \frac{x-h}{a} &= \cos t. \end{aligned}$$

Similarly, we rewrite the equation for y as

$$\frac{y-k}{b} = \sin t.$$

Using the Pythagorean Identity gives

$$\left(\frac{x-h}{a}\right)^2 + \left(\frac{y-k}{b}\right)^2 = \cos^2 t + \sin^2 t = 1.$$

> The implicit equation for an **ellipse** centered at (h, k) and with horizontal axis $2a$ and vertical axis $2b$ is:
>
> $$\frac{(x-h)^2}{a^2} + \frac{(y-k)^2}{b^2} = 1.$$

Exercises and Problems for Section 12.3

Exercises

For the ellipses in Exercises 1–4, find

(a) The coordinates of the center and the "diameter" in the x- and y-directions.

(b) An implicit equation for the ellipse.

1.

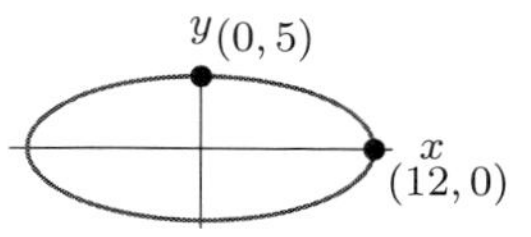

2.

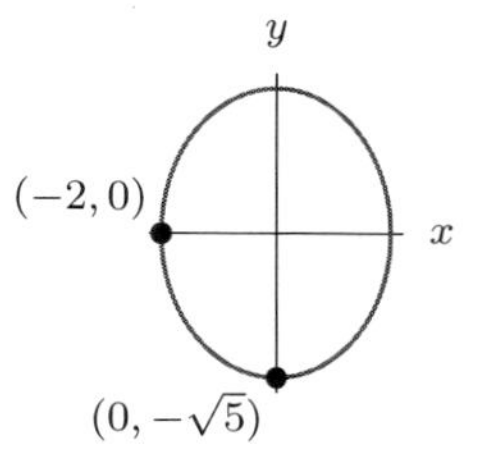

3.

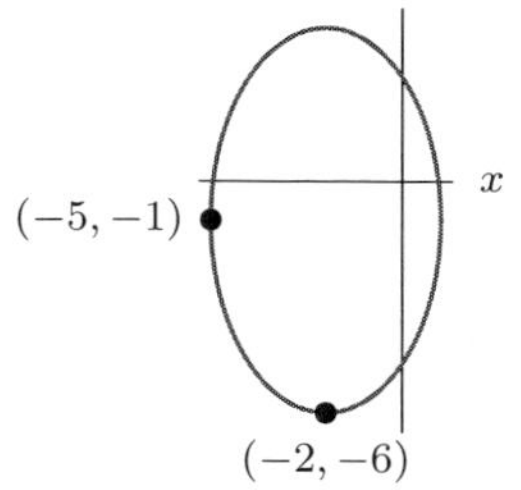

4.

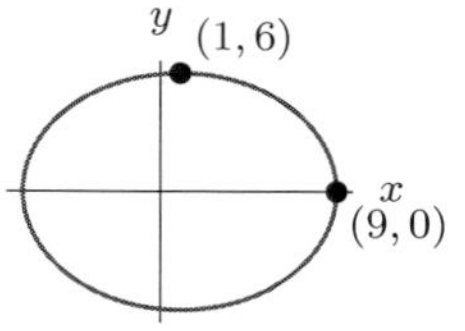

5. Parameterize the ellipse in Exercise 1 counterclockwise, starting at $(12, 0)$.

6. Parameterize the ellipse in Exercise 2 counterclockwise, starting at $(-2, 0)$.

7. Parameterize the ellipse in Exercise 3 clockwise, starting at $(-5, -1)$.

8. Parameterize the ellipse in Exercise 4 clockwise, starting at $(9, 0)$.

9. Compare the ellipse in Example 2 on page 530 with the ellipse given by

$$x = 7+2\cos(-s), \quad y = 4+5\sin(-s), \quad 0 \le s \le 2\pi.$$

10. Compare the ellipse in Example 2 on page 530 with the ellipse given by

$$x = 7+2\cos(2t), \quad y = 4+5\sin(2t), \quad 0 \le t \le 2\pi.$$

Problems

By completing the square, rewrite the equations in Problems 11–16 in the form

$$\frac{(x-h)^2}{a^2} + \frac{(y-k)^2}{b^2} = 1.$$

What is the center of the ellipse? The values of a and b?

11. $\dfrac{x^2+4x}{9} + \dfrac{y^2+10y}{25} = -\dfrac{4}{9}$

12. $\dfrac{x^2-2x}{4} + y^2 + 4y + \dfrac{13}{4} = 0$

13. $4x^2 - 4x + y^2 + 2y = 2$

14. $4x^2 + 16x + y^2 + 2y + 13 = 0$

15. $9x^2 - 54x + 4y^2 - 16y + 61 = 0$

16. $9x^2 + 9x + 4y^2 - 4y = \dfrac{131}{4}$

17. For each of the following ellipses, find the center and lengths of the major and minor axes, and graph it.

(a) $\dfrac{(x+1)^2}{4} + \dfrac{(y-3)^2}{6} = 1$

(b) $2x^2 + 3y^2 - 6x + 6y = 12$

For positive a, b, the equation of an ellipse can be written in the form

$$\frac{(x-h)^2}{a^2} + \frac{(y-k)^2}{b^2} = 1.$$

In Problems 18–19, rank h, k, a, b, and the number 0 in ascending order (i.e. increasing order). Assume the x- and y-scales are equal.

18.

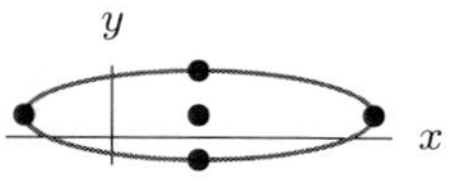

19.

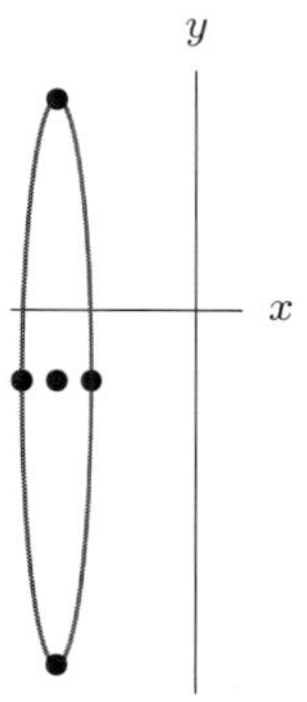

20. Show, by completing the square, that the equation $Ax^2 - Bx + y^2 = r_0^2$ represents an ellipse, where A, B, and r_0 are constants and $A > 0$. (See Problem 21.)

21. Consider the following equation in polar coordinates, where r_0 is a positive constant and where $0 \le \epsilon < 1$ is a constant:

$$r = \frac{r_0}{1 - \epsilon \cos\theta}, \qquad 0 \le \theta < 2\pi.$$

(a) Show, by converting to Cartesian coordinates, that this equation describes an ellipse. You may use the results of Problem 20.

(b) What are the minimum and maximum values of r? At what values of θ do they occur?

(c) Graph the ellipse for $r_0 = 6$ and $\epsilon = 0.5$, labeling the points from part (b) as well as the y-intercepts, that is, the points at $\theta = \pi/2$ and $\theta = 3\pi/2$. What is its center?

(d) Find a formula in terms of r_0 and ϵ for the length of the horizontal axis of the ellipse

(e) The constant ϵ is known as the *eccentricity* of the ellipse. Describe in words the appearance of an ellipse of eccentricity $\epsilon = 0$. What happens to the appearance of the ellipse as the eccentricity gets closer and closer to $\epsilon = 1$?

22. If you look at a circular disk, such as a quarter, head on, you see a circle. But if you tilt the disk away from your line of sight, as shown in Figure 12.24, the disk appears elliptical. The length of the apparent ellipse's horizontal axis does not change, but the length of the vertical axis decreases.

(a) Find a formula for ℓ, the length of the ellipse's vertical axis, in terms of θ, the angle of tilt with respect to the line of sight.

(b) What does your formula say about the appearance of the coin after being tilted through an angle of $\theta = 0°$? $\theta = 90°$? $\theta = 180°$? Does this make sense?

(c) Find a formula for the ellipse formed by tilting a disk of radius r through an angle of θ. Assume that the disk is centered at the origin and that the disk is being titled around the x-axis, so that the length of its horizontal axis does not change.

(d) The equation $x^2/16 + y^2/7 = 1$ describes an ellipse. If we think of this ellipse as a tilted disk, then what is the disk's radius, and through what angle θ has it been tilted?

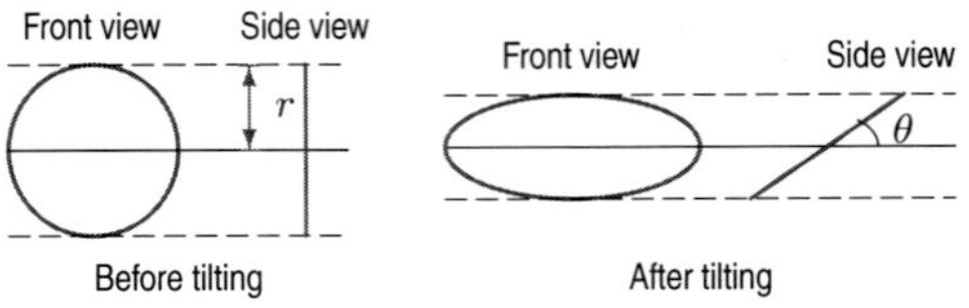

Figure 12.24: At left, front and side views of a circular disc of radius r. At right, the same disk as it appears after being tilted through an angle θ relative to the line of sight.

23. A fuel-efficient way to travel from earth to Mars is to follow a semi-elliptical orbit known as a *Hohmann transfer orbit*.[2] The spacecraft leaves the earth at the point on the ellipse closest to the sun, and arrives at Mars at the point on the ellipse farthest from the sun, as shown in Figure 12.25. Let r_e be the radius of earth's orbit, and r_m the radius of Mars' orbit, and let $2a$ and $2b$ be the respective lengths of the horizontal and vertical axes of the ellipse.

(a) Orienting the ellipse as shown in the figure, with the sun at the origin, find a formula for this orbit in terms of r_e, r_m, and b.

(b) It can be shown that $b^2 = 2ar_e - r_e^2$. Given this and your answer to part (a), find a formula for b in terms of r_m and r_e.

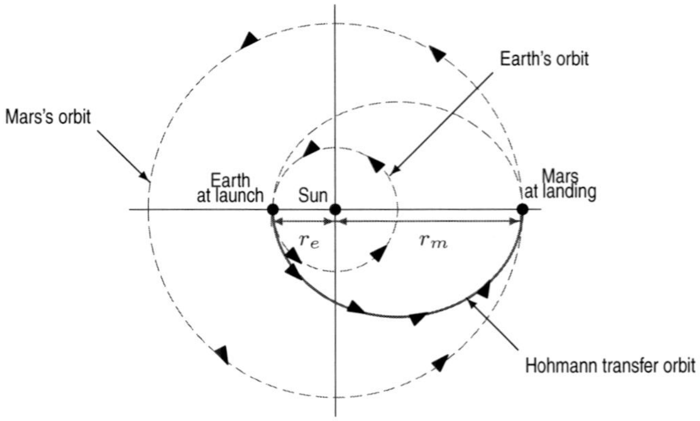

Figure 12.25

[2]*Scientific American*, March, 2000. Note that in an actual Hohmann transfer, the spacecraft would begin in low earth orbit, not from the earth's surface. Timing is critical in order for the two planets to be correctly aligned. Calculations show that at launch, Mars must lead earth by about 45 degrees, which happens only once every 26 months. Note also that the orbits of earth and Mars are actually themselves elliptical, though here we treat them as circular.

12.4 HYPERBOLAS

We now use another form of the Pythagorean identity to parameterize a curve. With $\tan t = \sin t/\cos t$ and $\sec t = 1/\cos t$, on page 282 we showed that

$$\sec^2 t - \tan^2 t = 1.$$

We use this identity to investigate the curve described by the parametric equations

$$x = \sec t \qquad \text{and} \qquad y = \tan t.$$

By eliminating the parameter t, we get

$$x^2 - y^2 = 1.$$

This implicit equation looks similar to the equation for a unit circle $x^2 + y^2 = 1$. However, the curve it describes is very different. Solving for y, we find that

$$y^2 = x^2 - 1.$$

For large values of x, the values of y^2 and x^2 are very nearly equal. For instance, when $x = 100$, we see that $x^2 = 10{,}000$ and $y^2 = x^2 - 1 = 9999$. Thus, for large values of x,

$$y^2 \approx x^2,$$
$$y \approx \pm x.$$

The graph of $y = \pm x$ is in Figure 12.17 on page 526; it looks like an X.

What happens for smaller values of x? Writing the equation as $y^2 = x^2 - 1$, shows us that x^2 cannot be less than 1; otherwise y^2 would be negative. See Table 12.1 and Figure 12.26.

Table 12.1 *Points on the graph of the hyperbola, $x^2 - y^2 = 1$*

x	y	Points	x	y	Points
-3	± 2.83	$(-3, 2.83)$ and $(-3, -2.83)$	3	± 2.83	$(3, 2.83)$ and $(3, -2.83)$
-2	± 1.73	$(-2, 1.73)$ and $(-2, -1.73)$	2	± 1.73	$(2, 1.73)$ and $(2, -1.73)$
-1	± 0	$(-1, 0)$ and $(-1, -0)$	1	± 0	$(1, 0)$ and $(1, -0)$
0	Undefined	None			

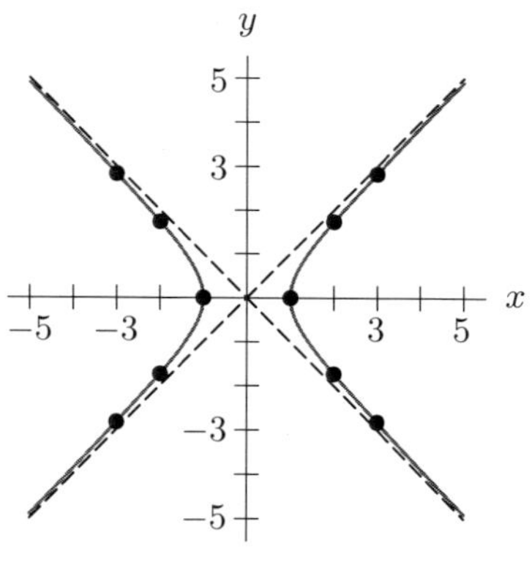

Figure 12.26: A graph of $x^2 - y^2 = 1$. The X-shaped graph of $y^2 = x^2$ has been dashed in

This graph is an example of a *hyperbola*. By analogy to the unit circle, it is called the *unit hyperbola*. It has two branches, one to the right of the y-axis and one to the left. As x grows large in magnitude, either toward $+\infty$ or $-\infty$, the graph approaches the asymptotes, $y^2 = x^2$. To see how the two branches of a similar hyperbola are traced as t increases from 0 to 2π, see Exercise 1.

A General Formula for Hyperbolas

We can shift and stretch the unit hyperbola using the equations $x = h + a\sec t$ and $y = k + b\tan t$. This has the effect of centering the hyperbola at the point (h, k).

> The parametric equations for a **hyperbola** centered at (h, k) and opening to the left and right are
> $$x = h + a\sec t \qquad y = k + b\tan t \qquad 0 \le t \le 2\pi.$$
> We usually take $a > 0$ and $b > 0$.

We eliminate the parameter, t, from the parametric equations to obtain the implicit equation

$$\frac{(x-h)^2}{a^2} - \frac{(y-k)^2}{b^2} = 1.$$

Notice that this equation is similar to the general equation for an ellipse.

Example 1 Graph the equation $\dfrac{(x-4)^2}{9} - \dfrac{(y-7)^2}{25} = 1.$

Solution Since $h = 4$ and $k = 7$, we have shifted the unit hyperbola 4 units to the right and 7 units up. We have $a^2 = 9$ and $b^2 = 25$, so $a = 3$ and $b = 5$. Thus, we have stretched the unit hyperbola horizontally by a factor of 3 and vertically by a factor of 5.

To make it easier to draw hyperbolas, we imagine a "unit square" centered at the origin. This unit square helps us locate the vertices and asymptotes of the hyperbola $x^2 - y^2 = 1$. See Figure 12.27. Stretching and shifting the hyperbola causes this square to be transformed into a rectangle centered at the point $(h, k) = (4, 7)$ with width $2a = 6$ and height $2b = 10$. See Figure 12.28. The rectangle enables us to draw the X-shaped asymptotes as diagonal lines through the corners. The vertices of the transformed hyperbola are located at the midpoints of the vertical sides of the rectangle.

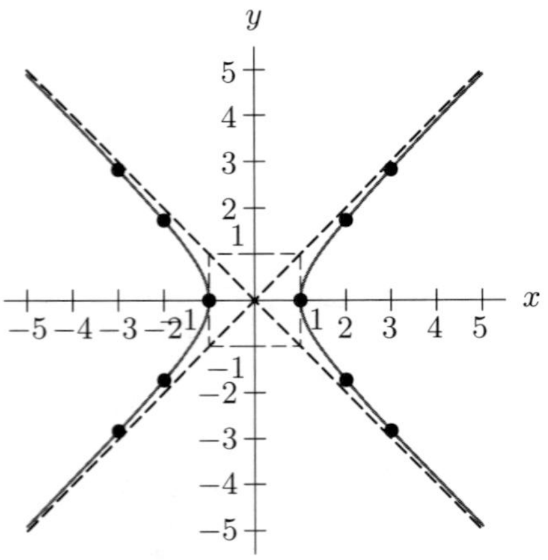

Figure 12.27: The hyperbola $x^2 - y^2 = 1$, with the unit square dashed in

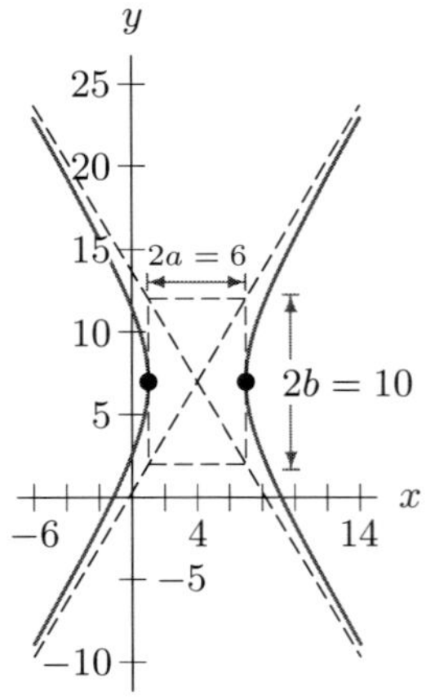

Figure 12.28: A graph of $\frac{(x-4)^2}{9} - \frac{(y-7)^2}{25} = 1$. A rectangle of width $2a = 6$ height $2b = 10$ has been dashed in, centered at the point $(h, k) = (4, 7)$

Note that the ellipse $\frac{(x-4)^2}{9}+\frac{(y-7)^2}{25}=1$ would exactly fit inside the rectangle in Figure 12.28. In this sense, a hyperbola is an ellipse turned "inside-out."

Example 2 Graph the equation $\dfrac{(y-7)^2}{25}-\dfrac{(x-4)^2}{9}=1.$

Solution This equation is similar to that in Example 1, except that the terms have opposite signs. The graph in Figure 12.29 is similar to the one in Figure 12.28 except that it opens up and down instead of right and left.

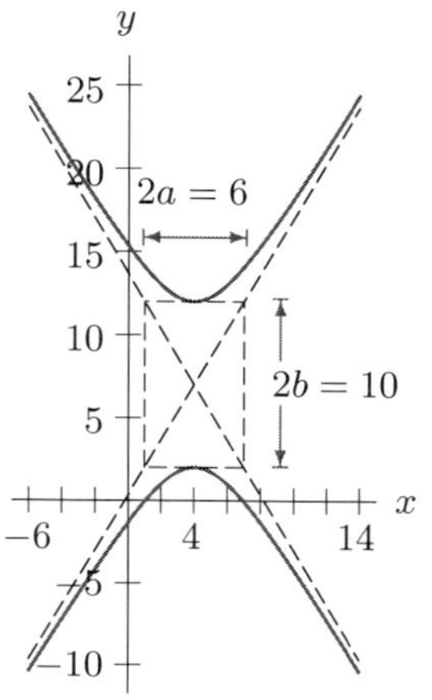

Figure 12.29: A graph of $\frac{(y-7)^2}{25}-\frac{(x-4)^2}{9}=1$. This graph is similar to the one in Figure 12.28 except that here the hyperbola opens up and down instead of right and left

In summary:

> The implicit equation for a **hyperbola**,
>
> $$\frac{(x-h)^2}{a^2}-\frac{(y-k)^2}{b^2}=1,$$
>
> describes a hyperbola that opens left and right. Its asymptotes are diagonal lines through the corners of a rectangle of width $2a$ and height $2b$ centered at the point (h,k). The graph with equation
>
> $$\frac{(y-k)^2}{b^2}-\frac{(x-h)^2}{a^2}=1$$
>
> has a similar shape, except that it opens up and down.

Exercises and Problems for Section 12.4

Exercises

For the hyperbolas in Exercises 1–4, find

(a) The coordinates of the vertices of the hyperbola and the coordinates of its center.

(b) The equations of the asymptotes.

(c) An implicit equation for the hyperbola.

1.

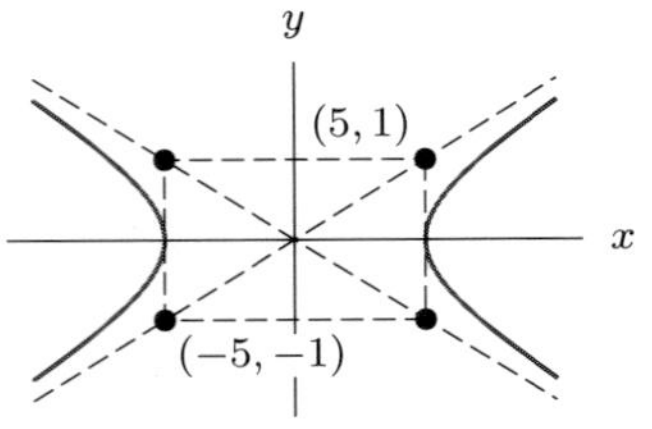

2.

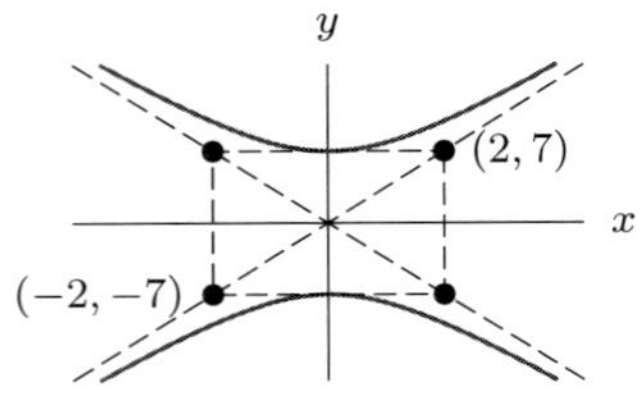

3.

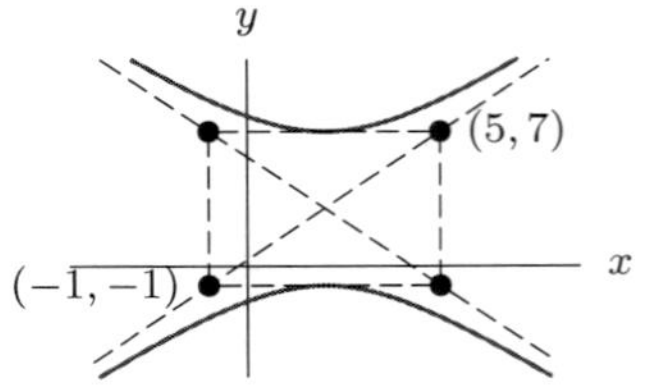

4.

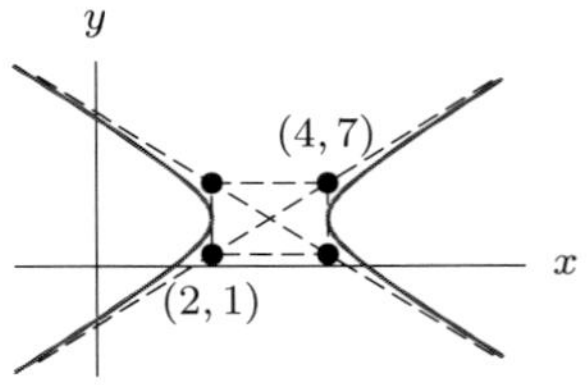

5. Parameterize the hyperbola in Exercise 1. What t values give the right half?

6. Parameterize the hyperbola in Exercise 2. What t values give the upper half?

7. Parameterize the hyperbola in Exercise 3. What t values give the lower half?

8. Parameterize the hyperbola in Exercise 4. What t values give the left half?

Problems

For positive a, b, the equation of a hyperbola can be written in one of the two forms

I. $\dfrac{(x-h)^2}{a^2} - \dfrac{(y-k)^2}{b^2} = 1$

II. $\dfrac{(y-k)^2}{b^2} - \dfrac{(x-h)^2}{a^2} = 1$

In Problems 9–10, which form applies? Rank h, k, a, b, and the number 0 in ascending order (i.e. increasing order). Assume the x- and y-scales are equal.

9.

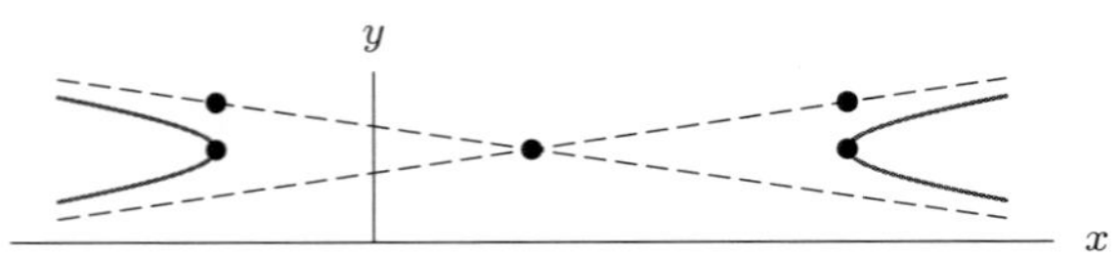

10.

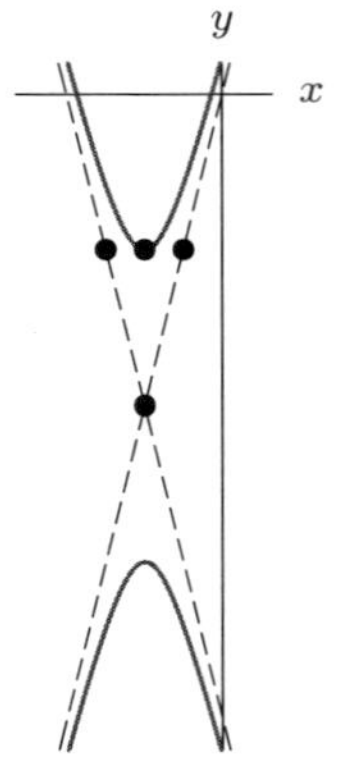

11. For each of the following hyperbolas, find the center, vertices, and asymptotes, and graph it.

(a) $\dfrac{(x+5)^2}{6} - \dfrac{(y-2)^2}{4} = 1$

(b) $x^2 - y^2 + 2x = 4y + 17$

By completing the square, rewrite the equations in Problems 12–17 in the form

$$\frac{(x-h)^2}{a^2} - \frac{(y-k)^2}{b^2} = 1 \text{ or } \frac{(y-k)^2}{b^2} - \frac{(x-h)^2}{a^2} = 1.$$

What is the center, and does the hyperbola open left-right or up-down? What are the values of a and b?

12. $\dfrac{x^2-2x}{4} - y^2 + 4y = \dfrac{19}{4}$

13. $\dfrac{y^2+2y}{4} - \dfrac{x^2-4x}{9} = \dfrac{43}{36}$

14. $x^2 + 2x - 4y^2 - 24y = 39$

15. $9x^2 - 36x - 4y^2 + 8y = 4$

16. $4x^2 - 8x = 36y^2 - 36y - 31$

17. $9y^2 + 6y = 89 + 8x^2 + 24x$

18. By scattering positively charged alpha particles off of atoms in gold foil, Earnest Lord Rutherford demonstrated that atoms contain a nucleus of concentrated charge. The scattered alpha particles followed hyperbolic trajectories, as shown in Figure 12.30, and Rutherford was able to measure the angle θ. However, the value of θ alone was not enough for Rutherford to determine whether the nuclear charge is negative or positive. A negatively charged nucleus would exert an attractive force on the alpha particles, and the positively charged nucleus would exert a repulsive force, but in either case, the trajectory would be a hyperbola specified by the angle θ. Note that in Figure 12.30, the incoming particle would pass through the origin if it were not scattered by the nucleus at $(d, 0)$.

(a) The hyperbola is $x^2/a^2 - y^2/b^2 = 1$. Find a formula for b in terms of a and θ.

(b) It turns out that the charge in the nucleus is positive, and that the alpha particles undergo a repulsive force. What must be true about the parameter a?

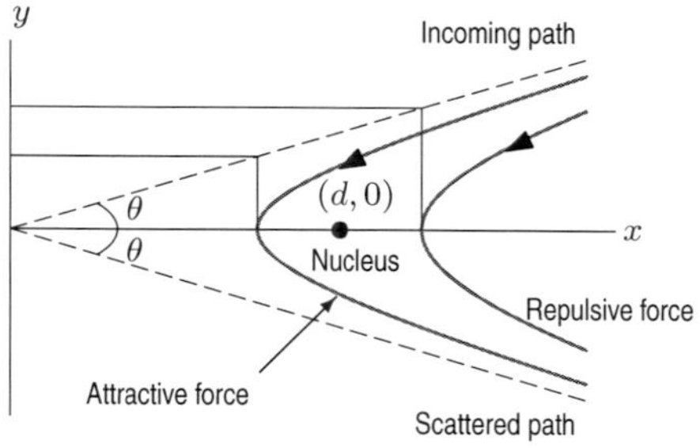

Figure 12.30

12.5 GEOMETRIC PROPERTIES OF CONIC SECTIONS

Having defined functions numerically, graphically, and algebraically, in this section we define the conic sections using geometry. The geometric properties of the conic sections lead to many applications in astronomy and engineering.

Using Geometry in Mathematical Models: Orbits

For centuries, natural philosophers tried to explain the motion of the planets. Over time, accounts based on the gods for whom the planets are named gave way to increasingly mathematical accounts. The seventeenth-century German astronomer Johannes Kepler believed nature could be explained mathematically, although today his approach would be considered mystical. He believed the planets orbited the sun within a geometrical framework of spheres nested within a cube, a pyramid, and other polyhedra.[3] After twenty years of attempting to fit a circle to the orbit of Mars, Kepler realized its orbit is not a circle, but an ellipse.

Other scientists have used geometry to build mathematical models. For instance, just as Kepler abandoned the perfect circle for the more complicated ellipse, Einstein abandoned perfectly flat Euclidean geometry in favor of the more complicated Riemannian geometry in his development of general relativity.

Orbits are Conic Sections

Although Kepler discovered that an orbit can take the shape of an ellipse, it was Newton's laws of motion and gravity that provided the explanation. Newton showed that the orbit of any object

[3] See http://en.wikipedia.org/wiki/Johannes_Kepler for more information. Page last accessed February 8, 2006.

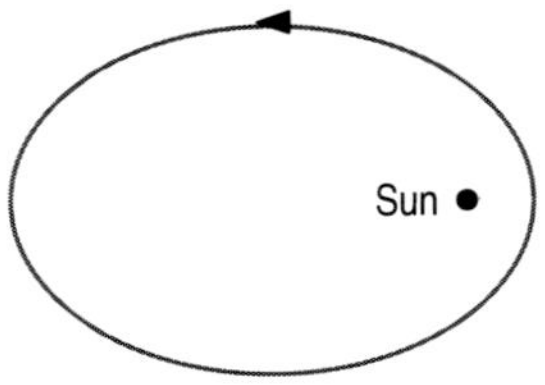

Figure 12.31: Halley's comet follows an elliptical orbit around the sun, like the one shown here

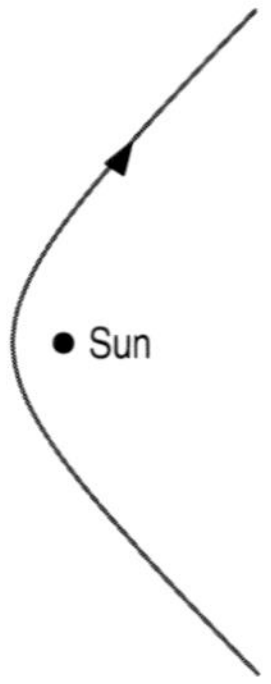

Figure 12.32: Some comets follow a hyperbolic trajectory, deflected by the sun but never to return

that cannot escape the gravitational pull of another object is an ellipse.[4] Planets in the solar system follow an elliptical orbit around the sun, satellites orbiting the earth follow an elliptical orbit around the earth, and many comets follow elliptical orbits through the solar system. The most famous of these is Halley's comet, which passes within sight of earth once approximately every 76 years.[5] See Figure 12.31.

In contrast, when a body is not held in an elliptical orbit by another object, its trajectory is hyperbolic.[6] Comets that follow hyperbolic orbits through the solar system have paths that are deflected as they pass the sun, but they do not return. See Figure 12.32.

Slicing a Cone

The ancient Greeks studied the curves formed when a plane cuts a cone. These curves are called *conic sections*, where *section* is another word for a cut (as in dissection).

Although conic sections get their name from slicing cones, the Greek mathematician and astronomer Apollonius of Perga gave alternative definitions of the conics using distances in the plane. His work on conics is considered one of the wonders of mathematics, and he wrote it without benefit of either Cartesian coordinates or algebra.

Geometric Definitions of Conics

In a perfectly circular orbit, the sun would be at the circle's center. We have seen two ways of defining a circle. On page 527, a circle of radius r centered at point (h, k) is defined parametrically by the equations

$$x = h + r\cos t \qquad y = k + r\sin t \qquad 0 \le t \le 2\pi.$$

On page 527, the same circle is defined by the equation

$$(x-h)^2 + (y-k)^2 = r^2.$$

[4]Unless, of course, the two objects collide.

[5]Halley's comet passed within sight of earth in 1986, and is predicted to return in 2061.

[6]In principle an orbit can be parabolic but in real systems such orbits are essentially impossible.

Circles: A Geometric Definition

These definitions of a circle are algebraic: they specify a set of points whose x- and y-coordinates satisfy certain equations. We can also give a geometric definition using the distance between points, not functions like squares, sines, and cosines. The geometric definition of the circle follows:

> A *circle* of radius r is the set of points in the plane a distance r from a given point.

Notice that unless we restrict our points to the plane, our definition gives a sphere of radius r.

Ellipses: A Geometric Definition

In an elliptical orbit, like that of Halley's comet, the sun is not located at the center, but at a point called a *focus*. Any ellipse (other than a circle) has two foci. These points are used to give the geometric definition:

> An *ellipse* is the set of points in the plane for which the sum of the distances to the foci is constant. [7]
> - Each of the two points is called a *focus* or *focal point* of the ellipse.
> - The two focal points lie on the major axis, and are equally spaced about the minor axis. The constant sum of distances is the length of the major axis.

No matter which point we pick on the ellipse, the sum of the distances from that point to the two focal points is the same. See Figures 12.33 and 12.34.

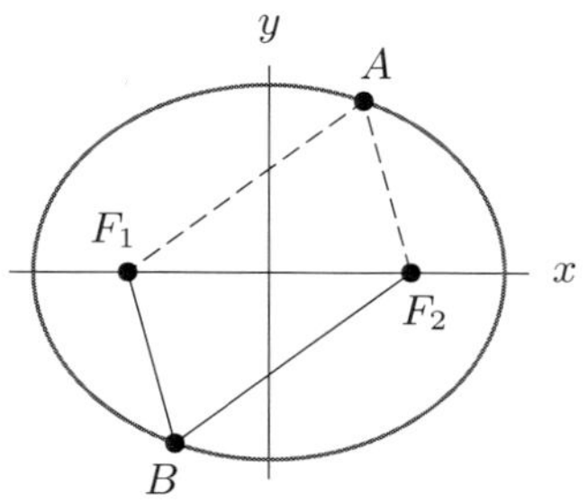

Figure 12.33: Sum of the distances from any point on the curve, such as A or B, to the foci, F_1 and F_2, is constant

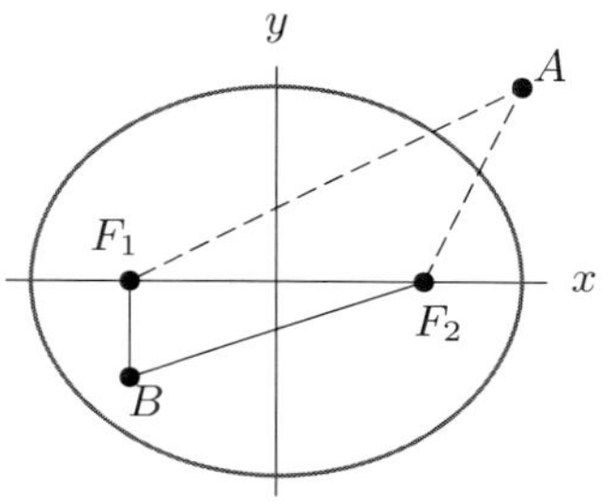

Figure 12.34: Sum of distances from points A and B, which do not lie on ellipse, to the foci, F_1 and F_2, is not constant

Example 1 Pluto's distance from the the sun ranges from 29.7 astronomical units (AU) at its closest point to 49.3 AU at its farthest. Center Pluto's elliptical orbit at the origin with the major axis on the x-axis.

(a) What is the length of the major axis?
(b) What are the coordinates of the two focal points?
(c) What is the length of the minor axis?
(d) Give an equation for this ellipse in rectangular coordinates.
(e) Give parametric equations for this ellipse.

[7] If the constant equals the distance between the two points, we get a degenerate ellipse.

Solution (a) In Figure 12.35, Pluto is closest to the sun at $(a, 0)$; Pluto is farthest from the sun at $(-a, 0)$. We see that

$$\text{Length of major axis} = 29.7 + 49.3 = 79 \text{ AU}.$$

(b) The two focal points are at $(c, 0)$ and $(-c, 0)$ in Figure 12.35. Half the length of the major axis is 39.5, so $a = 39.5$. Since a is 29.7 units from the closest focus, $c = 39.5 - 29.7 = 9.8$. The coordinates of the two focal points are $(9.8, 0)$ and $(-9.8, 0)$.

(c) When Pluto is at the point $(a, 0)$, the sum of its distances to the focal points is 79. Since the value of this sum is the same at every point on the ellipse, at the point $(0, b)$, the sum of the distances to the focal points is also 79. By symmetry, the distance from $(0, b)$ to the focus $(9.8, 0)$ is $79/2 = 39.5$. Using the Pythagorean Theorem, we have:

$$\begin{aligned}(9.8)^2 + b^2 &= (39.5)^2 \\ b^2 &= (39.5)^2 - (9.8)^2 \\ b &= 38.3.\end{aligned}$$

We see that

$$\text{Length of minor axis} = 2 \cdot 38.3 = 76.6 \text{ astronomical units.}$$

(d) In Section 12.3, we saw that the equation in rectangular coordinates for an ellipse centered at the origin is

$$\frac{x^2}{a^2} + \frac{y^2}{b^2} = 1.$$

For Pluto's orbit, we have $a = 39.5$ and $b = 38.3$, so the equation is

$$\frac{x^2}{(39.5)^2} + \frac{y^2}{(38.3)^2} = 1.$$

(e) Parametric equations for this ellipse are:

$$\begin{aligned}x &= 39.5 \cos t \\ y &= 38.3 \sin t.\end{aligned}$$

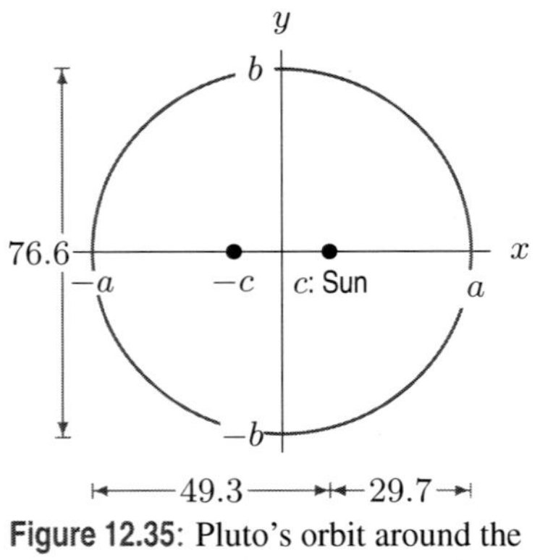

Figure 12.35: Pluto's orbit around the sun appears circular, but is actually an ellipse

In part (e) of Example 1, we give parametric equations for the orbit of Pluto. It would be natural to assume t stands for time. However, Kepler discovered that a planet travels fastest in its orbit when closest to the sun, and slowest when it is farthest away. Thus t cannot represent time, or the speeds would be the same at the closest and farthest points.

The Circular Appearance of Planetary Orbits

Notice in Example 1 that the length of the major axis in Pluto's orbit is not much longer than the length of the minor axis, making its orbit in Figure 12.35 look circular. Since Pluto's orbit is the most "squashed" of all the planetary orbits, it is easy to understand why Kepler assumed the orbit of Mars to be circular.

Example 2 By finding a formula for c in terms of a and b, determine the focal points $(\pm c, 0)$ for the ellipse where a and b are positive and $a > b$:

$$\frac{x^2}{a^2} + \frac{y^2}{b^2} = 1.$$

Solution The graph of this ellipse is in Figure 12.36. From the geometric definition of an ellipse, the sum of the distances to the two focal points is the same at every point on the curve. Thus

$$\begin{array}{c}\text{Sum of distances to}\\ \text{focal points from } (a,0)\end{array} = \begin{array}{c}\text{Sum of distances to}\\ \text{focal points from } (0,b)\end{array}.$$

On the left side of this equation, the distance from $(a, 0)$ to $(c, 0)$ is $a - c$ and the distance from $(a, 0)$ to $(-c, 0)$ is $a + c$, so

$$\begin{array}{c}\text{Sum of distances to}\\ \text{focal points from } (a,0)\end{array} = (a - c) + (a + c) = 2a.$$

Notice that, as before, the sum of distances to the foci is the length of the major axis, $2a$.

Using the distance formula, we see that the distances from $(0, b)$ to $(c, 0)$ and from $(0, b)$ to $(-c, 0)$ are both $\sqrt{b^2 + c^2}$, so

$$\begin{array}{c}\text{Sum of distances to}\\ \text{focal points from } (0,b)\end{array} = 2\sqrt{b^2 + c^2}.$$

Putting this together and solving for c gives

$$\begin{aligned} 2a &= 2\sqrt{b^2 + c^2} \\ a &= \sqrt{b^2 + c^2} \\ a^2 &= b^2 + c^2 \\ c^2 &= a^2 - b^2 \\ c &= \sqrt{a^2 - b^2}. \end{aligned}$$

Notice that applying the Pythagorean Theorem to the triangles in Figure 12.36 also gives $a = \sqrt{b^2 + c^2}$.

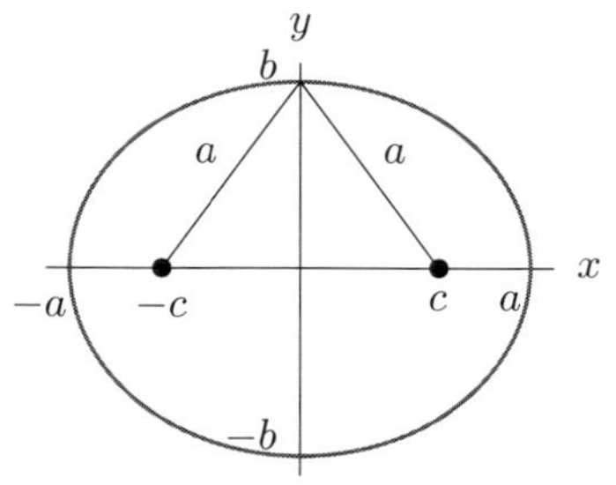

Figure 12.36: Locating the focal points

We have the following result:

Location of the Focal Points of an Ellipse
The focal points of the ellipse $x^2/a^2 + y^2/b^2 = 1$ with $a > b > 0$ are $(\pm c, 0)$ where

$$c = \sqrt{a^2 - b^2}.$$

The focal points lie on the major axis.
If $b > a$, the focal points are $(0, \pm c)$ with $c = \sqrt{b^2 - a^2}$.

Reflective Properties of Ellipses

The ellipse has a reflective property, which is used in applications involving light, sound, and shock waves. See Figure 12.37.

Reflective property of an ellipse
A ray that originates at one focus of an ellipse is reflected off the ellipse to the other focus.

This property of ellipses is used in lithotripsy, a medical procedure for treating kidney stones. The patient is placed in an elliptical tank of water with the kidney stone at one focus. High energy shock waves generated at the other focus are concentrated on the stone, breaking it into small bits. The reflective property is also used in the construction of "whispering rooms." A person standing and whispering at one focus in an elliptical room can be heard at the other focus, but not at places in between. John Quincy Adams reportedly discovered this property while in the House of Representatives, and situated himself at a focal point in order to eavesdrop on other House members.

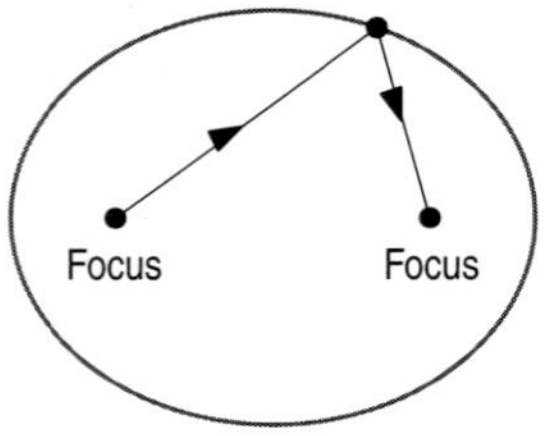

Figure 12.37: Reflective property of an ellipse

Example 3 An elliptical whispering gallery has a major axis of 100 m and minor axis of 40 m. Where should two speakers position themselves so as to most easily hear each other?

Solution Imagine the room centered at the origin with major axis on the x-axis. Then the positive x-intercept is $a = 50$ and the positive y-intercept is $b = 20$. The focal points are located at $c = \pm\sqrt{a^2 - b^2} = \pm\sqrt{50^2 - 20^2} = \pm 45.8$. The speakers should position themselves approximately 4.2 m in from the ends of the major axis.

Equivalence of the Algebraic and Geometric Definitions of the Ellipse

In Section 12.3, an ellipse was defined using rectangular coordinates as

$$\frac{x^2}{a^2} + \frac{y^2}{b^2} = 1.$$

In this section, we defined an ellipse is the set of points in the plane for which the sum of the distances to two focal points is constant. We now show that these two definitions are equivalent in the case in which the focal points are $(\pm c, 0)$ and the intercepts are $(\pm a, 0)$, with $0 < c < a$. Some of the details are omitted: see Problem 43.

Consider the set of points (x, y) such that the sum of the distances to two focal points $(\pm c, 0)$ is constant. We have

$$\text{Sum of distances from } (x,y) \text{ to focal points} = \text{Sum of distances from } (a,0) \text{ to focal points}$$
$$\sqrt{(x-c)^2+y^2}+\sqrt{(x+c)^2+y^2} = 2a$$
$$\sqrt{(x-c)^2+y^2} = 2a-\sqrt{(x+c)^2+y^2}.$$

Square both sides and simplify to obtain:

$$-4cx = 4a^2 - 4a\sqrt{(x+c)^2+y^2}.$$

Isolating the square root gives

$$a\sqrt{(x+c)^2+y^2} = a^2+cx.$$

We square both sides and simplify again to obtain:

$$a^2x^2+a^2c^2+a^2y^2 = a^4+c^2x^2.$$

Let $b = \sqrt{a^2-c^2}$. Then $b > 0$. Since $c^2 = a^2 - b^2$, simplifying again gives

$$-a^2b^2+a^2y^2 = -b^2x^2$$
$$b^2x^2+a^2y^2 = a^2b^2.$$

Dividing through by a^2b^2 gives the equation for an ellipse in rectangular coordinates

$$\frac{x^2}{a^2}+\frac{y^2}{b^2} = 1.$$

Hyperbolas

An ellipse is the set of points for which the sum of the distances to two points is constant. If the *difference* in the distances to two points is constant, we have a hyperbola. The two points are again called the focal points, or foci. See Figure 12.38.

> A *hyperbola* is the set of points in the plane for which the difference of the distances to two points is constant.[8] Each of the two points is called a *focus* or *focal point* of the hyperbola.

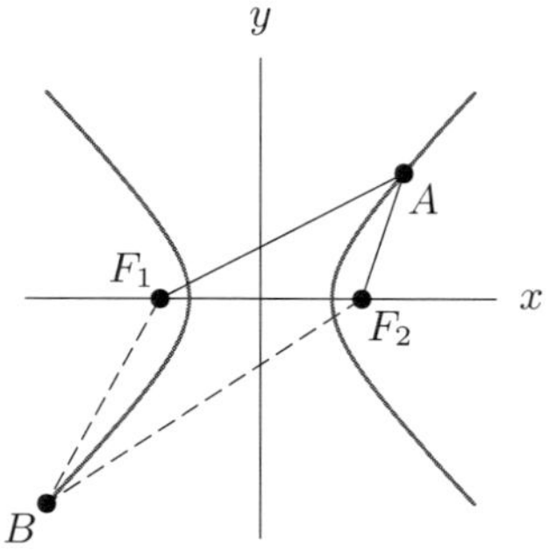

Figure 12.38: Difference between distances from any point on the curve, such as A or B, to the foci, F_1 and F_2, is constant

[8]We always subtract the shorter distance from the longer to get a positive constant. If the constant equals the distance between the two points, we get a degenerate hyperbola.

Hyperbolic orbits describe two objects that pass by each other but in which neither object "captures" the other. Hyperbolas have two parts, or *branches*, as in Figure 12.38, but a hyperbolic orbit only uses one branch. In the hyperbolic orbit of a comet, for example, the comet comes in on almost a straight line, gets bent around the sun due to the gravitational pull, and goes out on almost a straight line. The straight lines that hyperbolas approach as the curve gets farther from the focus are the *asymptotes* of the hyperbola.

We saw in Section 12.4 that the equation for a hyperbola that opens left and right and is centered at the origin is

$$\frac{x^2}{a^2} - \frac{y^2}{b^2} = 1.$$

For this hyperbola, the asymptotes are

$$y = \pm\frac{b}{a}x,$$

and the focal points can be shown to be $(\pm c, 0)$ with

$$c = \sqrt{a^2 + b^2}.$$

Location of the Focal Points of a Hyperbola
The focal points of the hyperbola given by equation $x^2/a^2 - y^2/b^2 = 1$ are $(\pm c, 0)$ where

$$c = \sqrt{a^2 + b^2}.$$

We have said that a hyperbola is the set of points for which the difference of the distances from two focal points is constant. Section 12.4 gave an equation for a hyperbola. The justification that these two definitions of a hyperbola are equivalent is similar to the justification for ellipses on page 542. See Problem 44.

Example 4 (a) Find the focal points and the asymptotes for the hyperbola given by

$$\frac{x^2}{9} - \frac{y^2}{16} = 1.$$

(b) Graph the hyperbola.
(c) For each of the two x-intercepts, find the difference of the distances to the two focal points.

Solution (a) We have $a = 3$ and $b = 4$. Since $c = \sqrt{a^2 + b^2} = \sqrt{9 + 16} = 5$, the two focal points are $(\pm 5, 0)$. The asymptotes are the two lines $y = \pm(b/a)x = \pm(4/3)x$.
(b) See Figure 12.39.
(c) The intercept $(3, 0)$ is distance 2 from one focus and distance 8 from the other focus. The difference of the distance to the two focal points is 6. Similarly, the difference of the distances to the intercept $(-3, 0)$ is also $8 - 2 = 6$.

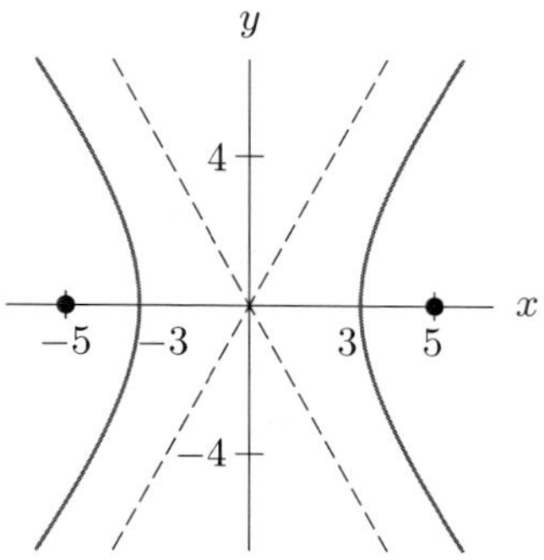

Figure 12.39: The hyperbola $x^2/9 - y^2/16 = 1$

Application of Hyperbolas: Constant Time Difference

LORAN (LOng RAnge Navigation) is a navigational system that uses low frequency radio transmitters to determine the location of a ship or aircraft. If two radio transmitters are transmitting signals, the time difference between the receipt of the signals by the ship or aircraft is measured. The ship or aircraft must be somewhere on the hyperbola with this constant difference. To determine the exact location, a third transmitter is needed. The constant time difference between this third transmitter and one of the first two gives a second hyperbola. The intersection of these two hyperbolas determines the location.

Sonic booms also involve hyperbolas. The cone created by the circular shock wave as its radius increases intersects the plane of the earth in a hyperbola.

Reflective Properties of Hyperbolas

Hyperbolas have a reflective property similar to that of ellipses. See Figure 12.40.

Reflective property of hyperbolas:
A beam aimed at the far focus behind the hyperbola is reflected off the curve to hit the focus in front of the hyperbola. Alternately, a beam originating from the focus in front of the hyperbola is reflected off the curve so that it appears to have originated from the focus behind the hyperbola.

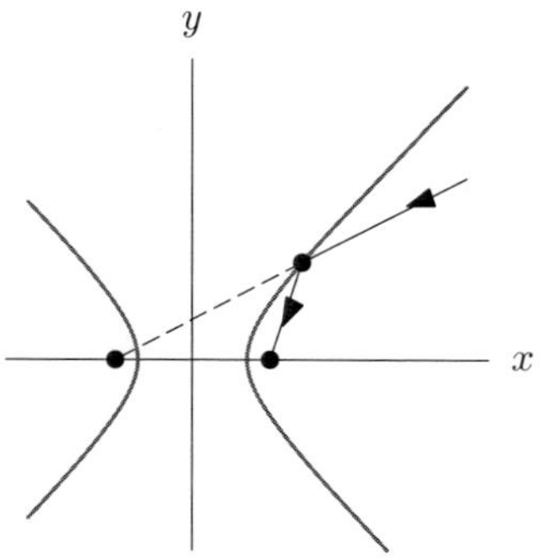

Figure 12.40: Reflective property of a hyperbola

Example 5 A hyperbolic mirror has equation

$$\frac{x^2}{16} - \frac{y^2}{9} = 1.$$

Find the equation of the line along which a beam of light could be sent out from point $A = (15, 50)$ to reflect off the hyperbola and arrive at the focus closest to point A.

Solution The focal points for this hyperbola are $(\pm c, 0)$ with $c = \sqrt{a^2 + b^2} = \sqrt{16 + 9} = 5$. The focal point closest to A is $(5, 0)$ and the focal point farthest away is $(-5, 0)$. Since we want the beam of light to reflect off the hyperbola and arrive at $(5, 0)$, we aim the beam at the other focal point, $(-5, 0)$. The line from point A to $(-5, 0)$ has slope

$$m = \frac{50 - 0}{15 - (-5)} = \frac{50}{20} = 2.5.$$

The equation of the line is

$$y = 2.5(x - (-5)) = 2.5x + 12.5.$$

A beam of light sent along the line $y = 2.5x + 12.5$ toward the point $(-5, 0)$ is reflected off the hyperbola to the point $(5, 0)$.

Parabolas

Similar to an ellipse and a hyperbola, a parabola can be defined in terms of the distance from a focal point. For a parabola, however, the definition involves one focal point, and a line called the *directrix*.

> A *parabola* is the set of points in the plane for which the distance from a point is equal to the distance from a fixed line. The point is called the *focus* and the line is called the *directrix* .

The focus lies on the axis of symmetry of the parabola and the directrix is perpendicular to the axis of symmetry. See Figure 12.41. If the focus is on the positive y-axis at $(0, c)$ and the directrix is the horizontal line is $y = -c$, Example 6 shows that the geometric definition gives a parabola.

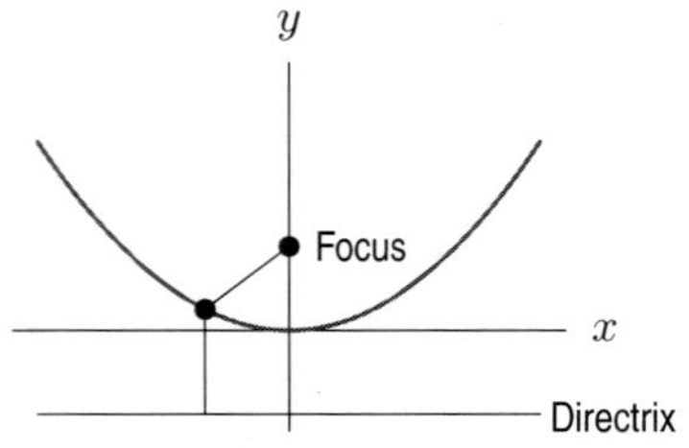

Figure 12.41: The focus and directrix of a parabola. Even though they may not look it, the two line segments are the same length

Example 6 For c a positive constant, find the equation for the set of points of equal distance from the point $(0, c)$ and a horizontal line $y = -c$..

Solution Let (x, y) be any point satisfying

$$\begin{aligned}
\text{Distance from } (x,y) \text{ to the point } (0,c) &= \text{Distance from } (x,y) \text{ to the line } y=-c \\
\sqrt{x^2+(y-c)^2} &= y+c \\
x^2+(y-c)^2 &= (y+c)^2 \\
x^2+y^2-2cy+c^2 &= y^2+2cy+c^2 \\
x^2 &= 4cy \\
y &= \frac{1}{4c}x^2.
\end{aligned}$$

Thus the set of points is a parabola through the origin.

Summarizing Example 6, we have the following result:

> **Location of the Focus and Directrix of a Parabola**
> The focus of the parabola $y = ax^2$ is the point $(0, c)$ where $c = 1/(4a)$.
> The directrix is the line $y = -c$.

Example 7 (a) Find the focus and the directrix of the parabola $y = 3x^2$.
(b) Show that the point $(1, 3)$ on the parabola is equidistant from the focus and the directrix.

Solution (a) We have $a = 3$ so the focus is at the point $(0, 1/12)$ and the directrix is the line $y = -1/12$.
(b) We have

$$\begin{aligned}
\text{Distance from } (1,3) \text{ to the focus } (0,1/12) &= \sqrt{1^2+(3-1/12)^2} \\
&= \sqrt{\frac{12^2+35^2}{12^2}} \\
&= \frac{37}{12}.
\end{aligned}$$

The distance from $(1, 3)$ to the directrix $y = -1/12$ is the difference in the y-values, since the directrix is a horizontal line:

$$\begin{aligned}
\text{Distance from } (1,3) \text{ to directrix } &= 3+\frac{1}{12} \\
&= \frac{37}{12}.
\end{aligned}$$

Thus the distance from the point $(1, 3)$ to the focus equals the distance from the point $(1, 3)$ to the directrix. This property holds for all points (x, y) on the parabola.

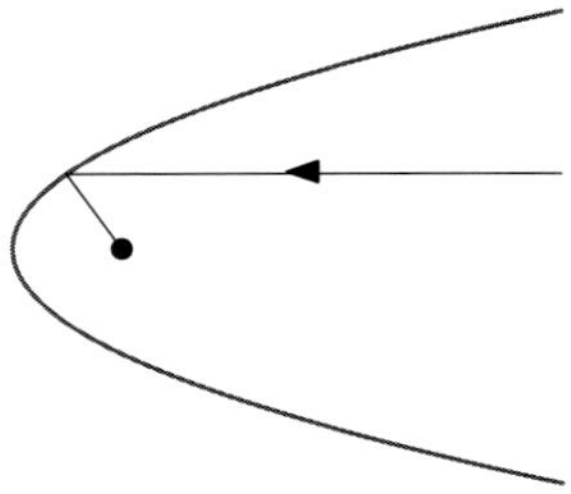

Figure 12.42: Reflective property of a parabola

Reflective Properties of Parabolas

Parabolas also have the reflective property illustrated in Figure 12.42.

> **Reflective property of parabolas**
> All beams parallel to the axis of symmetry reflect off the parabola to the focus.

Archimedes discovered this reflective property of parabolas in the third century BC, and legend has it that he constructed parabolic mirrors to concentrate the sun's rays at a single point and set fire to enemy targets.[9]

A solar furnace produces heat by focusing sunlight using parabolic mirrors. In fact, the word *focus* comes from Latin for "fireplace." Giant parabolic mirrors are used in telescopes to focus light and radio waves from outer space.

The reflective property of parabolas also works in reverse. A light placed at the focus of a parabolic mirror sends out light rays that are reflected out in a beam parallel to the axis of symmetry of the parabola. Car headlights have parabolic reflectors with the lightbulb at the focus.

Example 8 The parabolic reflector behind a car head light is 16 cm across and 3 cm deep. Where should the bulb be positioned?

Solution The shape of the parabolic reflector is made by rotating a parabola $y = ax^2$ about the y-axis; see Figure 12.43. Since the rim of the reflector lies on the parabola, the point $(8, 3)$ satisfies the equation $y = ax^2$ so $3 = a(8^2)$ and $a = 3/64$. The focus is located at:

$$c = \frac{1}{4a} = \frac{1}{4(3/64)} = \frac{16}{3}.$$

The bulb should be placed $16/3$ cm from the vertex of the parabola along the axis of symmetry.

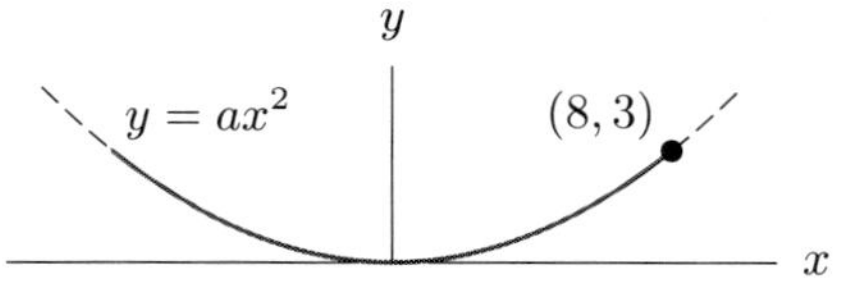

Figure 12.43: Car headlight

[9]The Discovery Channel's *Mythbusters* program has questioned the feasibility of such a strategy, though experiments conducted at MIT have demonstrated its potential (see http://web.mit.edu/2.009/www/lectures/10_ArchimedesResult.html, page last accessed February 27, 2006). However, the MIT experiments suggest that a number of carefully arranged small, flat mirrors might have been more practical than one large parabolic mirror.

Exercises and Problems for Section 12.5

Exercises

1. A circle is centered at the origin and includes the point $(-1, 3)$. Find the radius.

2. A circle is centered at the origin and includes the point $(1, 2)$. At what points does the circle intersect the vertical axis?

3. An ellipse with horizontal and vertical axes is centered at the origin and includes the point $(3, 2)$. Which of the following points are also on any such ellipse?

(a) $(-3, 2)$ **(b)** $(3, -2)$
(c) $(-3, -2)$ **(d)** $(2, 3)$

4. An ellipse is centered at the origin and has one focal point at $(0, 2)$. Find the other focal point and say whether the major axis is horizontal or vertical.

5. For the ellipse $\frac{x^2}{9} + \frac{y^2}{16} = 1$, find the focal points.

In Exercises 6–10, identify the conic and identify the major axis if an ellipse or else the line of symmetry through the vertex or vertices.

6. $\frac{x^2}{5} + \frac{y^2}{3} = 1$ **7.** $\frac{x^2}{5} - \frac{y^2}{3} = 1$

8. $\frac{x^2}{2} - \frac{y^2}{3} = -1$ **9.** $\frac{x}{2} + \frac{y^2}{3} = 1$

10. $y^2 = 1 + \frac{x^2}{3}$

11. A hyperbola with foci on one of the axes is centered at the origin and includes the point $(1, 2)$. Which of the following points are also on any such hyperbola?

(a) $(-1, 2)$ **(b)** $(1, -2)$
(c) $(-1, -2)$ **(d)** $(2, 1)$

12. A hyperbola is centered at the origin and has one focal point at $(0, -3)$. Find the other focal point and the axis through the vertices.

13. For the hyperbola $x^2 - y^2 = 1$, find the focal points.

14. A parabola is centered at the origin, opens downward, and includes the point $(-1, -2)$. Which of the following points are also on any such parabola?

(a) $(1, 2)$ **(b)** $(1, -2)$
(c) $(-1, 2)$ **(d)** $(2, 1)$

15. Find the equation of a parabola centered at the origin and having a focal point at $(0, -3)$.

16. Find the equation of a parabola centered at the origin and having a focal point at $(-1, 0)$.

Problems

For Problems 17–19, find the vertex and focal point of the parabola.

17. $(x + 2)^2 = 5(y - 1)$ **18.** $y^2 = \frac{1}{4}(x + 2)$

19. $y = x^2 - x + 1$

20. In Figure 12.44, where F is the focus, find the equation of the parabola and its directrix.

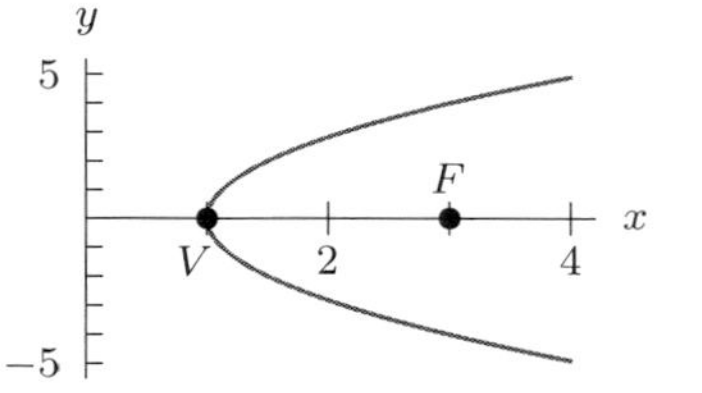

Figure 12.44

21. In Figure 12.45, find the equation of the parabola, its focus, and its directrix.

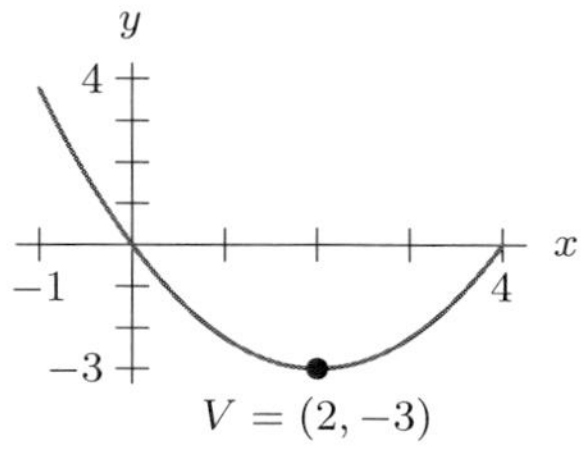

Figure 12.45

For Problems 22–24, find the center and focal points of the ellipse.

22. $\frac{x^2}{10} + \frac{y^2}{10} = 1$ **23.** $(y - 3)^2 + 25x^2 = 25$

24. $x^2 + 2y^2 + 4x - 12y - 3 = 0$

25. In Figure 12.46, find the equation of the ellipse and give its focal points.

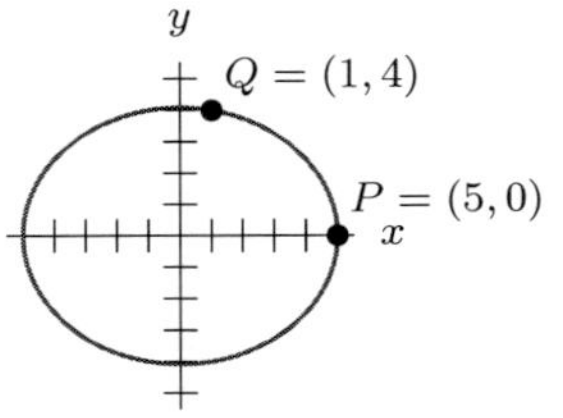

Figure 12.46

26. In Figure 12.47, find the equation of the ellipse. The foci are at $F_1 = (0, 1)$ and $F_2 = (0, 3)$.

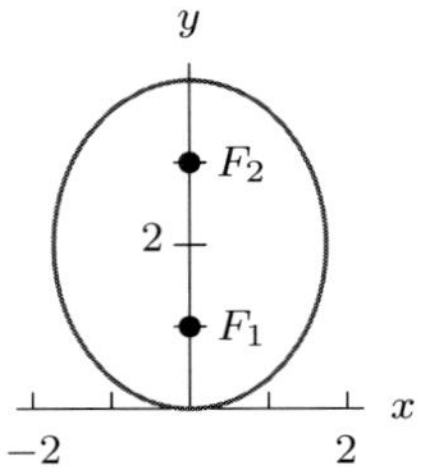

Figure 12.47

For Problems 27–29, find the vertices and focal points of the hyperbola.

27. $\dfrac{(y+2)^2}{25} - \dfrac{(x-2)^2}{4} = 1$

28. $x^2 - \dfrac{y^2}{2} = 8$

29. $4y^2 - x^2 + 56y - 8x + 100 = 0$

In Problems 30–31, find the equation and focal points of the hyperbola.

30.

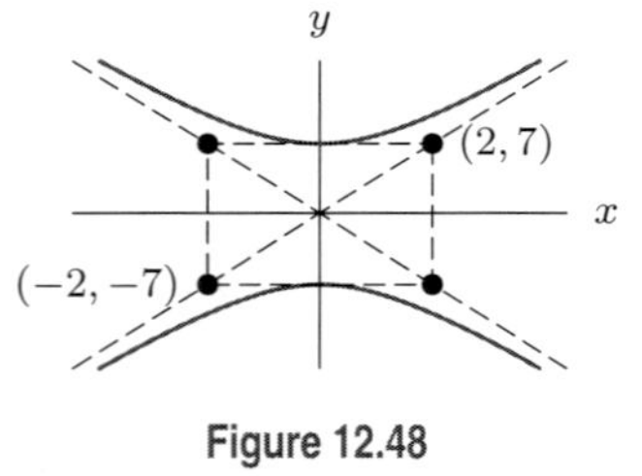

Figure 12.48

31.

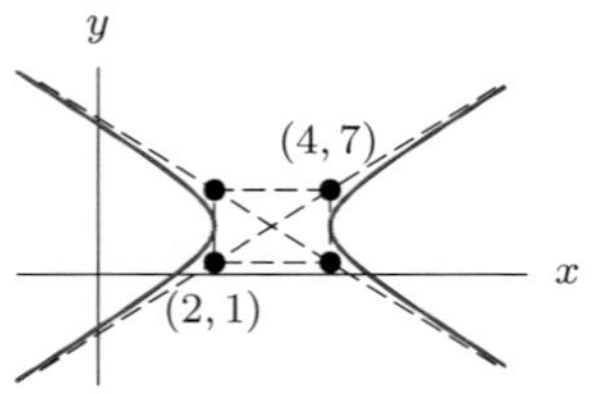

Figure 12.49

32. On an elliptical billiard table, a ball shot in any direction starting at a focal point and bouncing off a rail passes through the other focal point. What happens if the ball continues and bounces off a rail again?

33. Each end of a string is tacked to the points $(-2, 0)$ and $(2, 0)$ on centimeter grid paper. The string is 10 cm long. A pencil pulls the string taut and rotates to sweep out an ellipse. Find the equation.

34. A laser beam is aimed straight down into an upward facing hyperbolic mirror. If the beam is not aimed at the center then is it possible for a reflecting ray to pass through the near focal point?

35. A laser beam is aimed straight down into an upward facing parabolic mirror. Is it possible for a reflecting ray to pass through the focal point?

36. Satellite television receivers use a dish with an arm extending over the center of the dish, as shown in Figure 12.50. The shape of the dish is made by rotating a parabola, $y = ax^2$ about the y-axis. (This three-dimensional shape is called a paraboloid.) If a receiver dish is 24 inches across and 4 inches deep, how far above the center of the dish should the receiving end of the arm be positioned so that it is at the focus to receive the signal?

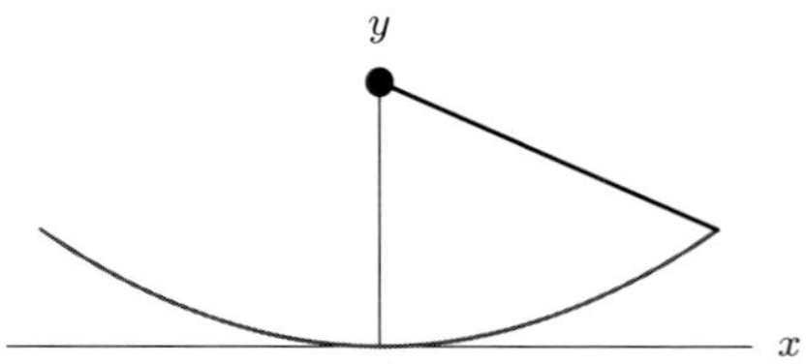

Figure 12.50

37. Satellite television receivers use a dish with a receiving arm extending over the center of the dish. The depth of modern dishes has decreased. If the depth of a dish with a given diameter is halved, what is the effect on the distance of the focal point from the center of the dish?

38. A searchlight in Los Angeles uses a 6 ft wide parabolic mirror that is 9 inches deep, where should the lamp be positioned to create a 6 ft wide beam to the sky?

39. The earth moves in an elliptical orbit with the sun as a focus. The nearest it gets to the sun is 146 million km and the farthest is 152 million km.

(a) The sun is at a focal point. Find the length of the major axis.

(b) Find an equation for earth's orbit if the sun is at the origin, with the major axis on x-axis.

40. Halley's Comet moves in an elliptical orbit with the Sun at a focus. The nearest it gets to the sun is 88 million km and the farthest is 5250 million km.

(a) Find the length of the major axis.
(b) Find an equation for its orbit if the sun is at the origin, with the major axis on x-axis.
(c) Give the parametric equations for the orbit of Halley's Comet.

41. A lamp with a shade casts a shadow described by a conic equation. With the bottom of the upright shade just touching the wall, consider an xy-axis on the wall, with the horizontal x-axis at the height of the lightbulb and the vertical y-axis touching the bottom of the shade. The equation for the shadow above the shade is

$$\frac{y^2}{(R_b A_t/R_t)^2} - \frac{x^2}{R_b^2} = 1$$

where R_b is the bottom radius of the shade, R_t is the top radius of the shade, and A_t is the vertical distance from the bulb to the level of the top rim. The equation for the shadow below the shade is

$$\frac{y^2}{A_b^2} - \frac{x^2}{R_b^2} = 1$$

where R_b is the bottom radius of the shade, and A_b is the vertical distance from the bulb to the level of the bottom rim.[10] See Figure 12.51.

(a) Identify the conic from the equations.
(b) Given that a specific lamp shade is 8 inches across the top, 16 inches across the bottom and 14 inches high, find the two implicit equation for its upper and lower shadows if the bulb is centered in the lamp.
(c) Solve each of the two equations for y and give an explicit equation in terms of x for each shadow.

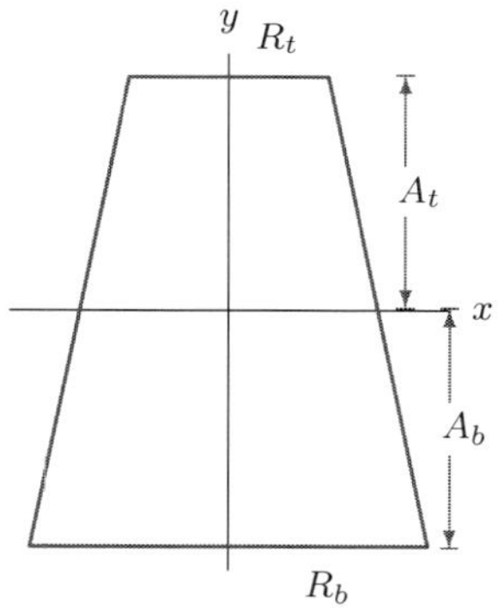

Figure 12.51

[10]Adapted from *The Physics Teacher*, March 2001.

42. The use of the LORAN navigation system is currently diminished by the use of the GPS system, but LORAN continues being used as a backup system to GPS. The technique for finding your position is to measure the time for a wave to be sent and returned from three transmitters. Using a pair of transmitters as foci it is possible to find a hyperbolic curve on which your position coordinates lie. Using a second pair of foci, a second curve is found and your position is the intersection of the two hyperbolic curves.

We use the property that the difference between the distance from two foci is constant for a hyperbola. However, for convenience in this exercise, we replace time measure with a distance measure. Suppose that you are at position P in Figure 12.52 and that the three transmitters lie along the vertical axis as shown.

(a) Find the equation of a hyperbola that passes through P and has foci T_1 and T_2.
(b) Find the equation of a hyperbola that passes through P and has foci T_1 and T_3.
(c) Solve for y in the equations you found in parts (a) and (b). In each case choose the branch that passes through P.
(d) Use a calculator or computer to graph the branches of the hyperbolas from part (c).

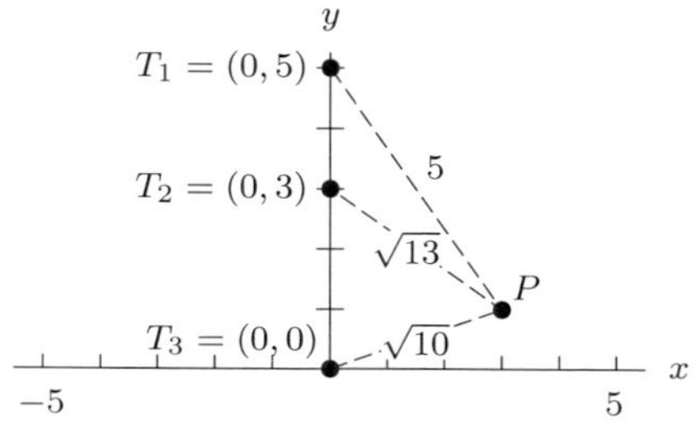

Figure 12.52

43. On page 542, we outline an argument to show that the set of points (x, y) satisfying the condition "the sum of the distances to two focal points $(\pm c, 0)$ is constant" is an ellipse. Fill in all the details of this argument. Assume that the x-intercepts are at $(\pm a, 0)$, where $a > c > 0$.

44. Show that the set of points (x, y) satisfying the condition that "the difference of the distances to two focal points $(\pm c, 0)$ is constant" is a hyperbola. Assume the intercepts are at $(\pm a, 0)$ with $0 < a < c$. [Hint: The justification in this problem is similar to the justification for ellipses on page 542.]

12.6 HYPERBOLIC FUNCTIONS

There are two combinations of the functions e^x and e^{-x} used so often in engineering that they are given their own names. They are the hyperbolic sine, abbreviated sinh (pronounced "cinch") and hyperbolic cosine, abbreviated cosh (rhymes with "gosh"). They are defined as follows:

Hyperbolic Sine and Hyperbolic Cosine

$$\cosh x = \frac{e^x + e^{-x}}{2} \qquad \sinh x = \frac{e^x - e^{-x}}{2}$$

Properties of Hyperbolic Functions

Figures 12.53 and 12.54 show graphs of $\cosh x$ and $\sinh x$ together with graphs of $\frac{1}{2}e^x$ and $\frac{1}{2}e^{-x}$.

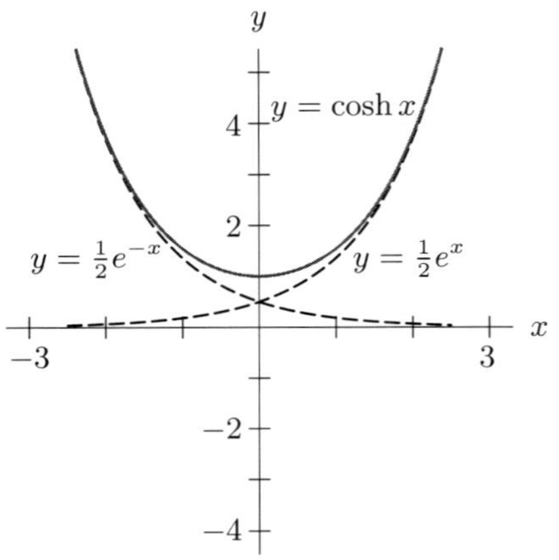

Figure 12.53: Graph of $y = \cosh x$ with multiples of e^x and e^{-x}

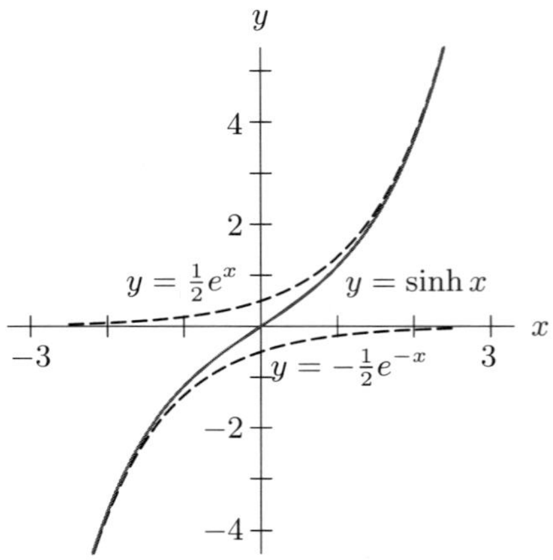

Figure 12.54: Graph of $y = \sinh x$ with multiples of e^x and e^{-x}

The graphs suggest that $\cosh x$ is even and $\sinh x$ is odd and that $\cosh x$ has a y-intercept of 1, whereas $\sinh x$ has a y-intercept of 0:

$$\cosh 0 = 1 \qquad \sinh 0 = 0$$
$$\cosh(-x) = \cosh x \qquad \sinh(-x) = -\sinh x$$

To show that the hyperbolic functions really do have these properties, we use their formulas.

Example 1 Confirm that (a) $\cosh 0 = 1$ (b) $\cosh(-x) = \cosh x$

Solution (a) Substituting $x = 0$ into the formula for $\cosh x$ gives the y-intercept:

$$\cosh 0 = \frac{e^0 + e^{-0}}{2} = \frac{1+1}{2} = 1.$$

(b) Substituting $-x$ for x gives

$$\cosh(-x) = \frac{e^{-x} + e^{-(-x)}}{2} = \frac{e^{-x} + e^{x}}{2} = \cosh x.$$

Thus, $\cosh x$ is an even function.

Example 2 Describe and explain the behavior of $\cosh x$ as $x \to \infty$ and then as $x \to -\infty$.

Solution From Figure 12.53, we see that as $x \to \infty$, the graph of $\cosh x$ resembles the graph of $\frac{1}{2}e^x$. Similarly, as $x \to -\infty$, the graph of $\cosh x$ resembles the graph of $\frac{1}{2}e^{-x}$. Since $e^{-x} \to 0$ as $x \to \infty$ and $e^x \to 0$ as $x \to -\infty$, we can predict these results algebraically:

$$\text{As } x \to \infty, \qquad \cosh x = \frac{e^x + e^{-x}}{2} \to \frac{1}{2}e^x.$$

$$\text{As } x \to -\infty, \qquad \cosh x = \frac{e^x + e^{-x}}{2} \to \frac{1}{2}e^{-x}.$$

Figures 12.53 and 12.54 suggest that the graph of $\cosh x$ is always above the graph of $\sinh x$. In the following example, we use the formulas to confirm this.

Example 3 Show that $\cosh x > \sinh x$ for all x.

Solution The difference between $\cosh x$ and $\sinh x$ is given by

$$\cosh x - \sinh x = e^{-x}.$$

Since e^{-x} is positive for all x, we see that

$$\cosh x > \sinh x \quad \text{for all } x.$$

Identities Involving cosh x and sinh x

The hyperbolic functions have names that remind us of the trigonometric functions because they have some properties that are similar to those of the trigonometric functions. Consider the expression

$$(\cosh x)^2 - (\sinh x)^2.$$

This can be simplified using the fact that $e^x \cdot e^{-x} = 1$:

$$\begin{aligned}(\cosh x)^2 - (\sinh x)^2 &= \left(\frac{e^x + e^{-x}}{2}\right)^2 - \left(\frac{e^x - e^{-x}}{2}\right)^2 \\ &= \frac{e^{2x} + 2e^x \cdot e^{-x} + e^{-2x}}{4} - \frac{e^{2x} - 2e^x \cdot e^{-x} + e^{-2x}}{4} \\ &= \frac{e^{2x} + 2 + e^{-2x} - e^{2x} + 2 - e^{-2x}}{4} \\ &= 1.\end{aligned}$$

Thus, writing $\cosh^2 x$ for $(\cosh x)^2$ and $\sinh^2 x$ for $(\sinh x)^2$, we have the identity

$$\cosh^2 x - \sinh^2 x = 1.$$

This identity is reminiscent of the Pythagorean identity $\cos^2 x + \sin^2 x = 1$. Extending the analogy, we define

Hyperbolic Tangent

$$\tanh x = \frac{\sinh x}{\cosh x}.$$

Parameterizing the Hyperbola Using Hyperbolic Functions

Consider the curve parameterized by the equations

$$x = \cosh t, \quad y = \sinh t, \quad -\infty < t < \infty.$$

The parameter, t, can be eliminated using the identity $\cosh^2 t - \sinh^2 t = 1$. This gives

$$x^2 - y^2 = \underbrace{\cosh^2 t}_{x^2} - \underbrace{\sinh^2 t}_{y^2} = 1,$$

which is the implicit equation for the unit hyperbola in Figure 12.56:

$$x^2 - y^2 = 1.$$

Notice that since $\cosh t > 0$, we have $x > 0$, so the parameterization gives only the right branch of the hyperbola. At $t = 0$, we have $(x, y) = (\cosh 0, \sinh 0) = (1, 0)$, so the curve passes through the point $(1, 0)$. As $t \to \infty$, both $\cosh t$ and $\sinh t$ approach $\frac{1}{2}e^t$, so for large values of t we see that $\cosh t \approx \sinh t$. Thus, as t increases the curve draws close to the diagonal line $y = x$. Since $\sinh x < \cosh x$ for all x, we know that $y < x$, so the curve approaches this asymptote from below. See Figure 12.55.

Similarly, as $t \to -\infty$, we know that $x \to \frac{1}{2}e^{-t}$ and $y \to -\frac{1}{2}e^{-t}$, which means that $y \approx -x$. Thus, as $t \to -\infty$, the curve draws close to the asymptote $y = -x$. In this case, the curve approaches the asymptote from above, because the y-coordinate is slightly larger (less negative) than the x-coordinate.

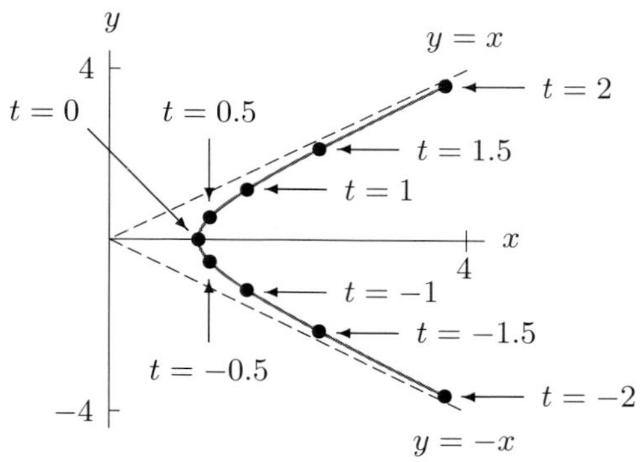

Figure 12.55: The hyperbola parameterized by $x = \cosh t,\ y = \sinh t,\ -\infty < t < \infty$

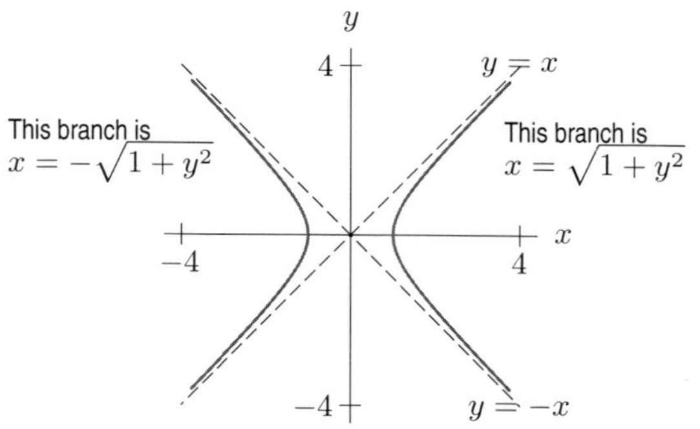

Figure 12.56: The hyperbola defined implicitly by $x^2 - y^2 = 1$

In addition to the parameterization, we can write an equation for each branch of the hyperbola. Solving the equation $x^2 - y^2 = 1$ for x, we obtain two values of x for every value of y:

$$\begin{aligned} x^2 - y^2 &= 1 \\ x^2 &= 1 + y^2 \\ x &= \pm\sqrt{1+y^2}. \end{aligned}$$

The right branch has equation $x = \sqrt{1+y^2}$; see Figure 12.55. The left branch is the graph of $x = -\sqrt{1+y^2}$, which is a reflection across the y-axis of the graph of $x = \sqrt{1+y^2}$. The two branches make up a single hyperbola.

Exercises and Problems for Section 12.6

Exercises

1. Show that $\sinh(-x) = -\sinh(x)$.

2. Show that $\sinh 0 = 0$.

3. Using the method of Example 2 on page 553, describe and explain the behavior of $\sinh x$ as $x \to \infty$ and as $x \to -\infty$.

In Exercises 4–13, write a parameterization for one branch of the hyperbola using hyperbolic functions.

4. $y^2 - x^2 = 1, \quad y > 0$

5. $y^2 - x^2 = 1, \quad y < 0$

6. $x^2 - y^2 = 1, \quad x < 0$

7. $(x-2)^2 - (y-3)^2 = 1, \quad x > 2$

8. $\dfrac{(y+1)^2}{9} - \dfrac{(x-1)^2}{4} = 1, \quad y < -1$

9. $\dfrac{(x-1)^2}{4} - \dfrac{(y+1)^2}{9} = 1, \quad x > 1$

10. $9(y+3)^2 - 4(x+1)^2 = 1, \quad y < -3$

11. $4(x+3)^2 - (y-1)^2 = 1, \quad x > -3$

12. $9(y+3)^2 - 4(x+1)^2 = 36, \quad y > -3$

13. $12(x-1)^2 - 3(y+2)^2 = 6, \quad x > 1$

Problems

In Problems 14–17, complete the square to write a parameterization for one branch of the hyperbola in terms of hyperbolic functions.

14. $x^2 + 2x - 4y^2 + 8y = 7, \quad x < -1$

15. $x^2 - 2x = y^2 - 4y + 4, \quad x > 1$

16. $25 + 2x^2 - 12x = 4y^2 + 4y, \quad y > -\frac{1}{2}$

17. $y^2 - 4x^2 + 8x = 12, \quad y < 0$

18. This problem concerns the parameterization of the unit hyperbola $x^2 - y^2 = 1$ given by

$$x = \sec t \quad \text{and} \quad y = \tan t \quad \text{for} \quad 0 \le t \le 2\pi.$$

Show on a graph of the hyperbola which section of the hyperbola is traced out and in which direction for

(a) $0 \le t < \pi/2$ **(b)** $\pi/2 < t \le \pi$
(c) $\pi \le t < 3\pi/2$ **(d)** $3\pi/2 < t \le 2\pi$

19. Show how to use hyperbolic functions to parameterize the hyperbola

$$\frac{(x-h)^2}{a^2} - \frac{(y-k)^2}{b^2} = 1.$$

20. Is there an identity analogous to $\sin 2x = 2\sin x \cos x$ for the hyperbolic functions? Explain.

21. Is there an identity analogous to $\cos 2x = \cos^2 x - \sin^2 x$ for the hyperbolic functions? Explain.

22. Consider the family of functions $y = a\cosh(x/a)$ for $a > 0$. Sketch graphs for $a = 1, 2, 3$. Describe in words the effect of increasing a.

23. Show that $\sinh(ix) = i\sin x$.

24. Find an expression for $\cos(ix)$ in terms of $\cosh x$.

25. Find an expression for $\sin(ix)$ in terms of $\sinh x$.

CHAPTER SUMMARY

- **Parametric Equations**
  Describing motion; parameterizing curves.
- **Implicitly Defined Curves**
  Implicit and parametric equations.
  Circle: $(x-h)^2 + (y-k)^2 = r^2$

  $$x = h + r\cos t; y = k + r\sin t.$$

  Ellipse: $$\frac{(x-h)^2}{a^2} + \frac{(y-k)^2}{b^2} = 1$$

  $$x = h + a\cos t; y = k + b\sin t.$$

  Hyperbola: $$\frac{(x-h)^2}{a^2} - \frac{(y-k)^2}{b^2} = 1$$

  $$x = h + a\sec t; y = k + b\tan t.$$

- **Geometric and Reflective Properties of Conic Sections**
  Circle: Distance to a fixed point is constant
  Ellipse: Sum of distances to two fixed points is constant; Ray from one focus reflects to the other.
  Hyperbola: Difference of distances to two fixed points is constant; Ray aimed at one focus will be reflected to the other.
  Parabola: Distance to a focus equals distance to the directrix line; Ray from a focus reflects to a line parallel to the axis of symmetry.
- **Hyperbolic Functions**
  Formulas: $\cosh x = \dfrac{e^x + e^{-x}}{2}$; $\sinh x = \dfrac{e^x - e^{-x}}{2}$.
  Properties; identities: $\cosh^2 x - \sinh^2 x = 1$; parameterizing the hyperbola.

REVIEW EXERCISES AND PROBLEMS FOR CHAPTER TWELVE

Exercises

Write a parameterization for the curves in Exercises 1–8.

1. A circle of radius 3 centered at the origin and traced clockwise.

2. A circle of radius 5 centered at the point $(2, 1)$ and traced counterclockwise.

3. A circle of radius 2 centered at the origin traced clockwise starting from $(-2, 0)$ when $t = 0$.

4. The circle of radius 4 centered at the point $(4, 4)$ starting on the x-axis when $t = 0$.

5. An ellipse centered at the origin and crossing the x-axis at ± 5 and the y-axis at ± 7.

6. A hyperbola crossing the x-axis at $x = \pm 1$ and with asymptotes $y = \pm 2x$.

7. An ellipse centered at the origin, crossing the x-axis at ± 3 and the y-axis at ± 7. Trace counterclockwise, starting from $(-3, 0)$.

8. A hyperbola crossing the y-axis at $(0, 7)$ and $(0, 3)$ and with asymptotes $y = \pm 4x + 5$.

In Exercises 9–16, decide if the equation represents a circle, an ellipse, or a hyperbola. Give the center and the radius (for a circle) or the values of a and b (for an ellipse or hyperbola) and whether it opens left-right or up-down (for a hyperbola).

9. $x^2 + (y-3)^2 = 5$

10. $7 - 2(x+1)^2 = 2(y-4)^2$

11. $\dfrac{x^2}{4} - \dfrac{(y-1)^2}{9} = 1$

12. $(x+1)^2 = 1 - \dfrac{(y+2)^2}{8}$

13. $9(x-5)^2 + 4y^2 = 36$

14. $63 + 9x^2 = 7(y-1)^2$

15. $6 + 2(x+\frac{1}{3})^2 - 3(y-\frac{1}{2})^2 = 0$

16. $\dfrac{4-2(x-1)^2}{(y+2)^2} = 1$

17. For the ellipse $\dfrac{x^2}{25} + \dfrac{y^2}{4} = 1$, find the focal points.

18. For the hyperbola $y^2 - x^2 = 5$, find the focal points.

19. For the parabola $y = 5x^2$ find the focal point and directrix line.

Problems

Write a parameterization for the curves in Problems 20–22.

20. The horizontal line through the point $(0, 5)$.

21. The circle of radius 1 in the xy-plane centered at the origin and traversed counterclockwise.

22. The circle of radius 2 centered at the origin, starting at the point $(0, 2)$ when $t = 0$.

By completing the square, put the equations in Problems 23–26 into one of the following forms

$$\frac{(x-h)^2}{a^2} + \frac{(y-k)^2}{b^2} = 1$$

or

$$\frac{(x-h)^2}{a^2} - \frac{(y-k)^2}{b^2} = 1$$

or

$$\frac{(y-k)^2}{b^2} - \frac{(x-h)^2}{a^2} = 1$$

and decide if it represents a circle, an ellipse, or a hyperbola. Give its center, radius (for a circle), or the values of a and b (for an ellipse or a hyperbola), and whether it opens left-right or up-down (for a hyperbola).

23. $x^2 + 2x + y^2 = 0$

24. $2x^2 + 2y = y^2 + 4x$

25. $6x^2 - 12x + 9y^2 + 6y + 1 = 0$

26. $\dfrac{x^2 - 4x}{2y - y^2} - 1 = 0$

27. On a graphing calculator or a computer, plot $x = 2t/(t^2+1)$, $y = (t^2-1)/(t^2+1)$, first for $-50 \le t \le 50$, and then for $-5 \le t \le 5$. Explain what you see. Is the curve really a circle?

28. Plot the Lissajous figure given by $x = \cos 2t$, $y = \sin t$ using a graphing calculator or computer. Explain why it looks like part of a parabola. [Hint: Use a double angle identity.]

29. A planet P in the xy-plane orbits the star S counterclockwise in a circle of radius 10 units, completing one orbit in 2π units of time. A moon M orbits the planet P counterclockwise in a circle of radius 3 units, completing one orbit in $2\pi/8$ units of time. The star S is fixed at the origin $x = 0$, $y = 0$, and at time $t = 0$ the planet P is at the point $(10, 0)$ and the moon M is at the point $(13, 0)$.

(a) Find parametric equations for the x- and y-coordinates of the planet at time t.
(b) Find parametric equations for the x- and y-coordinates of the moon at time t. [Hint: For the moon's position at time t, take a vector from the sun to the planet at time t and add a vector from the planet to the moon].
(c) Plot the path of the moon in the xy-plane, using a graphing calculator or computer.

30. A hyperbolic mirror is shown in Figure 12.57, and a beam needs to be directed from point P at $(8, 8)$ to point Q at $(0, 2)$. Because of the barrier shown in the figure, the direct path is obstructed. The hyperbola is defined by $3y^2 - x^2 = 3$.

(a) Find the foci of the hyperbola.
(b) Find the equation of the line through P along which a beam should be sent for it to be reflected to the point Q.

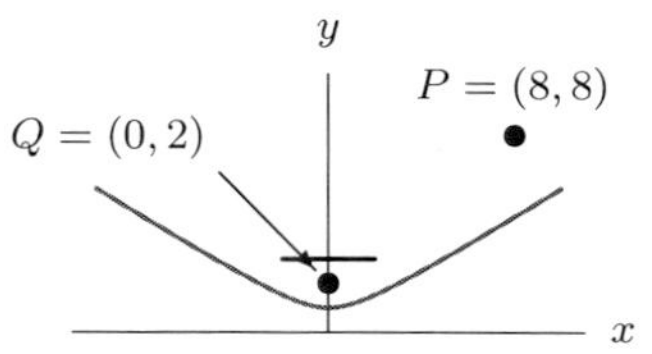

Figure 12.57

CHECK YOUR UNDERSTANDING

Are the statements in Problems 1–33 true or false? Give an explanation for your answer.

1. A moving particle with motion $x = 2t, y = 3t, 0 \le t \le 100$ is moving in a straight line.
2. A moving particle with motion $x = 2t^2, y = 3t^2, 0 \le t \le 100$ is moving in a straight line.
3. The motions $x = t, y = t$ and $x = t^3, y = t^3$ are on the same path.
4. If the motion of an object is described by $x = \sin(t/2)$ and $y = \cos(t/2), 0 \le t \le 4$, then when $t = \pi$ the object is located at the point $(1, 0)$.
5. If the motion of an object on the unit circle is described by $x = \sin(t/2)$ and $y = \cos(t/2), 0 \le t \le 4$, then the object is moving counterclockwise.
6. An object with motion $x = 2t, y = 3t+1$ passes through the point $(10, 16)$.
7. The graph of $x^2 - y^2 = 0$ is a circle.
8. The graph of $(x + 3)^2 + y^2 = 4$ is a circle of radius 4 centered at $(-3, 0)$.
9. The circle $x^2 + 8x + y^2 + 10y = 100$ has center $(4, 5)$.
10. The circle $x^2 + 8x + y^2 + 10y = 100$ has radius 10.
11. The parametric equations $x = t^3, y = t^6$ describe a parabola.
12. The graph of $y^2 - 4x^2 = 0$ is a hyperbola.
13. There is only one parameterization of the circle.
14. If the parameter t is eliminated from the parametric equations $x = 2t + 1$ and $y = t - 1$, the resulting equation can be written as $y = \frac{1}{2}(x - 3)$.
15. If a circle is centered at the point $(1, -1)$ and its radius is $\sqrt{5}$, then its equation is $(x - 1)^2 + (y + 1)^2 = 5$.
16. The equation $4x^2 + 9y^2 = 36$ represents a circle.
17. The equations $x = 2\cos t, y = \sin t$ parameterize an ellipse.
18. The horizontal axis of the ellipse $9x^2 + 4y^2 = 1$ is longer than its vertical axis.
19. The graph of $\dfrac{(y-k)^2}{b^2} - \dfrac{(x-h)^2}{a^2} = 1$ is a hyperbola that opens up and down.
20. The asymptotes of the hyperbola $x^2/4 - y^2/9 = 1$ are given by the equation $x^2/4 - y^2/9 = 0$.
21. The hyperbolic sine is defined as $\sinh x = \dfrac{e^x + e^{-x}}{2}$.
22. As x approaches infinity, $\cosh x$ approaches $\frac{1}{2}e^x$.
23. An identity that relates $\sinh x$ and $\cosh x$ is $\cosh^2 x + \sinh^2 x = 1$.
24. The function $\cosh t$ is periodic with period 2π.
25. The function $\cosh x$ is an even function of x.
26. The function $\cosh x$ grows faster than any power of x as $x \to \infty$.
27. $\cosh 0 = e$.
28. $\sinh x < \cosh x$ for all x.
29. $\sinh \pi = 0$.
30. $\cosh(\ln 2) = 5/4$.
31. If you know the lengths of the major and minor axes of an ellipse, then you can determine the distance between the two foci.
32. The distance between the vertex of a parabola and its focus is the same as the distance from the vertex to the directrix of the parabola.
33. Each asymptote of a hyperbola passes through one of its focal points.

Chapter 12

ADDITIONAL EXAMPLES AND EXERCISES

12-A CONIC SECTIONS

In this section we will discuss some of the basic geometric properties of parabolas, ellipses, and hyperbolas. These curves play an important role in calculus and also arise naturally in a broad range of applications in such fields as planetary motion, design of telescopes and antennas, geodetic positioning, and medicine, to name a few.

Some students may already be familiar with the material in this section, in which case it can be treated as a review. Instructors who want to spend some additional time on precalculus review may want to allocate more than one lecture on this material.

Conic Sections

Circles, ellipses, parabolas, and hyperbolas are called ***conic sections*** or ***conics*** because they can be obtained as intersections of a plane with a double-napped circular cone (Figure 12-A.1). If the plane passes through the vertex of the double-napped cone, then the intersection is a point, a pair of intersecting lines, or a single line. These are called ***degenerate conic sections.***

Definitions of the Conic Sections

Although we could derive properties of parabolas, ellipses, and hyperbolas by defining them as intersections with a double-napped cone, it will be better suited to calculus if we begin with equivalent definitions that are based on their geometric properties.

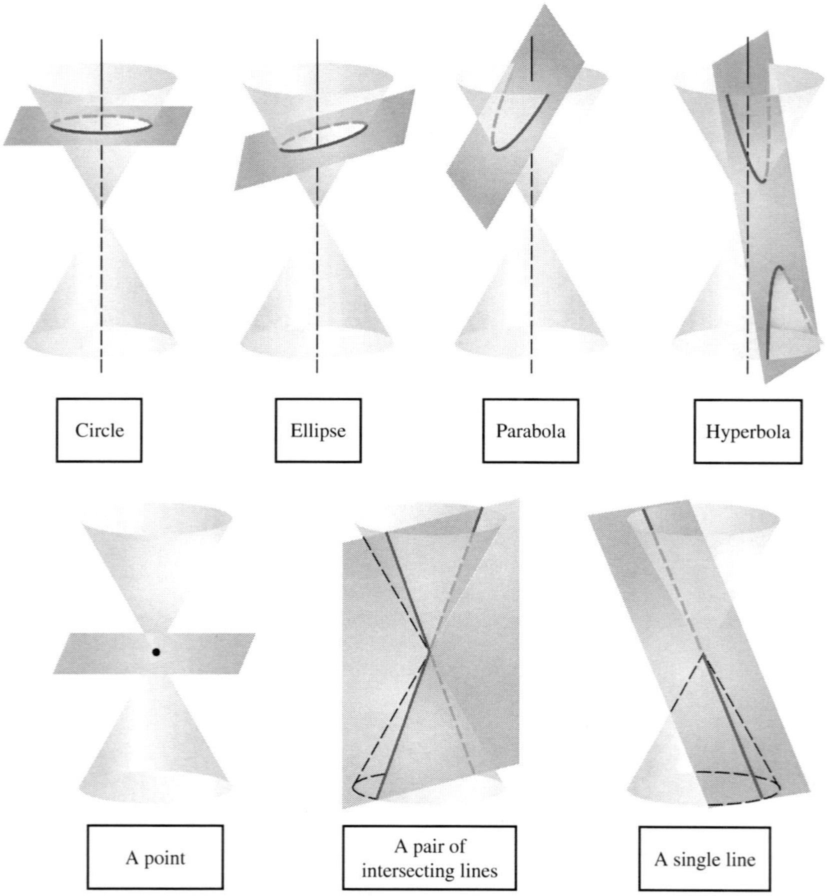

Figure 12-A.1

12-A.1 DEFINITION. A ***parabola*** is the set of all points in the plane that are equidistant from a fixed line and a fixed point not on the line.

The line is called the ***directrix*** of the parabola, and the point is called the ***focus*** (Figure 12-A.2). A parabola is symmetric about the line that passes through the focus at right angles to the directrix. This line, called the ***axis*** or the ***axis of symmetry*** of the parabola, intersects the parabola at a point called the ***vertex.***

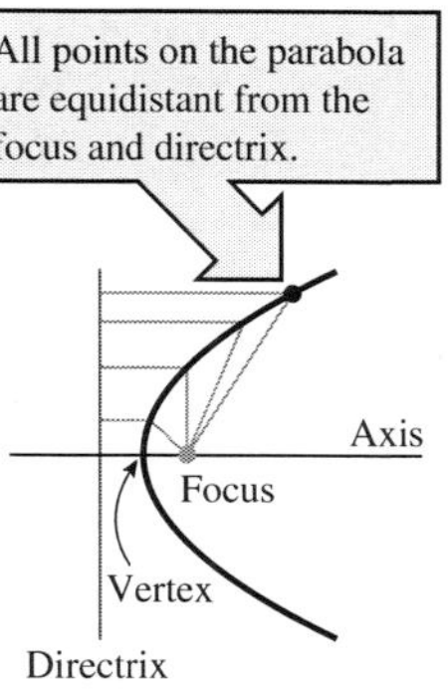

Figure 12-A.2

12-A.2 DEFINITION. An ***ellipse*** is the set of all points in the plane, the sum of whose distances from two fixed points is a given positive constant that is greater than the distance between the fixed points.

The two fixed points are called the ***foci*** (plural of "focus") of the ellipse, and the midpoint of the line segment joining the foci is called the ***center*** (Figure 12-A.3*a*). To help visualize Definition 12-A.2, imagine that two ends of a string are tacked to the foci and a pencil traces a curve as it is held tight against the string (Figure 12-A.3*b*). The resulting curve will be an ellipse since the sum of the distances to the foci is a constant, namely, the total length of the string. Note that if the foci coincide, the ellipse reduces to a circle. For ellipses other than circles, the line segment through the foci and across the ellipse is called the ***major axis*** (Figure 12-A.3*c*), and the line segment across the ellipse, through the center, and perpendicular to the major axis is called the ***minor axis.*** The endpoints of the major axis are called ***vertices.***

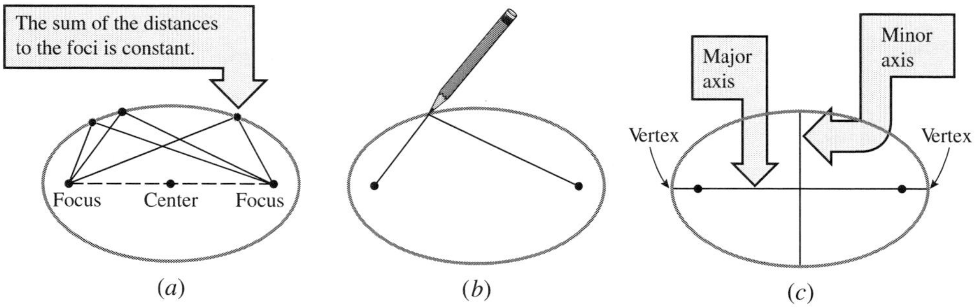

Figure 12-A.3

> **12-A.3 DEFINITION.** A ***hyperbola*** is the set of all points in the plane, the difference of whose distances from two fixed distinct points is a given positive constant that is less than the distance between the fixed points.

The two fixed points are called the ***foci*** of the hyperbola, and the term "difference" that is used in the definition is understood to mean the distance to the farther focus minus the distance to the closer focus. As a result, the points on the hyperbola form two ***branches,*** each "wrapping around" the closer focus (Figure 12-A.4*a*). The midpoint of the line segment joining the foci is called the ***center*** of the hyperbola, the line through the foci is called the ***focal axis,*** and the line through the center that is perpendicular to the focal axis is called the ***conjugate axis.*** The hyperbola intersects the focal axis at two points called the ***vertices.***

Associated with every hyperbola is a pair of lines, called the ***asymptotes*** of the hyperbola. These lines intersect at the center of the hyperbola and have the property that as a point P moves along the hyperbola away from the center, the vertical distance between P and one of the asymptotes approaches zero (Figure 12-A.4*b*).

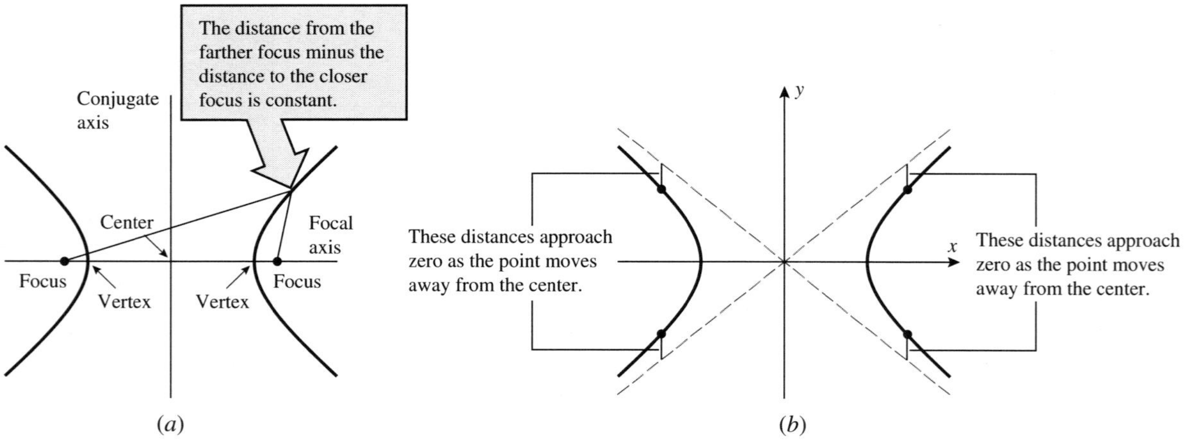

Figure 12-A.4

Equations of Parabolas in Standard Position

It is traditional in the study of parabolas to denote the distance between the focus and the vertex by p. The vertex is equidistant from the focus and the directrix, so the distance between the vertex and the directrix is also p; consequently, the distance between the focus and the directrix is $2p$ (Figure 12-A.5). As illustrated in that figure, the parabola passes through two of the corners of a box that extends from the vertex to the focus along the axis of symmetry and extends $2p$ units above and $2p$ units below the axis of symmetry.

The equation of a parabola is simplest if the vertex is the origin and the axis of symmetry is along the x-axis or y-axis. The four possible such orientations are shown in Figure 12-A.6. These are called the ***standard positions*** of a parabola, and the resulting equations are called the ***standard equations*** of a parabola.

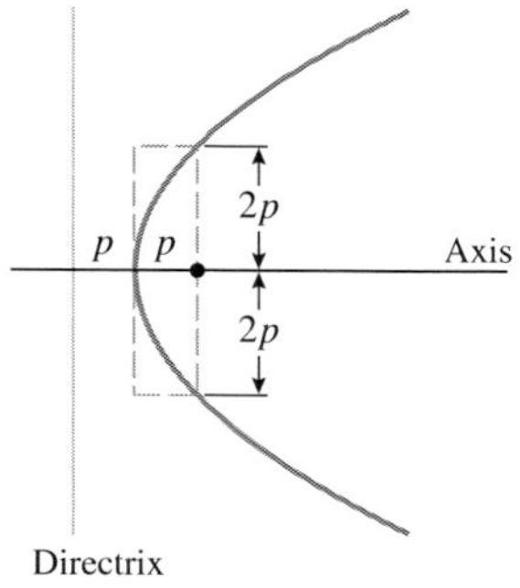

Figure 12-A.5

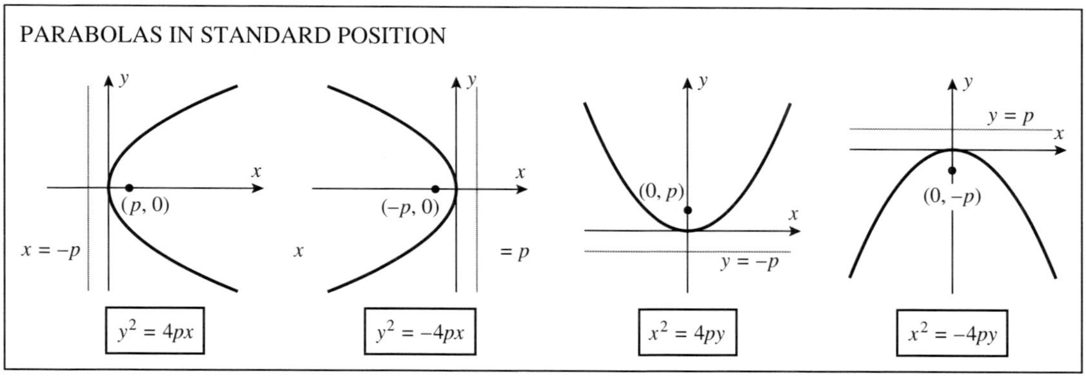

Figure 12-A.6

To illustrate how the equations in Figure 12-A.6 are obtained, we will derive the equation for the parabola with focus $(p, 0)$ and directrix $x = -p$. Let $P(x, y)$ be any point on the parabola. Since P is equidistant from the focus and directrix, the distances PF and PD in Figure 12-A.7 are equal; that is,

$$PF = PD \tag{1}$$

where $D(-p, y)$ is the foot of the perpendicular from P to the directrix. From the distance formula, the distances PF and PD are

$$PF = \sqrt{(x-p)^2 + y^2} \quad \text{and} \quad PD = \sqrt{(x-p)^2} \tag{2}$$

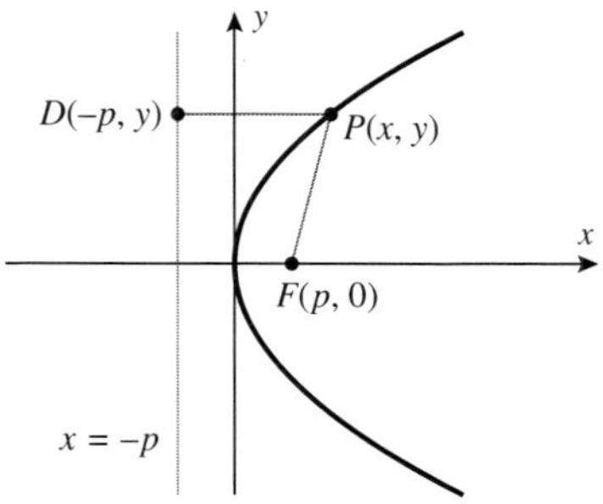

Figure 12-A.7

Substituting in (1) and squaring yields

$$(x - p)^2 + y^2 = (x + p)^2 \tag{3}$$

and after simplifying

$$y^2 = 4px \tag{4}$$

The derivations of the other equations in Figure 12-A.6 are similar.

A Technique for Sketching Parabolas

Parabolas can be sketched from their *standard equations* using four basic steps:

Sketching a Parabola from Its Standard Equation

Step 1. Determine whether the axis of symmetry is along the x-axis or the y-axis. Referring to Figure 12-A.6, the axis of symmetry is along the x-axis if the equation has a y^2-term, and it is along the y-axis if it has an x^2-term.

Step 2. Determine which way the parabola opens. If the axis of symmetry is along the x-axis, then the parabola opens to the right if the coefficient of x is positive, and it opens to the left if the coefficient is negative. If the axis of symmetry is along the y-axis, then the parabola opens up if the coefficient of y is positive, and it opens down if the coefficient is negative.

Step 3. Determine the value of p and draw a box extending p units from the origin along the axis of symmetry in the direction in which the parabola opens and extending $2p$ units on each side of the axis of symmetry.

Step 4. Using the box as a guide, sketch the parabola so that its vertex is at the origin and it passes through the corners of the box (Figure 12-A.8).

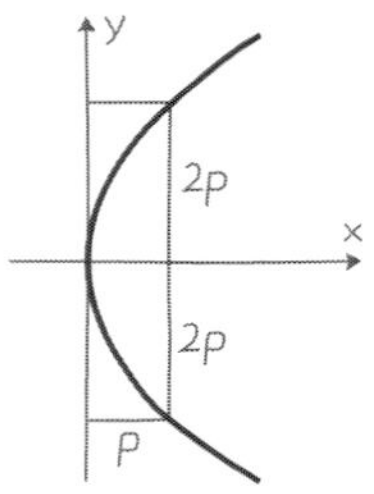

Figure 12-A.8

Example 1 Sketch the graphs of the parabolas

$$\text{(a) } x^2 = 12y \qquad \text{(b) } y^2 + 8x = 0$$

and show the focus and directrix of each.

Solution (a) This equation involves x^2, so the axis of symmetry is along the y-axis, and the coefficient of y is positive, so the parabola opens upward. From the coefficient of y, we obtain $4p = 12$ or $p = 3$. Drawing a box extending $p = 3$ units up from the origin and $2p = 6$ units to the left and $2p = 6$ units to the right of the y-axis, then using corners of the box as a guide, yields the graph in Figure 12-A.9.

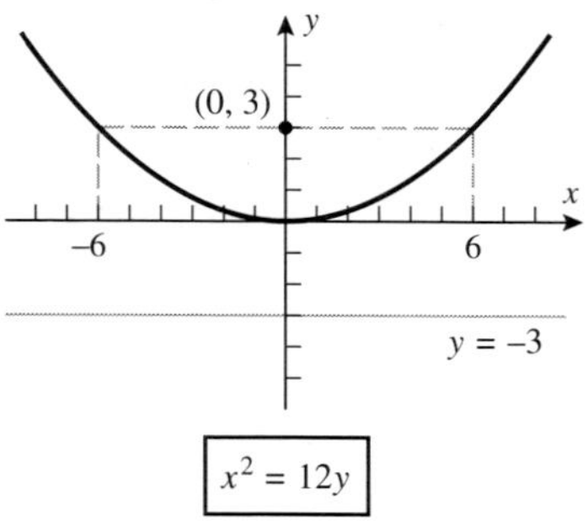

Figure 12-A.9

The focus is $p = 3$ units from the vertex along the axis of symmetry in the direction in which the parabola opens, so its coordinates are (0, 3). The directrix is perpendicular to the axis of symmetry at a distance of $p = 3$ units from the vertex on the opposite side from the focus, so its equation is $y = -3$.

Solution (b) We first rewrite the equation in the standard form

$$y^2 = -8x$$

This equation involves y^2, so the axis of symmetry is along the x-axis, and the coefficient of x is negative, so the parabola opens to the left. From the coefficient of x we obtain $4p = 8$, so $p = 2$. Drawing a box extending $p = 2$ units left from the origin and $2p = 4$ units above and $2p = 4$ units below the x-axis, then using corners of the box as a guide, yields the graph in Figure 12-A.10.

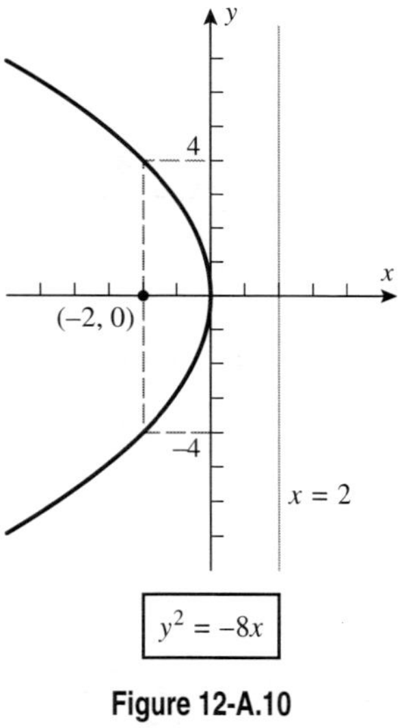

Figure 12-A.10

Example 2 Find an equation of the parabola that is symmetric about the y-axis, has its vertex at the origin, and passes through the point (5, 2).

Solution Since the parabola is symmetric about the y-axis and has its vertex at the origin, the equation is of the form

$$x^2 = 4py \quad \text{or} \quad x^2 = -4py$$

where the sign depends on whether the parabola opens up or down. But the parabola must open up since it passes through the point (5, 2), which lies in the first quadrant. Thus, the equation is of the form

$$x^2 = 4py \tag{5}$$

Since the parabola passes through (5, 2), we must have $5^2 = 4p \cdot 2$ or $4p = \frac{25}{2}$. Therefore, (5) becomes

$$x^2 = \frac{25}{2}y$$

Equations of Ellipses in Standard Position

It is traditional in the study of ellipses to denote the length of the major axis by $2a$, the length of the minor axis by $2b$, and the distance between the foci by $2c$ (Figure 12-A.11). The number a is called the ***semimajor axis*** and the number b the ***semiminor axis*** (standard but odd terminology, since a and b are numbers, not geometric axes).

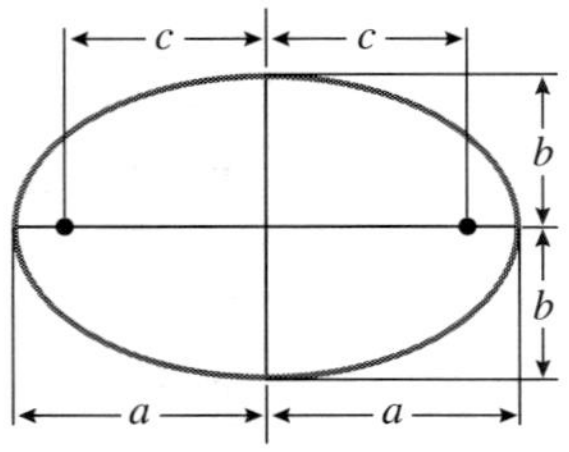

Figure 12-A.11

There is a basic relationship between the numbers a, b, and c that can be obtained by examining the sum of the distances to the foci from a point P at the end of the major axis and from a point Q at the end of the minor axis (Figure 12-A.12). From Definition 12-A.2, these sums must be equal, so we obtain

$$2\sqrt{b^2 + c^2} = (a - c) + (a + c)$$

from which it follows that

$$a = \sqrt{b^2 + c^2} \tag{6}$$

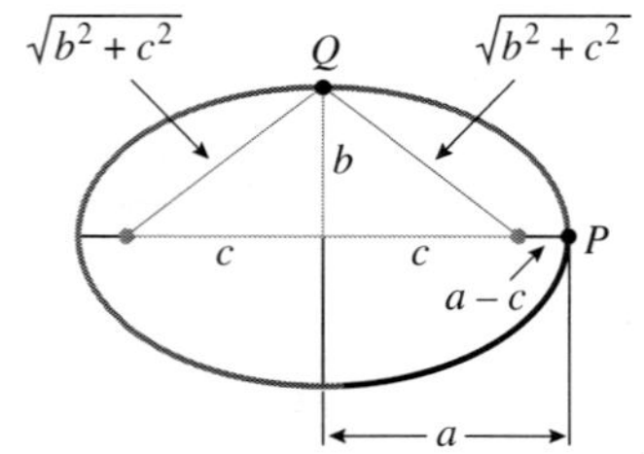

Figure 12-A.12

or, equivalently,

$$c = \sqrt{a^2 - b^2} \tag{7}$$

From (6), the distance from a focus to an end of the minor axis is a (Figure 12-A.13), which implies that for *all* points on the ellipse the sum of the distances to the foci is $2a$.

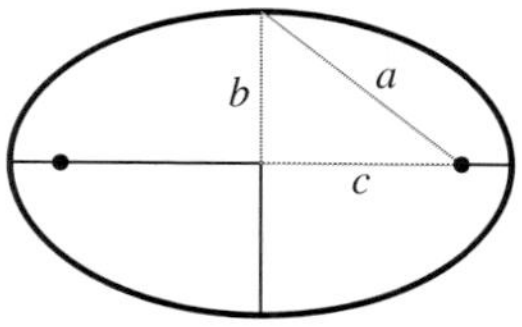

Figure 12-A.13

It also follows from (6) that $a \geq b$ with the equality holding only when $c = 0$. Geometrically, this means that the major axis of an ellipse is at least as large as the minor axis and that the two axes have equal length only when the foci coincide, in which case the ellipse is a circle.

The equation of an ellipse is simplest if the center of the ellipse is at the origin and the foci are on the x-axis or y-axis. The two possible such orientations are shown in Figure 12-A.14. These are called the ***standard positions*** of an ellipse, and the resulting equations are called the ***standard equations*** of an ellipse.

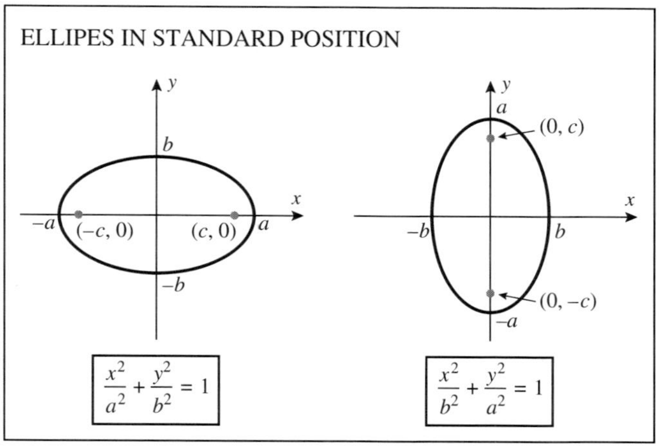

Figure 12-A.14

To illustrate how the equations in Figure 12-A.14 are obtained, we will derive the equation for the ellipse with foci on the x-axis. Let $P(x, y)$ be any point on that ellipse. Since the sum of the distances from P to the foci is $2a$, it follows (Figure 12-A.15) that

$$PF' + PF = 2a$$

so

$$\sqrt{(x+c)^2 + y^2} + \sqrt{(x-c)^2 + y^2} = 2a$$

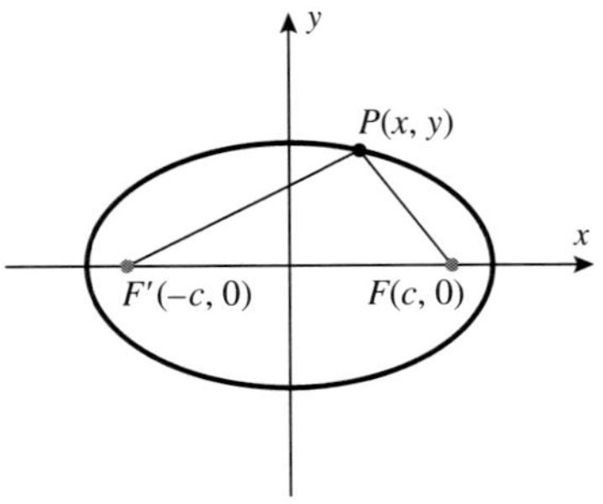

Figure 12-A.15

Transposing the second radical to the right side of the equation and squaring yields

$$(x + c)^2 + y^2 = 4a^2 - 4a\sqrt{(x - c)^2 + y^2} + (x - c)^2 + y^2$$

and, on simplifying,

$$\sqrt{(x - c)^2 + y^2} = a - \frac{c}{a}x \tag{8}$$

Squaring again and simplifying yields

$$\frac{x^2}{a^2} + \frac{y^2}{a^2 - c^2} = 1$$

which, by virtue of (6), can be written as

$$\frac{x^2}{a^2} + \frac{y^2}{b^2} = 1 \tag{9}$$

Conversely, it can be shown that any point whose coordinates satisfy (9) has $2a$ as the sum of its distances from the foci, so that such a point is on the ellipse.

A Technique for Sketching Ellipses

Ellipses can be sketched from their *standard equations* using three basic steps:

Sketching an Ellipse from Its Standard Equation

Step 1. Determine whether the major axis is on the x-axis or the y-axis. This can be ascertained from the sizes of the denominators in the equation. Referring to Figure 12-A.14, and keeping in mind that $a^2 > b^2$ (since $a > b$), the major axis is along the x-axis if x^2 has the larger denominator, and it is along the y-axis if y^2 has the larger denominator. If the denominators are equal, the ellipse is a circle.

Step 2. Determine the values of a and b and draw a box extending a units on each side of the center along the major axis and b units on each side of the center along the minor axis.

Step 3. Using the box as a guide, sketch the ellipse so that its center is at the origin and it touches the sides of the box where the sides intersect the coordinate axes (Figure 12-A.16).

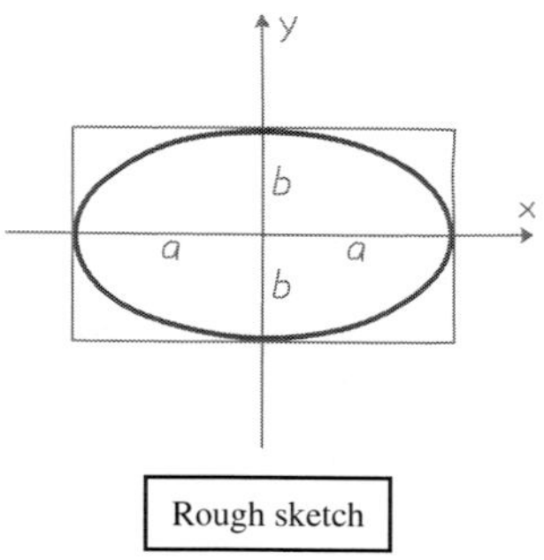

Figure 12-A.16

Example 3 Sketch the graphs of the ellipses

$$\text{(a)}\ \frac{x^2}{9} + \frac{y^2}{16} = 1 \qquad \text{(b)}\ x^2 + 2y^2 = 4$$

showing the foci of each.

Solution (a) Since y^2 has the larger denominator, the major axis is along the y-axis. Moreover, since $a^2 > b^2$, we must have $a^2 = 16$ and $b^2 = 9$, so

$$a = 4 \quad \text{and} \quad b = 3$$

Drawing a box extending 4 units on each side of the origin along the y-axis and 3 units on each side of the origin along the x-axis as a guide yields the graph in Figure 12-A.17.

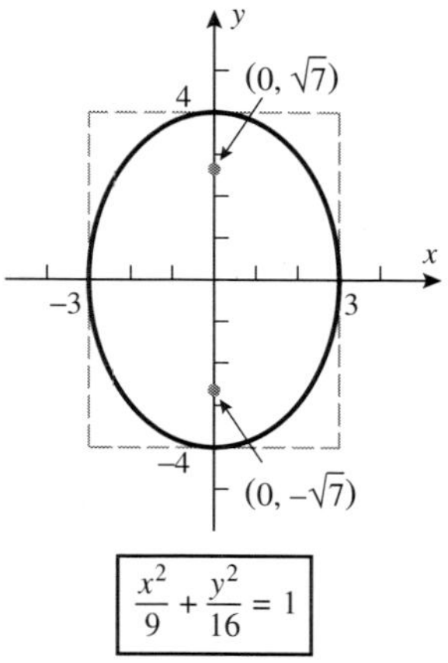

Figure 12-A.17

The foci lie c units on each side of the center along the major axis, where c is given by (7). From the values of a^2 and b^2 above, we obtain

$$c = \sqrt{a^2 - b^2} = \sqrt{16 - 9} = \sqrt{7} \approx 2.6$$

Thus, the coordinates of the foci are $(0, \sqrt{7})$ and $(0, -\sqrt{7})$, since they lie on the y-axis.

Solution (b) We first rewrite the equation in the standard form

$$\frac{x^2}{4} + \frac{y^2}{2} = 1$$

Since x^2 has the larger denominator, the major axis lies along the x-axis, and we have $a^2 = 4$ and $b^2 = 2$. Drawing a box extending $a = 2$ units on each side of the origin along the x-axis and extending $b = \sqrt{2} \approx 1.4$ units on each side of the origin along the y-axis as a guide yields the graph in Figure 12-A.18.

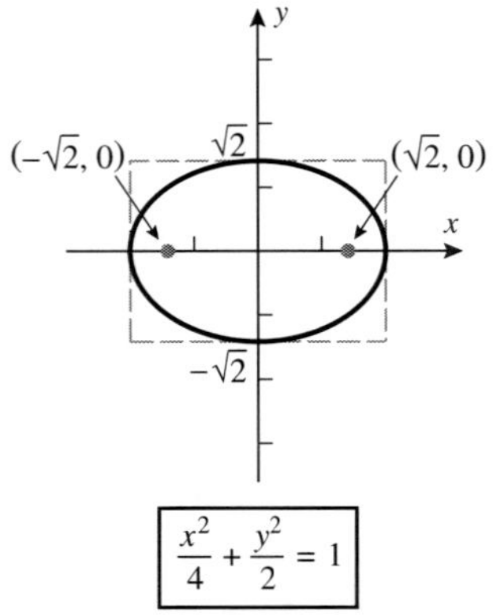

Figure 12-A.18

From (7), we obtain

$$c = \sqrt{a^2 - b^2} = \sqrt{2} \approx 1.4$$

Thus, the coordinates of the foci are $(\sqrt{2}, 0)$ and $(-\sqrt{2}, 0)$, since they lie on the x-axis.

Example 4 Find an equation for the ellipse with foci $(0, \pm 2)$ and major axis with endpoints $(0, \pm 4)$.

Solution From Figure 12-A.14, the equation has the form

$$\frac{x^2}{b^2} + \frac{y^2}{a^2} = 1$$

and from the given information, $a = 4$ and $c = 2$. It follows from (6) that

$$b^2 = a^2 - c^2 = 16 - 4 = 12$$

so the equation of the ellipse is

$$\frac{x^2}{12} + \frac{y^2}{16} = 1$$

Equations of Hyperbolas in Standard Position

It is traditional in the study of hyperbolas to denote the distance between the vertices by $2a$, the distance between the foci by $2c$ (Figure 12-A.19), and to define the quantity b as

$$b = \sqrt{c^2 - a^2} \tag{10}$$

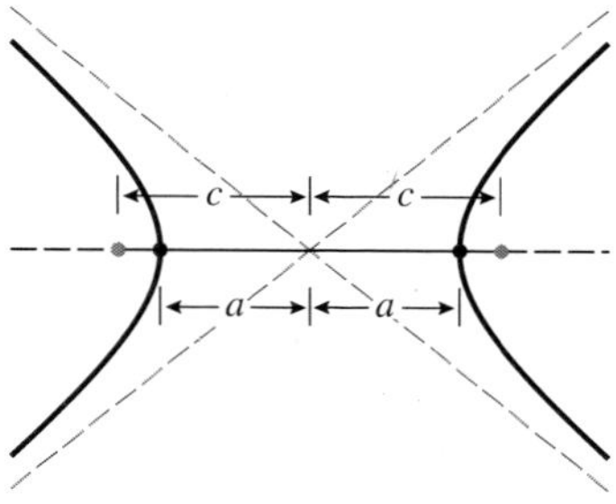

Figure 12-A.19

This relationship, which can also be expressed as

$$c = \sqrt{a^2 + b^2} \tag{11}$$

is pictured geometrically in Figure 12-A.20. As illustrated in that figure, and as we will show later in this section, the asymptotes pass through the corners of a box extending b units on each side of the center along the conjugate axis and a units on each side of the center along the focal axis. The number a is called the ***semifocal axis*** of the hyperbola and the number b the ***semiconjugate axis.*** (As with the semimajor and semiminor axes of an ellipse, these are numbers, not geometric axes.)

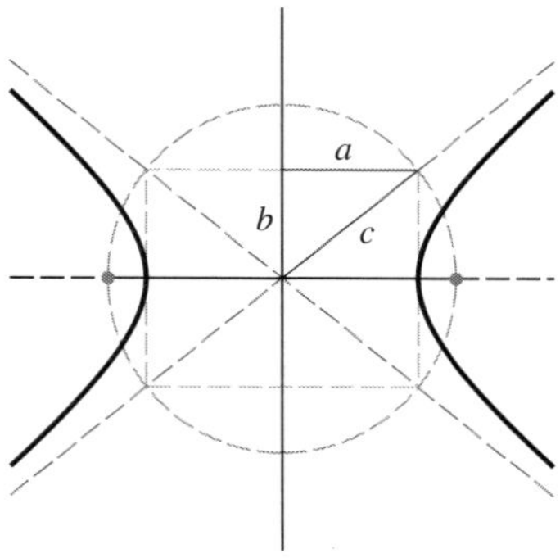

Figure 12-A.20

If V is one vertex of a hyperbola, then, as illustrated in Figure 12-A.21, the distance from V to the farther focus minus the distance from V to the closer focus is

$$[(c - a) + 2a] - (c - a) = 2a$$

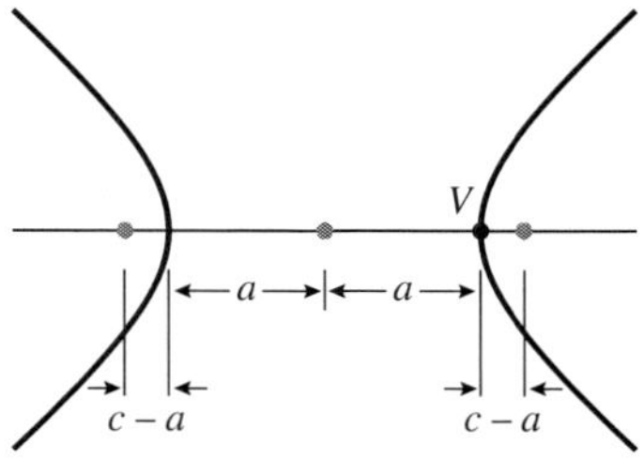

Figure 12-A.21

Thus, for *all* points on a hyperbola, the distance to the farther focus minus the distance to the closer focus is $2a$.

The equation of a hyperbola has an especially convenient form if the center of the hyperbola is at the origin and the foci are on the x-axis or y-axis. The two possible such orientations are shown in Figure 12-A.22. These are called the ***standard positions*** of a hyperbola, and the resulting equations are called the ***standard equations*** of a hyperbola.

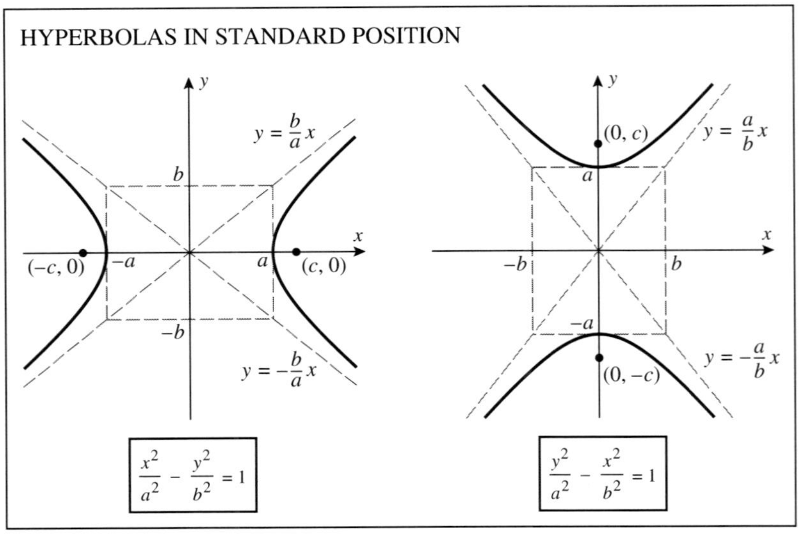

Figure 12-A.22

The derivations of these equations are similar to those already given for parabolas and ellipses, so we will leave them as exercises. However, to illustrate how the equations of the asymptotes are derived, we will derive those equations for the hyperbola

$$\frac{x^2}{a^2} - \frac{y^2}{b^2} = 1$$

We can rewrite this equation as

$$y^2 = \frac{b^2}{a^2}(x^2 - a^2)$$

which is equivalent to the pair of equations

$$y = \frac{b}{a}\sqrt{x^2 - a^2} \quad \text{and} \quad y = -\frac{b}{a}\sqrt{x^2 - a^2}$$

Thus, in the first quadrant, the vertical distance between the line $y = (b/a)x$ and the hyperbola can be written (Figure 12-A.23) as

$$\frac{b}{a}x - \frac{b}{a}\sqrt{x^2 - a^2}$$

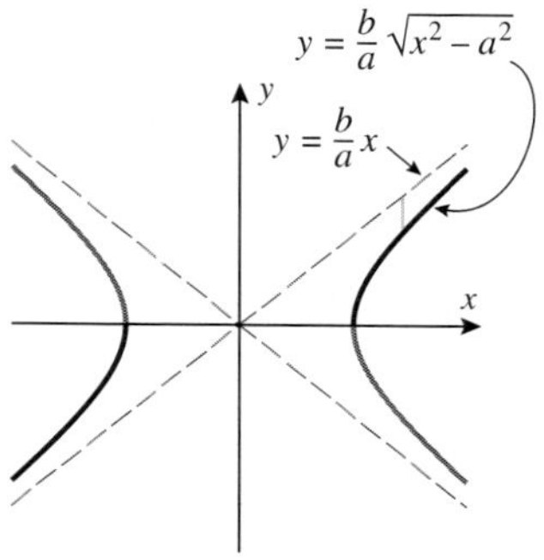

Figure 12-A.23

But this distance tends to zero as $x \to +\infty$ since

$$\lim_{x\to+\infty}\left(\frac{b}{a}x - \frac{b}{a}\sqrt{x^2 - a^2}\right) = \lim_{x\to+\infty}\frac{b}{a}(x - \sqrt{x^2 - a^2})$$

$$= \lim_{x\to+\infty}\frac{b}{a}\frac{(x - \sqrt{x^2 - a^2})(x + \sqrt{x^2 - a^2})}{x + \sqrt{x^2 - a^2}}$$

$$= \lim_{x\to+\infty}\frac{ab}{x + \sqrt{x^2 - a^2}} = 0$$

The analysis in the remaining quadrants is similar.

A Quick Way to Find Asymptotes

There is a trick that can be used to avoid memorizing the equations of the asymptotes of a hyperbola. They can be obtained, when needed, by substituting 0 for the 1 on the right side of the hyperbola equation, and then solving for y in terms of x. For example, for the hyperbola

$$\frac{x^2}{a^2} - \frac{y^2}{b^2} = 1$$

we would write

$$\frac{x^2}{a^2} - \frac{y^2}{b^2} = 0 \quad \text{or} \quad y^2 = \frac{b^2}{a^2}x^2 \quad \text{or} \quad y = \pm\frac{b}{a}x$$

which are the equations for the asymptotes.

A Technique for Sketching Hyperbolas

Hyperbolas can be sketched from their *standard equations* using four basic steps:

Sketching a Hyperbola from Its Standard Equation

Step 1. Determine whether the focal axis is on the x-axis or the y-axis. This can be ascertained from the location of the minus sign in the equation. Referring to Figure 12-A.22, the focal axis is along the x-axis when the minus sign precedes the y^2-term, and it is along the y-axis when the minus sign precedes the x^2-term.

Step 2. Determine the values of a and b and draw a box extending a units on either side of the center along the focal axis and b units on either side of the center along the conjugate axis. (The squares of a and b can be read directly from the equation.)

Step 3. Draw the asymptotes along the diagonals of the box.

Step 4. Using the box and the asymptotes as a guide, sketch the graph of the hyperbola (Figure 12-A.24).

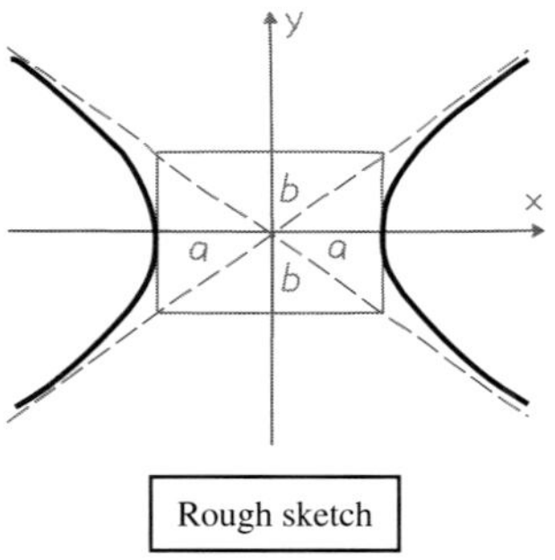

Figure 12-A.24

Example 5 Sketch the graphs of the hyperbolas

$$\text{(a) } \frac{x^2}{4} - \frac{y^2}{9} = 1 \qquad \text{(b) } y^2 - x^2 = 1$$

showing their vertices, foci, and asymptotes.

Solution (a) The minus sign precedes the y^2-term, so the focal axis is along the x-axis. From the denominators in the equation we obtain

$$a^2 = 4 \quad \text{and} \quad b^2 = 9$$

Since a and b are positive, we must have $a = 2$ and $b = 3$. Recalling that the vertices lie a units on each side of the center on the focal axis, it follows that their coordinates in this case are $(2, 0)$ and $(-2, 0)$. Drawing a box extending $a = 2$ units along the x-axis on each side of the origin and $b = 3$ units on each side of the origin along the y-axis, then drawing the asymptotes along the diagonals of the box as a guide, yields the graph in Figure 12-A.25.

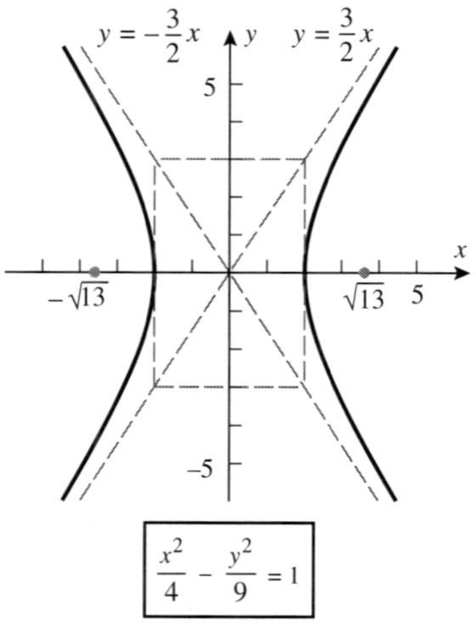

Figure 12-A.25

To obtain equations for the asymptotes, we substitute 0 for 1 in the given equation; this yields

$$\frac{x^2}{4} - \frac{y^2}{9} = 0 \quad \text{or} \quad y = \pm\frac{3}{2}x$$

The foci lie c units on each side of the center along the focal axis, where c is given by (11). From the values of a^2 and b^2 above we obtain

$$c = \sqrt{a^2 + b^2} = \sqrt{4 + 9} = \sqrt{13} \approx 3.6$$

Since the foci lie on the x-axis in this case, their coordinates are $(\sqrt{13}, 0)$ and $(-\sqrt{13}, 0)$.

Solution (b) The minus sign precedes the x^2-term, so the focal axis is along the y-axis. From the denominators in the equation we obtain $a^2 = 1$ and $b^2 = 1$, from which it follows that

$$a = 1 \quad \text{and} \quad b = 1$$

Thus, the vertices are at $(0, -1)$ and $(0, 1)$. Drawing a box extending $a = 1$ unit on either side of the origin along the y-axis and $b = 1$ unit on either side of the origin along the x-axis, then drawing the asymptotes, yields the graph in Figure 12-A.26. Since the box is actually a square, the asymptotes are perpendicular and have equations $y = \pm x$. This can also be seen by substituting 0 for 1 in the given equation, which yields $y^2 - x^2 = 0$ or $y = \pm x$. Also,

$$c = \sqrt{a^2 + b^2} = \sqrt{1 + 1} = \sqrt{2}$$

so the foci, which lie on the y-axis, are $(0, -\sqrt{2})$ and $(0, \sqrt{2})$.

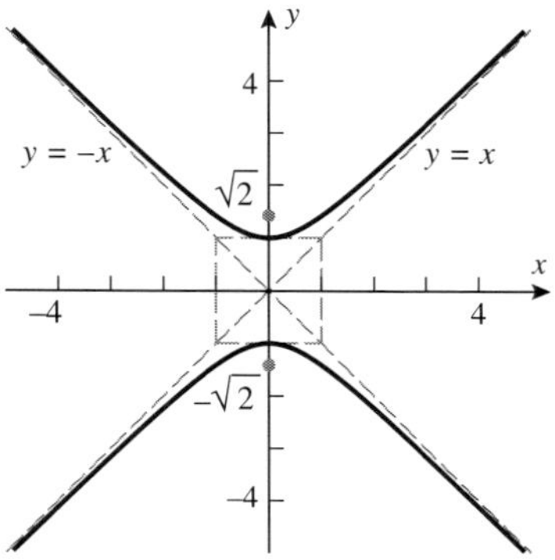

Figure 12-A.26

> A hyperbola in which $a = b$, as in part (b) of Example 5, is called an ***equilateral hyperbola.*** Such hyperbolas always have perpendicular asymptotes.

Example 6 Find the equation of the hyperbola with vertices $(0, \pm 8)$ and asymptotes $y = \pm\frac{4}{3}x$.

Solution Since the vertices are on the y-axis, the equation of the hyperbola has the form $(y^2/a^2) - (x^2/b^2) = 1$ and the asymptotes are

$$y = \pm\frac{a}{b}x$$

From the locations of the vertices we have $a = 8$, so the given equations of the asymptotes yield

$$y = \pm\frac{a}{b}x = \pm\frac{8}{b}x = \pm\frac{4}{3}x$$

from which it follows that $b = 6$. Thus, the hyperbola has the equation

$$\frac{y^2}{64} - \frac{x^2}{36} = 1$$

Translated Conics

Equations of conics that are translated from their standard positions can be obtained by replacing x by $x - h$ and y by $y - k$ in their standard equations. For a parabola, this translates the vertex from the origin to the point (h, k); and for ellipses and hyperbolas, this translates the center from the origin to the point (h, k).

Parabolas with vertex (h, k) and axis parallel to x-axis

$$(y - k)^2 = 4p(x - h) \quad \text{[Opens right]} \tag{12}$$

$$(y - k)^2 = -4p(x - h) \quad \text{[Opens left]} \tag{13}$$

Parabolas with vertex (h, k) and axis parallel to y-axis

$$(x - h)^2 = 4p(y - k) \quad \text{[Opens up]} \tag{14}$$

$$(x - h)^2 = -4p(y - k) \quad \text{[Opens down]} \tag{15}$$

Ellipse with center (h, k) and major axis parallel to x-axis

$$\frac{(x - h)^2}{a^2} + \frac{(y - k)^2}{b^2} = 1 \quad [b \le a] \tag{16}$$

Ellipse with center (h, k) and major axis parallel to y-axis

$$\frac{(x - h)^2}{b^2} + \frac{(y - k)^2}{a^2} = 1 \quad [b \le a] \tag{17}$$

Hyperbola with center (h, k) and focal axis parallel to x-axis

$$\frac{(x - h)^2}{a^2} - \frac{(y - k)^2}{b^2} = 1 \tag{18}$$

Hyperbola with center (h, k) and focal axis parallel to y-axis

$$\frac{(y - k)^2}{a^2} - \frac{(x - h)^2}{b^2} = 1 \tag{19}$$

Example 7 Find an equation for the parabola that has its vertex at (1, 2) and its focus at (4, 2).

Solution Since the focus and vertex are on a horizontal line, and since the focus is to the right of the vertex, the parabola opens to the right and its equation has the form

$$(y - k)^2 = 4p(x - h)$$

Since the vertex and focus are 3 units apart, we have $p = 3$, and since the vertex is at $(h, k) = (1, 2)$, we obtain

$$(y - 2)^2 = 12(x - 1)$$

Sometimes the equations of translated conics occur in expanded form, in which case we are faced with the problem of identifying the graph of a quadratic equation in x and y:

$$Ax^2 + Cy^2 + Dx + Ey + F = 0 \qquad (20)$$

The basic procedure for determining the nature of such a graph is to complete the squares of the quadratic terms and then try to match up the resulting equation with one of the forms of a translated conic.

Example 8 Describe the graph of the equation

$$y^2 - 8x - 6y - 23 = 0$$

Solution The equation involves quadratic terms in y but none in x, so we first take all of the y-terms to one side:

$$y^2 - 6y = 8x + 23$$

Next, we complete the square on the y-terms by adding 9 to both sides:

$$(y - 3)^2 = 8x + 32$$

Finally, we factor out the coefficient of the x-term to obtain

$$(y - 3)^2 = 8(x + 4)$$

This equation is of form (12) with $h = -4$, $k = 3$, and $p = 2$, so the graph is a parabola with vertex $(-4, 3)$ opening to the right. Since $p = 2$, the focus is 2 units to the right of the vertex, which places it at the point $(-2, 3)$; and the directrix is 2 units to the left of the vertex, which means that its equation is $x = -6$. The parabola is shown in Figure 12-A.27.

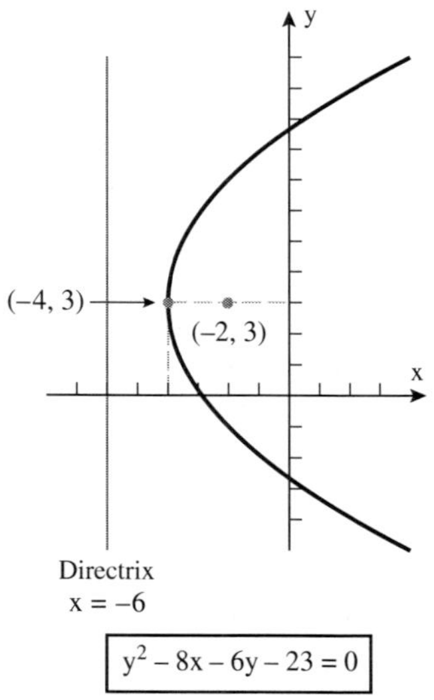

Figure 12-A.27

Example 9 Describe the graph of the equation

$$16x^2 + 9y^2 - 64x - 54y + 1 = 0$$

Solution This equation involves quadratic terms in both x and y, so we will group the x-terms and the y-terms on one side and put the constant on the other:

$$(16x^2 - 64x) + (9y^2 - 54y) = -1$$

Next, factor out the coefficients of x^2 and y^2 and complete the squares:

$$16(x^2 - 4x + 4) + 9(y^2 - 6y + 9) = -1 + 64 + 81$$

or

$$16(x - 2)^2 + 9(y - 3)^2 = 144$$

Finally, divide through by 144 to introduce a 1 on the right side:

$$\frac{(x - 2)^2}{9} + \frac{(y - 3)^2}{16} = 1$$

This is an equation of form (17), with $h = 2$, $k = 3$, $a^2 = 16$, and $b^2 = 9$. Thus, the graph of the equation is an ellipse with center (2, 3) and major axis parallel to the y-axis. Since $a = 4$, the major axis extends 4 units above and 4 units below the center, so its endpoints are (2, 7) and (2, −1) (Figure 12-A.28). Since $b = 3$, the minor axis extends 3 units to the left and 3 units to the right of the center, so its endpoints are (−1, 3) and (5, 3). Since

$$c = \sqrt{a^2 - b^2} = \sqrt{16 - 9} = \sqrt{7}$$

the foci lie $\sqrt{7}$ units above and below the center, placing them at the points $(2, 3 + \sqrt{7})$ and $(2, 3 - \sqrt{7})$.

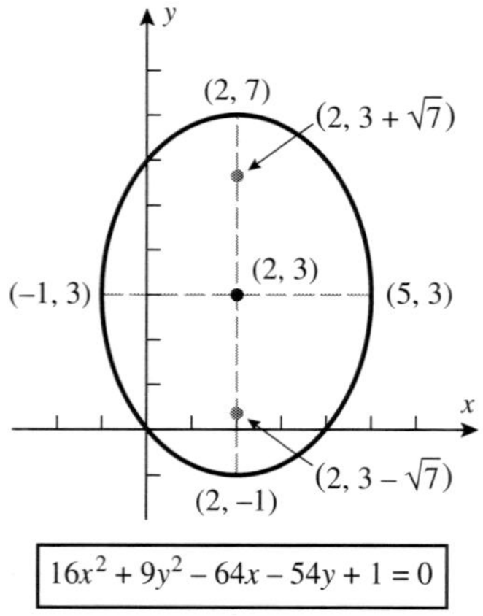

Figure 12-A.28

Example 10 Describe the graph of the equation

$$x^2 - y^2 - 4x + 8y - 21 = 0$$

Solution This equation involves quadratic terms in both x and y, so we will group the x-terms and the y-terms on one side and put the constant on the other:

$$(x^2 - 4x) - (y^2 - 8y) = 21$$

We leave it for you to verify by completing the squares that this equation can be written as

$$\frac{(x-2)^2}{9} - \frac{(y-4)^2}{9} = 1 \tag{21}$$

This is an equation of form (18) with $h = 2$, $k = 4$, $a^2 = 9$, and $b^2 = 9$. Thus, the equation represents a hyperbola with center (2, 4) and focal axis parallel to the x-axis. Since $a = 3$, the vertices are located 3 units to the left and 3 units to the right of the center, or at the points $(-1, 4)$ and $(5, 4)$. From (11), $c = \sqrt{a^2 + b^2} = \sqrt{9 + 9} = 3\sqrt{2}$, so the foci are located $3\sqrt{2}$ units to the left and right of the center, or at the points $(2 - 3\sqrt{2}, 4)$ and $(2 + 3\sqrt{2}, 4)$.

The equations of the asymptotes may be found using the trick of substituting 0 for 1 in (21) to obtain

$$\frac{(x-2)^2}{9} - \frac{(y-4)^2}{9} = 0$$

This can be written as $y - 4 = \pm(x - 2)$ which yields the asymptotes

$$y = x + 2 \quad \text{and} \quad y = -x + 6$$

With the aid of a box extending $a = 3$ units left and right of the center and $b = 3$ units above and below the center, we obtain the sketch in Figure 12-A.29.

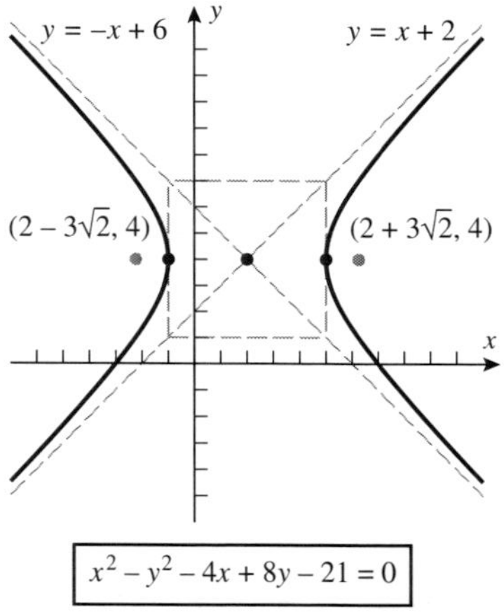

Figure 12-A.29

Reflection Properties of the Conic Sections

Parabolas, ellipses, and hyperbolas have certain reflection properties that make them extremely valuable in various applications. In the exercises we will ask you to prove the following results.

12-A.4 THEOREM (*Reflection Property of Parabolas*). *The tangent line at a point P on a parabola makes equal angles with the line through P parallel to the axis of symmetry and the line through P and the focus* (Figure 12-A.30*a*).

12-A.5 THEOREM (*Reflection Property of Ellipses*). *A line tangent to an ellipse at a point P makes equal angles with the lines joining P to the foci* (Figure 12-A.30*b*).

12-A.6 THEOREM (*Reflection Property of Hyperbolas*). *A line tangent to a hyperbola at a point P makes equal angles with the lines joining P to the foci* (Figure 12-A.30*c*).

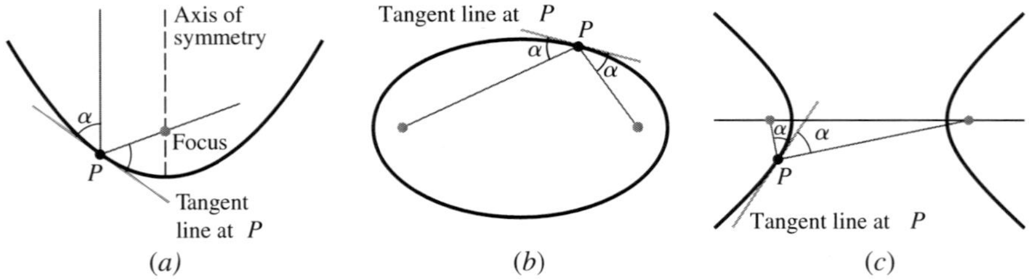

Figure 12-A.30

Applications of the Conic Sections

Fermat's principle in optics implies that light reflects off of a surface at an angle equal to its angle of incidence. (See Exercise 65 in Section 4.5.) In particular, if a reflecting surface is generated by revolving a parabola about its axis of symmetry, it follows from Theorem 12-A.4 that all light rays entering parallel to the axis will be reflected to the focus (Figure 12-A.31*a*); conversely, if a light source is located at the focus, then the reflected rays will all be parallel to the axis (Figure 12-A.31*b*). This principle is used in certain telescopes to reflect the approximately parallel rays of light from the stars and planets off of a parabolic mirror to an eyepiece at the focus; and the parabolic reflectors in flashlights and automobile headlights utilize this principle to form a parallel beam of light rays from a bulb placed at the focus. The same optical principles apply to radar signals and sound waves, which explains the parabolic shape of many antennas.

Visitors to various rooms in the United States Capitol Building and in St. Paul's Cathedral in Rome are often astonished by the "whispering gallery" effect in which two people at opposite ends of the room can hear one another's whispers very clearly. Such rooms have ceilings with elliptical

Incoming signals are reflected by the parabolic antenna to the receiver at the focus.

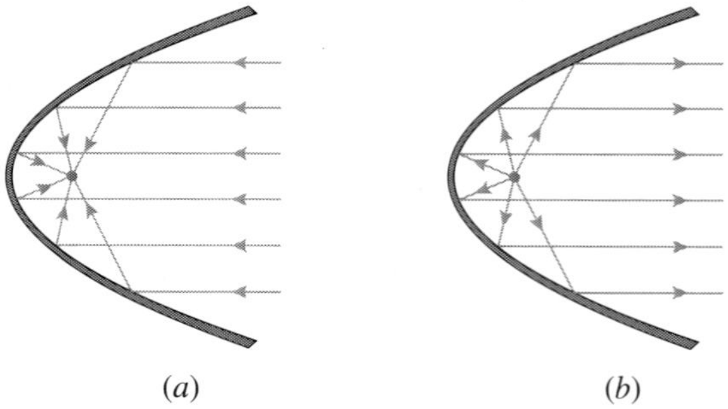

Figure 12-A.31

cross sections and common foci. Thus, when the two people stand at the foci, their whispers are reflected directly to one another off of the elliptical ceiling.

Hyperbolic navigation systems, which were developed in World War II as navigational aids to ships, are based on the definition of a hyperbola. With these systems the ship receives synchronized radio signals from two widely spaced transmitters with known positions. The ship's electronic receiver measures the difference in reception times between the signals and then uses that difference to compute the difference $2a$ between its distances from the two transmitters. This information places the ship somewhere on the hyperbola whose foci are at the transmitters and whose points have $2a$ as the difference in their distances from the foci. By repeating the process with a second set of transmitters, the position of the ship can be approximated as the intersection of two hyperbolas (Figure 12-A.32). (The modern "global positioning system" is based on the same principle.)

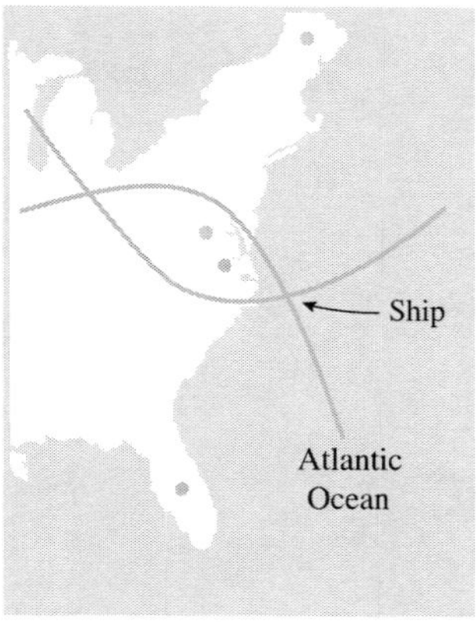

Figure 12-A.32

QUICK CHECK EXERCISES 12-A

1. Identify the conic.

(a) The set of points in the plane, the sum of whose distances to two fixed points is a positive constant greater than the distance between the fixed points is ______.

(b) The set of points in the plane, the difference of whose distances to two fixed points is a positive constant less than the distance between the fixed points is ______.

(c) The set of points in the plane that are equidistant from a fixed line and a fixed point not on the line is ______.

2. **(a)** The equation of the parabola with focus $(p, 0)$ and directrix $x = -p$ is ______.

(b) The equation of the parabola with focus $(0, p)$ and directrix $y = -p$ is ______.

3. **(a)** Suppose that an ellipse has semimajor axis a and semiminor axis b. Then for all points on the ellipse, the sum of the distances to the foci is equal to ______.

(b) The two standard equations of an ellipse with semimajor axis a and semiminor axis b are ______ and ______.

(c) Suppose that an ellipse has semimajor axis a, semiminor axis b, and foci $(\pm c, 0)$. Then c may be obtained from a and b by the equation $c =$ ______.

4. **(a)** Suppose that a hyperbola has semifocal axis a and semi-conjugate axis b. Then for all points on the hyperbola, the difference of the distance to the farther focus minus the distance to the closer focus is equal to ______.

(b) The two standard equations of a hyperbola with semi-focal axis a and semiconjugate axis b are ______ and ______.

(c) Suppose that a hyperbola in standard position has semi-focal axis a, semiconjugate axis b, and foci $(\pm c, 0)$. Then c may be obtained from a and b by the equation $c =$ ______. The equations of the asymptotes of this hyperbola are $y = \pm$______.

EXERCISE SET 12-A ~ Graphing Utility C CAS

Focus on Concepts

1. In parts (a)–(f), find the equation of the conic.

(a)

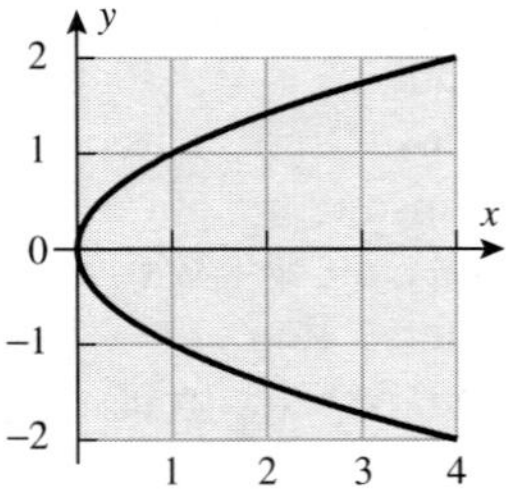

(b)

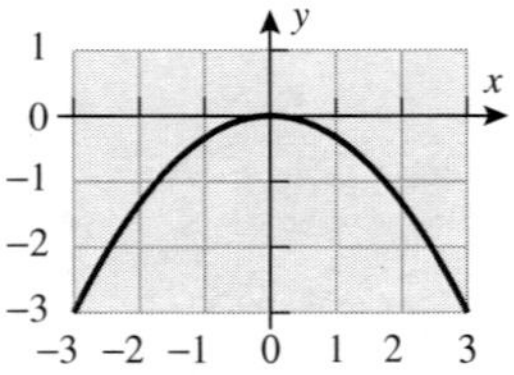

(c)

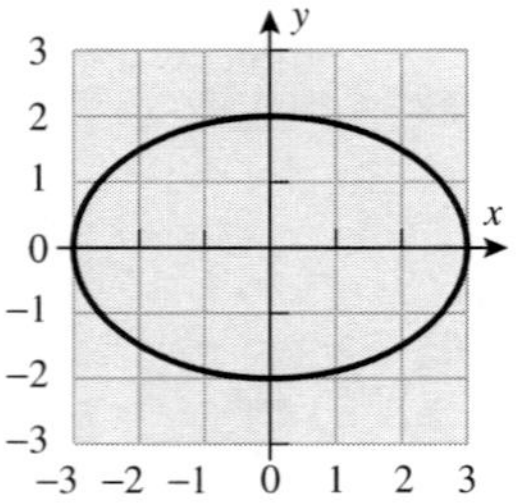

(d)

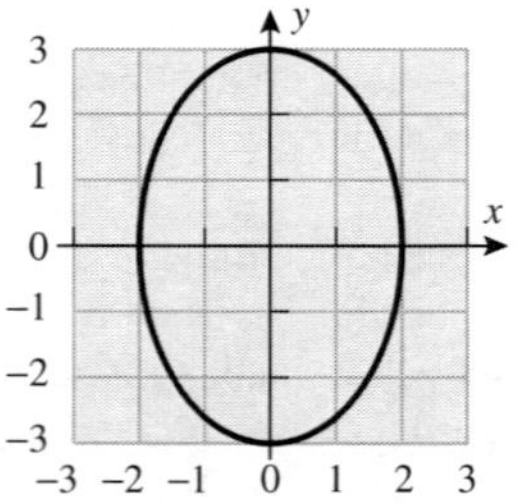

(e)

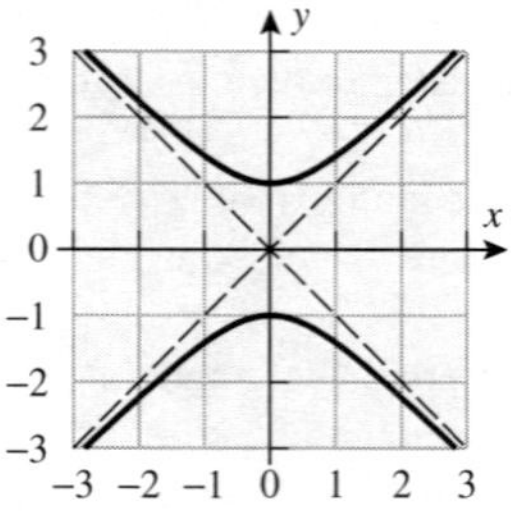

(e)

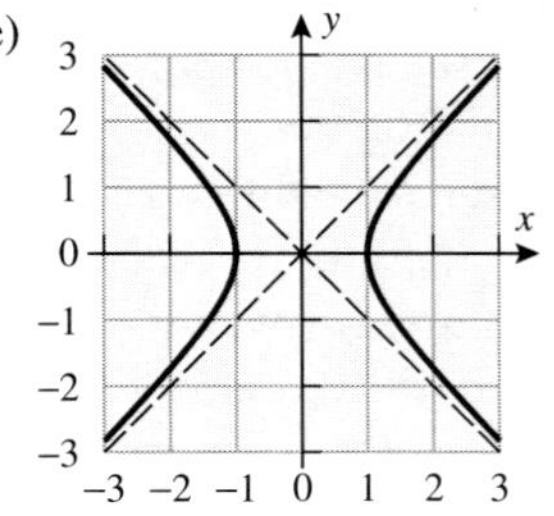

2. **(a)** Find the focus and directrix for each parabola in Exercise 1.
(b) Find the foci of the ellipses in Exercise 1.
(c) Find the foci and the equations of the asymptotes of the hyperbolas in Exercise 1.

3–8 Sketch the parabola, and label the focus, vertex, and directrix.

3. **(a)** $y^2 = 4x$ **(b)** $x^2 = -8y$

4. **(a)** $y^2 = -10x$ **(b)** $x^2 = 4y$

5. **(a)** $(y-3)^2 = 6(x-2)$ **(b)** $(x+2)^2 = -(y+2)$

6. **(a)** $(y-1)^2 = -12(x+4)$ **(b)** $(x-1)^2 = 2(y-\frac{1}{2})$

7. **(a)** $x^2 - 4x + 2y = 1$ **(b)** $x = y^2 - 4y + 2$

8. **(a)** $y^2 - 6y - 2x + 1 = 0$ **(b)** $y = 4x^2 + 8x + 5$

9–14 Sketch the ellipse, and label the foci, the vertices, and the ends of the minor axis.

9. **(a)** $\dfrac{x^2}{16} + \dfrac{y^2}{9} = 1$ **(b)** $9x^2 + y^2 = 9$

10. **(a)** $\dfrac{x^2}{25} + \dfrac{y^2}{4} = 1$ **(b)** $4x^2 + y^2 = 36$

11. **(a)** $16(x-1)^2 + 9(y-3)^2 = 144$
(b) $9(x+2)^2 + 4(y+1)^2 = 36$

12. **(a)** $(x+3)^2 + 4(y-5)^2 = 16$
(b) $\frac{1}{4}x^2 + \frac{1}{9}(y+2)^2 - 1 = 0$

13. **(a)** $x^2 + 9y^2 + 2x - 18y + 1 = 0$
(b) $4x^2 + y^2 + 8x - 10y = -13$

14. **(a)** $9x^2 + 4y^2 - 18x + 24y + 9 = 0$
(b) $5x^2 + 9y^2 + 20x - 54y = -56$

15–20 Sketch the hyperbola, and label the vertices, foci, and asymptotes.

15. **(a)** $\dfrac{x^2}{16} - \dfrac{y^2}{9} = 1$ **(b)** $9y^2 - x^2 = 36$

16. **(a)** $\dfrac{y^2}{9} - \dfrac{x^2}{25} = 1$ **(b)** $16x^2 - 25y^2 = 400$

17. **(a)** $\dfrac{(x-1)^2}{9} - \dfrac{(y+2)^2}{4} = 1$

(b) $4(y-3)^2 - 9(x-2)^2 = 36$

18. **(a)** $\dfrac{(y+4)^2}{3} - \dfrac{(x-2)^2}{5} = 1$

(b) $16(x+1)^2 - 8(y-3)^2 = 16$

19. **(a)** $x^2 - 4y^2 + 2x + 8y - 7 = 0$
(b) $16x^2 - y^2 - 32x - 6y = 57$

20. **(a)** $4x^2 - 9y^2 - 16x - 54y - 29 = 0$
(b) $4y^2 - x^2 - 40y - 4x = -60$

21–26 Find an equation for the parabola that satisfies the given conditions.

21. **(a)** Vertex (0, 0); focus (3, 0).
(b) Vertex (0, 0); directrix $x = 7$.

22. **(a)** Vertex (0, 0); focus (0, −3).
(b) Vertex (0, 0); directrix $y = \frac{1}{4}$.

23. **(a)** Focus (0, −3); directrix $y = 3$.
(b) Vertex (1, 1); directrix $y = -2$.

24. **(a)** Focus (6, 0); directrix $x = -6$.
(b) Focus (−1, 4); directrix $x = 5$.

25. Axis $y = 0$; passes through (3, 2) and $(2, -\sqrt{2})$.

26. Vertex (5, −3); axis parallel to the y-axis; passes through (9, 5).

27–32 Find an equation for the ellipse that satisfies the given conditions.

27. **(a)** Ends of major axis $(\pm 3, 0)$; ends of minor axis $(0, \pm 2)$.
(b) Length of major axis 26; foci $(\pm 5, 0)$.

28. **(a)** Ends of major axis $(0, \pm\sqrt{7})$; ends of minor axis $(\pm 1, 0)$.
(b) Length of minor axis 8; foci $(0, \pm 3)$.

29. **(a)** Foci $(\pm 1, 0)$; $b = \sqrt{2}$.
(b) $c = 2\sqrt{3}$; $a = 4$; center at the origin; foci on a coordinate axis (two answers).

30. **(a)** Foci $(\pm 3, 0)$; $a = 4$.
(b) $b = 3$; $c = 4$; center at the origin; foci on a coordinate axis (two answers).

31. **(a)** Ends of major axis $(0, \pm 6)$; passes through (−3, 2).
(b) Foci (−1, 1) and (−1, 3); minor axis of length 4.

32. **(a)** Center at (0, 0); major and minor axes along the coordinate axes; passes through (3, 2) and (1, 6).
(b) Foci (2, 1) and (2, −3); major axis of length 6.

33–38 Find an equation for a hyperbola that satisfies the given conditions. (In some cases there may be more than one hyperbola.)

33. **(a)** Vertices $(\pm 2, 0)$; foci $(\pm 3, 0)$.
(b) Vertices $(\pm 1, 0)$; asymptotes $y = \pm 2x$.

34. **(a)** Vertices $(0, \pm 4)$; foci $(0, \pm 5)$.
(b) Vertices $(0, \pm 2)$; asymptotes $y = \pm\frac{2}{3}x$.

35. **(a)** Asymptotes $y = \pm\frac{3}{2}x$; $b = 4$.
(b) Foci $(0, \pm 5)$; asymptotes $y = \pm 2x$.

36. **(a)** Asymptotes $y = \pm\frac{3}{4}x$; $c = 5$.
(b) Foci $(\pm 3, 0)$; asymptotes $y = \pm 2x$.

37. **(a)** Vertices (0, 6) and (6, 6); foci 10 units apart.
(b) Asymptotes $y = x - 2$ and $y = -x + 4$; passes through the origin.

38. **(a)** Foci (1, 8) and (1, −12); vertices 4 units apart.
(b) Vertices (−3, −1) and (5, −1); $b = 4$.

39. **(a)** As illustrated in the accompanying figure, a parabolic arch spans a road 40 ft wide. How high is the arch if a center section of the road 20 ft wide has a minimum clearance of 12 ft?
(b) How high would the center be if the arch were the upper half of an ellipse?

40. **(a)** Find an equation for the parabolic arch with base b and height h, shown in the accompanying figure.
(b) Find the area under the arch. 20 ft

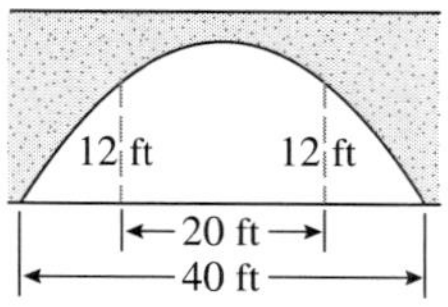

Figure Ex-39

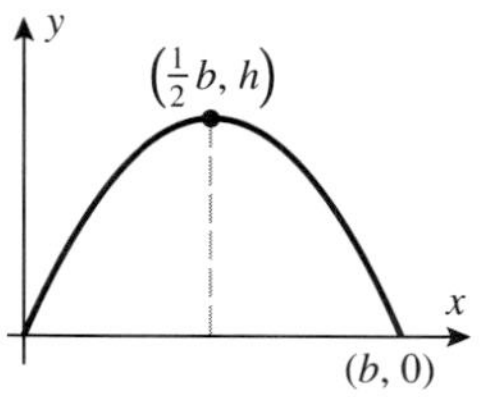

Figure Ex-40

41. Show that the vertex is the closest point on a parabola to the focus. [*Suggestion:* Introduce a convenient coordinate system and use Definition 12-A.1.]

42. As illustrated in the accompanying figure, suppose that a comet moves in a parabolic orbit with the Sun at its focus and that the line from the Sun to the comet makes an angle of 60° with the axis of the parabola when the comet is 40 million miles from the center of the Sun. Use the result in Exercise 41 to determine how close the comet will come to the center of the Sun.

43. For the parabolic reflector in the accompanying figure, how far from the vertex should the light source be placed to produce a beam of parallel rays?

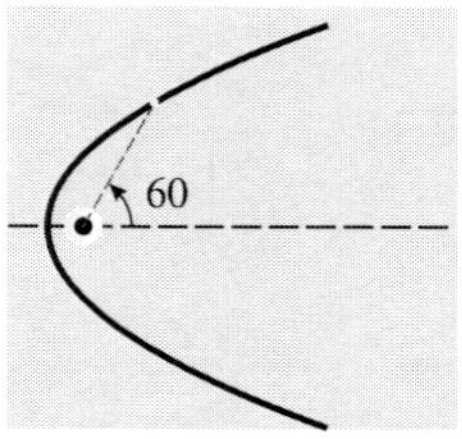

Figure Ex-42

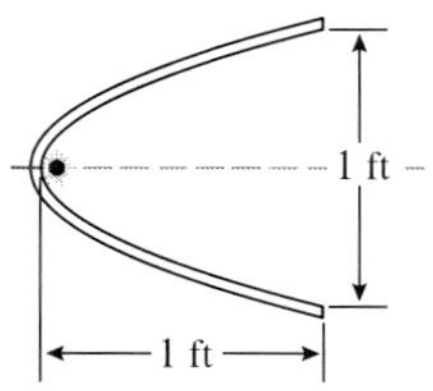

Figure Ex-43

ANSWERS TO ODD NUMBERED PROBLEMS

Section 1.1

1 2.9

3 $0, 4, 8$

5 $m = f(v)$

7 (a) (I), (III), (IV), (V) (VII), (VIII)
(b) (i) (V) and (VI)
(ii) (VIII)
(c) (III) and (IV)

9 (a) 40
(b) 2

11 (a) w
(b) $(-4, 10)$
(c) $(6, 1)$

13

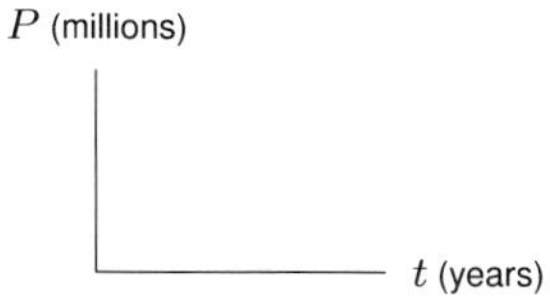

15

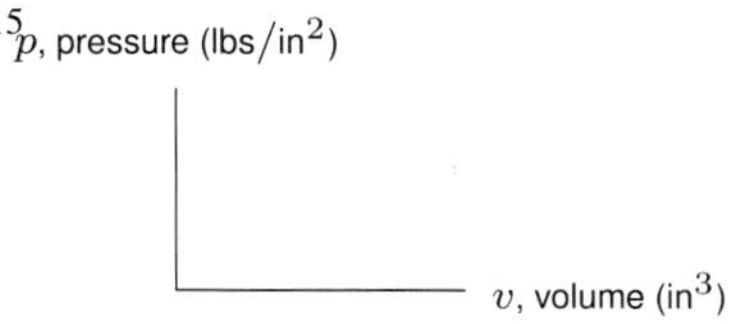

17 (a) $f(s) = 0.1s$
(b) $f(5) = 0.5$
(c) $s = 50$

19 (a) 100.3 m. own phones in 2000
(b) 20 m. own phones a years after 1990
(c) b m. own phones in 2010
(d) n m. own phones t years after 1990

21 $200 \leq s \leq 1000$

23 (a) 69°F
(b) July 17 and 20
(c) Yes
(d) No

27 (a) Yes
(b) No
(c) 8 female senators in 104th Congress
(d) 14 female senators in 108th Congress

29

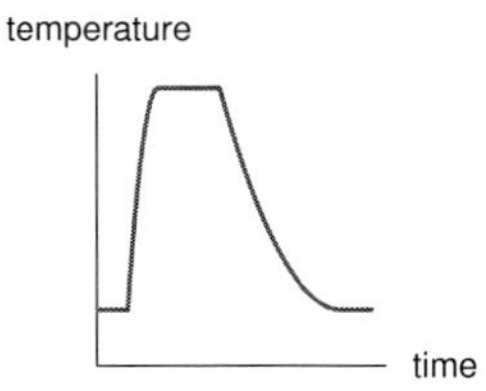

31 $S = 6\pi r^2$

33 (a) (ii)
(b) (i)
(c) (v)
(d) (iv)
(e) (ii)

35 $T(d) = d/5 + (10 - d)/8$

Section 1.2

1 $G(3) - G(-1) > 0$

3 0.513

5 0.513

7 (a) Negative
(b) Positive

9 Decreasing

11 (a) $300, -150$ people/yr
(b) $167, -83$ people/yr
(c) $235, -118$ people/yr

13 Between 3000 and 4000 years ago

15 (a) 2
(b) Increasing
(c) Increasing everywhere

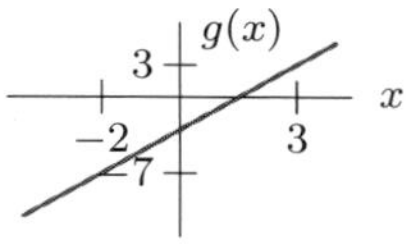

17 (a) $1/2$
(c) $1/4$

19 (a) (i) $1/2$
(ii) $1/2$
(iii) $1/2$
(b) Always $1/2$

21 (a) 9
(b) $(n - k)/(m - j)$
(c) $6x + 3h$

23 (a) 10
(b) 30, 30, 55, 29
(c) No; $\Delta G/\Delta t$ not constant

25 (a) $0°\text{C/meter}$
(b) $-0.008°\text{C/meter}$
(c) $0.009°\text{C/meter}$.

Section 1.3

1 Not linear

3 No

5 Yes

7 Vert int: 54.25 thousand; Slope: $-2/7$ thousand/yr

9 Vert int: $-\$3000$; Slope: $\$0.98$/item

11 $P = 18{,}310 + 58t$

15 (a) \$5350, \$5700, \$6750, \$8500, \$12,000

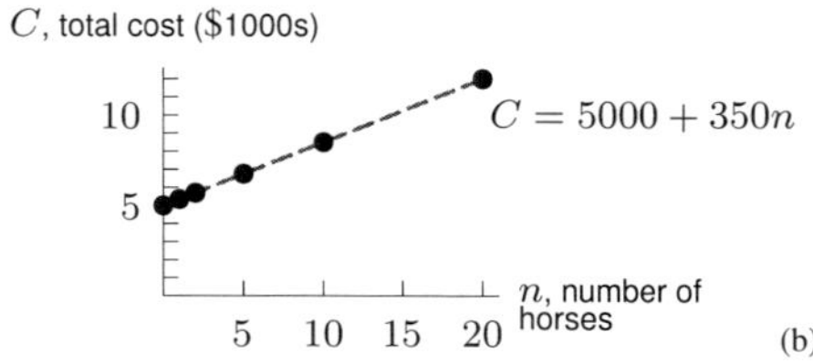

(b) $C = 5000 + 350n$
(c) \$350/horse

17 (a) Radius and circumference
(c) 2π

19 (a)

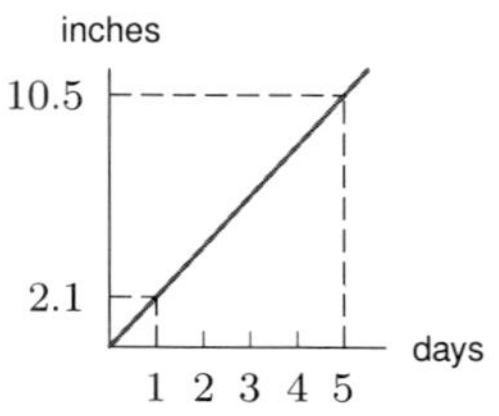

(b)

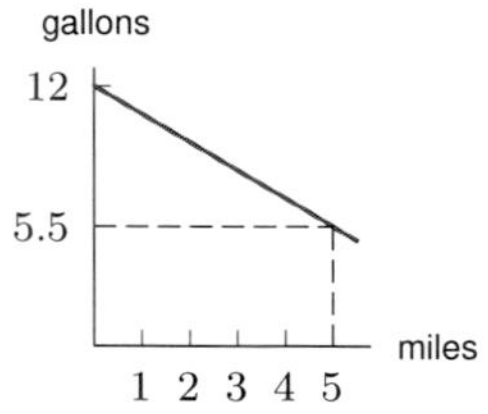

21 (a) No
(b) Looks linear

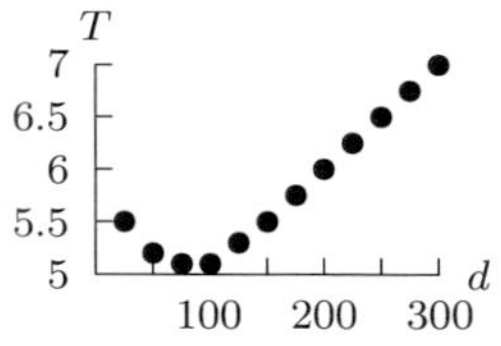

(c) $\Delta T/\Delta d = 0.01°\text{C/meter}$

23 (a) $T = \$1900$
(b) $C = 7$
(d) Twelve credits
(e) Fixed costs that do not depend on the number of credits taken

25 (a) $m = -0.62, A = 16.2$
(b) $f(15) = 15.27$
(c) 0.93 acres
(d) $t = 67.7$

29 No

31 (a)

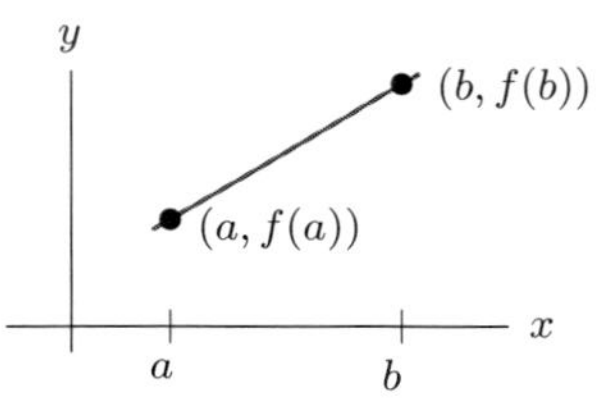

(b) $(f(b) - f(a))/(b - a)$

Section 1.4

1 $y = 4/5 - x$

3 $y = 180 - 10x$

5 $y = -0.3 + 5x$

7 $y = -40/3 - 2/3x$

9 $y = 21 - x$

11 Yes; $F(P) = 13 + (-1/8)P$

13 Yes; $C(r) = 0 + 2\pi r$

15 Yes; $f(x) = n^2 + m^2 x$

17 $y = 8 + 3x$

19 $y = (11 + 2x)/3$

21 $y = 0.03 + 0.1x$

23 $f(x) = 3 - 2x$

25 $q = 2500 - 2000p$

27 $y = 459.7 + 1x$

29 $u = (1/12)n$

31 $y = -4 + 4x$

33 $h(t) = 254 - 248t$

35 $y = \frac{16+5\sqrt{7}}{2+\sqrt{7}} - \frac{3}{2+\sqrt{7}}x$ or
$y = (1 + 2\sqrt{7}) + (2 - \sqrt{7})x$

37 (a) 1000, 990.2, 980.4, 970.6, 960.8
(b) v decreasing at constant rate
(c) Slope: -9.8 meter/sec^2
v-intercept: 1000meters/sec
t-intercept: 102.04 sec

39 (b) $p = 12 - s$
(c)

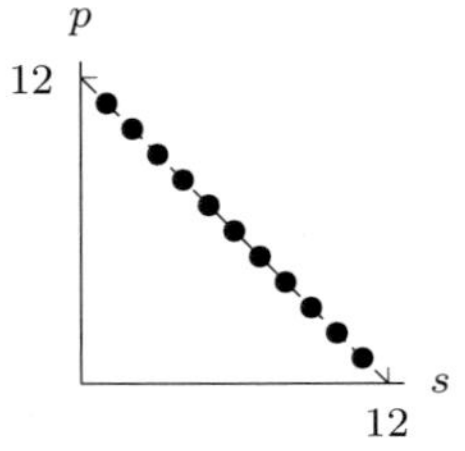

41 (a) $q = 170 - 50p$

43 (a)

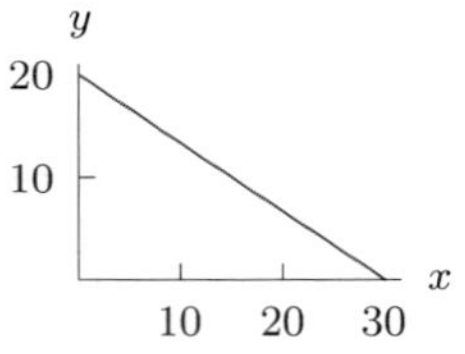

(b) $f(x) = 20 - (2/3)x$
(d)

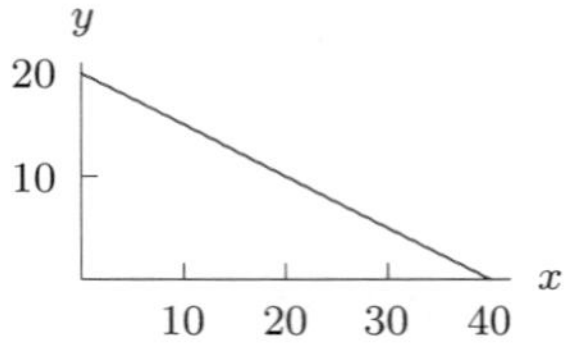

45 (a) $r = 0.005H - 0.03$
(b) $S = 200$

Section 1.5

1 (a) (ii)
(b) (iii)
(c) (i)

3 (a) $y = -2 + 3x$
(b) $y = -1 + 2x$

5 (a) $y = (1/4)x$
(b) $y = 1 - 6x$

7 A, h
B, f
C, g
D, j
E, k

9 Parallel

11 Parallel

13 Neither

15 $y = 24 - 4x$

17 (a) $y = 9 - \frac{2}{3}x$
(b) $y = -4 + \frac{3}{2}x$
(c)

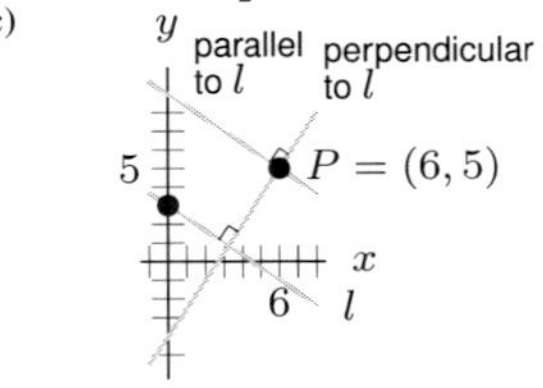

19 (b) $y = 0$

21 (d) Steepness appears to decrease as vertical height of window increases

23 Slope $= -2/5$
$y = -(2/5)x + 3$
(answers may vary)

25 $y = (3/2)x$

27 $\beta = 5$

29 $y = -2x - 3$

31 (a) Company A: $20 + 0.2x$
Company B: $35 + 0.1x$
Company C: 70
(b)

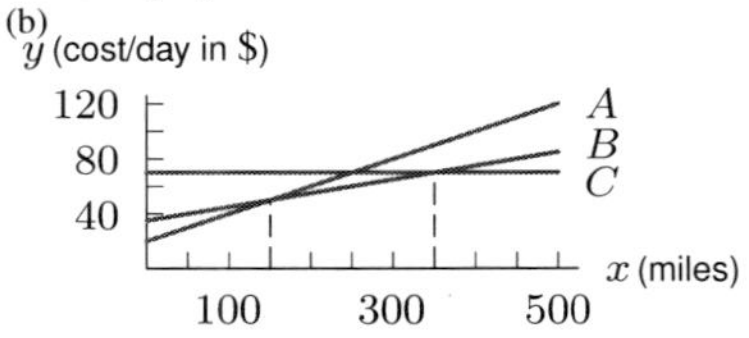

(c) Slope: mileage rate
Vertical intercept: fixed cost/day
(d) A for $x < 150$
B for $150 < x < 350$
C for $x > 350$

33 (a) $y = -7065.9 + 3.65t$
(b) 307.1 million tons

35 (a) $m_1 = m_2$ and $b_1 \neq b_2$
(b) $m_1 = m_2$ and $b_1 = b_2$
(c) $m_1 \neq m_2$
(d) Not possible

Section 1.6

1 (a) $r = 1$
(b) $r = 0.7$
(c) $r = 0$
(d) $r = -0.98$
(e) $r = -0.25$
(f) $r = -0.5$

3 (a) and **(b)**

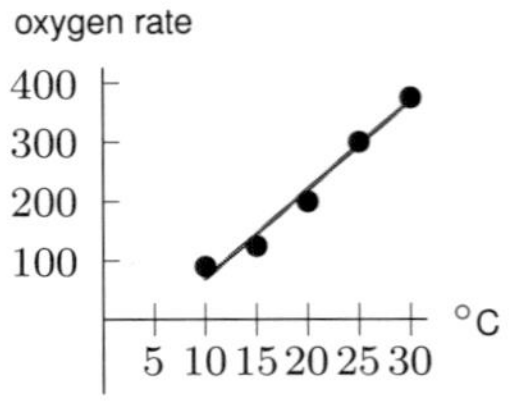

(c) $y = 15x - 80$
(e) Strong positive correlation

5 (a) and **(b)**

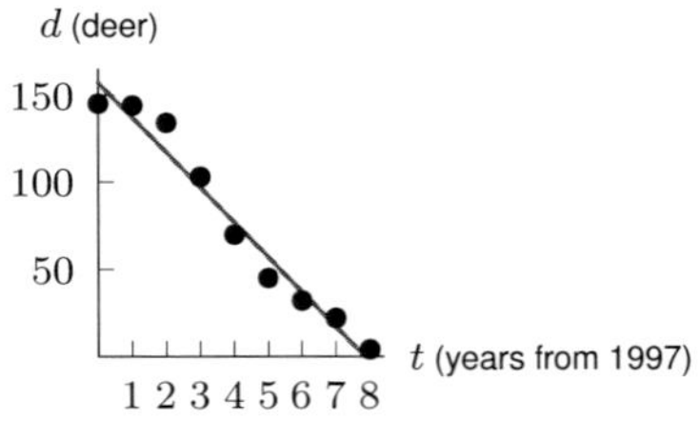

(c) $y = -20x + 157$
(e) Strong negative correlation.

7 (a) A column of zeros

Chapter 1 Review

1 Neither

3 Both

5 Both

7 (a) 9
(b) $\dfrac{n-k}{m-j}$
(c) $6x + h$

9 Yes

11 $f(t) = 2.2 - 1.22t$

13 (a) $y = 7 + 2x$
(b) $y = 8 - 15x$

15 Neither

17 Perpendicular

19 $f(x) = -12.5 - 1.5x$

21 $h(t) = 12{,}000 + 225t$

23 $y = 6 - (3/5)x$

25 Parallel line:
$y = -4x + 9$
Perpendicular line:
$y = 0.25x + 4.75$

27 $\pi(n) = -10{,}000 + 127n$

29 (a) $r_m(0) - r_h(0) = 22$

(b) $r_m(9) - r_h(9) = -2$
(c) $r_m(t) < r_a(t)$ for $t = 5$ to $t = 9$

31 (a) Yes
(b) No

33
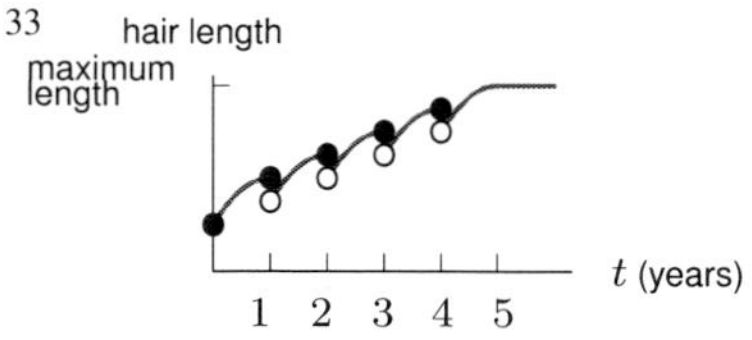

35 $0.945P$

37 (a) 10.71 gallons
(b) 0.25 gallons
(c) 55 mph

39 (a) $x + y$
(b) $0.15x + 0.18y$
(c) $(15x + 18y)/(x + y)$

41 (a) $\$11{,}375$
(b) $\$125$
(c) $\$5$

43 $C(n) = 10{,}500 + 5n$

45 (b) (i)
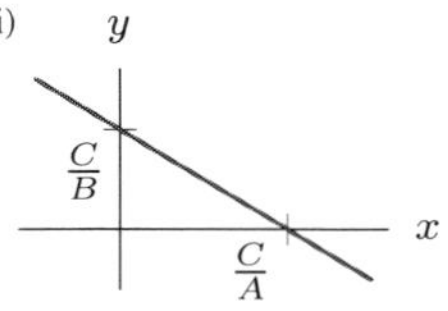

(ii)
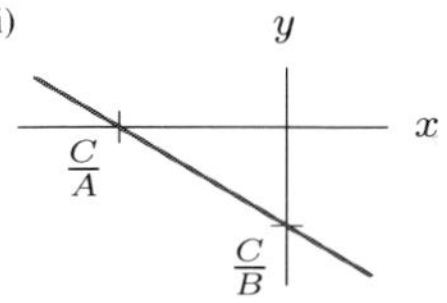

(iii)
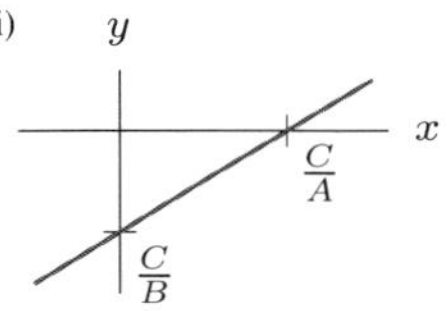

Ch. 1 Understanding

1 False
3 True
5 True
7 True
9 True
11 True
13 True
15 True
17 False
19 False
21 False
23 True
25 True
27 False
29 True
31 False
33 False
35 True
37 False
39 False
41 True
43 False
45 True
47 True
49 False
51 True
53 True

Chapter 1 Tools

1 $x = 5$
3 $z = 11/2$
5 $y = 26$
7 $x = 9$
9 $t = 45/13$
11 $x = 20$
13 $y = -17/16$
15 $r = 10$
17 $s = -0.115$
19 $r = C/(2\pi)$
21 $F = (9/5)C + 32$
23 $a = 2(h - v_0t)/t^2$
25 $x = (c - ab)/(2a)$
27 $v = (3w - 2u - z)/(u + w - z)$
29 $x = -a(b + 1)/(ad - c)$
31 $y' = 4/(y + 2x)$
33 $x = 4,\ y = -1$
35 $x = 4,\ y = 3$
37 $x = -55, y = 39$
39 $x = 13.5; y = -12$
41 $x = 3/2, y = 3/2$
43 $x = 3, y = 6$
45 $A = (17, 23);\ B = (0, 40);\ C = (-17, 23)$
47 $A = (3, 21)$
49 $B = (7, 4), A = (11, 0)$
51 $A = (2, 5), B = (5, 8)$

Section 2.1

1 $f(-7) = -9/2$
3 $-3/31$; 1
5 32; $\sqrt[3]{9/4}$
7 (a) 1
(b) $-1/2$
9 (a) 6
(b) 2, 3
11 2/3
13 -8
15 (a) $1/(x + 3)$
(b) $(1/x) + (1/3)$
17 (a) $2, 0, -2$
(b) $x = -1$
19 100
21 6π cubic inches
23 $p(-1) = 1; -p(1) = -3$; Not equal
25 (a) $t = 6$
(b) $t = 1, t = 2$
27 (a) (i) $1/(1 - t)$
(ii) $-1/t$
(b) $x = 3/2$
29 (a) 24
(b) 10
(c) -7
(d) 0
(e) 20
31 (a) -1
(b) $x = \pm 3$
(c) 0
(d) -1
(e) $3, -3$
33 (a) 48 feet for both
(b) 4 sec, 64 ft
35 (a) $s(2) = 146$
(b) Solve $v(t) = 65$
(c) At 3 hours
37 (b) No
(c) $f(0) = 0$
$f(-1) = 1$
$f(-2) = -1$
$f(0.5)$: undefined

Section 2.2

1 $f(x) \leq -(1/2)$ or $f(x) \geq (1/2)$
3 $-4 \leq f(x) \leq 5$
5 Domain: all real x
Range: all real $f(x)$
7 Domain: all real x
Range: $f(x) \leq 9$
9 Domain: $x \leq 8$
Range: $f(x) \geq 0$
11 Domain: $x, x \neq -1$
Range: $f(x) < 0$
13 D: $1 \leq x \leq 7$; R: $2 \leq f(x) \leq 18$
15 Domain: $x \geq 3$ or $x \leq -3$
Range: $q(x) \geq 0$
17 Domain: all real numbers
Range: all real numbers
19 D: all real numbers; R: all real numbers ≤ 7
21 D: all real numbers; R: all real numbers ≥ 2
23 D and R: all real numbers
25 $y = \frac{1}{x-3} + \sqrt{x}$
27 D: $0 \leq t \leq 12$
R: $0 \leq f(t) \leq 200$
29 Domain: integers $0 \leq n \leq 200$
Range: $0, 4, 8, \ldots, 800$
31 (a) 162 calories

(c) (i) Calories = $0.025\times$ weight

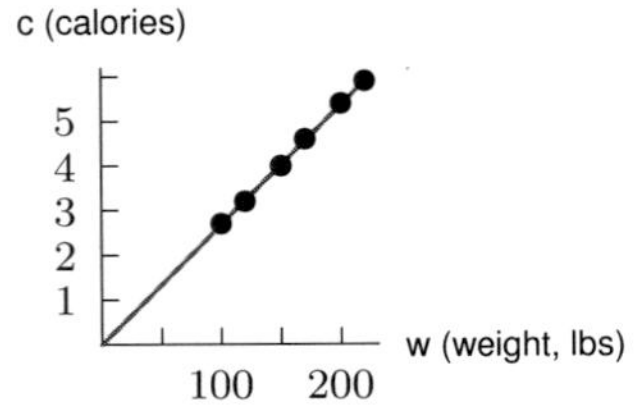

(ii) (0,0) is the number of calories burned by a weightless runner

(iii) Domain $0 < w$; range $0 < c$

(iv) 3.6

33 D: all real numbers $\geq b$;
R: all real numbers ≥ 6

35 (a) $p(0) = 50$
$p(10) \approx 131$
$p(50) \approx 911$
(c) $50 \leq p(t) < 1000$

Section 2.3

1

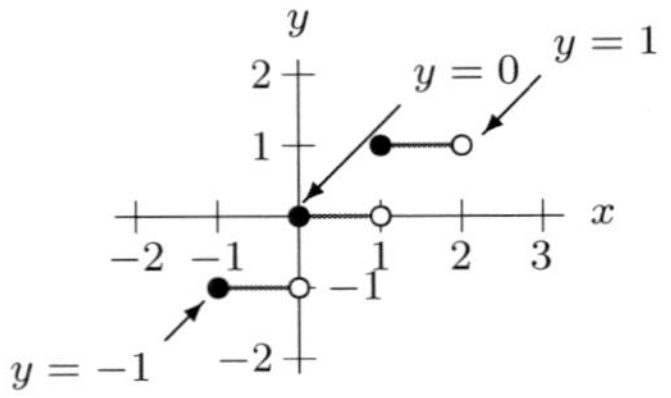

3

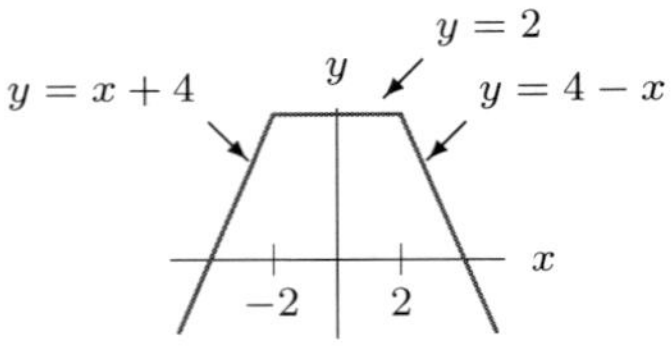

5 Domain: all reals;
Range: $G(x) < 0$ and $G(x) \geq 3$

7 $y = \begin{cases} 5 - x & \text{for } x < 3 \\ -1 + (1/2)x & \text{for } x \geq 3 \end{cases}$

9 $y = \begin{cases} 4 - \frac{1}{2}x & \text{for } 1 \leq x \leq 3 \\ -9 + 2x & \text{for } 5 \leq x \leq 8 \end{cases}$

11 (a) Yes
(b) No
(c) $y = 1, 2, 3, 4$

13 (c) Domain: all x, $x \neq 0$
Range: -1 and 1
(d) False, $u(0)$ is undefined

15 (a)

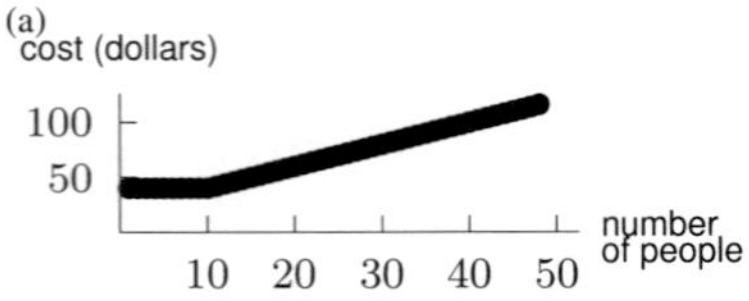

(b) Integers from 1 to 50
Even integers from 40 to 120

17 (a) $n(L) = 2L + 10$
(b) Domain: $L \geq 5$
Range: $n(L) = 20, 21, 22, 23, \ldots$

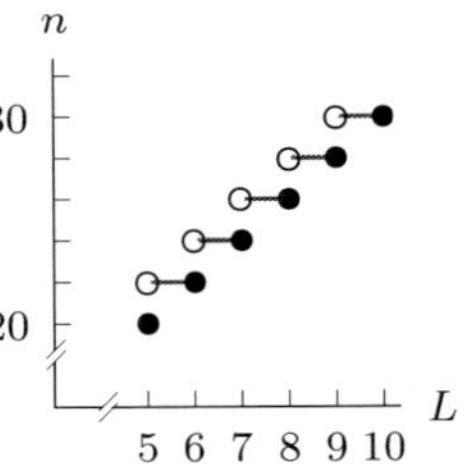

(c) Domain: All real numbers
Range: All real numbers

19 (a) \$1.12
\$1.26
(b) Domain: All positive integers
Range: All positive multiples of 0.14

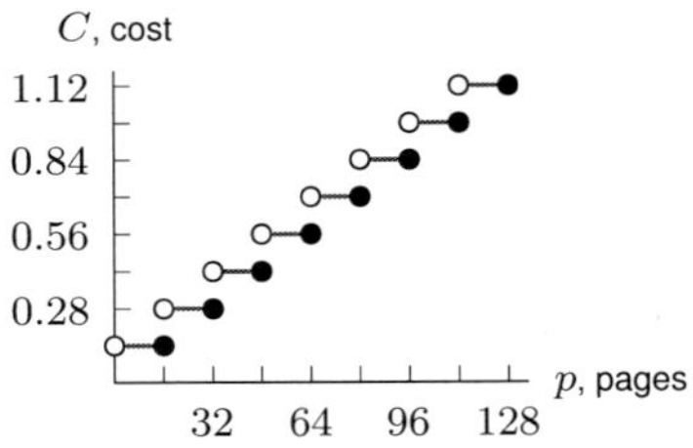

Section 2.4

1 Area in sq cm at time t

3 Acceleration in m/sec^2 at wind speed w

5 10

7 5

9 $4x^2 + 12x + 10$

11 $x^4 + 2x^2 + 2$

13 Year pop is P; years

15 Days for N inches snow; days

17 Interest rate for \$$I$ interest; %/year

19 $f^{-1}(Q) = (Q - 3)^{1/3}$

21 $f^{-1}(P) = (P + 2)/14.$

23 (a) b
(b) a
(c) a
(d) b

25 $f^{-1}(f(A)) = A$
$f(f^{-1}(n)) = n$

27 $g^{-1}(7) = 1, g^{-1}(12) = 2, g^{-1}(13) = 3, g^{-1}(19) = 4, g^{-1}(22) = 5$

29 $f^{-1}(V) = \sqrt[3]{3V/4\pi}$

31 (a) 12, perimeter for $s = 3$
(b) 5; side for $P = 20$
(c) $f^{-1}(P) = P/4$

33 (a) $s = f(A) = +\sqrt{\frac{A}{6}}$
(b) $V = g(f(A)) = \left(\sqrt{A/6}\right)^3.$

35 (a) 5000 loaves cost \$653
(b) 620 loaves \$80
(c) \$790 for 6300 loaves
(d) 1200 loaves for \$150

37 $f(t) = 4\pi(50 - 2.5t)^3/3$

39 (a) Domain: $0 \leq t \leq 5$, range: $365 \leq C(t) \leq 375$
(b) $C(4)$ is the concentration in 2002
(c) $C^{-1}(370)$ is number of years after 1998 when the concentration was 370 ppm

41 (a) $P = f(s) = 4s$
(b) $f(s + 4) = 4(s + 4) = 4s + 16$
(c) $f(s) + 4 = 4s + 4$
(d) Meters

Section 2.5

1 Concave down

3 Concave up

5 Concave up

7 Concave up

9 Rates of change: 4.35, 4.10, 3.80; Concave down

11 Possible graph:

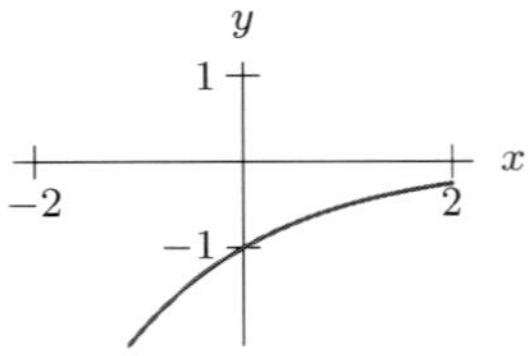

13 Increasing;
concave up

15 Increasing;
concave up then down

17 Increasing;
concave up then down

19 (a) F, IV
(b) G, I
(c) E, II
(d) H, III

21 (a) From A to F
(b) From O to A
(c) From D to E
(d) From F to I

Section 2.6

1 Yes; $f(x) = 2x^2 - 28x + 99$

3 Yes; $g(m) = -2m^2 + \sqrt{3}m + 42$

5 Not quadratic

7 Yes; $T(n) = (\sqrt{3} - 1/2)n^2 + \sqrt{5}$

9 $x \approx -0.541$ and $x \approx 5.541$

11 (a) $x \approx 0.057, x \approx 1.943$
(b) No real solutions

13 $x = 2, 3/2$

15 $x = 2, x = -1$

17 $x = -1/3$

19 $x = (-1 \pm \sqrt{6})/5$

21 $x = (23 \pm \sqrt{1821})/(-34) \approx 1.932$ or -0.579

23 No zeros

25 4.046 sec

27 Rates of change: $0, -4, -8$; Concave up

29

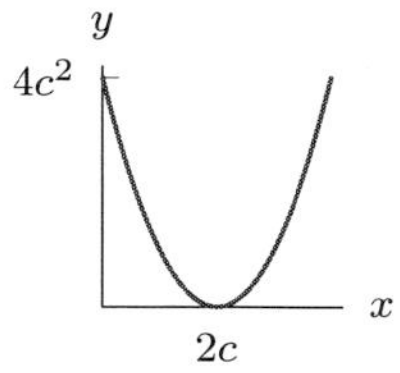

31 (a) 4 meters per second
(b) 2 seconds
(c) Concave up

33 (a) Similar shape, end behavior, Different intercepts, vertices
(b) Differences in intercepts become less significant
(c) Graphs look more like each other as the scale causes the distances between intercepts to decrease

35 (a) $y = 0.01x^2 + 2x + 1$
(b) $y = 2.03x + 0.98$
(c) 0.02
(d) 23.52
(e) $-0.791 \le x \le 3.791$

Chapter 2 Review

1 54

3 (a) $h(1) = b + c + 1$
(b) $h(b+1) = 2b^2 + 3b + c + 1$

5 $l = 0$ and $l = 7$

7 Domain: all real numbers
Range: $h(x) \ge -16$

9 Domain: $4 \le x \le 20$
Range: $0 \le r(x) \le 2$

11 (a) $-3(x^2 + x)$
(b) $2 - x$
(c) $x^2 + x + \pi$
(d) $\sqrt{(x^2 + x)}$
(e) $2/(x + 1)$
(f) $(x^2 + x)^2$

13 $3x^3 - 4$

15 $f^{-1}(y) = (y + 7)/3$

17 (a) 1.5
(b) 2.2
(c) 2.2
(d) 1.5

19 Time, yrs, at which pop is P mil

21 $(0, 2)$

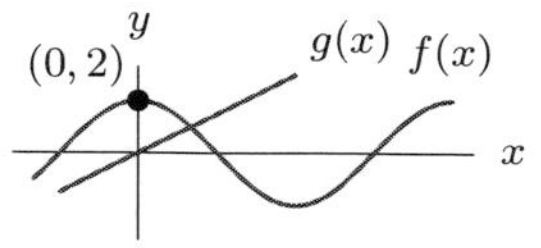

23 Intersect at $x = 2$

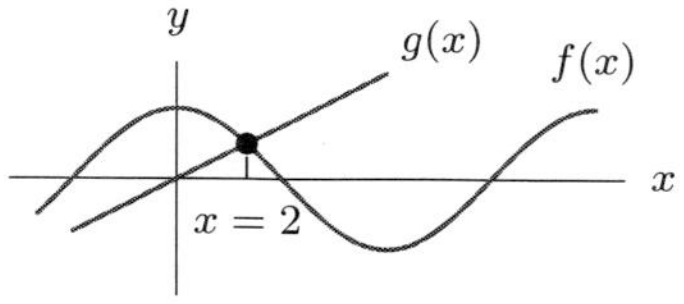

25 $a/2$

27 $a/(a - a^2 + 1)$

29 (a) 5000 loaves cost \$653
(b) 620 loaves \$80
(c) \$790 for 6300 loaves
(d) 1200 loaves for \$150

31 (a) $-17.778°$C
(b) $32°$F
(c) $37.778°$C
(d) $212°$F

33 $f^{-1}(T) = gT^2/(4\pi^2)$;
Length for period T

35 (a) 7000
(b) 8500; 4 weeks after the beginning of the epidemic
(c) $w = 1, w = 10$
(d) $1.5 \le w \le 8$

37 (a) 2
(b) Unknown
(c) 5

39 (a) $f(1) = 2$
$g(3) = 4$
(c) $f(5) = 14$
$f(-2) = -7$
$g(5) = 16$
$g(-2) = 9$
(d) $f(x) = -1 + 3x$
$g(x) = (x - 1)^2$

41 (a) 2 lbs cost \$4.98
(b) 0.5 lb costs \$1.25
(c) \$0.62 buys 1/4 lb
(d) \$12.45 buys 5 lb

43 (a) (i) 6
(ii) 5
(iii) Not defined
(b) (i) $50 \le s \le 75$
(ii) $76 \le s \le 125$

47 (a) $A = 2\pi r^2 + 710/r$
(b)

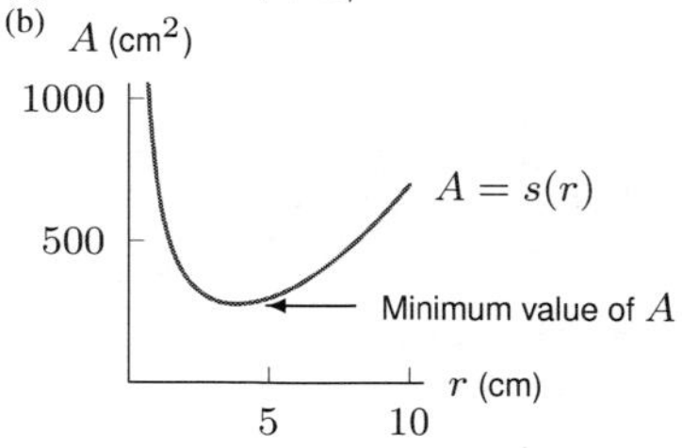

(c) Domain: $r > 0$
Approximate range: $A > 277.5$ cm^2
(d) ≈ 277.5 cm^2
$r \approx 3.83$ cm
$h \approx 7.7$ cm

Ch. 2 Understanding

1 False
3 False
5 False
7 False
9 True
11 True
13 False
15 False
17 True
19 True
21 True
23 False
25 True
27 True
29 True
31 True
33 False
35 True
37 True
39 True
41 True
43 True
45 False
47 False
49 False
51 False

Chapter 2 Tools

1 $3x + 6$

3 $6x - 14$

5 $12x + 12y$

7 $2x^2 + 5x$

9 $-50r^2 - 60r^2s$

11 $5xz - 10z - 3x + 6$

13 $x^2 + 4x - 12$

15 $3x^2 - 2x - 16$

17 $96wy + 84y - 40w - 35$

19 $x - 7$

21 $4x^2 + 11x - 20$

23 $Pp^2 - 6Ppq + 9Pq^2$

25 $4x^2 - 24x + 43$

27 $2^u + u2^{2u}$

29 $3(y + 5)$

31 $2(2t - 3)$

33 $u(u - 2)$

35 $3u^2(u^5 + 4)$

37 $7rs(2r^3s - 3t)$

39 Cannot be factored

41 Cannot be factored
43 $(x-3)(x+1)$
45 $(x+3)(x-1)$
47 $(3x-4)(x+1)$
49 $(x+7)(x-4)$
51 $x(x+3)(x-1)$
53 $(x+2y)(x+3z)$
55 $(ax-b)(ax+b)$
57 $(B-6)(B-4)$
59 Cannot be factored.
61 $(t-1)(t+7)$
63 $(a-2)(a^2+3)$
65 $(d+5)(d-5)(c+3)(c-3)$
67 $(r+2)(r-s)$
69 $xe^{-3x}(x+2)$
71 $(s+2t+2p)(s+2t-2p)$
73 $(x-3+2z)(x-3-2z)$
75 $(r-2)(\pi r+3)$
77 $x=-6$ or $x=-1$
79 $w=(-5/2)$ or $w=2$
81 $x=(7/16)$
83 $y=170$
85 $x=(37/2)$
87 $t=3\pm 2\sqrt{3}$
89 $g=3, -2,$ or 2
91 $p=3, p=-3, p=-1/2$
93 $t=0, t=\pm 8$
95 $x=4, x=-3/4$
97 $n=5, -2,$ or 2
99 $y=-2\pm\sqrt{6}$
101 $x=\pm 1$
103 $q=\pm\sqrt{2}$
105 $x=-1/8$
107 $v=700\pi$
109 $l=gT^2/4\pi^2$
111 $x=3, x=-4$
113 $x=3$ and $y=-3$, or $x=-1$ and $y=-3$
115 $x=(3\pm\sqrt{63})/2,$
$y=(-3\pm\sqrt{63})/2$
117 $x=2.081, y=8.013$
$x=4.504, y=90.348$
119 $(4,3), (-3,-4)$

Section 3.1

1 1.03 (per year)
3 0.99 (per day)
5 550
7 495
9 411.8
11 $a=1750; b=1.593; r=59.3\%$
13 $a=79.2; b=1.002; r=0.2\%$
15 (a) III
(b) I
(c) II
(d) VI
(e) V
(f) IV
17 20%; 2%.
19 $P=2200(0.968)^t$
21

Yr	2006	2007	2008	2009	2010
$	73	80.30	88.33	97.16	106.88

23 $Q=726(0.94374)^t$

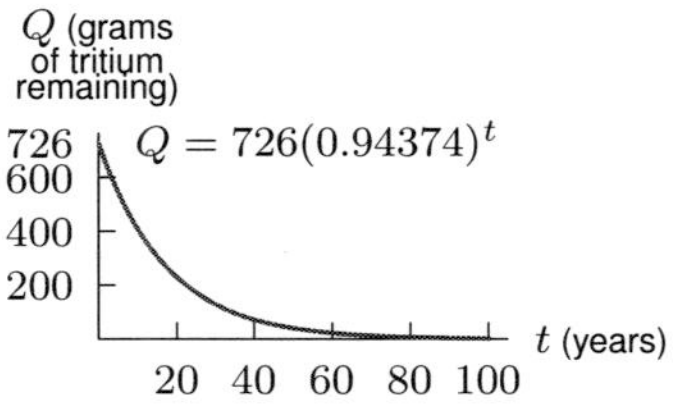

25 $109,272.70
27 (a) $P=7.50(1.035)^t$
(b) ≈ $14.92
29 (a) $C=100(0.84)^t$
(b) 41.821 mg
31 $f(n)=P_0(0.8)^n$

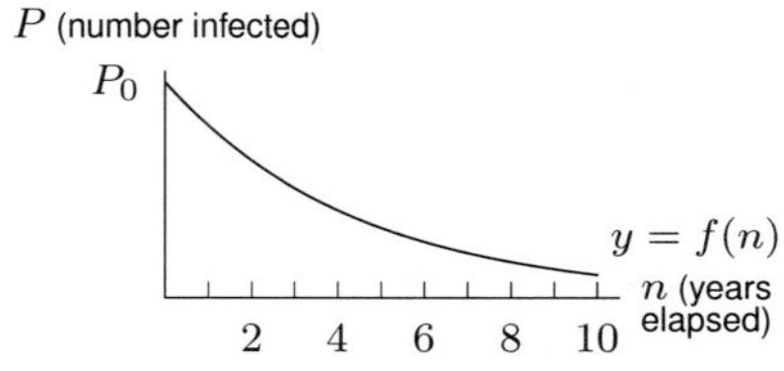

33 $L=420(0.15)^n$
35 (a) $573.60 per month
(b) $440.40 per month
(c) $645.00 per month
(d) $12,852
(e) $55,296
37 $b(b^4-1)$
39 (a) $R=Nr$
(b) $A=R/P=Nr/P$
(c) $N_{new}=1.02N$
$r_{new}=1.03r$
(d) $R_{new}=1.0506R$; 5.06%
(e) $A_{new}=(0.9728)A$; average revenue falls by 2.7%
41 b_0
43 t_0 decreases

Section 3.2

1 B, C, D exponential
3 (a) $P=100+10t$.
(b) $P=100(1.10)^t$.
(c)

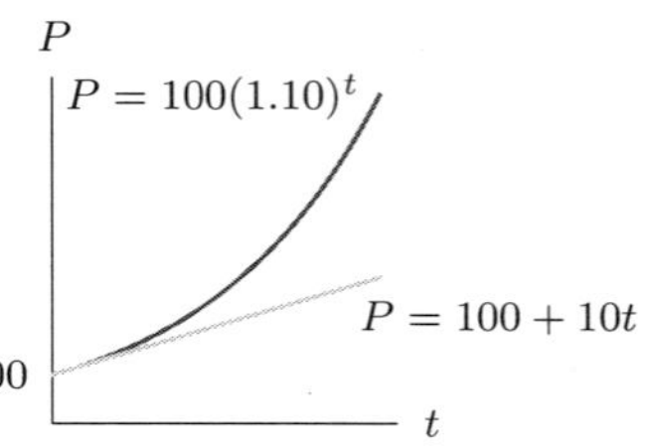

5 $f(x)=16{,}384(1/4)^x$
7 (a) $g(x)$ is linear
(b) $g(x)=2x$

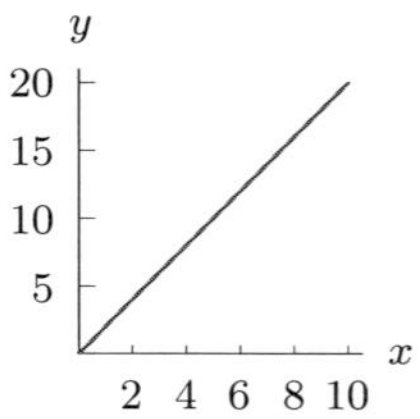

9 (a) $i(x)$ is linear
(b) $i(x)=18-4x$

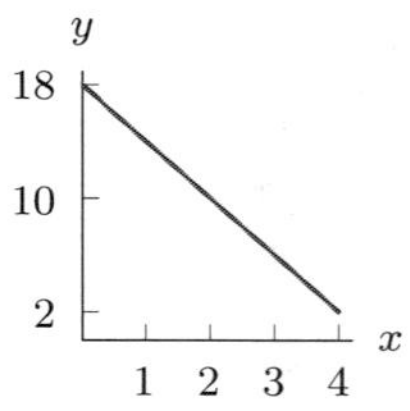

11 $h(x)=3(5)^x$
13 $g(x)=2(4)^x$
15 $g(x)=14.20(0.6024)^x$
17 (a) $f(x)=\frac{31}{8}x+\frac{49}{4}$
(b) $f(x)=5(2)^x$
19 $y=(1/2)^x$
21 $y=\frac{1}{5}(3)^x$
23 $y=2(0.8)^x$
25 $p=20(1.0718)^x; q=160(0.8706)^x$
27 Exponential: $q(x)=82.6446\cdot 1.03228^x$
29 Exponential; $g(t)=1024\cdot 0.5^t$
31 (a) $P(t)=2.58+0.09t$, increases by 90,000 people per year
(b) $P(t)=2.68(1.026)^t$, increases by 2.6% per year
33 (a) $P=1154.160(1.20112)^t$
(b) $1154.16
(c) 20.112%
35 (a) $S=91(1.52)^t$
(b) Increasing by 52%/yr
(c) No
37 $N=10(1.13)^t$; 13%/yr
39 (a) $V=(61{,}055)(0.916)^t$
(b) $V=61{,}055-4012.19t$
(c) Linear
41 (a) Penalty A: $291 million
Penalty B: $5.369 million
(b) $A(t)=(1+10t)10^6$ dollars
$B(t)=(0.01)2^t$ dollars
(c) $t\approx 35.032$ days

Section 3.3

1 (b)

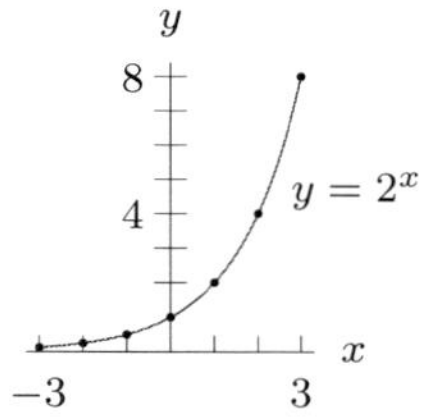

3 $h(x)$ top; $g(x)$ middle; $f(x)$ bottom

5 D

7 D

9 $x = 7.158$

11 $w = 1.246$

13 Zero

15 (a) $P = 651(0.9925)^t$
(b) $603{,}790$
(c) $t = 22.39$

17 (a) $y = (10{,}000) \cdot (0.9)^t$
(b) ≈ 5905 cases
(c) ≈ 21.854 years

19 (a) All
(b) b
(c) b, a, c, p
(d) $a = c$
(e) d and q

21 y_0 decreases , $y_0 > 0$

23 t_0 decreases

27

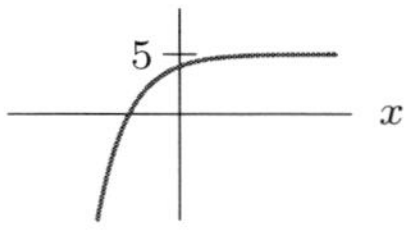

29

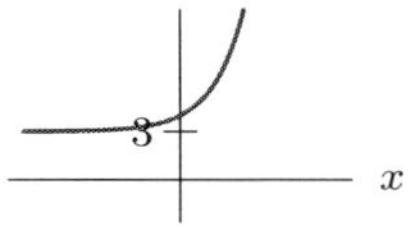

31 (a) 5
(b) -3

33 (a) 0
(b) 0
(c) 0.7

35 $y = 8$

39 (a)

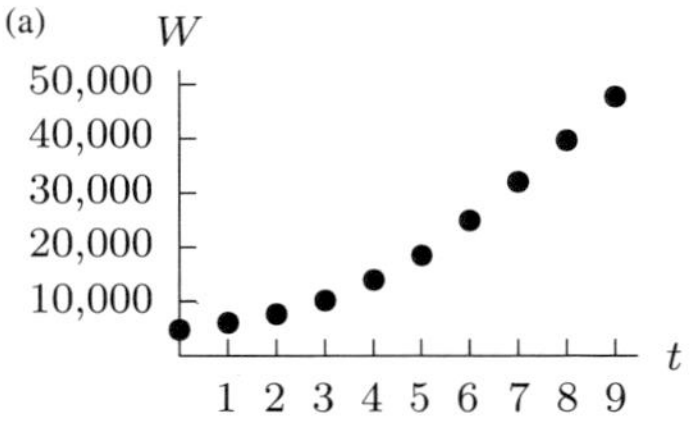

(b) $W = 4710(1.306)^t$; answers may vary
(c) 30.6%/yr

41 Domain: all t
Range: $Q > 0$

43 (a)

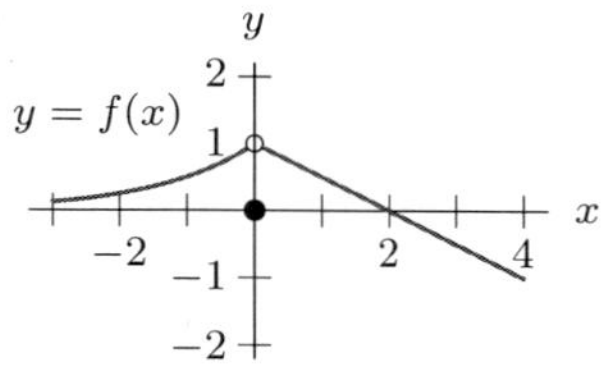

(b) $f(x) < 1$
(c) $(0, 0)$ $(2, 0)$
(d) As $x \to +\infty$, $f(x) \to -\infty$
As $x \to -\infty$, $f(x) \to 0$
(e) Increasing for $x < 0$, decreasing for $x > 0$

Section 3.4

1 Bottom to top:
$y = e^x, y = 2e^x, y = 3e^x$

3 (a)=(II); (b)=(III); (c)=(IV); (d)=(I)

5 $f(x) = e^{-x}$
$g(x) = e^x$
$h(x) = -e^x$

7 (a)=(I); (b)=(II); (c)=(III); (d)=(IV)

9 46.210 years

11 0

13 2

15 $a > 0, k > 0$

17 (a) $\$4157.86$
(b) $\$4234.00$

19 (a) $P(t) = 25{,}000e^{0.075t}$
(b) 7.788%

21 (a) $A = 50e^{-0.14t}$
(b) 12.330 mg
(c) 2016

23 (a) $P = 77.2e^{0.016t}$
(b) 84.979 million tons
(c) Middle of 2013

25 (a) Increasing, 2.7169239, $2.7181459, 2.7182682, 2.7182805$
(b) Approximately 10^7
(c) $(1 + 1/10^{16})^{10^{16}}$ appears to be 1

27 (a)

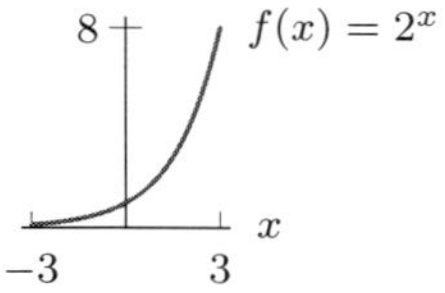

(b) 0.69
(c) 1.10
(d) $e \approx 2.72$

Section 3.5

1 (a) $\$505$
(b) $\$505.02$
(c) $\$505.03$
(d) $\$505.03$

3 (a) $\$525$
(b) $\$525.62$
(c) $\$525.64$
(d) $\$525.64$

5 A is continuous, B is annual
Initial deposit is $500

7 (a) $\$7401.22$
(b) $\$7459.12$

9 5.127%

11 (a) Nom: 1% Eff: 1%
(b) Nom: 1% Eff: 1.004%
(c) Nom: 1% Eff: 1.005%
(d) Nom: 1% Eff: 1.005%

13 (a) Nom: 3% Eff: 3%
(b) Nom: 3% Eff: 3.034%
(c) Nom: 3% Eff: 3.045%
(d) Nom: 3% Eff: 3.045%

15 (a) (i) 6.14%
(ii) 6.17%
(iii) 6.18%
(iv) 6.18%
(b) 1.0618
The highest possible APR is 6.18%.

17 (i) (b)
(ii) (a)
(iii) (c)
(iv) (b), (c) and (d)
(v) (a) and (e)

19 From best to worst: B, C, A

21 (a) The effective annual yield is 7.763%
(b) The nominal annual rate is 7.5%

23 $\$27{,}399.14$

25 5.387%

Chapter 3 Review

1 Yes; $g(w) = 2(1/2)^w$

3 Yes; $f(x) = (1/4)9^x$

5 Yes; $q(r) = -4(1/3)^r$

7 Yes; $Q(t) = 2^t$

9 Not exponential

11 (a) Starts at 200, grows at 2.8%/yr
(b) Starts at 50, shrinks at continuous rate of 17%/yr
(c) Starts at 1000, shrinks by 11%/yr
(d) Starts at 600, grows at continuous rate of 20%/yr
(e) Starts at 2000, shrinks by 300 animals/yr
(f) Starts at 600, grows by 50 animals/yr

13 $y = 10(1.260)^x$

15 $y = (1/2)(1/3)^x$

17 $y = 4(0.1)^x$ or $y = 4(10)^{-x}$

19 (a) (ii)
(b) (i)
(c) (iv)
(d) (ii)
(e) (iii)
(f) (i)

21

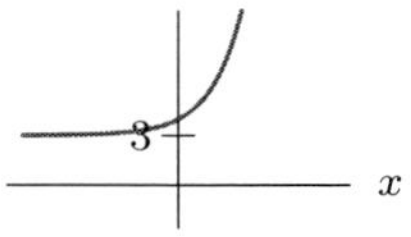

23

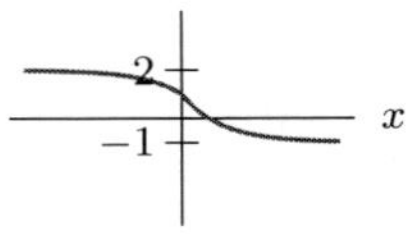

25

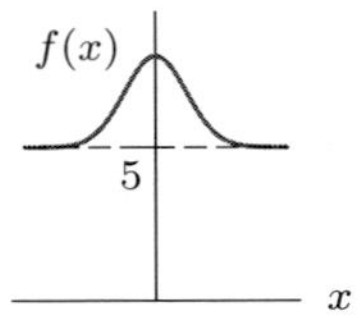

27 0

29 5.1

31 ∞

33 $Q = 70.711(0.966)^t$

35 $f(x) = 2(1/3)^x$

37 $Q = 0.7746 \cdot (0.3873)^t$

39 $Q = 1149.4641(0.935)^t$

41 $V = 2500e^{0.042t}$

43 Linear, $Q(t) = 7 + 0.17t$

45 Neither

47 (a) $N = 178.8 + 0.84t$
(b) $N = 178.8(1.0046)^t$

49 (a) Initial balance = $\$1100$
Effective yield = 5%
(b) Initial balance = $\$1500$
Effective yield $\approx 5.13\%$

51 $a > c$

53 $x_1 < x_3 < x_2$

55 (a) 10.409%
(b) 10.503%
(c) 42.576%
(d) 48.770%

57 (a) Exponential
(b) $P(t) = 20{,}000(1.0414)^t$

59 (a) $a = 12, \quad b = 0.841$
(b) $C = 11.914(0.840)^t$

61 (a) $15.269(1.122)^t$
(b) 108,066
(c) Not useful

63 (a) $P(t) = 0.755\,(1.194)^t$

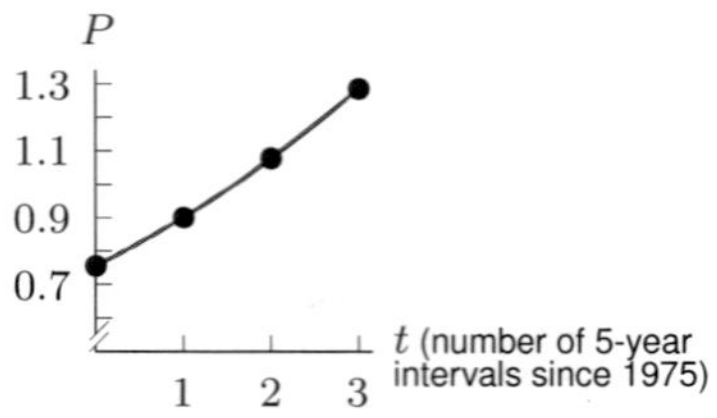

(b) 20 years; 2135

65 (a) $f(x) = p_0(2{,}087{,}372{,}982)^x$
(b) 8.55
(c) BAC of 0.051

Ch. 3 Understanding

1 True

3 True

5 False

7 True

9 True

11 False

13 False

15 True

17 True

19 False

21 True

23 False

25 True

27 False

29 True

31 False

Chapter 3 Tools

1 64

3 121

5 1

7 5

9 1,000

11 1

13 4

15 16

17 49

19 -1

21 -18

23 2100

25 4

27 8

29 1024

31 -5

33 -6

35 4

37 $1/9$

39 $1/25$

41 $1/125$

43 0.5

45 y^4

47 $x^{5/2}y^2$

49 $7w^{9/2}$

51 $|r|$

53 r^2

55 $8s^{7/2}$

57 $4\sqrt{3}u^5v^6y^{5/2}$

59 $6s^{3/2}t^4v^4$

61 3^{x+1}

63 $70w^{5/6}$

65 e^x

67 e^{kt+4}

69 $9x^5$

71 e^{2y}/y^4

73 $64A$

75 $4u^4v^2w^4$

77 $25(2b+1)^{20}$

79 -8

81 Not a real number

83 $1/512$

85 Not a real number

87 $x = 0.585$

89 $x = 1.842$

91 $x = \pm 0.1$

93 $(2.5, 31.25)$

95 False

97 False

99 True

101 False

103 True

105 $x = r + s$

107 $x = 5/a$

109 $x = 3/a$

111 $x = b/a$

Section 4.1

1 $19 = 10^{1.279}$

3 $26 = e^{3.258}$

5 $P = 10^t$

7 $8 = \log 100{,}000{,}000$

9 $v = \log \alpha$

11 $(\log 11)/(\log 2) = 3.459$

13 $(\ln 100)/(0.12) = 38.376$

15 $(\log(48/17))/(\log(2.3)) = 1.246$

17 $2\log(0.00012)/(-3) = 2.614$

19 (a) 0
(b) -1
(c) 0
(d) $1/2$
(e) 5
(f) 2
(g) $-1/2$
(h) 100
(i) 1
(j) 0.01

21 3.47712

23 (a) 0
(b) $\sqrt{2} - 1$
(c) $\frac{31}{30}$
(d) 27
(e) $\frac{1}{4}$
(f) 1

25 (a) $2x$
(b) $3x + 2$
(c) $-5x$
(d) $\frac{1}{2}x$

27 (a) False
(b) True
(c) False
(d) False

29 (a) 10^p

(b) 10^{3q}
(c) $p+3q$
(d) $p/2$

31 (a) 25; 7.5%
(b) $t \approx 19$
(c) $t = (\log 4)/(\log 1.075) = 19.169$

33 $S(x) = 42.0207(0.8953)^x$

35 $\log(7/4)/\log(1.171/1.088)$

37 $\log(17/3)/\log 2$

39 $(\ln 7 - 5)/(1 - \ln 2)$

41 $(\log c - \log a)/\log b$

43 $(\ln Q - \ln P)/k$

45 $\frac{1}{4}(\ln(30/58) - 1)$

47 $(\ln 50 - \ln 44)/(0.15 - \ln 1.2)$

49 $-2, 1/3, -1/3$

51 (a) $(\ln(0.5))/2 = -0.347$
(b) $(\ln(b/3))/3$

53 $0 < A < 1 < A^2B^2 < 100 < B^2$

Section 4.2

1 $a = 230, r = 18.2\%, k = 16.72\%$

3 $a = 0.81, r = 100\%$, and $k = 69.31\%$

5 $a = 12.1, r = -22.38\%, k = -25.32\%$

7 $a = 5.4366, b = 0.4724, r = -52.76\%, k = -3/4$

9 $Q = 4e^{1.946t}$

11 $Q = 4e^{2.703t}$

13 $Q = 4 \cdot 1096.633^t$

15 $Q = (14/5)1.030^t$

17 $y = 25(1.0544)^t$, 5.44%/yr, 5.3%/yr

19 $y = 6000e^{-0.1625t}$, -15%/yr, -16.25%/yr

21 $t \approx 3.466$

23 About 26 years

25 About 12.3 years

27 6.301 minutes

29 $Q = 28.9101(1.0074)^t$

31 (a) 5.726%
(b) 5.568%

33 3.053%/yr

35

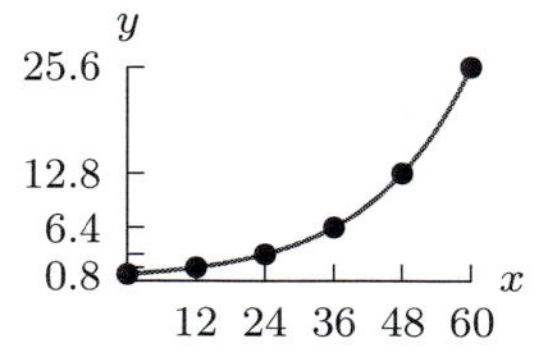

37 $10^{(\log 5)/(4 \log 3)}$

39 No solution

41 2

43 (a) 1.3%
(b) 5.4 million
(c) 84.509 years
(d) 1192.455 years ago; No

45 (a) $P = 3(3.218)^t$
(b) 0.940 hours

47 (a) 13.9% per year
(b) 7.9 years

49 (a) $o(t) = 245 \cdot 1.03^t$
(b) $h(t) = 63 \cdot 2^{t/10} \approx 63 \cdot (1.072)^t$
(c) 34.162 years

51 5092.013 years ago

53 2.291 hours

55 Yes, after 108.466 years

57 (a) 190°F; 100°F; 77.5°F
(b) 1.292 hours; 2.292 hours

59 t_0 decreases

61 s must increase

Section 4.3

1 $y = 0, y = 0, x = 0$

3 $A: y = 10^x$, $B: y = e^x$
$C: y = \ln x$, $D: y = \log x$

5 (a) 0
(b) $-\infty$

7

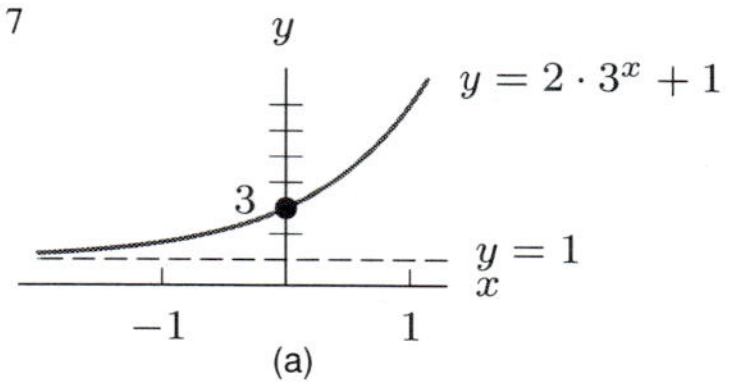

(a)

9

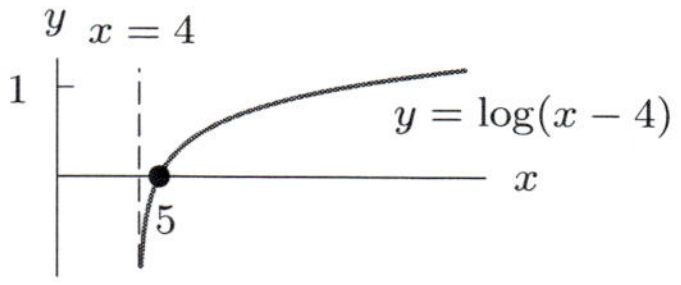

(c)

11 Vertical asymptote at 3, Domain $(3, \infty)$

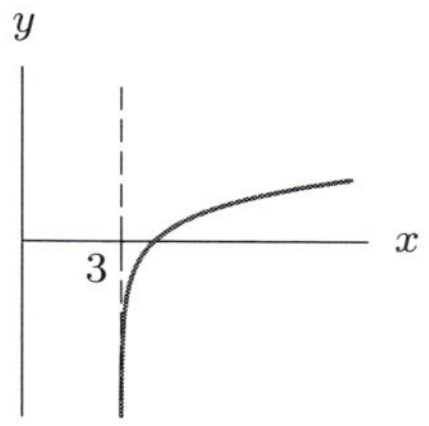

13 10^{-13} moles/l

15 5.012×10^{-9} moles/l

17 1 mole/l

19 (a) is (III)
(b) is (IV)
(c) is (I)
(d) is (I) or (II)

23 $y = b^x, 0 < b < 1$

25 $y = \ln x$

27 $y = -b^x, b > 1$

29 $x > 0$

31 $x > 3$

33 (a) $1.01 \cdot 10^{-5}$ moles/liter
(b) 4.996

35 37

37 (a) $M_2 - M_1 = \log(W$ ...
(b) 40

Section 4.4

1 Log

3 Linear

7 (a) $y = -3582.145 + 236.314x$; $r \approx 0.7946$
(b) $y = 4.797(1.221)^x$; $r \approx 0.9998$
(c) Exponential is better fit

9 $10^{-3.65}$ million years ago

17 Yes

19 (a) $y = 14.227 - 0.233x$
(b) $\ln y = 2.682 - 0.0253x$
(c) $y = 14.614e^{-.0253x}$

21 (a)

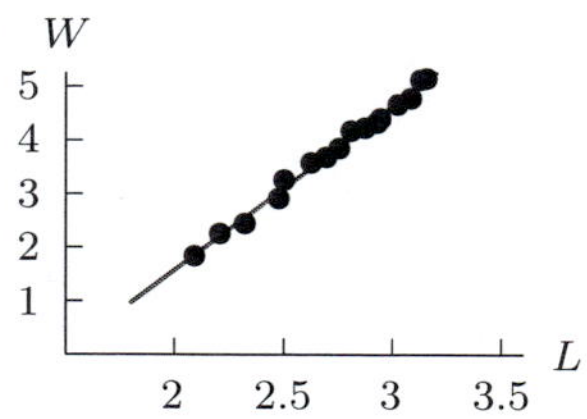

(b) $W = 3.06L - 4.54$
(c) $w = 0.011\ell^{3.06}$

23 (a) Log function

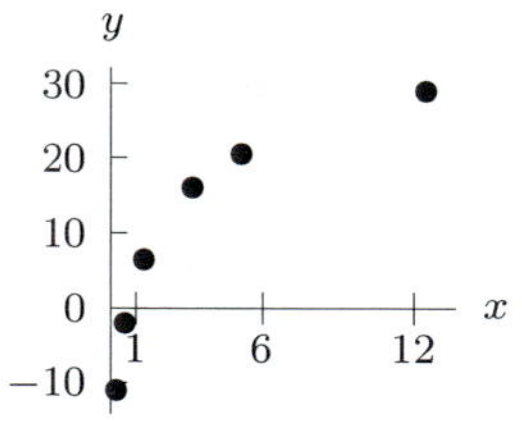

(b)

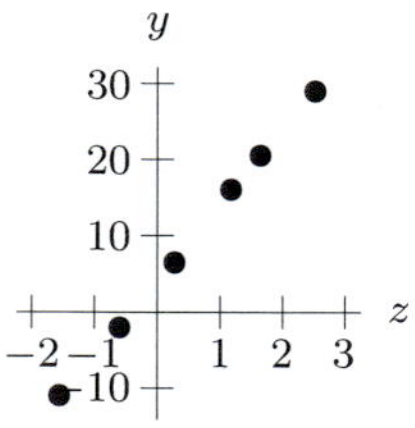

(c) Linear; $y = 4 + 9.9z$
(d) $y = 4 + 9.9 \ln x$
(e) $x = 0.67e^{0.1y}$; Exponential function of y

Chapter 4 Review

1 $Q = 12e^{-0.105t}$

3 $Q = 14e^{-0.208t}$

5 $Q = 7(0.0000454)^t$

7 $(\ln 50)/(\ln 3) = 3.561$

9 $(\ln 0.3)/(-0.2) = 6.020$

11 $a = 0.05, b = 1.13, r = 0.13, k = \ln 1.13$

13 (a) $3\%, 2.96\%$
(b) $3.05\%, 3\%$
(c) Decay: $6\%, 6.19\%$
(d) Decay: $4.97\%, 5.1\%$
(e) $3.93\%, 3.85\%$
(f) Decay: $2.73\%, 2.77\%$

15 $\ln(1328/400)/\ln 1.112$

17 $\ln(28/55)/(-0.223)$

19 $x = \ln(55/5)/\ln 1.1$

21 $x = 0.25$

23 $(1/0.049) \cdot \ln(25/13) \approx 13.345$

25 $x = 1000$

27 $2(x+1)$

29 $\ln(AB)$

31 $g(t) = 97.1187(0.9762)^t$

33 $p(t) = 5000(1.0234)^t$

35 $x < -2$ and $x > 3$

37 In 1969

39 (a) $W(t) = 50(1.029)^t$, $C(t) = 45(1.032)^t$
(b) 36.191 years
(c) 274.287 years

41 (a) 1412 bacteria
(b) 10.011 hours
(c) 1.005 hours

43 (a) 13.333%
(b) 2.534%
(c) 2.503%

45 (a) $Q(t) = 2e^{-0.04t}$
(b) 3.921%
(c) After 51.986 hours
(d) 54.931 hrs after second injection

47 (a) $\ln 8 - 3 \approx -0.9206$
(b) $\log 1.25/\log 1.12 \approx 1.9690$
(c) $-\dfrac{\ln 4}{0.13} \approx -10.6638$
(d) 105
(e) $\frac{1}{3}e^{3/2} \approx 1.4939$
(f) $e^{1/2}/(e^{1/2}-1) \approx 2.5415$
(g) -1.599 or 2.534
(h) 2.478 or 3
(i) 0.653

49 Predicted (yrs):
70, 35, 14, 10, 7
Actual (yrs):
69.661, 35.003, 14.207, 10.245, 7.273

51 100 months

53 About 12:53 pm

55 (a) Use $e < 3 < 4$
(b) Use $e < 4 < e^2$

Ch. 4 Understanding

1 True

3 False

5 True

7 False

9 True

11 True

13 True

15 True

17 True

19 False

21 False

23 False

25 True

27 False

29 False

Chapter 4 Tools

1 0

3 -4

5 3

7 0

9 2

11 $\log 100{,}000 = 5$

13 $\log 3 = 0.477$

15 $\ln 0.135 = -2$

17 $10^{-2} = 0.01$

19 $e^{-1} = x$

21 $\log 2 + \log x$

23 Cannot be rewritten

25 $\log(x^2+1) - 3\log x$

27 Cannot be rewritten

29 $\log(x+y) + \log(x-y)$

31 $(2\ln x)/\ln(x+2)$

33 $\ln x^5$

35 $\log(\sqrt{x}y^4)$

37 $\log(2/5)$

39 $\ln(((x^2)y + 4xy)/2z)$

41 $(5x)^{-1}$

43 $2\sqrt{x}$

45 $t^2/2$

47 1

49 $(1/2)\ln(x^2+16)$

51 $1/2$

53 $x = (\log 9)/(\log 4) \approx 1.585$

55 $x = (\log 3)/(\log 5) \approx 0.683$

57 $x = \ln 8 \approx 2.079$

59 $x = -(\ln 9)/5 \approx -0.439$

61 $x = (\log 3)/(5\log 12 - 2\log 15) \approx 0.157$

63 $x = 93/2$

65 $x = (10^{3/2} - 17)/9 \approx 1.625$

67 $x = (e^5 - 4)/3 \approx 48.138$

Section 5.1

1 (a) $-3, 0, 2, 1, -1$
One unit right
(b) $-3, 0, 2, 1, -1$
One unit left
(c) $0, 3, 5, 4, 2$
Up three units
(d) $0, 3, 5, 4, 2$
One right and three up

3
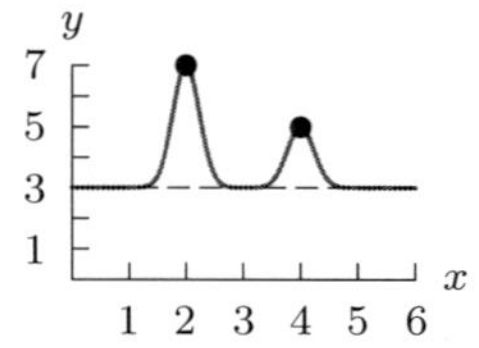

5
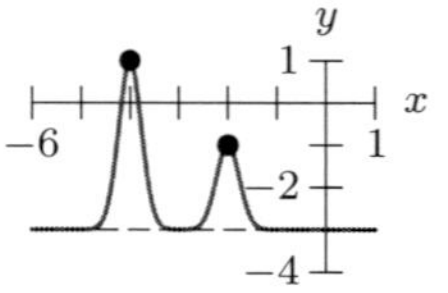

9 (a) $(3, 1)$
(b) $(-2, -4)$
(c) $(6, -6)$

11 $-50 \le R(s) - 150 \le 50$

13 $(1/2)n^2 + n + 1/2$

15 $(1/2)n^2 - 3.7n + 6.845$

17 $(1/2)n^2 + 2\sqrt{2}n + 4$

19 $(1/2)n^2 - 17n - 29/2$

21 3^{w-3}

23 $3^{w+\sqrt{5}}$

25 $3^{w-1.5} - 0.9$

27 (a) $g(x) + 3$
(b) $g(x-2)$

29 (a) Population 100 people larger than original
(b) Population same as 100 years earlier

31 $x = 0$ and $x = 3$

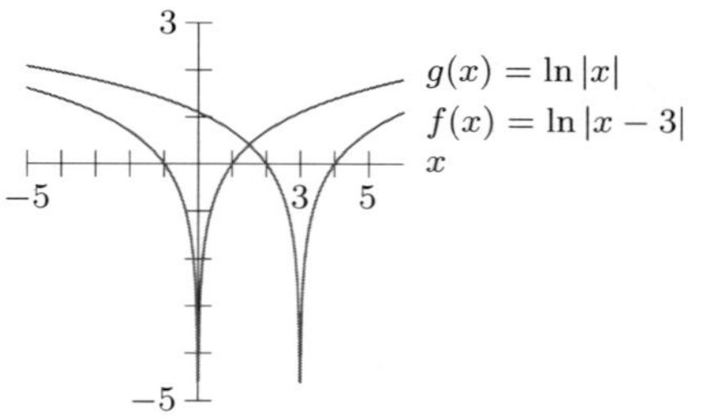

33 Shift up 3

35 Shift left 4

37 Shift left b, down a

39 (a) $T(d) = S(d) + 1$
(b) $P(d) = S(d-1)$

41 (a) $H(t+15) = 68 + 93(0.91)^{t+15}$
$H(t) + 15 = 83 + 93(0.91)^t$

(b)

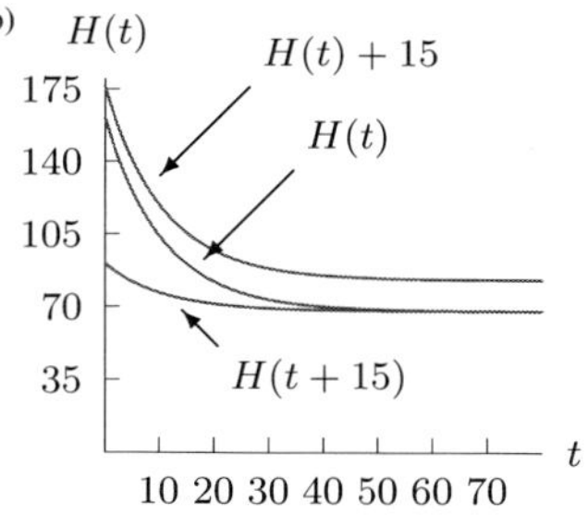

(d) $H(t+15)$ and $H(t)$ approach $68°$F

43 $H(t) = 280(0.97)^t + 70$

45 $-\ln 5$

Section 5.2

1 (a) $(-2, -3)$
(b) $(2, 3)$

3 -7

5 $y = -e^x$

9 Reflected across x-axis;
$-g(x) = -(1/3)^x$

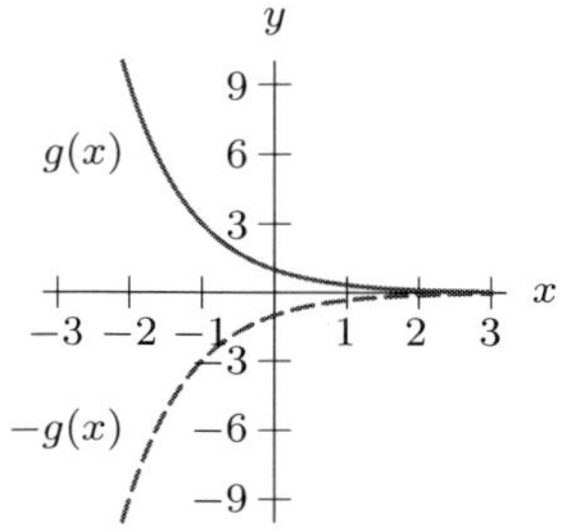

11 $-m(n) = -(n)^2 + 4n - 5$

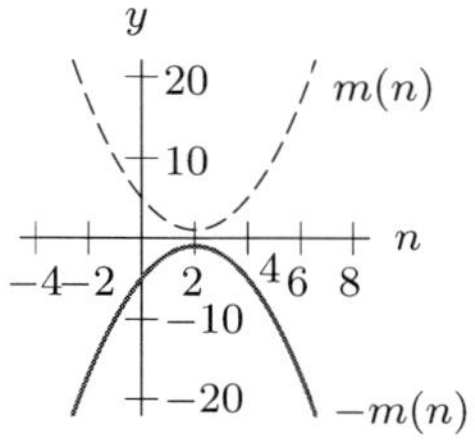

13 $-m(-n) + 3 = -n^2 - 4n - 2$

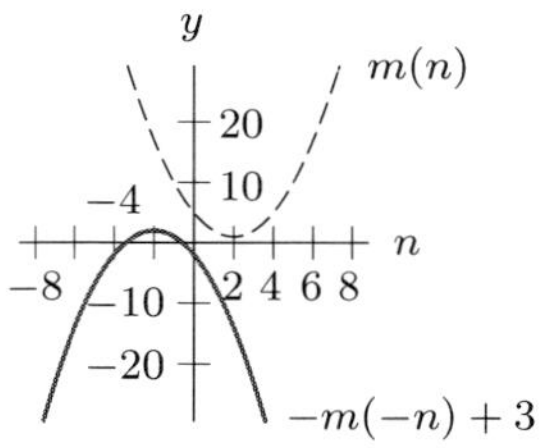

15 $-k(w) = -3^w$

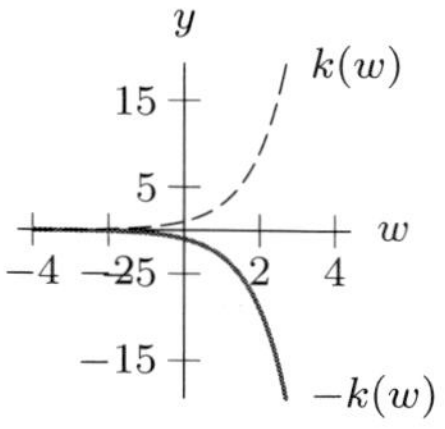

17 $-k(w-2) = -3^{w-2}$

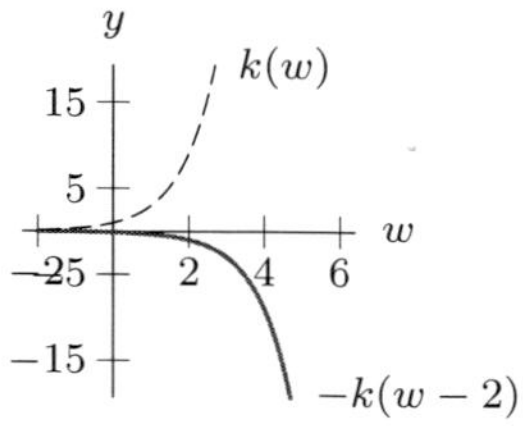

19 $-k(-w) - 1 = -3^{-w} - 1$

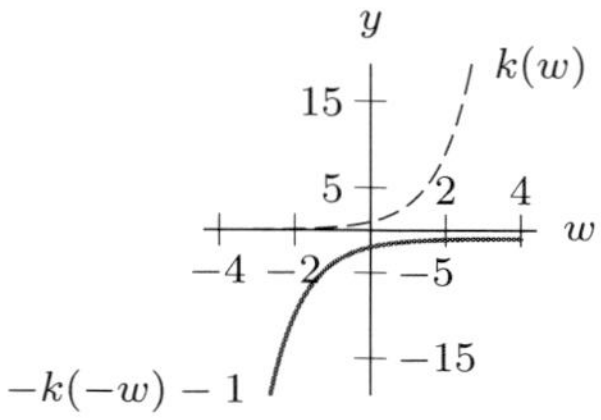

21 Odd

23 Neither

25 (a) $y = 2^{-x} - 3$

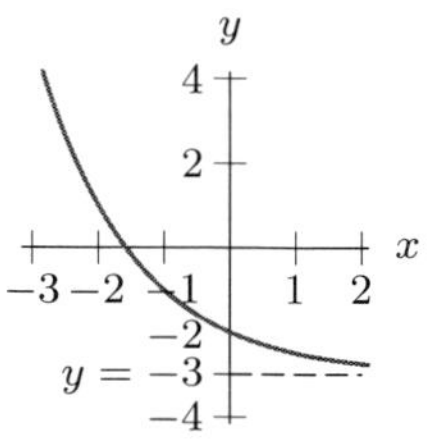

(b) $y = 2^{-x} - 3$

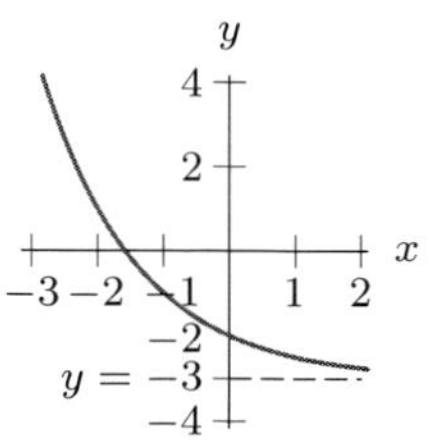

(c) Yes

27 Reflections across x-axis

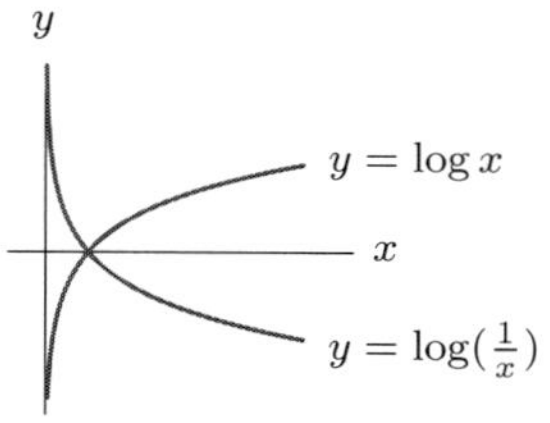

29 (i) c
(ii) d
(iii) e
(iv) f
(v) a
(vi) b

31 (a)

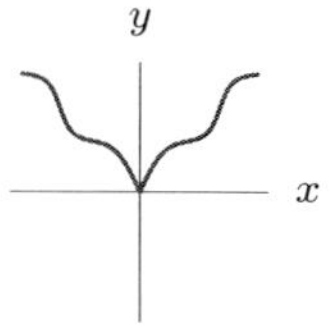

(b)

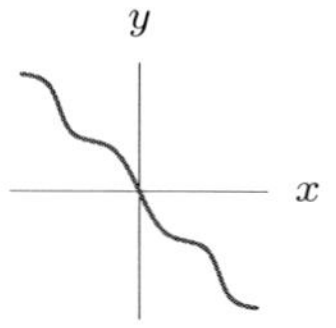

(c)

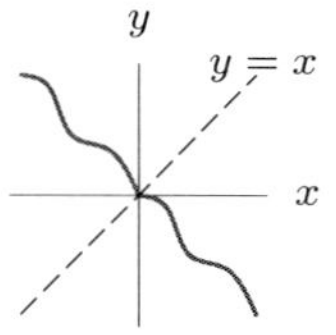

33 $y = x$. $y = -x + b$, where b is an arbitrary constant

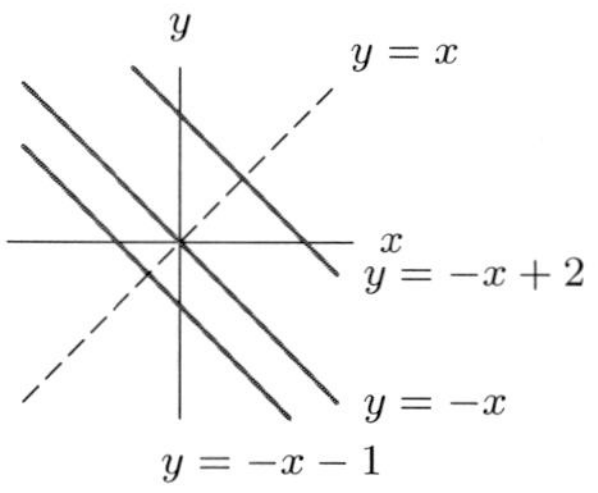

37 If $f(x)$ is odd, then $f(0) = 0$

41 Yes, $f(x) = 0$

Section 5.3

1 $y = 10f(x - 2)$

3 $-0.25 \leq 0.25C(x) \leq 0.25$

5

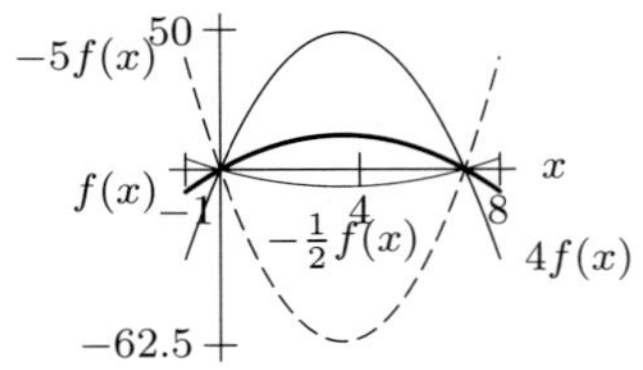

7

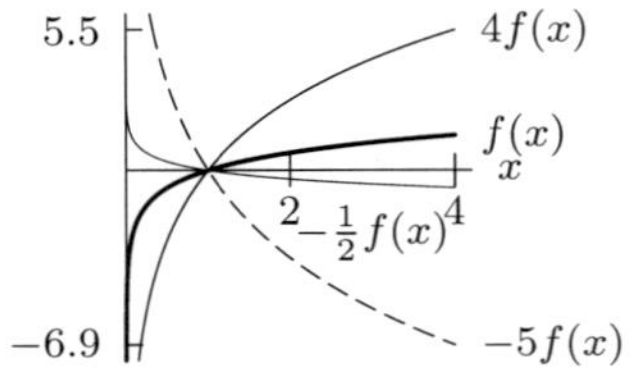

9 (d) All three

11

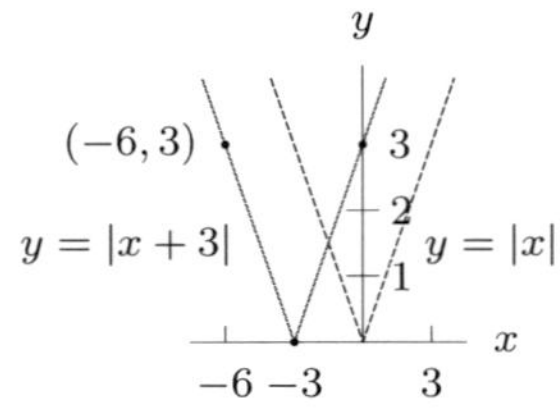

13

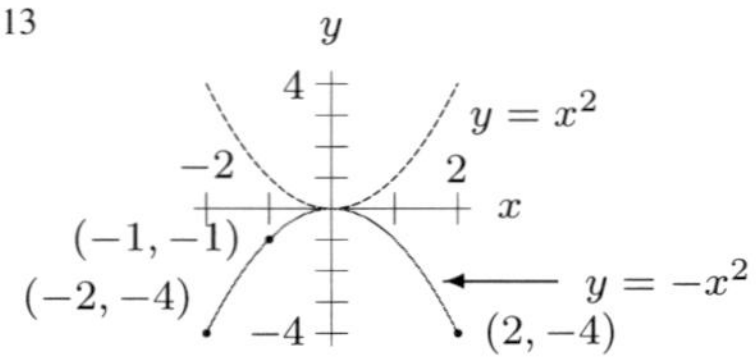

15

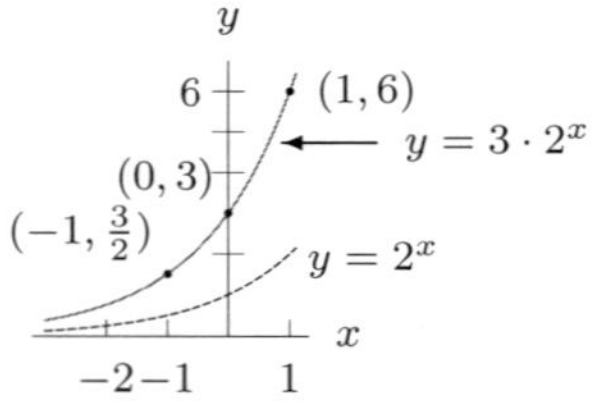

17 (i) i
(ii) c
(iii) b
(iv) g
(v) d

19 35 mph

21

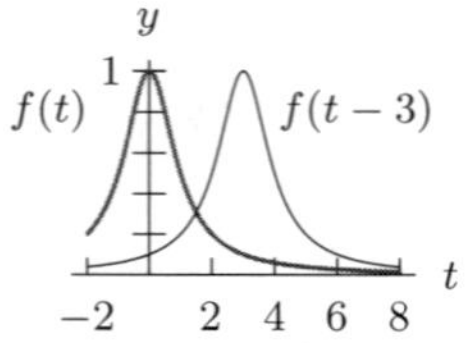

23

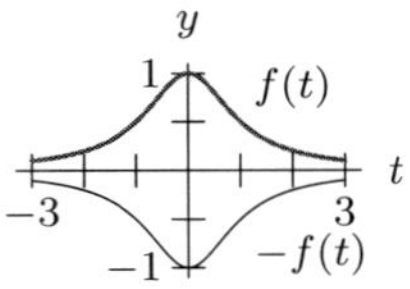

25 (a) (iii)
(b) (i)
(c) (ii)

29

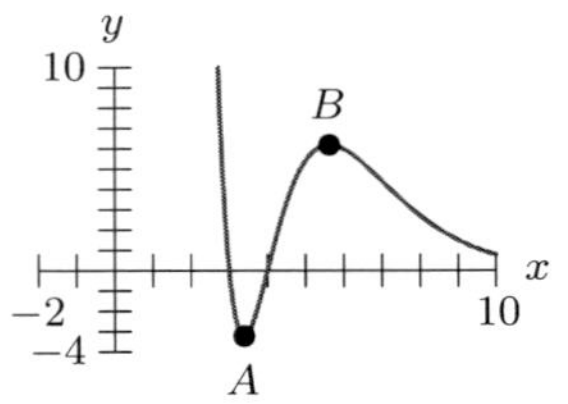

31

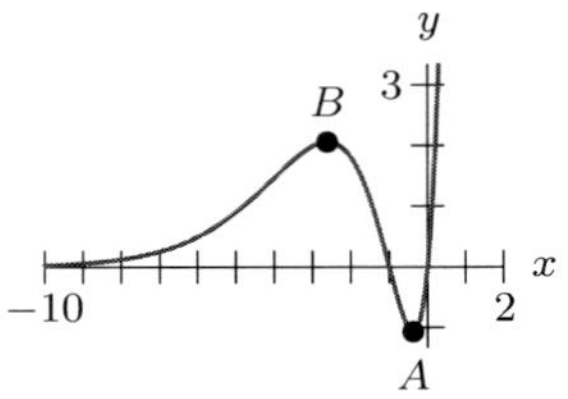

33

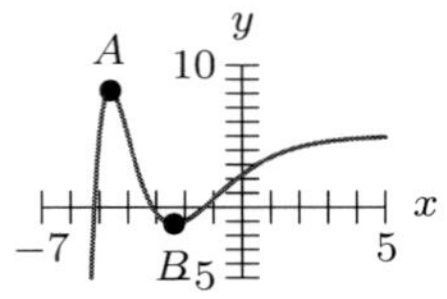

35 (a) $y = -2f(x)$
(b) $y = f(x) + 2$
(c) $y = 3f(x - 2)$

Section 5.4

1 $(1, 3)$

3 Same function values for $x = -6, -4, -2, 0, 2, 4, 6$

5

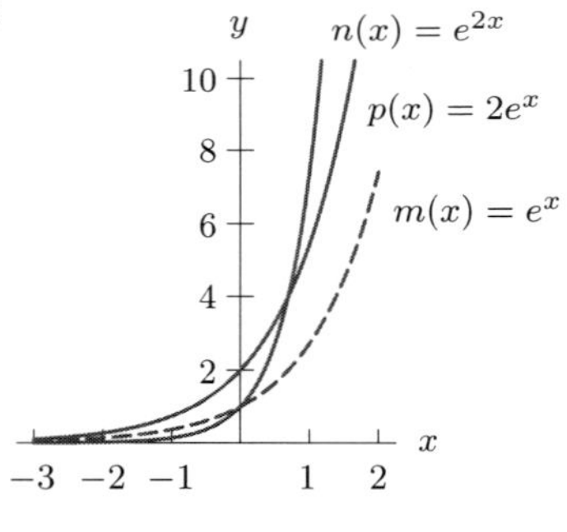

7

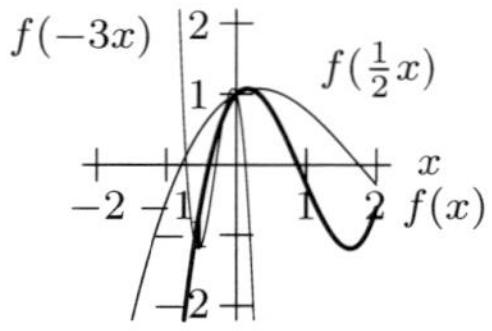

9

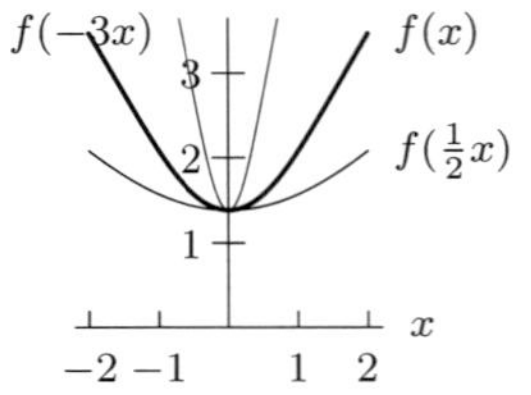

11 2/3

13 $(a/d, b + c)$

15

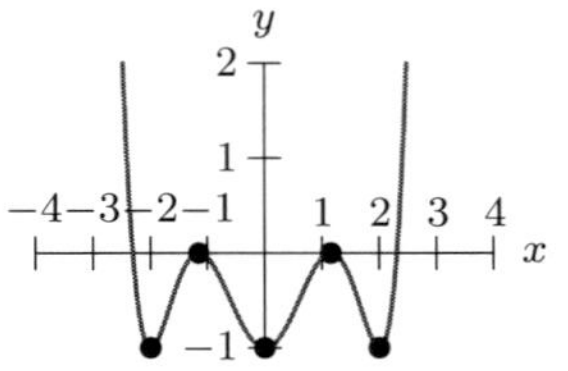

17 $r(t)$: half the level
$s(t)$: half the rate

19 (a) III
(b) II
(c) I
(d) IV

21 $y = -f(-\frac{1}{2}x)$

23

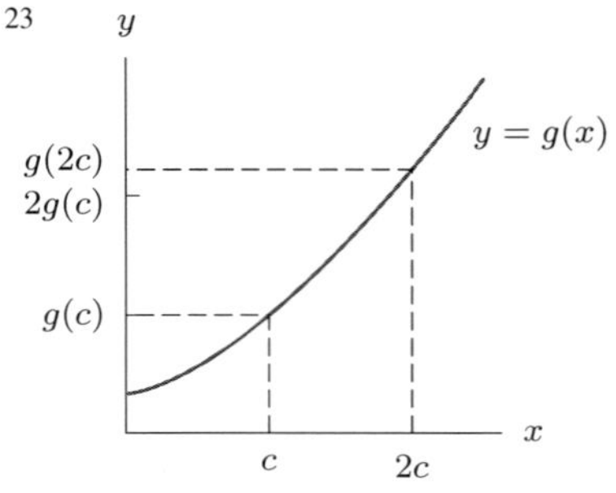

25

d	20	45	70	95
$h(d)$	5.5	5.2	5.1	5.1
d	120	145	170	195
$h(d)$	5.3	5.5	5.75	6

27

d	25	50	75	100
$n(d)$	8.25	7.8	7.65	7.65
d	125	150	175	200
$n(d)$	7.95	8.25	8.63	9

29

d	25	50	75	100
$q(d)$	10.25	9.8	9.65	9.65
d	125	150	175	200
$q(d)$	9.95	10.25	10.63	11

Section 5.5

1 $(1, 2)$; $x = 1$; opens upward

3 (a) $a = 1, b = 0, c = 3$
Axis of symmetry: y-axis
Vertex: $(0, 3)$
No zeros
y-intercept: $y = 3$

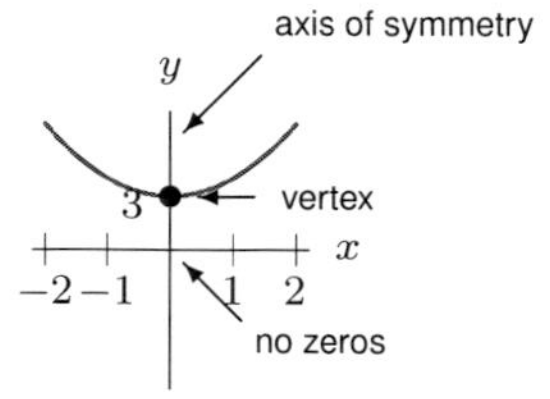

(b) $a = -2, b = 4, c = 16$
Axis of symmetry: $x = 1$
Vertex: $(1, 18)$
Zeros: $x = -2, 4$
y-intercept: $y = 16$

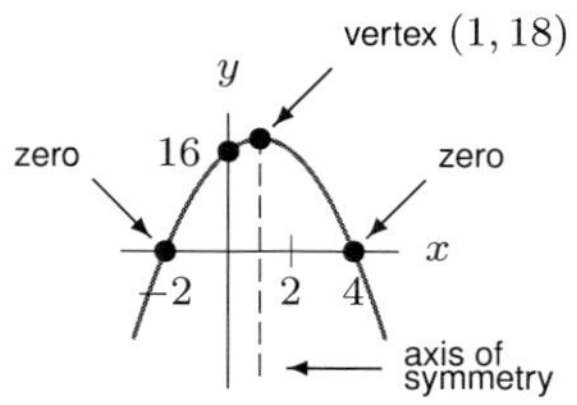

5 Vertex is $(-5, 106)$
Axis of symmetry is $x = -5$

7 $k = 4$

9 $y = -(3/16)(x - 4)^2 + 7$

11 $y = (1/3)x^2 - (2/3)x - 1$

13 $y = -(5/12)x^2 - (5/3)x + 5$

15 $f(x) = (x + 4)^2 - 13$;
Vertex: $(-4, -13)$; axis: $x = -4$

17 $y = -3(x - 2)^2 + 5$

19 $y = (1/4)(x - 4)^2 + 2$

21 $y = (-2/49)(x - 4)^2 + 2$

23 $y = (7/4)(x + 2)^2$

25 Shift $y = x^2$ right by 5 units to get $y = (x - 5)^2 = x^2 - 10x + 25$

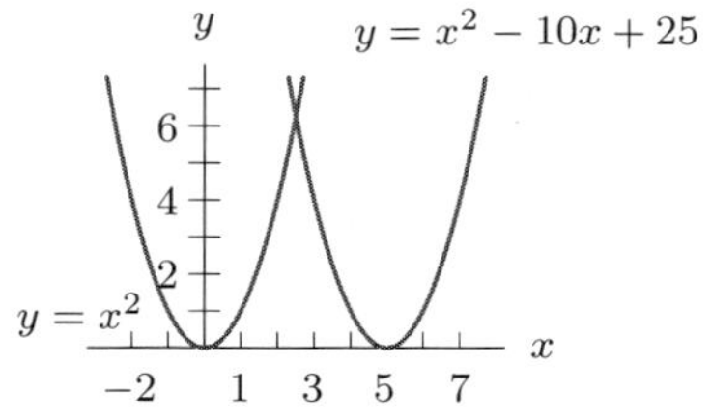

27 (a)

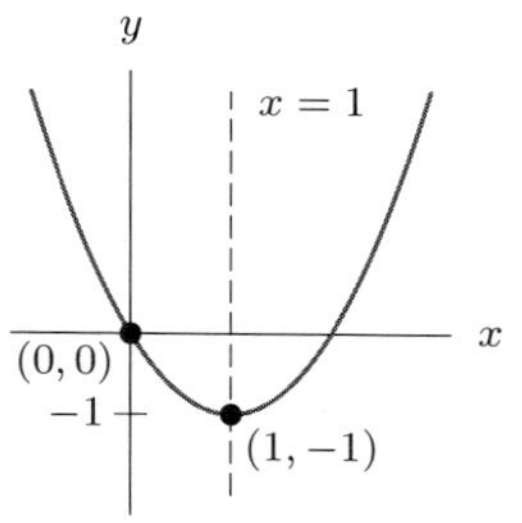

(b) $y = (x - 1)^2 - 1$ or $y = x^2 - 2x$
(c) Range: $y \geq -1$
(d) The other zero is $(2, 0)$

29 Yes. $f(x) = -(x - 1)^2 + 4$

31 (a)

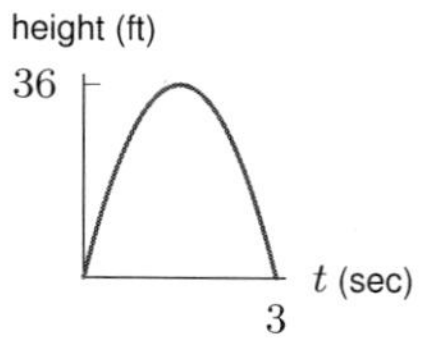

(b) $t = 0, 3$; being thrown; hitting the ground
(c) 1.5 seconds
(d) 36 feet

33 12.5 cm by 12.5 cm; $k/4$ by $k/4$

35 (b) Maximum height: $t = T/2$

Chapter 5 Review

1 (a) 4
(b) 1
(c) 5
(d) -2

3 (a) $(6, 5)$
(b) $(2, 1)$
(c) $(1/2, 5)$
(d) $(2, 20)$

5 Odd

7 Neither

9 Even

11 (a) $f(2x) = 1 - 2x$
(b) $f(x + 1) = -x$
(c) $f(1 - x) = x$
(d) $f(x^2) = 1 - x^2$
(e) $f(1/x) = (x - 1)/x$
(f) $f(\sqrt{x}) = 1 - \sqrt{x}$

13 $y = (x + 1)(x - 3)$

15 $y = -(x - 2)^2$

17

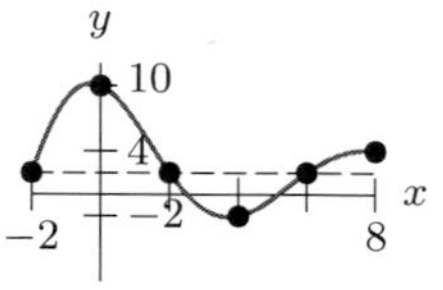

19 $y = f(t + 4) - 8$

21 (b) Decreasing

23 (a) VI
(b) V
(c) III
(d) IV
(e) I
(f) II

25 (a) $-16t^2 + 23$
$-16t^2 + 48t + 2$
(b)

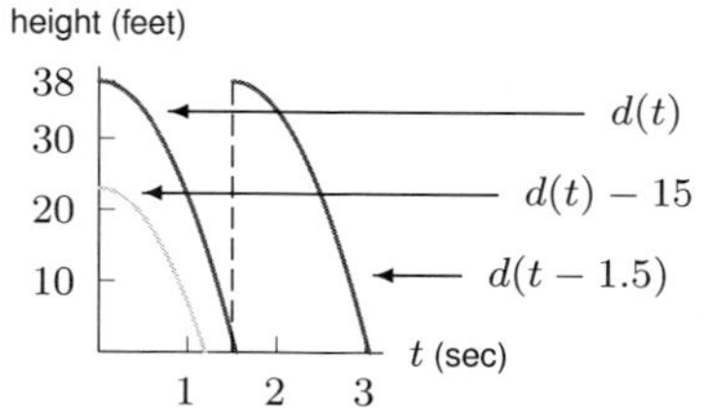

(d) 1.541 secs
1.199 secs
(e) 3.041 secs

27 $y = -(x + 1)^3 + 1$

29 $y = (1/2)h(x + 6) + 1$

31 $p \approx 15, q \approx 7190$

35

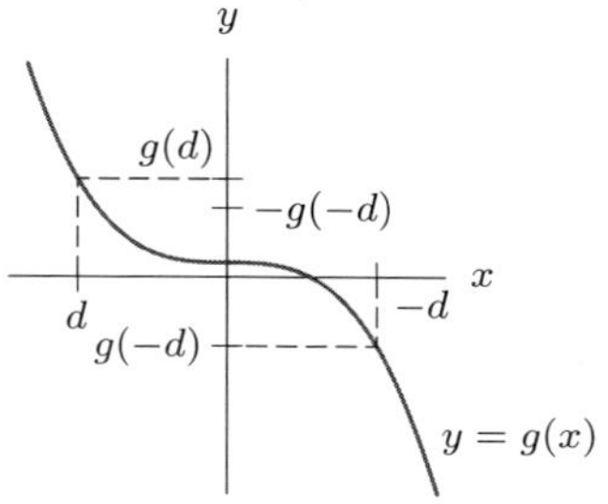

37 $y = 3h(x)$

39 $y = -h(2 - 2x)$

41 1/2 second

Ch. 5 Understanding

1 True

3 True

5 True

7 False

9 True

11 False

13 False

15 True

17 True

19 False

21 False

23 True

25 True

27 False

29 True

Chapter 5 Tools

1 $(x + 4)^2 - 16$

3 $(w + (7/2))^2 - (7/2)^2$

5 $(s + 3)^2 - 17$

7 $(a - 1)^2 - 5$

9 $(c+3/2)^2-37/4$

11 $4(s+1/8)^2+31/16$

13 $(x-1)^2-4$

15 $-(x-3)^2+7$

17 $(-3,-6)$

19 $(-4,18)$

21 $(1/2,-23/4)$

23 $(1,-2)$

25 $(7/4,-25/8)$

27 $r=4,2$

29 $p=1\pm\sqrt{7}$

31 $d=2,-1$

33 $s=-5/2\pm\sqrt{27}/2$

35 $r=3/14\pm\sqrt{177}/14$

37 $n=6,-2$

39 $k=-1/3,-3/2$

41 $z=-2\pm\sqrt{10}$

43 $r=4,-2$

45 $n=-5,1$

47 $z=-2,\pm\sqrt{3}$

49 $u=(3\pm\sqrt{5})/5$

51 $y=1\pm\sqrt{7}$

53 $w=3,2,-2$

55 $m=(-5\pm\sqrt{3})/7$

Section 6.1

1 Yes

3 No

5 No

7 Yes

9 4

11 3

13

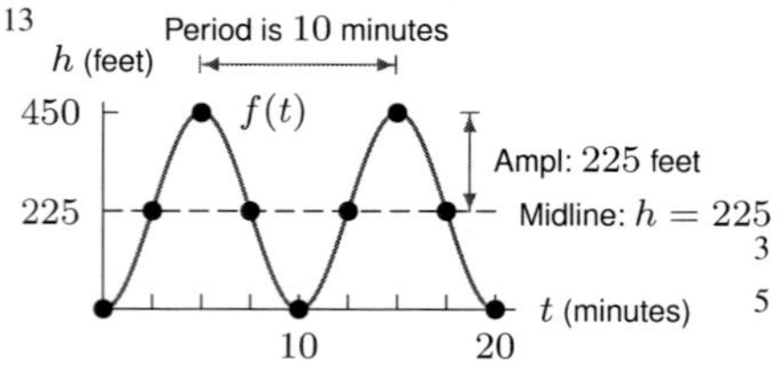

15

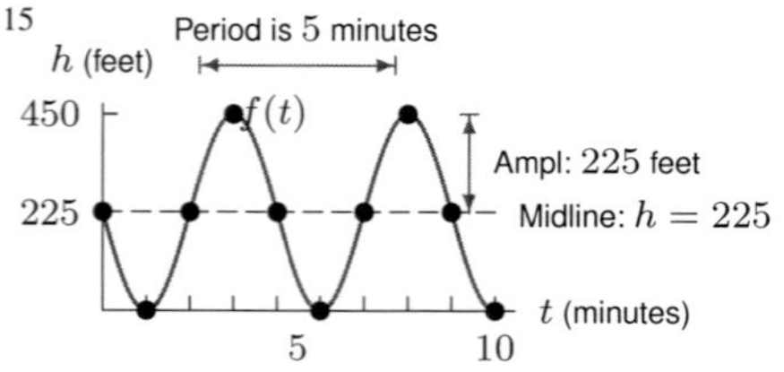

17

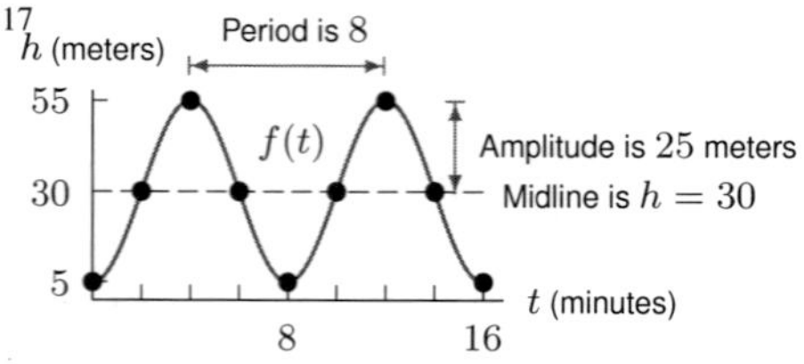

19 3 (or 9) o'clock; rising; 4 minutes; 30 meters; 5 meters; 11 minutes

21 3 (or 9) o'clock; upward; 5 minutes; 40 meters; 0 meters; 11.25 minutes

23 Midline: $y=10$;
Period: 1;
Amplitude: 4;
Minimum: 6 cm;
Maximum: 14 cm

25 Graph is same except starts at a peak

27 (b) Period: $1/60$ seconds;
Amplitude: 155.6 volts;
Midline: $V=0$

29 (a) Periodic
(b) Not periodic
(c) Not periodic
(d) Not periodic
(e) Periodic
(f) Not periodic
(g) Periodic

31 Midline: $h=2$;
Amplitude: 1;
Period: 1

Section 6.2

1 (a) $(-0.174, 0.985)$
(b) $(-0.940, -0.342)$
(c) $(-0.940, 0.342)$
(d) $(0.707, -0.707)$
(e) $(0.174, -0.985)$
(f) $(1, 0)$

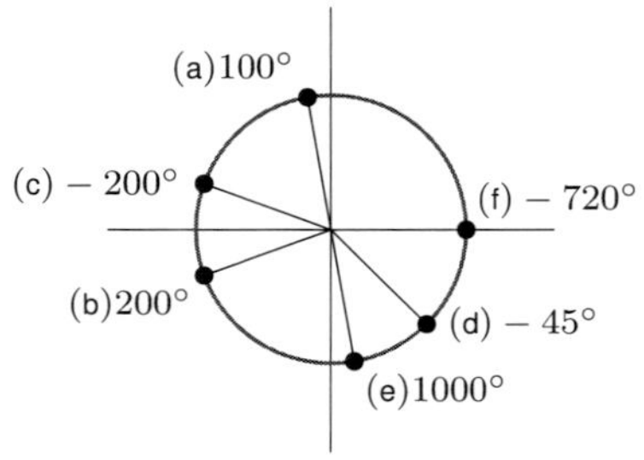

3 $-720°$

5 $S=(-0.707,-0.707)$, $T=(0,-1)$, $U=(0.866,-0.5)$

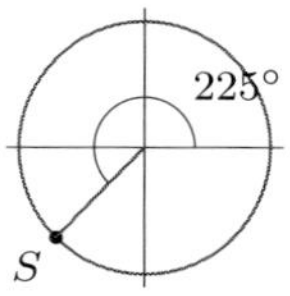

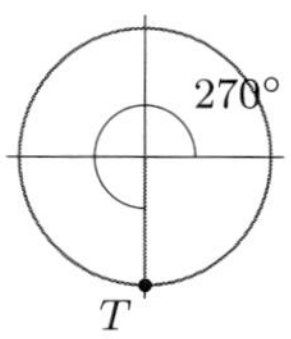

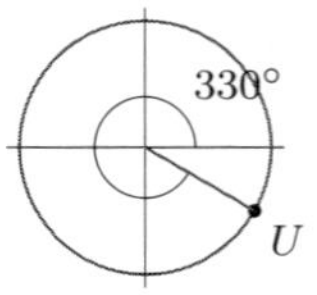

7 $A=(0.866,0.5)$, $B=(-0.707,0.707)$, $C=(0.866,-0.5)$

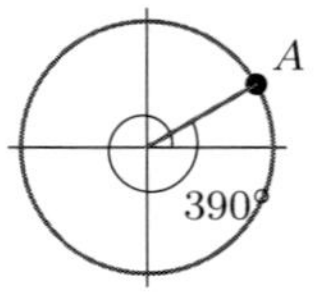

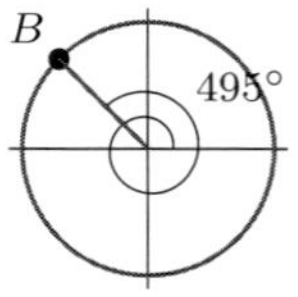

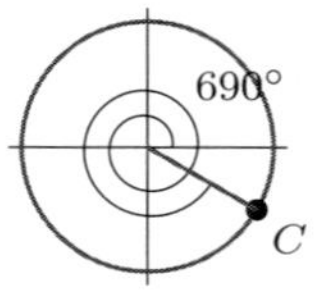

9 $S=(-3.536,-3.536)$
$T=(0,-5)$
$U=(4.330,-2.5)$

11 $(0,3.8)$

13 $(-3.8,0)$

15 $(0,3.8)$

17 $(3.687,-0.919)$

19 $(3.8\sqrt{2}/2, 3.8\sqrt{2}/2)$ or $(2.687, 2.687)$

21 $(-3.8\sqrt{2}/2, -3.8\sqrt{2}/2)$ or $(-2.687,-2.687)$

23 $(3.742,-0.660)$

27 (a) $307°$
(b) $127°$

29 (a)

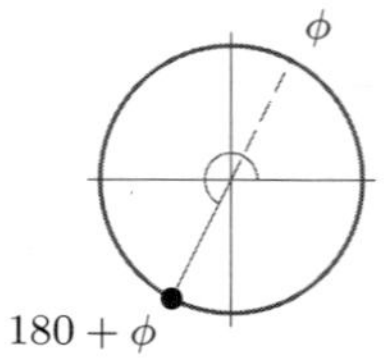

(b)

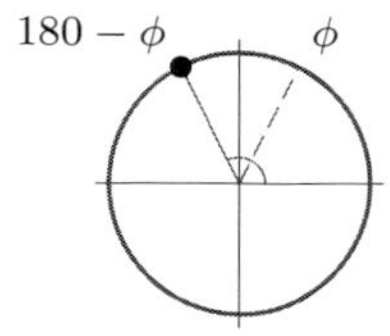

(c)

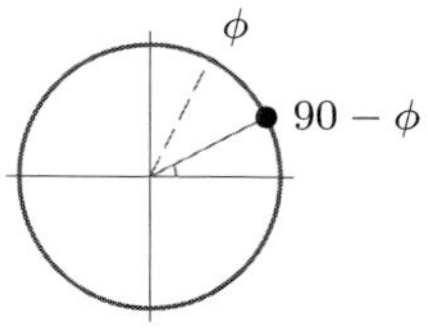

(d)

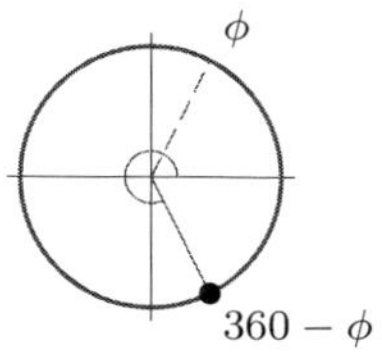

33 $4/3$ minutes

35 (a) 90°
(b) 90°
(c) 180°
(d) 180°
(e) A

Section 6.3

1 $\pi/3$

3 1.7453 radians

5 $5\pi/6$

7 $-3\pi/2$

9 630°

11 $16{,}200/\pi \approx 5156.620^\circ$

13 $8100/\pi \approx 2578.310^\circ$

15 (a) I
(b) II
(c) II
(d) III
(e) IV
(f) IV
(g) I
(h) II
(i) II
(j) III

17 -4π

19 8.54π

21 $6.2\pi/4 \approx 4.869$

23 $6.2a\pi/180$

25 5π feet

27 $\pi/9$ radians or 20°

29 $r = \sqrt{65}$; $\theta = 0.5191$ rad $= 29.7449^\circ$; $s = 4.185$; $P = (7, 4)$

31 $r = 12$; $\theta = 1.3$ rad $= 74.485^\circ$; $s = 15.6$; $P = (3.2100, 11.5627)$

33 $\theta = 0.4$ rad $= 22.918^\circ$; $P = (0.9211r, 0.3894r)$

35 (a) Negative
(b) Negative
(c) Positive
(d) Positive

37 $\sin\theta = 0.6$; $\cos\theta = -0.8$

39 $(-0.99, 0.14)$

41 $\pi/6$ feet

43 3998.310 miles

45 $t \approx 0.739$

Section 6.4

1 Mid: $y = 0$; Amp: 1

3 Mid: $y = -4$; Amp: 7

5 Mid: $y = -2$; Amp: 3

7 Mid: at 185 cm; Amp: 15cm

9 (a) (i) $0 < t < \pi$ and $2\pi < t < 3\pi$
(ii) $-\frac{\pi}{2} < t < \frac{\pi}{2}$ and $\frac{3\pi}{2} < t < \frac{5\pi}{2}$
(iii) $-\pi < t < 0$ and $\pi < t < 2\pi$
(b) $t = 0, 2\pi$

11 They are equal

13 $\sqrt{3}/2$

15 $\sqrt{3}/2$

17 $g(x) = \cos x$, $a = \pi/2$, $b = 1$

19 $f(x) = \sin(x + \frac{\pi}{2})$
$g(x) = \sin(x - \frac{\pi}{2})$

23 (a) $1/\sqrt{2}$
(b) $1/2$
(c) $-\sqrt{3}/2$
(d) $-1/2$
(e) $1/\sqrt{2}$

25 $(-5\sqrt{3}, -5)$

27 (a) (i) p
(ii) s
(iii) q
(iv) r

(b)

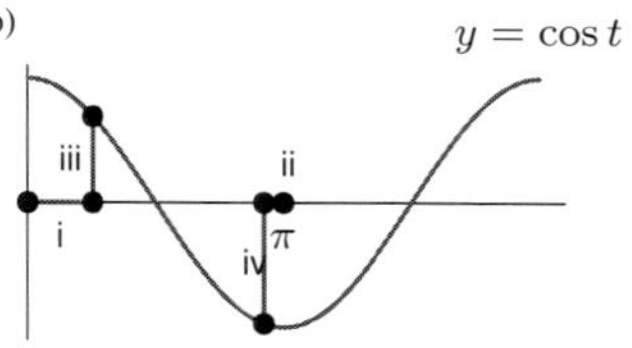

29 (a) $m = (\cos b - \cos a)/(b - a)$
(b) $-6(1 + \sqrt{2})/(13\pi)$

Section 6.5

1 Mid: $y = 0$; Amp: 6; Per: 2π

3 Mid: $y = 1$; Amp: $1/2$; Per: $\pi/4$

5 Hor: $-4/3$; Phs: -4

7 Both f and g have periods of 1, amplitudes of 1, and midlines $y = 0$

9 Period: 6; Amp: 5; Mid: 0

11 $h(t) = 4\sin(2\pi t)$

13 $g(t) = -2\cos(t/2) + 2$

15 $y = 4000 + 4000\sin((2\pi/60)x)$

17 $y = -2\sin(\pi\theta/6) + 2$

19

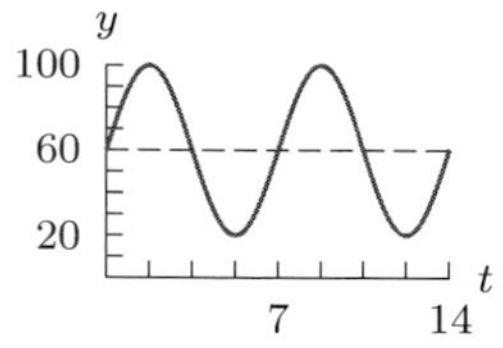

21

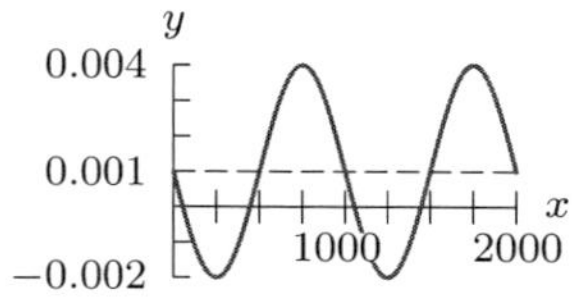

23 $1/4$, $g(x) = 3\sin((\pi/4)x - \pi/2)$

25 $f(x) = \sin x$, $a = \pi/2$, $b = \pi$, $c = 3\pi/2$, $d = 2\pi$, $e = 1$

27 Amplitude: 20
Period: $3/4$ seconds

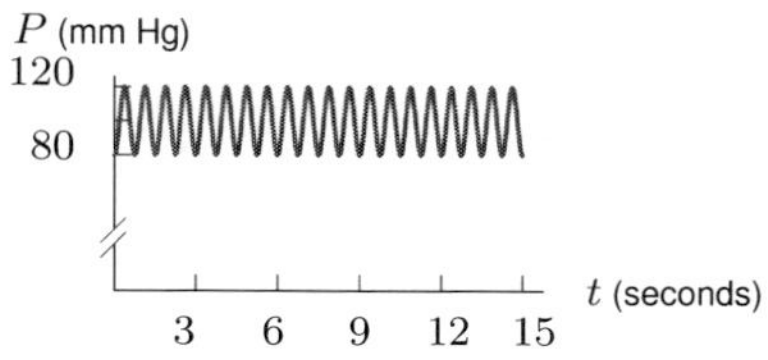

29 $f(t) = 14 + 10\sin(\pi t + \pi/2)$

31 $f(t) = 20 + 15\sin((\pi/2)t + \pi/2)$

33 (a) 12°/min
(b) $\theta = (12t - 90)^\circ$
(c) $f(t) = 225 + 225\sin(12t - 90)^\circ$
(d) Amp = Midline = 225 feet
Period = 30 min

35 (a) $P = f(t) = -450\cos(\pi t/6) + 1750$
(c) $t_1 \approx 1.9$; $t_2 \approx 10.1$

37 $y = 3f(x)$

39 $y = -f(2x)$

41 Amplitude: 41.5;
Period: 12 months

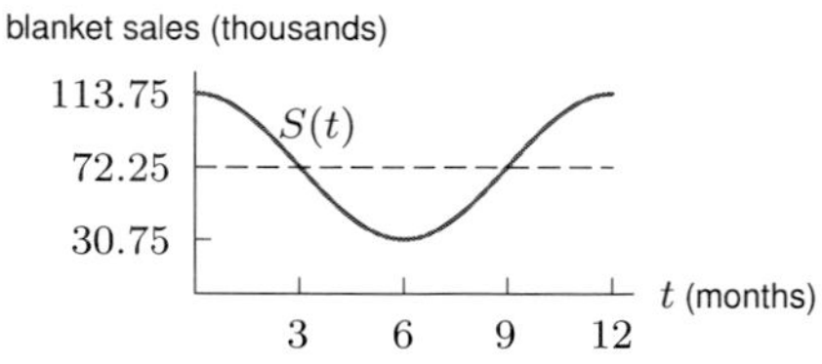

43 $f(t) =$
$-100\cos(\pi t) + 100$ (for $0 \le t \le 1$)
$10\cos(4\pi t) + 190$ (for $1 < t \le 2$)

45 (a)

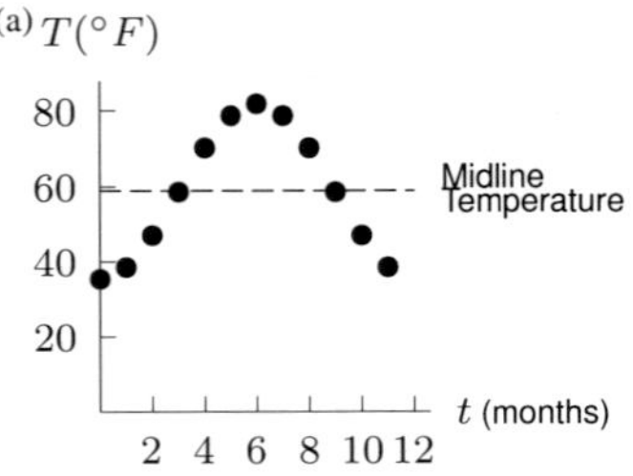

(b) $23.2°$; 12 months
(c) $T = f(t) =$
$-23.2\cos((\pi/6)t) + 58.6$
(d) $T = f(9) \approx 58.6°$

47 $f(t) = 3\sin((\pi/6)(t - 74)) + 15$

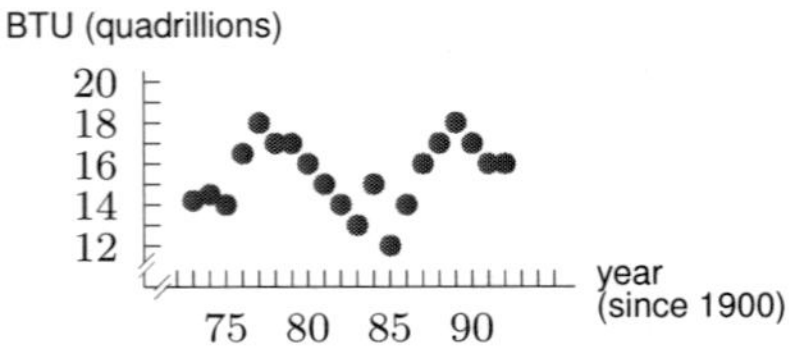

Section 6.6

1 0, 1, 0

3 1

5 −1

7 −1

9 1

11 $-\sqrt{3}$

13 $-\sqrt{2}$

15 $2/\sqrt{3}$

17 $\sec\theta = 2$
$\tan\theta = \sqrt{3}$

19 $\sec\theta = 3/\sqrt{8}$
$\tan\theta = 1/\sqrt{8}$

21 $f(\theta) = (1/2)\tan\theta$

23 (a) $\sin\alpha = -\sqrt{22}/5$,
$\tan\alpha = \sqrt{22/3}$
(b) $\sin\beta = -4/5$,
$\cos\beta = -3/5$

25 $\cos\theta = \sqrt{1 - y^2}$

29 $\sin\theta = \sqrt{x^2 - 16}/x$,
$\tan\theta = \sqrt{x^2 - 16}/4$

31 $\cos\theta = 9/\sqrt{x^2 + 81}$,
$\sin\theta = x/\sqrt{x^2 + 81}$

33 $y = y_0 + (\tan\theta)(x - x_0)$

35 No

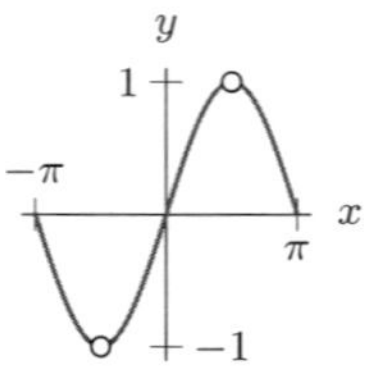

37 $u = -5\cos 2$
$v = 5\sin 2$
$w = 5\sqrt{2(1 - \cos 2)}$

Section 6.7

1 1.570

3 1.330

5 −1.447

7 $3\pi/2$

9 π

11 $\pi/4, 5\pi/4$

13 $\pi/3, 4\pi/3$

15 (a) 1.88, 4.41
(b) 1.88, 4.41

17 $\pi/6$

19 $\pi/3$

21 $\pi/3$

23 0.850

25 (a) $-1/2$
(b) $\sqrt{2}/2$
(c) $-\sqrt{2}/2$
(d) $-\sqrt{3}/2$

27 $\theta = 0.708, 2.434$

29 $t = 1.813, 4.473$

31 0.340, 2.802

33 0.152, 2.989, 3.294, 6.131

35 1.914, 4.653

37 0.305, 2.837

39 4.069, 5.356, 10.352, 11.639

41 $\theta = \pi/6 + 2\pi k, 11\pi/6 + 2\pi k$, k an integer

43 $\theta \approx 1.893$

45 $t = \pi/6,\ 5\pi/6,$
$7\pi/6$, or $11\pi/6$

47 $t = \pi/2,\ 3\pi/2,$
$\pi/6$, or $5\pi/6$

49 $\approx 39.806°$

51 (a) $f(t) = 40{,}000\cos\left(\frac{\pi}{6}t + \frac{\pi}{6}\right) + 60{,}000$
(b) $f(3) = \$40{,}000$
(c) Mid-March and mid-September

53 P: $x \approx 0.819$;
Q: $x \approx 3.181$

55 (a) $\pi/3$
(b) π
(c) ≈ 0.1

57 (a) $t_1 \approx 0.161$ and $t_2 \approx 0.625$.
(b) $t_1 = \arcsin(3/5)/4$ and
$t_2 = \pi/4 - \arcsin(3/5)/4$

59 Statement II is always true;
statement I is not always true

61 (a) $d = \sqrt{2rx + x^2}$
(b) $d = 25{,}238.776$ meters

Chapter 6 Review

1 (i) is B; (ii) is C; (iii) is A

3 (a) I
(b) I and III
(c) II and IV
(d) IV
(e) III

5 $7\pi/4$

7 $\pi^2/30 \approx 0.329$

9 $32{,}400/\pi \approx 10{,}313.240°$

11 8π

13 32.8π

15 $6.2 \cdot 17\pi/180 \approx 1.840$

17 12.4

19 Mid: 3; amp: 1; per: 2π

21 Mid: 7; Amp: 2; Per: 2

23 Amplitude: 20
Period: 1/2
Phase shift: 0
Horizontal shift: 0

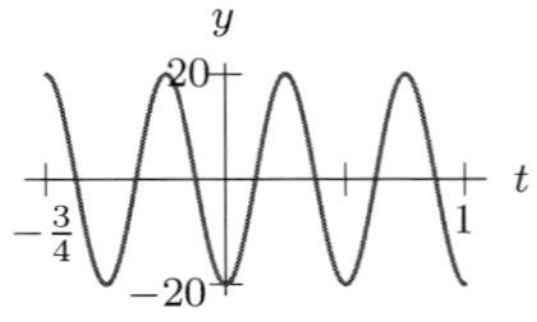

25 Amplitude: 3;
Period: $1/2$;
Phase Shift: -6π;
Horizontal Shift: $-3/2$ (left)

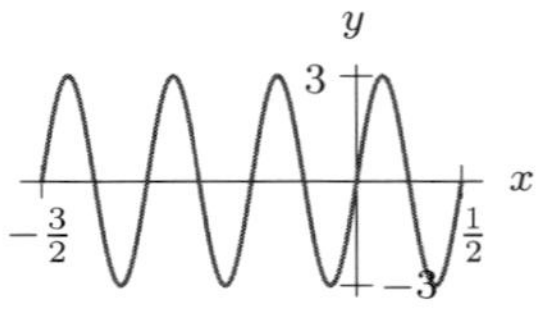

27 Amp: 30; mid: $y = 60$; per: 20

29 Amp: 50; mid: $y = 50$; per: 64

31

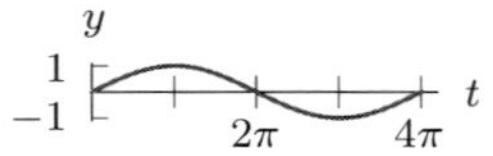

33

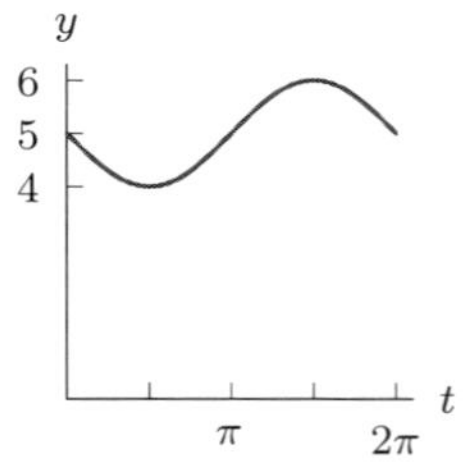

35 −1

37 $\sqrt{3}$

39 0.412

41 0.979

43 (a) $y = 600 - 300\cos(2\pi x/80)$
(b) $x = 14.5279, 65.4721, 94.5279$

45 $f_1(x) = 6\cos((1/2)(x - 3\pi)) + 2, f_2(x) = -6\cos((1/2)(x - \pi)) + 2, f_3(x) = 6\sin((1/2)(x - 2\pi)) + 2, f_4(x) = -6\sin((1/2)x) + 2$; answers may vary

47 $f_1(x) = 5\cos((\pi/6)(x+2))+3, f_2(x) = -5\cos((\pi/6)(x - 4)) + 3, f_3(x) = 5\sin((\pi/6)(x - 7)) + 3, f_4(x) = -5\sin((\pi/6)(x - 1)) + 3$; answers may vary

49 $\pi/6,\ 7\pi/6$

51 $2.897,\ 6.038$

53 $0,\ \pi,\ 1.107,\ 4.249$

55 69.115 miles

57 Outer edge: 3770 cm/min;
Inner edge: 471 cm/min

59 0.1345 radians

61 $f(t) = -900\cos((\pi/4)t) + 2100$

63 $y = 30\sin(105t - \pi/2) + 150$

Ch. 6 Understanding

1 True

3 False

5 True

7 True

9 False

11 False

13 True

15 False

17 True

19 True

21 False

23 True

25 True

27 True

29 False

31 True

33 False

35 True

37 True

39 True

41 False

43 False

45 True

47 False

49 True

51 True

53 True

55 False

57 False

59 False

61 True

63 True

65 True

67 True

69 True

71 False

73 True

75 True

77 False

79 False

81 False

83 False

85 True

87 False

Chapter 6 Tools

1 (a) $\tan\theta = 2$
(b) $\sin\theta = 2/\sqrt{5}$
(c) $\cos\theta = 1/\sqrt{5}$

3 (a) $5/\sqrt{125}$
(b) $10/\sqrt{125}$
(c) $10/\sqrt{125}$
(d) $5/\sqrt{125}$
(e) $1/2$
(f) 2

5 (a) $\sqrt{45}/7$
(b) $2/7$
(c) $\sqrt{45}/2$

7 (a) $8/12$
(b) $\sqrt{80}/12$
(c) $8/\sqrt{80}$

9 (a) $\sqrt{117}/11$
(b) $2/11$
(c) $\sqrt{117}/2$

11 $r = 7\sin 17°$; $q = 7\cos 17°$

13 $r = 6/\cos 37°$; $q = 6\tan 37°$

15 $r = 9/\tan 77°$; $q = 9/\sin 77°$

17 $c = 34.409$; $A = 35.538°$, $B = 54.462°$

19 $B = 62°$; $a = 9.389$; $b = 17.659$

21 Height $= 46.174$ ft;
Incline $= 205.261$ ft

23 $h = 400$ feet; $x = 346.410$ feet

25 74.641 feet

27 $d = 35000/\tan\theta$ feet

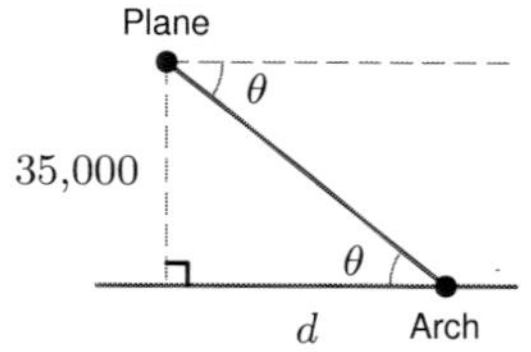

29 $d = 3\tan\phi$ miles

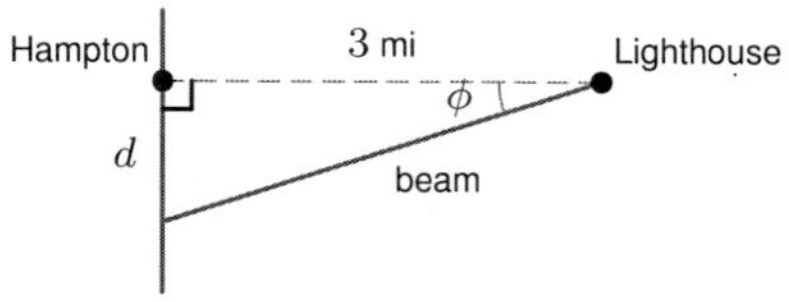

31 $d \approx 15.877$ feet

Section 7.1

1 $A = 25.922°$, $B = 37.735°$, $C = 116.343°$

3 $b = 31.762$, $A = 38.458°$, $C = 60.542°$

5 $c = 10.954$, $A = 54.010°$, $B = 45.990°$

7 $c = 7.2605$; $A = 21.4035°$; $B = 126.597°$

9 $a = 15.860$, $b = 2.569$, $C = 66°$

11 $a = 10.026$, $b = 6.885$, $C = 61°$

13 $a = 2.079$, $b = 3.090$, $B = 18°$

15 $a = 1.671$, $b = 4.639$, $B = 166°$

17 $a = 13.667$, $A = 90.984°$, $C = 17.016°$

19 $a = 12.070$, $A = 135.109°$, $C = 27.891°$
or
$a = 3.231$, $A = 10.891°$, $C = 152.109°$

21 $b = 0.837$ m, $c = 2.720$ m; $\gamma = 143.7°$

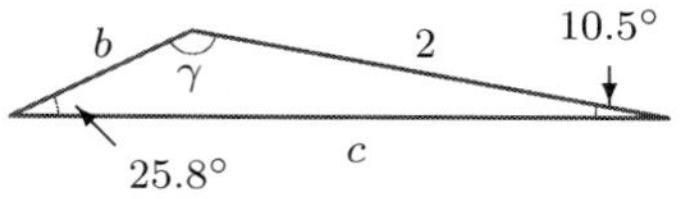

23 $\alpha \approx 41.410°$, $\beta \approx 82.819°$, $\gamma \approx 55.771°$

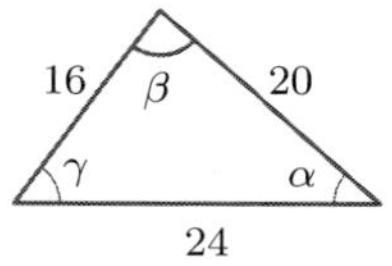

25 (a) $\sin\theta = 0.282$
(b) $\theta \approx 16.374°$
(c) $12.077\ \text{cm}^2$

27 Length of arc ≈ 0.174533 feet;
Length of chord ≈ 0.174524 feet

29 B closer by 2.387 miles

31 396.004 miles

33 $(18.876, 10.071)$

37 (a) First; 3.062 feet closer
(b) 157.279 feet to home
113.218 feet to third

39 158.926 feet

Section 7.2

1 $\sin x$

3 $2\sin\alpha$

5 $\cos t - \sin t$

7 0

9 $\cos^2\theta + \sin^2\theta = 1$; $\cos 2\theta = \cos^2\theta - \sin^2\theta = 2\cos^2\theta - 1 = 1 - 2\sin^2\theta$

17 $\pi/2, 7\pi/6, 11\pi/6$

19 $0, 3\pi/4, \pi, 7\pi/4, 2\pi$

21 (a) (x, y)

23 (a) $(x, -y)$

25 Not an identity

27 Not an identity

29 Not an identity

31 Identity

33 Identity

35 Identity

37 Not an identity

39 Not an identity

41 (a) $\theta = 60^\circ$, 180°, and 300°
(b) $\theta = \frac{7\pi}{6}, \frac{3\pi}{2}, \frac{11\pi}{6}$

43 (a) $\sqrt{1-y^2}$
(b) $y/(\sqrt{1-y^2})$
(c) $1-2y^2$
(d) y
(e) $1-y^2$

45 $\sin 2\theta = \frac{2x}{9}\sqrt{9-x^2}$

47 (a) $2x\sqrt{1-x^2}/(2x^2-1)$
(b) $2x/(1+x^2)$

49 $\sin 4\theta = 4(\sin\theta\cos\theta)(2\cos^2\theta - 1)$

Section 7.3

1 $10\sin(t-0.644)$

3 $\sqrt{2}\sin(t+3\pi/4)$

5 $\sin 15^\circ = \cos 75^\circ = (\sqrt{6}-\sqrt{2})/4$
$\cos 15^\circ = \sin 75^\circ = (\sqrt{6}+\sqrt{2})/4$

7 $\sqrt{6}/2$

9 $(\sqrt{6}+\sqrt{2})/4$

11 (a) 1.585
(b) 0.053
(c) 1.216
(d) -0.069

19 $\cos 3\theta = 4\cos^3\theta - 3\cos\theta$

Section 7.4

1 All integral multiples of π

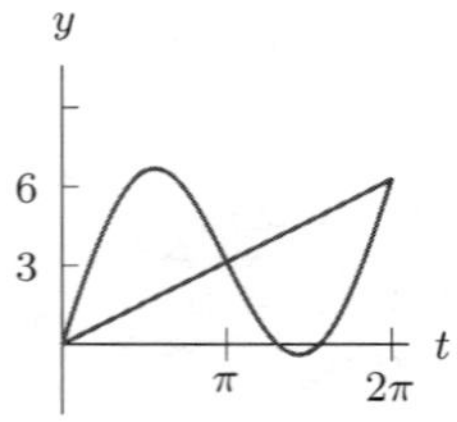

3 (a) $m = 2.5$; $b = 20$; $A = 10$
(b) Roughly in January and December
(c) Roughly between May and September

5 (a) $y = 1$
(b) f oscillates faster and faster between -1 and 1 as t increases.
(c) ≈ 0.540
(d) $t_1 = \ln(\pi/2)$
(e) $t_2 = \ln(3\pi/2)$

9 (a) $h = f(t) = 25 + 15\sin(\pi t/3) + 10\sin(\pi t/2)$
(b) $f(t)$ is periodic with period 12

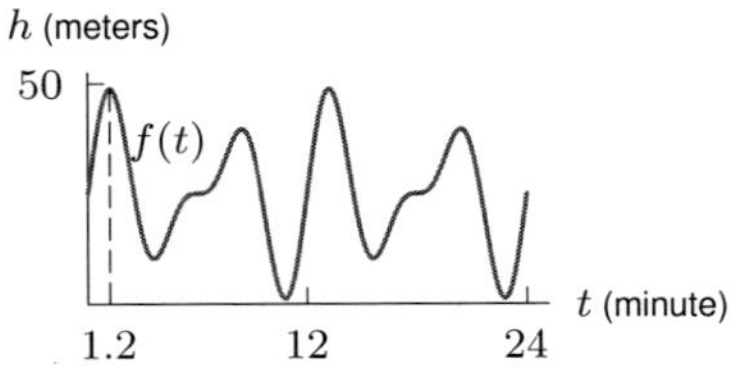

(c) $h = f(1.2) = 48.776$ m

Section 7.5

1 IV

3 II

5 III

7 I

9 IV

11 90° to 180°

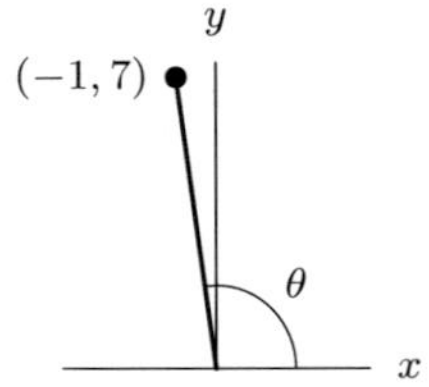

13 180° to 270°

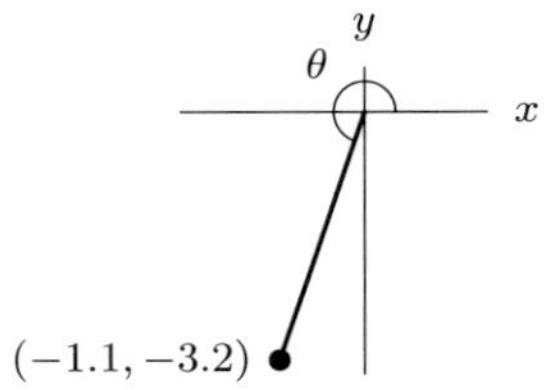

15 $(1, \pi)$

17 $(2, 5\pi/6)$

19 $(-\sqrt{6}/2, -\sqrt{6}/2)$

21 $(-\sqrt{3}, 1)$

23 $x^2 + y^2 = 6x$

25 $y = x^2 - 2x$

27 $r = \sqrt{5}$

29 $r = 1/\sqrt{2\cos\theta\sin\theta}$

31 $H: x = 3,\ y = 0;\ r = 3,\ \theta = 0$
$M: x = 0,\ y = 4;\ r = 4,\ \theta = \pi/2$

33 $H: x = 3/2,\ y = 3\sqrt{3}/2;\ r = 3,\ \theta = \pi/3$
$M: x = 0,\ y = 4;\ r = 4,\ \theta = \pi/2$

35 $H: x = -1.5,\ y = -3\sqrt{3}/2;\ r = 3,\ \theta = 4\pi/3$
$M: x = 0,\ y = 4;\ r = 4,\ \theta = \pi/2$

37 $H: x \approx -2.974,\ y \approx 0.392;\ r = 3,\ \theta = 172.5\pi/180$
$M: x = 4,\ y = 0;\ r = 4,\ \theta = 0$

39 $0 \le r \le 2$ and $-\pi/6 \le \theta \le \pi/6$

41 (b)

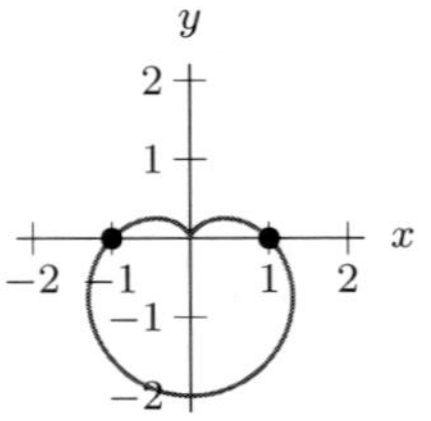

(c) Cartesian:
$(\sqrt{3}/4, 1/4)$;
$(-\sqrt{3}/4, 1/4)$ or polar:
$r = 1/2, \theta = \pi/6$ or $5\pi/6$

(d)

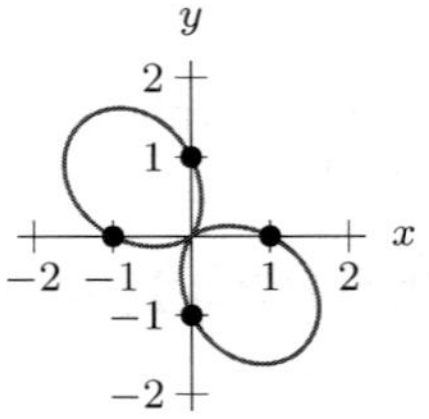

43 Looks the same

45 Rotated by 90° clockwise

47 $\pi/4 \le \theta \le 5\pi/4$;
$0 \le \theta \le \pi/4$ and $5\pi/4 \le \theta \le 2\pi$

Section 7.6

1 $5e^{i\pi}$

3 $0e^{i\theta}$, for any θ.

5 $5e^{i4.069}$

7 $-5+12i$

9 $-3-4i$

11 $-\frac{1}{2} + i\frac{\sqrt{3}}{2}$

13 $\frac{\sqrt{3}}{2} + \frac{i}{2}$

15 (a) $e^{i\pi/2}$

17 $\sqrt{2} + i\sqrt{2}$

19 $\frac{\sqrt{3}}{2} + \frac{i}{2}$

21 $\sqrt{2}/2 + i\sqrt{2}/2$

23 $\sqrt{2}\cos\frac{\pi}{12} + i\sqrt{2}\sin\frac{\pi}{12}$

25 $2.426 + 4.062i$

27 $A_1 = 2 - i$
$A_2 = -2 + 2i$

29 $p = -b$ and $q = \sqrt{c - b^2}$

Chapter 7 Review

1 (a) $a = 4$; $c = 2$; $B = 60^\circ$
(b) $A \approx 73.740^\circ$; $B \approx 16.260^\circ$; $b = 7$

3 3.004, 5.833; 31°

5 14.188, 9.829; 105°

7 $\cos\theta$

9 $2\cos\phi$

11 $-\sqrt{2}/2$

13 $(\sqrt{2}-\sqrt{6})/4$

15 I

17 I

19 $(1.571, 0)$

21 $(0, 0)$

25 $\cos\theta = 3/\sqrt{14}$

27 (a) $\cos\theta = -\sqrt{48}/7$
(b) $\theta \approx 2.998$ radians

29 $\sin(\ln(xy)) \approx 0.515$

35 $(\tan^2 x)(\sin^2 x) = \tan^2 x - \sin^2 x$

37 (a)

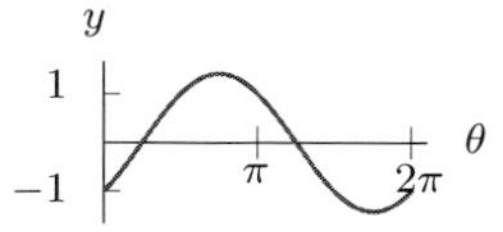

(b) $\sqrt{2}\sin(\theta - \pi/4)$
$\sqrt{2}\cos(\theta - 3\pi/4)$

39 $\theta = 101.068^\circ$

43 Height: 541.723 ft;
Distance between: 718.891 ft

45 (b) 12 o'clock $\rightarrow$ $(x, y) = (1, 1)$ and $(r, \theta) = (\sqrt{2}, \pi/4)$,
3 o'clock $\rightarrow$ $(x, y) = (2, 0)$ and $(r, \theta) = (2, 0)$,
6 o'clock $\rightarrow$ $(x, y) = (1, -1)$ and $(r, \theta) = (\sqrt{2}, -\pi/4)$,
9 o'clock $\rightarrow$ $(x, y) = (0, 0)$ and $(r, \theta) = (0$, any angle)

Ch. 7 Understanding

1 True

3 False

5 True

7 True

9 True

11 True

13 False

15 True

17 True

19 True

21 False

23 False

25 False

27 False

29 False

31 True

33 False

35 True

37 True

39 True

41 False

43 False

45 True

47 True

49 False

51 True

53 True

55 True

57 True

Section 8.1

1 $r(0) = 4, r(1) = 5, r(2) = 2, r(3) = 0, r(4) = 3, r(5)$ undefined

3 $\pi/2, \pi/3, \pi/4, \pi/6, 0$

5 2^{x^2}; 4^x

7 $(x^2 - 6x + 10)/(x^2 - 6x + 9)$

9 $x + 1$

11 $(x - 3)/(10 - 3x)$

13 $\sin x$

15 $x^3 + 3x + 1$

17 $5 + 1/x$

19 Length in terms of time

21 Time in terms of temp

25 (a) 4
(b) 1
(c) 4
(d) 0

27

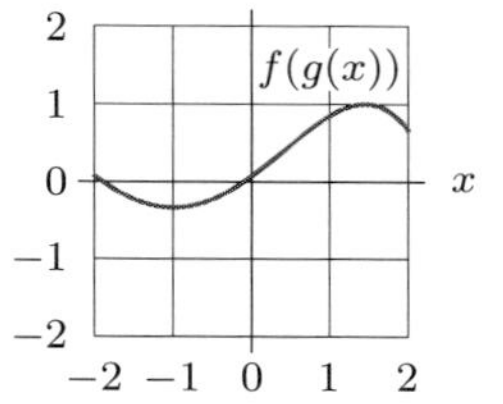

29

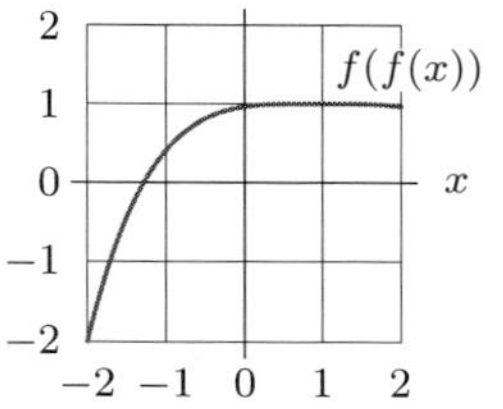

33 (a)

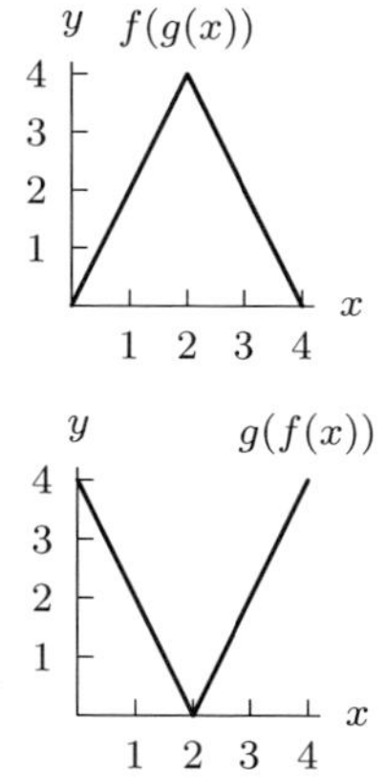

(b) $0 < x < 2$
(c) $2 < x < 4$

37 $1/2$

39 $2x + h + 1$

41 $-1/(x(x + h))$

43 $f(x) = 2x$

45 $f(x) = \ln x$

47 (a) $v(x) = -\sqrt{x}$
(b) $u(x) = e^{-x}$

49 (a) $v(x) = 2^x$
(b) $u(x) = 2^{1-x}$

51 (a) $v(x) = 2\cos x$
(b) $u(x) = e^{2x}$

53 $u(x) = \sin x, \quad v(x) = x^2$

55 $u(x) = e^x + x, \quad v(x) = \sin x$

57 $u(x) = 2/x, v(x) = 1 + \sqrt{x}$

59 $u(x) = 8 - 2x, v(x) = |x|$

61 (a) $f(x) = 114.64x$
$g(x) = 0.829x$
$h(x) = 0.00723x$
(b) \$1000 trades for yen, then for 828.8472 euros

Section 8.2

1 Not invertible

3 Not invertible

5 Not invertible

11 Yes, $f(f^{-1}(x)) = f^{-1}(f(x)) = x$

13 Yes, $f(f^{-1}(x)) = f^{-1}(f(x)) = x$

15 $h^{-1}(x) = \sqrt[3]{x/12}$

17 $k^{-1}(x) = \frac{1}{2}\ln(x/3)$

19 $n^{-1}(x) = 10^x + 3$

21 $h^{-1}(x) = (x/(1-x))^2$

23 $f^{-1}(x) = (4x^2 - 4)/(x^2 - 7)$

25 $f^{-1}(x) = 1/(e^x - 1)$

27 $q^{-1}(x) = (3 + 5e^x)/(e^x - 1)$

29 (a) $p(q(t)) = q(p(t)) = t$; inverses
(b) Line $y = t$

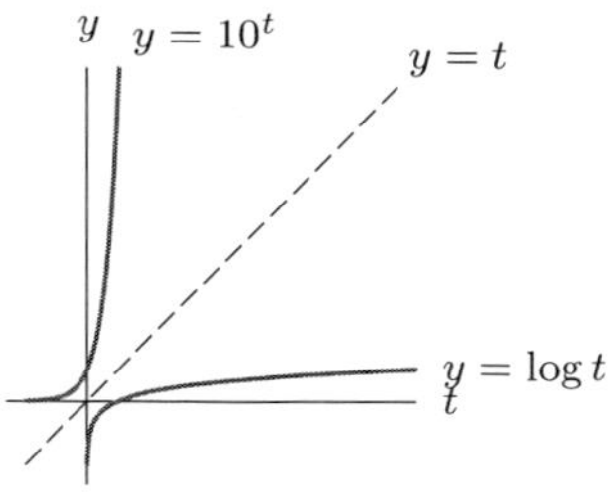

31 $f^{-1}(P) = 50\ln(P/10)$

33 $f^{-1}(N) = I_0 10^{N/10}$

35 $x = (\ln 3/\ln 2) - 5$

37 $x = e^{1.8} - 3$

39 $x = (19 - \sqrt{37})/2$

41 $f^{-1}(3) < f(3) < 0 < f(0) < f^{-1}(0) < 3$

43 $1 - t^2$

45 (a) $A = \pi r^2$

(b)

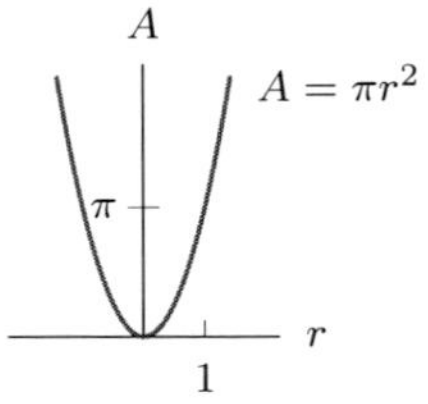

(c) $r \geq 0$

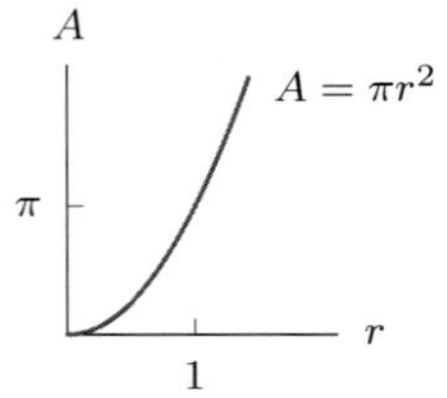

(d) $f^{-1}(A) = \sqrt{A/\pi}$

(e)

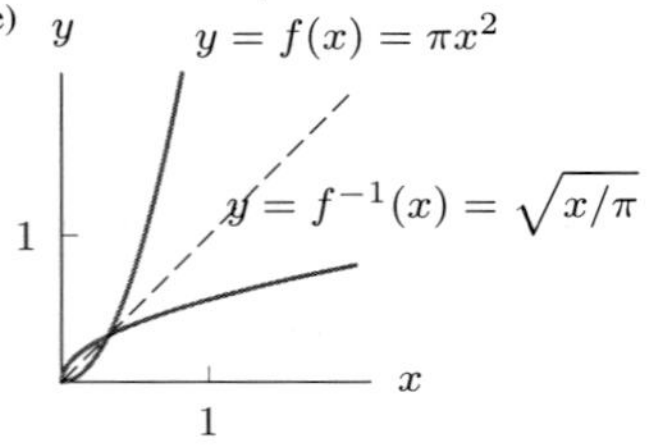

(f) Yes

47 (a) $P(t) = 150(1.1)^t$

(b) $P^{-1}(y) = (\log(y) - \log(150))/(\log(1.1))$

(c) 10.3 years

49 (a) 36 m

(b) 6 seconds

(c) One possibility: $3 \leq t \leq 6$

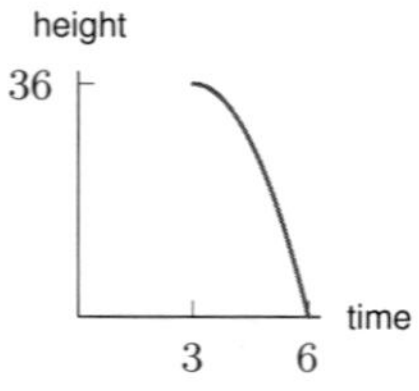

(e)

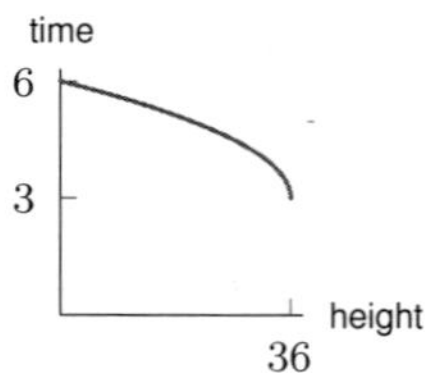

51 (a) A (wrong) guess is 0.99 ml

(b) $x = C^{-1}(0.98)$

(c) 50 ml

59 $f^{-1}(L) = -\frac{1}{k}\ln(1 - L/L_\infty)$
$f^{-1}(L) =$ Age of fish of length L
Domain: $0 \leq L \leq L_\infty$

Section 8.3

1 (a) $f(x) + g(x) = 3x^2 + x + 1$
(b) $f(x) - g(x) = -3x^2 + x + 1$
(c) $f(x)g(x) = 3x^3 + 3x^2$
(d) $f(x)/g(x) = (x+1)/(3x^2)$

3 (a) $f(x) + g(x) = 2x$
(b) $f(x) - g(x) = 10$
(c) $f(x)g(x) = x^2 - 25$
(d) $f(x)/g(x) = (x+5)/(x-5)$

5 (a) $f(x) + g(x) = x^3 + x^2$
(b) $f(x) - g(x) = x^3 - x^2$
(c) $f(x)g(x) = x^5$
(d) $f(x)/g(x) = x$

7 $3x^2 + x$

9 $2x\sqrt{x+2}$

11 $3x/2 - 1/2$

13 $f(x) = e^x(2x+1) = 2xe^x + e^x$

15 $h(x) = 4e^{2x} + 4e^x + 1$

17 $\sin x + x^2$

19 $(\sin x)/x^2$

21 $\sin^2 x$

23 (a)

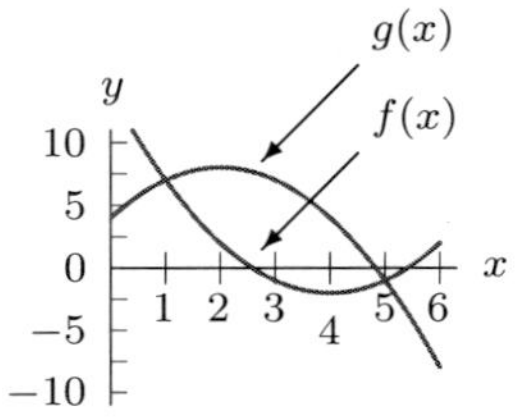

(c)

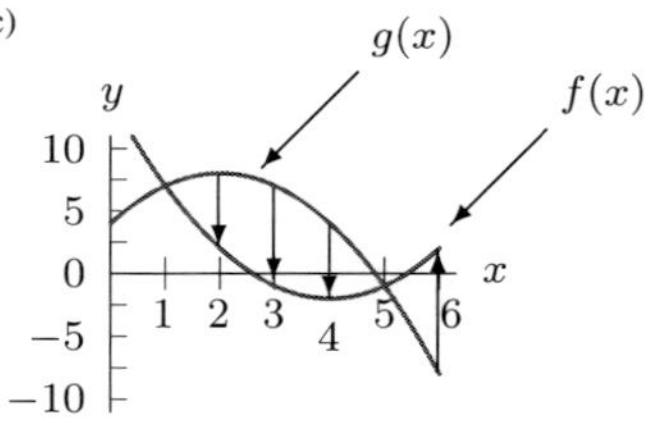

(f) $f(x) = x^2 - 8x + 14$;
$g(x) = -x^2 + 4x + 4$;
$f(x) - g(x) = 2x^2 - 12x + 10$

(g) Yes

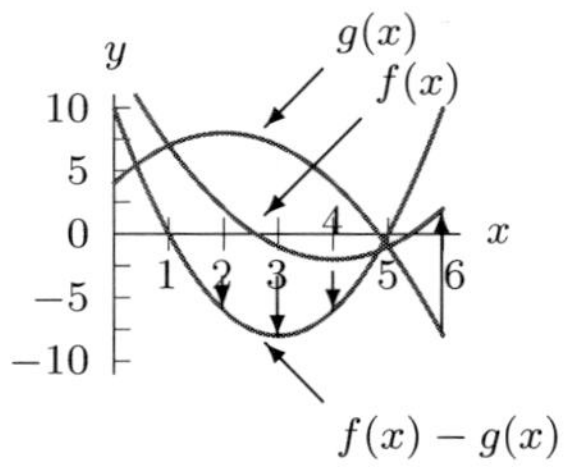

25 (a)

t (yrs)	0	1	2
$f(t)$ ($)	109,366	111,958	111,581
$g(t)$	13.61	13.68	13.03
t (yrs)	3	4	5
$f(t)$ ($)	114,931	119,162	122,499
$g(t)$	13.42	13.73	13.63

27 (a)

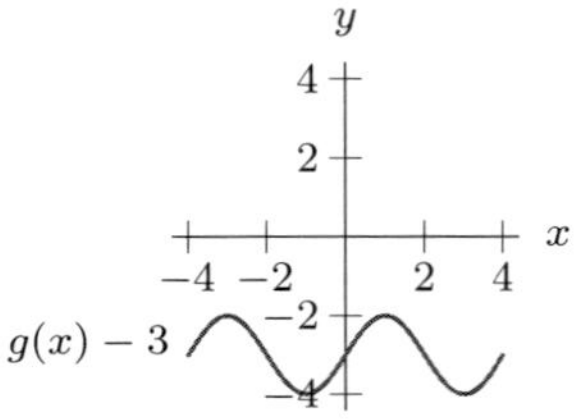

(b)

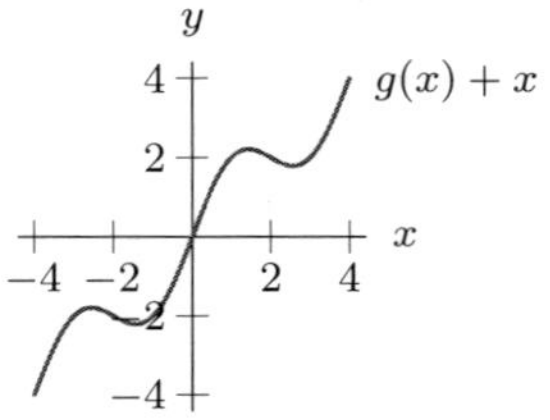

29

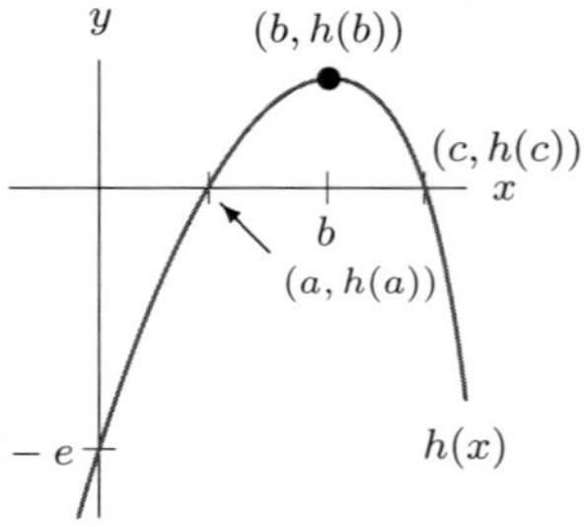

31

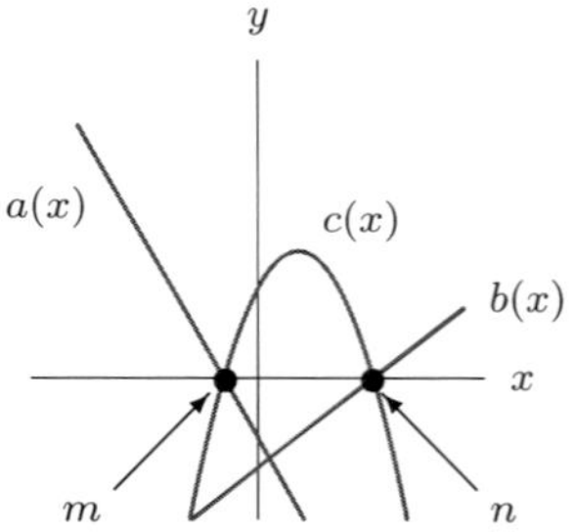

33 (a) $h(x) = x^2 + x, h(3) = 12$
(b) $j(x) = x^2 - 2x - 3, j(3) = 0$
(c) $k(x) = x^3 + x^2 - x - 1, k(3) = 32$
(d) $m(x) = x - 1$ for $x \neq -1, m(3) = 2$
(e) $n(x) = 2x + 2, n(3) = 8$

35 (a) $f(\theta) = (\tan\theta)/(1 + \tan\theta)$
(b)

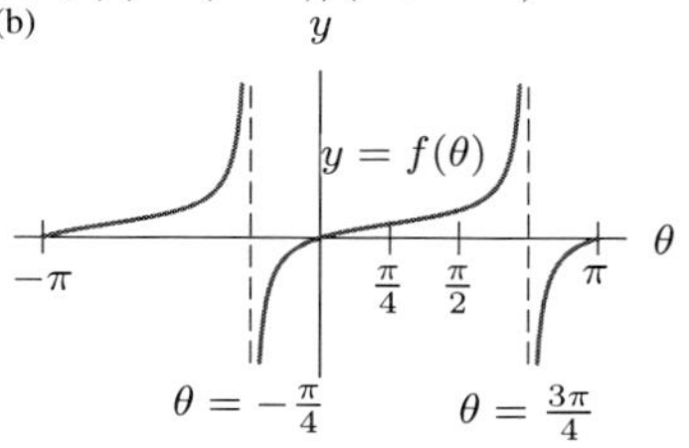

(c) $f(\theta)$ is periodic

37 40

39 (a) Even
(b) Odd
(c) Neither

41 (a) $A(t) = 316.75(1.004)^t$
(b) $V(t) = 3.25\sin 2\pi t$; other answers possible
(c)

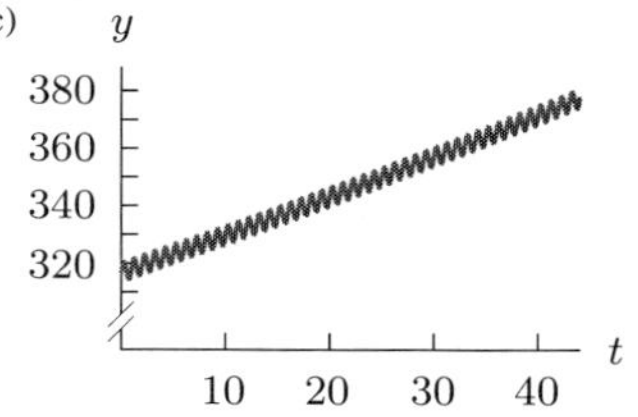

43 (b) $p(t) = (f_{\text{CA}}(t) \cdot g_{\text{CA}}(t) + f_{\text{FL}}(t) \cdot g_{\text{FL}}(t))/(f_{\text{US}}(t) \cdot g_{\text{US}}(t))$

Chapter 8 Review

1 $\sqrt{27x - 6}$

3 $3x$

5 $108x^2 - 2$

7 $2^{x/(x+1)}$

9 $2e^x - 1$

11 $4x - 3$

13 $\sqrt{x}e^{2x-1}$

15 (a) $f(2x) = 4x^2 + 2x$
(b) $g(x^2) = 2x^2 - 3$
(c) $h(1 - x) = (1 - x)/x$
(d) $(f(x))^2 = (x^2 + x)^2$
(e) $g^{-1}(x) = (x + 3)/2$
(f) $(h(x))^{-1} = (1 - x)/x$
(g) $f(x)g(x) = (x^2 + x)(2x - 3)$
(h) $h(f(x)) = (x^2 + x)/(1 - x^2 - x)$

17 $x/(1 + e^{2x})$

19 $x^{3/2}\tan 2x$

21 $\tan((3x - 1)^2/2) - 27x^{3/2}$

23 $f^{-1}(x) = (x + 7)/3$

25 $h^{-1}(x) = (2x + 1)/(3x - 2)$

27 $g^{-1}(x) = e^{(5+7x)/(1-2x)}$

29 $f^{-1}(x) = (\arccos x)^2$

31

t	u	v	w
0	2	3	1
1	3	4	2
2	1	1	4
3	4	2	0
4	0	0	3

33 $g(x) = x^2, h(x) = x + 3$

35 $g(x) = x^2 + x, h(x) = 3x$

37 65

39

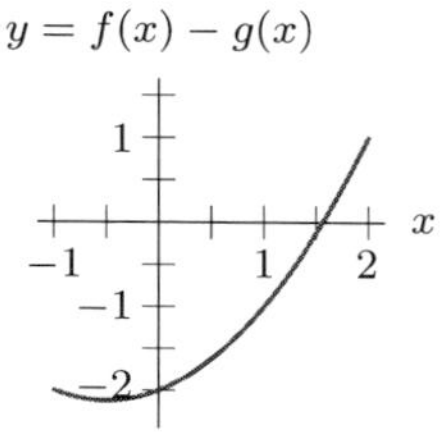

41

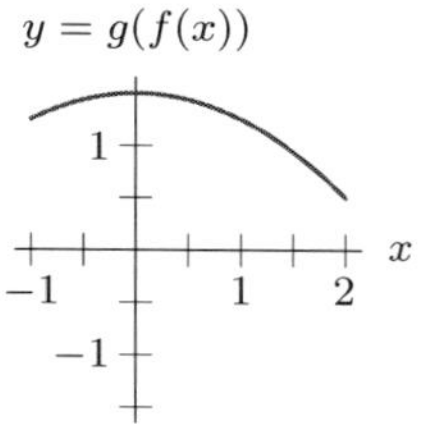

43 (a) II
(b) I
(c) None
(d) None
(e) IV
(f) None
(g) None

45 (a) $x \neq 3$
(b) $x \leq 2$

47 $r^{-1}(x) = (x + 1)/(2 - 2x)$

49 $u(x) = 0.5\sqrt{x}$

51 (a) $f(t) = 800 - 14t$ gals
(b) (i) 800 gals
(ii) 57.143 days
(iii) 28.571 days
(iv) $14t$

53 (a) $N(20{,}000) = 380$
(b) $N(x) = \frac{1}{250}x + 300$
(d) $N^{-1}(500) = 50{,}000$

55 Ace estimates 2000 ft^2 of office space costs \$200,000.

57 $f(2x) < 2f(x)$

59 Space estimates 1500 ft^2 of office space costs \$200,000

61 Ace seems to be a better value

63 Decreasing

65 Increasing

Ch. 8 Understanding

1 False

3 True

5 True

7 True

9 True

11 False

13 False

15 True

17 False

19 False

21 False

23 False

25 True

27 True

29 True

31 False

33 True

35 True

37 True

Section 9.1

1 Yes; $g(x) = (-1/6)x^9$

3 Yes; $Q(t) = (1/8)t^{-3/2}$

5 Yes; $T(s) = 6es^{-5}$

7 Odd

9 Odd

11 $y = 0.7x^{1.252}$

13 $y = 2x^{1.194}$

15 $k = 3/2$; $y = (3x)/2$; $x = 5.33$

17 $k = 5$; $c = 5d^2$; $c = 125$

19 $j(x) = 2x^3$

21 $g(x) = -\frac{1}{6}x^3$

23 (a) 0
(b) 0

27 (a) $x^{-3} \to +\infty, x^{1/3} \to 0$
(b) $x^{-3} \to 0, x^{1/3} \to \infty$

29 (a) No
(b) No new deductions
(c) All points satisfying the equation $y = -3x^5$

31 (a) $f(x) = x^{1/n}$
$g(x) = x^n$
(b) $(1, 1)$

33 $c = 81.67x$; 326.68

35 $y = 1350/n$; 14

37 $d = 0.1x$; 32.5 miles

39 (a) $T = kR^2D^4$
(b) Increases by factor of 4
(c) Increases by factor of 16
(d) Reduce to 44.4%

41 Decrease by 29.289%

43 (a) 20 lbs; 1620 lbs
(b) 3/10

45 (a) $h(x) = -2x^2$
(b) $j(x) = \frac{1}{4}x^2$

Section 9.2

1 Yes, 1

3 Yes, 2

5 No

7 $y \to \infty$; like $4x^4$

9 $y \to \infty$; like $2x^9$

11 (a) ∞
(b) ∞

13 -16.543

15 $y = (-1/8)(x+2)(x-2)^2(x-5)$;
$y = (-1/20)(x+2)(x-2)(x-5)^2$

17 $y = \frac{1}{2}x - 1$

19 (a)

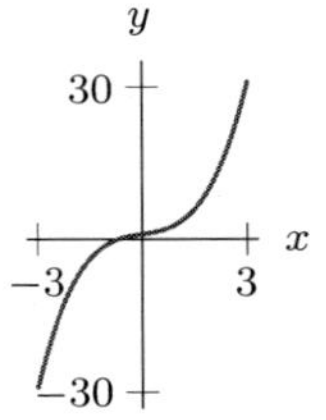

(b) $f(0.5) = 1.625$;
$f^{-1}(0.5) \approx -0.424$.

21 (a)

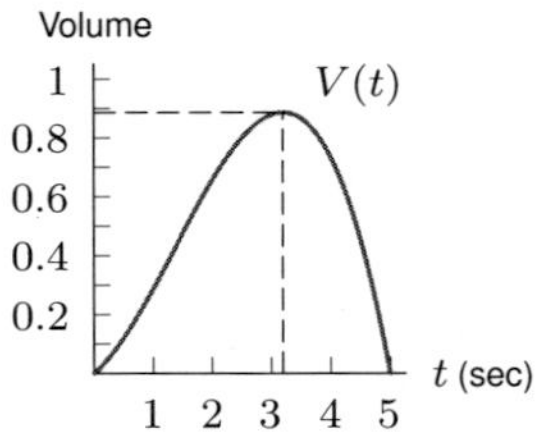

(b) $V \approx .886$ at $t \approx 3.195$
(c) $(0,0)$ and $(5,0)$;
Lungs empty at beginning and end

23 (a) 500 people
(b) May of 1908
(c) 790; February of 1907

25 Yes

27 (a) False
(b) False
(c) True
(d) False

29 (c)

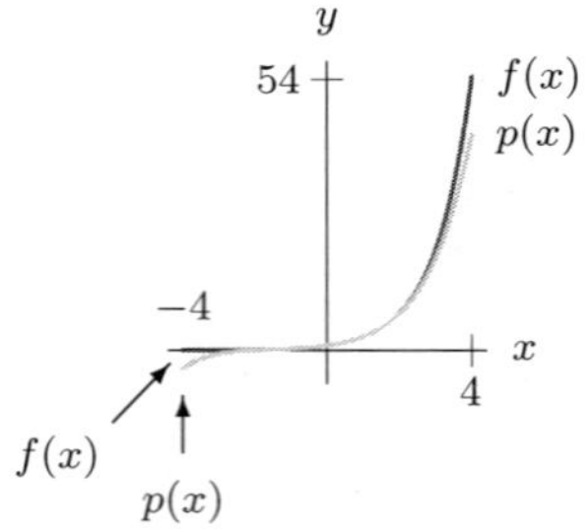

Section 9.3

1 $-3, 2, -7$

3 $0, -4, -3$

5 $x(x+2)(x-2)(x-4)$

7 $h(x) = x(x+2)^2(x-3)$

9

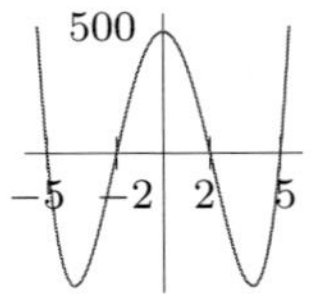

11 (a) $1/2, 1/3, 7$, and 9
(b) No
(c) $0 \le x \le 10, -500 \le y \le 1500$ and $0 \le x \le 1, -2 \le y \le 2$

13 $p(x) = (x+15)(x-1)^2(x-3)$

15 C

17 $f(x) = x^2 - 2x$

19 $f(x) = -\frac{1}{2}(x+3)(x-1)(x-4)$

21 $p(x) = x^2 + 2x - 3$

23 $y = \frac{1}{2}(x + \frac{1}{2})(x-3)(x-4)$

25 $y = 2x(x+2)(x-2)$

27 $y = 7(x+2)(x-3)^2$

29 $y = -\frac{1}{3}(x+2)^2(x)(x-2)^2$

31 $g(x) = -\dfrac{1}{4}(x+2)^2(x-2)(x-3)$

33 $g(x) = -\frac{1}{3}(x+2)(x-2)x^2$

35 $j(x) = (x+3)(x+2)(x+1) + 4$

37 $x = -2$ or $x = -3$

39 $x = \pm\frac{1}{2}$

41 $x = (3 \pm \sqrt{33})/4$

43 (a) $V(x) = x(6-2x)(8-2x)$
(b) $0 < x < 3$
(c)

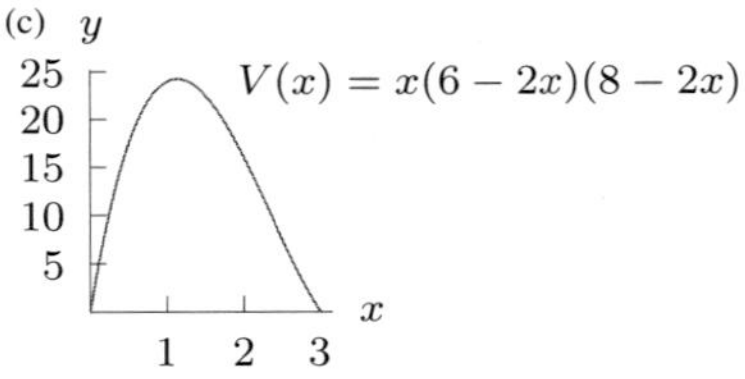

(d) ≈ 24.26 in^3

45 7.83 by 5.33 by 1.585 inches

47 (a)

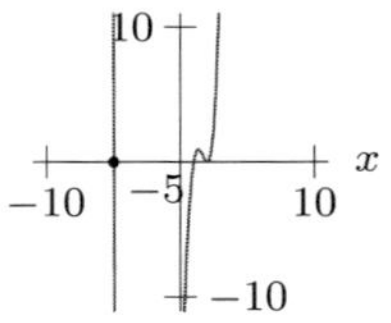

(b) No
(c) 3; $-6 \le x \le 3, -3 \le y \le 3$
(d) $f(x) = (x+5)(x-1)(x-2)^2$
(e) 3; No

49 (a) Never true
(b) Sometimes true
(c) Sometimes true
(d) Never true
(e) True
(f) Never true

Section 9.4

1 Rational; $(x^3 + 2)/(2x)$

3 Not rational

5 Rational; $(x^2 - 5)/(x - 3)$

7 (a) 2
(b) $5/6$

9 $y = 1$

11 $y = 4$

13 (a) $n > m$
(b) $n = m$

15 $f^{-1}(x) = (4x+4)/(5x+3)$

17 (a) $g(x) = (x^3+2)/(2x^2)$

19 (a) $f(x) = (3+x)/(12+x)$
(b) (i) 28%
(ii) 25%
(iii) $\approx 18.2\%$
(iv) 6
(v) -3
(c)

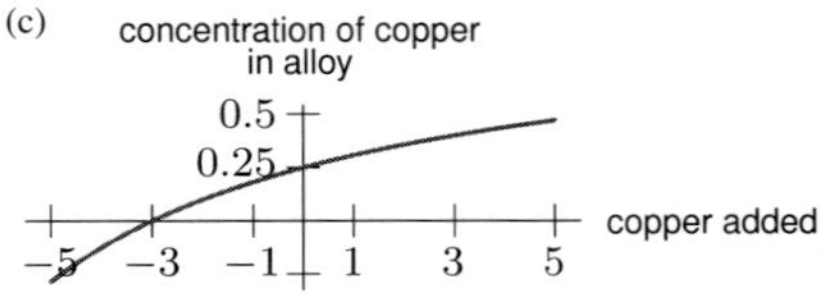

(d)

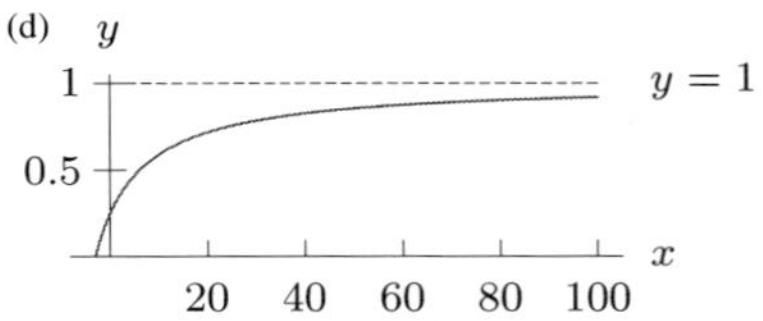

21 (a) (i) $C(1) = 5050$
(ii) $C(100) = 10{,}000$
(iii) $C(1000) = 55{,}000$
(iv) $C(10000) = 505{,}000$
(b) (i) $a(1) = 5050$
(ii) $a(100) = 100$
(iii) $a(1000) = 55$
(iv) $a(10000) = 50.5$
(c) $a(n)$ gets closer to \$50

23 No

Section 9.5

1 Zero: $x = -3$;
Asymptote: $x = -5$;
$y \to 1$ as $x \to \pm\infty$

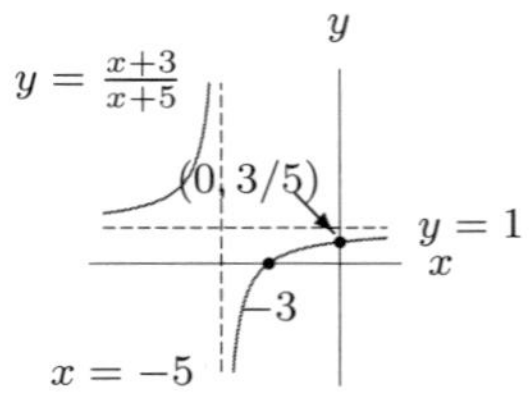

3 Zeros: $x = 4$;
Asymptote: $x = \pm 3$;
$y \to 0$ as $x \to \pm\infty$

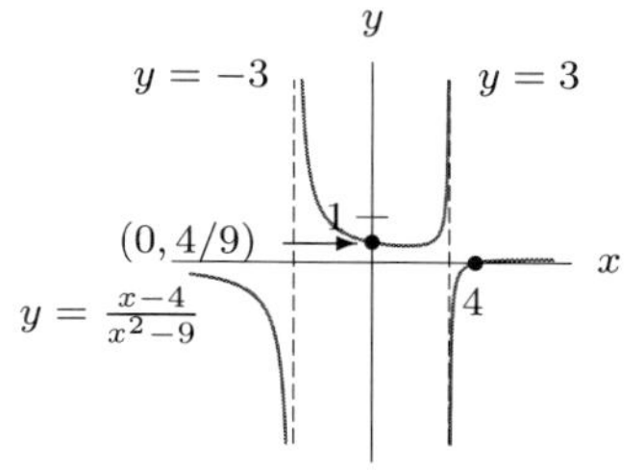

5 x-int: $x = 2$
y-int: $y = 1/2$
Horiz asy: $y = 1$
Vert asy: $x = 4$

7 x-int: $x = \pm 2$
y-int: None
Horiz asy: $y = 0$
Vert asy: $x = 0, x = -4$

9 (c) Horizontal: $y = 0$
Vertical: $x = 3$

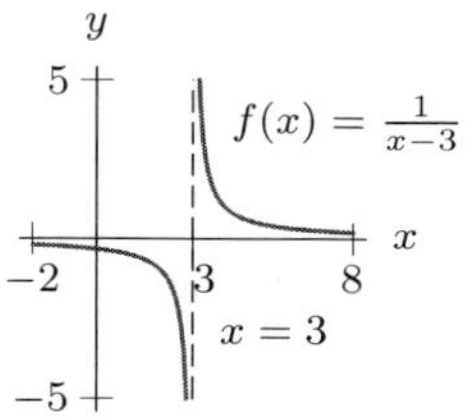

11

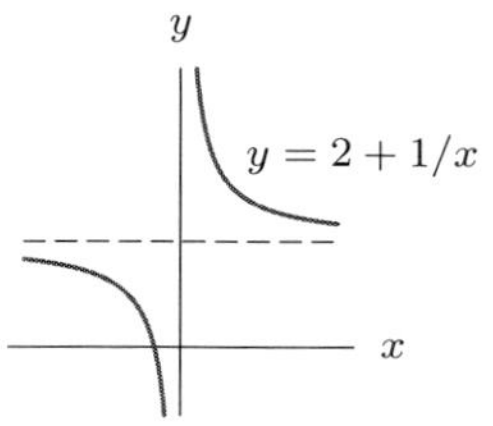

13 (a) $-\infty$
(b) $+\infty$

15 (a) 2, 2
(b) $\lim_{x\to -4^+} f(x) = -\infty$;
$\lim_{x\to -4^-} f(x) = \infty$

17 (a) (iii)
(b) (i)
(c) (ii)
(d) (iv)
(e) (vi)
(f) (v)

19 (a) Small
(b) Large
(c) Undefined
(d) Positive
(e) Negative

21 (a)

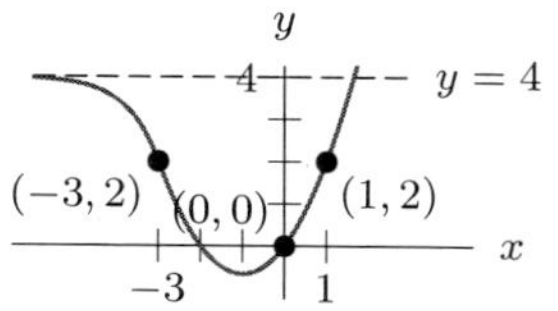

(b)

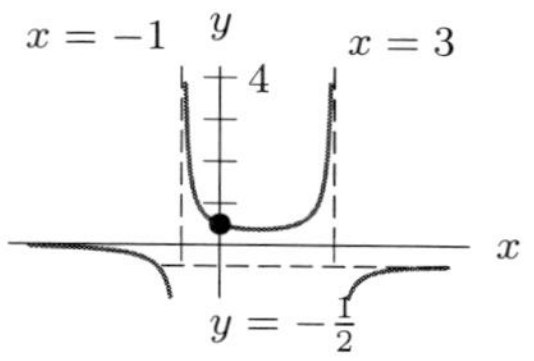

23 (a) $y = -1/(x+2)$
(b) $y = -1/(x+2)$
(c) $(0, -1/2)$

25 (a) $y = -1/(x-3)^2$
(b) $y = -1/(x^2 - 6x + 9)$
(c) $(0, -1/9)$

27 (a) $y = (1/x^2) + 2$
(b) $y = (2x^2 + 1)/x^2$
(c) None

29 (a) $1/x$
(b) $y = (1/x) + 2$

31 (a) $y = 1/x$
(b) $y = x/(2x-4)$

33 $y = \frac{1}{16}x^2(x+3)(x-2)$

35 $y = (x+2)(x-3)/((x-2)(x+1))$

37 $y = ((x+3)(x-1))/(x+2)^2$

39 $y = (x-2)/((x+1)(x-1))$

41 $g(x) = (x-5)/((x+2)(x-3))$

Section 9.6

1 Neither

3 $p(x) = 25^x$

5 $r(x) = 2(\frac{1}{9})^x$

7 A - (i)
B - (iv)
C - (ii)
D - (iii)

11 $y = 4e^x$

13 $y = ax^3$

15 $y = e^{-x}$

17 (a) $f(x) = 720x - 702$
(b) $f(x) = 2(9)^x$
(c) $f(x) = 18x^4$

19 (a) $f(x) = y = \dfrac{63}{4}x + \dfrac{33}{2}$
(b) $f(x) = 3 \cdot 4^x$
(c) $f(x) = \frac{3}{4}x^6$

21 A: $kx^{5/7}$; B: $kx^{9/16}$;
C: $kx^{3/8}$; D: $kx^{3/11}$

23 (a) $v = 40$
(b) $r(x) > t(x)$
(c) $t(x) > r(x)$

25 $y \to 0$ as $x \to \pm\infty$

27 $y \to \infty$ as $x \to \infty$
$y \to 0$ as $x \to -\infty$

29 $y \to 0$ as $t \to \infty$
$y \to \infty$ as $t \to -\infty$

31 $y \to 0$ as $t \to \infty$
$y \to 0$ as $t \to -\infty$

33 $y \to \infty$ as $t \to \pm\infty$

35 $y \to 1$ as $x \to \infty$
$y \to -1$ as $x \to -\infty$

37 (a) $h(r) = b \cdot r^{5/4}$;
$g(r) = a \cdot r^{3/4}$
(b) $a \approx 8$; $b \approx 3$

39 (a) $p_5(r) = 1000[(1+r)^5 + (1+r)^4 + (1+r)^3 + (1+r)^2 + (1+r) + 1]$;
$p_{10}(r) = 1000[(1+r)^{10} + (1+r)^9 + (1+r)^8 + (1+r)^7 + (1+r)^6 + (1+r)^5 + (1+r)^4 + (1+r)^3 + (1+r)^2 + (1+r) + 1]$
(b) 20.279%

Section 9.7

1 $f(x) = x^{\ln c/\ln 2}$

3 $h(x) = 2.35(1.44)^x$

5 (a) $f(x) = 201.353x^{2.111}$
(b) $f(20) = 112{,}313.62$ gm
(c) $x = 18.930$ cm

7 $y = \frac{3}{2}x$

9 $y = e^{0.4x}$

11 $y = x^{\frac{3}{2}}$

13 (a) FM

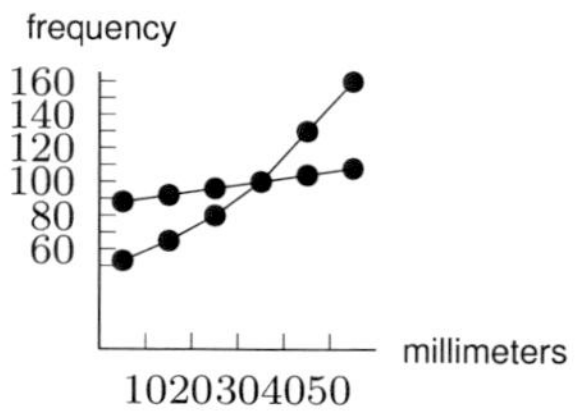

(b) AM
(c) $y = 86 + 0.4x$
(d) $y = 46.5e^{0.023x}$

15 (a) $N = -14t^4 + 433t^3 - 2255t^2 + 5634t - 4397$
(b) Good fit

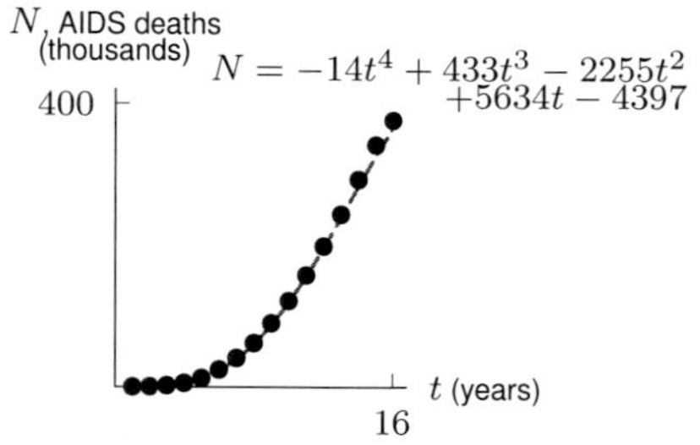

(c) Model fails for $t > 20$

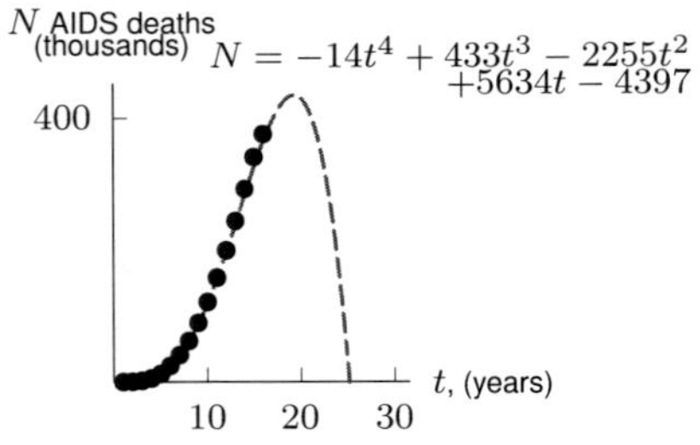

17 (a) $C(t) = 664.769(1.388)^t$
(b) 38.8% per year
(c) slower growth; concave down

19 (a)

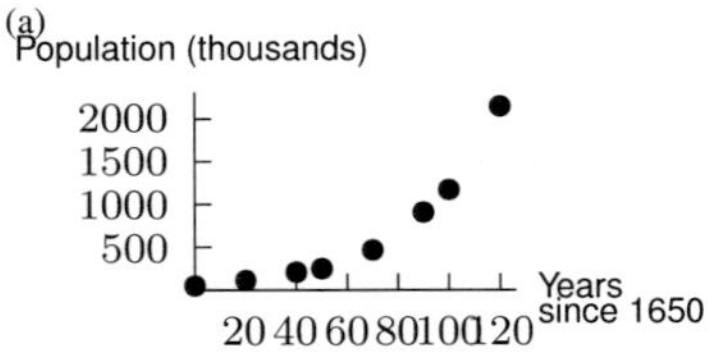

(b) $P(t) = 56.108(1.031)^t$, answers may vary
(c) 56.108 is 1650 population, 1.031 means 3.1% annual growth
(d) $P(100) = 1194.308$, Slightly higher
(e) $P(150) = 5510.118$, higher

21 (a) $N = 1271.2e^{0.3535t}$

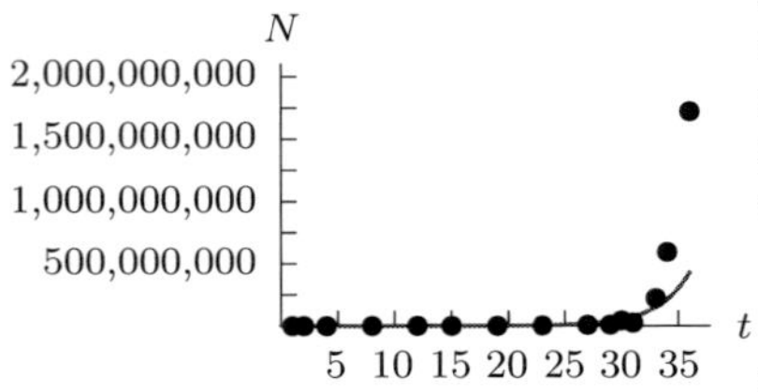

(b) About 1.96 years

23 (a) $y = 0.310t^2 - 12.177t + 144.517$
(b) $y = 3.01t^2 - 348.43t + 10{,}955.75$

25 (b) Points lie on a line

27 (a) Quadratic
(b) $y = -31.798x^2 + 3218.719x - 37{,}169.5$; answers may vary
(c) $38,392; answers may vary
(d) Age 10, −$8162, not reasonable; answers may vary

29 (a) $t = 8966.1H^{-2.3}$

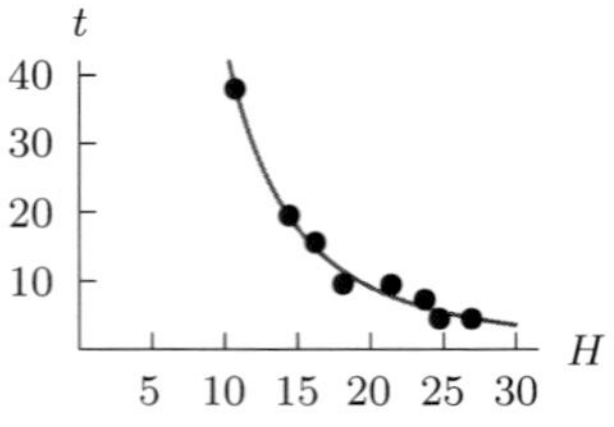

(b) $r = 0.0124H - 0.1248$

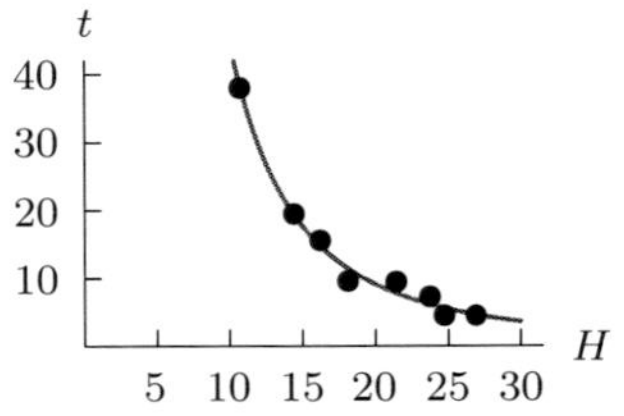

(c) $H = 0°$C; $H = 10.1°$C; model (b)

Chapter 9 Review

1 No
3 Yes; $y = x^2$
5 No
7 Even
9 $f(x) = (3/2) \cdot x^{-2}$
11 3rd degree
13 Degree: 4; Terms: 3; $x \to \pm\infty$: $y \to -\infty$
15 $6, 2, 3$
17 Rational; $(3x + 1)/6$
19 $y = 7/x^{-4}$
21 4
23 $3/4$
25 $-0.1 \le x \le 0.1, 0 \le y \le 0.011$
27 $-1.1 \le x \le 0.3, -0.121 \le y \le 0.117$
29 Graph (i): J;
Graph (ii): L;
Graph (iii): O;
Graph (iv): H
31 $f(x) = -(x + 1)(x - 1)^2$
33 $h(x) = \frac{1}{12}(x + 2)(x + 1)(x - 1)^2(x - 3)$
35 $g(x) = \frac{1}{6}(x + 1)(x - 2)(x - 4) + 4$
37 (a) Even
(b) Neither
(c) Odd
(d) Neither
(e) Neither
(f) Even
(g) Neither
(h) Even
39 (a) $b = 0$
(b) $b = d = 0$
41 (a) 147.628 million miles
(b) Yes
43 (a) 1.2 meters/sec
(b) 21% longer than the existing ship
45 (a) $f^{-1}(x) = 5x/(1 - x)$
(b) $f^{-1}(0.2) = 1.25$
(c) $x = 0$
(d) $y = -5$
47 $p(x) = \frac{7}{1080}(x+3)(x-2)(x-5)(x-6)^2$
49 $h(x) = (1/5)(x + 5)(x + 1)(x - 4) + 7$

53 (a)

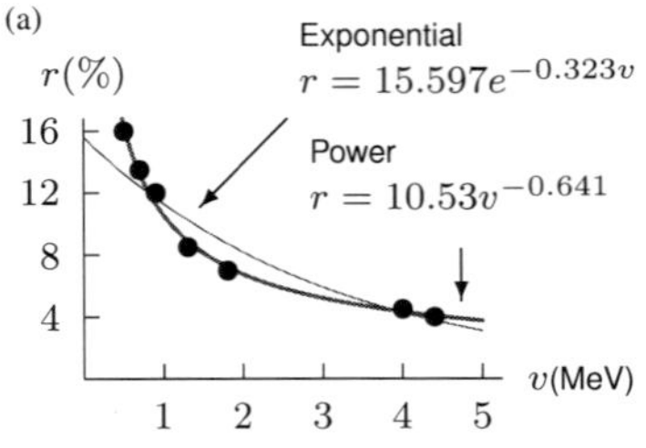

(b) High-energy photons
(c) $r = 15.597e^{-0.323v}$; $r = 10.53v^{-0.641}$ (best fit)
(d) Power function

55 (a) $I(x) = 0.01x$ (number infected); $N(x) = 0.99x$ (number not infected)
(b) $T(x) = 0.0098x$
(c) $F(x) = 0.0297x$
(d) $P(x) = 0.0395x$
(e) $0.01x/(0.0395x) \approx 0.253$
(f) No

Ch. 9 Understanding

1 False
3 True
5 True
7 False
9 True
11 True
13 False
15 True
17 False
19 False
21 True
23 False
25 True
27 True
29 True
31 True
33 True
35 False
37 False
39 True
41 True
43 False
45 False
47 False

Chapter 9 Tools

1 $41/35$
3 $(3 - 4x)/6x$
5 $-2(1 - 2y)/yz$
7 $(2 - 3x)/x^2$
9 $15/7$
11 $18/x^3$
13 $20/(x - 1)$
15 $(4y^3z - 3wx)/(x^2y^4)$
17 $(8(y + 4))/(y - 4)$

19 $x^3/2$
21 $(-27x+44)/((x+1)(3x-4))$
23 $2(11x+27)/((x-3)^2(x+5)^2)$
25 $(x^2+1)/(x-1)$
27 $\left((u+a)^2+u\right)/(u+a)$
29 $(1+e^x)/e^{2x}$
31 $(0.28+3M^3)/(4M)$
33 8
35 $x(x+2)$
37 $1/2$
39 $1/(a^2b^2)$
41 $-2x-h$
43 $1-(1/a)$
45 $p^3/(p^2+q^2)$
47 $(1-2\ln x)/\left(3x^3\right)$
49 $(-4x^2+2x+1)/(x^4\sqrt{2x-1})$
51 $1/3+1/\sqrt{x}$
53 $7/(p^2+11)+p/(p^2+11)$
55 $1/t^{5/2}+1/t^{3/2}$
57 $1+3/(q-4)$
59 $1+2/(2u+1)$
61 $1+1/e^x$
63 True
65 False
67 True

Section 10.1

1 Scalar
3 Scalar
5 $\vec{p}=2\vec{w}$
$\vec{q}=-\vec{u}$
$\vec{r}=\vec{u}+\vec{w}$
$\vec{s}=2\vec{w}-\vec{u}$
$\vec{t}=\vec{u}-\vec{w}$
7

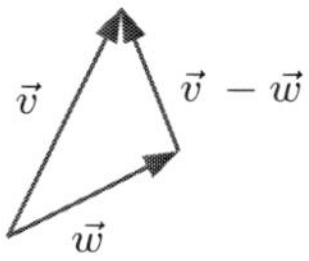

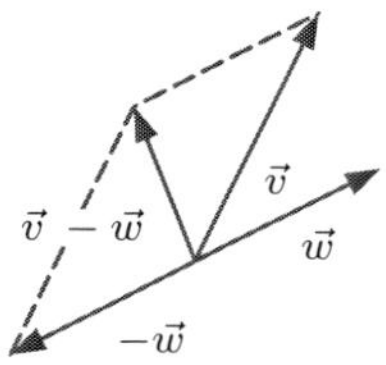

9

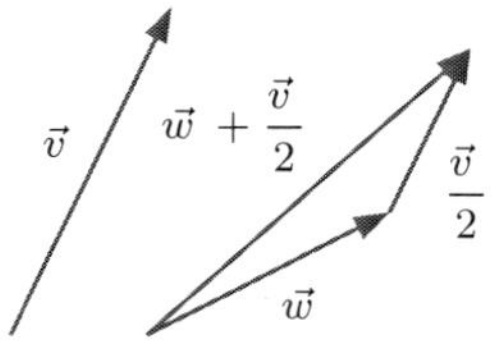

11

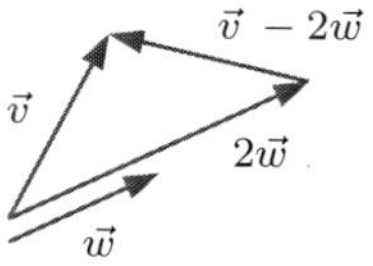

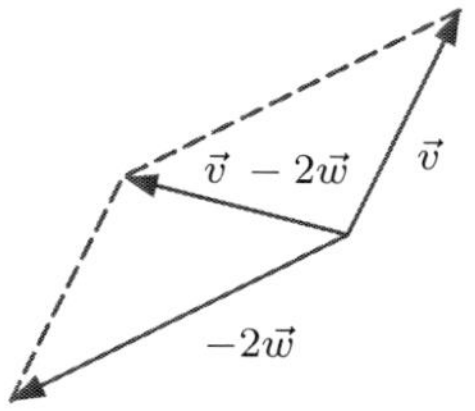

13 5.116 miles; 14.639° east of north
15 (a) 1.710 miles
(b) 5.848 miles
17 13,757 meters at an angle of $67.380°$ north of west and $19.093°$ from the horizontal
19 7.431 ft/sec and $16.588°$; 3.576 ft/sec and $36.387°$

Section 10.2

1 $10\vec{i}-4\vec{j}$
3 $3\vec{i}-7\vec{j}$
5 $-3\vec{i}-4\vec{j}$
7 $\vec{w}\approx-0.725\vec{i}-0.95\vec{j}$
9 $\sqrt{6}\approx 2.449$
11 7.649
13 (a) $50\vec{i}$
(b) $-50\vec{j}$
(c) $(25\sqrt{2})\vec{i}-(25\sqrt{2})\vec{j}$
(d) $-(25\sqrt{2})\vec{i}+(25\sqrt{2})\vec{j}$
15 $45°$ or $\pi/4$
17 $-140.847\vec{i}+140.847\vec{j}+18\vec{k}$
19 $21\vec{j}+35\vec{k}$
21 $\vec{u}$ and $\vec{w}$; $\vec{v}$ and $\vec{q}$.
23 3.982 mph; $51.116°$
25 $\vec{i}+4\vec{j}$
27 $-\vec{i}$

Section 10.3

1 $(6,7,9,11,14,18)$
3 $(3,5,0,-5,-17,-36)$
5 $(12.872,15.233,18.661,22.089,26.584,32.146)$
7 $(3.63,1.44,6.52,1.43,1.20,0.74)$
9 $(3.467,1.277,6.357,1.267,1.037,0.577)$
11 (79.000,79.333,89.000,68.333,89.333)
13 $3.378°$ north of east
15 (a) $\vec{v}=4.330\vec{i}+2.500\vec{j}$
For the second leg of his journey, $\vec{w}=x\vec{i}$

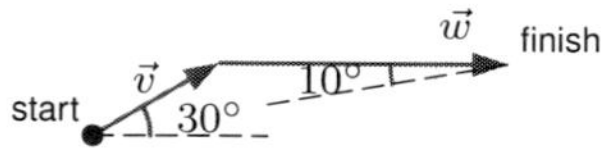

(b) $x=9.848$
(c) 14.397
17 (a) $\sqrt{34}$
(b) 17.6
(c) $(9.065, 15.086)$
19 (a)

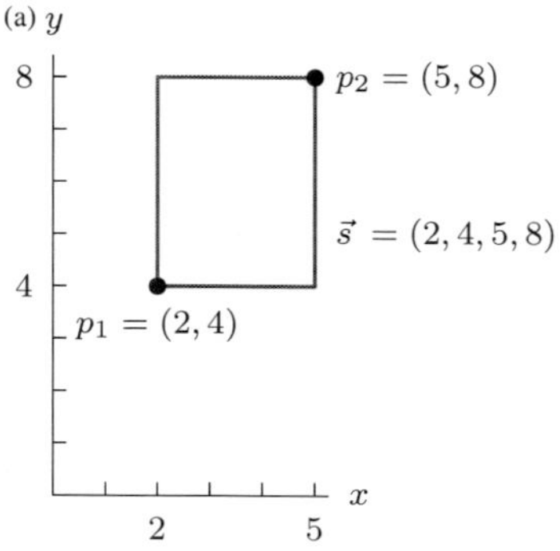

(b) (i) 1 unit right
(ii) 1 unit left
(iii) 2 right, 3 up
(iv) 1 right, 1 up
(v) k right, k up

Section 10.4

1 -7
3 -14
5 -22
7 -2
9 185
11 (a) 6 ft-lb
(b) -10 ft-lb
13 Parallel:
$3\vec{i}+\sqrt{3}\vec{j}$ and $\sqrt{3}\vec{i}+\vec{j}$
Perpendicular:
$\sqrt{3}\vec{i}+\vec{j}$ and $\vec{i}-\sqrt{3}\vec{j}$
$3\vec{i}+\sqrt{3}\vec{j}$ and $\vec{i}-\sqrt{3}\vec{j}$
15 Perpendicular: $t=2$ or -1
No values of t make $\vec{u}$ parallel to $\vec{v}$
17 No
23 Bantus
25 (a) $\sqrt{2}$
(b) 0.32

Section 10.5

1 (a) $\begin{pmatrix} 15 & 35 \\ 10 & -5 \end{pmatrix}$

(b) $\begin{pmatrix} -2 & 10 \\ 0 & -16 \end{pmatrix}$

(c) $\begin{pmatrix} 4 & 2 \\ 2 & 7 \end{pmatrix}$

(d) $\begin{pmatrix} -8 & -26 \\ -6 & 11 \end{pmatrix}$

(e) $\begin{pmatrix} 13 & 45 \\ 10 & -21 \end{pmatrix}$

(f) $\begin{pmatrix} k & -5k \\ 0 & 8k \end{pmatrix}$

3 (a) $\begin{pmatrix} 12 & 8 & 20 & 4 \\ 16 & 24 & 28 & 12 \\ 4 & 36 & 20 & 32 \\ 0 & -8 & 16 & 24 \end{pmatrix}$

(b) $\begin{pmatrix} -2 & -12 & -8 & -4 \\ -6 & -10 & 2 & -14 \\ -18 & -8 & -14 & -6 \\ -4 & -16 & -8 & -10 \end{pmatrix}$

(c) $\begin{pmatrix} 2 & -4 & 1 & -1 \\ 1 & 1 & 8 & -4 \\ -8 & 5 & -2 & 5 \\ -2 & -10 & 0 & 1 \end{pmatrix}$

(d) $\begin{pmatrix} 6 & -12 & 3 & -3 \\ 3 & 3 & 24 & -12 \\ -24 & 15 & -6 & 15 \\ -6 & -30 & 0 & 3 \end{pmatrix}$

(e) $\begin{pmatrix} 4 & 8 & 9 & 3 \\ 7 & 11 & 6 & 10 \\ 10 & 13 & 12 & 11 \\ 2 & 6 & 8 & 11 \end{pmatrix}$

(f) $\begin{pmatrix} 10 & -4 & 12 & 0 \\ 10 & 14 & 30 & -2 \\ -14 & 28 & 6 & 26 \\ -4 & -24 & 8 & 14 \end{pmatrix}$

5 (a) $(51, 15, 38)$
(b) $(-8, -11, 33)$
(c) $(70, 20, 22)$
(d) $(11, -6, 17)$
(e) 681
(f) $\begin{pmatrix} 24 & 60 & 84 \\ 48 & -72 & 36 \\ 192 & -60 & 0 \end{pmatrix}$

7 (a) $\mathbf{T} = \begin{pmatrix} 0.90 & 0 & 0 \\ 0.10 & 0.50 & 0.02 \\ 0 & 0.50 & 0.98 \end{pmatrix}$
(b) $\vec{p_1} = (1.8, 0.2, 0)$,
$\vec{p_2} = (1.62, 0.28, 0.1)$,
$\vec{p_3} = (1.458, 0.304, 0.238)$

9 (a) $\mathbf{T} = \begin{pmatrix} 0.97 & 0.05 \\ 0.03 & 0.95 \end{pmatrix}$
(b) $\vec{p}_{2006} = (214, 386)$,
$\vec{p}_{2007} = (226.88, 373.12)$.

11 (a) $\vec{v} = \begin{pmatrix} 11 \\ 19 \end{pmatrix}$
(b) $\vec{v} = \begin{pmatrix} 5 \\ 11 \end{pmatrix}$
(c) $\vec{v} = \begin{pmatrix} 2a+b \\ 3a+2b \end{pmatrix}$

13 (a) $\lambda_2 = -1$
(b) $\lambda_3 = -1$
(c) $\mathbf{A}\vec{v} = \lambda\vec{v}$, and $\mathbf{A}\vec{v}$ is parallel to $\vec{v}$

15 (a) $\begin{pmatrix} 3 & 5 \\ 2 & 4 \end{pmatrix}\begin{pmatrix} a \\ b \end{pmatrix} = a\begin{pmatrix} 3 \\ 2 \end{pmatrix} + b\begin{pmatrix} 5 \\ 4 \end{pmatrix}$
(b) $\vec{v} = \begin{pmatrix} -8.5 \\ 5.5 \end{pmatrix}$
(c) $\vec{v} = -8.5\vec{c_1} + 5.5\vec{c_2}$

Chapter 10 Review

1 $(3, 3, 6)$

3 $(-3, -2, 9)$

5 $(7, 8, -21)$

7 $(4, -2, 18)$

9 $-4.5\vec{i} + 8\vec{j} + 0.5\vec{k}$

11 13

13 6

15 $6\vec{i} + 6\vec{j} + 6\vec{k}$

17 $\vec{a} = \vec{b} = \vec{c} = 3\vec{k}$
$\vec{d} = 2\vec{i} + 3\vec{k}$
$\vec{e} = \vec{j}$
$\vec{f} = -2\vec{i}$

19 $\|\vec{u}\| = \sqrt{6}$
$\|\vec{v}\| = \sqrt{5}$

21 (a) Yes
(b) No

23 (a) $\vec{L} = (11, 7, 11, 7, 13)$
(b) $\vec{F} = (32, 36, 21, 8, 4)$,
$\vec{G} = (3, 3, 2, 0, 7)$

25 $F = g\sin\theta$

29 $0.4v\vec{i} + 0.693v\vec{j}$

31 (a) $\overrightarrow{AB} = 2\vec{i} - 2\vec{j} - 7\vec{k}$
$\overrightarrow{AC} = -2\vec{i} + 2\vec{j} - 7\vec{k}$
(b) $\theta = 44.003^\circ$

35 $\overrightarrow{AB} = -\vec{u}$; $\overrightarrow{BC} = 3\vec{v}$; $\overrightarrow{AC} = \overrightarrow{AB} + \overrightarrow{BC} = -\vec{u} + 3\vec{v}$; $\overrightarrow{AD} = 3\vec{v}$

37 $3\vec{n} - 3\vec{m}$;
$3\vec{m} + \vec{n}$;
$4\vec{m} - \vec{n}$;
$\vec{m} - 2\vec{n}$

Ch. 10 Understanding

1 False

3 False

5 False

7 False

9 True

11 True

13 False

15 True

17 True

19 False

21 True

23 False

25 True

Section 11.1

1 Arithmetic

3 Not arithmetic

5 Not arithmetic

7 Arithmetic, $a_n = -0.9 - 0.1n$

9 Geometric

11 Geometric

13 Geometric; $4 \cdot 3^{n-1}$

15 Geometric; $2(-2)^{n-1}$

17 Geometric; $1/(1.2)^{n-1}$

19 $10.8, 64.8, 4.8 + 1.2n$

21 $7.9, 57.4, 2.4 + 1.1n$

23 $1.661, 7(0.75)^{n-1}$

25 $486, 2 \cdot 3^{n-1}$

27 (a) $646.7, 650.580, 654.484, 658.411$
(b) $367.7, 396.748, 428.091, 461.911$
(c) 2012

29 (a) 17.960, 18.314, 18.675
(b) $17.960(1.0197)^n$
(c) 36.5 years

31 Arithmetic, $d > 0$

33 Arithmetic, $d < 0$

35 2, 7, 12, 17; $a_n = -3 + 5n$

37 3, 7, 15, 31;
$a_n = 2^{n-1} \cdot 3 + 2^{n-2} + 2^{n-3} + \cdots + 1$

39 (a) $256
(b) $d_n = 4^n$

Section 11.2

1 Not arithmetic

3 Not arithmetic

5 $a_1 = 2, d = 5$

7 $a_1 = 7, d = 3$

9 $a_1 = 6, d = 11$

11 $11^2 + 12^2 + 13^2 + \cdots + 21^2$

13 $-6 - 3 + 0 + 3 + 6 + 9$

15 $(-1)^0 2^1 + (-1)^1 2^2 + (-1)^2 2^3 + \cdots (-1)^6 2^7$

17 $\sum_{n=0}^{4} (10 + 3n)$

19 $\sum_{n=0}^{5} (30 - 5n)$

21 500,500

23 2625

25 -561

27 -111.3

29 -132

31 (a) (i) 24,717; 41,129; 62,248; total

number of deaths up to that year

(ii) 12,110; 16,412; 21,119; deaths in each year

(b) a_{n+1}; deaths in year $n+1$

33 0

35 20

37 784

39 $16n^2$

41 (a) $1, n$
(b) $n, n(n+1)/2$

43 (a) 297

Section 11.3

1 Yes, $a = 2$, ratio $= 1/2$

3 No. Ratio between successive terms is not constant

5 1,572,768

7 5.997

9 781.248

11 7.199

13 $\sum_{n=1}^{6}(-1)^{n+1}(3^n)$

15 $\sum_{n=0}^{5}(-1)^n 32(\frac{1}{2})^n$

17 (a) \$64,735.69
(b) \$65,358.46

19 \$26,870.37

21 (a) \$12,351.86
(b) Shortly before 20th deposit

Section 11.4

1 Yes, $a = y^2$, ratio $= y$

3 Yes, $a = 1$, ratio $= -x$

5 $y^2/(1-y)$, $|y| < 1$

7 $-4/3$

9 0.011

11 32/3

13 (a) $0.23 + 0.23(0.01) + 0.23(0.01)^2 + \ldots$
(b) $0.23/(1-0.01) = 23/99$

15 613/99

17 431/900

19 $Q_3 = 260.400$
$Q_{40} = 260.417$
Approaches 260.417 mg

23 \$926.40

25 (a) \$1000
(b) When the interest rate is 5%, the present value equals the principal.
(c) The value of the bond.
(d) Because the present value is more than the principal.

Chapter 11 Review

1 (a) $1 + 5 + 9 + 13 + 17$
(b) 45

3 $\sum_{n=0}^{10}(100 - 10n)$

5 Yes, $a = 1$, ratio $= -y^2$

7 $1/(1+y^2)$, $|y| < 1$

9 (a) $1, 4, 9, 16; n^2$
(b) $1, 3, 5, 7, 2n-1$

11 (b) $c_n = 300 + \frac{1}{2}(n-1)(3n-46)$
$c_{12} = 245$, $c_{50} = 2848$
(c) 19

13 8160, 8323, 8490, 8787, 9094, 9413, 9742, 10,083, 10,436, 10,801

15 (a) $\sum_{n=1}^{25} 81(1.012)^{n-1}$
(b) 2345.291 bn barrels

17 (a) Increase
(b) \$121,609.86
(c) \$2912.62

19 (a) 1365 people
(b) 1,398,101
(c) 6 stages

23 (a)

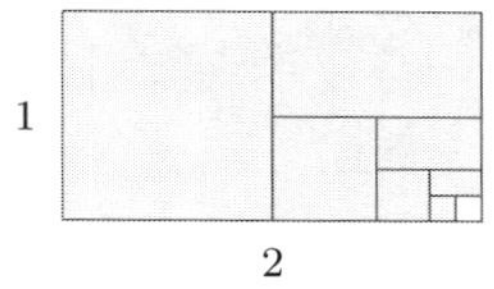

(b) $1 + \frac{1}{2} + \frac{1}{4} + \cdots + \frac{1}{2^{n-1}}$
(c) 2

Ch. 11 Understanding

1 True

3 True

5 True

7 True

9 True

11 False

13 True

15 False

17 False

19 False

21 True

23 False

25 False

27 True

29 False

31 False

33 False

Section 12.1

1 $y = 11 - 2x$

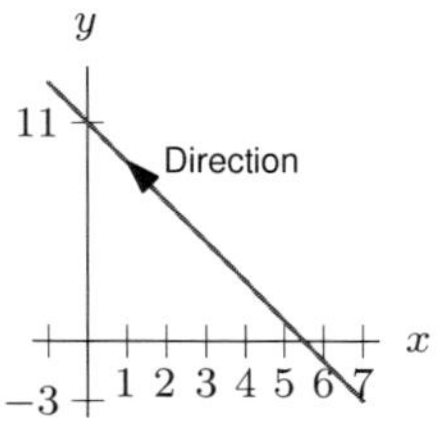

3 $y = 2x^2 + 1$

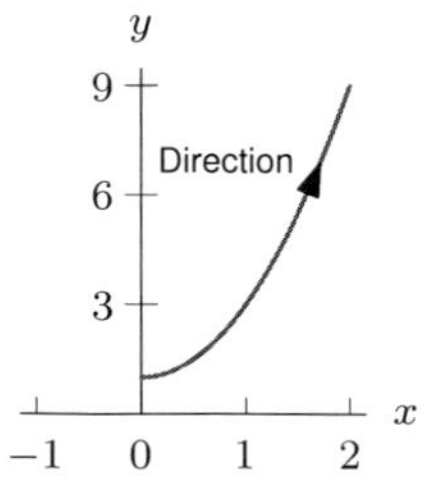

5 $y = (x+4)^2$

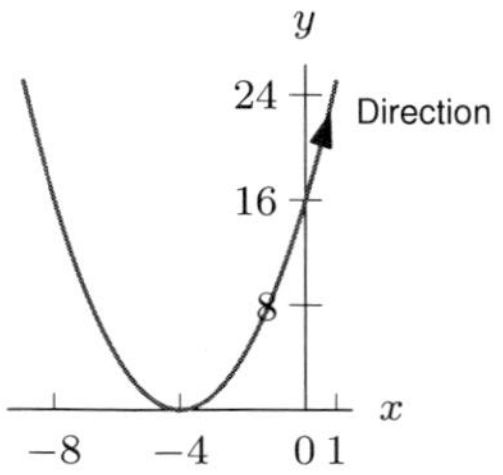

7 $y = x^{3/2}$

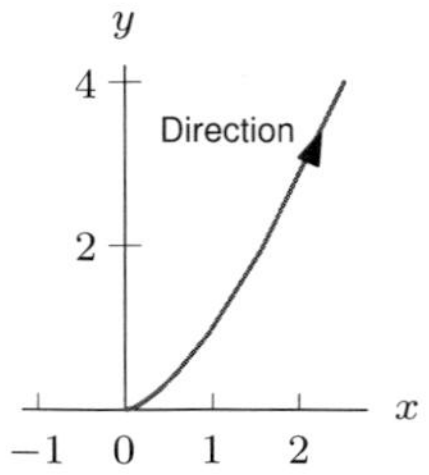

9 $y = (2/3)\ln x$, $x > 0$

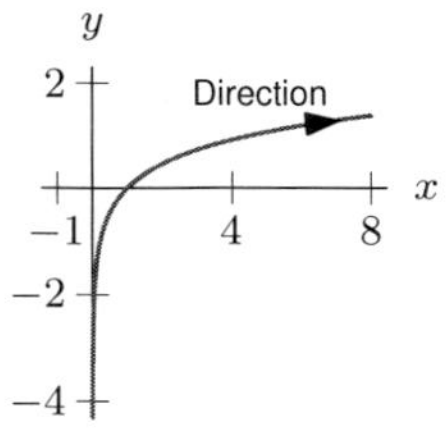

11 $(x^2/4) + (y^2/9) = 1$

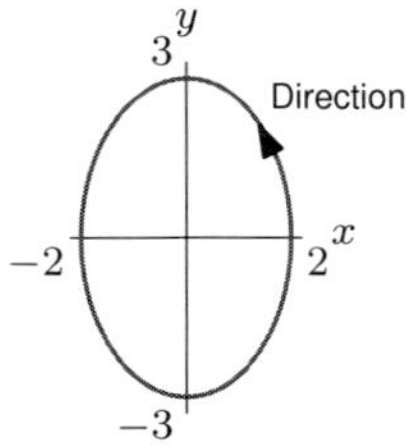

13 Lines from $(0,0)$ to $(2,0)$ to $(2,1)$ to $(0,1)$ to $(0,0)$

15 Lines from $(2,0)$ to $(1.5,1)$ to $(0.5,-1)$ to $(0,0)$ to $(0.5,1)$ to $(1.5,-1)$ to $(2,0)$

17 Clockwise for all t.

19 Clockwise: $t<0$,
Counter-clockwise: $t>0$.

23 (a) $x=t, y=t^2$
$x=t+1, y=(t+1)^2$
(b) $x=t, y=(t+2)^2+1$
$x=t+1, y=(t+3)^2+1$

25

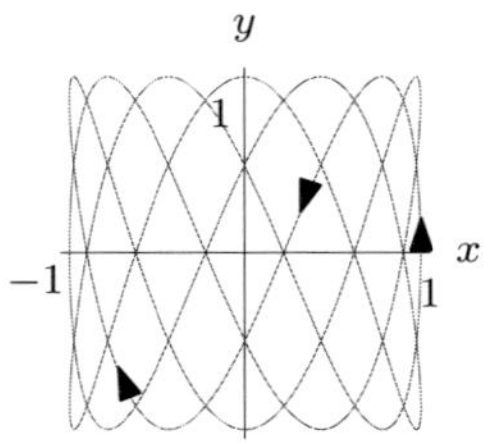

27

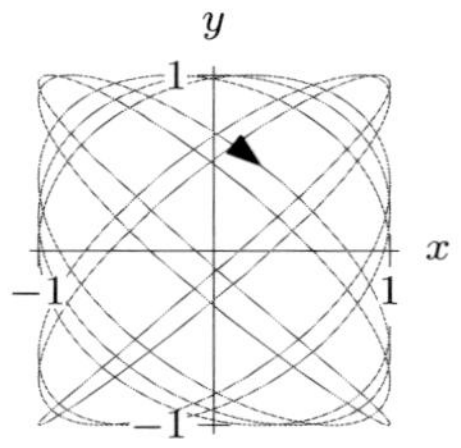

29 (a) Line $y=x$
(b) Circle, with starting point $(1,0)$ and period 2π
(c) Ellipse, with starting point $(1,0)$ and period 2π

31 $x=t, y=-4t+7$

33

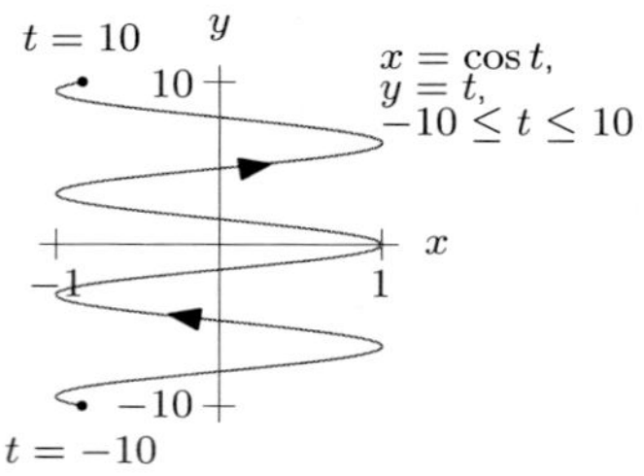

35 (a) $x=t, y=-16t^2+48t+6$
(b)

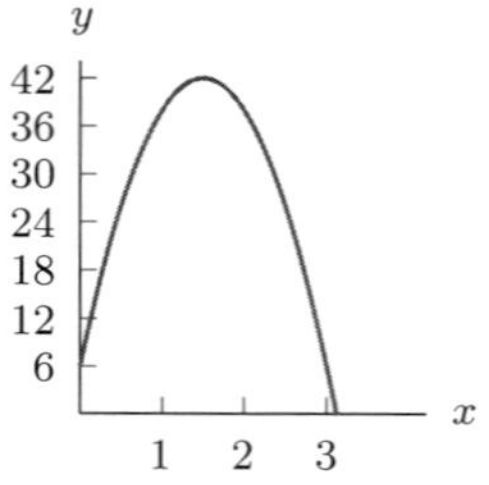

(c) 6 feet
(d) 3 seconds
(e) 42 feet

Section 12.2

1 $(0,0)\,;3/2$

3 $(-1,-3)\,;4$

5 $x=3\sin t, y=3\cos t, 0\le t\le 2\pi$

7 $x=-7\cos t, y=-7\sin t, 0\le t\le 2\pi$

9 $x=3+4\sin t, y=1+4\cos t, 0\le t\le 2\pi$

11 $x=-1-3\cos t,\ y=-2-3\sin t,\ 0\le t\le 2\pi$

13 (a) Center $(2,-4)$, radius $\sqrt{20}$
(b) Center $(-1,2)$, radius $\sqrt{11}$

15 Parabola:
$y=(x-2)^2, 1\le x\le 3$

17 $x=4(y-3)^2, 2\le y\le 4$.

19 Implicit: $xy=1, x>0$
Explicit: $y=1/x, x>0$
Parametric: $x=t, y=1/t, t>0$

21 Explicit: $y=\sqrt{4-x^2}$
Implicit: $y^2=4-x^2$ or $x^2+y^2=4, y>0$
Parametric: $x=4\cos t,\ y=4\sin t$, with $0\le t\le\pi$

23 (a) $x=t, y=1$
(b) $x=t+\cos t, y=1-\sin t$

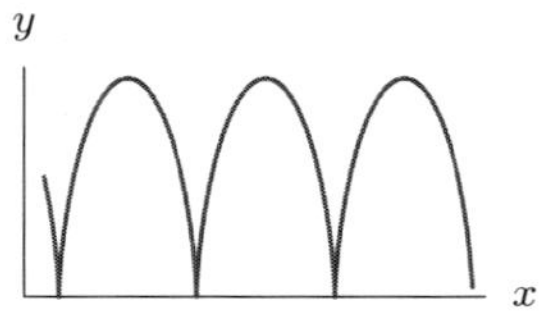

Section 12.3

1 (a) $(0,0)$; 24; 10
(b) $(x^2/144)+(y^2/25)=1$

3 (a) $(-2,-1)$; 6; 10
(b) $((x+2)^2/9)+((y+1)^2/25)=1$

5 $x=12\cos t, y=5\sin t, 0\le t\le 2\pi$

7 $x=-2-3\cos t,\ y=-1+5\sin t,\ 0\le t\le 2\pi$

9 Same ellipse; traced opposite direction

11 $((x+2)^2/9)+((y+5)^2/25)=1$; $(-2,-5)$; 3; 5

13 $(x-1/2)^2+(y+1)^2/4=1$; $(1/2,-1)$; 1; 2

15 $((x-3)^2/4)+((y-2)^2/9)=1$; $(3,2)$; 2; 3

17 (a) Center $(-1,3)$, major axis $a=\sqrt{6}$, minor axis $b=2$

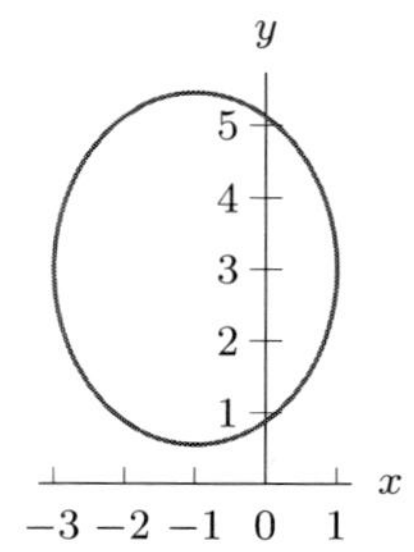

(b) Center $(3/2,-1)$, major axis $a=\sqrt{39}/2$, minor axis $b=\sqrt{13/2}$

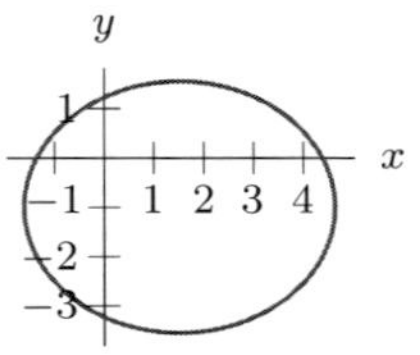

19 $h<k<0<a<b$

21 (b) Min$=r_0/(1+\epsilon)$
Max$=r_0/(1-\epsilon)$
(c) Center$=(8,0)$

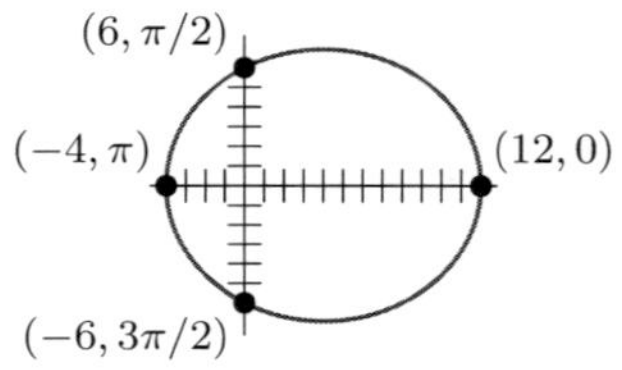

(d) $2r_0/(1-\epsilon^2)$

23 (a) $\left(\dfrac{2x-r_m+r_e}{r_e+r_m}\right)^2+\dfrac{y^2}{b^2}=1$
(b) $b=\sqrt{r_e r_m}$

Section 12.4

1 (a) $(5,0)$; $(-5,0)$; $(0,0)$
(b) $y=x/5$; $y=-x/5$
(c) $(x^2/25)-y^2=1$

3 (a) $(2,7)$; $(2,-1)$; $(2,3)$
(b) $y=(4x+1)/3$; $y=(-4x+17)/3$
(c) $((y-3)^2/16)-((x-2)^2/9)=1$

5 $x=5/\cos t, y=\tan t$;
Right half: $0\le t<\pi/2, 3\pi/2<t<2\pi$

7 $x=2+4\tan t, y=3+3/\cos t$;
Lower half: $\pi/2<t<3\pi/2$

9 I; $0<b<k<h<a$.

11 (a) Center $(-5,2)$; Vertices $(-5\pm\sqrt{6},2)$; Asymptotes $y=\pm(2/\sqrt{6})(x+5)+2$
(b) Center $(-1,-2)$; Vertices $(-1\pm\sqrt{14},-2)$; Asymptotes $y=\pm(x+1)-2$

13 $((y+1)^2/4)-((x-2)^2/9)=1$
$(2,-1)$; up-down; 3; 2

15 $((x-2)^2/4)-((y-1)^2/9)=1$
$(2,1)$; right-left; 2; 3

17 $((y+1/3)^2/8)-((x+3/2)^2/9)=1$
$(-3/2,-1/3)$; up-down; 3; $2\sqrt{2}$

Section 12.5

1 $\sqrt{10}$

3 (a), (b), (c)

5 $(0, \pm\sqrt{7})$

7 Hyperbola; x-axis

9 Parabola; x-axis

11 (a), (b), (c)

13 $(\pm\sqrt{2}, 0)$

15 $y = -(1/12)x^2$

17 $(-2, 1)$; $(-2, 9/4)$

19 $(1/2, 3/4)$; $(1/2, 1)$

21 $y = (3/4)(x-2)^2 - 3$; $(2, -8/3)$; $y = -10/3$

23 $(0, 3)$; $(0, 3 \pm \sqrt{24})$

25 $x^2/25 + y^2/(50/3) = 1$; $(\pm 5/\sqrt{3}, 0)$

27 $(2, -2 \pm 5)$; $(2, -2 \pm \sqrt{29})$

29 $(-4, -7 \pm \sqrt{20})$; $(-4, -7 \pm 10)$

31 $(x-3)^2 - (y-4)^2/9 = 1$; $(3 \pm \sqrt{10}, 4)$

33 $x^2/25 + y^2/21 = 1$

35 Yes

37 Distance doubles

39 (a) 298 million km
(b) $(x-3)^2/149^2 + y^2/148.970^2 = 1$

41 (a) Hyperbola
(b) $y^2/196 - x^2/64 = 1$; $y^2/49 - x^2/64 = 1$
(c) $y = 14\sqrt{1 + x^2/64}$; $y = -7\sqrt{1 + x^2/64}$

Section 12.6

3 $\sinh x \to (e^x)/2$ as $x \to \infty$
$\sinh x \to -(e^{-x})/2$ as $x \to -\infty$

5 $x = \sinh t$, $y = -\cosh t$, $-\infty < t < \infty$

7 $x = 2 + \cosh t$, $y = 3 + \sinh t$, $-\infty < t < \infty$

9 $x = 1 + 2\cosh t$, $y = -1 + 3\sinh t$, $-\infty < t < \infty$

11 $x = -3 + (\cosh t)/2$, $y = 1 + \sinh t$, $-\infty < t < \infty$

13 $x = 1 + (1/\sqrt{2})\cosh t$, $y = -2 + \sqrt{2}\sinh t$, $-\infty < t < \infty$

15 $x = 1 + \cosh t$, $y = 2 + \sinh t$, $-\infty < t < \infty$

17 $x = 1 + \sqrt{2}\sinh t$, $y = -2\sqrt{2}\cosh t$, $-\infty < t < \infty$

19 $x = h + a\cosh t$ and $y = k + b\sinh t$

21 Yes, $\cosh 2x = \cosh^2 x + \sinh^2 x$

25 $\sin(ix) = i\sinh x$

Chapter 12 Review

1 $x = 3\cos t$, $y = -3\sin t$, $0 \le t \le 2\pi$

3 $x = -2\cos t$, $y = 2\sin t$, $0 \le t \le 2\pi$

5 $x = 5\cos t$, $y = 7\sin t$, $0 \le t \le 2\pi$

7 $x = -3\cos t$, $y = -7\sin t$, $0 \le t \le 2\pi$

9 Circle; $(0, 3)$; $\sqrt{5}$

11 Hyperbola, $(0, 1)$; 2; 3; left-right

13 Ellipse, $(5, 0)$; 2; 3

15 Hyperbola, $(-1/3, 1/2)$; $\sqrt{3}$; $\sqrt{2}$; up-down

17 $(\pm\sqrt{21}, 0)$

19 $(0, 1/20)$; $y = -1/20$

21 $x = \cos t$, $y = \sin t$

23 Circle; $(-1, 0)$; 1

25 Ellipse; $(1, -\frac{1}{3})$; 1; $\sqrt{2/3}$

27 No, since $(0, 1)$ not on curve

29 (a) $x = 10\cos t$, $y = 10\sin t$
(b) $x = 10\cos t + 3\cos 8t$
$y = 10\sin t + 3\sin 8t$
(c)

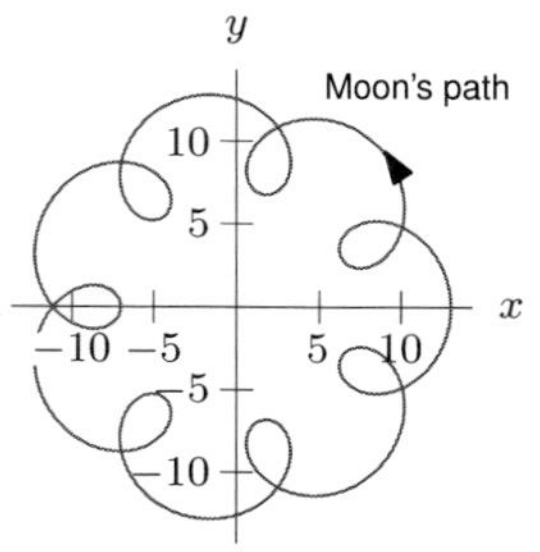

Ch. 12 Understanding

1 True

3 True

5 False

7 False

9 False

11 True

13 False

15 True

17 True

19 True

21 False

23 False

25 True

27 False

29 False

31 True

33 False

INDEX

FORMULA SUMMARY: ALGEBRA

Lines

Slope of line through (x_1, y_1) and (x_2, y_2):

$$m = \frac{y_2 - y_1}{x_2 - x_1}$$

Point-slope equation of line through (x_1, y_1) with slope m:

$$y - y_1 = m(x - x_1)$$

Slope-intercept equation of line with slope m and y-intercept b:

$$y = b + mx$$

Distance from (x_1, y_1) to (x_2, y_2):

$$\text{Distance} = \sqrt{(x_2 - x_1)^2 + (y_2 - y_1)^2}$$

Midpoint of a line segment whose ends are at (x_1, y_1) and (x_2, y_2):

$$\text{Midpoint} = \left(\frac{x_1 + x_2}{2}, \frac{y_1 + y_2}{2}\right)$$

Definition of Zero, Negative, and Fractional Exponents

$$a^0 = 1, \quad a^{-1} = \frac{1}{a}, \quad \text{and, in general, } a^{-x} = \frac{1}{a^x}$$

$$a^{1/2} = \sqrt{a}, \quad a^{1/3} = \sqrt[3]{a}, \quad \text{and, in general, } a^{1/n} = \sqrt[n]{a}.$$

$$\text{Also, } a^{m/n} = \sqrt[n]{a^m} = (\sqrt[n]{a})^m.$$

Rules of Exponents

1. $a^x \cdot a^t = a^{x+t}$ For example, $2^4 \cdot 2^3 = (2 \cdot 2 \cdot 2 \cdot 2) \cdot (2 \cdot 2 \cdot 2) = 2^7$.
2. $\dfrac{a^x}{a^t} = a^{x-t}$ For example, $\dfrac{2^4}{2^3} = \dfrac{2 \cdot 2 \cdot 2 \cdot 2}{2 \cdot 2 \cdot 2} = 2^1$.
3. $(a^x)^t = a^{xt}$ For example, $(2^3)^2 = 2^3 \cdot 2^3 = 2^6$.

Definition of Log

$y = \ln x$ means $e^y = x$; for example: $\ln 1 = 0$, since $e^0 = 1$.

$y = \log x$ means $10^y = x$; for example: $\log 10 = 1$, since $10^1 = 10$.

Rules of Logarithms

$$\log(AB) = \log A + \log B \qquad \ln(AB) = \ln A + \ln B$$

$$\log\left(\frac{A}{B}\right) = \log A - \log B \qquad \ln\left(\frac{A}{B}\right) = \ln A - \ln B$$

$$\log A^p = p \log A \qquad \ln A^p = p \ln A$$

Identities

$$\log 10^x = x \qquad \ln e^x = x$$

$$10^{\log x} = x \qquad e^{\ln x} = x$$

TABLE OF GREEK LETTERS

Name	Capital	Lowercase
Alpha	A	α
Beta	B	β
Gamma	Γ	γ
Delta	Δ	δ
Epsilon	E	ϵ
Zeta	Z	ζ
Eta	H	η
Theta	Θ	θ
Iota	I	ι
Kappa	K	κ
Lambda	Λ	λ
Mu	M	μ
Nu	N	ν
Xi	Ξ	ξ
Omicron	O	o
Pi	Π	π
Rho	P	ρ
Sigma	Σ	σ
Tau	T	τ
Upsilon	Υ	υ
Phi	Φ	ϕ
Chi	X	χ
Psi	Ψ	ψ
Omega	Ω	ω

TRIGONOMETRIC IDENTITIES

- **Tangent identity**:

$$\tan t = \frac{\sin t}{\cos t}$$

- **Periodicity**:

$$\sin(t + 2\pi) = \sin t \quad \cos(t + 2\pi) = \cos t \quad \tan(t + \pi) = \tan t$$

- **Pythagorean identity**:

$$\sin^2 t + \cos^2 t = 1$$

- **Double-angle formula for sine**:

$$\sin 2t = 2 \sin t \cos t$$

- **Double-angle formula for cosine** (expressed in three different ways):

$$\cos 2t = 1 - 2\sin^2 t$$
$$\cos 2t = 2\cos^2 t - 1$$
$$\cos 2t = \cos^2 t - \sin^2 t$$

- **Double-angle formula for tangent**:

$$\tan 2t = \frac{2 \tan t}{1 - \tan^2 t}$$

- **Negative angle identities**:

$$\sin(-t) = -\sin t \quad \cos(-t) = \cos t \quad \tan(-t) = -\tan t$$

- **Identities relating sine and cosine**:

$$\sin t = \cos\left(t - \frac{\pi}{2}\right) \qquad \cos t = \sin\left(t + \frac{\pi}{2}\right)$$

- **Sum-of-angle** and **difference-of-angle formulas for sine and cosine**:

$$\sin(\theta + \phi) = \sin\theta\cos\phi + \cos\theta\sin\phi$$

$$\sin(\theta - \phi) = \sin\theta\cos\phi - \cos\theta\sin\phi$$

and

$$\cos(\theta + \phi) = \cos\theta\cos\phi - \sin\theta\sin\phi$$
$$\cos(\theta - \phi) = \cos\theta\cos\phi + \sin\theta\sin\phi.$$

- **...nd difference of sine and cosine:**

$$\ldots + \cos v = 2\cos\frac{u+v}{2}\cos\frac{u-v}{2} \qquad \sin u + \sin v = 2\sin\frac{u+v}{2}\cos\frac{u-v}{2}$$

$$\cos u - \ldots = -2\sin\frac{u+v}{2}\sin\frac{u-v}{2} \qquad \sin u - \sin v = 2\cos\frac{u+v}{2}\sin\frac{u-v}{2}$$